Plant Physiology

New findings populate the enormous literature on plant physiology, almost on a daily basis. This text is a detailed introduction to the essential concepts of this rapidly advancing field of study; to important physiological aspects related to the functioning of plants. It covers a wide range of topics including water, absorption of water, ascent of sap, transpiration, mineral nutrition, fat metabolism, enzymes and plant hormones. Photosynthesis, respiration and nitrogen metabolism get discussed in separate chapters because their contribution towards food security, climate resilient farming and sustainable life needs highlighting. Unlike other books on the subject, this text lays due emphasis on the conceptual framework.

Our emphasis is on the concepts of water use efficiency (WUE) and nitrogen use efficiency (NUE), to lessen pressure and our dependence on our natural resources. A special feature of the book is a discussion on 'molecular mechanism of abiotic and biotic stresses'. Seminal contributions of an 'international group of plant physiologists' and 'implications of plant physiology in agriculture' would be of immense interest to the readers. Alongside its emphasis on theoretical concepts, this text details experiments relating to most topics/chapters. A structured approach including principle, procedure, discussion, results and observation, and precautions has been used to explain the experiments.

S. L. Kochhar taught courses in economic botany and plant physiology at Sri Guru Tegh Bahadur Khalsa College, University of Delhi, for more than four decades (1965–2007). His areas of interest include botany, crop science, tropical crops and plant physiology.

Sukhbir Kaur Gujral has been teaching in the Department of Botany at Sri Guru Tegh Bahadur Khalsa College, University of Delhi for more than three decades. Her areas of interest include plant physiology and cell biology.

Plant Physiology

Theory and Applications

2nd Edition

S. L. Kochhar

Sukhbir Kaur Gujral

CAMBRIDGE
UNIVERSITY PRESS

University Printing House, Cambridge CB2 8BS, United Kingdom

One Liberty Plaza, 20th Floor, New York, NY 10006, USA

477 Williamstown Road, Port Melbourne, VIC 3207, Australia

314–321, 3rd Floor, Plot 3, Splendor Forum, Jasola District Centre, New Delhi – 110025, India

79 Anson Road, #06–04/06, Singapore 079906

Cambridge University Press is part of the University of Cambridge.

It furthers the University's mission by disseminating knowledge in the pursuit of education, learning and research at the highest international levels of excellence.

www.cambridge.org
Information on this title: www.cambridge.org/9781108486392

1st edition published under Foundation imprint (now defunct) in 2017

2nd edition first published 2020

Printed in India by Nutech Print Services, New Delhi 110020

A catalogue record for this publication is available from the British Library

Library of Congress Cataloging-in-Publication Data

Names: Kochhar, S. L., author. | Gujral, Sukhbir Kaur, author.
Title: Plant physiology : theory and applications / S. L. Kochhar, Sukhbir Kaur Gujral.
Description: Second. | New York, N.Y. : Cambridge University Press, [2020] | Includes bibliographical references and index.
Identifiers: LCCN 2020001149 (print) | LCCN 2020001150 (ebook) | ISBN 9781108486392 (hardback) | ISBN 9781108707718 (paperback) | ISBN 9781108639736 (ebook)
Classification: LCC QK711.2 .K63 2020 (print) | LCC QK711.2 (ebook) | DDC 581.3--dc23
LC record available at https://lccn.loc.gov/2020001149
LC ebook record available at https://lccn.loc.gov/2020001150

ISBN 978-1-108-48639-2 Hardback

ISBN 978-1-108-70771-8 Paperback

Contents

Foreword

Plant Physiology is a rapidly advancing field of study where new findings are surfacing in the literature almost daily. The ways of teaching the subject at both undergraduate and postgraduate levels have undergone a sea change and there is now a far greater emphasis on the conceptual framework than there was some years ago. It has become increasingly difficult for students to keep abreast with the ongoing explosion of knowledge in this field.

This book on Plant Physiology – Theory and Applications is both authoritative and timely. The book focuses on in-depth analysis of wide-ranging topics included in the Plant Physiology syllabi of different universities in India and overseas. The textbook starts with water and concludes with experimental exercises, project works and physiological set-ups or demonstration experiments. The chapters dealing with photosynthesis, respiration, and nitrogen metabolism are important since an understanding of their role for sustaining life on earth is essential for food security and climate resilient farming. Experiments are underway to convert C3 plants like rice into C4 plants in order to improve their photosynthetic efficiency. Researches are underway to transfer nitrogen-fixing genes into cereal crops to lessen our dependence on nitrogenous fertilisers. I am confident that the students going through the book will be stimulated to undertake the innovative experiments which can help to shape the physiological rhythms of plants to the emerging era of global warming and climate change. The material is presented in a concise and lucid manner so that the readers can easily comprehend the conceptual complexities of recent achievements in the field. A discussion of Molecular Mechanism of Abiotic and Biotic Stresses would be of immense help to the students. In view of climate change, the authors have emphasised the concepts of water use efficiency (WUE) and nitrogen use efficiency (NUE). A feature of the book is the inclusion of Review Questions with answers at the end of each chapter. This would be of enormous help to the students preparing for their academic course work.

We are indebted to Dr S. L. Kochhar and Dr Sukhbir Kaur Gujral for their labour of love in presenting the secrets of life to young scholars in an easily understandable manner. I also compliment Cambridge University Press for publishing this book as a part of their commitment to make scientific knowledge available in an authentic manner at affordable prices to young university students. I hope the book will be read and used widely.

M. S. Swaminathan
Founder Chairman, M. S. Swaminathan Research Foundation
Ex-Member of Parliament (Rajya Sabha)

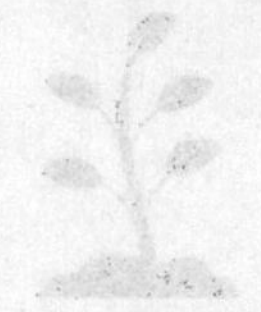

Preface to the Second Edition

The Foundation Imprint of our book, *Plant Physiology: Theory and Applications*, which was first published in 2017 by CUP, India, met with an overwhelming response and the stocks were exhausted within a year or so. In a large measure, the textbook succeeded very well in addressing the need for such a tome amongst students and scholars in the fields of botany and agricultural sciences. However, review inputs and advice from subject experts overseas and within the country, as well as suggestions offered to us by the editorial staff of the Cambridge University Press, led us to considerable soul-searching on how to enhance its value.

And so, it gives us great pleasure in presenting to the readers an expanded and updated version whereby the readers can assimilate knowledge without undue effort as they imbibe; this self-contained textbook should, therefore, have a much wider appeal. The material is presented in a concise and lucid manner so that the readers can develop better understanding of the conceptual complexities in the field. We hope it will prove to be even more useful than the previous ones.

The salient features of this Second Edition are as follows:

The present edition is divided into 7 units, with a total of 23 chapters in all. In addition to the accurate and authoritative textual content in each chapter, we have endeavoured to present, through models and flow charts, more information about growth and development such as mode of action, physiological role, biosynthesis and inactivation of auxins, gibberrelins, ethylene and abscisic acid. The role of recently discovered hormones such as jasmonates, polyamines, salicylic acid and nitric oxide and others has been emphasised. The mode of action of phytochrome-mediated responses based on their requirements such as LFRs, VLFRs and HIRs finds a special mention. Included in the new edition are chapters on 'Concepts of Metabolism', 'Carbohydrate Metabolism', 'Sulphur, Phosphorus and Iron Assimilation in Plants', 'Plant Photoreceptors' and 'Ripening, Senescence and Cell Death' as well the work on *Arabidopsis thaliana* as an experimental tool and a model system for research in genetics and molecular biology, detailing its advantages over other plant material. A special feature of the chapter on Stress Physiology is an invited article on 'Molecular Mechanism of Abiotic and Biotic Stresses' which is being reproduced in the original. We have also included here, a discussion of Reactive Oxygen Species (ROS) and the Asada-Halliwell or Ascorbate-Glutathione Pathway.

Other useful additions are the inclusion of numericals dealing with pH, the molecular movement of water between cells, and energy transformation during respiratory and fat metabolism. We have laid due emphasis on the concepts of water use efficiency (WUE) and nitrogen use efficiency (NUE) to lessen pressures and dependence on our natural resources. Furthermore, we have devoted one chapter to the seminal contributions of an international group of plant physiologists whose discoveries have impacted the very fundamentals of this field. We have also dealt with classical experiments that have shaped its development. Another addition that should be of immense importance to readers is the chapter on 'Implications of plant physiology in agriculture'. And as a special for Indian readers, is the chapter on the development and status of plant physiology in India. Yet another significant addition is the inclusion of experimental exercises, project work and physiological set-ups or demonstration experiments. By engaging themselves in conducting such experimentation, students would not only hone their practical skills in plant physiology but would also imbibe aspects of the subject with greater thoroughness – so vital for their semester course work. The fact is that we have attempted to integrate into this compendium, relevant practical details, such as Requirements, Principle, Methodology, Observation, Discussion and Conclusions, et al. This approach should enable the students to have an access and a deeper insight into experimental working and how it has led to significant findings.

Another salient feature is the inclusion of Review Questions with answers at the end of each chapter which should be of much help to the students, preparing for their examinations, (including the competitive ones). A glossary is provided to assist students in reviewing both the new and the familiar terms. The compilation is profusely illustrated to enhance the clarity and the utility of the book. At the end of units, we have interpolated many high-resolution coloured images pertaining to the subject matter, besides over a hundred well-labelled line diagrams. The online resource material includes discussions like 'Various Laboratory Techniques' and 'Microchemical Tests for Major Food Constituents'.

Preface to the First Edition

Our earlier book, *Comprehensive Practical Plant Physiology*, was first published in 2012 and was very warmly received by students and teachers alike. In a large measure, the book succeeded in generating a lot of interest amongst students and scholars alike, in the field of Botany and Agricultural Sciences. However, the readers have urged us to restructure the text and update the information so that the book becomes self-contained in itself, matching other leading titles in the field of plant physiology.

During the course of reorganisation, we were greatly influenced by the feedback reviews we received from the many experts based within the country and in Southeast Asia, as well as the suggestions offered to us by the editorial staff at Cambridge University Press, India. We are much pleased to present to our readers an altogether new book, in which we have ensured a continuous flow of information so that the readers can assimilate the knowledge without too much effort. The material is presented in a concise and lucid manner, so that the readers can comprehend the conceptual complexities, come to know about the recent achievements in the field, and share the joy that we feel for this subject.

Salient features of this edition are as follows:

In addition to the generalized and well-informed textual content in each chapter, we have attempted to highlight the important information through models and flow charts; such as the information in the chapter on 'growth and development' regarding various topic like, mode of action, physiological role, biosynthesis and inactivation of auxins, gibberellins, ethylene, and abscisic acid. The role of the recently discovered hormones such as jasmonates, polyamines, salicylic acid and nitric oxide, etc., has also been emphasised upon appropriately. The mode of action of phytochrome-mediated responses based on their requirements such as LFRs, VLFRs and HIRs, also finds a special mention. Included also in the discussion is the work on *Arabidopsis thaliana* as an experimental tool and model system for research in genetics and molecular biology, enumerating its advantages over other plant materials. We have also included a discussion on Reactive Oxygen Species (ROS) and Asada-Halliwell or Ascorbate-Glutathione Pathway in the chapter 'Stress Physiology and Secondary Metabolites'.

Another striking feature is the inclusion of numerical questions dealing with pH, molecular movement of water between cells and energy transformation during respiratory and fat metabolism. Furthermore, we have consolidated in a single chapter the seminal contributions of an international group of plant physiologists whose discoveries impacted our fundamental thoughts. We have not hesitated to include the classical experiments that had a revolutionary impact on the development of the subject. A chapter on the development and status of plant physiology in India will be of special interest to the Indian students. Another significant addition is the inclusion of Experimental Exercises, Project work and Physiological set-ups or Demonstration Experiments. This is an attempt to make students feel comfortable with their semester course work and hone their practical skills in plant physiology. In fact, we have tried to integrate the practical details, such as Requirements, Principle, Methodology, Observation, Discussion and Conclusions, etc., for the very same reason. This approach should enable the students to have not just an access but a deep insight into the various experimental working and findings.

Another salient feature of the book is the inclusion of Review Questions along with their answers at the end of each chapter, which would be of great help to the students preparing for their examinations. A glossary is provided to help students in reviewing both new and familiar terms. The book is profusely illustrated to enhance the clarity and utility of the book.

The online resource material includes chapters like, The Status and Development of Plant Physiology in India and Various Laboratory Techniques; and several appendices like, Microchemical Tests for Major Food Constituents, Concepts of Thermodynamics, The Laws of Thermodynamics, Redox Reactions, Chemosynthesis, ATP: The Energy Currency of the Cell, Coupled Reactions and Metabolic Pathways.

We wish to express our deep gratitude to Professor M. S. Swaminathan FRS, World Food Laureate, Founder Chairman and Mentor, M. S. Swaminathan Research Foundation, Chennai, not only for his kindness in writing the Foreword to the book but also for his great encouragement and continued interest in this project. We are also thankful to his Foundation for sending us pictures of *Rhizophora mucronata* – a component of the mangrove ecosystem.

We wish to offer our gratitude to Professor S. C. Maheshwari, well-known for his pioneering work on haploid from pollen, for permitting us to use pictures of some leading physiologists appearing in his article and also for sending his photograph to be included. The authors are highly indebted to Dr Govindjee, Professor Emeritus, University of Illinois, Urbana-Champaign, USA, for sending us his research contributions as well as his photograph. Additionally, we are grateful to Professor R. C. Sachar, formerly of Delhi University for sending us a write-up about his research contributions. Our sincere thanks are also for Professors A. K. Bhatnagar, formerly of Delhi University; Professors R. R. Singh and Nirmala Nautiyal of Lucknow University; Dr H. N. Krishnamoorthy, formerly of HAU, Hisar; Dr Alok K. Moitra, Executive Secretary, INSA, Delhi, for arranging pictures of some Indian physiologists. The authors express their gratitude to Dr P. S. Deshmukh (Emeritus Scientist), Dr M. C. Ghildyal, Professor Raj Kumar Sairam, Professor G. S. Sirohi and Dr Ajay Arora of the Division of Plant Physiology, IARI, New Delhi, for their help and cooperation.

Teaching Plant Physiology together at the college for over 25 years has strengthened our ties at a familial level and we both are indebted to our families, especially Urmil Kochhar and Jasbir Singh Gujral, without whose support and continued interest this book would not have materialised. We value our association with our former colleagues Dr G. N. Dixit and Dr S. Bala Bawa with whom we shared the joy of teaching plant physiology at the college.

Our special thanks are to Saurabh Kochhar, Monica Manchanda and Shalloo Chaudhary for their support in gathering information from multiple sources. Our unending gratitude reaches out to Dr Inderdeep Kaur, Associate Professor at Khalsa College, for her many helpful suggestions from time to time. Special thanks are also to Dr Jaswinder Singh, Principal of Khalsa College, for his continued interest in our project. We are also thankful to S. K. Dass for his help in providing photographic images.

We wish to express our deep gratitude to the editorial and production team at the Cambridge University Press, India, for their courtesy, cooperation and technical guidance about details, and for their meticulous efforts in bringing out this edition well in time and in the best possible format.

We would highly appreciate receiving suggestions from our colleagues in India and overseas for the improvement of this new edition.

Acknowledgements

We wish to express our deep gratitude to Professor M. S. Swaminathan, FRS, World Food Laureate; and Founder Chairman and Mentor of the M. S. Swaminathan Research Foundation, Chennai, India, not only for his kindness in writing the Foreword to the book but also for his great encouragement and continued interest in this project. We are also thankful to his Foundation for providing us pictures of *Rhizophora mucronata* – an important component of the mangrove ecosystem.

We wish to offer our gratitude to (Late) Professor S. C. Maheshwari, well-known for his pioneering work on haploid from pollen, for having permitted us earlier to use pictures of some leading physiologists, from his article and also for sending us his own photograph for inclusion. The authors are highly indebted to Dr Govindjee, Professor Emeritus, University of Illinois, Urbana-Champaign, USA, for sending us his research contributions as well as his photograph for insertion. We are further grateful to him for having permitted us to use a cartoon, from the book *The Quantum Requirements Controversy between Warburg and the Midwest-gang* which he had jointly authored with Nickelsen. We also wish to express our sincere thanks to Professor Jaswinder Singh of the Plant Science Department, McGill University, Macdonald Campus, Quebec, Canada for his many helpful suggestions. We are indeed grateful to Professor R. C. Sachar, formerly of Delhi University for sending us his piece about his research contributions. Our sincere thanks are also due to Professors R. R. Singh and Nirmala Nautiyal of Lucknow University; Dr H. N. Krishnamoorthy, formerly of HAU, Hisar, Haryana; and Dr Alok K. Moitra, Executive Secretary, INSA, Delhi for having arranged some pictures of Indian Physiologists. The authors express their gratitude to Dr P. S. Deshmukh (Emeritus Scientist), Dr M. C. Ghildyal, Professor Sai Ram, Professor G. S. Sirohi and Dr Ajay Arora of the Division of Plant Physiology, IARI, New Delhi for their help and cooperation.

Teaching Plant Physiology together at the college for over twenty-five years has strengthened our ties at the family level and we are indeed indebted to our families especially Mrs Urmil Kochhar and Mr Jasbir Singh Gujral, without whose support and constant encouragement, the book would not have materialised. Sincere thanks are also due to our children for their deep interest, support and genuine concern about its Cambridge edition. A special word of appreciation goes to our beloved grandchildren as they waited with keen anticipation over the years to see the book in print.

We indeed value our association with our former colleagues Dr G. N. Dixit and Dr (Mrs) S. Bala Bawa with whom we shared the joy of teaching plant physiology at college. Our thanks are due to Drs Inderdeep Kaur and Surinder Kaur, Associate Professor, Khalsa College, University of Delhi for their many helpful suggestions from time to time. Our special thanks are due to Dr Jaswinder Singh, Principal of Khalsa College for his continued interest in our project. Sincere gratitude is due to Saurabh Kochhar for his support in gathering information from multiple sources and his assistance with computer applications.

We wish to express our deep gratitude to the editorial and production teams of the Cambridge University Press, India, for their courtesy, cooperation and technical guidance, their attention to details, and for their meticulous efforts in bringing out this edition well in time and in the best possible format.

We would indeed appreciate receiving suggestions from our fellow botanists in India and overseas for the further improvement of this second edition.

Photo credits

The authors express their gratitude to the following for having provided us some of the high resolution coloured images used in this book: Dr Surendra Singh, former Associate Professor at MSJ Government Post Graduate College, Bharatpur, Rajasthan; Dr Prabir Ranjan Sur, former Scientist at Botanical Survey of India, Kolkata, West Bengal; Dr (Mrs) C. T. Chandra Lekha, Devamatha College Kuravilangad, Kottayam, Kerala; Dr B. K. Mukherjee, IIT, Kharagpur, a well-known Cosmos Healer and Author; Mr Thilak Makkiseril, Seafood Technologist, Cochin, Kerala; Dr Manju Choudhary, Associate Professor, Government College Jhunjhunu, Rajasthan; Mr Sushain Babu, Shree Haripal Shastri Smarak Mahavidyalaya, PheenaBijnor, Rohilkhand, Uttar Pradesh; Mr Sandesh Parashar, former Principal, Panchayat Training Centre, Kota, Rajasthan; Dr Md. Ahsan Habib, Assistant Professor and Head Department of Biology, Mohammadpur Preparatory School and College, Dhaka, Bangladesh; Mr Rabiul Hasan Dollar, Chapainawabganj, Bangladesh; Mr B. K. Basavaraj, Forest Department, Mangalore, Karnataka; Dr Preeti V. Phate, Assistant Professor and Head, Department of Botany, JSM College, Alibag, Raigad, Maharashtra; Miss Madhumati Yashwant Shinde, PG Department of Botany, Dattajirao Kadam Arts, Science and Commerce College, Ichalkaranji, Kolhapur, Maharashtra; Mrs Jyotirmayi Parija, Forest Officer, Government of Odisha; Mr Satya Prakash Mahapatra, Education Officer (Retd.), Government of India; Ms Christine Walters, Redwood National and State Parks (RNPS), Crescent City, California, USA; Mr Andrei Savitsky, Cherkassy, Ukraine; and Mr Bobby Verma, Government Brijindra College Faridkot, Punjab.

Delhi. S. L. Kochhar

Sukhbir Kaur Gujral

Some Common Abbreviations used in the Text

2,4,5-T	2,4,5-Trichlorophenoxyacetic acid
2,4-D	2, 4-Dichlorophenoxyacetic acid
4-CPA	4-Chlorophenoxyacetic acid
ABA	Abscisic acid
ABC proteins	ATP binding cassette proteins
ACC	1-aminocyclopropane-1-carboxylic acid
ACO	ACC oxidase
ACP	Acyl carrier protein
ACS	ACC synthase
ADH	Alcohol dehydrogenase
ATP	Adenosine triphosphate
BR_S	Brassinosteroids
BSA	Bovine serum albumin
CAC	Citric acid cycle
CAM	Crassulacean acid metabolism
cAMP	Cyclic adenosine monophosphate
COP gene	CONSTITUTIVE PHOTOMORPHOGENESIS gene
DAG	Diacyl glycerol
DCPIP	2,6-Dichlorophenol indophenol
DCMU	3,4-dichlorophenyl-1,1-dimethyl urea
DNP	2,4-Dinitrophenol
DRE	Dehydration response elements
EC	Enzyme commission
EDTA	Ethylenediaminetetraacetic acid
ETS	Electron transport system
EXPA	α-expansins
FCR	Folin-Ciocalteau reagent
Fd	Ferredoxin

FLC gene	FLOWERING LOCUS C gene
FT protein	FLOWERING LOCUS T protein
GA	Gibberellin
GA_3	Gibberellic acid
GAREs	GA response elements
GDH	Glutamate dehydrogenase
GLC	Gas–liquid chromatography
GOGAT	Glutamate synthase or glutamine-2-oxoglutarate-amino transferase
GPP	Gross primary productivity
GS	Glutamine synthetase
HIRs	High irradiance responses
HPLC	High performance (pressure) liquid chromatography
Hpt protein	Histidine phosphotransfer protein
HSF	Heat shock factor
HSPs	Heat shock proteins
IAA	Indole-3-acetic acid
IBA	Indole-3-butyric acid
IP_3	1,4,5-triphosphate inositol
IPP	Iso-pentenyl pyrophosphate
LAI	Leaf area index
LHb	Leghaemoglobin
LDH	Lactate dehydrogenase
LEA Proteins	Late embryogenesis abundant proteins
LFRs	Low fluence responses
LHC	Light-harvesting complex
NAA	Naphthalene acetic acid
NAD	Nicotinamide adenine dinucleotide
NADH	Nicotinamide adenine dinucleotide (reduced)
NADP	Nicotinamide adenine dinucleotide phosphate
NAM gene	NO APICAL MERISTEM gene
Nif genes	Nitrogen fixing genes
NiR	Nitrite reductse
NEDD	N-(1-naphthyl) ethylenediamine dihydrochloride
Nod genes	Nodulation genes
NPA	N-1-naphthylphthalamic acid
NR	Nitrate reductase
OAA	Oxaloacetate
OD	Optical density
OEC	Oxygen-evolving complex
OPCC	Oxidative pentose phosphate cycle
PAGE	Polyacrylamide gel electrophoresis
PAR	Photosynthetically active radiation
PC	Plastocyanin
PCD	Programmed cell death

PCK	PEP carboxykinase
PCMBS	p-chloromercuribenzene sulphonic acid
PCR	Photosynthetic carbon reduction cycle
PEP	Phosphoenolpyruvate
PEPcase	Phosphoenolpyruvate carboxylase (PEP carboxylase)
PFK	Phosphofructokinase
pmf	Proton motive force
PQ	Plastoquinone
PR-proteins	Pathogenesis-related proteins
PS	Photosystem
Q_{10}	Temperature coefficient
RCF	Relative centrifugal force
R_f	Relative front
RIA	Radio immunoassay
rpm	Revolutions per minute
RQ	Respiratory quotient
Rubisco	Ribulose-1,5-bisphosphate carboxylase/oxygenase
RUP proteins	REPRESSOR OF UV-B PHOTOMORPHOGENESIS proteins
SAGs	Senescence associated genes
SAM	S-adenosyl methionine
SAR	Systemic acquired resistance
SDS	Sodium dodecyl sulphate
SPAC	Soil–plant–air continuum (Soil–plant–atmosphere continuum)
TIBA	2,3,5-Triiodobenzoic acid
TLC	Thin-layer chromatography
TPP	Thiamine pyrophosphate
UVR8	Ultraviolet resistance 8
VAM	Vesicular-arbuscular mycorrhiza
VLFRs	Very low fluence responses
ZTL	Zeitlupe

Abbreviations for Units

%	per cent
A	Absorbance
Å	Angstrom
atm	Atmosphere
°C	Degree Celsius
cal	Calorie
cm	Centimeter
D	Dalton
[E]	Enzyme concentration
E	Extinction
E'^{o}	Redox potential
emf	Electromotive force
g	Gram
h	Hour
ln	Logarithm to the base e
K	Kelvin
kcal	Kilocalorie
kD	Kilodalton
kg	Kilogram
Kj	Kilojoules
Km	Michaelis–Menten constant
L	Litre
log	Logarithm to the base 10
m	Molal (concentration)
M	Molar (concentration)
min	Minute
ml	Millilitre
mm	Millimeter

N	Normal (concentration)
nm	Nanometer
ppm	Parts per million
R	Gas constant
s	Second
[S]	Substrate concentration
S	Svedberg unit
T	Absolute temperature in Kelvin
V	Volt

Unit I Water and Mineral Translocation in Plants

Chapter 1

Plant–Water Relations

1.1 Water – A Universal Solvent

Keywords

- Adhesion
- Aquaporins
- Chemical potential
- Cohesion
- Dialysis
- Electrochemical gradient
- Gated channel
- Hydration shell
- Hydraulic conductivity
- Hydrogen bond
- Ion channel
- Proton motive force
- Relative water content
- Tensile strength
- Transport proteins
- Water potential

Water is the most abundant resource for nearly all organisms. Covering over 70 per cent of the earth's surface and making up as much as 95 per cent of the matter of living organisms, it is virtually unique among all liquids. The water content of plants is in a continuous state of flux, based on the degree of metabolic activity, the water status of the surrounding air and soil, and a variety of other factors.

Water plays a crucial role in the physiological processes of plants; roles for which it is well-suited because of its special physical and chemical properties, some of which are:

- Water is an excellent solvent, making it a suitable medium for the absorption and translocation of mineral elements and other solutes necessary for normal plant growth and development.
- Most of the 'biochemical reactions', that are so characteristic of life, occur in water and water itself participates directly or indirectly in all the metabolic reactions.
- The 'thermal properties' of water, i.e., high specific heat, high latent heat of vaporization and high heat of fusion, ensure that water remains in the liquid form over the range of temperatures where most biological reactions take place.
- The combined properties of **'cohesion'** (capacity to bind strongly with itself), **'adhesion'** (capacity to bind strongly with other molecules containing oxygen such as glass and cell walls) and **'tensile strength'** (a measure of the maximum tension a material can

withstand before breaking) are especially significant for maintaining the continuity of water columns in plants.

- The 'transparency' of water facilitates penetration of sunlight into the aqueous medium of cells thus enabling photosynthesis and other physiological processes to function in the cells.

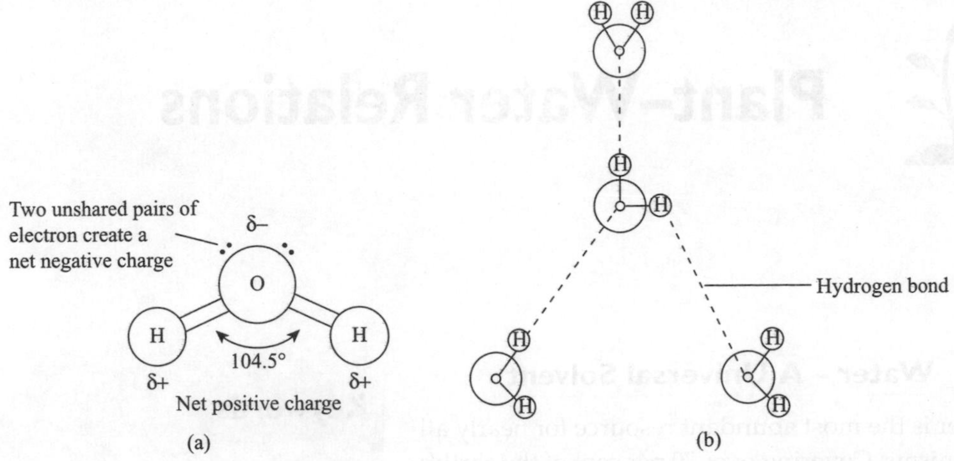

Fig. 1.1 (a) Structural configuration of a water molecule. Because of asymmetrical distribution of charge, the water molecule is polar. The bonding angle between the two hydrogen atoms is 104.5° rather than 180°. The oxygen atom shares one electron with each hydrogen atom. (b) Hydrogen bonds among the water molecules resulting from the electrostatic attraction between the partial positive charge on one molecule and the partial negative charge on the next.

These properties derive primarily from the **'polar'** structure of the water molecule (Figure 1.1a). The water (H_2O) molecule (molecular weight, 18) is composed of an oxygen atom covalently bonded to two hydrogen atoms that are attached asymmetrically to one side. The two O–H bonds form an angle of 104.5° rather than 180°. The oxygen atom being more electronegative than hydrogen, has a tendency to attract the electrons of the covalent bond thus resulting in a partial negative charge at the oxygen molecule and a partial positive charge at each hydrogen. Due to equal partial charges, the water molecule has no net charge.

The polarity and the tetrahedral shape of water molecules permit them to form **hydrogen bonds** that give water its unique physical properties (Figure 1.1b). Hydrogen bonding is a key feature that makes certain compounds soluble in water. Besides interactions among water molecules, hydrogen bonding is also accountable for the attraction between water and other molecules. However, when the groups responsible for hydrogen bonding are linked to a group other than water, they lose their solubility. This commonly occurs when hydrogen bonding exists within the same molecule, i.e., intra-molecular hydrogen bonding. For example: cellulose-a polymer of glucose found in plant cell walls. Hydrogen bonding is the basis for formation of **hydration shells** that develop around biological macromolecules such as proteins, lipids, nucleic acids and carbohydrates.

When water freezes below 0°C, it produces crystal 'lattice work', which we know as snow and ice. Ice is much less dense (a density of about 0.9179 g cm^{-3}) than liquid water (0.999 g cm^{-3}). This enables the ice to float on the lakes and ponds surfaces rather than sinking to the bottom where it might remain there throughout the year . This is extremely important for the survival of all types of aquatic life. If it were otherwise, it would have disastrous consequences, such as killing aquatic life. This unusual feature enables icebergs to float. The maximum density of water is 1.0 g cm^{-3} at 4 °C. This happens because molecules in the liquid state are tighter packed than in the crystalline state of ice.

The water expands when it freezes, resulting in the bursting of water pipes, and in plants, the inter- or intra-cellular risk of tissue damage increases manifold during extreme cold conditions.

The '**surface tension**' – the tendency of the surface to contract and resemble an elastic membrane–is high for water (Figure 1.2). And so, water droplets tend to form spheres as the water molecules at the surface are subjected to unequal forces and are pulled in towards the centre. Because of surface tension, the molecules tend to occupy the least possible space. In a deeper part of a mass of water, the molecules of water are attracted with equal force in different directions while at the air–water interface the water molecules, however, are attracted downward and sideways by other water molecules. Here, all of the hydrogen bonds in water face downward, causing the water molecules to cling together, resembling an elastic membrane. Some insects and spiders can glide on water surface due to the surface tension property of water, which is stronger than the force that one foot brings to bear.

The various physiological activities of the cell take place in an aqueous medium. They actually operate in dilute aqueous solutions, suspensions and colloidal phases. A study of their properties is, therefore, necessary for a better understanding of the different physiological processes.

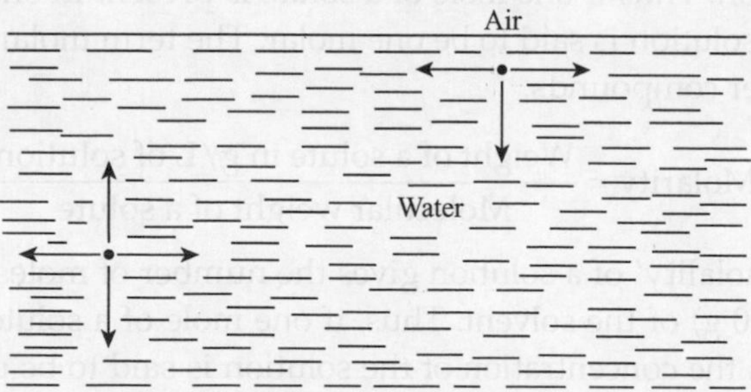

Fig. 1.2 Diagrammatic representation explaining surface tension and forces in a liquid. The particle is pulled downward because of unbalanced forces. A particle deeper in the liquid is attracted by forces from all directions.

1.1.1 Solution

A 'solution' is a homogenous mixture composed of two or more components. A component that is present in greater proportion is termed the 'solvent' while the other component that is present in smaller amounts is called the 'solute'.

A solution may exist in gaseous, liquid or solid state. It is convenient to distinguish solutions in terms of the physical states of the mixing components. The various types of solutions with examples are given in Table 1.1.

Most significant types of solution are: (i) solids-in-liquids; (ii) liquids-in-liquids; and (iii) gas-in-liquids.

Table 1.1 Various types of solutions.

Solute	Solvent	Examples
Gas	Gas	Air
Gas	Liquid	CO_2 in aerated water
Gas	Solid	H_2 in platinum or palladium
Liquid	Gas	Moisture in air during mist
Liquid	Liquid	Alcohol in water
Liquid	Solid	Jellies, mercury in silver
Solid	Gas	Camphor in air
Solid	Liquid	Sugar in water
Solid	Solid	Solid solutions, alloys, etc.

Terminology connected with the concentration of solutions

The concentrations of solutions are expressed in different units.

Molarity (M): The 'molarity' of a solution gives the number of moles of the solute present per litre of the solution. Thus if one mole of a solute is present in one litre of solution, the concentration of the solution is said to be one molar. The term molar is generally used for salts, sugars and other compounds.

$$\text{Molarity} = \frac{\text{Weight of a solute in g/L of solution}}{\text{Molecular weight of a solute}}$$

Molality (m): The 'molality' of a solution gives the number of moles of the solute present in one kilogram (1000 g) of the solvent. Thus, if one mole of a solute is dissolved in 1000 grams of the solvent, the concentration of the solution is said to be one molal. Molality is used in situations, where it is desirable to express the ratio of solute to solvent molecules, for example, in discussions of osmotic pressure.

$$\text{Molality} = \frac{\text{Weight of a solute in g/kg of solvent}}{\text{Molecular weight of a solute}}$$

Normality (N): In volumetric work, we express the concentrations in terms of normality of the solutions.

The normality of a solution gives the number of gram equivalents of the solute per litre of the solution. Thus, if one gram equivalent of a solute is dissolved per litre of the solution, the concentration of the solution is said to be one normal. A normal solution is generally used for acids and bases.

$$\text{Normality} = \frac{\text{Amount of a solute in g/L of solution}}{\text{Equivalent weight of solute}}$$

Per cent Solution (%): The concentration of substances that do not have a uniformly defined composition, such as proteins, nucleic acids, and such, is expressed in weight per unit volume rather than moles per unit volume. Thus, a 5 per cent solution would imply 5g of solute dissolved in solvent to give a final volume of 100ml of the solution. When either solute or solvent is taken by weight or volume, then per cent solution results in v/v (volume/volume), v/w (volume/weight), w/w (weight/weight) or w/v (weight/volume).

ppm Solution (mg/L or µg/ml solution): ppm solution needs to be prepared when a solute is present in a very small quantity. It represents one gram of a solute per million grams of the solution or one gram of a solute per million ml of the solution.

$$\text{ppm} = \frac{\text{mass of the component}}{\text{total mass of the solution}} \times 10^6$$

or

$$\text{ppm} = \frac{\text{g or ml of solute}}{\text{g or ml of solution}} \times 10^6$$

When 1 mg of a solute is dissolved in water and the volume is made up to 1000 ml, it results in 1 ppm (parts per million) solution.

In the true solution, the molecules of the solute do not settle down and cannot be visibly distinguished from the solvent even under the highest power of microscope. A true solution is transparent and has a higher boiling point, and lower freezing point as well as water potential, as compared to pure water. The number of solvent and solute molecules is constant in a sample of given solution. If to a given amount of solvent maintained at a constant temperature, a solute is added gradually in increasing amounts, a stage will be reached when some of the solute will remain undissolved, no matter how vigorously we stir. The solution is then said to be **saturated**. The excess solute is in dynamic equilibrium with the solute that has gone in solution.

1.1.2 Factors affecting solubility

1. ***Nature of Solute and Solvent***: In general, if a solute and solvent have similar chemical characters, the solubility is very high and if they have entirely different chemical character, the solubility is extremely low. For example, sugars are largely soluble in water but not in benzene.
2. ***Temperature***: The solubility of a solute in a given solvent varies greatly with temperature. It may increase or decrease with rise in temperature. Most of the salts (e.g., KNO_3, NH_4Br) show marked increase in solubility with rise in temperature, whereas others (such as, NaCl) show only a small increase. There are only a few substances (e.g., anhydrous sodium sulphate, Na_2SO_4, caesium sulphate) which show decrease in solubility with rise in temperature.

'Standard solution' contains a known weight of the substance dissolved in a known volume of the solvent.

1.1.3 Suspension

A 'suspension' is a two-phase system where solute particles are not dispersed as molecules or ions. The larger solute particles (>0.1 µm in diameter) remain suspended throughout the liquid. The system is unstable and the particles settle down with the passage of time.

1.1.4 Emulsion

Unstable emulsions are prepared by vigorously shaking two immiscible liquids together. Small droplets (dispersed phase) of one of the liquids will be dispersed throughout the other (dispersion medium). An emulsion can be made stable by the addition of an emulsifying agent. These substances generally function either by decreasing the surface tension of the liquids or by forming a protective layer around the droplets, thus making it impossible for them to combine with each other. Milk is an example of emulsion containing butter fat dispersed in water, with casein acting as an emulsifying agent.

1.1.5 Acids, Bases and Salts

Acids are ionizable compounds that give up protons. Strong acids like HCl, H_2SO_4 and HNO_3 tend to ionize completely when in solution.

$$HCl \rightarrow H^+ + Cl^-$$

Weak acids are the organic acids and tend to ionize only partially when in solution. Thus, acetic acid ionizes to form acetate ions and protons, but some acetic acid still remains unionized.

$$\underset{\text{Acetic acid}}{CH_3COOH} \rightleftharpoons \underset{\text{Acetate ion}}{CH_3COO^-} + \underset{\text{Hydrogen ion}}{H^+}$$

Bases are ionizable compounds that tend to take up protons. Strong bases such as sodium hydroxide (NaOH) dissociate into its ions, when in solution.

$$NaOH \rightarrow Na^+ + OH^-$$

In contrast, ammonia is a relatively weak base. Occasionally, substances that yield OH^- ions are considered bases. For example: acetate (CH_3COO^-) is the conjugate base to acetic acid (CH_3COOH).

Salts are compounds formed from the conjugate base of an acid and the conjugate acid of a base, e.g. on mixing a solution of NaOH with acetic acid, the salt of sodium acetate is formed.

1.1.6 Electrolytes and Non-electrolytes

Electrolytes are those substances that can conduct an electric current and undergo ionization when dissolved in water. Acids, bases and salts are electrolytes. Some common strong electrolytes with complete ionization in plant tissues are the inorganic ions K^+, Na^+, Ca^{2+}, Mg^{2+}, Mn^{2+}, Zn^{2+}, Cl^-, SO_4^{2-} and PO_4^{3-}. Common weak electrolytes with incomplete ionization are the organic cations and anions, amino acids, organic acids and the complex charged polymers, such as proteins and nucleic acids. The behaviour of electrolytes was first discussed by Arrhenius in 1880.

Non-electrolytes are substances that do not undergo ionization on dissolving in water. Examples are: gases (oxygen and hydrogen) and most of the organic compounds such as sugars and alcohols and others. However, they do carry groups having positive or negative potential, such as, amino NH_2, imino NH, amide $CONH_2$, carbonyl CO, carboxyl COOH, hydroxyl OH and so on. These groups combine with water molecules which protect them from coming together by forming water shells.

1.1.7 Polar and Non-polar Compounds

Jacobs (1924) emphasised the importance of recognizing two classes of compounds, polar and non-polar, from the stand point of permeability. Polar compounds are compounds that have slight charges (+ve as well as –ve) within the compound. For example, water (H_2O), hydrogen fluoride (HF), hydrogen chloride (HCl), and ammonia (NH_3). Slight charges are due to electro-negativity differences between 0.3 and 1.4. In polar molecules, two atoms do not share electrons equally in a covalent bond. In addition to water, acids, bases and salts, there are many polar organic compounds such as sugars and alcohols.

Non-polar compounds do not have slightly negative and positive charges within the compound (in other words, the electrical charges are evenly distributed across the molecule), such as hydrocarbons, organic molecules: benzene, methane, ethylene, carbon tetrachloride, and the noble gases: helium, neon, argon, krypton, xenon and so on. Electronegativity differences of non-polar compounds lie between 0 and 0.2. In non-polar molecules, electrons are shared equally by the atoms of the molecules and there is no net electrical charge. Some molecules such as those of the fatty acids are mostly non-polar (the long hydrocarbon chain) but have polarity at one end (–COOH group).

1. Polar substances tend to be soluble in water and other polar solvents while non-polar substances are soluble in fats and organic solvents.
2. Polar compounds, with the exception of water, penetrate cells with difficulty, whereas non-polar compounds penetrate cells rapidly.
3. Polar compounds ionize readily and are highly reactive (except sugars and alcohols), whereas non-polar compounds do not ionize and are electrically neutral.

1.1.8 pH and Buffers

pH

The term 'pH' has been derived from the French word 'Pouvoir Hydrogene' (power of hydrogen) and is used to quantitatively measure the acidity or alkalinity of aqueous or other liquid solutions. The concept of 'pH' was introduced by a Danish chemist, Søren Peder Lauritz Sørenson (1909), to express concentration of the hydrogen ion (H^+) in an aqueous solution. The 'pH' can be defined as the 'negative logarithm' of the hydrogen ion concentration.

$$pH = -\log [H^+]$$

or

$$\text{mathematically speaking, } pH = \log \frac{1}{[H^+]}$$

Water dissociates into equal numbers of H^+ and OH^- ions and is therefore neither an acid nor a base. At ordinary temperature (22 °C), the actual concentration of H^+ (and of OH^-) ions in pure water is very small, i.e., 10^{-7} mol/L each of H^+(protons) and OH^- (hydroxyl) or [0.0000001 M]. The product of the two (H^+) × (OH^-) is 10^{-14} and this remains constant for all aqueous solutions. By knowing the concentration of H^+, the concentration of OH^- can be calculated or vice-versa.

$$H_2O \rightleftharpoons H^+ + OH^-$$

$$[H^+][OH^-]$$

$$[10^{-7}][10^{-7}] = 10^{-14} \text{(to multiply } 10^{-7} \text{ by } 10^{-7}\text{), exponents are added to give } 10^{-14}.$$

The pH of pure water, (with H^+ concentration of 1.0×10^{-7} M), will be 7 as calculated below:

$$pH = \log \frac{1}{[H^+]} = \log \frac{1}{1 \times 10^{-7}}$$

$$= \log [1 \times 10^7]$$

$$= \log 1.0 + \log 10^7$$

$$= 0 + 7$$

$$pH = 7$$

Hence, pure water with equal numbers of protons and hydroxyl ions will have a pH of 7. However, the pH of water is lowered down at higher temperature, e.g., at 50 °C, the pH of water is 6.55 ± 0.01.

The pH scale covers a range of values from 0 to 14, in which the midpoint is 7.0 (Figure 1.3). Solutions having pH less than 7.0 are acidic (having more H^+ ions), while those having pH greater than 7.0 are basic or alkaline (having more OH^- ions than H^+ ions). Thus a solution of 6.0 pH will have 10 times more H+ ions than that of a 7.0 pH solution. Similarly a solution of 8.0 pH will have 10 times less H^+ than that of a 7.0 pH solution.

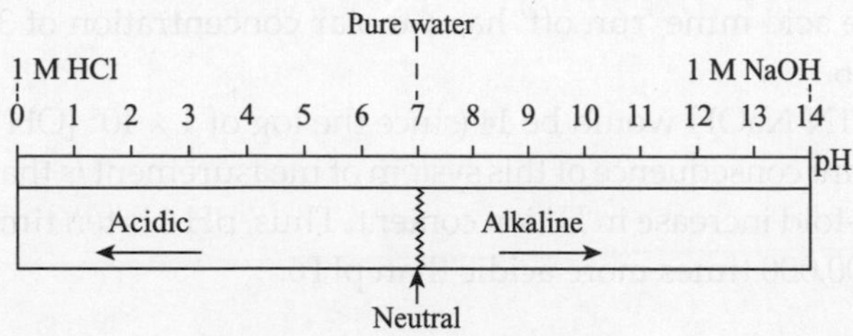

Fig. 1.3 The pH scale.

There is always an inverse relationship between the H^+ and OH^- ions concentration in a solution (Table 1.2). One needs to know only the value $[H^+]$ to determine $[OH^-]$ as well. Even a very strong acid of say pH 1 contains some OH^- ions (10^{-13}N) and a strong base of pH 14 has a few H^+ ions (10^{-14}N).

Table 1.2 Relationship between H^+ and OH^- ions concentration in a solution and the corresponding pH values.

H^+ mol/L	OH^- mol/L	Log $[H^+]$	$-$Log $[H^+]$ = pH
1×10^0	1×10^{-14}	0	0
1×10^{-1}	1×10^{-13}	−1	1
1×10^{-4}	1×10^{-10}	−4	4
1×10^{-7}	1×10^{-7}	−7	7
1×10^{-10}	1×10^{-4}	−10	10
1×10^{-13}	1×10^{-1}	−13	13
1×10^{-14}	1×10^0	−14	14

Thus, 1N HCl or any other strong acid that dissociates (100 per cent) shall have a pH of 0 because

$$pH = -\log [H^+] \text{ or } [H] \text{ would be equal to 1.}$$

$$= -\log [1]$$

$$= 0$$

Actually, the dissociation is not quite complete and 1.0 N HCl gives a pH of 0.10.
A 5N solution of hydrochloric acid, for example, would have a pH value of a little less than 0; correspondingly a 5N solution of sodium hydroxide would have a pH value a little greater than 14.

However, exceptionally acidic or basic solutions may have pH less than 0 or more than 14. For example, the acid mine 'run off' has a molar concentration of 3981 M, and the pH can be as low as –3.6.

While the pH of 1N NaOH would be 14 (since the log of $1 \times 10^{\circ}$ [OH^-] would be –14).

The most important consequence of this system of measurement is that each unit decrease in pH indicates a 10-fold increase in H^+ ion content. Thus, pH 6 is ten times more acidic than pH 7, and pH 1 is 1,00,000 times more acidic than pH 6.

Because the amount of water ionized to form H^+ and OH^- is very small, there will be essentially no reduction in the total amount of [H_2O] present.

A weak acid is one in which only a small fraction of the molecules are dissociated. An example is acetic acid, in which the 1N acid gives a pH of 2.37 because only 0.42 per cent of the molecules are dissociated.

$$[H^+] \text{ in 1N acetic acid} = \frac{0.42}{100} \times 1 = 0.0042 \text{ or } 10^{-2.37}$$

Thus pH = 2.37 because –2.37 is the log of 0.0042.

Many of the chemical and physical reactions occurring in plant cells are regulated by pH. Virtually all enzymatic reactions have rather narrow pH optima and many of the chemical reactions of cells either give up or take up protons. The pH drives many of the transport and energetic reactions in the cells. Thus, pH and its regulation in the cells are intimately tied to metabolism. The usual range of pH in plant fluids is from 5.2 to 6.2 (acidic). It is well known that the soluble pigments of the anthocyanin group change colour with pH. Thus, the floral colours of many plants are pH dependent.

Buffers

Buffers are the solutions that resist changes in pH by absorbing or reacting with either protons or hydroxyl ions. These are generally mixtures of either weak acids and their salts or weak bases and their salts. In biological systems, the vast array of amino acids and proteins operate as primary buffers.

A solution of acetic acid and sodium acetate will form an effective buffer. These compounds ionize in solution as follows:

$$CH_3COOH \rightleftharpoons CH_3COO^- + H^+$$

$$CH_3COONa \rightleftharpoons CH_3COO^- + Na^+$$

On addition of a small amount of NaOH, the OH^- ions released in solution are neutralized by the free H^+ ions in the buffer solution. This causes more of the acetic acid to ionize and thus restores the original H^+ ion concentration. As more NaOH is added, more acetic acid ionizes until all of the acetic acid in solution is ionized. At this point, any further addition of NaOH will cause an abrupt rise in pH (Figure 1.4).

When a small amount of HCl is added to the acetic acid–sodium acetate buffer, the released H^+ ions rapidly unite with free acetate ions to form undissociated acetic acid.

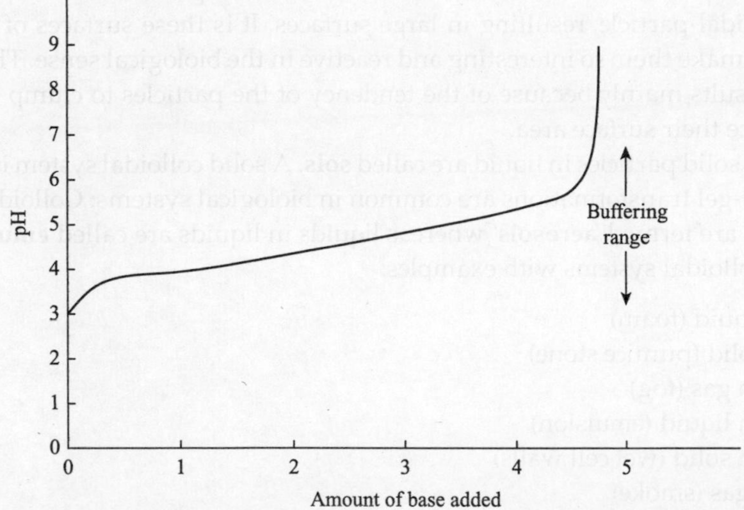

Fig. 1.4 Buffering range of a solution on addition of a base.

Therefore, there is no change in H^+ ion concentration. As more of HCl is added to the buffer, more H^+ ions will be available to combine with free acetate ions to form acetic acid, until all of the acetate ions have been converted. When this happens, any further addition of HCl will cause an abrupt drop in pH.

Buffer solutions are used to maintain specific pH of solutions in biochemical experiments. These are abundant in living plant cells and play vital roles in their existence. Enzymes generally function within a narrow pH range. A deviation of any magnitude impairs or completely inhibits their function. The living systems cannot tolerate any large increases or decreases in H^+ ion concentration. The biochemical reactions continuously produce acids and bases within cells. Moreover, we all consume foodstuff that are either acidic or basic; for example, a cola drink is a moderately strong acidic fluid with a pH of 3.0. In spite of such variations in the H^+ and OH^- ion concentrations, the pH of all living organisms is maintained relatively constant with the use of buffers solutions.

1.1.9 Colloids

If ordinary clay soil is agitated thoroughly with water and the resulting murky liquid uniformly brown in colour is allowed to stand, it clears rapidly. The larger particles settle out first followed by the smaller particles. Very minute soil particles still remain suspended in water and the stable, heterogeneous mixture that results is called a **colloid.** The suspended phase is called the **dispersed phase** (discontinuous phase) and the medium in which the dispersion takes place is called the **dispersion medium** (continuous phase). The dispersed particles range between 0.001 and 0.1 µm.

The term 'colloid', first suggested by T. Graham in 1861, derives its name from the Greek words 'Kolla' (= glue) and 'eidos' (= like). He used the term to describe glue-like preparations, such as solutions of certain proteins and liquid preparations of vegetable gums, such as gum arabic. But it is more than likely that there are multiple ions or molecules making up each colloidal particle, resulting in large surfaces. It is these surfaces of the colloidal particles that make them so interesting and reactive in the biological sense. The stability of the colloid results mainly because of the tendency of the particles to clump together and thereby reduce their surface area.

Colloids of solid particles in liquid are called **sols.** A solid colloidal system is called a **gel**. Reversible sol-gel transformations are common in biological systems: Colloids of solids or liquids in gas are termed '**aerosols**' whereas liquids in liquids are called **emulsions.**

Types of colloidal systems with examples:

1. Gas in liquid (foam)
2. Gas in solid (pumice stone)
3. Liquid in gas (fog)
4. Liquid in liquid (emulsion)
5. Liquid in solid (wet cell walls)
6. Solid in gas (smoke)
7. Solid in liquid (suspension)
8. Solid in solid (copper into zinc, if the bars of two metals are kept pressed with one another).

Depending on their properties, colloids can be lyophilic (show an affinity for solvent) or lyophobic (no affinity) or, if specifically aqueous, hydrophilic or hydrophobic. Hydrophilic colloids are those in which the particles have an affinity for water; hydrophobic colloids lack such affinity. It is believed that hydrophilic sols base their stability on water hydration of the particles and hydrophobic sols are stable because of water exclusion and repulsion of particles due to charged surfaces. Plant gums and mucilages, starch and proteins tend to form hydrophilic sols or colloids, whereas metallic sols are frequently hydrophobic.

Particles of colloidal dimension range in size from 1 to 200 nm. Colloidal particles are so tiny that they cannot be viewed even with the light microscope. The size of the colloidal particles is somewhere between the size of the particles that form true solutions and the size of particles that are found in unstable suspensions.

Properties

The most important properties of colloids are based on their large surface area and sol-gel transformations:

1. ***Surface area***: Several biochemical reactions occur on the surface of colloids. Thus surface area is an important part in determining the physiological behaviour of protoplasm.

2. ***Filterability***: Colloids are considered to be filterable even though they will pass through the usual laboratory filters. Ultra-filters (composed of biologically inert cellulose esters) of the proper dimensions allow for the separation of colloidal particles from those in true solution. This property is used in separating a colloid from a true solution by the process of dialysis.
3. ***Tyndall effect***: Colloids illuminated with a beam of light at right angles to the observer are seen to scatter or diffract light. The Tyndall effect (reported by M. Faraday in 1858), known to anyone who has observed dust in room air, is indicative of a colloid. The greater the index of diffraction, the greater will be the visual effect. Scattering of light entering a leaf is important for photochemical reactions in which light is absorbed by pigments.
4. ***Viscosity***: Colloids decrease in viscosity with increase in temperature. Hydrophilic sols increase exponentially in viscosity when the colloidal particles are increased. For example plant gums, mucilages and protein sols become highly viscous when concentrated.
5. ***Electrical properties***: The micelles of colloids are almost always charged because of the chemical groups associated with the particles. For example, the micelles of proteins have a net negative charge and therefore cations will always be attracted to the negative charge. The ionic double layer is one force giving stability to the colloid. Besides having an ionic double layer, hydrophilic colloids will also have a hydration shell associated with them resulting in even more stability.
6. ***Coagulation***: Coagulation can be brought about by removing the water layer and neutralizing the charge. Proteins will tend to clump at their isoelectric points when their net charge is zero or upon treatment with solvents that remove their water shells. These properties are frequently manipulated in protein purification in which unlike proteins are separated.
7. ***Brownian movement***: Colloidal particles show Brownian movements, first described by Robert Brown. The random motion of the particles is caused by their uneven bombardment by molecules of the dispersion medium. The smaller the suspended particles, the greater will be the amplitude of the Brownian movement. This is more active at higher temperatures because of greater kinetic activity and lower viscosity of the medium.

1.1.9.1 Protoplasm as a colloidal system

Physically, the protoplasm is a living, heterogenous polyphasic colloidal system, consisting of a thick or viscous (semi-fluid, jelly-like), translucent and odourless fluid matrix or ground substance surrounded by the cell wall. The protoplasm is a complex colloidal system of many phases where water is the dispersion medium (about 80 per cent) with dissolved inorganic ions and small molecules. The dispersed phase comprises mainly the large molecules of proteins (14 per cent), lipids (3 per cent), carbohydrates (1 per cent) and nitrogen-based derivatives of DNA and RNA and such, representing many phases. Much of the colloidal nature is due to the protein phase which forms the major part of the living protoplasm, about 79 per cent of the dry weight of protoplasm. Proteins are very large complex molecules which sometimes reach colloidal dimension. They are dispersed throughout the ground substance of the protoplasm where they are involved in such cell activities as respiration, digestion and secretion. Undoubtedly, the immense surface area provided by protein enzymes dispersed in the protoplasm is of extreme importance to the many enzyme–substrate reactions upon which all life depends.

Lipids constitute only a small fraction of the protoplasm as compared to the proteins. They are important structural components of the cell. Being insoluble in water (a polar solvent), they occur in the protoplasm as minute globules (or droplets) or form an emulsion. However, they are soluble in the non-polar organic solvents, including chloroform, ether, acetone and benzene.

The colloidal nature of protoplasm was suggested by Wilson Fisher (1894) and William Bate Hardy (1899).

Although protoplasm exhibits the properties of a 'sol', it sometimes becomes gel-like rigid. The change from sol to gel consistency takes place during cell division in the formation of nuclear fibres and during the formation of plasma membrane. The plasma membrane and cell walls are gel-like and so are the chromosomes and centrosomes.

1.2 Permeability

Permeability is a fundamental phenomenon for the functioning of living cells and for maintaining satisfactory intracellular physiological activities. Every cell and organelle needs some sort of physical barrier to keep its contents in and external materials out, as well as some means of controlling exchange between its internal environment and the extracellular environment. Ideally, such a barrier should be impermeable to most of the molecules and ions found in cells and their surroundings. At the same time, it must be readily permeable to water because water is the basic solvent system of the cell and must be able to flow in and out of the cell as required. As expected, the membranes that surround the cells and organelles satisfy these criteria admirably.

The three types of membranes found in plants are **permeable, semipermeable** and **impermeable**. The cellulosic cell wall is a permeable membrane, which allows both water and dissolved substances to pass through. The cutinized and suberized cell walls are impermeable membranes which allow neither water nor solutes to pass through. The cell membrane and vacuolar membrane (tonoplast) behave as semipermeable membranes allowing water to pass through but impermeable to most of the solutes. Although the term 'semipermeable' is often used, it is not descriptive of biological membranes. Semipermeable implies that the membrane is permeable to many substances but without specificity and the term fails to describe the true dynamic nature of such membranes. Instead, the term 'differentially permeable' is widely used to describe a membrane that regulates the passage of diverse materials in and out of the cell, organelles and vacuoles at different rates, depending on the relative solubility of those materials for the lipid and protein constituents.

1.2.1 Cell membrane as permeability barrier

Permeability may be passive if it obeys only physical laws, as in the case of diffusion. If a concentrated solution of soluble substance, e.g., sugar is placed in water, there will be a net movement of the solute along a concentration gradient. However, if the plasma membrane is interposed, the diffusion process is greatly modified and the membrane acts as a barrier to the movement of water-soluble molecules.

At the end of nineteenth century, Overton demonstrated that substances that are soluble in lipids pass more readily into the cell. Further Collander and Bärlund in their classic experiments with the cells of *Chara* demonstrated that the rate at which substances penetrate depends upon their solubility in lipids and the size or dimension of the molecule. The molecules that are more soluble in lipids pass across the lipid layer more quickly and further, with equal solubility in lipids, the smaller molecules traverse at a faster rate.

The permeability (P) of molecules across the membrane is expressed as follows:

$$P = \frac{KD}{t}$$

where K, is the partition coefficient; D, the diffusion coefficient (determined by the molecular weight); and t, is the thickness of the membrane.

Various theories or hypotheses have been proposed over a period of time to describe the permeability of the cell membrane: Retention pressure theory (Traube, 1867); Solution theory (Overton, 1895); Fluid-mosaic theory (Nathansohn, 1904); Sieve or ultrafiltration theory (Ruhland, 1912; Hoffman, 1925; Kuster, 1911, 1918); Colloidal theory (Sparth, 1916); Electro-capillary theory (Michaelis, Scarth and Lloyd).

The permeability attribute depends upon the chemical composition of the membrane and the molecular arrangement of its constituents. Our current understanding of membrane structure is the culmination of almost a century of studies, beginning with the recognition that lipids are an important membrane component and moving progressively from a lipid monolayer to a phospholipid bilayer (Figure 1.5). Once proteins were recognized as important components, Davson and Danielli (1935) proposed their 'Sandwich model', with the lipid bilayer surrounded on both sides by layers of proteins. Their model was acknowledged as the basis of the 'unit membrane' put forwarded by Robertson (1950) but was later discredited because it could not account for key features of membrane proteins. In its place, the **fluid-mosaic model** (S. J. Singer and G. L. Nicolson, 1972) has emerged as the generally accepted description of membrane structure (Figure 1.6). The membrane is envisioned as a mosaic of proteins discontinuously embedded in a bimolecular lipid layer. Two principal types of lipids – phospholipids[1], and sterols,[2] especially stigmasterol - exist in the plasma membrane of plant cells. The phospholipid bilayer structure gives the membrane three properties that are particularly significant. These are: being highly fluid, very stable, and impermeable to most molecules of biological interest.

---

[1] Phospholipids and transmembrane proteins with hydrophilic as well as hydrophobic groups are referred to as 'amphipathic molecules' (the Greek prefix *amphi* means 'of both kinds'). The term 'amphipathy' was coined by Hartley in 1936.

[2] Cholesterol is the major sterol in animal tissues.

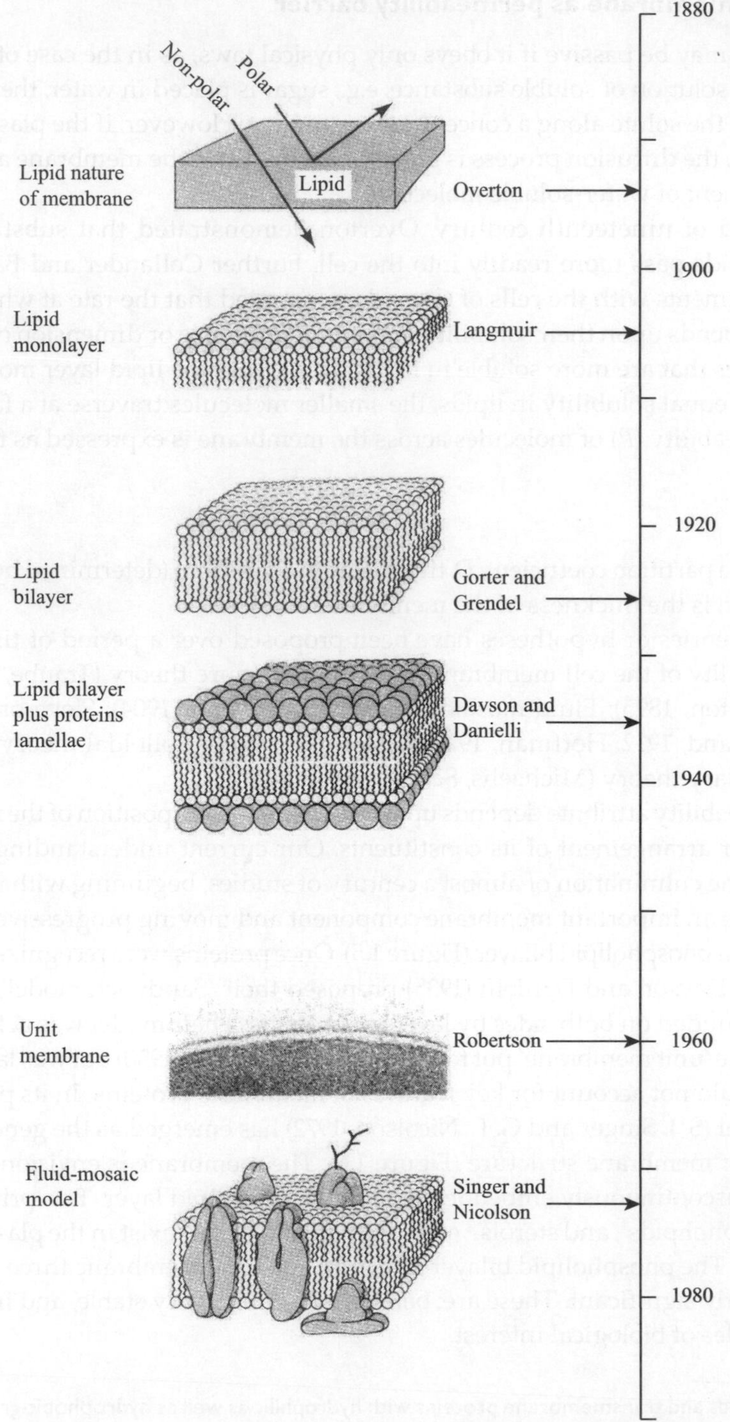

Fig. 1.5 Timeline for the development of our understanding of cell membrane structure.

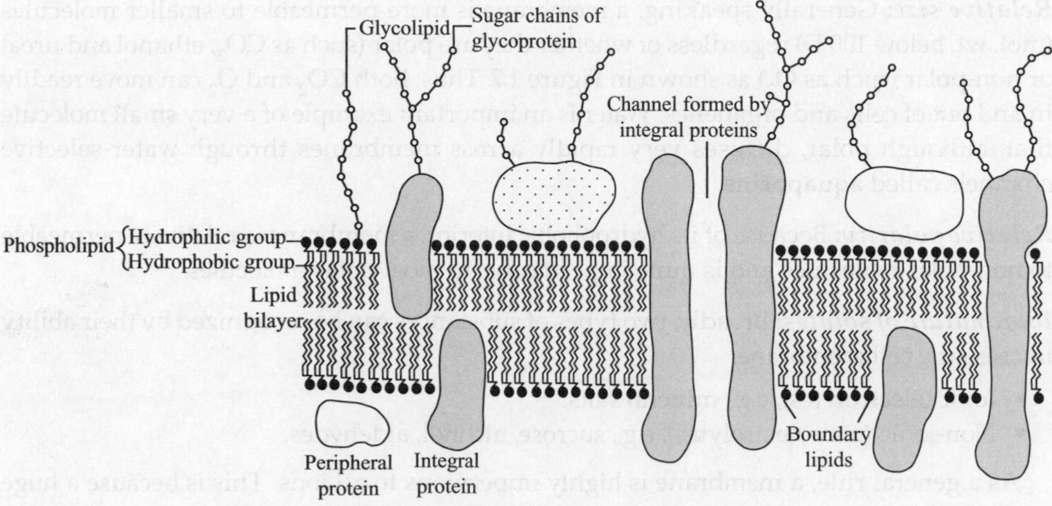

Fig. 1.6 The fluid-mosaic model of cell membranes.

This model recognizes two broad categories of membrane proteins that differ in their affinity or attraction for the inside of hydrophobic region of the membrane. **Integral membrane proteins** (intrinsic proteins) have one or more hydrophobic regions with an affinity for the hydrophobic region of the bimolecular lipid layer. These proteins are therefore intimately associated with the membrane and cannot be readily removed. In contrast, **peripheral membrane proteins** (extrinsic proteins) lack distinct hydrophobic segments and therefore do not enter into the lipid bilayer. Instead, they associate with the membrane surfaces through weak electrostatic forces, binding either to the polar heads of membrane lipids or to the hydrophilic portions of integral proteins that extend out of the membrane. Peripheral proteins are much more readily removed from the membrane than integral proteins and can usually be rendered soluble by aqueous salt solutions, without resorting to detergents. Proteins of the plasma membrane play a variety of roles in plant cells. They may serve as enzymes, or as anchors for structural elements of the cytoskeleton, or as, transport proteins for movement of specific ions and hydrophilic solutes, receptors for specific chemical signals that impinge on the cell from its environment and need to be transmitted inward to the cell.

The peripheral proteins and the parts of the integral proteins that occur on the outer membrane surface are attached to short-chain carbohydrates (oligosaccharides), thereby forming glycoproteins. Some of the phospholipids (specifically the hydrophilic region) at the outer membrane surface are linked to carbohydrates to form glycolipids.

1.2.2 Factors affecting permeability

Permeability of the membrane primarily depends upon the nature of the solute as well as of membrane. Three primary properties of the solutes must be considered: relative size, relative polarity and the ionic nature of solutes.

Relative size: Generally speaking, a membrane is more permeable to smaller molecules (mol. wt. below 100 D) regardless of whether they are polar (such as CO_2, ethanol and urea) or non-polar (such as O_2) as shown in Figure 1.7. Thus, both CO_2 and O_2 can move readily in and out of cells and organelles. Water is an important example of a very small molecule that, although polar, diffuses very rapidly across membranes through water-selective channels called **aquaporins.**

Relative polarity: Because of its hydrophobic interior, a membrane is relatively permeable to non-polar molecules and is quite impermeable to most polar molecules.

Ionic nature of solutes: Broadly, two types of substances can be recognized by their ability to cross the cell membrane:

- Ionic (electrolytes), e.g., mineral salts.
- Non-ionic (non-electrolytes), e.g., sucrose, alcohol, aldehydes.

As a general rule, a membrane is highly impervious to all ions. This is because a huge amount of energy (~40 kcal/mol) is needed to transport ions from an aqueous environment into a non-polar environment.

The impermeability of membranes to ions is very important for cell activity, because nerve cells, mitochondria and chloroplasts must maintain large ion gradients in order to function. Even the smallest ions (Na^+, K^+) are excluded from the hydrophobic interior of the membrane due to the charge on an ion and the sphere of hydration that surrounds the ion. Indeed, cells have ways for transferring ions such as Na^+ and K^+ as well as a wide variety of polar molecules across membranes that are otherwise not permeable to these substances. These membranes are equipped with **transport proteins** to serve this function. A transport protein is a specialized transmembrane protein that serves *either* as a hydrophilic **ion channel** (K^+, Ca^{2+} channels) through an otherwise hydrophobic membrane (Figure 1.7) *or* as a carrier that binds a specific solute molecule towards either side of the membrane accompanied by change in its conformation and thus transporting the solute across the membrane. Whether a channel or a carrier, each transport protein is specific for a particular molecule or ion. Diffusion through a channel depends on the size of the hydrated ion and the charge carried by it. Channels may exist in two different configurations – open and closed, i.e., **gated channels.** As a result, biological membranes can be best described as 'selectively permeable'.

1.2.3 Cell wall as permeability barrier

The presence of a wall around a plant cell poses a significant permeability barrier, though only for large molecules. In a typical cell wall, the diameter of the pores is about 5 nm, big enough to allow the passage of globular molecules with molecular weight up to 20,000 D. Not surprisingly, the hormones that serve as intercellular signals in plants are without exception small molecules, with molecular weight well below 1000 D.

Plasmodesmata are cytoplasmic channels (40–50 nm in diameter) through relatively large pores in the cell walls, connecting protoplasts of adjacent plant cells and allowing cytoplasmic exchange between the cells (Figure 1.8).

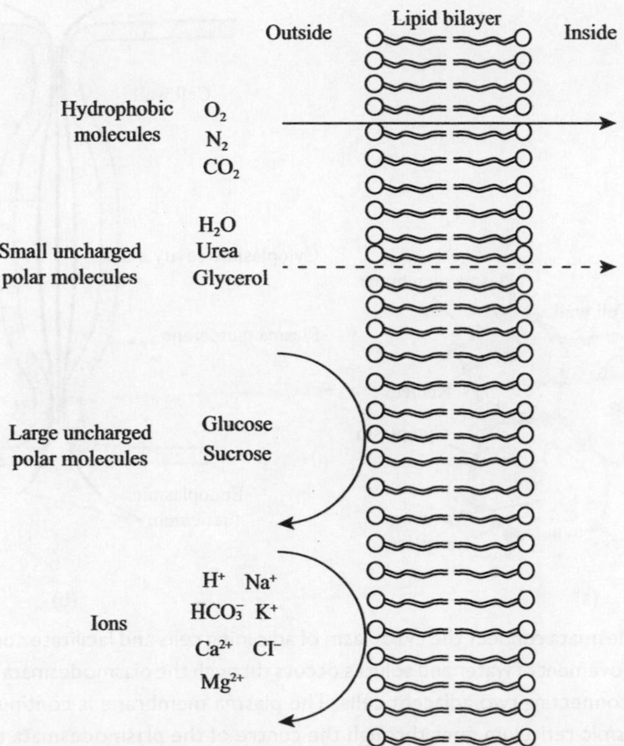

Fig. 1.7 **The relative permeability of the lipid bilayer to different types of molecules (hydrophobic molecules, small and large uncharged polar molecules, various types of ions, and such).**

Plasmodesmata can be primary or secondary; depending upon whether they are created during cell division or after the cell division is completed. The endoplasmic reticulum network of adjacent cells is connected, forming the desmotubule that runs through the centre of the channel. Proteins assemble on the external surface of the desmotubule and the internal surface of the plasma membrane; both the surfaces are linked by spokes, the filamentous proteins, which divide the cytoplasmic sleeve into microchannels. Surrounding the necks of the channel at either end are valve-like wall collars, composed of the polysaccharide callose, which serves to restrict the size of the pore.

Calcium appears to regulate movement of molecules through plasmodesmata. Intercellular transport of water and solutes can occur through plasmodesmata, which provide high-resistance connections between almost all living cells in plants. Only exceptions are stomatal guard cells and embryonic cells which lack plasmodesmata. The frequency of plasmodesmata varies widely from about 0.1 to 10 per square micrometer. Cells that are interconnected by plasmodesmata make up the symplasm and all space outside the plasma membrane is called the 'apoplast'. Not only small molecules but also macromolecules including virus and endogenous mRNA and proteins have been reported to move from one cell to another within the symplasm.

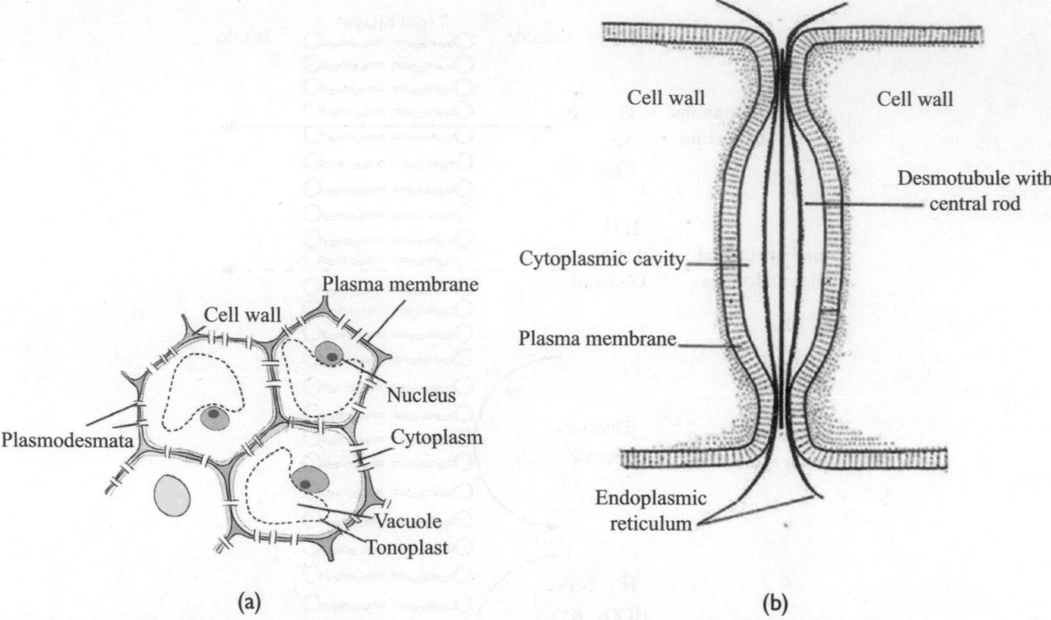

Fig. 1.8 (a) Plasmodesmata connect the cytoplasm of adjoining cells and facilitate communication between the cells. Symplastic movement of water and solutes occurs through the plasmodesmata. (b) Detailed structure of a plasmodesmata connecting two adjacent cells. The plasma membrane is continuous between the two cells and the endoplasmic reticulum goes through the centre of the plasmodesmata that forms the tubule.

The virus genome code for the '**movement proteins**' that facilitate movement of the virus particles by interacting with plasmodesmata, and this operates through one of the two mechanisms:

i. Viruses, such as cowpea mosaic virus and tomato spotted wilt virus, encode movement proteins that form a transport tubule within the plasmodesmatal channel that enhances the passage of mature virus particles through plasmodesmata.
ii. Movement proteins of tobacco mosaic can move between cells in leaves that are susceptible to the virus, where it recruits other proteins in the cell, thus increasing the size of the plasmodesmatal pore. So virus-sized particles can readily move through the plasmodesmata to a neighboring cell.

However, there is a restriction limit on the size of molecules that can be transported *via* the symplast (called **size exclusion limit**), which varies with cell type, environment, and developmental stage. The movement through plasmodesmata can be regulated or gated by altering the dimensions of wall collars, the cytoplasmic sleeve, and the lumen inside the desmotubule. In addition, adjacent plasmodesmata can form interconnections (branched plasmodesmata) that alter the size exclusion limit.

1.2.4 Plant vacuoles—multifunctional compartments

Vacuoles, the fluid-filled organelles or compartments surrounded by a single membrane, the **tonoplast,** occupy 30–80 per cent of the cell volume and enable plant cells to grow rapidly with the least amounts of energy and proteins. The number and size of the vacuoles in different plant cells vary greatly.

Plant vacuoles perform a multitude of functions such as:

i. **Plant cell expansion:** Cell expansion is controlled by the osmotic entry of water into the plant vacuoles followed by modified cell wall extensibility. Solutes (such as K^+, Na^+, Ca^{2+}, Cl^-, SO_4^{2-}, PO_4^{3-}, NO_3^-, amino acids, organic acids, and sugars) must be actively transported into the vacuole to maintain its osmotic concentration and also the turgor pressure of continuously growing cells. An electrochemical gradient across the vacuolar membrane, maintained by electrogenic H^+-ATPase pumps, provides the driving force for solute intake. Water movement across the tonoplast is mediated by tonoplast intrinsic proteins (TIPs)

ii. **Storage**: Plants store large amounts of proteins in their vacuoles, especially in seeds, which can be used in various metabolic pathways whenever required. Storage compounds in the vacuoles are also responsible for flavours of vegetables and fruits.

iii. **pH and ionic homeostasis**: Large vacuoles serve as reservoirs of protons and metabolically important ions, such as Ca^{2+}. Typically, the pH of plant vacuoles lies between 5.0 and 5.5 but vacuoles of lemon fruit and unactivated protein storage cells have pH of 2.5 and 7.0 respectively. Therefore, cells can regulate cytosolic pH, the enzyme activity, events such as assembly of cytoskeleton components and membrane fusion by regulating the efflux of protons and other ions into the cytosol.

iv. **Pigmentation**: Vacuoles of plant cells especially fruits and flower petals contain anthocyanin pigments which are used to attract pollinators and seed dispersers. In leaves, these pigments help in preventing photo-oxidative damage caused to the photosynthetic apparatus, by filtering out UV and visible light.

v. **Sequestration of toxic compounds**: Plants sequester heavy metals and metabolites like oxalate in vacuoles. Foreign compounds **(xenobiotics)** are transported from cytoplasm into vacuoles by an **ABC family of transporters** or specific cells develop vacuoles containing an organic matrix within which oxalate reacts with Ca^{2+} to form calcium oxalate crystals. Leaves are shed regularly because of accumulation of toxic compounds in leaf vacuoles.

vi. **Defence against microbial pathogens and herbivores**: Plant cells accumulate several toxic compounds in their vacuoles, which reduce feeding by herbivores and also destroy microbial pathogens. These compounds include:

- Phenolic compounds, alkaloids, cyanogenic glycosides, protease inhibitors
- Cell wall degrading enzymes, i.e., **chitinase and glucanase**, and defensive molecules such as **saponins** to destroy pathogenic fungi and bacteria
- Latexes and wound-clogging emulsions of hydrophobic polymers possessing insecticidal and fungicidal properties to serve as anti-herbivory agents.

vii. **Digestion**: Vacuoles have been shown to contain acid **hydrolases** such as proteases, nucleases, glycosidases and lipases, which degrade and recycle the cellular components. Such recycling is required for the retrieval of valuable nutrients during programmed cell death associated with development and senescence as well as for the normal turnover of cellular structures.

1.2.5 Transport channels

Transport channels in plants can be classified into two categories: **Ion channels and Aquaporins**

Ion channels

An ion channel forms a selective, water-filled pore in the membrane which permits the transport of a large number of ions very rapidly because of little interaction between the ion and the channel. Ion channels are ubiquitously distributed in the membranes of plant cells (Figure 1.9). Movement through ion channels is passive and is dictated by the electrochemical potential for the particular ion. The membrane potential has a strong influence on the diffusion of ions across the channels. Because the membrane potential of the plasma membrane is usually negative, cations tend to diffuse into the cytosol and anions tend to diffuse out. Ion channels show strong selectivity for either cations or anions based on size exclusion. Most plasma membrane anion channels allow the passage of anions such as chloride, nitrate and organic acids whereas vacuolar anion channels are specific for malate. Similarly cation channels are specific for divalent and monovalent cations. Ion channels play important regulatory roles e.g. they regulate osmotic concentration by allowing the flux of K^+ into and out of cells, and set the concentration of ion important in cellular signalling, i.e., cytosolic Ca^{2+}.

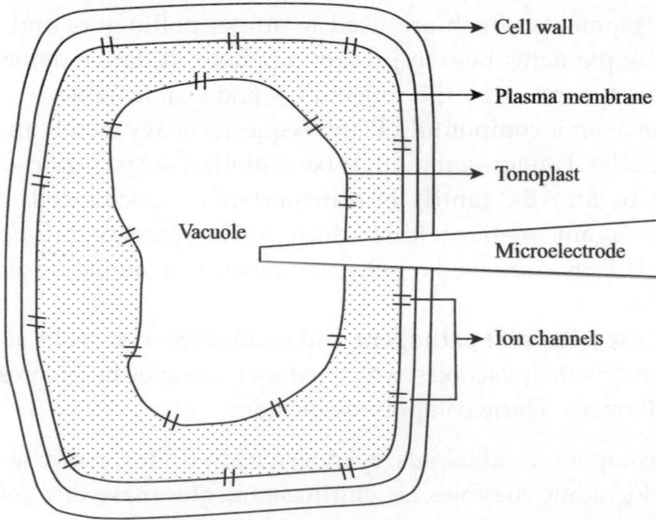

Fig. 1.9 Ion channels present in the plasma membrane and use of microelectrode to measure electrical properties of the cell.

The movement of ions through channels results in the flow of current, which can be used as a measure of the activity of a channel. Equation for Ohm's law can be used to model the movement of ions across the membrane.

$I = V/R$, where I = current, V = membrane potential, R = resistance of the membrane

The resistance of a membrane to a particular ion depends on its selectivity and the number of channels. No current will pass through an impermeable membrane, whereas membranes that are freely permeable will behave as a resistance-free circuit.

Channel opening and closing, termed as **gating**, is regulated by membrane potential. Voltage-gated K^+ channels are important in maintaining the cell's membrane potential. The plasma membrane has channels that allow either inward or outward movement of K^+, referred to as 'inwardly rectifying channels' or 'outwardly rectifying channels'. These channels operate like valves and are said to rectify. By opening and closing in response to voltage, these channels can keep membrane potential constant. Some ion channels can also be gated by ligands such as hormones, Ca^{2+}, G-proteins and pH and others by changes in cell's turgor pressure. Complex regulatory circuits can be built up by the interaction of these various regulators of ion channel activity; for example, the closing of stomatal guard cells in the presence of abscisic acid (ABA). The tonoplast has at least four voltage-dependent Ca^{2+} ion channels, and these may be regulated by binding of ligands or covalent modification. The presence of multiple Ca^{2+} channels gated by different signals allows for dynamic changes in cytosolic Ca^{2+} in response to a variety of stimuli.

Patch-clamp technique allows the study of single ion channels in selected cellular membranes (Figure 1.10). The microelectrode tip (1 μm diameter) is held in contact with the membrane outside an isolated protoplast and with little suction, a tight seal between the electrode and the membrane is formed. The small section of membrane in contact with the electrode is known as the '**patch**'. Measurements can be done either with the whole cell attached or the electrode can be pulled away from the cell and then the patch sticking to the electrode tip can be bathed in solutions of known composition. The external surface of the membrane is in contact with the microelectrode solution-appropriately termed as 'inside-out' configuration. With a suitable electrical circuit, the potential difference can be held across the patch at some predetermined value. Simultaneously, the circuit will monitor any current flowing through the membrane patch.

The patch-clamp technique exemplifies the major advancements made in the field of electrophysiology and provides insights into transport mechanisms across cellular membranes in plants.

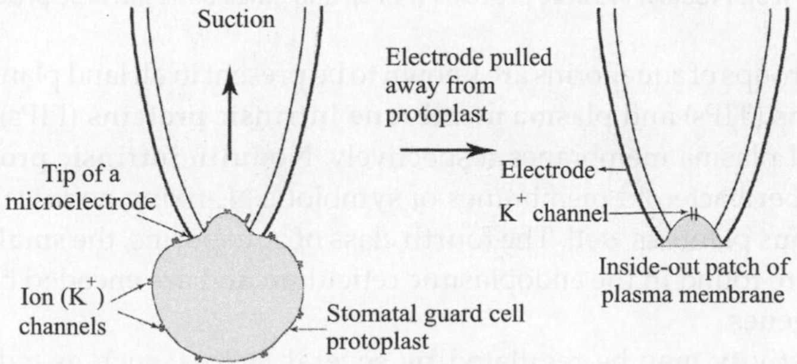

Fig. 1.10 Patch-clamp technique that is used to measure the flow of current through a single ion channel. A small section of membrane containing a individual ion channel can be isolated at the tip of the microelectrode. Sensitive amplifiers are used to measure the current carried by ions flowing through the channel.

Aquaporins

Aquaporins are specialized membrane proteins controlling the selective movement of water. Plant aquaporins were first isolated from tonoplast but they are also localized to the plasma membrane as well as to some other endomembranes, excluding the chloroplast and mitochondria (Figure 1.11). An aquaporin consists of four subunits (tetramers), each with a mass of ~30 kD. Each of the four subunits has six transmembrane domains. The interior and the exterior of each aquaporin channel consists of hydrophilic and hydrophobic amino acids that interact with water molecules and hydrophobic fatty acids of the lipid bilayer respectively (Figure 1.12).

Aquaporins are divided into four subgroups:

1. Tonoplast intrinsic proteins(TIPs)
2. Plasma membrane intrinsic proteins(PIPs)
3. Nodulin intrinsic proteins(NIPs)
4. Small basic intrinsic proteins(SIPs)

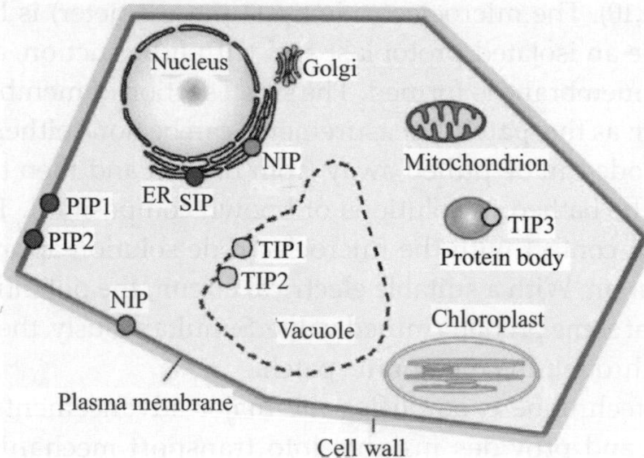

Fig. 1.11 Various types of aquaporins in a plant cell. Plasma membrane intrinsic proteins (PIPs), Tonoplast intrinsic proteins (TIPs), Nodulin intrinsic proteins (NIPs), and Small basic intrinsic proteins(SIPs).

The four subgroups of aquaporins are known to be present in all land plants. The **tonoplast intrinsic proteins** (TIPs) and **plasma membrane intrinsic proteins** (PIPs) are localized to the vacuolar and plasma membranes respectively. **Nodulin intrinsic proteins** (NIPs) are confined to the peribacteroid membranes of symbiotic N_2-fixing nodules and are present in non-leguminous plants as well. The fourth class of aquaporins, the **small basic intrinsic proteins** (SIPs) are found in the endoplasmic reticulum and are encoded by a small family of two or three genes.

Aquaporin activity may be regulated by several factors, such as cytosolic pH, Ca^{2+} concentration, nutrient and water stress, hormones, regulatory molecules, and so on. Thus, they can have a profound impact on the hydraulic conductivity of a membrane. Phosphorylation of aquaporins has also been shown to regulate their opening. Ca^{2+}-

dependent protein kinase may catalyze this phosphorylation, thereby suggesting a relation between Ca^{2+} signalling and the regulation of water flow between cells. They may also facilitate the movement of CO_2, NH_3, H_2O_2, boron and silicon.

Aquaporins are essential for osmoregulation in plant cells (intra- as well as inter-cellularly) and provide a least-resistance pathway for the movement of water thus enhancing its flow across the biological membranes.

Aquaporins are part of relatively large gene families in plants; for example, *Arabidopsis*, corn and rice have 35, 36, and 33 aquaporin genes respectively. There is high sequence homology (>97%) among members of each subgroup. The C- as well as N- terminal sequences of all aquaporins localize to the cytosol as do connecting loops II and IV. The I, III and V connecting loops of PIPs face the apoplast, whereas those of TIPs and SIPs face the lumen of the vacuole and endoplasmic reticulum, respectively. Four subunits (tetramer) make a functional aquaporin complex and an individual subunit of the tetramer forms a water channel (Figure 1.12). Post-translational modification of aquaporins occurs by phosphorylation and methylation of amino acids at the N- as well as C- terminals. Phosphorylation at multiple amino acids at the C terminus, and methylation of amino acids at the N terminus has been shown in PIP-type aquaporins.

1.2.5.1 Transport of molecules and ions across membranes

Three categories of transport of small organic molecules and inorganic ions have been identified:

Cellular transport involves the exchange of substances between the cell and its immediate environment. For example, the uptake of nutrients and other raw materials across the cell membrane and the removal of waste and secretory products.

Intracellular transport involves the exchange of substances through membranes of organelles within the cell. For example, traffic of molecules and ions in and out of organelles such as the nucleus, mitochondria, chloroplast, lysosome, peroxisome, Golgi complex and endoplasmic reticulum.

Transcellular transport involves the exchange of substances across the cell membrane, thereby accomplishing net transport across the cell, e.g., the cells of a plant root that are responsible for the absorption of water and mineral salts.

Transport of solutes across a membrane can be either passive or active, depending on the energy requirement of the process. Active transport is catalyzed by specialized membrane proteins, often referred to as '**pumps**'. Due to the energy input, active transport results in the generation of a 'concentration gradient' across the membrane. Active transport of H^+, K^+, Na^+ and Ca^{2+} across the membranes of the cells is quite important. Since ions have an electrical charge, the transport process and resulting concentration gradient can generate membrane potential across the membrane. The combined concentration gradient as well

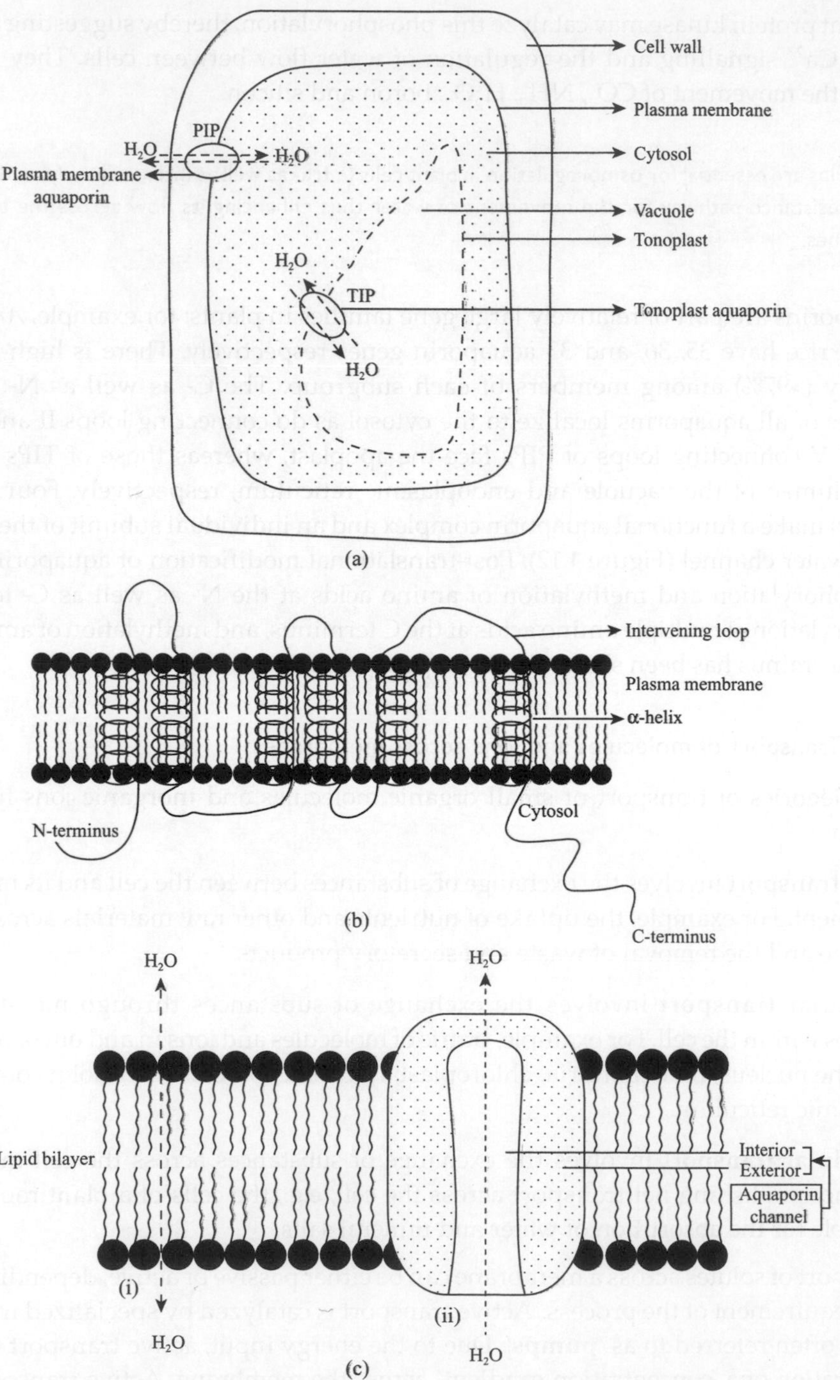

Fig. 1.12 (a) Aquaporins in the plasma membrane and vacuolar membrane in a plant cell. (b) Structure of a single subunit of a plant aquaporin. (c) Movement of water across the plasma membrane through (i) the lipid bilayer, and (ii) the aquaporin channel.

as the membrane potential, are termed as an 'electrochemical gradient'. The stored energy of such gradients is used to drive a variety of processes in the cells. In mitochondria and chloroplasts, for example, the stored energy of an electrochemical gradient of protons is used to drive the synthesis of ATP by their F-type ATPases during cellular respiration and photosynthesis, respectively. H^+ -ATPase complexes (ATPase-proton pumps), localized to the plasma membrane, vacuolar membrane, and other organelle membranes, behave as a proton-translocating carrier protein. The free energy of ATP hydrolysis is utilized to establish a proton gradient across the membrane. The proton gradient (ΔpH), along with transmembrane electrical potential (ΔE) contributes to a **proton motive force** (pmf), which is the primary source of energy for several plant activities such as active transport of solutes, regulation of cytosolic pH, stomatal opening and closing, sucrose transport during phloem loading and hormone-regulated cell elongation.

$$\text{pmf} = \Delta E - 59\ \Delta\text{pH}$$

1.3 Diffusion, Imbibition, Osmosis and the Water Potential Concept

Plant physiology, as we know, deals with different life processes operating within the cell, and between the cell and its environment. The former includes processes like photosynthesis, respiration, fat metabolism, nitrogen metabolism, and such, whereas the latter includes diffusion of gases, absorption and translocation of water and minerals, transpiration, and so on. The mineral elements necessary for plant growth and development, the compounds essential for energy transfer and storage, and the components of structural compounds all require water as a translocation and reaction medium. These materials are dissolved in water, and in this form, distributed throughout the plant. The processes of **diffusion, osmosis** and **imbibition** are intimately associated with the essential function of the translocation of water and solutes from site of origin to site of activity.

1.3.1 Imbibition

Imbibition (derived from the Latin word imbibere, 'to drink in') is a first step towards seed germination and water uptake by the root hairs. The process of adsorption of water by hydrophilic colloids of dry seeds (or non-living wood) is called 'Imbibition'. It is a purely physical phenomenon requiring no expenditure of metabolic energy. It can be best demonstrated by placing dry gram seeds in water. A noticeable swelling takes place after some time and is accompanied by a considerable increase in volume. Dry seeds are a rich reservoir of storage food such as starch and proteins, as well as cell wall materials in the form of cellulose and hemicellulose which being colloidal in nature imbibe a large quantity of water. Similarly, gum tragacanth, agar–agar and raisins when put in water, swell up considerably. Swelling of wooden doors and windows following torrential rainfall is also a common example of imbibition.

Imbibition does not require a selectively permeable membrane and involves the chemical and electrostatic attraction of water to cell walls, proteins, and other hydrophilic cellular materials. An **imbibition pressure**[3] of great magnitude is generated when an adsorbent or 'imbibant' (rich in hydrophilic colloids), in an enclosed space, is placed in a liquid to be imbibed. In cocklebur (*Xanthium strumarium*), it may go up to 1000 atmospheres. Dry wooden stakes, driven into a small crack in a rock and then soaked with water can develop enough pressure to split the rock. This process has been used in the past by Egyptians to build pyramids. The ancient witch-doctors used to perform crude surgical operation by splitting open the sutures of the bones of a skull by putting soaked seeds on the head and wrapping all over. Another example of imbibition pressure can be demonstrated in an inverted plaster of paris cone with sandwiched dry seeds. When kept in water the cone will break after some time due to imbibition by the seeds and their subsequent swelling (Figure 1.13).

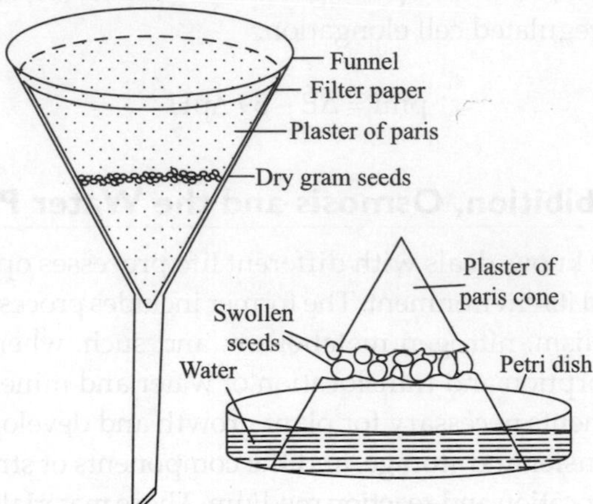

Fig. 1.13 Demonstration of imbibition pressure.

Conditions necessary for imbibition

1. A water potential gradient must be present between the surface of the imbibant and the liquid to be imbibed (water potential in dry seeds or wood is practically zero).
2. A certain amount of affinity or attraction must be present between the components of the imbibant and the liquid to be imbibed (e.g., dry seeds swell appreciably in water but not in ether).

Factors affecting imbibition

1. ***Temperature***: The rate and extent of imbibition increases with an increase in temperature because of increase in kinetic activity of the imbibing molecules.
2. ***Osmotic potential of the substance to be imbibed***: The rate of imbibition decreases with increasing solutes concentration in the medium as the water potential gradient becomes less steep.

[3] The old term 'imbibition pressure' has now been replaced by 'matric potential'.

Consequences of imbibition

1. Although the volume of an imbibant always increases but the final volume of the system as a whole decreases. Water molecules in free-state occupy more space; while in a bound form they undergo reorientation or realignment on to the surface of colloidal materials, and thus are packed tightly and occupy less volume.
2. There is always an increase in temperature following imbibition. Some of the kinetic energy possessed by these molecules gets lost because of tight adsorption of water molecules and is liberated in the form of heat.

Importance of imbibition

1. It was used for the widening of shoes in earlier days.
2. It was used for breaking open sutures of bones for performing crude operations as done earlier by 'witch-doctors'.
3. Dry wooden stakes, driven into small crack in rocks, when soaked in water can develop enough pressure to split open the rock. This method of quarrying was practiced by the ancient Egyptians who built pyramids from rock structures split open by wetting dry wood wedges.
4. It is the initial step toward germination of seeds and absorption of water by the root hairs.
5. It causes swelling of seeds and results in cracking of seed coat, followed by growth and development, leading to flower and fruit formation.
6. It is an innovative biotechnological approach requiring no sophisticated expensive laboratory facility to deliver foreign DNA molecules, using pollen tubes as conduit. Alien genes of DNA can be absorbed by root hairs where they might integrate with the host genome.
7. Seed imbibition allows uptake of plasmid DNA by embryos.

1.3.2 Diffusion

Cellular processes depend on the transport of molecules both to the cell and away from it. Diffusion is the natural tendency of systems to proceed towards increasing entropy, or disorder and the lowest possible state of energy. The driving force for the movement of a molecule either in solution or in the gas phase, is a gradient in its potential energy. Diffusion is most effective way for molecules and ions to move short cellular distances, but it is ineffective for long-distance movement.

Diffusion can be defined as the net movement of molecules from a region of (their) higher concentration into another region of lower concentration, i.e., down the concentration gradient, because of the random, translational kinetic motion of molecules (or ions/atoms).

All components of matter are in a state of motion, at temperatures above absolute zero (0 °K or –273.18 °C), i.e., they possess a certain amount of translational kinetic energy and will distribute themselves uniformly in the space that they occupy. The molecules move randomly in all directions and collide with one another. Taking the example of the air we breathe: it is primarily a mixture of N_2, O_2 and CO_2 molecules. These gas molecules are mixed uniformly in the atmosphere by constant movement in a random manner. Similarly, if a bottle containing ammonia (or any scent) is opened in one corner of the room, its odour can be detected at another corner as time passes. This is because the gas molecules of ammonia (or scent) move constantly and diffuse into the air molecules. Among the three states of matter, the gases offer the least resistance to diffusing molecules. At normal pressure and temperature, gas molecules are widely separated. That is why diffusion through gases or air is most rapid and while it is the slowest through solids.

Since diffusion depends on molecular movement, it must be quantitatively related to two components:

(i) Thermal agitation (the rate of motion of an average molecule)
(ii) Number of molecules/unit volume.

The free energy of a substance is dependent upon these two components and is termed as the **'chemical potential'** of the substance.

Diffusion $\propto \mu$, (μ; chemical potential of the diffusing substance).

At constant temperature, chemical potential is directly proportional to the concentration of the diffusing substance at the two ends. The greater the difference, the more rapid is the diffusion. However, the rate of diffusion must also depend on spatial factors. The greater the area across which the substance is diffusing, the greater is the number of molecules diffusing per unit time.

Further, small molecules are found to diffuse more rapidly than large molecules.

$$\text{Coefficient of diffusion, } D \propto \frac{1}{\text{molecular size}}$$

or,

$$\text{For small molecules, } D \propto \frac{1}{\text{mass}}$$

$$\text{For large molecules and colloidal particles, } D \propto \frac{1}{\text{radius}}$$

It is now obvious why the larger colloidal particles diffuse so much more slowly than the smaller molecules.

Diffusing particles also exert pressure, known as **diffusion pressure**. Actually it is a hypothetical term, which describes the potential ability of a gas, liquid or solid to diffuse from a region of its higher concentration to a region of its lesser concentration in a confined space. For example, in an inflated balloon, as the air gets concentrated, more molecules will be hitting the surface of balloon, resulting in an increased diffusion pressure. The walls of the balloon will stretch in order to compensate for the increase in pressure, giving visual evidence of the ability of a gas to exert pressure. If it is done excessively, the balloon bursts producing a bang because of the collision of gas molecules.

Law of Independent Diffusion

The direction of diffusion of a substance is determined entirely by the difference in diffusion pressure of that substance at the two ends and is completely independent of the diffusion pressure of the surrounding substances. For example, (i) in a balloon filled with nitrogen gas suspended in air, carbon dioxide from the air moves inside the balloon irrespective of high concentration of nitrogen gas; (ii) it is not desirable to sleep underneath a tree during night as more CO_2 diffuses out from the intercellular spaces of the leaf into the surrounding air, irrespective of oxygen concentration.

Factors affecting diffusion

1. ***Temperature***: A slight increase in temperature results in an increased rate of diffusion due to increasing kinetic energy and thus the velocity of diffusing particles.
2. ***Density of the diffusing substance***: According to **Graham's law of diffusion**, 'the rates of diffusion of gases are inversely proportional to the square roots of their densities'.

$$\frac{r_1}{r_2} = \frac{\sqrt{d_2}}{\sqrt{d_1}}$$

$$\text{For example, } \frac{r_H}{r_O} = \frac{\sqrt{16}}{\sqrt{1}} = 4$$

where r_1 and r_2 = diffusion rates of gases

d_1 and d_2 = densities of gases

This shows that the rate of diffusion of hydrogen is four times more than that of oxygen.

3. ***Medium of diffusing substance***: The more concentrated a medium is, the more slowly molecules will diffuse through it. For example, a gas diffuses more rapidly through a vacuum than through air. Similarly, a bromine gas vial when broken in a jar containing air or in an evacuated jar, bromine gas would fill the evacuated jar instantaneously.
4. ***Diffusion pressure gradient***: In general, the steeper the diffusion pressure gradient, the faster is the diffusion rate.

Fick's law

The process of diffusion occurs spontaneously and no additional energy needs to be provided. It occurs in all biological systems—within cells or across the cellular membranes. The rate at which a molecule diffuses is dependent on four factors: size of the molecule, magnitude of its concentration gradient, viscosity of the medium, and temperature. The relationship between these factors is represented by **Fick's law.**

The German scientist Adolf Fick (1850s) noticed that the rate of diffusing molecule is directly proportional to its concentration gradient ($\Delta Cs/\Delta x$).

According to Fick's first law:

$$J_s = -D_s \frac{\Delta C_s}{\Delta x}$$

Js; amount of substance 's' crossing a unit cross-sectional area per unit time - mol m^{-2} s^{-1}
Ds; the diffusion constant

$\Delta Cs/\Delta x$; the difference in the concentration of a substances between two ends separated by a distance x.

The negative sign in the above equation indicates that a substance moves down the concentration gradient in diffusion.

So a substance will diffuse rapidly when the concentration gradient (ΔCs)becomes more steep or the diffusion coefficient (Ds) increases. The importance of Fick's law is that one can calculate the time utilized by a molecule to diffuse a given distance. Small molecules in the gas phase diffuse rapidly, whereas large molecules, such as proteins and polysaccharides in solution diffuse slowly. The chemical potential and Fick's law describes the flux of an uncharged solute; which is governed only by its concentration.

Importance of diffusion in plants

Diffusion, mainly driven by concentration gradient, is a major factor governing the absorption of water as well as minerals through roots and their distribution throughout the plant. It is an important factor regulating the supply of CO_2 and O_2 during photosynthesis and respiration, respectively. It regulates loss of water vapour from the leaf intercellular spaces to the surrounding atmosphere during stomatal transpiration.

1.3.3 Osmosis

Plant and animal membranes are constructed in such a way that they are extremely permeable to water and are much less permeable to dissolved solutes. This is important because when a membrane separates two solutions having different solute concentration, there is an intrinsic tendency for water to move across the membrane towards the solution containing the greater solute concentration. In other words, there is a tendency for water to flow in a direction so as to equalize the solute concentration on each side of the membrane. The term for water movement in this manner is called 'osmosis', which may be thought of as a special type of diffusion. **Osmosis** is defined as the movement of water across a selectively permeable membrane from an area where its concentration is higher to an area where its concentration is lower.

The phenomenon of osmosis in non-living systems can be demonstrated by a U-tube glass apparatus having pure water and 5 per cent glucose solution separated by semi-permeable membrane. The water molecules will move from pure water to the 5 per cent glucose solution, till the concentration of water becomes equal on both the sides (Figure 1.14).

In such a physical system, there is only one force involved, i.e., difference between concentrations at two ends.

When an actively metabolizing plant cell is immersed in pure water or hypotonic solution,[4] water molecules tend to enter into the cell sap as a result of **endosmosis**. If the direction of movement is reverse, i.e., from cell sap to external medium as in hypertonic solutions, it is

[4] External solution is said to be hypotonic, hypertonic or isotonic with reference to cell sap if its osmotic pressure is less than, higher than or equal to osmotic pressure of the cell sap, respectively.

termed **exosmosis**. In an isotonic solution, the cell will not lose or absorb any water, thus maintaining the net volume of water constant.

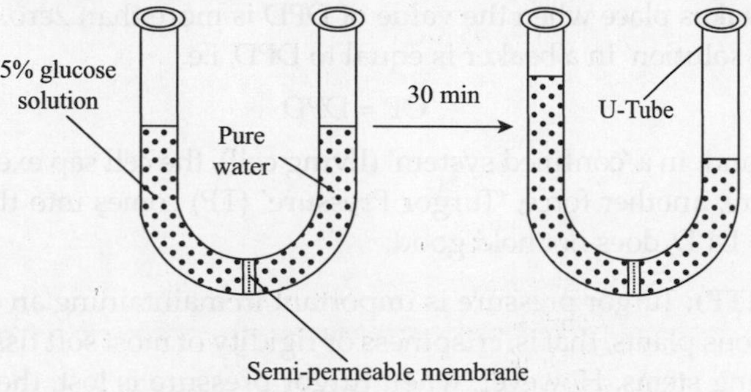

Fig. 1.14 Demonstration of osmosis.

Osmotic Pressure (OP): It is the maximum amount of pressure that can be developed in a solution separated from pure water by a semipermeable membrane.

$$OP = CRT$$

C = concentration of the solute
R = gas constant (0.082 L atm $K^{-1}mol^{-1}$)
T = absolute temperature (273.18 +°C).

This equation is applicable only in the case of non-electrolytes such as sucrose. In the case of electrolytes, such as NaCl, OP obtained is multiplied by its degree of dissociation.

Osmotic pressure (OP) of a solution is directly dependent upon the solute concentration. Therefore, it is a **colligative property**.[5]

Importance of osmosis in plants

1. The water uptake from the soil by the root hairs and its movement from one cell to another within the plant is partly controlled *via* osmosis.
2. This phenomenon is also involved in the movement of water from the non-living xylem elements into living cells.
3. The rigidity of plant organs is maintained through osmosis which helps in keeping the young stems erect and leaves expanded.
4. Growing points of root remain fully turgid due to osmosis and are, thus, able to move through the soil particles.
5. The osmotic uptake of water by the cells of the root and shoot generates the hydraulic pressure required for growth in these organs.
6. Many of the growth and turgor movements (opening as well as closing of stomata and flowers, nyctinasty of leaves, etc.) depend on this phenomenon.

[5] The other three colligative properties of solutions are freezing point depression, boiling point elevation and lowering of vapour pressure of the solvent.

Diffusion Pressure Deficit (DPD): The term 'diffusion pressure deficit' was put forth by Meyer (1938). The amount by which the diffusion pressure of a solution is lower than that of its pure solvent, is termed as DPD. It denotes the net pressure that causes water to enter the cell. Osmosis takes place when the value of DPD is more than zero. Osmotic pressure of an 'unconfined solution' in a beaker is equal to DPD, i.e.,

$$OP = DPD$$

On the other hand, in a 'confined system' (living cell), the cell sap exerts a high osmotic pressure. Therefore, another force, 'Turgor Pressure' (TP) comes into the picture and the relationship, OP = DPD, does not hold good.

Turgor Pressure (TP): Turgor pressure is important in maintaining an erect habit of non-woody or herbaceous plants, that is, crispiness or rigidity of most soft tissues of plants such as leaves and young stems. However, when turgor pressure is lost, they become limp or wilted. It is a unique property of a plant cell that is surrounded by a rigid cell wall. If a living plant cell is immersed in water, it swells due to absorption of water. As a result, pressure is developed in the vacuole that presses the protoplasm against the cell wall, and this is called 'turgor pressure'. A cell experiencing turgor pressure is said to be **turgid**, whereas a cell that experiences water loss to the point where turgor pressure is reduced to zero is said to be flaccid. Indeed, one of the first outward signs of water deficit in plants is the wilting of leaves due to loss of turgor in the leaf cells.

The rigid and elastic cell wall exerts an equal and opposite pressure against the expanding protoplasm, termed as 'wall pressure' (WP).

$$\text{At a given time, } TP = WP$$

TP is exerted by the cell vacuole to the protoplasm against the cell wall, while WP is exerted by the cell wall to the protoplasm against the vacuole.

The net force with which the water is drawn into the plant cell depends upon the difference between two entities, osmotic pressure and turgor pressure.

$$DPD = OP - TP\ (WP)$$

Thus, in a living system or in other actively metabolizing cells, there is an interplay of only two forces: OP and TP.

1.3.4 Plasmolysis

In a turgid plant cell, the cell sap presses the protoplasm against the cell wall. When an actively metabolizing plant cell is kept in a hypertonic solution, water moves out of the cell due to exosmosis with a consequent shrinkage in the volume of the vacuole. With the result, the protoplasm is no longer pressed against the cell wall and begins to retract and ultimately separates away from the cell wall, assuming a more or less spherical shape in the centre of the cell. This phenomenon is called **plasmolysis** (Figure 1.15). In a plasmolysed cell, the space between the cell wall and protoplasm is occupied by the external solution. The immediate effect of plasmolysis is the retardation of metabolic activities in cells, leading to permanent wilting of tissues and eventually death of the plant if it continues for a longer period. That is why irrigation of the soil with a concentrated solution of salt or any other solute or chemical fertilizers causes burning of the plants.

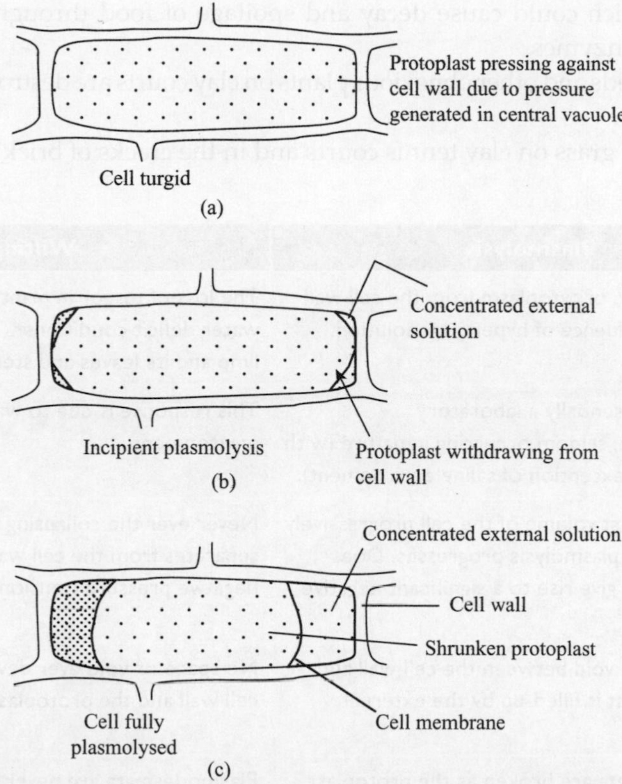

Fig. 1.15 The figures (a, b, c) show various stages of plasmolysis. In the fully plasmolysed state, the protoplast is rounded up in the centre of the cell. The space between the shrunken protoplast and the cell wall is occupied by the external hypertonic solution.

Importance of plasmolysis

1. The phenomenon of plasmolysis, first used by De Vries, is helpful in determining the approximate osmotic pressure of the plant cell.
2. It may be used to determine whether a cell is living or non-living. Plasmolysis never occurs in non-living cells such as cork or xylary elements.
3. It also proves that cytoplasm acts as a semipermeable membrane. If it were not so, the external bathing medium would just enter the vacuole, bringing about equilibrium between the vacuolar sap and the external solution. There will be no plasmolysis then.

So the phenomenon of plasmolysis provides important clues to the plant physiologists.

Applications of plasmolysis

Plasmolysis has some useful applications in our everyday life:

1. The salting of fresh meat and fish as in preserved or canned foods, and the addition of sugar to jams and fruit jellies – this is done to plasmolyse microbial spores of bacteria

and fungi which could cause decay and spoilage of food through the secretion of extracellular enzymes.

2. Unwanted weeds and other obnoxious plants on clay courts are destroyed by sprinkling salt on the soil.
3. The growth of grass on clay tennis courts and in the cracks of brick walls is similarly checked.

Plasmolysis	Wilting
1. The shrinkage of cytoplasm from the cell wall under the influence of hypertonic solution.	The loss of turgor in plant tissues under water deficit conditions*. The cells become limp and its leaves and stem wilt.
2. It remains essentially a laboratory phenomenon, seldom occurring in nature (with the possible exception of saline environment).	This response is due to water-stressed environment.
3. The protoplast volume of the cell progressively decreases as plasmolysis progresses. Does not normally give rise to a significant negative pressure.	Never ever the collapsing protoplast separates from the cell wall but significant negative pressure (tension) may develop.
4. The space or void between the cell wall and the protoplast is filled-up by the external solution.	No space or void ever develops between the cell wall and the protoplast.
5. Plasmodesmata are broken as the protoplast pulls away from the cell wall.	Plasmodesmata are never broken between the adjacent cells.

* Resulting from excessive loss of water than water uptake or actual deficiency of water in the soil.

1.3.5 Water potential

The best way to express spontaneous movement of water from one region to another is in terms of the difference in free energy of the water between two regions.

Free energy is the energy of the system available for doing some work at constant temperature. Free energy per mole of any substance in a chemical system is known as **chemical potential** (μ). As the chemical potential of a substance increases, its tendency to undergo chemical reactions and other processes (such as diffusion and osmosis) also shows an increasing trend. Chemical potential of a substance, at constant temperature and pressure, depends on the number of moles of substance present, its electrical charge and hydrostatic pressure.

In plant–water relations, the chemical potential of water is known as **water potential** (ψ_w). The term 'water potential' was introduced by Australian scientists, Slatyer and Taylor in 1960. Water potential is defined as the chemical potential of water divided by the partial molal volume of water. Thus it is a measure of the free energy of water per unit volume (J m^{-3}; equivalent to pressure units such as pascal, which is the common measurement unit for water potential).

$$\psi_w = \mu_w - \mu_w^{\circ} = RT \ln e/e^{\circ}$$

μ_w is the chemical potential of water at any point in a system; μ°_w is the chemical potential of pure water under standard conditions; *R*, is the gas constant (erg/mole/degree); *T*, the absolute temperature (°K); *e*, is the vapour pressure of the solution in a system at temperature, *T*; *e*°, the vapour pressure of pure water at the same temperature.

Water potential in a system is measured in bars [1 bar = 0.987 atm = 10^6 dynes/sq cm].

Pure water, at atmospheric pressure, has zero potential. Water potential of a solution or in a living cell is always less than zero because the addition of solutes to water increases the disorder or entropy of the system and thereby reduces the free energy of water, thus decreasing the water potential. The direction of water flow will be 'energetically downhill', i.e., from region of greater water potential (pure water) to the region of lesser water potential (solution). The water flow continues until the water potential of the two regions become equal.

In terms of the free energy concept, there is an interplay of three major forces –the osmotic potential (ψ_π), the pressure potential (ψ_p), and the matric potential(ψ_m).

∴ Water potential (ψ_w) is the algebraic sum of potentials of these components :

$$\psi_w = \underset{(-ve)}{\psi_\pi} + \underset{(+ve)}{\psi_p} + \underset{(-ve)}{\psi_m} \quad \text{(Values)}$$

Osmotic potential[6] (ψ_π) is defined as the amount by which the water potential is reduced as a result of the presence of solutes. The osmotic potential of a solution is directly proportional to the concentration of dissolved solute but completely independent of the specific nature of the solute. Osmotic potential is equal to osmotic pressure but carries a negative sign. It is expressed in bars.

Pressure potential (ψ_p) refers to turgor pressure exerted by the vacuole. It is usually positive and tends to maintain turgidity of cells. In young unvacuolated cells and slightly plasmolysed cells, ψ_p is zero. ψ_p becomes negative in the xylem.

Matric potential (ψ_m) refers to decrease in water potential as a result of adsorption of water to colloidal substances, i.e., soil, protoplasm, cell wall, etc. It is especially important in the initial stages of imbibition, soil–water relations, as well as succulents with high mucilage content. ψ_m is always negative and becomes negligible (<–0.1 bar) in vacuolated plant cells.

$$\therefore\ \psi_{soil} = \psi_\pi + \psi_m$$

Besides these three forces, two other forces namely, gravitational potential and electrical potential, also contribute to water potential.

[6] The term 'solute potential (ψ_s)' was used earlier in place of osmotic potential.

Gravitational potential (ψ_g) may become a significant factor in tall trees. As the water column increases in height, the water potential increases correspondingly. Gravitational force causes water to flow in the downward direction, unless it is opposed by an equal and opposite force.

Therefore, ψ_g depends on the height (h) of the water column above the reference-state water, the density of water (ρ_w) and the acceleration due to gravity (g), as shown below:

$$\psi_g = \rho_w g h$$

($\rho_w g$ has a value of 0.01 MPa atm^{-1})

Gravitational forces are generally ignored or cancel out in plant–water relations, especially when considering adjacent cells at the equal level.

The water potential upstream is greater than the water potential downstream by virtue of the position in a gravitational field, one of the factors affecting water potential. We know there is substantial free energy present because the flow of water can be used to turn a paddle wheel, which in turn can be used to drive a turbine for the production of hydroelectric power.

Electrical potential (ψ_e): Electro-osmosis is a process by which water moves along electrical gradients, and could be of some significance since water is a dipole and can readily move in an electric field. Potentials across living membranes, however, are only of the order of 100 mV. This voltage is probably insignificant as a force for water transport in plants and as a consequence, ψ_e is usually not considered.

The relationship of volume, water potential, osmotic potential, and pressure potential of a living cell with elastic cell wall

If a flaccid cell with ψ_p zero (the point 1.0 on the abscissa) is kept in pure water, water starts moving into the cell, thereby increasing its volume. Hence, the osmotic potential (ψ_π) becomes less negative but there is an increase in ψ_p (Figure 1.16).

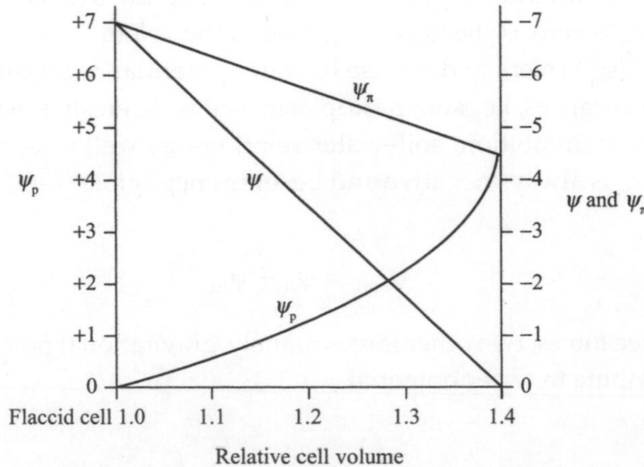

Fig. 1.16 Diagrammatic representation of relationship between volume changes and variations in ψ, ψ_p, ψ_π.

At the point where the osmotic potential equals the ψp (but opposite in sign), the ψ reaches zero and the cell is said to be fully turgid and there is no further increase in the volume of cell. The system would be in equilibrium with water outside and hence no more water will enter (i.e., no net water flow).

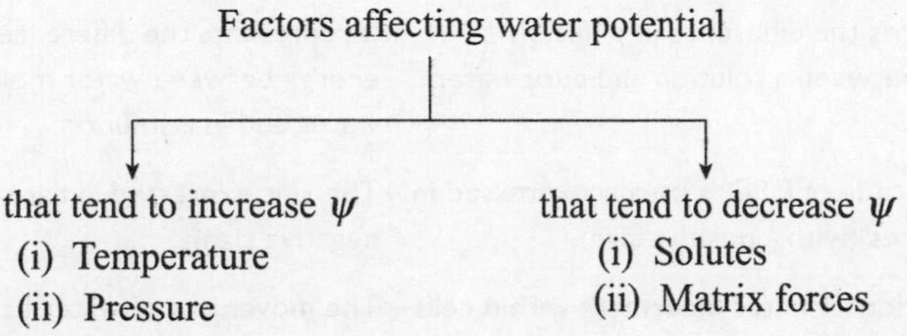

Significance of water potential in plants

1. Water potential is the quantity that regulates the direction of movement of water across cell membranes by osmosis.
2. Water potential is a measure of the water status of a plant. Water deficits lead to inhibition of various activities in plants such as cell division, wall and protein synthesis, solute accumulation, stomatal opening, photosynthesis, etc.
3. Water potential also indicates how 'wet' or 'dry' a plant is and thus provides a relative index of the water stress the plant is going through.

Overall, water potential is a good indicator of plant health and has a strong influence on crop yield. Water potentials can be measured either by psychrometer or pressure chamber.

Water potential is not the only parameter that determines how rapidly water moves into the cells, but the **hydraulic conductivity** (Lp) of the plasma membrane also influences how quickly cells equilibrate with their local environment. It is a measure of the ease with which water crosses the membrane and is expressed in units of volume of water per unit area/per unit time/per unit driving force ($m^3\ m^{-2}\ s^{-1}\ MPa^{-1}$).

The rate of water flow across a membrane (*J*v) is given by:

$$Jv = Lp \cdot \Delta\psi_w$$

where, Lp is the hydraulic conductivity and $\Delta\psi_w$, the driving force.

Diffusion pressure deficit (DPD)	Water potential (ψ)
1. The term 'DPD' was coined by Meyer in (1938) but was described originally under the name 'Suction Force' (Saugkraft) or 'Suction Pressure' (SP) by Renner (1915).	The term 'water potential (ψ)' was given by Slatyer and Taylor (1960).

Diffusion pressure deficit (DPD)	Water potential (ψ)
2. The diffusion pressure deficit is abbreviated as DPD.	Water potential is symbolized by the Greek letter or uppercase psi, ψ (pronounced 'sigh').
3. It represents the difference in diffusion pressure between a solution and pure water (solvent).	It represents the difference in free energy between water molecules in pure water and in a solution.
4. The magnitude of DPD is usually expressed in atmospheres (with a positive sign).	The ψ is expressed in bar units (with a negative sign).
5. The direction of water movement within cells is from low DPD to high DPD.	The movement of water between cells is from high ψ to low ψ, i.e., in an energetically downhill direction or from less negative ψ to more negative ψ.
6. The DPD of a cell is equivalent to: DPD = OP – TP	The water potential (ψ) is equal to the algebraic sum of three forces: $\psi = \psi_\pi + \psi_p + \psi_m$ (–ve) (+ve) (–ve)— (Values)

Water potential is expressed in terms of pressure units, most commonly mega-Pascals (MPa), although some other units of pressure are also used.

1 mega-Pascal (MPa) = 10 bars = 9.87 atm

1 bar = 0.987 atm

Review Questions

RQ 1.1. Prepare 1 molar solution of sucrose.

Ans. Dissolve 342.2 g (mol. wt.) of sucrose in about 500 ml of water and make the final volume to 1 L (1000 ml).

RQ 1.2. Prepare 1 molal solution of NaCl.

Ans. Dissolve 58.35 g (mol. wt.) of NaCl in 1000 ml of water.

RQ 1.3. Prepare 1N hydrochloric acid (HCl).

Ans. 36.5 g of HCl/L makes 1N solution. In terms of volume, $V = M/D = 36.5/1.16 = 31.4$ ml. [specific gravity of HCl is 1.16]. The purity is only 28 per cent; hence, $31.4 \times 100/28 = 112.1$ ml (approximately 115 ml will give strength a little more than 1N).

RQ 1.4. Prepare 1N H_2SO_4 and dilute it to 0.1N H_2SO_4.

Ans. 49 g of H_2SO_4 in 1 L of water = 1N H_2SO_4. Mol. wt. of H_2SO_4 is 98 and it has two replaceable hydrogen atoms; therefore, a normal solution of H_2SO_4 contains 49 g (98/2) of H_2SO_4 per litre.

In terms of volume, $V = M/D = 49/1.84 = 26$ ml [1.84 being the specific gravity of H_2SO_4]. As the purity of H_2SO_4 is about 8 per cent, about 28 ml of conc. H_2SO_4 dissolved in 1000 ml gives a solution of a little more than 1N H_2SO_4. 0.1N H_2SO_4 = 26/10 = 2.6 ml of concentrated H_2SO_4 added to a litre of distilled water will yield a solution of about 0.1 N.

RQ 1.5. Prepare (i) 10 per cent solution of NaCl (weight/volume) (ii) 5 per cent solution of glycerine (volume/volume).

Ans. (i) 10 g of NaCl with water to make the final volume 100 ml → 10 per cent solution of NaCl.
(ii) 5 ml of glycerine with 95 ml of water→5 per cent solution of glycerine.

RQ 1.6. Prepare (i) 100 ppm and (ii) 10 ppm solution of auxin (IAA).

Ans. For making 100 ppm sol: (i) 100 mg of IAA added to distilled water to make the final volume to 1 L (1000 ml).
100 mg IAA/1000 ml of solution.

For 10 ppm sol: (ii) Stock solution (S) = 100 ppm,
concentration of the required solution (R) = 10 ppm
Total volume required = 10 ml.

Amount of stock solution required

$$= \frac{\text{Concentration of required solution (R)}}{\text{Concentration of stock solution (S)}} \times \text{Total volume required}$$

$$\text{Amount of stock solution} = \frac{10}{100} \times 10 = 1 \text{ ml}$$

So 1 ml stock solution is added to 9 ml distilled water to make 10 ml of 10 ppm solution

(i) Distilled water should be used for the preparation of various solutions.
(ii) Acids/alkalies should be added to water but not the *vice versa.*

RQ 1.7. Why is the pH value only in between 0 and 14?

Ans.

$$H_2O \rightleftharpoons H + OH^-$$

$$K = \frac{[H^+][OH^-]}{H_2O}$$

$$K\,[H_2O] = [H^+] \times [OH^-]$$

$$= [H^{-7}] \times [OH^{-7}] = 10^{-14}$$

Hence, the pH values will range only in between 0 and 14.

RQ 1.8. Why in a given volume of pure water, are the number of H^+ ions exactly equal to the number of OH^- ions?

Ans. Pure water molecules have a 'slight' tendency to ionize, separating into H^+ and OH^- ions. The proportion of H^+ and OH^- ions is constant since water tends to dissociate into ions and this is exactly balanced by the tendency of the ions to combine again. Thus, even as some water molecules are ionizing, an equal number of others are forming covalent bonds.

RQ 1.9. What effect do you expect on pH when carbon dioxide gas or limestone is dissolved in distilled water?

Ans. The presence of dissolved CO_2 in distilled water may raise the hydrogen ion concentration so that it may reach 10^{-4} M (mol/L), i.e., pH 4. The hydroxyl ion concentration is then 10^{-10} M.

$$CO_2 + H_2O \rightarrow \underset{\text{Carbonic acid}}{H_2CO_3}$$

$$H_2CO_3 \rightarrow H^+ + HCO_3^-$$

If tap water contains much dissolved limestone ($CaCO_3$), the hydrogen ion concentration can reach close to neutral or even slightly basic (pH 7 to 8).

However, the pure water when exposed to the atmosphere will take up CO_2, some of which reacts with water to form carbonic acid, thereby reducing or lowering pH to about 5.7.

RQ 1.10. How is the pH of cytoplasm kept near 7?

Ans. The pH of cytoplasm is kept near 7 by buffers, and by pumps operating in the plasma membrane and vacuolar membrane (tonoplast) that transport H^+ into the vacuole and out of the cytoplasm into extracellular spaces (apoplast) (refer to RQ 4.46). The vacuole of a plant cell is a reservoir of organic acids, so that its pH is usually in between or in the region of 3.5–5.5, although in certain fruits like lemon it may reach as low as 1.0.

RQ 1.11. How many H^+ and OH^- do you expect in a solution of pH 3? Compare it with respective solutions having pHs of 1 and 2.

Ans. A solution having a pH of 3 will have 10^{-3} or 0.001 H^+ ions. It will also have 10^{-11} OH^- (0.00000000001). On the other hand, a solution of pH 1 and pH 2 will have 10^{-1} or 0.1 and 10^{-2} or 0.01 hydrogen ions respectively, while their OH^- concentration will have a corresponding value of figures 10^{-13} and 10^{-12}.

The negative logarithm of a number is the logarithm of its reciprocal. Therefore, the pH corresponding to a hydrogen ion concentration of 10^{-3} is 3.

$$pH = \log \frac{1}{10^{-3}} = \log 10^{-3} = 3.00$$

RQ 1.12. List the different methods that are used for estimating the pH of an aqueous solution.

Ans. The methods used for measuring pH of a solution are as follows:

- By using various indicator dyes, such as phenolphthalein and phenol red. The dyes ionize at a specific pH, producing coloured ions.
- By using litmus paper (that changes colour with change in pH) which gives approximate idea of pH.
- pH can be measured by an instrument called pH meter. In most pH meters, there is an ion selective electrode that is sensitive only to H^+ ions. The signal from this electrode is amplified and then compared to the signal originating from another electrode dipped in a solution of known pH strength.

One molar solution of HCl or other acid, producing 1.0 M H^+ has a pH of 0 (as log 10 of 1 is 0). On the contrary, 1.0 M NaOH or other alkali will produce 1.0 M OH^- and 1×10^{-14} M H^+ and the pH will be 14. It is noteworthy to mention that the pH scale is logarithmic, and not arithmetical. Thus, a solution of 6.0 pH will be 'ten times' more acidic than a solution having pH value 7.

RQ 1.13. Milk curdles heavily when lemon juice is poured into boiling milk. Comment.

Ans. Milk is a common colloid (to be more exact an emulsion), composed of butter and fat dispersed in water, with casein as an emulsifying agent. All the particles in a colloidal system bear similar charges and thus, repel each other. This keeps the particles dispersed.

Addition of lemon juice (or for that matter any acid) into boiling milk would increase the hydrogen (H^+) ion concentration. The +vely charged ions in the lemon juice cause the -vely charged milk particles to lose their charge. As a result, the dispersed particles in the milk would collide, aggregate, and finally settle out in the form of precipitates. That is why milk curdles heavily on addition of lemon juice that contains citric acid. The pH at which the positive charge exactly neutralizes the negative charge (no net charge) is termed as the '**isoelectric point**', or **pK**. At the pK, proteins do not have the protection from agglomeration provided by the presence of an electrical charge. The proteins will thus tend to clump together when the 'net charge' is **zero**.

RQ 1.14. Why are deltas often found at the mouth of a river? Explain.

Ans. Weathering of parent rock yields sand, silt, clay and soluble mineral nutrient ions. Colloidal clay consists of much finer particles (0.001–0.1 μ) than may be seen by the naked eye. However, they remain in a suspended form in water and will not settle out. Clay particles consist primarily of aluminium silicates, predominantly 'negatively charged' because of 'exposed carboxyl and hydroxyl groups'. Positively charged cations are preferably adsorbed on to clay particles.

The clay particles are carried downhill in a suspended form by the river water and as it meets the ocean or sea water, the 'charged cations' present in the saline water (predominantly NaCl, $NaNO_3$) cause the negatively charged clay micelle of river water to lose their charge and settle out. Destruction of the electric charge will cause the dispersed particles of a colloidal suspension to collide, aggregate, and finally settle down in the form of precipitate.

This phenomenon leads eventually to the formation of a delta – a fertile stretch of land which is often produced at the mouth of a river.

RQ 1.15. What is common and what is different between clay and humus? Why are they so important for plants?

Ans. Both clay and humus are hydrophilic colloids. Clay is inorganic material consisting of aluminium and silicon oxides while humus is organic material, derived from the dead and decomposed remains of plants and animals. The surface of the colloidal particles is negatively charged. The negative charge arises from exposed carboxyl and hydroxyl groups. As a result, cations like calcium, potassium, magnesium and others are attracted or preferentially absorbed. The roots secrete H^+, which due to great affinity for the negatively charged colloidal surface, tend to displace cations into the soil solutions.

So this colloidal fraction of clay and humus, owing to their large surface area per unit mass, becomes a principal nutrient reservoir for release into the soil solution.

RQ 1.16. Beetroot loses its red colour when boiled in water but does not do so in benzene. Why?

Ans. The anthocyanin or betacyanin pigments in beetroot are located in the vacuolar sap which is surrounded by tonoplast. However, the pigments are lost when beetroot is boiled in water as the membrane structure is completely lost because of denaturation of membrane proteins and it becomes permeable, allowing the pigments to diffuse. In benzene also, the permeability is lost but the solution remains colourless as the pigments are insoluble in benzene. However, the colour will appear if water is added to this benzene solution as the pigments are water soluble.

RQ 1.17. Although beetroot grows underground, it does not lose its red colour in the soil solution. Explain.

Ans. Beetroot does not lose its red colour in the soil solution as the plasma membrane and the vacuolar membrane or tonoplast (enclosing the anthocyanin) behave like differentially or selectively permeable membrane, preventing the diffusion of pigments in the soil solution.

RQ 1.18. Animal cells do not possess a cell wall, how then these cells are able to maintain strength and rigidity of tissues and organelles?

Ans. True. The strength and rigidity of animal tissues and organs is provided by the extracellular matrix consisting of collagen, glycoproteins and other compounds.

RQ 1.19. List the cell organelles that are bounded externally by (1) a single membrane and (2) a double or paired membrane.

Ans. Organelles with a single membrane: Microbodies such as Glyoxysomes and Peroxisomes. Organelles with a double membrane: Chloroplast, Mitochondria, Endoplasmic Reticulum and Nucleus with numerous pores.

RQ 1.20. Where do you find some unusual structural membranous features: (1) three-layered plasma membrane and (2) single lipid bilayer.

Ans. 1. Three layered (trilaminar) structure of plasma membrane: Erythrocytes
2. Single lipid layer: Liposomes (are being tested as vehicles to deliver drugs or DNA molecules within bodies).

RQ 1.21. What are amphipathic molecules?

Ans. The amphipathic molecules such as phospholipids have a hydrophilic (water-soluble) polar head and two (lipid-soluble) hydrophobic tails.

RQ 1.22. List the various functions attributed to proteins embedded in the membrane.

Ans. 1. Channel Proteins: form small opening for molecules to diffuse through.
2. Carrier Proteins (also called Transport Proteins): binding sites on the protein surface 'grab' certain molecules and pull them into the cells (Gated Channels).
3. Receptors Proteins: for attachment to suitable raw materials required for various functions in the cells. Acting as molecular triggers that set off cell responses (such as release of homones or opening of channel proteins).
4. Enzymatic proteins: carry out metabolic reactions.

RQ 1.23. Although the amount of CO_2 in the air is very low yet it enters the leaf tissues in spite of very high concentration of O_2 inside. Explain how?

Ans. The CO_2 level in the atmosphere is just only 0.03 per cent by volume yet it enters the leaf tissue because its molecular movement is controlled by its 'own concentration gradient' irrespective of concentration of other gases inside. After entering the leaf through stomata, it (CO_2) is immediately consumed during photosynthesis during the day and the CO_2 level remains low as compared to the air. Thus, the CO_2 gas will continue to enter as long as there is a positive gradient and it will not be affected by the oxygen concentration brought about by photosynthetic activity. This is in accordance with the 'principle of independent diffusion of gases'.

At night time there is build-up of CO_2 because of respiration and secondly because of lack of photosynthesis. CO_2 readily 'passes out' of the leaf tissue as its concentration outside is much less. That is why it is injurious to sleep under a tree at night. However, during daytime, O_2 level inside is much higher than outside, it will tend to diffuse outward from the leaf and helps to purify air. The molecular movement of oxygen is controlled by its own gradient, irrespective of very high concentrations of N_2 gas (which is most abundant, constituting 78 per cent of air).

RQ 1.24. What is plasmolysis? How does it differ from wilting?

Ans. Plasmolysis is the shrinkage of cytoplasm from the cell wall under the influence of 'hypertonic concentration' outside in the bathing medium. The protoplasm begins to retract from the cell wall and ultimately separates out, forming a more or less spherical mass. If it continues for a longer time, it will eventually lead to death of the cell. Plasmolysis *per se* does not occur in plant cells in the atmosphere.

Wilting is in response to loss of turgor pressure inside the leaf tissue, occurring during water stress conditions. There are three types of wilting: incipient, temporary and permanent. Here the cytoplasm never ever separates from the cell wall. A single severe wilting often results in 50 per cent or more reduction of yield.

When the plants begin to wilt there is sudden increase in the level of ABA in leaf cells, i.e., from about 20 μg/kg fresh weight to 500 μg/kg and the stomata close down.

RQ 1.25. Excessive use of chemical fertilizers is dangerous. Comment.

Ans. Chemical fertilizers, such as nitrogenous, phosphatic and potash (NPK) are used to increase productivity in our crops. However, if they are used in excessive amounts and indiscriminately, these have deleterious effects because they lead to plasmolysis of roots (especially the root hairs) and these will no longer be functional in absorbing water and minerals. Plants will soon wilt. Even the burning of plants (often turning brown) will be caused by residues of fertilizers. That is why these fertilizers are applied to the soil quite judiciously in appropriate quantities.

The cells of the plants especially roots would be plasmolysed and the plants would undergo wilting and finally death unless the excess salts are washed away by rain or irrigation.

RQ 1.26. What occupies the space between the cell wall and cytoplasm in a plasmolysed cell? Explain.

Ans. In a plasmolyzed cell, the space between the cell wall and the protoplasm is occupied by the 'bathing medium' or the 'external hypertonic solution' in which tissues were immersed. The stage at which initial pulling away of the plasma membrane from the cell wall begins is called **incipient plasmolysis** (the magnitude of turgor pressure here is zero). Here the cell wall is 'just left behind alone', and being permeable it will allow the external solution to move freely towards inside as there is no barrier to the solute movement, thus filling the space as the cytoplasm shrinks.

As a result of outward diffusion of water (exosmosis), the vacuole shrinks and the protoplasm and plasma membrane are pulled away from cell wall.

RQ 1.27. Explain the terms diffusion, chemical potential and water potential.

Ans. Diffusion is the unequal movement of the molecules (be it a gas, liquid or solute) from an area of higher concentration or high diffusion pressure to an area of low concentration or low diffusion pressure and is controlled by their kinetic activity. The molecules are constantly moving at random and occasionally colliding with one another. The rate of diffusion depends upon concentration gradient under conditions of constant pressure and temperature: (i) concentration difference at the two regions and (ii) the distance through which the diffusing molecules have to move. The smaller the distance between the two concentrations, the more the number of molecules that would cross over per unit time.

Quantitatively speaking, the process of diffusion has two aspects – the rate of motion of average molecules (thermal agitation) and the number of molecules per unit volume, both parameters controlling the free energy. This free energy of any substance in a chemical system is called 'chemical potential'.

Hence,

Diffusion $\alpha\ \mu$

(μ = chemical potential of the diffusing substance).

$\mu = RT$ lna Joules/g

R = 0.082 (gas constant)

T = temperature (degrees absolute)

a = 'effective' concentration or activity of the substance

ln = natural logarithm

At standard or constant temperature or pressure, the molecules diffuse from a region of high concentration or chemical potential to regions with lower chemical potential.

Thus, the water potential (ψ; pronounced psi) is the chemical potential of water and is equivalent to the algebraic sum of three forces operating or interplaying at the same time.

Water potential = osmotic or solute potential + pressure potential + matric potential

$$\psi = \psi_\pi + \psi_p + \psi_m$$

(–ve) (+ve) (–ve) values

Other forces like gravitational, electrical, etc., play a negligible role and are often overlooked.

Water will flow from a region of high ψ to low ψ, i.e., in an energetically downhill direction or from a region of less negative ψ to more negative ψ.

RQ 1.28. Tabulate *only* the water potential (ψ) values in different plant systems.

Ans. The range of different types of 'potentials' found in plant systems is as under:

ψ_w = water potential
zero for pure water
zero (at field capacity) or negative for cells

ψ_π = osmotic potential
always negative – becoming more or more negative with increasing concentrations or with more and more addition of solutes.

ψ_p = pressure potential
usually positive in living cells,
but negative in xylem cells which are under tension.
zero at incipient[7] plasmolysis
negative (in plasmolysed cells).

For details of ψ in soil, root hairs, stem, leaf and the atmosphere (see RQ 3.6, page 97)

RQ 1.29. The osmotic relations of a cell depend upon the state of activity or physical state of the cell. Explain by assuming the magnitude of OP and TP, according to the 'old concept' of DPD.

Ans. Assuming cells with OP = 10 units – 'kept in water', and at 'incipient stage of plasmolysis' or 'plasmolysed conditions'

1. In the case of fully turgid cell

OP = 10 and TP = 10
DPD = OP – TP
= 10 – 10
= 0 (no net movement of water)

Animal cells often burst if placed in pure water, a process called 'lysis', but plant cells do not or seldom burst, bursting occurring only under heavy downpour or torrential rains.

2. In the case of flaccid cell (showing slight or incipient plasmolysis)

OP = 10 and TP = 0
DPD = 10 – 0
= 10

Water will enter into the cell (endosmosis).

3. In the case of plasmolysed cell

OP = 10 and TP = –2 (negative turgor pressure)
DPD = 10 – (–2)
= 10 + 2
= 12 (endosmosis will occur)

---

[7] Incipient plasmolysis is defined as the point at which 50 per cent of the cells are plasmolysed.

RQ 1.30. The osmotic relation values in plant cells depend upon the physical state. Discuss the earlier example in terms of the 'new concept' of water potential (ψ).

Ans. **1.** In the case of fully turgid cell:

assuming osmotic potential value = 10

Water potential = osmotic potential + pressure potential

$$\psi_w = \psi_\pi + \psi_p$$
$$\psi = -10 + 10 \text{ bars}$$
$$= 0 \text{ bar (cell in a state of dynamic equilibrium)}$$

2. In the case of a flaccid cell:

$$\psi_\pi = -10 \text{ bars and } \psi_p = 0 \text{ bars}$$
$$\psi = -10 + 0$$
$$= -10 \text{ bars}$$

ψ negative as compared to water

3. In the case of a plasmolysed cell (in a state of tension):

$$\psi_\pi = -10 \text{ bars and } \psi_p = -2 \text{ bars}$$
$$\psi = \psi_\pi + \psi_p$$
$$= -10 + (-2)$$
$$= -10 - 2 = -12 \text{ bars (more negative unlike above cited two examples)}$$

RQ 1.31. The adjacent cells A and B have osmotic pressure 12 and 10 atmospheres, respectively. Their turgor pressures are 9 and 4, respectively. In which way will the water flow and why?

Ans.

A	B
OP = 12 TP = 9	OP = 10 TP = 4
DPD = OP −TP = 12 −9 = 3	DPD = OP −TP = 10 −4 = 6

Water will flow from cell A to B, i.e., from low DPD to high DPD.

RQ 1.32. Two adjacent cells A and B have osmotic potential 12 and 10 bars. Their respective pressure potentials are 9 and 4 bars. In which way will the water flow and how? Explain.

Ans.

A	B
$\psi_\pi = -12$ $\psi_p = +9$	$\psi_\pi = -10$ $\psi_p = +4$
$\psi = \psi_\pi + \psi_p$ $= -12 + 9$ $= -3$ less negative ψ	$\psi = \psi_\pi + \psi_p$ $= -10 + 4$ $= -6$ more negative ψ

Water will move from cell A to B, i.e., from less negative ψ (less concentrated) to more negative ψ (more concentrated).

RQ 1.33. Explain how the ψ gradient concept is useful in hydroelectric power generation?

Ans. Let us visualize a river flowing downstream. The water potential (ψ) upstream is more than the ψ of the water downstream on account of its position in a gravitational field, one of the factors controlling ψ (gravitational potential). Water will flow downhill and this can be used to turn a paddle wheel, which in turn can be used to drive a turbine for hydroelectric power generation.

However, this flow of water downhill can be prevented by constructing a barrier (or a dam). From there water could be pumped upward against the potential gradient but this would need energy. The water so diverted can be used for irrigation or for power generation elsewhere.

RQ 1.34. Where do you experience negative pressure potential in plant tissues?

Ans. Pressure potential is usually positive but tends to become negative in its magnitude under following situations:

- Xylem elements of the stem during ascent of sap (under tension)
- Substantial negative pressure potential may develop in wilted leaves or fully plasmolyzed tissues

Hence, the water potential of wilted or fully plasmolysed tissues becomes increasingly negative as it is the total sum of the negative osmotic potential (ψ_π) plus the negative pressure potential (ψ_p).

RQ 1.35. Where do you find the value of water potential (ψ) equivalent to zero.

Ans. Water potential has a zero value as in:

(i) Pure water

(ii) Fully turgid cells where ψ_π (osmotic potential) is equal to ψ_p (pressure potential) as given here:

$$\begin{aligned}\psi &= \psi_\pi + \psi_p + \psi_m\\ &= -10 + 10\\ &= 0\end{aligned}$$

(the value of ψ_m is negligible)

When turgor pressure of the cell is reduced to zero it is said to be flaccid as it happens at incipient plasmolysis – the protoplast exerts no pressure against the cell wall but neither is it withdrawn from the walls.

RQ 1.36. Under which situations do you expect the magnitude of the pressure potential to be zero?

Ans. The magnitude of pressure potential is zero in:

(i) Unvacuolated cells as occurring in meristematic regions

(ii) Cells at incipient plasmolytic stage

Chapter 2

Absorption and Translocation of Water

2.1 Absorption of Water

Keywords

- Acoustic detection technique
- Active transport
- Anaerobiosis
- Apoplast
- Asphyxiation
- Cavitation
- Embolism
- Passive transport
- Pressure bomb technique
- Root pressure

Water is the essential medium of life. Thus, Plants that are land-based (as opposed to aquatic ones) are faced with potentially lethal desiccation through water loss to the atmosphere. This problem is aggravated by the large surface area of leaves, that are exposed to high levels of radiant energy (sunlight) and their need to have an open pathway for CO_2 uptake. Thus, there is a conflict between the need for water conservation and the need for CO_2 assimilation.

The need to resolve this vital conflict determines much of the structure of plants that grow on land, namely: (i) the development of an extensive root system to absorb water from the soil; (ii) a low-resistance pathway through the tracheary elements to bring water to the leaves; (iii) hydrophobic cuticle lining the surfaces of the plant to reduce evaporation; (iv) microscopic stomata on the leaf surface to allow gaseous exchange; (v) guard cells to regulate the diameter and diffusional resistance of the stomatal opening.

Algae and simple land plants like mosses and lichens may absorb water through their entire surface but in the vascular plants, the absorption of water takes place mostly through the roots. The source of water supply, with few exceptions, is the soil. The principal source of soil water is rain.

2.1.1 Different types of water

- *Run-away water* (not available to the plant): After a heavy rainfall or irrigation, some of the water drains away along the slopes. This is called run-away water.

- *Gravitational Water* (also not available to the plant): Some of the water percolates downwards through the larger pore spaces between the soil particles under the influence of gravitational pull until it reaches the water table.
- *Hygroscopic Water* (again not available to the plant): Water gets adsorbed on the surface of soil colloids and held tightly by them.
- *Chemically combined Water* (again not available to the plant): A small amount of water is bound to the molecules of some soil minerals by strong chemical bonds.
- *Capillary Water* (the form that is available to the plant): The remainder of the water that fills the spaces between the non-colloidal smaller soil particles.

Capillary water plays a major role as it is readily adsorbed by the roots. The total amount of water held in the soil is called 'holard'.

Holard	=	Chesard	+	Echard
[Total amount of soil water]		[Water available to the plants]		[Water not absorbed by the plants]

Field capacity or water-holding capacity of soil: This is the amount of water held by the soil after the excess water has been drained by gravitation. The hygroscopic and the capillary water together represent the field capacity of the soil (or its water-holding capacity). Much of the rainwater is, however, retained by the soil particles against the force of gravity thus making the soil wet.

Wilting coefficient of soil or permanent wilting point (PWP) is the amount of moisture left in the soil and expressed as a percentage of the dry weight of the soil, after a plant has permanently wilted'.

2.1.2 Roots hairs facilitate water absorption by roots

The root system serves to anchor the plant in the soil and above all, to meet the tremendous water requirements of the leaves resulting from transpiration. Most of the water that a plant takes from soil enters through the younger parts of the root. The root hairs, located several millimeters above the root tip, provide an enormous surface area for water absorption.

Root hairs are microscopic prolongations of root epidermal cells that greatly increase the surface area of the root, thus providing greater capacity for absorption of ions and water from the soil (Figure 2.1). Water enters the root most readily in the apical part of the root that includes the root hair zone. More mature regions of the root often have an outer layer of protective tissue called an **exodermis** that contains hydrophobic material in its walls and is relatively impermeable to water. The cell wall of root hair is thin, delicate and permeable to both solute and solvent molecules. It is comprised of two distinct layers (Figure 2.2).

1. The outermost layer consists of pectic substances, which are hydrophilic in nature and possess great water-imbibing properties.
2. The inner layer is made up of cellulose.

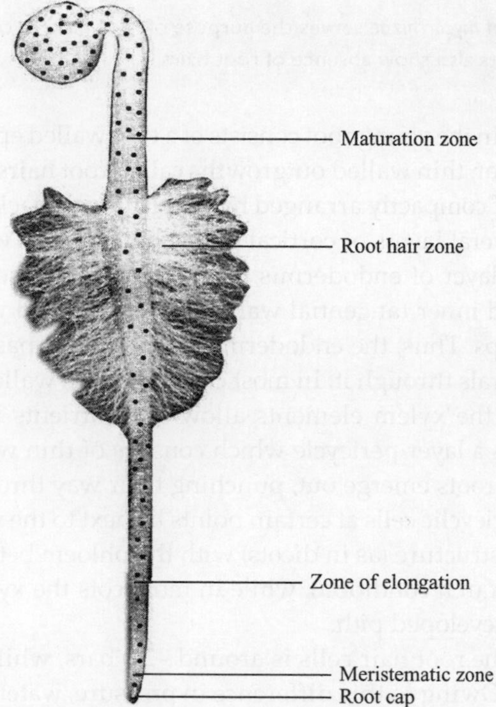

Fig. 2.1 A Young root showing various zones.

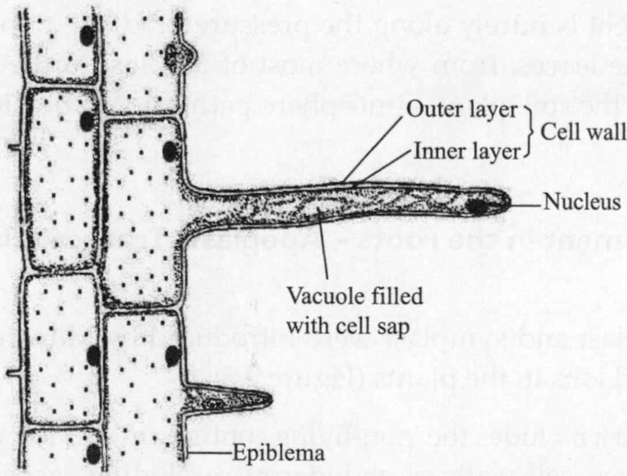

Fig. 2.2 Structure of a single root hair.

The root hairs remain active only for a short period, and as the root grows, most of them die and are sloughed off and new ones are continuously produced in the zone of elongation. Thus, as the plant grows the root hair zone is moving farther away from the base of the stem. The root hairs have sticky walls allowing them to adhere tenaciously to water films on soil particles.

Extensive development of *mycorrhizae* serves the purpose of root hairs in conifers and many forest trees. A few aquatic plants also show absence of root hairs.

The outermost layer in the young root consists of a thin walled epidermal layer (epiblema) from which arise tubular, thin walled outgrowths called root hairs. Below it, in some cases, there is a single layer of compactly arranged hypodermal cells lacking intercellular spaces. This is followed by several layers of cortical parenchyma tissue with intercellular spaces. Internal to this, lies a layer of endodermis, whose cells are closely packed and in many species their radial and inner tangential walls are impregnated with lignin and suberin, forming casparian strips. Thus, the endodermis obstructs the passage of water as well as the movement of minerals through it. In most cases, the thin walled passage cells (lacking thickenings) opposite the xylem elements allow the nutrients to pass through *via* the symplast. Inside this is a layer pericycle which consists of thin walled parenchyma cells, from where the lateral roots emerge out, punching their way through endodermis, cortex and epidermis. The pericyclic cells at certain points lie next to the xylem tissue arranged in a star or cross-shaped structure (as in dicots) with the phloem between the arms of xylem (radial bundles with exarch condition), while in monocots the xylem and phloem form a ring with a very well developed pith.

Water potential of the root hair cells is around –2.0 bars, while that of soil water is of the order of –0.5 bars. Owing to this difference in pressure, water will move from the soil (less negative water potential) into root hair cells (more negative water potential), that is from high DPD to low DPD, according to the old concept. A gradient of water potential exists between root hair cells, cortical cells and xylem elements. Once inside the elements of xylem, the movement is purely along the pressure gradient, passing through the root and stem and into the leaves, from where most of it is lost to the outer atmosphere by transpiration. Hence, the soil–plant–atmosphere pathway can be viewed as a continuum of water movement.

2.1.3 Water movement in the roots – Apoplast, Transcellular and Symplast Pathways

The concepts of apoplast and symplast were introduced by Münch (1930) to explain the transport of water and ions in the plants (Figure 2.3).

1. **Apoplast.** Apoplast includes the 'non-living continuum', i.e., interconnecting cell walls, intercellular spaces, cell walls of endodermis excluding casparian strips, cell walls of pericycle, xylem tracheids and vessels. Water moves through apoplast because of capillary action or free diffusion along the pressure gradient.
2. **Transcellular.** Transport of water occurs from cell to cell, passing from vacuole to vacuole.
3. **Symplast.** Symplast includes the 'living continuum', i.e., entire network of cell cytoplasm interconnected by plasmodesmata. Water moves through symplast because of osmosis.

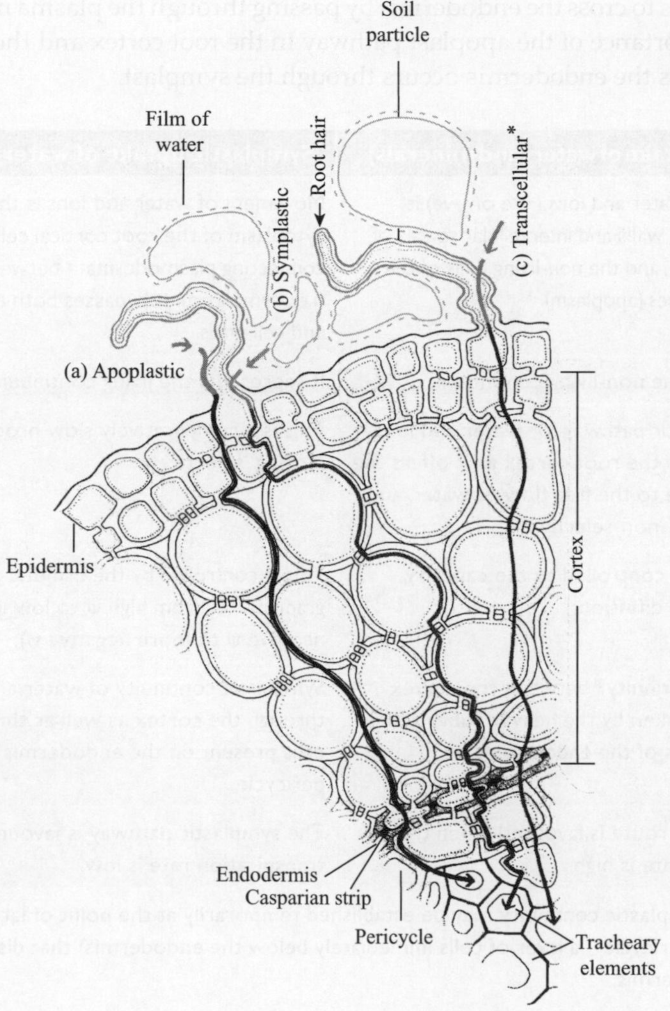

Fig. 2.3 Sectional view of the root showing apoplastic, symplastic (the vacuole is not a part of either pathway) and transcellular movement of water. The casparian strip creates a discontinuity in the apoplast. After crossing the endodermis, the water and ions may once again enter apoplastic and symplastic routes through the pericycle before reaching the xylem elements.

*This is yet another possible pathway for the movement of water called 'transcellular', i.e., from cell to cell, passing from vacuole to vacuole (rather than from protoplast to protoplast *via* plasmodesmata).

In practice, these pathways are closely associated. Apoplastic water is in constant equilibrium with water in the symplast and cell vacuoles. Rather, the apoplast appears to offer little resistance and probably accounts for a larger proportion of the water flow.

At the endodermis, movement of water through an apoplast pathway is hindered by casparian strips, bands of radial cell walls in the endodermis that are impregnated with the wax-like, hydrophobic substance suberin. Suberin acts as a barrier to water and solute movement. The casparian strip breaks the continuity of the apoplast pathway and forces water and solutes to cross the endodermis by passing through the plasma membrane. Thus, despite the importance of the apoplast pathway in the root cortex and the stele, the water movement across the endodermis occurs through the symplast.

Apoplastic uptake of water (and minerals)	Symplastic uptake of water (and minerals)
1. Movement of water and ions (+ve or –ve) is through the cell walls and intercellular spaces of the root cortex, and the non-living cells of the conducting tissues (apoplasm).	Movement of water and ions is through the cytoplasm of the root cortical cells and the connecting plasmodesmata between the cells (i.e., symplasm). It bypasses both the cell vacuoles and cell walls.
2. It represents the non-living continuum.	It represents the living continuum or component.
3. This is the major pathway for water and ions uptake through the root cortex as it offers the least resistance to the free flow of water, and ions that move 'non-selectively'.	This is a comparatively slow process with ions moving 'selectively'.
4. This process is controlled by the capillary action and free diffusion.	This is controlled by the osmotic or water potential gradient, i.e., from high ψ to low ψ (or from less negative ψ to more negative ψ).
5. Apoplastic continuity* between the cortex and stele is broken by the heavily suberized casparian strips of the endodermis.	Symplastic continuity of water and ions occurs through the cortex as well as through the passage cells present on the endodermis and then to the pericycle.
6. The apoplastic route is favoured when the transpiration rate is high.	The symplastic pathway is favoured when the transpiration rate is low.

* However, the apoplastic continuity may be established temporarily at the point of lateral root formation (arising from the pericycle – a layer of cells immediately below the endodermis) that disrupts the casparian strips of the endodermis.

2.1.4 Environmental factors influencing absorption of water by the roots

The absorption of water by the roots is controlled by the following external conditions:

1. *Soil temperature.* The rate of water absorption increases markedly with a rise in soil temperature up to a certain limit. Temperatures above 35 °C decrease water absorption by affecting the permeability of the plasma membrane. But soil temperatures below 27 °C reduces water absorption until it is insignificant at 0 °C. Cold soils are, therefore, 'physiologically dry'. Some of the most important causes of decreased absorption at low soil temperatures are as follows:
   (i) Increased viscosity of water.
   (ii) Decreased rate of root elongation.
   (iii) Alterations in the properties of the protoplasm.
   (iv) Reduced metabolic activity of the living cells of the roots.

2. *Soil air.* Water absorption by root cells is not a mere physical process but is a vital process that involves the expenditure of energy by cells through respiration. Oxygen, which is essential to plant respiration, is also essential for water absorption. In water-logged soils, water absorption is reduced because of **asphyxiation** of the roots. Soils lacking in oxygen become the hot beds of decomposition by anaerobic bacteria and the products of their activities, such as CO_2 and acids that are toxic to the roots, inhibit their development and metabolic activity. Poor aeration also causes decreased permeability of the roots to water. Water-logged soils are therefore, 'physiologically dry'.
3. *Available soil water.* Decrease in the soil water below the permanent wilting percentage is followed by considerable decrease in the absorption of water. If the soil water is increased appreciably beyond the field capacity, the soil aeration is affected, resulting in reduced absorption of water.
4. *Concentration of Mineral salts in the soil.* Absorption of water is also retarded by high concentration of salts in the soil water on account of high osmotic pressure (OP) of soil solutions. Alkaline soils and salt marshes are consequently physiologically dry for the plants. Therefore, high concentration of chemical fertilizers should be avoided. The best method is to tie the fertilizer in a muslin bag and to place the bag in the water channel at the time of irrigation.

Some morphological and anatomical features of the roots that allow maximum absorption of water from the soil:

1. The branching of the roots ensures that a large area of the soil can be tapped for water absorption.
2. The presence of the root cap is a safeguard against injury to the delicate parts of the root during its downward journey through the soil.
3. The abundant development of root hairs is an adaptation for increasing the water absorbing surface of the root.
4. The thin walls of the root hairs allow easy entry of water into the interior of the roots.
5. The anatomical architecture of the root is also an adaptation to the inward radial movement of water such as, absence of thick-walled cells in the cortex, thin-walled cells of pericycle, presence of thin-walled passage cells or unthickened pits in the tangential walls of cells and radial arrangement of vascular tissues.

2.1.5 Methods of water absorption

Any cell in an unsaturated condition may absorb water by virtue of its suction pressure when water is supplied to it. Leaves moistened with rain or dew may absorb water in this way. However, the amount of water absorbed is insignificant. A special type of absorption of water is found in many epiphytic flowering plants, especially in certain orchids. These plants have long, aerial adventitious roots that develop on the outside of their cortex a special water-absorbing tissue called the *Velamen*. Other epiphytes absorb water through hygroscopic hairs developed on their aerial parts. The absorbing surfaces of some conifers and many forest trees and shrubs are considerably modified by the presence of mycorrhizae.

Kramer (1949) recognized two distinct mechanisms:

- Active absorption (which may be either):
  (i) Osmotic
  (ii) Non-osmotic
- Passive absorption

2.1.5.1 Active absorption

Active uptake refers to water absorption in which driving forces originate in the root and are controlled by the living activity of cells. It occurs in early morning, late hours of evening and at night time.

Osmotic theory

The absorption of water from the soil occurs mainly through the root hairs. The vacuole of root hair contains an aqueous solution, containing minerals, sugars and organic acid. As a result, it has a high osmotic pressure (3–8 atm), whereas the soil solution is relatively dilute with low osmotic pressure (<1 atm). Thus, the capillary movement of water from the soil particles towards the root hairs is due to the cohesive force of water molecules. Water is then passed on to tracheary elements *via* cortex, endodermis and pericycle. Being inelastic the walls of xylem vessels do not exert turgor pressure; the whole of the osmotic pressure of xylem sap constitutes its diffusion pressure deficit (DPD).

In xylem cells, TP = 0, ∴ DPD = OP. Hence, the DPD gradient determines the movement of water from the soil to the xylem. Different layers of the cortex behave like a complex or multiple 'semi-permeable membrane'. Hence the rate of absorption of water depends upon the difference between osmotic pressure of the xylem sap and the soil solution, irrespective of the osmotic pressure of the intervening cortical cells.

The water is pushed upward from the roots into the stem through the cortical cells with forces originating in the innermost layer of cortex. The deeper layers of cortical cells push the water into the xylem elements of the roots and then to the stem with a force called **Root Pressure**. The magnitude of root pressure varies between 1 and 3 atmospheres under the most favourable environmental conditions.

Evidence

Root pressure ceases when the root cells are killed by the addition of poisonous substances to the soil. The high osmotic pressure of root cells which is necessary for water absorption results from absorption and accumulation of salts which is an energy requiring active process. Therefore, absorption of water indirectly requires metabolic energy.

Non-Osmotic theory

Some of the earlier workers like Bennet Clark (1936), Thimann (1951) and Bogen and Prell (1953) believed that there was a non-osmotic active water uptake mechanism in plants. Water was believed to be absorbed against a concentration gradient or at an accelerated rate. Such 'active' intake of water would require an expenditure of energy coming from the metabolic activity (e.g., respiration) of living root cells.

Evidences

1. Anaerobiosis (lack of O_2), low temperature and respiratory inhibitors (dinitrophenol, arsenic, azide, and such) which inhibit respiration of living cells also reduce water uptake.
2. Chemicals like auxins, which increase metabolic activity of cells, bring about an increased rate of water absorption.
3. The temperature coefficient (Q_{10}) of water absorption varies from 2 to 3, whereas for a purely physical process, the Q_{10} would be 1.2–1.3.

2.1.5.2 Passive absorption

According to theory of passive absorption, water is absorbed 'through' the roots of plants rather than pumped into the plant by them. The forces responsible for movement of water into the root originate in the shoot (transpiring cells) at the expense of solar energy and the root system, merely act as a physical absorbing system. In other words, water is simply pulled up or lifted from the bottom to the top.

The loss of water from the leaves by transpiration creates a tension in the xylem sap because of transpiration pull. This tension is transmitted downward to the root xylem creating a difference in the pressure potential between the root xylem and the soil solution, which forces the water movement from the soil into the xylem across the cortical cells of the root as a continuous stream.

Active absorption of water	Passive absorption of water
1. It is controlled by the activity of parenchyma cells of the roots where movement of water is from high ψ to low ψ (or, from less negative ψ to more negative ψ). The various forces originate in the roots (such as root pressure).	The forces (such as transpiration pull) that are responsible for uptake of water originate in the transpiring cells (leaves) owing to solar energy. The living cells of roots and stem play no role.
2. Absorption of water is 'by' the roots, and the water is 'pushed' or 'pumped' from below to the top.	Absorption of water is 'through' the roots of the plant, and it is truly 'pulled' upward from below as a continuous column.
3. It ceases when the living root cells are killed by high temperature or poisons.	It continues even if the living cells of the stem or roots are killed, i.e., does not depend upon living root cells.

Active absorption of water	Passive absorption of water
4. It occurs mainly in the early hours of morning before sunrise, evening, and at night time when there is no or little transpiration.	It occurs during day time when plants are actively transpiring.
5. It accounts for only about 5 per cent of the total water absorbed.	The bulk of water absorption (i.e., about 95 per cent) is through this mechanism.
6. It can be osmotic (or sometimes non-osmotic) requiring expenditure of metabolic energy that is in fact indirectly used to absorb and accumulate ions in the roots to build-up high osmotic pressure.	The movement of water is passive through the roots and stem, needing no expenditure of energy.
7. Root pressure and other related phenomenon such as guttation and bleeding are the main evidences. Here, the water oozes out from the stump when cut.	The water in the xylary or tracheary elements is under tension (or under negative pressure) and the water when supplied or applied at the cut 'stump' will be sucked in.

According to P.J. Kramer (1969), the active absorption of water is of negligible importance in the water economy of most or perhaps all plants.

2.2 Translocation of Water

Water in plants can be considered a continuous hydraulic system, connecting the soil water with the water vapours transpired from the leaves into the atmosphere. However, before the absorbed water is transpired, it must be conducted upwards to the transpiring surfaces against the force of gravity. The distance to be covered is sometimes very large, especially in tall trees and exceeds 300 feet as in the Australian eucalypts (*Eucalyptus amygdalina*) and some conifers. This upward translocation of water is called the 'ascent of sap' (sap includes water as well as minerals).

There are two aspects to the issue of translocation, namely, the pathways or channels governing the ascending stream of water and the forces that are involved in raising water to the top of 200–300 ft tall trees.

2.2.1 Pathway of the water stream

The upward translocation of sap in the plant occurs through the lumen of xylem elements. The tracheids and vessel elements are structured to be good conduits for water transport. This can be readily demonstrated by a simple experiment. A cabbage leaf or a twig of balsam (*Impatiens balsamina*) with a semi-transparent stem is excised under water and then placed in a weak solution of basic fuchsin. After a while, the presence of red streaks in the veins of the leaves indicate the involvement of xylem elements in the ascent of sap. Transverse sections of the stem or petiole cut at different levels also reveal that only the xylem (tracheids and vessels) has turned red while the phloem remains colourless.

It may be further confirmed from the ringing or girdling experiment conducted by the Italian scientist Malpighi[1] (1679). Bark and wood were cut in separate trials. The plant with the xylem cut and phloem intact will wilt indicating that upward transport of water is through the xylem. However, when the bark is cut and the xylem left intact, the plant remains fresh and the downward translocation of nutrients stops, leading to swelling above the girdle and even formation of adventitious roots because of accumulation of sugars (Figure 2.4 a, b).

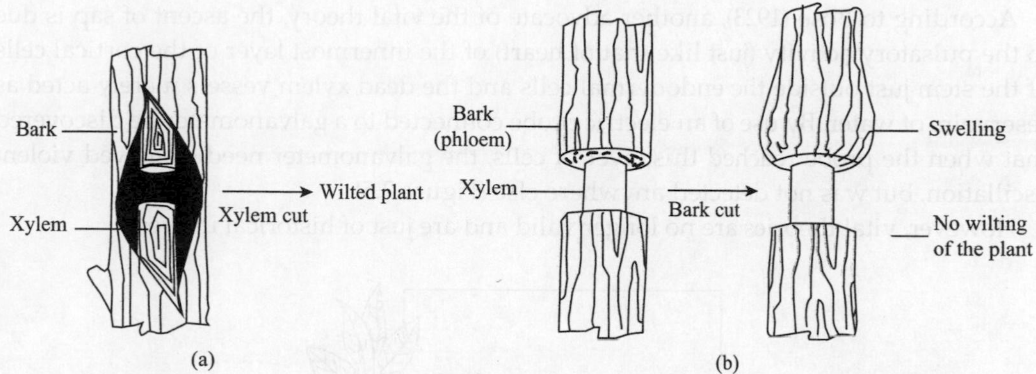

Fig. 2.4 Girdling experiment demonstrating: (a) the involvement of xylem (b) the involvement of phloem in translocation of water and nutrients respectively.

Besides the vertical movement of water through the xylem lumen, lateral transport also occurs from cell to cell. A water column within a single conduit along the length of a tall tree would not be very stable, and it is the lateral connections between the cells which give stability to the conducting system.

From time to time, various theories have been propounded to explain the ascent of sap in plants. These fall into two categories: Vital and Physical theories. According to the vital theories, ascent of sap takes place through the activities of living cells. However, as ascribed in physical theories, it occurs through dead cells due to certain physical forces like capillarity, imbibition, atmospheric pressure, root pressure, transpiration pull, and so on.

2.2.2 Vital theories

Godlewski (1884) proposed that the upward movement of water was due to the pumping action of the living cells of xylem parenchyma and medullary rays brought about by periodic changes in their osmotic concentration. With an increase in the osmotic pressure in these cells, water gets withdrawn from the neighbouring vessels. This leads to a decrease in their osmotic pressure and water is thus, again pumped into the vessel above. In this way water moves up step by step. This theory was refuted by the experiments of Strasburger (1893) who demonstrated that water continued to rise in plants even after their living cells had been killed by exposure to high temperature or through immersion in solution of poisonous substances. He performed a simple experiment that seemed to eliminate the necessity of

[1] Marcello Malpighi is also famous for discovering the Malpighian tubules of the kidney.

involving living parenchyma cells of the xylem in water transport. He decapitated the trunk of 22 meter tall oak tree at the base and placed the cut stump in a solution of picric acid. The picric acid solution was transported to the leaves along with the sap, killing the parenchyma cells on its way. Transpiration and upward picric acid movement occurred until the acid solution reached the leaves. Acid destruction of the leaves stopped transpiration and further upward movement of the water. This yielded virtually all the information required for Dixon and Joly to formulate the cohesion hypothesis for the ascent of sap.

According to Bose (1923), another advocate of the vital theory, the ascent of sap is due to the pulsatory activity (just like that of heart) of the innermost layer of the cortical cells of the stem just outside the endodermal cells and the dead xylem vessels merely acted as reservoirs of water. By use of an electric probe connected to a galvanometer he discovered that when the probe reached this layer of cells, the galvanometer needle showed violent oscillation, but was not detected anywhere else (Figure 2.5).

However, vital theories are no longer valid and are just of historical interest.

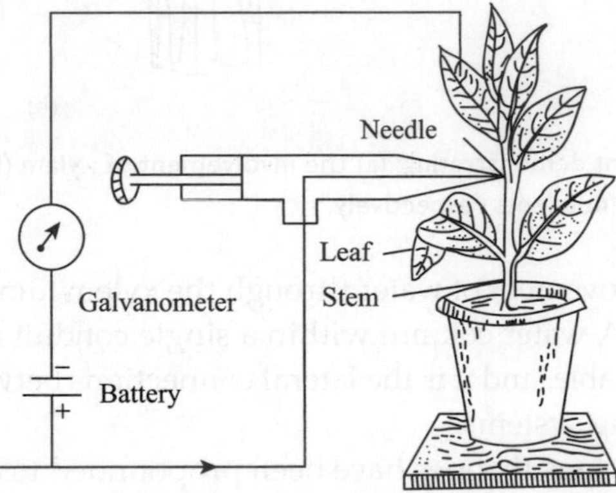

Fig. 2.5 By using an electric probe, Sir J. C. Bose proposed that the ascent of sap is due to the pulsatory activity of the innermost layer of cortical cells of the stem.

2.2.3 Physical theories

2.2.3.1 Capillarity theory (Boehm, 1809)

Xylem ducts are considered to consist of very narrow capillary tubes and ascent of sap occurs because of capillary force in much the same way as oil rises in the wick of an oil lamp. The smaller the capillary bore, the greater would be the rise of water. Capillarity is due to interaction of several forces, i.e., adhesion, surface tension and the force of gravity acting on the water column. The chief objections to this theory are factors such as absence of free surface or meniscus in the vessels, the presence of cross walls and the low magnitude of the force that is insufficient to raise water to greater heights. Moreover, the vessels possessed by the tallest trees do not have a narrow bore as expected. The capillary bore of xylem vessels or tracheids is approximately 40 µm in diameter which can raise water by capillary force only up to 40 cm. Further, for capillary action to operate, one end of the capillary tube should be in touch with water whereas in nature, the xylem ducts of the root are not in contact with soil water.

2.2.3.2 Imbibitional theory (Sachs, 1878)

'Imbibition' is the term for processes such as when a sponge absorbs water. According to this theory it is imbibitional force of the cell walls of the xylem that is responsible for the ascent of sap. However, it has been established beyond doubt that the translocation of water occurs through the xylem lumen.

2.2.3.3 Atmospheric pressure theory

Here, it is contended that transpiration creates a vacuum in the plant with the result that water is forced upwards from below to compensate for the pressure. However, the atmospheric pressure needs a free surface at the lower end. But it does not exist due to the presence of several layers of living cortical cells that lie between xylem cells and the soil water. Secondly, atmospheric pressure would be able to lift water only to a maximum of 30 feet.

2.2.3.4 Root pressure theory (Stocking, 1956)

He defined root pressure[2] as 'a pressure developing in the tracheary elements of xylem as a result of the metabolic functions of roots'. Although root pressure is purely a matter of diffusion pressure gradients, it is, however, maintained by the activity of living cells. Therefore, it is referred to as an active process. However, the movement of water up the stem as a result of root pressure is due to osmotic mechanisms that are created through active absorption of salts by the roots. Guttation and bleeding are the best examples of root pressure.

Objections

1. The magnitude of root pressure is about 2 atmospheres while a pressure of about 20 atmospheres is needed to raise water to the topmost part of tall trees. In addition, it does not take into account the friction encountered in the passage of water through xylem ducts.
2. Root pressures of any magnitude are conspicuously absent from most of the conifers, which are among the tallest of trees.
3. Water continues to rise upwards even in the absence of root pressure.
4. Root pressure shows seasonal fluctuations – highest during spring when transpiration is slow and lowest in summer when transpiration is relatively very rapid. So root pressure is less when requirement for water is more.

2.2.3.5 Cohesion hypothesis

Cohesion theory is the most widely accepted theory for ascent of sap through the plants. This was proposed simultaneously by two groups of scientists Dixon and Joly (1894) in UK and Askenasy (1895) in Germany. Dixon and Joly got the recognition since Askenasy's

[2] The term 'Root Pressure' was coined in 1727 by Stephen Hales, the father of plant physiology.

work was published later. It was later elaborated by Dixon (1914, 1924) and supported by the works of Renner (1911, 1915), Curtis and Clark (1951), Bonner and Galston (1952) and Kramer and Kozlowski (1960).

Principle: Water is assumed to form a continuous column extending from the growing shoots and leaves to the roots. Evaporation of water from the mesophyll cells of the leaf provides the driving force for water movement and thus increases the suction force that leads to withdrawal of water from the dead tracheary elements. This puts the column of water under a strain or tension. The xylem cells merely act as inert tubes but are of considerable importance because they prevent the collapse of the water column and also prevent the entrance of air bubbles. The column resists breaking because of the strong cohesive and adhesive forces between the water molecules and with the cell walls respectively and thus the water is pulled upward from the roots as a 'continuous column'. The living cells of the stem do not play any role.

Water covers the leaf mesophyll cell surface in the form of a thin film, clinging to cellulose and other hydrophilic surfaces. Evaporation from the intercellular spaces causes the water–air interface to withdraw into the small spaces between cellulose microfibrils and at the junctions of leaf mesophyll cells. As the water withdraws, the resulting surface tension pulls water from the nearby cells. The water column being continuous, this tension is transmitted through the column, ultimately down to the roots and soil water (Figure 2.6), therefore popularly known as the 'cohesion-tension' hypothesis also.

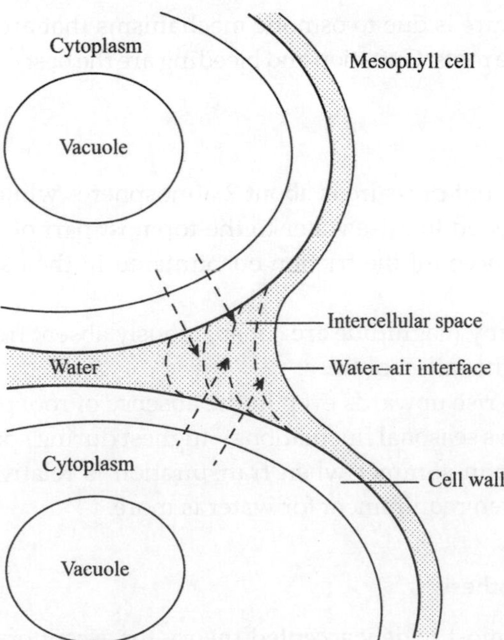

Fig. 2.6 An outline diagram of two neighbouring cells from a leaf along with intercellular space showing negative pressure in the water column, which is ultimately transmitted to the roots and soil water.

Pre-requisites

For this mechanism to be effective, the following conditions must be fulfilled:

- Parenchyma cells at the top of a plant must have an osmotic concentration in excess of that necessary to raise the water to the height and to overcome the frictional resistance.
- Cohesive strength of water must be sufficient to withstand the pull exerted upon it.
- Xylem conduits must prevent the entrance of air bubbles, which would otherwise expand and break the water column.
- Water conducting xylem conduits must be rigid enough to prevent collapse when the contents are under tension.

Evidences

- Osmotic concentrations of the cells in the upper part of a plant (which are between 10–20 atmospheres of pressure) are more than adequate to raise the water to the top.
- Cohesive strength of water has been found to be approximately 200 atm, which is sufficient to overcome the frictional and gravitational forces encountered in vertical rise of water in a plant.
- Although air bubbles are found to occasionally enter tracheids and vessels, the wet cell walls seem to be adequate in effectively preventing such bubbles from spreading into neighbouring elements.
- The strongest evidence in favour of Cohesion hypothesis has come from the work of Thut (1932) working at Cornell University (USA), who demonstrated the movement through the stem when water is actually under tension. By using leafy twigs of several woody plants, he demonstrated a rise of mercury above atmospheric pressure ranging up to 101.8 cm, when atmospheric pressure was mostly below 75 cm. This proves that water can be pulled up as a cohesive column.
- Dendrographic measurements by Kramer and Kozlowski (1960) also revealed that the diameter of the tree trunks decreased significantly during the day when transpiration rate was high. This was due to the strain caused on the water column by tension. Direct measurement of tension in xylem vessels was made possible by the introduction of the **Pressure Bomb Technique** by Scholander et al. (1965).

Limitations – cavitation and embolism

Despite the strong cohesive forces, however, water columns under tension in the xylem are somewhat unstable and if the atmospheric pressure is reduced, air enters at the wound site. Water tends to vaporize, filling the entire tracheid (**cavitation**) followed by the release of dissolved gases in the form of bubbles (**air embolism**), thus blocking water movement, i.e., cavitation precedes the air embolism.

Before 1972, cavitation was thought to be a rare phenomenon but now it is considered to be a regular event faced by tropical as well as temperate plants. Ideas began to change in 1974, when John Milburn at the University of Glasgow discovered the **acoustic detection**

technique. The basic fact is that the sudden release of pressure during cavitation causes the cell walls to vibrate, producing an audible click (Figure 2.7).

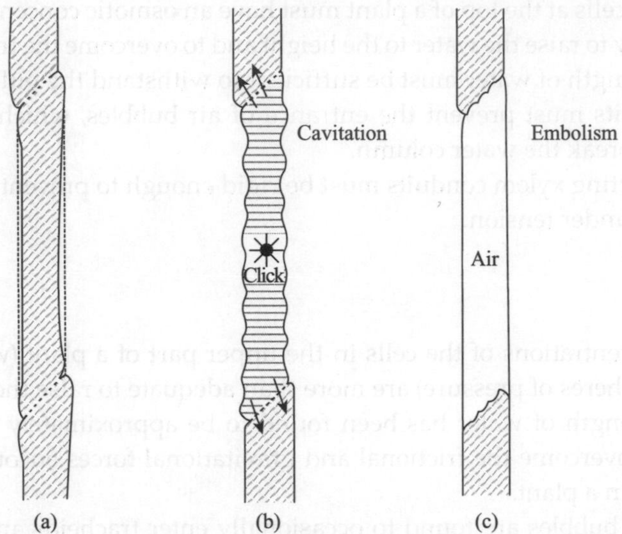

Fig. 2.7 Representation of cavitation and embolism formation in a xylem conduit. (a) Under extreme tension, the walls are strained inwards from the dotted line. (b) A cavitation 'bubble', filled with water vapour, forms and the strained walls vibrate, detectable as an audible 'click'. Water exits to adjoining conduits (arrows). (c) The evacuated conduit gradually becomes air-filled (embolism) due to gas coming out of solution from sap in the walls, etc.

With suitable amplifying equipment, the clicks can be heard and counted. Each click is known to represent the formation of air bubbles (or embolisms), usually confined to a single vessel element because the pits which link up adjacent cells are lined by a 'pit membrane' and through surface tension effects, this prevents the bubble from spreading. Bordered pits act as more sophisticated valves that close when the pressure on one side rises. Another feature of xylem that minimizes the effects of embolism is the ease with which water can move sideways through the pits, thus bypassing the blocked conduit (Figure 2.8).

Causes of cavitation and embolism

- Water stress developing with high rates of transpiration and low xylem pressures, particularly in herbaceous plants cause tiny bubbles of air to squeeze through pores in the thin pit membranes, thus separating an accidentally embolised conduit from its neighbour. The bubbles act as seeds or nuclei for further cavitation and the bigger the pores, the more likely is bubble formation.
- Sudden increase in temperature in the early morning – when the sun rises, dissolved gases are likely to appear in the form of gas bubbles inside the xylem ducts just like as the bubbles form when a glass of cold water becomes warm.

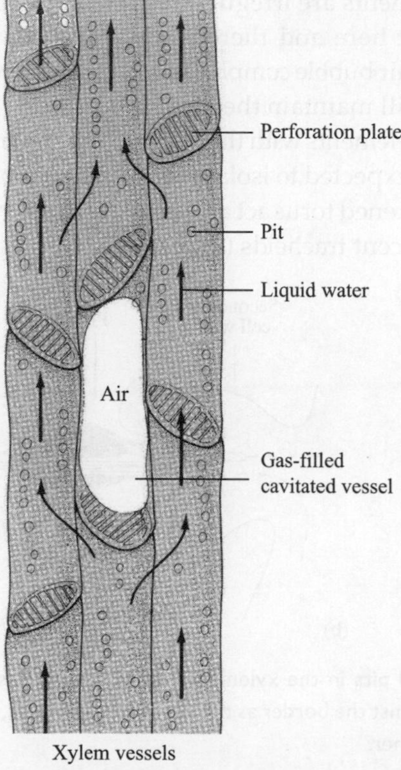

Fig. 2.8 Diagram showing sideways flow of water through pits in xylem elements, thus bypassing the blocked vessels.

- Freezing of xylem sap in winter – the dissolved gases become less soluble in water and tend to come out in the form of bubbles as we see in an ice slab or cubes. During thawing, these small bubbles expand and nucleate cavitation.
- Pathogen-induced embolism has been identified recently as the first major damage caused to the xylem by the Dutch Elm disease fungus. The pathogens might release oxalic acid into xylem sap. Since the oxalic acid is known to lower surface tension, it greatly facilitates air-seeding at pit membranes.
- Shaking and jarring action of wind, mechanical disturbances, i.e., natural calamities like tornado, windstorm and ionizing radiations can also disrupt the continuity of water column and stop water transport.

The impact of cavitation and embolism on long-term survival of plants could be dangerous if there were no ways of their removal or minimizing their effects.

Plants have been able to overcome the problem of cavitation to a great extent by the following means:

- The xylem elements form an intricate (anastomosing) meshwork within the plant system and even if a part of the elements are air-plugged, the continuity of water will be still maintained through the other elements of the meshwork.

- Walls of the xylem elements are irregular. So the air bubbles never occupy the entire space, crevices are left here and there in the cell and water column may still be continuous. Even if the air bubble completely ruptures the water column, water imbibed by the cell walls will still maintain the continuity.
- Wet cell walls of xylem elements with their special pit-closing membranes (with torus in gymnosperms) can be expected to isolate occasional bubbles from the water columns.
- Bordered pits with thickened torus act as 'self-regulating' valves, preventing the escape of bubbles into the adjacent tracheids (Figure 2.9).

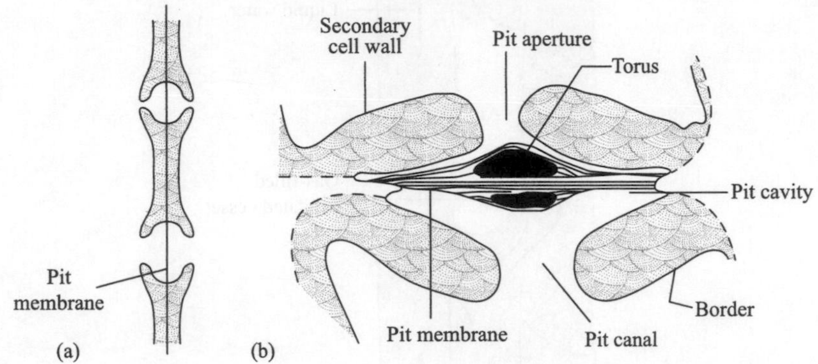

Fig. 2.9 Structure of bordered pits in the xylem. Bordered pits act as more sophisticated valves. The thickened torus is pushed out against the border as the gas bubble expands, thus preventing it from spreading from one xylem element to another.

- Following water stress, when adequate water supply returns, xylem conduits transport enough water to the shoot to enable the cambium to form new xylem cells (sapwood) for the shoot.
- Although the plants are subjected to natural stresses like strong winds and gales but they become least susceptible to jarring action because of their deep-seated location (forming the central core) and lignified walls of the xylem elements.
- Plants may also avoid long-term damage by repairing the embolism during night time, when transpiration rate is at a minimum. Decreased tension in the xylem sap allows the gas to re-dissolve in the xylem solution.

As the embolism is usually confined to a single tracheid or vessel; so, all the xylem conduits seldom cavitate at the same time; so the water conducting system as a whole remains functional.

Review Questions

RQ 2.1. List out the different types of water in soil. Which one is of paramount importance to plants?

Ans. The different types of water in the soil are categorized as follows:

1. Run-away water.
2. Gravitational water.

3. Hygroscopic or imbibed water – it is held tightly on the surface of soil colloids such as clay and humus.
4. Capillary water – it fills the spaces between non-colloidal smaller soil particles or forms thin films around them. It is of greatest importance to plants.
5. Chemically combined water.
6. Water vapours – epiphytic bromeliads, particularly *Tillandsia*[3] species.
7. Dew-water that has been condensed from the air.
8. Mist from the sea – *Welwitschia mirabilis* growing in the Nambibian desert.

RQ 2.2. Absorption of water is independent of absorption of minerals. Comment!

Ans. Absorption of water is independent of the absorption of mineral salts as (i) they are absorbed from different regions of the roots, while the water is absorbed by the root hairs; the minerals are, however, taken up from the meristematic zone and the zone of cell elongation at the root tip; (ii) water is absorbed entirely by osmotic forces, water molecules moving from region of high ψ in the soil to regions of low ψ in the roots. The cells of the roots are fully permeable to water. 'Active absorption' occurs in the early morning, evening and at night time when there is low transpiration. However, during daytime when transpiration rate is high, absorption of water is mainly through 'passive absorption'. On the other hand, absorption and accumulation of minerals is against the concentration or electrochemical gradient, requiring expenditure of metabolic energy (active transport), and (iii) absorption of mineral salts by the roots is a selective process. The roots take up ions vigorously when no absorption of water is occurring. Different mechanisms for salt uptake have been proposed from time to time.

RQ 2.3. Why are cold soils physiologically dry?

Ans. Cold soils do contain water which is rendered unavailable to the plants. Some of the causes of decreased absorption of water are as follows:

- As the temperature of the soil decreases, the viscosity of water increases correspondingly. This not only retards the movement of water from the soil to the roots but also slows down the water movement through the roots.
- At low temperatures, the viscosity of protoplasm increases, becoming less permeable to water. As a result the movement of water across the roots is considerably slowed down.
- At low temperatures, the roots fail to encroach upon deeper and farther regions of the soil which affect the water uptake.
- At low temperatures, there is reduced metabolic activity of the living cells of the roots which affects the energy level which may be directly or indirectly required for water absorption.

RQ 2.4. When crushed ice is placed over the surface of the soil in a potted plant, it soon wilts. Explain.

Ans. When crushed ice is placed over the soil in a potted plant the soil temperature decreases. Such cold soils would affect or retard the uptake of water (see question RQ 2.3 given above). However, the rate of transpiration far exceeds the rate of water uptake, leading to loss of turgor in the leaves and stem, which will then cause wilting.

RQ 2.5. Plants show sign of wilting if the fields are heavily flooded or irrigated. Comment.

Ans. Water absorption by root cells is not merely a physical phenomenon (involving imbibition and osmosis) but also is a vital process requiring energy that is released during respiration. Oxygen is essential for respiration as well as for water absorption.

---

[3] These species are without roots and can obtain water from mist as well as from dew, e.g., *T. purpurea* and *T. usneoides*.

Naturally, enough O_2 is present in the soil air. However, water-logging or flooding rapidly depletes the available O_2 in the soil and the roots are asphyxiated. Under such conditions, root growth and metabolism are both retarded. Absorption of water lags behind the rate of transpiration and the plants undergo severe wilting.

Soils lacking O_2 become the hotbeds of decomposition by anaerobic bacteria. The end-products of fermentation such as H_2S and ammonia are injurious or toxic to root cells, thus inhibiting their development and metabolic activity. Furthermore, the hormonal levels change in flooded plants. Ethylene (C_2H_4), for example, often increases, while gibberellins and cytokinins usually decrease, contributing to abnormal growth patterns.

Lack of oxygen (anaerobiosis, as it is called) results in anaerobic respiration (glycolytic fermentation in roots) that greatly reduce the supply of ATP which is necessary for the uptake of minerals that help in maintaining osmotic balance within root cells. This is necessary for the absorption of water.

RQ 2.6. Imbibition and other physical forces are the first step towards absorption of water by plants growing under lowly transpiring conditions. Comment!

Ans. The absorption of water takes place chiefly through the root hairs. As the cell walls of the root hairs are covered by pectic material, they absorb water through imbibition. From this point onward, water passes into the root hairs by osmotic forces. As the films of capillary water are removed by the root hairs, the other films of water from the adjacent soil particles are drawn in, which in turn draws water from the deeper layers. It should, however, be emphasized that this capillary movement of water towards the root hairs is due to the cohesive force between water molecules. In this way, the water from considerable long distances is brought into contact with root hairs.

RQ 2.7. List evidences supporting active absorption of water.

Ans. The following indirect evidences in favour of the theory of 'active absorption' of water are:

1. The factors that inhibit respiration also inhibit the absorption of water.

2. Low temperature, lack of O_2 and the respiratory inhibitors which inhibit respiration of living cells, also inhibit water absorption.

3. Auxins that increase the respiration rate also bring about an increase in water absorption.

4. Poisons like KCN reduce respiration, which in turn decrease the absorption of water.

These four evidences given above suggest that respiration rate and absorption of water are closely linked and are interdependent.

RQ 2.8. When do the mechanisms of active and passive absorption of water operate in plants?

Ans. Active absorption of water occurs under mesophytic or lowly transpiring conditions associated with positive root pressure. This occurs in the early morning, evening and at night time. The water is pushed up from below by forces such as root pressure developing in the water-absorbing organs.

On the other hand, passive absorption that accounts for about 98 per cent of water uptake takes place during day time when the plants are actively transpiring. Here, water is pulled upward by the forces developing in the transpiring cells of the leaves.

RQ 2.9. List the important functions of roots.

Ans. Roots have the following important functions.

- Anchoring the plants firmly in the soil.
- Providing a place for storage of carbohydrates and other organic molecules.
- Roots are sites of synthesis of alkaloids and some hormones. The synthesis of cytokinins is largely in the apical meristem of the roots from where they are transported up in the water stream. The biogenesis of nicotine in tobacco occurs in the roots but it accumulates in the leaves.

- Absorb and transport mineral salts and water upward to the stem.
- For vegetative propagation in certain species.

RQ 2.10. What special functions do the endodermal cells of the roots perform? Explain.

Ans. In many species of flowering plants, the endodermal cells are lignified, suberized or partially cutinized on the radial and tangential walls, thus obstructing the movement of water and minerals. The endodermal cells with casparian strips act as a barrier that prevents diffusion along the walls between cortex and stele, i.e., apoplastic movement. Here, the movement of water and solute is through the symplast, and only the ions that the plants require are 'selectively' pumped in and eventually released into the xylem for onward and upward transport.

Firstly, casparian bands[4] also prevent the free return of water back into the cortex, which would otherwise affect O_2 diffusion and the tissue would suffocate. Secondly, the impermeable endodermal cell walls also function to prevent the salts pumped into the tracheary elements from diffusing back into the cortex and then to the exterior through interconnecting plasmodesmata in the cell walls of adjacent cortical cells, thus forming an osmotic barrier.

RQ 2.11. Respiratory inhibitors or anaerobiosis inhibit the uptake of water. Explain.

Ans. Poisoning the respiration with respiratory inhibitors (such as azides, cyanide, arsenates, dinitrophenol) and 'anaerobiosis' (depriving the tissues of oxygen – called 'anoxia') greatly reduces or stops water uptake. This statement is true but their role is indirect.

Respiratory inhibitors will 'stop' the flow of electrons *via* electron transport carriers (such as from NADH, $FADH_2$ to cytochromes b, c, a, a_3) during oxidative phosphorylation occurring in the mitochondria. Hence, 'energy' in the form of ATP is 'not' released. However, this energy is not 'directly' involved in the absorption of water but is used up in the absorption and accumulation of salts in the root cells, thus maintaining the osmotic balance which is necessary for the active absorption of water.

Similarly, oxygen is essential for the terminal oxidation in the electron transport chain during oxidative phosphorylation. Lack of oxygen (anaerobiosis) will prevent the unloading of NADH to NAD. NAD is essential for the continuation of glycolytic and Krebs cycle reactions (or it will come to an abrupt halt unless the 'H_2' component is unloaded; NAD occurs in minute amount in the cell).

Absence of oxygen around the roots inhibits aerobic respiration in roots. Such soils become the hotbeds for the activities of anaerobic bacteria and their by-products like hydrogen sulphide, ethylene, ammonia and other products of fermentation, especially ethyl alcohol, lactic acid and malic acid that are potentially harmful to the roots, thus affecting root growth and metabolism. Poor aeration causes decreased permeability of roots to water. Therefore, the absorption of water is inhibited and the plants undergo wilting as the rate of transpiration far exceeds the rate of uptake of water.

> Irrigation of soil beyond a certain limit (flooding) has a retarding effect on water uptake due to depleted supply of O_2 in the soil (i.e., the air is dispelled out from the soil under flooded conditions).

Under such waterlogged conditions, the anaerobic soil bacteria reduce nitrate and sulphate into ammonia and hydrogen sulphide, respectively. These gases are toxic, and even in traces would

[4] The casparian band is principally composed of suberin, a complex mixture of hydrophobic, long chain fatty acids and alcohol.

kill the plants. Moreover, the mineral salts like nitrate, sulphur, iron and manganese become reduced and are rendered unavailable to the plants. Mineral uptake which is an active process becomes reduced because not much respiratory energy is available. This alters the osmotic balance, thus affecting water uptake although plants are growing literally in standing water. Due to impaired water absorption, plants will show symptoms of wilting.

RQ 2.12. Cavitation precedes embolism in xylem conduits during water transport. Comment.

Ans. Under extreme tension (low pressure), water columns in the xylem become unstable and the water is likely to undergo vaporization locally, filling the xylem conduits with 'water vapour'. Water molecules in the vapour phase have very low cohesion and submicroscopic bubbles are formed first. This process is called 'cavitation'. The air then escapes out in the form of bubble or 'embolism', blocking water movement. The small bubbles act as seeds or nuclei for further expansion of bubbles. The resulting large gas bubble forms an obstruction.

The sudden release of pressure during cavitation causes the xylem walls to vibrate, producing an audible 'click' and with suitable amplifying equipment the click can be heard and counted.

Air embolism is no longer a rare event, occurring in both tropical and temperate plants.

RQ 2.13. Outline the historical background of the transpiration pull theory.

Ans. Transpiration pull theory was proposed by Dixon (Irish botanist) and Joly (a physicist colleague) in 1894, and Askenasy in Germany in 1895. The theory was elaborated by Dixon (1914, 1924) in greater details and that is why it is popularly known as Dixon's cohesion–adhesion hypothesis or transpiration pull theory. It has been supported by workers like Renner (1911, 1915), Curtis and Clark (1951), Bonner and Galston (1952) and Kramer and Kozlowski (1960).

RQ 2.14. What is acoustic detection technique? In which ways is it useful in the studies of ascent of sap?

Ans. Until the mid-1970s, cavitation was considered to be relatively a rare event but in 1974, John Milburn and Johnson at the University of Glasgow devised an acoustic technique for the detection of cavitation in plants. If a plant is put under severe water stress (i.e., deprived of water), cavitation starts to occur in individual xylem tube and this can actually be heard as 'clicks' by using sensitive microphones and amplifiers placed next to the stem. Each click is believed to correspond to a bubble formation in a single vessel member. And so it was realized that cavitation was not a rare event but quite common.

Melvin Tyree and his collaborators from the University of Toronto (1989) have calculated the reduction of water flow through the xylem of corn plant (*Zea mays*) and maple (*Acer saccharum*). Such flow reductions are potentially a serious matter as it influences the productivity or yield. We should look for ways how to repair or circumvent the damage.

RQ 2.15. The water molecules do not snap away from each other due to tension in the xylem. Comment!

Ans. The water column in the xylem elements is under great tension (i.e., shows negative pressure), its magnitude may be as high as −142.9 bars but it resists being pulled apart because of strong cohesive force between water molecules (more than adequate; i.e., about 200 bars or 197.38 atmospheres) and the adhesive force between water molecules and the walls of tracheary elements. Water simply moves passively upwards through the xylem conduits as a continuous column.

RQ 2.16. In multi-storied buildings, it is necessary to use a booster pump to lift water to the top. How is it possible to lift water in tall trees, measuring as much as 400 ft high? Explain.

Ans. It is absolutely essential to install a booster pump to lift water in our multi-storied buildings. However, in tall trees, the function of booster pump is taken over by the transpiring cells of the leaves (working as booster pumps powered by solar energy). Transpiration results in the development of transpiration pull in the leaves and this pull is transmitted through the petiole down to the stem, and then all the way down to the roots. The water is simply pulled upward or lifted to the top as a continuous column. It resists breaking because of the strong cohesive force between water molecules, and the adhesive force between water molecules and the walls of tracheary elements. In fact the living tissues of the leaves here act as a booster pump.

RQ 2.17. Compare the energy requirement needed to translocate and evaporate water.

Ans. To raise 1 ml of water to 100 m height would require about 0.5 cal, while to evaporate 1 ml at 20 °C would require 592.5 cal which is 1185 times more. Thus, the energy requirement to raise the water is insignificant as compared with that necessary to evaporate it.

> At sea level, the atmosphere pressure is sufficient to raise water in a narrow tube to a height of about 10.4 m or 34 ft and it is clearly inadequate for raising water in 'taller' trees.

RQ 2.18. Explain where and when stem pressure is independent of the root system?

Ans. Positive xylem pressure occurs in the sap flow in the sugar maple tree (*Acer saccharum*) of Northeastern United States and Canada. This is generally seen on warm days following a cold night. Cold nights lead to the hydrolysis of starch reserves in the xylem parenchyma into sugars which are then actively transported into the vessel elements of the xylem.

Warm temperatures cause the release of CO_2 from the sap, thus creating a pressure in the xylem and as a consequence, water and sugars are forced up the trunk, just before spring growth starts.

RQ 2.19. Explain how the solubility of gases is affected by temperature and air pressure?

Ans. It is interesting to note that the solubility of gases in liquids decreases as temperature increases. For example, at 760 mm pressure, 0.04889 L of oxygen will dissolve in 1 L of water at 0 °C and only 0.01761 L at 80 °C. In fact, boiling is one common laboratory practice for removing gases from a liquid. However, the solubility of gases (with the exception of highly soluble gases) in liquids increases with an increase in pressure. The dissolution of carbon dioxide in carbonated beverage is accomplished under a pressure of about 5 atmospheres and the bottled beverage is then capped. When the cap is removed, the pressure above the solution drops to 1 atm, and the gas escapes in the form of bubble – this phenomenon of release of gas bubbles from a liquid is known as 'effervescence'.

> Dissolved gases in the xylem sap are likely to form air bubbles when the temperature increases by several degrees after the sunrise or air pressure is reduced when the water column in the xylem tubes is under tension (or negative pressure). Similarly, dissolved gases in a bottled cold drink form bubbles when air pressure is reduced such as while sipping the drink with a straw.

RQ 2.20. What is root pressure? Discuss briefly its magnitude and significance.

Ans. Root pressure is a vital phenomenon, depending upon the activity of living cells of the roots. During 'active absorption', water is forced into the xylem elements by the surrounding cortical cells of the roots with a certain force which induces a pressure sufficient enough to raise the water to many feet in the xylem. This pressure is called 'root pressure'.

Root pressure can be positive or negative depending upon environmental conditions. Positive root pressure is exhibited by plants growing under well watered and lowly transpiring conditions while negative root pressure is often found under xeric and highly transpiring conditions.

The magnitude of root pressure is usually small (1–3 atmospheres) under favourable conditions, and the higher magnitudes reported by Molisch and White are of rare occurrence. The xylem is under positive pressure as a result of osmotic forces generated by the roots. Guttation is a consequence of positive root pressure. A large quantity of fluid is forced out of the vein endings through hydathode.

Root pressure is a potent force that 'pushes' the water from below upward in small herbaceous plants rather than a 'pull' from above (as in passive absorption).

Root pressure, although usually small, may function as a refilling mechanism in herbaceous plants such as grape vine (*Vitis*) in spring, forcing air out of embolized xylem conduits causing it (gases) to enter the solution.

RQ 2.21. Why is Strasburger's experiment of 1893 on the ascent of sap so significant? What conclusions do you draw from here?

Ans. Eduard Strasburger, a German physiologist (a pioneer investigator of mitosis and meiosis in plants), decapitated a 22-m-tall oak tree at the base and then the cut stump was placed in a huge trough containing a solution of picric acid or $CuSO_4$.

The picric acid solution was transported to the leaves along with the xylem sap, killing living parenchyma cells of the stem on its way. Transpiration and upward movement of picric acid continued until the acid solution reached up to the living cells of the leaves. Acid destruction of the leaves stopped transpiration and hence the upward movement of water.

This simple experiment provided virtually all clues or all the information that prompted physiologists to formulate the hypothesis regarding water uptake.

1. Since the roots were removed, it eliminated the role of root pressure as the major force.
2. It further eliminated the role of living cells of the stem in the ascent of sap, as they were killed by picric acid, and water simply moved through the non-living tracheary elements.
3. Thirdly the water must have been pulled upward by the forces, resulting from transpiration. Water flow along with dye stopped once the leaves were killed.

Chapter 3

Transpiration

Keywords

- Antitranspirants
- Bleeding
- Foliar transpiration
- Guttation
- Hydathode
- Transpiration flux

In plant-water relations studies, the dominant processes are the uptake of huge quantities of water from the soil, its translocation through the plant and its subsequent loss from the aerial parts to the surrounding atmosphere in the form of water vapour through transpiration. Out of the total water absorbed by the plant during a growing season, about 99 per cent is lost to the atmosphere in the form of water vapour and only 1 per cent is retained by the plant. Out of this 1 per cent, roughly 0.9 per cent is retained as free water within the tissue and about 0.1 per cent enters into the plant's metabolism as a reactant in chemical reactions (e.g., photosynthesis). The process of water loss in the form of vapour from the leaf surfaces of the plant is called **transpiration.**

The loss of water is alarmingly high. The daily water loss of a large, well-watered tropical plant such as palm may be as high as 500 litres. A corn plant may lose 3–4 L/day and a medium-sized elm tree may lose more than a ton of water every day.

Transpiration is not a purely physical process like evaporation (loss of water from the surface other than the plant, such as soil) but is a vital phenomenon controlled by the cells. Most of the transpiration takes place from the leaves and is known as **foliar transpiration** that includes types like: stomatal, cuticular and lenticular.

3.1 Foliar Transpiration

Stomatal transpiration. Stomata are the primary portals of water loss during transpiration. It accounts for 90–95 per cent of water loss from leaves.

Cuticular transpiration. It accounts for 5–10 per cent of total water loss and is mainly dependent upon the thickness of the cuticle. Although this layer (consisting of a waxy substance cutin, and other hydrophobic polymers) generally can retard water loss, but, it is still permeable to small quantities of water vapour.

Lenticular transpiration. About 0.1 per cent of the total loss of water vapour occurs through lenticels present on fruits and woody stems.

Plants also lose water in liquid form *via* 'guttation' and 'bleeding'.

3.2 Guttation

Guttation is quite common in plants like garden nasturtium, colocasia, oat leaves, and such. (Figure 3.1). It is the oozing of drops of water from the uninjured tips or edges of the leaf where a major vein terminates. It occurs under conditions of active absorption of water and low water loss (in moist, humid conditions) just before dawn. The water of guttation contains a variety of dissolved solutes like carbohydrates, nitrogenous compounds, organic acids and mineral salts. On evaporation, these solutes get concentrated on the leaf margins and may cause injury to the leaf. Guttation takes place through **hydathodes** or **water stomata** (large stoma-like structure with a pore at the tip, which always remains open) as a result of the root pressure developing in xylem elements (Figure 3.2).

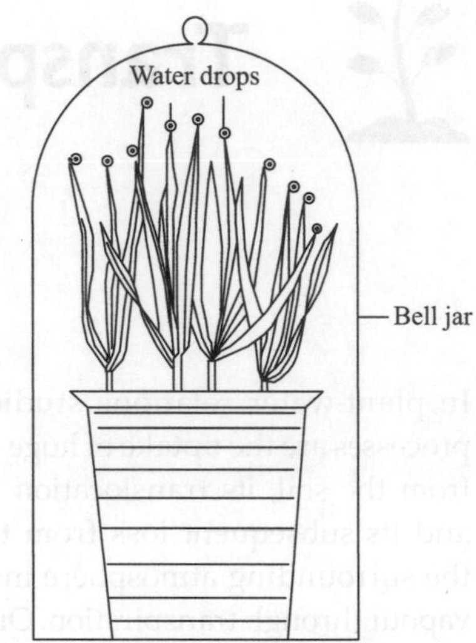

Fig. 3.1 Guttation in oat leaves.

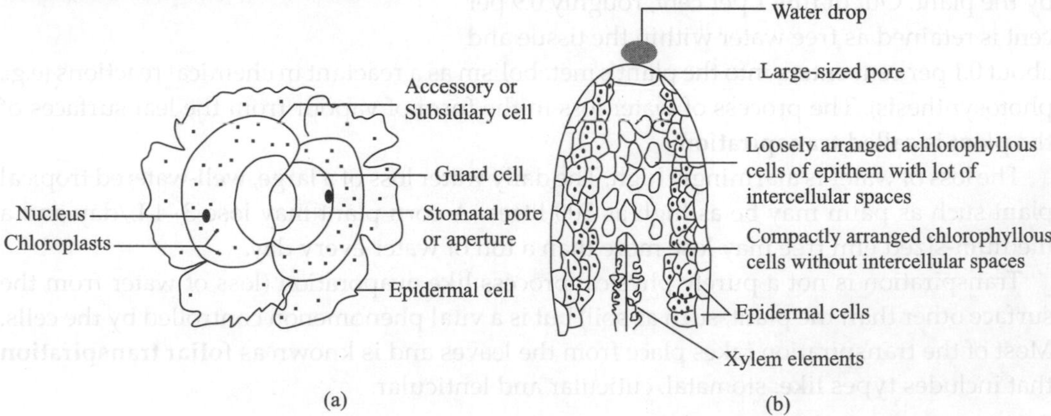

Fig. 3.2 Structure of (a) a stomata and (b) a hydathode.

3.3 Bleeding

Bleeding is the term for exudation of liquid water with the dissolved substances from the injured parts of the plant. The duration of bleeding and the amount of sap exuded shows great variations in different plants. Bleeding is of great economic importance. The exuded sap from sugar maple (*Acer saccharum*) yields sugar. The toddy palm (*Caryota urens*) in South India yields an alcoholic drink called 'toddy'. The exudation of latex in para rubber tree (*Hevea brasiliensis*) is yet another example.

Transpiration	Guttation
1. It is the loss of water in the form of water 'vapour' from the aerial parts of the plant. It does not contain dissolved solutes.	It is the loss of water in the form of liquid from the leaves. It contains dissolved substances like sugars, minerals, organic acids and amino acids, enzymes, proteins. The water evaporates as the day passes by, leaving behind a crust.
2. It occurs all through the day.	It occurs in the wee or early hours of morning before sunrise.
3. It is of three types, mainly stomatal (through the stomates) or cuticle (cuticular) or lenticles (lenticular).	It occurs through hydathodes or 'water stomata' present at the leaf endings or where the principal vein ends.
4. Stomata show opening and closing.	The terminal pore of hydathode remains permanently open (i.e., incapable of opening and closing).
5. Controlled by water potential (ψ) gradient in the mesophyll cells and is at the expense of solar energy.	Controlled by positive 'root pressure' originating in the root cortex. The water is literally forced out of leaf through the opening.
6. It is a rapid process, e.g., *Withania somnifera.*	It is relatively a slower process, e.g., many grasses, garden nasturtium (*Tropaeolum*) leaves, *Colocasia,* potato, tomato.

The term 'guttation' was first used by A. Burgerstein in 1887 and is derived from the Latin word 'gutta', meaning drop.

3.4 Stomata

3.4.1 Size and distribution

Stomata are present on exposed parts of virtually all plants except algae and fungi. They are to be found on the leaves, stems, floral parts and fruits. On an average, there are about 10,000 stomata/cm^2 of leaf surface.

The size of the stomatal aperture measures about 20 μm long and 10–20 μm wide when fully open. The aggregate area of the stomatal pores rarely exceeds 1–2 per cent of the entire leaf surface. Depending upon distribution of stomata, three categories of leaves have been recognized:

(i) ***Amphistomatous:*** stomata are confined to upper and lower surface of the leaves (e.g., corn, oat, grasses); (ii) ***Epistomatous:*** stomata are found only on the upper leaf surface (e.g., floating hydrophytic leaves); (iii) ***Hypostomatous:*** stomata are present only on the lower leaf surface (e.g., woody plants).

3.4.2 Morphology of stomata

The stomatal apparatus is composed of two guard cells surrounding a stomatal aperture, two or more adjacent subsidiary cells and a sub-stomatal cavity. Guard cells are kidney-shaped (in dicots) and have a large well-defined nucleus, small chloroplasts, mitochondria and other cellular inclusions. Guard cells show great diversity in their morphology, but there are two major types: one is typical of grasses, while the other is found in mosses, ferns, gymnosperms, and other flowering plants. In grasses (monocots), guard cells have a characteristic dumbbell shape, with inflated ends containing their cytoplasmic contents. The inflated ends of both the guard cells are joined by highly thick walls leaving a long slit or pore in between. Most other plants especially dicots, have an elliptical or kidney shaped guard cells with the pore at their centre. The ventral (inner) walls of guard cells adjacent to the aperture are thickened and inelastic, whereas dorsal (outer) walls distal to the aperture are thin and elastic (Figure 3.3). A characteristic feature of guard cells is the presence of distinct wall thickenings along the inner and outer margins of the ventral wall, which are modified into specialized structures called **ledges** (one or two), providing protection to the stomatal throat (Figure 3.4). The outer ledge helps to check the entry of water from the guard cells into the sub-stomatal cavity, which can otherwise adversely affect exchange of CO_2 and O_2 gases. The guard cells are suspended over the sub-stomatal cavity by the subsidiary cells, which act as storehouses for K^+ ions and water.

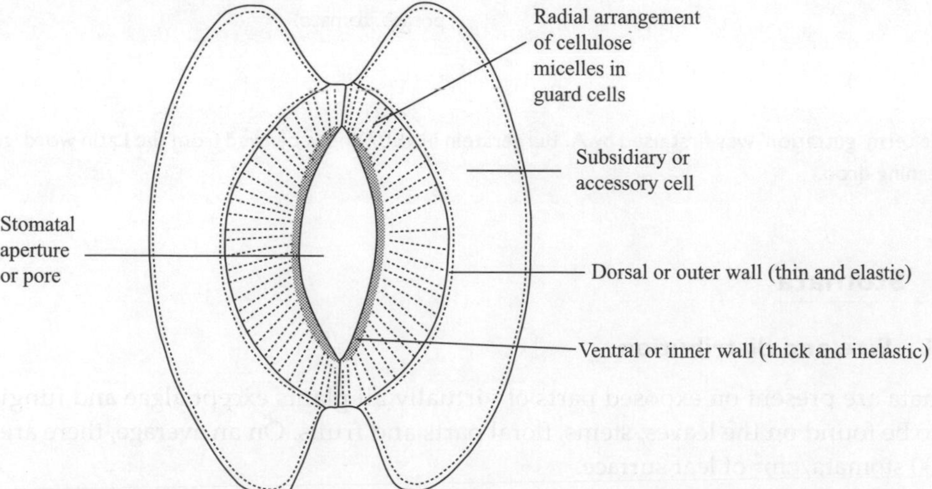

Fig. 3.3 Diagram of a stomatal apparatus showing radial arrangement of cellulose micelles (microfibrils) in the guard cells. The subsidiary or accessory cells act as storehouses or reservoirs of water and potassium ions. Radial micelles restrict the expansion of cell walls in transverse direction.

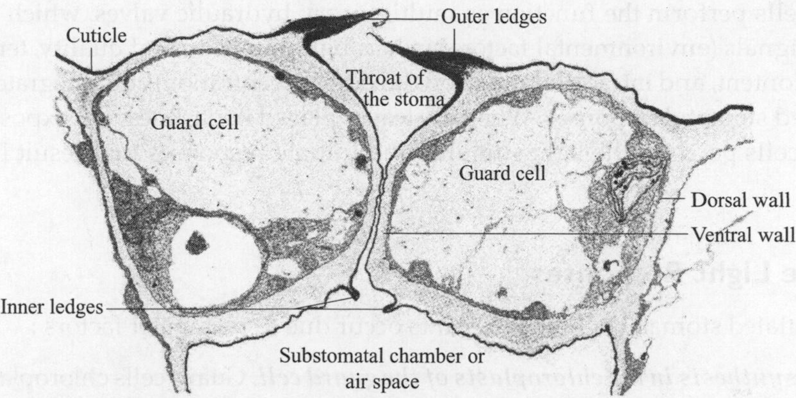

Fig. 3.4 Cross-section through guard cells. The well developed outer ledges are heavily cuticularized (electron-dense region) while the inner ledges are often small or absent.

The alignment of **cellulose microfibrils**, which strengthen cell walls of plants, determines the cell shape and plays a central role in the stomatal mechanism. In cylindrically shaped cells, cellulose microfibrils are oriented in a transverse manner to the longitudinal axis of the cell, resulting in cell expansion in the direction of its long axis.

Cellulosic microfibrilar organization is different in two types of guard cells (Figure 3.5). In graminaceous plants, the dumbbell-shaped guard cells behave like beams with bulbous ends. The orientation of the cellulose microfibrils is such that as the inflated ends of the guard cells increase in volume followed by separation of guard cells from each other, thus leaving a wide pore between them. In kidney-shaped guard cells, cellulose microfibrils radiate out from the stomatal pore. As a result, the inner wall is much stronger than the outer wall. This causes the guard cells to stretch apart and the stomatal pore to open.

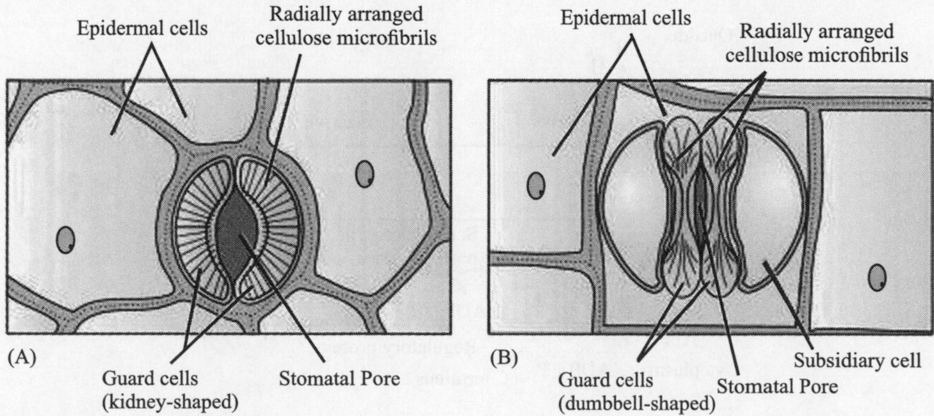

Fig. 3.5 Orientation of cellulose microfibrils in the cell walls of guard cells. Microfibrils are radially arranged in the cell walls of both (A) kidney-shaped guard cells as in dicots (B) dumbbell-shaped guard cells as in monocots, especially in grasses.

Guard cells perform the function as multisensory hydraulic valves, which are able to sense the signals (environmental factors such as light intensity and quality, temperature, leaf water content, and intracellular carbon dioxide concentration) and integrate them into well-defined stomatal responses. When the leaves placed in darkness are exposed to light, the guard cells perceive the light stimulus and initiate responses that result in stomatal opening.

3.5 Blue Light Responses

Light-stimulated stomatal opening in plants occur due to two major factors :

- ***Photosynthesis in the chloroplasts of the guard cell.*** Guard cells chloroplasts usually contain large starch grains. Starch content in guard cell chloroplasts decreases in the morning at the time of stomatal opening and increases at the end of the day when stomata close.
- ***Specific response of stomata to blue light.*** The stomatal response to blue light is localized in the guard cell and is quite rapid as well as reversible. Blue/UV-A wavelengths are perceived by **phototropins.** Phototropins are plasma membrane-associated protein kinases responsible for controlling stomatal aperture as well as regulating phototropism, light-induced chloroplast movement, hypocotyl and leaf expansion. Light absorption by phototropins causes their autophosphorylation and induces signal transduction pathways that lead to the redistribution of PIN auxin efflux carriers necessary for directional growth and to the changes in plasma membrane ion fluxes that regulate guard cell turgor.

Blue light activates H^+-ATPase at the guard cell plasma membrane, H^+ are pumped out of the cell thus lowering the pH of the apoplastic space surrounding the guard cells and generating an electrochemical-potential gradient that drives ion uptake (Figure 3.6).

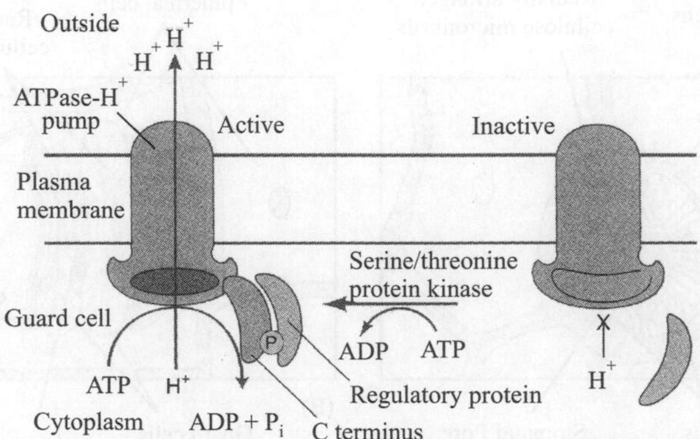

Fig. 3.6 Regulation of stomatal movement by blue light. On absorption of blue light, guard cell H^+-ATPase gets activated. A regulatory protein binds to the enzyme in phosphorylated form. H^+ are pumped out of the cell, and this is balanced by entry of K^+ into the guard cell.

Two important properties of blue-light responses are:

(i) Prolonged existence of blue light response after the light-off signal. Blue light induces rapid conversion of an inactive state of its photoreceptor to an active state, and this is followed by the slow reversion to the inactive form again when the blue light is switched off.
(ii) Significant lag time (approx 25 sec) between the exposure of the light signal and the onset of the response. This time period might be required for the signal transduction process to occur from the receptor site to the proton-pumping ATPase site and subsequently to establish the proton gradient.

Blue light regulates the osmotic balance of guard cells: At the time of stomatal opening, K^+ ions concentration in guard cells increases several-fold, which are electrically balanced by Cl^- and malate^{2-} ions. During stomatal opening, K^+ and Cl^- ions are transferred into guard cells *via* secondary active transport driven by the proton-motive force, generated by H^+-ATPase pump. The guard cells uptake Cl^- ions from the apoplast through a proton-chloride symporter. In the cytoplasm of the guard cell, blue light also promotes starch degradation into malate anion. Thus both components of the electrochemical proton gradient generated by blue light-dependent proton pumping play key roles in the uptake of ions for stomatal opening. Accumulation of sucrose and K^+ and its counter ions within guard cells results in opening of the stomata.

Blue light stimulated stomatal opening by manipulation of zeaxanthin: Manipulation of the carotenoid zeaxanthin, a component of the xanthophyll cycle, in guard cells allows regulation of stomatal opening in response to blue light. The carotenoid **Zeaxanthin** acts as a blue-light photoreceptor in chloroplasts of the guard cells and is believed to be directly involved in blue light-stimulated stomatal opening.

Evidence

(i) The absorption spectrum of zeaxanthin strictly follows the action spectrum for blue light-stimulated stomatal opening.
(ii) Blue light-stimulated stomatal opening is inhibited, if zeaxanthin accumulation in guard cells is blocked by the reducing reagent dithiothreitol (DTT).

Light quality can change the activity of pathways associated with osmoregulation and the stomatal movements. Stomatal opening is associated primarily with K^+ uptake while stomatal closing is associated with a loss of K^+ and a decrease in sucrose content. The blue-light response of guard cells is reversed by green light in the natural conditions. From spectroscopic studies it is quite clear that green light is very effective in the isomerisation of zeaxanthin and this changes the orientation of the molecule within the membrane, a transition that would be very effective as a transduction signal.

A carotenoid-protein complex in cyanobacteria, the **orange carotenoid protein (OCP)**, shows blue/green reversibility and functions as a light sensor. The OCP provides a molecular model for blue-light sensing by zeaxanthin in guard cells.

It has been concluded that **DCMU,** an inhibitor of photosynthetic electron transport chain, also inhibits stomatal opening, indicating the involvement of photosynthetic process in stomatal mechanism. The inhibition is however only partial, meaning that the second mechanism i.e. specific stomatal response to blue light is equally responsible.

3.6 Transpiration: A Two-Stage Process

1. The evaporation of water from the wet cell walls of mesophyll cells into sub-stomatal cavity.
2. The diffusion of water vapour from the sub-stomatal cavity into the external atmosphere through the stomatal pore (aperture).

Transpiration flux,[1] (the driving force for transpiration), is the difference in concentration of water vapour between a leaf and the external atmosphere.

$$\text{Transpiration flux} = \frac{\text{Magnitude of the driving force}}{\text{Resistance in the pathway}}$$

$$= \frac{C_\ell - C_a}{R_S + R_a}$$

where C_ℓ and C_a are the water vapour concentrations in sub-stomatal cavities in the leaf and in the external atmosphere, respectively; *Rs* is the stomatal resistance and *Ra* is the resistance of the boundary layer external to the leaf.

The greater the resistance ($R_s + R_a$), the lesser would be the transpiration flux; hence, there will be a lower rate of transpiration. There is a direct correlation between transpiration flux and rate of transpiration.

3.6.1 Mechanism of stomatal opening and closing

The regulation of stomatal aperture is of utmost importance in controlling the rate of transpiration. Generally, the stomata open in light and close in darkness.

Light	Dark
Increase in osmotically active substances in the guard cells. ↓	Decrease in osmotically active substances in the guard cells. ↓
Water flows into guard cells from adjacent epidermal cells because of high osmotic pressure. ↓	Water flows out of guard cells into adjacent epidermal cells because of low osmotic pressure. ↓

[1] The term 'saturation deficit' has been replaced by transpiration flux now. 'Saturation deficit' is the difference between two variables, i.e., amount of water present in the air and the amount required to saturate it.

Guard cells become turgid. ↓	Guard cells become flaccid. ↓
Outer thin walls of guard cells become stretched while the inner thick walls being inelastic fail to stretch and are pulled apart and become concave. ↓	Inner walls of guard cells become straight and approach each other. ↓
Results in gap between the guard cells. ↓	Stomatal pore becomes narrow. ↓
Leads to opening of stomata.	Leads to closing of stomata.

Hugo von Möhl had proposed in 1856 that chloroplasts present in guard cells photosynthesize in light resulting in production of carbohydrates and thus help to increase the osmotic pressure of guard cells, leading to stomatal opening.

3.6.2 Starch–Sugar hypothesis (Classical hypothesis)

This hypothesis was first formulated by J.D. Sayre (1926) and later supported by Scarth (1932) and Small and Clarke (1942). Stomata are sensitive to changes in pH. Generally, a high pH (alkalinity/basicity) favours opening and low pH (acidity) favours closing of stomata.

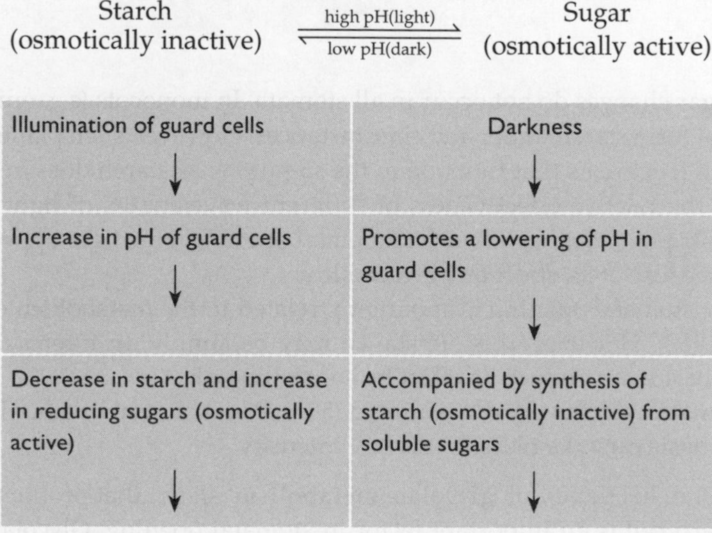

$$\underset{\text{(osmotically inactive)}}{\text{Starch}} \underset{\text{low pH(dark)}}{\overset{\text{high pH(light)}}{\rightleftharpoons}} \underset{\text{(osmotically active)}}{\text{Sugar}}$$

Illumination of guard cells ↓	Darkness ↓
Increase in pH of guard cells ↓	Promotes a lowering of pH in guard cells ↓
Decrease in starch and increase in reducing sugars (osmotically active) ↓	Accompanied by synthesis of starch (osmotically inactive) from soluble sugars ↓

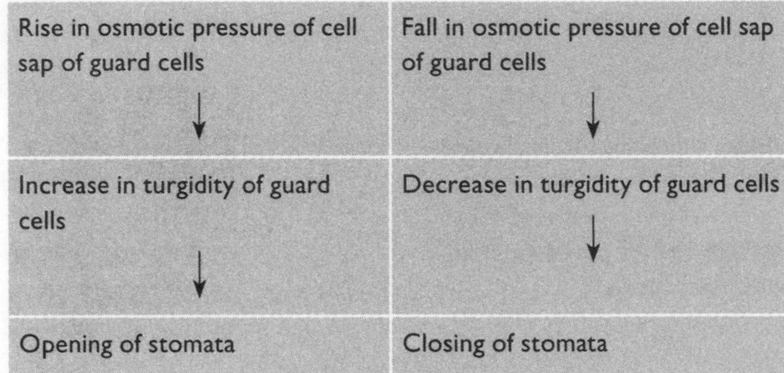

However, conversion of starch into sugar does not bring about significant changes in osmotic pressure of guard cells.

Yin and Tung (1948) discovered the presence of a reversible enzyme, starch phosphorylase in the chloroplasts of guard cells that favors the degradation of starch at pH 7.0 and starch synthesis at pH 5.0.

$$\text{Starch} + \text{Pi} \underset{\text{pH 5.0}}{\overset{\text{pH 7.0, starch phosphorylase}}{\rightleftharpoons}} \text{Glucose-1-phosphate}$$

According to Scarth, CO_2 through carbonic acid (H_2CO_3) is the major controlling force for pH change. The increase of pH during daytime is directly caused by the decreased carbon dioxide concentration in the cell sap (that is, low carbonic acid content means less of H^+ ions thus leading to alkalinity of pH 7.0 or above). At night time, there is a build-up of CO_2, firstly due to lack of photosynthesis and secondly because of respiration that also goes on at night. Due to high level of CO_2, there is greater production of H_2CO_3, i.e., more of H+ ions, thus leading to acidity (pH 5.0 or less).

Criticism

1. Starch-Sugar changes do not occur in all stomata. In monocots (e.g. onion), the guard cells do not form starch under any circumstances. However, such plants contain large amounts of fructosans that function in the same way as starch does in other plants.
2. Although the relative effectiveness of different wavelengths of light is, in general, the same for photosynthesis, but for stomatal opening, blue light appears to be more effective because of its effect on permeability.
3. In tobacco, stomatal opening is apparently related to the metabolism of glycolic acid (Zelitch, 1963). However, this substance may be simply an intermediate in sugar synthesis or it may produce its effect by controlling pH.
4. While stomata open in very dim sunlight (50–100 lux) or even in moonlight, not much photosynthesis can take place at this low intensity.

Zelitch (1963) in his theory of **'glycolate metabolism'** stated that production of glycolic acid in the guard cell is an important factor in stomatal opening. Glycolate is produced under low concentration of CO_2, thus raising the osmotic pressure and participating in the production of ATP required for the opening of stomata.

According to the modified scheme proposed by F.C. Steward (1964), glucose-1-P is converted to glucose and inorganic phosphate (Pi) – with both of them being osmotically active; they play an equally important role in stomatal opening (Figure 3.7).

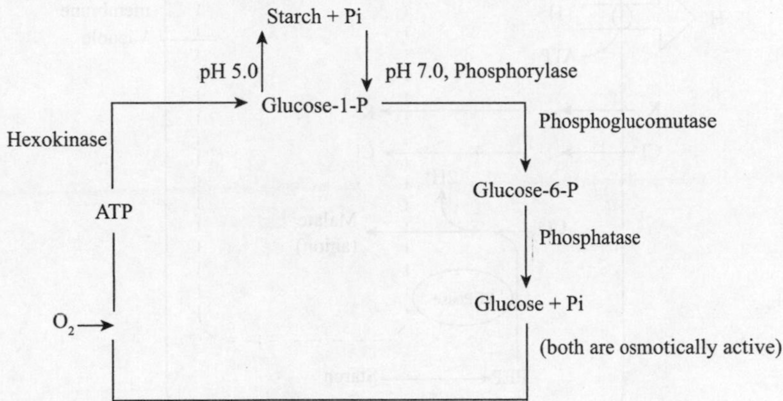

Fig. 3.7 Modified scheme proposed by Steward for stomatal opening and closing.

3.6.3 Modified classical theory

In 1967, J. Levitt modified the 'classical theory' by proposing that carbonic acid is 'too weak an acid' to affect pH changes in the guard cells where conversion of 'starch-sugar' (or *vice versa*) occurs during day and night time.

Removal of CO_2 during photosynthesis would lead to 'low CO_2 concentration' in the guard cells thus favouring 'decarboxylation' (breakdown of organic acids), that would mean less of H^+ ions, i.e., alkalinity or high pH. Under alkaline pH, starch is converted into soluble sugar that would lead to low ψ, thus withdrawal of water from adjacent subsidiary cells and the stomata open.

On the other hand, during night time there is a build-up of CO_2 in the guard cells. High CO_2 concentration in the guard cells promotes 'carboxylation' (formation of organic acids) which leads to low pH (acidic reaction) that would favour conversion of sugar to 'insoluble starch', followed by outward movement of water (exosmosis) into subsidiary cells. Thus the stomata close down.

3.6.4 Accumulation of K^+ ions in guard cells

Accumulation of K^+ in guard cells of plants was first noticed by Macallum in 1905 and was confirmed by Iljin (1915) and Lloyd (1925) but the main credit for identifying the role of K^+ in stomatal movement goes to Japanese workers (Imamura, 1943; Yamashita, 1952; Fujino, 1959, 1967; Fisher, 1968).

Fujino (1967) suggested that active transport of K^+ into the guard cells and out of them leads to stomatal opening and closing. Recent analysis with a very sensitive tool, the electron beam microprobe,[2] has shown that K^+ fluxes do occur in stomatal movements (Figure 3.8). Fujino's observations have been supported by Sawhney and Zelitch (1969) and Humble and Raschke (1971).

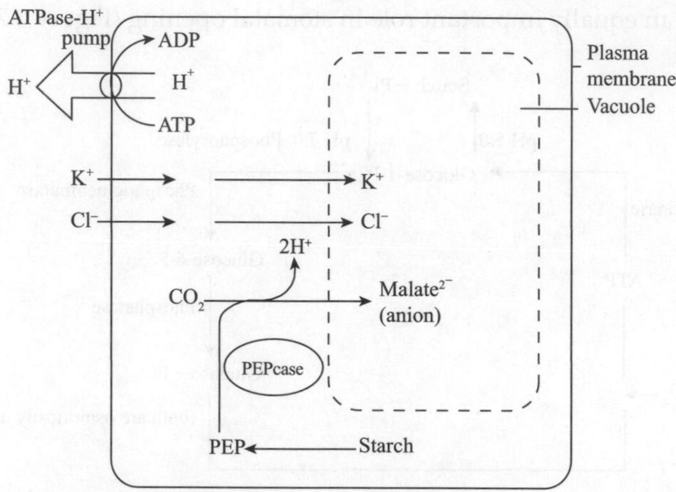

Fig. 3.8 The general model showing flow of ions in the guard cells during stomatal opening. K^+ uptake is controlled by H^+-ATPase pump present on the plasma membrane. Transport of ions (K^+, Cl^-, malate^{2-}) in the guard cell vacuole lowers its water potential, activates the osmotic uptake of water resulting in an increase in turgor pressure.

Noggle and Fritz (1976) have summarized the sequence of events occurring during light-induced stomatal opening as follows:

Light → Malic acid production → Dissociation into hydrogen and malate ions → [Influx of potassium ions (and efflux of hydrogen ions) through the agency of a hydrogen–potassium ion exchange pump in the plasma membrane]

→ Transport of potassium malate into vacuoles → Osmosis of water from adjacent epidermal cells → Turgor pressure increases → Stoma opens

3.6.5 Proton transport theory

Proton transport theory advanced by Jacob Levitt (1974) is a combination of Scarth's classical pH theory and active K^+ ion transport theory.

---

[2] Electron microprobe analyzers are used to measure distribution and amount of K^+ in a single cell/even in a part of the cell (*in situ*).

The complete scheme of Levitt during day time is as follows:

(a) All chloroplast – containing cells (mesophyll and guard cells)

1. Cytoplasm $\xrightarrow[\text{H}^+\text{ transport}]{\text{light induced}}$ chloroplasts
2. pH of cytoplasm $\rightarrow$ pH of chloroplast
   (8-9) (5)
3. $CO_2 \xrightarrow[\text{ATP}]{\text{RuBP}} (CH_2O)$ in the chloroplast
4. $[CO_2]_{intercell} \xrightarrow{\text{decreases}} < 0.01\%$
5. Cytoplasm: $CO_2 \xrightarrow{\text{pH 8-9}} HCO_3^-$

(b) Guard cells[3] only

6. Cytoplasm: $HCO_3^- + PEP \xrightarrow{\text{PEPC}} R(COOH)_2 + P$
7. Cytoplasm: $R(COOH)_2 \rightleftharpoons R(COO^-)_2 + 2H^+$
8. Cytoplasm: $K^+ \underset{\text{efflux}}{\overset{\text{influx}}{\rightleftharpoons}} H^+ (H^+, K^+\text{–ATPase})$
9. Chloroplast: $(CH_2O) + Pi \rightarrow PEP$
10. $R(COO^-)_2 + 2K^+ \rightarrow K^+$ malate (osmotic pressure increases)
11. $H_2O \xrightarrow{\text{endosmosis}}$ turgor pressure increases
12. Stomata opens

The potassium ions play a three-fold role in the opening of stomata

i. K^+ ions help in transport of maximum H^+ into chloroplast.
ii. K^+ ions maintain high cytoplasm pH to permit the continued conversion of PEP and HCO_3^- to organic acid.
iii. K^+ ions along with organic acid raise the osmotic potential of guard cells to cause stomatal opening.

3.6.6 Environmental factors affecting stomatal opening and closing

I. Light

In most plants, stomata remain open during the day time and closed during darkness. Stomatal opening early in the day is associated with a peak of K^+ and is thought to be triggered by blue light. Later in the day, photosynthetically active light stimulates and maintains stomatal opening through the synthesis and accumulation of sucrose.

[3] Guard cells are unique in having PEP carboxylase in their cytoplasm besides ribulose bisphosphate carboxylase in the chloroplast.

Blue light photoreceptors in guard cell plasma membranes activate the plasma membrane H^+ pump

↓

Which, in turn, causes acidification of the cell wall solution and a decrease in the plasma membrane potential (**hyperpolarization**)

↓

Hyperpolarization opens K^+ and Cl^- channels, leading to accumulation of K^+

↓

The resulting increase in water potential, as a result of solute accumulation, causes water to diffuse into the guard cells by osmosis, increasing turgor pressure

↓

This causes stomata to open

2. Abscisic acid

ABA causes stomatal closure by triggering Cl^- efflux through Cl^- channels in the plasma membrane of guard cells

↓

This results in less negative membrane potential (**depolarization**)

↓

ABA inhibits the activity of plasma membrane H^+ pump, leading to further depolarization

↓

The loss of solutes occurs because of opening of outwardly rectifying K^+ channels

↓

Water leaves the guard cells by osmosis, causing turgor pressure to fall

↓

This causes stomata to close.

3. Soil moisture

Stomata close when the soil is dry and root cells experience water stress. ABA is synthesized by root cells and is quickly translocated to the leaves through xylem where it induces stomatal closure.

4. Intracellular CO_2 concentration

A high level of CO_2 concentration causes stomatal closure. Mechanism of stomatal closure is similar to that of ABA. It triggers the opening of Cl^- channels, leading to the efflux of Cl^- ions and is followed by membrane depolarization. Depolarization, in turn, triggers the opening of outwardly rectifying K^+ channels and subsequent K^+ efflux.

5. Humidity

Low levels of humidity also cause stomata to close. Relative humidity of the air affects stomatal opening independently of its effect on guard cells.

6. Oxygen and respiratory inhibitors

Anaerobiosis will prevent stomatal opening. Many respiratory inhibitors such as phenolic compounds, azide, and arsenate also inhibit stomatal opening.

3.6.7 Midday stomatal closure

It is a temporary event existing for an hour or so.

The midday closure of stomata on a hot, sunny day is because of the following reasons:

1. The rate of transpiration far exceeds the rate of water uptake, thus resulting in loss of turgidity in guard cells, which thus becoming flaccid and the stomatal aperture closes.
2. Under such water stress conditions, the level of ABA within the guard cells rises significantly, leading to efflux of K^+ ions from the guard cells into the subsidiary cells while there is influx of H^+ ions from the adjacent subsidiary cells into guard cells. In the absence of potassium malate, the water potential of guard cells becomes less negative in comparison to subsidiary cells. Water moves outward leading to guard cells flaccidity and then closure.
3. Accumulation of intracellular CO_2 at high temperatures (i.e., 30–34 °C) because of increased respiration results in closure of stomata (Heath and Orchard, 1957). Stomata close when the internal concentration of carbon dioxide of the leaf is high.
4. Hall and Kaufmann (1975) attribute midday closure to decrease in humidity of the external air surrounding the leaf. This has a direct effect in reducing the elasticity of the walls of guard cells due to greater outward diffusion of water vapour from the walls of guard cells.

3.7 Plant Antitranspirants

Conservation of water is of prime importance for efficient agriculture. Antitranspirants are the compounds applied to the surface of the leaves in plants to retard transpiration. The objective of using them is to check huge losses of water *via* transpiration by artificial means in the cultivation of high-valued field crops or in seedling transplantation in nurseries. It is of course desirable that antitranspirants be cheap, readily available, non-toxic, stable with long-lasting effects on stomatal mechanism. These should also help in improvement of quality and productivity, as well as the disease resistance parameters in crops.

Antitranspirants are of four types:

1. ***Stomatal closing type:*** These antitranspirants such as, phenyl mercuric acetate (PMA), abscisic acid (ABA), and those with high CO_2 concentration; minimize the rate of transpiration by closure of stomata. With stomatal closure, diffusion of carbon dioxide gas into the leaf reduces, thus resulting in lower crop yields. Although use of PMA as an antitranspirant is quite common, but it is poisonous for fruits and vegetables. It also damages foliage by blocking the phosphorylation part of photosynthesis. CO_2 is another effective antitranspirant. A slight increase in CO_2 concentration from 0.03 to 0.05 per cent brings about stomatal closure partially, but beyond that it adversely affects the physiological processes such as photosynthesis and respiration by complete stomatal closure. Inhibitors of photosynthetic electron transport chain such as triazine, atrazine, simazine (herbicides) can also be used as antitranspirants at the lower concentrations. Generally, these types of antitranspirants are used for ornamentals, ground covers and other such utility plants.
2. ***Film forming type:*** These compounds (such as low viscosity waxes, colourless plastics, and silicone oils) form a thin film on the surface of leaves, fruits, root stocks, propagules and others; and retard the water loss from them but at the same time allow diffusion of carbon dioxide into the leaves. So they prevent excessive water loss during storage and transportation for which CO_2 uptake is not an important consideration. The disadvantages of using them are that they are effective only at low temperatures and also they obstruct the stomatal movement thus preventing gaseous exchange.
3. ***Reflectance type:*** This type of chemical such as kaolin and lime water reflect the radiation falling on the leaves, thus decreasing their temperature. These check the water loss without any effect on carbon dioxide assimilation. Kaolin does not affect any metabolic activity and usually forms a thin white film on the leaf when sprayed (2–5%).
4. ***Growth retardants:*** These chemicals such as, cycocel (CCC) suppress leaf area development as well as shoot growth, promote root growth, and induce stomatal closure thus making the plants drought-resistant.

Antitranspirants are generally employed for reducing transpiration loss to horticultural plants. Several strategies to improve productivity have been adopted by farmers and plant physiologists. The work on stomatal physiology to reduce water loss by spraying anti-transpirants has given encouraging results.

3.8 Factors Affecting Transpiration

The rate of transpiration is affected by a number of factors, both external and internal.

External Factors

1. ***Atmospheric humidity***: The rate of transpiration depends upon the capacity of the atmosphere to take up more moisture and this in turn, depends upon the difference between the amount of water vapour actually present in the air (absolute humidity)

and the amount of water required to completely saturate it (relative humidity). The difference between the two variables is called the 'saturation deficit'. The more the saturation deficit, the more rapid is the transpiration and *vice versa*.

2. ***Temperature***: Temperature affects transpiration indirectly through its effect on the saturation deficit of the air. An increase in temperature leads to decrease in relative humidity of the air, thus increasing the rate of transpiration.
3. ***Wind***: Wind affects the atmospheric humidity directly. Dry winds lower the amount of air moisture by removing moist air from the **boundary layer** (a layer of undisturbed air on the the leaf surface) and thus increase the rate of transpiration. The boundary layer is a zone immediately close to the leaf wherein air velocity is modified by the leaf itself. It consists of two layers – laminar sub-layer and turbulent layer. Diffusion of water vapour and other gases occurs readily through both the layers (Figure 3.9).

   In still air, the thick boundary layer prevents the loss of water vapour from the leaf. With an increase in wind velocity, the boundary layer at the surface of the leaf becomes thin and the vapour pressure gradient steepens thereby increasing the transpiration rate. This relationship, however, is true at lower wind velocities. As wind velocity increases further, it tends to close the stomata because of (i) cooling the leaf surface (ii) causing sufficient desiccation (dehydration). Either of these two factors tends to lower the rate of transpiration. Therefore, high wind velocity will have less effect on transpiration rate than expected on the basis of thickness of boundary layer alone.

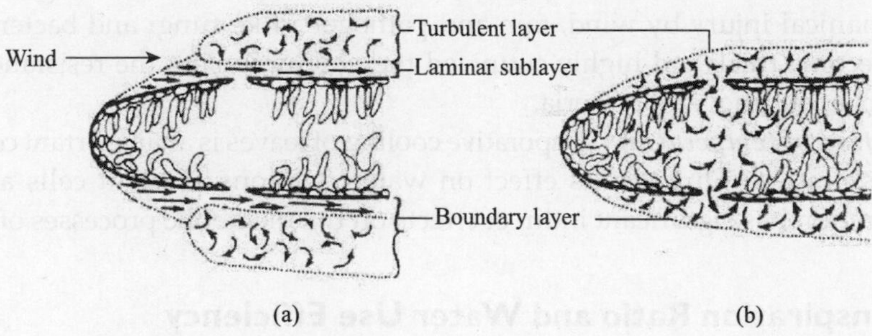

Fig. 3.9 Schematic drawings of a leaf boundary layer. (a) boundary layer as depicted based on the laminar boundary layer theory (b) boundary layer as it exists in nature.

   Besides wind speed, anatomical and morphological characteristics of the leaf (such as leaf size and shape, orientation of the leaves relative to the wind direction, the presence of leaf hairs i.e. trichomes and sunken stomata, and such) also determine the thickness of the boundary layer.

4. ***Light***: The rate of transpiration increases with increasing light intensity because of increase in temperature of the leaves and opening of stomata.
5. ***Available soil water***: Transpiration is directly linked to availability of soil water. High concentration of salts in the soil water reduces transpiration, probably by increasing the osmotic pressure of plant cells.

Besides external factors, some internal factors like stomatal apparatus of the plant, water content of mesophyll cells and structural peculiarities of the leaves also affect the rate of transpiration.

3.9 Significance of Transpiration

Water is of vital importance to the plant and yet out of the total amount of water absorbed, 99 per cent is lost in the form of water vapour. Transpiration has been described as a necessary evil. Though the leaf structure with its thin-walled mesophyll cells, intercellular spaces and the stomata is so built as to make maximum absorption of CO_2 by leaf cells and gaseous exchange between the leaf and its surroundings, but this leaf architecture at the same time makes water loss inevitable.

Transpiration is regarded as being advantageous to the plant in many ways:

1. ***Circulation of water within the plant*:** Transpiration establishes a continuous water stream from the roots to the uppermost parts of the plant, thus increasing circulation of water within the plant resulting in its growth and development.
2. ***Absorption and translocation of minerals*:** Transpiration streams bring about a far more rapid transport of mineral nutrients by the mass flow of water through the dead xylem ducts than would be possible by a cell to cell diffusion of the salts.
3. ***Development of mechanical tissues*:** Increased transpiration also favours the development of mechanical tissues in the plants, making them more tough and resistant to mechanical injury by wind, rain and pathogens like fungi and bacteria. The cell walls become thick and highly cutinized thereby increasing the resistance of plants to pathogenic fungi and bacteria.
4. ***Regulation of temperature*:** Evaporative cooling of leaves is an important consequence of transpiration. Through its effect on water relations of plant cells and tissues, transpiration has significant indirect effects on the metabolic processes of the plant.

3.10 Transpiration Ratio and Water Use Efficiency

3.10.1 Transpiration ratio

Transpiration ratio is defined as the amount of water transpired by the plant divided by the amount of CO_2 assimilated through photosynthesis. It is a parameter to assess the effectiveness of plants in balancing loss of water coupled with adequate intake of CO_2 for photosynthesis.

Causes of high transpiration ratio

(i) The concentration gradient responsible for water loss is approximately 50 times more as compared to that of the intake of CO_2. This might be because of the low concentration of CO_2 in air (0.036%) and relatively high concentration of water vapour within the leaf.

(ii) The diffusion rate of CO_2 is about 1.6 times slower through the air than that of water. This is because the CO_2 molecule being bigger than the water molecule, has a lower diffusion coefficient.

(iii) In order to be assimilated in the chloroplast, CO_2 must diffuse through the plasma membrane and the chloroplast membrane through cytoplasm, which contribute to the resistance of the CO_2 diffusion passage.

Transpiration ratios of C3, C4, and CAM plants are as follows:

- C3 plants have a transpiration ratio of 400 (hence, water use efficiency is 0.0025 or 1/400; approximately 400 water molecules are lost for one carbon dioxide molecule assimilated through photosynthesis).
- C4 plants have a transpiration ratio of 150 (comparatively less water gets transpired per molecule of CO_2 assimilated. This is because C4 photosynthesis results in a lower CO_2 concentration in the intercellular air space, thus creating a larger driving force for the uptake of CO_2 and allowing these plants to operate with smaller stomatal apertures and thus lower transpiration rates.
- CAM plants have even lower transpiration ratio, approximately 50. Because their stomata open at night, so transpiration is much lower at night.

The reciprocal of transpiration ratio is termed as **water use efficiency.**

3.10.2 Water use efficiency (WUE)

Globally, modern day agriculture is the largest single user of fresh water, consuming over 90 per cent of fresh water. An irrigated or lowland rice utilizes nearly 45 per cent of the total fresh water accounting for about/almost 2 to 3 times of that consumed by other cereals. With the world's population imminently set to increase by the end of 2050, the additional food needed to feed humans in the future will put a vast strain on our depleting fresh water resources. Raising water-use efficiency of both irrigated and rain-fed or water stressed arid or semi-arid regions/areas of crop production is of prime importance.

Many nations are now solely dependent on groundwater resources that face depletion by the 2020s at the current rate of consumption. And if preventive action is not immediate, in less than 30 years, Great Britain (UK) might face acute water shortages in view of present projections. Globally, many parts will face water deficits by 2050 unless we take urgent and concrete steps to change the existing systems. We will lack sufficient water supply to fulfil our requirements. The obvious reasons are our recent climate changes (nations facing hotter and drier summers) and our population explosion. A change of mindset is essential towards our conservation strategies to find solutions for problems that mankind is facing. To offset extreme shortages of water, demand and greed would have to be curtailed and replaced by steps like controlling seepage by strengthening permanent drainage systems, cutting down our personal needs, and the expansion of future water supplies. A **twin-track strategy** would be the right approach; reducing our water requirements while expanding our sources of water supply.

Ways to improve water-use efficiency

1. Reduce conveyance losses by substituting open irrigation channels or canals over a period of time with large diameter prefabricated piped irrigation network or closed conduits supplying water to one or more farms.

   Drip irrigation is the most viable water and nutrient delivery system for growing crops. Water and nutrients are delivered across the field through 'a filter' into a special drip pipes, with emitters located at intervals. Water and nutrients are uniformly delivered directly, through a special slow release emitters, to plant's root system in the right quantities and right time into the soil. Farmers can produce higher yields with a huge saving on water (no evaporation, no runoff, very little leaching and avoiding contamination of ground water) and fertilizers, energy, and even crop protection products. Very efficient solar-powered drip irrigation kits are available.
2. Reduce direct transpiration by skipping midday sprinkling at the time of irrigation
3. Reduce water loss caused by overland flow and seepage by regulating irrigation scheduling
4. Reduce transpiration by weeds, keeping inter-row strips dry and apply weed control measures by using selective herbicides where needed.
5. Proper irrigation scheduling
6. Drought-tolerant crops suitable for dry farming in arid and semi-arid regions
7. Reduce evaporation by using compost, mulch and cover crops from soil by keeping the inter-row strips dry
8. Establishing alternative sources of water irrigation, such as mega water desalination plants, and even use of slightly brackish water (though not good for drinking for human consumption) for irrigation
9. Integrating waste water being recycled with most of it being diverted for agricultural use.
10. Mandatory for building establishments or farmland to install rainwater harvesting systems
11. Waste water generated by big consumers should be cleaned, refined, and reused for horticulture, thus avoiding use of fresh water. Singapore has become self-sufficient in water by recycling all sewage water. Israel recycles around 94 per cent of water it uses.

Professor Mark van Loosdrecht of Environmental Biotechnology at Delft University of Technology, the Netherlands, and Bruce Rittman, Regents' Professor of Environmental Engineering and Director of Biodesign Swette Center of Environmental Biotechnology at the Biodesign Institute, Arizona State University, USA, have been awarded the 2018 **Stockholm Water Prize** for revolutionizing water and waste water treatment by using microbiological-based technologies. They have demonstrated the possibilities to remove harmful contaminants from water for safe drinking, cost-effective waste water treatments, minimizing the energy footprints/reduce energy use, and even recover chemical and nutrients for recycling.

Pearl Millet—The Super Crop

In view of shortages of water, **millets** (Sorghum, Pearl Millet and Finger Millet as well as minor millets like foxtail millet, little millet, kodo millet, proso millet and barnyard millets) are increasingly being used in the food basket of our rural and urban population. Millets can grow equally well even in dry areas, receiving just only 20 cm of rainfall, thus overcoming drought like conditions. Only 200 litres of water is required to grow one kilogram of millet, as compared to almost 9,000 litres of water for growing one kilogram of rice or wheat. Their production cost is much less unlike rice or wheat, and millets are also much healthier for human consumption .They require less water for growth and can also withstand extreme temperatures, critical for farmers in view of climatic changes now taking place. Millets, the yesteryear's dietary food of a vast majority of people living in the semi-arid Asian regions, especially India, could become major **climate-resilient crops** .Millets are capable of withstanding the adverse impact of climate change as compared to most other food crops. They can be grown in almost any kind of soil-ranging from sandy to varying degrees of acidity. Hardly they require any irrigation or any kind of fertilizers. Millets are generally resilient to pests. Millets lack gluten and have less glycemic index, thus are ideal for people suffering from diabetes and celiac disorders. Millets have better tillering capacity than maize and sorghum. They have a better forage value too.

Pearl millet is composed of quality protein (9–13%) and compared to other cereals, with a better balanced amino acid profile, vitamins, minerals, crude fibres, antioxidants, easy digestible fats (7–8%). It may contain as much 74 per cent unsaturated fatty acids of total fat with higher content of nutritionally important Omega-3 fatty acids. It is a rich source of micronutrients such as iron, zinc, calcium, magnesium, phosphorus, folic acid, riboflavin and significantly rich in soluble and insoluble dietary fibres. Whole grain pearl millet and its bran are rich sources of phenolic compounds (highest among cereals) which are natural antioxidants and are associated to reduced risk of chronic diseases related to oxidative stress. Being a 'powerhouse of nutrients', pearl millet is notified as one of the millets under 'Nutri-Cereals' for production, consumption and trade, identified by the Agriculture Ministry, Government of India. **Millets can be the wonder crop, providing food, nutrition, and livelihood and security, while coping with the adverse impact of climate change.**

Review Questions

RQ 3.1. Discuss briefly the significance of structural peculiarities in the microanatomy of guard cells.

Ans. There are three structural features in the guard cells that help in stomatal movement:

1. Differential behaviour of the two walls of the guard cells – the outer (or dorsal) walls are thin while the inner (or ventral) walls are quite thick (less elastic) than the outer.
2. Radial micellation – cellulosic micelle converging from the outer dorsal walls to the inner walls of the guard cells (or fanning out from the inner wall to the outer wall), thus making the inner walls very thick resulting in the ability to extend greatly (when turgid) parallel to the length of the guard cells rather than across the width. Moreover, the thicker inner walls check the entry of water into the stomatal aperture of the two guard cells.
3. The large and quick changes in the cell volume of guard cells would not be possible if they were connected to their neighbouring cells by plasmodesmata through which solutes and water could escape. The guard cells lack plasmodesmata.

RQ 3.2. Is it possible to regulate the opening and closing of stomata at will? If it is so, then explain how?

Ans. Generally, stomata are open in light (photoactive) and close in darkness (scotoactive). The only exception is in CAM plants that show inverted stomatal scycle.

However, stomata can be made to close during day time by exposing the plants to vapours of glacial acetic acid or by spraying abscisic acid or exposing them to huge CO_2 concentrations. Similarly, stomata can be made to open at night time or darkness by floating the epidermal peels in KCl solution or by exposing the parent plant to vapours of ammonia. In the dark, stomata will open at low concentrations of CO_2.

Stomatal opening, like photosynthesis, is affected by blue and red wavelengths.

The blue and red lights activate proton efflux by the guard cells and promote malate synthesis – both essential for opening stomata. Red light (660 nm) is the most effective in regulating stomatal opening, followed by blue light (445 nm).

RQ 3.3. Plants growing in arid regions have developed extraordinarily survival mechanism. Comment.

Ans. Plants adapted to arid regions have undergone modifications – morphological, anatomical and physiological – to conserve as much water as possible to tide over water stress conditions.

1. Such plants are characteristically dwarf with small thick, heavily cuticularized leaves (microphylly).
2. Tend to have fewer stomata per unit area than do mesophytes; some with sunken stomata in crypts or pits in the leaf surface, guarded by trichomes, thus minimizing the water loss through increased boundary layer resistance.
3. Leaves covered with a thick felt of hairs, which reflect radiations from the leaf surface, thereby reducing the heat load, i.e., thermoregulation.
4. Well-developed root system, allowing absorption of water and minerals from the deeper layer of the soil.
5. Presence of multiple epidermis and water-retentive mucilage cells.
6. Most remarkably, the stomata of some xerophytes (CAM plants) open during night when the transpiration rate is low and closed during day time when the conditions are most conducive to transpiration.
7. During water stress conditions, there is a dramatic increase in the level of abscisic acid (ABA) in guard cells that leads to efflux of K^+ ions from the guard cells into the adjacent subsidiary cells and influx of protons from the subsidiary cells into guard cells, followed by water loss and stomatal closure.

RQ 3.4. Explain why is transpiration regarded as a necessary evil?

Ans. Transpiration has been very rightly described as a necessary evil by Curtis (1926). The leaf anatomy with well-saturated, thin-walled mesophyll cells with intercellular spaces and stomata is primarily built for absorption of carbon dioxide and exchange of gases during photosynthetic and respiratory processes. Such a gaseous exchange is essential for the synthesis of carbohydrates and other organic compounds that are vital for growth and development (or for the well-being of plants). Incidentally, such a leaf structure or architecture will not be able to prevent the loss of water which has become inescapable or inevitable. This loss of water is alarmingly very high. A medium-sized elm tree loses nearly 1 ton of water every day.

Then the question arises why the plants are expending so much energy in developing ways and means for increased absorption of water, out of which only 1 per cent is used or enters into the plant's metabolism and the remaining 99 per cent is lost through transpiration. The energy so spent on absorbing water merely goes as wastage. That is why it is considered 'an evil'. Even under unfavourable conditions, the deciduous trees have to shed their foliage to conserve water even at the cost of cutting off their photosynthetic activity.

RQ 3.5. An increase in the rate of transpiration results in an increase in the rate of mineral salt absorption. Comment.

Ans. It was earlier thought that the increased absorption of water as a result of transpiration brings into the plant an increased amount of mineral salts from the soil. This view has been found to be based upon illogical reasoning as it overlooks the basic fact that absorption of mineral salt is independent of the absorption of water. Both water and minerals are absorbed from different regions of the roots, the former taking place through the root hairs while minerals are absorbed from the meristematic zone and the zone of cell elongation near the root tip. Furthermore, the mechanisms involved in both the cases are quite distinct.

Although there is no direct relation between the amount of transpiration and the amount of solute absorbed, once the minerals have been absorbed, their further distribution to different plant parts is much increased by the rate of transpiration. This is reflected in an increase in the percentage of ash content of the plant.

RQ 3.6. What do you mean by the term 'soil-plant-air continuum (SPAC)'? Has it any regulatory effect in controlling the rate of transpiration?

Ans. Water flow through the plant system is under the influence of ψ gradient. The net movement is always in the direction of decreasing ψ, that is, ψ is highest in the soil, somewhat lower in the root cells, still lower in the stem, and lowest in leaf mesophyll and sub-stomatal cavity. The greatest drop in ψ occurs at the leaf–air interface where there is a change from liquid to gaseous vapours of the atmosphere. Theψ of air may decrease to well below –1000 bars, provided the relative humidity of the air is ~50 per cent at 22°C.

Due to the very sharp change in ψ at the leaf–air interface, it is assumed that greatest control of transpiration is exerted at the stomatal region:

Soil water	–	– 0.5 bars
Root	–	– 2.0 bars
Stem	–	– 5.0 bars
Leaf	–	– 15 bars
Air	–	– 1000 bars

Estimated values for water potential in a hypothetical soil–plant–air system.

RQ 3.7. List out different measures that might prove useful in conserving water in crops grown in dry regions.

Ans. Conservation of water is of paramount importance for our agricultural crops. Their productivity is seriously limited by the availability of water. Scientists are toying with idea of controlling transpiration without affecting photosynthetic activity. Antitranspirants are materials or chemicals used for plants to check transpiration as well as to conserve water. There are mainly 'two' types of antitranspirants:

1. Colourless plastics, silicone oils, and low viscosity waxes when sprayed form a thin film on the surface of the leaves that prevents loss of water as it is impermeable but at the same time permit exchange of O_2 and CO_2.

2. Antitranspirants such as abscisic acid – a plant hormone (ABA) and phenylmercuric acetate (PMA - used also as a fungicide) cause closure of stomata. Although these reduce transpiration but they also inhibit the process of photosynthesis and respiration.

- High concentration of CO_2 also leads to closure of stomata, thus preventing water loss.
- Lack of oxygen (anaerobiosis) and respiratory inhibitors (like azides, arsenate and phenolic compounds) or inhibitors of cyclic photophosphorylation prevent opening of stomata, possibly by blocking oxidative phosphorylation and photophosphorylation that provide ATP, necessary for operating H^+–K^+ ion pumps in the plasma membrane of guard cells.

Sealants prevent water loss without interfering with CO_2 uptake.

RQ 3.8. Enumerate the different factors affecting stomatal opening.

Ans. The rate of transpiration is affected by a number of factors that can be classified as external (physical or environmental) and internal:

External	Internal
1. Humidity of air	Water content of mesophyll cells of the leaves.
2. Temperature	Stomatal frequency (the number of stomata per unit area).
3. Light	Stomatal peculiarities (such as sunken stomata, cuticle, waxy surfaces and epidermal hairs).
4. Wind	Stomatal and air boundary resistance.
5. Atmospheric pressure	Root/shoot ratio.
6. Available soil water	

RQ 3.9. Droplets of water often seen on the tips of grass blades are commonly mistaken as 'dew' by the general public. Are they dew drops or something else? Explain.

Ans. To a layman, these droplets of water so often found in the early morning in a grassy lawn appear to be 'dew' but in reality this is a guttated water, oozing out of a special type of stomata called 'water stomata' or 'hydathode' (Figure 3.10). This water of guttation contains some dissolved substance, like varying proportion of carbohydrates, nitrogenous compounds, organic acids and mineral salts.

The liquid is forced out of the tracheary elements into loose intercellular spaces of the 'epithem', then into an air cavity and finally out of the large-sized pore that always remains open.

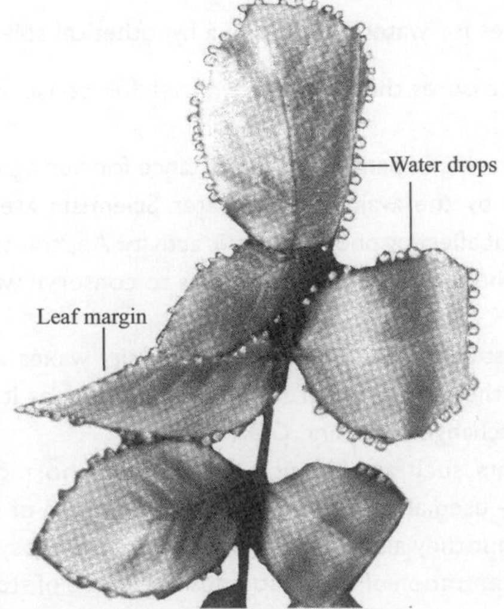

Fig. 3.10 Droplets of water along the margins of leaves, lost through guttation.

RQ 3.10. Why is it so that many fruits, rootstocks and certain other propagules are routinely sealed with waxes? Explain.

Ans. Many fruits, rootstocks and certain other propagules (meant for propagation) are routinely sealed with waxes to prevent excessive loss of water during storage and transportation over long distances. In such cases, CO_2 uptake is not as important a consideration as in saplings of herbaceous annuals.

RQ 3.11. Discuss briefly the analogy of the leaf epidermis to a multiperforate septum.

Ans. Stomata occur on both surfaces of the leaves of many plants, and there are on the average 10,000 stomata per square centimeter. Water vapour diffusing out through the small stomatal pore (about 10 μm in diameter) spread out in all directions from the margins of the pore, forming 'vapour caps' or 'diffusion shells' (Figure 3.11). When stomata are far enough apart, i.e., acting independently, the diffusion shells are practically spherical but when they are 'too close by', the vapour shell that forms over each pore overlap and eventually coalesce. Coalescence, results in a single vapour shell over the entire surface (representing boundary layer) of the leaf, and thus comparable to 'multiperforate physical' systems. In such cases, the rate of diffusion of water vapour is not proportional to the circumference or perimeter[4] of the pore. Thus, the 'widely-spaced' small pores are extremely efficient in terms of diffusion of vapours, per unit area.

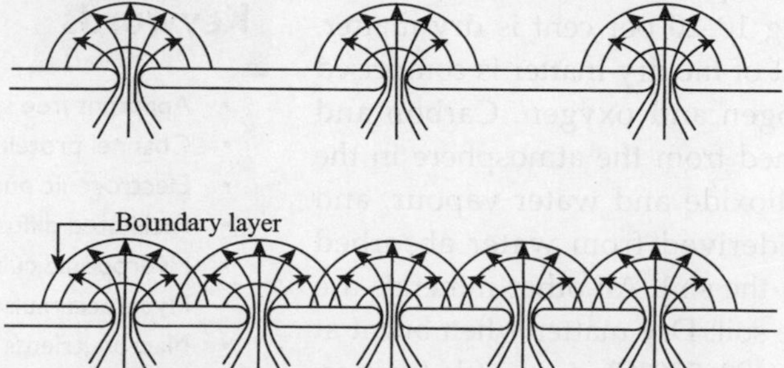

Fig. 3.11 When the stomata are far apart, the water vapour coming out through the small pores spread out in all directions from the margins forming 'vapour caps' or 'diffusion shells' that are practically spherical. When the stomata are close-by, the diffusion shells are not fully formed due to interference from the vapour caps of adjacent stomata. In such cases, the diffusion rate of water vapour does not correspond to the circumference of the pore.

[4] 'The rate of diffusion of water vapours from the surface of the leaf is proportional to the circumference or perimeter rather than their areas' was proposed by Brown and Escombe, while working in the Jodrell Laboratory in Kew Gardens around 1900.

Chapter 4

Mineral Nutrition and Absorption

Keywords

- Apparent free space
- Channel proteins
- Electrogenic pump
- Facilitated diffusion
- Hydroponic culture
- Hyperaccumulators
- Macronutrients
- Micronutrients
- Nutrient film technique
- Osmoregulation
- Salt respiration
- Transmembrane potential

Plants contain 80–90 per cent of water by weight and the remaining 10–20 per cent is dry matter. Nearly 95 per cent of the dry matter is comprised of carbon, hydrogen and oxygen. Carbon and oxygen are obtained from the atmosphere in the form of carbon dioxide and water vapour, and hydrogen being derived from water absorbed by the roots from the soil. All other minerals are obtained from the soil. Dry matter when burnt at high temperature (400–500°C) in a muffle furnace, leaves behind non-volatile residue, commonly called plant ash, i.e., metal oxides. The ash content of different plants and tissues varies from 1 to 4 per cent of fresh weight – being the least in aquatic plants and the highest in plants growing in saline or dry soil. Succulent tissues and fleshy fruits are poor in mineral proportion, whereas the leaves contain a relatively high proportion. A careful chemical analysis of the ash reveals that it is made up of as many as sixty different mineral elements found in the plants.

For the actual recognition and the dependence of plants on mineral elements present in the soil, credit must be given to de Saussure. In his book *Recherches chimiques sur la végétation*, published in 1804, de Saussure clearly demonstrated that the inorganic minerals present in the ash of plants are obtained from the soil through the root system. He further stated that mineral elements, including nitrogen, derived from the soil were important for the growth and development of plants. But the essentiality of the inorganic constituents of plant ash to the general welfare of the plant was not recognized until it was reported by Liebig in 1840.

The earliest kinds of experiments to study mineral nutrition and soil improvement were done in the field. In 1843, the oldest existing Rothamsted Experimental Station, Harpenden, in Hertfordshire, UK was established by Sir John Lawes and John Gilbert. Together they established the first protocols for the extensive experimentation that followed on mineral nutrition of plants and were able to successfully convert insoluble rock phosphate to soluble phosphate(superphosphate).

In 1860, Julius Sachs, a leading German botanist, reported for the first time that plants could be grown to maturity in a well-defined nutrient media without the presence of soil. Towards the end of the century, the agricultural use of NPK fertilizers was well established in Europe.

Serious attempts to determine experimentally the mineral content of plants were made by Sachs and Knop as far back as 1830. Using solution cultures, they were able to show that various elements (C, H, O, N, P, K, Ca, S, Mg, Fe) were required for the normal growth and development of plants. Elements required in trace amounts remained undetected. By 1939, the necessity of the trace elements (Mn, Zn, B, Cu and Mo) had been detected in various plants.

4.1 Techniques Used in Nutritional Studies

The analysis of plant ash and the use of solution and sand cultures are the techniques used for the study of plant nutrition in laboratories throughout the world.

Ash analysis: A reasonably reliable means of detecting the mineral element content of a plant is to subject the plant to high temperature (600 °C) and then analyze its ash content. Only the mineral elements are present in the ash, all other organic compounds having been decomposed and passed off in the form of gases. Ash is dissolved in dilute hydrochloric acid and elements are detected and estimated by chemical analytical methods or by absorption spectrophotometer.

Solution cultures. Careful studies of mineral nutrition using aqueous solution were first conducted by Hoagland, Arnon, Stout and their co-workers. In 1938, Hoagland and Arnon published their now widely used 'Hoagland's solution' for the aqueous culture of plants. This nutrient solution includes all the essential elements in a ratio conducive to good growth for most plants.

Over the years, a number of nutrient solutions have been devised for growing plants in the absence of soil. Solution cultures **(Hydroponic cultures)**[1] provide an excellent means for controlling the quantity and relative proportions of mineral salts given to a plant in anyone experiment (Figure 4.1). Two other reasons for using solution cultures in mineral nutrition studies are the excellent solvent characteristics of water and the relative ease with which water can be freed of most contaminating influences. In early studies of plant nutrition, the nutrient reagents used presented a major source of contamination. These reagents had to be purified by various means before trace element deficiencies could be demonstrated.

An ample supply of oxygen to the root system is achieved by vigorous bubbling of air through the solution.

[1] **Hydroponics** is the technique of growing plants in a soil-less, nutrient-enriched watery medium. With this method, physiologists are able to study root functions and plant nutritional requirements.

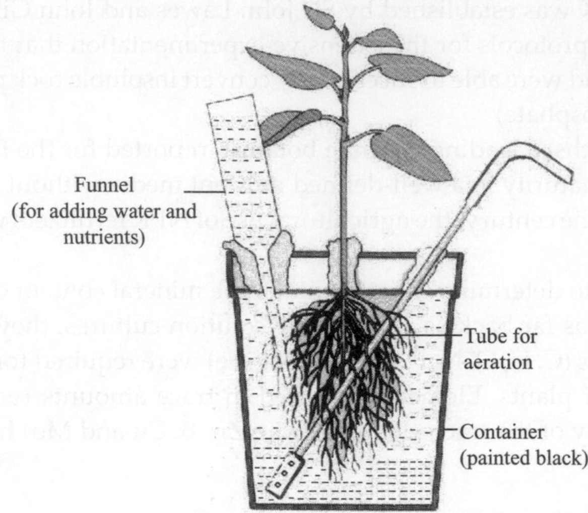

Fig. 4.1 Hydroponic culture.

The hydroponics technique is now widely used in North America for the year-round commercial production of vegetables such as lettuce, tomato, sweet bell-pepper, asparagus, broccoli, marijuana, cucumber and many more.

Sand cultures: There are certain shortcomings in using simple solution culture techniques to study the mineral nutrients requirements of plants. Such shortcomings include: (i) selective depletion of ions, (ii) changes in solution pH that might occur with the continuous absorption of nutrients by the plant roots. In order to avoid such problems, plants are grown in a non-nutritive, inert medium such as acid-washed quartz sand, gravel, perlite or vermiculite. Perlite is amorphous volcanic glass rock and vermiculite is a hydrous silicate mineral of the mica family.

Nutrient film techniques: In most commercial hydroponic techniques, some variation of the nutrient film operation are being used such as plant roots are continuously bathed with the nutrient solution, often recirculated in a thin layer by using a pump. The advantages of nutrient film techniques are: (i) they provide a medium for proper aeration of the roots (ii) they allow continuous monitoring of both the pH and the nutrient uptake of the solution.

Aeroponics: In this technique, the roots of growing plants are suspended in air along with a continuous spray of the nutrient solution. With this approach, it becomes easy to manipulate the gaseous environment around the roots but its use is not widespread because it requires higher concentrations of nutrients to maintain plant growth.

Ebb-and-flow systems: The nutrient solution goes up at regular intervals to submerge plant roots and then retreats, thereby exposing the roots to damp atmospheric conditions. But this again requires higher concentration of nutrients as compared to techniques described earlier.

Comparison of above mentioned techniques has been shown diagrammatically in Figure 4.2.

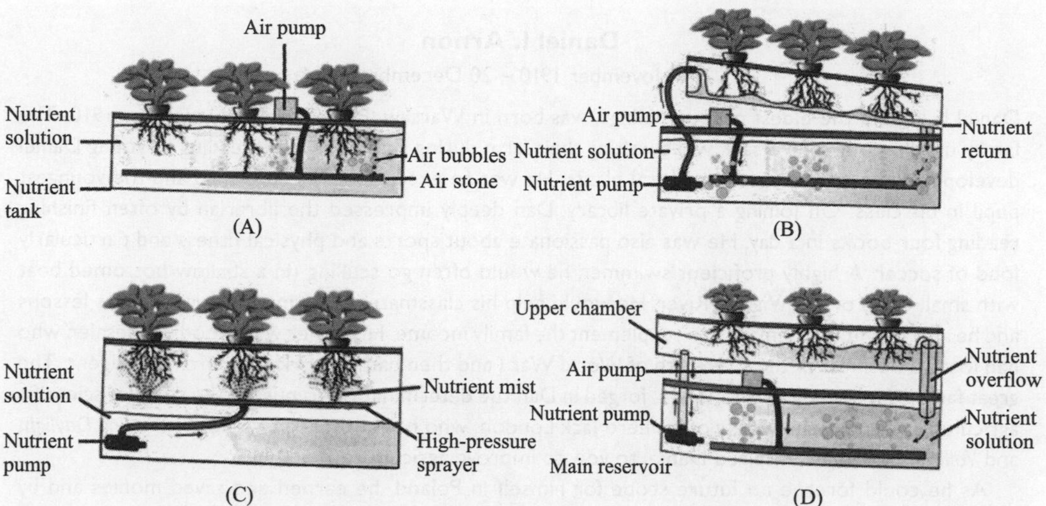

Fig. 4.2 Comparison of various solution culture techniques. (A) The **Hydroponic technique**; plant roots are submerged in a nutrient solution without the soil. Air stone[2] is used to pump air to maintain enough supply of oxygen to the root system of the plant (B) The **Nutrient film technique**; a pump drives the nutrient solution along the bottom of a tilted trough in the form of a thin film and then back to the nutrient recovery chamber (C) The **Aeroponic growth system**; a high-pressure sprayer pump is used to spray continuously nutrient solution on plant roots enclosed in a tank (D) The **Ebb-and-flow system**; a solution pump periodically fills the upper chamber with nutrient solution and then drains the solution back into the main tank.

Rhizotron

Another widely used technique employs 'Rhizotron' – a laboratory constructed underground and equipped with close-fitting insulated transparent panels in the form of glass or plastic installed against the soil profile that allows for non-invasive (or non-destructive) *in-situ* viewing, measuring and photographing of the root system as it grows. The studies include morphological and architectural parameters of root growth (that is, information on the number of the roots, root diameter and the distribution of roots, etc.). The researchers are thus able to collect digital images in the visible and infra-red range and by applying computational techniques are able to resolve root architecture and its dynamics. The use of chemical indicator dyes and fluorescent tags allows further probe into rhizosphere physiology.

Rhizotrons are in use at Botanic Gardens at Kew, at the USDA Northern Research Station at Houghton, Michigan (USA), at The Treborth Botanic Garden, University of Bangor (Wales, UK), at the East Malling Research Station, England, and at the University of Guelph, Canada.

4.2 Criteria of Essentiality

Daniel Arnon and Perry Stout of the University of California, USA (1939), established three criteria for essentiality of the mineral nutrient. These criteria are listed as follows:

[2] An air stone (also called an aquarium bubbler) is traditionally a piece of limewood or porous stone whose purpose is to gradually diffuse air into the nutrient solution of the tank, eliminating noise.

Daniel I. Arnon

(14 November 1910 – 20 December 1994)

Daniel I. Arnon, the eldest of four siblings, was born in Warsaw, Poland, on 14 November 1910. The family used to stay in Warsaw, while visiting their farm during summer breaks leading to young Daniel developing keen interest in agricultural plants. He was an exceptionally gifted child and the youngest pupil in his class. On joining a private library, Dan deeply impressed the librarian by often finishing reading four books in a day. He was also passionate about sports and physical fitness and particularly fond of soccer. A highly proficient swimmer, he would often go sculling (in a shallow bottomed boat with small oars) on the Vistula River. He would help his classmates by giving them gymnastics lessons and he also taught mathematics to supplement the family income. His father was a food wholesaler, who had lost his business in the aftermath of World War I and then had to work as a purchasing agent. The great famines following World War I, forged in Dan the determination to pursue the study of scientific agriculture. Reading the works of his hero Jack London, who had authored two books *Burning Daylight* and *Valley of the Moon*, inspired Daniel to vow to improve agriculture.

As he could foresee no future scope for himself in Poland, he earned and saved monies and by the age of 18 decided to go to New York (USA) where he took a job to save enough for purchasing a 'Greyhound' bus ticket for California where he was accepted by the University of California. He completed his Bachelor's degree from the University of California, Berkeley in 1932 and later his PhD in Plant Physiology in 1936 under Professor Dennis R. Hoagland, a pioneer in the field of mineral nutrition of plants . There they devised the basic principles for a nutrient media, famously known as Hoagland's solution (still in vogue worldwide for growing plants artificially in the laboratory). He discovered independently the essentiality of molybdenum (Mo) for the growth of plants and of vanadium (V) for algal cultivation. The molybdenum work had agronomic applications - addition of small amounts of Mo to deficient soils leads to significant increase in crop productivity in many parts of the world. Later, he became Professor of Cell Physiology and Biochemistry at the same University . During WW II, Arnon was given the rank of Major in the U.S. Army (1943–46) where he used his expertise on water culture of plants to grow crops in gravel and nutrient-rich water, while stationed on uncultivable Ponape Island in the western Pacific, where there was no arable land available. Dan spent his entire professional career at Berkeley, except for his brief stint in the US Army, and sabbatical leaves in England, Sweden, Germany, Switzerland and Pacific Grove, California. He was elected President of the American Society of Plant Physiologists and was the moving force behind the publication of the 'American Society of Plant Physiology and Plant Molecular Biology'.

After leaving military service, Arnon started researching on chloroplasts and their involvement in photosynthesis. He demonstrated that the chloroplasts utilize sunlight energy within the cells to produce ATP, the universal energy currency. He discovered the process of cyclic photophosphorylation in which only ATP is synthesized, and noncyclic photophosphorylation where the liberation of oxygen is accompanied by the generation of NADPH, the biological reductant. Additionally, Arnon and his team in 1955 was the first to replicate the complete process of photosynthesis in the laboratory (i.e. outside living cells) wherein they produced sugar and starch from CO_2 and H_2O. A pathbreaking discovery in photosynthesis was achieved with isolated and intact chloroplasts from spinach leaves . Like most far-reaching discoveries, Arnon's work was not recognized initially by many authorities in the field.

Another pathbreaking event occurred in the year 1962, when Daniel discovered the role of red iron-sulphur protein (which he called ferredoxin) – a universal part of the photosynthetic mechanism. He demonstrated the catalytic role of ferredoxin in both cyclic and noncyclic photophosphorylation. During the 1970s Arnon worked tirelessly on cytochrome enzymes system in chloroplasts/cyanobacteria.

Arnon's death came suddenly on 20th December 1994, at the age of 84, of complications resulting from cardiac arrest. His wife, Lucile Soule, had already passed away in 1986. He is survived by three daughters, two sons, and eight grandchildren.

1. An essential element is needed for completion of the life cycle of the plant – right from seed germination, flowering to seed setting.
2. The element should be involved directly in playing a key role in plant metabolism. Its effect must not be indirect.
3. The element should not be replaced by any other element with similar properties. As an example: although sodium has properties similar to potassium but it cannot replace potassium entirely as plant nutrient. So, potassium can be considered an essential mineral nutrient.

4.3 Essential Elements

Table 4.1 shows the list of essential elements that are further classified into major elements (macronutrients) and minor elements (micronutrients). Macronutrients are required in large amounts (>10 mmol/kg of dry weight), whereas micronutrients are needed in relatively small amounts (<10 mmol/kg of dry weight).

Table 4.1 The essential nutrient elements of higher plants and their concentration considered adequate for normal growth.

Element	Chemical symbol	Available form	Concentration in dry matter (mmol/kg)
Macronutrients			
Hydrogen	H	H_2O	60,000
Carbon	C	CO_2	40,000
Oxygen	O	O_2, CO_2	30,000
Nitrogen	N	NO_3^-, NH_4^+	1,000
Potassium	K	K^+	250
Calcium	Ca	Ca^{2+}	125
Magnesium	Mg	Mg^{2+}	80
Phosphorous	P	HPO_4^-, HPO_4^{2-}	60
Sulphur	S	SO_4^{2-}, SO_2	30
Micronutrients			
Chlorine	Cl	Cl^-	3.0
Boron	B	BO_3^{3-}	2.0
Iron	Fe	Fe^{2+}, Fe^{3+}	2.0
Manganese	Mn	Mn^{2+}	1.0
Zinc	Zn	Zn^{2+}	0.3
Copper	Cu	Cu^{2+}	0.1
Nickel	Ni	Ni^{2+}	0.05
Molybdenum	Mo	Mo_4^{2-}	0.001

Except for boron, all these elements are also essential for human beings. But unlike plants, our diet must also provide us with fluorine, iodine, cobalt, selenium, chromium, and sodium. We obtain fluorine by adding it to our drinking-water and iodine is obtained by adding it to salt or by eating a large amount of seafood. Our lives depend on sodium and if we lose too much by sweating, we can quickly die. Plants, however, have no need for it at all.

Essential elements in plants serve the following major functions:

1. ***Osmotic function:*** The mineral content of plant cells is generally 10–1000 times higher in concentration than the surrounding soil. Hence, water enters the cells by osmosis and builds up turgor pressure, which in turn is responsible for maintaining the shape and size of leaves.
2. ***Nutritive function:*** It takes many elements to constitute a cell. C, H, and O are the components of cellulose and lignin, constituting the bulk of the dry weight of the plant material, and therefore, known as the 'framework elements'. N, P, and S are components of proteins and nucleic acids. Since they make up the protoplasm, they are therefore known as the 'protoplasmic elements'. Calcium is a component of the cell wall, and magnesium is a component of chlorophyll.
3. ***Catalytic function:*** Micronutrients, such as Zn, Mn, Mo, B, Fe, generally act as cofactors for enzymes in many metabolic processes in plants.
4. ***Balancing function:*** Many elements like Ca, Mg, K, counteract the toxic effect of other minerals by maintaining an ionic balance.

4.3.1 Mobile and immobile elements

An important diagnostic aspect is whether symptoms appear in young leaves or older leaves. This is related to the mobility of the essential element. **Boron, calcium, iron, sulphur, and copper** are **immobile elements.** Once incorporated into plant tissue, they neither return to the phloem nor can they move to younger parts of the plant.

The elements **chlorine, magnesium, nitrogen, phosphorus, potassium, sodium, zinc,** and **molybdenum** are considered to be **mobile elements**. They can translocate to younger tissue, even after they have been incorporated into previous plant tissue. After the soil becomes exhausted of one of these elements, mature leaves are sacrificed by the plant. Such mobile elements are recovered and translocated to growing regions.

4.3.2 Foliar spray

Most plants can absorb mineral nutrients such as iron, manganese and copper that are sprayed directly on leaves. In cases, foliar application of mineral nutrients may be more effective than applying in the soil, where, soil colloids tend to absorb these ions, rendering them inaccessible to the root system.

Nutrient absorption by leaves is most efficient when the nutrient solution is applied to the leaf in the form of a thin layer, which often requires that the nutrient solutions be supplemented with surfactant chemicals, such as the detergent **Tween 80** (polysorbate 80, polyoxyethylene sorbitan monooleate) or newly developed **organo-silicon surfactants,** that reduce surface tension. Foliar sprays should be avoided on warm days, since high evaporation rates may lead to accumulation of salts which can cause burning or scorching of the leaves.

4.3.3 Physiological effects and deficiency symptoms associated with essential elements

Various physiological effects and deficiency symptoms associated with essential elements have been listed in Tables 4.2 and 4.3 respectively.

Table 4.2 Physiological effects of various essential elements.

Macronutrients	Available form	Physiological effects
Nitrogen (N)	Inorganic nitrate ion (NO_3^-), nitrite ion (NO_2^-), NH_4^+ ions	• Constituent of proteins, nucleic acid, indole-3-acetic acid, cytokinins and chlorophyll
Phosphorus (P)	Phosphoric acid (H_3PO_4)	• Essential component of phospholipids present in membranes
		• Found as phosphate ester (including sugar-phosphates) which play an important role in photosynthesis and intermediary metabolism
		• Component of nucleoproteins, DNA, RNA, phosphorylated sugars, and organic molecules such as ATP, ADP, UDP, UTP, CDP, CTP, GTP and GDP (required for energy transfer by phosphorylation)
Potassium (K)	Potash or potassium carbonate (K_2CO_3)	• An activator for many enzymes catalyzing reactions of photosynthesis and respiration
		• Important role in regulating the osmotic potential of the guard cells (osmoregulation) and ionic balance
		• Controls many plant movements such as opening and closing of stomata, sleep movements, e.g., nyctinastic movements in *Oxalis*, and Touch-me-not (*Mimosa pudica*)
		• Maintains electroneutrality in plant cells
		• Prevents lodging in plants

Contd.

Contd.

Macronutrients	Available form	Physiological effects
Calcium (Ca)	Divalent cation (Ca^{2+})	• Component of middle lamella layer of cell wall in the form of calcium pectate
		• Plays role in spindle formation during cell division
		• Essential for the physical integrity and normal functioning of membranes
		• Calmodulin–calcium complex
		• Acts as second messenger in several hormonal and environmental responses (intracellular signal transduction)
		• Important element in regulating the activity of α-amylase in barley
Magnesium (Mg)	Divalent cation (Mg^{2+})	• Constituent of porphyrin moiety of the chlorophyll molecule
		• Required to stabilize the ribosome structure
		• Activator of critical enzymes of photosynthesis (RuBP carboxylase and PEP carboxylase) and respiration (enolase and phosphoglyceric kinase)
		• Important for reactions involving ATP, where it helps to link the ATP molecule to the enzyme's active site.
Sulphur (S)	Divalent sulphate anion (SO_4^{2-})	• Component of amino acids such as cysteine, cystine, methionine and is associated with their tertiary structure or folding of proteins.
		• A constituent of the vitamins (biotin and thiamine) and CoA – an important component in respiration and fatty acid metabolism. • Important in electron transfer reactions in the form of iron–sulphur proteins (ferredoxin) during photosynthesis and nitrogen fixation. • Important component of mustard oil (thiocyanates and isothiocyanates), responsible for the pungent flavours of mustard, cabbage, turnip, horseradish, etc.
		• Participation in the structure of lipoic acid, TPP (thiamine pyrophosphate), glutathione.

Contd.

Contd.

Micronutrients	Available form	Physiological effects
Iron (Fe)	Ferric (Fe^{3+}) ion or Ferrous (Fe^{2+}) ion	• A part of the catalytic group for oxidation–reduction enzymes, haeme-containing cytochromes and non-haeme iron-sulphur proteins (Rieske proteins and ferredoxin) associated with photosynthesis, nitrogen fixation and respiration.
		• A constituent of oxidase enzymes, such as catalase and peroxidase.
		• Required for the chlorophyll synthesis and chloroplast protein in leaves.
		• Participates in the structure of dinitrogenase enzyme (in prokaryotes) – a crucial enzyme for molecular nitrogen (N_2) fixation to ammonia.
Boron (B)	Boric acid (H_3BO_3)	• Essential for the structural integrity of the cell wall and cell membrane.
		• Stimulates pollen tube germination and elongation.
		• Plays major role in translocation of sugars in the plants.
		• Participates in nucleic acid synthesis.
		• Role in cell division and cell elongation especially in primary and secondary roots.
		• Influences Ca^{2+} utilization.
Molybdenum (Mo)	Molybdate ion (MoO_4^{2-})	• Critical role in nitrogen metabolism–required for the functioning of nitrate reductase (reduction of nitrate to nitrite) and dinitrogenase (conversion of dinitrogen to ammonia).
		• Component of sulphite oxidase, xanthine dehydrogenase and aldehyde oxidase.
		• Required for the synthesis of ascorbic acid.
Manganese (Mn)	Divalent cation (Mn^{2+})	• Required as a cofactor for enzymes, especially decarboxylases and dehydrogenases; component of superoxide dismutase.
		• Participates in photosynthetic O_2 evolution in the form of oxygen-evolving complex: 'manganoprotein'.
		• Also cofactor for enzymes like oxidases, peroxidases, kinases, arginase, hydroxylamine reductase.

Contd.

Contd.

Micronutrients	Available form	Physiological effects
Chlorine (Cl)	Cl^- ion (ubiquitous in nature)	• A major counterion to diffusible cations, thus helps to maintain electrical neutrality across membranes.
		• Major osmotically active solute in the vacuole contributing to the ionic and osmotic balance of the cell.
		• Needed for cell division in leaves as well as shoots.
		• Essential for oxygen-evolving reactions of photosynthesis.
Copper (Cu)	Divalent cupric ion (Cu^{2+})	• Functions as a cofactor for several oxidative enzymes including cytochrome oxidase, ascorbic acid oxidase and polyphenol oxidase (responsible for browning of freshly cut apples and potatoes).
		• Component of superoxide dismutase, and detoxifies superoxide radicals(O_2^{*-}).
		• An integral part of plastocyanin and plastoquinone (photosynthetic electron transport carriers).
Zinc (Zn)	Divalent cation (Zn^{2+})	• An activator of many enzymes – alcohol dehydrogenase (ADH) catalyzing the reduction of acetaldehyde to ethanol; carbonic anhydrase (CA) catalyzing the hydration of CO_2 to bicarbonate, superoxide dismutase, carboxypeptidase, glutamic acid dehydrogenase (GDH), component of RNA polymerase.
		• Essential for the synthesis of tryptophan – precursor of phytohormone, indole-3-acetic acid. • Essential for the structure and function of transcription factors such as Zn finger, Zn cluster and Ring finger domains • Associated with important enzymes such as superoxide dismutase.
Nickel (Ni)	Ubiquitous in plant tissues	• Mobilization of nitrogen during seed germination. • Component of two important enzymes– urease and hydrogenase.

Table 4.3 Deficiency symptoms and diseases associated with various essential elements.

Macronutrients	Deficiency symptoms	
Nitrogen	In young plants, stunted growth and yellowish-green leaves.	
	Older leaves light green, followed by yellowing and shedding.	
	Accumulation of anthocyanins contributing purplish colour to the stems, petioles and the underside of leaves.	
	Shoots short, thin, growth upright and spindly.	
	Flowering and fruiting delayed or suppressed.	
Phosphorus	Young plants stunted.	
	Leaves exhibit intense green colour (darker than normal), sometimes turn dark green-purple because of accumulation of anthocyanins and may become necrotic.	
	Rapid senescence and death of older leaves.	
	Meristematic growth ceases in potato.	
	Poor nodulation in legumes.	
	Fruits ripen slowly, plants often dwarf at maturity.	
Potassium	Leaves with marginal chlorosis (yellowing at the tips and edges, usually in younger leaves) and necrosis appearing first on old leaves, usually wrinkled, or corrugated between veins.	
	Short and weak stems because of poor development of mechanical tissues; easily lodged due to increase in susceptibility to root-rotting fungi.	
	Dominance of apical buds is greatly diminished, so lateral buds become active resulting in bushy growth.	
	Decrease in starch and protein synthesis.	
Calcium	Leaves chlorotic, rolled, curled, deformed and necrotic. New leaves misshapen (malformed or distorted) or stunted.	
	Breakdown of meristematic tissues in stems and roots.	
	Roots poorly developed, lack fibres, 'slippery' to touch because of deterioration of middle lamella.	
	Symptoms appear near growing points of stems and roots.	
	Little or no fruiting.	

Contd.

Contd.

Macronutrients	Deficiency symptoms	
Magnesium***	Chlorosis, particularly in interveinal regions – thus giving a striped appearance to the magnesium-deficient leaf, especially in monocots. Lower leaves turn yellow from outside, going in; veins remain green.	
	Symptoms occurring first on old leaves.	
	Formation of brown spots and dying of tips.	
	Brittleness of leaves quite common.	
	Severely affected leaves may wilt and shed or may abscise without the wilting stage.	
	Necrosis occurs quite often.	
Sulphur	A generalized chlorosis of the leaf, including the tissues surrounding the vascular bundles.	
	Symptoms tend to occur initially in the younger leaves.	
	Stems often slender.	
	Stunted plants, sparse foliage.	
	Inhibition of terminal bud; emergence of lateral bud.	
Micronutrients	**Deficiency symptoms**	**Disease**
Iron****	Interveinal white chlorosis, appearing first on young leaves (young leaves are yellow and white with green veins. Mature leaves are normal).	
	Tendency for chlorosis of all aerial parts, often becoming necrotic.	
	In some cases, leaves may be completely bleached, margins and tips scorched.	
Boron	Terminal leaves necrotic, shed prematurely.	• 'Stem crack' in celery
	Shortening of internodes of terminal shoots (usually leading to rosetting)	• 'Heart rot' of sugar beet
	Blackening of apical meristems and death. General breakdown of meristematic tissue.	• 'Drought spot' of apples (pit-like spots)
	Short and stubby roots.	
	Stunted growth leading to dwarfing.	
	Flower development and seed production impacted.	

Contd.

Contd.

Micronutrients	Deficiency symptoms	Disease
	Cambial cells in the growing regions cease to divide but cell elongation continues; thus resulting in dislocation of xylem and phloem from their original position and are rendered non-functional.	
	The breakdown of phloem may result in accumulation of carbohydrates and increase in anthocyanin formation.	
Molybdenum	Light yellow interveinal chlorosis of leaves – appears first in older leaves; the rest of the plant is often light green.	'Whiptail disease' of cauliflower.
	Leaf blade may fail to expand.	
	Mottling of leaf followed by necrosis of leaf margins.	
	Failure of nodulation in legumes.	
	Seeds remain poorly filled.	
	Stunted growth.	
Manganese	Mottled chlorosis with veins green and a leaf web tissue yellow or white, appearing first on young leaves and may spread to old leaves; yellow spots and or elongated holes between veins.	'Grey speck' of cereal leaves
	Stems yellowish-green, often hard and woody.	'Crinkle leaf'
	Carotene reduced.	
	Deformities in legume seed.	
	Disintegration of thylakoid membranes of chloroplast.	
	Inhibits calcium translocation into the shoot apex.	
Chlorine	Exhibit reduced growth.	
	A general chlorosis of leaf.	
	Wilting of the leaf tips.	
	Stunted root growth.	
	Bronzing and necrosis basipetally in areas proximal to the wilting.	
Copper	Stunted growth.	'Summer die-back' of citrus.
	Distortion of young leaves.	'Reclamation' disease of cereals and legumes.

Contd.

Contd.

Micronutrients	Deficiency symptoms	Disease
	Wilting of terminal shoots, frequently followed by death.	
	Fading of leaf color, often seen	
	Carotene and other pigments reduced.	
Zinc	Leaves chlorotic and necrotic, young growth first affected.	'Little leaf' disorder of fruit trees.
	Rosetting, premature shedding.	'Mottle leaf' disease of apple.
	Leaves small and distorted with white necrotic spots.	
	Failure of seed setting.	
	Decline of auxin levels.	
	Stunted growth (shortened internodes)	
Nickel	Depressed seedling vigour.	
	Chlorosis.	
	Necrotic lesions in leaves.	
	Decreased urease activity in soybean leaves.	
	Decreased levels of hydrogenase activity in the nodules of soybean, which in turn reduces the efficiency of nitrogen fixation.	

*Excess of N stimulates vegetative growth, favouring a high shoot/root ratio and delays flowering in horticultural crops. Excess N inhibits the development of sclerenchymatous tissues, making the stems weak and tender.

** An excess of P stimulates growth of roots over shoots, thus reducing the shoot/root ratio.

*** Soil deficiencies are usually corrected by the addition of $MgSO_4$.

**** Remedial measures – (i) Spraying the plants with iron salt solutions and (ii) drilling a hole into trunk of affected orchard trees and inserting a gelatin capsule filled with an iron salt.

Excessive micronutrients, particularly copper and zinc, can inhibit growth of the roots because roots are the first organ to accumulate the mineral nutrients.

Besides 17 essential elements, Na, Si, Co and Se have been put in the category of **beneficial elements**.

4.3.4 Role of beneficial elements

Silicon

1. Silicon plays role in fending off fungal infections and in providing mechanical support to the cell walls thus preventing lodging, especially in grasses.

2. It helps in increasing growth and fertility of plants and in reducing the rate of transpiration.
3. It neutralizes phosphate deficiency and reduces the toxicity caused by iron and manganese.
4. It improves resistance to pathogens.

Sodium

1. Sodium is an important micronutrient in C4 and CAM plants for regeneration of PEP.
2. It increases the rate of cell enlargement.
3. It is believed to be involved in stomatal opening and maintenance of water balance.
4. It regulates nitrate reductase activity and transport of amino acids to the nucleus and therefore, controls the synthesis of nucleoprotein.

Cobalt

1. Cobalt is a part of vitamin B_{12} which is itself a part of an enzyme.
2. It is required by nitrogen-fixing bacteria for nodule formation.

Selenium

Selenium accumulator plants such as *Astragalus* are of great importance to ranchers.

Trace elements (or micronutrients)	Tracer elements (labelled or radioisotopes)
1. The minerals are required in extremely small quantities for normal growth, i.e., few parts per million (ppm) but are toxic at high concentrations.	These are 'labelled' or 'radioisotopes' of atoms of C, O, N, P which are artificially incubated into the plants in only very small amounts but injurious at high levels.
2. These minerals are required for growth and development and their deficiency causes physiological symptoms or disorders, and the plant is not able to complete its life cycle. The common symptoms are chlorosis, browning and necrosis.	These elements (in the form of $^{14}CO_2$, $H_2^{18}O$, $(^{15}NH_4)_2\ SO_4$ are used to trace biochemical steps involved in plant metabolism and can be detected by Geiger–Muller counter.
3. These elements act as activators or cofactors for different enzymes. Examples are Mo, Cu, Fe, Mn, Zn, and B.	These tracer elements are passed through intermediate reactions involved in plant metabolism, e.g., ^{14}C, ^{18}O, ^{15}N, ^{35}S.

Chlorosis	Etiolation
1. Chlorosis is a physiological disorder caused by the deficiency of mineral element (s).	The growth and development of plant when grown in darkness.
2. It is characterized by the yellowing of leaves or other parts due to loss or reduction of chlorophyll synthesis and is followed by the expression of yellow pigments – carotene and xanthophyll.	The leaves do not expand but remain small, rudimentary and pale yellow in colour.

Contd.

Contd.

Chlorosis	Etiolation
3. Chlorosis is followed by mottling of leaves and the appearance of dead, brown necrotic spots. The leaves may then twist and wither with premature fall.	No formation of chlorophyll but protochlorophyll accumulates in darkness.
4. The deficiency symptoms may be general, i.e., uniform all over the leaves or sometimes strictly local, a single branch or even a single leaf being affected, or sometimes along the veins or between the veins.	The internodes elongate many times more than normal.
5. Chlorosis may occur first in the young leaves (as in Fe, Mn, B, Ca) as these minerals are 'immobile' or in the lower leaves (as in N, P, K and Mo) as these are 'more mobile', moving from older to younger leaves.	The whole plant or the seedling is affected where the tender apical growing point is protected by the hook.
6. Overall growth is stunted with sparce branching, short internodes and marked decrease in leaf size.	The plants become very tall, narrow and spindly.
7. Flowering and fruiting is delayed or remain suppressed.	This response has a survival value in seedlings since it permits them to come out of the soil quickly and the energy is not wasted on the growth of the leaves.*

*When the seedling shoot emerges from the soil surface and is exposed to light, its development 'switches' from skotomorphogenesis (dark grown) to photomorphogenesis (light grown). This process is called de-etiolation. Once the seedling breaks through the surface of the soil and reaches the light, the hook protecting the growing point then opens, the stem thickens and its elongation rate slows down and the leaves expand and become green due to chlorophyll formation (i.e., conversion of 'protochlorophyll' to chlorophyll).

4.4 Iron Stress

Iron deficiency related problems in higher plants can be overcome by adopting two-pronged strategies for the solubilization and absorption of partially soluble inorganic iron (Figure 4.3).

4.4.1 Chelating agents

H^+-ATPase pumps in the cortical cells of the root serve to acidify the rhizosphere which can solubilize Fe^{3+} from the soil colloids. Fe^{3+} is then chelated by caffeic acid, which, is also secreted into the rhizosphere by the roots. The Fe^{3+}-chelate complex, subsequently moves to the surface of the root where Fe^{3+} is reduced to Fe^{2+} by inducible Fe^{3+} reductase enzyme. The Fe^{2+} thus produced is immediately translocated across the plasma membrane with the help of Fe^{2+} transporter. Some of the natural chelating agents including porphyrins (as in haemoglobin, cytochromes and chlorophyll) and caffeic acid are specific for iron uptake. Among the synthetic chelating agents, sodium salt of ethylenediaminetetraacetic acid (**EDTA**), diethylenetriaminepentaacetic acid (**DTPA**, or pentetic acid), or ethylenediamine-N,N'-bis (o-hydroxyphenylacetic) acid (***o, o-EDDHA***) are more common. However these being not very specific, can attach to a variety of cations (Fe, Cu, Zn, Mn, Ca, etc.).

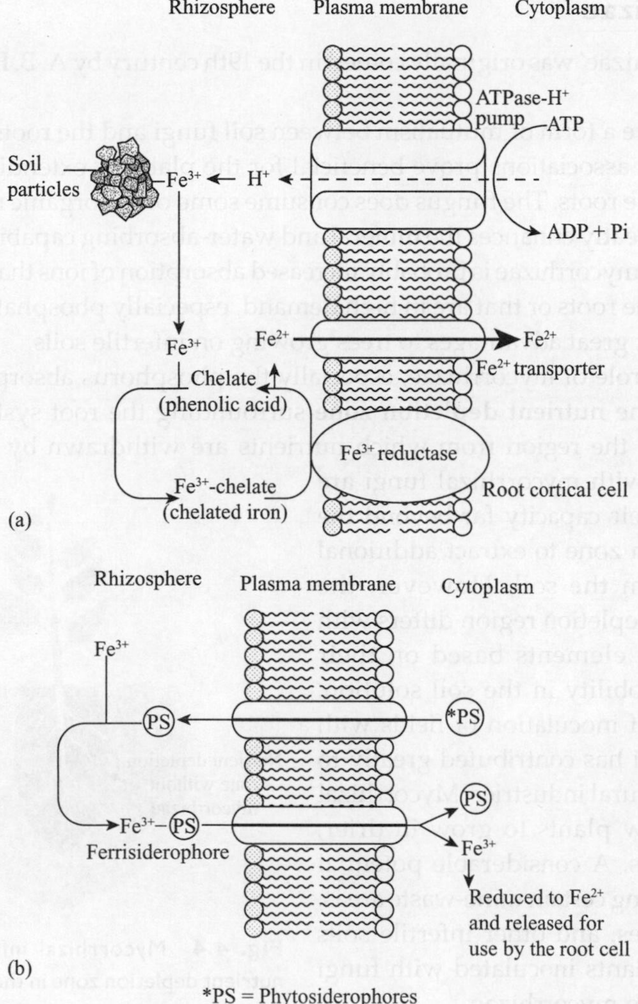

Fig. 4.3 Strategies for iron uptake in the higher plants under conditions of iron stress. (a) Role of chelating agents; (b) Involvement of phytosiderophores in the release of iron across the plasma membrane.

4.4.2 Phytosiderophores

Another strategy for uptake of iron under conditions of iron stress involves the use of siderophores. Siderophores, low-molecular-weight and iron-binding ligands are synthesized by bacteria, fungi, algae as well as roots of higher plants. Phytosiderophores are synthesized by the members of the Poaceae family (Gramineae) and are released only at the time of iron stress. Fe^{3+} present in the rhizosphere is solubilized by phytosiderophore secreted by the roots. The entire Fe^{3+} phytosiderophore complex is reabsorbed into the roots where Fe^{3+} is converted to Fe^{2+} and released for use by the plant. Phytosiderophores might be degraded afterwards or released from the root to transport more iron. **Avenic acid and mugineic acid** are common phytosiderophores released by roots of higher plants.

4.5 Mycorrhizae

The term 'mycorrhizae' was originally coined in the 19th century by A. B. Frank, the German botanist.

Mycorrhizae are a form of mutualism between soil fungi and the roots of higher plants. Such mycorrhizal associations prove beneficial for the plant by extending the volume of soil available to the roots. The fungus does consume some of the organic nutrients from the plant but it also greatly enhances the mineral and water-absorbing capabilities of roots. The greatest benefit of mycorrhizae is probably increased absorption of ions that normally diffuse slowly towards the roots or that are in high demand, especially phosphate, NH_4^+, K^+, NO_3^-. Mycorrhizae offer great advantages to trees growing on infertile soils.

The beneficial role of mycorrhizae, especially the phosphorus absorption, seems to be associated with the **nutrient depletion zone** surrounding the root system. The nutrient depletion zone is the region from which nutrients are withdrawn by the roots (Figure 4.4). Plant roots with mycorrhizal fungi are able to extend their capacity far beyond the nutrient depletion zone to extract additional phosphorus from the soil. However, the extension of the depletion region differs with various nutrient elements based on their solubility and mobility in the soil solution. The knowledge of inoculation of fields with mycorrhizal fungi has contributed greatly to forest and agricultural industries. Mycorrhizal associations allow plants to grow in drier, nutrient-lean soils. A considerable potential exists for populating certain mine-waste areas, landfills, roadsides, and other infertile soils by introducing plants inoculated with fungi capable of forming mycorrhizae.

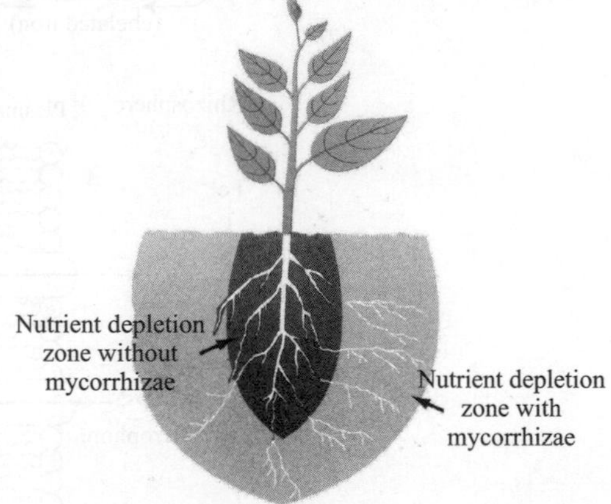

Fig. 4.4 Mycorrhizal infection extends the nutrient depletion zone in the soil surrounding the plant root system, from where nutrient elements are readily withdrawn.

Two types of mycorrhizae have been identified: (Figure 4.5).

Ectotrophic mycorrhiza (Ectomycorrhiza)	Endotrophic mycorrhiza (Endomycorrhiza)
The fungal mycelia surrounding the root produce a thick 'mantle'.	Do not form a thick 'mantle' of fungal hyphae surrounding the roots, both within the root itself and in the surrounding soil.
Some of the fungal hyphae penetrate the intercellular spaces of the cortical cells without entering the cell interior and here forms a meshwork of mycelium referred to as **Hartig net**.	After entering the root hairs or epidermis, the hyphae not only penetrate the intercellular spaces but also enter the individual cells* of the cortex where they can form oval shaped structures named vesicles and branched structures referred to as arbuscules – the site for exchange of nutrients between the fungal and the host cells.

The extent of fungal hyphae is so extensive that its whole mass is almost comparable with the root themselves.	The vesicular–arbuscular mycorrhizae (VAM) form only a small portion of fungal material constituting only 10 per cent or even less of the root weight.**
The mycorrhizal fungus infects only tree species, including woody gymnospermic and angiospermic plants.	The VAM occurs in association (mostly) with the roots of herbaceous angiosperms species.

*As the hyphae extend or enter cells, they do not destroy the plasma membrane or the vacuolar membrane ((tonoplast) of the host cells, but instead encourage these membranes to surround the mycelium, forming structures called arbuscules. A specific 'phosphate transporter' in the plant cell plasma membrane (different from the transporter that takes up phosphate directly from the soil) transfers the phosphate from the arbuscule into the cytosol of the cortical cell.

** Ironically, root-hair production either slows down or stops completely upon 'infection'. This decreases the water-absorbing capacity of root hairs but the hyphae take over the water-absorbing function.

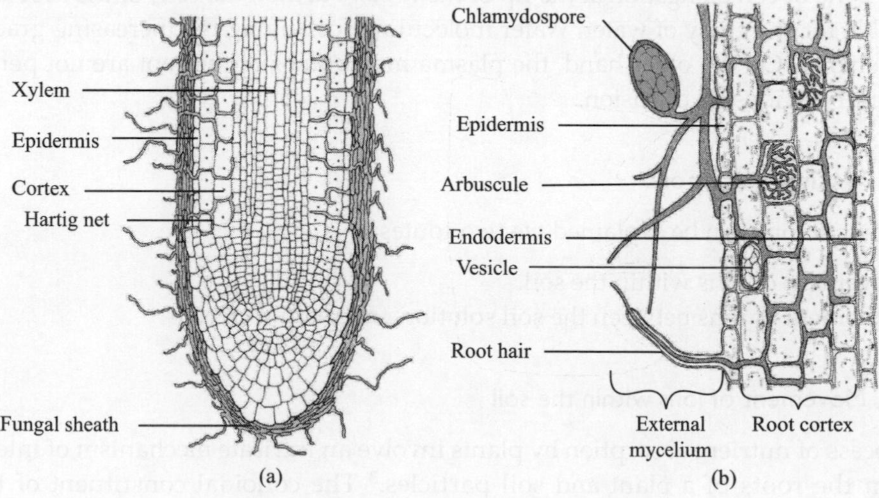

Fig. 4.5 Two types of mycorrhizae (a) Ectomycorrhiza and (b) Endomycorrhiza.

Harnessing **arbuscular mycorrhizal symbiosis** to optimize crop nutrition as fertilizers become increasingly expensive and will depend on understanding how the symbiotic partners interact to influence nutrient acquisition. Two major forms of arbuscular mycorrhizal colonization of the root cortex exist: ***Arum-type colonization*** characterized by the formation of intracellular, highly branched arbuscules in root cortical cells and ***Paris-type colonization*** characterized by the formation of intracellular hyphal coils in root cortical cells, some of which (called arbusculate coils) bear small arbuscule-like branches. Phosphate is transferred by the fungi directly to the root cortex of the plant *via* phosphate transporters, which are specifically expressed only in the plant membrane surrounding the arbuscules and coils in the root cortex.It has been reported that wheat plants inoculated with arbuscular mycorrhizal strains (*Funneliformis mosseae* and *Rhizoglomus irregulare*) experienced minimum yield losses caused by saline stress as compared to non-inoculated plants. This could be due to the enhanced water and phosphorus uptake, and more efficient photosynthesis. Mycorrhizal colonization could also increase total phenolic content and antioxidant composition in certain wheat cultivars.

4.6 Mechanism of Mineral Absorption

Absorption of minerals and absorption of water from the soil by the plant are two independent and distinct phenomena. These mineral salts are absorbed by the root system as ions but not as whole molecules. This is evident from the unequal accumulation of anions and cations of the same salt in the cell sap. Further roots absorb ions of some particular elements such as nitrogen, phosphorus, calcium and others and exclude some others like sodium, silicon and others. It means absorption of mineral salts by the roots is a selective process. Both water and minerals are absorbed from different regions of the root. While water is absorbed by the root hairs, minerals are taken up from the meristematic region and the zone of cell elongation at the tip of the roots. Further, the cells of the root are fully permeable for the entry of water. Water molecules move along the increasing gradient of kinetic energy. On the other hand, the plasma membranes of the root are not permeable for the entry of ions by diffusion.

4.6.1 Passive transport

Movement of ions can be explained *via* two routes:

1. Movement of ions within the soil.
2. Movement of ions between the soil solution and plant root cells.

4.6.1.1 Movement of ions within the soil

The process of nutrient absorption by plants involve an intricate mechanism of interaction between the roots of a plant and soil particles.[3] The colloidal constituent of the soil (comprising of clay micelles and humus) represents a highly specific region consisting primarily of negatively charged ions. The positively charged ions adsorbed onto the colloidal interphase constitute the chief reservoir of nutrient elements needed by the plants. Exchange of ions at colloidal matrix plays a principle role in providing the supply of nutrients to plants in a well-regulated manner.

Two theories have been proposed to explain the mechanism of ion exchange:

(a) ***Contact exchange theory:*** Ions do not dissolve in the watery medium before they get absorbed by the roots. Such ions adsorbed electrostatically to clay particles are not held so tightly or firmly, but oscillate within a given small oscillation volume of the space. As the oscillation volumes of two ions having similar charges overlap, ions are exchanged from one another (Figure 4.6a) (Jenny and Overstreet, 1939).

[3] The intimate relationship of roots, soil and microorganisms is called rhizosphere which affects mineral availability to plants.

(b) ***Carbonic acid exchange theory:*** Ions first dissolve in the soil solution. CO_2 evolved during root respiration dissolves in the soil solution to form carbonic acid (H_2CO_3), which being a weak acid ionizes into H^+ ions and HCO_3^- ions. H^+ ions of the soil solution get exchanged with K^+ ions adsorbed to clay micelles and ultimately may diffuse to the root surface in exchange for H^+ ions. The cation may also be absorbed, as ion pairs with bicarbonate (Figure 4.6 b).

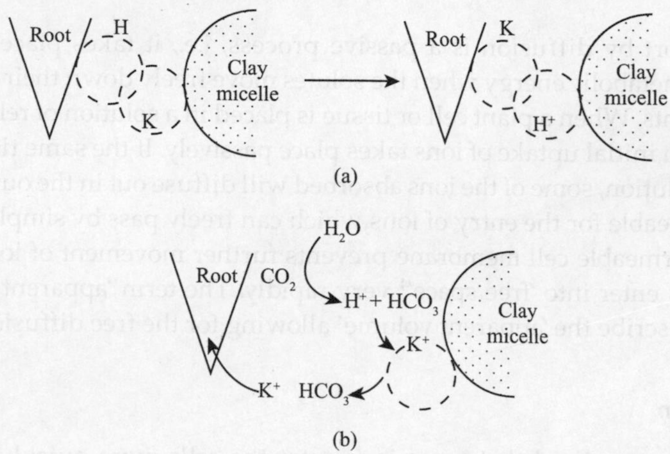

Fig. 4.6 Ion exchange between plant roots and soil particles: (a) contact exchange theory and (b) carbonic acid exchange theory.

Donnan equilibrium

According to this theory, certain 'fixed' or 'indiffusible' ions within the cells play an essential role in mineral absorption and these would require to be balanced by ions of the other charge. Assuming that a concentration of 'fixed' anions is present towards the inside of the membrane, more cations would be absorbed in addition to the normal exchange to maintain the equilibrium. Therefore, concentration of cations would be more in the internal solution as compared to the external solution.

When the product of the anions and the cations in the inner as well as the outer solution becomes equal, the **Donnan equilibrium** is attained according to the following equation.

$$\left[C_i^+\right]\left[A_i^-\right]=\left[C_o^+\right]\left[A_o^-\right]\left[\begin{array}{l}C_i^+ \text{ and } C_o^+ = \text{cations in the inner and outer solution}\\ A_i^- \text{ and } A_o^- = \text{anions in the inner and outer solution}\end{array}\right]$$

The Donnan equilibrium explains the accumulation of cations against a concentration gradient without the expenditure of metabolic energy.

Similarly, if there are 'fixed cations' towards the inner side of the membrane, the process will result in the accumulation of anions in the cell against a concentration gradient.

4.6.1.2 Movement of ions between the soil solution and plant root cells

Mineral nutrients must traverse the cell membrane before being absorbed into the plant root cells – thus making nutrient absorption fundamentally a cellular involvement.

Solutes may be transported across the cellular membranes by three modes: simple diffusion, facilitated diffusion, and active transport.

Simple diffusion

Cellular transport by diffusion is a passive process, i.e., it takes place without direct expenditure of metabolic energy when the solutes move freely down their electrochemical potential gradients. When a plant cell or tissue is placed in a solution of relatively high salt concentration, an initial uptake of ions takes place passively. If the same tissue is returned to a lower salt solution, some of the ions absorbed will diffuse out in the outer solution. The cell wall is permeable for the entry of ions which can freely pass by simple diffusion. The differentially permeable cell membrane prevents further movement of ions by diffusion. This means ions enter into 'free space'[4] very rapidly. The term 'apparent free space' was introduced to describe the 'apparent volume' allowing for the free diffusion of ions.

Facilitated diffusion

In the 1930s, it was noticed that some ions enter the cells more quickly than would be expected based on their diffusion through a lipid bilayer. So the term 'facilitated diffusion' was introduced to explain this rapid, assisted diffusion of solutes across the membrane with the aid of transport proteins in the membranes. Two entities - concentration gradient (for uncharged solutes) or electrochemical gradient (for charged solute molecules and ions) help in determining the direction of transport. Facilitated diffusion is **bidirectional**, i.e., net movement ceases across the membrane when the rate of transport through the membrane becomes equal in both the directions. Other characteristic features of facilitated diffusion are:

(i) specificity
(ii) passive nature
(iii) saturation kinetics

Two kinds of transport proteins are known to occur in the membranes – 'Carrier proteins' and 'Channel proteins' (Figure 4.7).

Carrier Proteins: These proteins transport certain specific ions and solutes (sugars and amino acids) across the membrane. Carriers are specific for a particular solute and can transport them in both the directions across the membrane (i.e. bidirectional). However, they facilitate the movement of solutes by physically binding to them, similar to enzyme–

[4] Free space refers to that part of the cell where ions can freely move by diffusion, e.g., cell wall and intercellular spaces.

substrate interactions. Binding of the solute towards the outside of the cell is followed by a conformational change in the carrier protein which then releases them on the other side of the membrane. Thus, the net movement always takes place from regions of high concentration to regions of lower concentration, similar to simple diffusion, the only difference is the presence of carriers which 'facilitate' the process.

***Channel Proteins*:** These are commonly visualized as charge-lined, water-filled channels (aquaporins) traversing or spanning the membrane. Channels are generally recognized by ion species (such as $K^+/Ca^{2+}/Cl^-$ channels) passing through them which, in turn, depends upon the 'hydrated size' of the ion and its associated charge. The associated water molecule must diffuse in conjunction with the ion. Ions to be transported do not bind to (or interact with) channel proteins. The direction of movement of ions depend upon (i) their relative concentration on both sides of the membrane and (ii) the difference in voltage across the membrane (the inside of the cell is generally negatively charged).

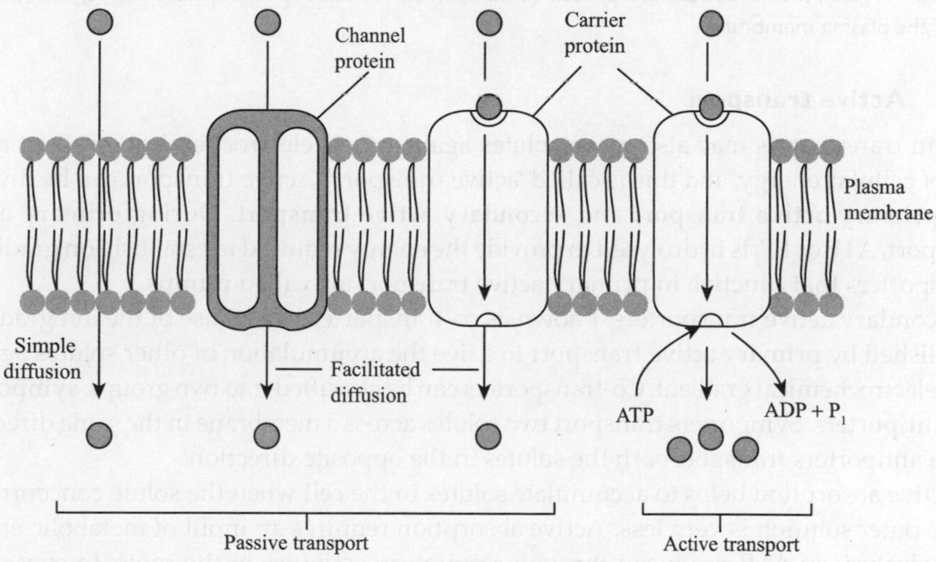

Fig. 4.7 **Two fundamental concepts of the ions (and solutes) exchange across the membranes – passive transport (simple diffusion and facilitated diffusion) and active transport.**

Several ion channels have been discovered in the membranes of plant cells. Each type of channel is specific for a particular ion such as calcium (Ca^{2+}), potassium (K^+) or chloride (Cl^-) and many others. Channel proteins are often 'gated' and may exist in two different conformations - open or closed (Figure 4.8). Solutes of suitable size and charge may pass across the membrane only when the channel 'gate' is open. Two kinds of gates are known to exist- an electrically gated channel opens in response to membrane potential and other channel opens only in the presence of the diffusing ion and may be regulated by signals such as light, hormones or other stimuli.

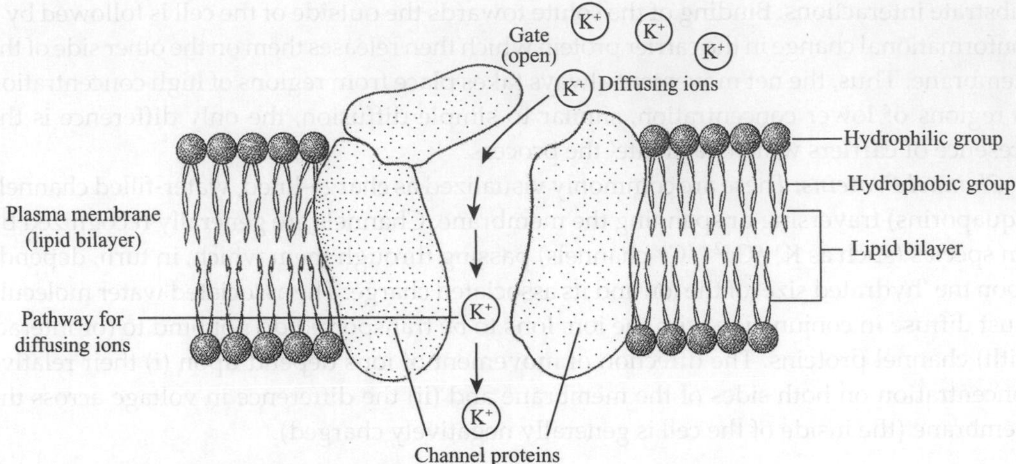

Fig. 4.8 A gated membrane channel protein (in an open conformation). K^+ions have been shown diffusing across the plasma membrane.

4.6.2 Active transport

Protein transporters may also move solutes against their electrochemical gradient at the cost of cellular energy, and this is called 'active transport'. Active transport can be divided into **primary active transport** and **secondary active transport**. During primary active transport, ATP or PP_i is hydrolysed to provide the energy required to establish ion gradients. Transporters that function in primary active transport are called **pumps**.

Secondary active transporters, known as co-transporters make use of the ion gradients established by primary active transport to drive the accumulation of other solutes against their electrochemical gradient. Co-transporters can be classified into two groups; **symporters** and **antiporters**. Symporters transport two solutes across a membrane in the same direction, while antiporters transport both the solutes in the opposite direction.

Active absorption helps to accumulate solutes in the cell when the solute concentration in the outer solution is very less. Active absorption requires an input of metabolic energy *via* hydrolysis of ATP, produced through respiratory activities of the roots. In contrast to simple and facilitated diffusion, the active transport is always unidirectional (either into or out of the cell) – and is assisted by carrier proteins.

There are a number of evidences for the existence of active absorption of minerals by plants, for example,

1. The rate of absorption of minerals is too rapid to be explained by passive absorption.
2. The diffusing solute is at a higher concentration within the cell, unlike lower concentration outside.
3. The rate and amount of solute uptake is directly proportional to the expenditure of metabolic energy. The decreased rate of ions absorption under anaerobic conditions or using metabolic inhibitors, such as cyanide or 2, 4-dinitrophenol, provide evidence in support of active transport mechanism.
4. The absorption of minerals increases with increase in temperature, up to a certain limit.

The degree of salt accumulation differs for different ions. Hoagland and Davies (1923), found 2,000 times as much potassium; 13 times as much calcium and 870 times as much phosphate in the cell sap of *Nitella* or *Valonia* as in the pond water in which the alga was growing. Hober (1945) also reported that fresh water alga *Nitella* accumulated K^+ ions to a level, 1000 times greater than the level of K^+ in the surrounding water. Obviously, the ions continued to be absorbed and retained even after the concentration of the ions inside the cell had gone much beyond the ions concentration in the external solution.

4.6.2.1 Carrier concept

The area between outer and inner space is impermeable to free ions and the movement across this is thought to require the participation of specific carriers. The carriers combine with ions in the outer space to form carrier – ion complex and release these ions into inner space where they aggregate (van den Honert, 1937).

1. carrier —kinase (ATP → ADP)→ carrier* (carrier activation)

2. carrier* —ion (+ or –)→ carrier* – ion (carrier – ion complex)

3. carrier* – ion —phosphatase→ carrier + ion (ion release)

The concept has derived great support from the three observations: isotopic exchange, saturation effects and specificity in the absorption of ions.

Based on the carrier concept, there are two possible mechanisms for salt absorption.

Lundegárdh's cytochrome theory

Lundegárdh and Burström (1933) observed that the rate of respiration increases when a plant is transferred from water to a salt solution. An increase in respiration over the normal one due to anion absorption is called 'anion respiration' (salt respiration).

Assumptions of Lundegardh's theory (1950, 1954) are as follows:

1. Anion absorption is independent of cation absorption. Only anions are actively transported across the membranes through the involvement of cytochrome system.
2. An oxygen gradient exists for oxidation at the external surface of the membrane and reduction at the inner surface.
3. Since both the salt respiration and the anion absorption are inhibited by cyanide or carbon monoxide, the anions are absorbed by a cytochrome pump in which cytochromes act as carriers.

According to him, the reactions brought about by the dehydrogenases on the inner surface result in the formation of protons (H^+) and electrons (e^-)

External medium	Cytochrome system	Internal medium
Salt	Anion transfer ⇌	Reduced substrate

As the electrons pass outward through the electron transport chain, there is corresponding inward passage of anions. The anions are picked up by the oxidized cytochrome oxidase and are transferred to the other members of the chain as they transfer the electron to the next component (Figure 4.9).

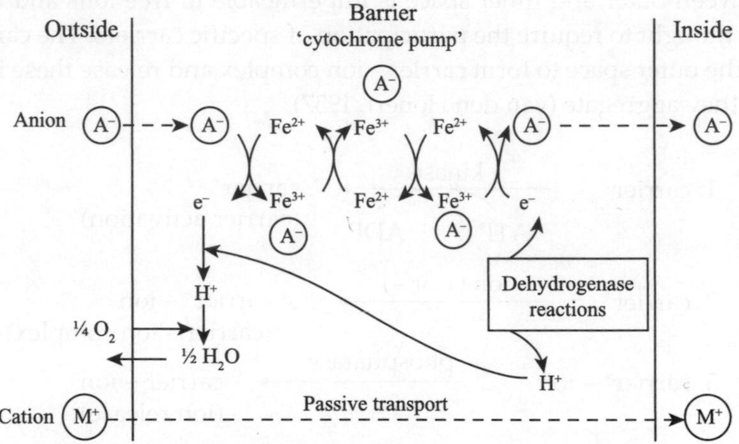

Fig. 4.9 Diagrammatic representation of the cytochrome theory of salt absorption.

The theory assumes that cations (M^+) move passively along the electrical gradient created by accumulation of anions (A^-) at the internal membrane surface.

Though this hypothesis explains the role of metabolic energy in the absorption of ions, it has been adversely criticized for assuming that there is a carrier only for anions and not for cations. The hypothesis also fails to explain the uptake of ions under anaerobic conditions following a period of active aerobic respiration.

Bennet–Clark's hypothesis (Lecithin hypothesis)

Bennet–Clark (1956) proposed an ATP carrier mechanism where protein associated with the phosphatide (lecithin[5]) acts as a carrier involved in the transport of ions across an otherwise impermeable cell membrane.

The carrier is 'amphoteric' and hence both cations and anions combine with it. In this transport, lecithin is synthesized and hydrolyzed in a cyclic manner – picks up ions on the outer surface and releases them into inner space on hydrolysis (Figure 4.10).

[5] Plants do not contain lecithin, choline or choline esterase, but have similar substances. Acetylcholine is one of the most common neurotransmitters in both invertebrates and vertebrates.

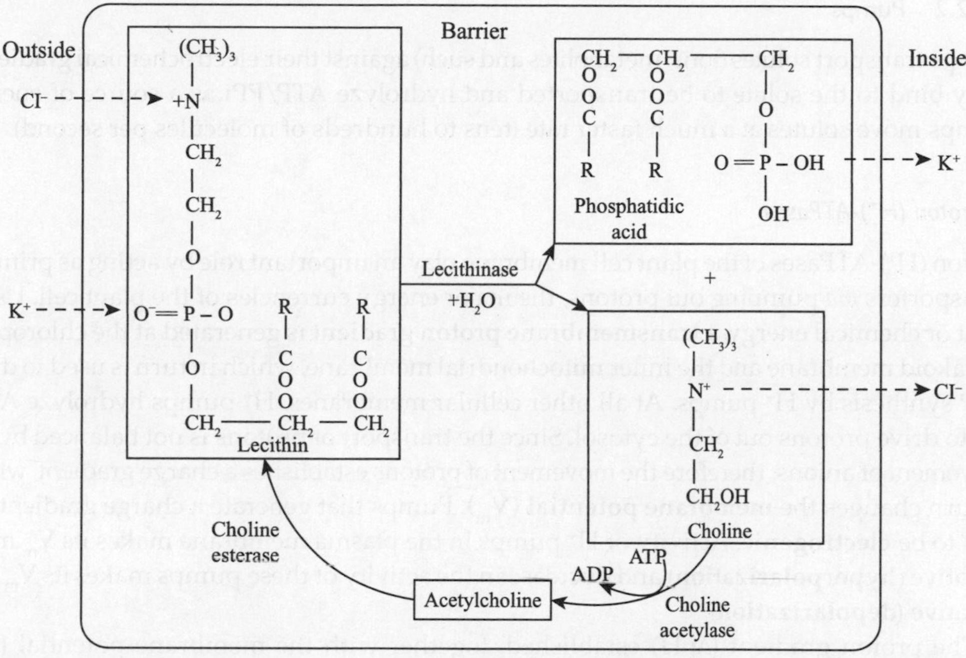

Fig. 4.10 Diagrammatic representation of the ATP carrier mechanism as proposed by Bennet–Clark.

The synthesis of at least one of the components of the phosphatide cycle or Lecithin carrier concept requires ATP.

Passive transport	Active transport
Movement of ions (both anions and cations) is from high chemical potential to low chemical potential, that is, along the chemical potential gradient.*	Movement of ions (+ve or –ve) is from low chemical potential to high chemical potential that is, against the potential gradient. Thus, the plant is able to absorb and accumulate ions within the tissue many times more than the surrounding soil solution.
It requires no expenditure of metabolic energy.	It requires expenditure of energy, derived from respiratory processes.
It is a slow process.	It is a much quicker process.
Involves phenomenon like mass flow, simple diffusion, facilitated diffusion (through channel proteins or aquaporins and carrier proteins), ionic exchange mechanism, and Donnan equilibrium.	Involves carriers (like proteins, lecithin and cytochrome pumps) located in or on the membrane – both plasma and vacuolar membranes, and 'ion pumps' mediated through ATPase (ATP-splitting enzyme).

* It takes into account the effects of both concentration and electrical forces of diffusing ions.

4.6.2.2 Pumps

Pumps transport solutes (ions, metabolites and such) against their electrochemical gradients. They bind to the solute to be transported and hydrolyze ATP/PPi as a source of energy. Pumps move solutes at a much faster rate (tens to hundreds of molecules per second).

(i) Proton (H^+)-ATPases

Proton (H^+)-ATPases of the plant cell membrane play an important role by acting as primary transporters *via* pumping out protons, the major energy currencies of the plant cell. Using light or chemical energy, a **transmembrane proton gradient** is generated at the chloroplast thylakoid membrane and the inner mitochondrial membrane, which in turn is used to drive ATP synthesis by H^+ pumps. At all other cellular membranes, H^+ pumps hydrolyze ATP/PP_i to drive protons out of the cytosol. Since the transport of protons is not balanced by the movement of anions, therefore the movement of protons establishes a charge gradient, which in turn changes the **membrane potential (V_m).** Pumps that generate a charge gradient are said to be **electrogenic**. Activity of H^+ pumps in the plasma membrane makes its V_m more negative (**hyperpolarization**) and decrease in the activity of these pumps makes its V_m less negative (**depolarization**).

The proton gradient (ΔpH) established, together with the membrane potential (ΔE), contributes to a **proton motive force (pmf).**

$$\text{pmf} = \Delta E - 59\ \Delta pH$$

The proton motive force is the primary source of energy for a variety of physiological activities such as secondary active transport of solutes, sucrose transport during phloem loading, turgor-related phenomena (such as nyctinastic and stomatal movements, plant and cell growth, salinity tolerance, etc.), hormone-mediated cell elongation, and regulation of cytoplasmic pH (7.0–7.3).

> The plasma membrane H^+-ATPase is a highly regulated enzyme with multiple physiological functions. The activity of this enzyme is the key to the transport activities of the plant cell and it is a major consumer of cellular ATP.

Implications of the H^+-ATPase pump

1. Only one ion species is transported in single direction – uniport system.
2. Because of the charge carried by the transporting ion, an electrochemical gradient is established across the membrane i.e., H^+-ATPase pump is electrogenic.
3. Since the transported ions are protons (H^+), the proton as well as the electrical gradients, are established by the H^+-ATPase pump across the membrane. Stored energy in the resulting electrochemical proton gradient can be coupled to cellular activities according to Mitchell's chemiosmotic hypothesis.

ATP–hydrolyzing pumps have been classified into three classes depending upon their structure, mechanism of action and inhibitor sensitivity

(a) ***F-type ATPases:*** These are found in the chloroplast thylakoid membranes, the mitochondrial inner membranes and the bacterial cell membranes. These permit the movement of protons across the membrane down their electrochemical proton gradient and use the energy associated with the gradient to drive ATP synthesis. These are large with multiple subunits and insensitive to the inhibitor vanadate.

(b) ***P-type ATPases (Plasma membrane H^+-ATPases):*** These are Mg-dependent, a single subunit, 100 kD protein with 10 membrane-spanning domains and are sensitive to vanadate. One H^+ ion is pumped out of the cytoplasm for each ATP hydrolyzed. Approximately 30–50 per cent of ATP produced by a cell is hydrolyzed by the plasma membrane H^+-ATPase.

    Plasma membrane H^+-ATPase pumps are encoded by a large multigene family and 11 genes have been identified in *Arabidopsis*. But it is regulated predominantly through enzyme activity rather than gene expression. The fungal toxin, fusicoccin, increases the activity of plasma membrane H^+-ATPase pumps and promotes the acidification of the plant cell walls, cell wall extensibility, and cell expansion resulting in growth and development.

(c) ***V-type ATPases (Vacuolar H^+-ATPases):*** These are present in almost all living organisms and are ubiquitously distributed among the endomembranes of plant cells except in the chloroplast and mitochondrial membranes. The V-ATPases on the membranes of the Golgi apparatus are associated with exocytosis and vesicle trafficking[6] in the secretory pathway. The H^+ pumping activity of the V-ATPase at the tonoplast causes vacuolar acidification (pH 5.5). The net positive charge of the vacuole lumen is balanced by the accumulation of organic anions (malate and oxalate) and inorganic anions (chloride).

In *Arabidopsis,* V-type ATPases are made up of 13 subunits, each of which can be encoded by any one of the 27 genes and are insensitive to vanadate. V-ATPases pump about three H^+ per ATP hydrolysed from the cytosol into the vacuole. The resulting proton gradient serves to drive anions and cations into the vacuole followed by turgor changes responsible for guard cell movements and nyctinastic responses. Hence they are electrogenic, contributing to the proton motive force and membrane potential (V_m) of the vacuolar membrane (tonoplast).

(ii) Ca^{2+} pumping ATPase

Ca^{2+} pumping ATPases are found on almost all endomembranes including the plasma membrane and their role is to maintain low concentration of Ca^{2+} in the cytosol (50–200 nM) by pumping Ca^{2+} out of the cell and into the vacuole and other organelles. Cytosolic Ca^{2+} must be kept low for the maintenance of many metabolic reactions and for Ca^{2+} to behave as

---

[6] Vesicle trafficking is the biological phenomenon by which vesicles transport material between different cellular compartments and between a cell and its environment.

a signalling molecule. The Ca^{2+} pump belongs to the P-type family of ATPases, having one polypeptide chain of about 110 kD with an extended N-terminal domain. ATP hydrolysis by this enzyme is accompanied by the transport of two Ca^{2+} ions across the membrane. Two broad classes of Ca^{2+}-ATPases identified in plant cells, are based on their ability to bind calmodulin, a cellular Ca^{2+} sensor. The calmodulin-binding class of Ca^{2+}-ATPases has the N-terminus domain (autoinhibitory) that binds to calmodulin which inhibits the pump and thus raises the cytosolic Ca^{2+} concentration. This is a good example of a negative feedback loop that maintains cytosolic Ca^{2+} homeostasis.

If levels of cytosolic Ca^{2+} are too high, calcium phosphate will precipitate, depleting the cytosolic phosphate pool. This pool is crucial for supporting a rapid turnover of ATP in cellular metabolism.

(iii) H^+-pyrophosphatase

Plants possess an H^+ pumping pyrophosphatase (H^+-PPase), a protein of about 80 kDa on the tonoplast membrane as well as in the Golgi, trans-Golgi network (TGN), the multivesicular body (MVB) and plasma membrane. The functional H^+-PPase is a homodimer.

Two kinds of H^+-PPases are found in plants:

- Type I H^+-PPases: stimulated by cytoplasmic K^+ and inhibited by Ca^{2+}ions
- Type II H^+-PPases: activated by Ca^{2+} and insensitive to K^+.

It is believed that H^+-pyrophosphatase evolved in plant cells to exploit the large pool of pyrophosphate generated during carbohydrate metabolism. By using pyrophosphate to pump H^+ at the tonoplast and other membranes, pyrophosphate concentrations are maintained at levels that do not adversely affect carbohydrate metabolism. Synthesis of ADP-G and UDP-G for starch and cellulose formation, respectively generate pyrophosphate, which in turn acts as a negative feedback regulator to slow their synthesis.

(iv) ABC transporters

ABC (ATP-binding cassette) transporters constitute one of the largest protein families found in all living organisms. These belong to P-type ATPases that hydrolyze ATP and are not electrogenic. ABC transporters are involved in multifarious activities in the cells:

(i) facilitate the transport of uncharged solutes (such as chlorophyll, anthocyanins, the auxin and antifungal compounds) out of the cytosol.
(ii) the transfer of waxes to the surfaces of leaf cells.
(iii) the sequestration of xenobiotics (such as synthetic herbicides) in the vacuole. The xenobiotic molecules are first modified by the addition of glutathione, which acts as an address label that targets the solute to an ABC transporter.

ABC transporters are found on the plasma and vacuolar membranes and are made up of two basic structural elements: integral, membrane-spanning domains and cytoplasmically oriented, nucleotide-binding folds that help in ATP hydrolysis.

4.7 Transmembrane Potential

A potential (or voltage) difference across the membrane is called its **transmembrane potential or Nernst potential.** This comes about due to unbalanced distribution of anionic and cationic charges on both sides of the membrane.

The relationship between transmembrane potential gradient and ion distribution across the membrane can be expressed quantitatively by the Nernst equation:

$$-z\Delta E_{nj} = 59\log\frac{C_j^i}{C_j^o} \left[\begin{array}{l} \Delta E_n = \text{Electrical potential difference for the ion } j \\ z = \text{Valency or charge for ion } j \\ \dfrac{C_j^i}{C_j^o} = \text{Ratio of molar concentration of ions inside and outside the cell} \end{array}\right.$$

This Nernst Equation can be employed to (i) find out the possibility of active or passive mode of transport of ions across the membrane and (ii) measure the transmembrane potential (Nernst potential) contributed by the ion, when its concentration is known.

4.8 Ion Antagonism

The effect of one ion upon another is known as **ion antagonism.**

Potassium ions (and other monovalent cations) tend to 'decrease' the cytoplasmic viscosity and increase membrane fluidity and permeability, while divalent ions such as Ca^{2+} act in a reverse manner. Hence, mixed salt solutions rather than single salt are usually supplied to the plants. For example, a plant when grown with roots in a dilute solution of NaCl will die soon because of deleterious effects. The same thing will happen if the same plant is grown in water containing only calcium chloride. However, when the same plant is grown in a solution containing the mixture of both these salts, no toxic effect is noticed. In such a solution, the harmful effect of sodium ions (Na^+) is counteracted by calcium (Ca^{2+}) ions and *vice versa*. Such an effect of one ion upon other is called 'ion antagonism'.

The role of phosphorus and nitrogen in plant metabolism seems to be interrelated. When available phosphates are low in the rooting medium, then the inorganic nitrogen compounds are rapidly absorbed and get accumulated in plant tissues. On the other hand, the absorption and accumulation of inorganic nitrogen compound is impeded or depressed if available phosphates are high in the rooting solution. Therefore, phosphatic fertilizers must be used judiciously. Otherwise, the nitrogen balance of the plant would be altered. Early maturation of plants often occurs when available phosphorus is high while the delayed maturity is occasioned by phosphorus deficiency.

Another example of absorption of ion inhibition results from ions having the same charge. This is because of mutual competition for the carrier site on the membrane. The most frequent competitions are between K^+ and Rb^+, Cl^- and Br^- and Ca^{2+} and Sr^{2+}. This is due to the fact that the carrier is unable to differentiate between closely related ions. Consequently, the relative concentration of ions will be the determining factor. Whichever of the two related ions is in higher concentration in the medium will be able to attach itself first to the carrier and thus would be transported across the membrane.

Some ions interfere with the uptake and transport of other ions because of competition for the specific membrane localized carrier site. Potassium, rubidium and caesium, of the same periodic group, have been observed to compete with one another for the same carrier site. For example, increase in the concentration of Rb^+ in the external solution decreases K^+ uptake, and *vice versa;* Cl^- and Br^- are also mutually antagonistic; and so is the case for the $Ca^{2+}$and Sr^{2+} ion pair.

4.9 Radial Path of Ion Movement through Roots

The root consists of three major regions:

(i) **Outermost region:** comprised of root epidermis and cortical cells.
(ii) **Intermediate region:** consisting of the endodermis with suberized casparian band.
(iii) **Innermost region:** consisting of vascular tissues

The uptake of ion starts with free diffusion into the apparent free space (AFS), which is equivalent to the apoplast outside the endodermal cells. The casparian strip checks further diffusion by the apoplastic mode through the endodermis layer into the vascular system. So the only possibility of ions to move through the endodermis is to enter the symplast aided by the carrier-mediated transport at the cell membrane. Symplastic connections (plasmodesmata) facilitate the passive movement of ions to xylem parenchyma cells from cell to cell. From there, the ions may be unloaded into the tracheary elements for long-distance transport to the leaves and other organs. Ions accumulate in the xylem against a concentration gradient, probably by an energy dependent, carrier-mediated transport process. The casparian strips also retard the loss of ions from the vascular tissue by blocking their diffusion down a concentration gradient. In some plants, **passage cells** of the endodermis may become a major port of entry for solutes into the vascular tissues.

Phytoremediation (plant-based bioremediation)

Contamination of the environment with heavy metals such as Zn, Cu, Hg, Ni, U, Cr and Cd radionuclides is a serious problem for human health and agriculture throughout the world. The presence of the metal hyperaccumulator species indicates that plants could be used to detoxify metal-contaminated soils by means of biological activity. **Hyperaccumulators** are the plants which are capable of building up and tolerating extremely high levels of trace elements in their tissues (more than 1 per cent of leaf dry matter). Hyperaccumnulators were first discovered more than 100 years ago and over 450 hyper-accumulators species have been discovered till date .These have potential uses in **bioremediation,** the reclamation of contaminated land by removal of toxic metals. The hyperaccumulation character has evolved between and within species as an adaptation to the abundance of particular elements in the environment. For example, all populations of *Thlaspi caerulescens*, a member of the Brassicaceae family, hyperaccumulate Zn, but when grown on the same substrate, populations from soils with high concentrations of trace metals generally accumulate less Zn than populations from soils with low concentrations.

Phytoremediation involves-Phytoextraction, Phytostabilization, and Phytodegradation

(i) ***Phytoextraction:*** removal of toxins from soil
(ii) ***Phytostabilization:*** complexation and immobilization of toxins within the soil
(iii) ***Phytodegradation:*** degradation of organic contaminants in the rhizosphere

One well-known heavy metal hyperaccumulator is *Noccaea caerulescens*, a member of family Brassicaceae that accumulates Zn (40,000 µg per gm dry weight) in their shoots. Other heavy metals including Cd are also reported to be accumulated in high concentrations in this species. The amazing physiology of heavy metal tolerance in this species makes it an interesting experimental plant choice, which can be used for deducing the mechanism of heavy metal hyperaccumulation, especially with the goal of transferring traits to species with a higher biomass. Thus, understanding the molecular process involved in toxicity has helped researchers develop better phytoremediation crops.

4.10 Mineral Toxicity

Aluminium (Al) toxicity is the major limiting factor in crop production on acidic soils. Aluminium is the most prevalent metal on the earth's surface. Most of the Al occurs as insoluble minerals above pH 5.0 but in more acidic conditions these minerals dissolve releasing several soluble monomeric Al species. Monomeric Al also forms complexes with various ligands, including sulphate, phosphate and carboxylates even when these groups are in macromolecules such as proteins and nucleotides. Stunted roots are the most obvious symptom of Al toxicity and this restricts the capacity of plants to access water and nutrients. Root apices are particularly sensitive and only a few millimetres need to be exposed to Al for growth to be inhibited. Al toxicity results from multiple interactions between Al and important biological ligands in the cell wall, the plasma membrane and cytosol, for example, Al can block Ca^{2+} and K^+ channels and inhibit Mg uptake which will result in mineral deficiency later. At the cellular level, Al induces oxidative stress, destabilizes phospholipid bilayers, disrupts cytoplasmic Ca^{2+} homeostasis, and disrupts components of the cytoskeleton.

Some plant species grow better on acid soils as compared to others. ***Al tolerance in plants*** relies on two types of mechanisms: ***exclusion mechanisms reduce Al uptake*** or minimize its interaction with root cells, and ***internal mechanisms enable plants to safely accommodate any Al*** that enters the cells. Al could be excluded from the symplasm by binding to cell walls, decreasing its permeability across the plasma membrane, chelating it with ligands released from the root, exporting it out of the cell or by increasing rhizosphere pH to reduce the local concentration of Al^{3+} near the root surface. Internal mechanisms could involve chelating Al in the cytosol or sequestering it to subcellular compartments such as the vacuole. Al exclusion in a different range of plant species, is based upon the efflux of organic anions such as malate, citrate, and oxalate from roots, which form strong complexes with Al^{3+} that serves to lower the free concentration of Al^{3+} and reduce its uptake by roots. Efflux of these anions is mediated by transport proteins from the **ALMT** (aluminium-activated malate transporter) and **MATE** (multidrug and toxic compound exudation) families.

Biotechnology provides opportunities of increasing food production on acidic soils in important crop species that possess little natural variants to breed for tolerance. Scientists are attempting to genetically engineer the trait of organic anion efflux in Al tolerance into model plant species. Two main strategies were employed to enhance organic anion efflux from roots: firstly through increased expression of genes involved in the synthesis of malate or citrate, while the second increased the capacity for organic anion transport out of the cells. Most success has been achieved by transforming plants with the ALMT and MATE-type tolerance genes.

4.10.1 Heavy metal homeostasis

Metal-binding proteins and metabolites are implicated in counteracting heavy metal toxicity. The maintenance of metal homeostasis is essential for viability. The range of mechanisms employed by plants to control the uptake and distribution of specific metal ions includes the capacity to detoxify non-essential metals and huge amounts of essential metals by converting them to complexes with low molecular weight or polymeric ligands. For example, **phytochelatins** are metal-binding polymers of different sizes with the general architecture (γ-Glu-Cys)nGly (n = 2 to 11), which are synthesized by plants in response to high concentrations of micronutrients such as Cu^{2+} or toxic cations such as Cd^{2+}. **Metallothioneins** are proteins with molecular mass ranging from 4 to 14 kD and have a high amount of cysteine, arranged in metal-binding motifs (Cys-Cys, Cys-X-Cys or Cys-X-X-Cys where X represents an amino acid). These regions of the polypeptide are able to coordinate with bivalent metal ions through the sulfhydral ligands of cysteine side-chains. Metallothioneins bind heavy metals such as Cd, and Cu, maintain Zn homeostasis, control ROS and participate in signal transduction pathways. Other metal-binding molecules with roles in nutrition, homeostasis and detoxification include **phytosiderophores, ferritins** and **metallochaperones.**

Research on the inorganic nutrients essential for crop plants has been of great potential value in agriculture and horticulture. The knowledge and techniques of plant breeding and plant nutrition are being used to select and develop cultivars of crop species that are better adapted for growth in nutrient-deficient environments. Hyperaccumulators, which concentrate trace elements, heavy metals or radionuclides at levels much higher than normal, are being identified and used to restore the soils around old smelters, metal finishers and nuclear weapon plants.

Review Questions

RQ 4.1. Arrange in descending order, the concentrations considered adequate for normal growth of plants. List out other micronutrients!

Ans.

Element	% dry weight
Carbon	45
Oxygen	45
Hydrogen	6
Nitrogen	1.5

Element	% dry weight
Potassium	1.0
Calcium	0.5
Magnesium	0.2
Phosphorus	0.2
Sulphur	0.1
Iron	0.01

Thus, carbon, hydrogen, and oxygen account for over 95 per cent of the total. These nine elements C, O, H, N, K, Ca, Mg, P and S make up the 'major elements' or 'macronutrients' because they are needed in relatively high concentrations. However, iron which accounts for approximately 100 ppm (0.01 per cent) lies on the borderline and is sometimes considered as also being a macronutrient.

Another six elements such as manganese, zinc, copper, boron, molybdenum and chlorine are required in extremely low concentrations, and are expressed in parts per million (ppm) rather than by dry weight. These trace elements are termed as 'micronutrients'.

Yet another element listed under trace elements is Nickel, an essential component of urease; the element is also involved in the degradation of arginine. Of the many elements that have been detected in plants tissues, only 16 are essential to higher plants. These may best be remembered by the phrase or mnemonic (formulated by Cyril Hopkins of the University of Illinois, USA) CHOPKNS Ca Fe Mg (read this as Hopkins Cafe, Mighty Good (stands for Mg); Mob comes (Zn) in cluster).

A few elements (e.g., cobalt, a constituent of vitamin B_{12}; selenium; silicon; sodium as in saltbush, *Atriplex* that accumulates very large quantities of NaCl-required by many other saline habitat plants, sugar beet, and blue green algae) have been shown to be essential for only a few plants, and therefore not included in this table. These are more appropriately called as 'beneficial elements'.

RQ 4.2. Discuss briefly the techniques used for growing plants for nutritional studies or for determining essentiality.

Ans. Mineral nutrition can be studied by using the following methods:

- Solution culture (nutrient solutions) or hydroponic culture: In this technique, the plants are grown with their roots submerged in a nutrient solution containing inorganic salts, without any soil (hydroponics or soil-less growth) and air is bubbled through the solution.
- Nutrient film technique: The nutrient solution is often recirculated in a thin layer, by using a pump.
- Aeroponic growth system: The plants are grown with their roots suspended over the nutrient solution which is whipped into a mist.
- Ebb-and-flow system: The nutrient solution periodically goes up to immerse plant roots and then recedes, exposing the roots to a moist atmosphere.

The earliest nutrient solutions were devised by Sachs, Knop and Dennis Hoagland.

RQ 4.3. Why fertilizers rich in phosphorus such as superphosphates or bone-meal are often used when transplanting perennials or propagules? In which way(s)is it helpful in agricultural crops?

Ans. An excess of phosphate in the soil stimulates the growth of roots over shoots, thus encouraging establishment of a strong root system which helps to absorb water and draw mineral nutrients effectively. The success of sugarcane cuttings during propagation depends upon the ease with which they can form roots and hence, phosphate-rich fertilizers (even growth hormones) promote rooting.

Application of phosphatic fertilizers reduces shoot growth but stimulates early flowering.

RQ 4.4. Why is excessive nitrogenous fertilizer used in forage crops?

Ans. Excess of nitrogenous fertilizer in the soil normally stimulates extensive growth of the shoot system, developing a high shoot over root ratio, and will often delay the process of flowering. This is especially important in forage crops where more leafy growth is important.

RQ 4.5. List out mineral(s) that are required for the synthesis or functioning of the following:

Ans. Nitrate reductase, dinitrogenase, glutamic acid dehydrogenase, carbonic anhydrase, carboxypeptidase, tryptophan, plastoquinone, plastocyanin, cytochrome a_3, catalase, PEP carboxylase, RuBP carboxylase, mustard oil glucosinolates, coenzyme A, phosphopyruvic transphosphorylase, phosphoglyceric transphosphorylase, enolase, alcohol dehydrogenase, α-amylase, and pyruvate kinase.

Enzyme/amino acid and others	Mineral(s)
1. Nitrate reductase	Molybdenum
2. Dinitrogenase	Iron and molybdenum
3. Glutamic acid dehydrogenase	Zinc
4. Carbonic anhydrase	
5. Carboxypeptidase	
6. Tryptophan (precursor of IAA)	
7. Plastoquinone	Copper
8. Plastocyanin	
9. Cytochrome a_3	Iron
10. Catalase	
11. PEP carboxylase	Magnesium*
12. RuBP carboxylase	
13. Mustard oil glucosinolates	Sulphur
14. Coenzyme A	
15. Phosphopyruvic transphosphorylase	Magnesium, potassium

Enzyme/amino acid and others	Mineral(s)
16. Phosphoglyceric transphosphorylase	Magnesium
17. Enolase	Magnesium
18. Alcohol dehydrogenase	Zinc
19. α-Amylase	Calcium
20. Pyruvate kinase	Potassium

*Mg^{2+} is an activator for both RuBP carboxylase and PEP carboxylase, two critical enzymes of CO_2 fixation. It is also required to stabilize ribosome structure.

RQ 4.6. Name the mineral responsible for (or that helps) in the following mechanisms:

Ans.

Osmoregulation	Potassium
Stomatal movement	Potassium
Translocation of food material	Boron
Cell wall stiffening	Calcium
Turgor movement in Touch-me-not	Potassium
Gravitropism	Calcium
Cysteine, cystine and methionine	Sulphur

RQ 4.7. Nitrogen is a constituent of many critical macromolecules. Name those macromolecules.

Ans. Nitrogen is a component of many essential molecules such as proteins, nucleic acids, some growth hormones (like indole-3-acetic acid, cytokinins), porphyrin ring structure of chlorophylls and cytochromes, and enzymes.

RQ 4.8. Sulphur is a component of many chemical compounds. Name those compounds.

1. Sulphur-containing amino acids – cysteine, cystine, methionine.
2. Component of vitamins such as thiamine and biotin.
3. Coenzyme A –it plays an important role in the respiratory process and fatty acid synthesis.
4. Component of iron–sulphur proteins (e.g., Rieske proteins and ferredoxin).
5. Constituent of allyl sulphide, mustard oil glycoside (sinigrin).

RQ 4.9. Mycorrhizal infection of the plant root depends much upon the nutrient status of the soil. Comment.

Ans. A well fertilized soil may tend to suppress mycorrhizal infection and the fungus continues to grow saprophytically. However, deficiency of nutrients such as phosphorus tends to promote infection, with the mycelium penetrating not only between the cortical cells but also enters within the cytosol of the cells. Here, it forms ovoid structures and branched structures designated as vesicles and arbuscules respectively – the sites for nutrient transfer between the fungus and the root cells. Mycorrhizal fungi assist in the uptake of phosphorus by extending their hyphae far away from the phosphorus depletion zone. Arbuscule hyphae are surrounded by an invaginated plasma membrane of the cortical cells.

RQ 4.10. Inoculation of fields with fungi capable of forming mycorrhizae is now a common practice in forestry and agriculture management. Comment.

Ans. The most well-authenticated advantage of mycorrhizae to plants is increased phosphate absorption, although absorption of other minerals and water is often increased. The greatest usefulness of mycorrhizae is probably increased intake of ions that normally move slowly into the roots or are more demanding, especially phosphates, ammonium ions, potassium, nitrates, etc. By doing so, we shall be able to resolve many problems associated with soil infertility and it would enable us to protect the ecosystem in native habitats.

Generally, only tender young roots become infected by the fungus. Ironically, root hair formation either slows down or stops altogether upon infection. This decreases the water absorbing power but the hyphae take over the absorbing function of root hairs.

RQ 4.11. Silicon is not an essential element yet it is important for certain plants. Comment.

Ans. Silicon is absorbed in the form of silicic acid, H_2SiO_4 [or $Si(OH)_4$] and it accumulates abundantly in the epidermal cells walls, as well as in primary and secondary walls of roots, stems and leaves of grasses (e.g., rice) and sedges. It accumulates in specialized epidermal cells called 'silica cells'. It is referred to as 'beneficial element'.

It seems to perform some functions like:

1. Reduces the rate of transpiration.
2. Prevents fungal infections.
3. Provides rigidity and limits compression of the tracheary or xylem elements, causing no or little lodging (bending down by wind action or rain).
4. Prevention of lodging or herbivory – an ecological requirement for some plants.

Silica is blamed for causing excessive wear on sheep's teeth.

Scouring rush or horsetail (*Equisetum arvense*) contain a high percentage of silicon; required for the growth of diatoms[7] whose cells walls are built of silica (SiO_2).

RQ 4.12. What is organic farming? In which ways is it better as compared to chemical fertilizers?

Ans. Organic farming is the practice of growing crops dependent only on the use of organic fertilizers, and avoiding any use of pesticides and herbicides. Organic fertilizers contain mineral nutrients in the form of complex organic molecules resulting from decay of wastes and residual products of plant and animal life, through microbial activity.

Organic farming has many advantages over the use of chemical fertilizers:

1. Improves the physical structure of the soil, making the soil easily cultivable and increases its capacity to hold air, water and mineral salts.
2. Can improve soil water retaining capacity, especially during drought, and helps increase drainage during wet weather.
3. Since the nutritional elements are locked in organic molecules, they must be released from them before these can be absorbed. Here, the nutrients are slowly released and become available to the plants over a period of weeks and months, thus reducing nutrient loss due to leaching.

[7] Diatoms are responsible for much of the photosynthesis in the oceans, i.e., about 2.5–3 times more carbon is fixed by photosynthesis in the oceans than on land. Diatoms frustules, deposited in the ocean floor when the cell dies, form the basis of many silicon-containing rocks and mineral deposits.

It is now realized that the crops grown on such manures are of high quality than those grown with chemical fertilizers and are healthier for human consumption. Also they are ecologically safe or less polluting on the environment.

Organic fertilizers include barnyard manure, green manure (obtained from legumes or other ploughed plants), compost, and sludge from sewage disposal plants.

Other items such as peat moss, vermiculite, and synthetic soil conditioners contain few or no minerals and are hence used primarily to improve soil structure.

RQ 4.13. Match the physiological disorders given in column I with minerals given in column II.

Ans.

Column I	Column II
1. Little leaf disorder	Mn
2. Grey speck disease of oats	B
3. Die-back of citrus	Cu
4. Whip tail disease of crucifers	Zn
5. Brown heart of turnip	Mo

Ans. 1. Zn; 2. Mn; 3. Cu; 4. Mo; 5. B

RQ 4.14. Why do plants growing in molybdenum-deficient soils show symptoms of nitrogen deficiency? Explain.

Ans. In the absence of Mo or when it is deficient in soil, the nitrates accumulate as they are not converted into nitrite by the enzyme 'nitrate reductase' which needs Mo as a cofactor. Thus, the synthesis of proteins is blocked and plant growth ceases.

$$NO_3 \xrightarrow[\text{Mo}]{\text{Nitrate reductase}} NO_2 \xrightarrow[\text{Cu and Fe}]{\text{Nitrite reductase}} NH_2OH \xrightarrow[\text{Mn}]{\text{Hydroxylamine reductase}} NH_3$$

Plants will show symptoms of nitrogen deficiency if any of the minerals such as Cu, Fe and Mn is deficient in the rooting medium.

RQ 4.15. Over-liming of the soil is injurious for plant growth. Explain.

Ans. Over-liming (or excessive application of calcium hydroxide to the soil), if done, in the soil would lead to an alkaline soil reaction which renders iron unavailable to plants by the formation of insoluble ferric hydroxide, $Fe(OH)_3$. This is not easily absorbed. Thus, the plants show symptoms of iron deficiency. Liming involves addition of calcium in various forms, often in mixtures. CaO (burned lime), $Ca(OH)_2$ (water-slaked or hydrated lime), or $CaCO_3$ (limestone, dolomite, or air-slaked lime).

RQ 4.16. List minerals that play an important role in the electron transport chain in photosynthesis and respiratory metabolism. What makes them so suitable?

Ans. Copper and iron are components of electron transport carriers – plastoquinone, plastocyanin, cytochromes, and ferredoxin - in the light reaction. While iron enters into the chemical composition of cytochromes (cyt *b*, cyt *c*, cyt *a* and cyt a_3) involved in the ETchain in mitochondria (oxidative phosphorylation). Copper exists in two forms (Cu^{3+} and Cu^{2+}), manganese in three forms (monovalent, divalent and trivalent) and iron occurs in Fe^{2+} (Ferrous state) and Fe^{3+}(Ferric state).

They get alternately oxidized and reduced during oxidation–reduction reactions taking place during ETS in photosynthesis and respiration.

RQ 4.17. Why is it so that plants growing in boron-deficient soil are often associated with a rosette habit and purplish coloration of the leaves? Explain.

Ans. **1.** In the absence of boron, shoot apices die. The degeneration of the terminal buds are accompanied by the lateral branches formation because of loss of apical dominance. The apical buds of lateral shoots, in turn, die resulting in further formation of branches. Thus, the plants will assume rosette formation or habit.

2. When boron is deficient, the cambial cells in the growing regions cease to divide but the cell elongation continues, resulting in the displacement of nearby tissues, especially the xylem and phloem which are rendered non-functional. The breakdown of phloem may result in the accumulation of carbohydrates and increase in anthocyanin formation which imparts a purplish colour to the leaves.

RQ 4.18. Why is phosphorus so important?

Ans. It is a common component of a vast number of organic compounds in plants as given under:

Glucose-6-phosphate, fructose-6-phosphate, phosphoglycerate, nucleic acids and phospholipids and phosphoamino acids (e.g., phosphoserine).

The central role of phosphorus is in the production of energy transfer compounds: ATP, ADP, Guanosine triphosphate (GTP), Guanosine diphosphate (GDP), Uridine diphosphate (UDP), Uridine triphosphate (UTP), Cytosine triphosphate (CTP), NAD, NADPH, pyridoxal phosphate, sugar phosphate, and phytic acid.

RQ 4.19. List two minerals that are associated with diseases in grazing animals, if consumed heavily.

Ans. **1.** Selenium-rich plants (or accumulators) –'Blind-Staggers' such as *Astragalus* and *Stanleya*.

2. Forage crops rich in molybdenum – Teart disease of livestock.

RQ 4.20. What are the ways in which the pH of the soil important for the availability and uptake of certain mineral ions by the plants? Explain.

Ans. The many ways where the soil reaction (i.e. pH) determines the availability and uptake of minerals by the roots are:

1. The rate of absorption of NO_3^- is generally at its maximum at low pH but it falls when the pH is on the higher side. The converse is true for NH_4^+ ions, which are rapidly absorbed at neutral pH but whose absorption is depressed as the pH falls.

2. Under alkaline soil reaction (high pH), iron is rendered unavailable to plants because of the formation of insoluble compounds [$Fe(OH)_3$]. However, in acidic soils, a large chunk of iron is in a soluble state (Fe^{2+}) and is therefore 'rapidly' absorbed by the plants and can accumulate in high concentrations which often have a toxic effect.

3. The availability of phosphorus in the soil depends upon its pH. At soil pH (<6.8), phosphorus occurs predominantly in the monovalent form of orthophosphate ($H_2PO_4^{3-}$) which is more rapidly taken up while at pH between 6.8 and 7.2 it mostly occurs as $H_2PO_4^{2-}$ in which state it is less readily absorbed. In basic conditions (pH>7.2), the trivalent PO_4^{3-} form predominates which cannot be absorbed.

(i) Phosphorus occurs in the soil primarily in the form of the polyprotic phosphoric acid (H_3PO_4) which consists of more than one proton, each having a different dissociation constant.

(ii) Phosphorus minerals have been used for centuries to increase crop yields. The manurial value of bones was recognized by Chinese and Welsh farmers about 2000 years ago. In the nineteenth century, the use of 'bones' as a fertilizer was widespread. The demand was so great that battle fields and graveyards of Europe were scoured for bones. By 1830, bone meal, bone dust, and bone charcoal were widely used as fertilizer. Peruvian guano (excrement of sea birds) then entered the European market. It was followed by superphosphate and its variants.

RQ 4.21. Why is foliar application of mineral nutrients often preferred over direct use in the soil? Comment.

Ans. Mineral nutrients (such as iron, manganese, copper, zinc and others like urea and ammonium salts) can be applied to the surface of leaves as sprays, a process known as 'foliar application' or 'foliar spray'. It has many advantages:

(i) It gives quicker and better results than the application of chemical fertilizer to the soil.

(ii) It is more economical and less wasteful as compared to direct use in the soil where not all of what has been applied may be available to the plant. They are absorbed by the soil and hence less available to the rooting media.

(iii) The practice is especially important in basic soils where micronutrients, for example, Mn, Cu, Zn and Fe are precipitated out into insoluble forms. Hence, it is considered essential to spray soluble salts of the nutrients directly to the foliage, especially in the case of fruit and forest trees and annual crops.

(iv) Foliar application of nutrients can reduce the 'lag time' (or time-lag) between application and absorption by the plants. The nutrient solution is absorbed through minute hairs (trichomes) or through rupture in the cuticle (breaks and cracks) and are able to reach the target site before any damage sets in or caused to the leaves. Hence, it is more efficient as compared to application through the soil where it takes a much longer time (minerals first reaching the site of absorption in the roots, their actual absorption and then upward translocation to the target site).

To promote the formation of a thin film on the surface of the leaf; the nutrient solution is often supplemented with surfactant chemicals such as oils that reduce surface tension and increase foliar absorption.

Foliar spraying should be done on a cool day or in the evening as it will not cause burning or scorching as it happens on a hot day where salts may accumulate on the leaf surface. Mineral salts in dilute concentrations are used alone or along with fungicides, insecticides, herbicides and others such as foliar sprays.

RQ 4.22. List out the commonly available N, P and K fertilizers.

Ans.

Nitrogen fertilizer	Formula
Ammonium sulphate	$(NH_4)_2SO_4$
Ammonium chloride	NH_4Cl

Ammonium nitrate	NH_4NO_3
Potassium nitrate	KNO_3
Urea*	$CO\ (NH_2)_2$
Calcium cynamide	$CaCN_2$
Potash fertilizer	
Potassium chloride (muriate of potash)	KCl (50–63% K_2O)
Sulphate of potash	KCl, $MgSO_4$ (48–52% K_2O)
Phosphate fertilizer	
Mono-ammonium phosphate	11% N, 48% P_2O_5
Di-ammonium phosphate	18% N, 46% P_2O_5
Bone-meal	

*The application of synthetic nitrogen fertilizers (urea) in agriculture-intensive areas in certain parts of the State of Punjab has led to higher levels of nitrate pollution and other toxic chemicals in ground water, thus affecting the quality of drinking water from nearby bore-wells. It can have adverse effects on plants, animals and humans, thus impacting the socio-economic and cultural aspects of its population.

RQ 4.23. Why symptoms of mineral deficiency are more pronounced in older leaves? List out the minerals involved.

Ans. The symptoms of mineral deficiency are more pronounced in older leaves as the minerals are quite mobile, being readily withdrawn from mature leaves and transported to the young tender leaves. Examples of highly mobile minerals are phosphorus, nitrogen, potassium and magnesium.

RQ 4.24. List two minerals that can prevent lodging in plants. Explain by giving reasons.

Ans. The phenomenon of lodging is caused by excessive elongation and softening of cells, especially in cereal crops where the tillers bend down due to the action of wind, rain or hail and snow storms. Two elements are known to prevent lodging – one is 'silicon' and the other is 'potassium'.

Silicon, although not an essential element, is known to accumulate in the walls of the epidermal cells and the xylem cells where it provides rigidity and limits compression (caused by bending in wind or rain). Further it makes the plants less susceptible to root-rotting fungi (or fewer fungal infections). Thus, silicon-rich cereal crops have less of the tendency to lodge, and so giving higher yield. This is, thus, more of an ecological aspect rather than a physiological or biochemical need.

Another element known to prevent lodging is 'potassium' which when present makes the stem stiffer and stronger and also improves resistance to root-rotting fungi.

RQ 4.25. What roles do the following minerals play in plant metabolism?

Ans.

Mineral	Constituent or Component (or Cofactor).
Nitrogen	Constituent of proteins, nucleic acids, indole-3-acetic acid, cytokinins and other organic molecules, e.g., chlorophyll.

Phosphorus	Essential component of lipoprotein membranes, nucleoproteins, DNA, RNA, phosphorylated sugars, ATP, ADP, UDP, UTP, CDP, CTP, GTP and GDP (required for energy transfer by phosphorylation).
Sulphur	Constituent of cysteine, cystine, methionine (amino acids), biotin and thiamine (vitamins), coenzyme A (much involved in respiration and fatty acid metabolism), mustard oil glycosides (thiocynates and isothiocynates) such as sinigrin, sinalbin, lipoic acid, TPP (thiamine pyrophosphate), glutathione.
Iron	Enters into the composition of cytochromes (cyt *b*, *c*, *a*, a_3), ferredoxin (playing an important role in electron transfer reactions of photosynthesis and nitrogen fixation), iron–sulphur proteins, many enzymes such as catalase, peroxidase, dinitrogenase (or nitrogenase).
Copper	Component of many respiratory enzymes such as ascorbic acid oxidase, cytochrome oxidase, tyrosinase, polyphenol oxidase (phenolase), superoxide dismutase, plastocyanin and plastoquinone (electron transport carriers).
Zinc	Component or activator of alcohol dehydrogenase (ADH), carbonic anhydrase (CA), glutamic acid dehydrogenase, carboxypeptidase, tryptophan (a precursor of IAA), component of zinc finger transcription factors.
Manganese	Acts as cofactor for decarboxylases and dehydrogenases, a part of the oxygen evolving complex (manganoprotein), enzymes like oxidases, peroxidase, kinases, arginase, hydroxylamine reductase.
Molybdenum	Required for the activity of nitrate reductase (that controls reduction of nitrate to nitrite) and dinitrogenase (conversion of dinitrogen to ammonia).
Magnesium	Constituent of chlorophyll, required to stabilize ribosome structure, activator of two critical enzymes of photosynthesis, i.e., ribulose bisphosphate carboxylase, phosphoenolpyruvate carboxylase; activation of enzymes involved in photosynthesis and respiration (phosphoglyceric transphosphorylase, enolase); helps to link the ATP molecule to the active site of enzyme in phosphorylation reactions.
Calcium	Component of cell walls, particularly middle lamella (calcium pectate, an integral compound), required during cell division (cytokinesis), calmodulin–calcium complex, as second messenger in many of the hormonal and environmental responses; maintenance of membrane permeability; intracellular signal transduction.
Boron	Needed for the structural integrity of plasma membrane and the cell wall matrix, stimulates pollen tube germination and elongation.
Chlorine	Required for oxygen-evolving reactions of photosynthesis, maintains electrical neutrality across membranes, a major osmotically active solute in the vacuole, for maintaining osmotic and ionic balance in cells.

RQ 4.26. Why is potassium so important in plant cell functions?

Ans. The importance of potassium lies in the fact that it is a major mineral element required in many cell functions, such as:

1. Potassium is an activator for a number of enzymes, involved in photosynthesis and respiration (i.e., used for 40 or more enzymes).
2. Important function in regulating the osmotic potential (osmoregulation) and ionic balance.
3. Controls many plant movements, such as opening and closing of stomata; sleep movements (nyctinastic movements in *Oxalis* and Touch-me-not (*Mimosa pudica*)).
4. Maintains electroneutrality in plant cells.
5. Prevents lodging in plants.

RQ 4.27. Outline the major deficiency symptoms associated with zinc, boron, molybdenum and copper.

Ans. **Zinc (Zn)**

1. Typically, zinc-deficient plants have 'shortened internodes', and, as a result, plants exhibit a rosette habit of growth having small (e.g., little leaf), stiff leaves at the tips of young shoots. Frequently, the shoots die off, leading to premature leaf fall.
2. Zn deficiency causes 'chlorosis' in the interveinal regions of the leaves, turning leaves pale green, yellow or even white (due to a decline in chlorophyll biosynthesis). The leaves may also remain small and distorted with white necrotic spots.

These disorders are associated with inadequate synthesis of hormone IAA. Zn acts as a cofactor in the synthesis of plant hormone IAA (from the amino acid tryptophan).

Zn is required as a cofactor for the enzyme carbonic anhydrase which catalyzes the hydration of CO_2 to produce carbonic acid (H_2CO_3). The enzyme has a functional role in CO_2 transport:

$$CO_2 + H_2O \xrightleftharpoons{Zn} H_2CO_3 \rightleftharpoons H^+ + HCO_3^-$$

Copper (Cu^{2+}, Cu^+)

Copper-deficient plants generally show dwarf growth, distortion of tender leaves, especially in citrus trees leading to loss of young leaves and even curling (called a 'dieback of citrus'). The initial symptoms of Cu deficiency is the formation of dark green leaves with some necrotic brown spots appearing first at the tips of young leaves (or reclamation disease – chlorosis) and then extending further along the leaf margins. The leaves may become twisted or malformed. Later, dead brown spots appear. In some tree species, copper deficiency may produce blisters or deep slits in the bark from which gums exude –a term called 'exanthema'.

The browning of freshly cut apples or cut potato surfaces is due to the activity of copper-containing enzymes called polyphenol oxidases or phenolases. Polyphenol oxidases, a large group of copper-containing enzymes, oxidize phenols to quinones (Figure 4.11).

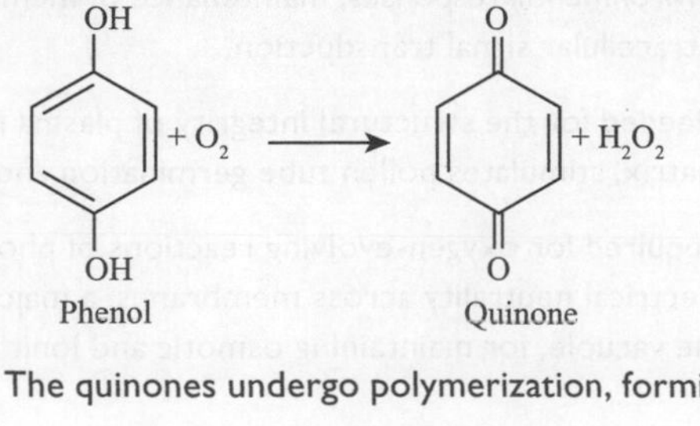

Fig. 4.11 The quinones undergo polymerization, forming tannins.

Boron (B)

In the absence of boron:

1. Meristematic activity is retarded and shoots and root apices die. Due to the disintegration of terminal buds, the lateral buds start developing into branches. The terminal buds of laterals in turn die, thus leading to more production of lateral shoots. This results in a rosette-like habit.
2. The roots become stunted and thickened and proper development of root tubercles in leguminous plant fails to occur.
3. The flower buds do not develop.
4. In boron-deficient plants, the cambial cells in the growing regions cease to divide but the cell elongation continues, thus resulting in the dislocation of xylem and phloem from their original position and are rendered non-functional. The breakdown of phloem may result in the accumulation of carbohydrates and increase in anthocyanin formation that imparts a purple colour to the leaves.
5. Stems become enlarged because of boron deficiency leading to various disorders like 'stem cracks' in celery, 'heart rot' of beets, 'drought spot' of apples.

Molybdenum (Mo)

The symptoms of Mo deficiency depend upon the form in which nitrogen fertilizers are applied. The plant grows satisfactorily when N is supplied in the form of ammonium salts or in organic form (urea).

1. In the absence of Mo, nitrates may accumulate because they are not converted to nitrite (NO_2^-) by 'nitrate reductase' and hence no ammonia is produced. The protein synthesis is thus blocked and plant growth ceases altogether. Overall, the plants show stunted growth.
2. In the absence of Mo, the root nodules in legumes fail to develop. The seeds remain poorly filled.
3. Deficiency symptoms are interveinal chlorosis, at first developing in the older leaves, progressing upward towards younger ones. Mottling of leaves is followed by necrosis of leaf margins because of nitrate accumulation.
4. Lack of Mo in crucifers (such as cauliflower, broccoli, etc.) induces a whip-tail disease, i.e., the partial or complete withering of leaf laminae, and the leaf thus looks like, 'a whip' and hence, termed 'whip-tail' disease. 'Yellowspot' disease of citrus is caused by Mo deficiency.

Excess of molybdenum causes chlorosis (golden yellow shoot in tomato). Forage crops rich in Mo seem to have a deleterious effect on animals (the 'teart disease' of livestock that was first reported in the UK).

Molybdenum requirement decreases or disappears when ammonium salts or organic nitrogen is fed to plants.

RQ 4.28. Why does chlorosis due to both magnesium and nitrogen usually develop in the older leaves while the reverse is seen during iron deficiency? Explain.

Ans. Interveinal chlorosis,(i.e., between the veins) occurs first in the mature leaves because magnesium and nitrogen are both 'highly mobile' minerals, that are moved from older to younger leaves. On the contrary, chlorosis develops first in the younger leaves as iron which is 'quite immobile', does not migrate freely from older to younger leaves.

RQ 4.29. In which ways are calcium (Ca^{2+}) ions important in regulating metabolic functions in plant cells?

Ans. The calcium ion (Ca^{2+}) is involved in controlling numerous physiological processes in plants:

- Cell division and elongation.
- Protoplasmic streaming.
- The secretion and activity of various enzymes.
- Hormone action.
- Tactic and tropic movements.
- Plays a role as 'second messenger' in biochemical mechanisms, e.g., Ca^{2+}-CaM complex that serves to activate enzymes such Ca^{2+}dependent ATPases in the plasma membrane, NAD kinase and protein kinase, etc.

Large amounts of calcium accumulate in the endoplasmic reticulum (ER), the mitochondria, and the large central vacuole but the cytosolic Ca^{2+}concentration is maintained low through membrane-bound calcium-dependent ATPases.

RQ 4.30. Who pioneered and refined the techniques for nutrient uptake studies? List the plant materials that have been used in experimental work.

Ans. The techniques for the study of mineral uptake were pioneered by Hoagland and Boyer in the 1930s and refined later by Epstein in the 1960s.

The plant materials often used were roots or tissue of excised barley (*Hordeum vulgare*) and slices of storage organs i.e. potato, carrot, or beet. Tissue slices or roots are grown under conditions that encourage depletion of nutrient elements. These so called 'low-salt' roots are excised and placed in cheesecloth 'teabags' that can be quickly and easily used for experimental work.

RQ 4.31. Outline some basic concepts of salt uptake.

Ans. Studies of ion uptake by living plant tissues or by isolated excised roots have shown the following basic ideas:

1. Roots take up ions vigorously even when little absorption of water is occurring, i.e., the process of absorption of mineral salts is independent of the absorption of water. Furthermore, the sites where they occur are different. Unlike the absorption of water that takes place through the root hairs, the absorption of minerals occurs through the epidermal cells of the meristematic region and the zone of elongation near the root tip (or from the entire root surface according to some researchers).
2. Absorption of mineral salts by the roots is a selective process.
3. Roots can take up ions 'against' the electrochemical gradient, although in the initial phase, diffusion of ions can occur only 'down or along' the diffusion or chemical potential gradient.
4. The mineral elements are almost always absorbed in their ionized form, and so these ions must first traverse the plasma membrane to enter the cytoplasm, and then the membrane surrounding the vacuole or cellular organelle present in the cytoplasm.
5. Ionic movement, whether passive or active, probably occurs at different sites in the membrane. Small lipid-soluble molecules move through the membranes more readily than do larger ones *via* small lipid channels traversing across the membrane. However, water-soluble inorganic ions too can move through the membrane *via* aqueous protein channels, appropriately called permease. Polypeptides and proteins with appropriate channelling properties (called ionophores) for specific ions have been isolated from bacteria and fungi. Some uncharged or non-ionic molecules such as sucrose move through the membranes in conjunction with ions usually H^+. This process is called co-transport or symport.

RQ 4.32. What do you mean by the term 'free space'? What is its significance?

Ans. The term 'free space' refers to the part of the root or tissues where ions can freely move in and out by diffusion. Just as the entry of ions into free space is easy and fast, similarly, ions can also exit out easily. Once these ions have crossed the barrier of cellular membranes, and reach the cytoplasm or vacuole, these cannot easily come out.

This free space (FS) is not, in fact, free. Neither any such free space can be seen under the microscope nor can it be exactly measured but a close approximation is the apparent free space (AFS) which on the basis of experimental data is represented by the ratio of total solutes in tissue divided by the solute concentration of the external bathing solution:

$$AFS = \frac{\text{Total solute in tissue}}{\text{Solute concentration of the external bathing solution}}$$

Approximate measurements indicate that the AFS of the roots is in the range of 6 to 10 per cent of the total tissue.

AFS of roots is essentially the cell walls and intervening intercellular spaces (i.e., equivalent to the apoplastic area) of the epidermal and cortical cells. It does not include vacuole, cytoplasm and the cellular membrane system of plasma membrane and tonoplast.

The AFS stops, almost certainly, at the endodermis where in most of the roots, the radial and transverse walls are heavily cutinized and suberized, 'filling the spaces' between the cellulose microfibrils.

RQ 4.33. List some of the earlier balanced nutrient solutions.

Ans. Some of the earliest balanced nutrient formulations developed in the past are as follows:

1. Arnon and Hoagland's solution
2. Knop's solution
3. Sach's solution
4. Knudson's medium
5. White's medium

RQ 4.34. List mechanisms where absorption of mineral ions is 'along' the concentration gradient without any involvement of metabolic energy.

Ans. There are three mechanisms where absorption and transfer of minerals do not require expenditure of metabolic energy:

1. Diffusion
   - (i) Simple
   - (ii) Facilitated
2. Ionic exchange mechanism
   - (i) Cation exchange theory
   - (ii) Carbonic acid exchange theory
3. Donnan equilibrium

RQ 4.35. Why is *Chara* or *Nitella* used for ionic relations studies instead of angiosperms?

Ans. In mature cells of angiospermic plants, the central vacuole often occupies 90 per cent or more of the cell's volume and the cytosol forms a thin peripheral layer around the cells. Hence, the cytosol of most angiosperm cells is difficult to assay chemically.

Certain green algae such as *Chara* and *Nitella* are ideal choices because their cells are several inches long and contain an appreciable volume of cytosol. In *Nitella*, the individual cells can exceed 10 cm in length and 2 mm in width. From each node of the cell, single-celled branches are formed.

RQ 4.36. Comment upon the significance of H^+-pumping ATPase.

Ans. In plants, H^+-pumping ATPase is the primary electrogenic transporter in both the membranes; the plasma membrane and the vacuolar membrane.

When protons (H^+) are effluxed from the cytosol to the external medium (i.e., extracellular spaces or vacuole) by the H^+-ATPase, both a pH gradient and a large interior or inside negative membrane potential is created (of the order of –116 mV). This electrochemical H^+gradient is termed as proton motive force.

This negative interior electrical potential would result in the inward diffusion of positively charged ions such as K^+ but would repel negative ions such as chloride (Cl^-). Some uncharged molecules such as sucrose move through the membranes in conjunction with an ion, usually H^+. This process, called 'cotransport' (symport) is particularly important in controlling the movement of sucrose in loading and unloading in and out of the phloem, respectively. Loading is believed to involve 'cotransport' of sucrose and H^+ ions through specific permeases in response to pH and electrochemical gradients. Similarly, secretion of sucrose out of phloem cell (unloading) occurs through permeases in response to pH and electrochemical gradient.

RQ 4.37. List three historical models that attempt to explain the uptake of ions or solutes?

Ans. Among the earlier models, the following three are important:

1. Van den Honert Carrier concept(1937)
2. Lundegardh's Cytochrome Theory of 'Ion Pumps' (1950,1954)
3. Bennet-Clark Phosphatide cycle/Lecithin Carrier concept(1956)

RQ 4.38. Define Inner Space.

Ans. Inner space is the space in a tissue or cell to which ions penetrate through the involvement or mediation of metabolic energy. Where outer space ends, the inner space begins. The area or barrier between outer and inner space is important to free ions.

RQ 4.39. The pH of the cytoplasm is kept close to 7.0. How is this maintained and what useful purpose does it serve to the plant?

Ans. Most mature plant cells have a large central vacuole which may occupy as much as 80–90 per cent of the cell volume. The cytoplasm together with all other organelles forms a thin film pressed against the cell wall. The vacuole is surrounded by a differentially permeable membrane called the tonoplast and contains an aqueous solution of salts, sugars and organic molecules, enzymes and other waste products of cell metabolism such as pigments and tannins.

The pH of the vacuolar sap of most plants lies between 3.5 and 5.5, but in some plants, it is as low as 1.0, whereas that of cytoplasm is near 7.0. So there is large difference in H^+ion concentration between the two. The tonoplast contains pumps energized by ATP that transports H^+ions from the cytosol to the vacuole, thus maintaining electroneutrality in the cytoplasm. Such a control is important for the proper activity of most of the enzymes that are essential for regulating metabolism.

RQ 4.40. How is the cytosolic Ca^{2+} concentration kept low? What is its significance?

Ans. The plasma membrane also contains calcium pumping ATPase (Ca^{2+}-ATPase) by which the calcium (Ca^{2+}) is pumped out of the cytoplasm back into the cell wall or extracellular spaces (causing stiffening of cell walls). This helps to keep the Ca^{2+}concentration low in the cytoplasm, which is essential to prevent precipitation of phosphates and to keep Ca^{2+}-dependent signalling mechanisms functioning in a proper manner.

RQ 4.41. In which way(s) is the ion exchange ability between the roots and soil nutrient useful? Explain.

Ans. In roots, H^+ ions are frequently exchanged for potassium (K^+). The K^+ is absorbed by the root system from the soil solution or from soil colloids surface to which it is adsorbed while the simultaneous exchange or export of H^+ ions makes the external medium more acidic. This can help in the breakdown process of soil particles.

RQ 4.42. Not all transport ATPases are electrogenic. Comment.

Ans. K^+/H^+-ATPases of bacteria are electrically neutral. This ATPase drives K^+ and H^+ against their electrochemical gradients,[8] but because equal number of K^+ and H^+ ions are moving in opposite directions, their charges cancel out.

RQ 4.43. Why are *Chara* or *Nitella* such favourites with plant physiologists engaged in plant nutritional studies? Explain.

Ans. *Chara* and *Nitella*, both members of green algae family Characeae, grow naturally in fresh or brackish water. They have giant internodal cells that are several times longer than wide, consisting of a large vacuole lined on the outside by a thin layer of cytoplasm with many nuclei. These cells are often 3 or more cm long, and therefore large enough to be handled singly. By snapping the ends of several cells and gently extruding the contents of their vacuoles, a small quantity of vacuolar sap uncontaminated with cytoplasm can be obtained, thus permitting direct chemical analysis (or by snapping the giant internodal cells, the vacuolar contents can be recovered and analyzed chemically).

> *Valonia*, a giant single-celled marine alga, is more or less spherical with a thin peripheral sheet of coenocytic cytoplasm, enclosing a large central cavity or vesicle full of aqueous solution. It has the remarkable ability to selectively accumulate certain ions. Although the sea water contains a high concentration of sodium but this does not accumulate within the cells. On the contrary, potassium ions which are present in low concentration in the sea water are known to accumulate manifold, representing the principal cation within the algal cells.

RQ 4.44. List only the metabolic inhibitors controlling salt uptake.

Ans. The different metabolic inhibitors are

- Oxidase inhibitor
- Azides (N_3^-)
- Carbon monoxide (CO)
- Cyanides (CN^-)

RQ 4.45. Discuss the strategies plants have developed for iron uptake under conditions of iron stress.

Ans. In response to iron stress conditions in the soil, plants have undergone several morphological and biochemical changes in the roots.

1. The formation of specialized transfer cells in the root epidermis.
2. Enhanced proton secretion into the soil surrounding the roots.
3. The release of strong ligands or chelating agents by the roots such as caffeic acid.
4. Simultaneously there is induction of reducing enzymes in the plasma membrane. of the roots epidermis.

Acidification of the rhizosphere promotes chelation of Fe^{3+} with caffeic acid, which then moves to the root surface where iron gets reduced to Fe^{2+} state at the plasma membrane. Reduced Fe^{2+} iron causes the ligand to release the iron which is immediately absorbed by the plants before it gets precipitated into insoluble ferric from.

[8] The electrochemical-proton gradient (or the proton motive force).

Another strategy for iron uptake by organisms (like bacteria, fungi and algae) involves the synthesis and release of chemicals of low-molecular-weight, iron-binding ligands called 'siderophores' (in higher plants phytosiderophores) that are synthesized and released by the plants roots under conditions of iron stress. Unlike the first type, the entire ferrisiderophore (siderophore-iron complex) is taken into the root cells where the iron is subsequently released (see Figure 4.3).

Mugineic acid and avenic acid are two phytosiderophore ligands.

RQ 4.46. What are chelating agents? Discuss briefly their importance in plant nutrition.

Ans. Iron deficiencies are common because of the propensity of Fe^{3+} (Ferric state) to form insoluble iron hydroxide as shown below:

$$2Fe^{3+} + 6OH^{-} \rightarrow 2Fe(OH)_3 \rightarrow Fe_2O_3 \cdot 3H_2O$$

(the reddish-brown oxide)

This problem is particularly severe in neutral and alkaline soils. This problem of iron deficiency can usually be overcome by supplying chelated iron, by adding directly to the soil or as a foliar spray (a chelate holds or binds iron or other atoms as if in a claw.)

A chelate (from the Greek, chele or claw-like) is a stable soluble product formed between a metal ion and organic molecule called a chelating agent or ligand that contains negatively charged carboxyl groups or electron-donating nitrogen groups that can bind the iron forming coordinate bond. Due to its high affinity for most metal ions, the possibility for formation of insoluble precipitates is reduced and at the same time, the metal ion can be easily withdrawn from the chelate during uptake by the plant.

One of the more common synthetic chelating agent is EDTA (sodium salt of ethylenediaminetetraacetic acid). Among the natural chelating agents are haemoglobin, cytochromes, and chlorophyll (all with a porphyrin ring) and a variety of phenolic acids. These are, however, not highly specific and will bind to a range of cations, including iron, copper, zinc, manganese and calcium.

Citric acid and tartaric acid were earlier used along with iron. Even each molecule of ATP and ADP is chelated with one Mg^{2+}.

Unit II Metabolism and Bioenergetics

Chapter 5

Concepts of Metabolism

5.1 Metabolism

Keywords

- Anabolism
- Autotrophy
- Catabolism
- Enthalpy
- Entropy
- Heterotrophy
- Homeostasis
- Intermediary metabolism
- Metabolic flux
- Metabolites

Plants can synthesize most of the organic matter on earth by assimilating inorganic elements (carbon and nitrogen), from the environment into organic molecules, driven by energy from sunlight. Every year, plants incorporate through photosynthesis nearly 100 billion metric tons of carbon, which is approximately 15 per cent of the total carbon dioxide (CO_2) present in the atmosphere. Respiration by plants and heterotrophic organisms converts the same amount of carbon present in organic compounds back into CO_2. These processes are important not only for the plants, but also for animals, fungi, and bacteria, which receive their carbon and nitrogen solely from organic compounds and thus completely rely on plant sources to obtain nutrition. Plants acquire the major elements (C, H, O and N) chiefly as carbon dioxide, water, and nitrate along with other minerals elements in lesser amounts from the soil. This mode of nutrition, as obtained from inorganic compounds, is called **autotrophy** and photosynthetic bacteria, green algae, and vascular plants are examples of **autotrophs**. On the other hand, for those organisms that get their carbon and nitrogen exclusively from organic compounds, their mode of nutrition is called **heterotrophy**. Multicellular animals, carnivorous plants, fungi, and most of the microorganisms are **heterotrophs**. Carbon, oxygen, and nitrogen are continuously recycled between the heterotrophs and autotrophs present in the biosphere, using the sun's energy as the driving force and this recycling process depends on a proper balance between their activites. These cycles of matter are controlled by huge energy flows within the biosphere, starting with the harvesting of sun's energy by photoautotrophs and its utilization in producing organic compounds including carbohydrates, which are then consumed as sources of energy by heterotrophs. Although carbon, oxygen, and nitrogen keep on recycling constantly, a part of the energy involved is being continuously changed into non-usable form such as heat.

Metabolism can be defined as 'the sum total of all the chemical reactions occurring in a cell or organism', and takes place through a sequence of reactions catalyzed by enzymes, thus constituting **metabolic pathways** and the metabolic intermediates are known as **metabolites**. Every successive step of the metabolic pathway results in a small and specific chemical change (elimination/inclusion/shifting of a specific molecule or the functional group). The collective action of all the metabolic pathways that is involved in interchange of reactants, metabolic intermediates, and products of low molecular weight (generally less than 1000 D) is termed as **intermediary metabolism**. There are two principal types of metabolites - primary and secondary. **Primary metabolites** are the compounds that are present in many organisms and are directly involved in growth, development, and reproduction of the organisms, e.g. carbohydrates, proteins, lipids, vitamins and so on; **secondary metabolites** are the end products of primary metabolism that include alkaloids, steroids, antibiotics, phenolics, essential oils and such, which provide protection to plants against herbivores and microbial infection, used as medicines, pigments, flavourings, and help in attracting pollinators (like bees and butterflies) or seed dispersal agents.

The term 'metabolism' includes the **catabolism** as well as the **anabolism** (coined by Gasket in 1886):

Catabolism involves the degenerative phase of reactions in which complex macromolecules such as carbohydrates, lipids, and proteins are degraded to simple products, which are then used as building materials for synthesis of cellular components required for cell structure and function. Catabolic pathways lead to decrease in atomic order (increase in entropy) and are exergonic (energy-releasing). Released energy is then stored as ATP and reduced coenzymes e.g. NADH, NADPH, and $FADH_2$, which are required to drive anabolic pathways.

On the other hand, **anabolism** is the biosynthesis of large and complex macromolecules (e.g., carbohydrates, lipids, proteins, and nucleic acids) from small, simple precursors. Anabolic reactions usually involve a significant increase in atomic order (decrease in entropy) and are endergonic (energy-consuming). Energy is supplied in the form of ATP molecules and the reduced coenzymes such as NADH, NADPH, and $FADH_2$. Anabolic pathways are associated with the growth and repair processes characteristic of living organisms.

Overall, catabolic pathways are known to be convergent and anabolic pathways are divergent in nature. Metabolic pathways can be **linear** or **branched** or **cyclic**. Branched pathways yield several useful final compounds from one precursor molecule or convert multiple starting substances into one product. In cyclic pathway, one of the starting compounds of the pathway is reproduced in a sequence of reactions and re-enters the same metabolic pathway (Figure 5.1).

> The catabolic and anabolic reactions that proceed in cells are accompanied by energy changes and it is the study of these changes that constitutes the field of **bioenergetics**.

Metabolic pathways listed in Table 5.1 can be regulated from inside as well as from outside the cell:

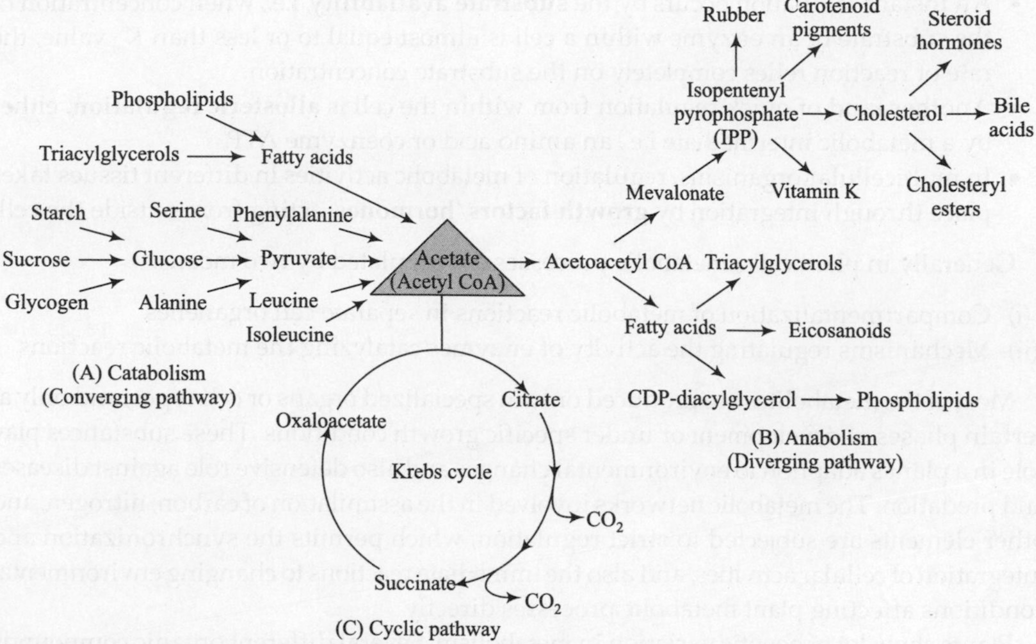

Fig. 5.1 Three types of metabolic pathways. (A) Catabolic or converging; acetate, a major intermediate of metabolism, is the catabolic product of multiple fuel molecules. (B) Anabolic or diverging; acetate behaves as the raw material for a vast array of products. (C) Cyclic; one of the initial compounds, oxaloacetate (OAA), is reproduced and again enters the Krebs cycle.

Table 5.1 The most common metabolic pathways in cells.

Pathway	Reaction
Catabolic	
Glycolysis	catabolism of monosaccharides to pyruvate
Fermentation	catabolism of monosaccharides to ethanol and CO_2 in the absence of O_2
Krebs cycle	oxidation of pyruvate and its catabolism to CO_2 and water in the presence of O_2
Pentose phosphate pathway	catabolism of pentoses
β-oxidation	breakdown of fatty acids
Anabolic	
Calvin cycle	fixation of CO_2 into carbohydrates
Glyoxylate cycle	synthesis of carbohydrates from fatty acids
Gluconeogenesis	synthesis of glucose from simple organic acids
Starch synthesis	incorporation of sugars into polysaccharides
Fatty acids synthesis	biosynthesis of fatty acids and triglycerides using acetyl CoA and glycerol
Amino acids synthesis	mechanism for formation of amino acids to be incorporated into proteins

- An instant regulation occurs by the **substrate availability**, i.e., when concentration of the substrate of an enzyme within a cell is almost equal to or less than K_m value, the rate of reaction relies completely on the substrate concentration.
- Another kind of quick regulation from within the cell is **allosteric regulation**, either by a metabolic intermediate i.e., an amino acid or coenzyme ATP.
- In multicellular organisms, regulation of metabolic activities in different tissues takes place through integration by **growth factors/hormones**, acting from outside the cell.

Generally in plant cells, metabolic processes are regulated by two means:

(i) Compartmentalization of metabolic reactions in separate cell organelles
(ii) Mechanisms regulating the activity of enzymes catalyzing the metabolic reactions

Most of the metabolites are produced only in specialized organs or cell types, and only at certain phases of development or under specific growth conditions. These substances play role in a plant's adaption to environmental changes and also defensive role against diseases and predation. The metabolic networks involved in the assimilation of carbon, nitrogen, and other elements are subjected to strict regulation, which permits the synchronization and integration of cellular activities, and also the immediate reactions to changing environmental conditions affecting plant metabolic processes directly.

Plants show lot of genetic variation in metabolism. Several different organic compounds have been found in plants, which include compounds associated with the metabolic pathways that absorb nutrients from the environment, in energy metabolism, and in biosynthetic pathways that provide the basic constituents of the plant cell: proteins, membranes, and cell walls.

Plants display a very flexible and strong metabolism. **Metabolic flux,** the overall rate of metabolites flow in a pathway, is attained by regulation of metabolism by the **pacemaker enzymes** that are involved in catalysis of rate-limiting steps of metabolic pathways. Understanding of regulation of such enzymes at the level of gene expression or protein degeneration would be of immense help in creating plants with transformed metabolism – **metabolic engineering**. Several complex enzymes of the metabolic pathway exist as multienzyme complexes (known as **metabolons)** as means of metabolic channelling. **Metabolic channelling** allows shuttling of reaction intermediates directly from one enzyme to another in a metabolic pathway, thus reducing their depletion rate due to diffusion. Every cellular compartment offers a favourable environment for particular metabolic pathways to operate at the optimum level. Specific transporters present in the organelle membranes facilitate and regulate the exchange of metabolites between various compartments within a cell.

Functions of metabolic pathways

1. Synthesis of an end-product required by the cell or organism; this product may be used as part of the structure of the cell or, in cases where it is secreted from the cell, it may be incorporated into extracellular elements e.g. pectins of the middle lamellae of plant cell walls. The end product may also function in the cell as a regulatory agent

for other reactions e.g. hormones. End products may also be stored as reserves, such as starch, glycogen, and certain lipids.

2. A metabolic pathway may function to provide energy-rich compounds such as ATP, GTP, and UTP and these may be used in other energy-requiring reactions.
3. Intermediates of a metabolic pathway may be drawn upon and used as substrates for other metabolic sequences. For example, many of the Krebs cycle intermediates are withdrawn and utilized for biosynthesis of other compounds (amino acids, lipids etc). Pathways involving metabolic reactions that serve in both a catabolic and an anabolic function are called **amphibolic pathways**. Amphibolic pathways which may have their intermediates diverted to other reaction sequences frequently include specialized reactions, known as **anaplerotic reactions**, that replenish these intermediates. For example, the glyoxylate pathway forming the Krebs cycle intermediates (succinate and malate), condensation of CO_2 with pyruvate to form oxaloacetate.

Reaction sequences and their intracellular location have been revealed using following techniques:

(i) ***Marker and Tracer techniques***: A frequently used technique for identifying the steps in a pathway is to follow the metabolism of a substrate molecule in which one of the atomic positions has been labelled with a radioactive isotope. Such a labelled substrate is said to be a tracer because specialized techniques are available to identify these labelled compounds. The tracer is then made available to the cell and metabolism allowed to proceed. After a short period of time, metabolism is halted by either rapidly dropping the temperature of the cells or by adding metabolic inhibitors. The cells are then broken open and potential chemical intermediates of the metabolic pathway are isolated.

(ii) ***Enzyme techniques***: All the metabolic reactions in cells are catalyzed by enzymes, it should be possible to identify an enzyme for each metabolic reaction. In enzyme studies, the presence of a particular enzyme can be demonstrated by measuring the rate of dissolution of the substrate molecule of that enzyme, or the rate of appearance of a specific end product. Different cellular fractions obtained after centrifugation can be tested for enzyme activity to establish the native location of the reactions in intact cells.

(iii) ***Use of enzyme inhibitors***: Numerous enzyme inhibitors are known that block specific reaction steps . The use of an inhibitor for one of the reactions of a metabolic pathway provides a means for identifying the intermediate before the block. Inhibitors can also be used to test for suspected intermediates that come after the block.

(iv) ***Mutant cells***: A mutant cell that lacks the genetic information for producing a specific enzyme may serve as a test organism for studying the steps of the metabolic pathway involving that enzyme.

To sum up, metabolism provides amazing and valuable insights into life, with numerous applications in the fields of medicine, agriculture, and biotechnology.

5.1.1 Metabolite pools in plants

There are three important metabolite pools in plants (Figure 5.2):

(i) Hexose phosphate pool
(ii) Pentose phosphate pool
(iii) Triose phosphate pool

Metabolites can be added to or withdrawn from the pool to serve the needs of various biochemical and metabolic pathways. Within a leaf, the direction of metabolite flow through the pools is based on the requirements and activities of the cell and changes during day and night. During day time, triose phosphates synthesized by the Calvin cycle in photosynthetic leaf tissue (source) leave the chloroplast, enter in the cytoplasmic hexose phosphate pool and get changed to sucrose, which then moves *via* phloem into non-photosynthetic tissues (sink). In sink tissues (such as roots, tubers, and bulbs) sucrose is converted to starch for long-term storage or utilized as a source of energy through glycolytic process. In developing plants, hexose phosphates are also drawn off from the pool and used for the cell wall synthesis. During night time, starch undergoes degradation in glycolysis and oxidative phosphorylation to supply energy for both source and sink tissues.

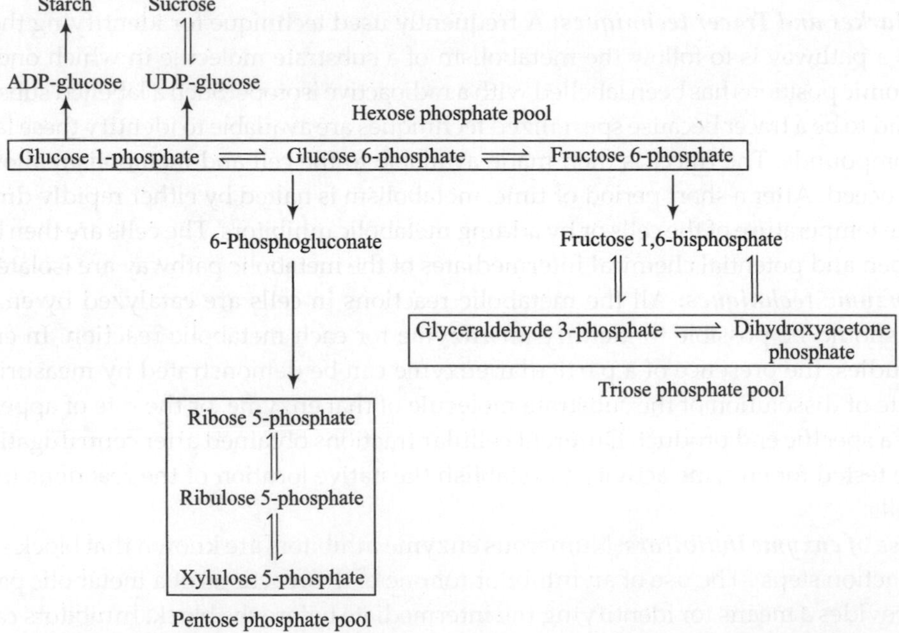

Fig. 5.2 Metabolite pools in plants- hexose phosphate, pentose phosphate and triose phosphate. The compounds in each pool are readily interconvertible. As soon as one compound of the pool is temporarily exhausted, a new equilibrium is rapidly attained to restore it. Transport of phosphorylated sugars between various compartments within a cell involves specific transporters in membranes of the organelles.

5.1.1.1 The hexose phosphate pool: a major crossroads in plant metabolism

Three interconvertible metabolites belonging to the hexose phosphate pool (glucose-1-phosphate, glucose-6-phosphate, and fructose-6-phosphate) are kept in equilibrium by the action of enzymes phosphoglucomutase and glucose-6-phosphate isomerase.

$$\text{Glucose-1-phosphate} \xrightleftharpoons{\text{Phosphoglucomutase}} \text{Glucose-6-phosphate}$$

$$\text{Glucose-6-phosphate} \xrightleftharpoons{\text{Glucose-6-phosphate isomerase}} \text{Fructose-6-phosphate}$$

Both reactions are readily reversible and remain close to equilibrium *in vivo.*

Carbon can enter and leave the hexose phosphate pool in a number of different ways.

- It can enter the pool during photosynthesis, when the triose phosphates produced are converted to hexose phosphates
- It can leave the pool for starch and sucrose synthesis
- It can leave the pool for cell wall formation
- At night, when no photosynthesis occurs, or in heterotrophic tissues, carbon can enter into the hexose phosphate pool through phosphorylation of free hexoses derived from the degradation of sucrose or starch
- Other major drains on the hexose phosphate pool in the absence of photosynthesis are the respiratory pathways of glycolysis and oxidation reactions of the pentose pathway, which supply energy and reducing power for the cell, respectively.
- An additional pathway for the production of hexose phosphates observed in germinated oilseeds is gluconeogenesis, by which storage lipids are converted to carbohydrates.

The hexose phosphate pools occur in different subcellular compartments; the cytosol and the plastid stroma. Equilibration between these pools depends on whether they are linked by the activities of metabolite transport proteins. However, the capacity for the transport of hexose phosphates across the cellular membranes differs. Most leaf chloroplast have low capacity for hexose phosphate transport, and their hexose phosphate pools are used in metabolic pathways distinct from those in the cytosol.

5.2 Concepts of Thermodynamics

Living organisms are highly ordered and structured systems where metabolic functions are carried out speedily and with great precision. The cell is an extremely organized biochemical machine inside which scores and sometimes hundreds of reactions may be occurring simultaneously. Energy[1] is needed to enable all chemical reactions to proceed. Organisms have developed mechanisms for storing chemical energy and using it to drive energy-consuming reactions. The fundamental laws of physics and chemistry apply to living organisms and explain, at a molecular level, how organisms operate. The information necessary to establish proper reactions acting on proper materials is 'stored in the genes', both in the nucleus and in plastids and mitochondria.

However, after death, decay is the process by which complex molecules of an organism break down and thus become more disorganized or disordered and scattered. This disorder or randomness is known as **Entropy**. It generally refers to a form of energy which cannot perform work. This increase in entropy cannot be prevented.

[1] Two forms of energy exist – the form which is available to do work (free energy) and the form which is not, is called entropy, i.e., generally referring to a form of energy which cannot perform work.

Our Sun is the ultimate source of all energy on this planet. The chlorophyll of green plants absorb this light which is then utilised in carbohydrate synthesis from simple molecules like carbon dioxide (CO_2) and water (H_2O). These complex sugars are the raw materials for the synthesis of more complex organic molecules like proteins and fats which together with carbohydrates serve as the source of energy for themselves and for heterotrophs, including animals. In different parts of the cell, the macromolecules may be broken into smaller molecules so that the carbon, hydrogen and oxygen atoms may be rearranged, substituted, deleted or supplemented to form a vast array of entirely different molecules, releasing energy[2] that is necessary for cell functioning. Some of the metabolic intermediates play key roles in providing large carbon pools from which other molecules are made.

More on energy: Energy can come in many forms, but here it will be in terms of heat, light and chemical energy which concerns us most. The last category of these is especially important. The energy of a molecule is made up of several components. First and foremost, there is the energy of the electrons arising from their position relative to the nucleus of a given atom (**electronic**). In addition, there are forms of energy such as **rotational** (atoms moving around each other), **vibrational** (atoms moving toward or away from each other), **nuclear** (based upon the state of atomic nuclei e.g. radioactive tracer chemicals), and **translational kinetic** (the driving force behind the straight-line movement of molecules of gases, liquid and solutions).

Energy can exist in multiple forms such as mechanical energy, heat, sound, electric current, light or radioactive radiation, so there are several ways for measuring energy. The most convenient is expressed in terms of heat because remaining other forms of energy can be changed into heat. The unit of heat normally used in biological studies is the kilocalorie (kcal), which is equivalent to 1000 calories (cal). One **calorie** is the heat required to raise the temperature of one gram of water by one degree celsius (°C). However, it should not be confused with calories for a term connected to diet and nutrition - the Calorie with a capital C.

Another energy unit, often used in physics, is the joule (one joule is equal to 0.239 calories).

Before discussing the basic concepts of thermodynamics (the study of energy, and heat related changes) it is pertinent to define some terms.

Energy is the capacity to do work, either actively (kinetic energy – the energy of motion) or stored for later use (potential energy) – the stored energy which has the capacity to do so, e.g., a boulder perched on a hill-top has potential energy, as it begins to roll downhill, some of its potential energy is converted into kinetic energy.

Free Energy: The amount of energy actually available to break chemical bonds that tend to hold the atoms in a molecule together, and subsequently form other chemical bonds in any system.

In a molecule within a cell, where pressure and volume usually do not undergo change, the free energy is symbolized by a word **G** (for Gibbs' 'free energy'). **G** is equal to the energy

[2] Some of the biologically important energy-releasing compounds are NADH, FADH, Acetyl-CoA and phosphorylated compounds. NADPH, built-up during light reaction, provides about 80 per cent of the energy required to fix CO_2 into carbohydrates. While the energy released during the oxidative phosphorylation of NADH (or FADH), for example, is used to build-up ATP from ADP plus Pi (energy currency of cells that powers many reactions).

contained in a molecule's chemical bond (called **enthalpy** and designated **H**) minus the energy unavailable because of disorder or randomness (called **entropy** and designated by the symbol **S**) times the absolute temperature, **T**, in degrees Kelvin (K = °C + 273):

$$G = H - TS$$

Chemical reactions break some bonds in the reactants and form new bonds in the products, thus resulting in changes in free energy. When a chemical reaction occurs under conditions of constant temperature, pressure, and volume – as it happens in most biological reactions, the change in free energy (given the symbol ΔG) is simply equal to:

$$\Delta G = \Delta H - T\Delta S$$

or

$$\Delta H = \Delta G + T\Delta S$$

(ΔH representing the chage in total heat energy, also called enthalpy and represented by the symbol H)

The change in free energy, or ΔG, (after American physicist Josiah Willard Gibbs, 1839–1903), is a fundamental property of chemical reactions. In some reactions, the ΔG is positive. This means that the products of the reaction contain **more** free energy than the reactants, the bond energy (H) is higher **or** the disorder (S) in the system is lower. Such reactions do not proceed spontaneously. They require extra energy for the reaction to proceed, and are thus **endergonic** (energy-consuming).

Energy is consumed (i.e., ΔG is positive)

In other reactions, the ΔG is negative, meaning thereby that the products contain less free energy than the reactants and the excess energy is released. Here the bond energy is lower or the disorder is higher, or both. Such reactions tend to proceed spontaneously as the difference in disorder (TΔS) is **greater** than the difference in bond energy between reactants and products (ΔH). Such reactions are **exergonic** (energy-releasing).

Energy is released (i.e., ΔG is negative)

5.3 Laws of Thermodynamics

A set of universal laws 'The Laws of Thermodynamics' govern all energy changes in the universe, right from nuclear reactions to the buzzing sound of a bee. Free energy is the thermodynamic parameter that determines the direction in which physical and chemical changes must happen and their ability to perform work.

First law of thermodynamics

It states that energy can change from one form to another (from potential to kinetic, for example) but it can *neither* be created *nor* can be destroyed. The total amount of energy in the universe remains constant. To be more precise, in all energy exchanges and conversions, the total energy of the system and its surroundings after the conversion is equal to the total energy before the conversion or the potential energy of reactants (the initial state) is equal to the potential energy of products (the final state) plus the energy released in the process (or reactions).

Second law of thermodynamics

It states that disorder (the correct term being entropy) in the universe is continuously increasing. Disorder and randomness is the natural scheme of things in our universe, and in all energy exchanges and conversions, if no energy enters or leaves the system under study, the potential energy of the final state will always be less than the potential energy of the initial state. A process in which the potential energy of the final state is less than that of initial state is one that releases energy i.e., exergonic ('energy output').

The laws of thermodynamics[3] apply only to closed systems, that is, to systems where or in which no energy is leaving or entering.

To explain further the laws of thermodynamics, let us take the classic example of aerobic respiration, where complete oxidation of glucose molecule yields carbon dioxide and water:

$$\underset{\text{Reactants}}{\text{Glucose + Oxygen}} \longrightarrow \underset{\text{Products}}{\text{Carbon dioxide + Water}}$$

To accommodate the 'first law of thermodynamics' it should be asserted that the left-hand side of the reaction exactly balances the right-hand side in terms of energy. Yet this reaction occurs without any outside energy source, and 'the second law' predicts that the products must somehow contain less intrinsic energy than the reactants. Are the laws contradicting each other? No, not really, because during the reaction some energy is released in various forms such as heat.

In biological systems, no significant temperature differences exist within a cell or different organs. Under these circumstances, the energy is not liberated as such but is distributed in the reacting molecules. In many of these, energy-liberating reactions are coupled with energy-consuming reactions. In most of these coupled reactions, the overall process proceeds spontaneously.

To maintain the biological order, 'homeostasis' in its broadest sense, a constant external supply of material and energy is required to overcome the tendency towards increasing disorder. Living cells or organisms tend to reduce their own 'entropy', that is, they reduce randomness or disorganization in their components and thus build-up and maintain their highly organized state by expending energy.

On the contrary, non-living systems are generally progressing towards a state of increasing randomness or disorganization (increasing entropy). The universe as a whole is constantly becoming less orderly, its disorder (entropy) is constantly increasing. Atomic reactions that

---

[3] The third law of thermodynamics states that the absolute entropy of most substances is zero at absolute zero (0°K or – 273.18°C)

generate sunlight cause greater disorder (solar flares) in the sun than sunlight causes order in the living organisms. Continents drift, galaxies explode and dead organisms decay. Our model universe has come to an end of its life! In a similar way the 'real' universe is, according to the second law of thermodynamics, irrevocably 'running down'. Energy is **not** being lost (first law) but useful (potential) energy is being converted into a non-useful form. Entropy is increasing.

Let us take the example of water. The change from ice to liquid water and the change from liquid water to water vapours are both endothermic processes, a considerable amount of heat is absorbed from the surroundings as they happen. The key factor in these processes is the increase in entropy. During the change from ice to water, a solid is being turned into a liquid, and some of the hydrogen bonds responsible for holding the water molecules together in crystal lattices of ice are being broken. As the liquid water turns to vapour, the remaining hydrogen bonds are ruptured as the individual water molecules separate, one by one. In each of the cases, the disorder or randomness of the system has increased (increasing entropy).

5.4 Redox Reactions

The processes of both photosynthesis and respiration include many oxidation–reduction (redox) reactions. Oxidation is the removal (loss) of one or more electrons from a compound while reduction is the addition (gain) of one or more electrons to a compound. Once the electron is transferred, a proton may follow, the net result being the addition of hydrogen during reduction and its removal during oxidation. In both the processes, the electron transport system (ETS) consists of a series of oxidation–reduction reactions, where oxidation of one compound is coupled with reduction of the another compound catalyzed by the same enzyme or the enzyme complex.

Let us consider the example of reduction of 3-phosphoglyceric acid (PGA) to glyceraldehyde-3- phosphate (GAP) by NADPH (the reduced form) during CO_2 fixation.

$$\underset{\text{(Phosphoglyceric acid)}}{\text{PGA}} + \text{NADPH} + \text{H}^+ \rightleftharpoons \underset{\text{(Glyceraldehyde-3- phosphate)}}{\text{GAP}} + \text{NADP}^+$$

Such a redox reaction may be conveniently split-up into two half-reactions, one involving the donation, and the other representing the acceptance of electrons as given below:

$$\text{NADPH} \rightleftharpoons \text{NADP}^+ + \text{H}^+ + 2\text{e}^-$$

$$\text{PGA} + 2\text{e}^- + 2\text{H}^+ \rightleftharpoons \text{GAP}$$

Thus the oxidation of NADPH to NADP is linked to the reduction of PGA to GAP. A reduced/ oxidized pair such as $NADPH/NADP^+$ is known as a **redox couple.** This mechanism of an oxidation–reduction reaction often involves the transfer of protons. The positively charged protons balance the negative charge of the acquired electrons, thus maintaining electroneutrality.

In biochemical reactions of photoynthesis, the important point in redox couples is more in their tendency to accept electrons from or donate electrons to another couple. This tendency is known as **redox potential.**

Other examples that could be cited are in Pigment Systems I and II or in the ETS of mitochondria.

5.5 ATP: The Energy Currency of the Cell

Adenosine triphosphate (ATP) is the most commonly used energy intermediate in the biological world.

ATP is a complex molecule containing the aromatic base *adenine,* the 5-C sugar *ribose,* and a chain of three *phosphate groups* linked to each other by **phosphoanhydride bonds** ('high-energy' bonds) and to the ribose by a **phosphoester bond**. The compound formed by the linking of adenine and ribose is called *adenosine.* Adenosine may be present in the cell in the unphosphorylated form or with one, or two, or three phosphate groups linked to C atom 5 of the ribose, thereby forming adenosine monophosphate (AMP), adenosine diphosphate (ADP), and adenosine triphosphate (ATP), respectively (Figure 5.3).

Hydrolysis of ATP to ADP and Pi can be catalyzed by ATPase present in the cell, with a standard free energy charge ($\Delta G^{0\prime}$) of about -7.3 kcal/mol (exergonic reaction). On the other hand, the synthesis of ATP by phosphorylation of ADP is a highly endergonic reaction, with a $\Delta G^{0\prime}$ of about +7.3 kcal/mol.

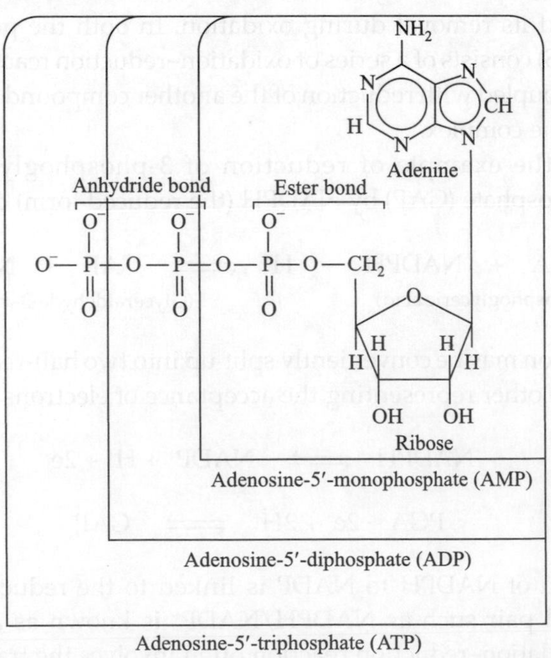

Fig. 5.3 The chemical structure of ATP, ADP and AMP. The phosphoanhydride bonds between the phosphate groups are important in energy transfer reactions.

5.6 Coupled Reactions

When a reaction occurs in more than one step and at least one of the steps is exergonic (energy-yielding), the energy released by that step of the reaction can drive all other steps forward to complete the reaction. Such reactions are called **coupled reactions.**

For Example

$$\text{Glucose + ATP} \xrightarrow{\text{Glucokinase}} \text{ADP + Glucose-6-phosphate}$$

$$\Delta G^0 = -4.0 \text{ kcal/mol}$$

$$\Delta G^0 = \Delta G_1^0 - \Delta G_2^0 = -7.3 + 3.3 = -4.0 \text{ kcal/mol}$$

This reaction can be written in two steps:

(i) ATP + H_2O $\longrightarrow$ ADP + Pi $\qquad \Delta G_1^0 = -7.3$ kcal/mol

(ii) Glucose + Pi $\longrightarrow$ Glucose-6-phosphate + H_2O $\qquad \Delta G_2^0 = +3.3$ kcal/mol

The exergonic reaction (i) is thus coupled to the endergonic reaction (ii) by glucokinase, and some of the free energy yielded by the exergonic reaction is consumed to drive the endergonic reaction.

Most of the major metabolic pathways, mentioned in Table 5.1, are common to all living organisms or cells. Some pathways are especially active, receiving as substrates the products of several other, less active pathways. By using these pathways, the cell is able to interconvert essentially all of the major compounds. Within the cell, many pathways are operative at the same time, and one pathway may influence both the rate and direction of reactions of another pathway. Intermediates in the degradation of carbohydrates can be diverted to lipid synthesis or to the formation of nitrogenous compounds (such as nucleotides and amino acids). Lipids present in microbial and plant cells can be transformed into carbohydrates and nitrogen compounds. Similarly, nitrogen compounds, once de-aminated, can be converted into lipids and carbohydrates. All these compounds may be further degraded, their catabolism acting as sources of reducing power, such as NADH and NADPH, to be used in cellular anabolic reactions.

Review Questions

RQ 5.1. Discuss the importance of compartmentation in the control of metabolic pathways.

Ans. Compartmentation of the plant cell into discrete organelles with distinct sets of internal conditions, increases the potential for metabolic flexibility and diversity.

(i) Firstly, it allows metabolic reactions requiring very different conditions to occur simultaneously in the same cell.

(ii) Secondly, it allows pathways and reactions that take place in more than one compartment to proceed with opposite net fluxes and at different rates in the same cell at the same time.

(iii) Thirdly, it provides a mechanism by which one metabolite can be used in two different processes without those processes being in direct competition.

RQ 5.2. Define entropy. What happens to entropy in a live against a dead plant?

Ans. Entropy is a term that reflects disorder in the universe. A living plant absorbs diffusely scattered molecules of CO_2, H_2O, and minerals and organises them (and thereby makes them more orderly) into organic molecules/cells/tissues/organs, thus exhibiting decrease in entropy. After death, molecules in an organism become more scattered and disordered leading to increase in entropy.

RQ 5.3. Distinguish between photoautotrophs and heterotrophs. How do photoautotrophs obtain energy? Can a plant be heterotrophic when a seedling and photoautotrophic when older?

Ans. Photoautotrophs are organisms that capture energy from sunlight to build their molecules using CO_2, H_2O, nitrates, sulphates, and other minerals. For example, green plants, algae, cyanobacteria and photosynthetic bacteria.

Heterotrophs are organisms that obtain energy from the oxidation of organic molecules such as carbohydrates, proteins, and fats. For example, animals, protozoa, parasitic plants, fungi, non-photosynthetic prokaryotes.

Young seedlings are heterotrophic when germinating underground and survive on reserves in cotyledons and endosperm. Seedlings become photoautotrophic only when they come out of soil and start photosynthesizing.

RQ 5.4. Describe the three methods of phosphorylation.

Ans. The three methods of phosphorylation of adenosine diphosphate (ADP) to adenosine triphosphate (ATP) are:

(i) ***Photophosphorylation*:** It involves sunlight as energy source in photosynthesis. Since animals, fungi, and non-chlorophyllous plant tissues lack the necessary pigments and organelles, they cannot perform this function. This process occurs only in chloroplasts.

(ii) ***Substrate-level phosphorylation*:** Compounds with high-energy phosphate groups are produced in some metabolic pathways, that transfer their phosphate group to ADP, thus synthesizing the ATP molecule. This process occurs in the cytosol, and these reactions do not involve oxygen.

(iii) ***Oxidative phosphorylation*:** In the terminal stages of respiration, ADP is phosphorylated to ATP. This process occurs in mitochondria and involves oxidations with oxygen.

RQ 5.5. Although ATP is an essential molecule, it constitutes only a tiny fraction of the plant body. Justify.

Ans. Although ATP is an essential molecule, it represents but a small fraction of the plant body. Every molecule of ATP is an energy carrier that shuttles between energy-releasing as well as energy-consuming modes. ATP is hydrolyzed to ADP and P_i in metabolic reactions, but the phosphate can be reattached with a high-energy bond through photosynthetic or respiratory reactions. Hence, each ATP molecule is recycled and reused repeatedly thousands of time per second.

RQ 5.6. What are coupled reactions, and how does ATP function as an intermediate between exergonic and endergonic reactions?

Ans. Coupled reactions are those in which endergonic reactions are linked to and driven by exergonic reactions that provide a surplus of energy. The result is that the net process is exergonic and thus able to proceed spontaneously.

The ATP molecule serves as the intermediate between the exergonic and endergonic reaction in coupled reactions.

$$ATP + H_2O \longrightarrow ADP + P_i + \text{energy}$$

When one phosphate group is removed from ATP by hydrolysis, ADP is produced with the release of free energy (7.3 kcal). This reaction is highly exergonic. Removal of a second phosphate group produces AMP and releases an equivalent amount of free energy.

$$ADP + H_2O \longrightarrow AMP + P_i + energy$$

So the energy yield is 14.6 kcal/one mole of ATP hydrolyzed.

On the other hand, sucrose synthesis is strictly endergonic, requiring an input of 5.5 kcal/one mole of sucrose synthesized.

$$Glucose + Fructose + energy \longrightarrow Sucrose + H_2O$$

However, when coupled with the hydrolysis of ATP, sucrose synthesis is actually exergonic. The balance 9.1 kcal is used to drive the reaction forward irreversibly and is eventually released as heat. This coupled reaction is an example of sucrose synthesis in sugarcane.

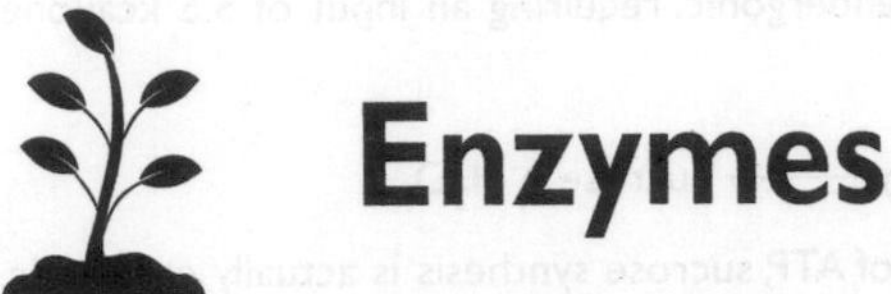

Chapter 6

Enzymes

Keywords

- Activation energy
- Active site
- Allosteric enzymes
- Apoenzymes
- Coenzymes
- Cofactor
- Competitive inhibition
- Holoenzymes
- Isozymes
- K_m
- Metalloenzymes
- Multienzyme complexes
- Prosthetic group
- Turnover number

6.1 Historical Background

The use of enzymes by mankind dates back to the Greek civilization, that first used enzymes in the process of fermentation to produce wine. After the discovery of the catalytic process in the early nineteenth century, a Swedish chemist, Brazelius (1836) suggested that the numerous chemical reactions in living organisms might depend upon the presence of catalysts within the tissues. In 1857, Louis Pasteur, a French scientist demonstrated the involvement of 'living, intact' yeast cells in the process of alcoholic fermentation and proposed the term 'ferments' for these biocatalysts. The term 'enzyme' was later coined by Kuhne (1878) for soluble ferment of yeast or bacteria. However, a significant breakthrough in the study of enzymes was made when the Buchner[1] brothers (1897) in Germany accidentally discovered that the 'squeezed out' juice from non-living yeast extract when mixed with sugar could bring about fermentation. Yeast juice is now known to be a mixture of at least twelve different catalysts. The name 'enzyme' was coined for the postulated catalyst in juice.

The chemical nature of enzyme remained uncertain until Sumner (1926) purified and crystallized first, the enzyme 'urease' from Jackbean (*Canavalia ensiformis*) and further discovered that it was proteinaceous in nature. Thereafter, hundreds of enzymes have

[1] E. Buchner gave the name 'Zymase' to the complex enzyme of the yeast juice.

been separated in a pure or semi-pure state and all have proved to be proteins except for **ribozymes**.[2]

6.2 Characteristics of Enzyme-catalyzed Reactions

- *High rates of reaction*: Enzyme-catalyzed reactions have typically 10^6–10^{12} times higher rates as compared to uncatalyzed reactions. Most of enzymes have the capability to convert thousands of substrate into product molecules every second.
- *High specificity*: Enzymes have the capablility to recognize extremely minute and specific differences in substrate as well as product molecules and can even distinguish between mirror images of any molecule (stereoisomers or enantiomers) e.g., D-Galactose and L-Galactose.[3]
- *Mild reaction conditions*: Enzymatic reactions usually take place at atmospheric pressure, relatively low temperatures and within a narrow range of pH (approximately 7.0) except for certain protein-degrading enzymes in vacuoles which function at pHs near 4.0, or enzymes present in thermophilic bacteria that can survive in hot sulphur springs (at 100 °C).
- *Opportunity for regulation*: The presence of a specific enzyme and its activity in plant tissues is subjected to regulation by controlled gene expression and protein turnover. Besides this, enzyme activity is also put to regulatory control by several activators as well as inhibitors. These opportunities for regulation are crucial to maintain balance of complex and competing metabolic reactions.

Enzymes exist either as a single unit (monomeric) or aggregates (oligomeric) of several subunits. Each subunit may have an active site where substrate molecules bind during enzymatic reaction.

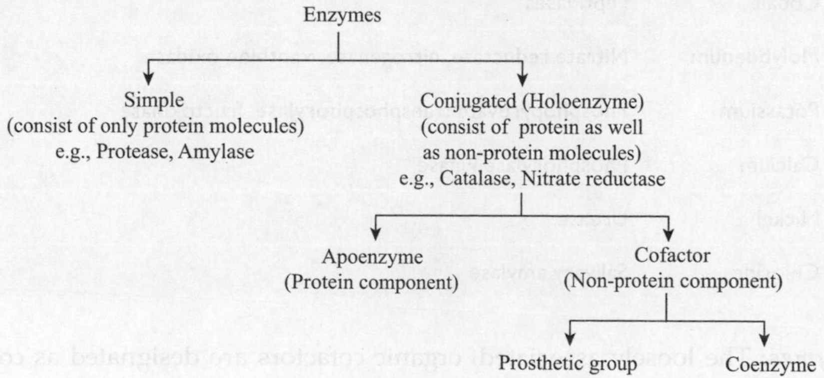

[2] Ribozymes are RNA molecules having catalytic properties of enzymes (Tom Cech, 1981). The earlier notion that all enzymes were proteins does not hold true any longer.

[3] Most of the naturally occurring sugars are in D-configuration form. The only exception is L-Galactose, a constituent of agar.

Neither the apoenzyme nor the cofactor alone has capability of catalytic operation, if they are not associated together.

6.3 Cofactors

Cofactors are classified into prosthetic groups and coenzymes depending upon their association with the enzymes.

Prosthetic groups: Prosthetic groups are organic compounds, permanently bound to the apoenzyme, e.g., haeme – a permanent part of peroxisomal enzymes, peroxidase and catalase; flavin adenine dinucleotide (FAD) – prosthetic group of succinic dehydrogenase, marker enzyme of inner mitochondrial membrane.

In the case of **metalloenzymes** (Table 6.1), metals are generally referred to as activators and these play an essential role in plant nutrition, e.g., zinc is the prosthetic group of proteolytic enzyme, carboxypeptidase.

Table 6.1 Some enzymes requiring metal activators.

Metals	Enzymes
Magnesium	Phosphatases, kinases, ATPase
Copper	Tyrosinase, respiratory proteins of invertebrate animals
Zinc	Dehydrogenases, carboxypeptidases, carbonic anhydrase
Iron	Cytochromes, catalase, peroxidase
Manganese	Peptidases, some enzymes of the nitric acid cycle
Cobalt	Peptidases
Molybdenum	Nitrate reductase, nitrogenase, xanthine oxidase
Potassium	Phosphopyruvate transphosphorylase, fructokinase
Calcium	Phosphorylase kinase
Nickel	Urease
Chloride	Salivary amylase

Coenzymes: The loosely associated, organic cofactors are designated as coenzymes. Coenzymes generally act as donor/acceptor of atoms that are either incorporated into or withdrawn from the substrate molecule.

Coenzymes have important role in oxidation–reduction reactions. These can be easily separated from the protein component under laboratory conditions but catalytic properties of the enzyme are greatly reduced.

Some of the important coenzymes are given in Table 6.2.

Table 6.2 Important coenzymes.

Nicotinamide adenine dinucleotide (NAD+)*
Nicotinamide adenine dinucleotide phosphate (NADP+)*
Adenosine triphosphate (ATP)
Coenzyme A (CoA)
Flavin mononucleotide (FMN)
Flavin adenine dinucleotide (FAD)
Vitamins of B-complex group, e.g., Thiamine (vit B_1), Riboflavin (vit B_2), Pantothenic acid (vit B_5), Pyridoxine (vit B_6), Niacin, Biotin (synthesized in plants but not in mammals)

*The reduced coenzymes are better written as NADH + H^+/NADPH + H^+ because only proton is released when these are oxidized.

Table 6.3 Various types of vitamins* with their sources, functions and effects when deficient.

Vitamin	Function	Source	Deficiency symptoms
Vitamin A (retinol)	Healthy skin and eyes	Milk products, fish, egg, liver, green and yellow vegetables	Night blindness, flaky skin and keratinization of cornea
	B-complex vitamins		
B_1 (thiamine)	Coenzyme in CO_2 removal during cellular respiration	Meat, grains, legumes, yeast, nuts	Beriberi, weakening of heart, edema, anorexia
B_2 (riboflavin)	Part of coenzymes FAD and FMN	Many different kinds of foods	Inflammation and breakdown of skin, eye irritation, nervous and muscular disorder
B_3 (niacin)	Part of coenzymes NAD and NADP	Liver, lean meats, grains	Pellagra, inflammation of nerves, mental disorders
B_5 (pantothenic acid)	Part of coenzyme A, regulation of metabolism	Many kinds of food	Rare: fatigue, loss of coordination
B_6 (pyridoxine)	Coenzyme during amino acid metabolism	Cereals, vegetables, meat	Anaemia, convulsions, irritability, nervous and muscular disorders
B_7 (Biotin)	Coenzyme in fat and amino acid synthesis	Meat, vegetables	Rare: depression and nausea
B_9 (Folic acid)**	Coenzyme in amino acid and nucleic acid metabolism	Green vegetables	Anaemia, diarrhoea

Contd.

Contd.

Vitamin	Function	Source	Deficiency symptoms
B_{12} (cyanocobalamin)**	Coenzyme during nucleic acids synthesis, nerve functions	Red meats, dairy products	Pernicious anaemia
Vitamin C (ascorbic acid)	Important for forming collagen, cement of bones teeth, connective tissues of blood vessels, resists infection	Fruits e.g., citrus, tomato, pineapple, Indian gooseberry (Amla), etc., green leafy vegetables	Scurvy, swelling and bleeding of gums, breakdown of skin, loss of appetite, delayed healing
Vitamin D (cholecalciferol)	Helps in absorption of calcium and promotes bone formation	Dairy products, fish, egg yolk, cod liver oil, and exposure to sun	Rickets and dental caries; osteomalacia (softening of bones in adults), bone deformities
Vitamin E (tocopherol)	Helps in protecting oxidation of fatty acids and cell membranes	Margarine, green leafy vegetables, wheat and nut fruits	Sterility, muscular degeneration
Vitamin K	Helps in clotting of blood	Green leafy vegetables, soybean, tomatoes, and carrots	Severe bleeding

* Vitamins are essential nutrients, responsible for maintaining good health in humans and animals, and generally they serve as enzyme cofactors. Broadly speaking, vitamins are classified into two categories: water-soluble (vitamins of B complex and vitamin C) and fat-soluble (vitamins A, D, E, and K). Vitamins A (carotenes), C, and E are important antioxidants, known to scavenge from the human system 'free radicals' that cause oxidative damage to cell membranes and nuclei, leading to ageing, cancerous growths and several neurodegenerative diseases and cardiovascular disorders, etc.

** Deficiency of folates or vitamin B_{12} (cyanocobalamin) due to dietary insufficiency or poor absorption (such as in alcoholism) leads to serious anaemias. Vitamin B_{12} is not present in plants, so strict vegetarians run the risk of anaemia.

Role of cofactors

1. In some cases, the cofactor completes the active site of the enzymes or modifies it in such a manner that substrate binding can occur.
2. Cofactors act as a donor of electrons or atoms to the substrate and following the reaction return to their former state.
3. Cofactors may serve as temporary recipients of either one of the reaction products/ electron/proton, again being recycled to their former state sometime after the main reaction is completed.
4. Finally, the cofactor, together with the side chains of residues at the active site, may serve to polarize the substrate and prime it for catalytic alteration.

Similarities between Coenzymes and Prosthetic Groups

- They help enzymes (or other proteins) to function, but they have no catalytic properties of their own.
- They are not themselves proteins (not composed of amino acids) and are invariably much smaller than the enzymes with which they are associated.
- They may undergo a temporary change during a reaction, but they are normally restored afterwards.
- They are required in small amounts.

Enzyme molecules are huge in size, e.g., catalase has a molecular weight of approximately 2,50,000 D while urease weighs more than 4,80,000 D. Owing to their large size, enzymes have a very low diffusion rate and make a colloidal system in water. The large colloidal nature of enzymes provide greater surface areas to the reactants which come into closer contact with one another, thus facilitating chemical reactions.

Most enzymes are intracellular **(endoenzymes),** i.e., they act within the cell in which they are produced while others found in fungi, bacteria and insectivorous plants are extracellular (**exoenzymes**), i.e., they diffuse out of the cell and act in some outside medium.

Enzymes, based on their occurrence in cells, fall into two categories:

1. **Constitutive enzymes:** Enzymes maintained at a constant level because the structural genes for these enzymes are continuously expressed, e.g., enzymes of glycolysis.
2. **Inducible enzymes:** Enzymes found lacking in the cells but their structural genes can be activated by the presence of specific inducer molecules in the cell, e.g., nitrate reductase (NR), isocitrate lyase, malate synthase.

Mobile Cofactors

The energy intermediates ATP, ADP, NADH, and NAD^+ as well as closely related intermediates such as GTP, GDP, NADPH, and $NADP^+$ are mobile cofactors when viewed in the context of metabolic pathways. Mobile cofactors can be thought of as links between or within metabolic sequences; connecting different pathways in parallel. The mobile cofactors are in limited supply, and so they must be regenerated by an external pathway in amounts sufficient to allow the pathway to proceed. Glycolysis exemplifies cofactor balance using both ATP and NADH. However, bound cofactors-often called prosthetic groups-are attached to to the enzyme, and must be regenerated along with the rest of the enzyme in the course of a single catalytic cycle. Thus, they cannot connect separate pathways.

6.4 Enzyme Classification

Given the specificity of enzymes and the large number of reactions that occur within a cell, it is not surprising that thousands of different enzymes have been identified. This enormous diversity of enzymes and enzymatic functions led to a variety of ways for naming enzymes:

1. Name of the enzyme indicated their respective substrates: e.g., amylase (acting on amylose), protease (acting on proteins), ribonuclease (acting on RNA) and so on.
2. Others were named to describe their functions: e.g., succinic dehydrogenase, PEP carboxylase, malate dehydrogenase.
3. Still other enzymes have names that provide little indication of either substrate or function: e.g., trypsin, catalase, lysozyme and so on.

The prevalence of common names for enzymes and the resulting confusion eventually prompted the **International Union of Biochemistry** (IUB) to appoint an Enzyme Commission (EC) in 1961 to devise a rationale for the nomencleature of enzymes based on reaction types and mechanisms (Table 6.4).

Characteristic features of the IUB system of classification

1. All the enzymes have been classified into six major classes, each of them having 4–13 subclasses.
2. Each enzyme name comprises of two parts: the first name is suggestive of the substrate and the second ending in suffix-ase, that describes the kind of catalyzed reaction.
3. Each enzyme has a systematic four-digit code number (EC). The first three numbers describe the major 'class', 'subclass' and 'sub-subclass' while the fourth number representing the 'serial number' is assigned to the enzyme when it was added to the list.
4. Additional information, if required, may be given in parenthesis.

Example: EC2.7.1.1. indicates a class 2 (transferase), subclass 7 (transfer of phosphate group from ATP), sub-subclass 1 (an alcohol functions as acceptor of the phosphate group), and the fourth digit 1 represents hexokinase (glycolytic enzyme), e.g., EC2.7.1.1.ATP.D-hexose-6-phosphotransferase (hexokinase).

$$\text{D-Hexose} + \text{ATP} \rightarrow \text{D-Hexose-6-phosphate} + \text{ADP}$$

6.5 Mechanism of Enzyme Action

The rates of reactions in an uncatalyzed cell at normal cellular temperature are very low as the reactant molecules lack adequate kinetic energy to exceed the activation energy barrier. In general, reactions can be accelerated by two means:

(i) Providing the reacting molecules with excess energy in the form of heat. With an increase in temperature, a greater number of reactants will gain sufficient energy of activation to the intermediate form and get converted to the product.
(ii) Lowering down the activation energy barrier by using a catalyst or an enzyme (Figure.6.1).

Activation energy can be defined as 'the extra energy needed to break existing bond(s) and initiate a chemical reaction'. For example, the activation energy required for

Table 6.4 The major classes of enzymes.

Class	Reaction type	Enzyme name	Example
1. Oxidoreductases	Oxidation–reduction reactions	Alcohol dehydrogenase (EC 1.1.1.1) (oxidation with NAD^+)	CH_3-CH_2-OH (Ethanol) $\rightleftharpoons$ ($NAD^+ \rightarrow NADH + H^+$) $CH_3-C(=O)-H$ (Acetaldehyde)
2. Transferases	Transfer of functional groups from one molecule to another	Glycerokinase (EC 2.4.3.2) (phosphorylation)	$HO-CH_2-CH(OH)-CH_2-OH$ (Glycerol) $\longrightarrow$ (ATP → ADP) $-HO-CH_2-CH(OH)-CH_2-O-PO_3^{2-}$ (Glycerol phosphate)
3. Hydrolases	Hydrolytic cleavage of one molecule into two molecules	Carboxypeptidase A (EC 3.4.17.1) (peptide bond cleavage)	$-NH-CH(R_{n-1})-C(=O)-NH-CH(R_n)-C(=O)-O^-$ (C-terminus of polypeptide) $\xrightarrow{H_2O}$ $-NH-CH(R_{n-1})-C(=O)-O^-$ (Shortened polypeptide) $+ H_2N^+-CH(R_n)-C(=O)-O^-$ (C-terminal amino acid)
4. Lyases	Removal of a group from, or addition of a group to, a molecule with rearrangement of electrons	Pyruvate decarboxylase (EC 4.1.1.1) (decarboxylation)	$CH_3-C(=O)-C(=O)-O^- + H^+$ (Pyruvate) $\longrightarrow$ $CH_3-C(=O)-H + CO_2$ (Acetaldehyde)
5. Isomerases	Movement of a functional group within a molecule	Malate isomerase (EC 5.2.1.1) (*cis-trans* isomerization)	$^-OOC(H)C=C(H)COO^-$ (Malate) $\rightleftharpoons$ $^-OOC(H)C=C(H)COO^-$ (Fumarate)
6. Ligases	Joining of two molecules to form a single molecule	Pyruvate carboxylase (EC 6.4.1.1) (carboxylation)	$CH_3-C(=O)-C(=O)-O^- + CO_2$ (Pyruvate) $\longrightarrow$ (ATP → ADP + Pi) $-C-C(=O)-CH_2-C(=O)-C(=O)-O^-$ (Oxaloacetate)

decomposition of H_2O_2 in the absence of catalase enzyme is 18,000 cal/mole; as compared to, 6,400 cal/mole in the presence of catalase.

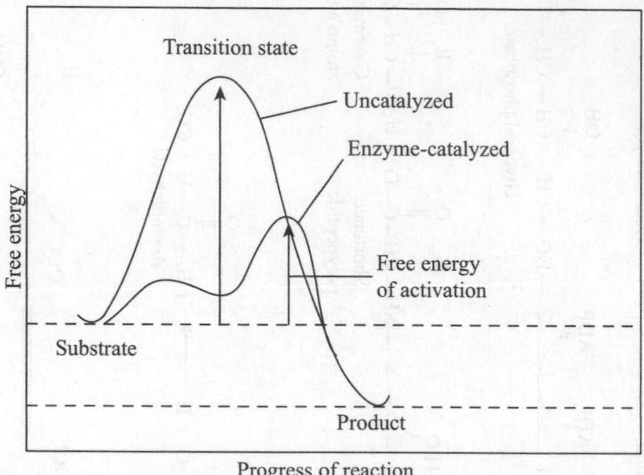

Fig. 6.1 Enzymes increase the reaction rate by lowering the activation energy.

Active site

Every enzyme contains somewhere within its tertiary configuration, a characteristic cluster of amino acids forming one or more active sites, where the substrate molecule binds and the actual catalytic event occurs. These occupy only a small part (approximately 5 per cent) of the enzyme's overall volume. The active site is usually a cleft or groove or pocket, with chemical and structural peculiarities that accommodate the entire substrate with high specificity. Generally, active sites are formed from the largest cleft on the protein surface, while the smaller clefts might be involved in binding regulatory molecules

Structurally, the active site can be considered to be composed of four main components (Figure 6.2):

1. **The Binding site,** (sites or residues) which attracts and positions the substrate and cofactors, and bind it to the enzyme.
2. **The Catalytic site,** where the reactions occur through bond-breaking and bond-forming.

Active site = Catalytic residues + Binding residues

3. **Structural residues,** which keep the enzyme's active site in proper configuration to allow it to perform its function effectively.
4. **Non-essential residues,** which are present close to the surface of an enzyme and can be replaced or eliminated without loss of any function.

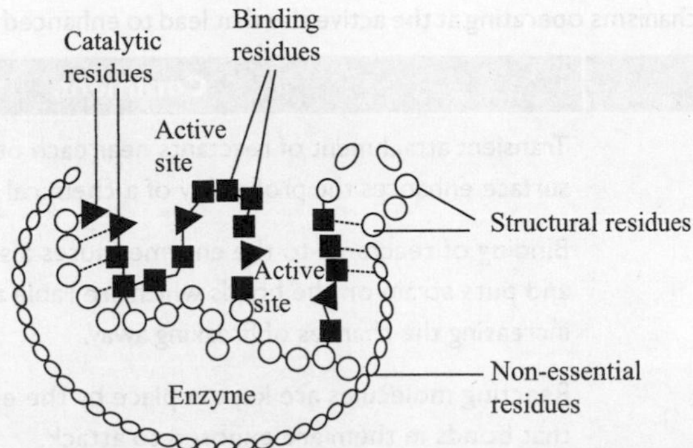

Fig. 6.2 The enzyme molecule comprises of essentially four categories of amino acid residues. The catalytic residues or sites, binding residues, structural residues and non-essential residues – the first two together form the active site.

The binding site is usually made up of a number of amino acids, whereas the catalytic site is composed of only a small number of amino acids. The amino acids constituting the active site are not usually contiguous to each other along the primary sequence of the protein. Rather they are brought together in proper conformation by the specific three dimensional (3-D) folding of the polypeptide chain. Only a few amino acids (cysteine, serine, histidine, aspartate, glutamate and lysine) participate in binding of substrate to the active site and may serve as donor/acceptor of protons, e.g., the active site of enzyme citrate synthase contains about 16 amino acids – 8 of these form temporary bonds with citric acid (substrate) and another 8 form bond with CoA.Therefore, the structure of the active site is responsible for the catalytic activity as well as specificity of the enzyme. It is generally believed that the substrate molecule fits into the active site of an enzyme with great specificity, somewhat like the way a key fitting into the lock. Nevertheless, both the geometric configuration of the active site and the 3-D structure of the substrate molecule controls binding to the amino acids at the binding site of an enzyme.

The three dimensional (3-D) structure of an enzyme is of paramount importance to all the functions performed by an enzyme, i.e., substrate binding, chemical catalysis, regulation and stability. Knowing how enzyme function is important in order to understand the overall cellular metabolism, the effect of drugs/toxins on metabolism, and genetic diseases. Several tools/techniques including **X-ray crystallography as well as nuclear magnetic resonance (NMR)** have been employed to elucidate enzyme structure and hence, enzyme function. **Chicken egg lysozyme** was the first one to be crystallized and have its 3-D structure successfully studied in 1965 by D. C. Phillips.

Large enzyme proteins tend to fold into several small self-contained structural and functional units i.e. **domains**, generally designated as the 'units of evolution'. These can often be exchanged between proteins without disturbing the folding pattern of other portions of the protein. Thus, within a single protein molecule, different combinations of domains can create novel functions.

Molecular basis of enzyme action: In molecular terms, several mechanisms, operate concurrently at the active site of an enzyme; and each one of them leads to reduction in the activation energy; and therefore, an increase in the reaction rate (Table 6.5).

Table 6.5 Molecular mechanisms operating at the active site that lead to enhanced rates of chemical reaction.

Method	Comment
1. Proximity effects	Transient attachment of reactants near each other on an enzyme surface enhances the probability of a chemical reaction.
2. Strain effects	Binding of reactants to the enzyme causes a slight distortion in them and puts strain on the bonds which are liable to be separated, thus increasing the chances of breaking away.
3. Orientation effects	Reacting molecules are kept in place by the enzyme in such a way that bonds in them are exposed to attack.
4. Micro-environmental effects	Hydrophobic amino acids establish a water-free region in which non-polar reactant molecules can combine readily.
5. Acid–base catalysis	Both acidic as well as basic amino acids of the enzyme speed up the electron transfer to and from the reactant molecules.

Two hypothesises have been proposed to describe the mechanism of enzyme action: The **Lock and Key hypothesis** and the **Induced-Fit hypothesis.**

6.5.1 Lock and Key hypothesis

Emil Fischer (1894) introduced the 'lock and key' hypothesis of enzyme action. He proposed that enzymes possess a well-defined 3-D structure that precisely fits the substrate molecule just as the key fits a lock. For a key to operate, it must be supplied with the appropriate lock and the same principle holds good for enzyme and substrate combination to yield products (Figure 6.3). This is one explanation for the precise specificity of enzymes towards their respective substrates.

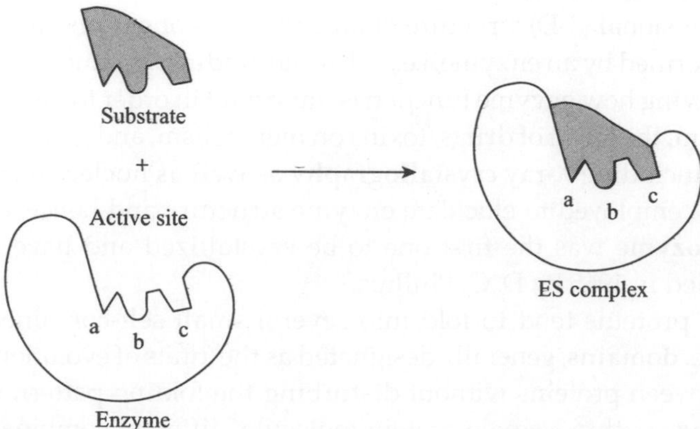

Fig. 6.3 Diagrammatic representation of the Lock and Key mechanism of enzyme action (Fischer, 1894).

$$\underset{\text{Enzyme}}{E} + \underset{\text{Substrate}}{S} \xrightleftharpoons{\text{Fast}} \underset{\text{Complex}}{ES} \xrightarrow{\text{Slow}} \underset{\text{Enzyme}}{E} + \underset{\text{Product}}{P}$$

For catalysis to take place within a complex, a substrate must lock up or fit accurately into an active site of an enzyme

↓

Side chains of the amino acid in an enzyme end up in close vicinity to specific bonds in the substrate

↓

The amino acid side chains have a chemical interaction with the substrate molecule, generally exerting strain and deforming a specific bond

↓

Consequently lowering the activation energy needed to break bond

↓

The substrate, now a product, then dissociates from the enzyme

6.5.2 Induced Fit hypothesis

According to Daniel Koshland (1958), a slight rearrangement of chemical groups occurs in both enzyme and substrate upon formation of an ES complex. Enzymes are therefore best regarded as rather 'flexible' molecules whose shape can change slightly under the influence of electrical charges present on the substrate (Figure 6.4).

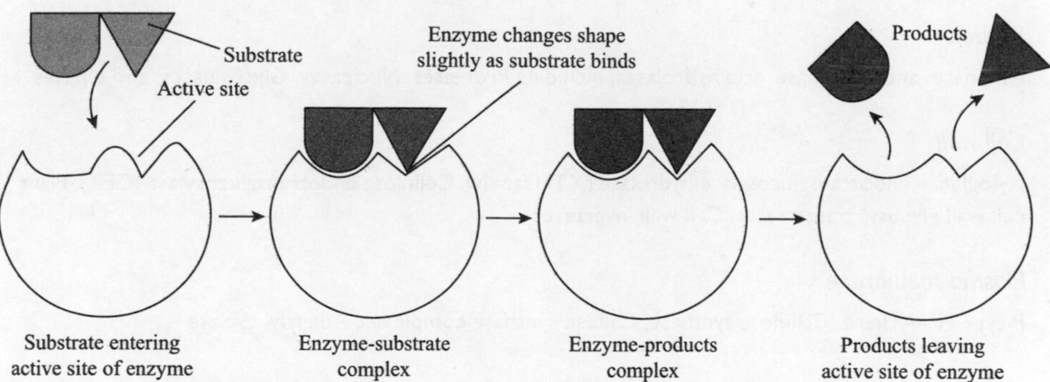

Fig. 6.4 Diagrammatic representation of Induced Fit hypothesis of enzyme action (Koshland, 1958).

The active site of enzyme and the substrate initially have different shapes but become complementary on substrate binding

↓

Enzyme molecule undergoes a conformational change and this places the catalytic residues in position to alter the bonds in the substrate

↓

Following which the products are released

↓

Active site returns to its initial state

Key Plant Enzymes Located in Cellular Compartments

Mitochondria

Enzymes of Krebs Cycle (Citrate synthase, Aconitase, Isocitrate dehydrogenase, 2-Ketoglutarate dehydrogenase complex, Succinyl-CoA synthetase, Succinate dehydrogenase, Fumarase, Malate dehydrogenase, Adenine nucleotide translocase) (For details, see chapter 9)

Chloroplast

Enzymes of Calvin cycle (C3 pathway), C4 pathway, and CAM pathway (For details, see chapter 7)

Peroxisomes

Glycolate oxidase, Catalase (For details, see chapter 7)

Glyoxysomes

Isocitrate lyase, Malate dehydrogenase (For details, see chapter 11)

Endoplasmic reticulum (ER)

UDP-glucose: glycoprotein glucosyltransferase, lipid metabolic enzymes such as Diacyl glycerolkinase, Phosphatidylinositol (PI) 4-phosphate 5-kinase, Choline-phosphate, Cytidylyl transferase, and PI 4-kinase

Vacuole

Chitinase, and Glucanase, acid hydrolases, including Proteases, Nucleases, Glycosidases, and Lipases

Cell wall

Xyloglucan endotransglucosylase/hydrolase (XTH family), Cellulose endotransglucosylase (CET), Plant cell wall glycosyl transferases, Cell wall invertases

Plasma membrane

P-type H^+-ATPase, Cellulose synthase, Callose synthase complexes, Adenylyl cyclase

6.6 Isozymes (Isoenzymes)

Isozymes are multiple forms of an enzyme (C. L. Markert and R. L. Hunter, 1959) that catalyze similar chemical reactions, but usually display different physical or kinetic parameters, such as isoelectric point, pH optimum, substrate affinity or effect of inhibitors. Isozymes are usually coded by different genes, and as a consequence they have different amino acid sequences and thus, different protein structures. These may be tissue- or organ-specific, and are subject to regulation by a host of chemical and environmental factors. Isozymes can be detected by gel electrophoresis.

- *Lactate dehydrogenase* (LDH), catalyzing the fermentation of pyruvate to lactate, has five isozymes – each with a molecular weight of 13,400 D and consisting of four polypeptide chains of two subunits: M and H (M4, M3H, M2H2, MH3, H4). In skeletal muscle and liver, the key isozyme consists of four M chains (M_4), and in heart, the principal isozyme is made up of four H chains (H_4). Different tissues may have different proportions of the five possible types of LDH isozymes. Difference in LDH isozyme content of tissues, can be used as an indicator to assess the heart damage (timing as well as extent) during myocardial infarction. Shortly after a heart attack, the level of total LDH increases in the blood. The pattern of LDH isozymes in the blood is suggestive of the tissue that released the isozymes and hence, can be utilized to diagnose myocardial infarction, infectious hepatitis and muscular diseases. The switch in H_4/MH_3 ratio, combined with increased concentration of creatine kinase, another enzyme present in the heart, provides strong evidence of myocardial infarction. Multiple LDH isozymes show different V_{max} and K_m values, particularly for pyruvate.

- *Malate dehydrogenase* that catalyzes conversion of oxaloacatate to malate and *vice versa*, exists in three organelle-specific forms. These forms exist in different subcellular compartments (microbodies, mitochondria and cytosol) and function in a different metabolic pathway but catalyze the same reaction. Hence, these play an important role in metabolic regulation.

By and large, the location of multiple isozymes of a particular enzyme, reflect the following:

- Distinct developmental stages in embryonic and adult tissues; e.g., the embryonic and adult liver have a characteristic isozyme distribution of LDH.
- Distinct sites and functional (metabolic) roles for isozymes within the same cell; e.g. isozymes of isocitrate dehydrogenase (in the cytoplasm and the mitochondrion), and of malate dehydrogenase (in the cytosol, the mitochondrion, and the microbodies).
- Distinct metabolic patterns in separate organs; e.g. isozymes of glycogen phosphorylase in the liver and the skeletal muscle possesses distinct regulatory properties, which reflects the different roles of glycogen breakdown in both tissues.
- Distinct responses of isozymes to allosteric modulators; e.g. isozymes of hexokinase in the liver and the other tissues exhibit difference in their sensitivity to inhibition by glucose-6-phosphate.

There is also some evidence that isozymes give some evolutionary stability to organisms. Mutations can occur without dramatic detrimental effects. Overall, isozymes are ubiquitous in plants and important to metabolism, growth and development.

6.7 Multienzyme Complex

A multienzyme complex is an aggregate of several enzymes that catalyze a number of reactions in a sequential manner. The end product of one particular reaction becomes the substrate for the next reaction occurring in the sequence. The intermediate products are transferred from enzyme to enzyme within the complex.

Three well-known examples of multienzyme complex are those involved in:

1. Oxidative decarboxylation of pyruvic acid during respiration - *Pyruvate dehydrogenase* (PDH)

$$\text{Pyruvate} + \text{CoA} + \text{NAD}^+ \xrightleftharpoons{\text{PDH}} \text{Acetyl CoA} + \text{NADH} + \text{H}^+ + \text{CO}_2$$

PDH – a multienzyme complex of pyruvate dehydrogenase, dihydrolipoyl transacetylase, dihydrolipoyl dehydrogenase, requires two coenzymes (CoA, NAD^+). Every complex possesses several copies of each of the three enzymes – total of 60 protein subunits working in concert, like a tiny factory.

2. Oxidative decarboxylation of α-ketoglutaric acid during critic acid cycle – *α-ketoglutarate dehydrogenase*

$$\alpha\text{-ketoglutarate} + \text{CoA} + \text{NAD}^+ \xrightleftharpoons{\alpha\text{-ketoglutarate dehydrogenase}} \text{Succinyl CoA} + \text{NADH} + \text{H}^+ + \text{CO}_2$$

3. Synthesis of fatty acids–*Fatty acid synthase*

$$8\ \text{Acetyl CoA} + 7\text{ATP} + 14\ \text{NADPH} \xrightarrow{\text{Fatty acid synthase}} \text{Palmitate} + 14\text{NADP}^+ + 8\ \text{CoA} + 6\ \text{H}_2\text{O} + 7\ \text{ADP} + 7\ \text{Pi}$$

In higher plants and bacteria, the fatty acid synthase is a complex of at least six enzymes and coenzyme called **acyl carrier protein (ACP)** (Figure 6.5). All the six enzymes are organized in a single polypeptide chain to facilitate the successive steps in the fatty acid synthesis.

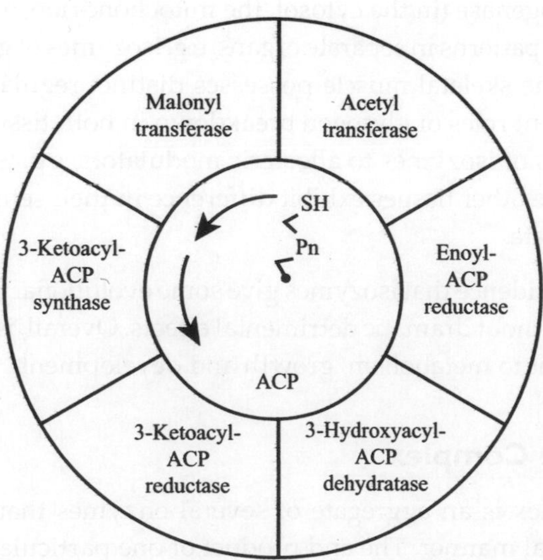

Fig. 6.5 Fatty acid synthase—a multienzyme complex involved in fatty acid synthesis.

ACP is a small protein containing a polypeptide that is 77 amino acids long with a prosthetic group of the vitamin, pantothenic acid. The enzyme complex is a dimer of two identical multi-enzymatic polypeptides and has a molecular weight of about 500,000 D.

Advantages:

1. All the reactions occurring within the complex can be regulated as a single unit.
2. As the reactant never exits the complex at the time of its migration through various steps of reactions, the probability of undesirable side reactions is negligible.
3. Within a complex, the end product of one chemical reaction can be transferred to the next enzyme in a sequence, without being released to diffuse away.

6.8 Enzyme Regulation

Enzymes must be regulated to adjust their activity levels to cellular needs. There are various control mechanisms in cells which help in enzyme regulation.

6.8.1 Substrate-level regulation

Substrate-level regulation includes the effect of substrate and product concentrations on the reaction rate. According to the Michaelis–Menten equation, an increase in concentration of the substrate results in higher rates of reaction. Conversely, an increase in product concentration reduces the speed at which substrate is converted to product.

6.8.2 Allosteric regulation

The term allosteric derives from the Greek word for another shape (state), which indicates that all enzymes capable of allosteric regulation exist in two different interchangeable states. In one state, the enzyme has a high affinity for its substrate, whereas in the other state, it has low affinity for its substrate. Such enzymes with two interconvertible forms are called **'allosteric enzymes'**. These enzymes are multi-subunit proteins with several catalytic and regulatory/allosteric subunits. Each catalytic subunit contains an active site to which the specific substrate binds, and every regulatory subunit has one or more allosteric sites to which the specific effector molecule binds, which may either inhibit or stimulate the enzyme, depending on which form of the enzyme (low affinity or high affinity) is favoured by effector binding. The binding of effector molecules at the allosteric site will cause changes in conformation of the enzyme molecule that may in turn alter the rate (kinetics) of the enzyme reaction (Figure 6.6). Allosteric enzymes normally have sigmoid kinetics rather than hyperbolic one.

The allosteric enzymes are the first enzymes in a series of enzymatic reactions of metabolic pathway. The final product of the pathway gets attached to the allosteric/regulatory site of the enzyme and inhibits its catalytic activity quite effectively. This sort of inhibition is called the **'feedback inhibition'** and the phenomenon has been termed **allosteric transition by** Jacob and Monod. Feedback inhibition is one of the most common mechanisms used by cells to ensure that the activities of reaction sequences are adjusted according to cellular needs.

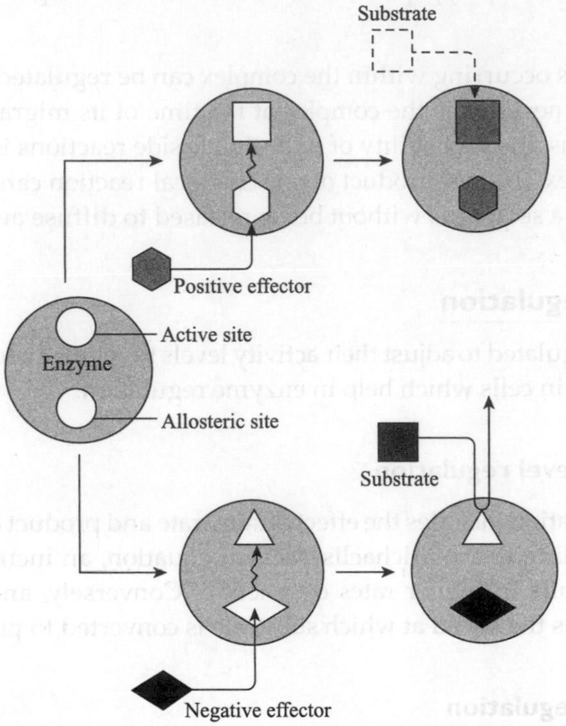

Fig. 6.6 A simple model of allosteric enzyme function.

The following reaction illustrates the conversion of the amino acid threonine to another amino acid isoleucine, catalyzed by a group of five enzymes, including threonine deaminase.

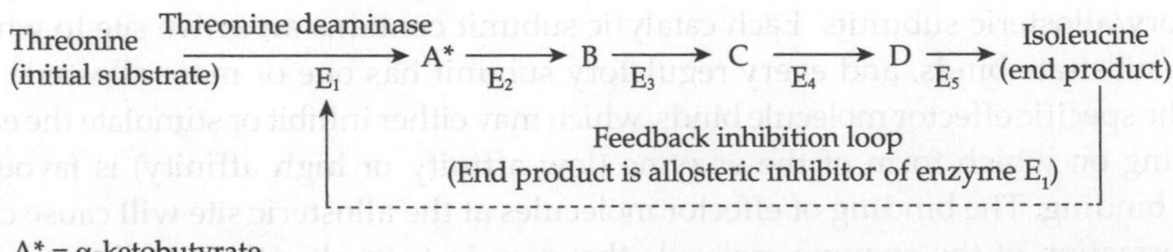

In this reaction, the first enzyme of the pathway is **threonine deaminase**, which is regulated by the concentration of isoleucine produced within the cell. If isoleucine is consumed further in protein synthesis, its concentration will remain low in the cell. The pathway is activated to produce more isoleucine; thereby meeting the ongoing need for this amino acid. If the need for isoleucine decreases, its concentration will increase leading to decrease in activity of threonine deaminase and hence to a reduced rate of isoleucine synthesis.

Many other enzymes in plant metabolism are regulated by allosteric mechanisms:

1. **Pyruvate dehydrogenase**, the enzyme that transforms pyruvate to acetyl CoA prior to oxidation in the Krebs cycle, is inhibited by ATP and citrate.
2. **Phosphofructokinase**, a key regulatory enzyme in the glycolytic pathway of respiration, is activated by AMP and inhibited by ATP.
3. **Aspartate transcarbamylase (ATcase)**, the key enzyme in pyrimidine biosynthesis, catalyzes the condensation of aspartic acid and carbamyl phosphate to cytidine triphosphate (CTP) with N-carbamyl aspartate as the first intermediate. The first reaction catalyzed by ATcase is the rate-determining step, and is inhibited by CTP when it is not being consumed further in the cell's metabolism. In contrast, ATP behaves as an allosteric activator, enhancing the activity of aspartate transcarbamylase for its substrates, thus resulting in enhanced enzyme activity.

6.8.3 Covalent modification

Enzyme regulation through covalent modification involves the enzyme whose activity is modified by the addition/removal of specific chemical groups i.e., phosphate, methyl, acetyl, and derivatives of nucleotides. The effect of modification is to either activate or to inactivate the enzyme, or at least to adjust its activity upward or downward.

1. *Phosphorylation/Dephosphorylation*: Phosphorylation involves the reversible addition of the phosphate group from ATP to the OH^- group of a serine, threonine, and tyrosine residue in the polypeptide, catalyzed by protein kinase. On the other hand, dephosphorylation occurs by the removal of a phosphate group from a phosphorylated protein, catalyzed by protein phosphatase.

   For example, the regulation of glycogen phosphorylase, an enzyme found in skeletal muscle cells; this enzyme breaks down glycogen by successive removal of glucose units as glucose-1-phosphate.

$$\text{Glycogen} \xrightarrow[\text{Pi}]{\text{Glycogen phosphorylase a}} \text{Glucose-1-phosphate + Glycogen chain with one less glucose group}$$

Glycogen phosphorylase b (inactive) ⇄ Glycogen phosphorylase a (active)

Forward: 2ATP → 2ADP, Phosphorylase kinase
Reverse: $2\,H_2O$ → 2Pi, Phosphorylase phosphatase

OH OH (phosphorylase b) — P P (phosphorylase a)

In addition to regulation by the phosphorylation/dephosphorylation mechanism, glycogen phosphorylase is also an allosteric enzyme, inhibited by glucose and ATP and activated by AMP.

2. *Proteolytic cleavage*: This kind of covalent modification involves the one-time, irreversible removal of a segment of the polypeptide chain by an appropriate proteolytic enzyme, e.g., activation of pancreatic zymogens (procarboxypeptidase, trypsinogen and

chymotrypsinogen). Pancreatic proteases are synthesized in their inactive form 'zymogen', and must be cleaved proteolytically to yield active enzymes. Otherwise, if pancreatic proteases are secreted in their active states, they would cause problems for the cells of the pancreas themselves.

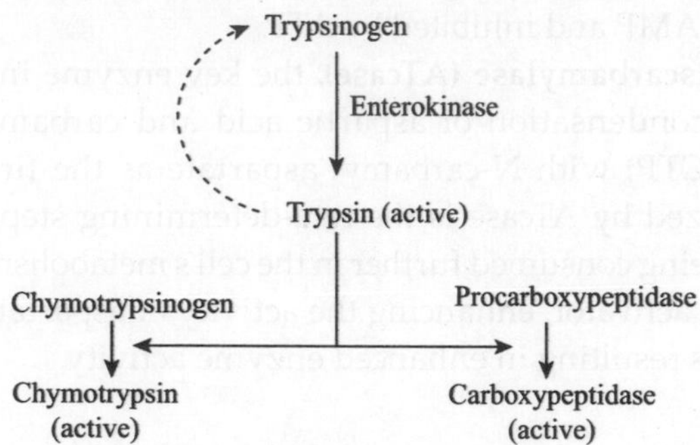

Other examples of reversible covalent modifications that are employed to modulate the activity of certain enzymes are **adenylylation** (the transfer of adenylate from ATP as seen in glutamine synthetase) and **ADP-ribosylation** (the transfer of an ADP-ribosyl moiety from NAD^+ as seen in RNA polymerase).

6.9 Michaelis Constant (K_m)

Each enzyme catalyzed reaction reveals a characteristic *Km* value. *Km* is defined as the concentration of the substrate necessary to produce half the maximum velocity. A small *Km* value means that only a small substrate concentration is needed to attain maximum velocity. It is an indicator of the affinity of an enzyme for a particular substrate (the smaller the *Km*, the greater the affinity). Leonor Michaelis and Maud Menten (1913) proposed this equation to explain the kinetics of enzyme action.

$$V_0 = \frac{V_{\max}[S]}{K_m + [S]}$$

V_0 = Initial reaction velocity,
V_{max} = Maximum reaction velocity,
$[S]$ = Substrate concentration

The Michaelis–Menten expression is very useful in explaining the behaviour of many enzymes in response to the variation in the substrate concentration (Figure 6.7).

Significance: Determination of *Km* has several practical applications:

1. The concentration of substrate required for maximum velocity can be determined.
2. Whether two enzymes from different sources are identical, can be also determined.
3. It indicates the degree of affinity for an enzyme with several substrates, and whether the enzyme can act on more than one substrate.
4. The significance of a reaction in metabolism within a cell can be established.

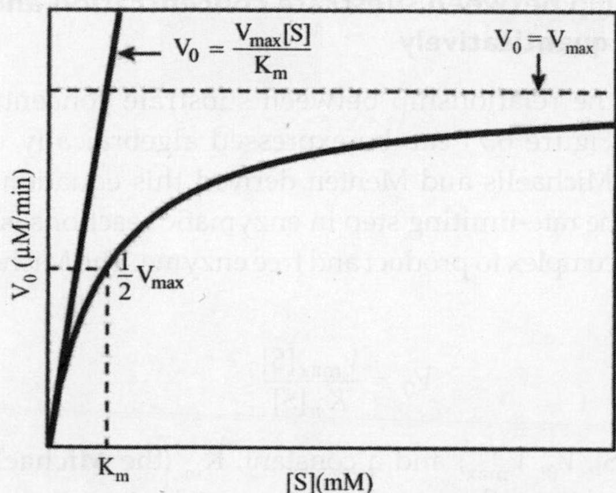

Fig. 6.7 Diagram representing the effect of substrate concentration on the rate of enzymatic reaction.

The determination of K_m and V_{max} from these hyperbolic curves is difficult. However, Lineweaver and Burk (1934) proposed a **double reciprocal plot** for direct calculation of K_m and V_{max} from linear graphs (Figure 6.8). These are helpful in evaluation of enzyme inhibitors.

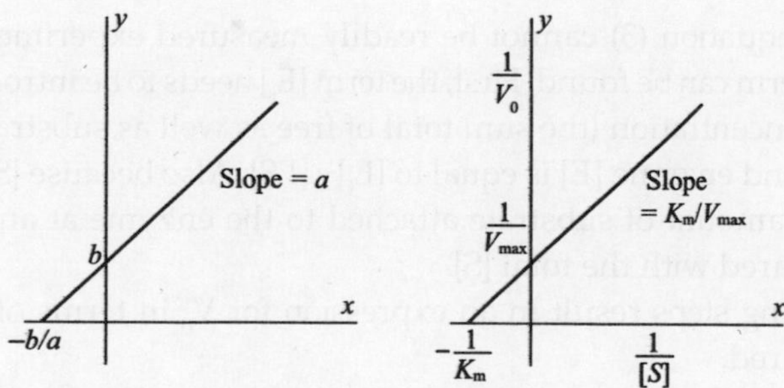

Fig. 6.8 **Lineweaver-Burk *double reciprocal plot* of $1/V_0$ and $1/[S]$ can be used to construct a line that may easily be extrapolated backward to find the K_m value for the enzyme.**

V_{max}: It is the maximum velocity attainable during the course of enzymatic reaction. The higher the V_{max}, the more is the product generated per minute. V_{max} is a useful parameter to calculate what is known as the **turnover number**, provided the concentration of enzyme is known.

$$\text{Turnover number} = \frac{\text{Moles of product formed per minute}}{\text{Moles of enzyme}}$$

Turnover numbers are rarely less than 10^4. The highest is for carbonic anhydrase, each molecule of which converts about 3.6×10^7 CO_2 molecules to H_2CO_3 per minute.

$$H_2O + CO_2 \xrightarrow[\text{in erythrocytes}]{\text{Carbonic anhydrase}} H_2CO_3$$

6.9.1 The relationship between substrate concentration and reaction rate can be expressed quantitatively

The graph depicting the relationship between substrate concentration, [S] and initial reaction velocity, V_0 (Figure 6.7) can be expressed algebraically using the **Michaelis–Menten equation (1)**. Michaelis and Menten derived this equation beginning with their basic hypothesis that the rate-limiting step in enzymatic reactions is the breakdown of the enzyme-substrate (ES) complex to product and free enzyme. The Michaelis–Menten equation is as follows:

$$V_0 = \frac{V_{\max}[S]}{K_m[S]} \tag{1}$$

All these terms – [S], V_0, V_{max}, and a constant, K_m (the **Michaelis constant**) – can be readily measured experimentally.

The derivation begins with the two basic steps i.e., the formation and breakdown of ES.

$$E + S \underset{k_{-1}}{\overset{k_1}{\rightleftharpoons}} ES \xrightarrow{k_2} E + P \tag{2}$$

V_0 can be calculated by the breakdown of ES to give product, which is determined by [ES]

$$V_o = k_2\,[ES] \tag{3}$$

Because [ES] in equation (3) cannot be readily measured experimentally, an alternate expression for this term can be found. First, the term $[E_t]$ needs to be introduced, representing the total enzyme concentration (the sum total of free as well as substrate-bound enzyme). Then, free or unbound enzyme [E] is equal to $[E_t]$ - [ES]. Also because [S] is generally much more than $[E_t]$, the amount of substrate attached to the enzyme at any given time is not significant as compared with the total [S].

Thus, the following steps result in an expression for V_0 in terms of parameters which can be easily measured.

Step 1: The rates of ES formation and its breakdown are calculated by the steps directed by the rate constants k_1 (formation) and $k_1 + k_2$ (breakdown to reactants and products, respectively),

$$\text{Rate of ES formation} = k_1([E_t] - [ES])[S] \tag{4}$$

$$\text{Rate of ES breakdown} = k_{-1}[ES] + k_2[ES] \tag{5}$$

Step 2: It has been assumed that the rate of formation of ES is equal to the rate of its breakdown. This is called the **steady-state assumption**. The expressions in equations (4) and (5) can be equated for the steady state, giving

$$k_1([E_t] - [ES])[S] = k_{-1}[ES] + k_2[ES] \tag{6}$$

Step 3: Solving equation (6) for [ES]. Multiplying out the left side and simplifying the right side

$$k_1[E_t][S] - k_1[ES][S] = (k_{-1} + k_2)[ES] \tag{7}$$

Adding the term $k_1[ES][S]$ to both sides of the equation and further simplifying

$$k_1[E_t][S] = (k_1[S] + k_{-1} + k_2)[ES] \tag{8}$$

By solving equation (8) for [ES]:

$$[ES] = \frac{k_1[E_t][S]}{k_1[S] + k_{-1} + k_2} \tag{9a}$$

This can now be simplified further, combining the rate constants into one expression:

$$[ES] = \frac{[E_t][S]}{[S] + (k_{-1} + k_2) / k_1} \tag{9b}$$

The term $(k_{-1}+k_2)/k_1$ is termed as the **Michaelis constant, K_m**. Substituting this into equation (9)

$$[ES] = \frac{[E_t][S]}{K_m + [S]} \tag{10}$$

Step 4: Expressing V_o in terms of [ES]. Substituting the right side of equation (10) for [ES] in equation (3)

$$V_0 = \frac{k_2[E_t][S]}{K_m + [S]} \tag{11}$$

Further simplification of equation (11): Because the maximum velocity occurs when the enzyme is fully saturated (i.e., when $[ES] = [E_t]$), V_{max} can be equated with $k_2[Et]$. Substituting this in equation (11) gives the Michaelis–Menten equation:

$$V_0 = \frac{V_{max}[S]}{K_m + [S]} \tag{11}$$

This is the Michaelis–Menten equation, the rate equation for a one-substrate enzyme-catalyzed reaction. It depicts the quantitative relationship between the initial velocity V_o, the maximum velocity V_{max}, and the initial substrate concentration [S], all related through the Michaelis constant K_m.

The special case when V_0 is exactly one-half of V_{max}. Then

$$\frac{V_{max}}{2} = \frac{V_{max}[S]}{K_m + [S]} \tag{12}$$

On dividing by V_{max},

$$\frac{1}{2} = \frac{[S]}{K_m + [S]} \tag{13}$$

Solving for K_m,

$$K_m + [S] = 2[S]$$

Or,

$$\mathbf{K_m = [S]}, \text{ when } V_0 = \frac{1}{2} V_{max} \tag{14}$$

This is a very useful, practical definition of K_m; K_m is equivalent to the substrate concentration at which V_o is one-half of V_{max}.

The Michaelis–Menten equation (equation 1) can be algebraically altered into versions that are helpful in the determination of K_m and V_{max} practically, and also in the analysis of inhibitor action.

6.10 Enzyme Inhibition

Enzyme inhibition is important for several reasons:

1. It plays a vital role as a control mechanism in cells.
2. It is important in understanding the mode of action of drugs, poisons, antibiotics, pesticides, and such.
3. It is useful to enzymologists as a tool in the studies of reaction mechanism or properties of enzymes.

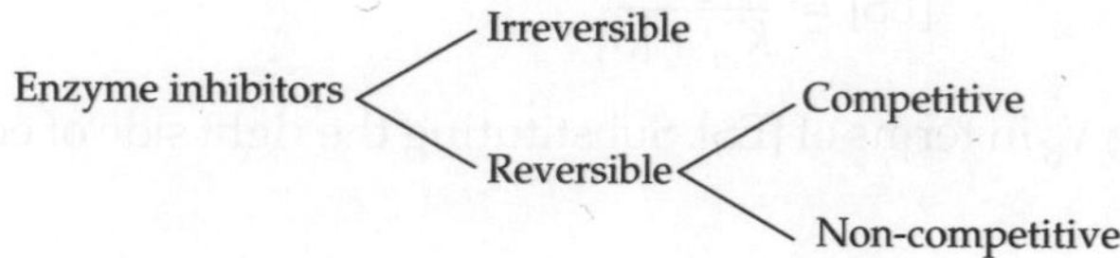

6.10.1 Irreversible inhibitors

Irreversible inhibitors such as, alkylating agents, nerve gas poison[4] (diisopropyl fluorophosphate), organophosphates, and pesticides, bind to the enzyme in a covalent manner resulting in irrevocable loss of catalytic activity. These bind to acetylcholine esterase, an enzyme crucial to the transmission of nerve impulses, thereby leading to rapid paralysis of vital function. Many natural toxins, e.g., alkaloid **physostigmine**, a potent inhibitor of acetylcholine esterase, are toxic to animals. Penicillin owes its antibiotic activity to its action as an irreversible inhibitor of a principal enzyme, **glycopeptide transpeptidase** involved in the formation of bacterial cell walls.

Plants also use irreversible inhibitors as part of their range of defences against predators and pathogens. An example is **mimosine**, an alkaloid originally isolated from the sensitive plant, *Mimosa pudica*. It is toxic to browsing animals because it irreversibly inhibits enzymes of DNA synthesis in susceptible cells.

6.10.2 Reversible inhibitors

Reversible inhibitors bind to the active site of an enzyme non-covalently in a dissociable manner.

1. Competitive inhibitors compete with the substrate for access to the active site of an enzyme because of their resemblance to substrate molecules. However, a higher substrate concentration can reverse the effect of competitive inhibition.

---

[4] Nerve gas (sarin) caused the death of several people when it was released by terrorists in the Tokyo subway in 1995.

V_{max} is not altered by the presence of a competitive inhibitor, but the K_m is elevated (Figure 6.9).

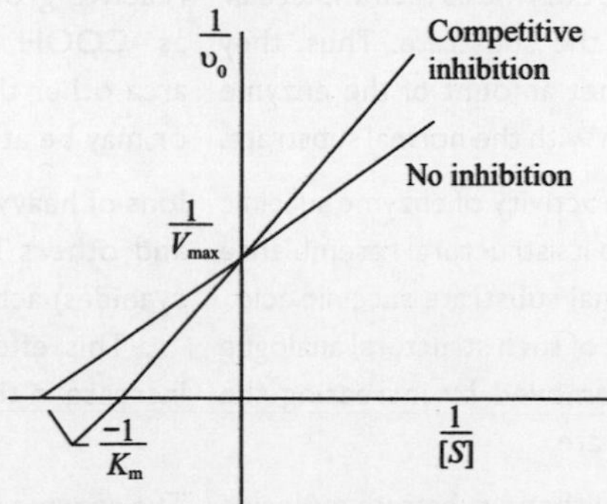

Fig. 6.9 Effect of a competitive inhibitor on normal enzyme kinetics. V_{max} is not altered, but the K_m value is increased. At low substrate concentrations, the effect of the inhibitor on the reaction rate is marked but is completely reversed as substrate concentration is increased and 1/[S] approaches 0.

1. Examples of competitive inhibition are as follows:

   (i) Competitive inhibition of succinate dehydrogenase (SDH) by malonate

$$\begin{array}{c} COO^- \\ | \\ CH_2 \\ | \\ CH_2 \\ | \\ COO^- \\ \text{Succinate} \end{array} \xrightarrow[\text{FAD} \;\to\; \text{FADH}_2]{\text{SDH}} \begin{array}{c} COO^- \\ | \\ CH \\ \| \\ CH \\ | \\ COO^- \\ \text{Fumarate} \end{array}$$

$$\begin{array}{c} COO^- \\ | \\ CH_2 \\ | \\ COO^- \\ \text{Malonate} \\ \text{(a structural analogue of succinate)} \end{array} \xrightarrow{\text{SDH}} \text{No reaction}$$

   (ii) Competitive inhibition of Rubisco by O_2

$$\text{RuBP} + CO_2 \xrightarrow{\text{Rubisco}} 2 \times \text{3-PGA (photosynthesis)}$$

$$\text{RuBP} + O_2 \rightarrow \text{3-PGA} + \text{p-glycolate (photorespiration)}$$

   CO_2, for example, is the substrate for RuBP carboxylase in C_3 plants but the enzyme activity depends a lot upon the CO_2/O_2 ratio: High CO_2/O_2 ratio promotes carboxylase activity while high O_2/CO_2 promotes oxygenase activity.

   (iii) Other examples of competitive inhibitors

Substrate	Competitive inhibitor
p-amino benzoic acid	Sulphanilamide
Hypoxanthine	Allopurinal

2. Non-competitive inhibitors bind to the enzyme surface at a location other than the active site, thus inhibiting catalytic activity. Increasing the concentration of substrate will not restore the original rate of reaction.

Competitive inhibition	Non-competitive inhibition
1. The inhibitor molecules compete with the substrate for the active sites of the enzyme as their molecular structure is similar to the substrate. Thus, they decrease the total or net amount of the enzyme available for the reaction with the normal substrate.	The inhibitor molecules combine or alter the reactive groups of enzyme molecule such as –COOH, –SH, $-NH_2$, etc., situated at an area other than the active sites of enzyme; or may be attached to the ES complex.
2. Malonic acid inhibits the activity of enzyme succinic dehydrogenase owing to its structural resemblance with the enzyme's normal substrate succinic acid. The inhibitory effect of such structural analogue may be reversed or removed by increasing the concentration of substrate.	Ions of heavy metals such as silver, mercury and others like Cu^{2+}, Zn^{2+}(even CO and cyanides) act as inhibitor. This effect can never be reversed by increase in the substrate concentration.
3. *The enzyme may bind either a substrate molecule or an inhibitor molecule, but not both at the same time. $E + S \rightleftharpoons ES \longrightarrow E + P$ $E + I \rightleftharpoons EI$ (I = Inhibitor)	The enzyme may bind the inhibitor molecule, the substrate molecule or both the inhibitor and substrate together. $E + S \rightleftharpoons ES \longrightarrow E + P$ $E + I \rightleftharpoons EI$ $ES + I \rightleftharpoons ESI$ $EI + S \rightleftharpoons ESI$ (I = Inhibitor)

*However, in uncompetitive inhibition, inhibitor binds only with enzyme-substrate and not with the free enzyme. This type of inhibition increases with an increase in substrate concentration. As a result, V_{max} is reduced, while K_m value increases.

Carbon monoxide, cyanides as well as ions of heavy metals, i.e., mercury, lead, silver, act as non-competitive inhibitors and alter reactive groups (–COOH, –SH, $-NH_2$, and such) of the enzyme molecules. In effect, the presence of the inhibitor prevents some percentage of the enzyme present from participating in normal catalysis. As a result, V_{max} is depressed, even though the K_m value remains the same (Figure 6.10).

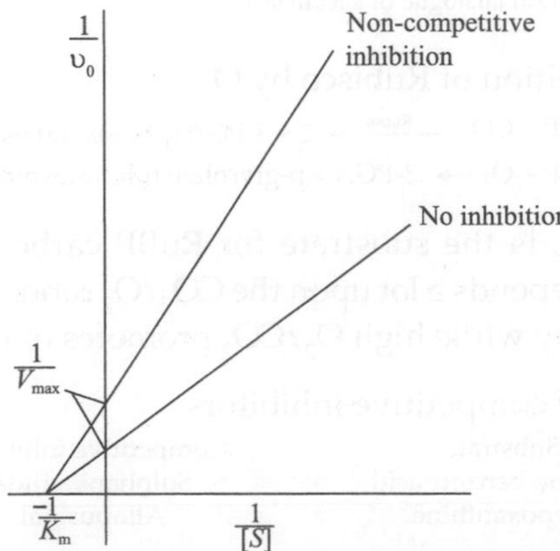

Fig. 6.10 Effect of a non-competitive inhibitor on normal enzyme kinetics. The K_m value is not altered, but V_{max} is lowered. Non-competitive inhibition can be reduced but not reversed by increasing the substrate concentration.

6.11 Factors Affecting Enzymatic Reactions

1. Temperature

In living organisms, the rate of reaction generally becomes double for every 10°C increase in temperature (temperature coefficient, $Q_{10} = 2$), within the normal range of environmental temperatures. An increased thermal activity because of increase in temperature causes a greater reaction rate but eventually leads to denaturation of enzymes. Thus, the usual temperature-dependent enzyme catalysis curve is bell-shaped (Figure 6.11). Enzymes vary in their temperature optima, with maximum activity at about 35 °C but plants from cold habitats/hot environments may have enzymes with vastly different thermal stabilities, e.g., enzyme **Taq polymerase** required in the polymerase chain reaction, is present in a bacterium that can survive at high temperatures in thermal springs, and thus, is suited to function optimally at high temperatures.

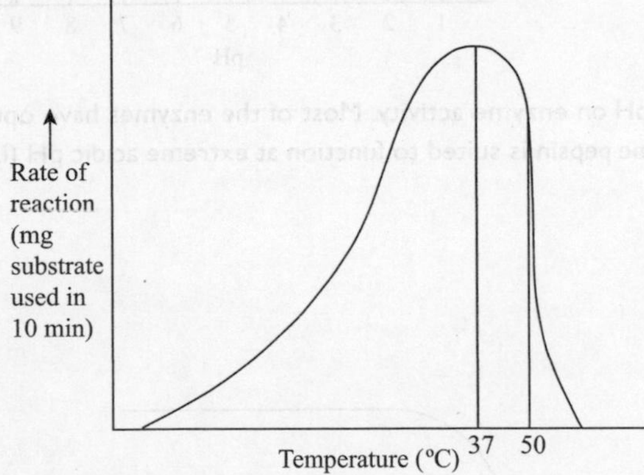

Fig. 6.11 Effects of temperature on the rate of enzymatically controlled reaction.

2. pH and ionic strength

Small deviation in pH has a marked impact on the structural configuration and ionic strength of enzymes because of various functional groups (–SH, –NH, –OH, –COOH) associated with them. In addition, many substrates, products, effectors and cofactors are also altered by pH and ionic strength. Much larger changes in pH cause denaturation of the proteinaceous enzyme itself, due to interference with weaker non-covalent links building its 3-D structure. A typical pH-dependent rate curve is bell-shaped (Figure 6.12). Different enzymes have different pH optima, with the maximum activity between pH of 6.0 and 8.0.

3. Substrate concentration

With increased substrate concentration, the rate of enzymatic reactions will increase because the maximum active sites of enzyme will be occupied and the enzyme is said to be working at maximum efficiency, all other factors being constant (Figure 6.13). Any further increase in substrate concentration will have no impact on the reaction because of the limiting effect of enzyme concentration thereby showing hyperbolic kinetics.

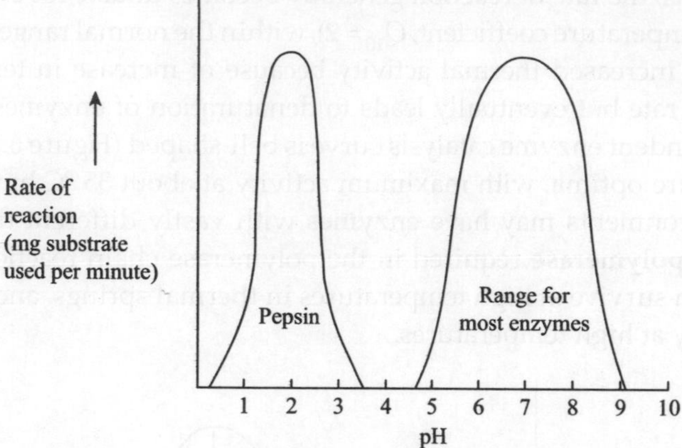

Fig. 6.12 The effects of pH on enzyme activity. Most of the enzymes have optimum activity around pH 7.0, but our digestive enzyme pepsin is suited to function at extreme acidic pH (i.e. 2.0) of the stomach.

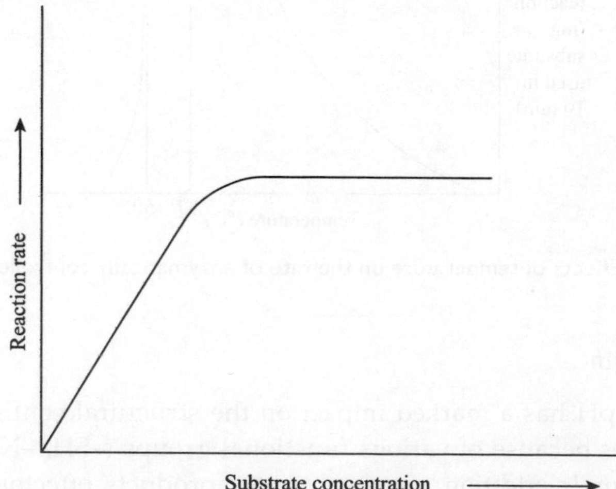

Fig. 6.13 Effect of substrate concentration on the rate of an enzyme-catalyzed reaction.

4. Enzyme concentration

The rate of an enzymatically controlled reaction increases linearly with the enzyme concentration under constant conditions. This constant rise in the rate of the reaction halts at a point and becomes constant because of the limiting effect of the substrate concentration (Figure 6.14).

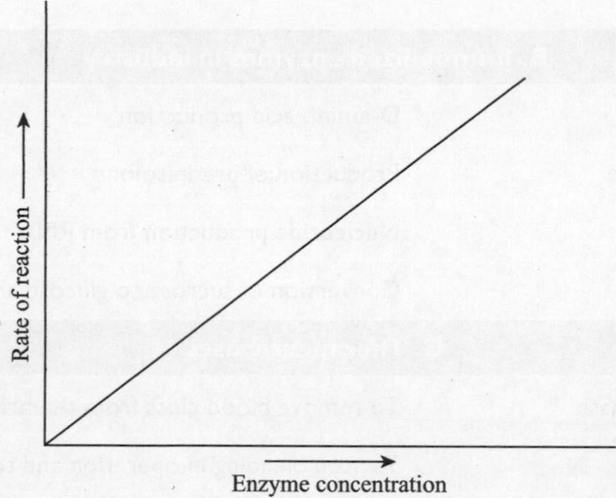

Fig. 6.14 Effect of enzyme concentration on the rate of an enzyme-catalyzed reaction.

6.12 Uses of Enzymes

Enzymes have many applications in industry, the medical sciences, pollution abatement, and genetic engineering. Some of the chief enzymes along with uses are presented in Table 6.6.

Table 6.6 Some important enzymes and their uses.

A. Industry	
1. Amylase	Textile industry, to destarch cloth before dyeing
2. Proteolytic enzymes	Leather industry, to digest collagen and connective tissues to remove hair and bating
3. Proteolytic enzymes	Detergent industry, the enzyme alcalase is used in detergents to remove stains from cloth
4. Endoxylanase (cellulase free)	Paper industry, to remove lignin-xylan complexes from pulp
5. Papain (Papaya), Bromelain (Pineapple), Pepsin, Trypsin	Meat processing, softens the meat by hydrolyzing peptide bonds
6. Renin, Catalase and Lipase	Cheese industry, renin converts Ca-casein to Ca-paracaseinate. Lipase adds flavour
7. Pectinase	Canned juice and wine, hydrolyse pectin making the juice and wine clearer
8. Glucose isomerase	Soft drink industry, converts hydrolyzed maize starch to high fructose syrup

Contd.

Contd.

B. Immobilized enzymes in industry	
1. L-amino acid oxidase	D-amino acid production
2. Δ-steroid hydrogenase	Production of prednisolone
3. Ribonuclease	Nucleotide production from RNA
4. Invertase	Conversion of sucrose to glucose and fructose
C. Therapeutic uses	
1. Streptokinase, Urokinase	To remove blood clots from the brain and artery
2. Thrombin	To stop bleeding in operation and tooth extraction
3. Asparaginase	Treatment of leukemia by reducing asparagine level of the blood plasma
4. Amylases, Lipases and Proteases	In digestive disorders
5. Papain (Papaya), Ficin (Fig)	Used in deworming of humans and animals
6. Resin (secreted by the kidney)	Diagnosis of hypertension
D. Enzymes as laboratory tools	
1. Restriction endonucleases	Cleave DNA at specific sites used in recombinant DNA technology and DNA fingerprinting
2. Terminal Nucleotidyl transferases	Add homopolymer tails to DNA fragment ends used in recombinant DNA production
3. Ligases, DNA polymerases, Alkaline phosphatases	Recombinant DNA production
4. Taq polymerase	In polymerase chain reaction cloning
5. DNA polymerase, Apyrase, Luciferase	DNA sequencing
6. Alkaline-phosphatase, Peroxidase	Enzyme linked immunosorbent assay
E. Enzymes as indicators of diseases (I) High enzyme levels	
1. α-Glutamyl transpeptidase	Alcoholism
2. 5′-nucleosidase	Hepatitis
3. Lactate dehydrogenase	Heart attack, liver disease
4. Acid phosphatase	Prostrate cancer
5. Alkaline phosphatase	Rickets, obstructive jaundice

Contd.

(II) Low enzyme level	
6. Amylase	Liver disease
7. Pseudocholinesterase	Viral hepatitis, liver cancer
F. Enzymes in diagnostic kits	
1. Peroxidase + glucose oxidase	Determination of glucose level in blood
G. Enzymes in pollution abatement	
Hydroxylation enzymes (from plasmids of *Pseudomonas* sp.)	Degradation of highly toxic detergents, plastics, chemicals (such as benzene, toluene, xylene)

Review Questions

RQ 6.1 What are isozymes? Explain, by giving an example.

Ans. Most enzymes exist in 'multiple forms' which catalyze the same reaction. These multiple forms are termed isozymes.

Such enzymes, despite being specific for substrates, are present in more than one molecular configuration. Let us take the example of malic dehydrogenase which is organelle-specific, occurring in the cytosol, mitochondria and microbodies (peroxisomes), each with the same catalytic function. Different isozymes occur in different compartments and each one is coded for by a different nuclear genome. The kinetic properties of isozymes are frequently different because the prevailing conditions are different for each of the situations.

Clement Markert and others in 1959 are credited with the detailed investigations of isozymes.

RQ 6.2. In which ways are enzymes different from hormones?

Ans.

Enzymes	Hormones
1. Nearly all enzymes are proteinaceous in nature (but for ribozymes).	Hormones may be a protein (as in insulin), terpene (GA, ABA) or a gas (as in ethylene, nitric oxide).
2. Enzymes catalyze biochemical reactions and increase the rate of reaction.	Hormones do not catalyze biochemical reactions.
3. Enzymes remain unchanged quantitatively and qualitatively at the end of chemical reaction they catalyze.	Hormones are used up or consumed during the course of reaction.

Contd.

Contd.

Enzymes	Hormones
4. Enzymes are of high molecular weight and produce reversible changes.	Hormones are of low molecular weight and produce irreversible changes.
5. Enzymes are generally intracellular, produced at the target site, that is, they are not translocated.	Hormones are synthesized in one part of an organ or organism and are transported to another by diffusion or transport channels where they elicit responses.
6. Enzymes catalyze reaction in the same cell in which they are produced.	In hormones, the site of synthesis (effector) and the site of reaction (target) are different.
7. Enzymes are very specific in their mode of action.	Hormones are not specific in their action.

RQ 6.3. What is the main feature of allosteric control that makes it such a tremendously important concept?

Ans. It is that an allosteric effector need not have any structural relationship to the substrate of the enzyme. This means that completely distinct metabolic systems can interact in a regulatory fashion.

RQ 6.4. What are the salient features of intrinsic and extrinsic regulation?

Ans. Intrinsic regulation is usually allosteric and can apply to a single cell. It keeps the metabolic pathway in balance. However, it cannot determine the overall direction of metabolism of a cell – whether, for example, it will store glycogen and fat or release these. These are determined by extrinsic controls, such as hormones for instance which direct the activities of cells to be in harmony with the physiological needs of the body.

RQ 6.5. Pyruvate dehydrogenase (PDH) is a major enzyme of the respiratory pathway. In general, products of the reaction inhibit the reaction. There are three mechanisms of control involved; what are these?

Ans. (i) Direct allosteric controls
(ii) Controls on a kinase that phosphorylates and inactivates PDH, and
(iii) Controls on protein phosphatase that reverses the phosphorylation.

RQ 6.6. Phosphofructokinase is the chief regulatory enzyme of glycolysis. What is the major allosteric effector for this enzyme and how is its level controlled in liver tissue?

Ans. Fructose-2,6-bisphosphate is the major allosteric effector for phosphofructokinase and cAMP increases its level.

RQ 6.7. K_m values of two enzymes are provided below, which of the two enzymes has greater affinity for its substrate and why?

Ans. i. Carbonic anhydrase 2.6×10^{-2}M
ii. Triose isomerase 4.7×10^{-4}M

Triose isomerase has greater affinity for its substrate.

The lower the K_m value of an enzyme, the more its affinity for its substrate.

RQ 6.8. In some multienzyme systems, the end product of the metabolic pathway is kept into balance with the cell's needs by feedback inhibition. Explain.

Ans. In a few metabolic pathways, the regulatory enzyme is specifically inhibited by the final product of the pathway that accumulates when its amount is far greater than the cell's requirements. By inhibiting the regulatory enzyme (the first enzyme in the sequence), the following enzymes function at decreased rates due to exhaustion of their substrates. Such regulation is termed **feedback inhibition**. When more of this end product is needed by the cell, these molecules will dissociate from the regulatory enzyme and activity of the enzyme will increase again. In this way the cellular concentration of the end product of the pathway is kept into balance with the cell's requirements.

RQ 6.9. What are the essential features that distinguish enzymes that undergo allosteric control?

Ans. These are multimeric enzymes that are controlled by effector molecules which bind at specific sites remote from the enzyme's active site. In doing so, these stabilize one of several possible conformations of the enzyme.

RQ 6.10. Do both the enzyme and substrate undergo a change when they interact?

Ans. Yes, the concept of 'induced-fit' of an active site to a substrate emphasizes the adaptation of the active site to fit the functional groups of the substrate. A poor substrate or inhibitor does not induce the correct conformational response in the active site.

RQ 6.11. What changes in kinetic parameters would you expect on adding (a) a competitive and (b) a non-competitive inhibitor to reaction mixtures?

Ans. (a) An apparent increase in K_m but no change in V_{max}.

(b) No change in K_m but a decrease in V_{max}.

RQ 6.12. Name the compounds that are vitamins for humans and that serve as cofactors within the cells.

Ans.

Vitamin	Cofactor	Usage Example
Thiamine (B_1)	Thiamine PP	Pyruvate dehydrogenase complex
Riboflavin (B_2)	FAD, FMN	Succinate dehydrogenase
Niacin (B_3)	NAD^+, $NADP^+$	Lactate dehydrogenase
Pantothenate (B_5)	Coenzyme A	Carnitine acyltransferase
Pyridoxine (B_6)	Pyridoxal P	Transaminase
Biotin (B_7)	Biotin	Pyruvate carboxylase
Folate (B_9)	Dihydrofolate, Tetrahydrofolate	Serine hydroxymethyl transferase
Cobalamin (B_{12})	Cobalamin	Methylmalonyl mutase

RQ 6.13. What are the two main ways by which the activities of enzymes may be reversibly modulated?

Ans. The two main methods of reversibly modulating enzyme activity are:

(i) By allosteric control (for details, see page 183)

(ii) By covalent modification of the enzyme, the chief mechanism for this being phosphorylation/dephosphorylation (for details, see page 185).

RQ 6.14. Protein degradation is a kind of cellular housekeeping. Comment.

Ans. A plant cell, at any given time, may contain upward of 10,000 or more proteins, many of these are no longer required for use because of their defunct nature, caused by errors during synthesis or may have been deformed by excessive heat and stresses. Thus the proteins are continually degraded by proteolytic enzymes such as proteases. Protein degradation is kind of housekeeping in which the unwanted proteins are split into their component amino acids to be recycled. The process of degradation goes on in several cellular components, including chloroplasts, nuclei, mitochondria and vacuoles. However, the process occurring in the cytosol is best understood (Figure 6.15). The mechanism of protein destruction is achieved by two main constituents – a small, highly conserved protein, 'ubiquitin', and a large, oligomeric enzyme complex, the 'proteasome'[5]. Through a series of interactions with the multi-enzymes complex, ubiquitin is conjugated with the target protein to be degraded. The ubiquitin-protein complex is then delivered to 20S core of the proteasome for degradation. After delivery, the ubiquitin separates out to be reused once again at the targeted protein. The proteasome is made up of of four stacked discs that produce a hollow cylinder. The regulatory protein (19S part) of the proteasome that binds to the ends of the stacked discs directs the ubiquitin-protein complex into the hollow core of the stacked discs where the protein-degrading active site is present.

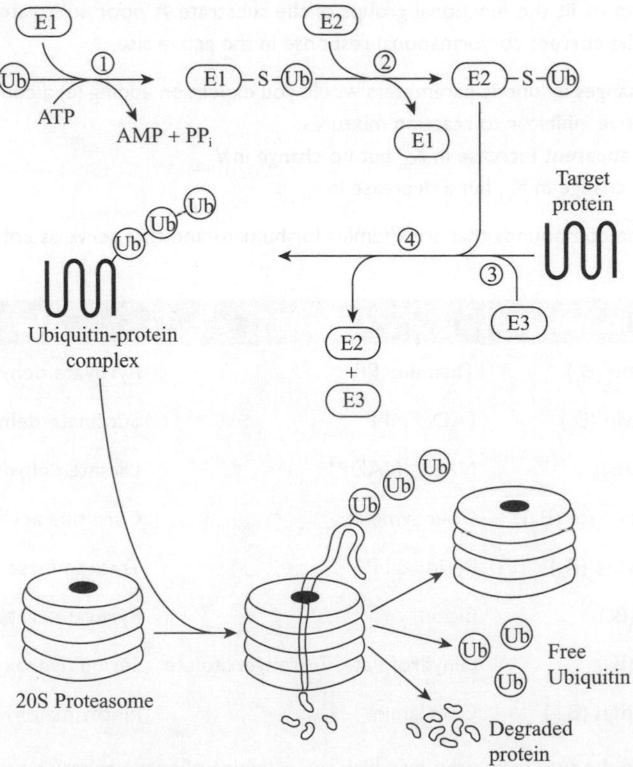

Fig. 6.15 Generalized view of the cytoplasmic pathway of protein destruction. ATP is needed for the initial activation of ubiquitin – activating enzyme (E1). E1 transfers ubiquitin to ubiquitin – conjugating enzyme (E2). Ubiquitin ligase (E3) serves as a mediator for the final transfer of ubiquitin to a target protein, which may be then ubiquitinated with multiple ubiquitin units. The ubiquitinated protein is thus targeted to the 20S proteasome where it is degraded.

[5] The proteasome consists of two parts - a 20S core proteasome and a 19S regulatory complex.

Chapter 7

Photosynthesis

Keywords

- Absorption spectrum
- Action spectrum
- Assimilatory power
- Autoradiography
- Crassulacean acid metabolism
- Emerson effect
- Fluorescence
- Hill reaction
- Kranz anatomy
- Photophosphorylation
- Photorespiration
- Photosystem
- Red drop
- Rubisco activase
- Solarization

Photosynthesis is by far the most significant biological process on the planet earth. It is through this phenomenon that all the useful organic matter available on earth has been produced. The leaves in the higher plants may be regarded as machinery engineered to undertake photosynthesis efficiently even under very hostile environments. Photosynthesis is not a monopoly of green plants but it also occurs in cyanobacteria and photosynthetic bacteria.

Photosynthesis can be defined as the anabolic process during which complex energy-rich organic molecules/compounds are synthesized by organisms from CO_2 and H_2O using solar energy. Every day, the radiant energy of the sun that bombards the earth equals nearly a million Hiroshima-sized atomic bombs. Out of the total supply of solar energy received by the earth, only 1–2 per cent is utilized by photosynthesis.

In higher and other non-flowering plants, the photosynthetic reactions occur in the 'chloroplast – an incredible thermodynamic machine'.

Photosynthesis occurs in three steps (Figure 7.1):

1. To harness solar energy by the chloroplasts
2. Using this solar energy to produce ATP and NADPH in the light reaction
3. Using the ATP and NADPH to power the synthesis of complex organic molecules using atmospheric CO_2 during the dark reaction

The following equation summarizes the overall process:

$$6CO_2 + 12H_2O \xrightarrow[\text{Chlorophyll}]{\text{673 Kcal of light energy}} \underset{\text{Carbohydrates}}{C_6H_{12}O_6} + 6O_2 + 6H_2O$$

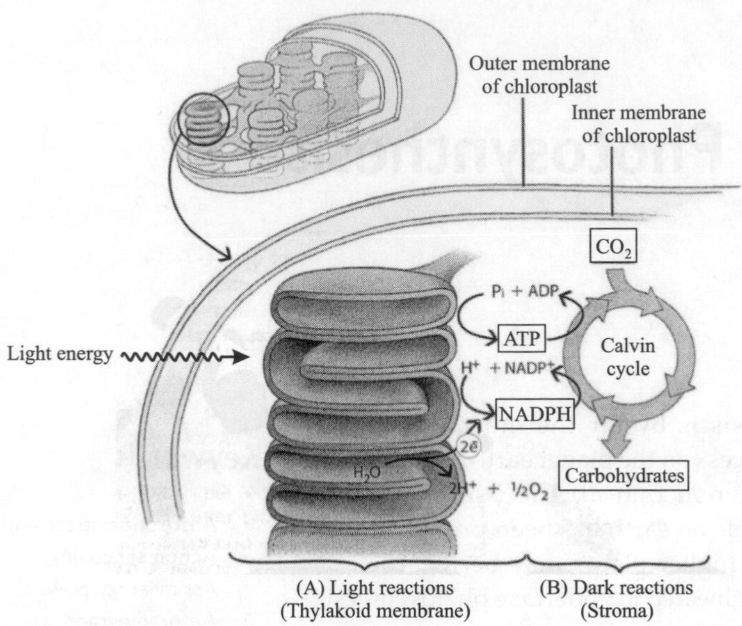

Fig. 7.1 Overview of photosynthesis in the chloroplasts. Photosynthesis is a two-step process: the light reaction and the dark reaction. (A) In the light reactions, solar energy absorbed by chlorophyll molecules in the thylakoid membrane is used to synthesize ATP. Simultaneously, water molecules split into oxygen gas and hydrogen atoms and electrons are accepted by $NADP^+$ and H^+ to produce NADPH. (B) In the dark reaction, CO_2 is fixed into carbohydrates in the stroma of the chloroplast, using the reducing power, ATP and NADPH, generated in the light reactions.

7.1 Historical Perspective

The beginnings of photosynthetic research can be traced back to approximately 300 years. The idea that water is an important reactant emerged from the experiments of a Dutch alchemist, Van Helmont (1648). The first account of photosynthesis appeared in the writings of Stephen Hales, an English naturalist and the 'Father of plant physiology'. Hales (1727) postulated that plants receive a large part of their nutritional requirements from the air but the role of light was not recognized at that time. Joseph Priestley (1772) demonstrated that plants have the capacity to purify air. His experiments excited Jan Ingenhousz who in 1773 noticed that photosynthesis will occur in the presence of sunlight and only the green parts of the plant purified air. But neither of them was aware of the chemical nature of pure or impure air. It was Antoine Lavoisier, a French chemist, who in 1785, identified the 'pure' component of air as 'oxygen' and the 'impure' as 'carbon dioxide'. Later Theodore de Saussure (1804), using eudiometery and gas analysis techniques, confirmed the equivalence of release of oxygen to consumption of carbon dioxide during photosynthesis.

Robert Mayer (1845) deduced that the energy used by plants and animals in their metabolism was ultimately derived from the sun and through photosynthesis it was converted from the radiant to the chemical form.

Theodore Engelmann (1883) conducted an experiment using filamentous alga, *Cladophora*, to determine the most effective wavelengths of light for photosynthesis (Figure 7.2). The efficiency of the different wavelengths of the visible light based on the evolution of oxygen was measured by him. Localized concentration of aerobic bacteria suggested that maximum oxygen was evolved in the blue and red regions of the visible spectrum.

Julius von Sachs (1887) was the first to discover that chloroplasts are the organelles in which carbon dioxide is fixed and oxygen is released. He also noticed that starch was the first visible product of the photosynthesis process.

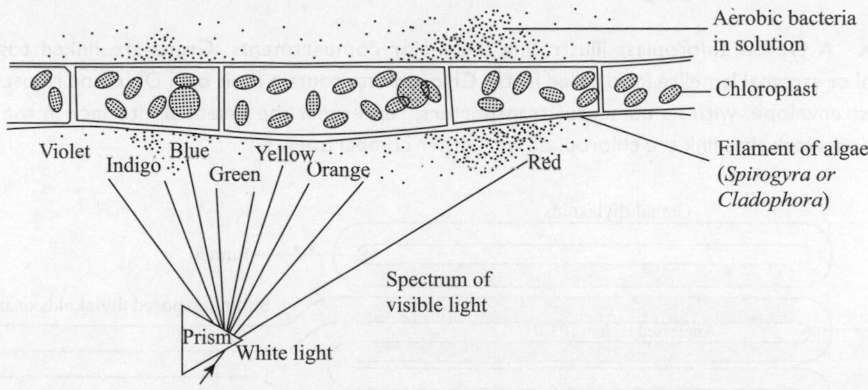

Fig. 7.2 Engelmann's experiment showing that the peak release of oxygen occurs at the blue and red wavelengths of the visible spectrum where there is localized concentration of aerobic bacteria. In this simple but elegant experiment, the German botanist Engelmann exposed a filamentous alga to white light that had been passed through a prism, thus exposing various portions of the alga to different wavelengths of light (i.e., monochromatic).

7.2 Chloroplast as Solar Harvesting Enterprise

Chloroplast was first described by a German botanist, von Möhl in 1837. A single leaf cell may contain as many as 50 chloroplasts, each with a 4–6 μm diameter and surrounded by a double membrane. Chloroplast consists of system of **lamellae** – grana lamellae and stroma lamellae. The inner membranes of chloroplast are folded extensively into sacs or vesicles known as **thylakoids** and often many thylakoids are stacked one above the other, in flattened columns to form **granum** (plural, grana). The space within the thylakoid membrane is called the **lumen**. Lamellae that connect the adjacent grana are termed as **stroma lamellae (frets)**. Surrounding the thylakoid membrane system is a semi-fluid substance called '**stroma**' (Figure 7.3A). Within the thylakoid membranes, photosynthetic pigments are organized in a network called '**photosystem**', capable of capturing photons of light (Figure 7.3B). The photosystem functions as a huge antenna, amplifying the power of individual pigment molecules to capture light.

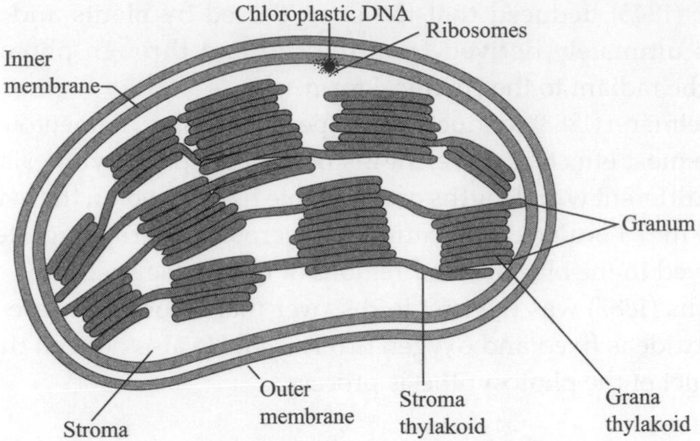

Fig. 7.3A A typical chloroplast illustrating the major compartments. Grana are linked together by intergranal or stromal lamellae (also called frets). Chloroplasts contain their own DNA and ribosomes. The chloroplast envelope, with its numerous transporters, represents the essential interface in the complex metabolic network that link the chloroplast with other compartments.

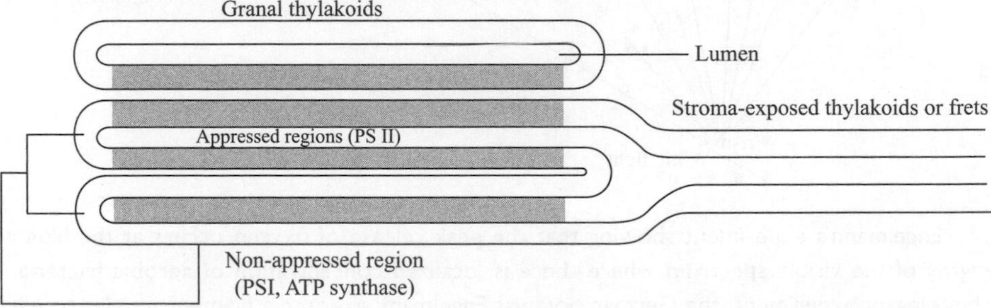

Fig. 7.3B A diagram showing the location of photosystem II and photosystem I (along with ATP synthase) in the appressed and non-appressed parts of a granum thylakoid, respectively.

> **Genetics, Assembly and Evolution of Photosynthetic Systems**
>
> - Chloroplasts have their own DNA, mRNA, and semi-autonomous protein synthesizing machinery. Most chloroplast proteins are synthesized in the cytoplasm and coded by nuclear genes. Chloroplast genes exhibit non-Mendelian or maternal patterns of inheritance.
> - Biosynthesis and breakdown pathways of chlorophyll are quite complex.
> - Chloroplasts have come about through the symbiotic relationship between a cyanobacterium and a non-photosynthetic eukaryotic cell–endosymbiosis.

Willstätter and Stoll (1912) studied the nature and composition of photosynthetic pigments in the chloroplasts.

Pigments in green plants have been categorized into two groups:

1. Vital (essential) pigments – Chl *a* ($C_{55}H_{72}O_5N_4Mg$)
2. Accessory pigments – Chl *b*, *c*, *d*, *e*, carotenoids, phycobilins

The **chlorophyll** molecule has a cyclic tetrapyrrolic structure – a porphyrin head with an isocyclic ring containing a magnesium atom at its centre. The phytol chain of the chlorophyll molecule extends from one of the pyrrole rings into the thylakoid membrane. The structure of the chlorophyll molecule roughly resembles a 'tennis racket' having a large flat head (the porphyrin ring) and a long handle (the phytol tail). The angle between head and tail is 45° (Figure 7.4). Chlorophyll molecules preferentially absorb the blue (400–500 nm) and red (600–700 nm) wavelengths of the electromagnetic spectrum and reflect the green wavelengths (500–600 nm) (Figure 7.5). Hence, leaves look green in color.

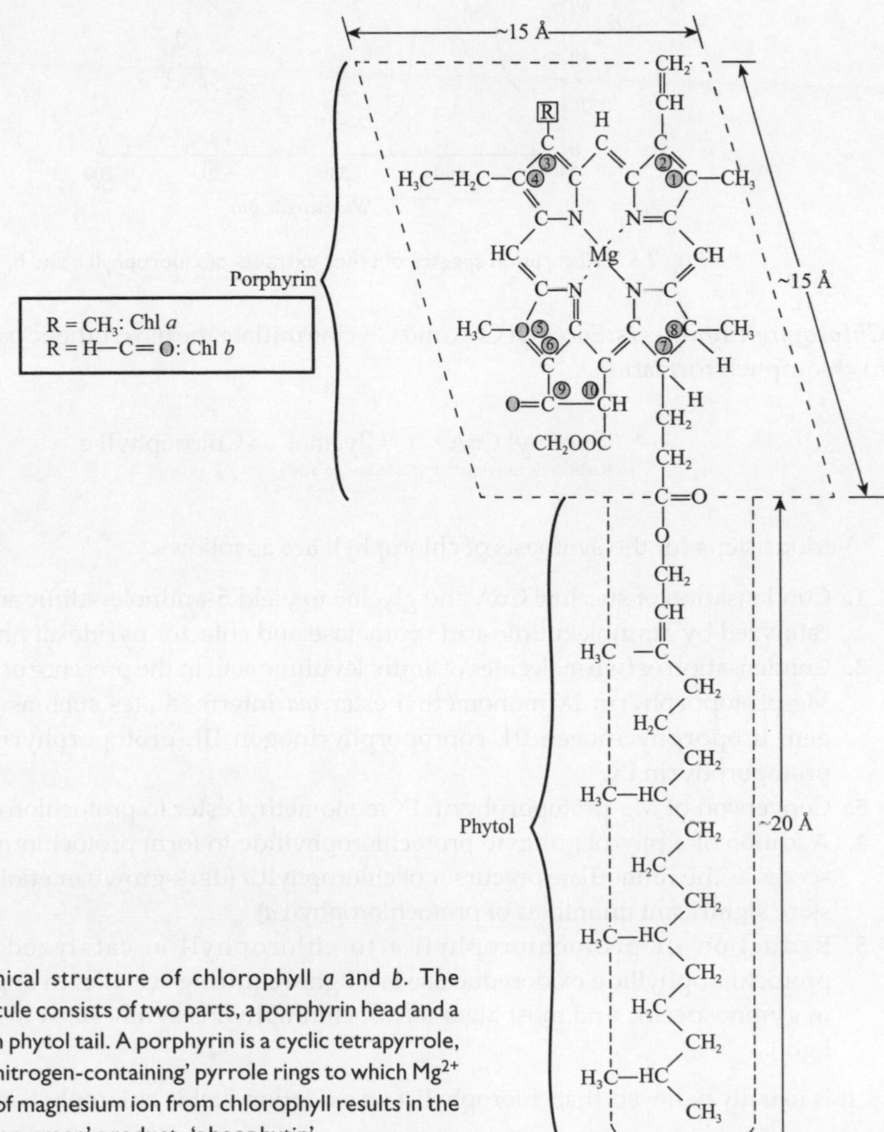

Fig. 7.4 Chemical structure of chlorophyll *a* and *b*. The chlorophyll molecule consists of two parts, a porphyrin head and a long hydrocarbon phytol tail. A porphyrin is a cyclic tetrapyrrole, made up of 'four nitrogen-containing' pyrrole rings to which Mg^{2+} is chelated. Loss of magnesium ion from chlorophyll results in the formation of a 'non-green' product, 'pheophytin'.

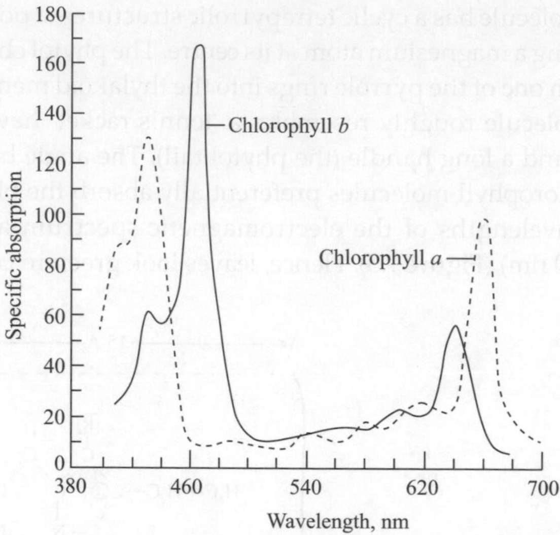

Fig. 7.5 Absorption spectra of ether extracts of chlorophyll *a* and *b*.

Chlorophyll synthesis: Succinyl CoA and Glycine initiate the biosynthetic pathway leading to chlorophyll formation.

$$\underset{\text{(a Krebs cycle intermediate)}}{\text{Succinyl CoA}} + \underset{\text{(an amino acid)}}{\text{Glycine}} \rightarrow \text{Chlorophyll } a$$

Various steps for the synthesis of chlorophyll are as follows:

1. Condensation of succinyl CoA and glycine to yield 5-aminolevulinic acid in the light, catalyzed by aminolevulinic acid synthetase and cofactor pyridoxal phosphate.
2. Condensation of two molecules of aminolevulinic acid in the presence of Mg to produce Mg-protoporphyrin IX monomethyl ester *via* intermediates such as porphobilinogen, uroporphyrinogen III, coproporphyrinogen III, protoporphyrinogen IX and protoporphyrin IX.
3. Conversion of Mg-protoporphyrin IX monomethyl ester to protochlorophyllide.
4. Addition of a phytol group to protochlorophyllide to form protochlorophyll *a*, which serves as the immediate precursor of chlorophyll *a* (dark-grown or etiolated seedlings store significant quantities of protochlorophyll *a*).
5. Reduction of protochlorophyll *a* to chlorophyll *a*, catalyzed by NADPH: protochlorophyllide oxidoreductase is a light-requiring reaction in angiosperms, but in gymnosperms and most algal forms, chlorophyll is synthesized in the absence of light).

It is usually believed that chlorophyll *a* on oxidation yields chlorophyll *b*.

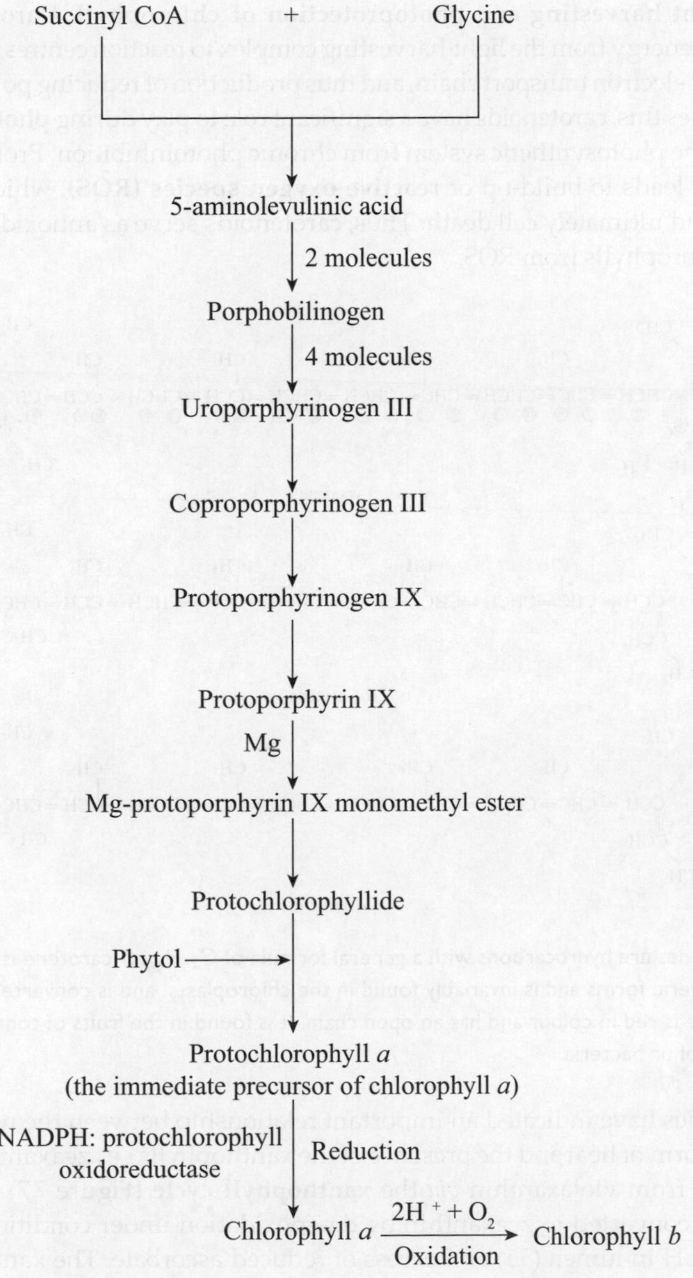

Flowsheet of biosynthetic pathway for the synthesis of chlorophyll *a* and chlorophyll *b*, beginning with succinyl CoA and glycine.

Carotenoids (comprised of carotene and xanthophyll pigments) are hydrocarbons with two aromatic rings at the ends linked by a long aliphatic chain (Figure 7.6). These absorb effectively in the blue region (400–500nm) of the spectrum. Carotenoids perform two major functions: **light harvesting** and **photoprotection of chlorophyll**. Carotenoids transfer absorbed light energy from the light-harvesting complex to reaction centres, which is used in photosynthetic electron transport chain, and thus production of reducing power, i.e., ATP and NADPH. Besides this, carotenoids have a significant role to play during **photoprotection,** i.e., protection of the photosynthetic system from chronic photoinhibition. Prolonged exposure to excess light leads to build-up of **reactive oxygen species (ROS),** which may result in cell damage and ultimately cell death. Thus, carotenoids serve as antioxidants, preventing damage to chlorophylls from ROS.

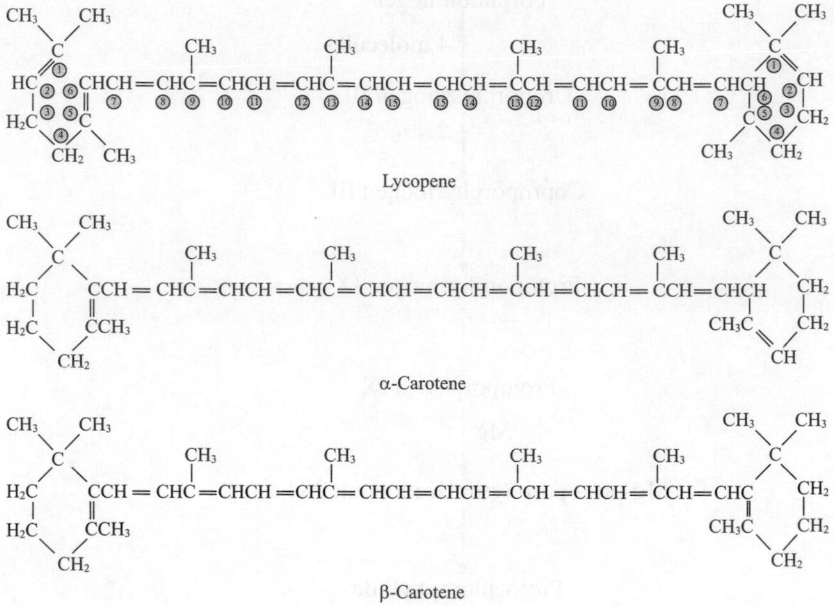

CHCH=CHC=CHCH=CHC=CHCH=CHCH=CCH=CHCH=CCH=CHCH

Lycopene

CCH=CHC=CHCH=CHC=CHCH=CHCH=CCH=CHCH=CCH=CHCH

α-Carotene

CCH=CHC=CHCH=CHC=CHCH=CHCH=CCH=CHCH=CCH=CHC

β-Carotene

Fig. 7.6 Carotenes are hydrocarbons with a general formula of $C_{40}H_{56}$. β-carotene is the most abundant of the three isomeric forms and is invariably found in the chloroplasts, and is converted into vitamin A by animals. Lycopene is red in colour and has an open chain. It is found in the fruits of tomato and chillies and also in purple sulphur bacteria.

Recent studies have indicated an important relationship between the removal of excess energy in the form of heat and the presence of the xanthophylls i.e., zeaxanthin. Zeaxanthin is synthesized from violaxanthin *via* the **xanthophyll cycle** (Figure 7.7). Violaxanthin is enzymatically converted to zeaxanthin by de-epoxidation under conditions of high light intensity, low pH in lumen (5.1) and excess of reduced ascorbate. The xanthophyll cycle is reversible in dim light conditions and nearly neutral pH (7.5).

Phycobilins are red (phycoerythrin) or blue (phycocyanin) pigments, occurring in red- and blue-green algae, respectively.

Photosynthetic bacteria contain two other forms of chlorophyll: bacteriochlorophyll *a* (in purple sulphur bacteria) and chlorobium chlorophyll (in green sulphur bacteria).

Pigments	Absorption peaks
(i) Chl *a*	449 nm, 660 nm
(ii) Chl *b*	453 nm, 642 nm
(iii) Carotenoids	440 nm, 470 nm
(iv) Phycobilins	
(a) Phycoerythrin	490 nm, 546 nm, 576 nm
(b) Phycocyanin	618 nm
(v) Bacteriochlorophyll	358 nm, 772 nm
(vi) Chlorobium chlorophyll	650 nm, 660 nm

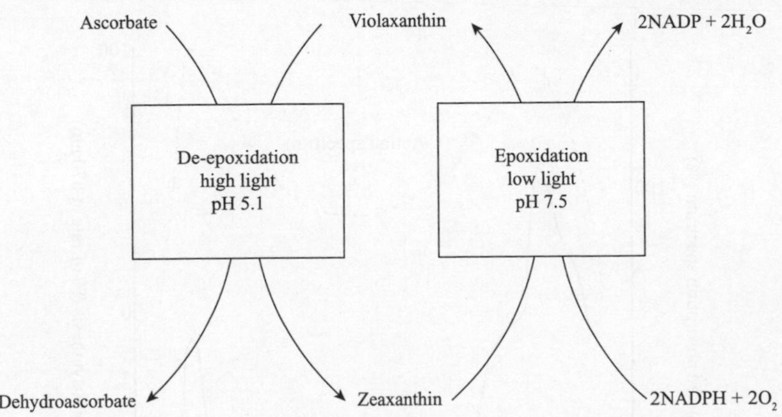

Fig. 7.7 Xanthophyll Cycle: De-epoxidation is favoured under high irradiance while epoxidation is favoured under limiting light. The xanthophyll violaxanthin is converted to Zeaxanthin (through an intermediary compound called antheraxanthin) whenever chloroplasts absorb excess light during peak irradiance. Zeaxanthin acts as a key facilitator of the dissipation of excess energy. Conversely, when light is not intense, Zeaxanthin is disengaged from energy dissipation and converted back to violaxanthin, thereby permitting efficient utilization of light energy in photosynthesis.

7.3 Absorption and Action Spectrum

'Absorption spectrum' is a graph depicting absorption as a function of wavelength, whereas **'action spectrum'** is a graph depicting the effectiveness of light in inducing a particular process (Figure 7.8). The two spectra provide useful information about the pigments involvement in the photochemical process.

Since the action spectrum for the photosynthesis process closely matches the absorption spectrum of the pigments responsible, it is believed that chlorophyll serves as the primary absorber of the light energy and carotenoids and phycobilins act as accessory pigments.

Absorption spectrum	Action spectrum
1. A graph plotted between light absorption and wavelength. Chlorophyll *a* shows maximum absorption peaks at 449 nm and 660 nm, whereas chlorophyll *b* absorbs strongly at 453 nm and 642 nm. Virtually no absorption occurs in the green region (about 500--560 nm) which is mostly transmitted or reflected.	A graph plotted between the wavelength of light absorbed and the photosynthetic yield (measured in terms of CO_2 fixation or O_2 evolved). There is a close parallelism between the amount of light absorbed, and the photosynthetic rate.
2. Absorption spectra of each molecule (pigment) is so unique that it can tell something about its structure or its precise identification. For example, different cytochromes have different absorption spectra. NADH and NADPH have absorption peaks at 340 nm.	Chlorophylls heavily absorb exactly those wavelengths that are most effective in photosynthesis. The light reaction is driven by 'short' wavelength and 'longer' red wavelengths (Emerson Enhancement Effect).

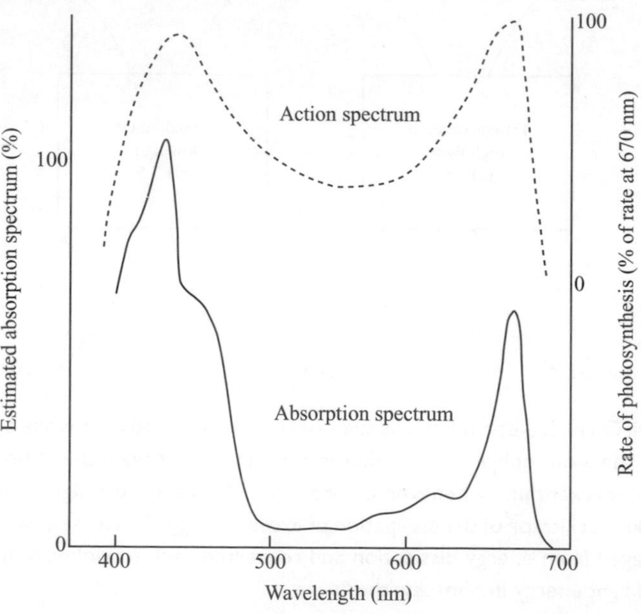

Fig. 7.8 Comparison of the action spectrum for leaf photosynthesis with the absorption spectrum of a leaf extract, containing primarily chlorophyll. The action spectrum shows peaks in the blue and red regions of the visible spectrum that closely follow the prinicipal absorption peaks for the pigment extract at the same wavelengths.

Absorption spectrum of chlorophylls and carotenoids (and other pigments) can be determined by using a spectrophotometer. Absorption peaks vary slightly depending upon the nature of organic solvents in which pigments are extracted.

The Sun – A Giant Thermonuclear Reactor

The sun is a thermonuclear device placed safely 93 million miles away from the earth, in which hydrogen atoms undergo fusion in a complex cycle of reactions carried out under very high temperature and pressure to produce helium (Figure 7.9).

$$4H^+ \rightarrow 2He$$

During this solar reaction, some mass is converted to energy (per Einstein's equation, $E=mc^2$), and this is radiated through space. A small fraction of this energy hits the earth and gets absorbed by pigments in the chloroplast and stored in the form of carbohydrates during the process of photosynthesis.

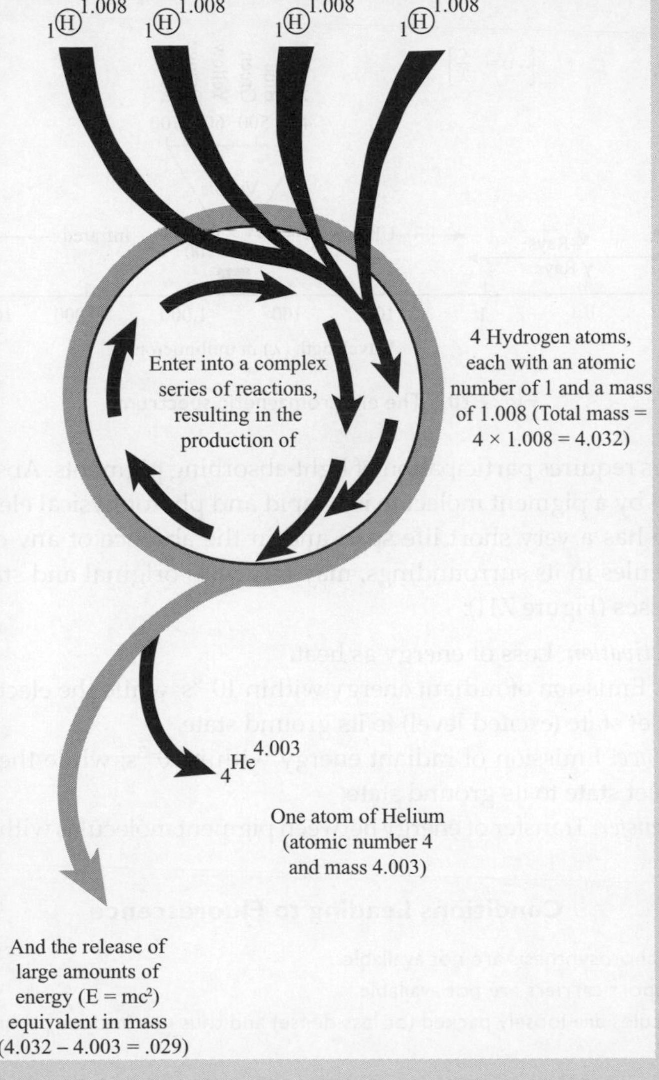

1(H)1.008
1(H)1.008
1(H)1.008
1(H)1.008
Enter into a complex series of reactions, resulting in the production of
4 Hydrogen atoms, each with an atomic number of 1 and a mass of 1.008 (Total mass = 4 × 1.008 = 4.032)
4He4.003
One atom of Helium (atomic number 4 and mass 4.003)
And the release of large amounts of energy (E = mc²) equivalent in mass (4.032 – 4.003 = .029)

Fig. 7.9 The sun is a giant thermonuclear reactor, which works by fusing hydrogen nuclei into helium.

7.4 The Light Reaction (Photochemical Phase)

Early in the twentieth century, Max Planck, the German physicist, postulated the 'quantum theory of light' and established that radiant energy occurs in quanta (singular, quantum), also called as photons. The energy content of these quanta is directly proportional to frequency of the radiation.

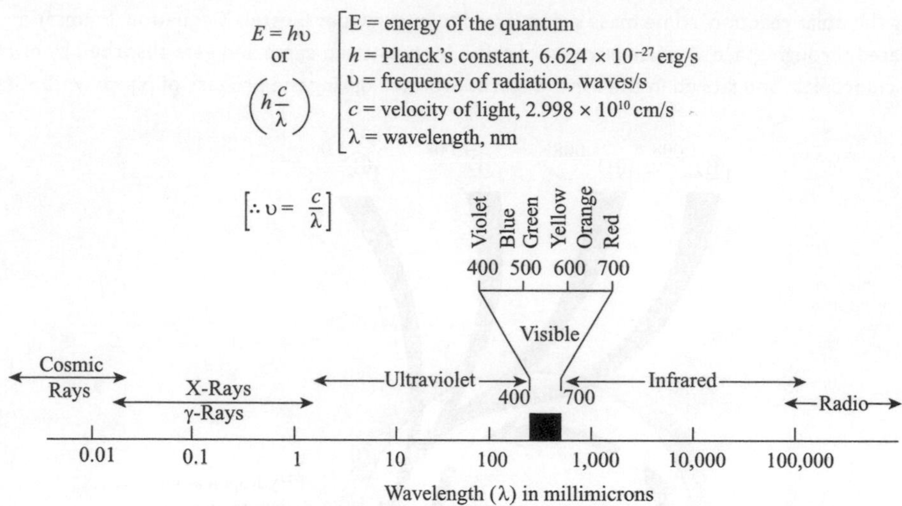

Fig. 7.10 The electromagnetic spectrum.

Photosynthesis requires participation of light-absorbing pigments. Absorption of visible light (Figure 7.10) by a pigment molecule is a rapid and photophysical electronic event. An excited molecule has a very short life span and in the absence of any chemical reaction with other molecules in its surroundings, may return to original and stable ground state by several processes (Figure 7.11):

(i) *Thermal deactivation*: Loss of energy as heat.
(ii) *Fluorescence*: Emission of radiant energy within 10^{-9}s; while the electron returns from its first singlet state (excited level) to its ground state.
(iii) *Phosphorescence*: Emission of radiant energy within 10^{-3}s; while the electron returns from its triplet state to its ground state.
(iv) *Resonance transfer*: Transfer of energy between pigment molecules within a photosystem.

Conditions Leading to Fluorescence

1. Reactants for photosynthesis are not available.
2. Electron transport carriers are not available.
3. Pigment molecules are loosely packed (or less dense) and thus not in close proximity.

The twentieth century saw the application of physical and chemical principles to unravel the mechanism of photosynthesis during this period. Blackman (1905) conducted temperature-coefficient experiments.

$$\text{Temperature coefficient } (Q_{10}) = \frac{\text{rate of reaction at } t+10\,°\text{C}}{\text{rate of reaction at } t\,°\text{C}}$$

For light reactions, $Q_{10} = 1$; and for dark reactions, $Q_{10} = 2$ or 3.

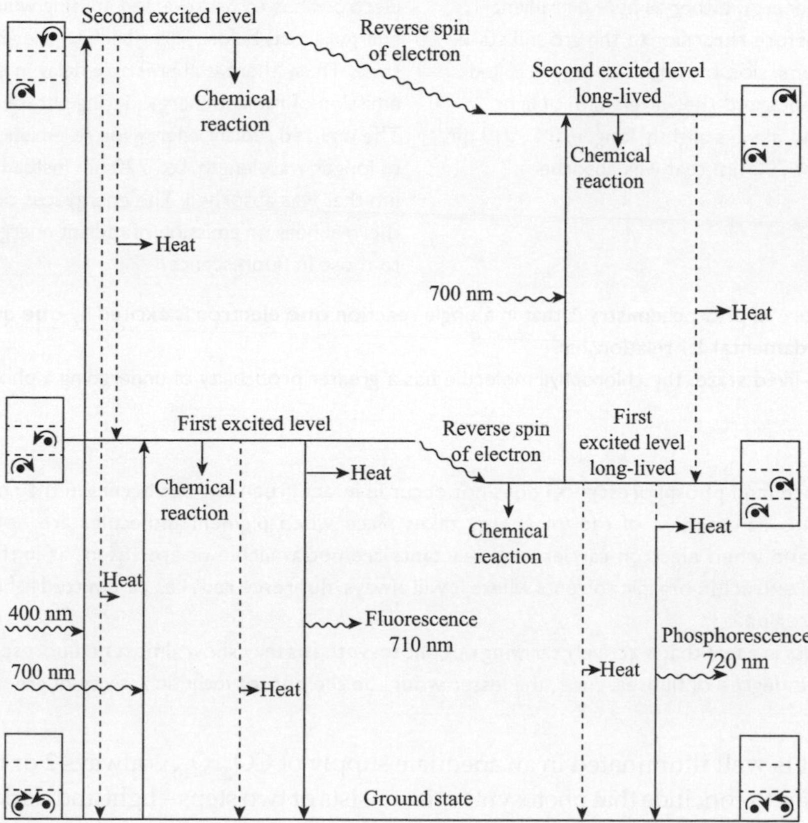

Fig. 7.11 Effects of two energetic photons of wavelengths, 400 and 700 nm as observable in a hypothetical pigment molecule in photosynthesis.

Fluorescence	Phosphorescence
1. Absorption of a photon* of red light of 700 nm (equivalent to 40.9 Kcal) excites the chlorophyll molecule to the 'First excited state' (having a half-life of 10^{-8} to 10^{-9} s) whereas a photon of blue light of 400 nm (equivalent to 71.5 Kcal) will raise one of the electrons of the pair to a still higher energy level or orbit, i.e., the 'Second excited state' (with a life of 10^{-11} to 10^{-12} s). The electrons are 'still spinning in opposite directions' (one right handed and the other left handed).	Absorption of a photon of light by chlorophyll molecules, either red or blue wavelengths will send one of the electrons of a pair to the 'First long-lived level'** or 'Second long-lived level' (or triplet state), respectively, with a half-life of 10^{-2} to 10^{-3} s. The spin of one of the electrons of the pair might be reversed and the two electrons would have a 'parallel' spin in the same direction (both electrons are moving right handed).

Contd.

Contd.

Fluorescence	Phosphorescence
2. In its short-lived excited state, the chlorophyll molecule is very unstable and 'very quickly' loses its energy, either as heat or radiant energy before returning to the ground state. This re-emission of radiant energy is called fluorescence and the wavelength of light emitted is 'always slightly longer' (i.e., 710 nm instead of 700 nm that was absorbed).	The electrons in the long-lived state 'do not emit energy so quickly' because the spin on one of the electrons has to be 'reversed' (during which some energy is lost) before falling back to the ground state. Thus, 'there is appreciable delay in the re-emission of radiant energy' during phosphorescence. The emitted radiant energy (or re-emission) is still of longer wavelength, i.e., 720 nm instead of 700 nm that was absorbed. The energy loss occurs by thermal heat or emission of radiant energy similar to those in fluorescence.

* A basic feature of photochemistry is that in a single reaction **one** electron is excited by **one** quantum, i.e., there is a fundamental 1:1 relationship.

** In the long-lived state, the chlorophyll molecule has a greater probability of undergoing a photobiological reaction.

Fluorescence (and phosphorescence) does not occur in intact green leaf but occurs in the chlorophyll extract. This re-emission of radiant energy takes place when pigment molecules are not in close proximity and when electron carriers and reactants are not available or are absent as in the case of chlorophyll extract in organic solvents where it will always 'fluoresce red', i.e., re-emitted light is always in the red region.

Pigments in a leaf that is actively carrying out photosynthesis may show almost no fluorescence. The greater 'the degree of fluorescence', the lesser would be the photosynthetic activity of green tissue.

If a plant is well illuminated in an adequate supply of CO_2, Q_{10} is always 2 or more. This led Blackman to conclude that photosynthesis consists of two steps – light and dark reactions:

$$\xrightarrow[\text{Photochemical (Rapid)}]{\text{Step I (Light)}} \xrightarrow[\text{Biochemical (Slow)}]{\text{Step II (Dark)}} \text{Photosynthesis}$$

Warburg (1919) through his experiments on 'intermittent light' discovered that if light was provided to a plant in short intense flashes, interrupted by dark periods, the photosynthetic yield per second of illumination was greater as compared to that in continuous light.

While working on photosynthetic bacteria, van Niel (1930s) established that hydrogen sulphide (H_2S) was used as the hydrogen donor during assimilation of CO_2 but not accompanied with release of oxygen.

$$2H_2S + CO_2 \xrightarrow{\text{Light}} 2S + (CH_2O) + H_2O$$

In green plants, water (H_2O) was found to be the donor of hydrogen accompanied with release of oxygen.

$$2H_2O + CO_2 \xrightarrow{\text{Light}} O_2 + (CH_2O) + H_2O$$

Bacteria Use a Photosynthetic Reaction Centre and Electron Transport System Similar to That in Plants

Much of our knowledge of capturing light energy and photosynthetic reaction centre complexes originally came from studies on photosynthetic bacteria. In the early 1980s, Hartmut Michel, Johann Deisenhofer and Robert Huber were able to crystallize a reaction centre complex from a purple bacterium, *Blastochloris virdis*, and determined its molecular structure by X-ray crystallography. For their exciting and ground-breaking contributions, all three shared the Nobel Prize in 1988.

Finding a similarity between the above equations, van Niel proposed a general equation for photosynthesis,[1]

$$2H_2A + CO_2 \xrightarrow{\text{Light}} 2A + (CH_2O) + H_2O$$

where A, stands for elemental sulphur in photosynthetic bacteria or oxygen in green plants.

Hill (1937) found that isolated chloroplasts, when illuminated, produced oxygen from water. Hydrogen acceptors (oxidants) present in the medium were reduced by the chloroplasts. The oxidants were called 'Hill reagents' (ferricyanide, benzoquinone, dichlorophenolindophenol, etc.) and the reaction as the '**Hill reaction**'.

$$2H_2O + 2O{=}C_6H_4{=}O \xrightarrow[\text{Isolated chloroplasts}]{\text{Light}} 2HO{-}C_6H_4{-}OH + O_2$$

Benzoquinone → Hydroquinone

Using the heavy isotopic form of oxygen (^{18}O) in water supplied to the plant, Ruben and Kamen (1941) found that during photosynthesis, the released oxygen came from the water molecule and not from carbon dioxide. Experiments of Hill and Ruben and Kamen led to the idea that the 'role of light in photosynthesis is essentially to split water and generate electrons to reduce CO_2.'

In an experiment with isolated chloroplasts, Arnon (1951) discovered that hydrogen released by photolysis of water molecule, was accepted by NADP which got reduced to NADPH. Vishniac and Ochoa in early 1950s verified Arnon's observations. In 1954, Arnon further noticed that chloroplasts could also synthesize ATP in the light from ADP and P_i (inorganic phosphate).

$$ADP + P_i \xrightarrow[\text{Photosynthetic phosphorylation}]{\text{Light}} ATP$$

7.4.1 The Emerson enhancement effect

The foundation of the current Z-Scheme known to exist in all higher plants, came from the studies of Emerson and Arnold, who extended Warburg's studies and found that the quantum requirement for evolution of a single oxygen molecule was 8, but not 4 as reported

[1] Photosynthesis in which the product A is other than oxygen, such as sulphur (photosynthetic prokaryotes) is termed as **anoxygenic** while in green plants where the product A is oxygen is referred to as **oxygenic**.

by Warburg. The requirement of 8 quanta for the removal of 4 electrons (from oxidation of two molecules of water) suggested that two photosystems were involved for the passage of each electron from water to $NADP^+$.

Blinks (1950) noticed that when photosynthetic cells were shifted from longer wavelengths (>680nm) to shorter wavelengths (<680nm), the result was a brief increase in the rate of photosynthesis (**Blinks effect**).

Later, Emerson (1957) observed a significant reduction in quantum yield[2] at wavelengths more than 680 nm, an area of spectrum occupied by the red absorption band of chlorophyll, referred to as '**Red drop**', although Chl *a* showed maximum absorption in that region (Figure 7.12). When he continued his experiments to decode the mystery of the red drop, he noticed that the efficiency of photosynthesis could be restored by a simultaneous application of wavelengths less than 680 nm. 'The effect of two superimposed beams of light on the rate of photosynthesis exceeds the sum effect of both the beams used individually'. This photosynthetic enhancement is termed as the '**Emerson enhancement effect**' (Figure 7.13).

By early 1960s, it became apparent that photochemical reactions require two distinct pigment systems called photosystem I and photosystem II (PS I and PS II), each possessing a specific group of pigment molecules (Rabinowitch and Govindjee, 1964).

PS I (>680 nm) – Chl *a* 683, carotenoids, P_{700}

↓

(associated with the production of reductant, NADPH)

PS II (<680 nm) - Chl *a* 673, Chl *b* 670, phycobilins in red and blue-green algae, carotenoids

↓

(associated with the production of oxygen and ATP)

Photosystem = Antennae + Reaction Centre (Figure 7.14)

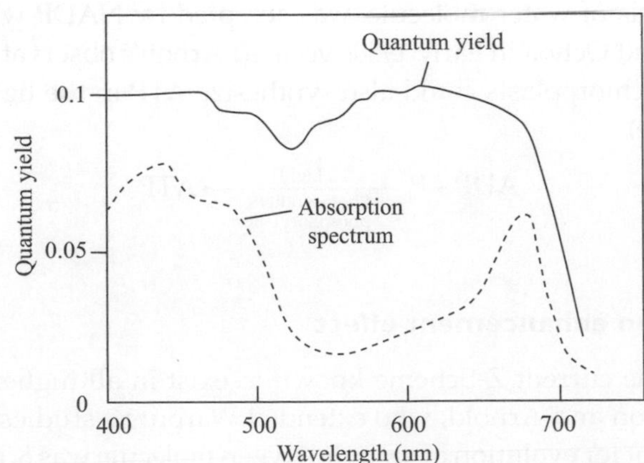

Fig. 7.12 Graphic representation of the 'red drop' phenomenon in the green alga *Chlorella*.

[2] Quantum yield is defined as the number of oxygen molecules evolved per light quantum absorbed.

Robert Emerson
(3 November 1903 – 4 February 1959)

Robert (Bob) Emerson was born on 3rd November 1903 in New York City; a scion of a distinguished family. In physical appearance, Bob was a typical New Englander: being tall, lean, long-headed and long-striding. His wife Claire Garrison also came from a distinguished Boston family and eventually, they were to be blessed with three sons, and a daughter. Robert Emerson's father, Dr Haven Emerson, was Head of the New York City Public Health Services, a position he occupied for many years. He was dedicated, strong-willed and hard-working, and stern person. His children never came to to terms with many aspects of his way of functioning and would end up rebelling against his ideology, but he left on them his imprint and influence. They would deeply respect his work ethics. In his eighties, Bob's father was involved not only in his professional work, but spent every moment of his leisure time, working in the garden of his estate on Long Island.

So, just like his father, Bob Emerson too, was a dedicated, hard-working man with strong convictions. After working for late hours, doing manometeric studies in the laboratory, he would always walk back home on foot, and would engage in gardening, until darkness forced him indoors. Emerson's backyard was the most systematically cultivated plot of land in Champaign County, styled in the Japanese fashion of careful and maximal utilization of every bit of ground. Passionate about perfection in human endeavour, including his own work, (be it an experimental work in photosynthesis, writing a research paper, wood carving or gardening and a range of activities). Emerson was a skilled glassblower and a specialist in artistically designed wood carpentry. Impeccable integrity and precision work were the tenets of his scientific career.

One of the missions of his life was to teach the virtues or values of integrity, hard work and thrift to his children, as well as to his students. In his belief, wars and violence were unjustifiable (he was a pacifist to the core) and believed in democratic socialism. He taught his children not to fight back when provoked or attacked in a school or in the neighbourhood. He felt strongly against economic disparities and racial injustice, and was always on the side of downtrodden. The only sport that Emerson really indulged in, was ice skating. He was active in the affairs of the University of Illinois skating rink and the figure skating club that was attached to it. He and his wife, who was an even more accomplished figure skater than Bob, were lauded as a lovable couple on ice.

Thrift was one of the virtues ingrained in him and he disliked any wasteful spending. He would purchase expensive items only in moments of great urgency. He scorned new cars, fitted with lots of latest features and scores of technical gadgets, preferring instead, old, simple, but strong vehicles built to withstand far longer distances on hard-surfaced roads. Emerson distrusted aeroplanes and would counsel others against flying. The Emerson family lived mainly on the produce of their backyard, despising store-bought items such as fruits, vegetables and chicken and such. He was teetotaller (except for an occasional glass of beer) and a non-smoker. The couple were early risers and early to go to bed.

He began studying animal physiology at Harvard with the aim of following in his father's footsteps and to follow a professional career in medicine. He obtained his Master's degree in Zoology from Harvard in 1925. However, under the influence of W.J.V. Osterhout's lectures in plant physiology, his interest turned from animals to plants. He went to Munich, Germany with the intention of researching on chlorophyll formation in plants under the Nobel Laureate, Richard Willstatter, famous for his work on chlorophyll synthesis. There, he soon found himself in conflict with the University administration because of the prevailing anti-semitic culture of both students and faculty members. He was advised to go to the Kaiser Wilhelm Institute of Biochemistry in Berlin-Dahlem, where another Nobel Prize recipient, Dr Otto Warburg, was engaged in research on quantitative studies in photosynthesis. After spending two years in Warburg's laboratory, Emerson received his doctorate in Botany from the University of Berlin in 1927. Thereafter, Emerson returned to Harvard in 1927 as a National Council Research Fellow, working during summer in the Harvard Tropical Laboratory in Cuba.

The fact was that botany was a subject which he had not studied as a major in his Under-Graduate courses. For Emerson, it was a great embarrassment that he possessed a PhD degree in Botany - a subject about which he knew almost nothing. This he candidly admitted shortly after he was honoured with the Stephen Hales Prize, instituted by American Society of Plant Physiologists in 1949, and shortly before he was elected to the US National Academy of Sciences membership in 1953.

Thomas Hunt Morgan invited him to join the Biology Division at the California Institute of Technology (CalTech); where he worked from 1930 to 1937, and again for a year in 1941 and 1945. From 1942 to 1945, he worked on producing rubber from a Mexican desert shrub called guayule (*Parthenium argentatum*) for the American Rubber Company. In 1947, he moved to the Botany Department of the University of Illinois at Urbana-Champaign (UIUC) where he worked for his remaining part of his life. The School of Integrative Biology at the University of Illinois has instituted Robert Emerson Memorial Award, given annually to a graduate student in the School of Integrative Biology for exceptional academic brilliance and research potential.

In the course of his photosynthetic research, Bob Emerson made multiple revolutionary discoveries as under:

1. Discovery of the Photosynthetic Unit in 1932 that included quantification of the ratio that it takes 2400 chlorophyll molecules to reduce just one molecule of oxygen (Emerson and William Arnold; the latter was then an Under-Graduate Student).
2. Demonstrated in 1939 that between 8–12 quanta of light are required to produce one molecule of oxygen (Emerson with physicist collaborator, Charleton Lewis) while working at the Carnegie Plant Biology Department at Stanford, USA.
3. After returning to CalTech in 1940, he discovered the roles of many accessory pigments during his action spectrum studies in photosynthesis.
4. The Red drop in Quantum Yield of Photosynthesis was discovered at Carnegie Institute, Stanford 1943).
5. The Emerson enhancement effect: In this most notable work, he laid the foundation of the most acceptable mechanism of two light reactions running in a series or, two distinct pigment systems or reaction centres. This was discovered jointly during 1957–59 by Robert Emerson and William Arnold in Natural History Building (NHB) at UIUC.

Under the aegis of the newly established 'Photosynthesis Research Project' at UIUC, Emerson invited his mentor Otto Warburg in the summer of 1948 so that they could work together to resolve the raging controversy between them about quantum requirements.

The idea of getting together was that the two researchers could examine and if required tweak their experimental protocols and thereby work out why there were disconcerting differences in their conclusions.

Both were expert in manometry. Also, both used the same technique and the same plant material. Emerson had been trained by Warburg. Yet they obtained results that differed by a factor of two or three. During their deliberations, however, Otto Warburg took a rigid stand on the Quantum Requirements, unlike the American Plant Physiologists.

However, this attempt at collaboration between guru and acolyte came to naught. Warburg's stay in Urbana proved unfruitful, and nothing was settled. The visit ended with bitterness on both the sides.

Warburg increasingly portrayed himself as the victim of conspiracy on the part of the Americans, whom he took to calling the 'Midwest-Gang'. This epithet would include, Robert Emerson, Eugene Rabinowitch (at Illinois) and James Frank and Hans Gaffron (of the University of Chicago, Illinois) since the last three were supporters of Emerson. This became a classical example of drawn-out international feuds between scientists in Germany vs. those in the US.

An interesting cartoon that beautifully depicts the controversy on the minimum quantum requirement of photosynthesis between the two giants Otto Warburg and Robert Emerson (his former student) appeared in the 1948 newsletter of the then Botany Department of UIUC which was also cited by Karin Nickelsen and Govindjee in their book 'The Maximum Quantum Yield Controversy: Otto Warburg and the Middlewest-Gang'.

On the left side, Emerson is shown wearing his Lab apron hinting that a minimum of 12 quanta is needed for the evolution of 1 oxygen molecule, whereas Warburg on the right in well dressed attire, is pointing out the number 4, instead. In the centre of the two opponents is Eugene Rabinowitch, an authority on photosynthesis, but also an affliaite of the Middlewest-Gang (as perceived by Warburg), holding a briefcase containing all the published papers, attempting to bring about a reconciliation or compromise between the two opposing sides. The dispute was finally resolved after the death of both men and it is now accepted that that Emerson was correct. It is now mostly accepted that 8–12 quanta of light are required for producing one molecule of oxygen.

At the height of his promising research career, Emerson met a sudden death on 4th February 1959 when the American Airline Flight 320 carrying him touched down La Guardia airport runway, missed it and then plunged straight off into the East River, New York. Originally booked on another flight, he was unluckily transferred into boarding the ill-fated flight.

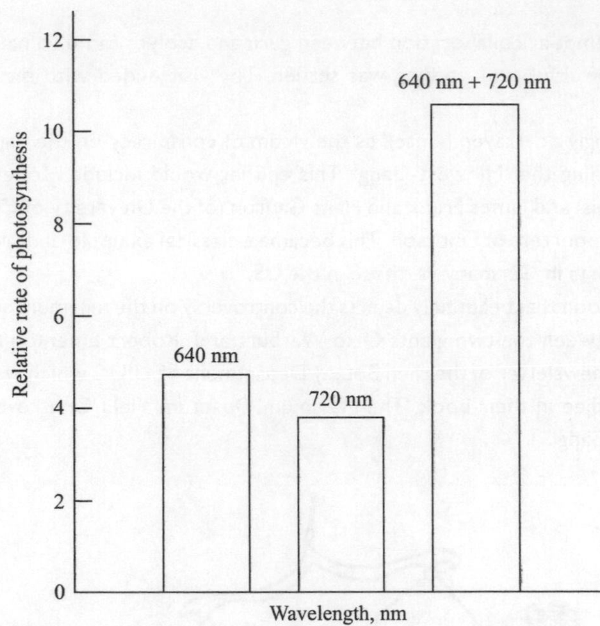

Fig. 7.13 An illustration of the 'Emerson enhancement effect'.

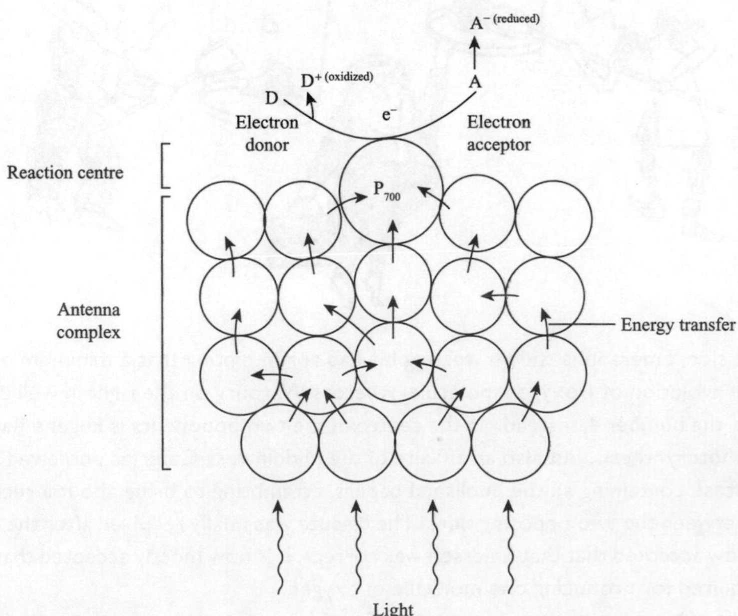

Fig. 7.14 A photosystem containing antennae molecules and a reaction centre (P_{680} and P_{700} for Photosystem II and Photosystem I, respectively).

Antennae molecules (~300 molecules of chlorophyll and carotenoid) capture incoming radiation and transfer the excitation energy to the reaction centre where oxidation–reduction or redox reactions take place. The reaction centre of chlorophylls (P_{700} and P_{680} in PSI and PSII, respectively) or '**daddy molecules**' are effective energy sinks; being the longest wavelengths (thus the least energy-absorbing) chlorophyll in the photosystem. The main advantage of combining a single reaction centre with multiple antennae molecules is to enhance efficiency in gathering light energy and its utilization. Plants growing under low light intensities have 'augmented' antennae complexes, light-harvesting complexes (LHC I and LHC II), in association with their respective photosystems to facilitate harvesting of available photons and keep the reaction centre operating at near optimal rates (Figure 7.15).

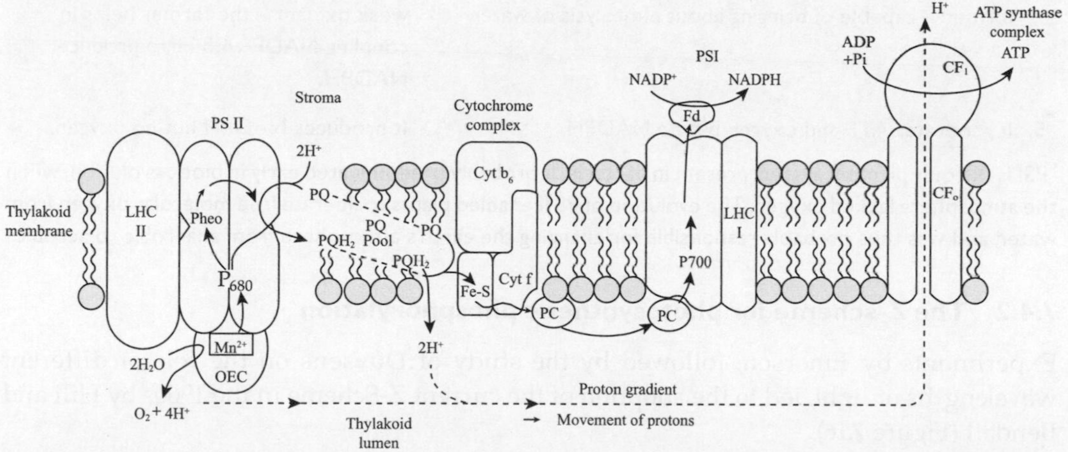

Fig. 7.15 The photosynthetic electron transport chain representing the arrangement of photosystem I as well as photosystem II along with LHC I and LHC II, cytochrome complex and ATP synthase complex in the thylakoid membrane. The plastoquinone, cytochrome b_6f complex, and plastocyanin have uniform distribution throughout the thylakoid membrane, while the major multiprotein complexes of the photosynthetic apparatus display lateral heterogeneity, that is they are not uniformly distributed throughout the membrane. These protein complexes, when damaged by high irradiance, move in a lateral manner from granal thylakoids to the non-appressed, stromal thylakoids for disassembly and reassembly.

Pigment System II	Pigment System I*
1. It is a larger, heavier light-harvesting complex (LHC) consisting of ~250 chlorophylls, ~50 carotenoids, a molecule of light trapping reaction centre of P 680 or Chl *a* 680, a tyrosine residue (a primary electron donor), a primary electron acceptor Q and associated electron carriers – a plastoquinone, two cyt b_{559}, one cyt b_6, plastocyanin, four Mn molecules and chloride ions. The core of the reaction centre is associated with proteins D1 and D2; proteins that regulate the movement of electrons ejected from P_{680} to plastoquinone.	It is a smaller, lighter LHC complex consisting of light trapping reaction centre P_{700} (that absorbs maximally at 700 nm), one cytf, one plastocyanin, two cyt b_{563}, FRS (ferredoxin reducing substance), ferredoxin and NADP. A lattice of 2 major proteins (Psa A and Psa B), help in holding the pigments close to each other. It is richer in iron and copper.

Contd.

Contd.

Pigment System II	Pigment System I*
2. It occurs in the stacked or appressed region of grana thylakoids.	It occurs in the stroma thylakoids and non-appressed or non-stacked region of thylakoid.
3. The ratio of Chl *a* over Chl *b* is low, i.e., rich amount of Chl *b*.	It has high ratio of Chl *a* over Chl *b*, i.e., richer in Chl *a*.
4. It generates a strong oxidant and weak reductant – the former is capable of bringing about photolysis of water.	It produces a strong reductant and weak oxidant – the former helps in coupling NADP with H^+ to produce NADPH.
5. It generates ATP and oxygen but no NADPH.	It produces NADPH but no oxygen.

*PSI is the only pigment system present in bacteria. It probably predominated early in biotic evolution, when the atmosphere lacked oxygen. The evolution of PS II enabled plants to liberate free molecular oxygen from water, and was thus probably responsible for changing the earth's atmosphere from anaerobic to aerobic.

7.4.2 The Z-scheme for photosynthetic phosphorylation

Experiments by Emerson, followed by the study of Duysens on the role of different wavelengths of light, led to the adoption of the current Z-Scheme in the 1960s by Hill and Bendall (Figure 7.16).

Photosynthetic phosphorylation is of two kinds – **cyclic** (involving PS I only) and **non-cyclic** (involving both PS II and PS I)

Cyclic photophosphorylation: The chlorophyll molecules of photosystem I (PS I) get excited upon absorption of photons. The extra energy absorbed by a chlorophyll molecule is transferred to one of its electrons which is raised to a higher energy level. As a result of the loss of the electron, the chlorophyll molecule assumes a positive charge.

$$\text{Chl} \xrightarrow{h\nu} \text{Chl}^*$$

$$\text{Chl}^* \longrightarrow (\text{Chl})^+ + e^-$$

The excited electron is then accepted by an electron acceptor, passes through a chain of cytochromes and finally back to the chlorophyll molecule. The union of the electron to the chlorophyll molecule neutralizes its positive charge and the molecule then returns to the ground state, ready to absorb another photon. Excess energy lost by electron is stored in the form of ATP.

In cyclic photophosphorylation[3], the electron travels in a cyclical manner and ultimately returns to the same chlorophyll molecule from which it came, i.e., the electron donor and the electron acceptor is the same chlorophyll molecule.

3 Cyclic photophosphorylation occurs when the amount of available $NADP^+$ is low, or PS II is absent, or monochromatic light 'beyond' 680 nm is given to the plants in the laboratory.

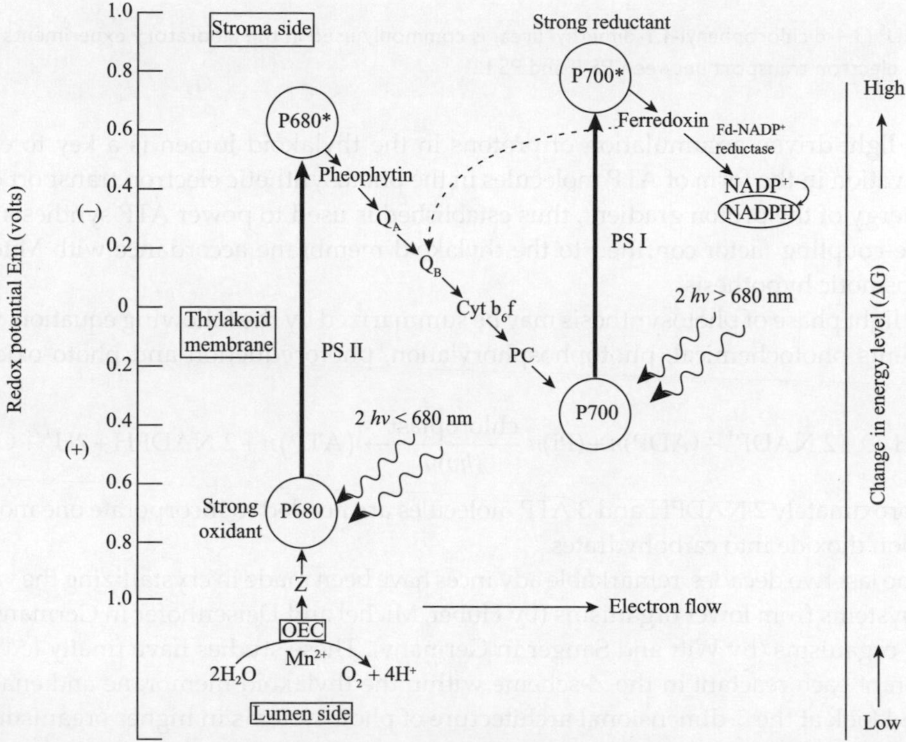

Fig. 7.16 The Z-scheme for photosynthetic phosphorylation. The Z-scheme (named because of its shape) depicts the flow of electrons through redox carriers in photosystem II and photosystem I from H_2O to $NADP^+$. Photons excite the reaction centre chlorophylls; P680 for PSII andP 700 for PSI, with the ejection of an electron. The excited PS II reaction centre chlorophyll, P680[4] transfers an electron to pheophytin. P680 oxidized by light is re-reduced by Z, which has received electrons from oxidation of water. Pheophytin transfers electrons to the plastoquinones (Q_A and Q_B) and the cytochrome b_6f complex, which in turn, transfers electrons to plastocyanin (PC), a copper-containing protein, and then reduces $P700^+$ (oxidized P700). A series of membrane bound, iron-sulphur proteins are involved in transferring electrons to soluble ferredoxin(Fd). The soluble flavoprotein, ferredoxin-$NADP^+$reductase (FNR), then reduces $NADP^+$ to NADPH, which ultimately causes CO_2 reduction by the Calvin-Benson cycle.

The dashed line indicates cyclic electron flow around PSI.

***Non-cyclic photophosphorylation*:** Electron transport starts with incoming excitation energy at photosystem II (PS II) reaction centre, i.e., P_{680}. An electron 'hole' created in PS II by photons is filled up by continuous supply of electrons from the photolysis of water. Excited electrons travel through carriers like plastoquinone, cytochromes, plastocyanin, ferredoxin, and are finally withdrawn by NADP to produce NADPH. There is a redox potential decrease of about 0.43 V from PS II to PS I, sufficient to produce ATP.[4]

In non-cyclic photophosphorylation, the emitted electrons from PS II, after passing through electron transport chain never come back to the same pigment molecule, i.e., the electron donor and the electron acceptor are two different molecules.

[4] Production of ATP and NADPH is collectively referred to as 'assimilatory power' or 'reducing power'.

DCMU[5] (3,4-dichlorophenyl-1,1-dimethyl urea) is commonly used in the laboratory experiments to block electron transport between PS II and PS I.

The light-driven accumulation of protons in the thylakoid lumen is a key to energy conservation in the form of ATP molecules in the photosynthetic electron transport chain. The energy of the proton gradient, thus established is used to power ATP synthesis using ATPase-coupling factor confined to the thylakoid membrane accordance with Mitchell's chemiosmotic hypothesis.

The light phase of photosynthesis may be summarized by the following equation, which represents photochemical, photophosphorylation, photoreduction and photo-oxidation events.

$$2H_2O + 2\,NADP^+ + (ADP)n + (Pi)n \xrightarrow[(hv)n]{\text{chloroplast}} (ATP)n + 2\,NADPH + 2H^+ + O_2$$

Approximately 2 NADPH and 3 ATP molecules are needed to incorporate one molecule of carbon dioxide into carbohydrates.

In the last two decades, remarkable advances have been made in crystallizing the various photosystems from lower organisms (by Huber, Michel and Deisenhofer in Germany) and higher organisms (by Witt and Sanger in Germany). These studies have finally led to the location of each reactant in the Z-scheme within the thylakoid membrane and enabled a detailed look at the 3-dimensional architecture of photosystems in higher organisms.

Cyclic photophosphorylation	Non-cyclic photophosphorylation
1. Only photosystem I is functional.	Both photosystem II and photosystem I operate simultaneously in a serial manner, i.e., one after the other.
2. The electrons ejected from PS I migrate in a cyclical manner – the electron donor and acceptor is the same chlorophyll molecule, i.e., P_{700} (reaction centre of PS I).	The ejected electrons from water (donor), after passing through a chain of electron transport carriers of PS I and PS II are finally accepted by NADP to produce NADPH (a biological reductant for dark reaction).
3. Oxygen is not evolved.	Photolysis of water occurs and oxygen is evolved as by-product.
4. One molecule of ATP is evolved during electron transfer from Fd, Cyt b_6, Cyt_f, PC.	One molecule of ATP is given out during electron transfer from Cyt·b/f to PC.
5. It occurs predominantly in photosynthetic bacteria.	This system is found dominantly in green plants.
6. The mechanism does not show any inhibition by DCMU.	DCMU acts as an inhibitor of PS II.
7. Ferredoxin reductase is bypassed, therefore NADPH is not produced and assimilation of CO_2 is slowed down.	NADPH is produced.

[5] DCMU is generally used as an effective herbicide in the fields.

7.4.3 Chemiosmotic mechanism of phosphorylation

Peter Mitchell (1966) proposed the chemiosmotic theory to explain the synthesis of ATP during photosynthetic phosphorylation (in chloroplasts) and oxidative phosphorylation (in mitochondria). The electrochemical gradient (or proton motive force)[6] established across the membrane during electron transport is a potential source of energy that binds ADP with inorganic phosphate, producing ATP.

In chloroplast grana, each thylakoid is a closed compartment, and during the light-driven reaction the electrons from PS II flow through a sequence of electron transport carriers such as Q (the primary electron acceptor), PQ, $b_{6/f}$ complex to plastocyanin, and then to PS I from where the electrons move uphill to be captured by X and finally to NADP, producing NADPH in the stroma region.

Plastoquinone is one of the electron carriers that takes up H^+ions from the exterior of the thylakoid membrane and releases these into the interior of the thylakoid lumen or sac or vesicle. For every two electrons moving across the transport system, two protons are transported inside by reduced PQ through the **'Q cycle'**. The release of protons from the photolysis of water ($2H^+ + ½ O_2$) further leads to the build-up of protons inside.

Thus, the outside (the stroma) becomes alkaline (this side becoming negatively charged – called the **N** side) while the interior of the thylakoid becomes acidic (this side which is called the P side - becoming positively charged) with respect to the exterior. The thylakoid membrane, like the inner mitochondrial membrane, is impervious to H^+ except in regions where coupling factors (F_1) or ATP synthase (ATPases) are embedded on the 'non-appressed' side of grana membrane. So, the protons move back out almost exclusively through the channels created by ATP synthase. The channels protrude out like knobs on the outer 'non- appressed' surface of the thylakoid membrane. As protons diffuse back out of the thylakoid through coupling factors, ADP is converted to ATP and is released into the stromal region inside the chloroplasts where the enzymes for CO_2 fixation are localized (Figure 7.17).

The 'proton' flow through the channels continues as long as their concentration inside the thylakoid exceeds that of the outside. Theoretically, one ATP molecule is synthesized for every $3H^+$ passing through the CF_1, unlike two protons flowing through mitochondria.

Mitchell's hypothesis was firmly supported by experiments carried out by Jagendorf and Uribe, 1966 (as given in chapter 21). In honour of his pioneering work, Peter Mitchell, an English biochemist, was awarded the Nobel Prize for chemistry in 1978.

---

[6] The proton motive force is the sum of proton chemical potential (or concentration gradient because of accumulation of H^+ions on one side of the membrane) and transmembrane electrical potential created by positively charged H^+ions (voltage difference across the membrane).

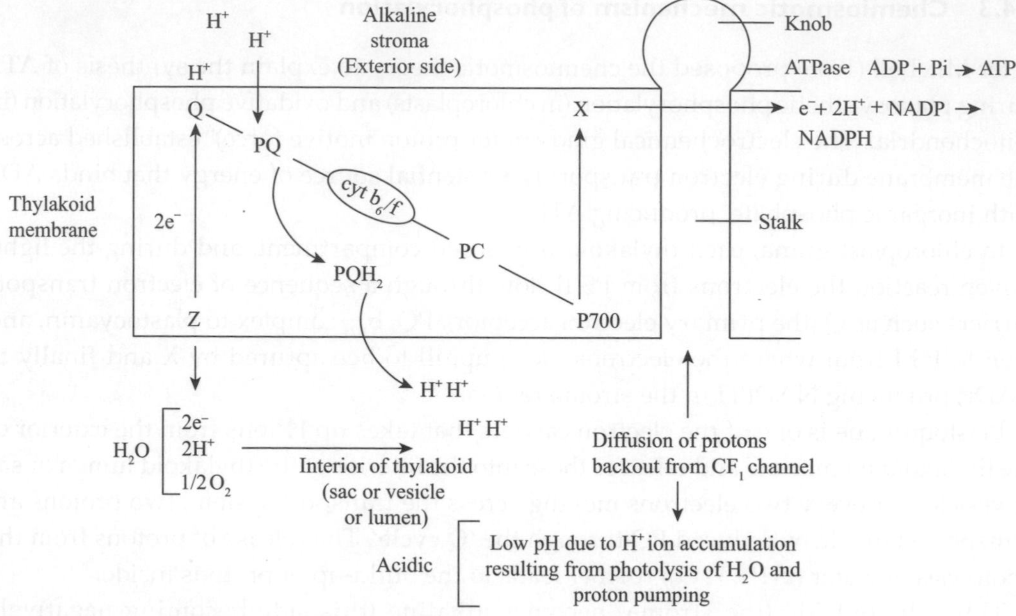

Fig. 7.17 A simplified scheme of chemiosmotic phosphorylation: Electron flow starts with water and ends up with NADP. Plastoquinone (a mobile factor) moves across the lipid phase of the thylakoid membrane. It gets alternately reduced and oxidized, ferrying protons to the thylakoid lumen to establish proton motive force (pmf), the same role which is played by ubiquinone in mitochondria. However, ATP synthesis is largely confined to the stroma – with its exposed membranes.

The Q Cycle in Photosynthesis

Pheophytin (a form of chlorophyll *a* in which the Mg^{2+} ion has been replaced by two hydrogen ions) is considered to be the first electron acceptor in PSII. Within a picosecond, the electron acquired by pheophytin instantaneously shifts to a plastoquinone molecule (to Q_A and then to Q_B). When PQ is reduced by PSII, it binds transiently to the DI protein (Q_B) as a semiquinone after it accepts first electron from Q_A.Following that, the Q_B semiquinone is changed to the fully reduced plastoquinol (PQH_2) after it has acquired another electron from Q_A, along with two protons picked up from the surrounding stroma. PQH_2 gets detached from the PSII complex and diffuses in a lateral manner through the membrane until it reaches the lumenal PQH_2 binding site of the cytochrome b_6/f complex where the translocation of H^+ across the thylakoid membrane occurs. There, two PQH_2 are reoxidized to PQ by the concerted action of the Rieske FeS–protein and the low reduction potential form of cytochrome b_6 (LP). Simultaneously, in this process, four H^+ are translocated (for each pair of electrons transported to PSI) across the thylakoid membrane from the stroma into lumen of the thylakoid. This aids in generating a proton gradient across the membrane that is utilised in ATPsynthesis. One of the PQ molecules goes back to the thylakoid PQ pool to be reduced again by PSII, while the other PQ molecule is shifted to the stromal binding site of the cytochrome b_6/f complex where it becomes reduced. This PQH_2 molecule is then set free from the stromal binding site and diffuses back across the membrane to repeat the process and undergoes recycling into the thylakoid PQH_2 pool. Electrons are transferred, one at a time, from cytochrome b_6/f complex to PSI *via* plastocyanin, the mobile electron carrier and a peripheral protein.

The Q cycle appears to present a paradox, in that the same products are cycled and remade while pumping protons into the lumen.

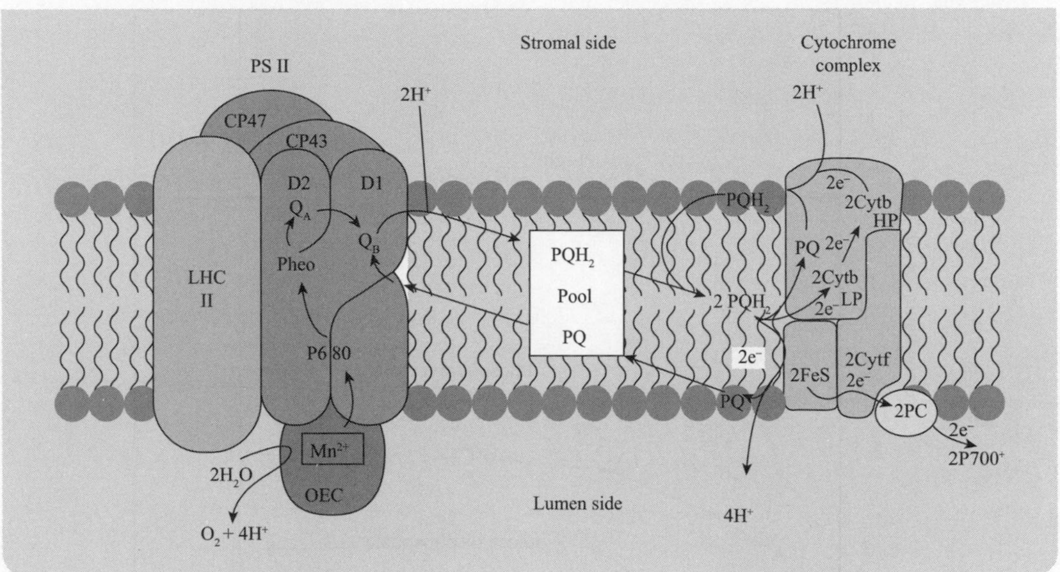

7.5 The Dark Reactions

Using photosynthetic green algae, *Chlorella/Scenedesmus,* Melvin Calvin and his collaborators identified 3-phosphoglyceric acid (3-PGA) to be the first stable compound following uptake of $^{14}CO_2$ during photosynthetic process. The first critical experiment was conducted in 1952 by Massini in Calvin's laboratory (Figure 7.18). He obtained evidence that ribulose bisphosphate was probably the acceptor molecule of CO_2 and that light was necessary for both its synthesis and the utilization of phosphoglycerate. An important experiment by Wilson and Calvin in 1954 confirmed the above findings (Figure 7.19).

Melvin Calvin and his associates exposed the photosynthetic algae (*Chlorella/Scenedesmus*) to $^{14}CO_2$ for 'only a few seconds'. They then found that the first compound to become radioactive was PGA. They further found that almost all the radioactivity in the PGA was located in the carboxyl group (–COOH) of PGA. Thus it appeared that this carboxyl group had been formed by the addition of $^{14}CO_2$ to the acceptor molecule as presented in the diagram below:

```
 CH2OP                                CH2OP
   |                                    |
  C=O         *CO2                  HO–C–H
   |            \  Rubisco              |  **
 H–C–OH    ─────────────────►         COOH
   |             +H2O                   +
 H–C–OH                               COOH
   |                                    |
 CH2OP                               H–C–OH
Ribulose-1,5-bisphosphate               |
                                      CH2OP
                              2 molecules of 3-PGA
```

*The carbon of CO_2 is labelled to show its point of entry into one of the two P-glycerate molecules. Rubisco produces the enediol intermediate through proton abstraction and electron rearrangement.

**The asterisks show how the –COOH group of one of the two molecules of PGA has been formed from CO_2.

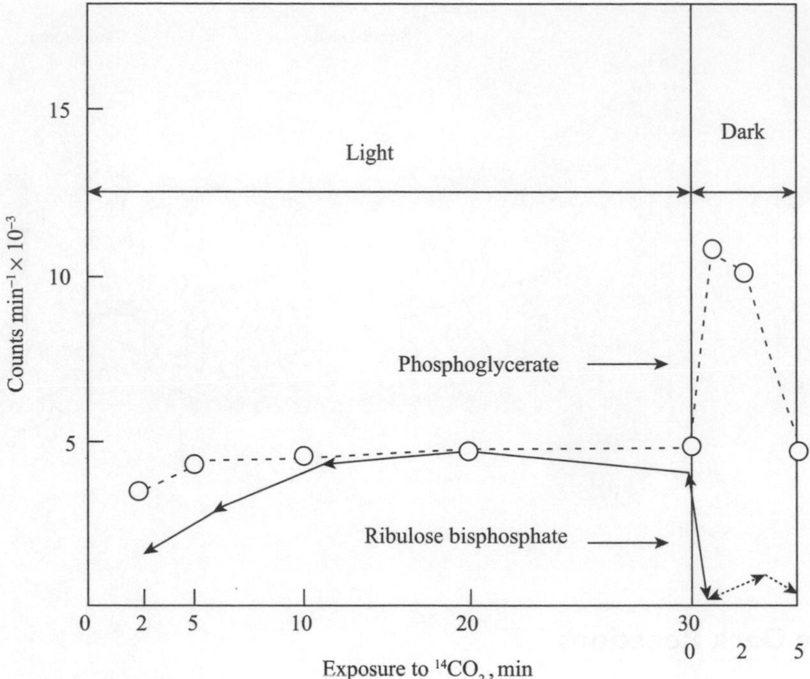

Fig. 7.18 The Massini experiment, in which light is removed in the presence of CO_2. Ribulose bisphosphate decreases and phosphoglycerate increases. This experiment suggested that ribulose bisphosphate was the CO_2 acceptor molecule and that light was necessary both for its synthesis and the utilization of the first product, phosphoglycerate.

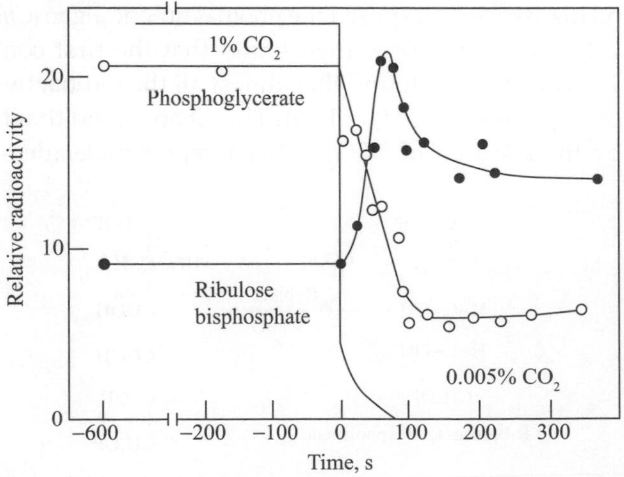

Fig. 7.19 The Wilson experiment, indicating that when CO_2 is removed in the presence of light, ribulose bisphosphate is no longer used but is continuously produced. Even though phosphoglycerate is no longer produced, it is used in the light.

7.5.1 Calvin–Benson pathway

The process of fixation of CO_2 was discovered in 1947 by Calvin, Benson[7], Bassham and their co-workers. The cylical pathway by means of which all photosynthetic eukaryotes ultimately incorporate CO_2 into carbohydrate is referred to as the **Calvin cycle** or **C3** pathway or **PCR** (Photosynthetic Carbon Reduction) pathway (Figure 7.20). Mapping the complex sequence of reactions involving the formation of organic carbon and its conversion to complex carbohydrates represented a major challenge in plant biochemistry. Calvin (1961) was awarded Nobel Prize for his efforts. The solution to the puzzle of carbon fixation was made possible in late 1940s because of two technological advances: the discovery of ^{14}C and the development of paper chromatography as an analytical tool.

Autoradiography: It is a special modification of the radioisotopic tracer technique and is most often used in conjunction with light and/or electron microscopy. Biological samples (usually tissue sections) containing the tracer are placed in contact with special photographic films or emulsions. The emitted rays expose the film producing grains, the numbers and distribution of which within a cell or tissue provide quantitative and qualitative information about the movement and/or localization of metabolic intermediates or products.

Calvin-Benson cycle can be discussed in three steps:

1. ***Carboxylation***: Condensation of CO_2 with Ribulose-1, 5-bisphosphate (RuBP) into two molecules of 3-phosphoglyceric acid (3-PGA) catalyzed by Ribulose-1,5-bisphosphate carboxylase/oxygenase (Rubisco).
2. ***Reduction***: Utilization of ATP and NADPH generated during photophosphorylation to reduce 3-Phosphoglyceric acid (3-PGA) into Glyceraldehyde-3-P (G-3-P)
3. ***Regeneration***: Consumption of extra ATP molecule to convert some of the triose phosphate (G-3-P) back to Ribulose-1,5-bisphosphate (RuBP).

3-Phosphoglycerate holds a pivotal position in the Calvin cycle. It may be transported out of the chloroplast and converted into various hexoses (including glucose, sucrose, fructosans (a polymer of fructose molecules) and cell wall carbohydrates) or may be directed to form starch within the chloroplasts *via* hexose phosphates or it may move into the metabolic pool.

[7] Melvin Calvin and his team of collaborators, with two major investigators being Andrew A. Benson and James Bassham spent some ten years in elucidating the photosynthetic pathway and identified a number of compounds that must be involved. But not all the evidences seemed to fit together as there were missing links in the envisaged cyclical pathway. Then in 1958, Calvin had a 'eureka moment'-the story is best described in his own words. 'One day I was sitting in the car parked in the red zone, while my wife was on an errand. While sitting at the wheel of the car, the recognition of missing compounds occurred just like that-quite suddenly. Suddenly also in a matter of seconds, the complete cyclic character of the path of carbon became apparent to me. It all occurred to me in the matter of 30s'. The missing link in the cycle that Calvin had just worked out was a compound called phosphoglyceric acid (PGA, a three-carbon compound) which is made from a compound Calvin already knew about called RuBP by the addition of carbon dioxide. The end product of photosynthesis glucose is generated in a chain of reactions that converts the PGA back to RuBP, to be used again. Calvin quoted, 'there is such a thing as inspiration, but one has to be ready for it. I don't know what made me ready at that moment, except I didn't have anything else to do but sit and wait. And perhaps that in itself has some moral'.

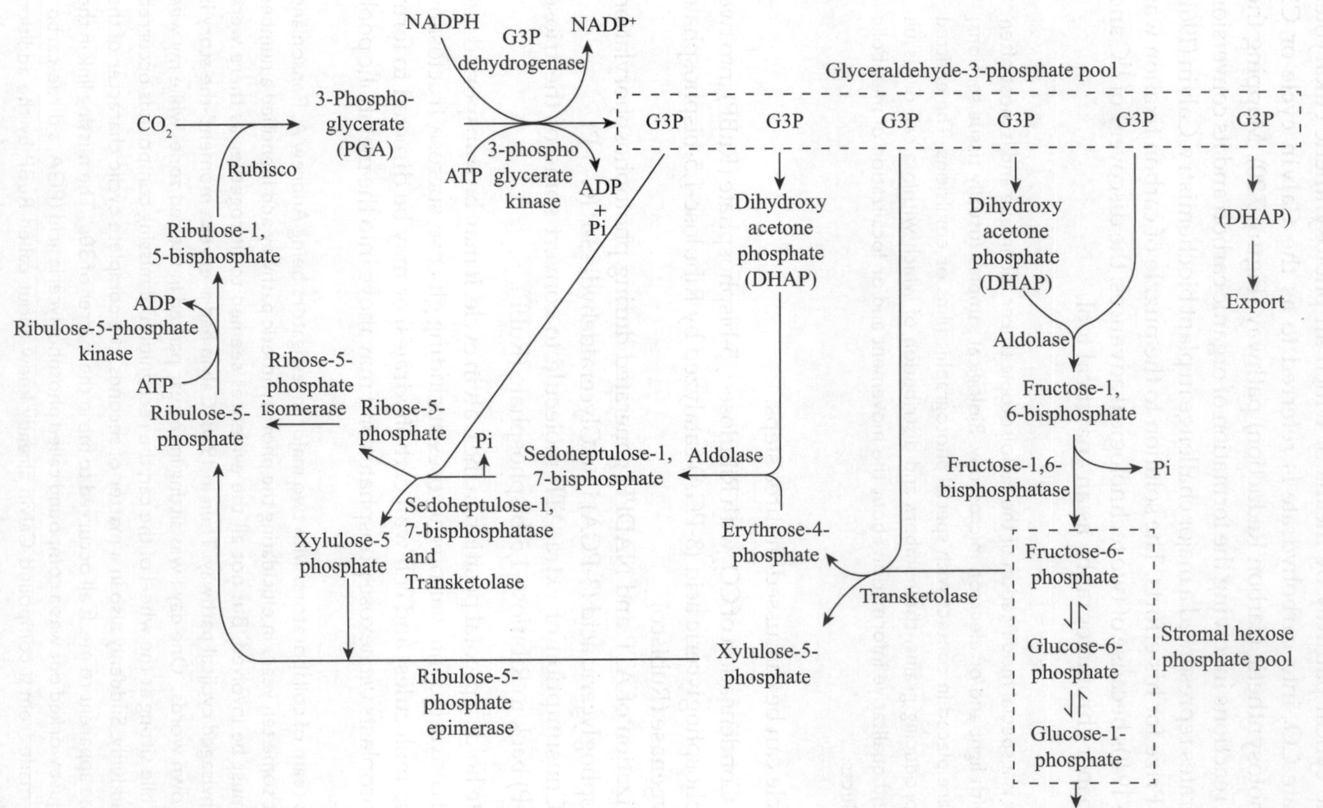

Fig. 7.20 The Calvin–Benson cycle or C3 pathway or photosynthetic carbon reduction (PCR) pathway or reductive pentose phosphate pathway. (i) Carboxylation phase – the carboxylation of three molecules of ribulose–1,5-bisphosphate by 3 molecules of CO_2 yields 6 molecules of 3-phosphoglycerate (3-PGA), catalyzed by Rubisco. (ii) Reduction phase – reduction of 3-PGA to Glyceraldehyde-3-phosphate (G-3-P) occurs in the presence of ATP and NADPH produced during light reactions, catalyzed by 3-PGA kinase and G-3-P dehydrogenase. (iii) Regeneration phase – among the 6 molecules of G-3-P, only one represents the net assimilation of 3 molecules of CO_2, while the other five undergo a series of reactions that fully regenerate the three starting molecules of RuBP.

The overall reaction of carbon fixation is as follows:

$$6RuBP + 6CO_2 + 12NADPH + 12H^+ + 18ATP \rightarrow C_6H_{12}O_6 + 6RuBP + 12NADP^+ + 18ADP + 18Pi$$

If we begin with 6 RuBP, then out of 12 PGA molecules generated, only two are available to enter into the pool for the synthesis of a glucose molecule and from it to other compounds such as starch, cellulose and so on. The remaining 10 molecules recycle through a sequence of reactions comprising 4-C, 5-C and 7-C intermediates to regenerate the five molecules of RuBP from 10 PGA.

7.5.2 Rubisco (Ribulose-1,5-bisphosphate carboxylase/oxygenase)

Rubisco, a dual functional enzyme, is found in all the green tissues of the leaves of C_3 or Calvin–Benson cycle plants and also in the parenchymatous bundle sheath cells surrounding the vascular tissues of C4 or Hatch and Slack cycle plants. It shows different physiological functions, depending upon the levels of CO_2/O_2. At high levels of CO_2 ($CO_2 > O_2$), the enzyme has carboxylase activity while in high concentrations of oxygen ($O_2 > CO_2$) it has oxygenase activity, producing one molecule each of phosphoglyceric acid (3C fragment) and phosphoglycolic acid (2C fragment), evolved during photorespiration in C3 plants.

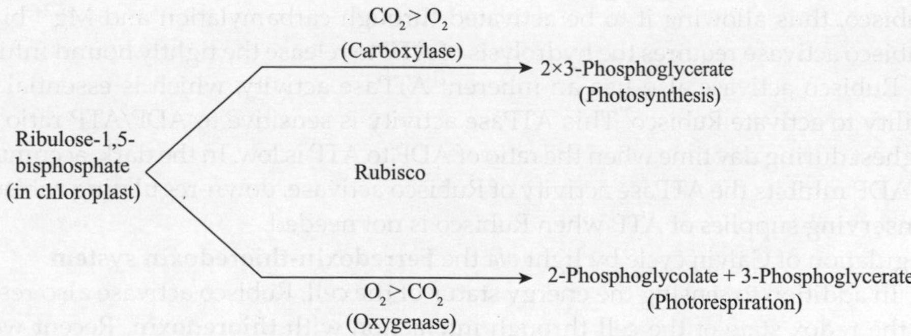

From this, it is evident that oxygen can replace CO_2 as a substrate for the enzyme, i.e., O_2 is a competitive inhibitor with respect to CO_2.

Rubisco, indeed the world's most abundant protein on the earth (~50 per cent of all the protein in the stroma), is present in green tissues of the leaf. Rubisco has been isolated and purified in crystalline form, from the leaves of several species (Lorimer and Andrews, 1981). It is composed of about 16 pairs of 2 kinds of subunits.

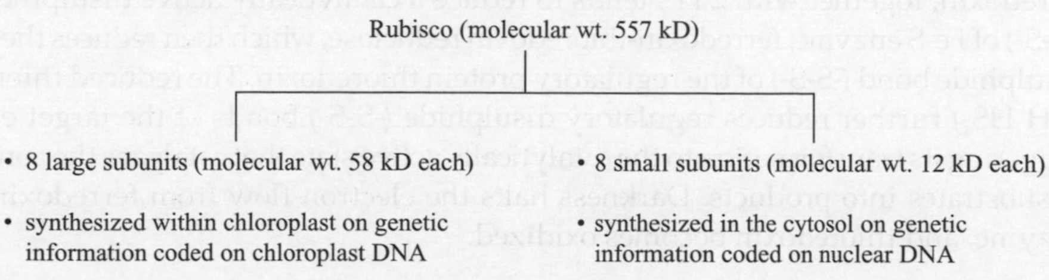

7.5.3 Regulation of the Calvin cycle

1. Regulation of catalytic activity of Rubisco by enzyme **Rubisco activase**.

   Calvin cycle activity is regulated by **Rubisco activase**, a protein that can modulate the activity of Rubisco. CO_2 functions as both an activator and a substrate for Rubisco. **Rubisco activase** complements the CO_2-dependent activation of Rubisco by lowering the inhibition caused by sugar phosphates; in other words, it functions by removing inhibitory sugar-phosphate compounds from the Rubisco active site, thus stimulating it's ability to fix carbon. This allows Rubisco to have great catalytic activity and promotes carbon fixation at otherwise suboptimal concentrations of CO_2.

   The catalytic activites of Rubisco-carboxylation and oxygenation require the formation of a lysal-carbamate (Rubisco-NH_2-CO_2-) by a molecule of CO_2 called 'activator CO_2'. The subsequent binding of Mg^{2+} to the carbamate stabilizes the carbamylated Rubisco (Rubisco-NH_2-CO_2-Mg^{2+}) and converts it to a catalytically competent enzyme. Another molecule of CO_2-substrate-CO_2- can then react with RuBP at the Rubisco active site, thereby releasing 3-phosphoglycerate (2 molecules). Phosphorylated sugars (such as xylulose 1,5-bisphosphate and the naturally occurring inhibitor 2-carboxyarabinitol 1-phosphate) and the substrate (ribulose 1,5-bisphosphate) prevent activation and inhibit catalysis by binding tightly to the uncarbamylated Rubisco and to the carbamylated Rubisco, respectively. Plants overcome this inhibition with the protein Rubisco activase, which removes the sugar phosphates from the uncarbamylated and carbamylated Rubisco, thus allowing it to be activated through carbamylation and Mg^{2+} binding. Rubisco activase requires the hydrolysis of ATP to release the tightly bound inhibitors.

   Rubisco activase also has an inherent ATPase activity, which is essential for its ability to activate Rubisco. This ATPase activity is sensitive to ADP/ATP ratio and is highest during day time when the ratio of ADP to ATP is low. In the dark, accumulation of ADP inhibits the ATPase activity of Rubisco activase, down-regulating rubisco and conserving supplies of ATP when Rubisco is not needed.

2. Regulation of Calvin cycle by light *via* the **Ferredoxin-thioredoxin system**.

   In addition to sensing the energy status of the cell, Rubisco activase also responds to the redox state of the cell through interaction with **thioredoxin**. Recent work by Archie Portis and co-workers shows that the ADP-induced inhibition of activity is greatly diminished in light by thioredoxin-mediated reduction of a disulphide bridge in Rubisco activase, allowing enhanced activity.

   The **Ferredoxin-thioredoxin system** couples the light signal sensed by thylakoid membranes to the activity of enzymes of the Calvin cycle in the chloroplast stroma (Figure 7.21). The activation of enzymes begins with the reduction of ferredoxin by the photosynthetic electron transport chain in the presence of light. In turn, reduced ferredoxin, together with 2 H^+, tends to reduce a catalytically active disulphide bond (-S-S-) of Fe-S enzyme, ferredoxin-thioredoxin reductase, which then reduces the unique disulphide bond (-S-S-) of the regulatory protein thioredoxin. The reduced thioredoxin (-SH HS-) further reduces regulatory disulphide (-S-S-) bonds of the target enzyme, triggering its transformation to the catalytically active state that catalyzes the conversion of substrates into products. Darkness halts the electron flow from ferredoxin to the enzyme, and thioredoxin becomes oxidized.

Proteomic studies have shown that the Ferredoxin-thioredoxin system regulates enzymes functional in numerous chloroplast processes other than carbon fixation. Thioredoxin also protects proteins against damage caused by accumulation of reactive oxygen species (ROS) such as hydrogen peroxide (H_2O_2), the superoxide anion (O_2^{*-}), and the hydroxyl radical (OH^{*-}).

In many plant species, the alternative splicing of a unique pre-mRNA yields two identical Rubisco activases that differ only at the carboxyl terminus: the long-form α (46 kD) and the short-form β (42 kD). The C-extension of the long-form α carries two cysteines that modulate the sensitivity of ATPase activity to the ATP:ADP ratio by thiol-disulphide exchange. In this way, regulation of Rubisco activase is linked to light through the **Ferredoxin-thioredoxin system.**

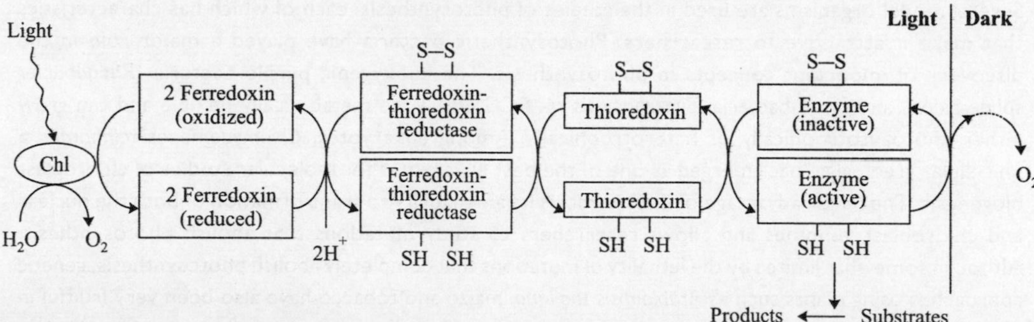

Fig. 7.21 The Ferredoxin–thioredoxin system links the light signal to the activity of Rubisco activase and four enzymes of the Calvin-Benson cycle in the chloroplast stroma. This mechanism uses ferredoxin, reduced by the photosynthetic electron transport system along with two chloroplast proteins, i.e., ferredoxin-thioredoxin reductase and thioredoxin to regulate activity of the enzymes (mentioned in text). Reduced ferredoxin reduces the disulphide bond (-S-S-) of ferredoxin- thioredoxin reductase, which in turn reduces the -S-S- bond of the thioredoxin. The reduced thioredoxin then reduces disulphide bonds of the target enzyme, transforming it to the active state that catalyzes conversion of substrates into products. In the dark, deactivation of thioredoxin-activated enzymes occurs.

Thus, Rubisco activase helps regulate the level of carbon fixation by Rubisco by responding to the level of light, the ADP/ATP ratio, and the redox state of the cell.

3. Light-activated ionic fluxes regulate enzymes of the Calvin cycle.

   When illuminated, the protons (H^+) flow from the chloroplast stroma into the thylakoid lumen is linked to the liberation of Mg^{2+}ions from the lumen or intrathylakoid space to the stroma. Light-dependent ionic fluxes lower the concentration of protons in the stroma, thus increasing pH from 7 to 8 followed by an increase in Mg^{2+} ion concentration by 2 to 5 mM. This activates enzymes of the Calvin cycle that function in the presence of Mg^{2+} and are more active at pH 8 rather than at pH 7 (e.g. Rubisco, fructose 1,6- bisphosphatase, sedoheptulose 1,7-bisphosphatase, and phosphoribulokinase). Darkness rapidly reverses the modifications of ionic composition of the chloroplast stroma.

4. Light regulates the assembly of chloroplastic enzymes into supramoleular complexes. The formation of supramolecular complexes with regulatory proteins has significant impact on the catalytic action of chloroplastic enzymes. For example, glyceraldehyde-3-phosphate dehydrogenase binds non-covalently to phosphoribulokinase as well as CP12- a protein of approximately 8.5 kD size, containing four conserved cysteines able to form two disulphide (-S-S-) bridges. Light regulates the stability of this ternary complex through the Ferredoxin-thioredoxin system. Reduced thioredoxin cleaves the S-S-linkages of both phosphoribulokinase and CP12, releasing glyceraldehyde-3-phosphate dehydrogenase and phosphoribulokinase in catalytically active conformations.

Other Model Organisms Used to Study Photosynthesis

Several model organisms are used in the studies of photosynthesis, each of which has characteristics that make it attractive to researchers. Photosynthetic bacteria have played a major role in the discovery of important concepts in photosynthesis. The anoxygenic purple bacteria (*Rhodobacter sphaeroides*), and cyanobacteria (*Synechocystis* sp. PCC 6803) are metabolically flexible and can grow either photoautotrophically or heterotrophically. Among eukaryotes, *Chlamydomonas reinhardtii*, a unicellular green alga, has emerged as one of the best eukaryotes for molecular studies of chloroplast biogenesis. The major advantage of *C.reinhardtii* is its amenability to transformation of both the nuclear and chloroplast genomes and allows researchers to study mutations that abolish photosynthesis. Although somewhat limited by the lethality of mutations that completely abolish photosynthesis, genetic approaches using plants such as *Arabidopsis thaliana*, maize and tobacco have also been very fruitful in photosynthesis research.

7.6 Photorespiration

Plants that exhibit Calvin cycle exclusively for CO_2fixation also display a competing process of light and oxygen-dependent evolution of CO_2referred to as **photorespiration** or the **PCO cycle** (Photosynthetic Carbon Oxidation cycle) (Figure 7.22).

This cycle also starts with Rubisco, which under excess of oxygen concentration (5–21 % and beyond) catalyzes the oxidation as well as carboxylation of RuBP. Thus, O_2 competes with CO_2 for binding at the catalytic site of Rubisco and often splits RuBP into phosphoglycolate (2-C compound) and 3-phosphoglycerate.The oxygenase activity of Rubisco is very significant because O_2 molecules in nature are many times more than CO_2. Photosynthesis can result in accumulation of even higher concentration of O_2 inside the leaf than is found in the atmosphere. Phosphoglycolate is transported from the chloroplast and oxidized through a sequence of reactions within peroxisomes to form glycine.

Glycine next moves to the mitochondrion where through a complicated reaction, the enzyme glycine decarboxylase, with its cofactor **tetrahydrofolate** (THF), catalyzes the NAD–dependent oxidative conversion of a molecule of glycine to one each of CO_2, NH_3 and CH_2-THF (methylene tetrahydrofolate). Glycine decarboxylase complex has four enzyme subunits, and pyridoxal phosphate, FAD and lipoamide cofactors, in addition to THF. Reaction of CH_2 – THF with a second glycine molecule, catalyzed by **serine hydroxymethyltransferase**, regenerates THF and produces serine which goes back to the

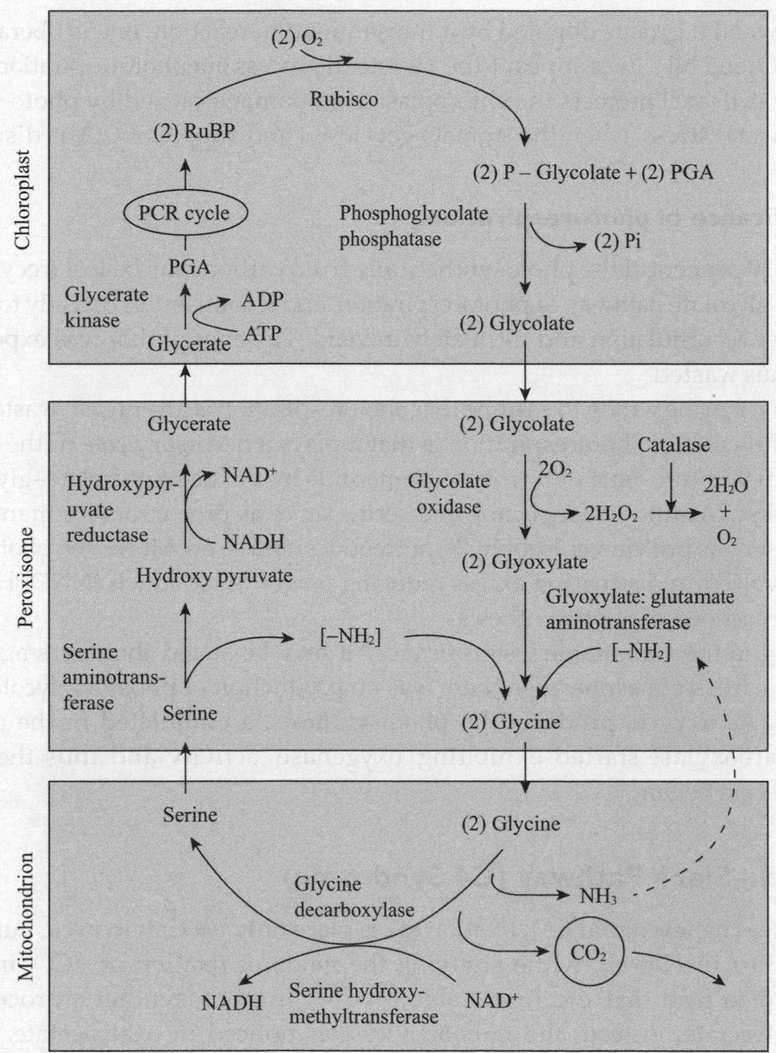

Fig. 7.22 Photorespiratory glycolate pathway or C_2 pathway or photosynthetic carbon oxidation (PCO) pathway.

peroxisome. Ammonia, a product of glycine decarboxylation in the mitochondria is efficiently reassimilated in the chloroplast *via* glutamine synthetase – glutamate α-ketoglutarate aminotransferase (GS-GOGAT) pathway through a series of reactions as shown in chapter 10.

Rubisco may seem inefficient since it suffers from photorespiration and low activity, but, it is one of a handful of enzymes, coupled with the Calvin cycle activity, that possesses autocatalytic ability, i.e., it regenerates one of its own substrates while having a synthetic capacity at the same time.

The pathway serves to recycle the remaining three carbon atoms out of four (from two molecules of glycine) through cell organelles. There is loss of one of them as CO_2. In addition, out of the two-NH_2 groups donated by a transamination reaction, one is liberated as NH_3. Release of CO_2 and NH_3 does appear to be a wasteful process but photorespiration is regarded a 'necessary evil' as it protects the chloroplast from damage caused by photo-oxidation at the time of water stress, when the stomata get closed and supply of CO_2 is disrupted.

7.6.1 Significance of photorespiration

As much as 30 per cent of the photosynthetically fixed carbon may be lost (recycled to CO_2) through the glycolate pathway of photorespiration, and it contributes directly to a reduction in dry matter accumulation and ultimately in yield. The energy that was expended to fix it (CO_2) is thus wasted.

However, it may be wrong to assume that photorespiration is an entirely wasteful process. The useful function of photorespiration is that it plays a necessary role in the metabolism or intraorganellar transport of nitrogen compounds by virtue of glycolate–glycine–serine transformation. Amino acids, glycine and serine serve as precursors for many important metabolites such as proteins, chlorophylls, nucleotides, and so on. Moreover, photorespiration might be involved in dissipating excess reducing power of too much (NADPH) generated in the light reactions of photosynthesis.

Returning to the evolutionary significance,[8] it may be stated that eon's ago there was no O_2 in the earth's atmosphere, so there was no production of phosphoglycolate by RuBP carboxylase. As oxygen, produced by photosynthesis, accumulated in the atmosphere, the RuBP carboxylase started exhibiting 'oxygenase' activity and thus the process of photorespiration began.

7.7 Hatch–Slack Pathway (C4 Syndrome)

Till 1965, it was believed that CO_2 fixation takes place only *via* Calvin cycle but Kortschak, Hartt and Burr in Hawaii, while studying the mode of fixation of $^{14}CO_2$ in sugarcane were baffled to find that the first stable product in photosynthetic process was not 3-phosphoglycerate; instead, the radioactivity was noticed in oxaloacetate, malate and aspartate. Further studies by Australian plant physiolgists, M. Hatch and R. Slack (1967) showed that this represented an additional pathway of CO_2 assimilation – **C4 pathway** and so, plants exhibiting this pathway are called C4 plants (e.g., corn, sorghum, amaranth, grasses, sugarbeet and so on).

C4 plants exhibit 'Kranz' (from the German, 'Wreath like') anatomy (Figure 7.23). Each vascular bundle in the leaves is covered by a layer of large, prominent parenchymatous cells – chlorophyllous parenchymatous cells known as **bundle sheath cells**, which in turn are surrounded by loosely organized spongy, undifferentiated mesophyll cells. Cells of the leaves of C4 plants possess dimorphic chloroplasts, mesophyll chloroplasts and bundle sheath chloroplasts (Figure 7.24).

---

8 Chloroplasts arose evolutionarily from cyanobacteria while mitochondria are evolutionary descendents of bacteria?

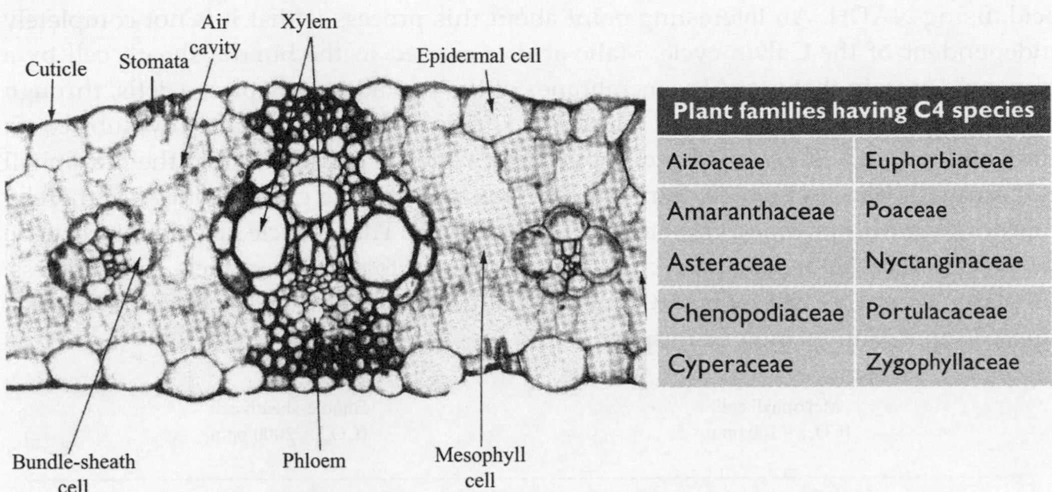

Plant families having C4 species	
Aizoaceae	Euphorbiaceae
Amaranthaceae	Poaceae
Asteraceae	Nyctanginaceae
Chenopodiaceae	Portulacaceae
Cyperaceae	Zygophyllaceae

Fig. 7.23 Kranz anatomy of the leaf from a C_4 plant.

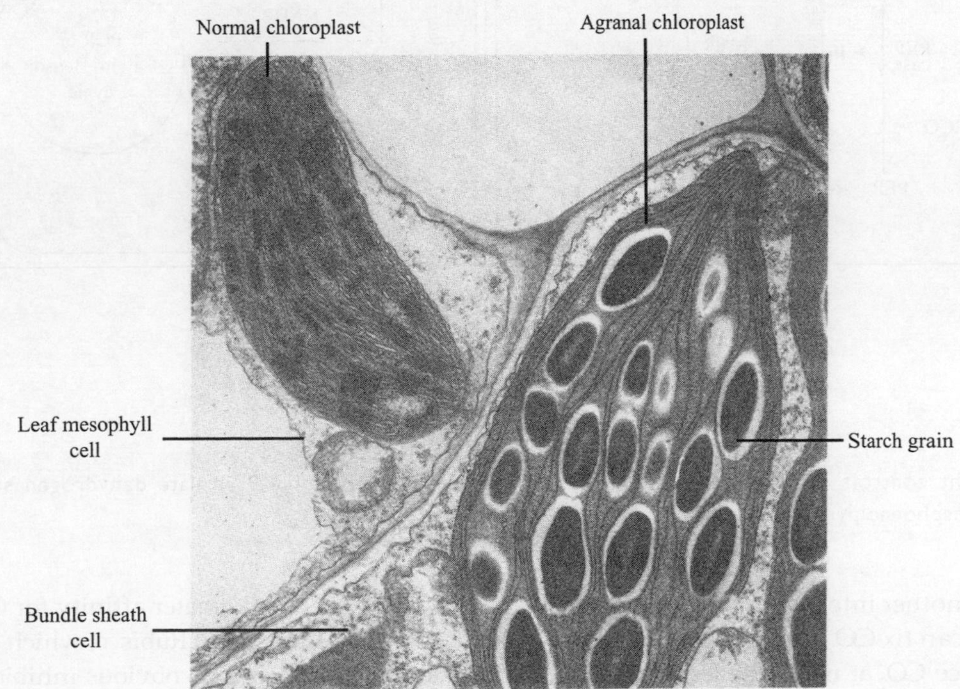

Fig. 7.24 Chloroplast dimorphism in C4 plants. Bundle sheath chloroplasts often have few grana, and in some cases none at all (agranal), while the chloroplasts of mesophyll cells are granal.

C4 plants capture CO_2 through phosphoenolpyruvate carboxylase (PEP carboxylase or PEPcase). Upon entry into the mesophyll cells, CO_2 is hydrated to bicarbonate ions (HCO_3^-) in a step catalyzed by carbonic anhydrase. CO_2 accepted by PEP is converted into a 4-carbon compound, oxaloacetic acid (OAA), which is immediately reduced to malic acid, using NADH. An interesting point about this process is that it is not completely independent of the Calvin cycle. Malic acid is carried to the bundle sheath cell by a transport protein that spans the membranes of the two adjacently placed cells, through a plasmodesmata. There it readily releases CO_2 to be refixed immediately by Rubisco *via* the Calvin cycle and gets oxidized to pyruvate, which is carried back to the mesophyll cell, through plasmodesmata. Pyruvate receives a phosphate group from ATP, thereby regenerating phosphoenolpyruvate (PEP) (Figure 7.25). The C4 cycle has, therefore, to be viewed not as an independent cycle, but as an adjunct to the Calvin cycle. Nonetheless, C4 plants do not fix CO_2 by both C3 and C4 pathways in 'any one cell', rather, there is a 'spatial separation' (separated in space).

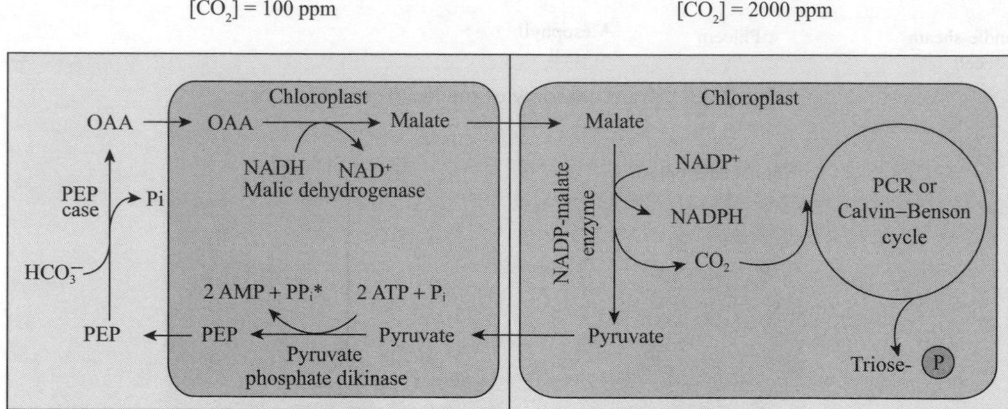

Fig. 7.25 C4 Pathway or Hatch–Slack Pathway.

$$\underset{(^{*}\text{PPi})}{\text{Pyrophosphate}} \xrightarrow{\text{Pyrophosphatase}} \underset{(2\,\text{Pi})}{\text{Inorganic phosphate}}$$

Light controls the functioning of key C4 cycle enzymes such as NADP-malate dehydrogenase, phosphoenolpyruvate carboxylase and pyruvate phosphate dikinase.

Another interesting aspect is that PEP carboxylase has a much greater affinity for CO_2 and can fix CO_2 at extremely low CO_2 levels (<10 ppm) as compared to Rubisco, which can reduce CO_2 at much higher CO_2 levels (50 ppm or above). It shows no obvious inhibition by O_2 with the net result that CO_2 initially captured over a large volume of leaf space is released into the relatively small volume of the bundle sheath. Consequently, the local CO_2 concentration in these cells is very high and refixed by Rubisco into the Calvin cycle. However, there is a price to pay for capturing atmospheric CO_2 at very low concentration. ATP is required to convert pyruvate to phosphoenolpyruvate (PEP) and perhaps also for the transport of organic acids to and from the mesophyll cells.

$$\text{Pyruvate} + \text{Pi} + \text{ATP} \xrightarrow[\text{(mesophyll cells)}]{\text{Pyruvate phosphate dikinase}} \text{PEP} + \text{AMP} + \text{PPi}$$

The location of bundle sheath cells close to vascular tissues also enables them to pass their final product of photosynthesis, i.e. sucrose, directly into the sieve tubes of phloem tissue for movement to other parts of the plant.

Significance of C4 photosynthesis appears to be its mechanism to reduce photorespiration and maximize CO_2 assimilation.

Overall, the C4 cycle is an adaptive feature to enable plants to survive better under environments of higher temperatures, intense solar radiation and moisture stress (as water is not the reactant in CO_2 assimilation in mesophyll cells).

C3 Plants (Calvin cycle)	C4 Plants (Hatch and Slack cycle)
1. Calvin cycle is present in all the green cells of the leaf.	Hatch–Slack cycle is present in the mesophyll cells while C3 cycle is restricted to the bundle sheath cells.
2. Ribulose 1,5-bisphosphate is the initial carbon dioxide (CO_2) acceptor.	Two CO_2 acceptors are present– phos-phoenolpyruvate in the mesophyll cells and RuBP in the bundle sheath cells.
3. The first stable compound is a 3-carbon compound, i.e., phosphoglyceric acid.	The first stable product is a 4-carbon compound, i.e., oxaloacetic acid/malic acid.
4. The leaves don't have 'Kranz anatomy'. The chloroplasts in the green cells are of one type. They are small with well-defined grana and have both, photosystems I and II.	The leaves have 'Kranz anatomy' (kranz = wreath, a German word). The vascular tissue is covered by a ring of large parenchyma cells – the bundle sheath. The chloroplasts are dimorphic. The cells of mesophyll have chloroplasts similar to those of C3 plants, accumulating very little starch while the chloroplasts of bundle-sheath (BS) cells are large and agranal* and contain large starch grains. They are centripetally arranged and lack PS II, and therefore dependent on the chloroplasts of the mesophyll for the supply of NADPH.
5. RuBP and other enzymes of the C3 pathway are found in the mesophyll cells.	RuBP and other enzymes of C3 pathway are present in the BS whereas the mesophyll cells contain the enzymes of C4 cycle.
6. The optimum temperature range is between 10 and 25 °C.	The temperature optima for C4 species is much higher than those for C3 plants. The optimum temperature range is between 30 and 45°C.**
7. Because of lesser affinity of RuBP for CO_2, they best operate at high CO_2 levels, i.e., 50 ppm.	PEP carboxylase has strong affinity for CO_2, which is why it can capture CO_2 at low concentration, i.e., 10 ppm or even lower.
8. Photorespiration is present that further reduces the photosynthetic yield. C3 plants are thus less efficient photosynthetically and have low productivity.	Photorespiration is absent, which is way C4 plants are photosynthetically efficient showing high productivity as they waste none of their photosynthetically fixed CO_2.

C3 Plants (Calvin cycle)	C4 Plants (Hatch and Slack cycle)
9. Oxygen has an inhibitory effect on photosynthesis.	Oxygen does not have any inhibitory effect.
10. In the C3 cycle, 18 ATP are needed for the production of one glucose molecule,.	In the C4 cycle, 30 ATP are used up for producing one glucose molecule.
Examples are pulses, most of the cereals, oil seeds, temperate and tropical tree species.	Examples are sugarcane, maize, sorghum, grain amaranth.

*Laetsch and Black (1968–69) have described the ultrastructure of the bundle sheath and mesophyll cells of C4 plants as well as C3 plants, and compared the structural differences between the two.

**C4 plants are especially well adapted to survive in low water regimes, high temperature and bright sunlight as prevailing in dry climates of tropics and subtropics that could be lethal to many C3 plants.

7.8 CAM Pathway (Crassulacean Acid Metabolism)

The dark uptake of CO_2 by succulents was observed as early as 1810 by De Saussure while studying a cactus. In 1815, Heyne noted that the succulent *Bryophyllum* accumulated organic acids at night. In the 1960s, there was renewed interest in Crassulacean acid metabolism because of the similarity of carbon metabolism between C_4 species and CAM species and the realization that CAM was an adaptive modification of basic photosynthetic metabolism. CAM pathway was named so, after the angiospermic family Crassulaceae in which it was first reported (Figure 7.26). CAM has been found in hundreds of species in 26 angiosperm families including the Cactaceae, Liliaceae, a few succulent members of Euphorbiaceae, epiphytic species of Orchidaceae and Bromeliaceae, non-epiphytic bromeliads (e.g. pineapple), in some pteridophytes and in a gymnosperm, *Welwitschia mirabilis*.

Plant families having CAM species	
Agavaceae	Labiatae
Asclepiadaceae	Liliaceae
Bromeliaceae	Orchidaceae
Cactaceae	Polypodiaceae
Crassulaceae	Vitaceae
Cucurbitaceae	Welwitschiaceae
Geraniaceae	

Fig. 7.26 *Kalanchoe blossfeldiana* – a member of family Crassulaceae (*Courtesy*: Mr Bobby Verma, Govt. Brijendra College, Faridkot, Punjab).

CAM is also a means of maintaining higher photosynthetic rates in high water-stressed environments. CAM plants exhibit an inverted stomatal cycle – open at night for CO_2 uptake, when moisture deficit is less and close during the day time, when moisture stress is more. CO_2 is stored overnight as organic acids (malate) in the vacuolar sap, again through the action of PEP carboxylase. Decarboxylation of malate to pyruvate during the day provides the required CO_2 for photosynthesis. Both carboxylation and decarboxylation occur in same cells, i.e., mesophyll cells but there is **temporal separation** (separated in time) of the two reactions (Figure 7.27). In CAM plants, closed cycle of carbon intermediates does not exist as there is in C_4 plants. Thus CAM is cyclic in time only.

Although the ability of a plant to perform CAM is genetically determined, it is also environmentally controlled. In general, CAM is favoured by hot days with high irradiance levels, cool nights, dry soils, and high salt concentrations.

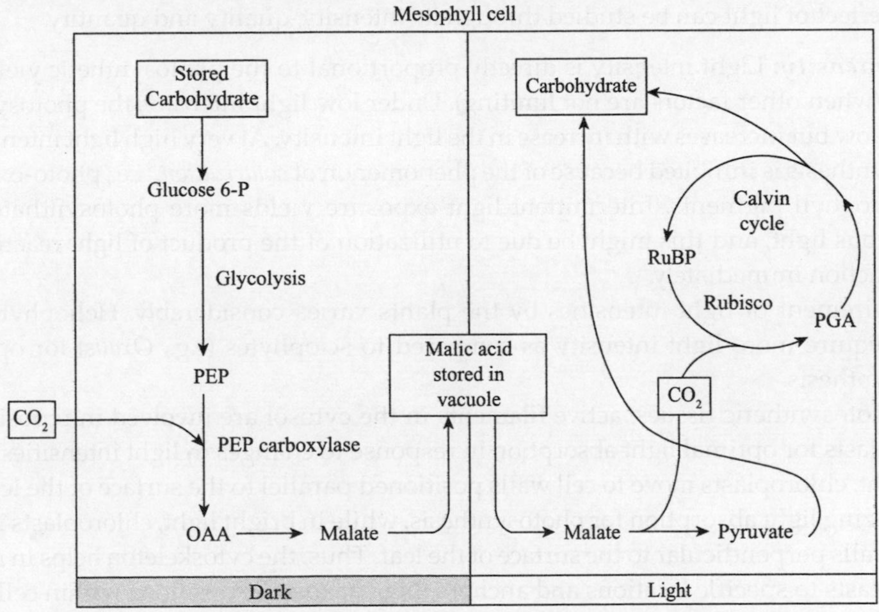

Fig. 7.27 Crassulacean acid metabolism (CAM) in the light and dark.

7.9 Factors Affecting Photosynthesis

Photosynthesis, like any other physiochemical process, is controlled by the environmental conditions in which it takes place. Sachs (1880) introduced the basic concept of three cardinal points, according to which there is a minimum, optimum and maximum value for each factor with regard to photosynthesis. Various external factors (such as CO_2 and O_2 concentration, light, temperature, available soil water, mineral elements, air pollutants) and internal factors (such as accumulation of end products, protoplasmic factors, chlorophyll content of leaf) affect the rate of the photosynthetic process. Before 1905, earlier physiologists, who conducted experiments to determine these cardinal points, were confronted with fluctuating optima. These investigators were actually treating the external factors affecting photosynthesis individually and not in relation to one another. These resulting contradictions were explained by Blackman in 1905 who proposed *'the law of limiting factors'* which is in fact an extension of' Liebig's law of minimum' on soil nutrition. According to Blackman, 'When a process is governed as to its rapidity by a number of separate factors, the rate of the process is limited by the pace of the slowest factor'. The 'slowest factor' refers to a factor which is present in quantity or intensity that is less than what is needed for the process.

External factors

Light

Light has direct as well as indirect effects on photosynthesis. Directly, it is involved in the photooxidation of water and excitation of chlorophyll molecules. Indirectly, it controls the stomatal movements, leading to diffusion of CO_2 inside the cell from outside.

The effect of light can be studied through its intensity, quality and quantity.

Light intensity: Light intensity is directly proportional to the photosynthetic yield up to a limit (when other factors are not limiting). Under low light intensity, the photosynthetic yield is low but increases with increase in the light intensity. At very high light intensity, the photosynthesis is inhibited because of the phenomenon of *solarization*,[9] i.e., photo-oxidation of chlorophyll pigments. Intermittent light exposure yields more photosynthates than continuous light, and this might be due to utilization of the product of light reaction into dark reaction immediately.

Requirement of light intensities by the plants varies considerably. Heliophytes (e.g., bean) require more light intensity as compared to sciophytes (e.g., *Oxalis*) for optimum photosynthesis.

In photosynthetic tissues, active filaments in the cytosol are involved in repositioning chloroplasts for optimal light absorption in response to changes in light intensities. Under dim light, chloroplasts move to cell walls positioned parallel to the surface of the leaf, thus maximizing light absorption for photosynthesis, while in bright light, chloroplasts migrate to cell walls perpendicular to the surface of the leaf. Thus, the cytoskeleton helps in moving chloroplasts to specific locations and anchors them at specific positions within cells.

Light quality: Photosynthetic pigments absorb the visible portion of the electromagnetic spectrum. In green plants, photosynthesis is maximum at the red wavelength, followed by the blue wavelength, while the green wavelength has the least effect. On the other hand, red algae show maximum photosynthesis in green light and brown algae in blue light. Blue light, because of having more energy, is able to reach submerged plants and pass through clouds.

Light quantity: The total quantity of light is the product of light intensity and exposure time; 10–12 h of continuous light is sufficient for optimum photosynthesis. In fact, any source of light of sufficient intensity containing rays within the visible part of the spectrum can promote photosynthesis.

[9] Exposure to high light irradiance may cause damage to photosystems, primarily the D1 polypeptides of PS II reaction centre.

The irradiance at which the photosynthetic rate equals the respiration rate in the plant is called as the ***Light compensation point.*** Below the light compensation point, plants respire faster than photosynthesis can assimilate CO_2 and this could be detrimental to the plant growth. Light compensation points of **heliophytes** (sun plants) vary from 10 to 20 µmol m^{-2} s^{-1}, whereas in case of **sciophytes** (shade plants), these values are much lower in the range of 1–5 µmol m^{-2} s^{-1}. This could be because of the low rate of respiration in shade plants.

Carbon dioxide

The chief source of CO_2 in land plants is the air which contains only 0.038 per cent (or 380 ppm)[10] per cent of the gas by volume. Therefore, it is generally the limiting factor in nature. Under standard prevailing conditions, an increase in the amount of CO_2 up to 1 per cent causes an increase in photosynthesis. However, higher concentrations have an inhibitory effect on photosynthesis because of stomatal closure.

At low CO_2 concentration and non-limiting light intensity, the rate of photosynthesis in a plant will be equal to the total amount of respiration. The ambient carbon dioxide concentration at which the rate of CO_2 uptake for photosynthesis is balanced by the rate of CO_2 evolution by respiration is referred to as the ***CO_2 compensation point***. For C_3 plants, the CO_2 compensation point varies from 20 to 100 µL CO_2 L^{-1} whereas for C_4 plants, the values fall in the range of 0–5 µL $CO_2 L^{-1}$.

The CO_2 compensation point reflects the balance between photosynthesis and respiration as a function of CO_2 concentration, whereas the light compensation point reflects that balance as a function of irradiance under constant CO_2 concentration.

Temperature

Temperature requirements for the optimum photosynthetic yield differ with the plant species and their environment. Usually, the photosynthetic rate shows an increasing trend with rise in temperature up to certain limit i.e., 40 °C when other environmental factors are at the optimal level, whereas inhibition of photosynthesis occurs at higher temperatures because of thermal deactivation of enzymes. In C3 plants, the most common inhibitory effects of high temperature on photosynthesis are due to an increase in the photorespiration rate. CO_2 fixation in C3 plants often decreases at 25–30°C. The C_4 plants, however, show proportional increases and attain an optimum photosynthetic rate above 30–35°C as photorespiratory rate is less in C4 plants.

Cold temperatures retard the rate of photosynthesis, both directly and indirectly. Directly, cold temperatures inhibit photosynthetic processes by deactivating the enzymes

[10] 380 ppm is a trend that is likely to continue and even accelerate. Projections by the international panel on climate change, based on a range of scenarios, anticipate a CO_2 level of 450–500 ppm by 2050.

participating in the dark reactions of photosynthesis; and indirectly, by the formation of ice within and outside the cell. The ice crystal formation creates drought-like conditions by draining water from the living cells. Ice formed inside a cell also causes mechanical injury, which upsets the architecture of the cell including chloroplasts as well as destroys the permeability properties of the membrane.

Water

Water is an essential raw material for photosynthesis. The important role of water is to supply electrons and protons for reduction of NAD^+ during photochemical reactions of photosynthesis. Water deficiency retards photosynthetic rate alongside other vital processes occurring in the entire living system, primarily because of reduced dehydration of the cytoplasm and closure of stomatal aperture. Dehydration of the cytoplasm leads to loss of its colloidal structure as well as impairment of enzymatic efficiency resulting in disruption of metabolic processes such as photosynthesis and respiration.

Oxygen

Increase in oxygen concentration beyond 21 per cent has been shown to inhibit photosynthesis in C3 plants and CAM plants. The inhibitory effect on photosynthesis by high O_2 concentrations is known as the **Warburg effect**. Oxygen favours a rapid respiration rate which oxidizes common intermediates and might also compete with CO_2 for hydrogen and gets reduced instead of CO_2. With rise in concentration of oxygen, the oxygenase activity of Rubisco increases, thus reducing the rate of CO_2 fixation and favouring photorespiration.

Mineral elements

Certain important mineral nutrients such as Mg, Fe, Cu, Cl, Mn and P act as cofactors for photosynthetic enzymes and thus, are closely associated with major reactions of photosynthesis. Deficiency of such minerals, ultimately lower the rate of photosynthesis.

Air pollutants

Air pollutants such as SO_2, NO_2, O_3, smog are known to decrease photosynthetic rates. Pollutants at very low level do not affect the photosynthetic rate but at higher levels they inhibit photosynthesis. Ozone, because of its strong oxidizing action, damages the permeability of membranes of photosynthesizing cells and cell organelles and thereby photosynthetic yields.

Besides the external factors discussed above, several internal factors such as photosynthetic pigments, protoplasmic factors, anatomy and age of the leaf, accumulation of photosynthetic products and plant hormones also influence the rate of photosynthesis.

Chemosynthesis

There are certain groups of **aerobic bacteria** which lack the chlorophyll pigment but can manufacture organic matter from CO_2 and H_2O. Instead of utilizing light energy they make use of the **chemical energy** which is generated during the oxidation of some inorganic substances by them. The synthesis of the organic matter by bacteria, using this chemical energy is termed as **chemosynthesis.** Common examples of chemosynthetic bacteria are as follows:

(i) Nitrifying bacteria

Nitrosomonas and ***Nitrosococcus*** oxidize ammonia to nitrite with release of chemical energy.

$$2NH_3 + 3O_2 \rightarrow 2HNO_2 + 2H_2O + \text{Chemical energy}$$

Nitrobacter oxidizes nitrite to nitrate to release chemical energy which is then utilized in the manufacture of food

$$2HNO_2 + O_2 \rightarrow 2HNO_3 + \text{Chemical energy}$$

(ii) Sulphur bacteria

Beggiatoa and ***Thiothrix*** oxidize hydrogen sulphide to sulphur with release of chemical energy that is consumed in food production.

$$2H_2S + O_2 \rightarrow 2H_2O + 2\,S + \text{Chemical energy}$$

(iii) Iron bacteria

Ferrobacillus, Leptothrix, etc., oxidize ferrous form of iron to ferric state, followed by the release of chemical energy.

$$4FeCO_3 + O_2 + 6H_2O \rightarrow 4Fe\,(OH)_3 + 4CO_2 + \text{Chemical energy}$$

*(iv) Carbon bacteria (e.g., **Bacillus oligocarbophilus**)*

$$2CO + O_2 \rightarrow 2CO_2 + \text{Chemical energy}$$

*(v) Hydrogen bacteria (e.g., **Bacillus pantotrophus**)*

$$2H_2 + O_2 \rightarrow 2H_2O + \text{Chemical energy}$$

Review Questions

RQ 7.1. Why photolysis of water occurs in pigment system II, but not in pigment system I? Explain.

Ans. The pigment system II uses a quantum of light to generate a strong oxidant Z (now known to be tyrosine) that is sufficiently strong to oxidize water.

$$2H_2O \rightarrow 4H^+ + O_2 + 4e^-$$

Unlike PS II, the quantum of light absorbed by PS I produces an oxidant that is too weak as an oxidizing agent to remove electrons from water.

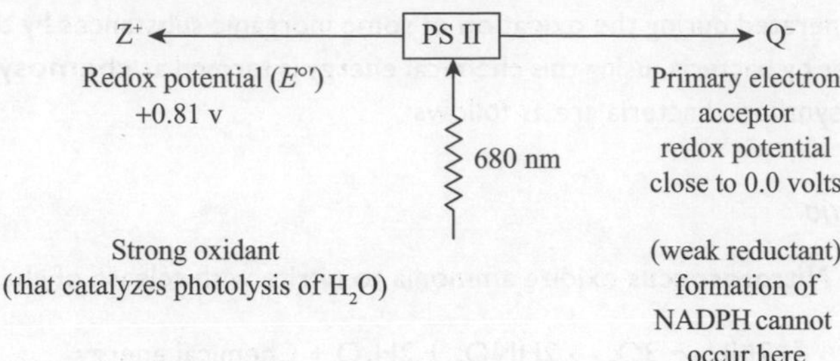

RQ 7.2. Why is it so, that the biological reductant NADPH is produced in PS I and not in PS II? Explain.

Ans. The pigment system I, after using a quantum of light (700 nm), generates a strong reductant X^- which can reduce ferredoxin (Fd) and thence reduce NADP to NADPH – the hydrogen atoms coming from the photolysis of water.

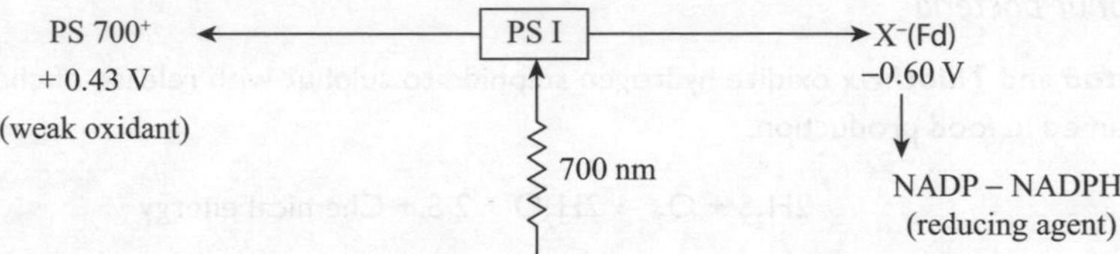

Unlike PSI, pigment system II, after absorbing a photon of light (680 nm) generates a weak reductant that is not capable of reducing ferredoxin and then NADP.

RQ 7.3. Why is the photosynthetic activity reduced or stopped when DCMU is sprayed on green plants? Explain.

Ans. On spraying dichlorophenyldimethyl urea (DCMU), a herbicide, the pigment system II is cut-off or inhibited and there will be no photolysis of water and hence no production of NADPH which is used for the fixation of CO_2 in the dark reaction. Hence, there will be no (or lower) production of hexose sugars.

However, ATP may be generated through cyclic photophosphorylation occurring in pigment system I.

RQ 7.4. What is 'Blinks effect'. How does it differ from 'Red drop' or long wavelength 'dropoff'?

Ans. Blinks in 1950 observed that when photosynthetic cells are shifted from red light of long wavelength (i.e.,700 nm) to a shorter wavelength of less than 670 nm, there was a brief increase in the rate of photosynthesis, unlike Emerson's 'Red drop'. Emerson in 1957 found that the rate of photosynthesis decreased when a plant was shifted from shorter wavelength (650 nm) to wavelengths longer than 680 nm. This decline in the photosynthetic rate under long wavelength of red light is known as 'Red drop' or long red wavelength 'drop off'. This is diametrically opposite to that of 'Blinks effect'.

RQ 7.5. List the chief pigments found in green leaves, photosynthetic bacteria, blue green algae (cyanobacteria), red algae and beetroot.

Ans.

1. Green leaves	Vital pigment: Chl *a* Accessory or antennae pigments: Chl *b*, carotene and xanthophyll (the last two collectively known as carotenoids)
2. Photosynthetic bacteria	Bacterioviridin – green sulphur bacteria Bacteriochlorophyll – purple sulphur bacteria
3. Blue-green algae or red algae	Phycocyanin (blue pigment) } Phycobilins Phycoerythrin (red pigment) }
4. Beetroot	Betacyanin (water soluble) or Betalains

RQ 7.6. Chlorophyll *a* exists in more than one form. Comment.

Ans. Chlorophyll *a* shows subtle variations, existing in three forms – Chl *a* 673, Chl *a* 685 and Chl *a* 700 (or P_{700}). These different forms are identified on the basis of small differences in absorption peaks. These variations in absorption spectrum are because of Chlorophyll *a* molecules being complexed with different proteins (not because of the substitutions on the porphyrin ring), that create a different kind of environment for chlorophyll molecules.

Chl a 673 and Chl a 685 were discovered by Butler in 1966 while Chl a 700 (P700) was discovered in 1966 by Kok and Clayton.

RQ 7.7. Water is the natural electron donor for photosynthesis. Comment.

Ans. The electron 'hole' created at the centre of Photosystem II by a photon of light is 'filled' up with electrons coming from water during the water-splitting process (photolysis). A strong oxidant (Z^+) is produced at the centre with a redox potential of about 0.81V. This strong oxidant is sufficiently strong to oxidize water, resulting in the generation of molecular oxygen and ejection of electrons:

$$2H_2O \rightarrow 4H^+ + 4e^- + O_2$$

RQ 7.8. How is the proton gradient established across the thylakoid membrane? Explain.

Ans. The proton gradient across the thylakoid membrane is established in the following manner:

1. By the vectorial transfer of protons from the outside of the membrane to inside of the thylakoid interior or sac by the carrier plastoquinone which is converted to PQH_2 (Plastoquinol).

2. By protons (H^+) produced from the photolyzed water molecules.

RQ 7.9. Enumerate briefly the role of antennae or accessory pigments.

Ans. The chloroplast pigments are classified into two main groupings (i) vital pigments represented by Chl *a* and (ii) accessory or antennae pigments such as Chl *b*, Chl *c*, carotene and xanthophylls (the last two are collectively called carotenoids).

The 'antennae' pigments do absorb light and ultimately transfer their energy by inductive resonance to the 'reaction centre' (or primary trap), i.e., to Chl *a* 700 or P_{700} of the photosystem I and the Chl *a* 680 (or P_{680}) of photosystem II. The carotenoids show strong absorption in the blue-violet end (390–430 nm) of the spectrum, with almost no absorption in the red end. They serve to protect the chlorophyll molecules from being destroyed by strong light (photo-oxidation).

Chemically, carotenoids consist of two carbon rings linked to by a chain of 16 carbon atoms with alternating single and double bonds (carotene with an empirical formula $C_{40}H_{56}$, as shown in Figure 7.6 and xanthophyll with $C_{40}H_{56}O_2$).

Flower colours[11] in the yellow, orange and red range are often due to carotenoids present in plastids of petal cells.

RQ 7.10. Why carrots, which are rich in β-carotene, enhance human vision? Explain.

Ans. β-carotene is a typical carotenoid present in carrot roots. It has two carbon rings connected by a chain of 18 carbon atoms with alternating single and double bonds.

On cleavage within the human system, a molecule of β-carotene produces two molecules (or equal halves) of vitamin A which on oxidation produces retinal, the pigment used in human vision.

β-carotene is converted into vitamin A by animals

$$\underset{\beta\text{-carotene}}{C_{40}H_{56}} + 2H_2O \rightarrow \underset{\text{vitamin A}}{2C_{19}H_{27}CH_2OH}$$

Carotene was isolated from carrot root tissues by Wackenroder in 1831.

RQ 7.11. What happens when the etiolated angiosperm seedlings are subjected to light? Explain the chemical basis.

Ans. When we subject the etiolated seedlings of angiosperms to light, protochlorophyllide is reduced to 'chlorophyllide *a*'. Subsequently, chlorophyllide *a*, in the presence of enzyme chlorophyllase, undergoes esterification with a phytol group to form 'chlorophyll *a*'.

Thus, seedlings turn green. However, in gymnosperms, some ferns, and many algae, chlorophyll *a* can be synthesized completely from protochlorophyllide in the dark, solely through enzymatic activity.

$$\text{Protochlorophyllide} \xrightarrow[\text{2H}^+\text{ (Photoreduction)}]{\text{Light}} \underset{\text{(Immediate precursor)}}{\text{Chlorophyllide } a} \xrightarrow[\text{Esterification of a phytol group}]{\text{+Phytol}} \text{Chlorophyll } a$$

The light required for the conversion of protochlorophyllide to chlorophyllide a appears to be an absolute necessity in angiosperms. This is not the case so in gymnosperm seedlings.

RQ 7.12. List only the features, characterizing etiolation. In which ways it is considered to have a survival value?

Ans. The seedlings kept in darkness show the following symptoms:

1. Chlorophyll does not form but protochlorophyllide or protochlorophyll accumulates.
2. The leaves do not expand but remain small and rudimentary.
3. The internodes elongate many times more than normal, forcing the seedlings to become very tall (i.e., rapidly elongating shoots) and spindly.
4. The tender apical growing point is protected by the recurved hook.

This growth habit has a great survival value to the plant, since it permits the seedling to reach light quickly. The tender apical growing point remains protected by the hook (a curving of the hypocotyl that is thought to protect the apical meristem from damage during growth through the soil), and the energy is not wasted on the growth of leaves.

[11] Many flower colours, including pinks, reds, blues and purples, are due to anthocyanins present in the vacuoles of petal cells. The colour depends on their molecular structure (particularly the position of OH groups in one of the rings), the pH of cell sap in the vacuole, and the formation of complexes with metal ions and other compounds such as 'flavones' and 'flavonols' also present in the vacuoles (visible as 'colours' to some pollinators, including bees but not to human eyes).

Once the seedling breaks through the soil surface and reaches the light, some of the phytochrome is converted to 'Pfr' state which brings about the morphogenetic changes. The hook then opens, the stem thickens, and its rate of elongation slows down, and the leaves expand and become green (a process known as 'de-etiolation'). Exposure to light causes conversion of yellowish 'protochlorophyll' or 'protochlorophyllide' to green chlorophyll.

RQ 7.13. How many ATP molecules are required to synthesize one glucose molecule in C3 and C4 plants? Explain.

Ans. In Melvin Calvin cycle (C3) plants, 18 molecules of ATP are used up to synthesize one glucose molecule

$$6\ CO_2 + 18\ ATP + 12\ NADPH \rightarrow C_6H_{12}O_6 + 18\ ADP + 18\ Pi$$

During this process, four electrons are removed from water while eight (8) photons are required (4 at PS I and 4 at PS II) to reduce one molecule of CO_2 or to generate one molecule of O_2.

However, in C_4 plants, 30 ATP molecules are required to produce one molecule of glucose.

RQ 7.14. C_4 plants are photosynthetically more efficient than C3 plants. Explain.

Ans.
1. C4 plants exhibit a division of labour where two groups of cells in the mesophyll and the bundle sheath work co-operatively to fix carbon dioxide, thus resulting in more photosynthetic yield.
2. The enzyme PEP carboxylase of mesophyll cells has greater affinity for CO_2, and it can fix or reduce CO_2 at low levels (even less than 10 ppm), whereas RuBP carboxylase present in bundle-sheath cells functions optimally as CO_2 is internally generated (CO_2-to-O_2 ratio is always high, i.e., $CO_2 > O_2$).
3. C4 plants do not photorespire appreciably and thus, they waste none of their photosynthetically fixed CO_2.
4. The sugars formed in Calvin cycle (i.e., in the bundle sheath cells) are directly 'loaded' into the phloem sieve tubes for transport to other parts of plants.
5. C4 plants can maintain higher rates of photosynthesis in habitats with little available water or at lower CO_2 levels, even when the stomata are partially closed to conserve water during periods of water stress.

RQ 7.15. A photon of blue light is more energetic than a quantum of red light. Explain Why?

Ans. The energy of one quantum[12] of blue light (440 nm) is about 64,000 calories, and that of red light (average wavelength 700 nm) is about 40,000 calories. It is true that longer the wavelengths of light, the lower is the energy level per the equation established by the German physicist Max Planck.

$E = h \times \upsilon$ (Greek nu)

E = energy of the quantum or photon

h = Planck's constant in units of Joule/seconds (6.6254×10^{-14} J/s)

υ = frequency of the radiation

Frequency (υ) is related to wavelength (λ-Greek lambda) through the fundamental expression:

$$C = \lambda \times \upsilon \quad \text{or} \quad \upsilon = \frac{C}{\lambda}$$

Velocity of light = wavelength × frequency

Hence, $$E = h \times \upsilon \quad \text{or} \quad E = h \times \frac{C}{\lambda} \qquad \left[\because \upsilon = \frac{C}{\lambda}\right]$$

[12] Quanta (singular, quantum) of light are also called photons.

The energy of the quantum is directly related to frequency ($E = h\upsilon$) but is inversely related to wavelength.

Clearly, the longer the wavelength of light, the lower is the frequency and the less energetic the quantum.

A quantum of red light (40 kcal/Einstein) is as effective as a quantum of blue light (70 Kcal/Einstein) but the extra energy of blue quantum is wasted, on the contrary. Thus, a quantum of UV light (<390 nm) has more energy than a quantum of blue light, which in turn is more energetic than a quantum of red light.

Having a very short life, a quantum is not capable of carrying out a photochemical reaction. It loses a part of energy and returns to the first excited singlet state.

RQ 7.16. Why chlorophyll solutions or leaves appear green to eyes?

Ans. Chlorophylls absorb strongly or heavily in the blue (400–450 nm) and red regions (600–700 nm) of the visible spectrum. Virtually no absorption occurs in the green-yellow region. Green rays (about 500–560 nm) are mostly transmitted or reflected and are least effective photosynthetically. Hence, the chlorophyll solution and leaves appear green to the eyes.

RQ 7.17. Who first ever measured the action spectrum in chlorophyllous plants?

Ans. By using a filamentous algae (such as *Spirogyra or Cladophora*), Engelmann in 1883 found that the oxygen seeking or aerobic (or aerotactic) bacteria[13] (*Bacterium termo*) congregated in areas illuminated by red or blue light – the two regions of the spectrum where chlorophyll pigments absorb heavily and generate the most of oxygen (Figure 7.2).

RQ 7.18. In which ways is chlorophyll *a* different from chlorophyll *b*?

Ans. All the different forms of chlorophylls have the same basic chemical composition but differ from one another in minor details.

Chlorophyll *a*	Chlorophyll *b*
1. It is a blue green pigment, with the empirical formula $C_{55}H_{72}O_5N_4Mg$.	It is a greenish yellow-black green pigment with the empirical formula $C_{55}H_{70}O_6N_4Mg$
2. Chl *a* has a methyl (–CH_3 group) attached to carbon number 3.	Chl *b* has an aldehyde (–CHO or formyl group) at carbon 3.
3. Chl *a* shows maximal absorption peaks at 449 nm and 660 nm.	Chl *b* absorbs strongly at 453 nm and 642 nm, i.e., within the range produced by Chl *a*.

RQ 7.19. The anatomy of the leaves of many C4 plants differs from that in C3 plants. Comment.

Ans. In C4 plants, the vascular bundle is surrounded by a layer of large parenchyma cells, the bundle sheath which is, in turn, surrounded by smaller mesophyll cells–an arrangement called **Kranz anatomy** (Kranz from the German word for wreath or border). Chloroplasts in the two types of cells are morphologically distinct, those in the bundle sheath cells have very large starch grains but often lack grana (**agranal**), whereas those in the mesophyll cells contain well-defined grana but accumulate very little starch (Figure 7.25). Mesophyll cells have PEP carboxylase that catalyzes the addition of CO_2 to PEP while the bundle sheath cells have RuBP carboxylase.

In C_3 plants, the chloroplasts in all the green cells of the leaves are alike. They are small and have well-defined grana and have both the photosystems I and II. RuBP carboxylase is the crucial enzyme.

[13] The photosynthetic bacteria were first discovered by Engelmann in 1883.

RQ 7.20. If a C3 and a C4 plant are placed together under a bell jar in limited CO_2 supply, the C3 plant will be starved by C4 plant and will die? Comment.

Ans. This occurs because the C3 plant photorespires away some carbon already fixed in the form of CO_2 which is immediately used up by the C4 plant as the enzyme PEP carboxylase in the mesophyll cells has greater affinity, operating at even 10 ppm of CO_2. The CO_2 optima for C3 plants is much higher than this value. Thus, C3 plants can be said to 'photorespire themselves to death'.

RQ 7.21. List out two mineral ions that are required for photolysis of water in pigment system II.

Ans. Photolysis of water in PS II occurs in the presence of manganese and chloride ions. In the early 1940s, Warburg reported that chloride ions had a stimulating effect on the Hill's reaction probably by facilitating the release of O_2 from OH^- ions. Another ion, Ca^{2+}, is also essential for photolysis of water and operation of the PS II reaction centre where it perhaps plays some structural and regulatory role.

RQ 7.22. What do you understand by the term genetic autonomy or semi- autonomous characterization of chloroplasts? Comment.

Ans. Chloroplasts contain their own specific DNA (transmitted genetically through proplastids in the maternal 'egg' cytoplasm) and their own ribosomes which are smaller than cytoplasmic ribosomes. It appears that at least some of the chloroplast-specific proteins are coded by chloroplast DNA and synthesized by chloroplast ribosomes.

The enzyme RuBP carboxylase (Rubisco) is composed of two protein parts – the smaller part or subunit of 12 kD size, is coded by nuclear DNA, synthesized in the cytosol and imported into chloroplast. However, the synthesis of larger subunits of 58 kD, is under the control of chloroplastic genome and synthesized within the chloroplast. The total molecular weight of the active protein is about 557 kD with about 16 pairs each of the smaller and larger subunits with 80 sulphydral (SH^-) group.

The core complex of PS I contains about 15 proteins of which some are nuclear and others are chloroplast-encoded. Psa A and Psa B are by far the largest proteins, being encoded by chloroplast DNA.

RQ 7.23. Outline the steps where O_2 is consumed and CO_2 released during the photorespiratory process.

Ans. The glycolate pathway of photorespiration consumes O_2 at two different steps.

- The initial reaction, which produces phosphoglycolate and phosphoglycerate, consumes one molecule of oxygen. A second molecule of O_2 is consumed in the glycolate oxidase step. However, half is regenerated by catalase.

  1. Ribulose bisphosphate + $O_2 \xrightarrow{\text{(chloroplast)}}$ Phosphoglyceric acid + Phosphoglycolic acid

  2. Glycolate + $O_2 \xrightarrow{\text{peroxisome}}$ Glyoxylate + Hydrogen peroxide (H_2O_2)

  3. Hydrogen peroxide $\xrightarrow[\text{catalase}]{\text{(peroxisome)}}$ $H_2O + \frac{1}{2}O_2$

- The photorespiratory CO_2 is released during glycine–serine transformation in mitochondria[14] (this is where all the photorespiratory CO_2 is lost):

$$\text{Glycine} \rightarrow \text{Serine} + NH_3 + CO_2$$

Serine is then transported from mitochondria to peroxisome where it is converted to hydroxypyruvate and then to glycerate.

---

[14] Mitochondria, like chloroplast, are semiautonomous organelle, since they contain ribosomes, RNA, and DNA, which encode some of the mitochondrial proteins (between 15 and 20).

RQ 7.24. How is H_2O_2 removed from subcellular compartments?

Ans. The hydrogen peroxide from the cellular pool is removed by three enzymes: catalase, peroxidase and ascorbate peroxidase.

Catalases, which are localized in glyoxysomes and peroxisomes, scavenge most of the H_2O_2, while ascorbate peroxidase removes hydrogen peroxide in other subcellular compartments, particularly chloroplasts.

RQ 7.25. List three groups of plants that exhibit an inverted stomatal cycle or reverse stomatal rhythm.

Ans.
1. Members of Crassulaceae (*Crassula, Sedum, Kalanchoe* and *Bryophyllum*)
2. *Welwitschia* – a gymnosperm
3. Members of Cactaceae, Euphorbiaceae and Bromeliaceae

RQ 7.26. Name an agricultural crop that has Crassulacean acid metabolism (acronym CAM)

Ans. Pineapple (*Ananas comosus*) is the most important agricultural crop with a CAM pathway.

RQ 7.27. Name an ornamental plant exhibiting a CAM pathway.

Ans. The Ice plant (*Mesembryanthemum crystallinum*), a fleshy annual of the family Aizoaceae, exemplifies the CAM pathway. The fleshy leaves glisten in the sun.

RQ 7.28. List major crops that have C4 pathways.

Ans.
1. Sugarcane (*Saccharum officinarum*) – the most efficient converter of solar energy
2. Grain amaranth (*Amaranthus* spp.)
3. Maize (*Zea mays*)
4. Sorghum (*Sorghum vulgare*)
5. Tropical grasses

RQ 7.29. List out simply; the steps that are used to minimize photorespiration.

Ans.
- High [CO_2] levels
- Low [O_2] levels
- Low light intensity
- Use of inhibitors of photorespiration

RQ 7.30. How are C_4 plant species classified? List the groupings.

Ans. On the basis of the metabolite transported from the mesophyll cells to bundle sheath cells, and the kinds of enzymes within the cells, C_4 plant species can be grouped into three classes.

1. NADP-malate enzyme type
2. Phosphoenolpyruvate carboxykinase type
3. NAD-malate enzyme type

Chapter 8

Carbohydrate Metabolism

8.1 Sucrose and Starch Metabolism

Keywords

- ADP-glucose pyrophosphorylase
- Cellulose synthase
- Csl genes
- Fructosyl transferases
- Q-enzyme
- Starch synthase
- Sucrose-6-phosphate synthase

Sucrose and starch, the principal leaf storage products, are biosynthesized in two different compartments – sucrose in the cytoplasm of photosynthetic cells and starch in the chloroplast. Both the compounds accumulate in daylight and are mobilised to meet the metabolic demands during night time or at the time of minimum photosynthetic yield. The pathway for sucrose synthesis is highly regulated and synchronized with the pathway associated with starch synthesis.

The appropriation of carbon fixed by the Calvin cycle into either starch or sucrose synthesis is referred to as **carbon allocation.**

8.1.1 Sucrose biosynthesis

Sucrose is the most common type of carbohydrate present in the translocation stream (Figure 8.1). Plants such as wheat, oats and barley temporarily accumulate large quantities of sucrose in the vacuole. Sucrose (non-reducing sugar) is a disaccharide and each molecule is composed of two monosaccharides, glucose and fructose (reducing sugars). Instead of free glucose and fructose, phosphorylated forms of these sugars serve as precursors of sucrose.

Glucose — α-1,2-link — Fructose

Sucrose

Fig. 8.1 A sucrose molecule

Uridine diphosphate-glucose (UDP-G) is the most preferred precursor for sucrose biosynthesis in the cytoplasm of cells of the leaf.

Reactions involved in the pathway of sucrose formation are as follows:

1. $\text{UTP} + \text{Glucose-1-P} \rightleftharpoons \text{UDP-glucose} + \text{PPi}$
2. $\text{PPi} + H_2O \rightleftharpoons 2\text{Pi}$
3. $\text{UDP-glucose} + \text{Fructose-6-P} \xrightleftharpoons[Mg^{2+}]{\text{Sucrose phosphate synthase}} \text{Sucrose-6-P} + \text{UDP}$
4. $\text{Sucrose-6-P} + H_2O \xrightleftharpoons[Mg^{2+}]{\text{Sucrose phosphate phosphatase}} \text{Sucrose} + \text{Pi}$
5. $\text{UDP} + \text{ATP} \rightleftharpoons \text{UTP} + \text{ADP}$

Overall equation for sucrose synthesis is

$$\text{Glucose-1-P} + \text{Fructose-6-P} + 2\ H_2O + \text{ATP} \rightarrow \text{Sucrose} + 3\text{Pi} + \text{ADP}$$

Only one ATP molecule is essential for the biosynthesis of sucrose. Since three ATP molecules are consumed in the Calvin cycle for each carbon in each hexose of sucrose (a total of 36 ATPs), one additional ATP needed to form a glycosidic bond in sucrose is a small additional requirement.

Sucrose exported from the leaf cell to non-photosynthetic tissues may be oxidized quickly or stored in the vacuoles for a short period or gets transformed to starch for long-term storage in the chloroplast.

***Regulation of sucrose-6-phosphate synthase*:** Sucrose phosphate synthase (SPS) is a key regulatory enzyme in sucrose biosynthesis and catalyzes the transfer of a glucose molecule from UDP-glucose to fructose-6-phosphate to yield sucrose-6-phosphate and UDP.

SPS is regulated *via* reversible protein **phosphorylation** and **dephosphorylation**. For example, SPS kinase causes phosphorylation of the enzyme sucrose phosphate synthase, thus converting this to an 'inactive' form while SPS phosphatase enzyme dephosphorylates the enzyme, thus, activating it.

SPS is also allosterically stimulated by G-6-P and repressed by Pi. Repression of the SPS kinase by G-6-P, and of the SPS phosphatase by Pi, enhances the consequence of G-6-P and P_i on sucrose biosynthesis. Thus, large quantities of G-6-P formed due to high photosynthetic rate stimulate sucrose phosphate synthase, thereby producing sucrose phosphate. A high P_i concentration, resulting from slow conversion of ADP to ATP, inhibits sucrose phosphate production (Figure 8.2)

8.1.2 Sucrose catabolism

In plants, sucrose is transported from photosynthetic tissues (source) to non-photosynthetic tissues (sink), where it is metabolized for providing energy as well as synthesizing polymeric compounds (such as, starch, cellulose) or it is stored to be consumed subsequently as per requirements. Sucrose can be degraded enzymatically by either **sucrose synthases** or **invertases**.

(i) $\text{Sucrose} + \text{UDP} \xrightleftharpoons{\text{Sucrose synthase}} \text{UDP-glucose} + \text{Fructose}$

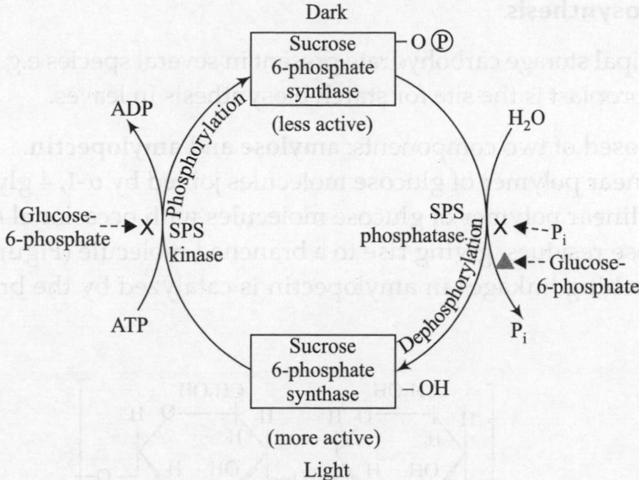

Fig. 8.2 Regulation of sucrose-6-phosphate synthase (SPS) *via* phosphorylation. SPS kinase phosphorylates serine residue in SPS and makes it less active. This inhibition is reversed by sucrose phosphate synthase phosphatase. Glucose-6-P represses the SPS kinase and also stimulates SPS allosterically. Pi inhibits the phosphatase and also sucrose phosphate synthase directly.

Sucrose synthase (SS) catalyzes reversible reaction and utilizes UDP for catalyzing sucrose breakdown, releasing UDP-glucose and fructose. UDP-glucose is converted to glucose-1-phosphate and UTP using PPi, catalyzed by enzyme **UDP-glucose pyrophosphorylase**. Sucrose synthase is present as soluble form in cytosol or is associated with plasma membrane in insoluble form. Both SS and UDP-glucose pyrophosphorylase can produce phosphorylated hexoses in an ATP-independent pathway and are thought to be important in amyloplasts of developing grains or in potato tubers.

(ii) $\text{Sucrose} + H_2O \xrightarrow{\text{Invertase}} \text{Glucose} + \text{Fructose}$

Invertases are the hydrolytic enzymes which hydrolyze sucrose to glucose and fructose, which must be phosphorylated using ATP in order to be utilized further. Invertases can be either acidic or alkaline. Acid invertases are present in both the vacuole as well as the apoplast and display acidic pH optima of 5.0, whereas alkaline invertases are localized in the cytosol as well as the plastid with neutral or alkaline pH optima of 7.5. Vacuolar invertases hydrolyze sucrose stored in the vacuoles accompanied by transport of free hexoses to the cytosol for further metabolism. Developing fruits and rapidly growing tissues, such as the elongation zone of roots and hypocotyls have been found to possesss high levels of vacuolar invertases. Vacuolar invertases are involved in cell expansion, sugar storage, and in regulation of cold-induced sweetening. The apoplastic invertases, found in plant reproductive tissues, are bound firmly to the cell wall matrix and hydrolyze apoplastic sucrose delivered to sink tissues. Cell wall invertases function in sucrose partitioning, plant development, and cell differentiation. Cytosolic invertases are important for plant growth, especially roots.

8.1.3 Starch biosynthesis

Starch is the principal storage carbohydrate present in several species e.g., spinach, tobacco and soybean. Chloroplast is the site for starch biosynthesis in leaves.

Starch is composed of two components: **amylose** and **amylopectin**.

Amylose is a linear polymer of glucose molecules joined by α-1, 4 glycosidic linkages. Amylopectin is a linear polymer of glucose molecules with occasional α-1,6 linkages for every 24–30 glucose residues, giving rise to a branched molecule (Figure 8.3). Formation of the α-(1,6) branching linkages in amylopectin is catalyzed by the branching enzyme **Q-enzyme**.

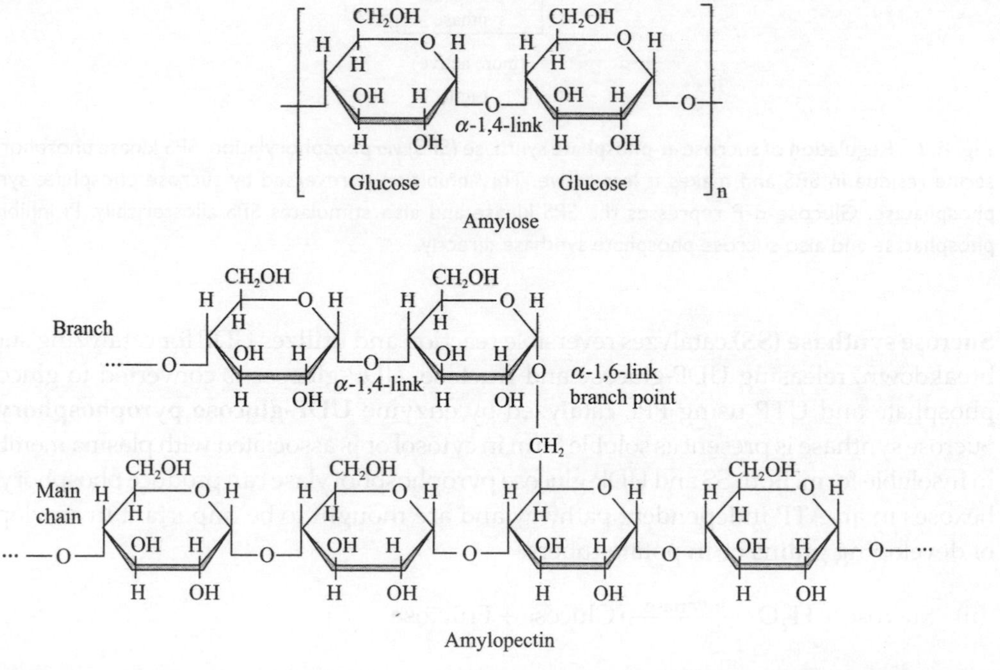

Fig. 8.3 Structural components of starch-amylose and amylopectin. Formation of α-(1,6) branching linkages in amylopectin

Reactions for starch biosynthesis are as follows:

1. $\text{ATP} + \text{glucose-1-P} \rightleftharpoons \text{ADP-glucose} + \text{PPi}$
2. $\text{PPi} + H_2O \rightleftharpoons 2\text{Pi}$
3. $\underset{(n\text{-glucose units}) \; (\alpha\text{-}(1\rightarrow4)\text{-glucan})}{\text{ADP} - \text{glucose} + \text{small amylose}} \xrightleftharpoons[K^+]{\text{Starch synthase}} \text{ADP} + \underset{(\text{with } n+1 \text{ glucose units}) \; (\alpha\text{-}(1\rightarrow4)\text{-glucosyl-glucan})}{\text{large amylose}}$

Starch biosynthesis occurs in plant plastids, using ADP-glucose as the preferred substrate. Starch synthesis occurs in chloroplasts and amyloplasts of the non-photosynthetic parts of plants for short- and long-term storage respectively. The process of starch synthesis is subjected to regulatory control by ADP-glucose synthesis. The major regulatory and rate-limiting enzyme in starch synthesis is **ADP-glucose pyrophosphorylase**, activated *via* allosterism by 3-phosphoglycerate (3-PGA) and repressed by P_i. Thus, the ratio of 3-PGA to P_i which increases with rise in photosynthesis during daytime, is a determining factor for starch synthesis (Figure 8.4).

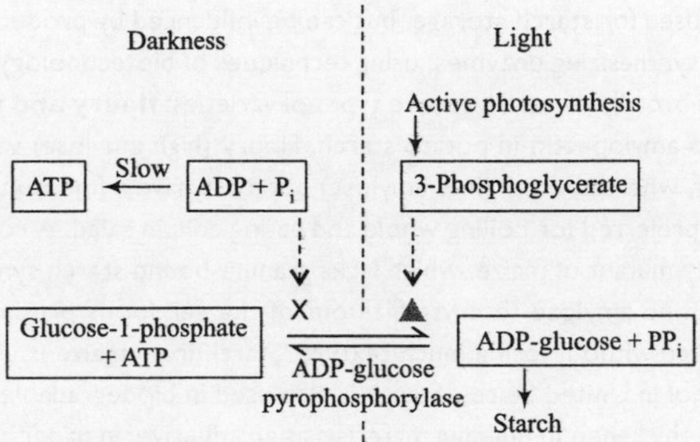

Fig. 8.4 Regulation of ADP-glucose phosphorylase, the key enzyme in starch synthesis. ADP-glucose phosphorylase is activated *via* allosterism by 3-phosphoglycerate (3-PGA) and repressed by P_i. The ratio of 3-PGA to P_i is the major controlling factor for starch synthesis at this step.

The two competing processes, storage as starch or its transport to the cytoplasm can be regulated by slight change in triose phosphate/Pi ratio and fructose-2,6-bisphosphate (F26BP), the regulator metabolite. The concentration of F26BP is inversely proportional to photosynthetic rate and thus, represses the generation of F-6-P, which is one of the the precursors of sucrose. F26BP affects, allosterically, the enzymes that determine the amount of F-6-P.

Starch synthase: The Starch synthase or SS enzyme is responsible for the synthesis of large amylose units of starch and is encoded by a family of five genes, classified as **GBBS (granule-bound starch synthase), SS1, SS2, SS3 and SS4**. GBBS is tightly associated with the starch granule and synthesizes amylose. The SS isoforms are soluble enzymes located in the plastid stroma or partly bound to the granule, and work with starch branching enzyme to synthesize amylopectin. There are two types of branching enzymes (BE) which differ in the length of chain they transfer. In addition to GBBS, the developing starch granule has association with multiple modifying enzymes (for example, amylases, phosphorylases and glycosyltransferases) which have the potential to restructure the growing granule as well as dismantle it during net degradation. By evolving multiple, specialized enymes, plants have developed the capacity to accumulate starches with a range of molecular architectures. This versatility is of commercial significance, since starches with different properties have a wide range of industrial uses. Deficiencies of specific isoforms of starch-synthesizing enzymes result in starch granules with different properties. Waxy variants of potato, and wrinkled pea seeds, are mutants low in amylose because they lack granule-bound starch synthase activity.

Starch: Its Roles in Food and Other Industries

Starch is the major nutritive component of our staple crops. Its unique properties affect the texture of the food we eat and make it useful for other non-food applications. The proportions of amylose to amylopectin, together with the size of starch granules, determine the properties of extracted starch that are important in both food and other industries. However, these characteristics differ in different species and the organ used for starch storage, but can be influenced by producing plants with altered amounts of the starch synthesizing enzymes, using techniques of biotechnology.

White potatoes are broadly divided into two types of varieties: **floury and waxy**, depending upon the ratio of amylose to amylopectin in potato starch. Floury (high amylose) varieties are best suited for baking and mashing, whereas waxy (low amylose) potatoes have a fine texture, retain their shape on cooking and so are preferred for boiling whole and eating cold in salad. A commonly used starch is obtained from the waxy mutant of maize, which lacks granule-bound starch synthase and so produces only amylopectin. With no amylose to crystallize out of the gel, foods produced from waxy starch can be frozen and thawed without losing much texture. Starch from maize is used extensively for the production of bioethanol in United States. Starch is also used in biodegradable plastics and packaging materials, in paints as a thickener, in building materials as an adhesive, in paper and textiles as a coating and sizing agent, and in pharmaceuticals and cosmetics as an inert carrying agent.

Deficiencies in starch synthesis are associated with the buildup of soluble sugars, as occurs in sweet corn varieties. Accumulation of sugars during low-temperature storage of potatoes can lead to the formation of unpalatable and carcinogenic products on frying or roasting. Low-temperature sweetening during dormancy is the result of the gradual degradation of starch and slow synthesis of sucrose using respiratory energy. Preferential inactivation of the enzyme phosphofructokinase during low-temperature storage leads to restriction of glycolysis, accumulation of hexose phosphates and increased sugar content.

8.1.4 Starch catabolism

Starch is degraded both by phosphorolytic and hydrolytic enzymes.

The **phosphorolytic mode of cleavage** requires **starch phosphorylase, glucosyltransferase** as well as **debranching enzyme. Debranching enzyme (R-enzyme)** acts on branch points i.e., α-(1,6)-glycosidic linkages, releasing linear glucan polymer, which are acted upon by starch phosphorylases at the non-reducing end, resulting in removing terminal glucose as glucose-1-phosphate, one at a time. Condensation of shorter glucan chains, to produce substrate for starch phosphorylase, is catalyzed by **glucosyltransferase**.

$$\alpha\text{-(Glucan)}_n + \text{Pi} \xrightarrow{\text{Starch phosphorylase}} \alpha\text{-(Glucan)}_{n-1} + \text{Glucose-1-phosphate}$$

The **hydrolytic mode of cleavage** includes **amylases** (exoamylases and endoamylases), **debranching enzymes** and **α-glucosidase** enzymes. Debranching enzyme is associated with hydrolysis of α-1,6-glycosidic linkages. **Exoamylase** (e.g., β-amylase) splits off two glucose molecules from the non-reducing end of starch molecule and **endoamylases** (e.g., α-amylase) catalyze hydrolysis of α-1,4-glycosidic linkages at the inerior of the starch molecule. Dextrins and maltoses, the degradation products of α-amylases, are further broken down to glucose molecules by α-glucosidase enzyme.

8.2 Cellulose Synthesis

Cellulose, a linear polysaccharide, is the most abundant organic compound on the earth. It is the basic component in the cell walls of plants, helping them to remain stiff and preventing the rupturing of cell when water enters into it by osmosis. It is composed of linear chains of thousands of (β)-linked D-glucose units, aggregated into bundles of at least 18 chains, that crystallize into microfibrils and ultimately integrate into the cell wall (Figure 8.5)

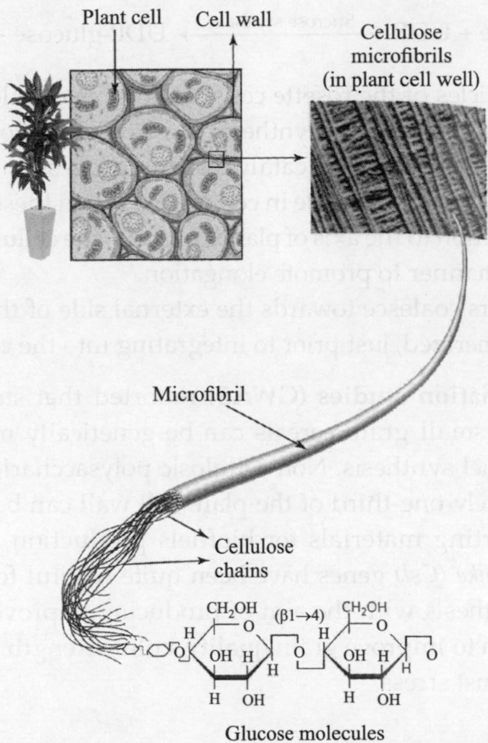

Fig. 8.5 Cellulose polymers, the basic units of plant cell wall.

The enzymatic machinery for synthesis of cellulose is more complicated as compared to synthesis of starch and sucrose. Freeze-fracture studies show that in plant cells, cellulose synthesis takes place in rosette-type cellulose synthesis complexes in the plasma membrane which contain multiple copies of cellulose synthase. Each complex (diameter about 30 nm) contains six large particles arranged in a regular hexagon. Much of the complex synthetic machinery is based on the genomic analysis of *Arabidopsis thaliana* and its structure resembles that of the bacterium *Rhodobacter sphaeroides*. The enzyme, **cellulose synthase,** exhibits glycosyl transferase activity in its cytoplasmic domain and creates a transmembrane channel through which the growing cellulose chain is extruded. Glucose units are transferred from UDP-glucose to the non-reducing end of the growing chain, making some polymers as long as 15,000 glucose units.

Steps for synthesis of cellulose

(i) Sucrose (produced during photosynthesis) combines with UDP in the presence of sucrose synthase to generate UDP-glucose to be used for cellulose synthesis per the following reaction:

$$\text{Sucrose + UDP} \xrightarrow{\text{Sucrose synthase}} \text{UDP-glucose + fructose}$$

(ii) Each of the six particles of the rosette consists of three cellulose synthase molecules, each one of them associated with synthesis of a single cellulose chain.

(iii) The large enzyme complex that catalyzes this process moves along the plasma membrane, directed by microtubule in cell cortex. When these microtubules lie in the perpendicular direction to the axis of plant's growth, the cellulose microfibrils are laid down in a similar manner to promote elongation.

(iv) 18 separate polymers coalesce towards the external side of the cell and solidify soon after they are polymerized, just prior to integrating into the cell wall.

Genome-wide association studies (GWAS) reported that stem cellulose content in bread wheat and other small grain cereals can be genetically manipulated to increase culm strength and biofuel synthesis. Non-cellulosic polysaccharides and hemicelluloses, constituting approximately one-third of the plant cell wall can be used as dietary fibres, food additives and starting materials for biofuels production. The knowledge of the wheat *cellulose synthase-like (Csl)* genes have been quite helpful for genetic manipulation of hemicellulose biosynthesis with the aim of producing improved wheat varieties with altered wall composition to improve grain quality, culm strength, biofuel production and increased tolerance against stress.

8.3 Fructans

Fructans are linear and branched polymers of sucrose. More than 36,000 plant species, belonging to members of the orders Asterales, Cyperales, Asparagales, Liliales and Poales, tend to accumulate carbohydrates as fructans. Fructans represent the polysaccharide reserves of onion bulbs, chicory and Jerusalem artichoke roots, and the vegetative tisues of temperate grasses (Figure 8.6). Besides this, bacterial strains of the genera *Bacillus, Streptococcus, Pseudomonas, Erwinia, Actinomyces* and such also synthesize fructans.

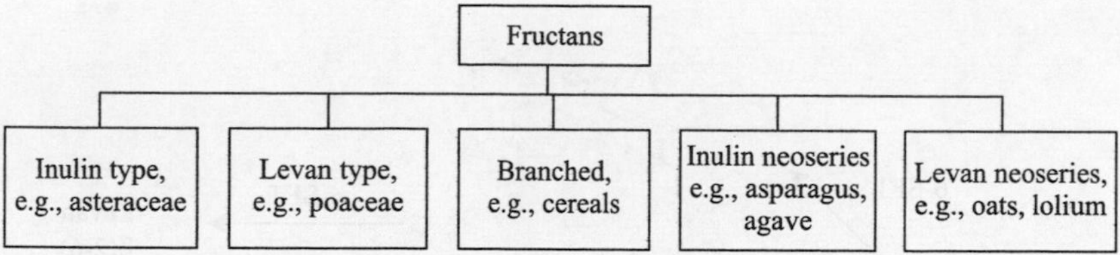

Fig. 8.6 Sources of different types of fructans

a. Fructans, the storage carbohydrate, exhibit the following characteristics. They are:
   i. Water-soluble
   ii. Synthesized and stored in cell vacuole
   iii. Osmotically active
   iv. Providers of stability to plasma membrane
   v. Behave as antistress metabolites

Fructans are derived from sucrose, to which fructose chains are joined by glycosidic bonds. **Kestose** (a trisaccharide) is the precursor of fructans, and has fructose residue joined to a sucrose molecule. Based on the bonding among sucrose and other fructose molecules, three kinds of kestose exist: 1**-kestose, 6-kestose, and neokestose**, and from them three types of fructans are obtained.

- 1-kestose types (inulin fructans); consist of chains of β(2→1)-linked fructose residues as in Dahlia tubers.
- 6-kestose types (levan fructans); consist of β(2→6)-linked fructose residues such as in grasses.
- Neokestose (found in wheat and barley); two polyfructose chains are connected with the sucrose. The first chain is joined to the glucose residue of sucrose by a β(2→6) glycosidic bond, while the second one is joined to fructose residue of sucrose by a β(2→1) glycosidic bond.

Fructans are biosynthesized from sucrose by the activity of **fructosyl transferases** in cell vacuoles of the storage tissues. Fructans as reserve compounds, are favoured over other polysaccharides, especially prior to flowering stage or quick seed development or exposure to abiotic stresses, such as low-temperature or water deficit . It is assumed that fructan synthesis might take place to regulate sucrose formation in cytoplasm. Fructans synthesis does not require activation of their precursor molecules and this is unlike starch and sucrose synthesis, that use UDPG and ADPG for activation of their synthesis. The thermodynamic requirement for fructosyl transfer is met by the breakage of the glucose-fructose bond ($\Delta G^{o\prime} = -29.3$ kJ mol^{-1}). Generally, sucrose acts as the starting material for fructan biosynthesis, with a glucose residue at the terminal end always. In some of the fructans, glucose may be eliminated by activity of endo-inulinase or α-glucosidase enzymes. Fructans are catabolized by sequential hydrolysis of fructose units from the terminal end, catalyzed by exo-hydrolytic enzymes. Reactions for fructan synthesis are given in Figure 8.7.

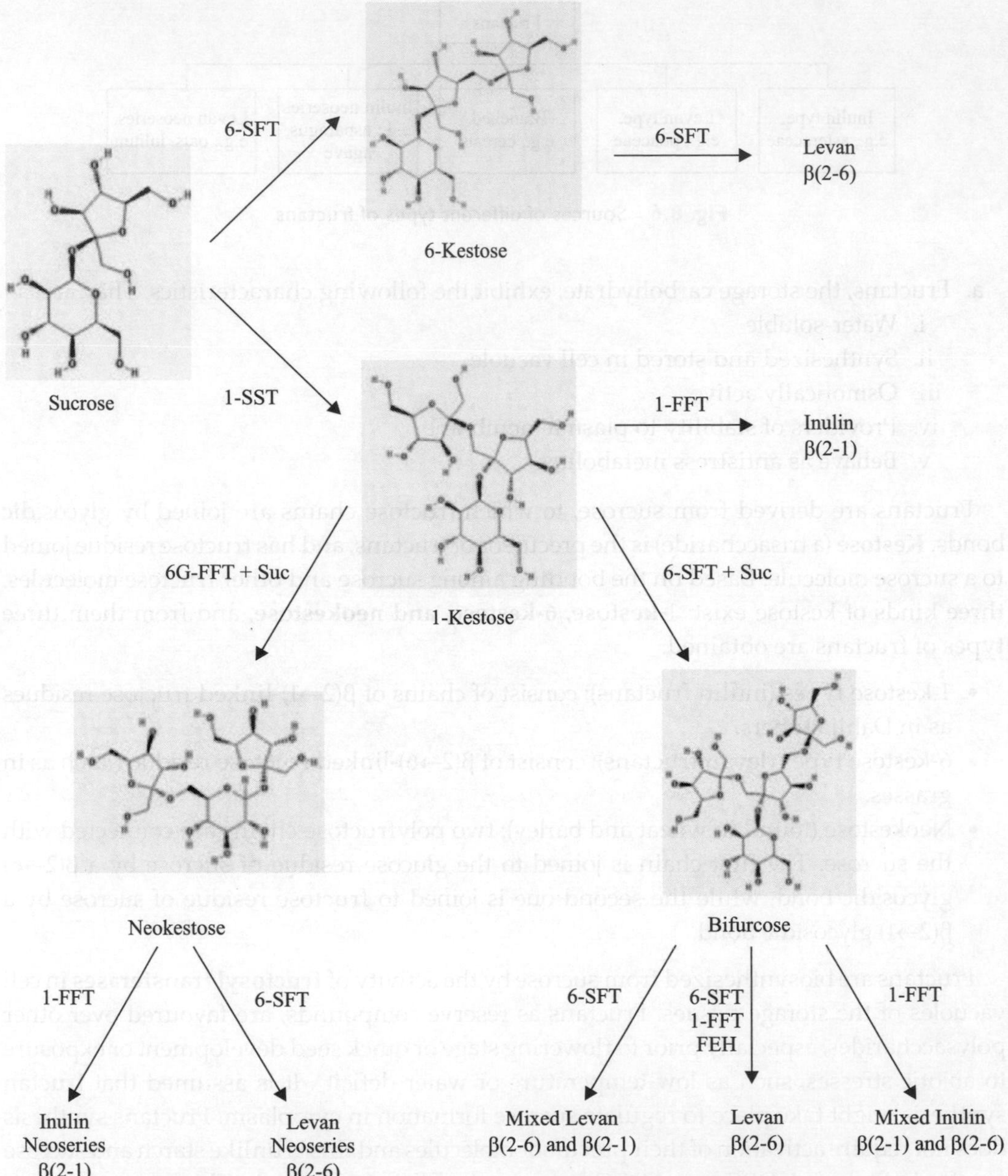

Fig. 8.7 Reactions that govern fructan synthesis

6-SFT: Sucrose: Fructan-6-fructosyl transferase
1-SST: Sucrose: Sucrose-1-fructosyl transferase
1-FFT: Fructan: Fructan-1-fructosyl transferase
FEH: Fructan 1-exohydrolase

Fructans can vary from small to large size. The smaller ones taste sweet with low calorific value and the larger ones have neutral taste with fat-like texture; so their use as a sweetener and low calorie fat has been adopted by the food industry.

8.4 Carbohydrate Metabolism in Plants and Animals

Plant cells function like animal cells in many respects:

- They perform the similar processes that produce energy in animal cells e.g.respiration
- They synthesize hexose compounds from 3- or 4-C compounds e.g. gluconeogenesis
- They oxidise hexose phosphates to pentose phosphates, producing NADPH e.g. oxidative pentose phosphate pathway
- They synthesize glucose polymer i.e. starch and its degradation to form hexoses.

However, carbohydrate metabolism in the plant cell is highly complicated compared to that in an animal cell. In addition to the carbohydrate transformations that the plant cells share with the animal cells, there are a number of functions that are exclusively performed by photosynthetic plant cells. These are:

- Assimilation of carbon dioxide into organic compounds (the carboxylation reaction);
- Using the final products of CO_2 assimilation to synthesize trioses, hexoses and pentoses (the Calvin-Benson cycle);
- Conversion of acetyl CoA produced by fatty acid breakown to 4-carbon compounds (the glyoxylate cycle);
- Conversion of 4-carbon compounds to hexose sugars (gluconeogenesis).

These processes, exclusively going on in the plant cell, and lacking in animal cells operate in different compartments; for example, the glyoxylate pathway (in glyoxysomes), the Calvin cycle (in chloroplasts), starch synthesis (in amyloplasts), and organic acid storage (in vacuoles). The inclusion and coordination of activities occurring among different compartments makes use of **specific transporters** in organelle membranes that transfer metabolites between organelles or into the cytoplasm. The regulation of these transporters has a marked influence on the degree to which the metabolic pools intermix. The pathways belonging to carbohydrate metabolism operating in plants exhibit extensive overlapping; these share metabolic pools of common intermediate compounds such as, phosphates of hexoses or pentoses or trioses, which are freely interchangeable through reactions which require little changes in standard free energy. Whenever any compound of the pool is exhausted, an original equilibrium is immediately reestablished to restore the depleted compounds. The extent and direction of metabolite flow through the pools keep changing based on changes in requirement of the plants and different tissue types.

During day time, triose phosphates generated in the plant leaf tissues ('source') *via* the Calvin cycle, exit the chloroplast to enter into the cytosolic hexose phosphate pool, where they are transformed to sucrose and eventually, transported to roots and tubers ('sink') *via* the phloem. Sucrose is either stored as starch within the sink tissues, or used as an energy source for both source and sink tissues through glycolytic as well as oxidative phosphorylation reactions during night time.

Review Questions

RQ 8.1. Outline how sucrose moves from leaves to non-photosynthetic parts of the plant.

Ans. Sucrose synthesized in the leaves is the source of carbon for all other cells in the plant and must be transported to non-photosynthetic parts of the plant i.e., sink organs (roots, meristems, young leaves, flowers, seeds and fruits, and vegetative storage organs such as tubers and rhizomes). The movement of sugars in the sieve elements of the phloem tissue occurs by a process called **pressure flow**. The pressure gradient in the phloem generated by the process of **phloem loading** causes the movement of phloem contents from the source to the sink organs.

RQ 8.2. Explain how sucrose is metabolized in non-photosynthetic cells to provide reducing power, ATP, and biosynthetic precursors.

Ans. Three basic components required for synthesis of a vast array of molecules essential for cell maintenance and growth are: reducing power (NADH or NADPH), energy (ATP), and precursor molecules. In a photosynthetic cell, the requirements for all these three components are met through photosynthesis, whereas in non-photosynthetic cells these three are provided by carbon compounds derived from sucrose imported from the leaves through phloem.

Sucrose entering the non-photosynthetic cells is initially metabolized to hexose phosphates (in cytosol), which are further metabolized by three interrelated metabolic pathways (glycolysis, the oxidative pentose phosphate pathway, and the Krebs cycle) to provide all of the reducing power, ATP, and precursor molecules for the cell.

RQ 8.3. Describe the main storage forms of carbohydrates in plants and where are they stored?

Ans. Sucrose entering the non-photosynthetic plant cells can be converted to storage compounds to be mobilized later, besides being used for biosynthesis and energy production.

(i) Sucrose can be either stored or can be used in the synthesis of **fructan** in the vacuole itself
(ii) It can be stored as **starch** granules in the plastids
(iii) It can be converted to storage **lipids** in oil bodies derived from the endoplasmic reticulum. Formation of storage lipids involves the synthesis of fatty acids in the plastid followed by export to ER for lipid synthesis.

The main form of carbohydrate in which carbon is stored varies with the cell types and the developmental stages.

RQ 8.4. Why must starch be hydrolyzed before it can be used as an energy source or transported through living systems?

Ans. Starch must be hydrolyzed to monosaccharides and disaccharides before being used as energy source or transported. The plant breaks down its starch reserves when it requires sugars for growth and development. We humans, hydrolyze these polysaccharides when our digestive systems break down the starch stored by plants in maize grains and potato tubers, making glucose available as a nutrient for our cells.

RQ 8.5. Why is it an advantage for a plant to store food energy as fructans rather than as starch? And, as oils rather than as starch or fructans?

Ans. Fructans (polymers of fructose) are generally the major storage polysaccharides in leaves and stems of some plants such as wheat, rye, and barley. These polymers are water-soluble and can be stored in much higher concentrations than starch. Further, it is advantageous for a plant to store oils rather than starch/fructans because oils (or fats) consist of a higher proportion of energy-rich carbon-hydrogen bonds than do carbohydrates and as a consequence, contain more chemical energy. On an average, fats yield about 9.1 kcal/g upon oxidation to release energy, compared with 3.8 kcal/g for carbohydrates.

RQ 8.6. Starch is never transported in plants yet potato tubers, which grow underground, are full of it. Explain.

Ans. Sucrose is the primary metabolite as well as the primary mobile product produced during photosynthesis, and is actively pumped into the sieve tubes of the smallest veinlets of the leaf, adjacent to xylem vessels (phloem loading). Loading is believed to involve cotransport of sucrose and H^+ ions through the specific permease in response to pH and electrochemical gradients. H^+ is later re-secreted through an ATP-utilizing proton transporter.

As sugar concentration increases in the sieve tubes, their water potential in turn decreases, and the water is thus withdrawn from the xylem vessels by osmosis. The turgor pressure thus built-up 'forces' the solution to 'move passively' through the sieve tubes, carrying it towards the root system and then onto the developing storage organs like tubers (or rhizomes and bulbs, etc.) that represent the 'consumption end or sink'. It is here that the soluble sugars are actively removed (phloem unloading) and are used up or else converted into insoluble starch through the enzymatic reactions as given below:

$$\text{Sucrose + UDP} \longrightarrow \text{UDP-glucose + Fructose}$$

As 'ADP-glucose' is preferred for starch synthesis, UDP-glucose is converted to ADP-glucose as follows.

$$\text{UDP-glucose + Pi} \rightleftharpoons \text{UTP + Glucose-1-phosphate}$$

$$\text{Glucose-1-phosphate + ADP} \rightleftharpoons \text{ADP-glucose + PPi}$$

The resulting ADP-glucose is then converted to starch by 'starch synthase'.

Sugar conversion or utilization guarantees the continued existence of a turgor pressure gradient from 'source' (the leaves) to 'sink' (the roots), thus facilitating the flow of metabolites. All movements of carbohydrates and even most of nitrogen compounds, whether upward or downward, occurrs in the phloem (Curtis).

RQ 8.7. Why is it that sucrose synthesis occurs 'exclusively' in the 'cytosol' of the photosynthetic cells? Outline the two routes of sucrose production in photosynthetic organisms.

Ans. Sucrose synthesis occurs in the 'cytosol' of photosynthetic cells as it can be effectively loaded from the smaller veins of the leaves into the sieve tubes of phloem. If sucrose was to be produced in the 'chloroplasts', it would be unable to exit as the inner membrane of the chloroplast which is impermeable to sucrose and thus would not allow it to enter the translocation stream.

There are two possible routes of sucrose synthesis in the photosynthetic cells but the principal pathway is regulated by the enzyme sucrose phosphate synthase as given in equation 1 below:

1. $$\text{UDP-glucose + Fructose-6-phosphate} \xrightleftharpoons{\text{Sucrose phosphate synthase}^{1}} \text{UDP + Sucrose-6-phosphate}$$

$$\text{Sucrose-6-phosphate} + H_2O \xrightleftharpoons{\text{Sucrose phosphate phosphatase}} \text{Sucrose + Pi}$$

---

[1] Sucrose phosphate synthase in some tissues can even use ADP-glucose but UDP-glucose is the most predominant form often used.

2. Another cytoplasmic enzyme capable of synthesizing sucrose is 'sucrose synthase'.

$$\text{Fructose} + \text{UDP-glucose} \xrightleftharpoons{\text{Sucrose synthase}} \text{UDP} + \text{Sucrose}$$

Sucrose synthesis in either of the both pathways, unlike starch biosynthesis, requires activation of glucose with the nucleotide UTP instead of ATP.

$$\text{Glucose-1-phosphate} + \text{UTP} \rightleftharpoons \text{UDP-glucose} + \text{PPi}$$

$$\text{PPi} + H_2O \rightleftharpoons 2\text{Pi}$$

Chapter 9

Respiration

Respiration is a catabolic, energy-releasing process in which sugars or other organic molecules are completely oxidized to carbon dioxide (CO_2) and water (H_2O). The potential energy locked within the carbohydrates as well as other organic compounds is transformed into kinetic energy by this process, which is similar to combustion. Combustion is a rapid release of energy at high temperatures, the energy being converted into heat and light, whereas respiration takes place at ordinary temperatures in living cells.

Keywords

- Alcoholic fermentation
- Amphibolic pathway
- Anaplerotic reactions
- Chemiosmotic hypothesis
- ETS-level phosphorylation
- Glycolysis
- Lactic acid fermentation
- Oxidative phosphorylation
- Pasteur effect
- Respiratory quotient
- Substrate-level phosphorylation
- Thermogenesis

Carbohydrates (glucose) are the principal substrates oxidized during respiration.

The overall reaction for complete oxidation of a glucose molecule:

$$\underset{\text{Glucose}}{C_6H_{12}O_6} + \underset{\text{Oxygen}}{6O_2} \rightarrow \underset{\text{Carbon dioxide}}{6CO_2} + \underset{\text{Water}}{6H_2O} + \text{Energy}$$

Oxygen being the ultimate electron acceptor, this reaction is highly exergonic, with the release of 686 kcal/mol under standard conditions.

Fats and proteins are comparatively richer sources of energy, the former yielding on an average 9.1 kcal/g and the latter approximately 6.7 kcal/g.

Comparisons between chemical equations of photosynthesis and respiration reveal that the two processes are diametrically opposite to each other. Photosynthesis builds up carbohydrates with the absorption of CO_2 and release of O_2; respiration breaks up carbohydrates with the absorption of O_2 and release of CO_2.

The destructive process of respiration goes on at all times and in all living cells, while the constructive process of photosynthesis proceeds only in sunlight and is confined to only green cells. Photosynthesis is a more rapid process as compared to respiration.

Respiration involves **glycolysis**, the **oxidative decarboxylation of pyruvate to acetyl CoA**, the **Krebs cycle (citric acid cycle)** and the **electron transport system**, which produces a gradient that drives oxidative phosphorylation (Figure 9.1).

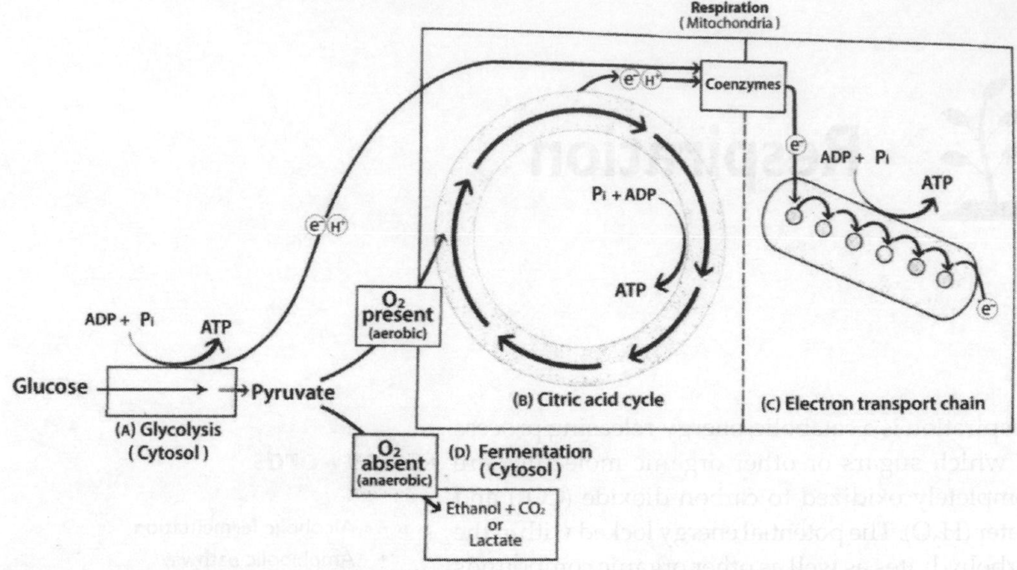

Fig. 9.1 Overview of respiration in mitochondria. The complete oxidation of glucose consists of three steps: (i) glycolysis (ii) the citric acid cycle (iii) the electron transport chain. (A) During glycolysis, glucose splits into pyruvate along with ATP production. (B) In the presence of O_2, pyruvate undergoes oxidative decarboxylation, producing acetyl CoA that enters into citric acid cycle (Krebs cycle), where additional ATP is synthesized. (C) In the electron transport system, the electrons are transferred to the oxygen via electron transport carriers, generating a proton gradient, which can drive more ATP synthesis. (D) In the absence of oxygen, pyruvate undergoes a fermentation process to produce ethanol and lactate, with no additional ATP production.

9.1 Glycolysis

Glycolysis ('glyco' meaning sugar and 'lysis' meaning splitting).

The glycolytic pathway is also referred to as **Embden–Meyerhof–Parnas (EMP) pathway** in the honour of three German biochemists who in 1940 elucidated the sequence of reactions from glucose to pyruvate. Glycolysis describes the sequential breakdown of glucose to pyruvate in a series of 10 reversible reactions, each catalyzed by a specific enzyme (Figure 9.2).

- Glycolysis takes place universally in all living organisms, from bacteria to the eukaryotic cells of plants and animals.
- Glycolysis is an anaerobic process that occurs in the cytosol.

Many anerobic organisms are entirely dependent on glycolysis. Glycolysis, the pathway with the highest carbon flux in most of the cells, is the only source of metabolic energy in a few mammalian tissues and cell types. Some of the plant tissues, such as starch-storing potato tubers and aquatic species, such as watercress (*Nasturtium officinale*), obtain a major part of their energy from glycolysis.

Biologically, glycolysis may be viewed as process since it, in all likelihood, came about before the origin of the aerobic environment and the cellular organelles.

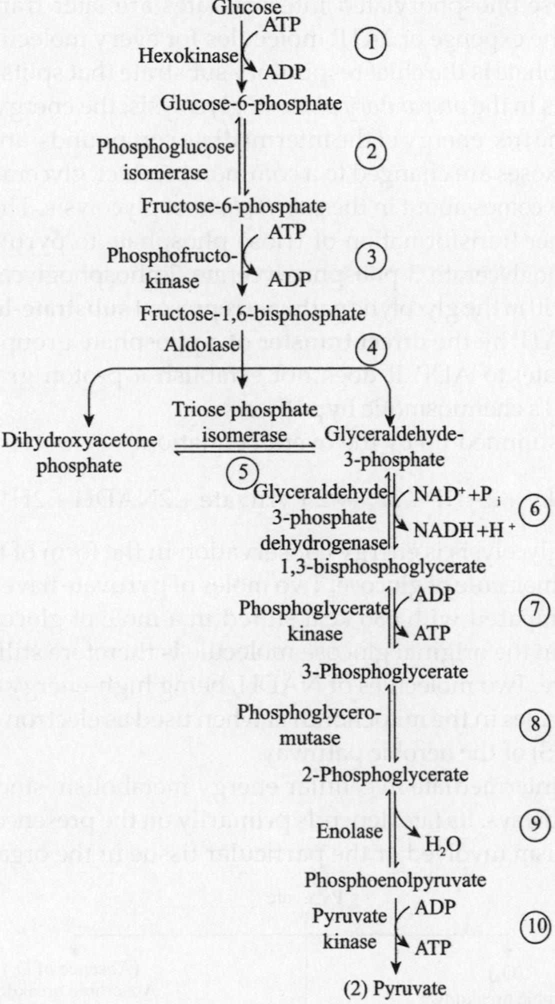

Fig. 9.2 The glycolytic pathway or Embden–Meyerhof-Parnas (EMP) pathway. For each molecule of glucose that passes through the preparatory phase two molecules of glyceraldehyde-3-phosphate are generated; both pass through the payoff phase and ultimately 2 molecules of pyruvate are formed.

Note: Steps 7 and 10 represent substrate-level phosphorylation. (Reactions 1–5 are Preparatory phase; Reactions 6–10 are Payoff phase).

There are two phases of Glycolysis as under:

1. ***Preparatory phase***: Production of triose phosphate (glyceraldehyde-3-phosphate) from glucose from a hexose-phosphate pool.
2. ***Payoff phase***: Oxidative conversion of triose phosphate into pyruvate and the associated reactions involving generation of ATP and NADH.

The first phase involves a sequence of reactions by which several types of glucose and fructose sugars obtained from storage carbohydrates are transformed to the common intermediate compounds such as triose phosphate through the cytoplasmic hexose-phosphate pool. These phosphorylated intermediates are later transformed to fructose-1,6-bisphosphate at the expense of 2ATP molecules for every molecule of hexose (glucose). Fructose-1,6-bisphosphate is the chief respiratory substrate that splits into two molecules of triose phosphate.Thus in the *preparatory phase* of glycolysis, the energy in the form of ATP is utilized, increasing the free energy of the intermediate compounds, and the carbon chains of all the metaolized hexoses are changed to a common product, glyceraldehyde-3-phosphate.

The gain in energy comes about in the *payoff phase* of glycolysis. The second step consists of reactions for further transformation of triose phosphate to pyruvate *via* intermediates such as 1,3-bisphosphoglycerate, 3-phosphoglycerate, 2-phosphoglycerate and phosphoenol pyruvate. Steps 7 and 10 in the glycolytic pathway represent **substrate-level phosphorylation**, i.e., the synthesis of ATP by the direct transfer of a phosphate group from a substrate (i.e., metabolic intermediate) to ADP. It does not establish a proton gradient and cannot be described by Mitchell's chemiosmotic hypothesis.

Glycolysis can be summed up by the overall equation:

$$\text{Glucose} + 2\text{NAD}^+ + 2\text{ADP} + 2\text{P}_i \rightarrow 2\ \text{Pyruvate} + 2\text{NADH} + 2\text{H}^+ + 2\text{ATP} + 2\text{H}_2\text{O}$$

The major role of glycolysis is energy conservation in the form of two molecules of ATP and NADH each per molecule of glucose. Two moles of pyruvate have a total energy content of about 546 kcal, compared with 686 kcal stored in a mole of glucose. About 80 per cent of the energy stored in the original glucose molecule is therefore still contained in the two molecules of pyruvate. Two molecules of NADH, being high-energy compounds can yield additional ATP molecules in the mitochondria when used as electron donors to the electron transport system (ETS) of the aerobic pathway.

Pyruvate is a key intermediate in cellular energy metabolism since it can be utilized in one of the several pathways. Its fate depends primarily on the presence of oxygen and partly on the specific organism involved or the particular tissue in the organism.

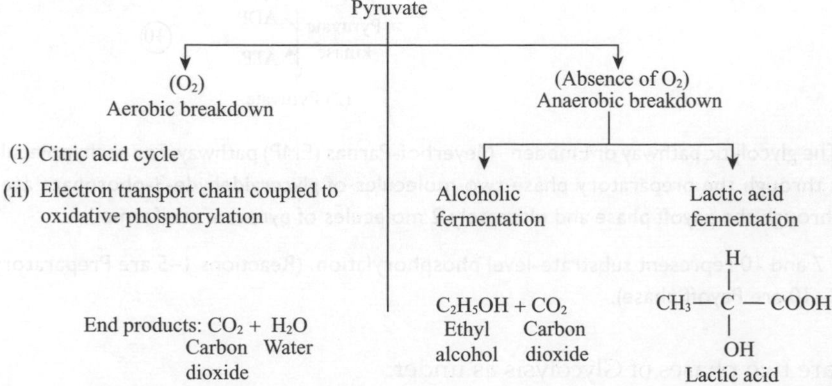

Although higher angiospermic plants are obligate aerobes, tissues or organs are occasionally subjected to anaerobic conditions. For example, roots growing in water saturated soil; when there is no oxygen to serve as terminal electron acceptor, mitochondrial respiration will come to a halt and metabolic activity will shift towards fermentation.

Before the cytosolic pyruvate reaches the mitochondrial matrix, it must cross both the outer and inner membranes of the mitochondria. The outer membrane is not a barrier for small molecules (those with a molecular weight of less than 10,000 D) which can readily pass through the '**porin**' channels traversing the outer membrane. Hence, pyruvate easily enters into the intermembranous space. On the other hand, the inner mitochondrial membrane is impervious to most of the bigger molecules, including pyruvate which requires a co-transport protein (**pyruvate transporter**) that can ferry pyruvate along with a proton to the matrix where subsequent oxidation occurs (Figure. 9.3).

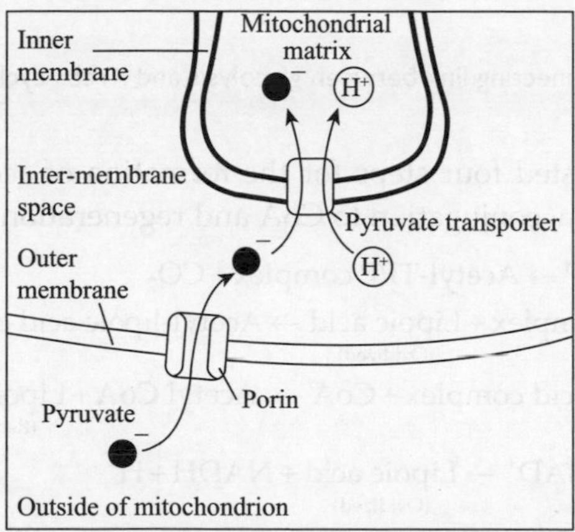

Fig. 9.3 Pyruvate transport by mitochondria. The outer membrane of mitochondria is not a barrier for small molecules (i.e., those less than 10,000 D) due to the presence of porin channels. Hence, pyruvate readily crosses into the inter-membrane space. Pyruvate is transported along with H^+ through a pyruvate transporter, embedded in the inner membrane. Because pyruvate has a -1 charge, no change in electrical charge of the matrix occurs as a result of this transport.

The two important mobile cofactor pairs in glycolysis are ATP/ADP and NAD^+/NADH. The first connects glycolysis to energy production, because net ATP is produced by the pathway and is used in other cellular reactions, such as the maintenance of the plasma membrane sodium gradient. The second is the redox cofactor pair, which is balanced if glycolysis converts glucose to lactate.

Glycolysis can be thought of as the crossroads of metabolism, because all of its intermediates intersect with other pathways.

9.2 Oxidative Decarboxylation of Pyruvate to Acetyl CoA

The pyruvate produced in the cytoplasm during glycolysis is transferred across the impermeable inner mitochondrial membrane into the mitochondrial matrix through a specific pyruvate translocase in a symport fashion alongwith a proton. Once inside the matrix, pyruvate undergoes decarboxylation (release of CO_2) in an oxidation reaction to form acetyl CoA. In the course of this exergonic reaction, a molecule of NADH is generated from

NAD^+ for every pyruvate molecule that is oxidized. Oxidation of 2 molecules of pyruvate, thus, produces 2 molecules of NADH and CO_2 each.

$$\text{Pyruvate} + \text{CoA} + \text{NAD}^+ \rightarrow \text{Acetyl CoA} + CO_2 + \text{NADH} + H^+$$

This reaction is catalyzed by a multienzyme complex, **pyruvate dehydrogenase** in association with five cofactors: thiamine pyrophosphate (TPP) – a derivative of vitamin B_1, Mg^{2+}, NAD^+, CoA and lipoic acid. The pyruvate dehydrogenase complex consists of 3 component enzymes, namely **pyruvate dehydrogenase, dihydrolipoyl transacetylase** and **dihydrolipoyl dehydrogenase**.

Acetyl CoA is the connecting link between glycolysis and Krebs cycle or citric acid cycle.

Gunsalus (1954) suggested four steps for the formation of acetyl CoA from pyruvate: decarboxylation, oxidation, conjugation to CoA and regeneration of lipoic acid.

$$\text{Pyruvate} + \text{TPP} \rightarrow \text{Acetyl-TPP complex} + CO_2$$

$$\text{Acetyl-TPP complex} + \underset{\text{(Oxidized)}}{\text{Lipoic acid}} \rightarrow \text{Acetyl-lipoic acid complex} + \text{TPP}$$

$$\text{Acetyl-lipoic acid complex} + \text{CoA} \rightarrow \text{Acetyl CoA} + \underset{\text{(Reduced)}}{\text{Lipoic acid}}$$

$$\underset{\text{(Reduced)}}{\text{Lipoic acid}} + \text{NAD}^+ \rightarrow \underset{\text{(Oxidized)}}{\text{Lipoic acid}} + \text{NADH} + H^+$$

$$\text{Finally: Pyruvic acid} + \text{CoA} + \text{NAD}^+ \rightarrow \text{Acetyl CoA} + CO_2 + \text{NADH} + H^+$$

Acetyl CoA, thus, formed is the starting material for the Krebs cycle or else is channelled into fatty acid synthesis.

9.3 Citric Acid Cycle[1] (Krebs Cycle)

The **citric acid cycle** was originally known as the **Krebs cycle** in honour of Sir Hans Krebs. Krebs postulated this metabolic pathway in 1937 and later in 1954 received a Nobel Prize in recognition of his brilliant work.

The Citric acid cycle (Figure 9.4) always begins with acetyl CoA, its substrate. Upon entering the Citric acid cycle, the 2-C acetyl group condenses with 4-C compound, oxaloacetate, to yield a 6-C compound, citrate. Through a series of enzyme-catalyzed reactions involving four oxidation steps (yielding NADH at step 3 and FADH at step 1), oxaloacetate is regenerated from citrate, accompanied by the release of 2 molecules of carbon dioxide. One ATP molecule is produced through **substrate-level phosphorylation** at step 6. α-Ketoglutaric acid is one of the key intermediate compounds of the Krebs cycle as it is involved in carbohydrate metabolism, lipid metabolism as well as during biosynthesis and degradation of amino acids.

[1] Krebs cycle is more commonly called the Citric acid cycle or tricarboxylic acid (TCA) cycle because it begins with the formation of citrate, which has three carboxylic acid ($-COO^-$) groups. The other important cycles in plants that similarly regenerate the primary acceptor molecules are the 'Glyoxylate pathway' that occurs in glyoxysomes during germination of fatty seeds and 'Reductive pentose phosphate pathway' that occurs in chloroplasts during carbon fixation in photosynthesis.

It is important to emphasize that the cycle must turn two times to oxidize the equivalent of one hexose sugar.

The overall equation for the citric acid cycle is therefore:

$$\text{Oxaloacetate} + \text{Acetyl CoA} + 3H_2O + \text{ADP} + P_i + 3\text{NAD}^+ + \text{FAD}^+ \longrightarrow$$
$$\text{Oxaloacetate} + 2CO_2 + \text{CoA} + \text{ATP} + 3\text{NADH} + 3H^+ + \text{FADH}_2$$

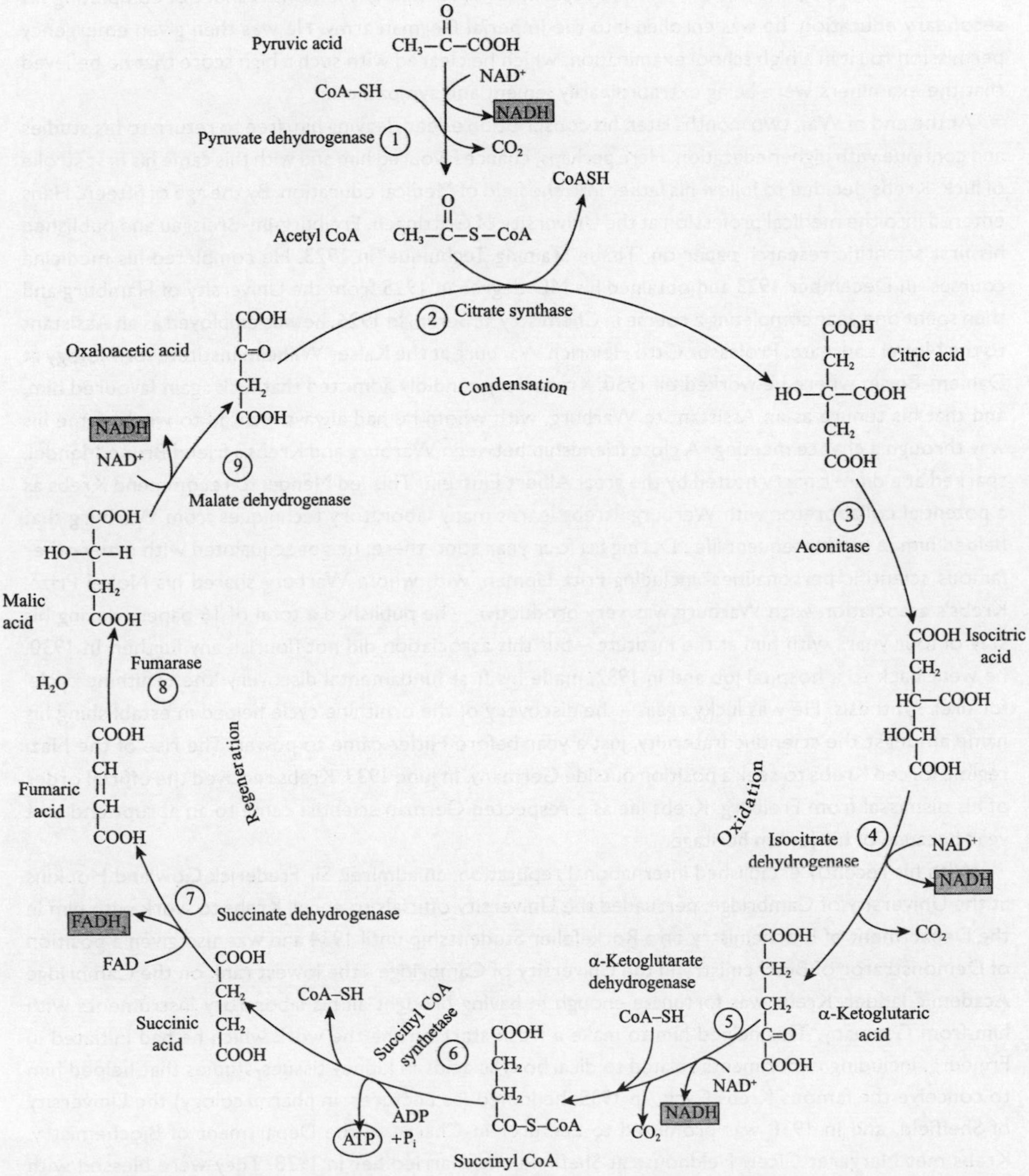

Fig. 9.4 Citric acid cycle or tricarboxylic acid (TCA) cycle or Krebs cycle. *Note*: Step 6 represents substrate-level phosphorylation.

Sir Hans Adolf Krebs

(25 August 1900 – 22 September 1981)

Sir Hans Adolf Krebs was born in Hildesheim (Germany) on 25th August 1900. Of Jewish-Silesian ancestry, he was the son of Georg Krebs MD, an ENT specialist, and Alma Krebs (nee Davidson). A shy young boy with a great thirst for knowledge, Krebs studied at the famous old Gymnasium Andreanum in his hometown . At the end of WWI, in September 1918, nearly six months short of completing his secondary education, he was enrolled into the Imperial German army. He was then given emergency permission to sit in a high school examination, which he cleared with such a high score that he believed that the examiners were being extraordinarily lenient and sympathetic.

At the end of War, two months later, his conscription ended, leaving him free to return to his studies and continue with higher education. Here perhaps, chance favoured him and with this came his first stroke of luck. Krebs decided to follow his father into the field of Medical education. By the age of fifteen, Hans entered into the medical profession at the University of Göttingen, Freiburg-im-Breisgau and published his first scientific research paper on 'Tissue Staining Technique' in 1923. He completed his medicine courses in December 1923 and obtained his MD degree in 1925 from the University of Hamburg and then spent one year completing a course in Chemistry at Berlin. In 1926, he was employed as an Assistant to the Nobel Laureate, Professor Otto Heinrich Warburg at the Kaiser Wilhelm Institute for Biology at Dahlem-Berlin where he worked till 1930. Krebs very candidly admitted that luck again favoured him, and that his tenure as an Assistant to Warburg, with whom he had always wanted to work, came his way through a chance meeting. A close friendship between Warburg and Krebs's friend Bruno Mendel, sparked at a dinner party hosted by the great Albert Einstein. This led Mendel to recommend Krebs as a potential collaborator with Warburg. Krebs learnt many laboratory techniques from Warburg that helped him in his subsequent life . During his four year stint there, he got acquainted with many other famous scientific personalities- including Fritz Lipman, with whom Warburg shared his Nobel Prize. Krebs's association with Warburg was very productive – he published a total of 16 papers during his stay of four years with him at the Institute – but this association did not flourish any further. In 1930, he went back to a hospital job and in 1932, made his first fundamental discovery 'the ornithine cycle' for urea synthesis. He was lucky again – the discovery of the ornithine cycle helped in establishing his name amongst the scientific fraternity, just a year before Hitler came to power. The rise of the Nazi regime forced Krebs to seek a position outside Germany. In June 1933, Krebs received the official order of his dismissal from Freiburg. Krebs life as a respected German scientist came to an abrupt end that year because of his Jewish heritage.

With his recently established international reputation, an admirer, Sir Frederick Gowland Hopkins at the University of Cambridge, persuaded the University officials to enroll Krebs to work with him in the Department of Biochemistry on a Rockefeller Studentship until 1934 and was also given a position of Demonstrator of Biochemistry in the University of Cambridge - the lowest rank on the Cambridge Academic ladder. Krebs was fortunate enough in having brought all his laboratory instruments with him from Germany. This helped him to make a fresh start on the the work which he had initiated in Freiburg, including experiments related to dicarboxylic acids in kidney tissues-studies that helped him to conceive the famous Krebs Cycle. In 1935, he joined (as Lecturer in pharmacology) the University of Sheffield, and in 1938 was promoted to Lecturer-in-Charge of the Department of Biochemistry. Krebs met Margaret Cicely Fieldhouse at Sheffield and married her in 1938. They were blessed with two sons, Paul and John, and one daughter, Helen.

In 1937, Krebs and his graduate student William A. Johnson attempted to publish their pathbreaking discovery in the the Journal 'Nature'. Surprisingly, their article which later turned out to be one of the most important biological discoveries of the 20th century, and that later won Krebs the 1953 Nobel Prize for Physiology or Medicine, was not accepted. The paper was eventually published in the Dutch Journal 'Enzymologia'. However, Nature was to send him an immediate congratulatory message to Krebs when he received his Nobel Prize. In 1945 Krebs was appointed Professor and Head and the Director of a Medical Research Council's (MRC) for Cell Metabolism, established in his Department. With his MRC funding, Krebs transformed the top floor of the old Scala cinema into a laboratory called 'Krebs's Empire'.

In 1954, Krebs moved out of Sheffield and was appointed Whitley Professor of Biochemistry at Oxford University, taking along with him the MRC unit. In 1967, Krebs retired at Oxford and promptly moved to the Nuffield Department of Clinical Medicine, at the Radcliffe Infirmary in Oxford, again taking his MRC unit with him. He had profoundly enjoyed and derived intellectual satisfaction from his research career. Admittedly, Krebs has been a hard taskmaster but he had also inspired the many people who came in touch with him, including some scientists who had begun their career with him as technicians and finally ended up becoming professors at other Universities. He is credited with the discovery of three cyclical pathways – the ornithine cycle, the glyoxylate cycle, most importantly the Krebs cycle. Krebs died in 1981 in Oxford, where he had spent thirteen years of his professional career (1954–1967).

The University of Sheffield established the 'Krebs Institute', in 1988. In 1990, the Federation of European Biochemical Societies instituted the Sir Hans Krebs Lecture and Medal. In 2008, a new building for the Department of Biochemistry at the University of Oxford was constructed, on which a plaque was placed on 20th May 2013 by the Association of Jewish Refugees. The plaque was unveiled by John Lord Krebs and the inscription reads: Professor Sir Hans Krebs FRS (1900–1981) Biochemist and discoverer of the Krebs Cycle, Nobel Prize winner 1953, worked here 1954–1967.

Krebs Cycle Intermediates – Word Origin

Many of the names for the Krebs cycle intermediates come from the organisms in which these compounds were first discovered. Citrate is of course derived from *Citrus* fruits. Aconitate is a compound from *Aconite* (monkshood), a member of the buttercup family. Succinate was first found in *amber* (fossilized resin); Fumarate in plants of the *Fumaria* genus (poppy family); Malate is an acid found in abundance in the *Malus* genus (crabapple, also spelled crab apple). While all of these organisms do have a Krebs cycle, the abundance of the compounds is due to specialized pathways, called **secondary metabolism**, which protect plants from predators.

9.4 Electron Transport System and Oxidative Phosphorylation

The term oxidative phosphorylation refers to two processes: **oxidation**, involving electron flow, and **phosphorylation**, the generation of ATP from ADP and Pi. Just like in intact cells, these processes are coupled in isolated mitochondria. Another kind of coupling also occurs in this process: coupling of electron and proton movement. As the electrons from NADH to O_2 move through the inner mitochondrial membrane, protons pass across the membrane from the interior of the mitochondrial matrix to the exterior of the inner membrane. This movement of protons is vectorial, because it has both a magnitude (the number of protons transported) and a direction (from the matrix to the cytoplasm).

The reduced compounds, NADH and FADH, generated from the the Krebs cycle represent reducing power that, using oxygen, are utilized to synthesize ATP. 'The synthesis of ATP by the electron flow through the carriers of electron transport system (ETS), with molecular oxygen as the terminal electron acceptor' is called **oxidative phosphorylation**. ETS is a sequence of electron carriers, which consists of NAD, FAD, ubiquinone and cytochromes (b, c_1, c, a, a_3) embedded in the inner mitochondrial membrane (Figure 9.5). Cytochromes are conjugated protein molecules with an iron-containing porphyrin ring or haeme group attached. Cytochromes pick up electrons on their iron atoms, which can be interchangeably reduced from ferric (Fe^{3+}) to ferrous (Fe^{2+}) forms. The respiratory poisons like cyanides, azides and even CO combine with the iron of cytochromes, thus preventing the conversion of $Fe^{+++} \rightarrow Fe^{++}$ transformation, i.e., stopping the flow of electrons with no release of ATP. Hence, CO deprives the flow of energy to the brain which chokes or throttles a person to death.

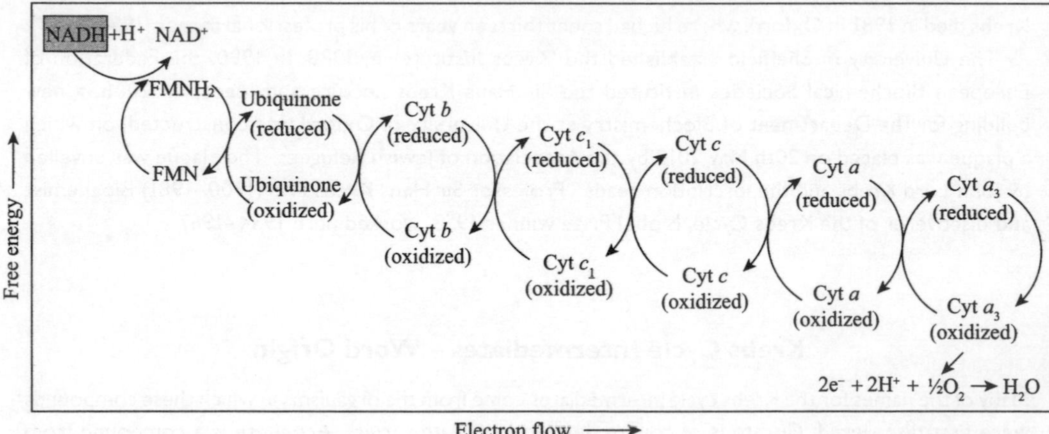

Fig. 9.5 Flow of electrons with decreasing (less electron negative) oxidation–reduction potential. Three ATPs are produced from NADH-one between NADH/FAD junction, the second ATP between Cyt b and Cyt c, and the third ATP between Cyt a, and Cyt a_3.

Interestingly, *Desulfovibrio* (anaerobic bacteria) uses sulphate as a terminal acceptor of electrons during respiration to produce ATP.

The pathway of connected redox reactions that comprise the respiratory chain, follows the order of increasing reduction potential (Table 9.1).

Table 9.1 Reduction potentials for mitochondrial electron transport chain mobile cofactors.

Half-Reaction	e° (mV)
NADH → NAD^+	– 0.32
Succinate → Fumarate	+ 0.03
$QH_2 \rightarrow Q$	+ 0.04
$Cyt_{c(red)} \rightarrow Cyt_{c(ox)}$	+ 0.23
$O_2 \rightarrow 1/2H_2O$	+ 0.82

The electron transport chain confined to the inner mitochondrial membrane comprises of following four transmembrane multiprotein complexes, joined by mobile electron carriers (Figure 9.6).

Complex I: NADH dehydrogenase
Complex II: Succinate dehydrogenase
Complex III: Cytochrome bc_1 complex
Complex IV: Cytochrome *c* oxidase

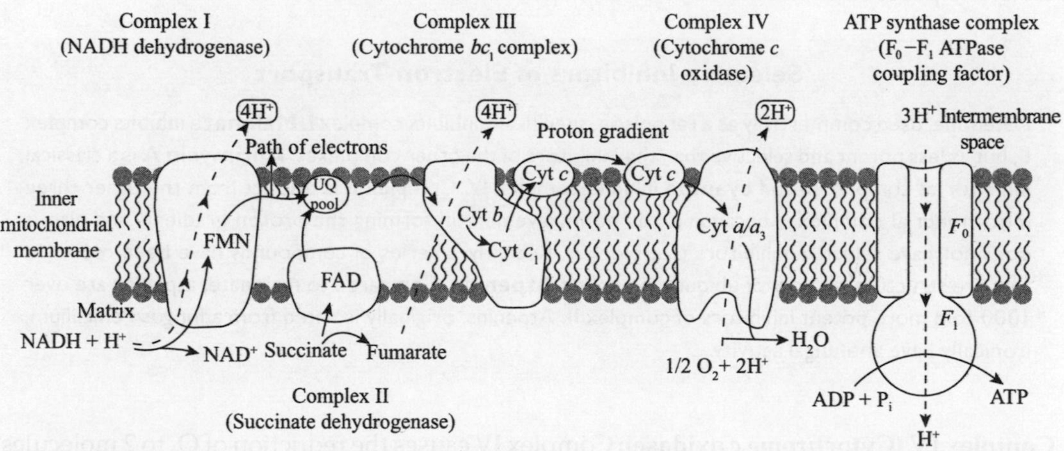

Fig. 9.6 Electron flow along the respiratory electron transport chain and synthesis of ATP in the inner mitochondrial membrane. The individual electron transport proteins are grouped into four transmembrane multiprotein complexes. Complex I, III and IV are engaged in proton pumping from matrix to intermembrane space.

Complex I (NADH dehydrogenase): Complex I oxidizes electrons generated from NADH in a matrix of the mitochondria during the Krebs cycle and in turn, transfers them to ubiquinone. Ubiquinone, a small electron and proton carrier, is present in the inner mitochondrial membrane and can diffuse within the hydrophobic layer of the membrane.

The electron carriers in complex I include FMN (flavin mononucleotide) as a cofactor and several iron–sulphur centres. For each pair of electrons moving through the complex, four H^+ are pumped out of the matrix into the intermembranous region.

Complex II (Succinate dehydrogenase): Complex II catalyzes the oxidation of succinate to fumarate in the Krebs cycle and transfers the electrons from $FADH_2$ and iron–sulphur centres to ubiquinone. Complex II does not involve any pumping of protons.

Complex III (Cytochrome bc_1 complex): Complex III oxidizes reduced ubiquinone (ubiquinol) and in turn transfers the electrons *via* an Fe–S centre, cytochrome *b* and a membrane-bound cytochrome c_1 to cytochrome *c* – a peripheral mobile electon carrier confined to the external surface of the inner mitochondrial membrane, that serves to transfer electrons between complexes III and IV.

Ubiquinone and cytochrome bc_1 complexes are identical to plastoquinone and the cytochrome b_6f complex in the photosynthetic electron transport chain.

Four protons per electron pair are pumped out from the matrix into the intermembrane region by complex III using a mechanism called **Q cycle** but only two out of four resulting electrons proceed down the electron transport chain to ultimately reduce O_2.

Selective Inhibitors of Electron Transport

Rotenone, used commercially as a rat poison, specifically inhibits complex I. **Malonate** inhibits complex II, but is less potent and selective than the inhibitors of the other complexes. **Antimycin A** is a classical inhibitor of complex III and **cyanide** inhibits complex IV. Complex II is distinct from the other three mitochondrial complexes because it has no involvement in forming the proton gradient, and also it does not have selective inhibitors that block it. Recently, a series of compounds have been reported that are structural analogs of ubiquinone, called **Atpenins**. Compared to malonate, atpenins are over 1000-fold more potent inhibitors of complex II. Atpenins, originally isolated from a fungus Penicillium, ironically have antifungal activity.

Complex IV (Cytochrome c oxidase): Complex IV causes the reduction of O_2 to 2 molecules of H_2O. This contains cytochromes *a,* and a_3 along with two copper centres (Cu_A and Cu_B). Two protons are pumped out of the matrix per electron pair.

During oxidative phosphorylation, electrons transfer from complexes I, III and IV to O_2 is linked to the production of ATP using ADP as well as Pi *through* the F_0–F_1 **ATP synthase complex**. For each ATP synthesized, $3H^+$ pass through F_0 component of the ATP synthase from intermembrane space to the matrix, down the electrochemical proton gradient. NADH leads to the production of three ATPs, while each FADH produces two ATPs. Rather than liberating energy in a single burst, electrons fall to low to lower and then to the lowest energy level (most electronegative electron acceptors) in steps, releasing locked up energy at each step.

The Q Cycle or Quinone Cycle in Mitochondria

This is actually a bi-cycle, because forms of Q and the electrons cycle through the membrane. During the Q cycle, ubiquinone or coenzyme Q (UQ) is reduced to ubiquinol (UQH_2) towards the matrix face of the mitochondrial inner membrane by electrons coming or migrating from complex I and complex II. In both cases, the reduction of UQ to UQH_2 by two electrons is followed by the uptake of two $2H^+$ from the mitochondrial matrix.

Firstly, the fully reduced Ubiquinol (QH_2) (at the site A of the membrane) dissociates, releasing protons to the cytosol and electrons are donated to an Fe-S protein (shown at B) and which are subsequently transferred to cytochromes c_1 and c, leaving the semiquinone Q^{*-} (a free radical anion and a negatively charged ion - which is mostly hydrophobic) at the cytosolic membrane. Unlike the other two 'mobile', entirely hydrophobic Q and QH_2 molecules, this molecule Q^{*-} is anchored to the membrane's water-phase edge. Next, (as shown at C) the electrons from the free radical anion are passed onto cytochrome b_{566}, forming Q. Then, these electrons are transferred to cytochrome b_{562} (at the site D). In the next step, Q can accept electrons from cytochrome b_{562} to form the second pool of Q^{*-} at the matrix membrane (at E) which together with another electron from a second b_{562} (shown at F) along with protons from the matrix produce QH_2 once again. QH_2 is mobile in the hydrophobic phase of the membrane and completes the cycle by migrating or diffusing back to the cytosolic phase and repeating the cycle.

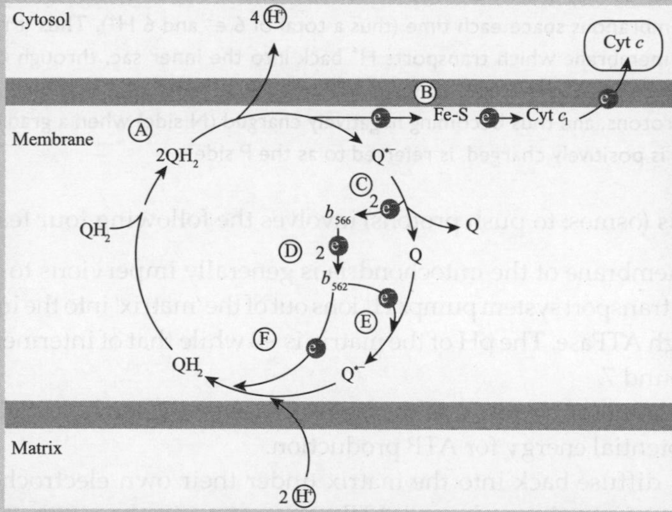

This presentation of the Q cycle when it was first proposed was considered highly controversial but it has gained much ground, explaining how protons can 'move across' the membrane and electrons can move through it. In fact, in this model, there are present in the mitochondria two pools of Q^{*-} as evidenced from the data gathered by 'paramagnetic resonance' – a technique that identifies and measures free radicals.

The net outcome of the operation of the Q cycle is the transfer of four H^+ from the mitochondrial matrix to the intermembranous space for every two electrons transferred from complex III to the cytochrome c.

A difference in redox potential between two consecutive carriers that exceeds 0.16 V is sufficient to couple ADP + Pi to form ATP in accordance with Mitchell's Chemiosmotic[2] hypothesis (Figure 9.7).

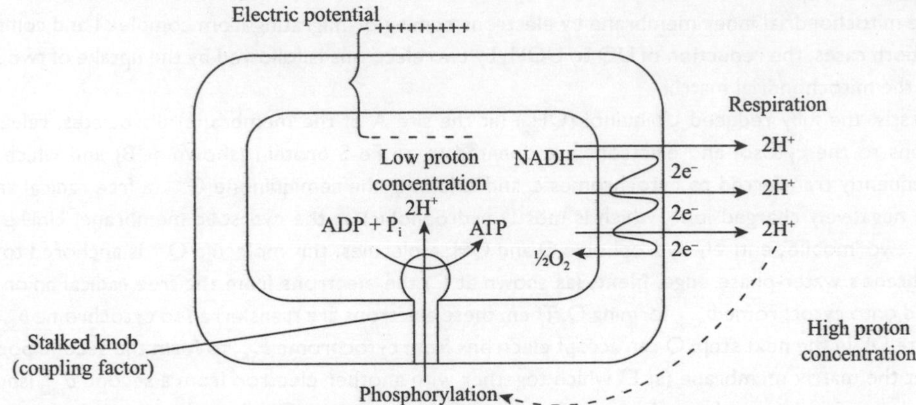

Fig. 9.7 Simplified diagrammatic sketch of the Chemiosmotic explanation for the generation of ATP in mitochondria. NADH gives up a pair of electrons which move back and forth through the mitochondrial membrane three times as they move from one carrier to the next one and finally to O_2, transporting 'two protons' to intermembranous space each time (thus a total of 6 e^- and 6 H^+). Thus it results in a high pH gradient across the membrane which transports H^+ back into the inner sac, through the coupling factor. The energy stored in the gradient is used to form ATP using ADP and Pi, catalyzed by ATPase. The space that is deficient in protons, and thus becoming negatively charged (N side) when a gradient is created. The opposite side which is positively charged, is referred to as the P side.

Chemiosmosis (osmos: to push protons) involves the following four features:

1. The inner membrane of the mitochondria is generally impervious to H^+.
2. The electron transport system pumps H^+ ions out of the 'matrix' into the intermembranous space through ATPase. The pH of the matrix is ~8 while that of intermembranous space remains around 7.
3. The accumulation of H^+ towards the 'cytosol' of inner mitochondrial membrane is a source of potential energy for ATP production.
4. The H^+ ions diffuse back into the matrix under their own electrochemical gradient through special membrane bound ATPase resulting in the synthesis of ATP in the matrix[3], under the control of 'proton-motive force'.

[2] Chemiosmotic power has other uses in living organisms: (1) It provides the power that drives the rotation of bacterial flagella; (2) It is involved in the formation of ATP using energy supplied to electrons by the sun in photosynthetic cells; (3) It can be used to power other transport processes, e.g., transport of phosphate and pyruvate through the inner mitochondrial membrane.

[3] A total of 10 protons are transported across mitochondrial membrane, for every hydrogen from NADH, while six (6) protons for every $FADH_2$ molecule. We also know, that 4 protons are required to synthesize 1 molecule of ATP, i.e., each NADH can make 2.5 ATPs (10/4) while each $FADH_2$ can produce 1.5 ATPs (6/4) and therefore the previous estimates of 3 ATPs per NADH and 2 ATPs for $FADH_2$ has been shown to be untenable but it is still being routinely taught.

The **proton-motive force (pmf)** across the inner mitochondrial membrane comprises of two major components: a chemical potential (ΔpH); reflecting the difference in proton (H^+) concentration, and an electrical potential (ΔE);reflecting the difference in charge between the matrix and the intermembranous space

$$\text{pmf} = \Delta E - 59\ \Delta pH$$

In contrast to the chloroplasts, ΔE is the major factor contributing to pmf generated by mitochondria.

The direction of proton gradient in $CF_1 - CF_0$ (chloroplast) is opposite to that of mitochondria with $F_1 - F_0$ complex; in the latter, electron flow (chain) starts with NADPH and ends up with water (in the opposite direction to what happens in chloroplasts).

The clinching evidence for chemiosmotic hypothesis comes from the use of uncouplers like 2,4-dinitrophenol (DNP)[4] which disrupts the pH gradient across the mitochondrial membrane, thereby preventing ATP generation. These make the inner membrane permeable to protons; putting a stop to the build up of a significantly high pmf to power ATP production. ATP synthesis will be restored when the pH gradient is imposed on the mitochondria.

Role of Uncouplers in Respiraton

Uncouplers are weak acids that transfer H^+ back into the matrix directly, avoiding the ATPase, thereby inhibiting ATP synthesis. These have a role in cold exposure or hibernation because the energy is utilized in generation of heat (as in thermogenesis) instead of synthesizing ATP. Salicylic acid is also an uncoupling agent and will slow down synthesis of ATP and increases body temperature if consumed in excessive amounts. The other powerful mitochondrial uncoupling agents include **CCCP** (Carbonyl cyanide *m*-chlorophenyl hydrazone) and **FCCP** (*p*-trifluoromethoxy carbonyl cyanide phenyl hydrazone). The collapse of the pH gradient generates heat.

Hibernating animals including human beings and newborn animals contain brown adipose tissue which is due to the high mitochondrial content. An endogenous protein called **thermogenin**, uncouples ATP synthesis but generates heat.

ATP synthase functions in a manner that is opposite to that of the proton pump, H^+-ATPase (Figure 9.8). ATP synthase uses the energy of protons moving down their gradient to produce ATP; on the other hand in a proton pump, ATP is used as an energy source to pump protons against their electrochemical gradient. The net yield from one molecule of glucose is 36 ATP; 6 from glycolysis, 6 from the oxidative decarboxylation of pyruvate to acetyl CoA and 24 from the Krebs cycle (Table 9.2).

ATP synthesized in the mitochondria is readily transported out of it, to be utilized elsewhere in the cell. It is achieved by two distinct translocator proteins confined to the inner

[4] During World War I, dinitrophenol was used as an explosive. This action, like the structurally similar trinitrotoluene (TNT), is the result of rapid release of its components into gases.

membrane – an adenine nucleotide transporter and an inorganic phosphate translocator. The former is involved in exchange of ATP and ADP on a one-for-one basis, whereas the latter exchanges Pi for OH^- ions.

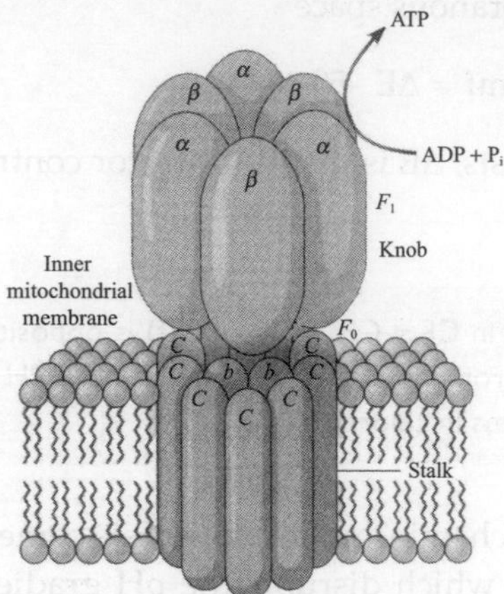

Fig. 9.8 A representation of the ATP synthase showing F_0 and F_1 components. The F_0 component is an integral part of the membrane protein complex, consisting of at least three different polypeptides. They create the channel through which H^+ move across the inner membrane. The F_1 component is a peripheral membrane protein complex towards the matrix, comprised of at least five different subunits. They possess the catalytic site for the converting ADP as well as Pi into ATP.

Table 9.2 A summary of the net yield of 36 ATP molecules from oxidation of one molecule of glucose.

	Cytosol	Matrix of mitochondrion	Electron transport and oxidative phosphorylation	
Glycolysis	2 ATP →	→	→	2 ATP
	2 NADH →	→	4 ATP (net yield) →	4 ATP
Pyruvate to acetyl CoA		2 × (1 NADH) →	2 × (3 ATP) →	6 ATP
Krebs cycle (Citric acid cycle)		2 × (1 ATP) →	→	2 ATP
		2 × (3 NADH) →	2 × (9 ATP) →	18 ATP
		2 × (1 $FADH_2$) →	2 × (2 ATP) →	4 ATP
				Total: 36 ATP

ATP synthase, the ubiquitous enzyme, plays the role of a turbine, allowing the proton gradient to power the synthesis of ATP.

During oxidative phosphorylation, several harmful **reactive oxygen species** (ROS) are produced in mitochondria which can play havoc in the cells by damaging enzymes, membrane lipids, and nucleic acids. The formation of ROS is favoured when mitochondria is under oxidative stress i.e. more electrons are available to enter the respiratory system than can be immediately passed through to oxygen. Although overproduction of ROS is quite detrimental, low levels of ROS serve as a signal reflecting the insufficient supply of oxygen (hypoxia) and coordinate mitochondrial oxidative phosphorylation with other metabolic pathways. ROS are deactivated by a group of protective enzymes, such as **superoxide dismutase and glutathione peroxidase**, thereby avoiding the harmful consequences of oxidative stress. Superoxides are produced due to reaction of O_2 with carriers of the electron transport chain. Detoxification of $O_2^{\cdot}$ is essential to avoid peroxidation of lipid membranes or oxidation of mitochondrial proteins and nucleic acids. A manganese-dependent **superoxide dismutase** in the mitochondrial matrix converts O_2^{*-} to hydrogen peroxide (H_2O_2), which is degraded to H_2O by a series of mitochondrial reactions that can be catalyzed by Trx-dependent peroxiredoxins, the ascorbate-glutathione cycle, or **glutathione dependent peroxidases** that have been reported in plant mitochondria from some species.

Besides this, cells have two other mechanisms of defence against reactive oxygen species:

(i) Regulation of pyruvate dehydrogenase (PDH); under hypoxia, PDH kinase causes phosphorylation of mitochondrial PDH, followed by its inactivation and a slow down of the supply of $FADH_2$ and NADH from the Calvin cycle to the mitochondrial respiratory chain.

(ii) Replacement of one subunit of Complex IV, COX4-1(cytochrome c oxidase), with another subunit, COX4-2, which is better adjusted to hypoxia.

Both the changes mentioned above, are mediated by HIF-1, **the hypoxia-inducible factor**. HIF-1 is a master regulator of O_2 homeostasis.

9.5 Regulation of Respiratory Pathways

This control is achieved by a variety of complementary mechanisms at several sites.

9.5.1 Regulation of the glycolytic pathway

Glycolysis plays a dual role in the cells: oxidation of glucose coupled with ATP synthesis and providing building blocks for the biosynthesis of cellular compounds. The transformation of one molecule of glucose to two molecules of pyruvate is highly controlled to fulfil these requirements. The glycolytic reactions are readily reversible except for the reactions 1, 3 and 10 (labelled in Figure 9.2) catalyzed by hexokinase, phosphofructokinase and pyruvate kinase.

- Hexokinase shows inhibition by G-6-P and it gets stored when phosphofructokinase is not active.
- Phosphofructokinase, the key control element in glycolysis, shows inhibition by large amounts of ATP as well as citrate and stimulation by AMP as well as fructose-2,6-bisphosphate.
- Pyruvate kinase, the other control site, shows allosteric inhibition by ATP as well as alanine and stimulated by fructose-1,6-bisphosphate. Hence, pyruvate kinase is at its peak activity when the energy charge is low and glycolytic intermediates get stored.

9.5.2 Regulation of the pyruvate dehydrogenase complex

Activity of the pyruvate dehydrogenase (PDH) complex is controlled by various mechanisms:

- Inhibition by products such as acetyl CoA and NADH.
- Feedback regulation by nucleotides such as GTP (inhibitor) and AMP (activator).
- Regulation by reversible phosphorylation – ATP-dependent phosphorylation of the pyruvate dehydrogenase complex diminishes, while dephosphorylation triggers its activity. PDH is phosphorylated by a regulatory kinase and dephosphorylated by a phosphatase.

9.5.3 Regulation of the Krebs cycle (Citric acid cycle)

Operation of the Krebs cycle is precisely regulated to fulfil the cellular requirement for ATP (Figure 9.9). The availability of $NAD^+$and FAD signals that the energy charge is low.

- Condensation of oxaloacetate and acetyl CoA to form citrate is an important regulatory step in the cycle. ATP is an allosteric inhibitor of the citrate synthase enzyme catalyzing this reaction.
- A second control point is isocitrate dehydrogenase. This enzyme is allosterically inhibited by NADH andATP and stimulated by ADP.
- A third regulatory site is α-ketoglutarate dehydrogenase, which shows inhibition by succinyl CoA and NADH.
- Another regulatory site is succinate dehydrogenase; promoted by high concentrations of succinate, phosphate, ATP, reduced ubiquinone and inhibited by oxaloacetate.

In short, the funnelling of 2-C units into the Krebs cycle and the operation of the cycle are lowered during the period, when the cell contains many ATP molecules.

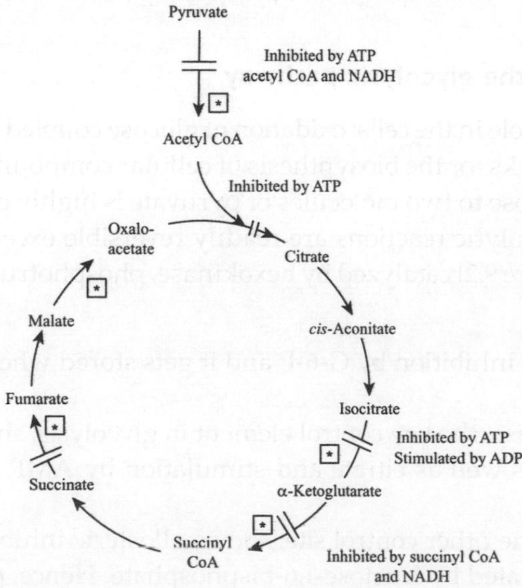

Fig. 9.9 Regulation of the Krebs cycle (Citric acid cycle) and the oxidative decarboxylation of pyruvate. ⊡ Steps requiring NAD^+ or FAD. \\ represents control point.

NAD	NADP
1. Known as nicotinamide adenine dinucleotide, also called Coenzyme I (previously designated as diphosphopyridine nucleotide – DPN).	Known as nicotinamide adenine dinucleotide phosphate, commonly known as Coenzyme II (earlier called as triphosphopyridine nucleotide – TPN).
2. NAD^+ is the oxidized form while its reduced form is written as NADH or more properly as NADH + H^+ (as shown below).	$NADP^+$ is the oxidized form while its reduced form is called NADPH or more appropriately as NADPH + H^+.
3. NAD has two phosphate groups in its molecule.	NADP has an additional phosphate group in its molecule, (i.e., having three phosphate groups) esterified to the 2'hydroxyl of the adenine part of the molecule.
4. NAD is most commonly used in most degradation reactions (i.e., both anaerobic and aerobic respiration), being converted to NADH. The energy conserved in this reduced molecule is released in the form of ATP during oxidative phosphorylation or in the electron transport chain.	NADP is used in synthetic reactions such as photosynthesis and fatty acid synthesis. NADPH (NADPH + H^+)* produced during light reaction of photosynthesis is involved in the fixation of CO_2 to carbohydrates which in turn provide a cellular energy reserve. NADP plays an important role in the 'hexose monophosphate shunt' of respiration, involved in nitrate reduction; and biosynthesis of proteins and nucleic acids.
5. NAD is usually found in its oxidized form (NAD^+).	NADP is usually present in its reduced form NADPH, thus acting as a storehouse of metabolic hydrogen (reducing agent).

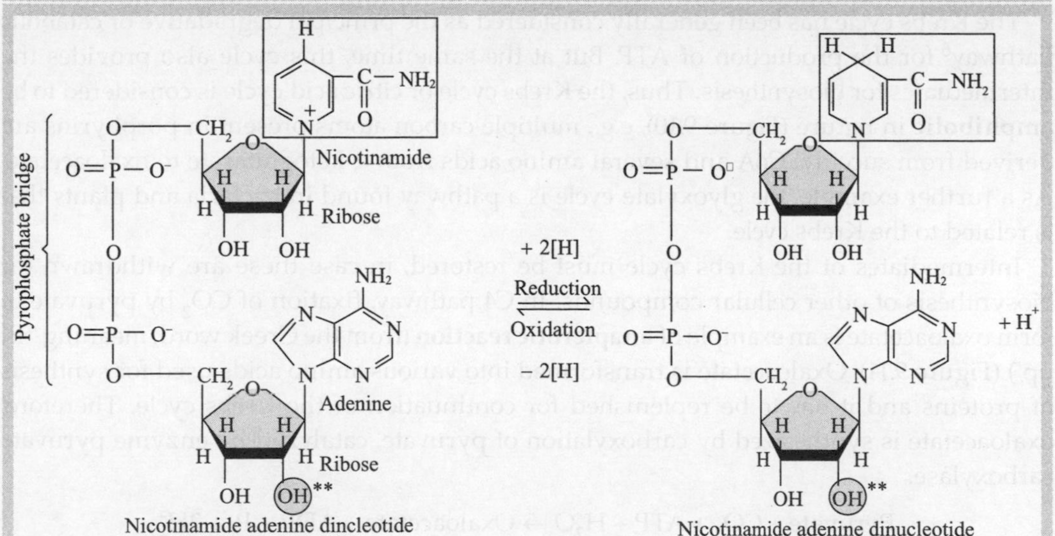

Pyrophosphate bridge; Nicotinamide; Ribose; Adenine; Ribose; + 2[H] Reduction; Oxidation – 2[H]; + H^+

Nicotinamide adenine dincleotide oxidized (NAD^+)

Nicotinamide adenine dinucleotide reduced (NADH)

* Chemists believe that NADP is already ionized, $NADP^+$, and so while two electrons have been taken up it has, to begin with just **one** hydrogen ion, and is in the form of NADPH + H^+. The addition of two electrons to NADP gives it a negative charge which is quickly balanced by the absorption of two hydrogen ions (H^+) forming NADPH and H^+.

$$NADP^+ + 2 \text{ electrons } (e^-) + 2H^+ \rightarrow NADPH_2 \text{ or more properly as } NADPH + H^+$$

** In $NADP^+$ and NADPH, 2' hydroxyl, (OH), of the adenine part of the molecule is replaced by phosphate group, (P).

Both these coenzymes are important in oxidation–reduction reactions ($E^{\circ\prime}$ is equivalent to −0.32 V) in which hydrogen transfer takes place. Both are biological reductants and electron carriers in redox reactions. NADH (or NADPH) and ATP, both together, represent the assimilatory power of the cell.

9.5.4 Regulation of oxidative phosphorylation

ADP content is very important to determine functioning of oxidative phosphorylation. NADH and FADH oxidation are both simultaneously linked to phosphorylation of ADP to ATP. This coupling is called respiratory control, and, can be interrupted by uncouplers such as 2,4-dinitrophenol (DNP[5]), which eliminate the proton gradient by transferring H^+ across the mitochondrial inner membrane.

9.6 The Krebs Cycle as a Second Crossroad of Metabolic Pathways

The Krebs cycle can be viewed as a metabolic hub in which, one or more of its enzymes play a part in other metabolic pathways. Just like an airline hub that connects numerous locations that by themselves have modest networks, the Krebs cycle intermediates are each connected to separate pathways in the cell. Thus, at any given moment, the pathway can be viewed as a separate entity, or as a portion of many others. It is important to remember that, as a distinct pathway, the Krebs cycle has just one substrate, acetyl CoA and one product, CO_2.

The Krebs cycle has been generally considered as the principal degradative or catabolic pathway[6] for the production of ATP. But at the same time, this cycle also provides the intermediates for biosynthesis. Thus, the Krebs cycle or citric acid cycle is considered to be **amphibolic** in nature (Figure 9.10), e.g., multiple carbon atoms present in porphyrins are derived from succinyl CoA and several amino acids from α-ketoglutarate to oxaloacetate. As a further example, the glyoxylate cycle is a pathway found in bacteria and plants that is related to the Krebs cycle.

Intermediates of the Krebs cycle must be restored, in case these are withdrawn for biosynthesis of other cellular compounds. In C4 pathway, fixation of CO_2 by pyruvate to form oxaloacetate is an example of **anaplerotic reaction** (from the Greek word, meaning 'fill up') (Figure 9.11). Oxaloacetate is transformed into various amino acids used for synthesis of proteins and it has to be replenished for continuation of the Krebs cycle. Therefore, oxaloacetate is synthesized by carboxylation of pyruvate, catalyzed by enzyme pyruvate carboxylase.

$$\text{Pyruvate} + CO_2 + \text{ATP} + H_2O \rightarrow \text{Oxaloacetate} + \text{ADP} + P_i + 2H^+$$

[5] Studies at Stanford University (USA) in early part of twentieth century showed that DNP was used as a 'diet pill' for obese patients, since it increases metabolic rate and hence, reduction in weight. However, its use was discontinued later because of certain complications.

[6] Degradative or catabolic pathways are said to be 'convergent' in nature, e.g., catabolism of proteins, polysaccharides and fats – all of which lead to formation of acetyl CoA, a pivotal molecule of cellular metabolism.

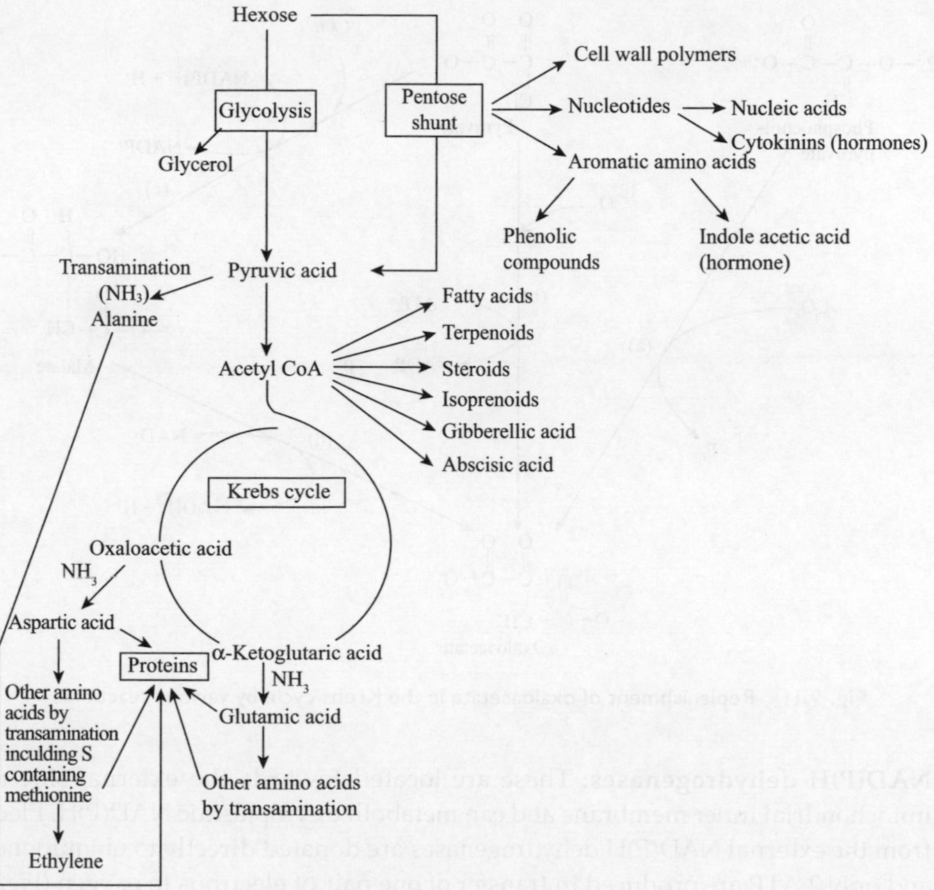

Fig. 9.10 A vast array of metabolic intermediates are produced during respiratory metabolism. As the Krebs cycle can function both in a catabolic mode and as a source of precursors for anabolic pathways, it is called an 'amphibolic pathway' (from the Greek amphi, meaning'both').

9.7 Alternative Electron Transport Pathways in Plant Mitochondria

Plants, fungi, and unicellular eukaryotic organisms, besides having the normal pathway for electron transfer coupled to ATP synthesis, have an alternative, uncoupled pathway that recycles excess NADH to NAD^+. During photorespiration, the major source of mitochondrial NADH is a reaction, where glycine is converted to serine

$$2\,\text{Glycine} + NAD^+ \rightarrow \text{Serine} + CO_2 + NH_3 + NADH + H^+$$

Plants must continue with this conversion even when there is no requirement of NADH for ATP generation. To reproduce NAD^+ from unwanted NADH, mitochondria in plants have evolved the process in which excessive energy conserved in NADH is eliminated in the form of heat (and this can prove very useful to the plant). Cyanide-resistant NADH oxidation is the hallmark of a unique pathway of electron–transfer in plants.

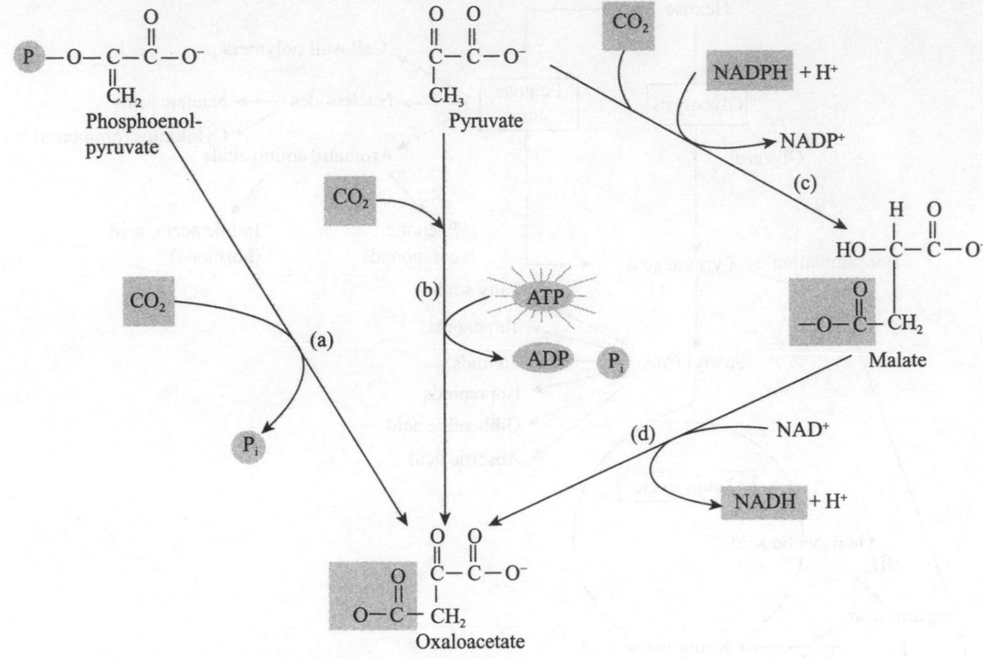

Fig. 9.11 Replenishment of oxaloacetate in the Krebs cycle by various reactions.

- **NAD(P)H dehydrogenases:** These are located towards the external side of the mitochondrial inner membrane and can metabolize cytoplasmic NAD(P)H. Electrons from the external NAD(P)H dehydrogenases are donated directly to ubiquinone pool and only 2 ATP are produced in transfer of one pair of electrons to oxygen (Fig. 9.12).

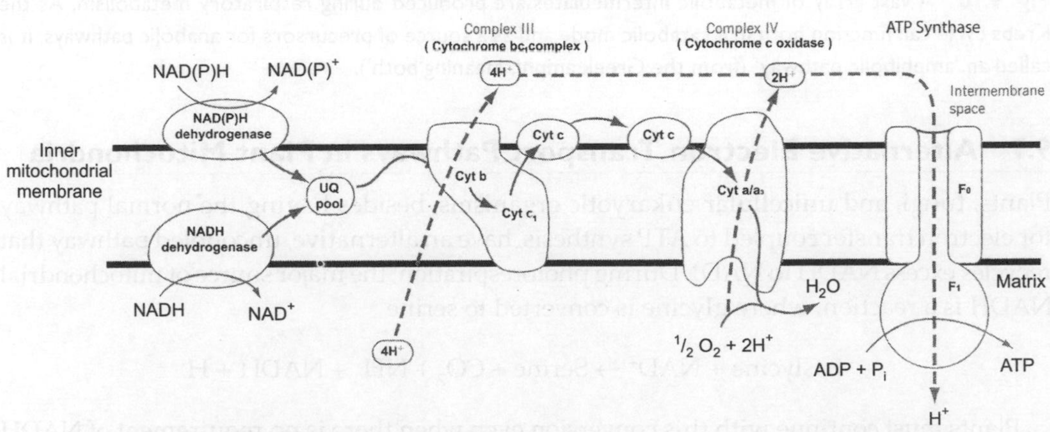

Fig. 9.12 External dehydrogenases in plant mitochondria. These do not traverse the mitochondrial inner membrane and hence, will not translocate protons as complex I does. Transfer of one pair of electrons to oxygen produces only 2 ATP molecules.

- **Rotenone-insensitive NADH dehydrogenases:** These are attached to the internal surface of the mitochondrial inner membrane, facing the matrix and oxidize only internal or matrix NADH. Electrons donated by the rotenone–insensitive dehydrogenases synthesize only 2 molecules of ATP per electron pair. These enzymes may function as a non-proton pumping bypass when complex I is overloaded (Figure 9.12).
- **Alternative oxidases:** These are present on the inner membrane, facing the matrix and accept electrons directly from ubiquinone, thereby reducing O_2 to H_2O. The alternative oxidase, unlike cytochrome c oxidase, is insensitive to inhibition by cyanide, CO, or NO.

9.7.1 Cyanide-resistant respiration

In many animals, inhibitors such as cyanide, carbon monoxide and azide completely inhibit oxygen uptake. However, in some plant tissues, respiration continues even in the presence of cyanide; and this is known as **cyanide-resistant respiration.** This involves the use of an **'alternative pathway'** of electron transport in the mitochondria and is catalyzed by **alternative oxidase**, which comprises of two identical subunits (homodimers) spanning the inner membrane. Electrons from the ubiquinone pool are accepted by alternative oxidase and finally transferred to O_2. Hence, the electrons processed by this enzyme bypass two locations of proton pumping (at cytochrome complexes III and IV), thereby disrupting the electrochemical gradient. When electrons are transferred through the alternative respiratory pathway, the free energy that would otherwise be conserved in the form of ATP gets transformed to heat (Figure 9.13). Cyanide-resistant respiration is, however, sensitive to inhibition by hydroxamic acid derivatives such as **salicylhydroxamic acid (SHAM).**

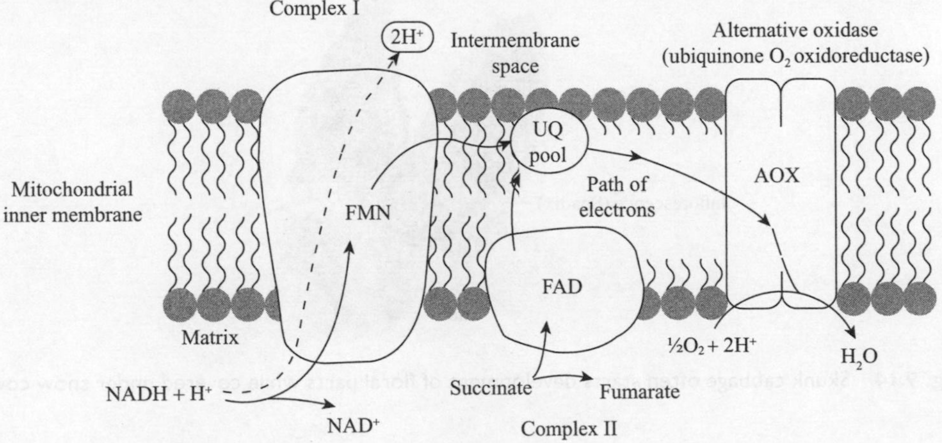

Fig. 9.13 The alternative respiratory pathway in plant mitochondria.

Physiological roles

1. ***Thermogenesis:*** Salicylic acid has been identified as the endogenous trigger, starting off the thermogenic function in the inflorescence spadices of voodoo lily (*Sauromatum guttatum*). Thus, a lot of heat is liberated, which raises the temperature by 10–15°C. This elevated temperature helps in the volatilization of compounds such as amines, ammonia, and indoles, that attract insects for pollination. Salicylic acid was later found to be also involved in plant pathogen defence.

   Secondly, the skunk cabbage (*Symplocarpus foetidus*) starts development of the floral parts while hidden under snow cover. When the inflorescence is ready to emerge, it produces enormous heat that can even melt snow, thus exposing the inflorescence to pollinators (Figure 9.14).
2. ***Energy overflow hypothesis:*** In most of the tissues, the alternative pathway[7] gets activated when the typical cytochrome pathway becomes overloaded and also when the respiratory activity exceeds the cell's demand for ATP. Therefore, this pathway permits the mitochondrion to regulate the relative rates of ATP generation and production of carbon skeletons to be used in synthetic reactions.
3. ***Prevention of overreduction of electron transport chain:*** Another function performed by the pathway is in the plants response to various types of stress conditions, which can inhibit mitochondrial respiration. In response to stress, the electron transport chain leads to surge in production of **reactive oxygen species (ROS)**. By withdrawl of electrons away from the ubiquinone pool, the alternative pathway stops the overreduction of respiratory electron transport chain. This limits the formation of reactive oxygen species and minimizes the harmful effects of stress on respiration.

Fig. 9.14 Skunk cabbage often starts development of floral parts while covered under snow cover.

---

7 The alternative oxidase activity increases during ageing and wounding such as in sliced potato tubers. Other factors like ethylene application, cold stress, high temperature, salicylic acid and salinity are known to trigger alternative oxidase.

9.8 Gluconeogenesis

Gluconeogenesis is a universal multistep pathway in which glucose is produced from non-carbohydrate precursors i.e. pyruvate, lactate, glycerol, oxaloacetate as well as certain amino acids. This takes place in all plants, animals, fungi, and microorganisms. Gluconeogenesis serves to convert lipids and proteins, stored in plant seedlings, to sucrose for distribution within the growing plant. Glucose and its derivatives are the starting compounds for formation of cell walls, nucleotides, coenzymes as well as several other essential compounds in plants.

Although gluconeogenesis and glycolysis pathways have many common steps, but they are not identical occurring in opposite directions. In gluconeogenesis, seven out of ten reactions are catalyzed by the enzymes also involved in glycolytic reactions, thus making them reversible (Figure 9.15).

Three irreversible reactions of glycolysis are skipped by reactions catalyzed by enzymes of gluconeogenesis. These are:

- Conversion of pyruvate to phosphoenol pyruvate *via* oxalocetate, catalyzed by pyruvate carboxylase and PEP carboxykinase
- Dephosphorylation of Fructose-1,6-bisphosphate (FBP), catalyzed by FBPase-1
- Dephosphorylation of Glucose-6-phosphate, catalyzed by G-6-phosphatase

The enzyme pyruvate carboxylase is activated by acetyl CoA, thus enhancing the gluconeogenesis rate when the cell has sufficient supply of other substrates i.e., fatty acids. Animals cannot transform acetyl CoA derived from fatty acids to glucose, whereas plants and microorganisms can do so.

Glycolysis and gluconeogenesis processes are regulated in a reciprocal manner to stop unwanted reactions simultaneously. Synthesis of one glucose molecule from pyruvate is comparatively expensive since it needs 4 ATP, 2 GTP, and 2 NADH.

9.9 Pentose Phosphate Pathway or Oxidative Pentose Phosphate Pathway

(An alternate route for glucose metabolism)

The pentose phosphate pathway (PPP) is sometimes called **the pentose shunt, the hexose monophosphate pathway** or the **phosphogluconate oxidative pathway** (Figure 9.16). The first enzyme of the pathway, glucose-6-phosphate dehydrogenase was discovered by Otto Warburg in 1931, and the complete pathway was subsequently elucidated by Fritz Lipmann, Frank Dickens, Bernard Horecker and Efraim Racker.

Pentose sugars are the major intermediate products of the oxidation of glucose, and hence, called the pentose phosphate pathway. This pathway exists in many organisms including plants and animals and shares several intermediates with glycolysis. All the reactions occur in the cytosol and require oxygen for the entire operation.

In the pentose phosphate pathway, glucose- 6-phosphate undergoes a dehydrogenation reaction to produe lactone, followed by its hydrolysis to form 6-phosphogluconate, and which

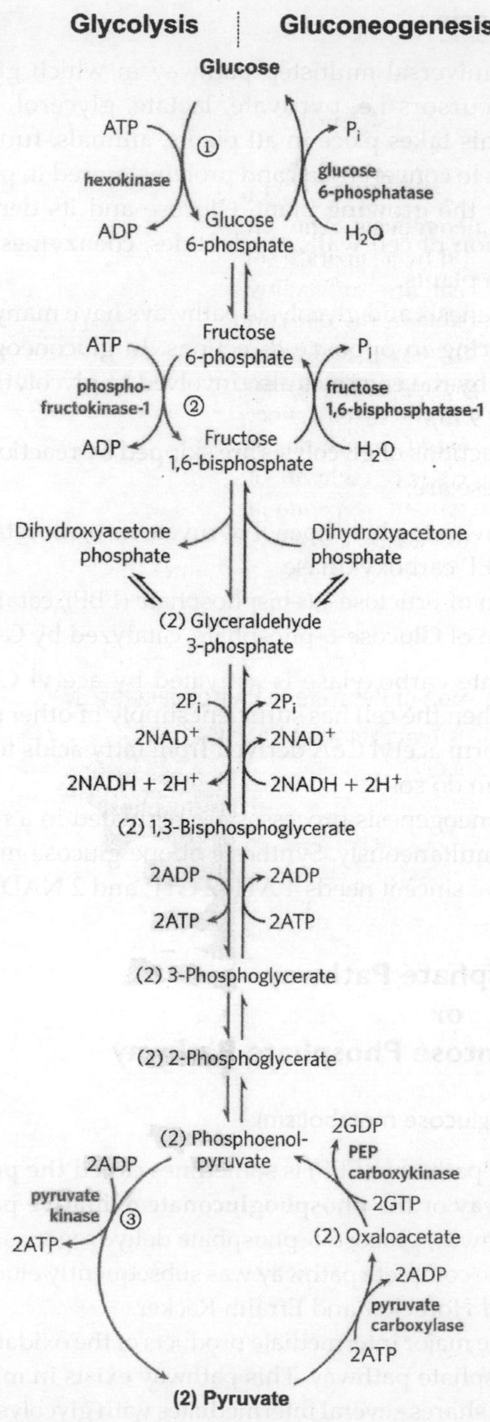

Fig. 9.15 Glycolysis and gluconeogenesis are not similar pathways going on in opposite directions. Out of 10 reactions of gluconeogenesis, 7 reactions are reverse to that of glycolysis. The remaining 3 glycolytic reactions (marked as 1, 2, 3) are essentially irreversible and do not function in gluconeogenesis.

in turn is oxidatively decarboxylated to ribulose-5-phosphate. In both the oxidative reactions, $NADP^+$ acts as acceptor of the electrons. Subsequent reactions lead to the generation of glyceraldehyde-3-phosphate and fructose-6-phosphate, to be oxidized *via* the glycolytic pathway. Erythrose-4-phosphate, Xylulose-5-phosphate, and sedoheptulose-7-phosphate are produced as intermediates in the interconversion reactions.

If an equivalent of the glucose molecule is oxidized to carbon dioxide and water *via* this direct oxidation pathway (six turns of cycle), then 12 molecules of NADPH would be formed. Because this mechanism bypasses the glycolytic cycle, it is also known as hexose monophosphate shunt.

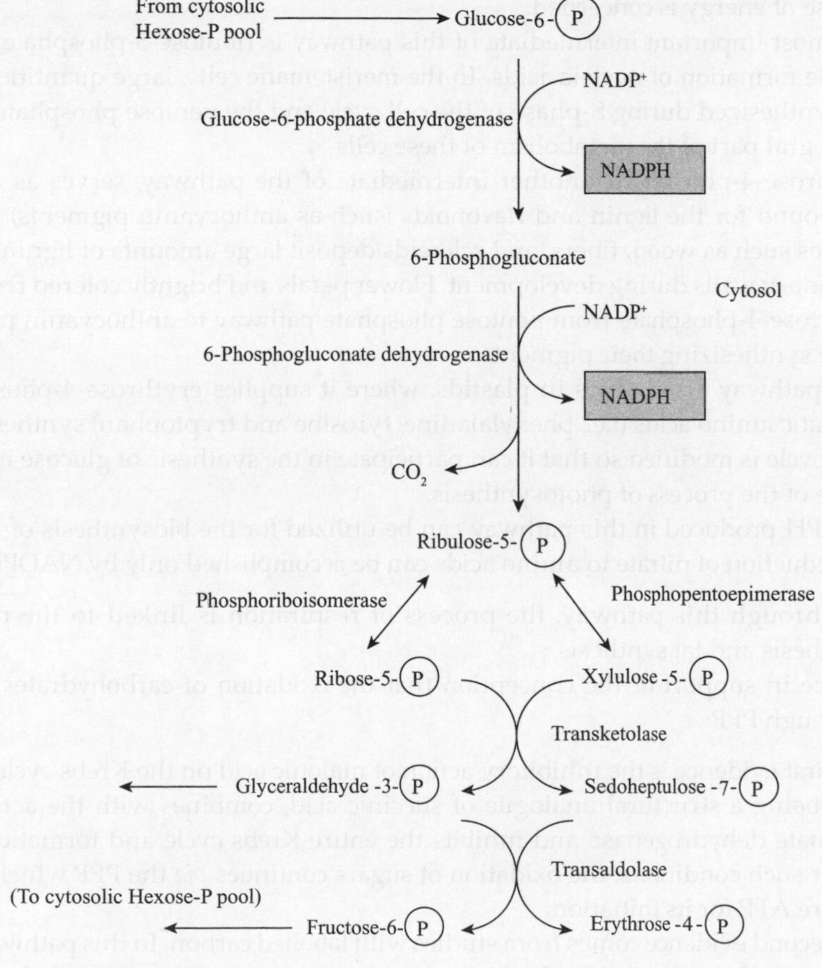

Fig. 9.16 The Pentose phosphate pathway or Oxidative pentose phosphate pathway.

Regulation of pentose phosphate pathway

The first reaction in the oxidative branch of this cycle, i.e., the oxidation of glucose-6-phosphate to 6-phosphogluconate, is a strictly irreversible and rate-determining step under physiological conditions, with $NADP^+$ as the regulatory factor. The non-oxidative branch of this cycle is regulated primarily by the substrate availability.

Significance

1. PPP is considered to be a means to produce 12 NADPH (36ATP) molecules to power the anabolic reactions in the cytosol, as efficient as glycolysis and Krebs cycle as far as release of energy is concerned.
2. The most important intermediate of this pathway is ribulose-5-phosphate, essential for the formation of nucleic acids. In the meristematic cells, large quantities of DNA are synthesized during S-phase of the cell cycle and the pentose phosphate pathway is integral part of the metabolism of these cells.
3. Erythrose-4-phosphate, another intermediate of the pathway, serves as a starting compound for the lignin and flavonoids (such as anthocyanin pigments) synthesis. Tissues such as wood, fibers, and sclereids deposit large amounts of lignin into their secondary walls during development. Flower petals and brightly colored fruits divert erythrose-4-phosphate from pentose phosphate pathway to anthocyanin production while synthesizing their pigments.
4. This pathway also occurs in plastids, where it supplies erythrose-4-phosphate for aromatic amino acids (i.e., phenylalanine, tyrosine and tryptophan) synthesis.
5. This cycle is modified so that it can participate in the synthesis of glucose in the dark phase of the process of photosynthesis.
6. NADPH produced in this pathway can be utilized for the biosynthesis of fats. Also, the reduction of nitrate to amino acids can be accomplished only by NADPH.

Thus, through this pathway, the process of respiration is linked to the process of photosynthesis and fat synthesis.

Evidence in support of the conception that the oxidation of carbohydrates proceeds partly through PPP:

1. The first evidence is the inhibitory action of malonic acid on the Krebs cycle. Malonic acid, being a structural analogue of succinic acid, combines with the active site of succinate dehydrogenase and inhibits the entire Krebs cycle and formation of ATP. Under such conditions, the oxidation of sugars continues *via* the PPP which does not require ATP for its initiation.
2. The second evidence comes from studies with labelled carbon. In this pathway, carbon number 1 of glucose will be the first to be released as CO_2 followed by carbon number 6. Such studies indicate that PPP accounts for as much as 10 per cent of the glucose metabolism.

In young growing cells, the glycolytic–citric acid cycle is the main pathway, whereas in older cells, PPP plays the major role.

9.10 Anaerobic Respiration

Anaerobic respiration, i.e., incomplete oxidation of hexose sugars, occurs in the cytosol in the absence or shortage of oxygen. It is an inefficient way of releasing energy as compared to aerobic respiration. It occurs in the initial stages of germination in number of seeds, e.g., pea and cereals and in many fruits, including grapes. The types as well as amounts of end products (ethyl alcohol, lactic acid, butyric acid, acetaldehyde and such) formed differ in different tissues and organisms.

Anaerobic conditions occur in plants and fungi growing in mud beneath stagnant water, especially in swamps and marshes. Rice seeds germinate and grow anaerobically until the shoots reach oxygenated water. Ethanol and lactate offer no good solutions to the problem of NADH accumulation. The pyruvate consumed is always present in adequate amounts, of course, because glycolysis itself produces it, but it is really expensive because many of its bonding orbitals have high-energy electrons. Furthermore, pyruvate could be used as a monomer for many types of synthesis. A lack of oxygen forces a cell to use a very valuable molecule as an electron dumping ground. Even worse, the products of anaerobic respiration, either lactate or ethanol, are toxic and if they accumulate in the tissues, they can kill the cells producing them.

Even in the presence of an oxygen-rich atmosphere and aerobic respiration, anaerobic respiration still has some selective advantage. For example, rice seedlings are capable of anaerobic respiration, so they germinate and establish themselves during times of floods when other plants or seedlings are suffocating. Although this metabolism is expensive for the rice, but by the time the flooding subsides, rice plants are already well rooted and out-compete other species whose seeds are just beginning to germinate.

9.11 Alcoholic Fermentation

Closely analogous to anaerobic respiration is the process of fermentation, in which organic substances are decomposed into simpler compounds due to the activity of microorganisms. The oldest and best-known example of fermentation is the alcoholic fermentation of sugars by yeasts (*Saccharomyces*), e.g., alcoholic fermentation by yeast[8] is an enzymatic process. Yeast is unable to survive alcohol concentration much above 10 per cent.[9]

---

[8] Chemically speaking, zymase secreted by yeast cells is an iron-containing multienzyme aggregate with two components – glycolase and carboxylase.

[9] Besides yeast, alcohol dehydrogenase (ADH) is present in the liver of human beings, where it is responsible for the oxidation of ethanol and other alcohols present there.

$$CH_3—\overset{O}{\overset{\|}{C}}—COOH \xrightarrow[\searrow CO_2]{\text{Pyruvate decarboxylase (TPP* as a coenzyme)}} CH_3—\overset{O}{\overset{\|}{C}}—H \xrightarrow[NADH + H^+ \;\; NAD^+]{\text{Alcohol dehydrogenase}^9 \text{ (Zn as an activator)}} CH_3—CH_2—OH$$

Pyruvic acid → Acetaldehyde → Ethanol

* TPP - Thiamine pyrophosphate

The net result of this anaerobic respiration is

$$\text{Glucose} + 2P_i + 2ADP + 2H^+ \rightarrow 2\text{ Ethanol} + 2CO_2 + 2ATP + 2H_2O$$

Many of the fermentation processes are of great importance in both households and industry, such as souring of milk, ripening of cheese, retting of flax, curing of tobacco and tea, tanning of leather, and so on. Brewers make beer preparations using ethanolic fermentation of the carbohydrates in barley, brought about by glycolytic enzymes in yeast. With continuous depletion of the fossil fuels reserves and the increasing demand for fuel used in internal combustion engines, there is increased interest in the employment of ethanol as a fuel substitute. The fermentation process can produce not only ethanol for fuel but also by-products (such as proteins) that can be utilized as animal feed.

9.12 Lactic Acid Fermentation

Lactate is generally produced from pyruvate reduction using NADH, in several microorganisms as well as in higher organisms when oxygen availability is in short supply, for example during intense muscular exercise. This step is catalyzed by the enzyme lactate dehydrogenase.

$$CH_3—\overset{O}{\overset{\|}{C}}—COOH + NADH + H^+ \xrightleftharpoons{\text{Lactate dehydrogenase}} CH_3—\underset{OH}{\underset{|}{\overset{H}{\overset{|}{C}}}}—COOH + NAD^+$$

Pyruvic acid ⇌ Lactic acid

The lactate produced by active skeletal muscles is carried to the liver tissue through the blood where it gets transformed to glucose during recovery periods from intense muscular activity. High amounts of lactate formed during vigorous muscle contraction, gets ionized in muscle and blood leading to acidification, which limits the period of vigorous exercise.

Just like in alcoholic fermentation, there is no involvement of net oxidation–reduction. NADH produced in the reaction involving oxidation of glyceraldehyde-3-phosphate is utilized during reduction of pyruvate to lactate/ethanol. This results in regeneration of NAD^+ that sustains the continuous functioning of glycolysis in the absence of oxygen. Since no ATP is produced in the above reactions, the net ATP gain from one molecule of glucose fermented is just two (from glycolysis).

9.13 A Summarized Overview of the Various Sites of Cellular Respiration alongside Diagrammatic Representation (Figure 9.17)

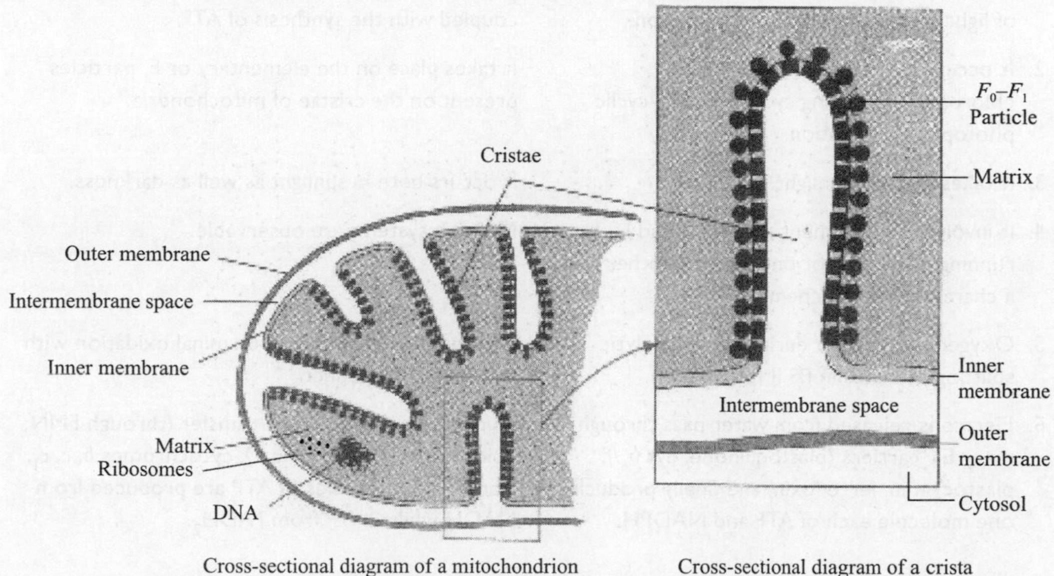

Fig. 9.17 Various sites of cellular respiration. The inner membrane of mitochondria encloses the mitochondrial matrix, which includes soluble enzymes, mitochondrial DNA and ribosomes.

Reactions	Site
Glycolysis	Cytosol
Pentose phosphate pathway	Cytosol
Oxidative decarboxylation of pyruvic acid	Cytosol
Alcoholic and lactic acid fermentation	Cytosol
Citric acid cycle (excluding succinate conversion to fumarate)	Mitochondrial matrix
Succinate $\xrightarrow{\text{Succinate dehydrogenase}}$ Fumarate	Mitochondrial inner membrane
Electron transport system	Mitochondrial inner membrane
Oxidative phosphorylation	ATP synthase complex (F_0–F_1 particles)

Photophosphorylation	Oxidative phosphorylation
1. The ability of green plants to make ATP from ADP and inorganic phosphate in the presence of light is called photophosphorylation.	The oxidation of NADH/$FADH_2$ (the reducing power of cells), in stepwise electron transfer, is coupled with the synthesis of ATP.
2. It occurs in the granal region of chloroplast, involving cyclic and non-cyclic photophosphorylation.	It takes place on the elementary or F_1 particles present on the cristae of mitochondria.
3. It takes place in sunlight.	It occurs both in sunlight as well as darkness.
4. It involves two pigment systems II and I – both running in a series or one after the other with a characteristic Z-scheme.	No such systems are observable.
5. Oxygen is given out during photocatalytic splitting of water in PS II (photolysis).	Oxygen is used up during terminal oxidation with water as end product.
6. Electrons released from water pass through many ET carriers (plastoquinone, cyt b_6/f, plastocyanin, ferredoxin) and finally producing one molecule each of ATP and NADPH.	During stepwise electron transfer (through FMN, ubiquinone or coenzyme Q, cytochromes b, c, c_1, a and a_3) 3 molecules of ATP are produced from NADH while 2ATP from $FADH_2$.

9.14 Other Substrates for Respiration

Besides glucose, fats and proteins can also be used as respiratory substrates. Both of them are catabolized to acetyl CoA, which is the substrate for the Krebs cycle.

Fats

A triglyceride molecule gets hydrolyzed to glycerol and three fatty acids. Then, beginning at the carboxyl terminal of the fatty acids, 2-carbon acetyl groups are successively eliminated as acetyl CoA by a process called β-oxidation.

Proteins

Proteins are similarly degraded to their amino acids accompanied by removal of the amino groups. Some of the residual carbon skeletons are transformed to Krebs cycle intermediates such as α-ketoglutarate, oxaloacetate and fumarate and thereby enter the cycle.

9.15 Respiratory Quotient

The ratio of molecules of CO_2 produced to molecules of O_2 consumed is called respiratory quotient (RQ).

$$RQ = \frac{\text{Number of molecules of } CO_2}{\text{Number of molecules of } O_2}$$

The RQ value of a respiring tissue provides valuable information about the nature of the substrate being oxidized and also the chemical transformation taking place during respiration. For example, when

1. RQ = 1 (respiratory substrate is carbohydrate)

$$\underset{\text{Glucose}}{C_6H_{12}O_6} + 6O_2 \rightarrow 6CO_2 + 6H_2O$$

$$RQ = \frac{6CO_2}{6O_2} = 1$$

2. RQ < 1 (respiratory substrate is either fat or protein)

$$\underset{\text{Tripalmitin}}{2C_{51}H_{98}O_6} + 145O_2 \rightarrow 102CO_2 + 98H_2O$$

$$RQ = \frac{102CO_2}{145O_2} = 0.7$$

$$\underset{\text{Triolein}}{C_{57}H_{104}O_6} + 80O_2 \rightarrow 57CO_2 + 52H_2O$$

$$RQ = \frac{57CO_2}{80O_2} = 0.7$$

Similarly, in the case of proteins, the value of RQ is about 0.5.

Since fats and proteins are more reduced than carbohydrates, more oxygen is required for their respiration, so RQ value is less than 1.

3. RQ > 1 (respiratory substrate is rich in organic acid)

$$\underset{\text{Oxalic acid}}{2(COOH)_2} + O_2 \rightarrow 4CO_2 + 2H_2O$$

$$RQ = \frac{4CO_2}{1O_2} = 4$$

$$\underset{\text{Tartaric acid}}{2C_4H_6O_6} + 5O_2 \rightarrow 8CO_2 + 6H_2O$$

$$RQ = \frac{8CO_2}{5O_2} = 1.6$$

$$\underset{\text{Malic acid}}{C_4H_6O_5} + 3O_2 \rightarrow 4CO_2 + 3H_2O$$

$$RQ = \frac{4CO_2}{3O_2} = 1.3$$

Organic acids (e.g. Krebs cycle intermediates) are highly oxidized than the carbohydrates and consequently need less oxygen for their oxidation, so RQ value is more than 1.

4. RQ = 0 (incomplete oxidation of carbohydrates in succulent plants, e.g., *Bryophyllum*, *Opuntia*)

$$\underset{\text{Glucose}}{2C_6H_{12}O_6} + 3O_2 \rightarrow \underset{\text{Malic acid}}{3C_4H_6O_5} + 3H_2O$$

$$RQ = \frac{0CO_2}{3O_2} = 0$$

In such plants, the respiration of carbohydrates occurs at night when the stomata are open and generally results in the production of organic acids. So RQ is always zero as oxygen is absorbed without any corresponding release of CO_2.

5. RQ = α (anaerobic respiration and fermentation)

Seeds in the early stages of germination show anaerobic respiration as very little oxygen is absorbed before the seed coats rupture and CO_2 is produced anaerobically. Hence, the RQ values become infinity in such a case.

$$\underset{\text{Glucose}}{C_6H_{12}O_6} \rightarrow 2CO_2 + \underset{\text{Ethyl alcohol}}{2C_2H_5OH}$$

$$RQ = \frac{2CO_2}{0O_2} = \alpha \text{ (infinite)}$$

Importance of RQ

The RQ value of a respiring tissue may provide valuable information to an investigator:

1. One can obtain a rough indication of type of substrate being oxidized.
2. RQ value also indicates whether oxidation process is complete or not, since intermediates show different RQ values.
3. Plants under starvation conditions will exhibit RQ values consistently below unity.
4. Plants with anaerobic respiration show RQ value approaching toward infinity.

9.16 Factors Affecting Respiration

Respiration process is affected by number of internal and external factors.

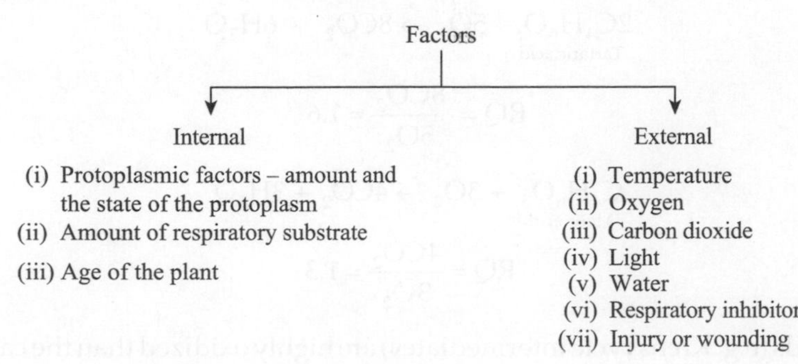

Internal factors

1. ***Protoplasmic factors***: The respiration rate is dependent upon the amount of protoplasm in the cells of the respiring tissues and the state of its activity (hydration of protoplasm, number of mitochondria per cell and pH status of protoplasm). Because of a large amount of active protoplasm, the rate of respiration is at its maximum in the meristematic cells as compared to older and mature cells or tissues. Developing buds generally display the highest respiration rate followed by vegetative organs. In mature vegetative organs, stems normally have the lowest respiratory rates, whereas in leaf and root, the respiration rate shows variation in different plant species and the conditions of plant growth such as availability of soil nutrients. In general, the rate of respiration varies among the different plant organs.
2. ***Respiratory substrate***: The rate of respiration of a given tissue is governed by the available amount of respirable material, especially sucrose. Respiratory substrates such as carbohydrates, proteins, lipids, have different ratios of carbon to oxygen atoms. Therefore the rate of CO_2 evolution to O_2 consumption (RQ) varies with the substrate oxidized. The etiolated leaves show a decline in the rate of respiration because of starvation or poor photosynthesis.
3. ***Age of the plant***: When a plant organ reaches maturity, its respiration rate declines slowly and the tissue undergoes senescence. An exception to this is the sudden increase in respiration, ultimately termed as **climacteric**, that follows the onset of ripening in several fruits (banana, apple, tomato, etc.) and senescence in detached flowers as well as leaves. Both the ripening process and the climacteric respiration are initiated by the endogenous synthesis of ethylene.

External factors

1. **Temperature:** The optimum temperature for respiration lies between 30°C and 40°C. Beyond 40°C, respiratory rate declines because of thermal denaturation of the enzymes participating in respiration process. Temperature coefficient, Q_{10}, usually varies from 2 to 2.5 within temperature range of 0–35°C.

$$Q_{10} = \frac{\text{Rate at}\,(t+10)\,°\text{C}}{\text{Rate at}\,t\,°\text{C}}$$

   It means that the rate of respiration increases by 2–2.5 times for a rise of 10 °C within this temperature range. Beyond 35 °C, the Q_{10} in most of the plants becomes less as availability of the substrate becomes limiting. This coefficient describes how respiration responds to short-term temperature changes, and it varies with plant development and external factors.
2. ***Oxygen:*** O_2 is one of the essential reactants for the aerobic respiration in higher plants. The air consists of significantly large amounts of oxygen (~21 per cent) for plant respiration. The amount of oxygen present in the air may be increased or decreased considerably without affecting the respiration. However, in an atmosphere of nitrogen (O_2 absent), respiration as measured in terms of loss of sugar and evolution of CO_2

increases rather than showing a decline because lack of oxygen promotes glycolysis **(Pasteur Effect)**.

The Pasteur Effect

The Pasteur Effect is an inhibitory action of oxygen (O_2) concentration on fermentation processes. In 1861, Louis Pasteur – an illustrious French microbiologist - concluded that 'aerating' yeast containing broth intermix promoted increased yeast cell growth, while conversely, the fermentation rate declines. However, in the absence of oxygen, yeast consumes more glucose than under oxygenic condition (in the presence of oxygen).

Yeast is an example of a facultative anaerobes, that is, it can generate energy by the usage of two distinct metabolic pathways. In the presence of molecular O_2, the product of glycolysis, pyruvate, is transformed to acetyl CoA which is later used in the Krebs cycle, producing 36 molecules of ATP per molecule of glucose. However, under anaerobic conditions, pyruvate is converted into ethanol and carbon dioxide (CO_2), and the energy generation efficiency is less (that is, 2 molecules of ATP per mole of glucose). In fact, under anaerobic conditions, yeast consumes more glucose than in the presence of oxygen, and thus, the rate of glucose breakdown is rapid, but the quantity of ATP generated is low. A switch from anaerobic to aerobic condition leads to decline in the rate of glucose breakdown in yeasts. But the increased rate of ATP and citrate synthesis during glycolysis and Krebs cycle slows down metabolism as both of them act as an allosteric inhibitor for phosphofructokinase - the enzyme that controls the conversion of fructose phosphate to fructose bisphosphate. On the contrary, high levels of ADP activate phosphofructokinase, committing more sugar to the catabolic pathways while high level of ATP inhibits or shuts down the catabolic pathway. So it is of great advantage for yeast to undergo the glycolytic cycle in the presence of oxygen, as greater number of ATP molecules are formed with low amounts of glucose.

The Pasteur Effect has some practical importance in storage of fruits and vegetables, especially apples. Here a goal of storage is to prevent extensive sugar loss and overripening. This is done by carefully decreasing O_2 to the concentration at which aerobic respiration is minimized but sugar breakdown by the anaerobic process is not stimulated.

3. ***Carbon dioxide:*** Increase in carbon dioxide concentration in the atmosphere lowers respiration and consequently inhibits seed germination. In poorly aerated soils, CO_2 accumulates due to respiration of microorganisms and affects the growth of plants. High CO_2 concentration and low O_2concentration would check the deterioration in the quality of fruits and vegetables by reducing the rates of respiration.
4. ***Light:*** It has an indirect effect on respiration and raises the temperature causing the stomata to open and brings about accumulation of respirable material by photosynthesis.
5. ***Water***: The rate of respiration increases with the amount of water due to the following reasons:
   - It maintains turgor of the respiring cells.
   - It brings about a hydrolytic conversion of reserve carbohydrates to soluble sugars.
   - It facilitates the action of respiratory enzymes.
   - It provides the medium in which oxygen diffuses into the respiring protoplasm.
6. ***Respiratory inhibitors***: Metabolic inhibitors like cyanide, azide, CO, fluoride have been known to decrease the rate of respiration. These choke a person to death because

of interruption in oxidative phosphorylation, thus no energy is released causing brain death.

7. ***Injury or wounding***: Wounding a plant organ leads to an increase in the respiration rate for a short period. For example, in potato tubers, wounding results in an increase in the sugar content of the injured portion, which accounts for the increased rate of respiration.

Review Questions

RQ 9.1. How is oxaloacetate for the Krebs cycle replenished?

Ans. Glycolytic and the Krebs cycle intermediates are drawn away to become precursors for the biosynthesis of cellular molecules. For the continuation of the Krebs cycle, a minimum amount of oxaloacetate must be maintained to permit its condensation with acetyl CoA. If not, the Krebs cycle will cease. The replenishment of oxaloacetate for the Krebs cycle is carried out by the following reactions:

- By the carboxylation of phosphoenolpyruvate:

$$\text{Phosphoenolpyruvate} + CO_2 \rightarrow \text{oxaloacetate} + P_i.$$

- By ATP-driven carboxylation of pyruvate:

$$\text{Pyruvate} + CO_2 + \text{ATP} \rightarrow \text{oxaloacetate} + \text{ADP} + P_i.$$

- If the amino acid aspartate is present, it can also be used as a source of oxaloacetate by the transamination reaction.

$$\text{Aspartate} \xrightleftharpoons{\text{Transamination}} \text{Oxaloacetate}$$

The carboxylation of pyruvate is an example of an *anaplerotic* reaction (from the Greek word, meaning 'to fill-up'). Oxaloacetate, the product of the pyruvate carboxylase reaction, is both intermediate in 'gluconeogenesis' and the 'citric acid cycle'.

Interestingly, in animals, PEP carboxylase is absent and its function is fulfilled by the biotinyl enzyme pyruvate carboxylase (not found in plants) which catalyzes the pyruvate + CO_2 + ATP → oxaloacetate + ADP + P_i.

RQ 9.2. Outline the three steps involved in the Krebs cycle where water molecules are consumed.

Ans. The three steps of the Krebs cycle where water molecules are consumed:

1. $$\text{Oxaloacetic acid} + \text{Acetyl CoA} + H_2O \xrightarrow{\text{Citrate synthase}} \text{Citric acid} + \text{CoA}$$

2. $$\text{Cis-aconitic acid} + H_2O \xrightarrow{\text{Aconitase}} \text{Isocitric acid}$$

3. $$\text{Fumaric acid} + H_2O \xrightarrow{\text{Fumarase}} \text{Malic acid}$$

RQ 9.3. Outline the biochemical steps in aerobic respiration where CO_2 molecules are generated.

Ans. The three steps where CO_2 molecules are released in aerobic respiration:

1. Pyruvic acid + CoA + NAD^+ $\xrightarrow[Mg^{++}]{\text{Pyruvate dehydrogenase}}$ Acetyl CoA + CO_2 + NADH

2. Oxalosuccinic acid $\xrightarrow[Mn^{++}]{\text{Oxalosuccinic dehydrogenase}}$ α-Ketoglutaric acid + CO_2

3. α-Ketoglutaric acid + CoA + NAD^+ $\xrightarrow{\alpha\text{-Ketoglutaric dehydrogenase}}$ Succinyl CoA + CO_2 + NADH

RQ 9.4. Why are niacin and riboflavin so important in our diet?

Ans. Niacin (nicotinic acid) and riboflavin (vitamin B_2)[10] are the starting material for the synthesis of dinucleotides like NAD/NADP and FAD, respectively. They are essential for respiratory processes such as glycolysis and the Krebs cycle for the continuous release of energy.

These nucleotides occur in very small quantities and their synthesis is important for human health, otherwise deficiency symptoms will appear because of inadequate quantities of NAD, NADP and FAD. This is normally maintained by oxidation of its reduced form by molecular oxygen. If the oxygen is in short supply, the unloading of NADH to NAD (or $FADH_2$ to FAD) will be slowed down and the process of ATP generation will be affected. It is likely that pyruvate may be pushed towards lactate or alcohol synthesis. In the long-term, lactic acid is toxic and must ultimately be removed. In most cases, muscular fatigue is linked to the formation of lactic acid.

RQ 9.5. With the help of an example, explain where energy during respiration is released in the form of heat? What significance does it have for the survival of the plant?

Ans. In some of the arum lilies, such as voodo lily (*Sauromatum guttatum*) and Lords-and-ladies (*Arum maculatum*), some of the stored energy of sugar molecules is lost as heat. The lilies have a tongue-shaped spike called a spadix, which has the stench of rotting meat. During the period of maximum odour production, the spadix temperature can be as much as 10–15°C above that of the surrounding air and the heat is dissipated in the atmosphere.

It seems that the heat generated in the spadix during 'cyanide-resistant' or 'alternative pathway' respiration is necessary for the volatilization of the extremely foul-smelling chemical compounds, that presumably help in attracting flies for pollination, without which the plant is unable to reproduce.

RQ 9.6. Why will plants generally die or wilt, if the roots and underground stems are deprived of oxygen as it happens during waterlogging or if hoeing of the soil is not properly done?

Ans. Plant organs, especially roots and underground stems, quite often become deprived of oxygen when the soil becomes waterlogged or hoeing of soils is not done.

In the absence or deficiency of oxygen (anaerobiosis as it is called), the reduced compounds NADH/$FADH_2$ are not re-oxidised to regenerate NAD^+/FAD that are essential for the continuation of EMP pathway and Krebs cycle and continued release of energy that is essential for the uptake of nutrients or minerals (active transport) from the soil and other vital processes. The resulting reduction or loss of energy release initially reduces the salt uptake (thus affecting the osmotic concentration in root cells) and then the process of water uptake. The plants thus undergo wilting as transpiration rate far exceeds water absorption. If this failure of energy supply continues, the whole plant will die due to accumulation of alcohol produced during anaerobic respiration.

RQ 9.7. Respiratory inhibitors choke or throttle a person to death. Comment.

Ans. During oxidative phosphorylation, electrons donated by NADH or $FADH_2$ are passed from one carrier to another present on the mitochondrial inner membrane. The final carriers are haeme-containing cytochromes (cyt *b*, *c*, c_1, *a*, a_3) in which electrons reduce the iron from Fe^{+++}(Ferric)

---

[10] Flavonoids and vitamin E in walnuts are known to destroy harmful free radicals that cause dementia in human beings. One should therefore, eat a few walnuts daily to improve mental skills!

to Fe^{++}(Ferrous) state, i.e., they are alternately reduced or oxidized. At different steps of ETS, energy is generated as ATP (i.e., 3 molecules for NADH and 2 molecules for every $FADH_2$).

However, the use of respiratory inhibitors (such as azides, malonates, etc.) obstruct or block the flow of electrons by combining with the iron of cytochromes, thus preventing $Fe^{+++} \rightarrow Fe^{++}$ transformation. Hence there is no release of energy.

Brain death occurs if a man is deprived of energy for more than few minutes, because of choking; eventually leading to death.

Carbon monoxide, a poisonous gas, combines with the haemoglobin of the blood to form carboxyhaemoglobin which will prevent the transport of O_2 in the blood, leading to brain death if a person, by accident, remains in an atmosphere of CO for some time. Like CO, cyanide (CN^-) and azide (N_3^-) interfere with the last electron carrier, cytochrome oxidase. When these inhibitors are present, electrons cannot pass from cyt *c* to oxygen. Hence, there is no release of ATP. Interestingly, heat generating respiration is not affected by cyanide. Hence, it is usually called *cyanide-resistant respiration* (a better name is thermogenic respiration).

RQ 9.8. Oxygen is not directly used up during glycolysis and Krebs cycle, yet it is essential for their continuation. Explain.

Ans. The process of glycolysis and the Krebs cycle consists of a series of reactions which do not use oxygen directly. However, some ATP is produced through substrate-level phosphorylation. But much energy is stored in the form of reduced compounds NADH, NADPH and $FADH_2$. During oxidative phosphorylation, the electron pairs are transported to molecular oxygen through a series of electron transport carriers and finally to cytochrome oxidase where NADPH or NADH is converted to NADP/NAD. Terminal oxidation is necessary as it results in the regeneration of NAD/NADP which then re-enters the reaction sequences of aerobic respiration (i.e., EMP pathway and Krebs cycle).

RQ 9.9. Potatoes that are grown in the mountainous regions produce larger-sized tubers than grown in the plains. Why so?

Ans. The optimum temperature for photosynthesis varies between 20°C and 25°C while that for respiration is about 35°C. This interrelationship between photosynthesis and respiration is important for determining crop yield. In cooler regions, the night temperature is low, reducing the breakdown of photosynthates, thus resulting in accumulation of surplus organic food material. Hence, the potato tubers will become large in size. On the other hand, high temperatures at night in the plains promote increased rate of respiration (high night temperatures are followed by a corresponding rise in the rate of respiration). There will be a greater breakdown of surplus carbohydrates. Hence, the tubers will be of smaller size as compared to those being grown at higher altitudes.

RQ 9.10. Why are fruits and vegetables kept under refrigeration during shipment and storage?

Ans. Low temperatures during storage and shipment reduce the rate of respiration to minimum, thereby preventing the breakdown of food materials. Hence, the fruits and vegetables remain fresh and maintain their flavour and quality better for longer period. Their shelf-life is also increased.

RQ 9.11. For every molecule of pyruvic acid, how many molecules are produced of NADH, NADPH, $FADH_2$, and ATP (through substrate-linked phosphorylation)? Correlate it in terms of the total number of ATPs generated.

Ans. During oxidative decarboxylation and the Krebs cycle:

		Number of ATPs
NADPH	1	3
NADH	3	9
$FADH_2$	1	2
ATP (through substrate-linked phosphorylation)	1	1
		15

RQ 9.12. For every molecule of glucose, how many molecules of the following are produced during glycolysis, oxidative decarboxylation and Krebs cycle?

Ans.

Pyruvic acid	2 molecules
Acetyl CoA	2 molecules
NADH	8
NADPH	2
$FADH_2$	2
ATPs (through substrate-level phosphorylation)	6 (−2*) = 4

*2ATPs are used during conversion of glucose to fructose-1,6- bisphosphate (the chief respiratory substrate) as given under:

1. Glucose + ATP $\rightarrow$ Glucose-6-phosphate + ADP
2. Glucose-6-phosphate $\rightleftharpoons$ Fructose-6-phosphate
3. Fructose-6-phosphate + ATP $\rightarrow$ Fructose-1,6-bisphosphate + ADP

RQ 9.13. Discuss the yield of NADH/NADPH, $FADH_2$ and ATP from the complete oxidation of glucose; the EMP pathway; and the Krebs cycle; along with the Net ATP yield.

Ans.

Step	NADH	$FADH_2$	ATP	NET ATP
Glycolysis	2	0	2	6
Oxidative decarboxylation	2	0	0	6
Krebs cycle	6	2	2	24
Total ATPs				36

Two molecules of NADH produced in the cytosol in glycolysis must be transported across mitochondrial membranes that require 1 ATP per molecule of NADH. The net ATP generation is decreased by two. Thus, the theoretical yield of ATP harvested from glucose by aerobic respiration totals 36 molecules (refer to Table 9.2).

Yield of ATP is computed on the basis of 3 molecules of ATP from NADH and 2 molecules from $FADH_2$.

RQ 9.14. Mention the sites along the electron transport system in the mitochondria where the ATP molecule is formed.

Ans. There are three 'sites' along the ETS where ATP is formed:

1. At the junction between NAD (or NADP) and FAD.

2. At the junction between Cytochrome *b* and cytochrome *c*.

3. At the junction between Cytochrome *a* and cytochrome a_3.

RQ 9.15. Why the pentose phosphate pathway is also called the hexose monophosphate shunt or the pentose monophosphate shunt?

Ans. In the pentose phosphate pathway,[11] glucose monophosphate undergoes direct oxidative decarboxylation without the intervention of glycolysis. Because this mechanism 'bypasses the glycolytic cycle' it is also called the 'hexose monophosphate shunt' or 'pentose monophosphate shunt'.

Substantial quantities of glucose are degraded by the PP pathway in most plants and in some animal tissues such as mammary glands.

RQ 9.16. Discuss briefly the significance of the pentose phosphate pathway.

Ans. **1.** The most important intermediate in this pathway is ribulose-5-phosphate and its isomeric form ribose-5-phosphate that provides 'ribose units' for RNA synthesis and its nucleotide, and several coenzymes like ATP, NAD, RNA and DNA.

2. The cytosolic NADPH[12] produced in this pathway can be utilized in the biosynthesis of fatty acids, cholesterol, steroids, carotene and other lipids or fats.

3. In addition, erythrose-4-phosphate in plants is a starting material of aromatic amino acids (phenylalanine, tyrosine and tryptophan). Tryptophan is the precursor of indole acetic acid (IAA) – a universal phytohormone.

RQ 9.17. Yeast cells are 'habitual drunkards'. Comment upon this statement.

Ans. While growing on a glucose-rich medium, yeast fungi (*Saccharomyces* sp.) are capable of decomposing glucose into ethyl alcohol and carbon dioxide as shown in the equation:

$$C_6H_{12}O_6 \xrightarrow{\text{Yeast cells}} \underset{\text{Ethyl alcohol}}{2C_2H_5OH} + 2CO_2 + \text{Energy}$$

This discovery was made by the German biochemist Eduard Buchner (1897).

During alcoholic fermentation, besides ethyl alcohol small quantities of other substances such as acetaldehyde, pyruvic acid and glycerol are produced.

Yeast fungi continue to grow in alcohol-rich medium (just like a drunkard) until its concentration reaches 10–15 per cent when the yeast cells begin dying.

[11] One important difference between the pentose phosphate pathway and glycolysis (and the Krebs cycle) is that in the former $NADP^+$ is always the electron acceptor, whereas in glycolysis it is NAD.

[12] NADP, according to some, is reduced to NADPH by the transfer of electrons from isocitric acid, leading to the formation of oxalosuccinic acid, catalyzed by 'isocitrate dehydrogenase'.

Alcoholic fermentation is controlled by the zymase enzyme system, which is secreted by the yeast cells. Zymase is an iron-containing multienzyme aggregate. Glycolase and carboxylase are two well-known components, requiring magnesium and inorganic phosphate.

The Vedas of India, which are said to be as old as 5000 years, contain references to alcoholic drinks such as 'Soma Ras'.

Starch cannot be fermented by yeast cells as they lack or do not produce amylase enzyme.

RQ 9.18. Comment on the statement that there is a physical separation of the three major processes involved in respiration in living cells.

Ans. The glycolysis reactions occur in the 'cytoplasm' of the cells, while those of the Krebs cycle occur in the 'mitochondrial matrix'. Lastly, the re-oxidation of these reduced molecules of NADH and NADPH (or $FADH_2$) takes place in the 'Elementary' or 'F_1 particles' present on the cristae of mitochondria.

RQ 9.19. No reaction either in glycolysis or the Krebs cycle involves oxygen. So why is oxygen so essential for the continuation of respiration?

Ans. The simple reason is that glycolysis and Krebs cycle function *only* when supplied continuously with the oxidized forms of the coenzyme –NAD^+ and FAD that are produced during terminal oxidation of ETS. If $NAD^+$and FAD are not regenerated, then both the cycles will stop and the process of fermentation will begin.

RQ 9.20. List only the principal functions of respiration.

1. Firstly, it releases the energy stored in sugar molecules and makes it available in the readily usable form of ATP.
2. Secondly, it provides a large number of intermediate chemical compounds of widely differing structure which are the precursors for the biosynthesis of a vast array of organic compounds necessary for the growth and development of a healthy plant.

The Krebs cycle is the central metabolic hub where one or more of its enzymes play a part in other metabolic pathways. The Krebs cycle intermediates are each connected to separate pathways in the cell.

RQ 9.21. In which tissues is the pentose phosphate pathway more active than others? Explain, why is it so?

Ans. The pentose phosphate pathway converts glucose into 4-carbon sugars (erythrose) and 5-carbon sugars (ribose) that are essential monomer units in many metabolic functions. The ribose-5-phosphate so produced can be shunted into nucleic acid metabolism, forming the basis of RNA (ribonucleic acid) and DNA (deoxyribonucleic acid) synthesis, especially during the S-phase of the mitotic cycle in the meristematic cells.

The four-carbon sugar (erythrose-4-phosphate) is the precursor for the biosynthesis of two important products – lignin and anthocyanin pigments.

Tissues such as wood, fibres and sclereids deposit large amounts of lignin during their secondary wall development. Much of erythrose phosphate comes from PP pathway. Flower petals and brightly coloured fruits divert erythrose-4-phosphate from the PP pathway to anthocyanin production.

Besides cytoplasm, the PP pathway also occurs in plastids, where it supplies eythrose-4-phosphate for the production of amino acids such as tyrosine, phenylalanine and tryptophan. Besides these 4- and 5-carbon sugars, the PP pathway also produces NADPH which is used in the reduction of nitrate (NO_3^-) to ammonia (NH_4^+).

RQ 9.22. In which way is the electron transfer during oxidative phosphorylation different from as seen in cyanide-resistant or insensitive plants? Explain.

Ans. During the electron transfer in oxidative phosphorylation, NADH is oxidized to NAD and ATP is produced. The electrons move from NADH to oxygen through $FADH_2$, ubiquinone and cytochromes, the terminal enzyme being cytochrome oxidase.

On the other hand, the cyanide-resistant (or insensitive) plants involve the use of an alternative pathway of electron transport in mitochondria. In this case, the terminal enzyme is 'alternative oxidase' – a quinol oxidoreductase that bypasses the cytochromes whereby no electrochemical gradient is formed and the energy is released in the form of heat. The production of ATP is significantly reduced.

Examples in this category are species such as *Sauromatum guttatum*, *Symplocarpus foetidus*, and *Arum maculatum*. Prior to the emergence of the inflorescence (spadix), the temperature inside may go up by 10–15 °C above ambient on account of elevated 'alternative oxidase pathway' of electron transfer. The heat so generated vaporizes compounds such as amines, ammonia, indole, etc., that attract pollinating insects.

It is believed that the staminate flowers of 'voodoo lilies' produce 'salicylic acid' (related to aspirin; acetyl salicylic acid) which then travels downward through the fleshy axis of the spadix until it reaches the base where it induces certain cells to undergo thermogenic respiration. Just prior to pollination, the tissue of the spadix undergoes an increase in oxygen utilization, termed a 'respiratory crisis'. This respiratory crisis is attributed almost entirely to an increase in **'alternative pathway respiration'**, heating up the tissues which give off foul smelling odours that attract pollinators.

It should be noted that these thermogenic plants have beautiful flowers and are plants of temperate climates where environmental temperature may not be sufficient enough to volatilize such chemicals for attracting insects on its own. Thus, this alternative oxidase pathway is an adaptive step to ensure pollination and hence, survival of these species.

RQ 9.23. Name two plants along with their botanical names where there is no proton gradient and no chemiosmotic production of ATP during respiration.

Ans. There are two common examples where a proton gradient is not there and so, no chemiosmotic production of ATP during respiration. The energy in NADH is converted entirely to heat instead of being conserved as ATP.

1. Voodoo lily (*Sauromatum guttatum*)
2. Skunk cabbage (*Symplocarpus foetidus*)

This heat generating respiration is usually called 'cyanide-resistant or cyanide-insensitive respiration' or more appropriately as thermogenic respiration.

RQ 9.24. In which way(s) is thermogenic respiration advantageous to the plants? Where does it occur? Explain.

Ans. In voodoo lily (*Sauromatum guttatum*), parts of the inflorescence become much warmer than the ambient air, thereby causing amines and other chemicals to volatilize, serving as chemical attractants for pollinators that help in the completion of their lifecycle.

Secondly, the skunk cabbage (*Symplocarpus foetidus*) begins development of floral parts while hidden under snow cover. When the inflorescence is ready to emerge, it produces enormous heat that melts the snow, thus exposing the inflorescence to pollinators.

Hibernating bears[13] stay warm just like newborn human babies, and almost like 'compost piles' which become warm because of heat loss during the respiration of the fungi and bacteria that help in compost decomposition.

RQ 9.25. Acetyl CoA is the connecting link between glycolysis and Krebs cycle. Comment.

Ans. All the reactions of glycolysis and the conversion of pyruvic acid to acetyl CoA through oxidative decarboxylation take place in the cytoplasm (or cytosol).

$$\text{Pyruvic acid} + \text{CoA} + \text{NAD}^+ \xrightarrow{\text{Pyruvate dehydrogenase}} \text{Acetyl CoA} + CO_2 + \text{NADH} + H^+$$

Acetyl CoA enters the mitochondrion and reacts with oxaloacetic acid (a 4-carbon acid) to form 6-carbon citric acid, catalyzed by a condensing enzyme, citrate synthase as shown in the equation below:

$$\text{Oxaloacetic acid} + \text{Acetyl CoA} \xrightarrow{\text{Citrate synthase}} \text{Citric acid} + \text{CoA}$$

The Krebs cycle begins with citric acid and the process goes on. The resulting CoA goes back to cytoplasm where it produces another molecule of acetyl CoA from pyruvic acid, thus serving as a connecting link between glycolysis and the Krebs cycle.

RQ 9.26. How much ATP does the catabolism of 6-carbon glucose molecule produce? Compare the energy release with a 6-carbon fatty acid (say hypothetically speaking).

Ans. A 6-carbon glucose molecule yields about 36 molecules of ATP during glycolysis and the Krebs cycle. During β-oxidation, 6-carbon fatty acid would be changed into 'three' molecules of acetyl CoA during the 'two rounds'. Each round needs one molecule of ATP to prime the process, but it also generates at the same time one molecule each of NADH and $FADH_2$ (or a total of 2 NADH and 2 $FADH_2$ molecules).

Thus, the total yield of ATP would be:

			ATP
1. 2 NADH (through oxidative phosphorylation)	= 2 × 3	—	+6
2. 2 $FADH_2$	= 2 × 2	—	+4
			+10
3. ATP molecule used during each 'priming'	–1 ATP	—	–1
			+9

The oxidation of each molecule of acetyl CoA in the Krebs cycle ultimately generates 12 molecules of ATP. Three molecules would produce energy equivalent to 36 ATP (3 × 12). Overall, the ATP yield of a 6-carbon fatty acid would be approximately 45 (36+9).

[13] Brown adipose tissue (or called just 'brown fat') has long been known to be present in animals that hibernate, and in small quantities in newborn humans. This fat is brown due to the abundance of mitochondria. These mitochondria contain a protein (UCP-1 or uncoupling protein-1). Under the appropriate stimulus, there is an increased 'uncoupling' activity. This generates heat which can lead to arousal of animals from hibernation.

Therefore, the respiration of a 6-carbon fatty acid yields about 20 per cent more ATP than the respiration of glucose.

Moreover, a fatty acid of that size would weigh less than two-thirds as much as glucose. So a gram of fatty acid contains twice more kilocalories as compared to a gram of glucose.

RQ 9.27. How much is the available energy of a glucose molecule stored as ATP? Explain.

Ans. One molecule of glucose contains about 686,000 calories of energy in the form of bonds.

$$C_6H_{12}O_6 + 6O_2 \rightarrow 6CO_2 + 6H_2O + 686 \text{ kcal}$$

The complete oxidation of one molecule of glucose results in the net gain of 38 (or 36 as a reference source) molecules of ATP. One ATP molecule when hydrolysed yields about 8,000 calories of energy. Thus, the total usable energy conserved from 38 ATP molecules will be $38 \times 8{,}000 \times 100 \div 686{,}000 = 304{,}000$ calories. Therefore, about 45 per cent energy of the glucose molecule is utilized during aerobic breakdown (i.e., glycolysis plus Krebs cycle) and the rest is lost as heat.

RQ 9.28. Write a descriptive note on substrate-level phosphorylation with examples.

Ans. The production of ATP by the direct transfer of a phosphate group from a substrate molecule to ADP is termed as substrate-level phosphorylation. It takes place during the following reactions of respiratory metabolism.

Glycolysis

1. 1,3-bisphosphoglycerate $\xrightarrow[\text{Mg}^{++}\text{, ADP}]{\text{Phosphoglycerate kinase}}$ 3-Phosphoglycerate + ATP

2. Phosphoenol pyruvic acid $\xrightarrow[\text{Mg}^{++}\text{, K}^{+}\text{, ADP}]{\text{Pyruvate kinase}}$ Pyruvic acid + ATP

Krebs cycle

3. Succinyl CoA + P_i + ADP $\xrightarrow{\text{Succinyl CoA synthetase}}$ Succinic acid + ATP + CoA

Reaction sequence involved:

Succinyl CoA + P_i $\longrightarrow$ Succinyl P + CoA

Succinyl P + GDP $\longrightarrow$ GTP + Succinic acid

GTP + ADP $\longrightarrow$ GDP + ATP

Finally: Succinyl CoA + P_i + ADP $\longrightarrow$ Succinic acid + ATP + CoA

RQ 9.29. Write an explanatory note on the malate–aspartate shuttle (the NADH shuttle) or an oxidation of extramitochondrial NADH.

Ans. NADH produced in the cytoplasm during glycolysis cannot move across the inner membrane of the mitochondria (the site of the electron transport chain) as it is impermeable. Thus some shuttle mechanisms have been proposed to help explain the phenomenon.

In the **malate–aspartate shuttle**, NADH in the cytosol powers the conversion of oxaloacetate to malate, which enters the mitochondrial matrix and powers the formation of a 'new' molecule of NADH which through oxidative phosphorylation releases 3 ATPs. Malate is converted to aspartate which is then transferred back out of the mitochondria into the cytoplasm where it gets converted back to oxaloacetate, thus repeating the cycle as shown in the diagram below (Figure 9.17):

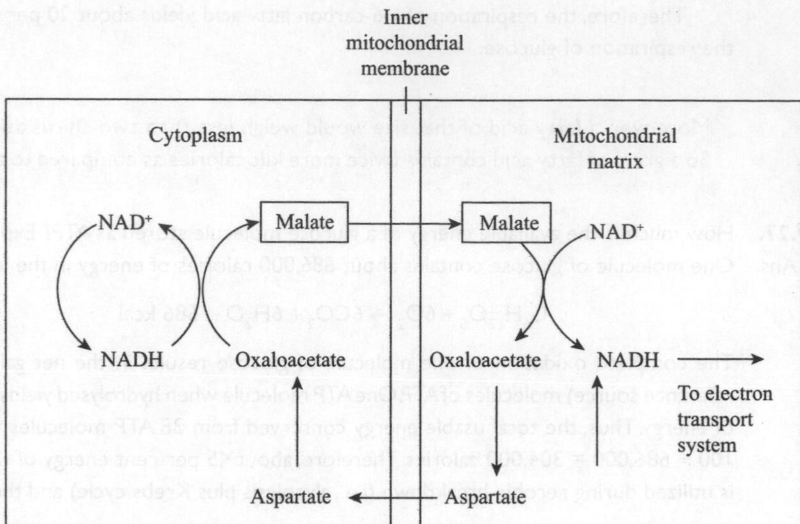

Fig. 9.17 The malate-aspartate shuttle allows reducing power to be transferred into mitochondria when NADH itself cannot cross the mitochondrial membranes because of impermeable nature.

In the second mechanism called the '**glycerol phosphate shuttle**', the cytosolic NADH reduces dihydroxyacetone phosphate (DHAP) to glycerol phosphate which then crosses over to the inner mitochondrial membrane to the matrix. There it is converted back to dihydoxyacetone phosphate, reducing FAD to $FADH_2$ in the process. Each cytosolic NADH results in the formation of one molecule of $FADH_2$ in the matrix which can drive the formation of 'only two' ATPs, unlike 3 obtained from NADH.

As a third possibility, the electrons from extramitochondrial NADH, in at least some plants, are transferred to 'ubiquinone' directly bypassing the first site of oxidative phosphorylation, thus generating two ATP molecules per external NADH, instead of three.

Chapter 10

Nitrogen Metabolism

Keywords

- Lectins
- Leghaemoglobin
- Nodulin
- Reductive amination
- Rhizosphere
- Symbiotic nitrogen fixation
- Transamination
- Zwitterion

Although the three elements carbon, hydrogen and oxygen constitute the bulk of dried plant material, nitrogen plays a vital role in plant metabolic processes. On the basis of dry weight, nitrogen is the fourth most prevalent mineral element in plants. It represents an integral component of protoplasm, including nucleic acids, proteins, growth hormones, porphyrins such as in chlorophyll and the haeme group of cytochromes, vitamin B and other primary and secondary plant metabolites.

The bulk of the air, 78 per cent by volume, comprises of molecular nitrogen (N_2 or dinitrogen), an inert, odourless and colourless gas. Despite its abundance, however, higher plants are not able to convert nitrogen into a biologically useful form. For this, plants must depend on prokaryote organisms.

The main source of nitrogen for the green plants is the supply of nitrogenous compounds present in the soil. Soils contain both organic and inorganic nitrogenous compounds. Organic nitrogen compounds are abundant in the soil humus and are produced by the microbial decay of plant and animal remains through ammonification and nitrification, e.g., urea.

Inorganic nitrogenous compounds in the soil are the nitrates, nitrites and ammonia compounds. Of these, the nitrate nitrogen is the most important for plants but this, like other anions, is leached out and therefore should be replenished, in order to maintain productivity.

10.1 Nitrate Assimilation

Most species of plants absorb their nitrogen in the form of nitrates (NO_3^-). Nitrogen (N_2) is present in a highly oxidized form in nitrates and in a highly reduced form in amino acids. Therefore, nitrates are to be first reduced to ammonia to be assimilated by the plant.

$$HNO_3 + 8H^+ \rightarrow NH_3 + 3H_2O$$

Thus, the transport of nitrate across root cell membrane is an energy-consuming process facilitated by a carrier protein. Once within the root, nitrates may accumulate in the vacuole or else be assimilated in the root cells or transported through the xylem to the leaves for further assimilation.

Reduction of nitrate to ammonia takes place in two stages:

1. The first step in the sequence is reduction of nitrate to nitrite, catalyzed by **nitrate reductase (NR)** enzyme

$$NO_3^- + 2H^+ + 2e^- \xrightarrow{NR} NO_2^- + H_2O$$

Nitrate reductase is a ubiquitous enzyme, predominantly localized in cytosol of prokaryotic and eukaryotic cells as well. It has been extracted and purified from many groups of plants and is comprised of two identical subunits having molecular weight of ~115 kD. NR is a metalloflavoprotein containing three major components:

- NAD(P)H – as an electron donor
- FAD – as a prosthetic group
- Mo – as an activator/cofactor.

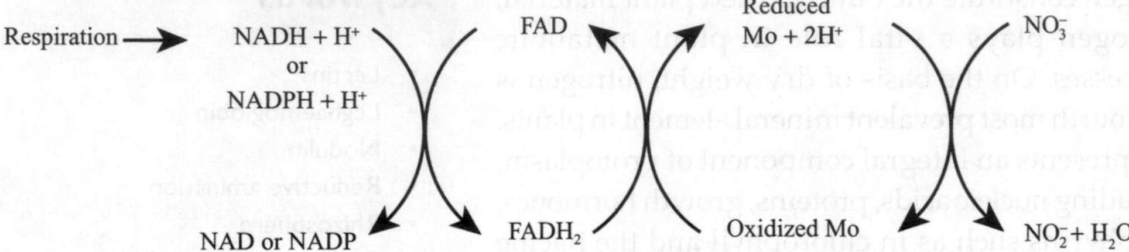

Fig. 10.1 Electron transport chain involved in nitrate reduction.

NADPH/NADH serve as electron donors in higher plants (ferredoxin – as an electron donor in cyanobacterial system). Electrons are passed from NADPH to FAD, producing $FADH_2$ from where the electrons are in turn passed to oxidized molybdenum, producing reduced Mo which then passes on the electrons to nitrate, and thus reducing it to nitrite (NO_2^-) (Figure 10.1).

The nitrate reductase enzyme in higher plants consists of two identical subunits (dimer), each having three prosthetic groups: FAD, haeme (or heme, as it is called in the United States), and a molybdenum (Mo) atom associated to an organic molecule known as **pterin**.

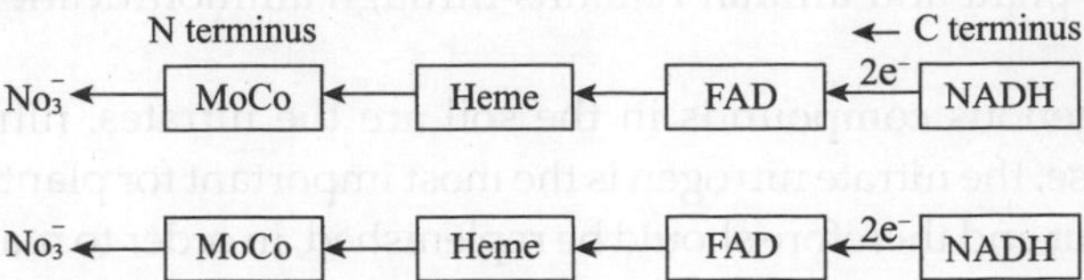

According to three-domain model of nitrate reductase dimer, the NADH complexes to the FAD-binding site of each subunit and involves transfer of two electrons from the carboxyl (C) terminus, by electron transfer components, to the amino (N) terminus. Nitrate is finally reduced at the molybdenum complex MoCo, present near the N-terminus.

Nitrate reductase is an example of **inducible enzyme**, i.e., it gets induced only in the presence of substrate, nitrate and it tends to disappear in the absence of nitrate. High intensity of light, nitrate and carbohydrate levels favour the maximum activity of this enzyme. On the other hand, nitrite accumulation inhibits its activity (**feedback inhibition**). Darkness and Mg^{2+} can also inactivate NR and such inactivation is faster than regulation by reduced synthesis or degradation of the enzyme. The nitrite accumulates if it is not further used for the formation of ammonia and this happens in plants suffering from Ca, Cu, Fe and Mn deficiency. Nitrate reductase activity can be reversibly controlled by red and far-red light, indicating the regulatory effect of phytochrome system.

2. The second step involves the reduction of nitrite to ammonia, catalyzed by the enzyme **nitrite reductase (NiR)**.

The nitrite (NO_2^-), that is the immediate product of nitrate reduction, is very reactive and potentially toxic ion as well. It quickly migrates to plastids (in roots) or chloroplasts (in foliage) where it gets reduced to an ammonium ion (NH_4^+) by the nitrite reductase enzyme system.

$$NO_2 + 6e^- + 6H^+ \xrightarrow{\text{NiR}} NH_4^+ + 2OH^-$$

The photosynthetic electron transfer is coupled to the reduction of nitrite to ammonium by nitrite reductase (NiR) enzyme system *via* ferredoxin. NiR is metalloflavoprotein and consists of two prosthetic groups, Fe_4S_4 and heme, participating in the reduction of nitrite to ammonium. Nitrite reductase is also an example of an inducible enzyme system.

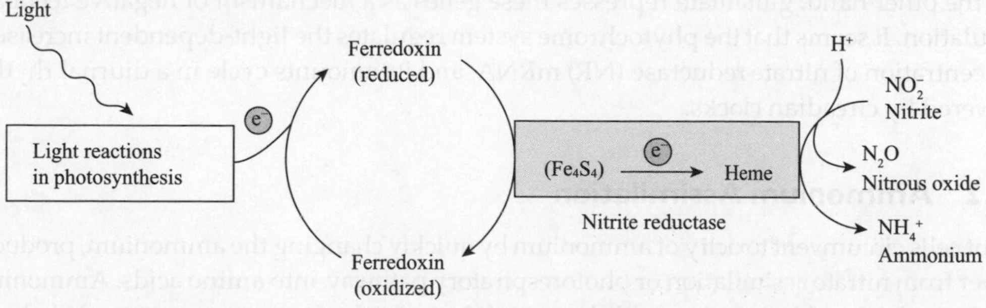

Elevated concentrations of nitrate or exposure to light stimulate the transcription of nitrite reductase mRNA, whereas accumulation of the final products in the reaction (such as asparagine and glutamine) suppress this induction.

Ammonia, produced during reduction of nitrates, is incorporated directly into the amino acids - the principal of these (in terms of quantity) being glutamic acid. Plants can make all the 20 amino acids that are essential for synthesis of proteins. These amino acids are the basic building blocks of proteins. Animals, on the other hand, typically lack the enzymes for some of the special pathways for amino acid biosynthesis, even though they can make simple amino acids like glutamic acid and alanine from carbohydrates and ammonium, but they must obtain several of other amino acid from their diet.

At high concentrations, ammonium (NH_4^+) is harmful to living tissues, but nitrate can be safely stored and translocated in plant tissues. However, if livestock or humans tend to eat plant material rich in nitrates, they may suffer from **methemoglobinemia**, a disorder during which nitrates are reduced to nitrites in the liver, which can combine with haemoglobin, thus making it unable to carry oxygen. Human beings as well as other animals may also convert nitrate into highly carcinogenic nitrosamines, or into nitric oxide, a potent signalling molecule associated with many biological processes. Some countries restrict the level of nitrates in plant materials sold for human use.

Nitrate Signalling

Nitrate acts as both nutrient as well as signal molecule and it has pronounced influence on growth and metabolism of the plant. Two types of symporters NRT 1 and NRT 2 (nitrate transporter 1 and 2) and one antiporter CLC (Chloride channel) are involved in nitrate transport in plants. NRT 1 and NRT 2 are associated with nitrate transport across plasma membrane and tonoplast, whereas CLC is associated with nitrate transport across tonoplast only.

Nitrate acts as a signal molecule that regulates its own uptake and assimilation. Nitrate regulates transcriptional responses, seed dormancy, leaf expansion and root architecture in plants. Nitrate triggers primary nitrate response within 10 minutes of addition of 100 nM nitrate, such as transcription of nitrate transporters, NR, NiR and production of the reducing equivalents required for nitrate assimilation. Nitrite also acts as a potent signal to upregulate gene expression needed for absorption as well as assimilation of nitrate. On the other hand, glutamate represses these genes as a mechanism of negative feedback regulation. It seems that the phytochrome system regulates the light-dependent increase in concentration of nitrate reductase (NR) mRNA, and its amounts cycle in a diurnal rhythm powered by circadian clocks.

10.2 Ammonium Assimilation

Plant cells circumvent toxicity of ammonium by quickly changing the ammonium, produced either from nitrate assimilation or photorespiratory pathway, into amino acids. Ammonium can be assimilated into amino acids by one of the several processes as mentioned below:

(i) The **GDH pathway** (or **Reductive Amination**); that generates glutamate by using NAD(P)H as a reducing agent.
(ii) **Transamination**; transferring of the amino group from glutamate to oxaloacetate to produce aspartate, catalyzed by the enzyme aspartate-aminotransferase.
(iii) The **GS-GOGAT** (or **Glutamate Synthase) pathway;** that generates glutamine and glutamate by using a reduced cofactor (ferredoxin or NADH).
(iv) **Amidation**; the formation of asparagine by transferring an amino group from glutamine to aspartate, catalyzed by the enzyme asparagine synthetase.

10.2.1 GDH (reductive amination) pathway

This method represents the principal route of entry of ammonia into amino acids. When roots are supplied with ammonium compounds labelled with heavy isotope of nitrogen (^{15}N), most of the labelled nitrogen is incorporated into glutamic acid. Glutamic acid plays a chief role in nitrogen metabolic functions of both plants as well as animals.

$$\underset{\substack{\alpha\text{-Ketoglutarate}\\ \text{(from Krebs cycle)}}}{\begin{array}{l} COOH \\ | \\ CH_2 \\ | \\ CH_2 + NH_3 \\ | \\ C{=}O \\ | \\ COOH \end{array}} \rightleftharpoons \underset{\alpha\text{-Iminoglutarate}}{\begin{array}{l} COOH \\ | \\ CH_2 \\ | \\ CH_2 \\ | \\ C{=}NH \\ | \\ COOH \end{array}} \underset{\substack{\text{Glutamate}\\ \text{dehydrogenase}\\ \text{(GDH)}}}{\overset{NAD(P)H + H^+}{\rightleftharpoons}} \underset{\text{Glutamate}}{\begin{array}{l} COOH \\ | \\ CH_2 \\ | \\ CH_2 + NAD(P)^+ + H_2O \\ | \\ C{-}NH_2 \\ | \\ COOH \end{array}}$$

The enzyme glutamate dehydrogenase (GDH) catalyzes a reversible reaction that involves *either* synthesis *or* deamination of glutamate. The NADH-dependent form of GDH is present within mitochondria, while the NADPH-dependent form is present in the chloroplasts of photosynthetic cells. Their primary function is the deamination of glutamate during the reallocation of nitrogen. Glutamate dehydrogenase requires zinc for its activity. This is one of the reasons why zinc is an essential element for plants.

Likewise, oxaloacetic acid (OAA) and pyruvic acid are converted into aspartic acid and alanine.

$$\underset{\text{Oxaloacetic acid}}{\begin{array}{l} COOH \\ | \\ CH_2 \\ | \\ C = O + NH_3 + NADH + H^+ \\ | \\ COOH \end{array}} \rightleftharpoons \underset{\text{Aspartic acid}}{\begin{array}{l} COOH \\ | \\ CH_2 \\ | \\ CHNH_2 + NAD^+ + H_2O \\ | \\ COOH \end{array}}$$

$$\underset{\text{Pyruvic acid}}{\begin{array}{l} CH_3 \\ | \\ C = O + NH_3 + NADH + H^+ \\ | \\ COOH \end{array}} \rightleftharpoons \underset{\text{Alanine}}{\begin{array}{l} CH_3 \\ | \\ CHNH_2 + NAD^+ + H_2O \\ | \\ COOH \end{array}}$$

10.2.2 Transamination

Once glutamic acid is available, other amino acids (e.g., aspartic acid and alanine) may be synthesized by the transfer of its amino group to the carbonyl group (-CO-) of a keto acid, catalyzed by the **transaminase** enzyme. The transfer of an amino group from the glutamate to the carbonyl group of oxaloacetate occurs in the presence of aspartate aminotransferase (glutamic-aspartic transaminase).

Aminotransferases are located in the cytosol as well as in various cellular organelles such as, chloroplasts, mitochondria, glyoxysomes, and peroxisomes. The aminotransferases present in chloroplasts play an important role in biosynthesis of amino acids such as glutamate, aspartate, alanine, serine, and glycine.

$$\underset{\text{Glutamic acid}}{\begin{matrix}COOH\\|\\(CH_2)_2\\|\\CHNH_2\\|\\COOH\end{matrix}} + \underset{\text{Oxaloacetic acid}}{\begin{matrix}COOH\\|\\CH_2\\|\\C{=}O\\|\\COOH\end{matrix}} \xrightleftharpoons[\text{transaminase}]{\substack{\text{Aspartate aminotransferase}\\ \text{or}\\ \text{Glutamic-oxaloacetic}}} \underset{^{1}\alpha\text{-Ketoglutaric acid}}{\begin{matrix}COOH\\|\\(CH_2)_2\\|\\C{=}O\\|\\COOH\end{matrix}} + \underset{\text{Aspartic acid}^{2}}{\begin{matrix}COOH\\|\\CH_2\\|\\CHNH_2\\|\\COOH\end{matrix}}$$

$$\underset{\text{Glutamic acid}}{\begin{matrix}COOH\\|\\(CH_2)_2\\|\\CHNH_2\\|\\COOH\end{matrix}} + \underset{\text{Pyruvic acid}}{\begin{matrix}CH_3\\|\\C{=}O\\|\\COOH\end{matrix}} \xrightleftharpoons[\text{transaminase}]{\substack{\text{Alanine}\\ \text{aminotransferase}\\ \text{or}}} \underset{\alpha\text{-Ketoglutaric acid}}{\begin{matrix}COOH\\|\\(CH_2)_2\\|\\C{=}O\\|\\COOH\end{matrix}} + \underset{\text{Alanine}}{\begin{matrix}CH_3\\|\\CHNH_2\\|\\COOH\end{matrix}}$$

All transamination reactions involve participation of cofactor, pyridoxal phosphate (Vitamin B_6). Using ^{15}N in one of the experiments, it has been found that ^{15}N is first incorporated into glutamic acid and very shortly afterwards into aspartate and alanine.

> Few amino acids found in plant tissues are generally of secondary origin, being produced from chemical transformation of amides and hydrolysis of proteins. Two of the protein-digesting enzymes **Papain**, from the latex of papaya (*Carica papaya*), and **bromelin**, from the fruit of pineapple[3] (*Ananas comosus*) hydrolyse proteins to amino acids.

Reductive Amination	Transamination
1. It is the condensation of NH_3 with a keto acid, by reduced NAD (NADH), giving rise to an amino acid.	It involves the transfer of an amino group from an amino acid to the carbonyl group of a keto acid, giving rise to a corresponding amino acid.
2. The dehydrogenase* enzyme catalyzes this chemical reaction, in the presence of NADH.	The enzyme catalysing the reaction is known as transaminase (aminotransferase). It requires pyridoxal phosphate, a derivative of pyridoxin (vitamin B_6) as their essential coenzyme.

[1] α-ketoglutarate or 2-oxoglutarate.

[2] Aspartate is an amino acid that participates in the malate-aspartate shuttle.

[3] A glass of pineapple juice is quite welcome before serving fleshy food (chicken, meat, etc.) to your honoured guest.

Reductive Amination	Transamination
3. This represents the principal route of entry of ammonia into amino acids. α-Ketoglutaric acid + NH_3 + NADH ⟶ Glutamic acid** + H_2O + NAD^+ Oxaloacetic acid + NH_3 + NADH ⟶ Aspartic acid + H_2O + NAD^+ Pyruvic acid + NH_3 + NADH ⟶ Alanine + H_2O + NAD^+	Some amino acids may be formed due to a transamination reaction. Glutamic acid + oxaloacetic acid → α-ketoglutaric acid + Aspartic acid

* Glutamate dehydrogenase is the chief enzyme of amino acid metabolism.
** Glutamate is the ionized form of glutamic acid.

10.2.3 GS-GOGAT pathway

- Glutamine synthetase (GS) enzyme helps to combine ammonium with glutamate to produce glutamine in the presence of ATP and Mg^{2+}/Co^{2+} as a cofactor. Plants have two kinds of GS, one in cytoplasm and the other in root plastids or shoot chloroplasts.
- GOGAT transfers the amide group from glutamine to α-ketoglutarate, generating two molecules of glutamate.

Through the **Glutamate synthase cycle** (shown below), ammonium concentration is kept low with the involvement of two enzymes: glutamine synthetase (GS) and glutamate synthase (GOGAT).[4]

$$NH_4^+ + \text{Glutamate} \xrightarrow[\text{GS}]{\text{ATP} \rightarrow \text{ADP + Pi}} \text{Glutamine (amide)}$$

GOGAT

Glutamate + NAD^+ ↓ Export

α-Ketoglutarate + NADH + H^+

Glutamate synthase cycle

The reduction of glutamine to glutamate occurs in plastids. There are two isoforms of GOGAT; the first accepts electrons from reduced ferredoxin and the second one from NADH. The ferredoxin-linked GOGAT predominantly occurs in leaves while the NADH-linked GOGAT occurring in roots.

Glutamine synthetase, a primary regulator of nitrogen metabolism, is present almost in all organisms. Besides its importance for ammonium assimilation in bacteria, it plays a pivotal

[4] The enzyme glutamate synthase (or in modern terminology, glutamine-oxoglutarate aminotransferase, GOGAT).

role in amino acid metabolism in mammals, transforming toxic free NH_4^+ to glutamine for circulation in the blood.

10.2.4 Amidation

In plants with a liberal supply of fertilizers, the amount of ammonia available to a plant is often in excess of its immediate requirements and accumulates in the form of amides. Amides store ammonia in a form in which it can be readily available when needed for the synthesis of amino acids through reductive amination and transamination and also in the synthesis of many metabolites, i.e., purine and pyrimidine.

The amides contain more nitrogen than their amino acids and their highly reactive $-CONH_2$ group can bring about amination of other organic acid. They also guard the protoplasm against the toxic effect of ammonia when it is in excess. The formation of amides converts the ammonia into the harmless amide group, especially under conditions of starvation when proteins are broken down with the release of ammonia.

$$\begin{array}{l} COOH \\ CH_2 \\ CH_2 \\ CHNH_2 + NH_3 + ATP \\ COOH \\ \text{Glutamic acid} \end{array} \xrightleftharpoons{\text{Glutamine synthetase}} \begin{array}{l} CONH_2 \\ CH_2 \\ CH_2 + H_2O + ADP + Pi \\ CHNH_2 \\ COOH \\ \text{Glutamine} \\ \text{(amide)} \end{array}$$

$$\begin{array}{l} COOH \\ | \\ CH_2 \\ | \\ CHNH_2 + NH_3 + ATP \\ | \\ COOH \\ \text{Aspartic acid} \end{array} \xrightleftharpoons{\text{Asparagine synthetase}} \begin{array}{l} CONH_2 \\ | \\ CH_2 + H_2O + ADP + Pi \\ | \\ CHNH_2 \\ | \\ COOH \\ \text{Asparagine} \\ \text{(amide)} \end{array}$$

Amides are the most important vehicles of nitrogen transport to the growing regions where they are utilized for the formation of amino acids through a process of transamination.

Asparagine was the first amide to be identified from *Asparagus*. It is the key compound of proteins as well as for nitrogen translocation and storage.

The main pathway for synthesis of asparagine requires the transfer of the amide nitrogen from glutamine to asparagine in the presence of **Asparagine synthetase (AS)**. This enzyme is present in the cytoplasm of leaves and roots, and in nitrogen-fixing root nodules.

$$\text{Glutamine} + \text{Aspartate} + \text{ATP} \rightarrow \text{Asparagine} + \text{Glutamate} + \text{AMP} + \text{PP}_i$$

Ammonia released by the symbiotic nitrogen-fixing prokaryotes must be immediately transformed into organic compounds (amides and ureides) in the root nodules or tubercles. The three principal ureides are **allantoic acid, allantoin,** and **citrulline**. Allantoin is synthesized from uric acid in peroxisomes, which produces allantoic acid in the endoplasmic reticulum (ER) . The precursor of citrulline is ornithine. These ureides are ultimately transported to the shoots *via* the xylem tissues and are rapidly degraded to ammonium that enters any of the assimilation pathway.

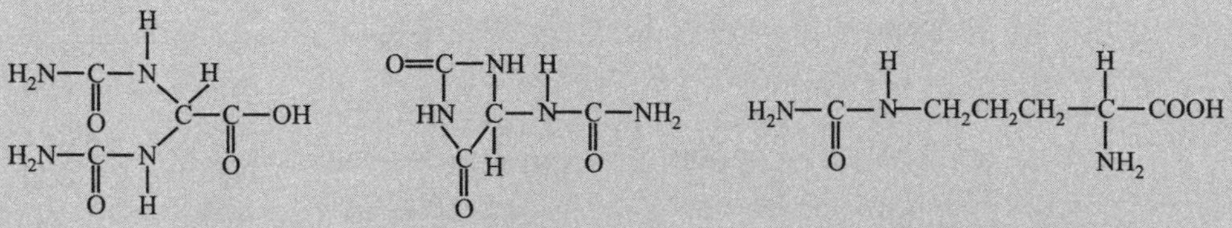

Allantoic acid Allantoin Citrulline

The three major ureide compounds that are used to transport nitrogen from sites of fixation to sites where their deamination will provide nitrogen for amino acid and nucleoside synthesis.

10.3 Amino Acids and Their Structure

Amino acids are small, water-soluble organic molecules, and universal in all living organisms.There are 20 standard amino acids, specified by the genetic code, that make up all proteins. All amino acids have the same basic structural organization, in which a single carbon atom (α-carbon) possesses both an amino group (–NH_2) and a carboxyl group (–COOH). At physiological pH, both the amino group and the carboxyl group present in an amino acid undergo ionization, thus possessing a negative as well as a positive charge. Such molecules are known as **Zwitterions.**

$$
\begin{array}{c} \mathrm{R} \\ | \\ {}^{+}\mathrm{H_3N{-}CH{-}COO^{-}} \end{array}
$$

Besides their primary role as protein precursors, amino acids perform other important functions in plants, including nitrogen metabolism, hormone biosynthesis, biosynthesis of alkaloids, flavonoids and isoflavonoids and adaptation to certain environmental stresses.

Most of the naturally occurring amino acids are of the L-amino type (Figure 10.2) and represent the backbone of proteins. These 22 amino acids are also present independent of protein, (in the free state) in the cytosol. Some exist as components of non-protein molecules and others are key components of intermediary metabolism such as glycine, alanine, serine and glutamic and aspartic acid and their amides.

In plants, more than 100 non-protein amino acids have also been isolated and identified. Some of these participate in nitrogen metabolism (ornithine, citrulline, canavine) or in energetics (phosphoserine), others are α-aminoadipic acid, ethanolamine, β-alanine, and L-homoserine. Gamma-aminobutyric acid (GABA) amino acid accumulates at the time of severe environmental (biotic as well as abiotic) stress in plants and is also involved in intermediary metabolism.

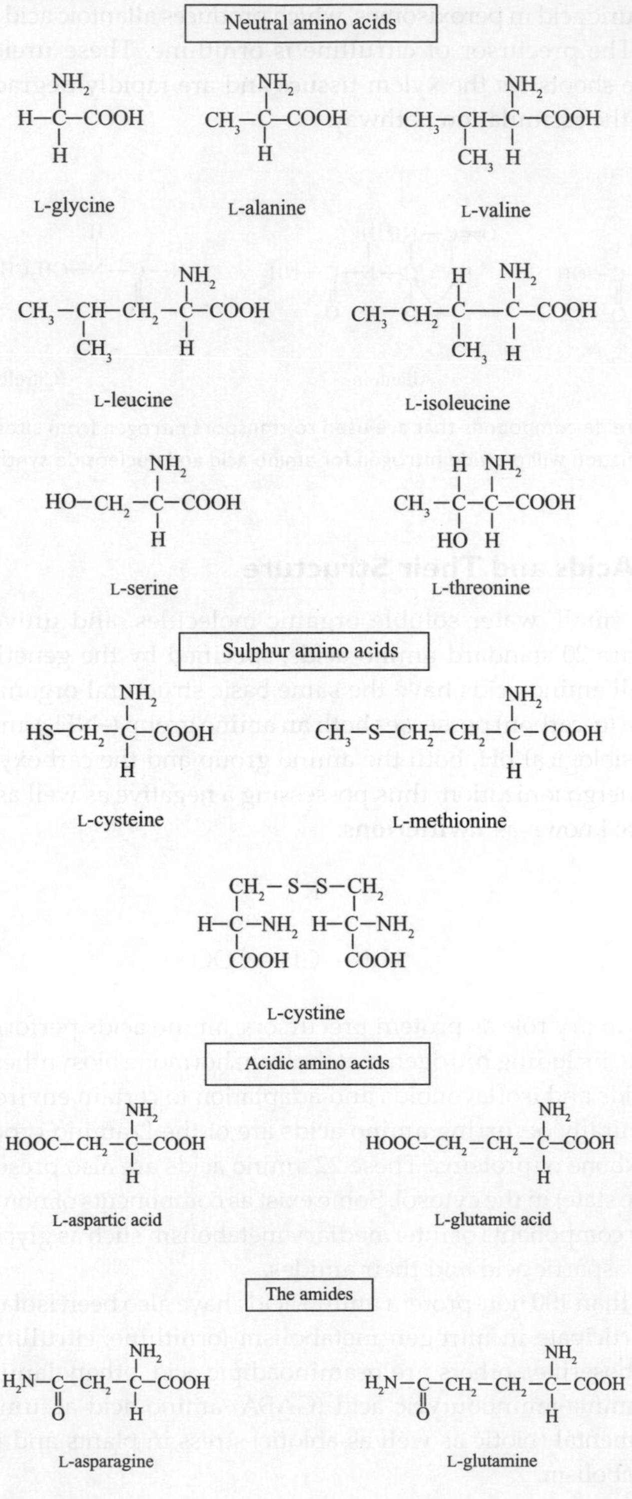

Neutral amino acids

L-glycine: $H{-}C(NH_2)(H){-}COOH$

L-alanine: $CH_3{-}C(NH_2)(H){-}COOH$

L-valine: $CH_3{-}CH(CH_3){-}C(NH_2)(H){-}COOH$

L-leucine: $CH_3{-}CH(CH_3){-}CH_2{-}C(NH_2)(H){-}COOH$

L-isoleucine: $CH_3{-}CH_2{-}C(H)(CH_3){-}C(NH_2)(H){-}COOH$

L-serine: $HO{-}CH_2{-}C(NH_2)(H){-}COOH$

L-threonine: $CH_3{-}C(H)(HO){-}C(NH_2)(H){-}COOH$

Sulphur amino acids

L-cysteine: $HS{-}CH_2{-}C(NH_2)(H){-}COOH$

L-methionine: $CH_3{-}S{-}CH_2{-}CH_2{-}C(NH_2)(H){-}COOH$

L-cystine: $CH_2{-}S{-}S{-}CH_2$; $H{-}C{-}NH_2$ $H{-}C{-}NH_2$; $COOH$ $COOH$

Acidic amino acids

L-aspartic acid: $HOOC{-}CH_2{-}C(NH_2)(H){-}COOH$

L-glutamic acid: $HOOC{-}CH_2{-}CH_2{-}C(NH_2)(H){-}COOH$

The amides

L-asparagine: $H_2N{-}C(=O){-}CH_2{-}C(NH_2)(H){-}COOH$

L-glutamine: $H_2N{-}C(=O){-}CH_2{-}CH_2{-}C(NH_2)(H){-}COOH$

Figure 10.2 *Contd.*

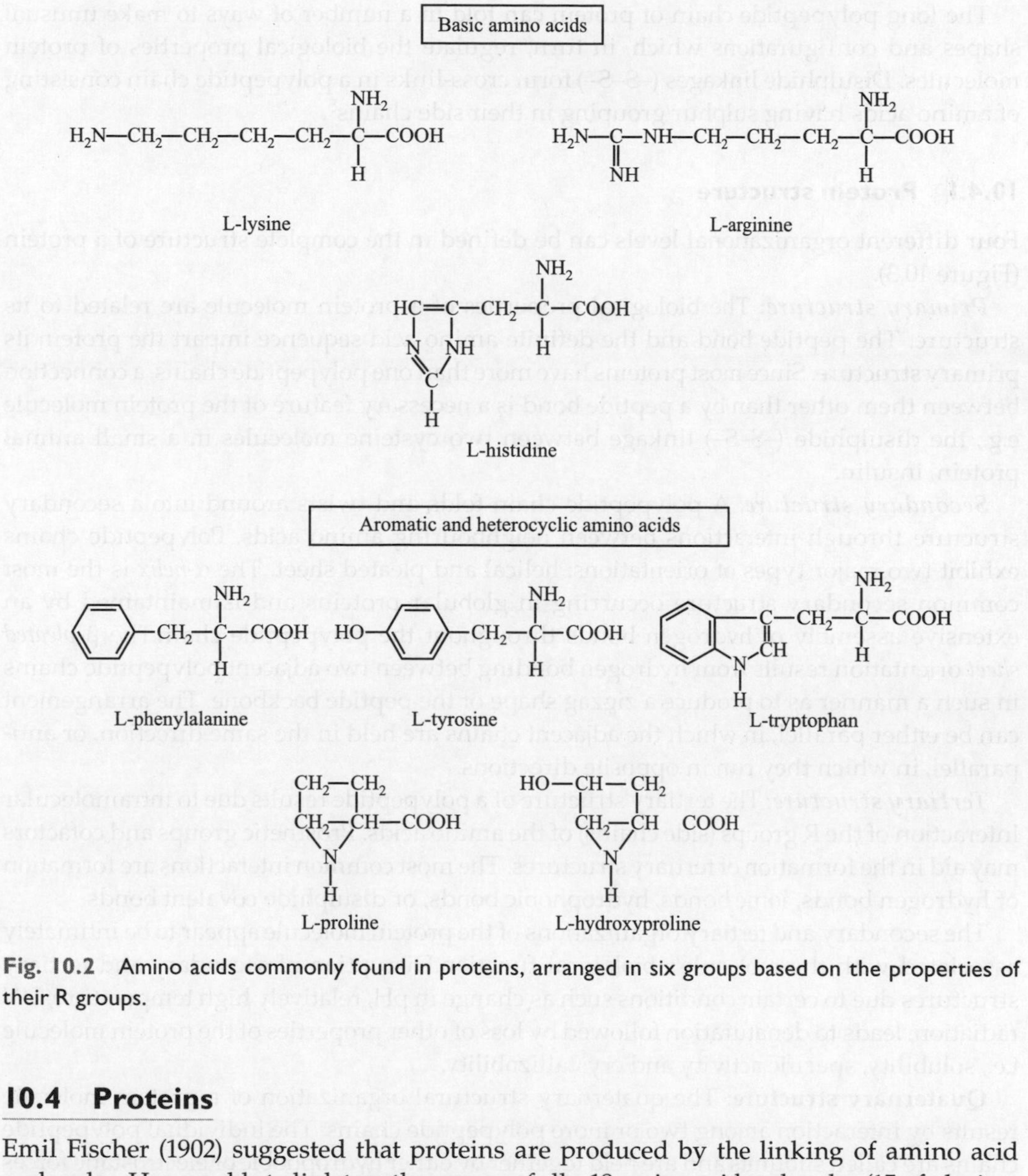

Basic amino acids

$H_2N{-}CH_2{-}CH_2{-}CH_2{-}CH_2{-}CH(NH_2){-}COOH$

L-lysine

$H_2N{-}C({=}NH){-}NH{-}CH_2{-}CH_2{-}CH_2{-}CH(NH_2){-}COOH$

L-arginine

L-histidine

Aromatic and heterocyclic amino acids

L-phenylalanine

L-tyrosine

L-tryptophan

L-proline

L-hydroxyproline

Fig. 10.2 Amino acids commonly found in proteins, arranged in six groups based on the properties of their R groups.

10.4 Proteins

Emil Fischer (1902) suggested that proteins are produced by the linking of amino acid molecules. He was able to link together successfully, 18 amino acid molecules (15 of glycine, 3 of leucine) into a synthetic polypeptide chain.

In a protein molecule, the amino acids are held together by covalent bonds among the carboxyl group (–COOH) of one amino acid and the amino group (–NH_2) of another amino acid by elimination of a water molecule. The bond linking the two amino acids (–CO–NH) is known as the **peptide bond**. When two amino acids are joined by one peptide bond, the resulting molecule is termed as a **'dipeptide'**. Three amino acids constitute a **'tripeptide'** and so forth. A complex chain with a large number of amino acids is referred to as **'polypeptide'**.

The long polypeptide chain of protein can fold in a number of ways to make unusual shapes and configurations which, in turn, regulate the biological properties of protein molecules. Disulphide linkages (–S–S–) form cross-links in a polypeptide chain consisting of amino acids having sulphur grouping in their side chains[5].

10.4.1 Protein structure

Four different organizational levels can be defined in the complete structure of a protein (Figure 10.3).

Primary structure: The biological properties of a protein molecule are related to its structure. The peptide bond and the definite amino acid sequence impart the protein its primary structure. Since most proteins have more than one polypeptide chains, a connection between them other than by a peptide bond is a necessary feature of the protein molecule e.g., the disulphide (–S–S–) linkage between two cysteine molecules in a small animal protein, insulin.

Secondary structure: A polypeptide chain folds and twists around into a secondary structure through interactions between neighbouring amino acids. Polypeptide chains exhibit two major types of orientations: helical and pleated sheet. The *α-helix* is the most common secondary structure occurring in globular proteins and is maintained by an extensive assembly of hydrogen bonds throughout the polypeptide chain.The *β-pleated sheet* orientation results from hydrogen bonding between two adjacent polypeptide chains in such a manner as to produce a zigzag shape of the peptide backbone. The arrangement can be either parallel, in which the adjacent chains are held in the same direction, or anti-parallel, in which they run in opposite directions.

Tertiary structure: The tertiary structure of a polypeptide results due to intramolecular interaction of the R groups (side chains) of the amino acids. Prosthetic groups and cofactors may aid in the formation of tertiary structures. The most common interactions are formation of hydrogen bonds, ionic bonds, hydrophobic bonds, or disulphide covalent bonds.

The secondary and tertiary organizations of the protein molecule appear to be intimately associated with the molecule's biological function.Disruption of secondary and tertiary structures due to certain conditions such as change in pH, relatively high temperature, UV radiation, leads to denaturation followed by loss of other properties of the protein molecule i.e., solubility, specific activity and crystallizability.

Quaternary structure: The quaternary structural organization of a protein molecule results by interaction among two or more polypeptide chains. The individual polypeptide chains are called subunits and are held together by either hydrophobic or electrostatic forces besides the involvement of hydrogen bonds in holding subunits together. Subunits making up the quaternary structure of proteins may be identical or dissimilar.

10.4.2 Protein classification

Proteins can be categorized into **simple proteins** and **conjugated proteins,** based upon their composition (Table 10.1)

---

[5] Sanger (1953) determined the amino acid sequence in the protein hormone insulin for the first time.

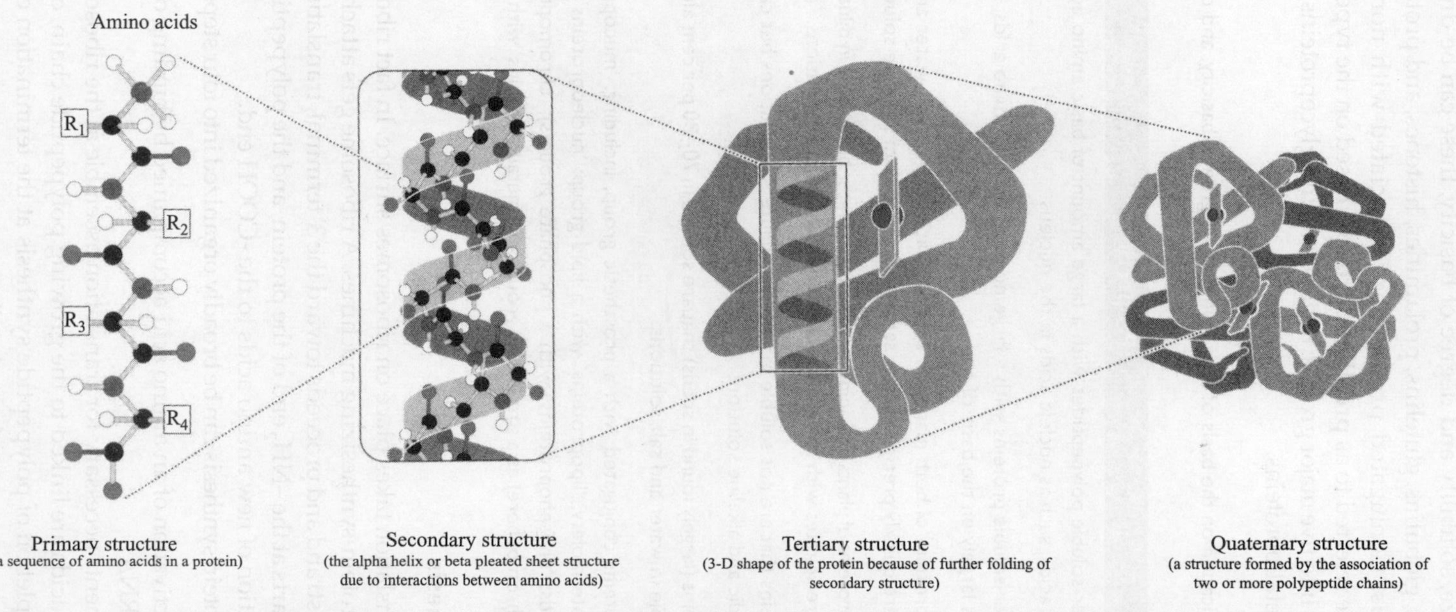

Fig. 10.3 Four different organizational levels of protein molecules: primary, secondary, tertiary and quaternary. R_1, R_2, R_3 and R_4 represent the side chains of four amino acids.

Simple proteins: Simple proteins, when undergoing hydrolysis, produce only amino acids. Based on their size, solubility and degree of basicity, these can be divided into six major groups: albumins, globulins, glutelins, prolamines, histones, and protamines.

Conjugated proteins: Conjugated proteins are associated with non-amino acid components, and may be referred to as **prosthetic groups**. Based on the type of prosthetic group, these are divided into five major groups: nucleoproteins, glycoproteins, lipoproteins, chromoproteins, and metalloproteins.

Table 10.1 Protein classification on the basis of size, solubility, degree of basicity, and conjugation with non-protein groups.

Proteins	Properties
Protamines	Small, water-soluble polypeptides with a large amount of basic amino acids; frequently occur with acids, such as nucleic acids in the nucleus.
Histones	Large, water-soluble proteins with a large number of basic amino acids; differ from the protamines largely on the basis of size.
Albumins	Common proteins of both plant and animals that are soluble in water and in dilute salt solutions; are quickly precipitated by saturated ammonium sulphate solutions.
Globulins	Common proteins of plants and animals, insoluble in water but soluble in dilute salt solutions; can form precipitates with half saturated ammonium sulphate solutions.
Glutelins	Plant proteins that are not soluble in water or dilute salt solutions but can be dissolved in both acidic and alkaline solutions.
Prolamines	Plant proteins (largely found in seeds) that are soluble in 70–80 per cent alcohol solutions but insoluble in water and salt solutions.
Conjugated proteins	Simple proteins conjugated with a prosthetic group, including: 'mucoproteins' with a carbohydrate moiety; 'lipoproteins' with a lipid group: 'nucleoproteins' conjugated to nucleic acids; 'phosphoproteins' with a phosphate group(s); 'chromoproteins' with a pigment (chromophore) as a prosthetic group; and metalloproteins' with a metal group.

10.4.3 Protein synthesis

Protein synthesis or translation takes place on ribosomes surface. In fact ribosomes can be considered as moving protein-synthesizing machines. A ribosome gets attached near the 5′ terminal end of m-RNA strand and proceeds toward the 3' terminal, translating the codons as it moves. Synthesis starts at the $-NH_2$ end of the protein, and the polypeptide undergoes elongation by incorporation of new amino acids to the -COOH end.

The mechanism of protein synthesis can be broadly organized into four steps (Figure 10.4):

1. tRNA charging – activation of an amino acid accompanied by binding of an activated amino acid to the tRNA
2. Initiation – components necessary for translation assemble at the ribosome surface
3. Elongation–amino acids are linked to the growing polypeptide chain, one at a time
4. Termination – completion of polypeptide synthesis at the termination codon and the release of translation components from the ribosome

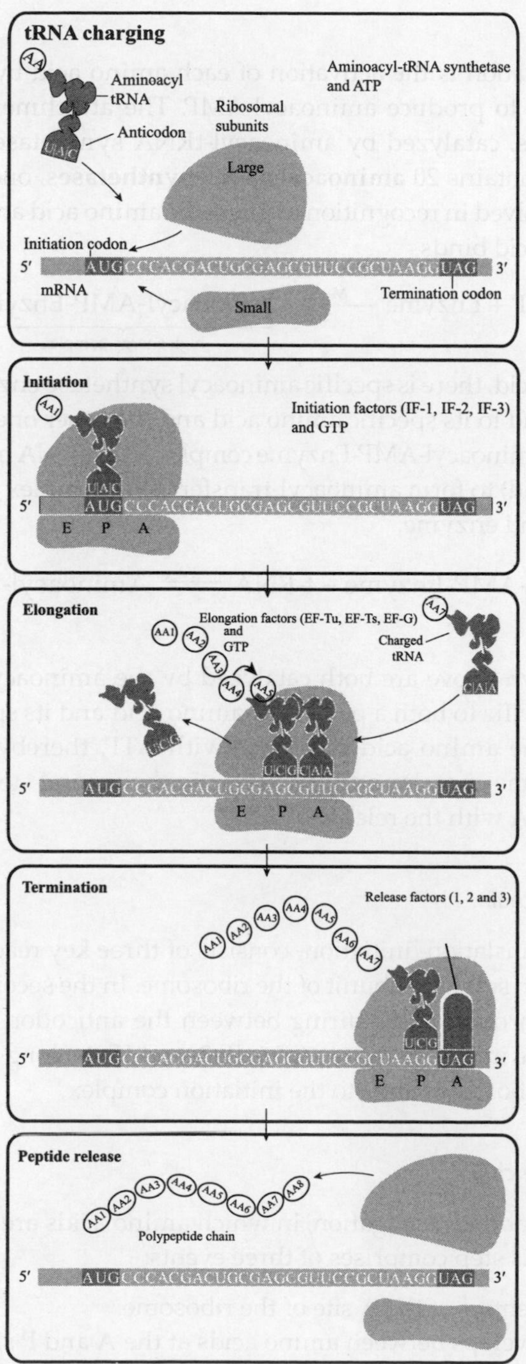

Fig. 10.4 Diagrammatic representation of four steps in the process of protein synthesis in *E.coli*. Translation consists of tRNA charging, initiation, elongation, and termination. During this process, amino acids (AA) are joined together in an order directed by mRNA to produce a polypeptide chain. Several initiation, elongation, and release factors participate in this process, and the necessary energy is provided by ATP and GTP.

tRNA charging

The first step of translation is the activation of each amino acid by ATP molecule in the presence of Mg^{2+} ions to produce aminoacyl-AMP. The attachment of activated amino acid to specific tRNAs, catalyzed by aminoacyl-tRNA synthetases, takes place in two reactions. Every cell contains 20 **aminoacyl-tRNA synthetases**, one for each amino acid. Each synthetase is involved in recognition of a specific amino acid and the tRNAs, to which the particular amino acid binds.

(i) Amino acid + ATP + Enzyme $\xrightarrow{Mg^{2+}}$ $\underbrace{\text{Aminoacyl-AMP-Enzyme}}_{\text{high energy complex}}$ + PPi

For every amino acid, there is specific aminoacyl synthetase enzyme with two binding sites – one can bind to its specific amino acid and the other one to an ATP molecule.

(ii) Condensation of aminoacyl-AMP-Enzyme complex with t-RNA molecule (that is specific for each amino acid) to form aminoacyl-transfer RNA complex with the simultaneous release of AMP and enzyme.

$$\text{Aminoacyl-AMP-Enzyme + t-RNA} \rightleftharpoons \text{Aminoacyl-tRNA complex + AMP + Enzyme}$$

The two steps shown above are both catalyzed by the aminoacyl synthetase enzyme, which is therefore specific to both a particular amino acid and its specific t-RNA.

In the first step, the amino acid combines with ATP, thereby producing activated aminoacyl-AMP-Enzyme complex and PPi, while in the second step, the amino acid gets transferred to the tRNA with the release of AMP.

Initiation of polypeptide chain

The second stage of translation- initiation- consists of three key reactions. In the first step, mRNA attaches itself to smaller subunit of the ribosome. In the second step, initiator tRNA binds to the mRNA through base pairing between the anticodon and the codon. These reactions need multiple initiation factors (IF-1, IF-2, and IF-3) along with GTP. Thirdly, the larger subunit of the ribosome binds to the initiation complex.

Elongation of polypeptide chain

The next step in translation is elongation, in which amino acids are linked to synthesize a polypeptide chain. This step comprises of three events:

(i) A charged tRNA enters at the A site of the ribosome.
(ii) A peptide link develops between amino acids at the A and P sites of the ribosome.
(iii) The ribosome moves to the next codon on mRNA.

Elongation steps need multiple elongation factors (EF-Tu, EF-Ts and EF-G) along with GTP.

After translocation, the aminoacyl or **A site** of the ribosome is unoccupied and in a position to accept the tRNA directed by the next codon. The elongation step repeats itself and during the course of all the events, the polypeptide chain remains associated to the tRNA at the peptidyl or **P site**.

Termination of polypeptide chain

Termination of the polypeptide chain occurs when the ribosome arrives at the termination codon (UAA, UAG and UGA). Release factors (R_1, R_2 and R_3) attach to the termination codon, thereby releasing the polypeptide chain from the final tRNA, and finally detaching tRNA and mRNA from the ribosome.

Table 10.2 shows the details of all the factors (initiation, elongation and release factors) participating in protein synthesis in *E.coli.*

Table 10.2 Components required in the four major steps in protein synthesis in *E. coli.*

Steps	Components
1. Activation of the amino acids	Amino acids tRNAs Aminoacyl-tRNA synthetases ATP Mg^{2+}
2. Initiation of the polypeptide chain	fMet-tRNA in prokaryotes (Met-tRNA in eukaryotes) Initiation codon in mRNA (AUG) mRNA GTP Mg^{2+} Initiation factors (IF-1, IF-2, IF-3) 30S ribosomal subunit 50S ribosomal subunit
3. Elongation	Functional 70S ribosomes Aminoacyl-tRNAs specified by codons Mg^{2+} Elongation factors (EF-Tu, EF-Ts and EF-G) GTP
4. Termination	Functional 70S ribosomes Termination codon in mRNA (UAA, UAG, UGA) Polypeptide release factors (R_1, R_2, R_3)

10.5 Nucleic Acids

Nucleic acids (both **DNA** and **RNA**) are large polymeric molecules composed of repeating units called **nucleotides**, bound together by sugar-phosphate linkages (Figure 10.5). The nucleotides are comprised of three components: (a) a nitrogenous base, (b) a pentose sugar, (c) phosphoric acid. There are two types of nitrogenous bases: purines and pyrimidines. Adenine and guanine represent the two purines commonly found, while thymine, cytosine, and uracil are the pyrimidines that are generally present. The pyrimidine thymine is found only in DNA, while the pyrimidine uracil is confined to the RNA molecule. DNA and RNA also differ in kind of sugar: deoxyribose in DNA, while ribose in RNA.

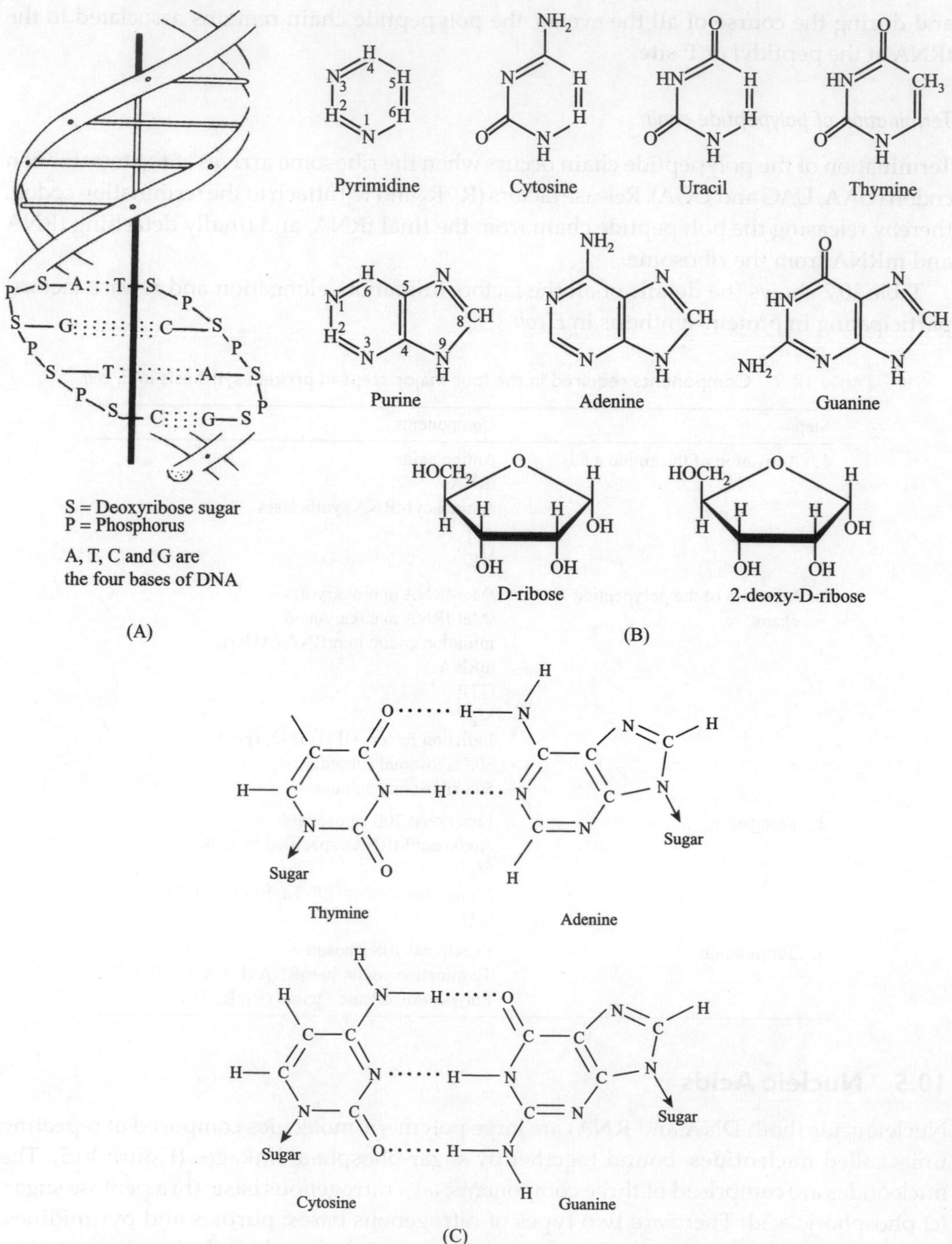

Fig. 10.5 (A) Double-helix structure of DNA. (B) Structures of nitrogenous bases and sugars present in plant nucleic acids. (C) Base pairing between nitrogenous bases (A = T; C ≡ G).

DNA is found in chromosomes, plastids, and mitochondria. RNA is found in chromosomes, nucleoli, cytoplasm (as transfer RNA), ribosomes (ribosomal and messenger RNA), plastids, and mitochondria. The biological functions of DNA and RNA are the transmission of hereditary characteristics and the biosynthesis of proteins, respectively.

Three types of RNA have been identified based on size and functions:

1. Ribosomal RNA (**rRNA**) is the largest RNA and is a constant component of ribosomes along with the proteins.
2. Messenger RNA (**mRNA**) is much smaller, found attached with ribosomes. The whole complex is called a **polysome** or **polyribosome**. The mRNA contains along its length a sequence of bases, complimentary to the base sequence in the DNA molecule. Each triplet of bases on the mRNA molecule codes for a specific amino acid and is called **codon**.
3. Transfer RNA (**tRNA**) is a relatively small molecule containing 70–100 nucleotides and is associated with protein synthesis (Figure 10.6). The tRNA picks up individual amino acid molecules from the cytoplasm and transports them to the ribosome where they are incorporated into the polypeptide chain. Each tRNA has a set of three unpaired bases called the **anticodon,** by which it binds itself to the corresponding codon bases on the mRNA strand *via* hydrogen bonds, ensuring that the amino acids get attached in the appropriate manner.

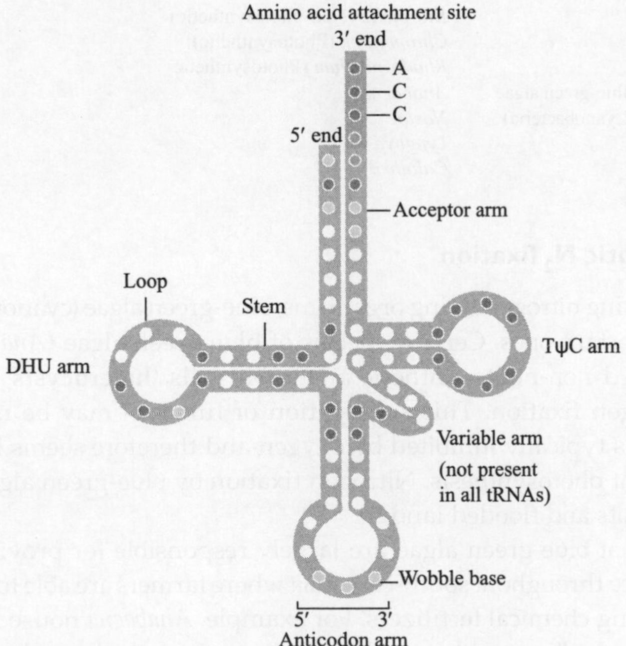

Fig. 10.6 Diagrammatic representation of structure of tRNA (clover-leaf model). The clover-leaf has four main arms: Acceptor arm, TψC arm, Anticodon arm, and DHU arm. The acceptor arm includes the 5‘ and the 3‘ terminals of the tRNA molecule. TheTψC arm is named after the bases of three nucleotides present in the loop of the arm; thymine (T), pseudouridine (ψ), and cytosine (C). The anticodon arm exists at the lowermost region of the tRNA. The DHU arm is designated so since it has modified base, dihydrouridine.

10.6 Molecular (Biological) Nitrogen Fixation

The mechanism of reduction of nitrogen to ammonia is called as **nitrogen fixation.** Approximately 10 per cent of the nitrogen is fixed yearly by nitrogen oxides in the atmosphere produced by lightning strikes and UV radiation, industrial combustion, power-generating stations, forest fires, automotive exhausts and such. Another 30 per cent of the total nitrogen fixed is because of industrial nitrogen fixation (Haber–Bosch process). This process combines nitrogen with hydrogen at elevated temperature (300–400°C) and pressure (300 atm). The balance 60 per cent nitrogen is fixed by living organisms through **biological nitrogen fixation.**

Biological nitrogen fixation is an exclusively prokaryotic domain because only prokaryotes (bacteria and blue-green algae[6]) possess the enzyme complex system known as **nitrogenase,** catalyzing the conversion of nitrogen to ammonia.

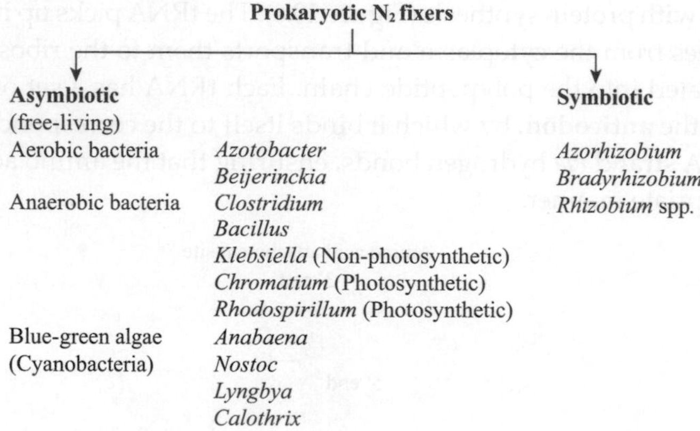

10.6.1 Asymbiotic N_2 fixation

Among the free-living nitrogen-fixing organisms, blue-green algae (cyanophycean members) are the most important ones. Certain species of blue-green algae (*Anabaena, Nostoc,* etc.) possess specialized non-photosynthetic anaerobic cells **'heterocysts'** with nitrogenase activity for nitrogen fixation. This segregation of function may be necessary because nitrogen fixation is typically inhibited by oxygen and therefore seems incompatible with normal green plant photosynthesis. Nitrogen fixation by blue-green algae is important in hydrophytic habitats and flooded land.

It is believed that blue-green algae are largely responsible for providing the nitrogen requirements of rice throughout south-east Asia where farmers are able to obtain maximum yields without using chemical fertilizers. For example, *Anabaena* housed in the leaves of a small aquatic fern, *Azolla,* provides combined nitrogen for the host plant. *Azolla* has been useful as green manure in the rice fields of south-east Asia where it is applied as a manure or

[6] The nitrogen-fixing photosynthetic blue-green algae are the most self-sufficient cells. Blue-green algae are believed to be the first microorganisms that colonized land during evolution.

co-cultivated alongside rice plants. Similar non-nodule forming associations occur between *Nostoc–Anthoceros, Nostoc–Cycas* (coralloid roots) and *Nostoc–Gunnera* (an angiosperm).

10.6.2 Symbiotic N_2 fixation

It is believed that symbiotic nitrogen fixing organisms account for substantial supply of nitrogen to the rhizosphere as compared to free-living bacteria. Rhizobia-legume is the most common type of symbiotic association and results in the development of large, multicellular structures called '**nodules** or **tubercles**' on the roots of host plant (peanut, pea, beans, clover, alfalfa and such). The host plant (legume) supplies the nodule microorganisms (rhizobia) with organic carbonaceous food while the latter supply the former with usable nitrogen, probably amino acids. Curiously neither the host plant nor the microorganism by itself can fix nitrogen; this ability comes about only when the two are associated.

Under waterlogged conditions, plants like *Aeschynomene* and *Sesbania rostrata* exhibit nodules on their aerial parts. While *Rhizobium* is restricted to members of *Fabaceae* and *Parasponia* of Ulmaceae, *Frankia* – a filamentous actinomycetes is associated with a wide variety of non-legume genera like *Alnus* and roots of *Casuarina.*

The root nodules of leguminous plants comprise of a large central region consisting of bacteroid-infected as well as uninfected cells surrounded by a comparatively narrow cortex. Vascular bundles, radiating from the site of attachment of the nodule to the root, are confined to the inner cortex. In soybean, both infected and uninfected cells play a role in the formation of ureides (derivatives of urea), the resulting end products of nitrogen fixation are exported from the nodule to the plant. Other legumes produce several nitrogenous compounds, the amino acid asparagine being one of the most common.

Leghaemoglobin is synthesized partly by the bacteroid (the haeme portion) and partly by the host plant (the globin portion) and is confined within the bacteroid-infected host tissue. It may constitute nearly 30 per cent of the host cell protein and imparts the nodule a distinct pink colour, upon exposure of cut surface to air. In structure and function, it resembles the haemoglobin of mammalian blood. It gets attached to oxygen and regulates the liberation of oxygen in the region of bacteroid. So leghaemoglobin generates the anaerobic conditions (in the bacteroid) in the aerobic environment of the host, required for the functioning of nitrogen reduction.

$$\text{LHb} + O_2 \rightleftharpoons \text{LHb}\cdot O_2$$

The association between *Rhizobium* and the host is not only at the morphological level but also at the molecular level; the globin portion is synthesized by the host while the haeme portion is synthesized by the *Rhizobium* and these two together form leghaemoglobin that gives a pink colour to the root nodule. The more the redness/ pinkness, the greater is its capacity to fix atmospheric nitrogen. Therefore, symbiotic nitrogen fixation specifically needs the co-ordinated expression of multiple genes of both the host and microsymbiont, controlling the formation of the globin protein and haeme respectively.

> Leghaemoglobin is thought to buffer the O_2 concentration within the nodule, allowing respiration without inhibiting the nitrogenase.

Mechanism: The series of events starting with bacterial infection of the root and terminating with development of mature, nitrogen-fixing nodules or tubercles has been divided into four principal steps (Figure 10.7):

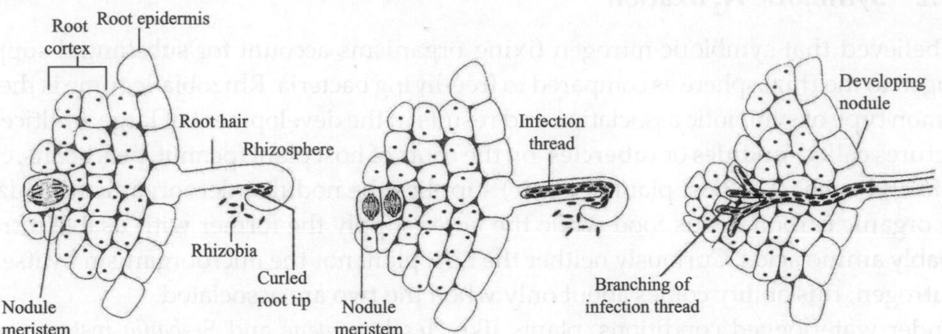

Fig. 10.7 Schematic presentation of the process of bacterial infection of the root leading to the development of nodules.

1. **Multiplication of the rhizobia bacteria in rhizosphere:** Leguminous plants root exudates, containing a variety of amino acid, sugars, organic acids and flavonoids, attract rhizobia bacteria which multiply in the vicinity of roots. These get attached to epidermal and root hair cells and may function as nutrients for them.
2. **Nodule initiation and its development in the root cortex zone:** Rhizobia secrete nodulation factors (**nod factors**) into the surrounding soil solution which stimulate initiation of nodule and its development accompanied by considerable changes in the growth and metabolism of the host roots like increased hair production, development of shorter and thicker roots, branching of the root hairs and their curling in the tip region. Prior to invading the host, rhizobia also discharge mitogenic signals that induce localized cell divisions in the cortex of roots and pericycle, forming the primary nodule meristem. In rhizobia–legume interactions, recognition seems to involve two kinds of compounds – complex polysaccharides and lectins[7].
3. **Invasion of root hair and the formation of an infection thread:** It is believed that rhizobia secrete enzymes i.e., pectinase, cellulase and hemicellulase, known to digest the root hair cell wall of the host and form an infection thread following invasion. The infection thread enters the specialized cells within the nodule and ultimately branches to penetrate several cortical cells. The tip of the infection thread thorugh its vesicles, releases *Rhizobium* cells packed in a peribacteroid membrane derived from the plasma membrane of the host cell. Bacteria keep on multiplying and the surrounding membrane undergoes an increase in surface area to accommodate their growing number by fusion with smaller vesicles.

[7] Lectins – proteins with a sweet tooth – are small non-enzymatic proteins produced by the host. Since these can bind with specific complex carbohydrates, they have proven very useful as an analytical and preparative tool in the laboratory.

4. **Release of bacteria and their differentiation into specialized nitrogen-fixing cells:** Very soon after release from infection thread, rhizobia bacteria stop dividing, grow bigger in size and develop into nitrogen-fixing endosymbiotic organelles referred to as **bacteroids** (Figure 10.8). The peribacteroid membrane covering the bacteroid is derived from the infected root nodule cells. The bacteroids are the seat of nitrogen fixation. As the nodule undergoes enlargement and maturation, the vascular links develop with the main vascular tissues of the root, which help to import photosynthetic carbon into the nodule and export fixed nitrogen from the nodule to the plant (Figure 10.9).

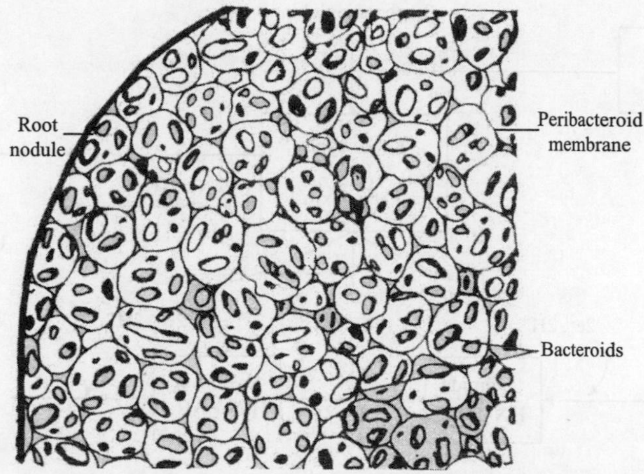

Fig. 10.8 Groups of bacteroids surrounded by peribacteroid membrane in a root nodule.

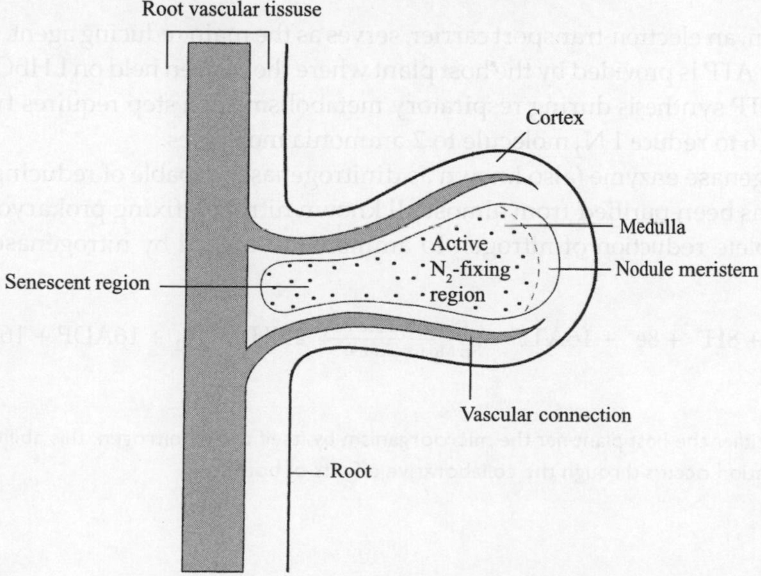

Fig. 10.9 Cross-section through a mature nodule.

Biochemistry: The multistep reduction is believed to proceed from molecular nitrogen to diimide, hydrazine and ultimately to ammonia (Figure 10.10).

$$\underset{\text{Nitrogen}}{N_2} \longrightarrow \underset{\text{Diimide}}{HN=NH} \longrightarrow \underset{\text{Hydrazine}^7}{H_2N-NH_2} \longrightarrow \underset{\text{Ammonia}}{2NH_3}$$

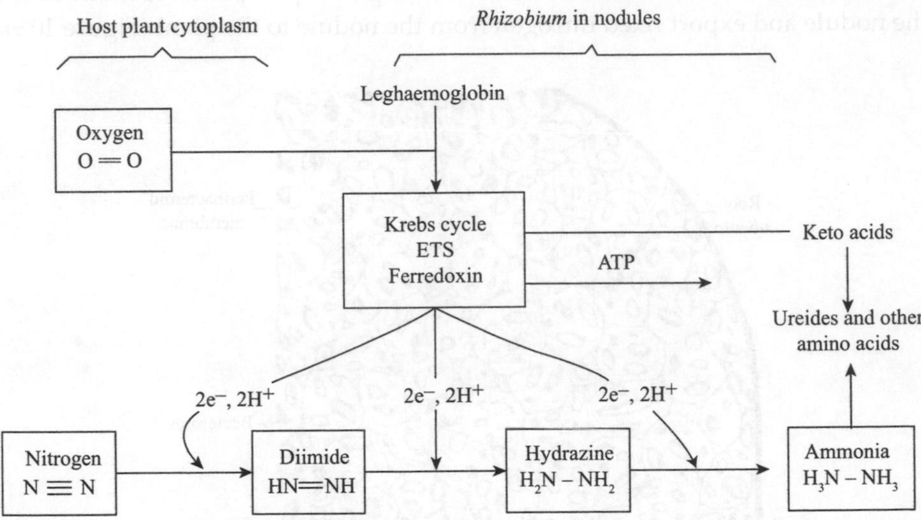

Fig. 10.10 Major reactions taking place during nitrogen fixation in *Rhizobium*, catalyzed by nitrogenase (an iron and iron-molybdenum protein complex). The reduction of N_2 (dinitrogen) to ammonia requires a strong reducing agent and ATP.

Ferredoxin, an electron-transport carrier, serves as the main reducing agent. In symbiotic associations, ATP is provided by the host plant where the oxygen held on $LHbO_2$ may be the source for ATP synthesis during respiratory metabolism.Each step requires two electrons for a total of 6 to reduce 1 N_2 molecule to 2 ammonia molecules.

The nitrogenase enzyme (also known as dinitrogenase), capable of reducing nitrogen to ammonia, has been purified from almost all known nitrogen-fixing prokaryotic systems.

The complete reduction of nitrogen to ammonia catalyzed by nitrogenase is outlined as below:

$$N_2 + 8H^+ + 8e^- + 16ATP \xrightarrow[\text{Fe, Mo (cofactors)}]{\text{Dinitrogenase}} 2NH_3 + H_2 + 16ADP + 16Pi$$

Curiously *neither* the host plant *nor* the microorganism by itself can fix nitrogen, this ability of nitrogen fixation occurs through the collaborative efforts of both.

[8] Hydrazine, a component of fuel launch, is colourless liquid with an ammonia like odour, toxic and used in reactors/satellite launch/propeller for shuttle launch.

The reduction of N_2 to NH_3 appears to be coupled to the evolution of H_2 from the reduction of two protons.

Dinitrogenase is a large, multimeric protein complex composed of two major functional components (Figure 10.11).

- The Fe Protein, the smaller one, is a dimer comprised of two identical subunits with a molecular weight of each subunit ranging between 24 to 36 kD. It contains four ferrous atoms linked to four sulphur atoms (Fe_4S_4).
- The Mo-Fe protein, the larger one, is a tetramer comprised of two pairs of identical subunits with a molecular weight of more than 220 kD. It has two molybdenum atoms bound to Fe_4S_4 clusters.

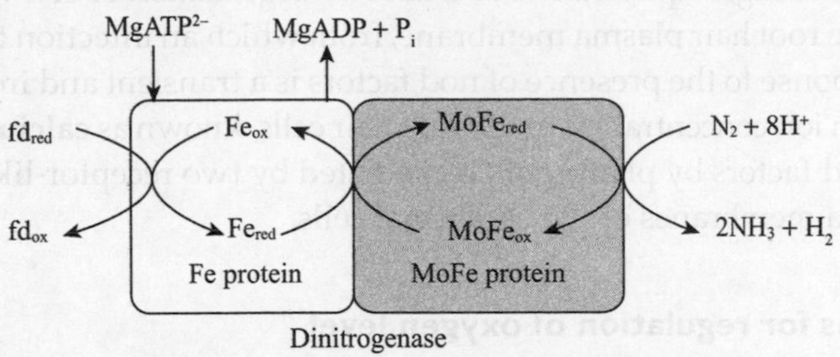

Fig. 10.11 Dinitrogenase reaction in the bacteroids.

Neither of the two components exhibits nitrogenase activity by itself. Like oxygen, nitrate and ammonia are strong inhibitors of N_2 fixation.

Biological reduction of nitrogen is quite expensive in terms of energy. A total of 25 ATP molecules for each molecule of N_2 fixed (16 ATP for each molecule of N_2 reduced + 9 ATP for reduced ferredoxin) are required as compared to 9 ATP for each molecule of CO_2 fixation (3 ATP for each molecule of CO_2 fixation + 6 ATP for reduced ferredoxin).Therefore, it is quite evident that fixation of nitrogen constitutes a significant drain on the energy and carbon reserves of the host plant.

One of the crucial problems being faced by nitrogen-fixing organisms is the sensitivity of the nitrogenase enzyme to molecular oxygen (O_2). It gets rapidly and irreversibly inactivated by oxygen concentration more than 10 nM.

10.6.3 Genetics of nif genes and nod factors

nif genes

In free-living nitrogen-fixing as well as symbiotic bacteria, the nitrogenase and other enzymes participating in nitrogen fixation to produce ammonia, at normal temperatures and atmospheric pressure, are encoded by bacterial *nif* genes and *fix* genes. In addition to the enzyme nitrogenase, *nif* genes also code for several regulatory proteins associated with fixation of nitrogen. For example, *nif* D and *nif* K genes code for the MoFe protein subunits; *nif* H and *nif* F genes encode the Fe protein and ferredoxin, while the remaining *nif* genes take part in the activation and processing of the enzyme complex.

Nod factors (NFs)

The nodulation process in legumes includes a complex array of signalling molecules, molecular recognition, and regulation. In the early stages of nodulation, a set of rhizobial nod (**nodulin**) genes is switched on by chemicals such as, flavonoids and betaines, released from roots of the host plant into the rhizosphere. The *nod A*, *nod B*, *nod C* are primary nodulation genes present in most of the rhizobia. The other nod genes (*nod P*, *nod Q*, *nod H*; or *nod E*, *nod F* and *nod L*) are host-specific and serve to determine the host range. Only the *nod D* gene is expressed constitutively and its final product controls the transcription of rest of the nod genes. *Rhizobia* bacteria, synthesize lipochitooligosaccharides (LCOs), known as **nod factors (NFs)** which behave as morphogenic signal molecules on legume host. Nod factors induce root hair curling, sequestration of rhizobia, degradation of cell wall, and bacterial accessibility to the root hair plasma membrane, from which an infection thread forms. One specific early response to the presence of nod factors is a transient and immediate increase in nuclear calcium ion concentration in the root hair cells, known as **calcium spiking**. Initial recognition of nod factors by plant roots is mediated by two receptor-like kinases (RLKs), present on plasma membranes of the epidermal cells.

10.6.4 Strategies for regulation of oxygen level

1. Most of the free-living nitrogen-fixing bacteria have conserved an anaerobic life style.
2. Certain species of blue-green algae have structurally isolated the nitrogen-fixing apparatus through **heterocysts** formation.
3. Supply of oxygen is carefully controlled by leghaemoglobin, O_2-carrying protein, in nodules of the leguminous plants.

Another problem faced by bacterial nitrogen-fixers is release of the hydrogen gas as a by-product, which is wasteful and exhausts energy that might otherwise be utilized for reduction of nitrogen. Most of the nitrogen-fixing bacteria consist of an enzyme **uptake hydrogenase**, which is dependent upon oxygen and can partly recover the energy loss due to production of hydrogen gas .Using biotechnological approach to increase the number of rhizobial strains possessing capacity for recycling of hydrogen, has the potential to enhance energy efficiency of molecular nitrogen fixation in agricultural crops.

The beneficial effects on the soil of growing leguminous plants have been recognized for centuries. Theophrastus who lived in the third century BC, wrote that the Greeks used crops of broad beans (*Vicia faba*) to enrich soils.In modern agriculture, it is common practice to rotate a non-leguminous crop, such as maize, with a leguminous crop, such as alfalfa, or maize with soybeans and then wheat. The leguminous plants are then either harvested for hay, leaving behind the nitrogen-rich roots or better still, simply ploughed under. A crop of alfalfa that is ploughed back into the soil may add nearly 300–350 kg of nitrogen for a hectare of soil. Approximately 150–200 million metric tons of reduced nitrogen per annum is put back into the earth's surface by such biological systems.

Both legumes and bacteria are being genetically screened for combinations that would result in increased nitrogen fixation in specific environments. Genetic engineering offers the possibility of transferring the genes necessary for nitrogen fixation (nif genes[9]) from one organism to another. Researchers, with the help of robotics, computers as well as advanced genetics, have tried to identify many gene regulators (**transcription factors**) in stem, shoots and leaves of the plants for proper utilization of nitrogen, which can check the environmental damage caused due to excess nitrogen present in the soil.

The production of **diazotrophic plants** through the transfer of *nif* genes straight away into the plant cell has been reviewed since long time back. The plastid offers an appropriate site for the insertion of *nif* genes because of the expression of chloroplastic genes in a prokaryotic-like manner, which permits translation of polycistronic messages. Efforts are now being made to insert nitrogenase genes into chloroplast genomes of non-leguminous plants.

Nitrogen fixation is an area of great interest in the current scenario due to its vast practical approach. Industrial production of ammonia used to synthesize fertilizers needs a huge and costly input of energy, and this has encouraged the scientists to create the recombinant (**transgenic) organisms** that can help to fix nitrogen. **Recombinant DNA techniques** might be employed to transfer the DNA, coding for the enzymes involved in nitrogen fixation, into non-nitrogen fixing microorganisms and plants. There are formidable challenges in engineering new nitrogen-fixing plants. The result of these experiments will be based on solving the issue of oxygen toxicity in the cells producing nitrogenase.

Nitrogen Fixation in Cereals

Basic human nutrition is derived from cereals such as maize, rice, wheat and sorghum. Since huge quantities of chemical fertilizers are required for their growth, it has always been recommended to encourage biological nitrogen fixation in these crops. Recent studies have confirmed the presence of a few rhizobia in cereal crops. Since the amounts of N_2 fixed by nitrogen-fixing bacteria are not adequate to sustain the plant's requirements, there is always need to enhance nitrogen fixation in cereals crops.

Use of Rhizobia as biofertilizers in agricultural crops has been known for long. The conceptualization of important avenues such as **usage of bradyrhizobial inoculates, development of the root nodular symbiosis and transfer of nitrogenase genes** can be explored to achieve this objective in cereal crops. Another approach is to produce cereals that encourage the growth of diazotrophs. So plants can be engineered to synthesize a certain specific compound or metabolite and thus generate a 'biased rhizosphere' to favour the growth of an introduced diazotroph that can consume the unique metabolite.

Efforts are being made to incorporate 'nif genes' into cereal crops so that they are able to convert atmospheric nitrogen into ammonia, thus lessening our dependence on the use of nitrogenous fertilizers as its production is quite energy-intensive.

10.7 Nitrogen Cycle

Although Earth's atmosphere consists of about 80 per cent of molecular nitrogen (N_2), only few plant species can reduce this atmospheric nitrogen into usable forms for living organisms. In the biosphere, the metabolic pathways of various species are interrelated to save and reutilize biologically available nitrogen in a complex **nitrogen cycle** (Figure 10.12).

[9] 17 'nif genes' have been identified from *Klebsiella pneumonieae*. Nitrogenase synthesis is directed by nif genes.

The first step in the cycle is reduction of atmospheric nitrogen to produce ammonia (NH_3 or NH_4^+), using nitrogen-fixing bacteria. Nearly all the ammonia reaching the soil is oxidized to nitrate by soil bacteria, *Nitrosomonas* and *Nitrobacter* (in two steps given below) that obtain their energy by the oxidation of ammonia to nitrite (NO_2^-) and then to nitrate (NO_3^-). This process is called **nitrification**.

$$NH_4^+ + 1\,½\,O_2 \rightarrow NO_2^- + H_2O + 2H^+$$

$$NO_2^- + ½\,O_2 \rightarrow NO_3^-$$

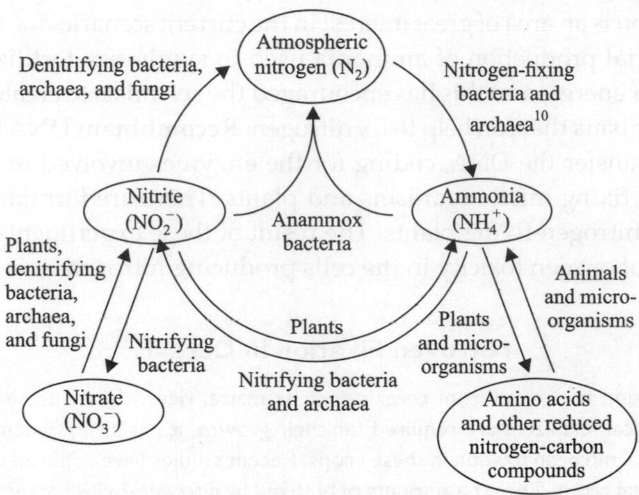

Fig. 10.12 The nitrogen cycle. The first reaction in the cycle envisages the fixation of atmospheric nitrogen by nitrogen-fixing bacteria to produce ammonia. Nearly all ammonia entering the soil is oxidized to nitrite and then to nitrate (nitrification). Plants readily reduce nitrite and nitrate to ammonia, which is incorporated into amino acids. Animals in turn, use plants as a source of amino acids. Microbial degradation of plant and animal remains returns ammonia to the soil. Denitrifying bacteria reduce nitrate and nitrite to N_2 under anaerobic conditions. The nitrogen cycle is short-circuited by anammox bacteria, which provide useful solution to waste-treatment problems.

Plants and many bacteria can absorb and promptly reduce nitrate and nitrite to ammonia by nitrate reductase and nitrite reductase respectively. Plants incorporate this ammonia into amino acids and animals, in turn, consume plants as a source of amino acids to synthesize their proteins. Death of plants and animals is accompanied by microbial decomposition of their proteins, returning ammonia back into the soil where nitrifying bacteria again transform it to nitrite and nitrate. An equilibrium is established between fixed nitrogen and atmospheric nitrogen by bacteria that reduce nitrate and nitrite to N_2 under anaerobic conditions (**denitrification**). These soil bacteria (*Thiobacillus denitrificans, Micrococcus*

[10] Archaea are single-celled microorganisms, structurally similar to bacteria. They are obligate anaerobes living in environments low in oxygen i.e., water and soil. Archaea form the third domain of life, since they are evolutionarily distinct from bacteria and eukaryotes.

denitrificans), instead of oxygen, use nitrate or nitrite as the ultimate electron acceptor in a sequence of reactions, creating a transmembrane proton gradient, which can drive ATP synthesis.

$$2\,NO_3^- + 10\,e^- + 12H^+ \rightarrow N_2 + 6\,H_2O$$

The nitrogen cycle is cut short by a process known as **anammox**, in which a set of bacteria encourage anaerobic oxidation of ammonia to N_2. Approximately 50 per cent–70 per cent of ammonia to nitrogen conversion in the biosphere may occur *via* this cycle, which went unnoticed until the 1980s. The obligate anaerobes, the **annamox bacteria**, provide a major challenge in waste treatment, thereby lowering the cost of ammonia removal by as much as 90 per cent and reducing the release of pollutants. A group of Dutch microbiologists identified these bacteria belonging to an unusual bacterial phylum, Planctomycetes. These bacteria solve the problem by sequestering hydrazine, a highly reactive molecule produced as an intermediate, in a specialized organelle - the **anammoxosome**. Its membrane is made up of lipids called **ladderanes**, which stack up into a thick and impermeable, hydrophobic layer, thus delaying the generation of hydrazine.

Human activity presents an increasing challenge to the global nitrogen balance, and to all life in the biosphere supported by that balance. Controlling the damaging effects of agricultural runoff and industrial pollutants will remain an important component of the continuing effort to expand food supply for a growing human population.

10.8 Nitrogen Use Efficiency (NUE)

Nitrogen (N), a macroelement, is essential for growth and developmentof a plant. It is a component of nucleic acids (DNA and RNA), proteins, enzymes and chlorophyll. Although it constitutes about 78 per cent of the total atmosphere, it is not readily available to most plants; its availibilty being restricted to only leguminous crops. During the last five decades, the usage of chemical (nitrogenous) fertilizers in our agricultural crops has led to a significant rise in productivity, but this has had a negative effect on the environment.

> Nitrogen use efficiency (NUE) is the ratio of the crop nitrogen uptake to the total input of nitrogen fertilizers. The greater the ratio, the better is the nitrogen use efficiency.

Producers and agronomists strive hard to optimize crop yields with minimum nitrogen inputs. Globally, nitrogen use efficiency (NUE) is nearly 33 per cent and the remaining 67 per cent represents $15.9 billion loss of N fertilizer which results from surface runoff, nitrate leaching, ammonia volatilization, nitrification and denitrification. Nitrogen in ammonium form, bound firmly to the negatively charged clay particles, can be interchanged with other cations (but it does not quickly leach out from the soil). On the other hand, nitrates present in the soil solution are easily leached from the soil. Nitrates are potentially harmful if present in drinking water consumed by animals and humans as well. Excess nitrates leads to massive algal bloom, leading to reduced oxygen levels (hypoxia), causing the death of fish and decline in aquatic life and clogging of waterways. After denitrification, nitrates discharge nitrous oxide into the atmosphere. This is one of the most potent greenhouse gases that degrades the ozone layer and contributes to global warming.

Over the years, the nitrogen use efficiency (NUE) of plants has been defined in many different ways, focussing on agronomical, ecological and breeding parameters. However, physiologically speaking, NUE is defined as the grain yield (dry mass productivity) per unit of available N (soil + fertiliser and it has 2 major components, such as nitrogen uptake efficiency (NUpE) and nitrogen utilisation efficiency (NUtE).

NUpE indicates plant ability to uptake nutrients from the soil and is defined as the amount of N in plant corresponding to available N in soil (soil + fertilizers).

NUtE indicates plant ability to convert the nitrogen taken from the soil into grain yield (dry mass) and is defined as grain yield corresponding to per unit N of plant.

Precisely referring to (1) nitrogen uptake efficiency (plant N uptake/available soil N, NUpE) and (2) nitrogen utilization efficiency (grain yield/total N uptake, NUtE), the products of NUpE and NUtE is NUE.

In brief, NUE = NUpE × NUtE.

In view of the high cost of nitrogenous fertilisers and its destructive impact on the environment, scientists are working on the concept of evolving crop varieties with improved NUE – the main thrust is to produce varieties with decreasing inputs of nitrogenous fertilisers through NUE. Genotypic variations in NUE, NUpE and NUtE occurs in different crops. More recently, varieties in rice and other crops are being genetically improved by altered expression of various nitrate transporter genes, facilitating more uptake of nitrates a crucial factor for NUE in future. Both agronomists and breeders will contribute to improved nitrogen use efficiency (NUE).

Nitrogen use efficiency in plants is regulated by interaction of several genetic and environmental factors. New strategies are urgently required for enhancing NUE, either by agronomic or by agricultural management practices or genetically modifying plants by biotechnological methods (such as, over expression of nitrate as well as ammonium transporters, and manipulation of the genes governing C and N metabolism).

Leaching of nitrogen requires proper management to make nitrates accessible to the plants and to minimize the nitrogen loss into streams. To improve the efficiency of utilization of nitrogenous fertilisers, a number of precision fertiliser application measures in the field are being practiced:

1. Deep placement of N fertilisers in a field before transplanting or sowing;
2. Application of different types of fertilisers, to be applied in split doses (or instalments) at proper times to minimise losses. Their mode of application and issues such as avoiding runoff all have an impact;
3. Use of slow and controlled-release nitrogenous fertilisers in the form of urea super granules or foliar sprays;
4. Use of organic manures, green manuring, legume intercropping systems;
5. Use of nitrification inhibitors such as dicyandiamide (DCD), acetone extract of neem cake and so on. (Meliacins in neem cake are responsible for inhibiting nitrification). Neem coated urea increases the efficiency of 10 per cent nitrogen usage as compared to uncoated urea and thus consumes 10 per cent less urea for the equivalent effect. The most common practice of broadcasting of nitrogenous fertilisers should be avoided as it leads to large N losses through ammonia volatilization.

Anhydrous ammonia (AA), a commonly used nitrogen fertilizer source, is generally applied to the ground prior to any crop being planted. Anhydrous ammonia is 82 per cent nitrogen by weight. Other types of fertilisers used to provide N, include dry urea pellets or prills (46 per cent nitrogen); liquid UAN (Urea ammonium nitrate) can be applied to soil to minimise leaching and volatilisation. Anhydrous ammonia is easily available and can kill some microbes. It is injected into the soil where the plant roots are present and turns into gas once in contact with air or wet soil.

Researchers are now looking for environmentally-safe as well as economical measures to improve NUE in plants. The production of varieties, inherently efficient in nitrogen utilization coupled with greater productivity and lower nitrogen inputs, is a major strategy on which breeders and biotechnologists are working on:

- Through genetic manipulation of amino acids translocation from the regions of source to sink.
- Using a nitrification inhibitor will help delay the denitrification process, preventing nitrogen loss.

10.9 Carnivorous Plants

Nitrogen is the primary mineral nutrient required by plants but its availability is limited in many terrestrial ecosystems. Carnivorous plants with modified/specialized leaves; can attract, capture, and digest insects using their secreted enzymes and then the available nutrients are absorbed by the plant for growth and development. Insectivory, the ability to trap and digest insects, has evolved in plants that grow in bogs (habitats poor in nitrates and ammonia). By digesting insects, plants obtain the nitrogen, organic compounds and minerals like potassium and phosphate, etc.

To date, quite a few studies have reported the discovery of distinct digestive enzymes in carnivorous plants e.g. **proteases, esterases, nucleases, phosphatases, glucanases, peroxidases, chitinases**. Proteases obtained from carnivorous plants are of industrial importance with myriad applications, such as in leather goods, detergents, meat tenderisers, food products, pharmaceutical as well as the waste processing industry. Special attention has been given to plant proteases (*versus* microbial and animal proteases) with commercial approach because of great stability in extreme conditions of temperature and pH, and wide substrate specificity.

Moreover, ethical issues that limit the uses of animal and microbe proteases in certain countries, necessitates a demand for plant proteases. Aspartic proteases are found in the plant leaf, flower, seed and also in the carnivorous plants. Proteases present in the digestive liquid of these plants are the only extracellular proteolytic enzymes of plant origin whereas most of the plant proteases (also known as proteinases or peptidases) are intracellular and confined to the vacuoles. Since the identification of **'nepenthesin'** protease in *Nepenthes* (pitcher plant), digestive enzymes from carnivorous plants have been the centre of attraction for genomics studies. Currently, recombination technology and protein purification tools have proved very helpful in the identification of enzymes from carnivorous plants.

Trap leaves in carnivorous plants can be classified as either **active traps**, that move during capture e.g. *Drosera* or **passive traps** incapable of movement e.g. *Nepenthes, Darlingtonia,* and *Sarracenia*

Nepenthes (Pitcher plant): The leaf lamina is highly modified i.e. tubular rather than flat and secretes a watery digestive fluid. The epidermis in the digestive region must be absorptive rather than impermeable. The throat of the pitcher contains numerous trichomes that point toward the liquid. Insects walking in the direction of the trichomes are thus led to their death.

Drosera (Sundew): It is a tiny plant with club shaped glandular trichomes on the upper surface of its leaves which secrete a sticky digestive liquid (Figure 10.13). After the insect is caught on a single trichome, the adjacent trichomes bend inward until the entire leaf blade curves around the insect. The hairs secrete at least six digestive enzymes, which together with bacteria-produced enzymes, especially chitinase, digest the insect. The nutrients released from the prey mix with the mucilage, which is then resorbed by the glands that secreted the digestive enzymes.

Fig. 10.13 Sundew (*Drosera spatulata*) (*Courtesy*: Dr C.T. Chandra Lekha).

Utricularia vulgaris (Bladderwort): It is a free-floating aquatic plant with tiny, flattened, pear-shaped bladders with a mouth guarded by a hanging door. Four stiff bristles are present near the lower free edge of the door. With the approach of a small insect or animal, the hairs distort the lower edge of the door, causing it to spring open. Water rushes into the bladder, carrying the insect inside and the door snaps shut behind it. Several enzymes secreted by the internal wall of the bladder and by the residential bacterial population digest the insect. Released organic compounds and minerals are taken up through the cell wall of the trap, while the undigested exoskeleton remains within the bladder.

Dionaea muscipula (Venus flytrap) : These have leaves that are held flat and this position is maintained because the motor cells along the upper side of the midrib are extremely turgid and swollen (Figure 10.14). When an insect walks across the trap, it brushes against trigger hairs. If two of these are stimulated, midrib motor cells lose water quickly, and the trap rapidly closes as the two halves of the lamina move upward. Long interdigitating teeth on the margin trap the insects and short glands secrete digestive fluid. After completion of digestion and absorption, the midrib motor cells fill with water, swell, and force the trap open, ready for a new victim.

Fig. 10.14 Venus flytrap (*Dionaea muscipula*) (*Courtesy*: Dr. Md. Ahsan Habib, Dhaka, Bangladesh).

Proteases from different carnivorous plant sources alongwith their applications are listed in Table 10.3.

Table 10.3 Proteases obtained from carnivorous plants.

Plant Source	Protease	Application/functional properties
Nepenthes	Nepenthesin I & II	Tool for digestion in H/D Exchange Mass Spectrometry
	Neprosin	Proteomic analysis / Histone mapping
		Gluten digestion
Papaya	Papain	Meat tenderiser
		Denture cleaner
		Detergent, healing burn wounds, textiles, cosmestics industry
	Caricain	Gluten-free food processing
Pineapple	Bromelain	Anti-inflammatory and anti-cancer agent
Fig (*Ficus carica*)	Ficin	Pharmaceutical industry
Kiwi fruit, Banana, Pineapple, Mango	Actinidin	Dietary supplement
Zinger	Zingipain	Anti-proliferative agent

Review Questions

RQ 10.1. What is zwitterion? How is its charge affected? Explain.

Ans. Amino acids consist of at least one carboxyl group (–COOH) and one amino group ($–NH_2$)[11]. An amino acid can carry either a +ve or a –ve charge, depending upon the pH of the solution. An amino acid might exist as a zwitterion, or a molecule consisting of both positive and negative charges at the same time. In this form, an amino acid is dipolar, by which we mean that it can function as an acid or a base (**amphoteric**).

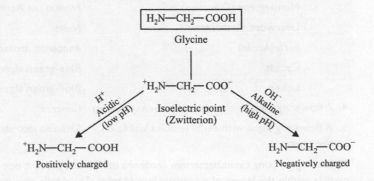

$H_2N—CH_2—COOH$

Glycine

$^+H_3N—CH_2—COO^-$

Isoelectric point (Zwitterion)

H^+ Acidic (low pH) → $^+H_3N—CH_2—COOH$ Positively charged

OH^- Alkaline (high pH) → $H_2N—CH_2—COO^-$ Negatively charged

[11] Some amino acids, such as arginine or lysine are more strongly basic because they contain extra amino (NH_2) groups; some, such as aspartic acid and glutamic acid, are more more strongly acidic because of extra carboxyl (COOH) groups.

The pH at which the 'zwitterion form' exists is usually referred to as the '**isoelectric point**.' Thus, the isoelectric form of amino acids exhibits a 'net zero' charge and does not migrate to either of the two poles in the electric field when subjected to electrophoresis.

RQ 10.2. What is a peptide bond? How is it formed?

Ans. Each amino acid has at least one acid or carboxyl (–COOH) group and one basic or amino (–NH_2) group in its molecule. The 'carboxyl group' of one amino acid molecule can combine with the 'amino group' of another amino acid molecule with the removal of a water molecule. In this enzymatically controlled reaction, the two amino acids are joined by a –CO–NH– bond called a **peptide bond.** Two amino acids bonded through such a peptide linkage is a **dipeptide**, three in a **tripeptide** while a long chain of several amino acids is known as **polypeptide**.

(amino group) NH_2 COOH (carboxyl group)

CH_2 — CO[OH + H]–NH–CH_2

(Glycine) (Glycine)

COOH

NH_2–[CONH]–CH_2 + H_2O

CH_2

Peptide bond

A dipeptide

RQ 10.3. Give **one** example each of free-living (asymbiotic) and symbiotic nitrogen-fixing organisms, together with their host.

Ans.

1. Asymbiotic	
Aerobic	*Azotobacter*
Anaerobic	*Clostridium*
Blue-green algae	*Nostoc*
2. Photosynthetic bacteria (anaerobic)	*Chromatium*
3. Symbiotic	
Legumes	*Rhizobium*
Non-legume (*Casuarina*)	*Frankia* (an Actinomycetes)
Liverwort (*Anthoceros*)	*Nostoc*
Fern (*Azolla*)	*Anabaena azollae**
Cycads	Blue-green algae (*Anabaena*)
Lichen	Blue-green algae (*Nostoc, Scytonema*)
4. A flowering plant (with a cyanophyceaen symbiont)	*Gunnera*
5. A flowering plant with aerial nodules visible on the stem.	*Sesbania rostrata (with Rhizobium)*

*The nitrogen-fixing cyanobacterium *Anabaena azollae* are present not in root nodules but in cavities within the leaves of an aquatic fern (*Azolla*). The *Azolla*-rice-duck agricultural system is practiced in rice paddies. The *Azolla-Anabaena* symbiosis provides fixed nitrogen to the rice plants while ducks graze on the *Azolla* and their faecal matter is rich in other nutrients required by the rice.

RQ 10.4. Name the microorganisms that show symbiotic relationship with the following:

Ans.

(i) A bacterium that produces aerial nodules on the stem of *Sesbania rostrata* and *Aeschynomene aspera*.	*Azorhizobium*
(ii) With non-legume genera like *Alnus, Casuarina* and *Myrica*	*Frankia* (an Actinomycetes)
(iii) Wetland rice ecosystem or agriculture	*Anabaena azollae*, cyanobacteria (free-living)

RQ 10.5. List the 'genes' controlling nitrogen fixation.

Ans. There are three classes of genes controlling N_2 fixation *in situ*. They are *nod genes*, *nif genes* and *fix genes*

1. *nif genes*: involved in the formation of functional nitrogenase enzyme.
2. *nod genes*: involved in the formation of root nodules.
3. *'fix' genes*: required for the maintenance of N_2 fixation by the nodule.

RQ 10.6. List only the essential requirements for biological (molecular) nitrogen fixation.

Ans. The following are the essential requirements for biological nitrogen (N_2) fixation:

1. Nitrogenase is the enzyme that catalyzes the conversion of molecular N_2 to NH_3.
2. Cellular mechanism to protect nitrogenase action from oxygen inhibition.
3. Presence of natural physiological electron donors, reduced ferredoxin (Fd_{red})
4. Provision for abundant supply of ATP (provided by the host).
5. Availability of Mo and Fe, integral constituents of nitrogenase; Mg^{2+} divalent cations.
6. Absence of fixed nitrogen sources like nitrate or ammonia.
7. Presence of an 'uptake hydrogenase' enzyme system.

RQ 10.7. Explain the O_2 protection mechanism for nitrogenase enzyme in cyanobacteria.

Ans. In cyanobacteria, such as *Nostoc* and *Anabaena*, the O_2 protection mechanism for nitrogenase is provided by **heterocyst** with a specialized structure, that has lost its capability of PS II activity during development because of which there is reduced oxygen tension.

Heterocysts are the sites of both synthesis and activity of nitrogenase enzyme. During their differentiation from ordinary vegetative cells, PSII activity is lost, resulting in the loss of O_2 production during photosynthesis. Secondly, during development, a glycolipid layer is newly laid in the wall of the heterocyst that constitutes a barrier to oxygen entry from the external environment.

> The loss of photosystem II activity and formation of a new glycolipid layer in the wall are the two apparent mechanisms for oxygen protection for nitrogenase.

RQ 10.8. Where are the following located in plants?

Ans.

Leghaemoglobin	Cytoplasm of the host cells
Nitrogenase	Bacteroid
Bacteroid	Host cells
Heterocyst	Cyanobacteria

RQ 10.9. Why is ammonium nitrate preferred over ammonium sulphate as a nitrogenous fertilizer? Explain.

Ans. If ammonium nitrate is added to the soil, both the basic and the acidic radicals are absorbed 'equally readily' with the result that none predominates and there is no appreciable change in the soil reaction. In the case of ammonium sulphate, the ammonium ions are readily absorbed while sulphate ions (basic radicals) are slowly absorbed. Such soils develop acidity, and extra 'liming' (application of calcium hydroxide) is needed to counteract this acidity.

Thus, ammonium nitrate (NH_4NO_3) is one of the best sources of quick acting fertilizers.

RQ 10.10. Urea is sometimes sprayed directly on the leaves of fruit trees or other crops. Comment.

Ans. Urea is one of nitrogenous wastes (the other being uric acid) excreted in large quantities in the urine. It is now synthetically produced. Urea can be absorbed by the plant roots and the leaves during foliar spray but it is possibly first hydrolyzed by the enzyme urease before its nitrogen is 'utilized or assimilated'.

$$CO(NH_2)_2 + H_2O \xrightarrow{\text{Urease}} 2NH_3 + CO_2$$

Ammonia in water will form ammonium ions, depending on the pH

$$NH_3 + H_2O \rightarrow NH_4^+ + OH^-$$

Urease is the 'first protein' demonstrated to have enzymatic activity. Nickel is an essential component of urease. In fact it was the first protein to be crystallized by Sumner (1926) from jackbean seeds (*Canavalia ensiformis*).

RQ 10.11. Name and explain reactions where the ammonium ions act as substrate.

Ans. **1.** α-Ketoglutaric acid + NADH + NH_3 $\xrightarrow[(Zn^{2+})]{\text{Glutamic dehydrogenase}}$ Glutamic acid + NAD^+ + H_2O

2. $\text{Glutamic acid} + NH_3 + ATP \xrightleftharpoons{\text{Glutamine synthetase}} \underset{\text{(amide)}}{\text{Glutamine}} + H_2O + ADP + Pi$

3. $\text{Aspartic acid} + NH_3 + ATP \xrightleftharpoons{\text{Asparagine synthetase}} \underset{\text{(amide)}}{\text{Asparagine}} + H_2O + ADP + Pi$

Glutamic or glutamate dehydrogenase and glutamine synthetase are the two primary enzymes involved in ammonia assimilation.

RQ 10.12. *Neither* the host plant *nor* the microorganism by itself can fix nitrogen. Comment.

Ans. The biological fixation of nitrogen in the root nodules of legume plants is controlled by the enzyme **nitrogenase** (now called **dinitrogenase**) that functions under anaerobic conditions. This anaerobic environment within the host cells is kept or maintained by leghaemoglobin (pinkish or reddish pigment).

$$LHb + O_2 \rightarrow LHb \cdot O_2$$

The globin portion of the 'leghaemoglobin' is formed by the host plant in response to infection by the bacteria (*Rhizobium*) and the 'haeme' portion is formed by the bacterial symbiont. Through the symbiotic relationship, the leghaemoglobin would be synthesized and neither the bacterium nor the host is capable of synthesizing leghaemoglobin which is essential for the enzymatic activity of nitrogenase.

RQ 10.13. All nitrogen fixation systems are poisoned by even slight traces of oxygen. How have legume root nodules and cyanobacteria been able to overcome this problem?

Ans. The anaerobic environment at the site of nitrogen fixation is achieved through means such as:

1. Legume root nodule: It is accomplished by leghaemoglobin (LHb), a reddish, iron-containing pigment.

$$LHb + O_2 \rightarrow LHb \cdot O_2$$

2. Nitrogen-fixing blue-green algae: The nitrogenase enzyme is localized in 'heterocysts', with specially thickened, non-photosynthetic, anaerobic cells with multilayered cell walls.

RQ 10.14. In which form is the fixed nitrogen transported from the root cells to different parts of the plant?

Ans. It is mainly in the form of four compounds – glutamic acid, glutamine, aspartic acid, and asparagine that fixed nitrogen is translocated from the root cells to remaining parts of the plant. In nodulated legumes, substituted urea derivatives (ureides such as allantoin, $C_4N_4H_6O_3$; allantoic acid, $C_4N_4H_8O_4$), as well as amides like asparagine ($C_4N_2H_7O_4$) have relatively high C:N ratios and are being transported upward *via* the xylem elements throughout the host plant.

RQ 10.15. Ammonia is toxic to plants cells. How have the plants overcome this problem of toxicity?

Ans. In plant cells, ammonia is not allowed to accumulate in substantial quantities as it reacts with α-ketoglutaric acid and oxaloacetic acid (by-products of Krebs cycle) to form glutamic acid and aspartic acid, and then with the addition of another NH_3 molecule, these are converted to glutamine and asparagine, respectively.

However, at high concentrations of ammonia (**The ammonia effect**), glutamic synthetase is converted to a 'form' that acts on 'nif genes', preventing transcription and ultimate synthesis of nitrogenase. Nitrogenase is synthesized when glutamine synthetase (GS) levels are high and that do not allow ammonia to accumulate. Thus, ammonia accumulation represses the synthesis of nitrogenase or dinitrogenase enzymes.

RQ 10.16. The enzyme nitrogenase needs the co-ordinated expression of several genes of the microsymbiont. Explain.

Ans. Nitrogenase is a large complex enzyme, composed of two major protein components:

1. Component I consists of both Mo and Fe and is called **Mo-Fe protein** (molecular weight of about 220 kD).
2. Component II contains just iron and is called **Fe-protein**. It is smaller with a molecular weight of about 24–36 kD and contains 4 iron atoms.

Components I and II are controlled by distinct genes of microsymbiont. *Neither* component I *nor* component II exhibit nitrogenase activity by themselves. The two together produce nitrogenase that is fully active.

RQ 10.17. List only the aromatic and sulphur-containing amino acids.

Ans.

1. Aromatic or heterocyclic amino acids	Tyrosine, tryptophan, phenylalanine, proline*
2. Sulphur-containing amino acids	Cysteine, cystine**, methionine

* Proline is actually an imino acid.

** Cystine is an oxidized form of cysteine. While homocysteine ($C_4H_9NO_2S$) is a non-protein α-amino acid, and a homologue of the amino acid cysteine. A high level of homocysteine in the blood (hyperhomocysteinemia) is a possible risk factor correlated with the occurrence of blood clots, heart attack and strokes.

RQ 10.18. Discuss the role of trace elements in the process of nitrate reduction to ammonia.

Ans. Nitrogen is principally absorbed by the roots as nitrate[12] (NO_3^-) and after entering the plant it must be reduced to ammonia before it is incorporated into the amino acids – the building blocks of proteins. The reduction of nitrate to ammonia is likely to involve such intermediates as nitrite (NO_2^-), hyponitrite (HNO), hydroxylamine (NH_2OH) and finally to NH_3.

$$NO_3^- \xrightarrow{+2e^-} NO_2^- \xrightarrow{+2e^-} HNO \xrightarrow{+2e^-} NH_2OH \xrightarrow{+2e^-} NH_3$$

These reactions take place in a series of steps and it is presumed that at each step, one pair of electrons is added on forming intermediate compounds. A total of four pairs of electrons are used up.

However, the presence of HNO has not been detected in the tissues of higher plants because it is unstable.

The different reactions can be summarized as follows:

$$NO_3^- + NADPH \xrightarrow[\text{FAD, Mo}]{\text{Nitrate reductase}^{13}} NO_2^- + NADP^+ + H_2O$$

$$NO_2^- + 2NADPH \xrightarrow[\text{Cu, Fe (cofactors)}]{\text{Nitrite reductase}} NH_2OH + 2NADP^+ + H_2O$$

$$NH_2OH + NADH \xrightarrow[\text{Mn as cofactor}]{\text{Hydroxylamine reductase}} NH_3 + NAD^+ + H_2O$$

Ammonia dissolves in water forming NH_4OH which in turn ionizes to form $NH_4^+ + OH^-$ ions.

$$NH_3 + H_2O \rightleftharpoons NH_4^+ + OH^-$$

The overall reduction of nitrate may be expressed by the following equation:

$$NO_3^- + 8e^- + 8H^+ \longrightarrow NH_4 + 2H_2O$$

Nitrate reductase is predominantly confined to the 'cytoplasm' while nitrite reductase is localized exclusively in chloroplasts.

RQ 10.19 What is' one-gene-one-polypeptide hypothesis? Is it tenable now?

Ans. While working on nutritional mutants in *Neurospora*, George Beadle and Edward Tatum of the Stanford University (USA) in 1941 concluded that each gene encodes the structure of enzymes. They called this relationship, 'one-gene-one enzyme hypothesis'. However, this has now been

---

[12] Nitrate from the soil reaches into root epidermal and cortex cells *via* nitrate transporters and is either reduced to ammonia for amino acid production and exported to the shoot mainly as glutamine and asparagine or translocated directly to the shoot as nitrate *via* xylem. Nitrate translocated to the shoot is converted to ammonium and later to amino acids in the chloroplast.

[13] The reduction of nitrate (NO_3^-) to ammonia (NH_3) can be accomplished only by NADPH, which can be generated by the pentose phosphate pathway or shunt. Although ATP is not used-up, the process is expensive energetically as the NADH, NADPH, and ferredoxin are consumed here, and are no longer available for ATP synthesis during oxidative phosphorylation.

modified because many enzymes contain multiple proteins or polypeptide subunits, each encoded by a separate gene, the relationship is today more commonly known as 'one-gene-one polypeptide'.

RQ 10.20 Give the singular contribution of Frederick Sanger.

Ans. The English biochemist Frederick Sanger in 1953 (the same year when Watson and Crick unravelled the structure of DNA) succeeded in determining the complete sequence of amino acids in the insulin, a small protein hormone. It was the first protein to be sequenced.

Chapter 11

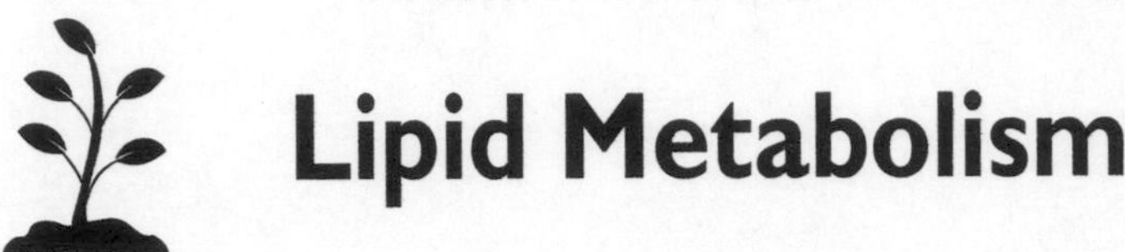

Lipid Metabolism

Keywords

- Acyl carrier protein
- Emulsification
- Fatty acid synthase
- Gluconeogenesis
- Phospholipid
- Triglyceride

Lipids constitute a heterogenous category of cellular components with low water solubility, but higher miscibility in organic solvents like acetone, benzene, chloroform and ether. Lipids play four primary roles in living organisms:

1. They are a constituent of certain structural framework of cells, particularly membrane-bound organelles (for example, glycerophospholipids and sphingolipids).
2. They may be stored within the cells as reserve sources of energy[1] (for example, triacylglycerols).
3. Fatty acids derivatives function as hormones (i.e., prostaglandins) and intracellular second messengers, e.g., diacylglycerol (DAG) and inositol-1,4,5-triphosphate (IP_3).
4. Several proteins are covalently linked by the fatty acids. For example, myristic (C14) and palmitic acid (C16) are directly attached to some proteins.

Like carbohydrates, lipids (fats) consist of only carbon, hydrogen and oxygen but in a varied proportion, with relatively very little oxygen. Lipids can be conveniently grouped into various categories – **triglycerides, phospholipids, sphingolipids, glycolipids, steroids, terpenes** and **waxes**.

Naturally occurring lipids are normally a mixture of **triglycerides** of different fatty acids that are generally configured as shown below:

$$\begin{array}{l} CH_2-COOR_1 \\ | \\ CH-COOR_2 \\ | \\ CH_2-COOR_3 \end{array}$$

[1] The complete oxidation of 1 g of fat or oil (which contains about 40 KJ of energy) can generate more ATP as compared to oxidation of 1 g of starch (with about 15.9 KJ).

R_1, R_2 and R_3 denote the carbon chains of various fatty acids.

11.1 Fatty Acids: Structure and Properties

A fatty acid comprises of a hydrocarbon chain with a terminal carboxyl group. Generally the fatty acids present in a biological system possess an even number of carbon atoms organized in an unbranched chain. Normally the chain length varies from 14 to 24 carbon atoms, with the most commonly occurring fatty acids consisting of 16–18 carbon atoms.

The fatty acids have been grouped into two categories – **saturated** and **unsaturated**. Saturated fatty acid contains all the carbon atoms present in its chain saturated with hydrogen atoms, with the general formula $CH_3(CH_2)_nCOOH$, where n is an even number. **Monounsaturated fatty acids (MUFA)** possess only one double bond in their structure, while **polyunsaturated fatty acids (PUFA)** contain two or more double bonds, all with **cis** configuration. However, during the oxidation of fatty acids, some **trans**–isomers are produced.

Oleate – the Origin of the Word

The term 'oleate' is derived from olive, a source of oleic acid. Triacylglycerols of the olive flesh consist mostly of oleic acid, an18:1 fatty acid. Since 1950s, margarine has been produced by the partial chemical saturation of plant oils, including oleic acid, and several others, such as peanut, soybean, and cotton seed oils. The high heat required for the process causes rearrangement of some *cis* double bonds to the *trans* configuration. Currently eating *trans* fats is associated with cardiovascular disease. More recently, Proctor and Gamble produced the eponymous olestra as a fat substitute. Olestra is not broken down in the intestine by esterases, and it is not incorporated as a dietary lipid, so it is calorie free. However people who consumed this, suffered intestinal disorders arising from undigested lipids passing through their intestinal tract. Olestra also decreases the absorption of fat-soluble vitamins.

The commonly occurring saturated fatty acids are palmitic acid (16:0) and stearic acid (18:0), the former being the most dominant in vegetable fats. The commonly occurring unsaturated fatty acids are oleic (18:1, one double bond), linoleic (18:2, two double bonds) and linolenic acids (18:3, three double bonds) (Table 11.1). The fatty acids characteristics are dependent upon the length of their chain and the number of double bonds. Fatty acids with shorter chain lengths have lower melting point than those with longer chains. Unsaturated fatty acids have lower melting point as compared to saturated fatty acids of the same chain length, while the comparable polyunsaturated fatty acids possess even lower melting temperatures. In fact, the triglycerides of unsaturated fatty acids are liquid (oils) and glycerides of saturated fatty acids with 12 or more carbon atoms are solid (fats) at ambient temperature. Examples of saturated fats are coconut oil, palm oil, palm kernel oil, cocoa butter and shea butter. Coconut oil, liquid in tropical climates, is solid in temperate regions. Triglycerides (Triacylglycerols) are synthesized by ester linkages between glycerol and three long-chain fatty acids (Figure 11.1).

Table 11.1 Important fatty acids with their empirical formulae.

Saturated acids		Unsaturated acids	
Acid	Empirical formula	Acid	Empirical formula
Capric	$C_{10}H_{20}O_2$	Oleic	$C_{18}H_{34}O_2$
Lauric	$C_{12}H_{24}O_2$	Linoleic	$C_{18}H_{32}O_2$
Myristic	$C_{14}H_{28}O_2$	Linolenic	$C_{18}H_{30}O_2$
Palmitic	$C_{16}H_{32}O_2$	Ricinoleic	$C_{18}H_{34}O_3$
Stearic	$C_{18}H_{36}O_2$	Erucic	$C_{22}H_{42}O_2$

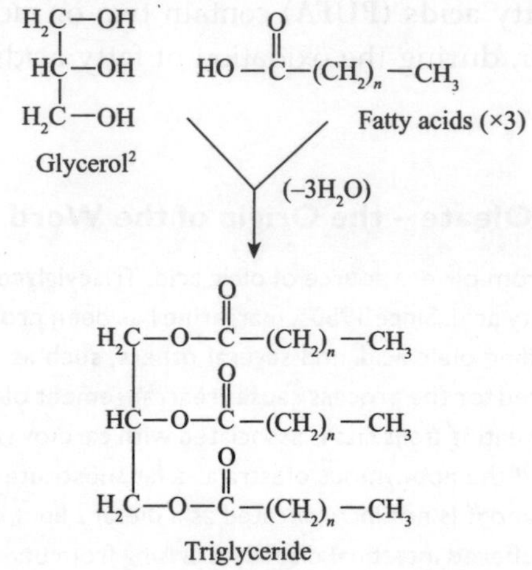

$H_2C—OH$
$HC—OH$ $HO—C(=O)—(CH_2)_n—CH_3$
$H_2C—OH$ Fatty acids (×3)
Glycerol[2]

$(-3H_2O)$

$H_2C—O—C(=O)—(CH_2)_n—CH_3$
$HC—O—C(=O)—(CH_2)_n—CH_3$
$H_2C—O—C(=O)—(CH_2)_n—CH_3$
Triglyceride

Fig. 11.1 The structure of a triglyceride molecule.

Triglycerides are hydrolyzed by c-AMP regulated lipase.

$$\text{Triglyceride} + 3H_2O \xrightarrow{\text{Lipase}} \text{Glycerol} + \text{Fatty acids (3)}$$

The site of fatty acids oxidation is the mitochondria, whereas glycerol undergoes phosphorylation in cytosol followed by oxidation to dihydroxyacetone phosphate which in turn is converted to its isomeric form, glyceraldehyde-3-phosphate.

[2] $H_2C—O\,NO_2$ / $HC—O\,NO_2$ / $H_2C—O\,NO_2$ The hydroxyl groups of glycerol on nitrosylation result in nitroglycerine – a compound that has both the explosive power to disintegrate rocks and the curative power to control blood vessel collapse in human beings. The explosion is caused by nitrogen (N_2) gas released during oxidation of nitroglycerine. However, nitroglycerine in the body is broken down to nitric oxide (NO) which relaxes and dilates blood vessels, helping in increased flow of blood.

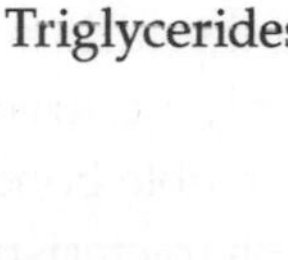

$CH_3-(CH_2)_{16}-C(=O)O^-$ **Stearic acid**

Hydrophobic Hydrophilic

$CH_3-(CH_2)_7-CH{=}CH-(CH_2)_7-C(=O)O^-$ **Oleic acid**

Saturated region (CH₃–(CH₂)₇) Saturated region ((CH₂)₇)

Hydrophobic Hydrophilic

$CH_3-(CH_2)_4-CH{=}CH-CH_2-CH{=}CH-(CH_2)_7-C(=O)O^-$ **Linoleic acid**

Saturated region (CH₃–(CH₂)₄) Saturated region ((CH₂)₇)

Hydrophobic Hydrophilic

The saturated fatty acid' stearic acid and the unsaturated fatty acids, oleic and linoleic acid.

Phospholipids refer to the lipids that have one or more phosphate groups. These are critical to the bilayered structure found in all membranes. Phospholipids contain a glycerol molecule esterified to two molecules of fatty acids, whereas the third hydroxyl group belonging to glycerol is occupied by a phosphate group and not the fatty acid. The length and extent of unsaturation of fatty acid chains in membrane phospholipids profoundly affect membrane fluidity and, in fact, can be regulated by the cell to control this crucial membrane property.

Phospholipids are modified triglycerides that are components of cellular membranes.

Sphingolipids are derived from sphingosine, an amino alcohol possessing a long, unsaturated hydrocarbon chain and are confined to membranes of animal cells.

Glycolipids (lipids attached to short chains of sugars) occur principally in the plasma membrane and in the photosynthetic tissues. These play vital roles in immunity, blood group specificity and cell–cell recognition.

Steroids share the common property of being non-polar and therefore hydrophobic, with the other three categories of lipids. Steroids stabilize cell membranes and also function as hormones. The most commonly occurring steroid in animals is ***cholesterol***, found primarily in the membranes. Cholesterol is also the precursor of the five major classes of steroid hormones, which include progesterone, estrogens, androgens, glucocorticoids and mineralocorticoids.

Terpenes possess certain fat-soluble vitamins (such as vitamins A, E and K), carotenoids and coenzymes (e.g., coenzyme Q, ubiquinone). Terpenes are synthesized from multiple numbers of a 5-C building block known as isoprene unit. The monoterpenes impart the characteristic odours and flavouring properties to plants (e.g., geraniol from geraniums, menthol from mint and limoneme from lemons).

Cutin and Suberin are unique lipids which are important structural components of cell walls of many plants. ***Waxes***[3], the most water-repellent of the lipids, are rarely found within living plant cells. These are the esters of fatty acids and long chain monohydroxy alcohols. Waxes occur predominantly as the protective coatings on the surfaces of leaves, stems, fruits and seeds and greatly limit the water loss through transpiration. Cutin, suberin and waxes are lipids that form barriers to water loss.

Lecithin

Lecithin, greek word for egg yolk, is commonly used by biochemists as synonym for phosphatidylcholine. In the food and pharmaceutical industries, a mixture of phospholipids extracted from animals or plants-largely containing phosphatidylcholine-is known as lecithin. Two common materials are 'egg yolk lecithin' and 'soybean lecithin'. While phosphatidylcholine is the major component of these materials, they contain other phospholipids (e.g. phosphatidylethanol amine) as well as lysophospholipids (i.e. phospholipids with just one fatty acid ester). **Lecithins are commonly used for the purposes of emulsification, for example, in drug delivery.**

11.2 Fatty Acid Oxidation

It was noticed even before the end of the nineteenth century that fatty acids are synthesized and oxidized in the cell by addition or subtraction of 2-C fragments, as most natural fatty acids possess an even number of carbon atoms. In 1904, Franz Knoop made a significant contribution and first elucidated the mechanism of fatty acid oxidation in plants. He postulated that fatty acids are degraded through oxidation at the β-carbon atom and the resulting 2-carbon fragments are successively released as acetyl CoA.

11.2.1 β-Oxidation

Fatty acid oxidation takes place in the cytoplasm of prokaryotic cells. In seed storage tissues of plants, the enzymes involved in *β*-oxidation are located exclusively in the glyoxysome or the peroxisome. In 1949, Eugene Kennedy and Albert Lehninger discovered that fatty acids are degraded in mitochondrial matrix of all other eukaryotes.

Activation of fatty acids: Subsequent studies proved that fatty acids undergo activation before they reach the mitochondrial matrix and the reaction occurs on the outer membrane of the mitochondria, where it is catalyzed by the enzyme **acyl CoA synthase** (fatty acid thiokinase).

$$\underset{\text{Fatty acid}}{R{-}C(=O){-}O^-} + ATP + HS{-}CoA \rightleftharpoons \underset{\text{Fatty acyl CoA}}{R{-}\overset{\overset{O}{\|}}{C}{-}S{-}CoA} + AMP + PPi$$

[3] Carnauba wax, used for car and floor polishes, is harvested from the leaves of the carnauba wax palm (*Copernicia cerifera*) from Brazil.

Paul Berg showed that the activation of a fatty acid occurs in two steps:

$$R{-}C(=O){-}O^- + ATP \rightleftharpoons R{-}C(=O){-}AMP + PPi$$

$$R{-}C(=O){-}AMP + HS{-}CoA \rightleftharpoons R{-}C(=O){-}S{-}CoA + AMP$$

Pyrophosphate, PPi is rapidly hydrolysed by an enzyme pyrophosphatase

$$PPi + H_2O \rightarrow 2Pi$$

In most of the biosynthetic reactions liberating PPi, pyrophosphatase activity makes the reaction more favorable energetically, thus rendering these reactions irreversible *via* hydrolysis of inorganic pyrophosphate.

Transport into mitochondria: Long-chain acyl CoA (more than 10 carbon atoms) do not quickly traverse the inner membrane of the mitochondria, but rather are transported across the inner mitochondrial membrane by **carnitine** in the form of acyl carnitine regulated by the enzyme carnitine acyl transferase I, confined to the cytosolic side of the inner mitochondrial membrane. Acyl carnitine then moves across the inner mitochondrial membrane through a translocase protein. The acyl group is transferred back to CoA, thereby forming acyl CoA towards the matrix face of the membrane catalyzed by enzyme acyl transferase II. Finally, carnitine goes back towards the cytosolic face through the translocase in exchange for an incoming acyl carnitine (Figure 11.2).

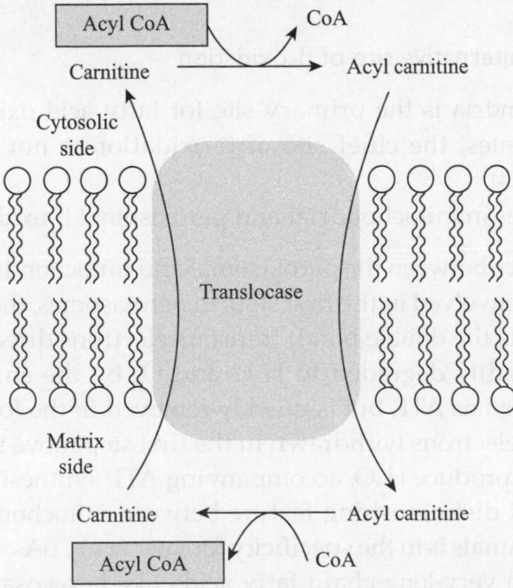

Fig. 11.2 The entry of acyl carnitine into the mitochondrial matrix is mediated by a translocase. Carnitine comes back to the cytosolic face of the inner mitochondrial membrane in exchange for acyl carnitine.

Degradation of fatty acids: A saturated acyl CoA is broken down in a sequential manner through four-step reactions: (i) oxidation by FAD (ii) hydration (iii) oxidation by NAD^+ and, (iv) thiolysis by CoA. The fatty acyl chain is cut short by two carbon atoms as a consequence of these reactions, and FADH, NADH and acetyl CoA are regenerated. David Green, Severo Ochoa and Feodor Lynen (1953) made important contributions to the elucidation of reactions of **β-oxidation pathway** (Figure 11.3).

The resulting fatty acid can be subjected once again to the similar degenerative pathway to obtain chunks of 2-C fragments as acetyl CoA. The acetyl CoA can then be oxidized in the Krebs cycle to CO_2 or else be used for the synthesis of fatty acids, steroids, isoprenoids, terpenoids, gibberellic acid, abscisic acid, etc. Overall, 3 ATP molecules are synthesized by 1 NADH, 2 ATP from 1 FADH and 12 ATP are generated by oxidation of acetyl CoA in the Krebs cycle.

The overall equation for palmitoyl CoA is given as below:

$$\text{Palmitoyl CoA} + 7\,\text{FAD} + 7\,\text{NAD}^+ + 7\,\text{CoA} + 7\,\text{H}_2\text{O} \rightarrow 8\,\text{Acetyl CoA} + 7\,\text{FADH}_2 + 7\,\text{NADH} + 7\,\text{H}^+$$

Thus, the net gain of ATP from the breakdown of palmitate is 129 (131 ATP produced – 2 ATP consumed for the activation of palmitate).

Regulation of β-oxidation pathway: The chief regulatory point of β-oxidation pathway is the availability of fatty acids. The primary source of free fatty acids in the bloodstream is from the degradation of triacylglycerol reserves in adipose tissue.This reaction is driven by triacylglycerol lipase which shows sensitivity toward hormones.The breakdown as well as the synthesis of fatty acids are coordinated in a controlled way to prevent a futile cycle.

11.2.1.1 Peroxisomes – alternative site of β-oxidation

The matrix of mitochondria is the primary site for fatty acid oxidation in animal cells. However, in plant tissues, the chief site of β-oxidation is not mitochondria but the peroxisomes (Figure 11.4).

Two major differences in mitochondrial and peroxisomal β-oxidation:

(i) The major difference between the peroxisomal and mitochondrial oxidation pathways is in the chemistry involved in the first step. In peroxisomes, the flavoprotein acyl CoA oxidase (that inserts the double bond) transfers electrons directly to oxygen, forming H_2O_2, which is readily degraded to H_2O and O_2 by the enzyme catalase. Energy released is not stored as ATP, but is quickly removed in the form of heat. Whereas in mitochondria, the electrons withdrawn in the first step move through the respiratory chain to oxygen to produce H_2O, accompanying ATP synthesis.

(ii) Another important distinguishing feature between mitochondrial and peroxisomal β-oxidation in mammals is in the specificity for fatty acyl CoAs; the peroxisomal system is more reactive on very-long-chain fatty acids like hexacosanoic acid (26:0) and on branched-chain fatty acids like phytanic acid and pristanic acid. Their degradation in the peroxisome makes use of a variety of auxiliary enzymes, unique to this organelle.

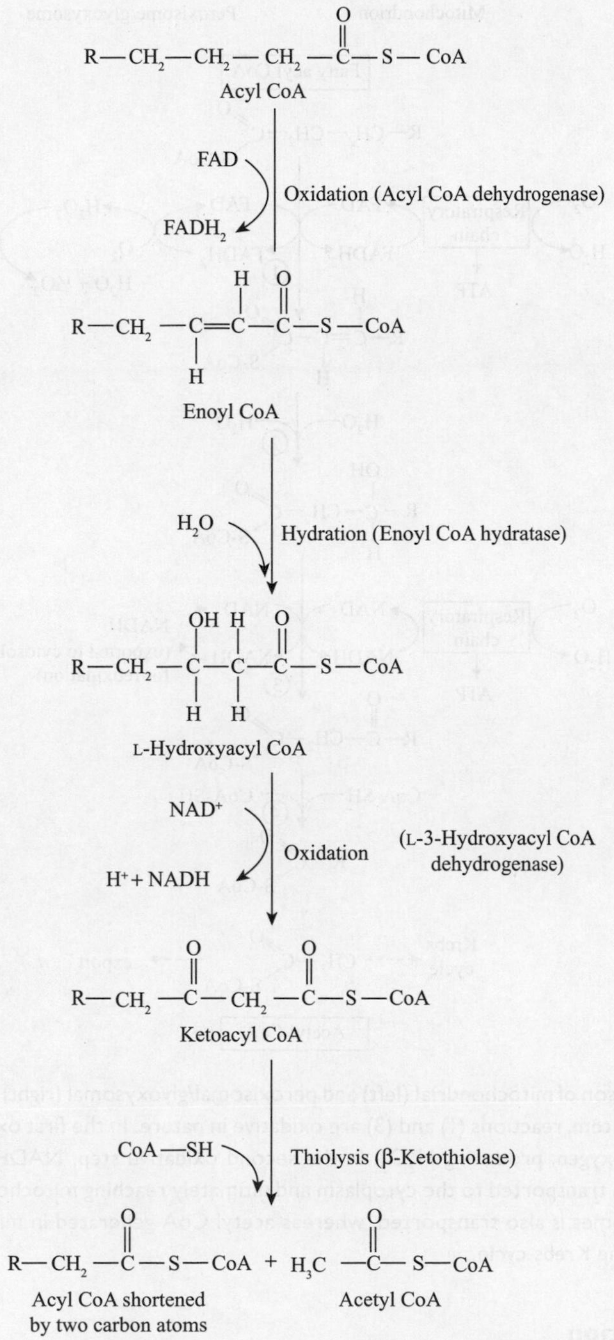

$R—CH_2—CH_2—CH_2—C(=O)—S—CoA$

Acyl CoA

FAD → $FADH_2$ Oxidation (Acyl CoA dehydrogenase)

$R—CH_2—C(H)=C(H)—C(=O)—S—CoA$

Enoyl CoA

H_2O Hydration (Enoyl CoA hydratase)

$R—CH_2—C(OH)(H)—C(H)(H)—C(=O)—S—CoA$

L-Hydroxyacyl CoA

NAD^+ → $H^+ + NADH$ Oxidation (L-3-Hydroxyacyl CoA dehydrogenase)

$R—CH_2—C(=O)—CH_2—C(=O)—S—CoA$

Ketoacyl CoA

CoA—SH Thiolysis (β-Ketothiolase)

$R—CH_2—C(=O)—S—CoA + H_3C—C(=O)—S—CoA$

Acyl CoA shortened by two carbon atoms

Acetyl CoA

Fig. 11.3 Sequence of reactions during breakdown (β-oxidation) of fatty acids involving oxidation, hydration, oxidation and thiolysis.

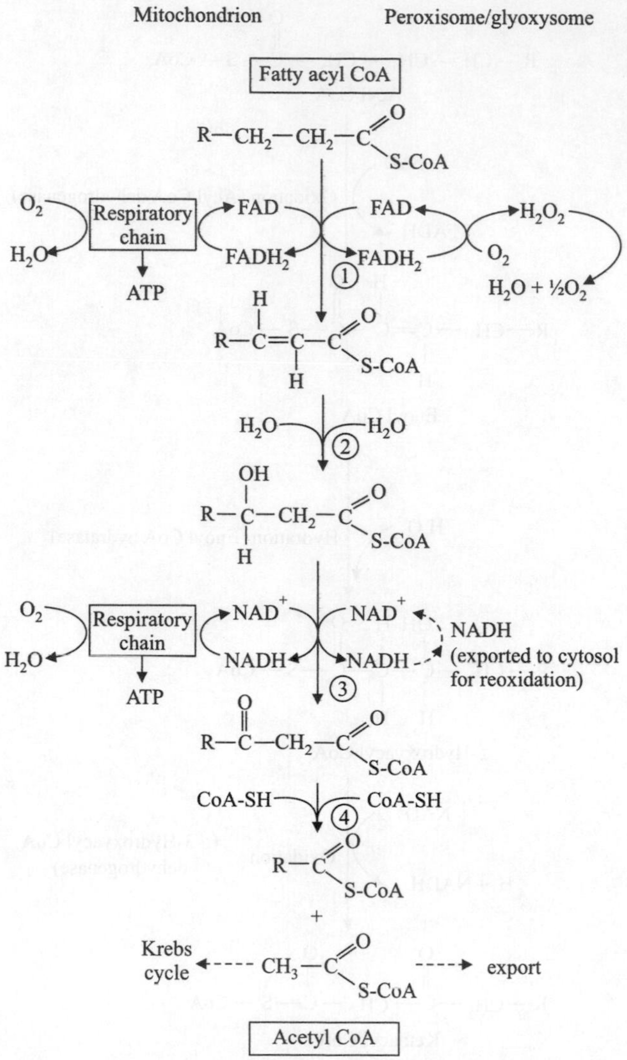

Fig. 11.4 A comparison of mitochondrial (left) and peroxisomal/glyoxysomal (right) β-oxidation pathways. In the peroxisomal system, reactions (1) and (3) are oxidative in nature. In the first oxidative step, electrons traverse directly to oxygen, producing H_2O_2. In the second oxidative step, NADH produced cannot be reoxidized, and thus is transported to the cytoplasm and ultimately reaching mitochondria. The acetyl CoA generated in peroxisomes is also transported, whereas acetyl CoA generated in mitochondria undergoes further oxidation in the Krebs cycle.

11.2.2 α-Oxidation

In germinating plant seeds, a special pathway of fatty acid oxidation occurs, referred to as α-oxidation (Stumpf, 1970) in which the carboxyl carbon of the fatty acids is released in the form of CO_2 and the α-carbon atom is oxidized to an aldehydic group at the cost of hydrogen peroxide (Figure 11.5). This reaction is catalyzed by fatty acid peroxidase. The hydrogen peroxide required is supplied *via* direct oxidation of reduced flavoproteins by O_2. The fatty aldehyde formed is oxidized to the corresponding carboxylic acid. This two-enzyme reaction sequence is then repeated on the shortened free fatty acid. As the fatty acid peroxidase attacks only fatty acids having 13–18 carbon atoms, this pathway cannot lead to a complete breakdown of long chain fatty acids. The aldehydes produced by α-oxidation may alternatively undergo reduction to yield long-chain alcohols, which occur in large amounts in plant waxes.

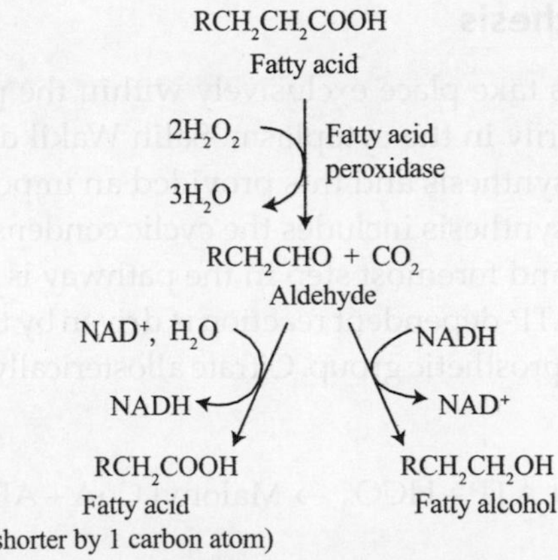

Fig. 11.5 The α-oxidation pathway in plants.

Fatty acids are synthesized and degraded by different mechanisms. Some of the important features of the biosynthetic pathway of fatty acids are:

1. Synthesis occurs in the cytoplasm, as compared to degradation which takes place in matrix of the mitochondria.
2. Intermediate products of fatty acid synthesis are covalently attached to the–SH groups of an acyl carrier protein (ACP), while the intermediates in fatty acid degradation are linked to coenzyme A.
3. The enzymes involved in fatty acid synthesis in higher plants are joined in a single polypeptide chain known as 'fatty acid synthase'. The degrading enzymes do not appear to be linked.
4. The growing fatty acid chain increases by the addition of 2-C units, derived from acetyl CoA, in a sequential manner. The activated donor of 2-C units is malonyl-ACP in the elongation step, powered by the evolution of carbon dioxide.
5. The reducing agent in fatty acid synthesis is NADPH.
6. Elongation caused by the fatty acid synthase complex comes to a halt upon synthesis of palmitate (C_{16}). Further elongation and the incorporation of double bonds are achieved by other enzyme systems.

It is believed that α-oxidation pathway is a method to synthesize acids with an odd number of carbon atoms. For instance, if a 18-carbon fatty acid is broken down *via* α-oxidation to a 13-carbon fatty acid, which then undergoes β-oxidation in order to produce 3-carbon propionic acid. Moreover, α-oxidation may function as a bypass of a substitution group that could prevent the operation of the β-oxidation sequence.

The breakdown products of certain membrane lipids, such as jasmonic acid (derived from linolenic acid), can function as signalling molecules in plant cells and also regulates the development of anthers and pollen during plant growth.

11.3 Fatty Acid Synthesis

The synthesis of fatty acids take place exclusively within the plastids in plants, whereas in animals, it occurs primarily in the cytoplasm. Salih Wakil discovered that bicarbonate is needed for fatty acid biosynthesis and this provided an important clue for determining the pathway. Fatty acid biosynthesis includes the cyclic condensation of 2-C units, derived from acetyl CoA. The first and foremost step in the pathway is the carboxylation of acetyl CoA to malonyl CoA. This ATP-dependent reaction is driven by the **acetyl CoA carboxylase** enzyme containing a biotin prosthetic group. Citrate allosterically stimulates this committed step in fatty acid synthesis.

$$\text{Acetyl CoA} + \text{ATP} + \text{HCO}_3^- \rightarrow \text{Malonyl CoA} + \text{ADP} + \text{Pi} + \text{H}^+$$

This reaction is completed in two stages:

1. $\text{Biotin-enzyme} + \text{ATP} + \text{HCO}_3^- \rightleftarrows \text{CO}_2\sim \text{Biotin-enzyme} + \text{ADP} + \text{Pi}$
2. $\text{CO}_2 \sim \text{biotin-enzyme} + \text{acetyl CoA} \rightarrow \text{Malonyl CoA} + \text{Biotin-enzyme}$

Substrates are attached to acetyl CoA carboxylase enzyme and the end-products are given out in a specific sequence.

Roy Vagelos (1963) demonstrated that the intermediate compounds in fatty acid synthesis in *E. coli* are joined to an acyl carrier protein. The enzyme system that regulates the production of saturated long fatty acid chain from acetyl CoA, malonyl CoA and NADPH is referred to as **fatty acid synthase.**

The fatty acid synthases of eukaryotic cells, as compared with those of *E. coli* are well-defined **'multienzyme complexes'** (refer to Figure 6.5). The one from yeast has a mass of 2200 kD and is ellipsoid with a length of 250 Å and a cross-sectional diameter of 210 Å. It is composed of two types of polypeptide chains with the subunit composition $\alpha_6\beta_6$. The α chain (185 kD) has the acyl carrier protein, 3-Ketoacyl-ACP synthase (condensing enzyme) and the 3-Ketoacyl-ACP reductase, whereas the β chain (175 kD) contains four enzymes such as acetyl transacylase, malonyl transacylase, 3-Hydroxyacyl-ACP dehydratase, and enoyl-ACP reductase. The flexible phosphopantetheinyl unit of ACP moves the substrate from one active site to another in this complex.

Acetyl-ACP is produced from acetyl CoA, while malonyl-ACP is generated from malonyl CoA. Acetyl-ACP and malonyl-ACP undergo condensation to form acetoacetyl-ACP, accompanied by CO_2 evolution from the activated malonyl unit. This is followed by reduction, dehydration and second reduction steps using NADPH as the reductant. The butyryl-ACP so produed takes part in a second round of elongation, beginning with the addition of a 2-C segment from malonyl-ACP (Figure 11.6).

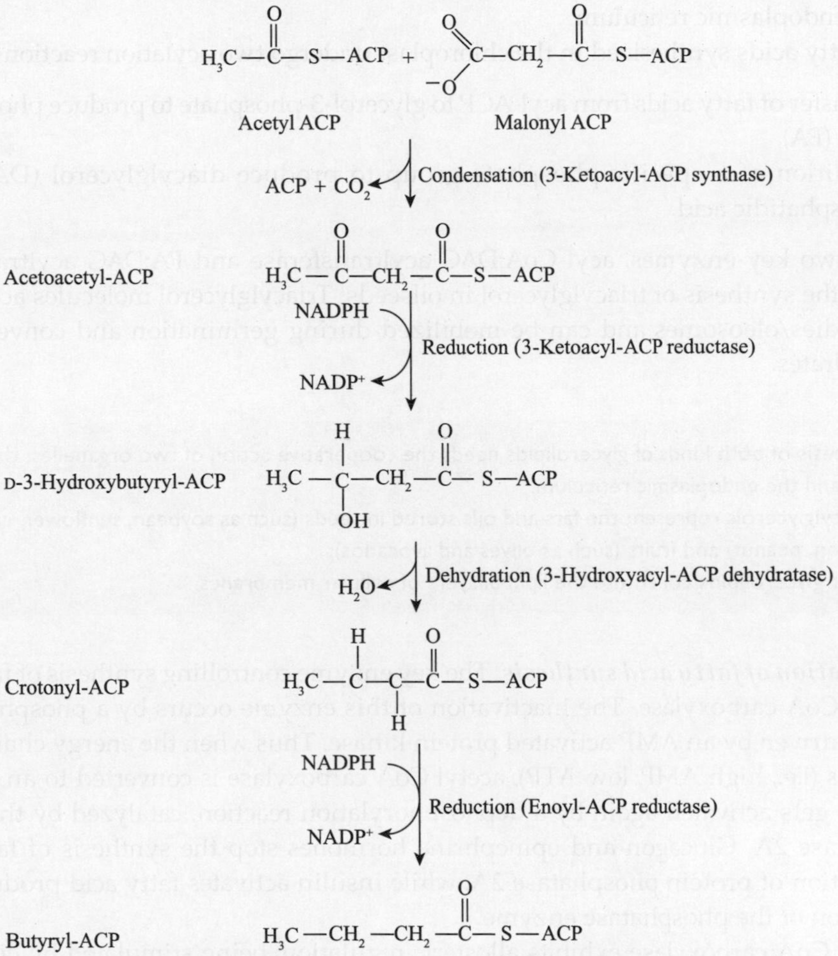

$$H_3C-\overset{O}{\overset{\|}{C}}-S-ACP \quad + \quad {}^{-}O-\overset{O}{\overset{\|}{C}}-CH_2-\overset{O}{\overset{\|}{C}}-S-ACP$$

Acetyl ACP   Malonyl ACP

↓ Condensation (3-Ketoacyl-ACP synthase) → $ACP + CO_2$

Acetoacetyl-ACP $\quad H_3C-\overset{O}{\overset{\|}{C}}-CH_2-\overset{O}{\overset{\|}{C}}-S-ACP$

↓ Reduction (3-Ketoacyl-ACP reductase): NADPH → $NADP^+$

D-3-Hydroxybutyryl-ACP $\quad H_3C-\underset{OH}{\overset{H}{C}}-CH_2-\overset{O}{\overset{\|}{C}}-S-ACP$

↓ Dehydration (3-Hydroxyacyl-ACP dehydratase) → H_2O

Crotonyl-ACP $\quad H_3C-\overset{H}{C}=\underset{H}{C}-\overset{O}{\overset{\|}{C}}-S-ACP$

↓ Reduction (Enoyl-ACP reductase): NADPH → $NADP^+$

Butyryl-ACP $\quad H_3C-CH_2-CH_2-\overset{O}{\overset{\|}{C}}-S-ACP$

Fig. 11.6 Reaction sequence during the synthesis of long-chain fatty acids, involve 4 steps: condensation, reduction, dehydration and reduction.

Seven rounds of elongation ultimately produce palmitoyl-ACP, which is then hydrolyzed to palmitate.

$$8 \text{ Acetyl CoA} + 7\text{ATP} + 14 \text{ NADPH} \rightarrow \text{Palmitate} + 14 \text{ NADP}^+$$
$$+ 8 \text{ CoA} + 6 H_2O + 7\text{ADP} + 7 \text{ Pi}$$

Palmitic acid (16:0–ACP) is set free from the fatty acid synthase complex, but most of the molecules that are elongated to stearic acid (18:0–ACP) are efficiently converted into oleic acid (18:1–ACP) by a **desaturase** enzyme. Additional double bonds are added in oleic acid

(18:1) through a series of desaturase isozymes to generate linoleic acid (18:2) and linolenic acid (18:3). Desaturase isozymes, the integral membrane proteins, are confined to the chloroplast and the endoplasmic reticulum.

Synthesis of triglycerides (triacylglycerol/glycerolipids): This occurs in the plastids and the endoplasmic reticulum.

The fatty acids synthesized in the chloroplast undergo two acylation reactions:

(i) Transfer of fatty acids from acyl-ACP to glycerol-3-phosphate to produce phosphatidic acid (PA)
(ii) Addition of a specific phosphate group to produce diacylglycerol (DAG) from phosphatidic acid.

The two key enzymes, acyl-CoA:DAG acyltransferase and PA:DAG acyltransferase, regulate the synthesis of triacylglycerol in oilseeds. Triacylglycerol molecules accumulate in oil bodies/oleosomes and can be mobilized during germination and converted into carbohydrates.

Biosynthesis of both kinds of glycerolipids needs the cooperative action of two organelles: the plastids and the endoplasmic reticulum.

i. Triacylglycerols represent the fats and oils stored in seeds (such as soybean, sunflower, canola, cotton, peanut) and fruits (such as olives and avocados).
ii. Polar glycerolipids constitute the lipid bilayers of cellular membranes.

Regulation of fatty acid synthesis: The key enzyme controlling synthesis of fatty acids is acetyl CoA carboxylase. The inactivation of this enzyme occurs by a phosphorylation reaction, driven by an AMP-activated protein kinase. Thus when the energy charge of the cell is less (i.e., high AMP, low ATP), acetyl CoA carboxylase is converted to an 'inactive' state and gets activated again by a dephosphorylation reaction, catalyzed by the protein phosphatase 2A. Glucagon and epinephrine hormones stop the synthesis of fatty acids by inhibition of protein phosphatase 2A, while insulin activates fatty acid production by stimulation of the phosphatase enzyme.

Acetyl CoA carboxylase exhibits allosteric regulation; being stimulated by citrate and inhibited by palmitoyl CoA.

11.4 Lipid–Sugar Conversion: The Glyoxylate Pathway

During germination of seeds that store lipids, such as castor bean (Figure 11.7), linseed, cotton and cucurbits, stored triglycerides are mobilized to produce sucrose. Sucrose is then utilized as a source of energy for growth and development processes. The primary purpose of the glyoxylate pathway, first delineated by Krebs and Kornberg in *Pseudomonas*, is to allow plants and microbes to consume fatty acids or acetate in the form of acetyl CoA, as the only carbon source, especially for the synthesis of carbohydrates from fatty acids. This pathway was first studied in higher plants by H. Beevers (1967).

Fig. 11.7 Castor bean (*Ricinus communis*) plant showing palmate leaves, male flowers and warty capsular fruits that enclose within them, seeds characterized by mottled testa.

The stored triglycerides are mobilized first by lipase activity, producing glycerol and free fatty acids. The free fatty acids are subsequently oxidized to acetate through β-oxidation. Acetate by condensation with glyoxylate molecule is first converted to malate through a complex series of reactions known as the **glyoxylate pathway**. The malate is then converted to phosphoenolpyruvate (PEP) through oxaloacetate and ultimately to hexoses by reversal of the glycolytic process and then to sucrose. The sequential reactions of β-oxidation and the glyoxylate pathway take place within the seed in the single membrane-bound microbody, the glyoxysomes.

The glyoxylate pathway is depicted in Figure 11.8. Acetyl CoA first undergoes condensation with oxaloacetate (OAA) to produce citrate which is then transformed into isocitrate by the action of aconitase as it happens in the Krebs cycle. However, the breakdown of isocitrate occurs by a pathway wherein three reactions of the Krebs cycle are bypassed. Isocitrate is first cleaved by 'isocitrate lyase' to form succinate and glyoxylate. Glyoxylate combines with another molecule of acetyl CoA to produce malate, catalyzed by 'malate synthase'. The malate then gets oxidized to regenerate oxaloacetate by the action of a malate dehydrogenase enzyme. The oxaloacetate may undergo condensation with acetyl CoA to begin another round of the glyoxylate pathway.[4]

[4] Other important cycles in plants that similarly regenerate the primary acceptor molecules are the Krebs cycle that occurs in mitochondria during aerobic respiration and the Calvin-Benson cycle that takes place in chloroplasts during carbon fixation in photosynthesis.

The succinate produced moves out of the glyoxysomes into the mitochondria where it is converted through the conventional Krebs cycle reactions to oxaloacetate.

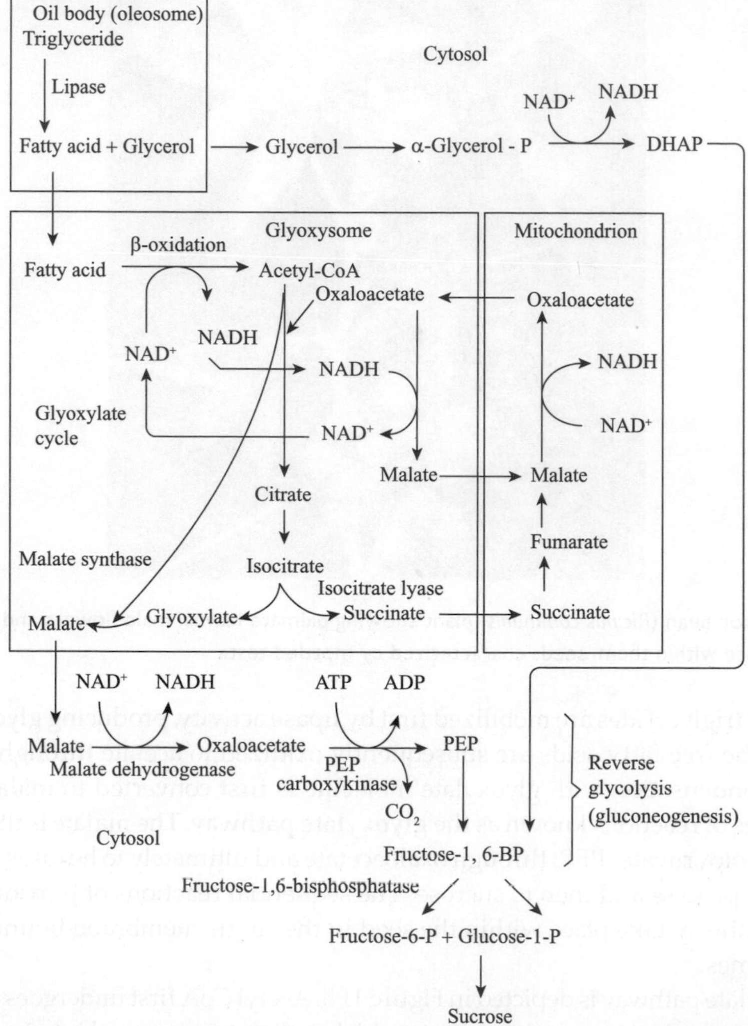

Fig. 11.8 Lipid catabolism, the glyoxylate cycle and gluconeogenesis.

Conversion of oxaloacetate to phosphoenolpyruvate (PEP), other glycolytic intermediates and ultimately to sucrose occurs in the cytosol by reversal of glycolysis (gluconeogenesis) utilizing ATP as a source of energy. The glyoxylate cycle is under allosteric regulation. Isocitrate lyase is strongly inhibited by PEP, a key intermediate in the biosynthesis of glucose from lipids.

Thus, the glyoxylate pathway is important to the germination and growth process of seedlings because it provides readily available carbohydrates from lipids.

The overall equation of the glyoxylate pathway:

$$2\text{Acetyl CoA} + \text{NAD}^+ + 2\text{H}_2\text{O} \rightarrow \text{Succinate} + 2\text{CoA} + \text{NADH} + \text{H}^+$$

Isocitrate lyase and malate synthase[5] are inducible enzymes: they are synthesized by plant cells only when they are needed.

11.5 Genetic Engineering of Lipids

1. ***Improvement of oil quality is a prime objective of agronomists:*** Rapeseed and many other species of the Brassicaceae family consist of huge amounts of long-chain (20–24 C) monounsaturated fatty acids.Although high in erucic acid (22:1) content, approximately 50 per cent of the fatty acids in rapeseed oil, made it more user-friendly for various industrial uses, but it prevented its widespread use as an edible crop. Extensive research was done in 1950s to produce cultivars of *Brassica napus* with low erucic acid and glycosinolate content, now known as **canola**.
2. ***Improvement of edible oils by metabolic engineering***: The 20- and 22-carbon omega-3 fatty acids in fish oils have health benefits that include likely reductions in heart disease, stroke, and other metabolic syndromes. Production of these fatty acids in seed oils is an attractive goal in oilseed engineering and has been achieved by the transgenic expression of a condensing enzyme (3-ketoacyl-ACP synthase) to catalyze fatty acid elongation. Many attempts have been done to develop monounsaturated oils i.e. to reduce the proportions of 16:0, 18:0, 18.2 and 18:3 in favour of 18:1. Further, oils low in 18:2 and 18:3 are much more stable, particularly in high-temperature food frying. Oil from genetically engineered soybean variety is similar in composition to olive oil and has improved health benefits due to reduction in omega-6 fatty acids and improved oxidative stability.
3. ***Increase in seed oil productivity through molecular genetic approaches:*** Expression of diacylglycerol acyltransferase and overexpression of the WRINKLED 1 transcription factors result in increased oil content in soybean or rapeseed . Similarly if oil palm (*Elaeis* spp.), which produces 4000 litres of oil per hectare per year with low inputs of agrochemicals, can be genetically engineered, it may be viable to produce biodiesel from a renewable source at a cost that is competitive with petroleum.
4. ***Industrial applications of fatty acids***: A few unusual fatty acids present in coconut, palm, rapeseed, castor, flax, jojoba, and soybean have non-food industrial applications such as production of soaps, detergents, paints, varnishes, lubricants, adhesives, plastics and so on.

Ricinoleic acid produced by castor bean is an extremely versatile natural product with industrial applications such as synthesis of nylon-11, hydraulic fluids, plastics, cosmetics and so much more. However, castor bean contains **ricin**, an extremely toxic lectin, as well as other poisons and allergens. Because of poorer yields, castor bean is a minor crop grown mainly in non-industrial states. Overexpression of a glycerolipid hydroxylase enzyme from castor bean can promote the synthesis of ricinoleic acid in transgenic plants. The successful

[5] Higher animals do not have isocitrate lyase and malate synthase enzymes. So, they are not able to transform acetyl CoA into new glucose molecules.

cloning of a gene that encodes the hydroxylase, responsible for 12-OH-18 : $1^{\Delta 9}$ synthesis from 18:$1^{\Delta 9}$ relied on a detailed knowledge of the biochemistry involved. As an alternative, the isolation of key genes directing the synthesis of a chain of fatty acid can provide a suitable means to genetically engineer agronomically suitable oilseed crops to produce the desired oil more easily and cheaply.

Similarly high lauric-content rapeseed can be successfully used in oilseed engineering.

Review Questions

RQ 11.1. **Assuming that one molecule of fatty acid containing 16 carbons (e.g., palmitic acid $C_{16}H_{32}O_2$) undergoes β-oxidation; then how many (i) acetyl CoAs, (ii) NADHs and (iii) $FADH_2$s will be produced? Calculate the number of ATPs produced.**

Ans.

		ATP
1. Acetyl CoAs	= 8	8 × 12 = 96
2. NADHs	= 7	7 × 3 = 21
3. $FADH_2$s	= 7	7 × 2 = 14
ATP (used for Priming)	= 1 or 2	
Total number of ATP produced	= 131 – 1 or 2 = 130 or 129 ATPs	

RQ11.2. **In the Krebs cycle, how many ATPs are produced per acetyl CoA used? Explain.**

Ans. The total ATPs produced are 12.

3 NADH	(3 × 3 ATP)	= 9
1 $FADH_2$	(1 × 2 ATP)	= 2
Through substrate-linked phosphorylation	1 ATP	= 1
	Total	= 12 ATPs

RQ 11.3. **Why is the breakdown of fatty acids known as β-oxidation? How does it differ from α-oxidation?**

Ans. It is 'β' carbon (second from functional group or the carboxylic acid group) that becomes oxidized, hence known as 'β-Oxidation'. The fatty acids are degraded stepwise to chunks of acetyl CoA (2-C fragments) that enters the Krebs cycle, thereby making the fatty acid chain shorter by two carbon atoms.

Under certain conditions or circumstances, the long-chain fatty acids can undergo oxidation to produce carbon dioxide, reducing the fatty acid chain length by one carbon atom (α-oxidation). The details of this degradation pathway were earlier worked out by Stumpf (1970).

RQ 11.4. **Why do we consume nuts rich in fats during winter months?**

Ans. Nuts such as peanut, almond, cashew or walnut are rich in fats, i.e., they are much more reduced than carbohydrates. More oxygen is used during their respiration than that needed for the respiration of carbohydrates. Consequently, more energy is released after consumption providing greater warmth to our bodies.

Oxygen requirements for tripalmitin and glucose:

$$\text{Tripalmitin: } 2C_{51}H_{98}O_6 + 145O_2 \rightarrow 102CO_2 + 98H_2O$$
$$C_6H_{12}O_6 + 6O_2 \rightarrow 6CO_2 + 6H_2O$$

Fats contain relatively little oxygen as compared to carbohydrates. They require greater quantities of oxygen for complete oxidation into CO_2 and H_2O.

RQ 11.5. Name the sites where fatty acids and fats are synthesized.

Ans. **1.** The principal site for fatty acid and fat synthesis is the smooth or agranular membranes of the smooth endoplasmic reticulum (SER) and microsomes. The different enzymes are present in the cytoplasmic matrix.

2. The second site is provided by the mitochondria.

3. The biosynthesis of fatty acids also occurs in 'chloroplasts stroma' and other plastids in the cytoplasm.

RQ 11.6. During the maturation of oilseeds, the increase in fat content is followed by a proportionate decrease in the carbohydrate content. Explain.

Ans. Fats are synthesized in the plant cells by the condensation of glycerol (1molecule) with fatty acids (3 molecules). Glycerol and fatty acids are obtained from carbohydrates as summarized in the diagram. The decrease in the carbohydrates content is because of its utilization as indicated below.

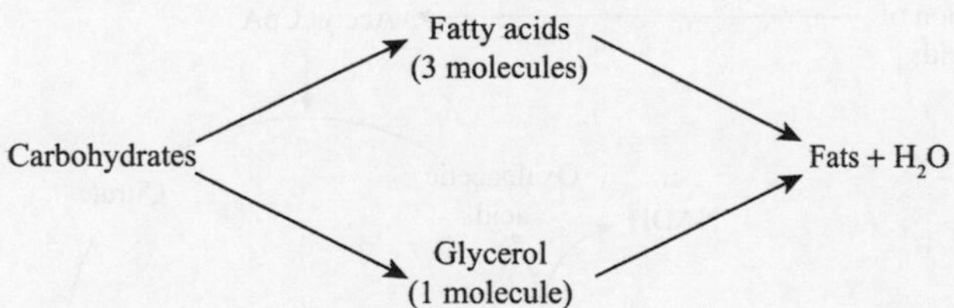

RQ11.7. Calculate the net gain in ATP when one molecule of glycerol undergoes oxidation to carbon dioxide and water *via* glycolysis and the Krebs cycle.

Ans. Glycerol is first phosphorylated and then oxidized to form glyceraldehyde-3- phosphate (GAP)

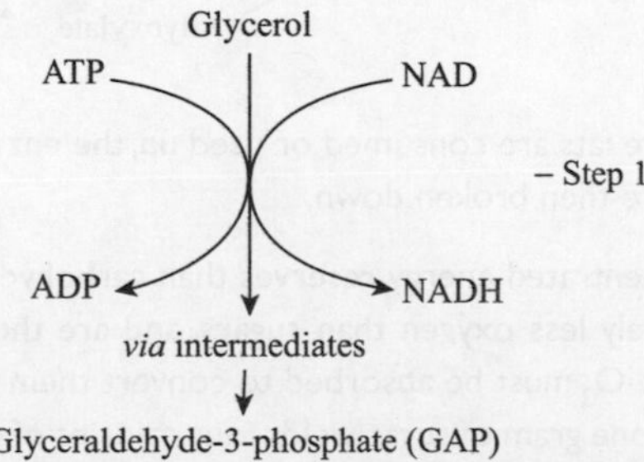

Net Gain of ATP		Total ATP
Step I	ATP (–1)	–1
	NADH (+3 ATP)	+3
Glycolysis	ATP (+2)	+2
	1 NADH (+3 ATP)	+3

Krebs cycle	15 ATP	+15
		ATPs =22* or ATPs =20

* Hypothetically 2 ATPs are generated when NADH is transferred from the cytoplasm to matrix of the mitochondria.

RQ 11.8. How are fats converted into carbohydrates during germination of oilseeds?

Ans. Acetyl CoA produced during β-oxidation of fats is converted into sucrose *via* the glyoxylate cycle occurring in glyoxysomes. Glyoxysomes possess the enzymes of the glyoxylate cycle by which insoluble fats are converted into soluble sugars which are translocated to the centres of growth. The details of the 'glyoxylate cycle' were worked out by Kornberg and Krebs (1957). Two special enzymes 'isocitrate lyase' (Isocitrase) and 'malate synthase' are mainly involved.

1. $\underset{\text{(in the Krebs cycle)}}{\text{Isocitric acid}} \xrightarrow{\text{Isocitrate lyase}} \text{Succinic acid} + \text{Glyoxylic acid}$

2. $\text{Glyoxylic acid} + \underset{\text{(resulting from β-oxidation)}}{\text{Acetyl CoA}} \xrightarrow[+H_2O]{\text{Malate synthase}} \text{Malic acid} + \text{Coenzyme A}$

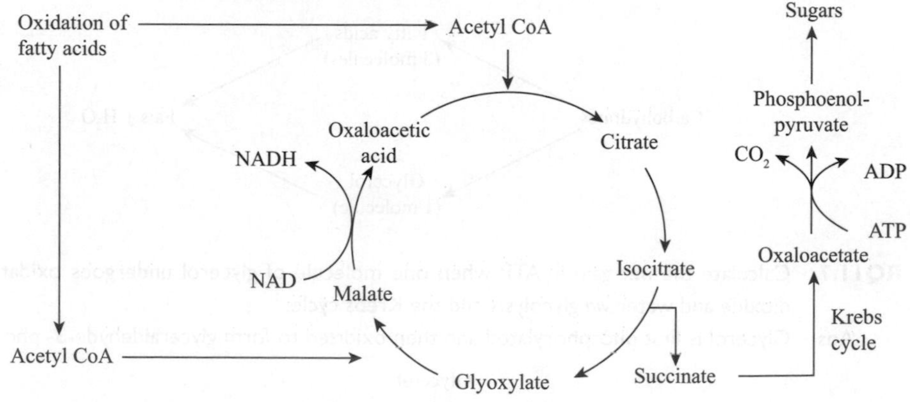

After the reserve fats are consumed or used up, the enzymes of the glyoxylate cycle are no longer needed and are then broken down.

RQ 11.9. Lipids are more concentrated energy reserves than carbohydrates. Comment.

Ans. Lipids contain relatively less oxygen than sugars, and are thus a more concentrated source of energy because more O_2 must be absorbed to convert them into carbon dioxide and water.

Combustion of one gram of sugar yields four calories of energy; while one gram of fat when burned or burnt liberates nine calories.

RQ 11.10. Plants are unable to transfer lipids from the endosperm or cotyledons to the roots and shoots of germinating seedlings. Then how do these overcome this problem?

Ans. Fatty seeds such as castor bean, watermelon and linseed convert their 'insoluble' stored fats into 'soluble' sugars, necessary for mobilization and translocation of carbon reserves to the centre of growth. The mechanism of conversion of fats to sugars involves several different cellular sites such as oleosome, glyoxysome, cytosol and mitochondria as shown in Figure 11.8.

The C_4 compound succinic acid, one of the products, released from the glyoxysomes, then enters the Krebs cycle in the mitochondria, producing oxaloacetic acid which is necessary to keep the glyoxylate cycle running because of two unique enzymes 'isocitrate lyase' and 'malate synthase'.

Malate produced from acetyl CoA during glyoxylate cycle is the substrate for carbohydrate biosynthesis in the 'cytosol'. The oxaloacetate so produced from malate undergoes decarboxylation by the enzyme phosphoenolpyruvate carboxykinase (PEPCK) to produce phosphoenolpyruvate, which in turn, by a reversal of glycolysis (gluconeogenesis), is converted to sucrose.

Likewise, α-glycerol-P produced from glycerol in the cytoplasm is first converted to DHAP, and then through reversal of glycolysis is converted to sucrose.

However, after the reserve fats are used up, the glyoxylate cycle enzymes are no longer needed and are broken down. This glyoxylate cycle does not operate in seeds that store starch.

> The glyoxylate cycle in the germinating fatty seeds ceases as soon as the fat reserves have been exhausted.

RQ 11.11. Why are oils and fats so important in our daily diet? Explain the physiological basis.

Ans. The reasons for adequate intake of fats in our daily diet are:

- Fats are highly concentrated reserves of energy which they release during their conversion into acetyl CoA (energy that is used up in the Krebs cycle plus NADH and $FADH_2$ representing the assimilatory power of the cell that provides ATP through oxidative phosphorylation).
- Function as solvents for fat soluble vitamins such as A (retinol), D (calciferol), E (α-tocopherol), and K, as well as sterols (constituents of hormones).
- Important constituents of biological membranes and cellular components.

RQ 11.12. How many ATP molecules are produced during complete oxidation of one molecule of stearic acid (with 18 carbon atoms)?

Ans. β-Oxidation of stearic acid involves sequential removal of 2-C fragments in the form of acetyl CoA. This sequence continues until the whole molecule is degraded into acetyl CoA (i.e., Cn fatty acid is degraded to *n*/2 molecules of acetyl CoA). Thus, one complete oxidation of stearic acid (with 18 C atoms) would produce large amount of ATP as given below:

$$\text{Stearic acid} + 2\,\text{ATP} + 8\,\text{NAD}^+ + 8\,\text{FAD} + 8\,\text{H}_2\text{O} + 9\,\text{CoASH} \rightarrow 9\,\text{Acetyl CoA} + 2\,\text{AMP} + 8\,\text{NADH} + \text{PPi} + 8\,\text{FADH}_2 + 8\,\text{H}^+$$

A summary reaction of ATP generated:

				ATP molecules	
8 NADH	=	8 × 3	=	24	
8 $FADH_2$	=	8 × 2	=	16	
				40	(i)
9 Acetyl CoA	=	9 × 12	=	108	(ii)
(Krebs cycle)				148	(i + ii)
ATP consumed during activation of fatty acid				–2	
		Total	=	146	

The first step in β-oxidation involves the activation of fatty acid in the presence of ATP (to prime the process) and the enzyme fatty acyl CoA synthase (thiokinase).

$$\text{Fatty acid} + \text{CoA} + \text{ATP} \xrightarrow[\text{Mg}^{++}]{\text{Fatty acyl CoA synthase}} \text{Fatty acyl CoA} + \text{AMP} + \underset{\text{(Pyrophosphate)}}{\text{PPi}}$$

The AMP molecule thus produced reacts with another ATP molecule under the enzymatic activity of 'Adenylate kinase' to produce 2ADP molecules.

$$\text{AMP} + \text{ATP} \rightleftarrows \text{ADP} + \text{ADP}$$

Thus, net gain in complete oxidation of one molecule of stearic acid: 148 –2 = 146 ATP (see above).

RQ 11.13. Calculate the energy yield, in terms of ATP, during the complete oxidation of palmitic acid (with 16 carbon atoms).

Ans. One molecule of palmitic acid will produce: 7 NADH, 7 $FADH_2$ and 8 molecules of acetyl CoA (and at the same time it will consume 2ATP in the first step of fatty acid activation).

				ATP molecules
7 NADH	=	7 × 3	=	21
7 $FADH_2$	=	7 × 2	=	14
8 Acetyl CoA	=	8 × 12	=	96
(during Krebs cycle)				131
ATP consumed during activation of fatty acid				–2
				129 ATP

One gram of fat or oil (~40kJ or 9.3kcal) can generate considerably more ATP than the oxidation of one gram of starch (~15.9 kJ or 3.97kcal).

Finally: $\text{Palmitic acid} + \text{ATP} + 7\ \text{NAD}^+ + 7\ \text{FAD} + 7\ H_2O + 8\ \text{CoASH} \rightarrow$

$$8\ \text{Acetyl CoA} + \text{AMP} + \text{PP}_i + 7\ \text{NADH} + 7\ H^+ + 7\ FADH_2$$

RQ 11.14. Explain the concept of multienzyme complexes with the help of examples.

Ans. In plant tissues, like animals, several enzymes collectively control the different sequential steps of a given reaction, and the reaction intermediates remain loosely bound with one another in non-covalently bonded assemblies called 'multienzyme complexes'. The product of one reaction becomes the substrate for the next enzyme, and the steps are so quick that none of these are released into intercellular pool, thus eliminating the possibility of unwanted side reactions.

Two well-studied systems[6] are 'pyruvate dehydrogenase' (during oxidative decarboxylation in respiration) and the 'fatty acid synthase complex'.

The long-chain saturated fatty acids (from 16 to 18 carbon-units) are synthesized by the repeated addition of two-carbon unit derived from acetyl coenzyme A (acetyl Co A). According to Green (1960), two enzyme complexes and five cofactors (ATP, Mn^{++}, biotin, NADPH and CO_2) are essential for the production of fatty acids. The reaction series involve several enzymes including

[6] The mitochondria and the chloroplast in the cell are other examples of multienzyme complexes. The zymase enzyme in yeast (*Saccharomyces*) is yet another example.

'acyl CoA carboxylase' (consisting of 3 protein subunits – biotin carboxylase, biotin carboxyl carrier protein-BCCP and a transcarboxylase) and the reactions are given as follows:

$$\text{Acetic acid} + \text{Coenzyme A} \rightarrow \text{Acetyl CoA (a highly reactive compound)}$$

$$\text{Acetyl CoA} + CO_2 + \text{ATP} \xrightarrow[\text{Enzyme} + Mn^{++}]{\text{Biotinyl}} \text{Malonyl CoA} + \text{Pi} + \text{ADP}$$

Malonyl CoA acts as a substrate for the enzyme system called 'fatty acid synthase complex' (a complex of at least 6 separate enzymes) and a heat stable coenzyme called acylcarrier protein - ACP.

$$\underset{(3C)}{\text{Malonyl CoA}} + \underset{(2C)}{\text{Acetyl CoA}} + 4\ \text{NADPH} \rightarrow \text{Butyryl CoA} + H_2O + 4\ \text{NADP} + CO_2$$

The butyryl CoA then reacts with another molecule of malonyl CoA (serving as the two-carbon donor) and the cycle is repeated till the chain length in the fatty acid reaches '16 or 18 carbons'.

There are obvious advantages to such a system. The complex would be considerably more efficient than a system in which each of the enzymes and substrates come together haphazardly to react. It is assumed that the separated enzymes are inactive, perhaps because they do not have the proper shape. Secondly, the control is more effectively exercised because blocking one reaction of the series blocks the entire system.

Chapter 12

Sulphur, Phosphorus and Iron Assimilation in Plants

Keywords

- Acid rain
- Biogeochemical cycle
- Ferredoxin reductase/oxidase (FRO)
- Iron-regulated transporter 1 (IRT1)
- Lipooligosaccharides
- Phytoalexins
- Safeners
- Sulphate permeases
- Vacuolar iron transporter 1 (VIT 1)
- Yellow stripe 1 (YS1)

12.1 Sulphur Metabolism

Sulphur is an essential macronutrient for all living organisms, and it occurs in nature in different states of oxidation as a component of inorganic and organic compounds. The chief inorganic form of sulphur, **sulphate**, is obtained predominantly from the weathering of parent rock material. Natural organic sulphur compounds include gases such as hydrogen sulphide, dimethylsulphide and sulphur dioxide, which are discharged into the air both by geochemical processes and by activity of the biosphere. In the gaseous phase, sulphur dioxide (SO_2) reacts with a hydroxyl radical and oxygen to form sulphur trioxide (SO_3) which upon dissolving in water produces a strong sulphuric acid (H_2SO_4), the major source of acid rain. A high concentration of sulphate is also found in the oceans (approximately 26mM or 2.8 g L-1). Plants have a capacity to metabolize sulphur dioxide absorbed in the gaseous phase through their stomata. A prolonged exposure of nearly more than eight hours to high atmospheric concentrations of SO_2 (i.e. more than 0.3 ppm) tends to damage plant tissues extensively due to formation of sulphuric acid.

12.1.1 Sulphate uptake and transport

Sulphate enters a plant from the soil solution, primarily through the roots by an active proton cotransport (H^+-SO_4^{2-} symporter). Sulphate uptake is generally inhibited by the presence of anions such as selenate, molybdate, and chromate anions, which can compete with sulphate for binding to the transporters. Sulphate uptake systems can be classified as either **sulphate permeases** or facilitated transport systems. Plant sulphate permeases are similar to fungal and mammalian H^+-SO_4^{2-} co-transporters. They consist of a single polypeptide chain with 12 transmembrane domains, characteristic of cation/solute transporters. A second mechanism, ATP-dependent transport, is exemplified by a system in cyanobacteria that includes a multiprotein complex of three cytoplasmic membrane components and a sulphate-binding protein in the periplasmic space.

Roots take up sulphate from the soil via electrogenic H^+-SO_4^{2-} symporters (SULTRs). Two high-affinity transporters are expressed in root epidermis and cortex. Although sulphate assimilation occurs in roots, most of the sulphate is transported to the shoots. In photosynthetic cells, sulphate can be translocated either into chloroplasts for assimilation, or into the vacuoles for storage. It has been found that leaves are usually more active in assimilation of sulphur as compared to roots, perhaps because photosynthesis supplies reduced ferredoxin, and photorespiration produces an amino acid serine, which might activate the synthesis of O-acetylserine. Sulphur assimilated in leaves is transferred *via* the phloem to the site of protein synthesis i.e. root and shoot apices, as well as fruits, generally as glutathione. **Glutathione** also functions as a signal molecule that regulates the transport of sulphate into roots and its assimilation by the shoot.

12.1.2 Role of sulphur in plants

1. In agricultural soils, amounts of methionine and cysteine have a significant impact on yield and quality of crops. Amino acids, **cysteine and methionine,** contain sulphur in their structure. In plants, the primary product of sulphate assimilation is cysteine, which in turn becomes the substrate to produce methionine. Cysteine is particularly essential for the structural and catalytic functions of the proteins. The thiol (-SH) groups of two cysteine amino acids upon oxidation yield a covalent disulphide linkages (-S-S-), which are important to establish tertiary and quaternary protein structures, and hence play both structural and regulatory roles in plants. These linkages can also be reversed by reduction, a property crucial for controlling activity of proteins. The redox state of the cysteines that can form disulphide linkages is ultimately controlled by cellular processes and reactions that generate biological reducing agents. The activities of some chloroplast enzymes functioning in carbon metabolism are regulated by reversible disulphide bond formation, which is mediated by proteins such as **thioredoxin** and **glutaredoxin**, and the nonprotein tripeptide, **glutathione**.
2. Several plant products, including coenzymes, lipids, and secondary metabolites, contain sulphur in their structure. Sulphur takes part in electron transport system through iron-sulphur (Fe-S) clusters. Several group-transfer coenzymes and vitamins, including coenzyme A, S-adenosyl-L-methionine (SAM), thiamine, biotin, lipoic acid, and S-methylmethionine consist of functionally important sulphur moieties. Sulphur is a part of the catalytic sites of many enzymes, including urease. Chloroplast membranes contain a sulpholipid, sulphoquinovosyl diacylglycerol (SQDG). Some **signalling molecules** contain sulphur as a principal component, including the sulphated **lipooligosaccharides** that act as rhizobial nod factors, and **turgorin**, a sulphated derivative of gallic acid involved in thigmotactic movement in the leaves of *Mimosa pudica*. Various peptide hormones, such as **phytosulphokines**, require sulphation of their tyrosine residues for their biological activity.

3. Sulphur containing compounds play a **defensive** role in **plant stress**. The major component of various defence systems is glutathione. In addition, many sulphur-containing **phytoalexins**, such as **camalexin,** are produced by members of the Brassicaceae in response to pathogen attack. Thioredoxin is emerging as a defence protein with certain pathogens. Some plants deposit elemental sulphur in their stems as a potent fungicide, but the best known sulphur-containing defence compounds are the secondary metabolites, for example, antiseptic **allins** (in onion and garlic) as well as **glucosinolates** (in Brassicaceae), which impart flavour to the corresponding vegetables coupled with health benefits, and anticarcinogen **sulphorafane** (in broccoli).

Earlier sulphur was considered to be a limitless nutrient for agriculture. But now the agricultural land must be supplemented with sulphate for maximum plant productivity. Sulphur deficiency initiates many changes in plant metabolism leading to reduction of crop yield as well as quality, and increased vulnerability to disease. Sulphur deficiency adversely affects the yield of oilseed *Brassica* species and the baking quality of wheat flour. Sulphur deficiency is also a cause of increased acrylamide contents in bakery products. Sulphur-deficient plants accumulate large quantities of free amino acids, thereby increasing the potential for acrylamide formation. Sulphur deficiency is associated with increased susceptibility of crops to attack by insects and fungi, consistent with the prominence of some sulphur-containing compounds and proteins in plants defence against pathogens. Several genetic and genomics technologies have been employed to study the molecular responses to sulphur deficiency, thus making plant sulphur physiology a perfect model for plant systems biology.

12.1.3 Biogeochemical sulphur cycle

The interconversion of oxidized and reduced sulphur species on earth is known as the **biogeochemical sulphur cycle**. Although the burning of fossil fuel is the greatest contributor to atmospheric sulphur content, a large amount of atmospheric sulphur is derived from marine algae, many of which produce dimethylsulphoniopropionate (**DMSP**, a tertiary sulphur analog of the quaternary nitrogen compounds known as betaines). DMSP has many functions, including roles as osmoprotectant, cryoprotectant, and repellent against planktonic herbivores. DMSP discharged from algae is degraded to dimethyl sulphide (DMS), vapours of which after being released into the atmosphere undergo oxidation to DMSO, sulphite, and sulphate. Atmospheric sulphate acts as a nucleus for formation of water droplets and is associated with the formation of clouds.

Plants and microorganisms reduce sulphate to sulphide for assimilation into cysteine. Some anaerobic bacteria use sulphate as a terminal electron acceptor much as aerobic organisms use oxygen. This process, known as **dissimilation**, generates H_2S. Oxidation of reduced sulphur to sulphate completes the cycle. Oxidation occurs through aerobic catabolism of sulphur compounds, carried out by animals, microorganisms, and plants or by bacteria that use reduced sulphur compounds as electron donors for chemosynthetic or photosynthetic reactions. Reduced sulphur is also oxidized geochemically when oxygen is present.

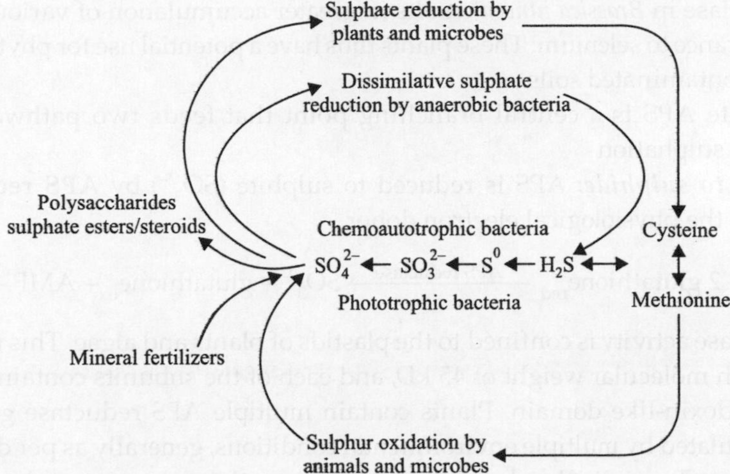

Fig. 12.1 The biogeochemical sulphur cycle. The global sulphur cycle includes microbial conversions among oxidized and reduced states. Sulphate is reduced either by plants and microorganisms to cysteine and methionine or by anaerobic bacteria to sulphur and H_2S. Biological oxidation of reduced sulphur to sulphate is carried out by chemotrophs and photosynthetic sulphur bacteria. Reduced sulphur is also oxidized geochemically to SO_2 in the presence of oxygen. Combustion of fossil fuels also adds to atmospheric SO_2. SO_2 enters the biological cycle again in water and soil as the dissolved form SO_3^{2-}.

12.1.4 Sulphur assimilation pathways

There are two principal sulphur assimilation pathways:

The reductive sulphate assimilation pathway

Sulphur gets reduced to sulphide, which is then incorporated into cysteine

(i) Sulphate assimilation is divided into three steps:

Inorganic sulphate activation: It occurs in the cytosol as well as plastids. Sulphate being chemically inert, requires activation by ATP, using the **ATP sulphurylase** (ATPS) enzyme to form adenosine 5′-phosphosulphate (APS) and the resulting compound consists of a phosphoric acid-sulphuric acid anhydride bond that enables the sulphuryl moiety to undergo modification in subsequent metabolic reactions.

$$SO_4^{2-} + MgATP \xrightleftharpoons{\text{ATP sulphurylase}} MgPP_i + APS$$

Plant ATP sulphurylase is a homotetramer of polypeptides with molecular weight of 52–54 kD, and its substrates bind in an ordered and synergistic manner, with MgATP binding before sulphate. Approximately 85–90 per cent of this enzyme activity is localized in the plastids and the remaining in the cytoplasm. Plant ATP sulphurylase is coded by small multigene families in all plant species. In most of the species, ATP sulphurylase gene expression is regulated by the demand for pathway end products. Overexpression of

ATP sulphurylase in *Brassica juncea* results in greater accumulation of various metals and increased tolerance to selenium. These plants thus have a potential use for phytoremediation of selenium-contaminated soils .

Intermediate APS is a central branching point that feeds two pathways; sulphate reduction and sulphation

Reduction to sulphide: APS is reduced to sulphite (SO_3^{2-}) by **APS reductase**, with glutathione as the physiological electron donor.

$$\text{APS} + 2\ \text{glutathione}_{red} \xrightarrow{\text{APS reductase}} SO_3^{2-} + \text{glutathione}_{ox} + \text{AMP} + 2H^+$$

APS reductase activity is confined to the plastids of plants and algae. This plant enzyme is a dimer with molecular weight of 45 kD, and each of the subunits contains a reductase and a glutaredoxin-like domain. Plants contain multiple APS reductase genes and the enzyme is regulated by multiple environmental conditions, generally as per demand of the reduced sulphur. Transcript levels as well as enzyme activity increase under conditions of sulphur starvation and are repressed when plants are treated with reduced forms of sulphur. Overexpression of APS reductase in plants increases the amounts of compounds containing reduced forms of sulphur and may result in cellular damage when the activity is too high.

Sulphite reduction is catalyzed by ferredoxin-dependent **sulphite reductase**. Plant sulphite reductase adds six electrons to free sulphite, forming sulphide (S^{2-}). Electrons are provided by reduced ferredoxin.

$$SO_3^{2-} + 6\ \text{Fdx}_{red} \xrightarrow{\text{Sulphite reductase}} S^{2-} + 6\ \text{Fdx}_{ox}$$

Plant sulphite reductase is a monomeric 65 kD hemoprotein found in plastids. Like NiR, sulphite reductase contains one siroheme and one iron-sulphur (Fe_4S_4) cluster. Unlike ATP sulphurylase and APS reductase, the enzyme activity and transcript levels of sulphite reductase are not affected appreciably by sulphur nutrition. Sulphite, being a toxic anion, would damage cells if accumulated, so its activity is maintained in excess. Significantly lower levels of sulphite reductase leads to a substantial reduction of plant growth.

Reduction of sulphate is highly active during day time as compared to night time and very energy-intensive. It is carried out in plastids of many cell types and but a preponderance of sulphate reduction occurs in green tissues.

Incorporation of sulphide into cysteine: Cysteine is synthesized by condensation of sulphide with **O-acetylserine (OAS)**, catalyzed by **OAS (thiol) lyase**. OAS used in this reaction is formed by the condensation of serine and acetyl CoA, in the presence of the enzyme, **serine acetyltransferase.**

$$\text{Serine} + \text{acetyl CoA} \xrightarrow{\text{Serine acetyltransferase}} \text{O-acetylserine} + \text{CoA}$$

$$\text{O-acetylserine} + S^{2-} \xrightarrow{\text{O-acetylserine (thiol) lyase}} \text{cysteine} + \text{acetate}$$

Both the enzymes, serine acetyltransferase and OAS (thiol) lyase, coded by small multigene families, are localized in all compartments i.e. the cytosol, plastids, and mitochondria. Cysteine functions as donor of reduced sulphur for production of other sulphur-containing metabolites, such as methionine and glutathione. Plants can detoxify many endogenously produced toxins and hormones, and xenobiotics such as herbicides, by forming conjugates with glutathione in the presence of **glutathione S-transferase**, which links the reactive thiol group of cysteine to the xenobiotic. The conjugates are then transported into vacuoles by an **ABC-type glutathione translocator**. Within the vacuole, detoxification is completed as glutathione conjugates are hydrolyzed to cysteine conjugates. Plant **glutathione reductases** are usually localized in the cytosol, where they catalyze the detoxification of xenobiotics and reduction of the organic hydroperoxides formed during oxidative stress. In addition, **glutathione reductases** facilitate transport of natural products into the vacuole and are involved in various stress-signalling pathways. This mechanism for herbicide detoxification has been applied successfully to agriculture by identifying the substances that stimulate expression of glutathione S-transferase and the vacuolar transporter, or that result in an increased rate of glutathione synthesis. These substances, termed **safeners,** are used to enhance crop resistance to herbicides.

The presence of **sulphite oxidase** enymes has been detected in plants recently particularly in peroxisomes. This enzyme catalyzes the oxidation of sulphite to sulphate, with molecular oxygen as an electron acceptor. Hydrogen peroxide (H_2O_2) thus generated can either be metabolized by catalase or can non-enzymatically oxidize another sulphite molecule.

$$SO_3^{2-} + H_2O + O_2 \xrightarrow{\text{Sulphite oxidase}} SO_4^{2-} + H_2O_2$$

The function of sulphite oxidase in plants is most probably protection against excess sulphite. But, a futile cycle between sulphate reduction to sulphite and reverse oxidation by sulphite oxidase is prevented by compartmentalization of the two processes in plastids and peroxisomes, respectively.

Alterative pathway: APS is further phosphorylated by APS kinase

In the first step, sulphate is activated by ATP sulphurylase, which adenylates the sulphate to generate the intermediate **APS**. In the second step, APS undergoes further phosphorylation by APS kinase to 3′-phosphoadenosine 5′-phosphosulphate **(PAPS)**, which is the molecular source for sulphate to be utilized in sulphated polysaccharides, glucosinolates, and other metabolites. PAPS is synthesized in the plastids and cytosol, the two compartments where APS is also produced

$$\text{APS} + \text{ATP} \xrightarrow{\text{APS kinase}} \text{PAPS} + \text{ADP}$$

Sulphate uptake and assimilation are regulated by demand, and at multiple levels from transcription to activity. This regulation integrates factors including growth, light, and sulphur availability, and these functions are an apealing target for systems analysis.

12.1.5 Sulphate assimilation is closely linked with carbon and nitrogen metabolism

Three major pathways of primary metabolism i.e. carbon fixation, nitrate assimilation, and sulphate assimilation are connected by synthesis of cysteine. These pathways are well coordinated. Sulphate uptake and APS reductase activity decrease during nitrogen and carbon starvation. Sulphate starvation results in decline in nitrate uptake and nitrate reductase activity, chlorophyll degradation, and reduced photosynthesis. APS reductase, just like nitrate reductase, displays a diurnal rhythm in activity and transcript levels, and it is induced by light and carbohydrates. Exchange through the sulphate reduction cycle is far greater during day time compared to night time and is increased by addition of reduced nitrogen sources and carbohydrates.

Phytohormones also affect sulphate assimilation. APS reductase, the key enzyme of sulphate assimilation, is upregulated by ethylene and salicylate. ABA has a negative effect on APS reductase, but promotes the synthesis of cysteine and glutathione. Cytokinins also regulate sulphur assimilation; the upregulation of sulphate transporters by sulphur starvation is partially dependent on cytokinin signalling. Jasmonic acid coordinates the upregulation of the genes of sulphate assimilation and glucosinolate synthesis. Thus, the regulation of sulphate assimilation is complex, involving and coordinating between many signals and effectors.

Sulphite can directly enter into the sulphur reduction pathway and undergo reduction to sulphide, followed by integration into cysteine, and eventually into other sulphur compounds. Alternatively, sulphite may also get oxidized to sulphate catalyzed by peroxidases or metal ions or superoxide radicals and ultimately reduced and assimilated again. Excess of sulphate is translocated into the vacuole.

The foliar uptake of hydrogen sulphide seems to depend upon the rate of its metabolism into cysteine and eventually into other sulphur compounds. There is strong evidence that O-acetyl-serine (thiol) lyase is directly involved in the active fixation of atmospheric hydrogen sulphide by plants.

Sulphur Metabolism in Plants and Air Pollution

The rapid economic growth, industrialisation and urbanisation are associated with a strong increase in energy demand and emissions of air pollutants including sulphur dioxide and hydrogen sulphide, which may affect plant metabolism. Sulphur gases are potentially phytotoxic, however, they may also be metabolized and used as sulphur source and even be beneficial if the sulphur fertilization of the roots is not sufficient.

Plant shoots form a sink for atmospheric sulphur gases, which can directly be taken up by the foliage (dry deposition). The foliar uptake of sulphur dioxide is generally directly dependent on the degree of opening of the stomates, since the internal resistance to this gas is low. Sulphur is highly soluble in the apoplastic water of the mesophyll, where it dissociates under formation of bisulphite and sulphite.

12.2 Phosphorus Metabolism

Elemental phosphorus (P) does not occur in free form on earth. It exists in living organisms, mineral deposits, soils and water, as phosphate (PO_4^{3-}, HPO_3^{2-}, $H_2PO_3^-$). Phosphate assimilation in plants differs from the reductive mode of nitrate and sulphate assimilation in that P remains in its oxidized state, entering into organic combination in the form of phosphate esters.

12.2.1 Biogeochemical phosphorus cycle

The various stages of phosphorus cycle are weathering, solubilisation, sequestration, leaching and precipitation. Unlike the nitrogen and sulphur cycles, the phosphorus biogeochemical cycle has no gas phase and only the oxidized form of phosphorus is incorporated in the biomolecules . Apart from small amounts in dusts and in acid rain as phosphoric acid, atmospheric P is of least significance in the overall balance. The largest reservoir of P is in sedimentary rock.

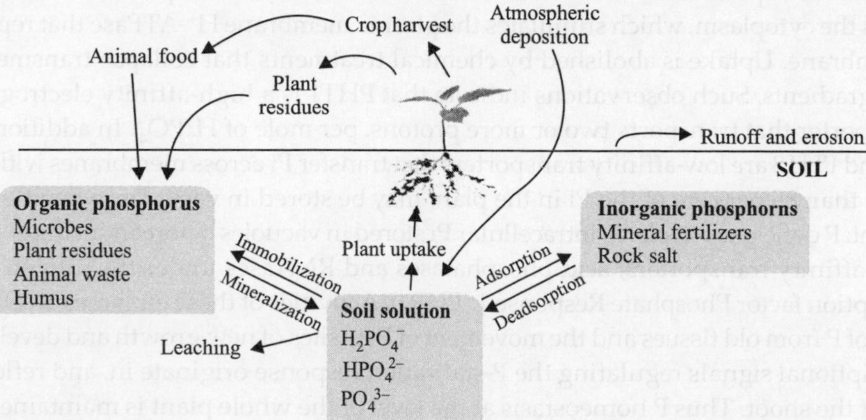

Fig. 12.2 The biogeochemical phosphorus cycle. Phosphorus is available in both inorganic and organic states in the soil. Phosphous is present in soil solutions as $H_2PO_4^-$, HPO_4^{2-}, and PO_4^{3-}. Phosphorus derived from animals as well as plants organic waste is recycled by them. Phosphates are not reduced and it is the oxidized form only which is integrated in the biomolecules.

The uptake of soil phosphate by plants is the basis of a terrestrial food chain eventually leading to herbivores and carnivores. P is returned to the soil in the form of urine and faeces, as well as by decomposition of plants and animals after death. Organic P is not directly usable for growth of the plants. Plants absorb P as soluble inorganic phosphate, designated Pi. Soil organic matter releases Pi through the process of mineralization. The P accessible to support plant growth can be limited because soil phosphate is readily leached, precipitated, adsorbed by organic and mineral materials, made unavailable by drought or rapidly depleted from the rhizosphere by roots. To obtain otherwise inaccessible P from the soil, roots may modify their structure and function, or manipulate the chemistry of the rhizosphere, or enter into mutualistic associations with mycorrhizal fungi.

12.2.2 Role of phosphorus in plants

Phosphorus is an essential component of phospholipids present in the membranes. It is found as phosphate ester (including sugar-phosphates) which play a principal role in photosynthesis and intermediary metabolism. It is also a component of nucleoproteins, DNA, RNA, phosphorylated sugars, and organic molecules such as ATP, ADP, UDP, UTP, CDP, CTP, GTP and GDP (required for energy transfer by phosphorylation).

12.2.3 Phosphate is actively accumulated by root cells

It is believed that root cells import phosphate against an electrochemical gradient, requiring an input of energy equivalent to at least 1 mol ATP hydrolyzed per mole of $H_2PO_4^-$ taken up. In plants, three types of inorgainc phosphate (Pi) transporters i.e. PHT1, PHT2 and PHT3 have been identified and studied in details. Pi uptake by roots is particularly dependent on high-affinity transporters of the PHT1 group. Uptake of Pi *via* PHT1 is accompanied by a transient depolarization of the cell membrane followed by a repolarization. Pi uptake acidifies the cytoplasm, which stimulates the plasma membrane H^+-ATPase that repolarises the membrane. Uptake is abolished by chemical treatments that collapse transmembrane proton gradients. Such observations indicate that PHT1 is a high-affinity electrogenic Pi/H^+ symporter that transports two or more protons, per mole of $H_2PO_4^-$. In addition to this, PHT2 and PHT3 are low-affinity transporters that transfer Pi across membranes within cells.

More than 85 per cent of the Pi in the plant may be stored in vacuoles, when P supply is sufficient. P deficiency depletes intracellular Pi stored in vacuoles before inducing expression of high-affinity transporters, acid phosphatases and RNAases under the influence of the transcription factor Phosphate Response 1 (PHR1). Activities of these enzymes accelerate the salvage of P from old tissues and the movement of Pi to sites of new growth and development. Transcriptional signals regulating the P-starvation response originate in, and reflect the P status of the shoot. Thus P homeostasis at the level of the whole plant is maintained by the flexibly-regulated coordination of Pi acquisition in response to Pi supply and the state of the internal phosphorus economy.

Plants modify the rhizosphere and form mycorrhizal associations to improve phosphorus availability. A major priority for crop breeders is to use traditional and molecular approaches to improve P utilization efficiency by modifying the capacity of crop plants to mine soil P. Soil P becomes unavailable to the plants because of low solubility of Pi in clay or calcareous soils and its conversion to organic forms by soil microbes.

Strategies evolved by plants to increase phosphorus accessibility

- In dry desert soils, some plants are able to release water from the roots at night, thereby improving Pi diffusion.
- Plants generally modify the rhizosphere by secretion of carboxylic acids (such as citrate, malate, and malonate), phytases, and phosphatases. Carboxylates are chelators, being polyanions bind to cations and prevent them from forming precipitates. Carboxylates form chelate complexes with metal cations that bind phosphate and can displace both forms of phosphorus i.e. inorganic and organic from the soil matrix, thus making it available to the plant. Carboxylate export occurs *via* an anion channel and is often accompanied by acidification of the rhizosphere. Secretion of organic acids in response to phosphorus deficiency is associated with coordinated changes in root metabolism

e.g., enhancement in activity of Krebs cycle enzymes (such as citrate synthase and malate dehydrogenase), and of the glycolytic enzyme PEP carboxylase. After the soil organic P (primarily in the form of esters of phosphate, including phytate) is rendered soluble by carboxylates, it must be hydrolyzed by acid phosphatases released by the roots for uptake by the plant. Roots of some plants also release considerable quantities of phytases.

- Several vascular plants have the potential to form mycorrhizal associations between their roots and fungal symbionts. Symbioses with mycorrhizal fungi improve the uptake of immobile Pi from the soil. The transporters of mycorrhizal fungi are of the PHT-like high-affinity Pi/H^+ symporter type. Absorbed phosphate is incorporated into fungal energy metabolism, nucleic acids and phospholipids or condensed into polyphosphate, the storage form of P in fungi. After translocation within the fungal hyphae, which forms arbuscules within root cells, phosphate is transferred to the host plant.

12.3 Iron Metabolism

Iron constitutes nearly 3 per cent of the earth's crust and ranks fourth major element in earth crust. Iron is obtained from the soil, where it is present primarily as ferric ion (Fe^{3+}) in oxides, such as $Fe(OH)^{2+}$, $Fe(OH)_3$, and $Fe(OH)_4^-$.

Iron is an essential element required by all living organisms as the structural component of heme proteins as well as iron-sulphur protein complexes. It is involved in electron transport system besides functioning as an activator of some of the enzymes.

12.3.1 Biogeochemical Iron Cycle

Soil pH has a great influence on the availability of iron to the plants (Figure 12.3). At neutral pH, the ferric ion is highly insoluble. Generally plants show iron deficiency symptoms even when iron is present in sufficient amounts in the soil because soils usually consist of low concentrations of Fe^{2+}, the soluble inorganic form of iron that is readily absorbed by plants. In aerated soils, acidophilic bacteria such as *Acidithiobacillus ferrooxidans* oxidise Fe^{2+} to Fe^{3+}, using O_2 as the terminal electron acceptor.

$$2Fe^{2+} + 1/2\, O_2 + 2H^+ \longrightarrow 2Fe^{3+} + H_2O$$

Oxidation of ferrous to ferric ions is used both for producing ATP as well as reducing CO_2 to carbohydrates.

$$4Fe^{2+} + HCO_3^- + 10H_2O + h\nu \longrightarrow 4Fe(OH)_3 + (CH_2O) + 7H^+$$

Ferric hydroxide is insoluble, and its formation is favoured at pH 5–8. The pH of aerobic soils range from neutral to basic, and therefore, iron is not available to plants. Iron is present in oxidized state or is complexed with organic material in soils having neutral and basic pH. However, in waterlogged soils, iron is available in a reduced form, Fe^{2+}. Plants can absorb Fe^{2+} since transporters for this are present in plasma membrane of the root epidermis.

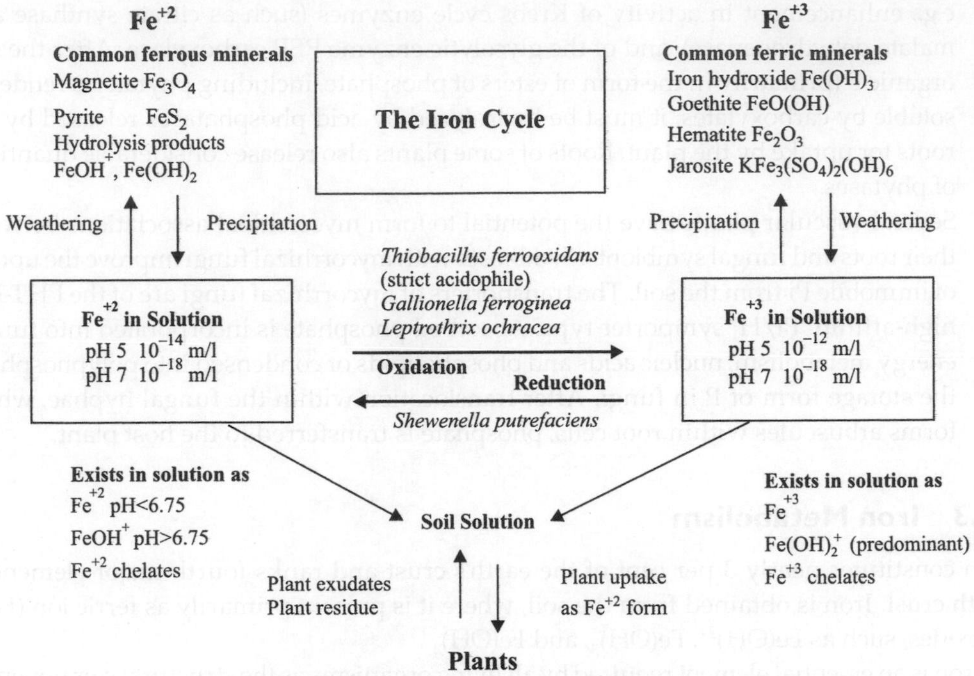

Fig. 12.3 The biogeochemical cycle of Iron

12.3.2 Role of iron in plants

Iron is an essential component of many of the catalytic groups for several redox enzymes e.g., heme consisting of cytochromes; and, non-heme iron-sulphur proteins (Rieske proteins and ferredoxin) associated with photosynthesis, nitrogen fixation and respiration.

It is also a key component of many oxidase enzymes, such as catalase and peroxidase.

Essential for the synthesis of chlorophyll and protein of the chloroplast in the leaves; it also participates in the structure of the dinitrogenase enzyme (in prokaryotes) - a crucial enzyme for molecular nitrogen (N_2) fixation to ammonia.

12.3.3 Iron uptake and transport in plants

Plants have evolved various strategies to take up iron from the soil, which include release of chelating organic compounds or phytosiderophores or protons by the roots, thereby facilitating release of iron from the organic complex forms, and reducing Fe^{3+} to Fe^{2+}state. This is accompanied by its absorption *via* transporters located on the root epidermis (refer to Figure 4.3). Transporters of Fe^{3+}-phytosiderophore complexes are integral membrane protein with 12 transmembrane domains coded by the gene **yellow stripe 1** (YS1). In many monocot plants (other than grasses), enzymes of the **ferredoxin reductase/oxidase (FRO)** family are synthesized that reduce Fe^{3+} to Fe^{2+} form, with cytosolic NADH or NADPH acting as the electron donor. Iron deficiency stimulates the activity of this enzyme as well as the export of protons by roots. The presence of **iron-regulated transporter (IRT1)** in root epidermis is associated with the absorption of Fe^{2+} ions. Coordinated action of ATPase-mediated H^+ pumps, FROs and IRT, is needed for iron absorption by the monocots and eudicots.

Once inside the cytoplasm, iron is either translocated to the shoots or redistributed intracellularly within the cells of roots or mesophyll, facilitated by transporters (i.e., mitochondrial ion transporters, vacuolar iron transporters) or is transported to chloroplasts. Mitochondria and chloroplasts are the principal sinks for iron because these two organelles house the electron transport chain. Plants show deficiency symptoms in case of unavailability of iron or excess of free iron which is toxic. To maintain iron homeostasis, it is either exported to the subcellular organelles or stored as **frataxin** in mitochondria or **ferritin** in chloroplasts or transferred into the vacuoles, facilitated by **vacuolar iron transporter (VIT 1)** as a complex of nicotianamine.

After the roots absorb iron cations, they oxidize it to a ferric form and translocate much of it to the leaves, where an iron cation undergoes an important assimilatory reaction. Free iron may combine with oxygen to produce highly damaging hydroxyl radicals. To prevent such damage, plant cells accumulate surplus iron in an iron-protein complex, ferritin. Studies on ferritin reveal that iron in this protein-bound form may be easily available to human beings, and foods rich in ferritin (such as soybean) may solve dietary anemia problems.

Plants serve as the prime conduits through which major nutrients (such as nitrogen, sulphur, phosphorus and iron) pass from 'inert geophysical domains' into 'dynamic biological domains' and so, the nutrient assimilation of plants, ends up playing a vital role in human diet.

Review Questions

RQ 12.1. Name the two activated sulphur compounds in plants. What is their role in sulphur assimilation?

Ans. The two activated sulphur compounds are adenosine 5'-phosphosulphate (APS) and 3'-phosphoadenosine 5'-phosphosulphate (PAPS). Both are used in the cell for esterification reactions of sulphate with organic substances. These are two analogues of ATP.

The APS pathway of sulphate reduction occurs in higher plants, algae and cyanobacteria; whereas PAPS pathway operates in organisms like yeast and *E.coli*.

RQ 12.2. What is meant by sulphur respiration?

Ans. Certain anaerobic bacteria like *Desulphovibrio* use sulphate as a terminal acceptor of electrons during respiration to generate ATP. This is termed as sulphur/sulphate respiration. Such a reduction of sulphate is known as **dissimilatory reduction**. This is in contrast to **assimilatory reduction** in which sulphate is reduced to a thiol group found in many coenzymes and in the amino acids-cysteine and methionine.

RQ 12.3. What is the main source of sulphur? Mention some of its plant sources as well. What is special about it?

Ans. Sulphur is the fifth most common element by mass on the Earth. Burning of fossil fuels such as coal, oil and natural gas are the main sources of sulphur dioxide emissions. Elemental sulphur is also found in pure form near hot springs and volcanic regions in many parts of the world, especially along the pacific Ring of Fire. Such volcanic deposits are currently being mined in Indonesia, Chile and Japan. Other sources of sulphur are the mineral pyrite and hydrogen sulphide gas distilled from natural gas wells, especially in the Middle East and Canada. Sulphur is also a common constituent of crude oil, from which it is extracted.

Garlic, onions, leeks and other vegetables in the *Allium* genus are among the best dietary sources of sulphur. These foods are rich sources of allicin, allinin, diallylsulphide, sulphides in general, as well as phytonutrients. The avocado is the fruit with the highest sulphur content, followed by kiwis, bananas, pineapples, strawberries, melon, grapefruit, grapes, oranges and peaches. Eggs are a great source of sulphur. Chicken eggs, particularly the yolks, are rich in it. Millets generally contain significant amounts of amino acids, particularly the sulphur-containing ones (methionine and cysteine). Sulphoraphane (sulphur rich compounds) are abundant in cruciferous vegetables like broccoli and cabbage and have been known to provide powerful health benefits. Raw broccoli has 10 times more sulphoraphane than cooked broccoli.

Sulphur is present in proteins, sulphate esters of polysaccharides, steroids, phenols, and sulphur-containing coenzymes. Sulphur is used in the vulcanization of natural rubber, as a fungicide, as a component of black gunpowder, in detergents and in the manufacture of phosphatic fertilizers. Sulphur is a vital component for all forms of life - a constituent of two amino acids, cysteine and methionine.

RQ 12.4. Explain the routes by which phosphorus is taken up from the soil .

Ans. There are two routes by which phosphorus can be acquired from the soil:-

(i) Phosphate in the soil is bound to clay particles and organic matter. A high-affinity **phosphate transporter** present in the plasma membrane of root epidermal cells takes up free phosphate from the soil. Free phosphate diffuses very slowly in the soil, and a zone of 1–5 mm beyond the root surface is usually depleted of phosphate. The root epidermal cells secrete phosphatases, organic anions (such as malate and citrate) and protons leading to acidification in the rhizosphere, which solubilizes bound phosphate from soil particles, thus enhancing phosphate availability.

(ii) Another means of phosphate acquisition by plants is **mycorrhizal association** (symbiotic association between plant and a fungus). Fungal hyphae take up phosphate from the soil surrounding the roots and transfer to arbuscules (fungal structures) inside the cortical cells of the root. Specific phosphate transporters in the cortical cell plasma membrane are involved in transfer of phosphate from the arbuscule into cytoplasm of root cortical cells.

RQ 12.5. Iron deficiency induces various morphological and biochemical changes. List out all of them.

Ans. Various changes induced by iron deficiency in the plants include:

(i) Root hair morphogenesis
(ii) Differentiation of rhizodermal cells into transfer cells
(iii) Yellowing of leaves
(iv) Ultrastructural disorganization of chloroplasts and mitochondria
(v) Increased synthesis of organic acids and phenolics
(vi) Activation of root systems responsible for enhanced iron uptake capacity.

RQ 12.6. Discuss the role of iron in plant growth and metabolism.

Ans. Iron is an essential micronutrient for almost all living organisms because of its redox properties. Although abundant in most well-aerated soils, the biological activity of iron is low because it primarily forms highly insoluble ferric compounds at neutral pH levels. It plays role in metabolic processes such as DNA synthesis, respiration, and photosynthesis. In plants, iron is essential for the synthesis of chlorophyll and the maintenance of chloroplast structure and function. That is why plants deficient in iron show chlorosis in the new leaves. Iron deficiency leads to interveinal chlorosis of the new leaves (I.e., leaves are yellow with green veins).

Iron plays a significant role in various physiological and biochemical pathways in plants. It serves as a prosthetic group for many critical enzymes and a component of cytochromes in the electron transport chain, thus required for a wide range of biological functions. Iron is a component of several pigments, and assists in nitrate and sulphate reduction, and energy production within the plant.

RQ 12.7. In which form is iron absorbed by plants? How do you add iron in plants?

Ans. Iron is absorbed as ferrous and ferric ions by plants. Plants usually absorb iron from the region of soil, where the roots form extensive ramifications. The region of soil (called the **rhizosphere)** is influenced by the plant root secretions as well as microorganisms present in the soil.

***Treatment of soil*:** Powdered or granular chelated iron is the best option for soil amendments. Sprinkle it around the root zone of the plant. Phosphorus overload can contribute to iron chlorosis, so if your supplements also contains fertilizers, make sure it's phosphorus free.

Chapter 13

Phloem Transport

Keywords

- Apoplast
- Mass flow
- Phloem loading
- Phloem unloading
- Sink
- Source
- Symplast
- Transcellular strands

Attempts to demarcate the two processes of transportation: of inorganic, as well as organic nutrients in plants, dates back to the 17th century. A plant anatomist, Malpighi conducted a experiment in which he separated a ring of bark (phloem) from the wood (xylem) of young stems by detaching the two in the region of vascular cambium - **'girdling'** or 'ringing' (Figure 13.1). Since the xylem remained intact, water and inorganic solutes kept on rising all the way upto the foliar region and the plant remained alive for a few days. However, girdled plants showed swelling of the bark in the area just above the girdle due to accumulation of photo-assimilates flowing downward. The downward stream also consisted of nitrogenous compounds and hormones, which caused cell enlargement above the girdle. Ultimately, the root system was subjected to starvation because of lack of nutrients and the girdled plants died away.

13.1 Evidences in Support of Phloem Transport

1. An ***analysis of phloem exudate*** obtained by making an incision into the phloem tissue provides evidence, supporting the fact that photoassimilates are translocated through the phloem.
2. ***Aphid technique:*** Aphids, constituting groups of small insects, feed on herbaceous plants by inserting a long mouth part (proboscis) deep into individual sieve-tube elements of the phloem. While aphids are feeding, they are anaesthetized with a gentle stream of CO_2 and the proboscis is carefully removed with a sharp blade. Meanwhile, the uncontaminated phloem liquid continues to ooze out from the cut proboscis for a long period (Figure 13.2). This demonstrates that phloem sap is under pressure. The

aphid technique has proved to be of great use in understanding the mechanism of phloem transport.

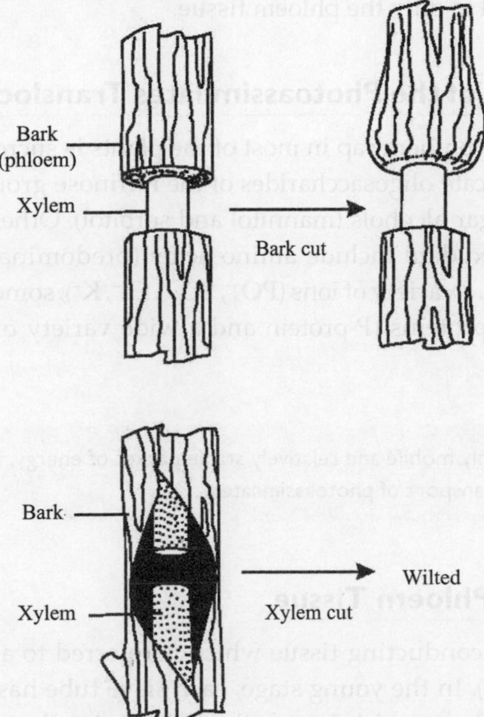

Fig. 13.1 Girdling experiment demonstrating the involvement of phloem and xylem in the translocation of inorganic nutrients and water, respectively.

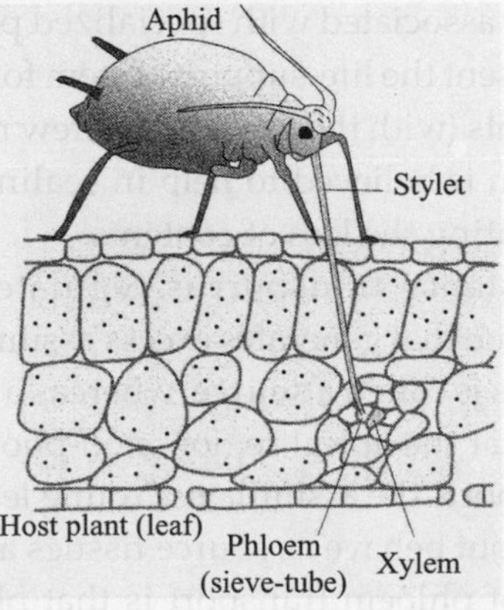

Fig. 13.2 Aphid feeding on the surface of the leaf of the host plant. The stylet is inserted deep into the sieve-tube element of the phloem.

3. ***Radioactive tracers:*** Evidence is also garnered by employing radioactive elements, specifically ^{14}C on the leaves of herbaceous and woody plants. The radioautographs that follow, reveal that radioactive photoassimilates being transported out from the leaves, are confined to only the phloem tissue.

13.2 Composition of the Photoassimilates Translocated in the Phloem

The major component of phloem sap in most of the plants is sucrose. But, a small number of plant families translocate oligosaccharides of the raffinose groups (raffinose, stachyose and verbascose) and sugar alcohols (mannitol and sorbitol). Other constituents identified in phloem sieve-tube exudate include amino acids (predominantly glutamic acid and aspartic acid), malic acid, a variety of ions (PO_4^-, SO_4^-, Cl^-, K^+), some plant hormones (auxin, gibberellin, cytokinin), proteins (P-protein and a wide variety of enzymes), and, RNAs, plant viruses, and such.

Sucrose, being a small, highly mobile and relatively stable packet of energy, is the most favoured means for long-distance transport of photoassimilates.

13.3 Anatomy of Phloem Tissue

Phloem tissue includes conducting tissue which is referred to as the **sieve elements** or **sieve tubes** (Figure 13.3). In the young stage, each sieve tube has a large central vacuole surrounded by the cytoplasm which forms a thin layer lining the cell wall and a prominent nucleus. At maturity, the nucleus degenerates, followed by the loss of ribosomes, Golgi apparatus and cytoskeleton. The mature sieve element contains modified but metabolically active protoplasm. All the sieve elements are interlinked *via* perforated ending walls of sieve plates.

Sieve elements are typically associated with specialized parenchymatous cells called the **companion cells,** which represent the life-support system for them. The protoplasts of sieve elements in angiospermic plants (with the exception of few monocots) are characterized by the P-protein bodies. P-protein is believed to help in sealing the sieve plate pores should they be wounded, thus preventing the loss of contents.

The direction of long-distance transport is regulated mainly by source-to-sink relationships. An organ or tissue that generates excess assimilates; more than what it needs for its own metabolic functions is called a **source,** whereas a **sink** is the organ (for example, meristems, developing leaves at the apical region, non-photosynthetic stem tissues, roots and storage tissues) which imports the assimilates. Young leaves function as sink organs in initial phase of development, but behave as source tissues as they reach maturity.

The underlying principle of phloem transport is that photosynthates are transported from a source towads a sink.

Sources	Sinks
Photosynthetic tissues:	Plant parts that are growing or developing food reserves:
• Mature green leaves	• Developing seeds
• Green stems	• Developing fruits
Storage organs (unloading their reserves):	• Growing leaves
• Storage tissues in germinating seeds	• Developing tap roots or tubers
• Tap roots or tubers at the onset of the growth season	• Roots that are growing and absorbing mineral ions using energy from cell respiration

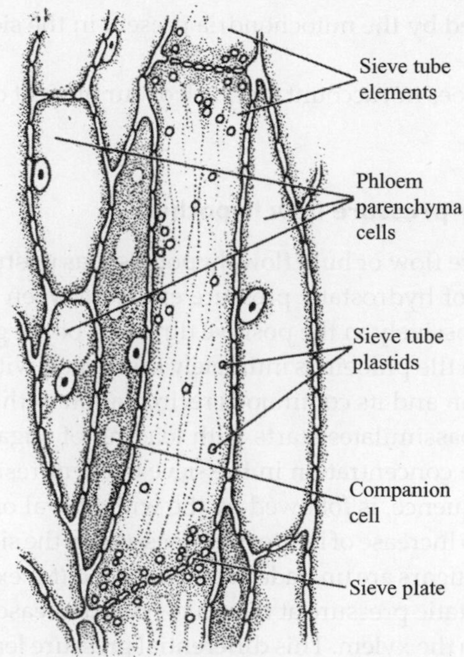

Fig. 13.3 A portion of the phloem tissue showing a sieve tube element, a companion cell and phloem parenchyma cells.

13.4 Mechanism of Phloem Transport

Various theories have been suggested to explain the transport of photosynthates in the phloem.

13.4.1 Protoplasmic streaming[1]

This theory formulated by Hugo de Vries (1885) assumes that solute particles are caught up within the circulating cytosol of the sieve elements and carried from one end of the cell to the other. These particles pass across the sieve plates by diffusing through cytoplasmic strands connecting one element to another.

Curtis (1935), through the streaming movement of the sieve tube cytoplasm, could account for the rapid movement of solutes, taking place simultaneously in both directions.

The main objection to this hypothesis is that it requires metabolic energy and the mature sieve tube, devoid of nucleus, is metabolically inactive. Canny (1962) and Thaine (1961–64), however, independently observed actual cytoplasmic streaming in sieve tube elements. Several cytoplasmic strands run through the length of the sieve elements sap and pass through pores of sieve plates between adjacent sieve tube elements forming continuous **transcellular strands**. Solute particles move in opposite directions in adjacent transcellular strands, thus explaining the bidirectional movement of solutes in the same sieve duct. Energy needed for this is provided by the mitochondria present in the sieve tube cytoplasm or by the companion cells.

However, this theory does not account for much of our present day knowledge of phloem transport.

13.4.2 Münch mass or pressure flow hypothesis

The mass flow (or pressure flow or bulk flow) hypothesis, as postulated by E Münch (1930), assumes that a gradient of hydrostatic pressure exists between the source and the sink. Metabolites are carried passively in the positive direction of the gradient. Translocation of photoassimilates through the phloem is intimately associated with the movement of water in the transpiration stream and its continuous recirculation within the plant.

Translocation of photoassimilates starts with loading of sugars into sieve elements at the source. Greater solute concentration in the sieve element results in lowering its water potential and as a consequence, is followed by the withdrawal of water from the adjacent xylary tissue. This leads to increase of hydrostatic pressure in the sieve elements at the source end. Simultaneously, the sugars are unloaded at the sink end (for example, in the root or stem storage cells). The hydrostatic pressure at the sink end is decreased as water exits the sieve elements and goes back to the xylem. This differential pressure leads to movement of water from the source to the sink and assimilates (sugars) are carried passively along (Figure 13.4).

This hypothesis can be conveniently demonstrated in a laboratory exercise by connecting two osmometers as shown in Figure 13.5. The main evidence in support of this mechanism is derived from the exudation of sap in many species when an incision is made in the phloem of the stem.

---

[1] Cytoplasmic streaming is the most important or significant contributor to the movement of water and solutes within plant cells. It is driven by activity of motor proteins of cytoskeleton and is fueled by ATP hydrolysis.

The objection to this hypothesis, which considers translocation as a purely passive process, is that the transport of assimilates is inhibited by metabolic inhibitors and temperature, thereby suggesting the involvement of metabolic energy. However, the impact of low temperature and metabolic inhibitors are either short-term or may cause disintegration of the P-protein, sealing the sieve plates. Requirement of energy for transportation within the sieve elements is therefore minimum and justifies the passive nature of the mass flow hypothesis. Moreover, absence of nucleus within the mature sieve tube indicates that metabolic energy is not involved in the process.

Swanson (1959) pointed out that the movement of sugars from the photosynthetic tissues of the leaves into the sieve tube elements takes place against a concentration gradient and is, therefore, an active process which requires metabolic energy. It has also been suggested that sugar (sucrose) may move into the sieve tubes as sugar phosphates with the involvement of ATP (Figure 13.6). Phosphorylation may facilitate the transport of sucrose across cell membranes or it may activate the sucrose molecule, thereby enabling it to unite with a carrier to form a complex which can readily move across cell membranes. The uptake of sugars from the sieve elements towards the sink is also considered to be an active process.

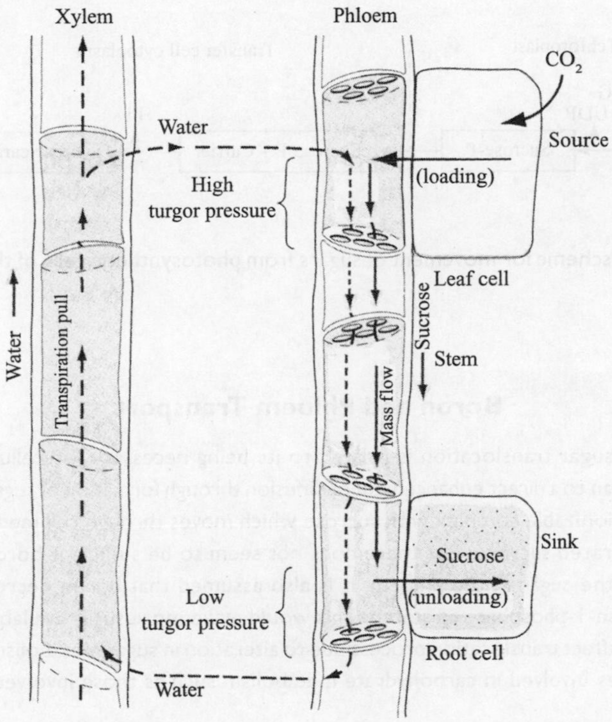

Fig. 13.4 Diagrammatic representation of functioning of the mass flow hypothesis. The pressure difference between the source and sink end of sieve-tube drives the bulk flow of solutes (sugars) within sieve tubes from source to sink.

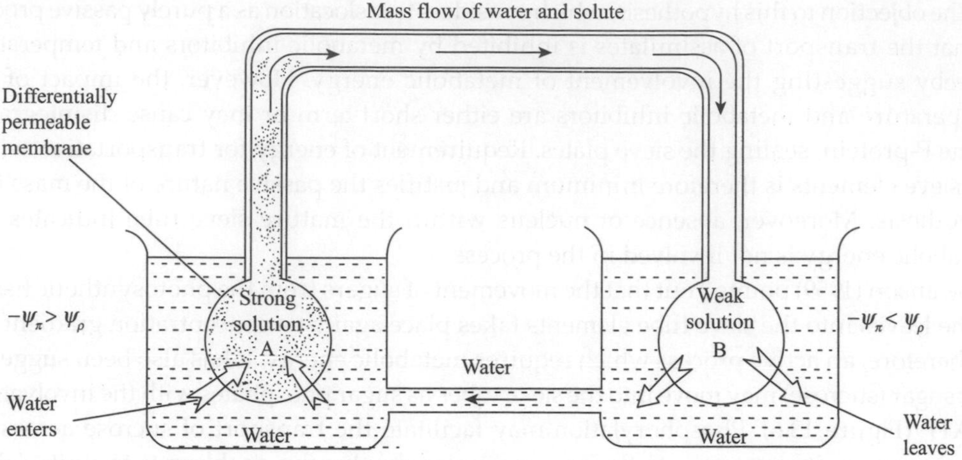

Fig. 13.5 A physical model for the mass flow hypothesis. A and B, are the two osmometers. The connecting links between the osmometers and water containers represent the sieve and xylem ducts respectively.

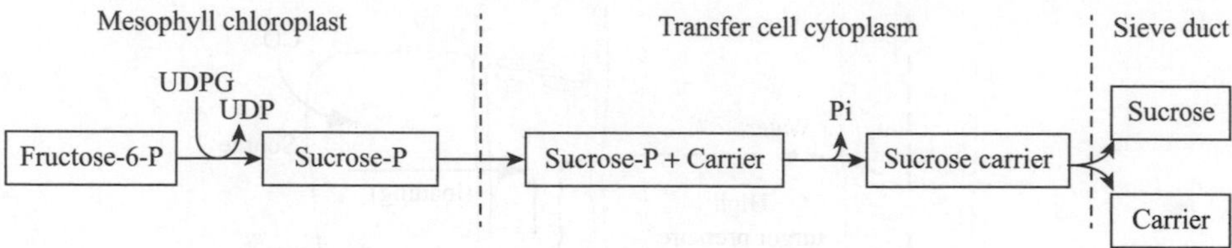

Fig. 13.6 Swanson's scheme for movement of sugars from photosynthetic cells of the leaves into the sieve elements

Boron and Phloem Transport

Effect of boron on sugar translocation is related to its being necessary for cellular activity in apical meristems, rather than to a direct enhancement of diffusion through formation of sugar-boron complexes.

Boron forms an ionizable complex with sucrose which moves through cell membranes much more readily than non-borated sucrose. But there does not seem to be sufficient boron in plants to form complexes with all the sugar in the phloem. It is also assumed that boron decreases the enzymatic conversion of glucose-1-phosphate to starch. This would make more sugar available for translocation. Overall, boron may affect translocation process due to alteration in sugar metabolism. It probably affects some of the enzymes involved in carbohydrate metabolism such as those involved in the synthesis of UDP-glucose.

Thus, the mass flow is not a purely passive process as was conceived by Münch. The bidirectional translocation of sugar from a photosynthetic leaf to the shoot apex and the root is explained by assuming that it takes place in separate sieve ducts.

13.4.3 Historical hypothesis

- *Interfacial flow hypothesis*

Van den Honert (1932) proposed that solutes move along the membranes interfaces such as the tonoplast or vacuolar membrane. The solutes are thought to be absorbed and distributed at the interface due to the decreased surface tension.

- *Diffusion hypothesis*

Translocation of solutes occurs from site of synthesis (high concentration) to the site of consumption (low concentration) by simple diffusion. This theory is not much convincing because the rate of translocation in phloem is much higher, ~40,000 times greater than what could actually be achieved by simple diffusion.

Secondly, phloem transport is dependent on metabolism and ceases if the phloem tissue is deprived of oxygen or is poisoned.

- *Activated diffusion theory*

This theory, proposed by Mason and Maskell (1936), assumes that the protoplasmic content of the sieve tubes accelerates the diffusion of solutes either by: 'activating' the diffusing molecules or by reducing the protoplasmic resistance to their diffusion process. However, this theory lacks any experimental evidence.

- *Electro-osmotic theory*

Spanner (1958) proposed that sugar molecules are tightly linked to K^+ in the sieve tube sap which are carried along with the movement of K^+ through the lumen of the sieve element. K^+ ion circulation can maintain electric potential at the sieve plate. Bowling (1968) found a potential difference of 4–48 mV in primary phloem of *Vitis vinifera* which was adequate for an electro-osmotic flow. However, this theory could not answer several problems of solute translocation.

- *P-protein theory*

According to this, P-protein present in the phloem seems to be contractile in nature and aids in transport of solutes by contraction and relaxation.

13.5 Phloem Loading and Unloading

13.5.1 Phloem loading

Translocation of assimilates from the photosynthesizing mesophyll tissues *via* the symplasm and plasmodesmata directly into the sieve element–companion cell complex (**phloem loading**) is an active process. Alternately, sucrose may be translocated through the membranes of mesophyll cell into the **apoplast** and from there, it would move across the membrane of the 'se-cc complex'and enter into the long-distance transport stream.

The **apoplastic model** for phloem loading received much support in the mid-1970s, based mainly on studies conducted in sugarbeet leaves. In apoplastic loading, the sugars enter into the apoplast quite near to the sieve element-companion cell (se-cc) complex, to be transported actively into the sieve elements with the aid of an energy-dependent, selective transporter confined to the plasma membranes of these cells (Figure 13.7).

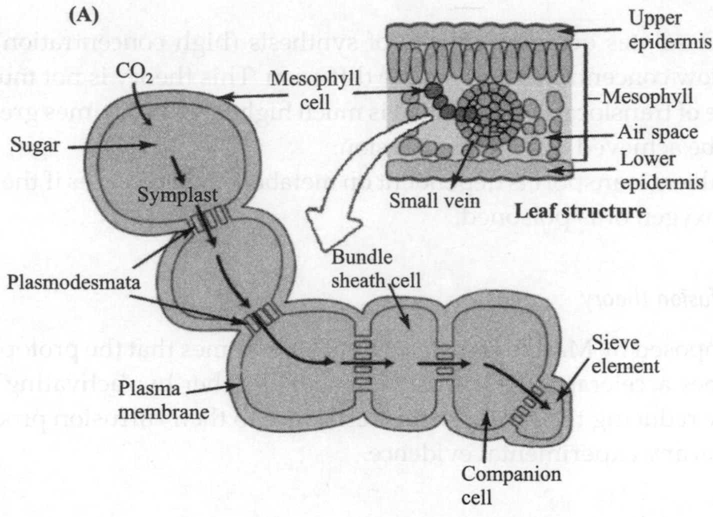

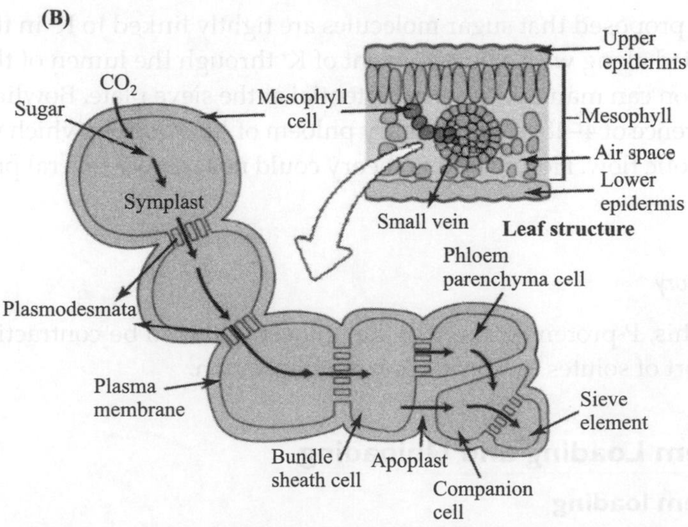

Fig. 13.7 Phloem loading in source tissues. (A) In symplastic phloem loading, sugars are transported from the photosynthetic mesophyll cells into the sieve elements *via* plasmodesmata. (B) In apoplastic phloem loading, sugars initially pass through the symplast but reach the apoplast prior to loading into the companion cells and further move into sieve elements through plasmodesmata.

Sucrose-proton (H^+) symporter is believed to regulate the movement of sucrose from the apoplast into the se-cc complex against a concentration gradient. Symport is a secondary active transport, utilizing the energy produced by the proton (H^+) pump. The energy dissipated by protons returning into the cell is linked to the intake of sucrose (Figure 13.8). Various sucrose-H^+ symporters have been identified in the phloem of spinach (*Spinacea oleracea*), potato (*Solanum tuberosum*), Greater plantain (*Plantago major*) and Thale cress (*Arabidopsis thaliana*). Two principal sucrose transporters, *SUT1* and *SUC2*, seem to be involved in phloem loading into the se-cc complex. Greater amounts of sucrose present in the se-cc complex as compared to adjoining cells suggests that sucrose is translocated actively against its chemical potential gradient.This is supported by the evidence that application of respiratory inhibitors to source tissues decreases ATP concentration as well as prevents loading of exogenous sugar.

Phloem loading by **symplastic pathway** is dominant in plants that (i) besides sucrose, translocate raffinose and stachyose in the phloem; (ii) possess intermediary cells in the minor veins of the leaves; (iii) have plenty of plasmodesmata leading into the minor veins. Examples are coleus, pumpkin, squash and melon.

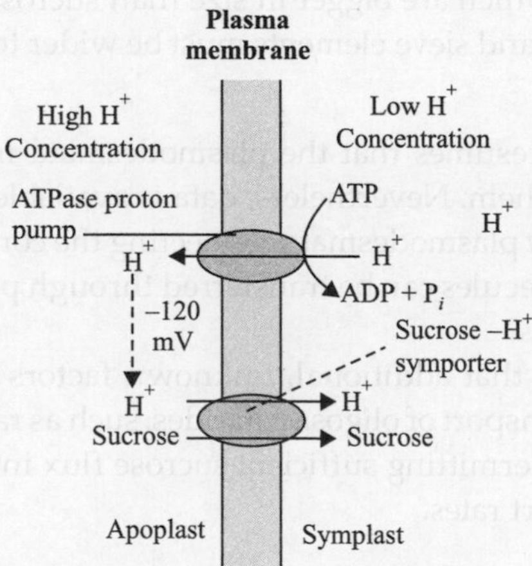

Fig. 13.8 Sucrose-H^+ cotransport. ATP-driven sucrose transport in apoplastic sieve element is linked to the uptake of H^+ *via* sucrose-H^+ symporter.

Two major questions arise regarding symplastic loading:

1. In several plant species, composition of sap in the sieve elements differs from the solute present in cells surrounding the phloem. This difference is indicative of the fact that some specific sugars are selected for translocation in the leaf (source).The association of sucrose-H^+symporters during apoplastic phloem loading explains a detailed process for selectivity, in contrast symplastic loading is dependent upon the diffusion of sugars from the mesophyll cells to the sieve elements *via* plasmodesmata. How can diffusion *via* plasmodesmata in symplastic loading be selective for certain sugars?
2. Analysis from various plants involving symplastic loading shows that sieve elements as well as companion cells have greater osmolyte quantities as compared to the mesophyll

cells. How can diffusion-driven symplastic loading explain the observed selectivity of translocated sugars and their accumulation against a concentration gradient?

The **polymer trap model** can answer these questions in species of cucurbits and coleus. This proposes that the sucrose produced within mesophyll cells diffuses from bundle sheath cells into the intermediary cells *via* plasmodesmata connecting them. In the intermediary cells, polymers of hexose sugars (such as, raffinose and stachyose) are produced from the transported sugars. Because of their large size, these polymers cannot diffuse back into the bundle sheath cells but can diffuse into the sieve elements and get carried away by mass flow. Sugar concentrations in the sieve elements of these plants can reach levels equivalent to those in plants that load apoplastically.

The polymer trap model is based on three assumptions:

1. Sucrose concentration should be higher in the mesophyll cells as compared to its concentration in the intermediary cells
2. Enzymes for the production of raffinose and stachyose should be preferably confined to the intermediary cells.
3. Plasmodesmata connecting the bundle sheath cells with the intermediary cells should eliminate molecules, which are bigger in size than sucrose. Plasmodesmata between the intermediary cells and sieve elements must be wider to allow passage of raffinose and stachyose.

The symplastic mode presumes that the plasmodesmata might restrict movement of bigger molecules through them. Nevertheless, data is available for passage of transporter protein as well as mRNA *via* plasmodesmata connecting the companion cells with the sieve elements. If such macromolecules can be transferred through plasmodesmata, why cannot the smaller oligosaccharides?

Recent models suggests that additional, unknown factors must be present to enable plasmodesmata to block transport of oligosaccharides, such as raffinose and stachyose, back into the mesophyll, while permitting sufficient sucrose flux into the intermediary cells to maintain observed transport rates.

Passive symplastic loading

It is quite clear that several tree species have a large number of plasmodesmata between the sieve element-companion cell complex and the adjoining cells, but do not have intermediary type companion cells and do not transport raffinose and stachyose. These plants have no concentrating step in the pathway right from the mesophyll cells into the sieve element-companion cell complex. Since a concentration gradient of sugars from the mesophyll cells to the sieve elements of phloem drives diffusion along this short-distance pathwy, the absolute levels of sugars in the leaves of these trees species must be high enough to maintain the required greater solute concentrations and the resulting high turgor pressures in the sieve elements.

Plasmodesmatal frequencies suggest that the passive loading mechanism is of ancient heritage in the angiosperms, while apoplastic loading and polymer trapping evolved later. Multiple loading mechanisms may allow plants to adapt quickly to abiotic stresses, such as low temperatures. Switching mechanisms may also reflect biotic stresses, such as viral infection.

13.5.2 Phloem unloading

Phloem unloading is the process whereby translocated sugars exit the sieve elements of sink tissues i.e., young roots, storage tubers, and reproducing bodies. After unloading, the sugars are translocated to cells within the sink through short-distance transport pathway. Finally, sugars are stored or assimilated in sink cells. An interesting aspect of phloem unloading is that, unlike the loading process, it lacks specificity.

Phloem unloading and short-distance transport can take place *via* symplastic or apoplastic pathways (Figure 13.9).

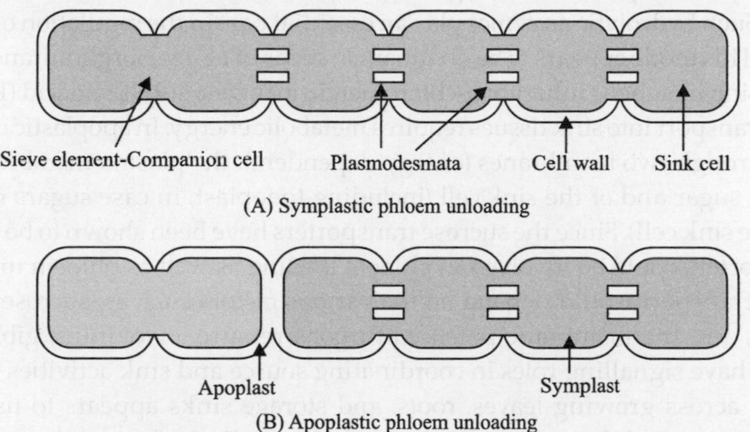

Fig. 13.9 Phloem unloading into the sink tissues. (A) Symplastic phloem unloading occurs entirely through plasmodesmata. The sieve element-companion cell complex behaves as one functional unit. (B) Apoplastic phloem unloading from the sieve element-companion cell complex takes place in the apoplast. The apoplastic step could be located at the site of unloading itself (as shown in figure) or farther away from sieve elements. Once the sugars have moved into the symplast of neighboring cells, the translocation pathway changes over to symplastic mode.

Within sink tissues, sugars are transferred from the sieve elements to the cells that accumulate or assimilate them. Other than developing vegetative tissues (i.e., root tips and young leaves), sinks also include storage tissues (such as, roots and stem) and reproductive and dispersal organs (such as, fruits and seeds). In sugar beet, sucrose actively accumulates in the vacuoles of sink cells in the root. By removing and consuming sugar, phloem unloading increases solute potential. Thus, water diffuses out of the sieve element-companion cell complex, causing the pressure at the sink end of a sieve element to be lower than that at the source end. The sugars may be translocated completely through the **symplast** *via* plasmodesmata in sink organs, or might move into the **apoplast** at some point of time. In young eudicot leaves (such as, sugarbeet and tobacco) as well as meristematic and elongation zones of primary root tips, both unloading process and the short-distance pathway seem to be entirely symplastic. In developing seeds, the apoplastic mode predominates because of absence of the symplastic links between the maternal and the embryonic tissues. Sugars leave the sieve elements through the symplastic mode and are transported from the symplast to the apoplast at some point withdrawn from the se-cc complex. Either the transport sugar i.e. sucrose can pass through the apoplast unaltered or can undergo hydrolysis in the apoplast itself by the action of enzyme acid invertase, strongly held to the cell wall. The hydrolysed products (such as, glucose and fructose) would make an entry into the sink cells and again combine to form sucrose which gets actively translocated into the vacuole to be stored.

$$\text{Sucrose} + H_2O \xrightarrow{\text{Acid invertase}} \text{glucose} + \text{fructose}$$

Sucrose hydrolysis within the apoplast, coupled with the irreversible action of the invertase enzyme, helps in sustaining the gradient and permits the unloading process to keep going. Such hydrolytic enzymes play an essential part in the regulation of transport by sink tissues. This mode appears to be dominant in seeds of maize, sorghum, and pearl millet.

Studies using metabolic inhibitor *p*-chloromercuribenzene sulphonic acid (**PCMBS**) have shown that transport into sink tissues requires metabolic energy. In apoplastic import, sugars must pass through two membranes (energy-dependent): the plasma membrane of the cell transporting sugar and of the sink cell (including tonoplast, in case sugars move into the vacuole of the sink cell). Since the sucrose transporters have been shown to be bidirectional, same transporters could be involved in phloem loading as well as phloem unloading. The direction of transport would depend on the various factors such as, sucrose gradient, the pH gradient, and the membrane potential.Turgor pressure, cytokinins, gibberellins and abscisic acid have signalling roles in coordinating source and sink activities

Transport across growing leaves, roots, and storage sinks appears to use symplastic phloem unloading and short-distance transport. Generally, in these sink organs, there is a requirement of metabolic energy for respiration and biosynthetic reactions, since transport sugars pass from the sieve elements to the sink tissues passively without crossing any membranes. In leguminous plants, sucrose unloading occurs into the apoplast using an ATP-dependent carrier. In contrast to phloem loading, there does not seem to be a common mechanism for phloem unloading into the growing embryo.

The mechanism of loading and unloading at the source and the sink respectively offer the motive power for long-distance transport and hence of great importance from agricultural viewpoint. An in-depth knowledge of such processes becomes the basis for technology geared towards increasing crop yield by enhancing the storage of photoassimilates in edible sink organs (such as cereal grains).

13.6 Photoassimilate Distribution: Allocation and Partitioning

The distribution of photoassimilates among plant metabolic pathways and its organs takes place at two stages: **allocation** and **partitioning**.

Allocation determines the instant metabolic fate of photoassimilates. These might be allotted for the immediate metabolic requirements of the leaf or accumulated to be assimilated during non-photosynthetic periods or exported out of the leaf to be partitioned among various competing sink organs. ***Partitioning*** is based on **sink strength,** comprising of two components - sink size and sink activity.

$$\text{Sink strength} = \text{Sink size} \times \text{Sink activity}$$

Sink size refers to the total mass of the sink organs, expressed as dry weight. **Sink activity** is the rate of assimilate intake, represented by per unit dry weight of sink per unit time. Determination of sink strength is quite helpful in analyzing the pattern of translocation of photoassimilates and is therefore closely related to crop productivity and yield.

Partitioning of assimilates between competing sinks is dependent upon three factors: **the placement of the sink with regard to the source, nature of vascular links between source and sinks, and sink strength.** Competition for the photosynthates among various sinks also determines the partitioning of transport sugar. Partitioning results in determination of growth patterns, which must be balanced between shoots and roots in such a way that the plant can respond to the challenges of a variable environment. Therefore, a secondary line of control has evolved between zones of supply and demand. Various factors such as turgor pressure, plant hormones, nutrients (potassium and phosphate ions), and transport sugars themselves, could be an important mode of communication between the source and the sink. Recent findings suggest that macromolecules (RNA and protein) may also play a role in photosynthate partitioning, perhaps by influencing transport through plasmodesmata.

Crop yield is essentially synonymous with sink size. But, **What is it that determines yield - the capacity of the sink or the strength of the source**? For a given crop, yield may be sink-limited under one set of conditions and source-limited under another. Another question is: **how is sink demand conveyed to the source, or source capacity signaled to a potential sink?** The answer depends on type of species and circumstances. When the sink becomes full, biosynthesis may close down, precursors back up in the translocation system, pool sizes of intermediates in source metabolism increase and assimilation is downregulated by feedback. In the case of events which reduce source output, such as defoliation, shading or disease, slowly developing and undersized sinks are formed. The nature of the signal between source and sink is also diverse. Nutrients, and products of photosynthesis (nitrate and sucrose) are mobile integrators of ***source-sink physiology***, acting both at the level of metabolism and through gene expression networks. **Hormones as well as mobile proteins i.e. flowering factor FT and antipathogenic signal systemin, also play a signalling role and are part of the source-sink communication system.**

Achievement of greater plant productivity is the major challenge on photosynthate allocation and partitioning. Total yield includes edible portions (grains and fruits) as well as inedible portions of the shoot. Plant breeders are constantly making efforts to enhance crop yield while modern plant physiology is playing its role in increasing crop yield based on a fundamental understanding of metabolism, development and partitioning. However, allocation and partitioning in the entire plant must be synchronized in such a manner that enhanced translocation to edible tissues does not take place at the cost of integral physiological phenomena.

The regulation of phloem loading and unloading as well as source-to-sink communication continues to be the promising avenues of exploration in times to come. Mobility in the phloem tissues is of special interest to the agrochemical industry for synthesizing **xenobiotic[2] compounds**. The degree of uptake and transportation of these compounds generally decides their usefulness as herbicides, growth regulators, fungicides or insecticides.

Two forces (**Diffusion** and **Bulk Flow**) drive transportation of water and solutes in plants. Diffusion can account for the movement of molecules over short distances, for example, water movement across membranes through osmosis. Bulk flow is pressure-driven transport of a fluid and its solutes as a unit. This differs from diffusion in which each molecular species moves down its own concentration gradient. An example of bulk flow is the flow of water in the pipes of a house or blood flow in an animal circulatory system. In vascular plants, for transport over distances exceeding several millimetres, the movement of solutes takes place by bulk flow.

Review Questions

RQ 13.1. Starch is never transported in plants yet potato tubers, which grow underground, are full of it. Explain.

Ans. Sucrose is the primary metabolite as well as the primary mobile product produced during photosynthesis, and is actively pumped into the sieve tubes of the smallest veinlets of the leaf, adjacent to xylem vessels **(phloem loading)**. Loading is believed to involve cotransport of sucrose and H^+ ions through the specific permease in response to pH and electrochemical gradient. H^+ ion is later re-secreted through an ATP-utilizing proton transporter.

As concentration of sugar increases in the sieve tubes, their water potential in turn decreases, and the water is thus withdrawn from the xylem vessels by osmosis. The turgor pressure thus built-up 'forces' the solution to 'move passively' through the sieve tubes, carrying it towards the root system and then onto the developing storage organs like tubers (or rhizomes and bulbs, and such-like) that represent the 'consumption end or sink'. It is here that the soluble sugars are actively removed **(phloem unloading[3])** and are used up or else converted into insoluble starch through the enzymatic reactions as given below:

$$\text{Sucrose} + \text{UDP} \rightarrow \text{UDP– glucose} + \text{Fructose}$$

As 'ADP-glucose' is favored for synthesis of starch, UDP-glucose is changed to ADP-glucose as follows.

---

[2] 'Xenobiotic compounds' can be defined as chemical substances that are not naturally produced in an organism (or biologically active molecules that are foreign/alien to an organism).

[3] Except for 'loading and unloading' of sucrose and other carbohydrates, the process of 'mass flow' is purely physical and does not need energy. The energy required for loading and unloading is supplied by the companion cells or parenchyma next to sieve tubes.

UDP-glucose + Pi $\rightleftharpoons$ UTP + Glucose-1-phosphate

Glucose -1-phosphate + ADP $\rightleftharpoons$ ADP – glucose + PPi

ADP-glucose formed changes later to starch using **'starch synthase'**.

Sugar conversion or utilization guarantees the continued existence of a turgor pressure gradient from 'source' (the leaves) to 'sink' (the roots); thus facilitating flow of metabolites. All movements of carbohydrates and even much of that of nitrogen compounds whether upward or downward, occurred in the phloem.

RQ 13.2. What role aphids have played in the studies of phloem translocation?

Ans. While feeding on herbaceous plants, aphids insert their stylet into a single sieve tube element, and the sugary phloem sap passes through aphid's intestines and is released in a chemically altered form as 'honeydrop'.

If a feeding aphid is anaesthetized and then its body is removed or severed, the stylet is left embedded in the sieve tube as 'microcanula' through which the sap continues to ooze out, indicating that the phloem contents are under pressure. The pure sieve-tube sap is directly collected from the stylet and then chemically analysed. The phloem sap contains 0.2–0.7 M sucrose plus other sugar alcohols (such as mannitol and sorbitol). In addition to sucrose, the sap contains small amounts of other sugars such as raffinose, stachyose and verbascose, traces of glucose and fructose, amino acids, amides, auxins, organic acids and minerals. More than 90 per cent of the dry weight of sieve tube sap contains sugars.

The aphid-stylus experiment on phloem transport was first conducted by Zimmerman at the Harvard Forest, Petersham, Massachusetts in 1961.

- Martin Zimmermann at the Harvard Forest measured the velocity of phloem translocation in ash trees (*Fraxinus*), i.e., it was about at the rate of 30–70 cm/h.
- Authored a book on '*Xylem Structure and Ascent of Sap*'. Springer, Berlin (1983).

RQ 13.3. What is the girdling or ringing experiment? What conclusions do you derive from such experiments?

Ans. Italian anatomist and microscopist Marcello Malphighi (of kidney malphighian tubule fame) is credited for this discovery of such 'girdling' or 'ringing' experiments.

- In the 'first' experimental set-up, a ring of bark, i.e., the phloem tissues (about 1 inch in length) outside the xylem from around a woody twig is removed without injuring the vascular cambium and the central xylem core.
- In the 'second' parallel set-up, by careful manipulation, the wood elements, i.e., xylem elements are scooped out a bit, leaving the bark with phloem and cortex intact. The girdling operation is carried out while the twig is still under water, and later both the set-ups are kept in a basal medium containing both the micro- and macro-nutrients.

It was discovered that in the former case, the leaves remain fresh and showed no sign of wilting. However, the portion of the twig just above the girdle becomes swollen due to the accumulation of food material as these are not being transported downward because the bark tissue with phloem has been removed (Figure 13.1).

In the second twig, where the xylem was cut and the bark (i.e., the phloem tissue) intact, the leaves undergo wilting (loss of turgor) in a short while, indicating that the translocation of water and minerals is through the xylem. This further shows that the outer tissues (i.e phloem) are not involved in the ascent of sap.

RQ 13.4. What will happen if the tissue outside the central xylem core of a rooted woody plant (with a single ring of collateral vascular bundle) is removed?

Ans. In such girdled plants, the roots die first as the 'operation' cuts off the food supply from the leaves to the roots. The roots cannot synthesize their own carbohydrates and other organic compounds, thus they die of starvation. As the roots gradually die, the upper part of the plant, above the girdle, will be deprived of water and minerals, with the result that the whole plant will die ultimately.

RQ 13.5. Girdling is of much practical value in the case of fruit trees. Explain.

Ans. While the fruits are developing, ringing or girdling of the leafy twigs in the basal region (i.e., removal of bark only) would prevent the downward flow of food material across the ring, with the result that all the food material synthesized in the leaves above the girdle is diverted to the fruits. Thus, the fruits would become much larger in size and of better quality. In practice, only a narrow ring is made without injuring the vascular cambium so that after sometime it would be able to regenerate the tissue outside and the wound would heal up.

RQ 13.6. Why is it that synthesis of sucrose, takes place 'solely' in the 'cytoplasm' of the photosynthetic cells? Outline the two routes of sucrose production in photosynthetic organisms.

Ans. Sucrose synthesis takes place in the 'cytoplasm' of photosynthetic cells because it can be effectively loaded from the smaller veins of the leaves into the sieve tubes of phloem. If sucrose was to be produced in the 'chloroplasts', it would be unable to exit as the inner chloroplastic membrane is not permeable to sucrose and thus would not allow it to reach the translocation stream.

There are two possible routes of sucrose synthesis in the photosynthetic cells but the principal pathway is regulated by the enzyme sucrose phosphate synthase:

1. $$\text{UDP-glucose} + \text{Fructose-6-phosphate} \xrightleftharpoons{\text{Sucrose phosphate synthase}^4} \text{UDP} + \text{Sucrose-6-phosphate}$$

$$\text{Sucrose-6-phosphate} + H_2O \xrightleftharpoons{\text{Sucrose phosphate phosphatase}} \text{Sucrose} + \text{Pi}$$

2. Another cytoplasmic enzyme capable of synthesizing sucrose is 'sucrose synthase'.

$$\text{Fructose} + \text{UDP-glucose} \xrightleftharpoons{\text{Sucrose synthase}} \text{UDP} + \text{Sucrose}$$

Sucrose synthesis in either of the both pathways, unlike starch biosynthesis, requires activation of glucose with the nucleotide UTP instead of ATP.

$$\text{Glucose-1-phosphate} + \text{UTP} \rightleftharpoons \text{UDP-glucose} + \text{PPi}$$

$$\text{PPi} + H_2O \rightleftharpoons 2\text{Pi}$$

RQ 13.7. Explain the empty-ovule (empty-seed) technique of phloem unloading in legume seeds.

Ans. The Empty-Ovule Technique (Figure 13.10) was developed by three groups of people, working independently and almost simultaneously in 1983 in three different countries–the United States (Thorne and Rainbow), the Netherlands (Wolswinkel and Ammerlaan), and Australia (John Patrick) to study the mechanism of phloem unloading into legume seeds. Later this technique was extended to non-leguminous seeds as maize and other species (Shannon et al., 1986).

---

[4] Sucrose phosphate synthase in some tissues can even use ADP-glucose but UDP-glucose is the most predominant form often used.

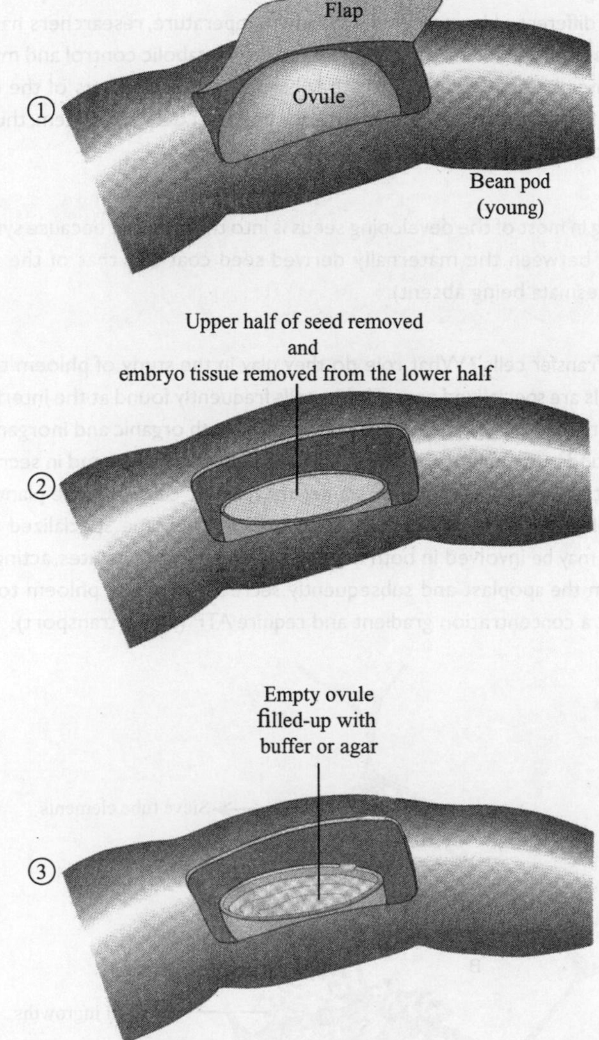

Fig. 13.10 The empty ovule technique used for studying phloem unloading in legumes. In step 1, a flap is cut into the walls of young bean pod; step 2, the distal half of the seed is surgically removed; step 3, the embryo tissues of the remaining half-seed are scooped out leaving behind only cup-shaped structure (the empty ovule) which is filled up with a buffer or agar to trap photoassimilates from the vascular tissue.

Incisions are made in the pod wall and the flap over a window is cut back, exposing a developing seed. The top (distal) half of the seed is decapitated. Then the embryo inside the lower half of the **'seed cup'** is carefully removed with a small spatula without damaging the maternal seed coat lining the endothelium. This practice exposes the maternal seed coat and the phloem 'unloading sites' meant for the nutrition of the embryo. The seed cup is then filled with 4 per cent warm agar which quickly solidifies or with a buffer solution (agar trap).

Phloem unloading of assimilates from the sieve tubes that now occurs, which would normally be absorbed by the developing embryo but now diffuse into the agar or buffer trap.

By using different metabolic inhibitors like cyanide, azide, PCMBS (it makes the source carrier ineffective), different pH ranges and even low temperature, researchers have conclusively proven that assimilate (or phloem sap) unloading is under metabolic control and may well involve carriers located in membranes. Furthermore, they have studied the effects of the environmental control of different factors on solute (photosynthate imports) delivery system, thus indirectly the whole process of phloem unloading.

Unloading in most of the developing seeds is into the apoplast because symplastic connection is lacking between the maternally derived seed coat and that of the developing embryo (plasmodesmata being absent).

RQ 13.8. What are 'Transfer cells'? What role do they play in the study of phloem transport ?

Ans. Transfer cells are specialized parenchyma cells frequently found at the interface region of symplast and apoplast (cell walls) where active transport of both organic and inorganic solutes is occurring. They are found in the xylem and phloem, in reproductive tissues, and in secretory glandular tissues (such as salt glands). Even, there are epidermal transfer cells. In some plants, the companion cells in the 'smallest' or 'minor' leaf veins (Figure 13.11) become specialized to form transfer cells where they may be involved in both loading and unloading of solutes, acting like a lens, scavenging solutes from the apoplast and subsequently secreting it out of phloem to specific neighbouring cells against a concentration gradient and require ATP (active transport).

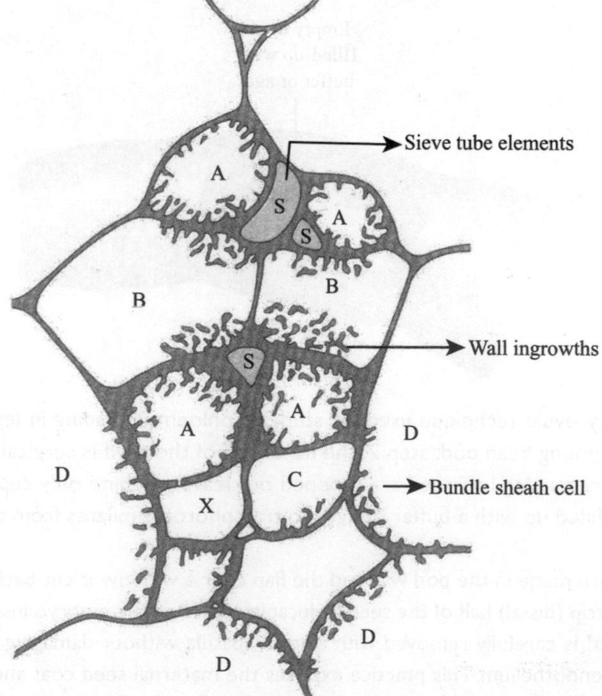

Fig. 13.11 A cross-section of a minor leaf vein showing four types of vascular transfer cells; (A) modified companion cells; (B) modified phloem parenchyma cells, both associated with sieve tube elements (S); (C) modified xylem parenchyma; and (D) modified bundle sheath cells – both associated with xylem (X). Seen are characteristic wall ingrowths in transfer cells.

Transfer cells are characterized by 'ingrowths'of cell walls that enhance the surface area of the plasma membrane, numerous plasmodesmata and dense cytoplasm with many mitochondria. The transfer cells have very distinctive ingrowths on the walls next to the parenchyma cells but not on the walls adjacent to the sieve elements where plasmodesmata are present.

Sucrose from leaf mesophyll cells (source) passes through plasmodesmata along a symplast route until it reaches the phloem and enters the apoplast. The apoplast transfer site could be a companion cell or sieve tube element or even a phloem parenchyma cell.

However, apoplastic transfer is a critical step because it is from the apoplast that active transport occurs and sucrose accumulates in phloem cells. A membrane carrier protein in the sucrose transport complex binds sucrose and a proton (H^+) at the outer surface, undergoes conformational change, and moves the sucrose and proton to the inner surface, where they are released.

Pate and Gunning (1972) proposed the involvement of special cells called 'transfer cells' in the movement of salts into the xylem elements.

Unit III Growth and Development

Chapter 14

Dormancy, Germination and Flowering

Keywords

- Critical photoperiod
- Cryptochrome
- Florigen
- High irradiance responses
- Photoblastic seeds
- Photoinductive cycle
- Photoperiodism
- Scarification
- Stratification
- Vernalization

The growth and development of a flowering plant is a complex phenomenon. It begins with seed germination, accompanied by a series of large number of morphological and physiological changes termed as **growth and development**. The plant is in part a consequence of its genetic composition and in part a consequence of the environment in which it grows and develops. To some extent, the genome appears to be preprogrammed, but the program is highly modified by environmental factors.

The terms ***Growth, Differentiation*** and ***Development*** are customarily used to describe different aspects of the events that a plant undergoes during its life span. *Development* is an expression of the genetic program that governs the activities and interactions of individual cells. Development is the sum total of two processes: growth and differentiation. *Growth* is a quantitative term, related to an increase in size and mass, while *differentiation* is a series of qualitative changes in the cell, tissue or organ. Many parameters (increase in number of cells/fresh weight/dry weight/length and width/area and volume) could be invoked to measure growth depending upon the need of a researcher. Whatever the measure, *growth is an irreversible increase in volume or size followed by the synthesis of new protoplasmic constituents.*

Growth comprises of three phases: (i) cell division, (ii) cell enlargement, (iii) cell maturation.

The orderly development of a complex multicellular organism is coordinated by internal and external controls. 'Internal controls' are expressed at both the intracellular and intercellular levels. Intracellular controls are primarily genetic, requiring a programmed sequence of gene expression. Intercellular controls are primarily hormonal, chemical messengers that permit cells to communicate with each other. 'External controls' are environmental such as light, temperature and gravity. Many environmental variables seem to act at least in part by altering gene expression or hormonal functions.

14.1 Growth Curve

The **sigmoid** (S-shaped) growth curve is characteristic for virtually growth in all living organisms. It can be classified into three phases: *Lag phase, Log phase,* and *Stationary phase* (Figure 14.1).

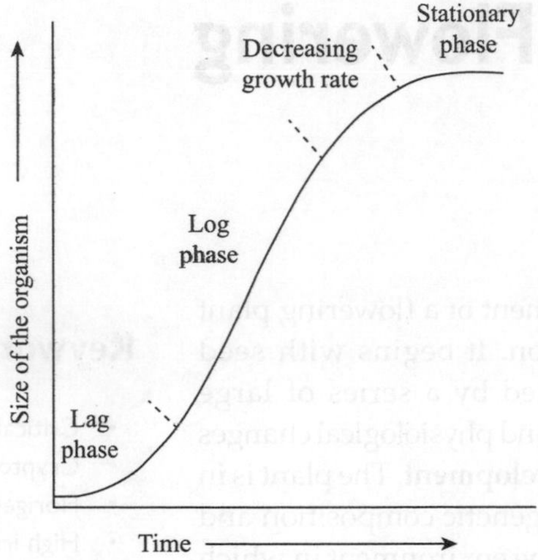

Fig. 14.1 The sigmoid growth curve in living organisms.

Lag phase

During this period, growth with time is rather slow but still occurs with an increasing rate. It extends between days 15 and 20. Initial plant growth is restricted by the quantity of food reserves present in the seed.

Log phase

Growth rate is proportional to time. Between days 20 and 90, the period of rapid increase in size was called the 'Grand Period of Growth' by von Sachs (1887). The log phase of growth is frequently called the 'phase of exponential growth' and high metabolic rates are maintained during this period.

Stationary phase

After the 'Grand period of Growth', a decreasing growth phase occurs, during which there is no further increase in size with time and is referred to as stationary phase of growth (senescence phase). The reduced growth could be due to several factors like competition for essential metabolites/growth substances/water or light, and the accumulation of inhibitors/ toxic materials or waste products.

14.2 Seed Germination

The life cycle of a flowering plant begins with the process of seed germination. The seed which represents a dormant stage-germinates, grows into a mature plant, shifts from vegetative to reproductive growth to form floral organs and seeds following pollination and fertilization.

Dry seeds remain inactive often for long periods. When the seed is placed in the soil under suitable conditions, the embryo becomes activated and resumes its activity, resulting in the production of a young seedling. The sequence of steps beginning with the uptake of water and leading to the cracking of the seed coat by the radicle or the shoot is called **Seed germination**.

The process of germination occurs in three phases based on the water uptake:

Phase I: The dry seed takes up water rapidly by the process of imbibition. Since water flows from the higher to the lower water potential, the water absorption by the seed ceases when the water potential between the seed and its environment becomes the same i.e., zero. In the dry seeds, the matric potential (ψ_m) also lowers the water potential (ψ) and creates the gradient. Imbibition ceases when ψ_m becomes less negative.

Phase II: The embryonic cells expand and the radicle emerges rupturing the seed coat. Metabolic activities including transcription and translation increase followed by cell wall loosening. During this phase, the osmotic potential (ψ_s) of the embryo gradually becomes more negative due to conversion of stored food reserves to osmotically active compounds.

Phase III: Water uptake resumes because of reduction in water potential as the seedling becomes established. The stored food reserves of the seed are completely mobilized that provide nutrients to the growing seedling until it becomes autotrophic.

The sequential steps taking place during the seed germination are as follows:

(i) Intake of water through imbibition
(ii) Hydration of the protoplasmic constituents
(iii) Synthesis and activation of enzymes such as amylases, lipases, proteases and such .
(iv) Increase in rate of respiration
(v) Increased production of nucleic acids and proteins
(vi) Synthesis and release of growth hormones from the embryo
(vii) Increase in cell division and cell enlargement
(viii) Hydrolytic breakdown of reserve food materials present in the endosperm/cotyledons
(ix) Consumption of soluble organic materials by the developing embryo
(x) Development of the radicle and plumule into the root and shoot respectively.

14.2.1 Mobilization of food reserves

Carbohydrates, proteins and lipids comprise the major food reserves in the cotyledons/ endosperm.

Starch stored in amyloplasts in the endosperm of the cereals[1] is degraded by α- and β-amylase.

$$\text{Starch} \xrightarrow{\alpha\text{-amylase}} \text{Oligosaccharides} \xrightarrow{\beta\text{-amylase}} \text{Maltose} \xrightarrow{\text{Maltase}} \text{Glucose}$$

The thickened cell walls of endosperm tissue in some seeds provide another source of carbohydrates for the growing seedling.

Proteins stored in storage vesicles are degraded to amino acids to be used for new protein synthesis in the seedling. In addition, **phytin** present in protein storage vesicles gets hydrolyzed by the enzyme phytase to release phosphate and other ions for use by the growing seedling.

Lipids stored in oleosomes of fatty seeds are also converted to simple sugars *via* the glyoxylate pathway and used for the nourishment of the seedlings.

Gibberellins are synthesized by the embryonic region and released into the aleurone layer of starchy endosperm through the scutellum. Living cells of the aleurone layer are activated to synthesize and secrete hydrolytic enzymes, including α-amylase, into the endospermic tissue during germination. As a consequence, the stored food reserves of the endosperm are broken down to smaller molecules such as sugars as well as amino acids, and are carried to the growing embryo *via* the scutellum.

There is enough evidence now for gibberellin-enhanced α-amylase synthesis at the gene transcriptional level. ABA inhibits the gibberellin-induced α-amylase production by preventing the transcription of α-amylase mRNA. The gibberellins receptor, gibberellins insensitive dwarf1 (GID1) promotes the degradation of negative regulators of α-amylase production, including DELLA proteins, thereby upregulating GA-MYB[2] proteins and α-amylase transcription. Within the aleurone cells, there are both Ca^{2+}-independent and Ca^{2+}-dependent gibberellin signalling pathway – the former leads to the production of α-amylase, while the latter regulates its secretion.

14.2.2 Factors affecting germination

Water: The water content of mature seeds is well below the threshold required for active metabolism. Seeds at the time of harvest contain only 8–10 per cent moisture and several hydrophilic colloids like starch, proteins and lipids. Therefore, these exert a matric potential (ψ_m) of –100 to –200 bars, often exceeding 1000 bars. Consequently, the dry seeds can take up large quantity of water even from relatively dry soil by the process of imbibition. Imbibition activates hydration of protoplasm and metabolic activities leading to germination. In addition, water uptake is required to generate the turgor pressure that powers cell expansion, the basis of vegetative growth and development.

---

[1] Cereal grains can be cut in two, and 'half-seeds' that lack the embryo make a convenient experimental system for studying the action of applied gibberellin (embryo is the source of bioactive gibberellin).

[2] MYB proteins are a class of transcription factors in all eukaryotes, including plants, implicated in GA signalling. GA-MYB activates α-amylase gene expression.

***Temperature*:** Germination is most rapid at an optimum temperature, being lower at the minimum and the maximum temperature. Generally, seeds of temperate plants have a lower optimum (20–25 °C) for seed germination than seeds of tropical species (30–35 °C). Most seeds can germinate at a constant temperature, both during the day and night with a few exceptions like *Oenothera*, *Apium*, etc.

***Oxygen*:** Seeds also require oxygen for germination and fail to germinate in an atmosphere of nitrogen and carbon dioxide, despite other conditions being favourable. However, the oxygen requirement varies from species to species. Increase in oxygen concentration beyond 21 per cent facilitates better diffusion through the hard seed coat and hence, better germination.

***Light*:** Seeds of most species can germinate equally well both in the dark and in light. Therefore, light is not considered to be necessary for germination. However, in a few species, seeds do not germinate until they are subjected to light, or after imbibition is complete and in still others light inhibits germination.

***Fire*:** Seeds of species adapted to global ecosystems such as grasslands, savannas, Mediterranean shrub-lands and boreal forests subject to regular burning are often dormant until they are exposed to heat and smoke from burning vegetation. Stimulation of seed germination by smoke from fires has been reported in approximately 170 species. For example, the Australian legume (*Acacia holoserica*) with a thick testa layer will not germinate readily until heat from a fire source forces the plug in the seed to come out. Recent work has shown that smoke from forest fires induce germination of yellow whispering bells (*Emmenanthe penduliflora*). Smoke has many components but nitrogen dioxide (NO_2) present in it is sufficient to trigger germination. A new class of growth-promoting chemicals derived from smoke, such as **karrikins** have been identified (Figure 14.2).

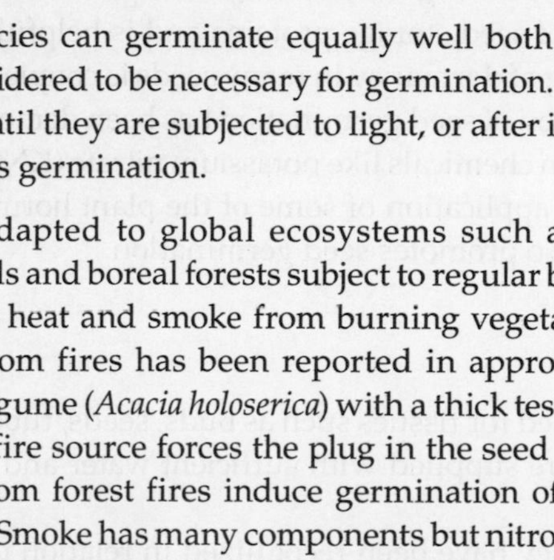

KAR_1 KAR_2

KAR_3 KAR_4

Fig. 14.2 Chemical structure of karrikins - bioactive germination promoters

Karrikins bear chemical similarity to strigolactones. The core structure of the bioactive karrikins is a pyran ring (comprising of five carbons and an oxygen atom) attached with a furan ring (made up of four carbons and one oxygen atom) bearing a keto group. Not only species of ecosystems regularly subjected to burning, but even *Arabidopsis* seeds respond to smoke-water. KAR_1 has been observed to induce expression of two key GA biosynthetic enzymes, GA3 oxidase 1 (GA3 ox 1) and GA3 oxidase 2 (GA3 ox 2), in dormant seeds of *Arabidopsis*. It is believed that karrikins work by interacting with GA/ABA signalling networks. There are even commercial 'smoke-water' preparations available for horticultural use.

14.2.3 Germination inhibitors and promoters

Many compounds may inhibit germination and usually act by blocking some process essential to germination. Natural germination inhibitors (ammonia, phthalides, ferulic acid, *para* ascorbic acid, cyanogenic compounds, coumarin and others) are present either in the embryo or in the seed coats. However, ABA is the most potent natural inhibitor. There is an increase in the level of ABA during initial seed development in many plant species which stimulates the induction of seed storage proteins and is helpful for inhibiting precocious germination. The breaking of dormancy in many seeds is coupled with decreasing levels of ABA in the seeds. Promotion of seed germination has been demonstrated in several species by the application of certain chemicals like potassium nitrate (KNO_3), thiourea and ethylene-chlorhydrin. Besides this, application of some of the plant hormones like gibberellic acid, cytokinin and ethylene also promotes seed germination.

14.3 Dormancy

The term 'dormancy' is used for tissues such as buds, seeds, tubers, and corms that cannot grow even though they are supplied with sufficient water and oxygen level at a suitable temperature.

Three types of dormancy have been recognized in relation to seasonal conditions and growth:

Ecodormancy is quiescence due to limitations in environmental factors. For example, the resting state of mature onion bulbs kept at low temperature.

Paradormancy is inhibition of growth by the influence of another part of the plant. For example, suppression of growth of lateral buds because of apical bud (apical dominance).

Endodormancy is inhibition residing in the dormant structure itself, e.g. seed dormancy and bud dormancy. In **developmental endodormancy** the quiescence of the structure is associated with physiological or structural immaturity, and in **seasonal endodormancy,** the intrinsic dormancy programme is executed in response to a sequence of environmental cues. During the seasonal cycle, a quiescent structure may pass through different modes of dormancy e.g. in *Salix,* the resting condition of buds shifts from paradormancy (in summer) to endodormancy (during winter) to ecodormancy (in spring).

14.3.1 Seed dormancy

Mature seeds typically have less than 0.1 g water/g dry weight at the time of shedding. Metabolic activities come to an end because of dehydration, with the seed entering into a quiescent (resting) phase. Quiescent seeds germinate upon rehydration, whereas dormant seeds require some signal or treatment for germination to occur. Most of the seeds can germinate immediately after maturation, e.g., maize and citrus; seeds of the para rubber tree, sugar maple and willow are viable only for a short time; some seeds are able to retain their viability for several hundred years[3] (*Mimosa glomerata*, *Albizzia julibrissin*[4]) while in many other plants, some of the viable seeds fail to germinate even if all the essential environmental conditions for growth are fulfilled. There is something that acts as a block to the processes leading to germination. The arrest in growth and development of seeds, buds and other plant parts under conditions seemingly suited for growth is referred to as '**dormancy**'.

Seed dormancy can be of different types based on the developmental timing of dormancy onset:

Primary dormancy: Newly dispersed, mature seeds fail to germinate under normal conditions. It is induced by ABA during seed maturation.

Secondary dormancy: Once primary dormancy has been lost, non-dormant seeds may acquire secondary dormancy if exposed to unfavourable conditions that inhibit germination over a period of time.

Seed dormancy may be due to various reasons, mainly: **Hard seed coat and immature embryo.**

(i) Hard seed coat

Seeds of many species of the families, particularly Fabaceae and Malvaceae remain dormant because of hard seed coat. The primary mechanisms of coat-imposed dormancy are inhibition of water uptake, mechanical limitations, interference with gaseous exchange, retention of inhibitory compounds and synthesis of inhibitor.

Mechanical constraint: In some seeds such as *Arabidopsis*, coffee, tomato, tobacco, the thick-walled endosperm may be too hard for the radicle to emerge out. For germination of such seeds, the cell wall of endosperm must be made weak by synthesis of cell wall-degrading enzymes.

The seeds of certain lupines possess unique hygroscopic valve, hilum, that limit the absorption of water, thus inhibiting germination before the seed coat is damaged by other methods.

Interference with gaseous exchange: Seed coat and other surrounding tissue limit the supply of oxygen to the embryo sometimes. The oxidative reactions involving phenolic compounds in the seed coat may consume large amounts of oxygen, thereby reducing oxygen availability to the embryo. In such seeds, dormancy can be overcome by exposures to an oxygen-enriched atmosphere.

Retention of inhibitors: Dormant seeds generally contain secondary metabolites, including phenolic acid, tannins and coumarins. The seed coat may impose dormancy by

[3] Indian lotus (*Nelumbo nucifera*) seeds have survived up to 1200 years because of their impermeable coats.

[4] During the Second World War (1942) when the British Museum Herbarium in London was damaged, 147 years old seeds of *Albizzia julibrissin* became soaked with water during the ensuing fire fighting operation and started germinating.

preventing release of inhibitors from the seed or they may enter into the embryo from the testa and prevent germination. Repeated rinsing of such seeds with water often promotes germination.

However, such dormant seeds promptly germinate under favourable conditions on removal of the seed coat – a phenomenon known as **Scarification**. This refers to any method that renders the seed coat permeable to water and oxygen or breaks the seed coat so that embryo expansion is not physically retarded.

Scarification is of two types: **Mechanical** and **chemical.**

Mechanical scarification: involves scratching the coat with a knife or shaking the seeds with some abrasive material (sand) to promote germination.

Chemical scarification ; involves the use of strong acid (H_2SO_4), organic solvents (acetone, alcohol) and sometimes even boiling water.

Scarification	Stratification
1. It is a treatment for breaking seed dormancy caused by hard testa layer that prevents uptake of water and exchange of oxygen.	The freshly harvested seeds of many temperate species remain dormant and do not germinate, if planted under moist conditions at 20 °C but will germinate, if kept under moist conditions at 0–5 °C for several weeks (chilling).
2. It involves rupturing, puncturing or weakening of the seed coats* by different methods such as: (a) mechanical injury (filing, rubbing with abrasives, chipping, etc.). (b) by treating with concentrated sulphuric acid for a very short duration, followed by thorough washing with water. (c) treatment with hot water. Examples: seeds of many species of Fabaceae and Malvaceae.	Dry seeds** cannot be stratified. Therefore, such seeds are allowed to imbibe water (to have a minimum amount of moisture) before exposing to low temperatures. During stratification, the inhibitory hormone (ABA) disappears but growth-promoting hormone (gibberellic acid) appears. The chilling requirement can be substituted by applying artificially the gibberellic acid to such seeds. Gibberellin-treated seeds can germinate without any chilling treatment. Examples: seeds of apple, walnut, rose, peach, *Picea, Pinus* and *Tilia*.

* However, in nature, dormancy, due to hard seed coats, is overcome by microbial decay of the seed coat while the seeds lie buried in the soil.

****Recalcitrant seeds** are those which die very quickly when subjected to the storage conditions at about 5 per cent humidity and about −20 °C or less (that are satisfactory for orthodox seeds). Examples of this particular type of seeds are many tropical fruit trees, many temperate fruits, a number of timber trees and other economically useful crops like rubber, oil palm, coffee, cacao and cola nut.

In nature, the process is accomplished by the acid and enzymatic actions of the digestive tracts of birds and other animals or abrupt changes in temperature or by the fungal action and other soil microflora or by exposure to fire by certain species.

'Stratification' is the term for the process of exposing seeds to low temperatures to break their dormancy; while the low temperature treatment that is given for floral initiation is called 'vernalization'.

(ii) Immature embryo

In most of the plants, when the seeds fall from the fruit, the embryo inside the seed is fully developed. However, in a few cases like *Fraxinus* and *Anemone*, the development of the embryo is incomplete when the seeds are dispersed. As a result, such seeds remain dormant and fail to germinate immediately. Rather, a prerequisite to germination for such seeds is a period of **after-ripening**, which occurs for some species during dry storage. For others, moisture and low temperatures are a necessary phenomenon known as **Stratification**. Natural stratification occurs when seeds shed in autumn are covered with cold soil, debris and snow. In artificial stratification, layers of seeds are alternated with layers of moist *Sphagnum* and sand and stored at low temperatures. Temperatures, generally lower than 10°C (ideally 5°C) are most suitable for breaking seed dormancy. The effectiveness of temperature is also a function of the length of the cold exposure .

The duration of after-ripening requirement may be as short as a few weeks as in barley or as long as five years as in *Rumex crispus*. During after-ripening, extensive transfer of compounds from storage cells to the embryo, sugar accumulation and digestion of various storage lipids takes place. Stratification may affect the disappearance of inhibitors like abscisic acid and the build-up of germination promoters such as gibberellins and cytokinins. Stratification has the added benefit that it synchronizes germination, which ensures that plants will mature at the same time.

Embryo dormancy can be broken by removal of the cotyledons, e.g., in European hazel and European ash. Immature embryos simply require additional time to enlarge under appropriate conditions before they can emerge from the seed, e.g., celery and carrot.

ABA/GA ratio is the primary determining factor of seed dormancy and germination according to the **hormone balance theory**. The amounts of the two hormones are regulated by their rates of synthesis vs. deactivation. According to a recent view, the balance between the concentrations of ABA and gibberellin in seeds is under developmental as well as environmental control. During the early phases of seed development, ABA sensitivity is quite high which favours dormancy over germination. Later in seed development, gibberellin sensitivity increases favouring germination. Besides ABA and gibberellin, ethylene and brassinosteroids also regulate seed dormancy. ABA inhibits ethylene biosynthesis, while brassinosteroids enhance it, thus highlighting the involvement of hormonal networks.

14.3.1.1 The relationship between embryo immaturity and capacity to germinate is influenced by ABA

Further embryo development and capacity to germinate requires adequate water and favourable temperatures. Under optimal conditions ash seeds take several months to complete maturation, whereas the immature seeds of marsh marigold are able to germinate after about 10 days. The relationship between dormancy and stage of maturity varies widely across different species. For example, mangroves proceed directly from embryogenesis to the seedling state whereas sycamore embryos are dormant throughout maturation and become responsive to chilling or removal of seed coat as they approach maximum weight and begin to dry out. In general, primary and middle phases of maturation show predominance of ABA action, initially secreted by maternal tissues. Isolating developing embryos from the influence of ABA allows precocious germination. So-called viviparous mutants exhibit this behaviour. There are at least 15 *vp* (viviparous) mutants in maize, each affecting a different step in ABA biosynthesis. Seeds of ABA-insensitive and ABA-deficient mutants of *Arabidopsis* also germinate precociously. It seems that embryo maturation and embryo dormancy are physiologically distinct processes under the control of separate though interacting regulatory systems.

14.3.1.2 Role of light and chemicals in dormancy

Light: It is not considered to be necessary for germination of seeds of most of the species. However, seeds of some species like lettuce, *Lactuca sativa* var. Grand Rapids, tomato and tobacco do not germinate till they are exposed to light. Such seeds are called '**positive photoblastic seeds**' and respond to light after imbibing 30–40 per cent moisture.Even exposure to very low light intensity for 1–2 minutes is sufficient to overcome dormancy. It has been observed that shorter wavelengths i.e., red light (660 nm) stimulates germination and longer wavelength i.e., far-red light (730 nm) prevents germination of seeds. The impact of red light is reversed by far-red light and *vice versa* and this seems to involve phytochromes. The behaviour of the seed depends upon the last treatment of radiation given. The presence of red light requirement for germination of lettuce (*Lactuca sativa*) seeds can be even substituted by the application of gibberellins and kinetin.

In a few species like *Phlox*, *Nemophila* and *Silene*, the seeds remain dormant when exposed to light but germinate only in the dark. These seeds are called '**negative photoblastic seeds**'.

All seeds requiring light exhibit coat-imposed dormancy, and the removal of outer layers/ tissues, specifically the endosperm permits the embryo to germinate even in the absence of light. The effect that light has on the embryo is thus to enable the radicle to penetrate the endosperm, a process facilitated in some plant species by enzymatic weakening of the cell walls in the region of micropyle, next to radicle.

The soil seed bank consists of buried dormant seeds that germinate when disturbance exposes them to light. Buried seeds in the soil seed bank may be considered to be in a dormant state and whether this dormancy is of the eco- or the endo- type depends on species, length of time in the soil and the presence of allelochemicals. Weed seeds (*Veronica arvensis*, *Matricaria recutita*) present in the soil seed bank may be released from dormancy by exposure to light. The light response shows red/far-red reversibility indicative of phytochrome control. In the *Arabidopsis* weed, seed coat removal, or altered testa tructure resulting from mutation, overcomes the light requirement which suggests that action of the phytochrome system in the intact seed is to stimulate a process that renders the coat permeable.

Chemicals: Various compounds e.g., respiratory inhibitors, sulfhydral compounds, oxidants and nitrogenous compounds have been shown to break seed dormancy. Nitrate, often along with light, is probably the most important chemical.

Another chemical agent that can break dormancy is nitric oxide (NO), a signalling molecule found in both plants and animals. Smoke produced during forest fires, contains multiple germination stimulants, but the most active of these is '**Karrikinolide**', which structurally resembles strigolactones. The chemical stimulants appear to break dormancy by altering the ABA/GA ratio, i.e., by down-regulating ABA synthesis or up-regulating gibberellin synthesis. Ethylene released by different germinating seeds such as oats, peanuts and clover is also effective in breaking dormancy.

Positively photoblastic seeds	Negatively photoblastic seeds
1. The seeds in which germination is stimulated by light.	The seeds where light inhibits germination but darkness promotes germination.
2. The shorter red light wavelength (660 nm) promotes seed germination, while the longer far-red radiations (730 nm) inhibit germination. If lettuce seeds are exposed alternately to '100 red and far-red' radiation treatments, it is the last radiation that will determine the germination response. If the last instalment given is red, then the seeds will show germination and if the last instalment is far-red, then germination is inhibited, e.g., *Lactuca sativa* var. Grand Rapids, *Nicotiana tabacum, Digitalis purpurea, Capsella bursa-pastoris*. The requirement of light can be substituted by gibberellic acid. Consequently, gibberellin-treated seeds can germinate in total darkness as well.	The red and far-red reversibility in the germination is not demonstrated (i.e., no such effect). e.g., Love-in-Mist (*Nigella damascena, Silene* sp., *Nemophila insignis*).

14.3.2 Bud dormancy

Bud dormancy is an important adaptive feature for woody species in cold climates. When a tree is exposed to the chilling temperature in winter, bud scales protect its meristems and inhibit bud growth. Dormant buds show a little respiratory activity with a significant loss of water and inability to grow. During their development, there is a build-up of storage compounds (such as starch, fats and proteins) in the bud primordia.

The onset of bud dormancy, induced by photoperiod, is a typical short-day response, coincident with decreasing temperature, leaf fall, decreased cambial activity and increased capacity to withstand low temperature or cold hardiness. A period of low temperature (near or just above freezing) is required to break bud dormancy. The amount of chilling required varies with species, cultivar, and even position of the buds on the trees. Ethylene can also break bud dormancy and ethylene treatment is sometimes used to promote bud sprouting in potato and other tubers. Bud dormancy prevents precocious bud growth, which ensures that the meristematic tissues within the bud are able to survive the impact of adverse conditions like winter. Premature growth of buds during the atypical warmer period in the winter, could have adverse impact on the survival of the plant.

Both the processes, bud dormancy and growth, are controlled by the balance between bud growth inhibitor (ABA) and growth–inducing substances (gibberellins and cytokinins).

14.3.3 Biological significance of dormancy

1. Temporary dormant period experienced by many cereal grains allows for their harvest, dry storage and ultimate use as food, otherwise these grains become useless for human consumption.
2. Dormant seeds[5] and organs in perennial plants resist unfavourable conditions for their development, particularly in temperate climates and thus maximize seedling survival by preventing germination.
3. Dormancy allows a greater opportunity for seed dispersal over greater geographic distances and for a seed to reach the optimum conditions for germination.

Plant dormancy, quiescence, and their breaking are under strong natural selection by the environment. However, the ability of certain weed seeds to lie dormant for many years in the soil has proven to be a great inconvenience.

14.3.4 Regulation of development and dormancy of resting organs

Resting structures pass through a common sequence of phases: **growth, transition into dormancy, dormancy maintenance, dormancy termination and resumption of growth**.

- The onset of dormancy is associated with arrest of the cell cycle in meristems. Endodormant buds of potato tubers are arrested in G1 phase and exhibit reduced rates of DNA, RNA and protein synthesis. Paradormancy is associated with arrest prior to the G1/S checkpoint. Release from cell cycle-related endodormancy is mediated by hormones and signalling mechanisms that control expression, assembly and activity of cyclin-dependent kinases (CDK) system. Closing down the cell cycle reduces the requirement for synthesis of chromosomal proteins which is reflected in low expression of histone genes during the dormancy phase and an increased activity in tissues prior to sprouting. Subsequent dormancy release involves the reactivation of cell cycle regulators followed by resumption of meristematic activity as well as DNA replication.
- Seasonal endodormancy is usually a response to declining day lengths and involves the interaction between the CO-FT module, the phytochrome and the circadian system. The CO-FT module has been shown to regulate bud set in *Populus*, tuberization in potato and bulb formation in onion. FT function in dormancy induction and release is suppressed by DAM (DORMANCY-ASSOCIATED MADS-BOX) proteins, whose genes are induced under short days.
- Dormant organs have specialized structures that develop in coordination with the transition to the dormant condition. For example, leaf morphogenesis is modified to form the scales of resting buds. During bud set, KNOX genes and genes regulating polarity and pattern formation in leaf primordia are up-regulated. Resumption

[5] In the Kew Gardens Millennium Seed Bank, UK, most of the stored seeds will remain viable for an estimated 200 years. The oldest verifiable record for seed germination is the sacred lotus, which gave a rate of 75 per cent germination after 1300 years of storage during which time some of the seeds had dispersed from China to the USA.

of growth and development following release from dormancy is associated with reinstatement of the regulatory systems of morphogenesis.

- Dormancy is controlled by the antagonistic actions of ABA and gibberellin. Meristem cells in the dormant state are insensitive to growth-promoting signals. Day length and temperature control of entry into and exit from dormancy are mediated through the GA-DELLA-UbPS network. ABA acts as a GA antagonist but does not seem to function in dormancy maintenance. Transcription of genes with functions in chromatin remodeling is stimulated by short days early in bud development. **Chromatin remodelling** is also required for appropriate expression of cell cycle-regulated genes.

14.3.5 Adaptive and evolutionary significance of the resting phase

- Most resting structures are propagules, means by which plants reproduce asexually. A clonally-propagated colony, a **genet**, consists of genetically-identical individuals propagated by rhizomes or stolons termed as **ramets**. Several aquatic species such as *Spirodela, Potamogeton, Hydrilla* and *Utricularia* propagate by **turions**- endodormant buds formed from modified shoot meristems under short days, disperse from the parent plant and grow into new individuals after overwintering. Turion formation is stimulated by ABA and inhibited by cytokinin.
- **Phenology,** the study of the timing of periodic life cycle events over the course of a year, contributes important evidence to the debate about climate change.
- The various life forms (**phanerophytes, chamaeophytes, hemicryptophytes, cryptophytes and therophytes**) have evolved through changes in integration of component developmental processes.
- The consumption of plant resting organs has influenced the course of human evolution.

14.4 Photoperiodism

The transformation of vegetative shoot apex into reproductive structure is a very important and extensively studied developmental phenomenon in plants. The process of flowering is controlled by two major environmental factors – light and temperature. The phenomenon of **photoperiodism** refers to response of plants to the duration of light. The early beginning of photoperiodism in plants goes back to Tournois (1912) and Klebs (1913), but it was left to two American researchers, Garner and Allard working in the U.S. Department of Agriculture, Maryland to demonstrate and develop, in 1920, the concept of photoperiodism. They demonstrated that a mutant strain of tobacco called 'Maryland mammoth' which remained vegetative during the season could be induced to flower by growing in a glasshouse by exposing it to short days.

Finally, Garner and Allard designed light-tight boxes to shorten the day length of the plants growing in summer and discovered that tobacco would flower if the day length was shorter than 14 hours. If grown on photoperiods of greater than 14 hours or in continuous light, no flowering occurred.

Based on the requirement for day length, the flowering plants are categorized into five types.

(i) ***Short-day plants (SDP):*** These plants flower when the length of the day is less than a certain critical photoperiod in a 24-hour cycle. For example, *Xanthium* flowers under any photoperiod which is less than and up to 15½ hour. Other examples of short-day plants include: Biloxi soybean, *Nicotiana tabacum, Pharbitis nil* and *Impatiens balsamina*. The absolute value of the critical photoperiod differs from plant to plant.

(ii) ***Long-day plants (LDP)***: These plants flower when the length of the day exceeds a certain critical photoperiod. Henbane plant (*Hyoscyamus niger*) can flower under any photoperiod which is more than 10½ h. It means it can flower with continuous light of 24 hours as well. Other examples of long-day plants are *Spinacea oleracea, Beta vulgaris, Plantago lanceolata*.

(iii) ***Day-neutral plants* (*DNP*)**: Flowering occurs after a period of vegetative growth, independently of the day length, i.e., flowering is typically under autonomous regulation. For example, *Pisum sativum, Lycopersicum esculentum, Mirabilis jalapa*.

(iv) ***Short-long day plants(SLDP)***: These plants flower only when exposed sequentially to short days followed by long days (i.e., June–July). For example, *Trifolium repens, Campanula medium, Echeveria harmsii*.

(v) ***Long-short day plants* (*LSDP*)**: Flowering in such plants occurs when exposed to long days followed by exposure to short days (i.e., September–October). For example, *Cestrum nocturnum, Bryophyllum daigremontianum, Kalanchoe* spp.

Short-day plants	Long-day plants
1. Flowering occurs when the day length exposure is less than the 'certain critical length'. The critical day length or photoperiod is the 'limit' below which they flower.	Flowering occurs when the day length exposure is greater than the 'certain critical length'. The critical photoperiod is the 'limit' above which they flower.
2. Short-day plants are better known as 'long- night plants' as here the continuity of the dark period is more important for flowering, and a 'night break' by a flash of red light (660nm), prevents flowering in SD plants.	In the Long-day plants, darkness continuity is not essential. Even a 'night break' will cause flowering in LD plants.
3. Pr form of phytochrome promotes flowering but Pfr form inhibits it, e.g., *Xanthium*.	Pfr triggers a series of reactions that finally lead to flowering (in other words, Pfr causes flowering of LD plants), e.g., *Hyoscyamus*.

14.4.1 Critical photoperiod

Critical photoperiod refers to the limit of the photoperiod up to which the short- day plants come to flowering. However, in long-day plants, critical photoperiod is the limit above which the plant flowers. The critical photoperiod varies from species to species. When exposed to 12 hours of light, both *Xanthium* and *Hyoscyamus* would flower although the former is a short-day plant (SDP) and the latter is a long-day plant (LDP). Therefore, it is clear that the same photoperiod can be short day for one species but long day for the other. Hence, no absolute duration of light can be fixed for both a short day and a long day.

A short-day plant flowers below its critical photoperiod and the long-day plant flowers above its critical photoperiod.

14.4.2 Photoperiodic induction

Photoperiodic induction is the phenomenon of conversion of leaf primordia into floral primordia under the influence of suitable inductive cycles in apical meristem. Number of photoinductive cycles varies from plant to plant. For example, *Xanthium* requires only a single photoinductive cycle for flowering, whereas *Glycine max, Salvia occidentalis* and *Plantago lanceolata* require 2–4, 17 and 25 cycles respectively for flowering. Although *Xanthium* will show response to a single inductive cycle but the initiation of floral primordia is more rapid when multiple cycles are given. Other plants may exhibit fractional induction, for example, *Plantago lanceolata* (LDP), if given 25 long days with an intervening break of 10 short days, will still show flowering. This indicates a summation of inductive photoperiods despite interruption with the non-inductive cycle. Once a plant has received the minimum number of photoinductive cycles, it will flower even if it is returned to non-inductive cycles.

A photoperiod is said to be inductive if it causes flowering in a plant and non-inductive if it does not lead the plant to flowering. Thus, short days are inductive for *Xanthium* but long days are non-inductive. Similarly, long days are inductive for *Hyoscyamus* but short days are non-inductive.

Night breaks: An initial observation by Hamner and Bonner in 1938 was that the short-day cocklebur plant (*Xanthium*) would not flower if the long night was interrupted with a brief light period. Thus, cocklebur grown in inductive photoperiods of less than 15½ hours will not flower when the continuity of the dark period is broken with a brief period of light (Figure 14.3). This interruption can be as short as 3 minutes and still be effective. The interruption of night by light prevented flowering in short-day plants (Figure 14.4). This was followed by experiments in which it was shown that long-day plants would flower if grown on short-days, non-inductive cycles and the long night was interrupted with light.

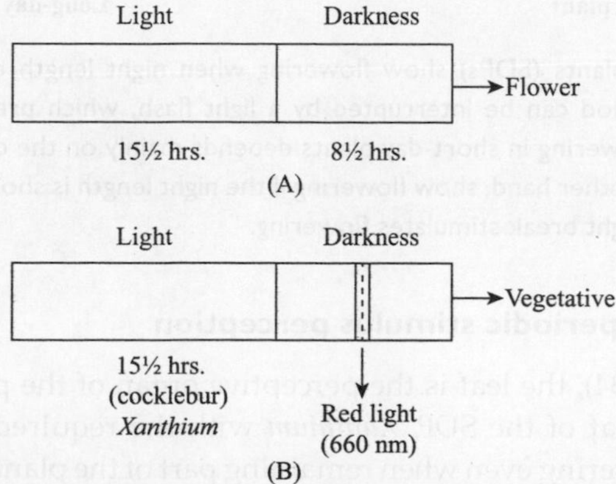

Fig. 14.3 **Significance of dark periods for flowering in SDP, *Xanthium*. The break of dark period by red light prevents flowering and the plants will remain vegetative.**

Besides this, the effect of a night break differs greatly depending upon the time when it is given. For both LDP and SDP, a night break has been found to be most effective when interrupted near the middle of a dark period of 16h.

The flowering response to night breaks confirms the role of circadian rhythms in photoperiodic timekeeping.

A light break of a long night period would promote flowering of long day plants and inhibit the flowering of short day plants. Both SDP and LDP behave diametrically opposite.

Further, it was noticed that an increase in length of the photoperiod leads to an increase in numbers of floral primordia, whereas the length of the dark period is important for flower induction but that of the photoperiod has a quantitative effect on flowering.

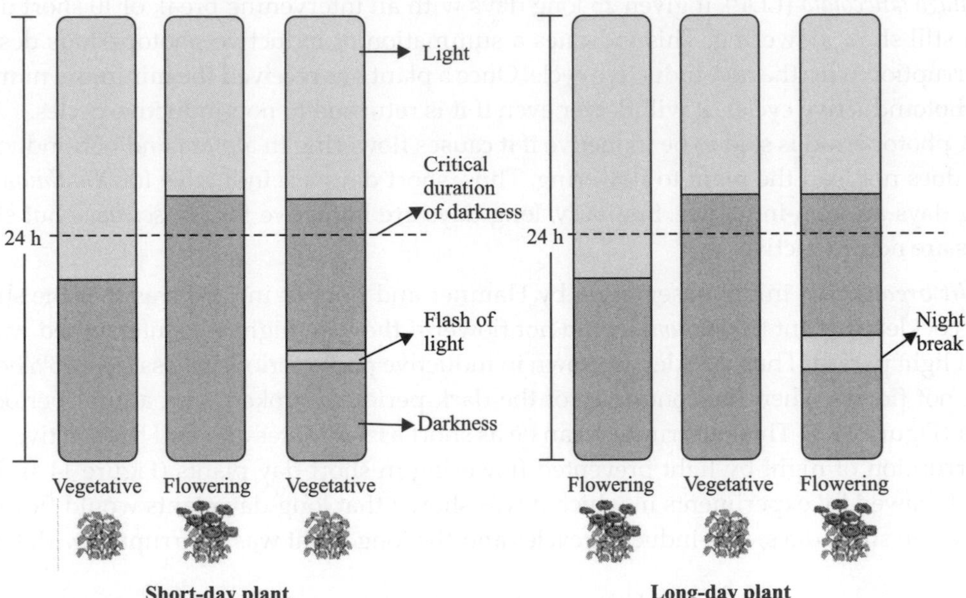

Fig. 14.4 (A) Short-day plants (SDPs) show flowering when night length exceeds a critical duration. Continuity of the dark period can be interrupted by a light flash, which prevents flowering and plants remain vegetative. Thus, flowering in short-day plants depends mainly on the duration of the dark period. (B) Long-day plants, on the other hand, show flowering if the night length is shorter than a critical duration. Shortening the night by a night break stimulates flowering.

14.4.3 Site of photoperiodic stimulus perception

According to Knott (1934), the leaf is the perceptive organ of the photoperiodic stimulus. Exposure of a single leaf of the SDP, *Xanthium* with the required short photoperiods is sufficient to initiate flowering even when remaining part of the plant is exposed to long days (Figure 14.5). While responding to photoperiods, the leaf transmits a signal that controls the transition to flowering in the apical region of the shoot. Photoperiodic induction can also occur in a leaf that has been detached from the plant. For example, in SDP *Perilla crispa*, an excised leaf exposed to short days can lead to flowering when grafted to non-induced plant maintained under long day conditions. The sensitive response of the leaf may vary with age. In *Perilla* and soybean, the youngest but fully expanded leaf was found to be the most sensitive.

With the knowledge gained through light-break experiments, two USDA scientists, Borthwick and Hendricks set out in the 1950s to ascertain the light quality that most strongly promoted the response. The response (action spectrum) in both SDP and LDP was most sensitive to red light having a wavelength maximum of 660 nm and reversal effect of this response was caused by light with a wavelength maximum of 730 nm.

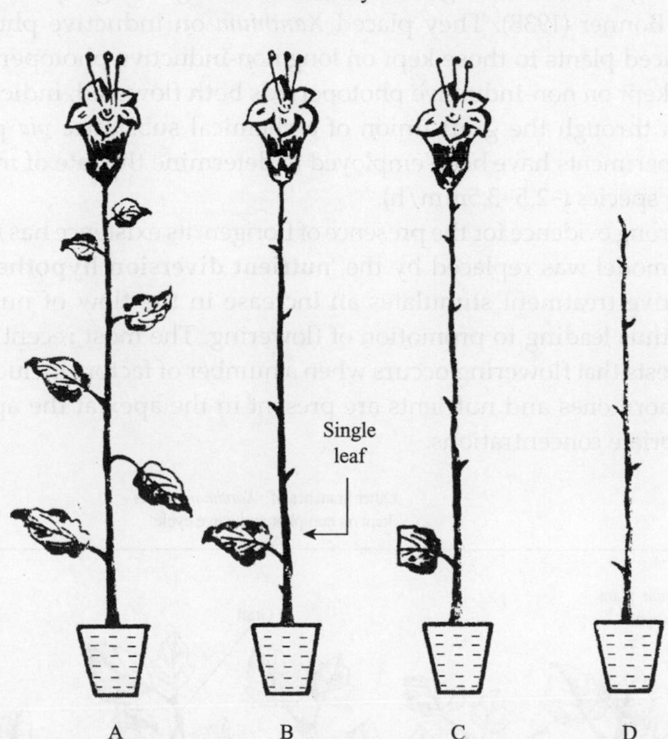

Fig. 14.5 An experiment to demonstrate that photoperiodic stimulus is perceived by the leaf (A). Flowering occurred in short-day plant, even if a single leaf is maintained under short-day conditions (B, C). However, the plants without leaves remained vegetative (D).

Studies in short day plants in late 1940s indicated that red light was most effective as a light-break, with a maximum effectiveness near 660 nm. Borthwicks and his colleagues reported that the inhibitory effect of red light on flowering in *Xanthium* was reversed with far-red. At about same time, photoreversibility of seed germination was noticed. Red, far-red photoreversibility of the light-break clearly envisages the role of phytochrome in controlling flowering in SDP and LDP. Based on recent investigation with phytochrome mutants in *Arabidopsis*, it has been proposed that phytochrome gene A (PHYA) is needed to stimulate flowering in LDP under specific conditions. On the other hand, phytochrome gene B (PHYB) seems to inhibit flowering.

14.4.4 Florigen: flowering stimulus

In 1936, Mikhail Chailakhyan, a Russian Scientist, proposed the existence of a universal flowering hormone, named **florigen**. It is translocated to distantly placed target tissue (shoot apex) along with photoassimilates in the phloem and stimulates the flowering response. The role of gibberellins needs a special mention here because the application of gibberellin to most long-day plants will cause them to flower when they are on a non-inductive cycle.

Evidence for florigen as a flowering hormone comes from grafting experiments conducted by Hamner and Bonner (1938). They placed *Xanthium* on inductive photoperiods and then grafted induced plants to those kept on long non-inductive photoperiods. When the grafted pair was kept on non-inductive photoperiods both flowered, indicating that there was transmission through the graft union of a chemical substance *via* phloem (Figure 14.6). Grafting experiments have been employed to determine the rate of movement of the stimulus in many species (~2.5–3.5mm/h).

Despite such strong evidence for the presence of florigen, its existence has not been proven and the florigen model was replaced by the **'nutrient diversion hypothesis'**. According to this, an inductive treatment stimulates an increase in the flow of nutrients into the apical meristem, thus leading to promotion of flowering. The most recent **'multifactorial hypothesis'** suggests that flowering occurs when a number of factors including promoters, inhibitors, phytohormones and nutrients are present in the apex at the appropriate time and in the appropriate concentrations.

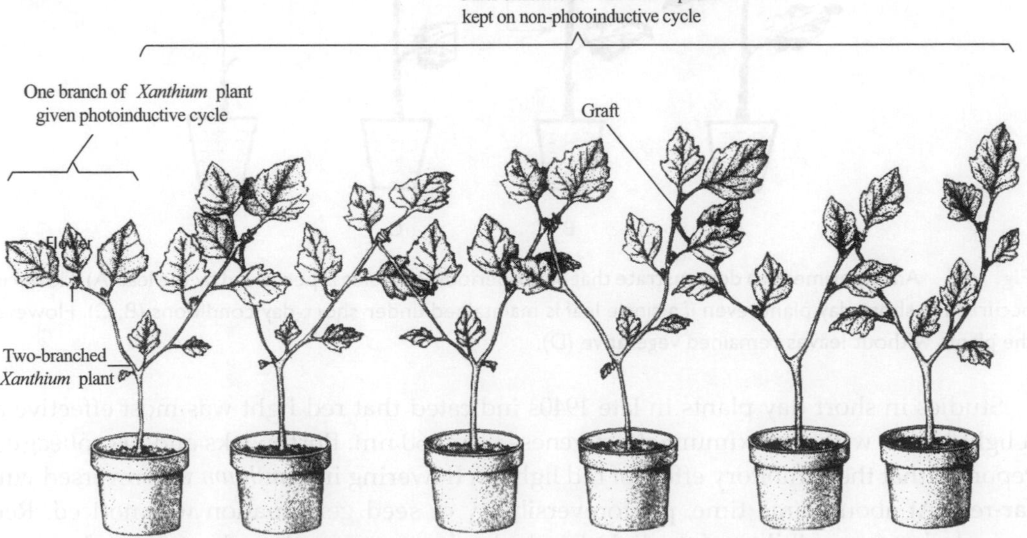

Fig. 14.6 Demonstration of a grafting experiment in SDP, *Xanthium* (two-branched), to indicate the transmission of floral stimulus from a single-induced leaf through rest of the plants.

Molecular studies with *Arabidopsis* (a facultative long-day plant) have thrown light on the nature of florigen. According to the **coincidence model**, gene CONSTANS (CO) is involved in producing a phloem-mobile transmissible floral signal, FLOWERING LOCUS T (FT). FT is a small globular protein with molecular weight of 20 kD. CO activated by phytochrome and cryptochrome responds to an inductive photoperiod by activating FT mRNA transcription and FT protein synthesis in phloem parenchyma cells of the leaves.

FT protein is carried through the phloem sieve tube elements from the leaves to the shoot apical meristem. It involves two critical steps: (i) the export of FT from companion cell to the sieve tube element. The endoplasmic reticulum-localized protein, FT INTERACTING PROTEIN 1 (FTIP1) is required for FT movement into the phloem translocation stream, which takes it to the meristem (ii) the activation of FT target genes at the shoot apex.

Once FT protein reaches the floral meristem, it enters into the nucleus and interacts with FLOWERING D (FD), a basic leucine zipper (bZIP) transcription factor to induce a group of flowering genes termed as **floral meristem identity genes** such as **APETALA 1** (AP1) in the floral meristem and **SUPPRESSOR OF CONSTANS 1** (SOC1) in the inflorescence meristem. This stimulates **LEAFY** (LFY gene) expression, which in turn, activates the expression of AP1 and FD, forming two positive feedback loops. CO protein undergoes degradation at different rates in the light period and the darkness. Light factor increases the stability of CO, thus permitting it to accumulate during the day and it quickly undergoes degradation in the darkness. Expression of CO gene and its downstream target gene FT appears to be highest in the phloem companion cells in the leaves. The signaling output of CO activity is regulated by the FT gene expression (Figure 14.7).

All pathways that control flowering converge to increase the expression of key floral regulators - FT in the vasculature and SOC1, LFY, and AP1 in the meristem. Expression of genes SOC1, LFY, and AP1, in turn activates downstream genes (such as AP3, PISTILLATA and AGAMOUS) essential for the development of floral organs. Both gibberellins and ethylene are involved in induction of flowering. Gibberellins appear to promote flowering in *Arabidopsis* by stimulating expression of the LEAFY gene, mediated by the transcription factor GA-MYB. GA-MYB is negatively regulated by DELLA proteins and is also modulated by micro RNA, which promotes the degradation of the GA-MYB transcript.

14.4.5 Biological significance of photoperiodism

1. The productivity of tubers, bulbs, corms and rhizomes can be enhanced by knowing favourable photoperiods for their formation.
2. Vegetative crops (radish, carrot, sugarcane, etc.) can be kept in the vegetative phase for a longer period of time. The concept of photoperiodism has been commercially exploited in sugarcane. In sugarcane, being SDP, continuity of the dark period is more important and if it is broken by a flash of light, plants can be kept vegetative. Flowering if allowed to occur causes a reduction of 10–20 per cent sugar content.
3. Annuals may be cropped 2 or 3 times a year.
4. Winter dormancy and autumn leaf fall can be prevented by increasing light hours.
5. Long-day treatment increases stolon formation in strawberry.

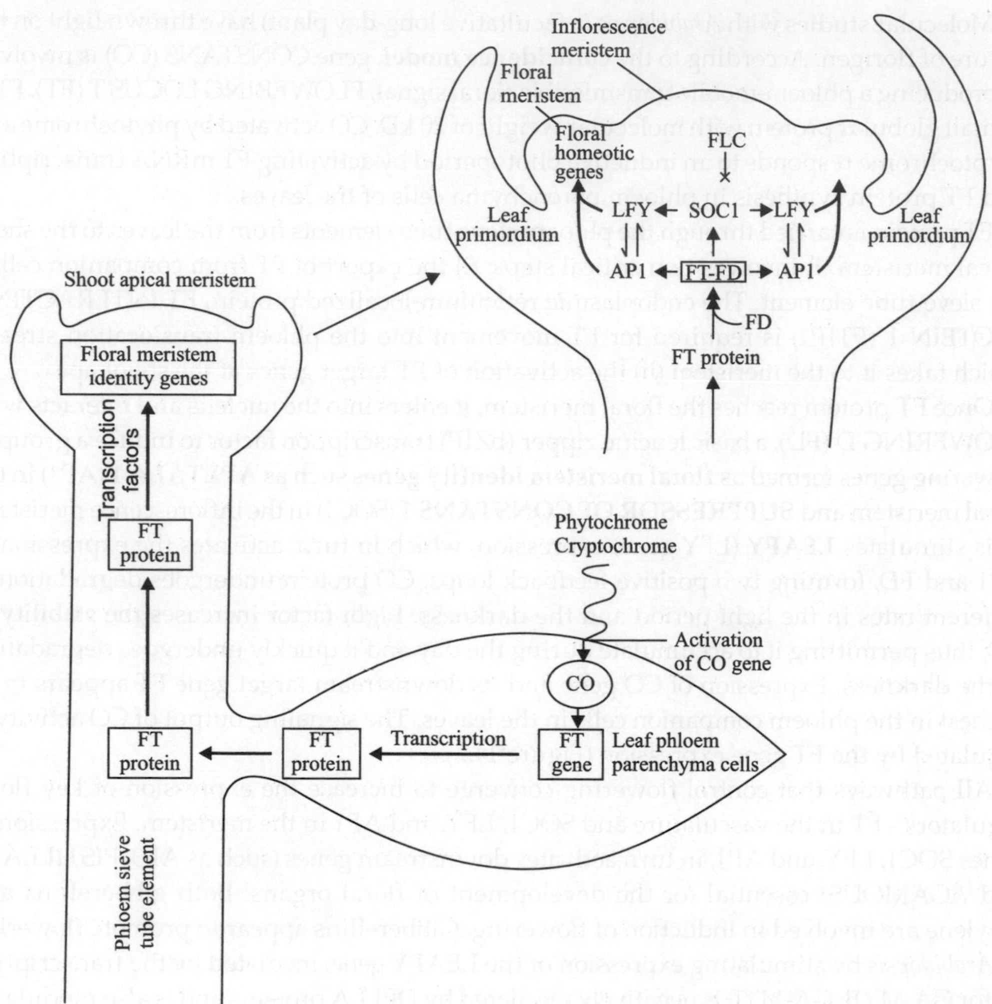

Fig. 14.7 A model to explain the transmission of flowering stimulus from leaf phloem parenchyma cells through the phloem sieve tube elements to the apical meristem of the shoot to induce flowering genes. FT protein after being unloaded from the phloem, forms a complex with the FD transcription factor in the meristem. The FT-FD complex stimulates SOC1 and AP1 genes in the inflorescence meristem and floral meristem respectively, resulting in activation of LFY gene expression. LFY and AP1, in turn, induce the expression of the floral homeotic genes. Flowering locus C (FLC) behaves as a negative regulator of gene SOC1 located in the meristem.

Since changing day length is the most reliable indicator of seasonal change, photoperiodism almost conclusively reflects the need for plants to synchronize their life cycle in accordance with the time of the year. Photoperiodism is helpful in ensuring that plants flower in their temporal niche, reducing competition with others as well as ensuring that reproductive development is completed before the onset of unfavourable wintry conditions.

Effects of photoperiod on geographical distribution of plants can have a direct effect on human health. Hay fever sufferers in most of the Western and Central Canada are spared from the troublesome consequences of highly allergenic pollen of plant ragweed (*Ambrosia artemisiifolia*), an annual short day plant with a critical photoperiod of about 14.5 hours.

14.4.6 Floral meristems and floral organ development

Morphological changes during transition from the vegetative to the reproductive phase are:

(i) During the vegetative period of growth, an apical meristem produces rosettes of leaves. At the onset of the reproductive phase, the vegetative meristem undergoes transformation into primary as well as secondary inflorescence meristem. Flowers arise from floral meristems that form on the flanks of the inflorescence meristem. In *Arabidopsis*, inflorescence meristem exhibits indeterminate growth, whereas floral meristem shows determinate growth.

(ii) Floral meristems increase in size as a result of the increased division of the cells in the central zone of the shoot apical meristem.

(iii) Floral meristems initiate differentiation of four types of floral organs; sepals, petals, stamens, and carpals, which are produced in concentric rings (whorls) around the flanks of the meristem.

14.4.7 Floral organ specification

Formation of the floral organs (sepals, petals, stamen, carpel) takes place in the apical meristem of shoot and several genes controlling floral morphogenesis have been identified in plant species. Both internal (autonomous) and external (environmental) signals enable plants to regulate precisely the time of flowering for reproductive development. **Photoperiodism** (response to changes in day length) and **Vernalization** (response to prolonged cold) are the two most important seasonal responses affecting floral development. Synchronized flowering favors crossbreeding and helps ensure seed production under favourable conditions.

The different kinds of floral organs are originated sequentially in four separate, concentric whorls. Two major categories of genes control floral development are:

- **Floral meristem identity genes** that code for transcription factors essential for the early induction of floral organ identity genes.
- **Floral organ identity genes** that are associated with the direct regulation of floral organ identity.

Specification of different floral structures is best understood in *Arabidopsis* because of availability of homeostatic mutants. Mutations in homeostatic floral identity genes bring about change in the types of organs developed in each whorl. The **ABC model** proposed in 1991 suggests that there are three overlapping fields (A, B and C) of gene expression to determine the organ identity in each whorl (Figure 14.8A).

- The A field is conceived as covering whorls 1 and 2
- The C field covers whorls 3 and 4
- The B field overlaps fields A and C in whorls 2 and 3, respectively.

A characteristic aspect of the ABC model is the antagonism between A and C, such that if expression of either is missing, the other will be expressed across the entire floral meristem.

- Field A acting individually specifies sepals
- Fields A and B act together to specify petals
- Fields B and C act together to specify stamens
- Field C acting individually specifies carpels

Field A and C activities mutually suppress each other and antagonism between these two is indicated by T-ended lines shown in Figure 14.8B. When one of these fields is missing, the other expands to cover all whorls.

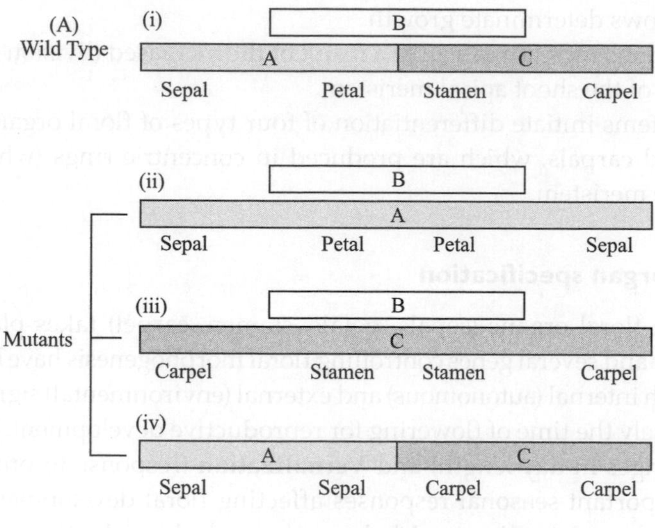

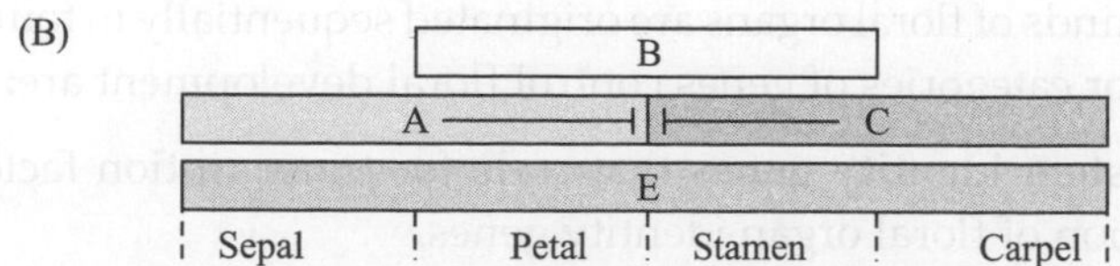

Fig. 14.8 (A) ABC model of flowering. (i) Wild-type: all three types of classes of genes A, B, and C are functional in the floral meristem. (ii) (iii) and(iv) Mutants: Loss of class C, A, and B genes results in expression of class A, C, and A+C functions respectively in the floral meristem. (B) ABCE model for floral development. Class E genes are required for floral organ identity.

Expression of the ABC genes throughout the plant is not adequate to impose floral organ identity onto a leaf developmental program. The original model has been expanded to include one more class of floral homeostatic genes, the class E genes, essential for the expression of class A, B and C genes. A, B, C and E genes are expressed in distinct zones of the floral apex as the flower primordium develops . The ABCE model is based on genetic experiments in *Arabidopsis* and *Antirrhinum.* Expression of class E floral meristem identity genes (e.g. SEPALLATA) is essential for expression of class A, B and C genes.

Many floral organ identity genes code for MADS domain-containing transcription factors that function as heterotetramers. The **Quartet Model** describes how these transcription factors might function collectively to specify floral organs. In addition to the class C and E genes, Class D genes are essential for normal development of the ovule, first discovered in *Petunia.* Ovule being present within the carpel, Class D genes, are not considered to be 'organ identity genes'.

14.5 Vernalization

In several plants, flowering is regulated either quantitatively or qualitatively by subjecting them to low temperatures - the process termed as vernalization. **Vernalization** is a method of preventing precocious reproductive development late in the growing season, thus making sure that seed formation does not start till the onset of the next growing season so that the seed will have adequate time to attain maturity.

Vernalization refers specifically to the promotion of flowering by exposure to low temperature. Coined by the Russian Lysenko in the 1920s, vernalization reflects the ability of a cold treatment to make a winter cereal imitate the behaviour of a spring cereal with respect to its pattern of flowering. The concept failed to get critical recognition from the scientific community until 1918, when Gassner demonstrated that low temperature requirement of winter cereals could be met at the time of seed germination. For his part, Lysenko got substantial notoriety for his conviction that the effect was an inheritable conversion of the winter strain to a spring strain. His position – a form of the thoroughly discredited Lamarckian doctrine of inheritance of acquired characteristics–was adopted as Soviet dogma in biology and remained so until the 1950s. The adoption of Lysenko's perspective as official dogma had a considerable effect on Soviet biology, imposing serious limitations to agriculture in the USSR for several years.[6]

One of the most extensive studies of vernalization was undertaken on the Petkus rye (*Secale cereale*) by Gregory and Purvis in 1930s.

There are two basic types of plants that respond to vernalization. Winter annuals such as winter wheat require a cold treatment after the seed has imbibed water or is in the seedling stage before they will flower. These plants will normally germinate in late winter and will come to flowering in the spring season or onset of summer.

Additionally, biennials and a few perennials require a cold treatment between seasonal growth prior to flowering. Biennials, for example, form rosettes the first year and during the winter, the rosette leaves may die, but the roots remain alive. Following the cold of the winter, the biennial will flower. The cold winter, i.e., the vernalization process is a requirement for flowering.

[6] The saga of vernalization is a classic example of what can happen when scientific ideas become entangled in political beliefs.

Site of perception of vernalization stimulus: Experiments reveal that the growing apex is the target tissue for cold treatment. Furthermore, the apex has to be growing with cell divisions to be receptive. If the apices are transplanted from vernalized plants to the tips of non-vernalized plants, flowering will occur as if they were vernalized. These grafting experiments led Melchers (1939) to advocate the existence of transmissible stimulus known as **vernalin**. Transmission requires a 'living' graft union and seems to be synchronized with the movement of photoassimilates between the donor and the receptor plants. Since gibberellin causes flowering in some biennials without the presence of cold treatment, it may be the hormone with vernalin activity.

***Temperature and age of the plant*:** The optimum temperature for vernalization differs based on the plant species and time duration of treatment . In Petkus rye, the effective range is –5 °C to +15 °C. Within these limits, vernalization is directly proportional to the duration of exposure. Within the effective range, the temperature optimum is normally greater for periods of lesser exposure. Like flowering, the vernalized state is nearly long-lasting in most plant species, bringing about the concept of an induced state. For example, vernalized *Hyoscyamus*, a LDP, can be kept under short days for nearly 10 months before losing the ability to respond to long-day treatment. On the contrary, all cold-requiring plants are capable of getting devernalized, i.e., reversal of vernalization treatment, if quickly followed by a high temperature treatment. Experiments have shown that after vernalization by cold treatment, grain could be devernalized by high temperature treatment during only one day at 35 °C. Vernalized seeds can also be devernalized by either drying them for many weeks or preserving them under anaerobic conditions for the time period subsequent to cold treatment.

Winter cereals may be vernalized immediately when the embryo has imbibed water and the germination has been induced. On the other hand, biennials must attain a certain minimum size before they can be vernalized. The low-temperature treatment seems to be effective only in the shoot apical meristem. Induced state established in a comparatively few meristematic cells can be maintained throughout the period of plant development.

With the use of flowering-time mutants, it has been confirmed that the phenomenon of vernalization behaves independently of the long-day, autonomous and gibberellin-dependent genetic pathways.

Both the autonomous pathway and vernalization minimize the expression of the gene FLOWERING LOCUS C (FLC). It seems that the autonomous pathway functions exclusively by regulating FLC expression, but vernalization can initiate flowering *via* two pathways: either by suppressing FLC expression or through some undiscovered FLC-independent mechanism.

It is known now that three genes (VRN 1, VRN 2 and VRN 3) determine the vernalization requirement in cereals.

Mechanism

According to Chailakhyan, vernalin is produced in some biennials and perennials at low temperatures.

In LDP, 'vernalin' is converted to GA. 'Anthesin' is already present in LDP. 'Anthesin' along with 'vernalin' cause flowering in LDP.

Vernalin + Anthesin → Gibberellin

However, in SDPs, vernalin is not converted to GA as they lack anthesin. Hence, flowering does not occur in SDPs.

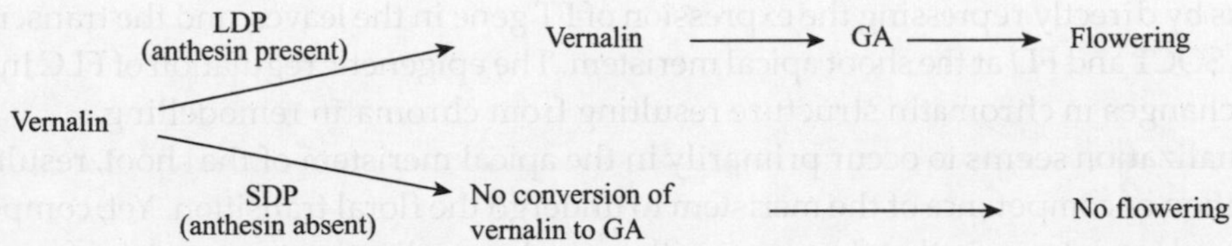

Purvis (1961) postulated the formation of a substance A from its precursor. A is converted to B after chilling. B being unstable is converted to compound D 'vernalin' at low temperature. At high temperature, B is converted to C through devernalization. At an appropriate photoperiod vernalin is converted to F 'Florigen', which induces flower formation.

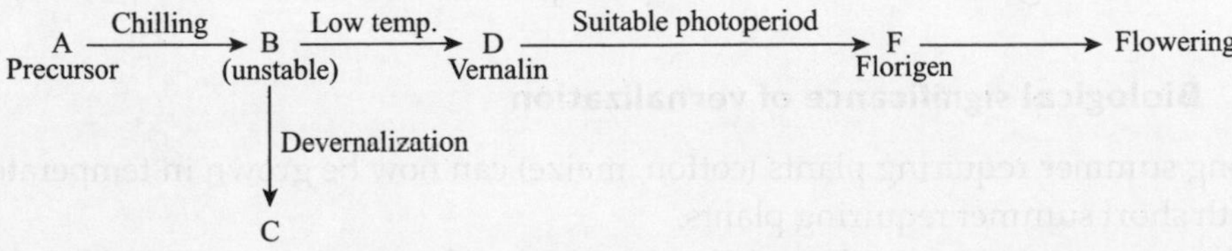

Adding to the complexity of vernalization is the apparent involvement of gibberellin in response to low temperatures. In 1957, Lang dramatically demonstrated that repeated application of GA_3 (10 µg) to the shoot apex would initiate flowering in non-vernalized plants of *Hyoscyamus* and various other biennials maintained under short day conditions. No such stimulation took place in *Xanthium* and other SDPs treated with gibberellin under non-inductive long days. Eventually, it has been suggested that gibberellin concentration is likely to escalate in response to low-temperature treatment in many cold-requiring plant species.

It appears apparent that there is a substance which gets converted to a precursor of 'vernalin' in the cold. Further in the cold, the precursor can then be converted to vernalin but at warm temperature, it reverts back to its original form.

$$X \underset{\text{Warm}}{\overset{\text{Cold}}{\rightleftharpoons}} X' \rightarrow \text{'Vernalin'}$$

14.5.1 Epigenetic changes in gene expression during vernalization

Active metabolic activity is needed during the cold treatment in vernalization; such as, cell division, DNA replication, sugars (source of energy) and oxygen to name some. In a few plant species, vernalization causes a stable change in the competence of the meristem to form an inflorescence. Stable changes in gene expression that do not involve alterations in the DNA sequence and can be passed on to descendent cells through mitosis or meiosis are known as **epigenetic changes.**

FLC is highly expressed in non-vernalized shoot apical regions and gets switched off epigenetically for the remaining plant's life cycle after vernalization. This allows flowering to occur in response to long days . However, the gene FLC is switched on again in the next generation. Thus in *Arabidopsis*, FLC gene is a major determinant of meristem competence and it works by directly repressing the expression of FT gene in the leaves, and the transcription factors SOC1 and FD at the shoot apical meristem. The epigenetic regulation of FLC involves stable changes in chromatin structure resulting from chromatin remodelling.

Vernalization seems to occur primarily in the apical meristem of the shoot, resulting in acquisition of competence of the meristem to undergo the floral transition. Yet, competence to flower does not ensure that flowering will occur. A vernalization treatment is often related to the need for a certain photoperiod, especially the cold treatment followed by long days results in flowering in early summer at high latitudes.

Vernalization is the process wherein repression of flowering is alleviated by a cold treatment provided to a hydrated seed or the growing seedling. In the absence of the cold treatment, plants remain vegetative or grow as rosettes, and they are not competent to respond to floral signals such as inductive photoperiods.

14.5.2 Biological significance of vernalization

1. Long summer requiring plants (cotton, maize) can now be grown in temperate areas with short summer requiring plants.
2. Winter crops can be sown in spring season so that they can be protected from freezing injury.
3. This process helps in production of more than one crop in a year because it serves to shorten the time to flower and hence the vegetative phase of the plant.

14.6 Biological Clock and Circadian Rhythms

Photoperiodism is associated to an internal time-duration measuring system known as the endogenous **biological clock**. Most of the organisms possess biological clocks which schedule events to occur on a daily basis – the so-called **circadian rhythms.** Many physiological and developmental processes show circadian rhythms. The first of these rhythms to be recognized was the leaf movements of *Mimosa* plants, opening and closing at a particular time each day. A circadian rhythm is comprised of a period of approximately 24 hour **(circa, about +diem, day).**

Circadian rhythms are **'cell autonomous'**, i.e., they are set largely independently in each cell of an organism. Circadian clocks regulate a large proportion of the protein-coding genes in the plants where such genes participate in many essential processes. Plants also use the circadian clock to measure the duration of day and night, thus sensing the season and influencing processes such as the onset of flowering or the start of leaf senescence.

Biological rhythms are universal in nature, from the beating of the heart to the rhythms of flowering plants.

Circadian rhythms are defined by three parameters:

(i) **Period** is the time between comparable points in the repeating cycle.
(ii) **Phase** is any point in the repeating cycle recognized by its relationship with the remaining cycle.
(iii) **Amplitude** is the distance between peak and trough. The amplitude of a biological rhythm can often differ, whereas the period remains unaltered.

An important characteristic of circadian rhythms is that the rhythmic activity continues when plants are moved from light-dark cycles to continuous dark or continuous light period. The rhythms that continue under constant conditions are called **free-running rhythms** (Figure 14.9).

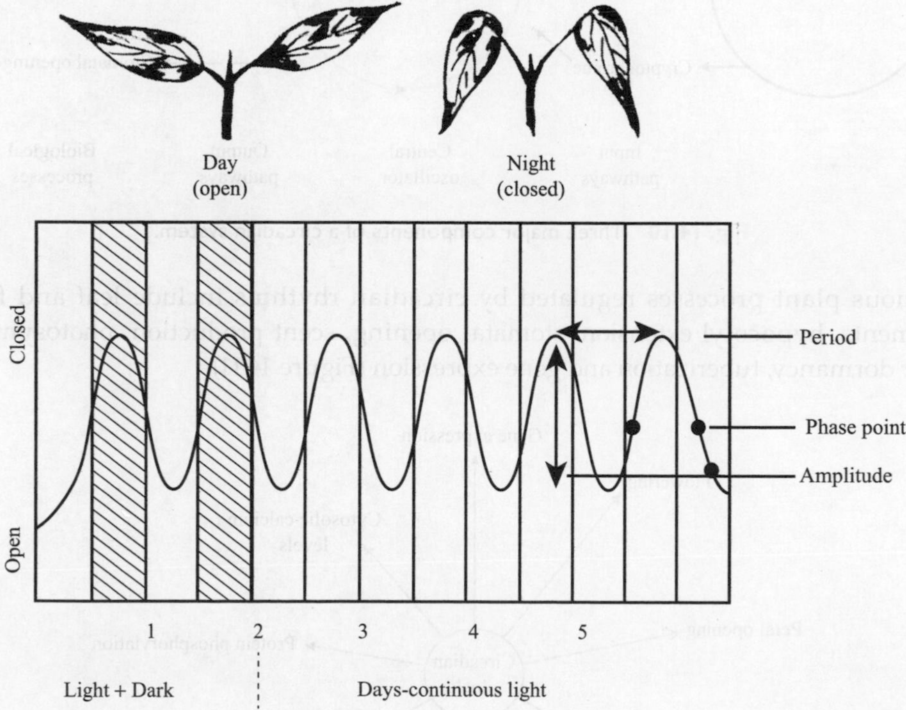

Fig. 14.9: Circadian plant rhythms: Graphic representation of leaf movements, under uniform environmental conditions. The peaks refer to the closed leaves during night time and troughs to the open leaves during day time.

14.6.1 Components of a circadian system

Three major components of a circadian system are as follows (Figure 14.10):

1. **Input pathways** synchronize (entrain) the oscillator to the daily day-night cycle. Light and temperature are the major environmental signals in plants, which entrain the oscillator through the phytochrome (PHY) and cryptochrome (CRY) photoreceptors. Coupling of a circadian rhythm to a regular external environmental signal is called '**entrainment**'. The signal that entrains the rhythms is termed as '**zeitgeber**'.
2. **Central oscillator** is the core of the circadian system and is the molecular machinery that generates the 24-hour time-keeping mechanism. In plants, the central oscillator is an 'auto-regulatory' negative feedback system.
3. **Output pathways,** controlled by the central oscillator, are individual clock-controlled biological processes.

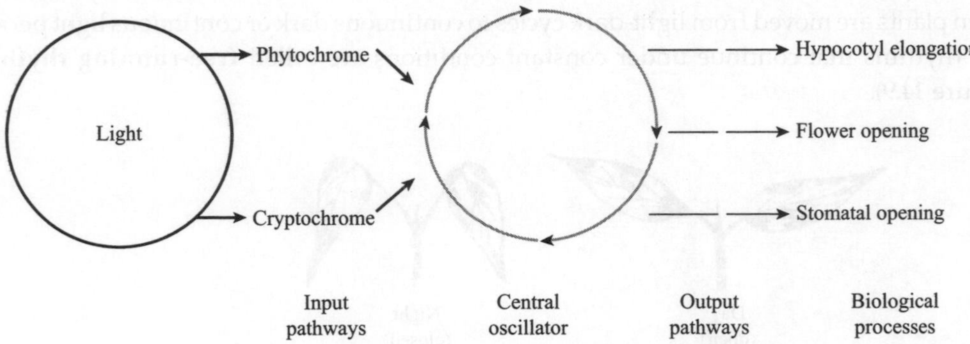

Fig. 14.10 Three major components of a circadian system.

Various plant processes regulated by circadian rhythms include leaf and flower movements, hypocotyl extension, stomatal opening, scent production, photosynthesis, winter dormancy, tuberization and gene expression (Figure 14.11).

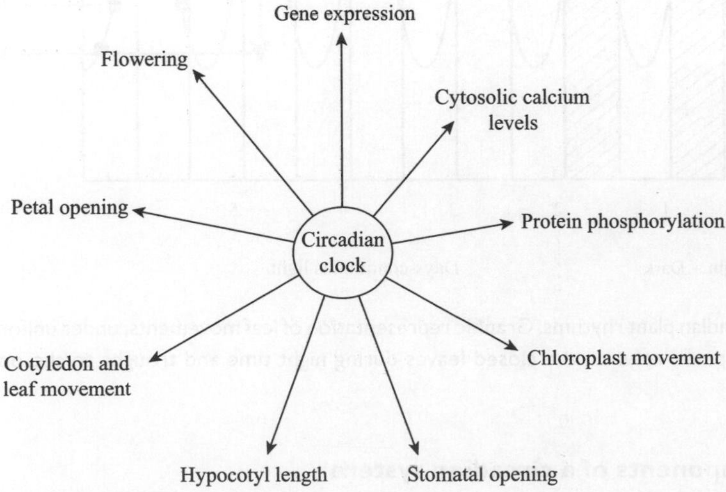

Fig. 14.11 Various plant processes regulated by circadian rhythms.

Three criteria are employed to distinguish between simple periodic phenomena and clock-driven rhythms. They are as follows:

(i) Clock-driven rhythms persist under constant conditions.
(ii) These are reset by environmental signals i.e., light and temperature.
(iii) These exhibit temperature compensation. Temperature compensation prevents temperature changes from affecting the period of the circadian clock.

Bünning (1936) first proposed that photoperiodism was linked to circadian rhythms. He suggested that the rhythm consists of two phases – the **photophile** (light-loving) and the **scotophile** (darkness-loving) – which alternated about every 12 hours. According to Bünning's hypothesis, light reaching the plant during the photophile phase would stimulate flowering, while the presence of light during the scotophile period would prevent flowering. Hillman later established that flowering takes place only when a dark period greater than the critical night coincides with the night phase of its circadian rhythm.

Circadian clocks are synchronized to the daily cycle at the molecular level through a mechanism of interacting feedback loops that result in oscillations in the expression of clock genes. The mechanism in *Arabidopsis* comprises of morning and evening loops in which some genes (e.g., circadian clock associated 1, CCA 1) are upregulated at dawn while expression of others (e.g., timing of CAB[7] expression1 and TOC1) increases at dusk. Many clock genes are light-responsive, thereby staying in sync with day length. Some components of the clock, **zeitlupe proteins**; ZTL, are post-transcriptional regulators of clock gene expression.

14.7 Photomorphogenesis

14.7.1 Etiolation

Seedlings of higher plants can adopt one of the two developmental pathways and the presence or absence of light determines which pathway is followed. When grown in the dark, seedlings exhibit **skotomorphogenesis** and become etiolated. **Etiolated** seedling have long, spindly hypocotyls, an apical hook, closed cotyledons, suppressed shoot apical meristem activities and non-photosynthetic proplastids, which causes the unexpanded leaves to have a pale-yellow colour. In contrast, seedlings grown in light, exhibit **photomorphogenesis**. Light-grown seedlings have shorter, thicker hypocotyls without apical hook, open cotyledons and expanded leaves with photosynthetically active

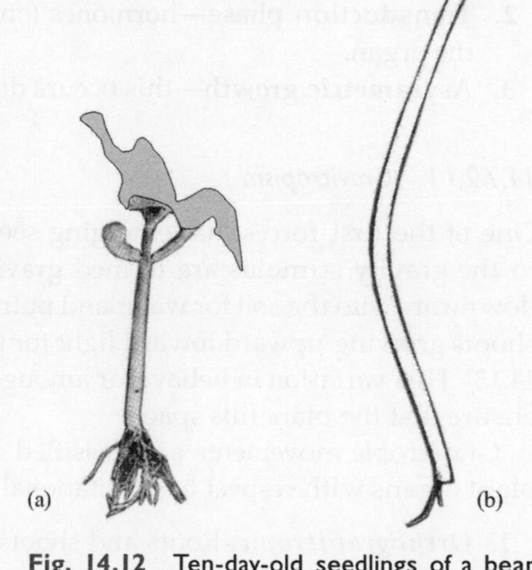

Fig. 14.12 Ten-day-old seedlings of a bean (*Phaseolus vulgaris*) plant. (a) Light-grown seedling; (b) dark-grown (etiolated) seedling.

[7] CAB – Chlorophyll a/b binding proteins

chloroplasts (Figure 14.12). When dark-grown seedlings are shifted to the light conditions, photomorphogenesis takes over and the seedlings are said to be **'de-etiolated'**. The difference between skotomorphogenesis and photomorphogenesis is the result of changes in the expression of specific genes i.e. their expression is activated or repressed by the availability or non-availability of light. Light thus acts as a signal to induce a change in the form of the underground seedling to the one that makes the plant to capture light energy efficiently. Phytochrome and cryptochrome are the photoreceptors that can induce photomorphogenic responses in plants. The transition from skotomorphogenesis to photomorphogenesis is surprisingly rapid. Gibberellins and brassinosteroids suppress photomorphogenesis in the dark and suppression is reversed by red light.

14.7.2 Plant movements

Plants show response to external stimuli such as gravity, light, touch by altering their growth and developmental patterns. Most of the plant movements, although relatively slow, perform important roles by orienting the plant organs for the absorption of water as well as nutrients, optimum perception of sunlight or getting nutrients (such as nitrogen) through the leaves. Plants exhibit two principal categories of movements i.e. **Tropic movements** and **Nastic movements.**

14.7.2.1 Tropic movements

Tropisms are directional growth responses in relation to external stimulus caused by asymmetric growth of the plant axis. Tropisms may be positive (growth toward the source of stimulus) or negative (growth away from the stimulus).

Tropic movements can be categorized into three phases:

1. **Perception phase**—the organ perceives a significantly higher stimulus on one side.
2. **Transduction phase**—hormones (one or more) become unevenly distributed across the organ.
3. **Asymmetric growth**—this occurs due to substantial cell elongation on one side.

14.7.2.1.1 Gravitropism

One of the first forces that emerging seedlings encounter, is gravity. Growth responses to the gravity stimulus are termed **gravitropism**. The primary roots of the plant grow downward into the soil for water and nutrients, thus tend to be **positively geotropic**, while shoots growing upward toward light for photosynthesis are **negatively geotropic** (Figure 14.13). This variation in behaviour among different levels of both shoot and root branches ensure that the plant fills space.

Gravitropic movements are classified as follows based on the orientation angle of the plant organs with respect to gravitational pull:

1. ***Orthogravitropic:*** Roots and shoots of the primary plant axis do orient themselves parallel to the direction of gravitational pull.
2. ***Diagravitropic:*** Organs such as stolons, rhizomes and some lateral branches grow at right angles to the pull of gravity.

3. ***Plagiogravitropic***: Lateral stems and lateral roots align themselves at same intermediate angle (between 0° and 90°).
4. ***Agravitropic***: Organs, probably tertiary roots, exhibit little or no sensitivity to gravity.

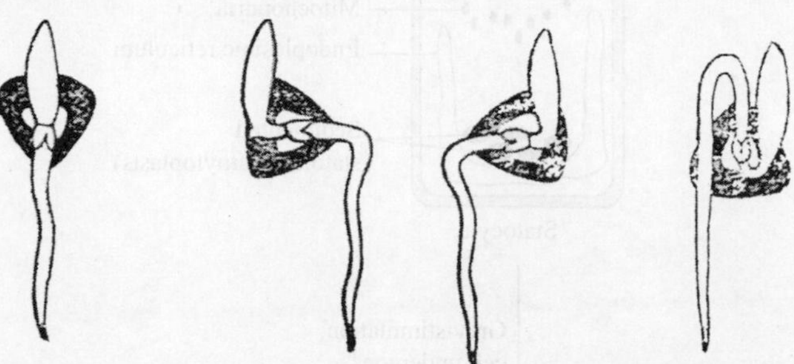

Fig. 14.13 Gravitropism in maize seedlings.

Gravitropic perception in the root is localized in the root cap, particularly in the central column (columella) of the root cap. Graviresponse signalling events initiated in the root cap must induce production of the chemical messenger that modulates growth in the elongation zone. Experimentally it has been established that the tip of the coleoptile could perceive gravity and redistribute auxin to the lower side. Tissues below the tip are able to respond to gravity as well.

Plant responses to gravity can be explained by the following hypothesis:

Starch–statolith hypothesis—Noll (1892) first suggested that plants might be able to sense gravity in a similar manner as seen in some of the animals. Haberlandt and Němec (1900) proposed the statolith theory to explain the plant responses to gravity. A **statolith** is a group of starch grains localized within a membrane, called an amyloplast. Statolith-containing cells are called **statocytes**. Starch is found abundantly in cell elongation zone and root cap.[8] Statolith can move freely in statocyte and persist even in conditions of extreme starvation. After receiving stimulus, statoliths move upside-down (from upper to the lower side) within the organs (Figure 14.14). Accumulation of statoliths on lower side exerts pressure on cytoplasm, which initiates a chain of events and leads to metabolic changes, including trigger of rechanneling of auxin flux and finally unilateral growth. Growth occurs more on the upper side and less on the lower side causing growth curvature.

According to Hasenstein and Evans model (1988), IAA synthesized in the shoot is transported downward (acropetally) in the roots *via* the stele and is distributed evenly on all sides of the root cap. Vertically, the diffusion of auxin from the root cap region is equal all over the circumference, with the result that a uniform rate of growth is maintained all around the root. Then, IAA is carried upward (basipetally) within the cortex to the zone of elongation, where it serves to regulate cell elongation (Figure 14.15a).

However, when the root is shifted from the vertical to horizontal orientation, gravitational forces cause sedimentation of the heavy starch grains, with nucleus and endoplasmic

[8] The root cap is a protective group of continually replaced cells, fitting snugly over the tip of the root. It is the sensitive region and when the root cap is removed it becomes practically insensitive to the gravity stimulus. It helps the roots to penetrate the soil.

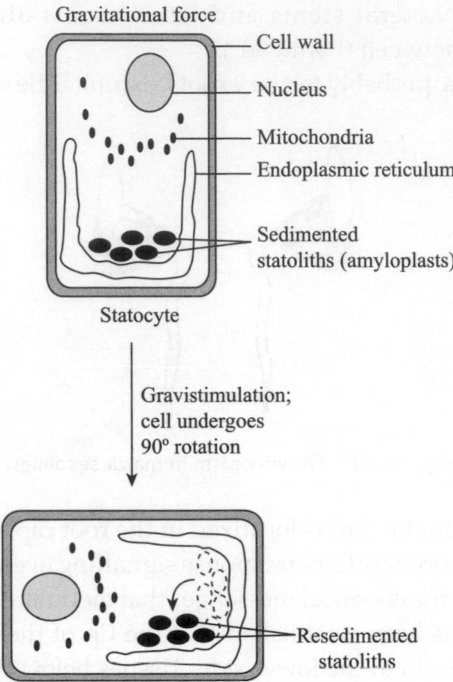

Fig. 14.14 Role of statoliths (amyloplasts) in gravitropism. When the orientation of statolith – containing cells, with respect to gravity, is altered, the statoliths resediment on the new bottom of the cell.

reticulum rising towards the top of the cell (Figure 14.15b). Such changes might be detected by the cell membrane and translated into hormonal signals which control the direction of growth of roots. Then a large amount of the auxin in the cap is transported basipetally in the cortical cells on the lower side of the root. The greater auxin concentration on the lower side of the root has an inhibitory effect on growth while the decreased auxin level on the upper side stimulates more growth and as a result the root bends in the zone of elongation.

Cholodny–Went hypothesis: Geotropism involves the lateral redistribution of auxin. According to the Cholodny–Went hypothesis, auxin in horizontally placed coleoptiles tip is transported laterally to the lower side, causing the lower side of the coleoptile to grow rapidly as compared to the upper side. So the negatively gravitropic organs (e.g., coleoptiles and shoots) would turn upward. In positively gravitropic organs such as root, the higher concentration of auxin is believed to inhibit elongation on the lower side as compared to the upper, causing the organ to grow downward. Exogenous application of auxin (10^{-6} M or more) effectively inhibits root growth- a concentration that normally promotes elongation of coleoptiles and shoots. Additionally, gravitropic curvature is prevented by inhibitors of auxin transport such as tri-iodobenzoic acid (TIBA) and N–(1-naphthylthalamic acid)(NPA).

Gravitropic response in roots also includes changes in second messengers (Ca^{2+}), pH gradients and inositol triphosphate. Role of Ca^{2+} in the gravitropic induction in shoots as well as roots is mediated by the calcium-binding protein **calmodulin**. Calcium has been shown to move toward the upper surface of gravistimulated shoots before the shoots curve upward

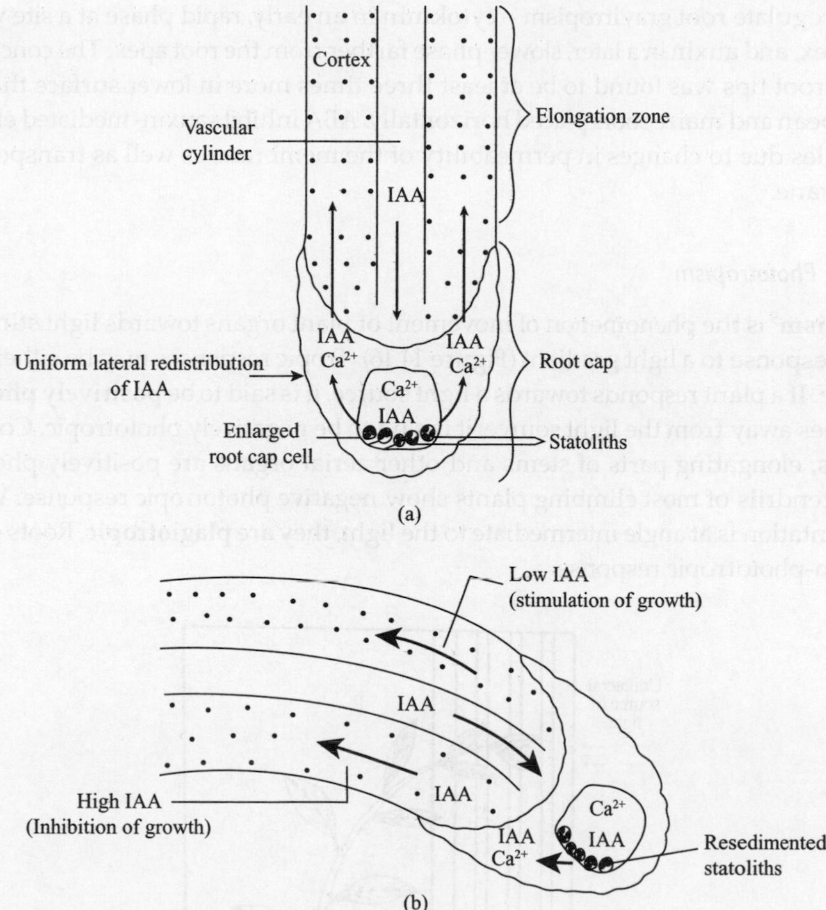

Fig. 14.15 A model proposed for the redistribution of calcium and auxin during gravitropism in roots. Note the orientation of the statoliths in the cap cells of the roots in the vertical (a) and in the horizontal direction (b).

and toward the lower surface of roots before the roots curve downward. According to the **tensegrity model** for gravitropism, local disruption of the actin meshwork in the columellar region of the root cap by the sedimenting amyloplasts leads to a transient increase in the cytosolic Ca^{2+} levels. This is due to the changes in the distribution of tension transmitted by the actin meshwork to calcium channels on the plasma membrane.

Changes in extracellular pH may also be an important signalling element that could modulate auxin responses by altering the chemiosmotic proton gradient.

Cytokinin has also been implicated in the gravitropic response of roots. In uprightly growing roots of *Arabidopsis*, cytokinin produced in the root cap is symmetrically distributed. When roots are positioned horizontally, the cytokinin is distributed asymmetrically, accumulating in the new lower side of the root cap. Such an asymmetrical distribution of cytokinin is believed to cause initiation of downward curvature. Both cytokinin and auxin appear to regulate root gravitropism – cytokinin in an early, rapid phase at a site very near the root apex, and auxin in a later, slower phase farther from the root apex. The concentration of ABA in root tips was found to be at least three times more in lower surface than upper surface of bean and maize roots placed horizontally. ABA inhibits auxin- mediated elongation in coleoptiles due to changes in permeability of the membrane as well as transport across the membrane.

14.7.2.1.2 Phototropism

Phototropism[9] is the phenomenon of movement of plant organs towards light stimulus, or, a growth response to a light gradient (Figure 14.16). Tropic responses may be either positive or negative. If a plant responds towards a light source, it is said to be **positively phototropic** and if it goes away from the light source, it is said to be negatively phototropic. Coleoptiles, hypocotyls, elongating parts of stems and other aerial organs are positively phototropic, while the tendrils of most climbing plants show negative phototropic response. When the leaves orientation is at angle intermediate to the light, they are **plagiotropic**. Roots generally exhibit non-phototropic response.

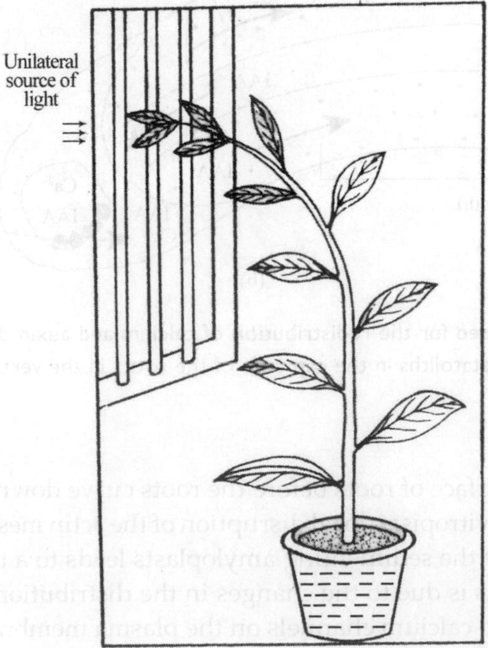

Fig. 14.16 A positively phototropic stem shown bending towards unilateral source of light.

[9] Seedlings of the arboreal, ground-germinating tropical vine, *Monstera gigantea* are attracted toward the darkest sector of the horizon. In nature, the potential host trees provide the dark sector. This response is termed as **Skototropism** (growth towards darkness).

Phototropic response to blue and UV-A light is regulated by a flavoprotein, **phototropin** (120 kD), located in the plasma membrane. Two flavoproteins such as phototropin 1 and phototropin 2, have been recognized as the photoreceptors for the blue-light signalling pathway in oat coleoptiles and *Arabidopsis* hypocotyls. Phototropism occurs due to the differential growth on the illuminated and the oppositely placed shaded regions of the responding organ (coleoptiles and stem).

According to the Cholodny–Went theory of the late 1920s, there is sideways redistribution of auxin as it flows down basipetally from the apical region. Accumulation of auxin on the unexposed side of the organ stimulates those cells to undergo more elongation than those present on the illuminated side. This differential growth is responsible for causing the curvature toward the light source. Moreover, a many-fold increase in the expression of auxin-regulated genes occurs on the unexposed side of hypocotyls. Auxin-regulated genes include two members of the α-expansin family of genes that are essential for cell wall extension.

Recent studies reveal that acidification of the apoplast plays role in phototropic growth. The apoplastic pH on the non-illuminated side of coleoptiles is more acidic as compared to the illuminated side. Acidic pH increases auxin transport by increasing both the rate of IAA entry into the cell and the chemiosmotic proton potential-driven auxin efflux mechanisms. Acidification is also expected to increase cell elongation. Thus, enhanced auxin transport combined with increased cell elongation on the shaded side contribute to bending toward unilateral light.

Although phototropins are the primary photoreceptors, **phytochromes** and **cryptochromes** also contribute to phototropic response.

14.7.2.1.3 Thigmotropism

Response of a plant or plant part to contact with a hard object is known as **thigmotropism**. For example, the coiling of tendrils in bur cucumber (*Cucumis anguria*) and pea plants.

Such reponses are so rapid, that some tendrils coil around a support twice or more in an hour's time. The coiling results from the fact that the cells in contact become slightly shorter while those on opposite side undergo elongation. In *Passiflora*, the tip of the tendril in the beginning moves freely in the air but it twines around a solid object as soon as it comes in contact, and then climbs upward.

14.7.2.1.4 Hydrotropism

Roots often penetrate cracked water pipes and sewer lines. In fact, roots have been found to ascend upward for considerably long distances in response to water leaks. Such growth movements are termed as **hydrotropisms**.

14.7.2.2 Nastic movements

Many plants and plant parts (leaves or flower petals) exhibit nastic movements. Nastic movements are non-directional that do not result in an organ being positioned toward or away from the stimulus direction. These movements may involve differential growth (such as epinasty, hyponasty and thermonasty) or because of turgor changes occurring in specialized motor cells located within the pulvinus (nyctinasty, seismonasty).

14.7.2.2.1 Nyctinastic movements

Nyctinastic movements (sleep movements) are common in the primary leaf of bean plants, *Coleus, Maranta* (prayer plant) and other house plants. Such responses are sensitive to blue light, status of phytochrome and endogenous rhythms. Nyctinastic responses depend on reversible changes in the turgor pressure in the pulvinus. The pulvinus is a bulbous structure present in the basal region of the petiole, the pinna or the pinnule in plants belonging to families Fabaceae and Oxalidaceae. The pulvinus contains a number of large, thin- walled motor cells, which brings about a shift in the position of the leaf by undergoing reversible changes in turgor. The oppositely placed sides of the pulvinus are termed as the **extensor** and **flexor** regions. During day time, large amounts of auxins are produced and are transported to the lower side of the petiole. In high auxin areas, K^+ and Cl^- ions get accumulated which leads to influx of water causing turgidity of extensor cells, thereby unfolding of leaves. At night, the auxin transport to pulvinus is inhibited. Reverse reaction takes place which leads to flaccidity in extensor cells and turgidity in the oppositely placed flexor cells. Turgid flexor cells force flaccid extensor cells to bend down, thereby folding leaves/leaflets.

The current model involves the interaction of phytochrome, the biological clock and inositol triphosphate (IP3), to provide an explanation of the leaf movements of nyctinastic plants. Light signal stimulates phytochrome, which enhances inositol phospholipid turnover. Activation by light leads to an increase in the level of the second messengers, inositol 1,4,5-triphosphate (IP3) and diacylglycerol (DAG). The second messenger triggers extrusion of calcium ions into the cytoplasm, and phosphorylation of proteins, which in turn promote the release of H^+ from the cell with simultaneous inward movement of K^+ into the cell. This is followed by uptake of water and motor cell swelling. An active transport pump extrudes Ca^{2+} which helps in restoration of Ca^{2+} homeostasis.

14.7.2.2.2 Seismonastic movements

Ordinarily, the leaves of the sensitive plant (*Mimosa pudica*) take up different positions by day and night but can also move extremely rapidly in response to mechanical, thermal, chemical and electrical stimuli. The so-called 'sensitive plant' or 'touch-me-not' (*M. pudica*) is a herbaceous perennial with doubly-pinnate compound leaves (Figure 14.17). Present at the base of the petioles, there are specialized swollen structures called 'pulvinus' (specialized motor organs located at the joint of the leaf) and similarly smaller 'pulvinules' at the base of leaflets (Figure 14.18). The bipinnate compound leaves have three 'pulvini'.

Fig. 14.17 Touch-me-not plant (*Mimosa pudica*) (*Courtesy*: Mr Sandesh Parashar)

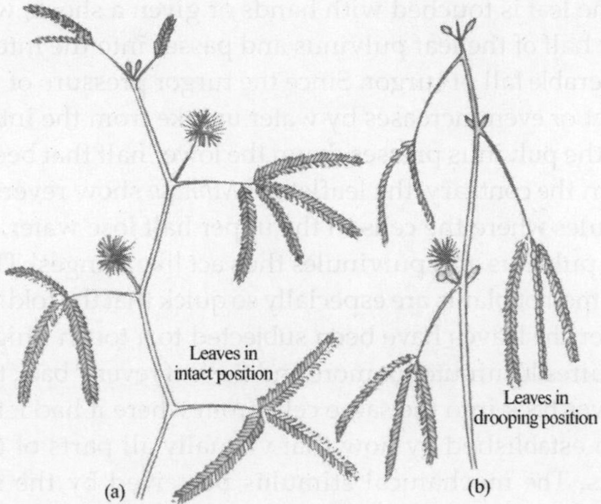

Fig. 14.18 Seismonastic movements in *Mimosa pudica.* (a) unstimulated leaves (in horizontal position) and (b) stimulated leaves (in drooping condition).

These seismonastic movements result from the differential loss of turgor on the two regions of the pulvinus. The lower half of the pulvinus consists of thin-walled cells with larger intercellular spaces while the upper half is composed of relatively thick-walled cells with fewer extracellular spaces. In the 'unstimulated' condition, the cells on both the regions of the pulvinus are fully turgid and the leaf is held in an erect or horizontal position (Figure 14.19).

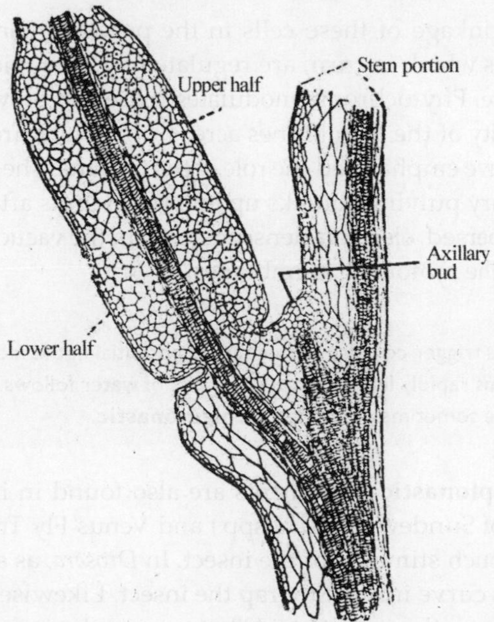

Fig. 14.19 Sectional view through a pulvinus of *Mimosa pudica* (a sensitive plant).

However, when the leaf is touched with hands or given a shock, water moves out from the cells of the lower half of the leaf pulvinus and passes into the intercellular spaces thus resulting in a considerable fall of turgor. Since the turgor pressure of the cells of the upper half remains constant or even increases by water uptake from the intercellular spaces, the upper turgid half of the pulvinus presses down the lower half that becomes flaccid and the leaf droops down. On the contrary, the leaflets of *Mimosa* show reversal in the differential turgor in the pulvinules where the cells in the upper half lose water, allowing the leaflets to turn upward. The pulvinus and pulvinules thus act like 'hinges'. The leaves and leaflets movements in touch-me-not plants are especially so quick that the folding takes place within 'a second or two' after the leaves have been subjected to a touch (thigmonastic). Recovery after stimulation requires 10 minutes or more and the leaf reverts back to its original position as water usually moves back into the same cells from where it had left.

In fact it has been established by now that virtually all parts of the *Mimosa* plant are sensitive to stimulus. The mechanical stimulus perceived by the sensory cells is first converted to electrical signals or impulse (called action potential) which travels rapidly until it reaches the 'motor cells' of the pulvinus which then undergo the volume change that leads to leaf/leaflets movement. It appears that phloem sieve tubes with their protoplasmic continuity can function as conduits for transmission of electrical signals. These electrical signals are, as in animals, an important means of communication between different tissues of the plant body. The electrical signal moves along the petiole at the rate of about 2 cm per second. The rate of transmission rises with rise in temperature and is dependent upon the living cells. The electrical signal gets converted into a chemical signal, with K^+ and Cl^- ions, followed by water moving from the cells in one half of the pulvinus to the intercellular spaces of the other half. Aquaporins facilitate a rapid outward movement of water.

The swellings or shrinkage of these cells in the pulvinus controls the up-and-down movements of the leaves which in turn, are regulated by the rapid entry and exit of ions, especially K^+ and malate. Phytochrome modulates these salt movements, presumably by affecting the permeability of the membranes across which ions are transported.

Some investigators have emphasized the role of the tannins. The vacuolar tannin mass of motor cells in the primary pulvinus breaks up into small units after stimulation, followed by changes in finely dispersed, electron-dense material in the vacuole. Some tannins appear to be excreted through the motor cell membranes.

Touch stimuli to the leaves trigger collapse of membrane potential in the motor cells (depolarization); potassium and chloride ions rapidly leave the cells and loss of water follows. Since seismonastic plants respond to touch, they are sometimes considered **thigmonastic**.

Thigmonastic or **Haptonastic** movements are also found in insectivorous plants, for example, in the leaves of Sundew (*Drosera* spp.) and Venus Fly Trap (*Dionaea* spp.) which are in response to the touch stimulus of the insect. In *Drosera*, as soon as an insect alights on the leaf, the tentacles curve inward to trap the insect. Likewise in *Dionaea*, the two leaf halves curve upward along the central midrib, trapping the insect in between. After the insect has been digested, the leaf parts return to their normal position.

14.7.2.2.3 Solar tracking

Several plants possess solar-tracking leaves with their blades positioned at right angles to the sun's rays throughout the day. Young plants of sunflower, for example, face the sun throughout the day **(heliotropic movements**[10]). Leaves often twist on their petioles and show perpendicular orientation to receive the maximum amount of light available.

Solar tracking is a blue-light response and the sensing of blue-light by the leaves takes place in specialized parts of the stem or leaf. Various plant species including alfalfa, cotton, soybean, beans, and lupine possess leaves which show capability of solar tracking.

14.7.2.2.4 Water conservation movements

Leaves of many grasses have specialized thin-walled **bulliform cells,** below parallel, lengthwise grooves on their surfaces. During periods of water stress, these cells lose their turgor followed by rolling of the leaves, which effectively reduces transpiration to 5–10 per cent of the normal.

14.7.2.2.5 Taxic movements

These movements involve either the entire plant or its reproductive cells and occurs in several groups of plants and fungi but not among flowering plants. Stimuli for taxic movements include chemicals, light, oxygen, and gravitational fields. For example, in ferns, the female reproductive structures produce a chemical that prompts a **chemotaxic** response in the male reproductive cells, i.e., the sperms swim toward the source of the chemical. Certain one-celled algae exhibit **phototaxic** responses, swimming either toward or away from a light source. Other similar organisms exhibit **aerotaxic** movements in response to changes in oxygen concentrations.

***Arabidopsis thaliana*: Another Model Organism**

Scientists all over the world are working to understand the molecular mechanism of plant growth and development, largely through investigations of a small, herbaceous weed of Eurasia, *Arabidopsis thaliana*, a member of the angiospermic family Brassicaceae. *A.thaliana* (mouse-ear or thale cress), a native of Europe, Asia and northwestern Africa, grows in barren fields, in disturbed lands and along roadsides throughout the temperate zones of northern hemisphere, forming vegetative rosette during winter. In the ensuing spring season, the stem elongates and flowers (Figure 14.20) .There are multiple reasons why *Arabidopsis* has become such an ideal experimental tool and model system for basic research in genetics and molecular biology. It is easy to grow and cross, and like Mendel's pea plant, it is able to self-fertilize,

[10] 'Phototropism' (tracking the light) term is used during the irreversible growth of the plant, whereas the term 'heliotropism' (tracking the sun) is used to describe the reversible movement of leaves and flowers in response to the position of the sun. However the primary mechanism remains the same i.e. adjustment of cell hydration in order to increase or decrease the turgor pressure within the individual cells.

making genetic studies convenient. It can be easily cultured in the laboratory with a short life span of 5 to 6 weeks from seed germination to fruit setting, a single plant producing thousands of seeds (up to 10,000 per plant). It has a comparatively small genome of about 125 Mb (Million of Base-Pairs), just the same size as present in the nematode *(Caenorhabditis elegans*) and the common fruit fly (*Drosophila melanogaster*) which is unlike our ornamental tulips with a genome of 24,704 Mb (containing over 170 times more DNA as compared to *Arabidopsis*).[11] In *Arabidopsis* the mutant lines are easily produced by treating the seeds with chemical mutagens or by irradiating the seeds. A large number of mutant lines are available for research, and the isolation and study of such mutants has already made great contributions to various areas of plant growth and development. Transgenic plants can be made easily using *Agrobacterium tumifaciens* as the vector to introduce foreign or alien genes. Although *Arabidopsis* is self fertile but it can easily be cross pollinated to do (i) genetic mapping work and (ii) to produce strains with multiple mutations

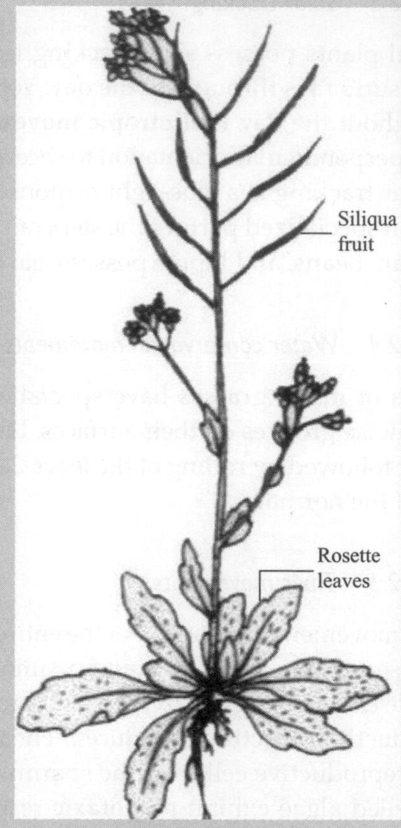

Fig. 14.20: An ideal experimental tool and model system for basic research in genetics and molecular biology.

Review Questions

RQ 14.1. Short-day plants are more correctly regarded as 'long night plants'. Explain and also discuss briefly the state of phytochrome that controls flowering in SDP and LDP.

Ans. In short-day plants, the continuity of the dark period is absolutely essential for flowering to occur. If the dark phase is broken in the middle by a flash of light, then the plants fail to flower. Hence, SD plants are more correctly known as 'long-night plants'. On the other hand, 'interruption of light period' in such plants by darkness had little or no effect on flowering. In short-day plants, most of the phytochrome at the end of day light is converted to a 'Pfr' form which inhibits flowering.

[11] The *Arabidopsis thaliana* was the first genome to be sequenced in entirety in 2000 by an international collaboration – the Arabidopsis Genome Initiative (AGI).

However, during the long dark period, this 'Pfr' form is slowly converted to a 'Pr' form which promotes flowering. However, when the dark phase is interrupted or broken by a brief flash of light in the middle, the Pr form is converted back to Pfr which inhibits flowering.

Like short-day plants, Pfr accumulates at the end of the day in long-day plants also, but it triggers a sequence of reactions which finally leads to flowering. In other words, Pfr causes flowering of long-day plants but Pr form inhibits it (the Pfr form is converted to the Pr form if the dark period is greater).

RQ 14.2. Enumerate 'three' methods by which flowering can be induced at will in the biennials.

Ans. In some of the biennial plants such as cabbage (*Brassica oleracea* var. *capitata*) and carrot (*Daucus carota*), the leaves are produced in rosettes as the internodes between them do not elongate. In these plants, flowering can be induced by exposure to appropriate 'long day photoperiods' or to cold or chilling.

The stem (or flowering axis) elongation (i.e., bolting) occurs due to an increase both in cell number and in cell elongation.

Exogenous application of gibberellins to such plants will induce bolting and flowering even without any appropriate exposure to long-day conditions or low temperature treatment.

RQ 14.3. Thigmomorphogenesis and solar tracking are familiar phenomena observed in plants. Explain.

Ans. In thigmotropism, plants may respond to mechanical stimuli (rubbing or bending of stems) by altering their growth patterns. Such plants typically are shorter and stockier than non-stimulated plants. Studies in *Arabidopsis thaliana* show that this phenomenon involves modifications in genes expression, encoding proteins associated with the calcium-binding protein calmodulin, suggesting a role for calcium in mediating growth response. Plants in a natural environment are, of course, subjected to similar stimuli, in the form of wind, rain drops, and rubbing by passing animals and machines.

On the other hand, the leaves and flowers of many plants follow or track the sun – a phenomenon known as solar tracking or heliotropism – during the course of the day, either maximizing or minimizing absorption of solar radiations. Heliotropic leaf movements are common in cotton, soybeans, sunflower, cowpeas, lupines and such.

RQ 14.4. In the perception of a stimulus, what are presentation time and threshold?

Ans. Presentation time is the length of time the stimulus must be present for the perceptive cells to react and complete transduction. After the stimulus has acted long enough to fulfil the presentation time, a response occurs even if the stimulus is removed.

Threshold refers to the level of stimulus that must be present during the presentation time to cause perception and transduction.

RQ 14.5. What is an **all-or-none response**? Give some examples of plants that flower with an all-or-none response.

Ans. In an all-or-none response, after the threshold and presentation time requirements are met, the stimulus is no longer important; the response is now completely internal. Individuals respond identically whether they received strong or weak stimuli, regardless of whether the stimulus was brief or long lasting. Examples are *Hibiscus syriacus*, poinsettia, chrysanthemum, and oats.

RQ 14.6. What is a **dosage-dependent response**? Name some plants that flower this way.

Ans. In dosage-dependent responses, the duration of the stimulus affects the amount and duration of the response. In species of this type, individuals that receive only minimum stimulation flower poorly, even if the plant is quite healthy. Those that receive stronger and longer stimulation produce more flowers e.g., turnip, a few varieties of cotton and potato, and *Cannabis sativa*.

RQ 14.7. Distinguish between nastic movements and tropic movements.

Ans.

Nastic movements	Tropic movements
1. These are non-directional and may be caused by diffused stimulus.	These are directional and caused by directional stimulus.
2. Direction of response is determined by organ and not by stimulus.	Direction of response is related to direction of stimulus.
3. These movements are reversible and plant organs revert back to original shape after some time of removal of stimulus.	These movements are irreversible and organs don't return to original shape after removal of stimulus.
4. These movements result due to turgor pressure changes.	Such movements result due to differential behaviour of growth within an organ or between two different organs.

Chapter 15

Plant Photoreceptors

The receptor molecules used by plants to detect sunlight are termed **photoreceptors.** Photoreceptors absorb a photon of a particular wavelength and utilize this energy, thus initiating a **photoresponse.** All the known photoreceptors (except UVR8) consist of proteins bound to non-protein light-absorbing prosthetic groups (**chromophores**). Protein structures of the different photoreceptors vary and are involved in regulation of downstream signalling. Other common aspects of photoreceptors include sensitivity to light (quantity, quality, and intensity) and the time duration of the photoperiod. Photoreceptors, including **phytochromes, cryptochromes** and **phototropins,** help plants regulate developmental processes over their lifetime by sensitizing them to incident light. They also initiate protective processes in response to harmful radiations.

Keywords

- Chromophore
- HIRs
- LFRs
- LOV-domain
- Phototropin
- Phytochrome-interacting factor
- Phytochromobilin
- VLFRs

The action spectra of light-regulated responses demonstrate that plants are highly sensitive to UV-B radiation (280–320 nm), UV-A radiation (320–380 nm), blue light (380–500 nm), red light (620–700 nm), and far-red light (700–800 nm). UV-B radiation is detected by the **UV RESISTANCE LOCUS 8 (UVR8)** protein; UV-A and blue light, by cryptochromes, phototropins, and the recently discovered **ZEITLUPE** (ZTL) family; and red and far-red light by phytochromes. Zeitlupes are light-responsive F-box proteins that target proteins functioning in the circadian rhythms and the photoperiodic regulation of flowering. Phytochrome and cryptochrome are soluble receptors that function primarily in the nucleus to regulate the activity of transcription factors. These two classes collectively regulate developmental responses to light intensity and also mediate responses to variations in the spectral quality of light that indicate shading by other plants.

15.1 Phytochrome

Historical background: The discovery of a ubiquitous plant pigment, the phytochrome, and the evidence that it regulates almost every aspect of a plant's response to light was a path-breaking discovery for plant science. The domino effects of their historic research have been stupendous, which prompted scientists worldwide focusing on phytochrome-related research. The United States Department of Agriculture (USDA), Beltsville Agricultural Research Center (BARC) in Beltsville, Maryland, USA, was recently nominated an American Chemical Society National Historic Chemical Landmark for the seminal work of USDA scientists in the discovery of phytochrome. The discovery team got motivated by the experimental work of H. A. Borthwick and M. W. Parker in the 1930s, which quantified the phenomenon of photoperiodism and along with S. B. Hendricks, a renowned soil chemist, searched for a chemical basis for non-photosynthetic processes of plants to light. In 1940–1960, they speculated about the presence of a photoreversible pigment regulating photoperiodism in flowering, seed germination, and several other photomorphogenic responses in plants. It was also noticed that the red light responses were reversed by far-red light. In 1959, the biophysicist W. Butler and biochemist H. Siegelman, using a specialized spectrophotometer, finally isolated the phytochrome pigment from dark-grown 'corn' seedlings. In 1964, H. Siegelman and E.M. Firir extracted phytochrome from oat seedlings in a highly purified state and also analyzed its structure. Surprisingly all these scientific stars assembled at a common place, which they rarely do, to complete the discovery.

In 1983, P. Quail and C. Lagarias purified the intact phytochrome molecule and subsequently, in 1985, H. Hershey and P. Quail published its first gene sequence. By 1989, molecular genetics studies revealed the existence of more than one type of phytochrome, e.g., the pea plant possessed two phytochromes (type I and type II predominant in etiolated seedlings and green plants respectively). Recently, a technique of genome sequencing has confirmed that *Arabidopsis* has five phytochrome genes (PHY A, B, C, D and E), the rice plant has only three (PHY A, B and C), and maize has six phytochromes (PHY A1, A2; PHY B1, B2; PHY C1, C2). All these phytochromes use considerably different protein parts, but possess common **phytochromobilin** as their light-harvesting chromophore. Phytochrome A is quickly destroyed in the Pfr form compared to other members of the family. In the late 1980s, it was established that Phytochrome A is destroyed by the ubiquitin system, a major intracellular protein degradation system identified in eukaryotes.

The chemical structure and domains of a plant phytochrome are shown in Figure 15.1 and Figure 15.2 respectively. Phytochrome is a large dimeric protein made up of two equivalent subunits. The monomer has a molecular mass of 124,000 D with a high proportion of polar amino acids and is water-soluble . It is believed to be a blue-green biliprotein and is covalently bound to an open chain tetrapyrrole chromophore molecule. It was established by **Borthwick and Hendricks** that phytochromes exist in two forms - **P_r (red light absorbing form)** and **P_{fr} (far-red light absorbing form)** as shown in Figure 15.3. Initial studies in SDPs in late 1940s suggested that red light was most effective as a light–break, with a maximum effectiveness near 660 nm. Borthwick and his colleagues also observed that red light inhibition of flowering in *Xanthium* could be reversed with far-red. At about same time, photoreversibility of seed germination was noticed. Thus, red/far-red photoreversibility of the light-break indicates the role of phytochrome in the control of flowering in SDP and LDP. Studies on phytochrome mutants in *Arabidopsis* have suggested that PHY A is essential to induce flowering of LDP under certain conditions, whereas PHY B seems to inhibit flowering. It has been confirmed now that the Pfr form induces flowering in long-day plants and Pr form in short-day plants (Figure 15.4).

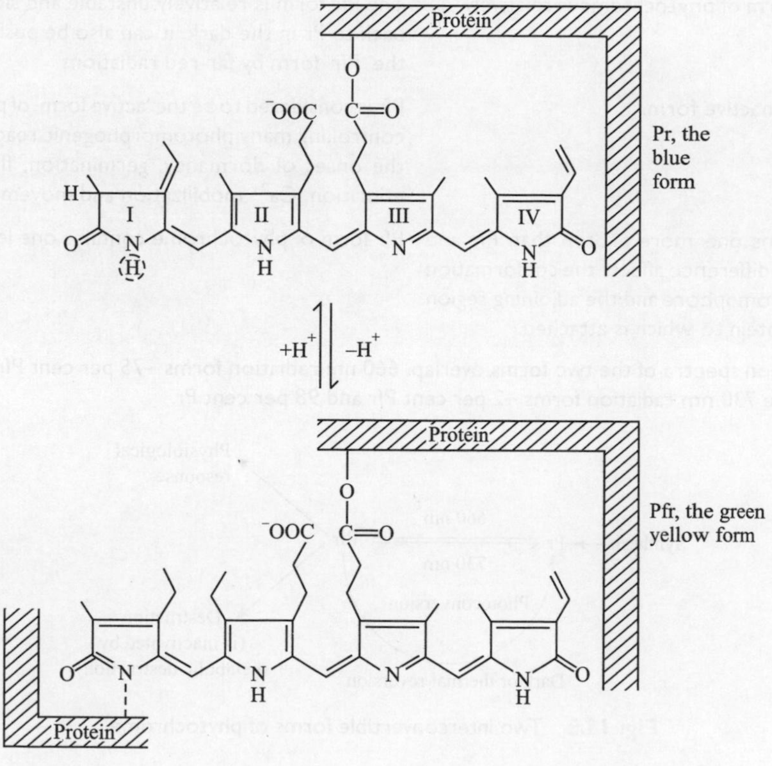

Fig. 15.1 The chemical structure of a plant phytochrome molecule.

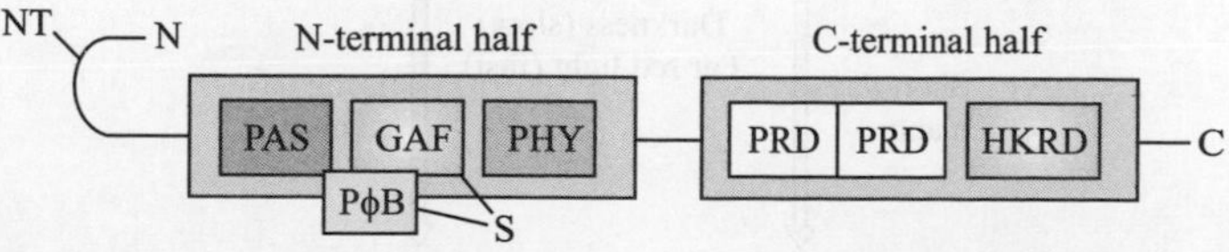

Fig. 15.2 Schematic representation of a plant phytochrome domain. The N-terminal moiety of phytochrome contains the PAS-GAF-PHY domains, which comprise the chromophore-binding, photosensory region of phytochrome. The C-terminal moiety of phytochrome contains two PRD domains and a HKRD domain. A hinge region separates the N-terminal and C-terminal moieties of the molecule. The chromophore phytochromobilin (PΦB) is attached to cysteine residues in the proteins by a thioester link (-S-). PAS domain (Per-Arnt-Sim domain); GAF domain (domain named after the proteins that are found in cGMP-specific phosphodiesterases; adenyl cyclases and FhlA); NT (N-terminal extension); PRD (PAS-related domain); HKRD (Histidine kinase-related domain).

P_r	P_{fr}
1. It is a 'red' light-absorbing form of phytochrome which to the human eye looks blue.	It is a 'far-red' light-absorbing form which is blue-green or greenish yellow.
2. Absorption of red light centred about 660 nm converts it to Pfr*.	Absorption of far-red light centred about 730 nm converts it to Pr*.
3. The Pr form of phytochrome is relatively stable.	The Pfr form is relatively unstable and slowly decays back to Pr in the dark, it can also be pushed back to the 'Pfr' form by far-red radiations.
4. Pr is the inactive form.	Pfr is considered to be the 'active form' of phytochrome, controlling many photomorphogenic reactions such as the onset of dormancy, germination, flowering, de-etiolation, Ca^{2+} mobilization and movements.
5. Pr contains one more proton than Pfr and this small difference affects the conformation of the chromophore and the adjoining region of the protein to which is attached.	Pfr form of phytochrome contains one less proton.

*The absorption spectra of the two forms overlap, 660 nm radiation forms ~75 per cent Pfr and 25 per cent Pr, while 730 nm radiation forms ~2 per cent Pfr and 98 per cent Pr.

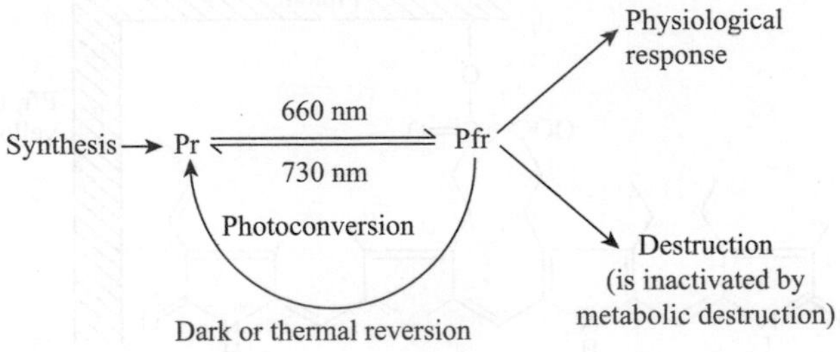

Fig. 15.3 Two interconvertible forms of phytochrome.

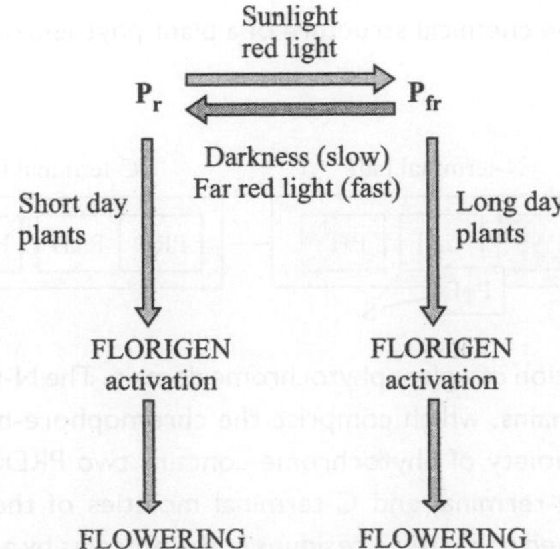

Fig. 15.4 Phytochrome is produced in Pr form during dark period. In dark-grown plants, phytochrome is found in a red light absorbing form known as Pr. Thus red light (660 nm) would convert the pigment to a far-red absorbing form known as Pfr (730 nm) and the conversion is fully reversible. The Pfr form promotes flowering in long-day plants and the Pr form in short-day plants.

Phytochrome-regulated responses are categorized into three groups based on their energy demands - **very low fluence responses (VLFRs), low-fluence responses (LFRs) and high-irradiance responses (HIRs).**

VLFRs

- These phytochrome responses can be induced by fluences as low as 0.0001 μmol m^{-2}. (Amount of light required to induce the responses is known as **fluence**. Total fluence = fluence rate x length of time of irradiation.)
- They become saturated at about 0.05 μmol m^{-2}
- Non-photoreversible
- Obey the ***law of reciprocity*** (reciprocal relationship between fluence rate and time duration)

Examples: Red-light induced germination of *Arabidopsis* seeds
Red-light stimulated growth of coleoptile in etiolated oat seedlings
Red-light induced inhibition of growth of mesocotyl in etiolated oat seedlings

LFRs

- These phytochrome responses cannot be induced until the fluence reaches 1.0 μmol m^{-2}
- They become saturated at about 1000 μmol m^{-2}
- Photoreversible
- Obey the law of reciprocity

Examples: Promotion of lettuce seed germination
Inhibition of hypocotyl elongation
Regulation of leaf movements
Promotion of de-etiolation
Seed germination

HIRs

- These require continuous exposure to light of relatively high irradiance. Response is directly proportional to the irradiance until the response saturates and additional light has no further effect.
- HIRs saturate at much higher fluences as compared to LFRs-at least 100 times more.
- Non-photoreversible and time-dependent
- Don't obey the law of reciprocity

Examples: Synthesis of flavonoids in various dicot seedlings
Inhibition of hypocotyl elongation in mustard, lettuce seedlings
Induction of flowering in *Hyoscyamus*
Production of ethylene in *Sorghum*

Mode of action

Phytochrome-mediated responses are linked to considerable changes in gene expression (Figure 15.5).

(i) Phytochrome - interacting factors (PIF) found in the nucleus inhibit the expression of phytochrome-dependent genes. PIF protein acts as a negative regulator and perhaps suppresses transcription by attaching to the promoter site of a photoresponsive gene. Most of the PIF proteins remain intact in the darkness, but are rapidly degraded in the light.

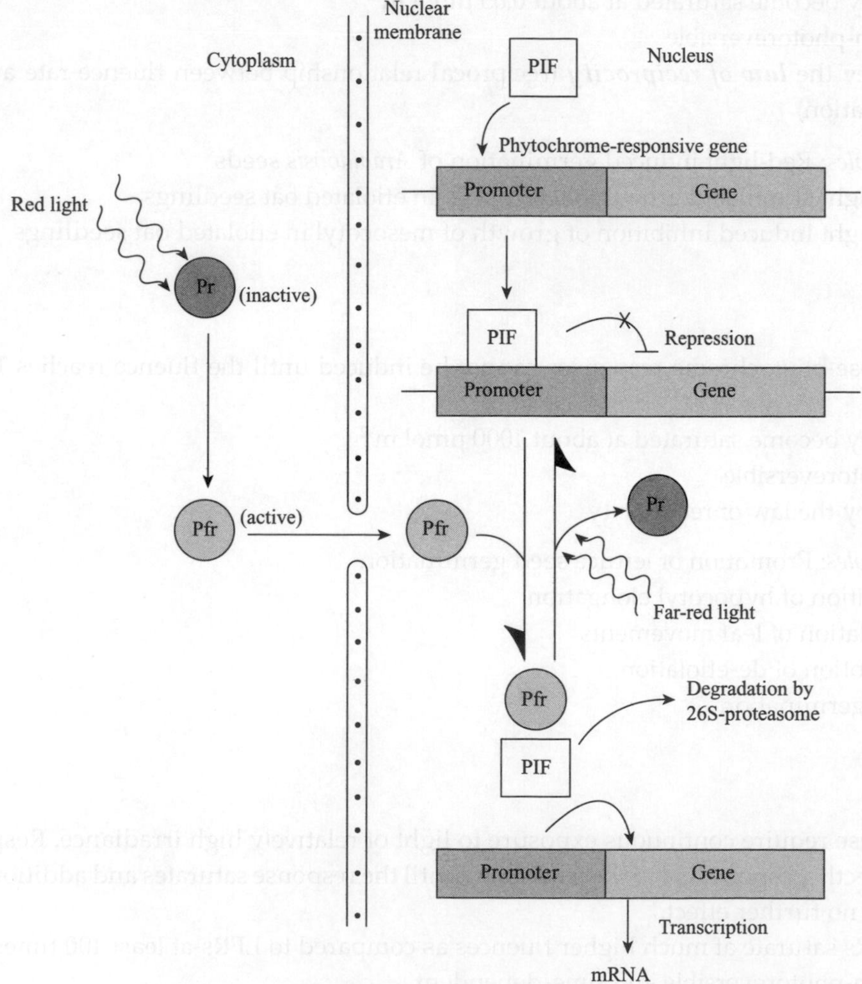

Fig. 15.5 A general model explaining the mode of action of phytochrome pigment. Phytochrome interacting factor (PIF) in the nucleus binds to promoter reigon of phytochrome-responsive genes and causes repression of transcription. P_{fr} form of phytochrome is transported from the cytoplasm to the nucleus and causes the degradation of PIF by 26S-proteasome system. This results in activation of gene transcription to mRNA. P_r form of phytochrome accumulates in the nucleus which allows PIF to reassociate with the promoter region and thereby causing gene repression again.

(ii) Phytochrome accumulates and remains in the cytosol as Pr (inactive) form in the dark. Upon irradiation with red light, Pr form is transformed to Pfr (active) form, which is then imported into the nucleus.
(iii) Pfr binds to PIF protein, which undergoes degradation by 26S-proteasome system, thereby activating transcription of phytochrome-responsive genes.
(iv) Far-red light will change the Pfr back to Pr.

15.1.1 Important phytochrome-mediated responses

The wide variety of developmental responses that are controlled by phytochrome are listed below:

1. Leaf expansion and stem elongation
2. Germination of many seeds
3. Straightening of the hook-shaped shoot of many seedling (unfolding of plumular hook)
4. Phototropism
5. Geotropism
6. Nyctinastic movements (day-night movements of leaves)
7. Orientation response of chloroplasts in certain algae (*Mougeotia*), i.e., the chloroplasts remain flat under low light intensities but rotate within the cell through an angle of 90°, or at higher intensities the chloroplast 'edge-on' to the light (section 15.1.1.2).
8. Regulate many biochemical phenomenon such as, formation of protochlorophyll as well as carotenoids in the leaves, and of anthocyanin and flavonoid pigments in flowers, stems and fruits.
9. Photoperiodic responses.
10. The cytosolic Ca^{2+} concentration (section 15.1.1.1)
11. Fern spore germination.
12. Sex expression
13. Regulating nitrate reductase activity
14. Pollen germination
15. Differentiation of stomata and tracheary elements
16. Tanada effect (section 15.1.1.3)

15.1.1.1 Calmodulin and cell wall stiffening

Calmodulin is a protein, present both in the cytosol and in organelles, such as plastids, mitochondria and nuclei. The Ca^{2+} level in the cytosol is usually very low (10^{-7} M). Each calmodulin molecule possesses four high-affinity calcium-binding regions. Calmodulin, by itself, has no regulatory role but calmodulin–calcium complex in the bound form can stimulate a large number of enzymes. Thus, it can act as a master switch controlling metabolic mechanisms in the cell. Calmodulin is the principal receptor of Ca^{2+} in plant and animal cells.

Red light (650–680 nm) converts Pr to Pfr, followed by Pfr-mediated increase in the cytosolic Ca^{2+} concentration (>10^{-6} M) which favours the activation of calmodulin, forming

Ca^{2+}-calmodulin complex.[1] This complex activates Ca^{2+}-ATPase located in the plasma membrane, which pumps Ca^{2+} out from the cytosol into the walls of the plant cells. Increased Ca^{2+} level within the cell walls in some way induces cell wall stiffening or inhibiting cell wall loosening.

This model (Figure 15.6) is speculative but it serves well to explain how Pfr-mediated signals are transmitted, activating Ca^{2+}-ATPase[2] system in the plasma membrane.

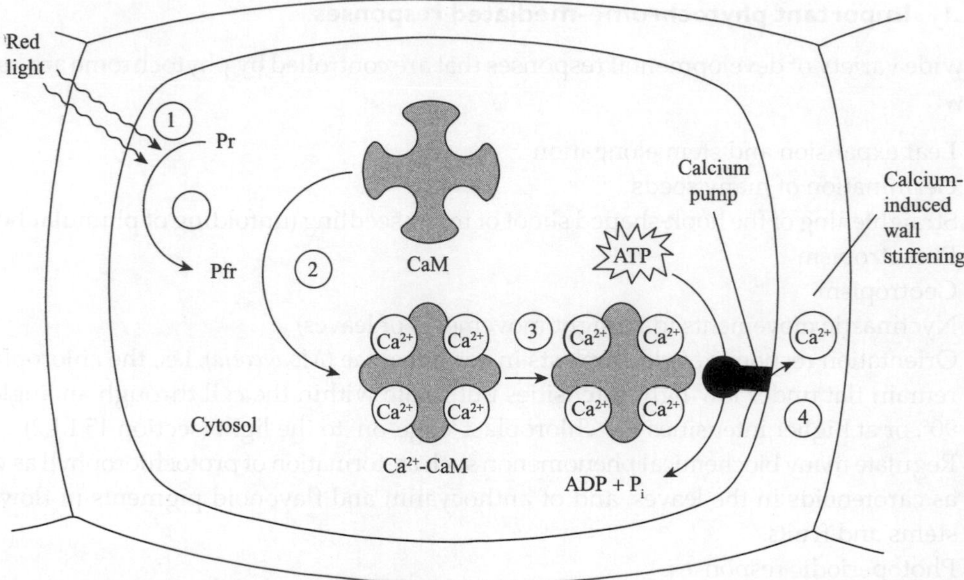

Fig. 15.6 A model for phytochrome regulation of cell wall stiffening showing the cascade of events initiated. The plasma membrane -bound Ca^{2+}-ATPase pumps Ca^{2+} out of cytosol into the wall of plant cells, leading to its stiffening.

15.1.1.2 *Chloroplast rotation in Mougeotia*

Mougeotia is a filamentous green alga of Chlorophyllaceae with a single flat, ribbon-shaped chloroplast lying in between two large vacuoles that are surrounded by a layer of cytoplasm. The chloroplast can 'rotate' about its long axis in response to different wavelengths of light.

By using a microbeam of red light, Wolfgang Haupt (1959) observed that chloroplast rotated perpendicularly to the direction of light (Figure 15.7), whereas no such orientation of chloroplast could be seen when exposed to a beam of far-red light. However, the impact of red light could be reversed if immediately exposed to far-red light. They further showed that the photoreceptor (phytochrome) involved in chloroplast rotation is localized near

[1] The calmodulin–calcium complex has a regulatory effect on activity of many enzymes such as protein kinases, NAD^+ kinase, phosphorylase kinase, adenylate cyclase, phosphodiesterase, guanylate kinase, and other processes like microtubule assembly, membrane phosphorylation and hormonal action.

[2] ATP phosphohydrolase is universally abbreviated as ATPase.

the plasma membrane surrounding the cytosol but *'not'* onto the chloroplast directly. The response is fast, occurring after a lag time as brief as a few minutes. Later studies have specifically revealed that phytochrome molecules are arranged in left-handed spirals around the cell.

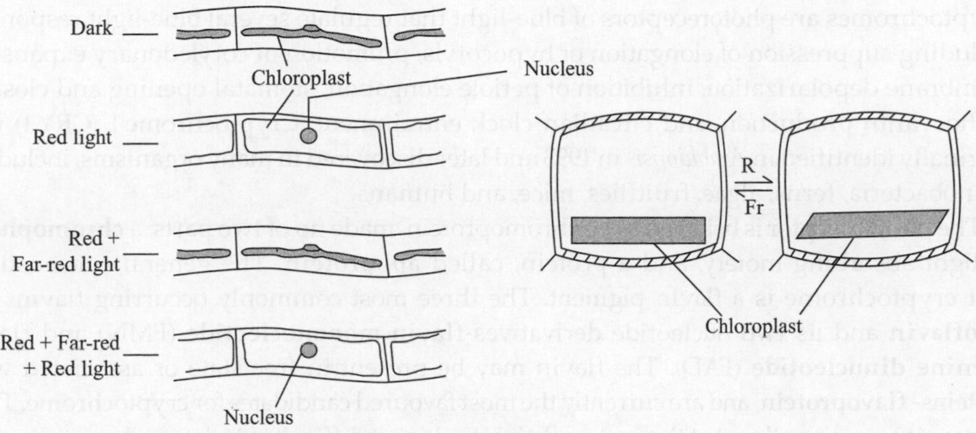

Fig. 15.7 An elegant experiment to illustrate the reversibility of red and far-red light on the orientation of the giant chloroplast* in the filamentous alga, *Mougeotia*. Red light causes the chloroplast to turn its 'broad face' lying parallel to the upper surface but the far-red light reverses the effect, and the chloroplast orientation is perpendicular to the upper surface. The phytochrome receptor is not in the chloroplast but is localized somewhere in the plasma membrane.

* Orientation of *Mougeotia* chloroplast may be helping it (or other aquatic plants) in utilizing any available light reaching the filament. The single chloroplast 'faces' red light source and becomes 'parallel' when exposed to a beam of far-red light.

15.1.1.3 Tanada effect

One of the most interesting experiments on photoreversibility of phytochrome showing immediate effect is known as the **'Tanada effect'**, discovered in 1968 by a Japanese scientist, Takuma Tanada. When dark grown, excised barley (*Hordeum vulgare*) root tips are floated in a glass beaker containing IAA, ATP, ascorbic acid and minerals like $MgCl_2$, $MnCl_2$, KCl and $CaCl_2$ (the **Tanada bath**) and are subjected to red light for a short period, it causes the root tips, within seconds, 'to stick to the walls' of the beaker which have developed a slight negative charge because of being washed with sodium monohydrogen phosphate. This means that red light exposed roots develop a slight positive surface charge because of H^+ extrusion from the cytosol to cellular wall by the activity of the ATPase located in the plasma membrane. The root tips remain attached to the glass walls until these are given a brief far-red exposure which rapidly separate out the root tips because of negative charge produced, thus neutralizing the positive charge. This shows that the red light results in creating a +ve charge on the root tips while the far-red light produces a negative charge.

This phenomenon of photoreversibility of red and far-red radiations also occurs in mung bean root tips (*Phaseolus aureus*).

15.2 Cryptochromes

Cryptochromes are photoreceptors of blue-light that regulate several blue-light responses, including suppression of elongation of hypocotyls, promotion of cotyledonary expansion, membrane depolarization, inhibition of petiole elongation, stomatal opening and closing, anthocyanin production, and circadian clock entrainment. Cryptochrome 1 (CRY 1) was originally identified in *Arabidopsis* in 1993 and later discovered in many organisms, including cyanobacteria, ferns, algae, fruitflies, mice, and humans.

The photoreceptor is believed to be chromoprotein, made up of two parts: a **chromophore**, or light absorbing moiety, and a protein, called **apoprotein**. The general consensus is that cryptochrome is a flavin pigment. The three most commonly occurring flavins are **riboflavin** and its two nucleotide derivatives-**flavin mononucleotide** (FMN) and **flavin adenine dinucleotide** (FAD). The flavin may be present in free state or associated with proteins- **flavoprotein**, and are currently the most favoured candidates for cryptochrome. They represent a very small part of the vast pool, thus making it difficult to isolate and unequivocally establish its physiological role. Cryptochrome possesses two chromophores - FAD and a pterin, resembling with microbial DNA repair enzyme, photolyase.

Most of the plant cryptochromes are 70–80 kD proteins with two distinct domains. Plant cryptochrome can be considered like a molecular light switch where absorption of blue photons at the N-terminal photosensory site alters the redox status of the bound FAD chromophore. This results in protein conformational changes at the C-terminus, which, in turn, initiates signalling by binding to specific partner proteins. Cryptochromes bind a flavin adenine dinucleotide (FAD) and the pterin derivative 5,10-methyltetrahydrofolate (MTHF) as chromophores (Figure 15.8). Dimerization of cryptochromes, mediated by the photolyase-like domain, may be important for their signalling. The coaction of cryptochrome, phytochrome and phototropins affects the developmental functions such as stem elongation inhibition, and regulation of flowering as well as the circadian clock.

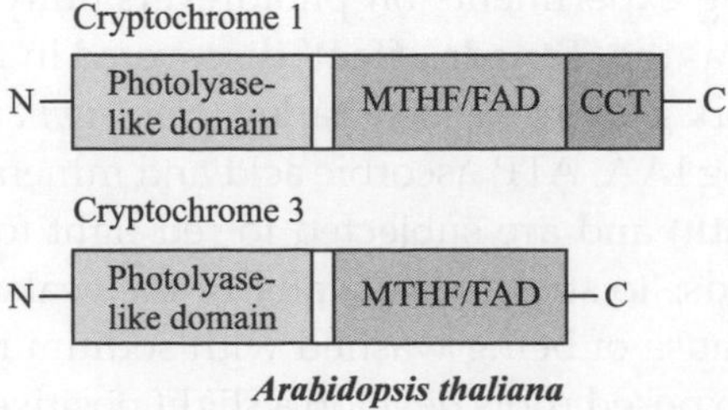

Fig. 15.8 The domain structure of cryptochromes (1 and 3) from *Arabidopsis thaliana* showing the photolyase-like domain, FAD-binding domain, and the cryptochrome C-terminal domain (CCT).The light-capturing cofactor 5,10-methyltetrahydrofolate (MTHF) and the catalytic cofactor flavin adenine dinucleotide (FAD) are linked to the protein *via* non-covalent association.

Arabidopsis contains three cryptochrome genes: CRY 1, CRY 2, and CRY 3. Cryptochrome homologs 1, 2, and 3 have different developmental impacts, and are strictly localized unlike phytochromes. While CRY 1 and CRY 2 are generally located in the nucleus, CRY 3 is confined to chloroplasts and mitochondria. CRY 2 protein tends to be damaged under blue light, whereas CRY 1 is much more stable. Unlike CRY 1, CRY 2 does not play any important role in stem elongation inhibition. In addition, CRY 1, and to a lesser extent CRY 2, are associated with the setting of the circadian clock in *Arabidopsis*, whereas CRY 2 has a key role in the initiation of flowering. Nuclear cryptochromes inhibit COP1-induced protein degradation. Blue light- induced phosphorylation of cryptochrome appears to be important in modulating its activity i.e. maintaining the C terminus of CRY 1 in an active conformation and in the case of CRY 2, promoting its degradation. In addition to controlling the levels of transcription factors, cryptochrome can also directly attach to and modulate the functioning of specific DNA-binding proteins

Mode of action: During dark, gene CONSTITUTIVE PHOTOMORPHOGENESIS 1 (COP1) in association with SUPPRESSOR OF PHYA 1(SPA1) and other factors, acts to damage transcription factors i.e. HY5, which induce the expression of genes required for photomorphogenesis. Upon activation by blue light, the nuclear cryptochrome 1 (CRY 1) forms a complex with SPA1 and COP1 that prevents them from acting, thus terminating the breakdown of HY5 and other transcription factors that promote photomorphogenesis (Figure 15.9).

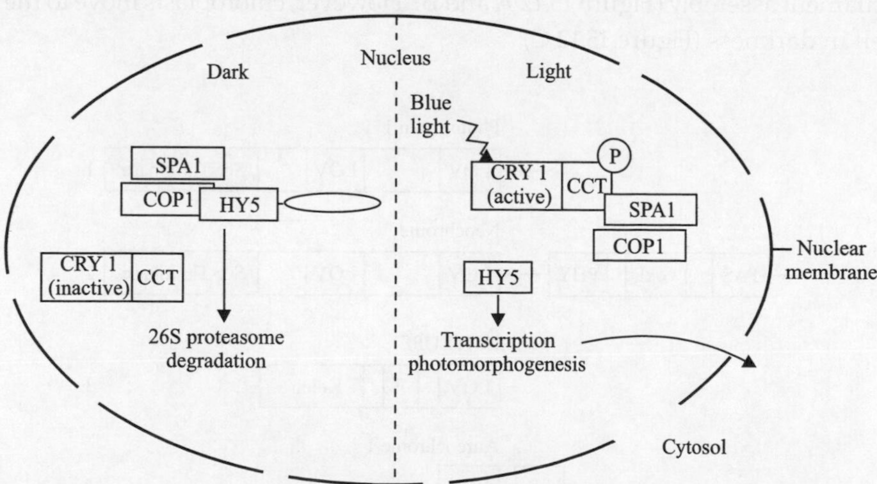

Fig. 15.9 Regulation of photomorphogenesis by interaction of CRY I with the COPI/SPAI complex. The cryptochromeI CRY I is inactive in the dark and COP I/SPA I complex degrades HY5, the transcription factor required for photomorphogenesis. Blue light activates CRY I and the activated CRY I forms a complex with COPI/SPAI, thereby preventing them from degrading HY5 protein and promoting photomorphogenic development.

15.3 Phototropins

Phototropins are plasma membrane-associated protein kinases responsible for regulating phototropism, light-induced chloroplast movement, control of stomatal aperture and regulation of hypocotyls and leaf expansion. Angiosperms contain two phototropin genes, PHOT1 and PHOT2. PHOT1, the primary phototropic receptor in *Arabidopsis,* regulates phototropism both in response to low and high fluence rates of blue light whereas PHOT2 mediates phototropism under high light intensities.

Each phototropin possesses two flavin mononucleotide (FMN) chromophores that can induce conformational changes. Phototropin consists of two light-sensing Light-Oxygen-Voltage (LOV) domains, LOV1 and LOV2, each bound to a chromophore flavin mononucleotide (Figure 15.10). LOV2 domain, in particular, is essentially required for blue light-induced kinase activation and autophosphorylation of the phototropin photoreceptor. The function of LOV1 domain might be in receptor dimerization. Blue light absorption by phototropins induces a conformational change that uncages their kinase domain, causing their autophosphorylation and induces signal transduction pathways that leads to the redistribution of PIN auxin efflux carriers necessary for directional growth and to the changes in plasma membrane ion fluxes that regulate guard cell turgor. Blue light, sensed by phototropins, triggers the activation of H^+-ATPase located in plasma membrane which ultimately regulates opening of the stomata (details explained in Figure 15.11). Phototropins mediate chloroplast accumulation and avoidance responses to weak and strong light through F-actin filament assembly (Figure 15.12 A and B). However, chloroplasts move to the bottom of the cell in darkness (Figure 15.12 C)

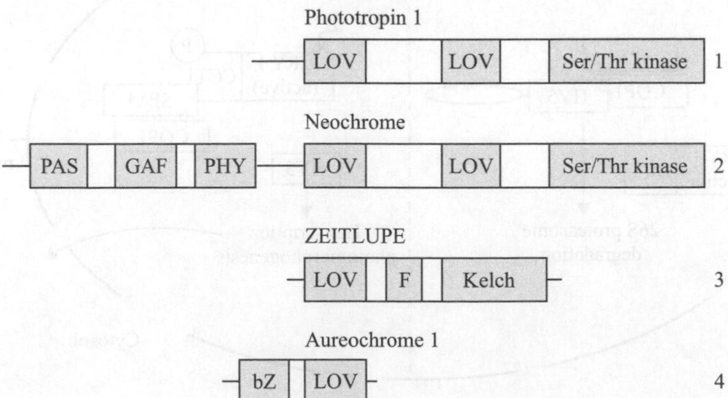

Fig. 15.10 The domain compositions of phototropin and related LOV-domain phototropins such as, neochrome, ZEITLUPE, and aureochrome. Photoreceptors 1 and 3 belong to *Arabidopsis*, 2 and 4 to *Adiantum* and *Vaucheria* respectively.

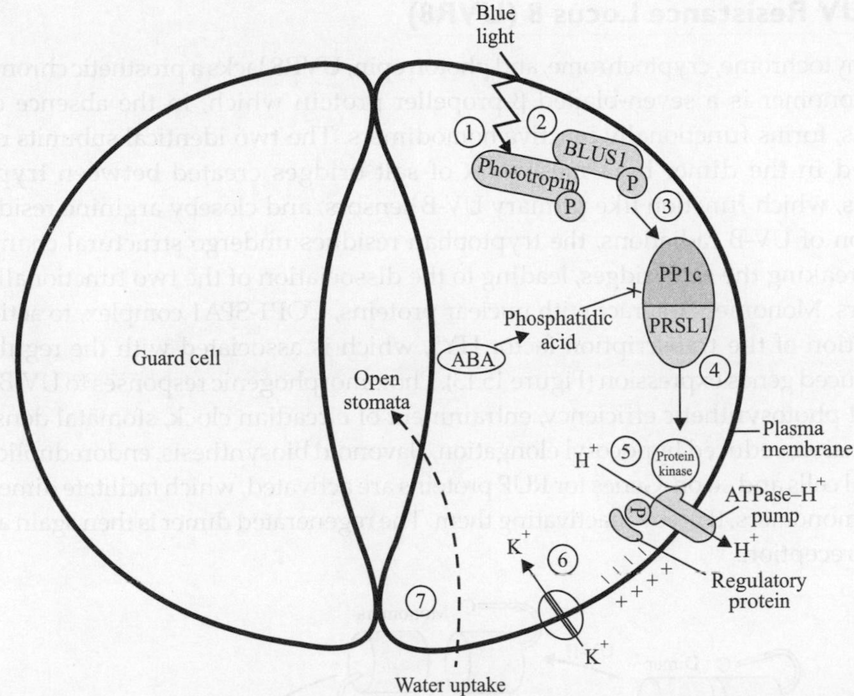

Fig. 15.11 Phototropin signal transduction pathway. (1) and (2) Phototropins get autophosphorylated on absorption of blue light and in turn, phosphorylate BLUE LIGHT SIGNALLING 1 (BLUS1). (3) and (4) BLUS1 controls a protein phosphatase1, PP1c (subunit PRSL1), which in turn controls the activity of protein kinase. (5) The protein kinase promotes the binding of a regulatory protein (14-3-3) to the H^+–ATPase pump confined to the plasma membrane. (6) and (7) Membrane hyperpolarization drives K^+ uptake leading to water absorption and finally opening of stomata. ABA triggers the production of phosphatidic acid, which blocks PP1 activity. PROTEIN PHOSPHATASE 1 (PP1) is a serine/threonine protein phosphatase composed of a catalytic subunit (PP1c) and a regulatory subunit, PP1 REGULATORY SUBUNIT2-LIKE PROTEIN1 (PRSL1).

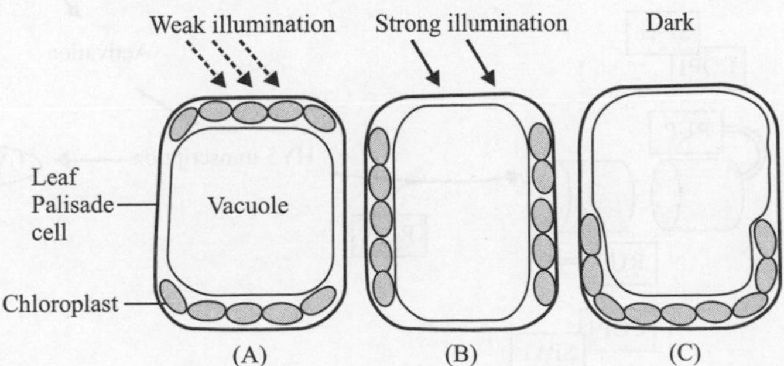

Fig. 15.12 Distribution patterns of chloroplasts within palisade cells of *Arabidopsis* leaf in view of the changing illumination conditions. (A) Under weak illumination, chloroplasts accumulate near the upper and lower sides of cells. (B) In strong illumination, the chloroplasts migrate to the lateral walls of cells. (C) In the dark, chloroplasts shift to the bottom surface of the cells. Redistribution of chloroplasts within the cells modulates light absorption and prevents photodamage.

15.4 UV Resistance Locus 8 (UVR8)

Unlike phytochrome, cryptochrome, and phototropin, UVR8 lacks a prosthetic chromophore. UVR8 monomer is a seven-bladed β-propeller protein which, in the absence of UV-B radiations, forms functionally inactive homodimers. The two identical subunits of UVR8 are joined in the dimer by a meshwork of salt bridges created between tryptophan molecules, which function like primary UV-B sensors, and closeby arginine residues. On absorption of UV-B radiations, the tryptophan residues undergo structural changes that help in breaking the salt bridges, leading to the dissociation of the two functionally active monomers. Monomers interact with nuclear proteins, COP1-SPA1 complex, to activate the transcription of the transcription factor HY5, which is associated with the regulation of UV-B induced genes expression (Figure 15.13). Photomorphogenic responses to UV-B include increased photosynthetic efficiency, entrainment of circadian clock, stomatal density, leaf cell expansion, reduced hypocotyl elongation, flavonoid biosynthesis, endoreduplication in epidermal cells and so on. Genes for RUP proteins are activated, which facilitate dimerization of UVR8 monomers, thereby inactivating them. The regenerated dimer is then again available for photoreception.

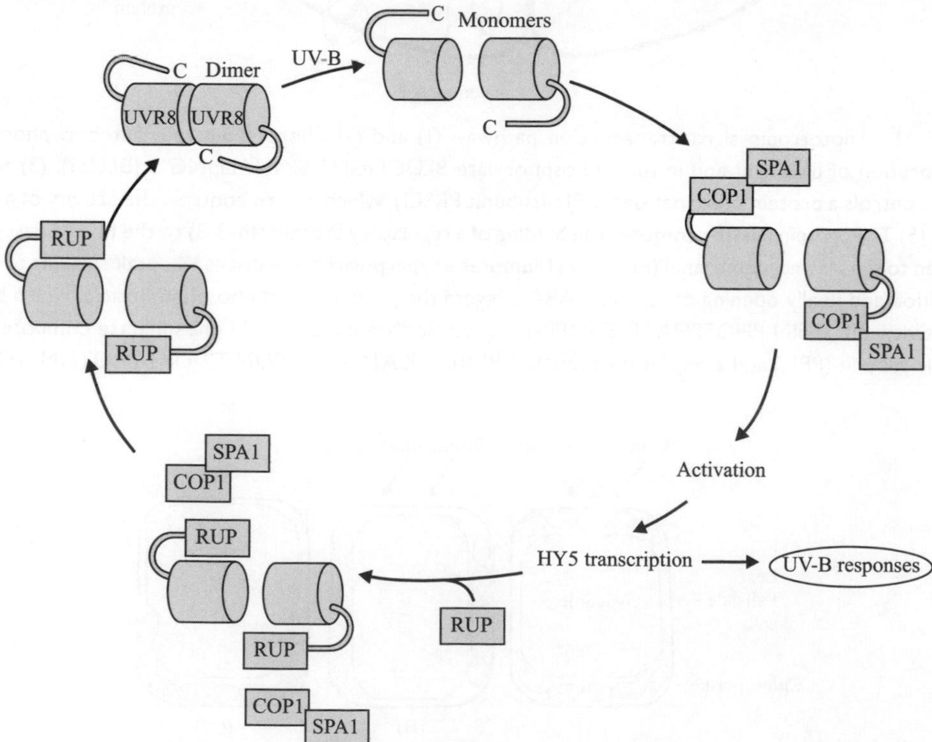

Fig. 15.13 The UVR8 signalling pathway. The photoreceptor UVR8 (dimer) absorbs UV-B radiations and forms monomers. The monomers then interact with the COP1/SPA1 complex to activate the transcription of UV-B responses. RUP proteins get induced and facilitate dimerization of UVR8 monomers as well as dissociation of COP1/SPA1 complex. The regenerated dimer is ready again for photoreception. RUP stands for REPRESSOR OF UV-B PHOTOMORPHOGENESIS 1 and 2.

Review Questions

RQ 15.1. Why are etiolated seedlings, used so often in phytochrome studies?

Ans. Etiolated seedlings have been used as favourite source material for phytochrome studies because of three reasons:

- Dark-grown seedlings grow quickly.
- They accumulate relatively large amounts of phytochrome pigment.
- The absence of chlorophyll makes them ideal for measuring the total phytochrome content and the relative proportion of Pr/Pfr in the tissues, with appropriate instruments used in spectrophotometry.

RQ 15.2. List the different physiological responses caused by blue light.

Ans. Blue light stimulates responses of plant development such as:

- Most effective wavelength for phototropism.
- It inhibits hypocotyl elongation once a seedling has emerged out of soil.
- Stimulates chlorophyll and carotenoid synthesis.
- Causes opening of stomata at dawn.

RQ 15.3. What is the chemical nature of photoreceptors? What are the different types of photoreceptors found in plants?

Ans. Photoreceptors are light-sensitive proteins involved in the sensing and subsequent response to light in a variety of organisms. Some examples are rhodopsin in the photoreceptor cells of the vertebrate retina, phytochrome in plants, and bacteriorhodopsin and bacteriophytochrome in some bacteria.

There are two types of photoreceptors that perceive different wavelengths of light (i) the blue light sensitive system and (ii) the red light sensitive system.

Blue light responses: Many plant responses are regulated by blue light such as phototropism, stomatal opening and chlorophyll synthesis. **Phototropins** are sensitive to blue light.

Red light responses: Many flowering plants use photoreceptor proteins, such as **phytochrome** or **cryptochrome**, to sense seasonal changes in night length or photoperiods. Phytochromes are a class of photoreceptors in plants, bacteria and fungi used to detect red/far-red wavelengths of light. They regulate the germination of seeds (photoblasty); the synthesis of chlorophyll; the elongation of seedlings; the size, shape, number and movements of leaves; and the timing of flowering in adult plants.

RQ 15.4. List the different kinds of photoreceptors of photosynthetic pigments.

Ans. There are seven main types of photoreceptors of photosynthetic pigments: Phycobilins, cryptochromes, UV-B receptors, flavonoids, betacyanins, chloroplasts and carotenoid pigments.

RQ 15.5. Where is phytochrome located in plants? What is the role of phyochrome in plants?

Ans. Phytochrome is a blue-green pigment found in most of the organs of seed plants, which regulates photomorphogenic aspects of plant growth and development, such as seed germination, stem elongation, leaf expansion, synthesis of certain pigments, chloroplast development, and flowering.

Chapter 16

Plant Hormones and Signal Transduction

Keywords

- Apical dominance
- Bioassay
- Bolting
- Climacteric fruits
- Crosstalk
- Second messengers
- Signal amplification
- Vivipary

Hormones are naturally occurring molecules, in plants and animals that function as messengers and signalling agents and thus impact growth and behaviour of the organism.

Plant hormones are classified as being of five major types: auxins, gibberellins, cytokinins, abscisic acid and ethylene (Table 16.1). Besides these plant hormones; various compounds such as salicylic acid, jasmonates, polyamines, strigolactones, brassinosteroids and nitric oxide[1] (NO) have been reported to act as signalling molecules in hormonal and defence responses.

16.1 Auxin: The Growth Hormone

16.1.1 Historical background

The way auxins were discovered is quite fascinating and deserves a special discussion because it was the first and foremost growth hormone to be discovered in plants (Figure 16.1).

In later years of his career, the great evolutionary biologist Charles Darwin became increasingly interested in the study of plants. In 1881, Darwin along with his son Francis authored a book called 'The Power of Movement of Plants' in which they highlighted a series of experiments on the growth response of plants to light – **Phototropism**. For their experiments, they used germinating oat (*Avena sativa*) and canary grass (*Phalaris canariensis*) seedlings. Both the Darwins found that if seedlings were exposed to light from one direction, they would bend strongly towards it. If the coleoptile tip was covered with foil,

[1] Yet another gas, hydrogen sulphide (H_2S), also produced by plants, can act as a potent signalling molecule and investigations are on to study its effect.

Table 16.1 Major plant hormones: their nature, occurrence, transport and physiological effects.

Hormone(s)	Chemical nature	Sites of biosynthesis	Transport	Physiological effects
Auxins Indole ring; Acetic acid side chain; CH_2—COOH Indoleacetic acid (IAA)	Indole-3-acetic acid is the principal naturally occurring auxin. Possibly synthesized *via* tryptophan-dependent and tryptophan-independent pathways.	Primarily in leaf primordia and young leaves and in developing seeds.	IAA is transported both polarly (unidirectionally) and nonpolarly.	Apical dominance; tropic responses; vascular tissue differentiation; promotion of cambial activity; induction of adventitious roots on cutting; inhibition of leaf and fruit abscission; stimulation of ethylene synthesis; inhibition or promotion (in pineapples) of flowering; stimulation of fruit development.
Cytokinins Kinetin (CH_2—NH)	N^6 adenine derivatives, phenyl urea compounds. Zeatin is the most common cytokinin in plants.	Primarily in root tips.	Cytokinins are transported *via* xylem from roots to shoots.	Cell division; promotion of shoot formation in tissue culture; delay of leaf senescence; application of cytokinin can encourage the development of lateral buds from apical dominance.
Ethylene $H_2C{=}CH_2$	Ethylene (C_2H_4) gas is synthesized from methionine. It is the only hydrocarbon, with a pronounced effect on plants.	In most tissues in response to stress, especially in tissues undergoing senescence or ripening.	Being a gas, ethylene moves by diffusion from its site of synthesis.	Fruit ripening (especially in climacteric fruits, such as apples, bananas and avocados); leaf and flower senescence; leaf and fruit abscission.

Hormone(s)	Chemical nature	Sites of biosynthesis	Transport	Physiological effects
Abscisic acid 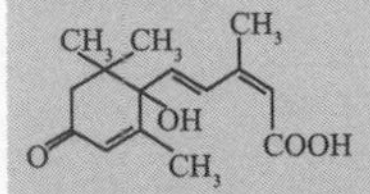	'Abscisic acid' is a misnomer for this compound, for it has little to do with abscission. ABA is synthesized from a carotenoid intermediate.	In mature leaves and roots, especially in response to water stress. May be synthesized in seeds.	ABA is exported from leaves in the phloem: from roots in the xylem.	Stomatal closure; induction of photosynthate transport from leaves to developing seeds; induction of storage protein synthesis in seeds: embryogenesis; may affect induction and maintenance of dormancy in seeds and buds of certain species.
Gibberellins 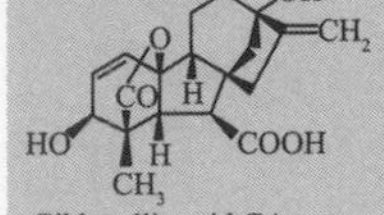 Gibberellic acid GA_3	Gibberellic acid (GA_3), a fungal product, is the most widely available. GA_1 is probably the most important gibberellin in plants. GAs are synthesized *via* terpenoid pathway.	In young tissues of the shoot and developing seeds. It is uncertain whether synthesis also occurs in roots.	GAs are probably transported in the xylem and phloem.	Hyperelongation of shoots by stimulating both cell division and cell elongation, producing tall, as opposed to dwarf plants; induction of seed germination; stimulation of flowering in long-day plants and biennials; regulation of production of seed enzymes in cereals.

it would not bend. The zone of coleoptile responsible for bending toward the light – 'the growth zone'–is many millimetres below the tip. Thus, they came to the conclusion that some kind of stimulus produced in the tip region travels to the growth zone and causes the non-illuminated (shaded) side to grow more quickly compared to the illuminated side. For 30 years, the Darwin's experiments remained the only source of information about this interesting phenomenon.

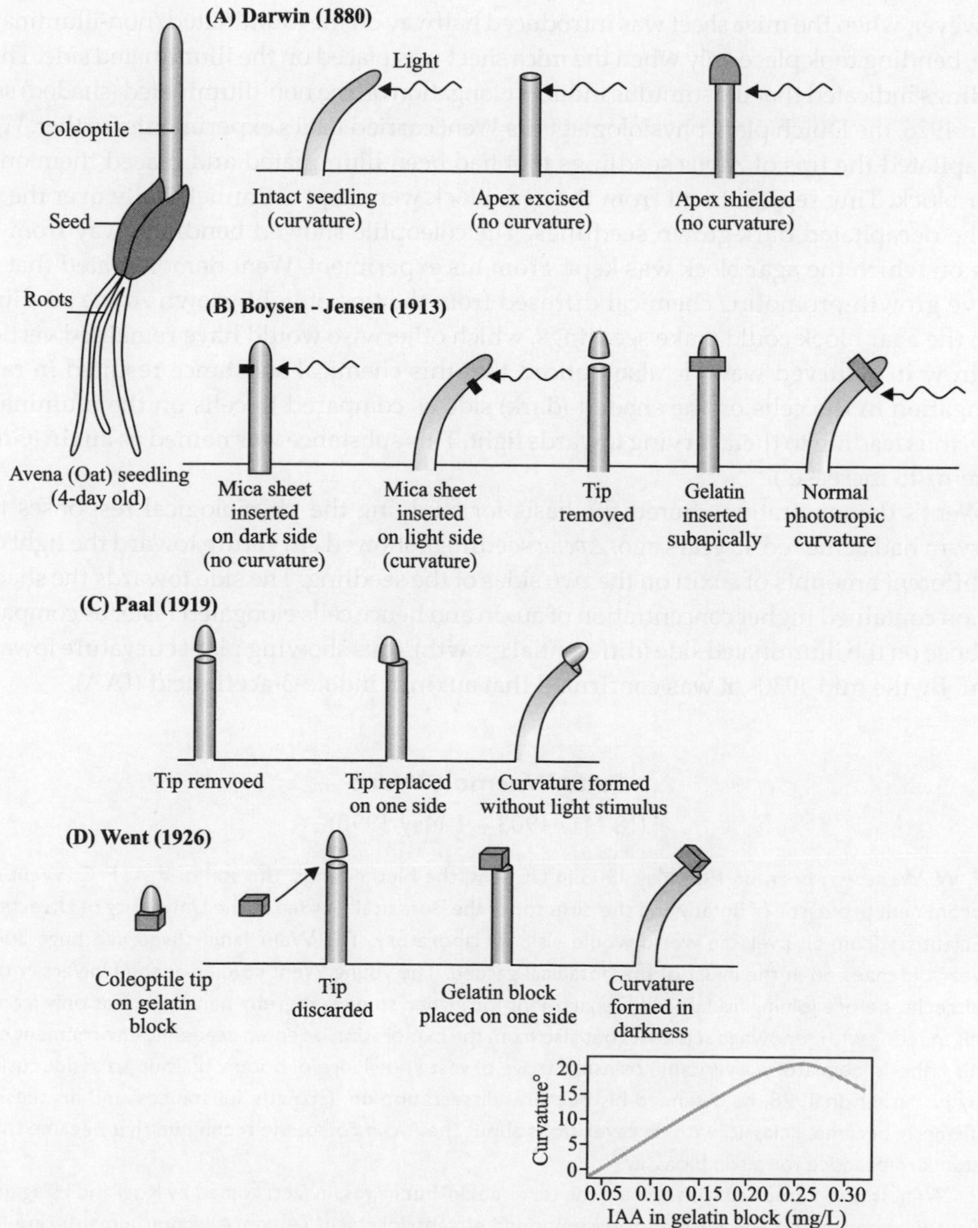

Fig. 16.1 Classical experiments which led to the discovery of auxins.

Later in 1913, the Danish plant physiologist Peter Boysen-Jensen and the Hungarian plant physiologist Arpad Paal (1919) independently proposed that the growth-promoting signal generated in the tip region was chemical in nature.

They decapitated the coleoptile of *Avena* seedling and in one experiment (a) inserted a small gelatin block between the stump and cut tip and in other (b) a thin sheet of mica. The coleoptile bending was observed in (a) but not in (b) because the growth stimulus passes across the gelatin block but not through a water-impermeable barrier such as the mica sheet. However, when the mica sheet was introduced halfway on the illuminated/non-illuminated side, bending took place only when the mica sheet was placed on the illuminated side. These findings indicated that the stimulus induces elongation on the non-illuminated (shaded) side.

In 1926, the Dutch plant physiologist Frits Went carried Paal's experiments further. Went decapitated the tips of *Avena* seedlings that had been illuminated and placed them on an agar block. Tiny segments cut from the agar block were kept asymmetrically over the top of the decapitated dark-grown seedlings. The coleoptile showed bending away from the side on which the agar block was kept. From his experiment, Went demonstrated that the active growth-promoting chemical diffused from the tips of light-grown *Avena* seedlings into the agar block could make seedlings, which otherwise would have remained vertical, to grow in a curved way. He also noticed that this chemical substance resulted in rapid elongation in the cells on the shaded (dark) side as compared to cells on the illuminated side, thus leading to their curving towards light. This substance was named as **auxin** (Greek auxein, 'to increase').

Went's demonstration offered the basis for studying the physiological responses that Darwin had achieved 45 years ago. *Avena* seedlings showed curvature toward the light due to different amounts of auxin on the two sides of the seedling. The side towards the shaded region contained higher concentration of auxin and hence cells elongated faster as compared to those on the illuminated side (differential growth), thus showing plant curvature towards light. By the mid-1930s, it was confirmed that auxin is indole-3-acetic acid (IAA).

Frits Warmolt Went
(18 May 1903 – 1 May 1990)

F. W. Went was born on 18th May 1903 in Utrecht, the Netherlands, the son of F. A. F. C. Went a prominent professor of Botany and the director of the Botanical Garden at the University of Utrecht. Scientists from all over the world would visit his laboratory. The Went family lived in a huge 300 year old mansion in the midst of the botanical garden. The young Went studied at the University of Utrecht, before joining his father's Department for higher studies; thereby benefiting not only from his meeting with renowned scientists but also from the exploration oriented academic environment of his father's laboratory, a veritable treasure trove of vast knowledge of botany built up so assiduously by his father. In 1928, he defended his doctoral dissertation on 'Growth Substances' and his thesis instantly became a classic, with its revelations about the *Avena* coleoptile technique that became the standard practice for auxin bioassay.

Went is often credited with coining the term 'auxin' but It was, in fact, coined by Kogl and Haagen-Smit who extracted and purified the compound auxentriolic acid (Auxin A) from human urine in

1931. Following that, Kogl isolated other compounds from human urine whose structure and function resembled with Auxin A, and one of them was named Indole-3-acetic acid (IAA).

After completing his PhD Went joined the Royal Botanical Garden at Buitenzorg (now Bogor) situated on the island of Java, which was then a part of the Dutch East Indies. He was accompanied by his wife Catharina and two childern Hans and Anneka. He worked there as a plant physiologist from 1928 to 1933. In 1933, he shifted to the California Institute of Technology to take the position of another young Dutch plant hormone researcher, Herman E. Dolk, who had lost his life in a tragic accident. At Caltech, he soon became the centre of attraction amongst an active group of plant hormone researchers, including Kenneth Thimann, James Bonner, Folke Skoog and Johannes van Overbreek, among others, all of whom later rose to prominence in their own right. Went's work on growth hormones reached culmination in 1937 with his (along with co-author Kenneth Thimann) book on 'Phytohormones', published by the Macmilan Company, New York. In 1937, he along with the Soviet botanist N. G. Cholodny proposed the Cholodny-Went Model for elucidating the phototropic and gravitropic responses of emerging shoots of monocot species.

With generous financial help from Lucy Mason Clark, he built at Caltech, many 'controlled-condition greenhouses' where plants could be grown under specific regimes of light, temperature, nutrition, humidity and air quality, in an insect-free environmental conditions. Later in 1949, with another donor Harry Earhart, he was able to construct a building at Pasadena (California), well equipped with air-conditioned greenhouses for plant growth. This state-of-the-art Earhart Plant Research Laboratory, along with its supporting engineering and machinery, was called the 'phytotron' by Went's colleagues at Caltech. It became the model for many similar projects and initiatives all over the world. He then shifted his interest to environmental impact on plant growth and development and played a significant role in the development of synthetic plant hormones, which laid the foundation of agricultural chemical industry.

Growing dissatisfied with his administrative burdens and the business of raising grants to support the phytotron, and the petty envy of his colleagues, he was becoming increasingly interested in molecular genetic approaches to the problems of biology. So, he quit his cherished phytotron as well as desert ecosystems and moved on. In 1958, after 25 years of service at Caltech, Went moved out of Pasadena and was appointed as a director of the famous Missouri (Shaw's) Botanical Garden at St. Louis. Drawing on his earlier vast experience in climate control at the Pasadena phytotron, Went oversaw design and construction of the Climatron. This majestic geodesic dome is open from inside, but cleverly manipulates the air movements and simulates all different climatic zones simultaneously. One of its impressive structures is a tunnel under Victoria Pool, named after *Victoria regia*, a water lily with a six-feet-wide leaves. The tunnel has curves so that at some places, visitors have a glimpse of only aquatic life while walking through the tunnel. After the inauguration of Climatron, Went's vision of a renowned MBG came into conflict with its Board of Trustees, and he eventually resigned from the post of a Director in 1963, but continued his next two years, teaching and doing research as Professor of Botany at Washington University, St. Louis .

In 1960, Went believed that the natural 'blue haze' of the Mountain Range that developed in US is owing to the minute particles emanating from the condensation of plant-derived terpenoids emissions from trees, dispersing light. Then in 1965, Went became the director of the Desert Research Institute in the city of Reno at the University of Nevada where he founded the laboratory of Desert Biology where he continued his research on desert plant ecosystems for the remaining part of his academic career. In 1967, Went spent 6 weeks aboard the research vessel 'Alpha Helix', which travelled up to Amazon River, but the Brazilian government confiscated the gas chromatograph equipments he was supposed to use for measurement of terpenoid emissions emanating from tropical plants.

F. W. Went died on 1st May 1990 at White Pine County, Nevada, USA, but left behind a legacy of insights for posterity that had fuelled research in various areas, like plant hormones, air pollution, desert ecology, plant-atmosphere interactions, and mycorrhizae, all of which had interested him at one time or other in his long and brilliant academic career.

16.1.2 Bioassays

Bioassay is the quantitative estimation of biologically active substances, reflected in the amount of their actions under standardized conditions on living organisms or live tissue for the known response.

Various bioassays for auxins

- Avena coleoptile curvature test
- Avena section test
- Split pea stem curvature test
- Cress root inhibition test

The *'Avena* Coleoptile Curvature test' is a sensitive measure of auxin activity.

A 3-day-old seedling of *Avena* is decapitated at the tip and the leaf inside is pulled up. The tip is placed on agar block for few hours. An agar block containing auxin is held eccentrically (i.e., on one side) of the decapitated stump. Auxin from the agar block is transported down the coleoptile and stimulates that side to grow more quickly compared to the opposite side, resulting in curvature. The degree of curvature is measured after 90 min.

The concentrations of hormones are measured in terms of the degree of bending angle, by comparing with the degree of angle produced by known IAA concentration (Figure 16.2). The greater the angle, the more is the hormone concentration.

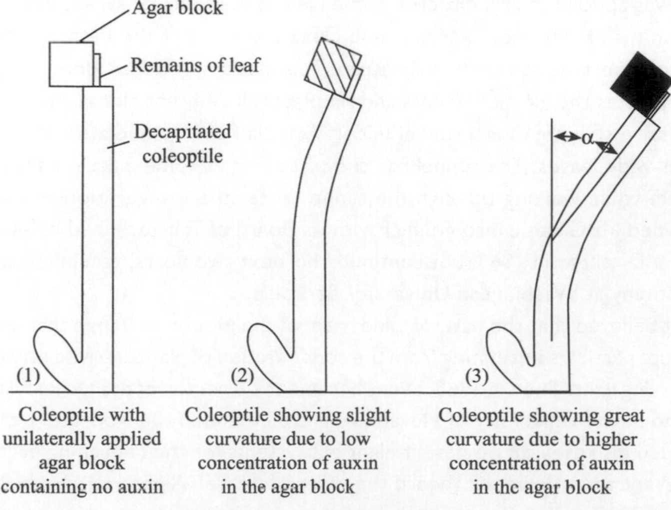

Fig. 16.2 The quantitative estimation of auxin concentration by means of the *'Avena* coleoptile curvature test'. The curvature angle (or the degree of bending) is directly proportional to the amount of auxin present in the agar block.

16.1.3 Distribution and transport

Auxins appear to be well distributed throughout the entire length of plant. These are synthesized in meristematic regions and other actively growing organs such as shoot and root tips, apical buds of growing stems, germinating seeds, young leaves, developing inflorescences and developing embryos. Indole-3-acetic acid (IAA) is the most widely distributed amongst natural auxin. Besides IAA, there are several other naturally occurring and synthetic auxins (Figure 16.3). The amount of IAA present is determined by various factors, such as the type and age of tissue, and its active state.

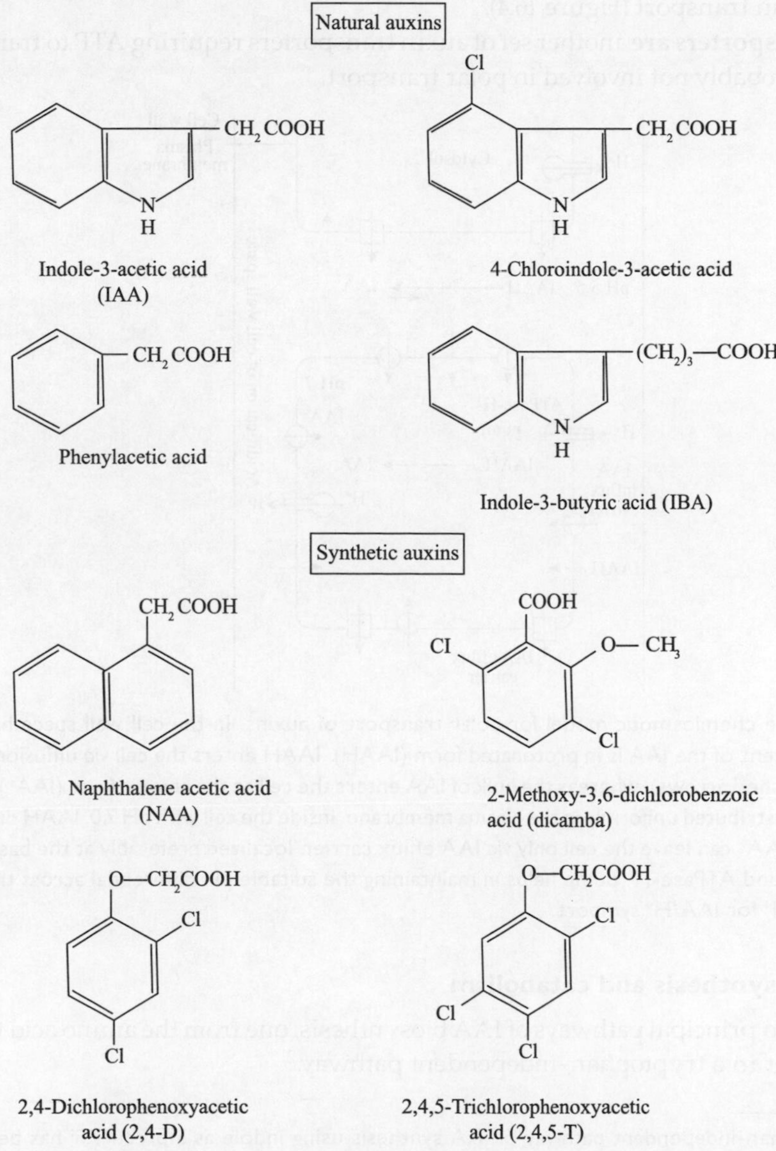

Fig. 16.3 Structural configurations of some naturally occurring as well as synthetic auxins.

Polar transport (basipetal) is a fundamental property of auxins and plays an important role in many developmental processes. Polar transport is an energy-consuming process dependent on the mitochondrial activity. It is believed to take place predominantly in the parenchymatous cells, associated with the vascular tissue. Polar auxin transport is mainly driven by gradients in protonated IAA (IAAH) in the apoplast and anionic form (IAA^-) in the symplast. IAAH readily enters cells by diffusion but IAA^- requires specific membrane transporters to enter or exit cells. **Auxin influx transporters** are present at the apical ends of cells and **auxin efflux carriers**, 'PIN proteins' are located at the lower ends. Differences in pH of the cell wall and cytoplasm (pH 5.5 and pH 7, respectively) are critical components of polar auxin transport (Figure 16.4).

ABC transporters are another set of auxin transporters requiring ATP to transport IAA. These are probably not involved in polar transport.

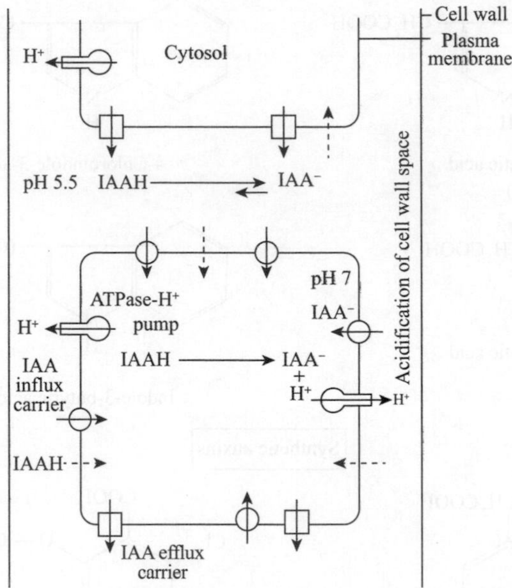

Fig. 16.4 The chemiosmotic model for polar transport of auxins. In the cell wall space having pH 5.5, nearly 20 per cent of the IAA is in protonated form (IAAH). IAAH enters the cell *via* diffusion through cell membrane (dashed arrows), whereas the bulk of IAA enters the cell as the anionic form (IAA^-) through IAA influx carrier distributed uniformly in the plasma membrane. Inside the cell with pH 7.0, IAAH dissociates into IAA^- and H^+. IAA^- can leave the cell only *via* IAA efflux carrier, localized preferably at the basal end of cell. Membrane-bound ATPase-H^+ pump helps in maintaining the suitable pH differential across the membrane and provides H^+ for IAA/H^+ symport.

16.1.4 Biosynthesis and catabolism

There are two principal pathways of IAA biosynthesis, one from the amino acid tryptophan and the other in a tryptophan–independent pathway.[2]

[2] A tryptophan-independent pathway of IAA synthesis using indole as a precursor has been described in many plant species, with indole-3-acetonitrile (IAN) and indole-3-acetamide (IAM) as possible intermediates. However, little is known about the enzymatic processes involved.

Synthesis of IAA occurs in three steps (Figure 16.5):

1. It begins with the removal of amino group on the tryptophan side chain to form indole-3-pyruvic acid (IPA), catalyzed by *tryptophan aminotransferase*.
2. IPA is decarboxylated to produce indole-3-acetaldehyde (IAAld) in the presence of *indole-3-pyruvate decarboxylase*.
3. IAAld undergoes oxidation to IAA in the presence of NAD-dependent *indole-3-acetaldehyde oxidase*.

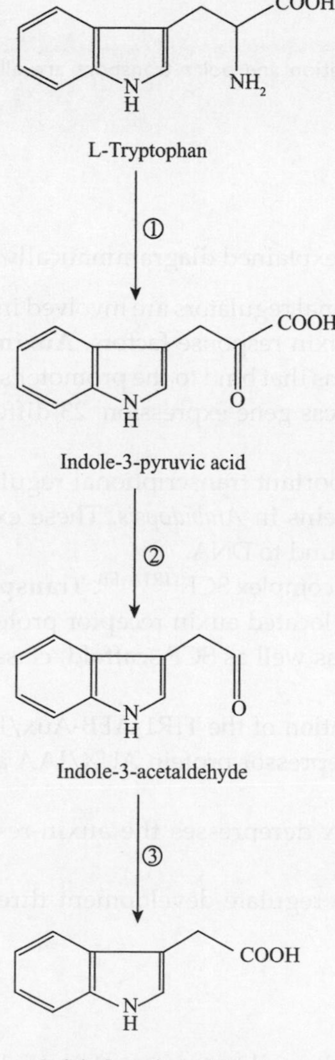

Fig. 16.5 Pathway for tryptophan-dependent biosynthesis of indole-3-acetic acid. The enzymes involved are (1) tryptophan aminotransferase; (2) indole-3-pyruvate decarboxylase; (3) indole-3-acetaldehyde oxidase.

Catabolism of IAA can also follow several routes, one that results in the formation of conjugates and others that bring about IAA degradation. IAA can be stored in the form of chemical conjugates (such as glycosyl esters) which will liberate active IAA on enzymatic hydrolysis. Glycosyl esters are the chief source of IAA during germination of seeds. Recent studies have indicated that conjugation of IAA with amino acids (such as alanine or aspartic acid) also results in irreversible deactivation. IAA is also deactivated by oxidation in the presence of IAA oxidase and peroxidase. IAA oxidation is an effective means of eliminating IAA after it has done its job. IAA is also readily degraded by acids, ultraviolet and ionizing radiations.

Biosynthesis, conjugation, degradation and polar transport are all mechanisms that control local concentration of auxins.

16.1.5 Mode of action

Mode of auxin action has been explained diagrammatically in Figure 16.6.

1. Two families of transcriptional regulators are involved in the auxin signalling pathway: AUX/IAA proteins and auxin response factors. **Auxin response factors (ARFs)** are short – lived nuclear proteins that bind to the promoter site of auxin – responsive genes and either activate or repress gene expression. 23 different ARF proteins have been identified in *Arabidopsis*.
2. **Aux/IAA** proteins are important transcriptional regulators. There are 29 such short – lived small nuclear proteins in *Arabidopsis*. These exert their regulatory effect by binding to ARF protein bound to DNA.
3. Auxin binds to its receptor complex $SCF^{TIR1/AFB}$. **Transport Inhibitor Response 1(TIR 1)**, a soluble and nuclear – located auxin receptor protein, is an F-box protein with a recognition site for auxin as well as SCF scaffold, classified as Auxin F-box binding protein (AFB).
4. Auxin stabilizes the formation of the TIR1/AFB-Aux/IAA heterodimer, followed by the ubiquitination of the repressor protein AUX/IAA and its degradation by the 26S proteasome.
5. Removal of the AUX/IAA derepresses the auxin-responsive gene and allows its transcription.
6. Therefore auxin seems to regulate development through derepression of auxin – responsive genes.

16.1.6 Physiological roles

Cell elongation: Auxins have the ability to promote cell elongation in excised coleoptile as well as stem segments. Rate of elongation is directly proportional to concentration of auxin applied. The auxin concentration that promotes stem elongation has an inhibitory effect on root cell elongation.

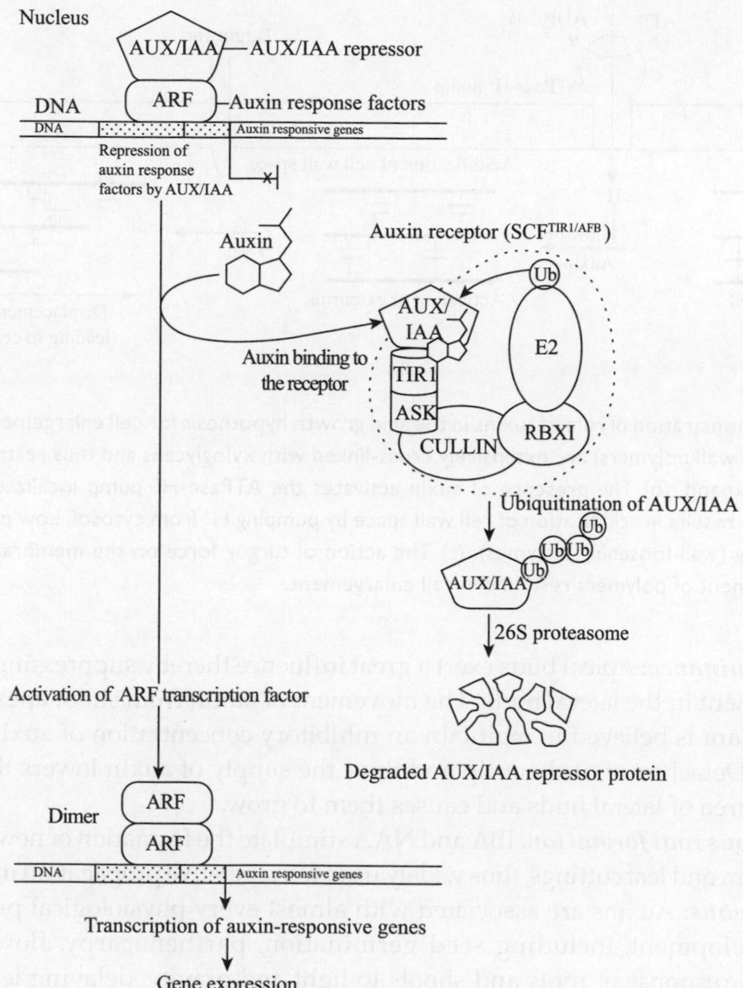

Fig. 16.6 A diagrammatic representation of mode of action in auxins . In the absence of auxin, AUX/IAA repressor protein inhibits ARF transcription factor. Auxins initiate an interaction between AUX/IAA and the TIRI/AFB component of an SCF$^{TIRI/AFB}$ complex. Auxin – activated SCF$^{TIRI/AFB}$ complexes bind ubiquitin molecules to AUX/IAA repressor protein, thus activating their degradation by the 26S proteasome. The degradation of AUX/IAA protein activates the ARF transcription factor. The ARF transcriptional activators attach to auxin response elements and activate the transcription of auxin- induced genes.

One of the downstream effects of auxin is enhanced plasticity of the plant cell wall, but it solely works on young cell walls lacking extensive secondary cell wall formation. A more plastic cell wall can stretch more while its protoplast is swelling during active cell growth. The **acid growth hypothesis** (Cleland and Hager, 1970) offers a model correlating presence of auxin to cell wall expansion (Figure 16.7).

Cell division in tissue and organ culture: Initiation of cell division by auxin is useful in formation of callus in tissue culture. High auxin to cytokinin ratio stimulates root formation and low auxin to cytokinin ratio promotes shoot formation.

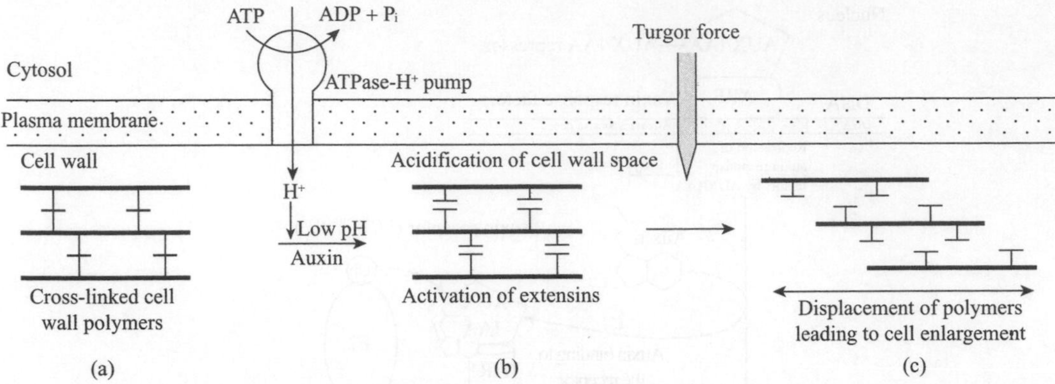

Fig. 16.7 Demonstration of role of auxins in the acid growth hypothesis for cell enlargement. (a) Cellulose microfibrils (cell wall polymers) are extensively cross-linked with xyloglycans and thus restrict the capacity of the cell to expand. (b) The presence of auxin activates the ATPase-H^+ pump localized in the plasma membrane. This results in acidification of cell wall space by pumping H^+ from cytosol. Low pH increases the expansin activity (wall-loosening enzymes). (c) The action of turgor force on the membrane and cell wall causes displacement of polymers resulting in cell enlargement.

Apical dominance: Apical buds exert a great influence thereby suppressing cell division and enlargement in the lateral buds. The movement of auxin from shoot apex towards the base of the plant is believed to maintain an inhibitory concentration of auxin around the lateral buds. Detachment of shoot tip and thus, the supply of auxin lowers the amount of auxin in the area of lateral buds and causes them to grow.

Adventitious root formation: IBA and NAA stimulate the formation of new adventitious roots from stem and leaf cuttings, thus widely used for vegetative propagation in horticulture.

Miscellaneous: Auxins are associated with almost every physiological process related to plant development, including seed germination, parthenocarpy, flower and fruit development, response of roots and shoots to light and gravity, delaying leaf abscission, prevention of lodging, shortening of internodes, sex expression, eradication of weeds using 2,4-D and 2,4,5-T and vascular differentiation.

16.2 Gibberellins: Regulators of Plant Height

16.2.1 Historical background

The discovery of gibberellins came about from more practical concerns. During the late 19th and early 20th centuries, Japanese rice farmers were concerned about a disease that greatly diminished their crop yield. The plants infected with the **bakanae disease** (commonly known as foolish seedling disease) displayed weak, spindly stems and produced little or no grain sometimes. In 1926, Japanese plant pathologist Eiichi Kurosawa became interested in developing methods for eradicating the disease and correlated its connection with the fungus, *Gibberella fujikuroi*. Kurosawa cultured *Gibberella* in the laboratory and isolated a compound that, if applied to rice plants, would produce symptoms of bakanae disease.

By 1938, Japanese researchers Yabuta and Sumiki had isolated and crystallized the active substance, which was named **gibberellin** after the generic name of the fungus.

Gibberellin did not receive any attention from Western plant physiologists until after the 1939–1945 war, when two groups (one led by Cross in England and other headed by Stodola in the United States) isolated and chemically characterized gibberellic acid from filtrates of the fungal cultures. During the same period, Japanese scientists isolated and crystallized three gibberellins (gibberellin A_1, A_2, A_3). Gibberellin A_3 proved to be similar to gibberellic acid. The first gibberellin to be obtained from higher plants was isolated from immature seeds of runner bean (*Phaseolus coccineus*) and found to be similar to GA_1. Although such chemicals were earlier considered to be only a novelty, they have since turned out to belong to a vast group of over 125 naturally occurring gibberellins, which differ slightly in their structural details based on the source from which they are isolated.

16.2.2 Bioassays

- Growth of lettuce hypocotyls
- Prevention of senescence in leaves as indicated by chlorophyll content
- Production of α-amylase by barley endosperm

16.2.3 Distribution and transport

Gibberellins are synthesized chiefly in the shoot apex and the young, developing leaves adjacently placed around the growing tips and sometimes in the roots of few plants. The highest concentrations of gibberellins are present in immature seeds and developing fruits. The endogenous level of active gibberellin controls its own synthesis by turning on or preventing the transcription of the genes that code for the enzymes of gibberellin synthesis and degradation. By this means, the concentration of active gibberellin is maintained within a narrow range, depending on the availability of precursors and the activity of enzymes responsible for gibberellin synthesis and degradation. Environmental factors such as photoperiod and temperature, and the auxin presence from the apex of the stem, also regulate gibberellin synthesis *via* transcription of the genes for enzymes involved in their synthesis.

Gibberellins seem to be translocated generally through the phloem in response to source–sink relationships.

16.2.4 Biosynthesis and catabolism

Gibberellins (GAs) belong to terpenoid compounds, consisting of isoprene units. GAs are synthesized from isopentenyl pyrophosphate (IPP) which is an end product of two different biosynthetic pathways, one located in the cytoplasm and the other in chloroplasts.

Biosynthesis of gibberellins occurs in three steps (Figure 16.8):

1. **Synthesis of isopentenyl pyrophosphate (IPP)**
   (a) ***Cytoplasmic pathway* – Mevalonic acid (MVA) pathway**

(i) The cytoplasmic pathway for IPP synthesis begins with the condensation of 3 molecules of acetyl CoA in a two-step reaction to form hydroxymethylglutaryl CoA, which in turn gets reduced to mevalonic acid.

(ii) Mevalonic acid is the primary precursor for terpenes and undergoes phosphorylation to form mevalonic acid pyrophosphate at the cost of two ATP molecules, followed by decarboxylation to form IPP.

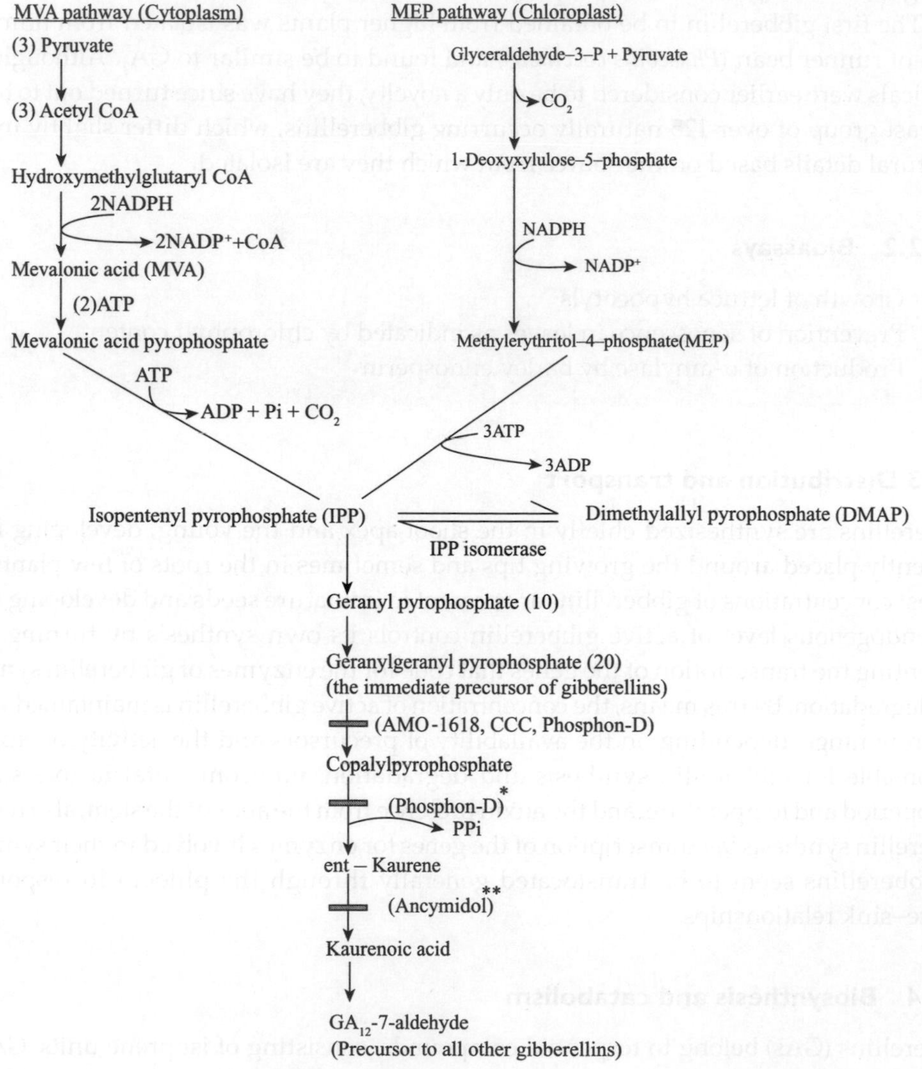

Fig. 16.8 Gibberellin biosynthesis in plants.

*Several anti-gibberellins such as AMO-1618, CCC, Phosphon-D and B-nine are commercially used in the horticulture industry for the production of ornamental plants.

**Ancymidol, a dwarfing agent, is commercially known as A-REST. The enzymes responsible for conversion of ent-kaurene to yield GA_{12}-7-aldehyde are NADPH-dependent, membrane bound cytochrome P450 mono-oxygenases.

(b) ***Chloroplastic pathway* – Methylerythritol-4-phosphate (MEP) pathway**

   (i) The chloroplastic pathway starts with the condensation reaction of glyceraldehyde-3-phosphate and pyruvic acid to yield deoxyxylulose phosphate, which is then reduced to MEP.

   (ii) MEP undergoes phosphorylation and oxidation to yield IPP.

2. **Isomerization of IPP to dimethylallyl pyrophosphate**
   IPP then isomerizes to dimethylallyl pyrophosphate (DMAP), catalyzed by IPP isomerase. IPP destined for ABA and GA synthesis and DMAP for the side chain of CKs are synthesized in the plastid, whereas the IPP required for brassinoids is synthesized along the mevalonic acid pathway in the cytosol.
3. The immediate diterpene precursor of GAs is geranylgeranyl pyrophosphate (GGPP) including other intermediates such as farnesyl pyrophosphate, copalyl pyrophosphate and ent-kaurene. Ent-kaurene can be readily converted to GA_{12}-7-aldehyde in the plants in the presence of cytochrome P450 monooxygenases.
   GA_{12}-7-aldehyde is the precursor to all other gibberellins.
   Gibberellin biosynthesis involves plastids, ER and cytosol simultaneously.
   Gibberellins are deactivated by 2β-hydroxylation or by conversion to inactive conjugates.

16.2.5 Mode of action

1. DELLA proteins represent the group of nuclear proteins that seem to behave as repressors in gibberellin signalling. GID1 (GA Insensitive Dwarf 1) receptor binds not only to gibberellin, and the F-box protein of SCF^{SLY1}, but also to DELLA protein to produce hormone–receptor-repressor complex (Figure 16.9).
2. The resulting complex leads to degradation of the DELLA repressor protein by the 26S proteasome pathway, thus relieving a DELLA-imposed repression and permitting expression of the gibberellin-responsive gene.

So, the function of gibberellin is similar to that of auxin – to derepress genes by accelerating the removal and ubiquitin-mediated degradation of the repressor protein.

16.2.6 Physiological roles

Bolting: Gibberellins promote hyper-elongation of intact stems, particularly in dwarf and rosette plants. A rosette form is an extreme example of dwarfism that results in compact growth habit, characterized by closely placed leaves. For example, environmentally-limited rosette plants (such as spinach, lettuce and cabbage) do not come to flowering in the rosette form. Just before initiation of flowering, these plants undergo **bolting** (excessive elongation of internodes as shown in Figure 16.10). Bolting is generally stimulated by an environmental signal (photoperiod/low temperature). Gibberellins have been found to be a limiting factor in the stem growth of rosette plants. Exogenous treatment of GA_3 can substitute for the low-temperature requirement in several plant species.

Mobilization of nutrient reserves during germination of cereal grains: In the germinating cereal grains, cells of the aleurone layer secrete hydrolytic enzymes, such as α-amylase and proteases, which can hydrolyze the stored carbohydrates and proteins in

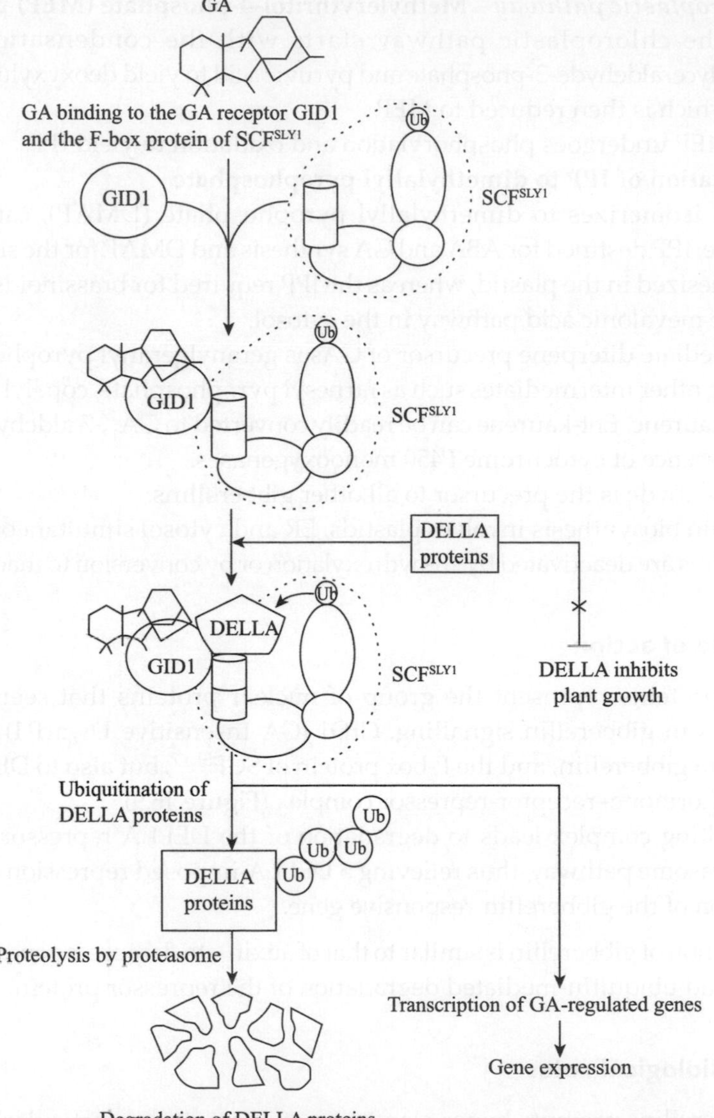

Fig. 16.9 A model showing the mode of action of gibberellins and the role played by DELLA proteins in GA signal transduction pathway. Gibberellic acid (GA) binds to its receptor GID1. The hormone–receptor complex then interacts with SLY1, the F-box protein of SCF^{SLY1} and together they bind DELLA proteins. This allows DELLA proteins to be ubiquitinated for subsequent proteolysis by the 26S proteasome. Degradation of DELLA proteins allows the transcription of GA-regulated genes. (e.g. α-amylase transcription).

the endospermic tissues. Germinating embryos transmit a signal i.e. gibberellin, to the aleurone cells, where it either stimulates or de-represses transcription of genes coding for the hydrolytic enzymes. These enzymes are released into the endosperm where they degrade starch and protein to meet the metabolic needs of the developing embryo (Figure 16.11).

Miscellaneous: Various other roles of gibberellins include regulation of transition from juvenile to adult phases, floral initiation, gender determination, hastening of seed germination (by substituting for the impact of low-temperature or light requirements). Gibberellins are employed commercially to space out developing grape flowers by increasing the internode length so that the fruits get more space to grow.

Fig. 16.10 The effect of gibberellins on bolting in a rosette lettuce plant.

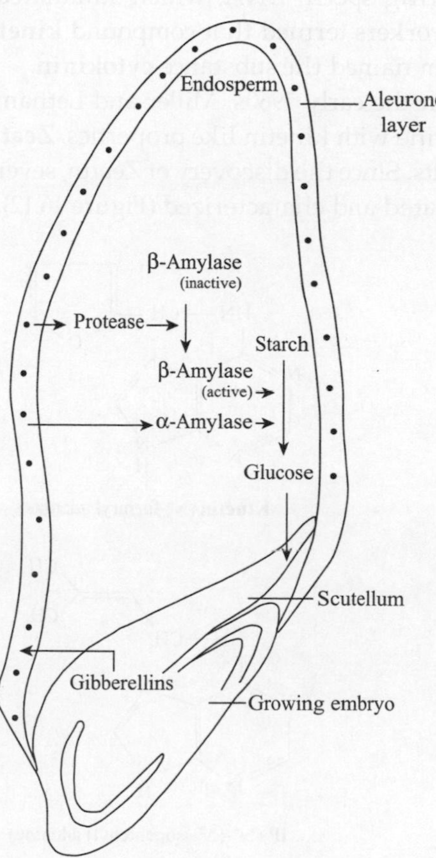

Fig. 16.11 Carbohydrate mobilization during germination of barley seed induced by gibberellins.

16.3 Cytokinins: Regulators of Cell Division

16.3.1 Historical background

The discovery of cytokinins came about because plant cells in culture would not divide. The first experimental evidence for chemical control of plant cell division was provided by Gottlieb Haberlandt of Austria in 1913, when he showed that phloem sap could cause non-dividing parenchymatous cells in potato tubers to transform into actively-dividing meristematic cells. Later some other cell division factors were identified in wounded bean pod tissue, extracts of *Datura* ovules and milky endosperm of coconut. Subsequent research has concentrated on the role of cytokinins in the differentiation of tissues from callus. In the 1940–50s, Skoog and his team at the University of Wisconsin discovered that extract of vascular tissue, coconut milk and yeast would all promote cell division in the presence of auxin. In 1956, Miller (then a post-doctoral student) in Skoog's laboratory isolated and crystallized a highly active adenine derivative N^6-furfuryl-aminopurine from autoclaved herring sperm DNA, which stimulated cell division in tissue culture. Miller and his coworkers termed that compound **kinetin**. Following that in 1965, Skoog along with his team named the substance **cytokinin**.

In the early 1960s, Miller and Letham independently communicated the isolation of a purine with kinetin like properties, **Zeatin**, from young, developing maize seed and plum fruits. Since the discovery of Zeatin, several other naturally occurring cytokinins have been isolated and characterized (Figure 16.12).

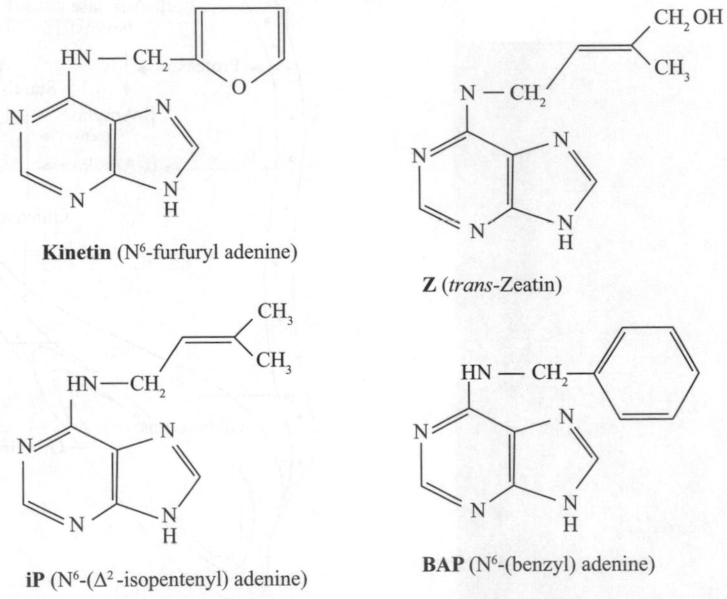

Kinetin (N^6-furfuryl adenine)

Z (*trans*-Zeatin)

iP (N^6-(Δ^2-isopentenyl) adenine)

BAP (N^6-(benzyl) adenine)

Fig. 16.12 The chemical structures of some of the naturally occurring and synthetic cytokinins.

16.3.2 Bioassays

- Retention of chlorophyll by ageing leaves after application of cytokinins
- Extent of enlargement of radish cotyledons

16.3.3 Distribution and transport

Cytokinins are synthesized in the root apical meristem and are translocated passively into the shoot *via* xylem tissues along with water and minerals transport stream. Developing fruits, young leaves and crown gall tissues are also important sites of cytokinin synthesis.

Cytokinins are also produced by plant associated bacteria, insects and nematode. Various cellular importers and exporters are needed for efficient mobilization and targeted translation of coupled high-affinity purine transport.

16.3.4 Biosynthesis and catabolism

Cytokinins are N^6 adenine derivatives with either an isoprenoid-related side chain or an aromatic side chain.

Biosynthesis of cytokinins occurs in the following steps (Figure 16.13):

(i) Biosynthesis of cytokinins starts with the condensation of isopentenyl pyrophosphate with an adenine nucleotide (AMP/ADP/ATP), catalyzed by adenine phosphate-isopentenyl transferase (IPT) to yield N^6-(Δ^2-isopentenyl)-adenosine-5′-monophosphate.
(ii) The nucleotide–phosphate group then splits off to produce the active forms; *trans*-Zeatin and isopentenyl adenine.

In *Arabidopsis*, a small gene family with seven members (AtIPT1, AtIPT3-AtIPT8) encodes the enzymes for synthesis of IPT. AtIPT genes are expressed in the specific root and shoot tissues suggesting that cytokinin synthesis takes place in all major organs. The tRNA is another possible source of cytokinins.

Cytokinins can be inactivated by glycosylation of the purine ring or of the side chain. In general, glucose is the conjugated sugar molecule. Glycosyl conjugation appears to be important for regulating cytokinin activity in some plant species. Various genes encoding cytokinin glycosyl transferases and glycosidases have been identified in plants. Alternatively, cytokinins undergo irreversible oxidative degradation of the N^6-side chain in the presence of enzymes cytokinin oxidases/dehydrogenases (CKX) possessing FAD as a cofactor, to yield adenine and an aldehyde.

16.3.5 Mode of action

The cytokinin signal is perceived and transduced by a multi-step phosphorelay system through a complex form of the two-component system (TCS) pathway.

1. Cytokinin receptors belong to a class known as histidine kinases. CRE 1 (cytokinin response 1) – the cytokinin receptor is an intracellular histidine kinase possessing three domains and two small membrane-spanning hydrophobic regions (Figure 16.14).
2. Binding with cytokinin initiates dimerization followed by autophosphorylation of the histidine kinase residue.
3. The signal chain involves a multistep transfer of the phosphoryl group *via* the phosphorelay network.
4. The final targets are A- and B-type response regulators (RRs) in the nucleus. B-type response regulators are transcription factors that activate the expression of genes responsible for cytokinin-regulated responses. On the other hand, A-type response regulators are not transcription factors and don't modulate gene expression but regulate cytokinin responses by affecting other metabolic aspects. When activated

by phosphorylation, the response regulators either induce transcription of cytokinin primary response genes or regulate other functions of cytokinin-related metabolism.

5. In the absence of cytokinin, CRE 1 reverses the process, unloads phosphoryl groups from HPTs and quickly inactivates the cytokinin response pathway.

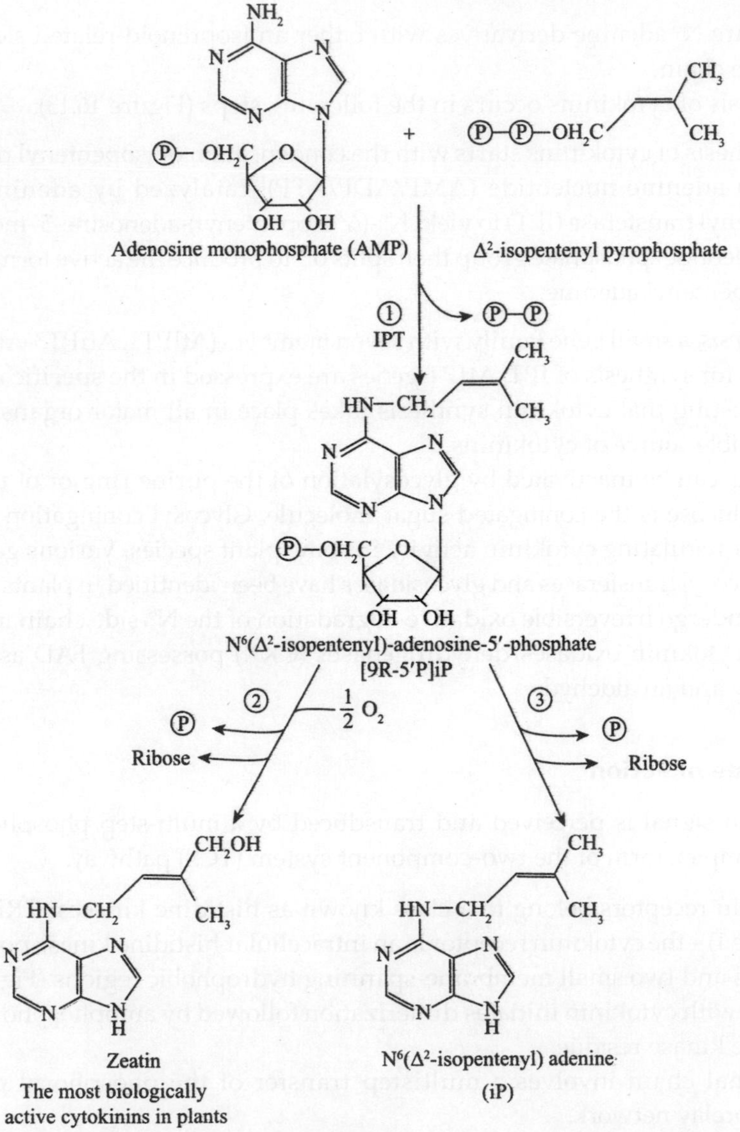

Adenosine monophosphate (AMP) + Δ^2-isopentenyl pyrophosphate

① IPT → Ⓟ–Ⓟ

N^6(Δ^2-isopentenyl)-adenosine-5′-phosphate [9R-5′P]iP

② $\frac{1}{2}O_2$ → Ⓟ, Ribose

③ → Ⓟ, Ribose

Zeatin — The most biologically active cytokinins in plants — N^6(Δ^2-isopentenyl) adenine (iP)

Fig. 16.13 A schematic pathway for the synthesis of isopentenyl adenine (iP) and *trans*-Zeatin (tZ) from adenosine monophosphate (AMP) and isopentenyl pyrophosphate. Reaction (I) shown in the pathway is the rate-determining step, which takes place in the presence of the enzyme adenine phosphate-isopentenyl transferase (IPT).

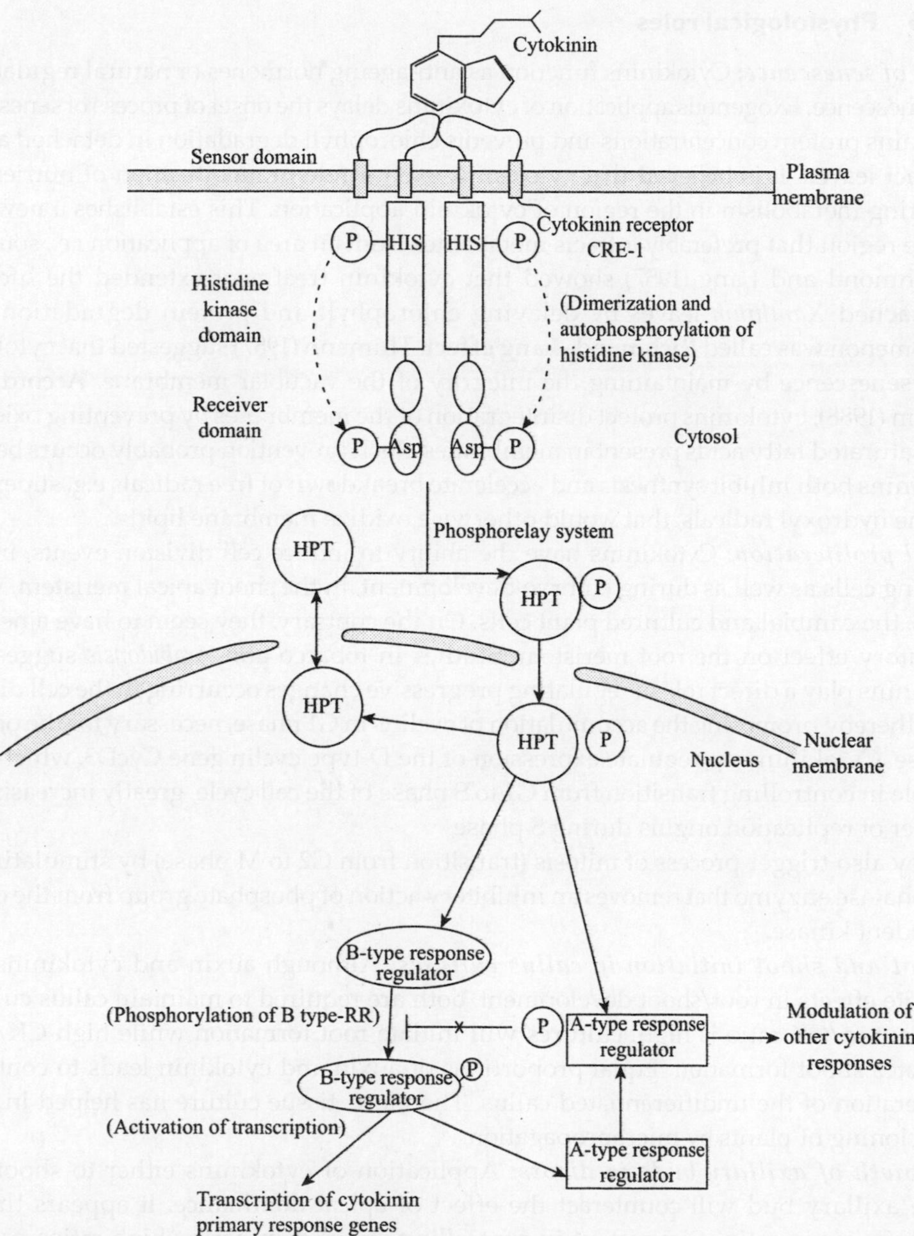

Fig. 16.14 A diagrammatic scheme for cytokinin signal transduction pathway. Cytokinin binding to the receptor CRE1 initiates dimerization and phosphorylation of a histidine kinase residue in each of two receptor molecules. The phosphate group is transferred firstly to an aspartate residue in the receiver domain and later to the histidine residue in a separate histidine phosphotransfer protein (HPT). The phosphorylated HPT moves into the nucleus where the phosphate group is transferred to either a B-type or A-type response regulators (RR). The activated B–type RR, in turn, activates transcription of cytokinin primary response genes, including the A–type RR which may positively or negatively modulate other cytokinin responses.

16.3.6 Physiological roles

Delay of senescence: Cytokinins function as anti-ageing hormones or natural regulators of leaf senescence. Exogenous application of cytokinins delays the onset of process of senescence, maintains protein concentrations and prevents chlorophyll degradation in detached as well as intact leaves. It is believed that cytokinins exert a role in mobilization of nutrients by activating metabolism in the region of cytokinin application. This establishes a new sink, i.e., the region that preferably attracts metabolites from an area of application i.e., source.

Richmond and Lang (1957) showed that cytokinin treatment extended the life span of detached *Xanthium* leaves by delaying chlorophyll and protein degradation. This phenomenon was called **Richmond–Lang effect**. Thimann (1987) suggested that cytokinins delay senescence by maintaining the integrity of the vacuolar membrane. According to Lesham (1988), cytokinins protect disintegration of the membranes by preventing oxidation of unsaturated fatty acids present in membranes. Such prevention probably occurs because cytokinins both inhibit synthesis and accelerate breakdown of free radicals e.g. superoxide and the hydroxyl radicals, that would otherwise oxidize membrane lipids.

Cell proliferation: Cytokinins have the ability to induce cell division events, in non-dividing cells as well as during embryo development, in the shoot apical meristem, young leaves, the cambial and cultured plant cells. On the contrary, they seem to have a negative regulatory effect on the root meristem. Studies in tobacco and *Arabidopsis* suggest that cytokinins play a direct role in regulating progressive changes occurring in the cell division cycle, thereby promoting the accumulation of cyclins in G1 phase, necessary for the onset of S phase. Cytokinins up-regulate expression of the D-type cyclin gene CycD3, which plays key role in controlling transition from G1 to S phase of the cell cycle, greatly increasing the number of replication origins during S phase.

They also trigger process of mitosis (transition from G2 to M phase) by stimulating the phosphatase enzyme that removes an inhibitory action of phosphate group from the cyclin-dependent kinase.

Root and shoot initiation in callus cultures: Although auxin and cytokinins have opposite effects in root/shoot development, both are required to maintain callus cultures. If the auxin/CK ratio is high, cultures will initiate root formation while high CK/auxin promotes shoot formation. Equal proportions of auxin and cytokinin leads to continued proliferation of the undifferentiated callus. Therefore, tissue culture has helped in large-scale cloning of plants by micropropagation.

Growth of axillary buds in dicots: Application of cytokinins either to shoot apex or the axillary bud will counteract the effect of apical dominance. It appears that the cytokinin/auxin ratio is important in controlling apical dominance; high ratios promote bud development while low ratios favour apical dominance. The phenomenon of Witches' broom is the perfect example of cytokinin–auxin antagonism in many plants. It is due to fungal infection that triggers accumulation of cytokinin through overproduction. The result is an unregulated release of axillary bud development.

Pathogenicity: Cytokinins are produced by many plant pathogenic bacteria that play a role in pathogenicity. The T-DNA harbours an IPT gene, which is expressed in the host cell, leading to cytokinin production. Other cytokinin-synthesizing bacteria are *Pseudomonas syringae*, which induces gall formation and *Rhodococcus fascians*, which brings about fasciation and abnormal growth resulting in **Witches' broom disease**. The root nodule forming and nitrogen-fixing symbiont *Rhizobium* species is also known to synthesize cytokinin.

***Formation of crown galls*:** Crown galls are the outcome of an infection of a plant with the bacterium *Agrobacterium tumefaciens* containing a typical bacterial plasmid, **Ti plasmid**, besides its own chromosomal DNA. A small segment of the Ti plasmid, known as the **T-DNA** (transferred DNA), gets integrated into the nuclear DNA of the host plant cell. T-DNA carries genes for the synthesis of cytokinins as well as auxins, along with unusual amino acid derivatives called **opines**. Both cytokinins and auxins collectively promote cell proliferation and neoplastic growth resulting in the development of a tumour called **'crown gall'**, while opines serve as nutrients for the invading bacterium.

***Chloroplast development and chlorophyll synthesis*:** Application of cytokinin to the etiolated leaves increases conversion of etioplasts into chloroplasts, particularly by promoting grana development and also enhances the rate of chlorophyll synthesis. It is believed that cytokinins increase the production of one or more proteins to which chlorophylls are bound, thus stabilizing its structure.

***Miscellaneous*:** Besides the roles discussed above, cytokinins are involved in maintaining a highly meristematic apical shoot, cell expansion in cotyledons and leaves, breaking dormancy of seeds, tuber formation, differentiation of vascular tissue, de-etiolation, transformation from vegetative to reproductive growth and such processes. Cytokinins also inhibit the development of lateral roots, while auxins on the other hand encourage their formation. Consequently, this relationship maintains the balance between cytokinins and auxin, which determines plant architecture. Cytokinins also regulate important physiological parameters that determine biomass formation and its distribution *via* central genes of primary metabolite pathways, including invertases, hexose transporters, key genes of phosphate and nitrogen signalling.

Cytokinins and Agricultural Biotechnology

Modulation of endogenous cytokinin content of plants or interfering with cytokinin signalling has a high potential for biotechnological applications in agriculture.

- Plants with increased cytokinins are more branched and senesce later.
- Moreover, cytokinins alter source-sink relations, a promising approach to regulate sink strength and improve yield attributes.
- Plants with reduced cytokinin content develop a larger root system leading to improved acquisition of minerals and water, factors which are often limiting for plant growth. Changes in cytokinin levels are positively correlated with levels of mineral nutrition, especially nitrogenous nutrients.

16.4 Abscisic Acid: A Seed Maturation and Anti-stress Signal

16.4.1 Historical background

Abscisic acid (ABA), a naturally occurring growth inhibitor, was discovered through independent investigators working on different physiological phenomena in different laboratories. In 1953, Bennet-Clark and Kefford isolated large amounts of 'inhibitor β' from axillary buds and the outer layer of dormant potato tuber which convinced them that it was associated with apical dominance and maintaining dormancy in potatoes. Meanwhile, other investigators confirmed the occurrence of inhibitors in buds and leaves that seemed to correlate with the onset of dormancy in woody plants. In 1964, Wareing at the University College of Wales at Aberystwyth, proposed the term 'dormin' for these endogenous dormancy-inducing chemicals. At about same time, Addicott and collaborators at University of California isolated natural substances, from senescing leaves of bean and from cotton and lupin fruits that accelerated abscission and were called 'abscission II'. By 1965, all these paths of research led to the discovery that a single hormone was involved in all of them. Finally, the name **abscisic acid** and abbreviation **ABA** were recommended by a panel of scientists to the International conference on plant growth substances held in Ottawa in 1967. The recommendation was accepted by the conference and the term 'abscisic acid' is now internationally used.

16.4.2 Bioassays

- Abscission of petioles from cotton seedling explants in sealed containers
- Inhibition of germination
- Interference with growth responses to the promoting hormones, auxin and gibberellin

Bioassays for ABA largely depend on its inhibitory properties.

16.4.3 Distribution and transport

ABA has been found to be a universal plant hormone in vascular plants, synthesized predominantly in leaves, fruits and root caps. ABA response varies with its concentration within the tissue and the sensitivity of the tissue to the hormone. ABA, being highly mobile, is translocated *via* both xylem and phloem (more abundant in the phloem sap) and moves rapidly out of the leaves to other parts of the plant, especially sink tissues (roots and developing seeds). Under conditions of water stress ABA is synthesized in the roots and rapidly exported to the leaves.

16.4.4 Biosynthesis and catabolism

ABA is a sesquiterpenoid (15-C) compound, obtained from the terminal region of carotenoids.

Biosynthetic pathway for ABA begins within chloroplast and ends in cytosol (Figure 16.15).

1. Glyceraldehyde-3-phosphate and pyruvate condense to synthesize isopentenyl pyrophosphate (IPP) in the chloroplast *via* MEP (methylerythritol-4-phosphate) pathway.
2. IPP produces a number of C_{10}, C_{20} and C_{40} terpenoids, including β-carotene, violaxanthin and xanthoxin.
3. Xanthoxin (C_{15}) is the end product in chloroplast which migrates to cytosol and forms ABA.

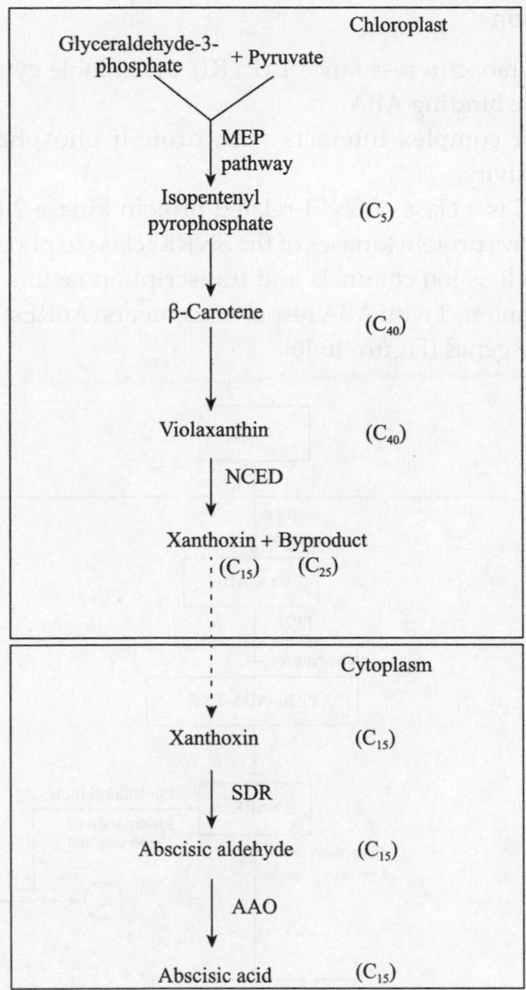

Fig. 16.15 A flowsheet showing the biosynthetic pathway for ABA (in the chloroplast and in the cytosol). Abscisic acid (ABA) biosynthesis from isopentenyl pyrophosphate (IPP) begins in the chloroplast with intermediates such as β-carotene and violaxanthin. Violaxanthin is then cleaved to xanthoxin in the presence of enzyme 9-cis-epoxycarotenoid dioxygenase (NCED). Xanthoxin, the end product in chloroplast, migrates in the cytoplasm and gets converted to ABA in two steps; xanthoxin → abscisic aldehyde in the presence of short chain dehydrogenase/reductase (SDR), and abscisic aldehyde → ABA catalyzed by abscisic acid oxidase (AAO).

ABA is degraded to phaseic acid *via* oxidation in the presence of cytochrome P450. Phaseic acid is further metabolized to dihydrophaseic acid (DPA) and 4′-glucoside of DPA. Conjugation of ABA to sugars is another route for removing active ABA from cells.

ABA synthesis and accumulation can fluctuate considerably in certain tissues during development or in response to changing environmental situations. All the processes of ABA synthesis, catabolism, compartmentalization and transport contribute to the concentration of active hormone in the tissue at any given point of development.

16.4.5 Mode of action

1. ABA receptors, pyrabactin resistance 1 (PYR1), are soluble cytoplasmic proteins that form dimers before binding ABA.
2. The ABA-receptor complex interacts with protein phosphatases type 2C(PP2C) inhibiting their activity.
3. The target of PP2C is a class of SNF1-related protein kinase 2 (SNRK2). Inhibition of PP2Cs activity allows protein kinases of the SNRK[3] class to phosphorylate and activate target proteins such as ion channels and transcription factors ABRE binding factors (ABFs) which then interact with ABA response elements (ABREs) in the promotor region of ABA-responsive genes (Figure 16.16).

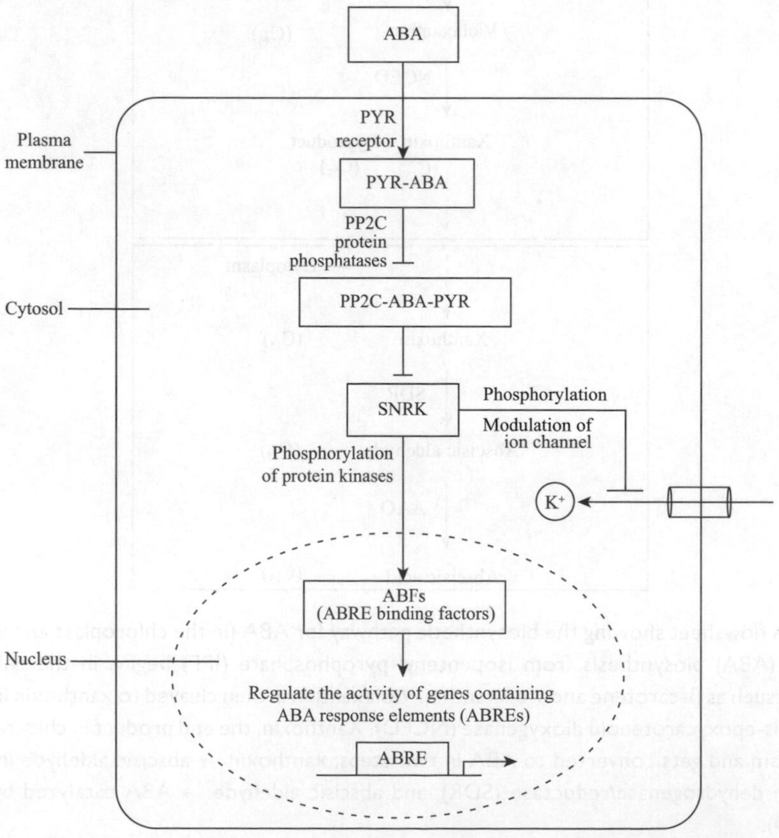

Fig. 16.16 A simple representation of the model of ABA signal transduction pathway. The soluble receptor, PYR I, binds ABA in the cytosol and then binds protein phosphatase PP2C, inhibiting its activity. Phosphorylation of protein kinases of SNRK class (downstream of PP2C), modulate the activity of K^+ ion channel, and of ABF transcription factor that regulate the activity of genes containing ABREs.

[3] SNRK (SNF I - related kinase): Protein kinases of SNRK class are positive regulators of the ABA signalling pathway.

16.4.6 Physiological roles

Regulation of embryo maturation and seed germination: ABA levels normally reach maximum when the embryo is at the late maturation phase. Role of ABA is to prevent precocious seed germination (**vivipary**) while it is still attached to the mother plant. ABA also promotes protein build-up in the later phase of soybean embryo development and prevents GA-induced α-amylase synthesis in cereal grains

Desiccation of the seed initiated by ABA may involve modulation of genes which code for proteins involved in desiccation tolerance.

ABA-induced stomatal closure: ABA content of monocot and dicot leaves can increase 50-fold within 4–8 h when subjected to water stress. It was found that water-stressed roots also form more ABA, which is transported through xylem to the leaves and causes stomata to close. ABA-induced stomatal closure occurs by inhibition of an ATP-dependent proton pump in the plasma membrane of guard cells. ABA shuts off K^+ influx, so K^+ and water leak out accompanied by reduction in turgor causing stomatal closure. According to recent view, effect of water deficit in leaves is linked to photosynthesis. Inhibition of photosynthetic electron transport and photophosphorylation in the chloroplast would impair proton accumulation within lumen of the thylakoid, accompanied by reduction in stromal pH. Simultaneously, pH of the apoplastic space adjacent to the mesophyll cells increases.

The resulting pH gradient promotes release of ABA from the mesophyll cell into the apoplastic space and from there it is carried to the guard cells along with transpiration stream (Figure 16.17).

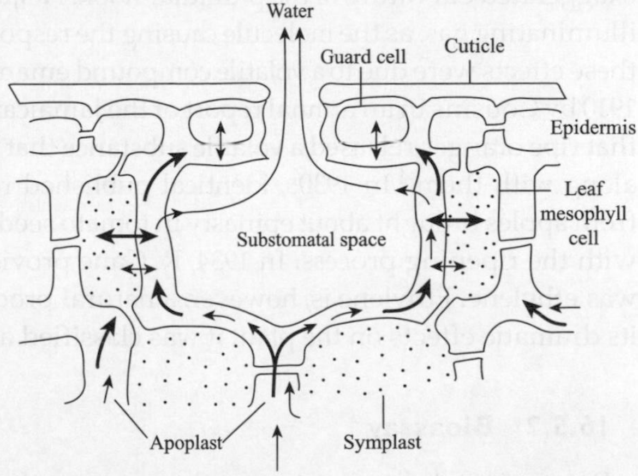

Fig. 16.17 Movement of abscisic acid in the apoplastic space. ABA synthesis takes place in the roots and is transported to the leaves and then *via* leaf mesophyll cells to stomatal guard cells (bold arrows) along with the transpiration stream (thin arrows).

Defence against salt and cold stress: ABA levels increase; when plants are subjected to stress caused by salt and chilling temperatures. Salt stress causes formation of several new proteins, especially a low molecular weight protein **osmotin**, which helps in protecting against the stress.

Bud dormancy: ABA triggers the development of winter buds which remain dormant throughout the winter. Like ethylene, abscisic acid may also prevent growth of dormant lateral buds. It seems that ABA can have some antagonistic effects of gibberellins by preventing growth and elongation of buds.

Miscellaneous: ABA increases the hydraulic conductivity of the root and enhances the root: shoot ratio at low water potentials. ABA is also associated with developmental responses such as heterophylly, initiation of secondary root, flowering, seed dormancy, promoting senescence, and inducing synthesis of storage protein in seeds.

16.5 Ethylene: The Gaseous Hormone

16.5.1 Historical background

Long before its role as a plant hormone was recognized, the simplest gaseous hydrocarbon ethylene ($H_2C = CH_2$) was known to defoliate plants when it leaked out from gaslights in street lamps. In 1886, Dimitry Nikolayevich Neljubow, a graduate student at the Botanical Institute of St. Petersburg in Russia, noticed that etiolated pea seedlings growing in the laboratory displayed symptoms that were later termed as the **'triple response'**: inhibition of hypocotyl and root cell elongation, a pronounced radial swelling of the hypocotyl, and exaggerated curvature of the plumular hook. Neljubow identified ethylene emanating from illuminating gas, as the molecule causing the response. One of the earliest study reports, that these effects were due to a volatile compound emanating out of plant tissue, was published in 1910 by Cousins in an annual report of the Jamaican Department of Agriculture. He noticed that ripe oranges released a volatile substance that would hasten ripening of bananas stored along with them.[4] In 1930s, identical published reports showed that volatile emanations from apples brought about epinasty in tomato seedlings and respiratory changes associated with the ripening process. In 1934, R. Gane provided evidence that the volatile substance was ethylene. Ethylene is, however, a natural product of metabolic activities and owing to its dramatic effects on the plant it was classified as a hormone.

16.5.2 Bioassay

Bioassay can be used to test for the presence of ethylene. Ethylene is readily assayed by enclosing tissues in gas-tight containers and sampling the surrounding atmosphere for ethylene using gas chromatography.

16.5.3 Distribution and transport

Ethylene is produced by almost every part of the higher plants, although the rate of production is dependent on the type of tissue and the stage of development. Generally, meristematic and nodal sites are the most active regions in ethylene production. However, ethylene biosynthesis also increases during leaf abscission, flower senescence, and fruit ripening. Any kind of wounding as well as physiological stresses i.e., biotic, and abiotic stresses such flooding, chilling, temperature, drought stress can trigger ethylene biosynthesis.

Ethylene plays a chief role in fruit development. At first, auxin, which is synthesized in appreciable amounts in pollinated flowers and developing fruits, triggers ethylene

[4] Ripe and rotten fruits hasten the ripening of other fruits stored close by. For example, bananas picked in Cuba and shipped by boat often reached in New York in an overripe and unmarketable state.

production; this in turn accelerates fruit ripening. When auxin is translocated down from the apical meristem of the shoot, it induces synthesis of ethylene in the tissues around the lateral buds and thus inhibits their growth. Ethylene also suppresses stem and root elongation, perhaps in a similar way. In addition to these effects, ethylene is associated with breaking of seeds and bud dormancy in some species, stimulating adventitious root formation in leaves and stems, inducing flowering in pineapple family and enhancing the rate of leaf and flower senescence.

16.5.4 Biosynthesis and catabolism

Ethylene (molecular weight, 28) is the simplest of all known gaseous hydrocarbon. It is lighter than air under physiological conditions, inflammable and rapidly undergoes oxidation. Lieberman and Mapson (1964) demonstrated that sulphur-containing amino acid, methionine, was readily converted to ethylene in plant tissues such as apple fruits. In 1977, Adams and Yang noticed that S-adenosyl methionine (SAM) was an intermediate compound during the production of ethylene from methionine. Furthermore in 1979, they reported the accumulation of 1-aminocyclopropane-1-carboxylic acid (ACC) under anaerobic conditions catalyzed by the enzyme ACC synthase. ACC in the presence of ACC oxidase and O_2, gets converted to ethylene. The production of ethylene is stimulated by various factors such as IAA, wounding as well as water stress. Ethylene formation is also activated by ethylene itself, a form of autocatalysis.

To maintain normal rates of ethylene formation, the sulphur released during ethylene synthesis is recycled back to methionine *via* the **Yang cycle** (Figure 16.18).

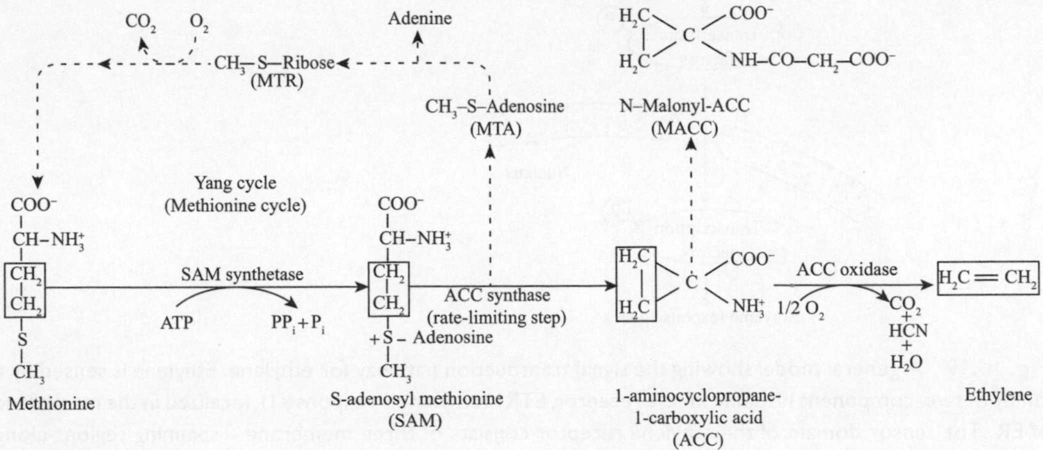

Fig. 16.18 A scheme for biosynthesis of ethylene from methionine in higher plants. The Yang cycle for recycling of sulphur is also shown in the figure.

Ethylene being a volatile gas is quickly released to the surrounding atmosphere by simple diffusion. Ethylene can be catabolized by oxidation to CO_2 or gets converted to ethylene oxide/ethylene glycol.

16.5.5 Mode of action

1. Ethylene is sensed by a family of two-component histidine kinase receptors (ETR1), associated with the endoplasmic reticulum membrane. The sensor domain of the ethylene receptor consists of three membrane-spanning regions and functions as a dimer (Figure 16.19).

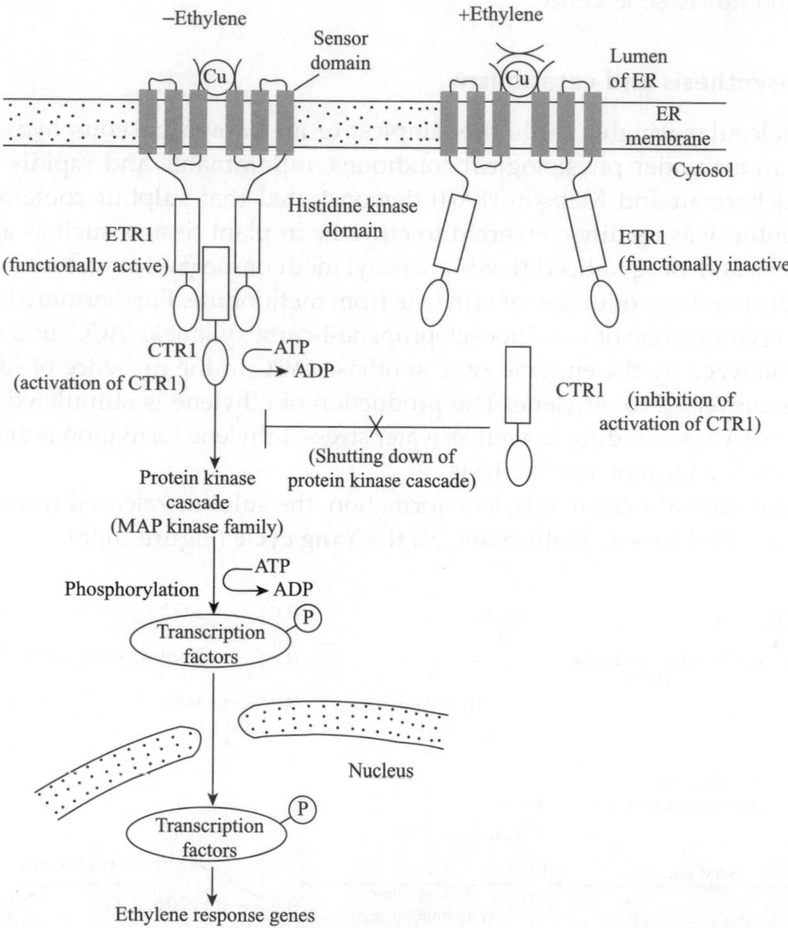

Fig. 16.19 A general model showing the signal transduction pathway for ethylene. Ethylene is sensed by a family of two-component histidine kinase receptor, ETR1 (ethylene – response 1), localized in the membrane of ER. The sensor domain of the ethylene receptor consists of three membrane – spanning regions along with copper as a cofactor (i). In the absence of ethylene, the receptor is functionally active and activates the serine-threonine kinase, CTR1 (constitutive triple response 1). CTR1 is the first component in a protein kinase cascade that targets one or more transcription factors. Phosphorylation activates transcription factors that get attached to the promoter sites of ethylene response genes in the nucleus and cause their transcription. This results in repression of the ethylene response pathway. (ii) The binding of ethylene to ETR1 dimer leads to its inactivation, which causes CTR1 to become inactive. The inactivation of CTR1 shuts down the protein kinase cascade, preventing phosphorylation of the transcription factors and turning off the ethylene-response genes.

2. In the absence of ethylene, the receptor and the subsequent signal chain are functional. The signal chain starts with a protein, **constitutive triple response 1** (CTR1) which physically associates with the histidine kinase domain of the receptor ETR1. This leads to phosphorylation of CTR1 as well as initiation of protein kinase cascade, accompanied by phosphorylation of one or more transcription factors and expression of ethylene response genes.
3. In the presence of ethylene, it binds with the receptor ETR1 and prevents the interaction of CTR1 with ETR1. This retards the initiation of the protein kinase cascade and the subsequent activation of the required transcription factors.
4. The receptor kinase system is constitutively active ('on') and is blocked ('turned off') by ethylene - unique to ethylene.

16.5.6 Physiological roles

Effect on plants in waterlogged soils and submerged plants: Waterlogged soils become hypoxic rapidly. Ethylene synthesis is inhibited because O_2 is required to convert ACC to ethylene. Ethylene already synthesized is trapped in the roots. It causes some of the cortical cells to synthesize cellulase which is partly responsible for degradation of cell walls. Cortex cells also lose their protoplasts and disappear forming the air-filled aerenchyma tissue.

Epinasty: The epinasty of petioles occurs because mature parenchymatous cells on the upper side of the petiole undergo elongation in the presence of ethylene, whereas those on lower side do not. Epinasty is a common type of nastic response to waterlogging of flood-prone plants such as tomatoes.

Suppression of stem and root elongation: Ethylene usually inhibits elongation of stems and roots (especially in dicots) which causes stems and roots to become thick by enhanced radial expansion of the cells. Root and stem thickening caused by ethylene is of survival value for dicot seedlings emerging from the soil.

Effect on flowering: The promotion of flowering in mangoes and bromeliads occurs by ethylene. An ethylene-releasing substance (ethrel/ethephon) rapidly breaks down in water at neutral or alkaline pH to form ethylene and it is used for fruit production in horticulture.

Induction of flower and leaf senescence: Numerous flowers undergo a climacteric rise in respiration and in ethylene production, which eventually causes senescence. Petals show wilting in response to an increase in permeability of the plasma membrane and tonoplast, accompanied by water as well as solutes loss to the cell walls and the intercellular spaces. Pollination also increases the rate of ethylene production, as ACC is translocated from the stigma causing ethylene release and senescence.

Fruit ripening: Ethylene is known for its effects on fruit ripening. During the development of **climacteric fruits**, there is regulated burst in the respiratory rate followed by ethylene formation. This is accompanied by the expression of genes responsible for development of fruit colour, flavour and texture.

Fruits can be catagorized as *climacteric* or *non-climacteric* on the basis of whether ripening is or is not accompanied by a rise in respiration. Another difference between these two groups is that ripening of climacteric fruit is followed by an increase in the synthesis of the PGR ethylene, known to play a key role in the ripening of climacteric fruit.

Miscellaneous: Effect of ethylene has also been studied in promotion of seed germination, inhibition of bud dormancy, diminished apical dominance, adventitious root formation, expression of flower sex in monoecious species, cell death and pathogen responses.

Inhibitors of the pathway of ethylene synthesis (silver salts) have been used to extend markedly the longevity of carnation flowers.

Ethylene gas volatilizes easily from the tissue and may affect other tissues or organs nearby. Therefore, ethylene-trapping mechanisms are employed during the storage of fruits, vegetables and flowers. Potassium permanganate ($KMnO_4$) is an effective absorbent of ethylene and can lower the concentration of ethylene in apple storage regions from 250 to 10 μl, markedly prolonging the storage life of the fruits.

Studies have indicated that ethylene plays a chief ecological role as well. Ethylene synthesis shows a rapid increase when a plant is subjected to ozone and other toxic chemicals, temperature extremes, drought, attack by pathogens or herbivores and other stresses. The increased production of ethylene can speed up the loss of leaves or fruits that have been damaged by such stresses. The synthesis of ethylene by plants attacked by herbivores or infected with pathogens may be a signal to activate the defence mechanism of the plants. This may include the production of compounds toxic to the pests.

16.6 Polyamines

The discovery of polyamines dates back to 1952, when Richards and Coleman found them in potassium-deficient barley. Their growth-regulating properties were first described in 1974 by Bagni and Serafini in sunflower. Polyamines were accepted as phytohormones in 1982 at the International Conference on Plant Growth Substances. Polyamines are small organic molecules with low molecular weight, having at least two or more amino groups. These are widely distributed in microbes, animals and plants (millimolar concentrations). Unlike phytohormones, these are almost immobile intracellularly. Polyamines have a high affinity for anionic groups in membranes and cell walls, including hydroxyl cinnamic acid as well as DNA, RNA, phospholipids and acidic proteins. These are also found as amide conjugates with *p*-coumaric, ferulic and caffeic acids.

Common polyamines are putrescine, spermine, spermidine, thermospermine, diaminopropane and cadaverine (abundant in legumes) (Figure 16.20).

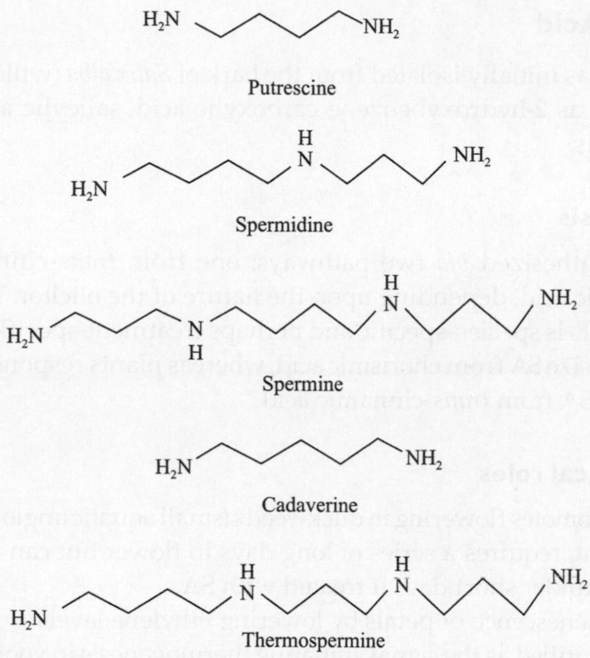

Putrescine

Spermidine

Spermine

Cadaverine

Thermospermine

Fig. 16.20 Chemical structures of some of the naturally occurring polyamines in plants.

16.6.1 Biosynthesis and catabolism

Putrescine, the smallest of polyamines, is synthesized from amino acids L-ornithine or L-arginine and serves as the precursor for the synthesis of spermidine. Spermine and thermospermine are synthesized from spermidine, when it condenses with S-adenosyl L-methionine. Polyamines and ethylene act as competitive inhibitors of each other's biosynthetic pathways because S-adenosyl methionine (SAM) is also a precursor of ethylene.

Polyamines are catabolized to inactive compounds by two classes of enzymes: polyamine oxidases and diamine oxidases, thereby generating H_2O_2.

16.6.2 Physiological roles

(i) Xylem differentiation in *Arabidopsis* and *Zinnia elegans*
(ii) Cell differentiation and division
(iii) Embryogenesis
(iv) Abiotic stress tolerance
(v) Flower development
(vi) Fruit ripening
(vii) Root initiation
(viii) Tuber formation and response to pathogen

Most of the responses to polyamines may be related to their effects on programmed cell death.

16.7 Salicylic Acid

Salicylic acid (SA) was initially isolated from the bark of *Salix alba* (willow), hence its name. Chemically known as 2-hydroxybenzene carboxylic acid, salicylic acid[5] has molecular weight of 138 daltons.

16.7.1 Biosynthesis

Salicylic acid is synthesized *via* two pathways; one from *trans*-cinnamic acid and the other from chorismic acid, depending upon the nature of the elicitor. The pathway for SA biosynthesis in plants is species-specific and perhaps treatment-specific. Plants exposed to abiotic stress synthesize SA from chorismic acid, whereas plants responding to the presence of pathogens make SA from *trans*-cinnamic acid.

16.7.2 Physiological roles

1. Salicylic acid promotes flowering in duckweeds (small aquatic angiosperms). Duckweed, a long-day plant, requires a series of long days to flower but can be induced to come into flowering under short days if treated with SA.
2. SA can retard senescence of petals by lowering ethylene levels.
3. SA has been identified as the signal, initiating thermogenesis in voodoo lily (*Sauromatum guttatum*). The thermogenic-inducing compound, initially named calorigen, was later identified as SA in the 1980s.

   Thermogenesis is a process that generates heat by uncoupled electron flow through the alternative oxidase pathway in mitochondria. Heat produced by spadix volatilizes pungent amines, which attract insect pollinators.
4. SA participates in plant responses to abiotic stress.
5. SA is also involved in plant defence against pathogens, especially viruses. Artificial inoculation of leaves with viruses causes SA levels to increase and these in turn activate disease-resistance genes. It can prevent local pathogen infection as well as regulate systemic acquired resistance (SAR), encouraging resistance at regions much away from the site of infection. SAR is also associated with the production of pathogenesis-related (PR) proteins which serve a protective function during pathogen attack.

16.8 Nitric Oxide

Nitric oxide (NO)[6] is a reactive gas produced by all living organisms. Although NO has been known to be produced by plants since 1980s but its role as a hormone has only recently been explored.

[5] Adding acetyl salicylic acid (aspirin) to a vase of cut flowers is a widely used remedy to delay their senescence.

[6] The 1998 Nobel Prize for the discovery of role of nitric oxide (NO) as a signalling molecule in the cardiovascular and nervous systems was jointly awarded to the three American pharmacologists, Drs. Robert F. Furchgott, Louis J. Ignarro, and Ferid Murad.

16.8.1 Biosynthesis

In plants, NO can be synthesized during nitrate assimilation when NO_3^- is reduced to NH_4^+, using the enzyme nitrate reductase. Bacteria generate NO during the process of denitrification and animals produce NO[7] from arginine using the enzyme NO synthase. But, NO synthases have not been found in plants.

Other biosynthetic pathways in plants can give rise to significant amounts of NO. Under conditions of limited oxygen availability, NO is produced by reduction of nitrite, which accepts electrons from cytochrome oxidase, Cyt *c*, Cyt P450 or other donors.

$$NO_2^- + 2e^- + 2H^+ \rightarrow NO + H_2O$$

A significant amount of NO can also be produced non-enzymatically from nitrite in the acidic condition found in plant cell walls.

$$NO_2^- + H^+ \rightarrow HNO_2 \xrightarrow{2RH} NO + H_2O + 2R \text{ (R = biological reductant)}$$

16.8.2 Physiological roles

1. Regulation of stomatal closure as a result of ABA signalling.
2. Lateral root initiation.
3. Delays flower senescence.
4. Stimulation of seed germination in sunflower, tomato, lettuce, etc.
5. Responses to a vast array of abiotic stresses (osmotic stress, salinity, flooding and temperature extremes). NO interacts with ROS produced during stresses- thus serving as antioxidant function.
6. Plant defence against pathogen attack by stimulating the production of H_2O_2.

NO interacts with melatonin[8] (N–acetyl-5-methoxytryptamine, a tryptophan derived molecule) as a long-range signalling molecule. When plants are infected by pathogenic bacteria, NO acts as a downstream signal leading to increased melatonin levels, which in turn induces the mitogen-activated protein kinase (MAPK) cascade and associated defence responses. The application of exogenous melatonin can also promote sugar and glycerol production, leading to increased levels of salicylic acid and NO. Melatonin and NO in plants can function cooperatively to promote lateral root growth, delaying ageing, etc.

There is a clearly a functional relationship between melatonin and NO during salt stress. There is some evidence that NO-induced-nitrosylation is involved in salt stress responses.

[7] However, direct formation of NO can take place in the human body when nitroglycerine is ingested where it is released through non-enzymatic reaction. NO is a vasodilator, helping to dilate blood vessels, thereby increasing the blood supply.

[8] Melatonin in plants controls a plethora of physiological processes, ranging from growth regulation to delaying leaf senescence ; promoting root generation, embryo morphogenesis, regulating circadian rhythm, photoperiodic flowering, resistance to pathogen invasion, and controlling the damage caused by substances such as heavy metals, salt and other chemicals, UV radiations, and temperature change. Melatonin is considered to have a multifunctional role, particularly in the control of ROS, reactive nitrogen species (RNS), harmful free radicals and other oxidative molecules in plant cells.

16.9 Strigolactones

Strigolactones (SLs) are terpenoid molecules synthesized from carotenoids in the plastids. SLs were originally isolated from cotton root exudates because roots of cotton and other plants are known to stimulate the seed germination of weeds from the Orobanchaceae family. SLs are present in plants at low concentration. More than 10 SLs have now been isolated from plants. Strigol was the first SL to be purified. It has a characteristic chemical structure that includes two lactone rings (ring C and D) linked *via* an enol ether bridge. Another one, **Orobanchol** has been isolated from red clover root exudates.

16.9.1 Biosynthesis

Strigolactones are synthesized in plastids from carotenoids. Carotenoids are cleaved by CCD7 and CCD8 to a product which is converted into active strigolactones in the cytosol by enzymes of the cytochrome P450 class. Signalling from SLs involves an F-box protein and leads to inhibition of shoot branching (Figure 16.21).

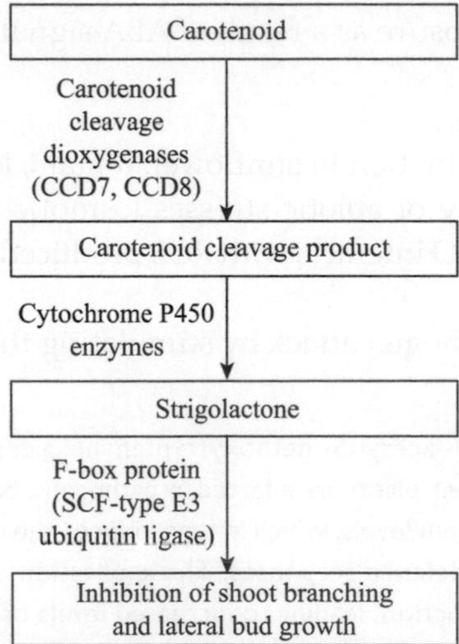

Fig. 16.21 A biosynthetic scheme for the synthesis of strigolactones (in the plastids) from carotenoids.

16.9.2 Physiological roles

1. Regulation of growth and development: These suppress shoot branching and stimulate cambial activity and secondary growth. SLs also reduce adventitious and lateral root formation and promote root hair growth.
2. Regulation of branching: These act as host recognition signals in arbuscular mycorrhizal fungi. These fungi form beneficial symbiotic associations with the roots of most of the plants. SLs released by the roots of host plants serve to attract arbuscular mycorrhiza to the root surface and thought to stimulate branching of fungal hyphae and facilitate phosphate uptake from soil.

3. Regulation of lateral bud dormancy. Very little is known about SL receptors or signal transduction except that SCF E3 ubiquitin ligases play an important role.

16.10 Jasmonates

Jasmonates (JAs) are a group of naturally occurring molecules that resemble mammalian prostaglandins. The role of jasmonates as plant hormones was discovered in the 1970s when these were reported to inhibit the growth of wheat, rice and lettuce seedlings. Jasmonic acid and methyl jasmonate are the most commonly occurring jasmonates in plants. Methyl jasmonate is one of the principal constituents of fragrance produced by jasmine (*Jasminum grandiflorum*).

16.10.1 Biosynthesis

Jasmonates are synthesized from α-linolenic acid released from membranes by lipase activity involving two organelles-the plastid and the peroxisome. JA conjugate with isoleucine (JA-Ile) is the functionally active form of jasmonate in thc cytosol. JA also acts as the precursor of volatile methyl jasmonate. Methyl jasmonate is biologically active and is thought to act as a volatile signal in plant defences against herbivory.

16.10.2 Mode of action

1. Jasmonate receptors are F-box proteins, components of SCF^{COL1}E3 ubiquitin ligase, that target a repressor of jasmonate-responsive genes (Figure 16.22).
2. Jasmonate forms a complex with the F-box protein COL1 that targets the transcription repressor JAZ for degradation by the ubiquitin–mediated proteasomes.
3. JAZ proteins repress the expression of defence proteins. Release of inhibition of JAZ repression allows the transcription of JA-responsive genes by transcription factors such as MYC2.
4. Transcribed genes include defence proteins.

16.10.3 Physiological roles

1. Inhibit seed and pollen germination.
2. Retard root growth.
3. Elicit the curling of tendrils.
4. Promote tuber formation.
5. Associated with the accumulation of storage proteins in seeds.
6. Methyl jasmonate is involved in signalling, especially in defence against herbivory (refer to the chapter on Secondary Plant Metabolites).

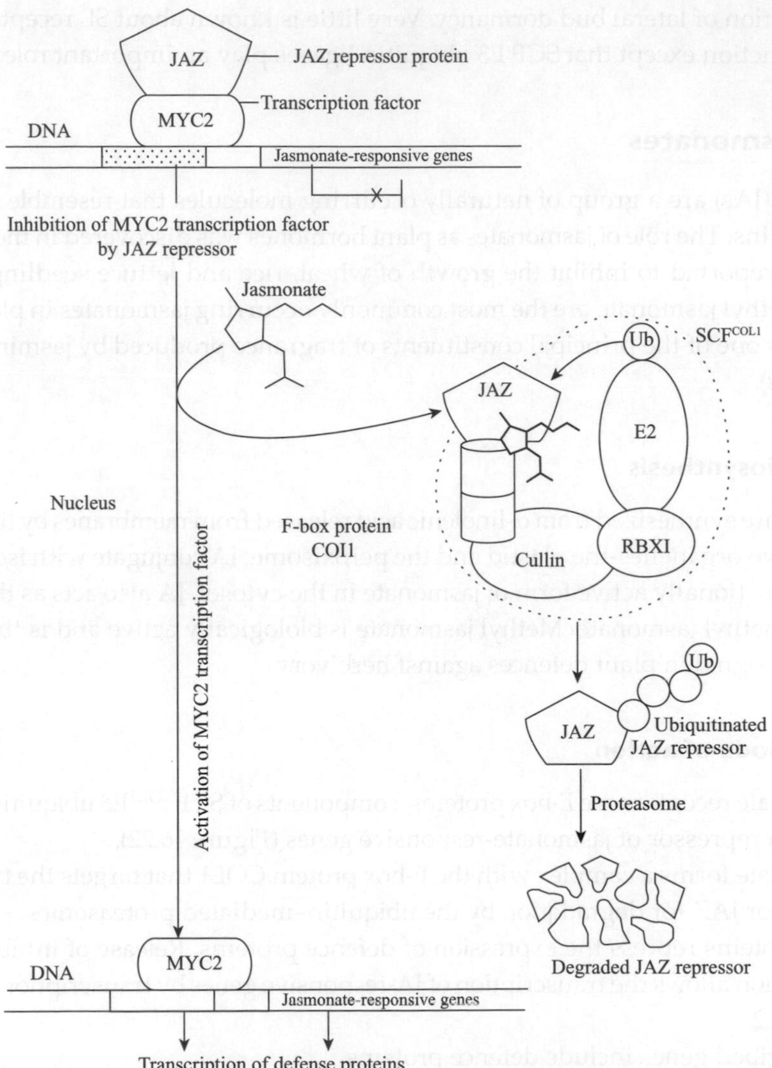

Fig. 16.22 A general model of the jasmonates signal transduction pathway resulting in the transcription of defence proteins. Jasmonate forms a complex with the F-box protein COI1 (CORONATINE-INSENSITIVE1) that targets the transcription repressor JAZ (JASMONATE ZIM-DOMAIN) for degradation by the proteasome. Release of inhibition of JAZ repression allows the transcription of jasmonate-responsive genes (including defence proteins) using transcription factor MYC2.

16.11 Brassinosteroids

The study of brassinosteroids (BRs) as growth regulators goes back to the early 1970s. In 1979, Grove and his colleagues identified the most active component of the native BRs in plants, **brassinolide** (BL). BL was first isolated from the germinating pollen of *Brassica napus*. Another BR **castasterone** was later isolated from insect galls of chestnut. Like other hormones, brassinolide is active in micromolar concentrations. Over 40 analogues of BL have now been isolated from nearly 60 plant species and almost from all types of tissues including pollen grains, roots, stems, leaves, seeds and flowers.

16.11.1 Bioassay

- Rice lamina inclination bioassay

16.11.2 Biosynthesis

BRs are polyhydroxylated plant sterols. Sterols are triterpene molecules (C_{30}) with distinct chemical structures containing C_{27}–C_{30} and are obtained from acetate *via* the mevalonic acid biosynthetic pathway (Figure 16.23).

There are two analogous pathways for the conversion of campesterol to brassinolide, depending on whether the oxidation at carbon-6 occurs early or late in the pathway. Campesterol is first converted to campestanol, which is then converted to castasterone through either of two pathways. All the enzymes converting campestanol to BL are cytochrome P450 monooxygenases located on the ER. BR function close to their regions of synthesis and do not participate in long–distance translocation.

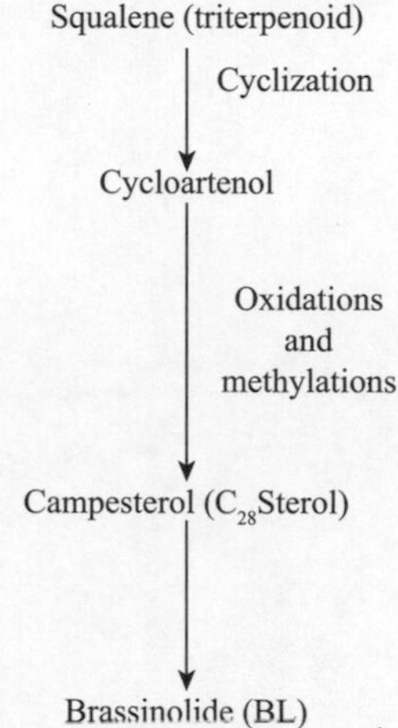

Fig. 16.23 Biosynthesis of brassinosteroids. The BR biosynthetic pathway starts with the sterol campesterol, which is ultimately derived from plant sterol precursor, cycloartenol.

16.11.3 Mode of action

1. Brassinosteroid sensing involves a plasma membrane-bound leucine-rich repeat (LRR)–receptor serine/threonine kinase that controls the phosphorylation and dephosphorylation of transcription factors (Figure 16.24).
2. **BRI 1 (brassinosteroid insensitive-1)**, an essential component of the BR signalling pathway, is the receptor for BRs (Brassinosteroids).

   Brassinolide gets attached to the extracellular domain of the BRI 1 receptor, a plasma membrane LRR – receptor kinase and thus activates the receptor. This results in increased autophosphorylation activity followed by its increased association with a second LRR – receptor kinase, **BAK 1(BRI 1 associated receptor kinase 1).**
3. The formation of the activated BRI 1/BAK 1 hetero – oligomer in the presence of BRs initiates a signalling cascade that leads to the BR – regulated gene transcription.
4. The next step in the pathway involves inactivation of the negative regulator **BIN 2 (brassinosteroid insensitive-2).**

   In the absence of BRs, BIN 2 phosphorylates nuclear protein **BZR 1 (brassinazole – resistant 1)**, thus inhibiting its transcriptional activity. The BR signalling chain dephosphorylates cytoplasmic BZR 1-P, which then moves into the nucleus and activates the transcription of BR-sensitive genes.
5. BR signalling may also inhibit the phosphorylating ability of BIN 2, thus ensuring the activation and nuclear localization of BZR 1.

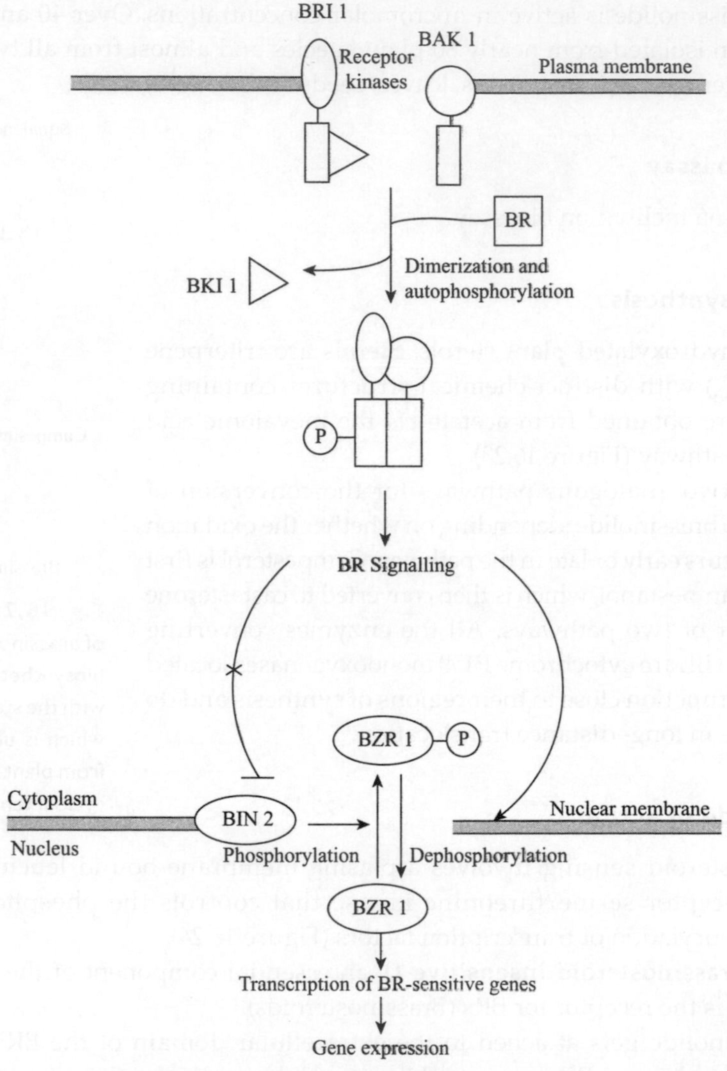

Fig. 16.24 A generalized scheme explaining the signal transduction for brassinosteroids. In the absence of BR, the two receptor kinases BRI 1 and BAK 1 exist as monomers in the plasma membrane. In the presence of BR, dimerization of the two receptors as well as autophosphorylation of kinase region of BRI 1 occurs by removal of inhibitory protein, BKI 1. The target of BR signalling chain is transcription factor such as BZR 1. Dephosphorylation of BZR 1-P occurs in the cytosol, which subsequently migrates into the nucleus and activates the transcription of BR- sensitive genes. On the other hand, BIN 2 protein phosphorylates BZR 1 causing it to move back in the cytoplasm, thereby inhibiting transcription. BR signalling may also increase transcription by blocking BIN 2.

16.11.4 Physiological roles

1. Increase in rate of stem and pollen tube elongation.
2. Increase in rate of cell division.

3. Seed germination.
4. Leaf morphogenesis.
5. Apical dominance.
6. Vascular differentiation.
7. Inhibitory effect on root elongation.
8. Increased senescence and cell death.
9. Mediation of responses to abiotic and biotic stress.
10. Increase in plastic extensibility of the cell wall by regulating genes coding for expansins and cellulose synthase.

Most of the responses regulated by brassinosteroids are also affected by auxin. It seems that considerable crosstalk exists between brassinosteroids and auxin signalling pathways.

16.12 Signal Perception and Transduction

Signals are environmental inputs that initiate one or more plant **responses** and the physical component that biochemically responds to that signal is referred to as **receptor**. Receptors are either proteins or pigments associated with proteins (in case of light receptors). Receptors are present throughout the cell - in plasma membrane, cytosol, the endomembrane system and nucleus - and are conserved across bacterial, plant, animal and plant kingdoms. Once receptors sense their specific signal, they must transduce the signal in order to amplify it and trigger the cellular response. Receptors often do this by modifying the activity of other proteins or by employing intracellular signalling molecule called **second messengers**, which in turn alter cellular processes, such as gene transcription.

All developmental signals, regardless of their nature, share the same sequence of signal perception, transduction, and response.

Signal-receptor ⟶ signal transduction ⟶ response

Many of the specific events and intermediate steps involved in plant signal transduction have now been identified, and these intermediates constitute the signal transduction pathways.

16.12.1 Signal perception

(A) The G Protein-coupled receptor

The G Protein system is a ubiquitous receptor system in plants and animals (Figure 16.25). G-proteins are a vast family of guanosine triphosphate (GTP) binding proteins. The two major components of the G-protein system are:

(i) G-protein itself with three distinct subunits (Gα, Gβ, Gδ), localized on the cytosolic face of the plasma membrane
(ii) A transmembrane protein known as the G protein-coupled receptor (GPCR)

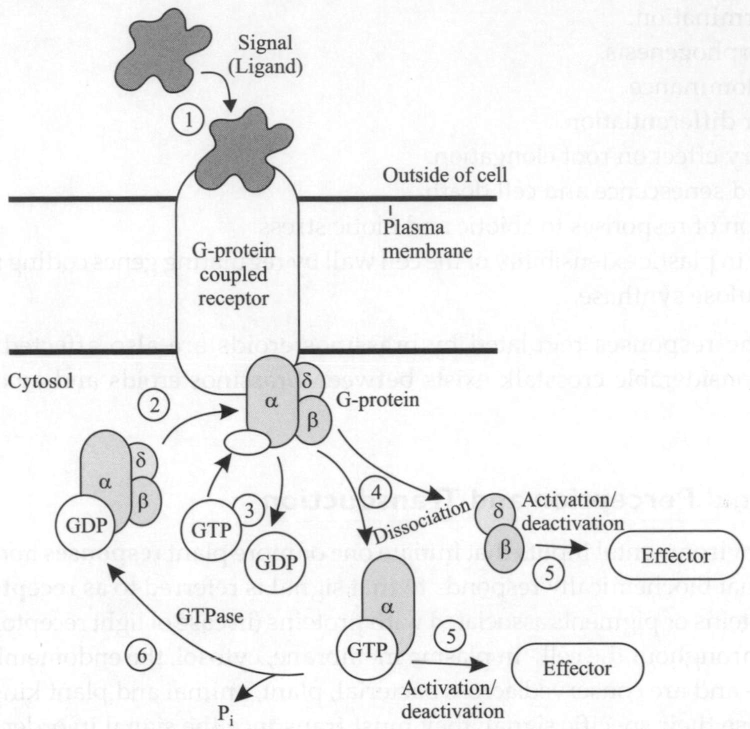

Fig. 16.25 Signal transduction by the G-protein coupled receptor system (GPCR). (1,2) The ligand molecule gets attached to GPCR and increases its affinity for the G-protein. (3,4) The G-protein exchanges its GDP for GTP and this brings about detachment of α subunit of the G-protein from its β δ dimer subunits. (5) Separation of both the subunits from the receptor is followed by binding to the effector molecule causing its activation or deactivation. (6) The in-built GTPase activity of α subunit hydrolyzes the GTP, thereby deactivating the α-subunit. This allows the two subunits to recombine.

A signal molecule (or the ligand) gets attached to the transmembrane GPCR, increasing its affinity for the G protein. The G-protein exchanges its GDP for GTP, which is responsible for separation of the Gα subunit from the Gβδ dimer followed by release of both the subunits from the receptor. Both of them may bind to an effector (such as, calmodulin, phospholipases, and protein and lipid kinases), thus activating or deactivating the effector. The intrinsic GTPase activity locked in Gα subunit brings about hydrolysis of GTP, which deactivates Gα and allows the two subunits to recombine. As long as the GPCR itself remains in active state, it can turn on multiple G proteins and amplify the stream of signals to the effector molecules.

The G-protein system acts as a biochemical on/off switch. In the absence of any signal, three subunits (α, β, δ) of the G-protein associate to form an inactive complex.

(B) F-box proteins

Plant F-box receptors/ubiquitin ligase systems are essential for many plant hormone receptor complexes. Eukaryotic E3 ubiquitin ligase complexes (both in the cytosol and nucleus), covalently attach ubiquitin to substrate proteins and tag them to be degraded by the 26S proteasome system. In the SCF (Skp, Cullin, and F-box protein) subfamily of E3 ligases, substrate identification is mediated by F-box proteins.

(C) Receptor kinases

When a protein behaves as a receptor and transduces the signal through phosphorylation of another molecule, it is termed as **receptor kinase.** Based on the kind of receptor kinase, a target protein can undergo phosphorylation at several amino acid residues (serine, threonine, tyrosine, or histidine) to alter its biological activity. Receptor kinases, although limited, but play important role in plants. For example, BR11 receptor kinase for brassinosteroid hormones plays an important role in development (refer to Figure 16.24). There are numerous receptor-like serine/threonine kinases (RLKs) in plants, which play a prominent role in plant-pathogen interactions.

(D) Two-component systems

Plant receptors that perceive the presence of the hormones cytokinin and ethylene; are derived from bacterial 'two-component' systems (Figure 16.26). Such systems are comprised of two functional components: a membrane-bound **sensor histidine kinase** (the 'input domain' and the 'transmitter domain'), which senses signal, and a downstream **response regulator** (the 'receiver domain' and the 'output domain'), whose activity is controlled by the sensor histidine kinase through phosphorylation.

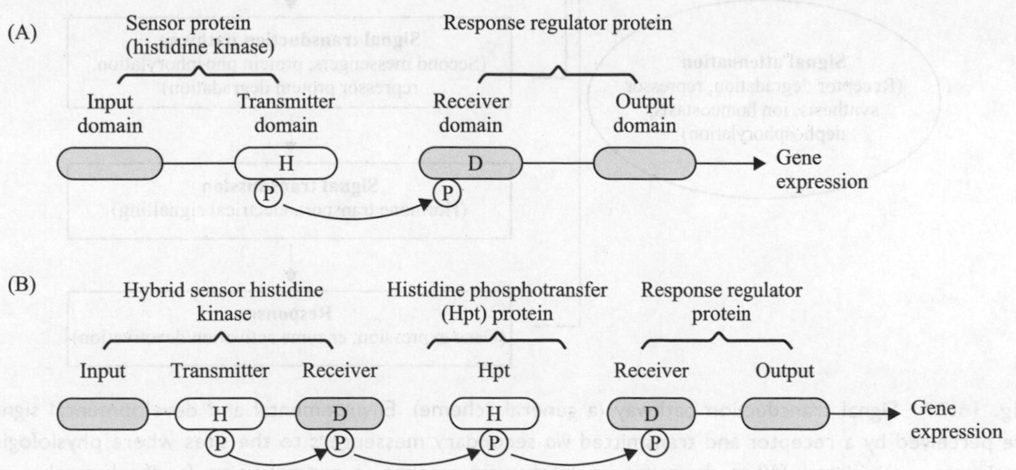

Fig. 16.26 Two-component signalling systems. (A) Simple two-component signalling system: It is found only in prokaryotes and consists of a sensor protein and a response regulator protein. (B) Phosphorelay two-component signalling system: It is found in both prokaryotes and eukaryotes and involves an intermediate histidine phosphotransfer protein (Hpt), called AHP in *Arabidopsis*. H, histidine; D, aspartate.

(E) Ion channels

Many plant receptors resemble those found in bacterial systems. For example, homologues of the bacterial mechano-sensitive ion channel are present in the plasma membrane and inner chloroplast membrane of plant cells. These ion channels act as stretch receptors (activated by stretching of cell membrane) and help cells and plastids adjust to osmotically-induced swelling.

16.2.2 Signal transduction

Signal perception is accompanied by a large number of biochemical events, referred to as **signal transduction or signalling.** All signal transduction pathways share common features. An internal stimulus is perceived by a receptor and transmitted *via* intermediate processes to the sites where physiological responses are initiated. When the response mechanism reaches an optimal point, feedback mechanisms attenuate the processes and reset the sensor mechanism (Figure 16.27).

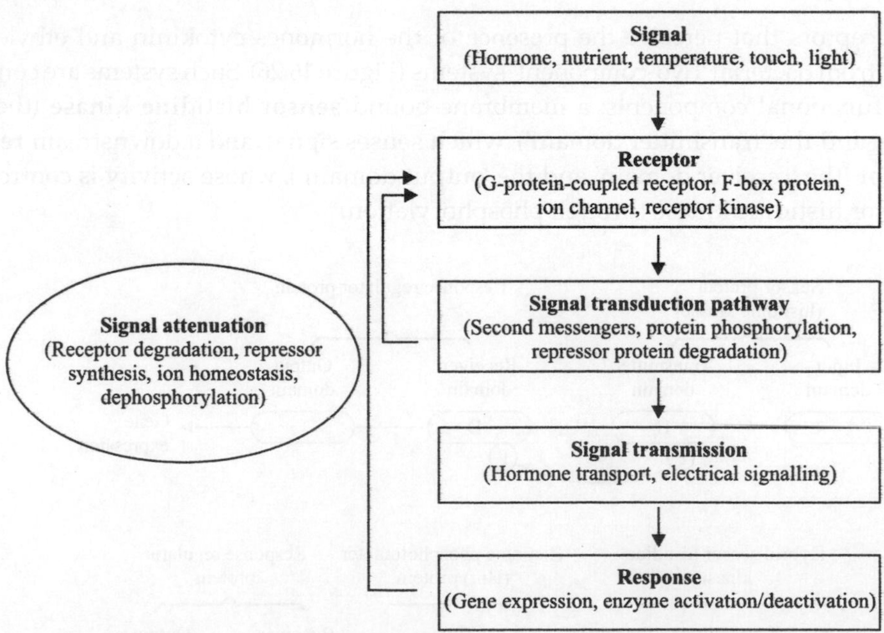

Fig. 16.27 Signal transduction pathway (a general scheme). Environmental and developmental signals are perceived by a receptor and transmitted *via* secondary messengers to the sites where physiological responses are initiated. When the response mechanism reaches an optimal point, feedback mechanisms (such as dephosphorylation, ion homeostasis, repressor synthesis, receptor degradation) attenuate the processes and reset the sensor mechanism.

Plant signal transduction mechanisms may be relatively fast e.g. immediate closing of leaflets of Touch-me-not plant (*Mimosa pudica)* upon being touched or extremely slow e.g. emission of volatile chemicals in plants attacked by insect herbivores to attract insect predators. Plant responses to environmental signals differ spatially. In a **cell autonomous response** to an environmental signal, both signal reception as well as response occur in the same cell such as opening of guard cells, where blue light activates membrane ion transporters to swell guard cells *via* the phototropin blue-light photoreceptors. In contrast, in a **non-cell autonomous response**, signal reception and the response occur in different cells such as formation of additional stomata when mature leaves are exposed to high light intensity in a process that requires transmission of information from one organ to another.

Signalling intermediates must be amplified intracellularly to prevent dilution of the signalling cascade and thus regulate their target molecules.

Protein kinase-based signalling

Protein kinases stimulate other proteins through regulation of their phosphorylation reactions and their action is balanced by phosphatases, which inactivate the proteins by eliminating the phosphate group. Protein kinases amplify weak signals and specific protein kinases act in series, thus creating a cascade. A signalling cascade composed of several kinases, is therefore able to regulate the activity of a large number of target proteins in response to relatively few ligand molecules originally binding to the receptor molecule localized in the plasma membrane. **The MAP (mitogen-activated protein) kinase cascade** has a key role to play in **signal amplification** in plants and other eukaryotic organisms (Figure 16.28). It is involved in many important signalling pathways, including those regulating hormones, abiotic stress, and defence responses.

Signals can also be amplified by **secondary messengers** i.e., **Ca^{2+}, cytosolic or cell wall pH, reactive oxygen species (ROS), modified lipids (phosphoglycerolipids and sphingolipids).** Ca^{2+} is the most ubiquitous second messenger in plants and other eukaryotes. Besides regulating various signalling pathways mentioned above, Ca^{2+} also regulates the signalling pathway associated with symbiotic interactions.

Secondary messengers are small molecules or ions that are readily produced or mobilized at relatively high levels after signal perception and can alter the activity of target signalling proteins, thus representing another strategy to enhance signals.

Phospholipid-based signalling

Phospholipids are an important class of secondary messengers, that are synthesized by the action of phospholipases. Four types of phospholipases (PL) are known: PL A_1, PL A_2, PL C, PL D and each of these enzymes brings about the hydrolysis of a particular bond in the phospholipid molecule.

Lipid-based signalling in plants by inositol triphosphate system has been illustrated in Figure 16.29 . The signal-receptor complex activates **phospholipase C** (PLC) involving G protein, which catalyzes the synthesis of two second messengers, inositol triphosphate (IP_3) and diacylglycerol (DAG), from the hydrolysis of membrane phospholipid phosphatidylinositol bisphosphate (PIP_2). IP_3 diffuses into the cytoplasm where it promotes the release of Ca^{2+} from intracellular stores *via* activation of calcium channels. This is immediately followed by phosphorylation of DAG to phosphatidic acid (PA), catalyzed by the plasma membrane-bound lipid kinase. Subsequently, PA diffuses into the cytosol where it might cause activation of ion channels, effector proteins, and enzyme systems.

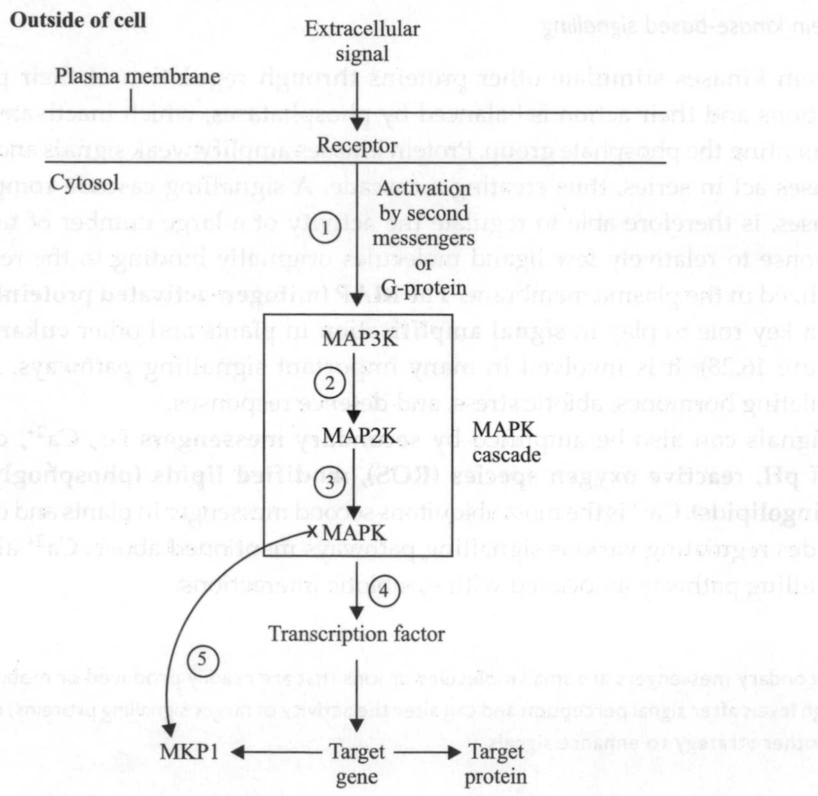

Fig. 16.28 The major components of a mitogen-activated protein kinase (MAPK) cascade. Extracellular signals are linked to transcription factor phosphorylation by MAPK cascade, which comprises of three kinase enzymes -MAP kinase, MAP kinase kinase (MAP2K), MAP kinase kinase kinase (MAP3K). (1) MAP3K can be activated directly by a signal-receptor complex/G-protein/second messenger. (2, 3) MAP3K activates MAP2K by phosphorylation, which further activates MAPK. (4) MAPK phosphorylates transcription factor that induces the response. (5) Expression of one of the genes MKP1, a protein phosphatase that can remove phosphate group from MAPK and block gene activation.

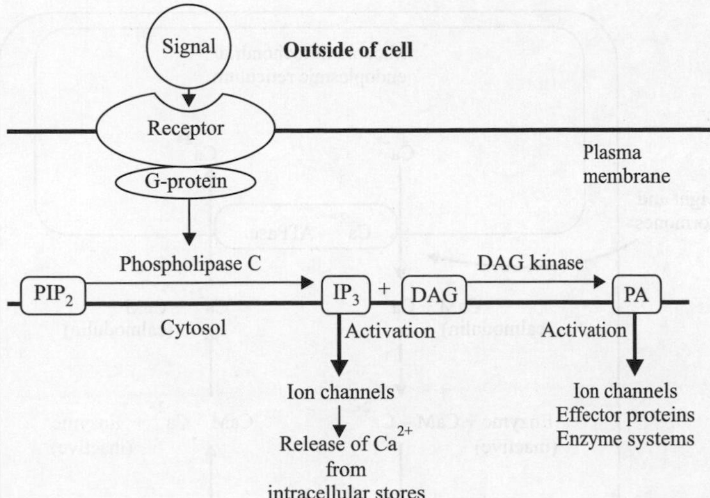

Fig. 16.29 Lipids as secondary messengers. The external signal activates plasma membrane- bound enzyme phospholipase C, which brings about hydrolysis of phosphatidylinositol bisphosphate (PIP_2) to form second messengers:-inositol triphosphate (IP_3) and diacylglycerol (DAG). IP_3 diffuses into the cytosol and activates calcium ion channel. DAG is immediately phosphorylated to phosphatidic acid (PA), which diffuses into cytosol and activates ion channels and effector proteins.

Calcium-based signalling

Calcium ions act as effective second messengers. The cytoplasmic calcium ion concentration is maintained at a low level because of membrane-bound Ca^{2+}-ATPases and both of them are influenced by stimuli such as light and hormones, etc. The chief Ca^{2+} receptor in plant and animal cells is **calmodulin**, a highly conserved, ubiquitous protein with a molecular mass 17–19 kD. Ca^{2+} present in cytosol combines with calmodulin to form an active complex, accompanied by conformational changes in calmodulin that helps in activating target proteins (Figure 16.30).

Transcriptional-based signalling

Some of the plant signals, such as phytochrome and hormones, may bypass plasma membrane receptors and directly intervene in gene expression. They diffuse into the nucleus and associate with specific **transcription factors**, which are small, DNA-binding proteins regulating the transcription of mRNA *via* attachment to specific regulatory DNA sequences in certain genes. They may either up-regulate or down-regulate the transcription of that particular gene. Some of the transcription factors possess two binding sites: one for DNA and other for a regulatory molecule. Binding of the signal molecule, such as a hormone, to the transcription factor initiates changes in the expression of the target gene.

6.12.3 Crosstalk among signal pathways

Signalling pathways are quite similar to interconnected web and established pathways are linked by multiple connections, a situation commonly referred to as **'crosstalk'**. Therefore,

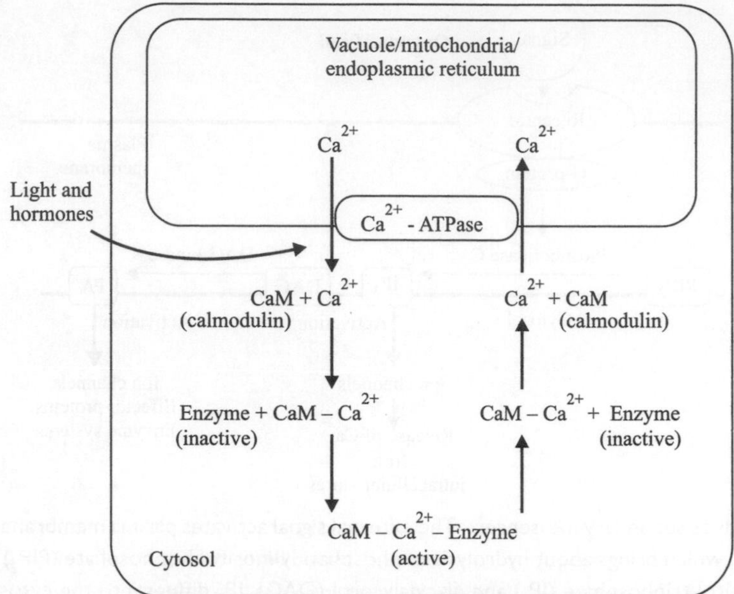

Fig. 16.30 Ca^{2+} ions function effectively as a second messenger. Stored intracellular Ca^{2+} diffuses into the cytosol and forms an active complex with calcium-binding protein, Calmodulin. Exchange of Ca^{2+} between the vacuolar sap and the cytoplasm is controlled by stimuli such as, light or hormones. Membrane bound Ca^{2+}-ATPase maintains low Ca^{2+} concentration in the cytosol.

biological crosstalk refers to processes in which one or more components of one signal transduction pathway affects another. Crosstalk includes interactions among different groups of signals, the branching and merging of transduction pathways, and common use of secondary messengers – all of them are involved in coordination of developmental signals. Various fast changing environmental signals must be integrated with a large number of control systems for execution of a finely tuned developmental program.

The maintenance of cellular homeostasis is the outcome of a complicated network of genetically controlled biochemical pathways. There seems to be overlapping among various signal transduction pathways associated with the physiological response of plants to all types of stress conditions, that inhibit photosynthesis directly or indirectly resulting in negative impact on plant survival and productivity.

Review Questions

RQ 16.1. Why is ethylene so widely used as a plant hormone in agriculture? Explain.

Ans. Ethylene is sprayed in the form of an aqueous solution of ethephon or 2-chloroethylphosphonic acid (known in the trade, as ethrel) and is quickly absorbed and translocated in plants. It gives off ethylene slowly by a chemical reaction. It has many commercial applications:

1. It hastens fruit ripening in banana, apple, kiwifruit, avocado and tomatoes, to name a few.

2. It helps in 'de-greening' citrus fruits.
3. It also synchronizes flowering and fruit set in some bromeliad, such as pineapples, making harvesting easier.
4. Used as a promoter of female sex expression in monoecious plants such as cucumber, squash and melon; It leads to a greater number of fruits per plant.
5. It accelerates abscission of fruits, and so may be ethylene is used to achieve thinning or fruit drop in cotton, cherries and walnuts.
6. Enhances latex flow for a longer period in para rubber trees (*Hevea brasiliensis*).

At home, ripening of unripe or immature banana fruits can be hastened by enclosing them tightly in a plastic bag so that the released ethylene gas is retained within the bag, and not allowed to diffuse. The dark flecks that appear on the banana peel are pockets of concentrated ethylene.

RQ 16.2. In which ways are the gibberellins important in the agricultural and horticultural industries? Explain.

Ans. Gibberellins are important in many ways in our agricultural and horticultural practices:

1. Gibberellins are often sprayed in very dilute concentration on vineyards to increase the number and size of grapes in a cluster of the 'Thompson' seedless variety. A single spray is often given in the early stage of development. causing the size to be doubled or tripled. However, the quality of the product remains unchanged.
2. In the brewing industry (malting barley), the gibberellins are added to barley seeds to promote even germination and to 'speed up' the conversion of starch to sugar. The early release of sugars from the endosperm tissue accelerates alcohol production. The entire process becomes more efficient, resulting in greater production of beer and hence greater revenues to the brewer.
3. Gibberellins, either alone or in conjunction with auxin, induce parthenocarpy (seedlessness) in apples. It also produces larger fruits.
4. Gibberellins are applied to celery to produce longer and crisper stalks.
5. They can break dormancy and can promote the germination of a wide variety of seeds, including 'Grand Rapids' variety of lettuce and can even substitute the red light.
6. Increased sugarcane yield because of 'elongation of sugarcane stalks' by GA. However, there is no change in sugar concentration.

The first gibberellin that was isolated and chemically characterized was named gibberellic acid (GA_3). More than 125 gibberellins are known to occur in different plants, including fungi. They differ slightly in their chemical structure but plants can easily recognize the various gibberellin molecules. They respond dramatically to one, and yet remain unaffected by another.

The two groups of scientists led by Stodola in Illinois (USA) and Brain at Imperial Chemical Industries (UK) identified the chemical nature of the hormone and named it gibberellic acid.

RQ 16.3. Cytokinins interact with IAA to control organ initiation in tissue culture studies. Comment.

Ans. The relative amounts of IAA and kinetin can determine whether a tissue culture remains undifferentiated (in the form of callus) or develop into structures like shoots or roots. Skoog and Miller in 1957 reported that high kinetin/auxin ratios promote formation of shoot buds in pith tissue culture (callus) of tobacco. However, when auxin concentration is higher than kinetin in the nutrient medium it will promote root formation. So therefore by maintaining a correct balance of IAA and kinetin in the nutrient medium, we can control organogenesis of 'either' roots or shoots, i.e., they can be tailored at will.

High cytokinin/auxin ratio causes bud initiation in some tissue cultures while low cytokinin/auxin ratios facilitate root initiation in callus culture.

Cytokinins are most abundant in growing fruits, seeds and roots. They are synthesized primarily in the root system and are then transported upward to the aerial parts through tracheary elements.

RQ 16.4. List the different measures that could be employed to increase the storage life of apples and other fruits.

Ans. Storage life of apples can be further increased dramatically by inhibiting ethylene biosynthesis through controlling atmospheric conditions such as:

1. Low temperature (just above freezing)
2. Low oxygen concentration
3. High CO_2 level or high ambient CO_2 concentration (3–5 per cent) to suppress respiration
4. High relative humidity
5. Continuous exchange of air to prevent a build-up of ethylene, i.e., ethylene is removed as soon as it is produced

Low pressure is applied 'to take away ethylene and oxygen from the storage chambers', thus slowing down the ripening process.

Richard Gane, in 1934, provided unflinching evidence that the volatile compound was ethylene, although a number of earlier reports published in the 1930s (Cousins, 1910) indicated that volatile products released from oranges could hasten the ripening process of other fruits stored close by!

RQ 16.5. Why is ethylene called the senescence hormone? How is ethylene produced?

Ans. Ethylene seems to be a prime controlling agent in many aspects of plant senescence such as:

1. The fading of flowers
2. The ripening of fruits
3. The abscission of leaves.

During senescence of flowers, fruits, and even leaves, the tonoplast or vacuolar membrane becomes slightly leaky so that methionine, the precursor of ethylene, leaks from the vacuole into the cytoplasm where it is converted to ethylene which causes degradative changes.

RQ 16.6. An orchid or any other flower remains fresh for a long time if it remains unpollinated. Why is it so?

Ans. If an orchid flower remains unpollinated it remains fresh for a long time. However, very soon after it is pollinated it starts to fade as pollination initiates the production of 'ethylene' which then causes the senescence of the flower petals.

Similarly, in morning glory (*Ipomoea tricolor*), the flowers open early in the morning, start to droop by mid-morning and by noon they form a curled-up mass of soggy petals. The curling of petals coincides with the onset of ethylene evolution which brings about turgor changes in the cells of the petal midribs which finally leads to curled-up mass of soggy petals.

RQ 16.7. Why are auxins so important in our agriculture, horticulture and for military use?

Ans. Auxins have been used commercially for various purposes such as:

1. To stimulate the production of adventitious roots in vegetatively propagated crops such as sugarcane (stem cuttings) or ornamental plants (leaf cuttings).
2. Induction of parthenocarpic fruits.
3. Induction of flowering in pineapples (actually caused by the auxin induced liberation of ethylene).
4. To promote the even growth of fruits in strawberry. If auxins are applied to 'deseeded' (seeds removed) young fruits, there will be normal growth and development. The developing seeds are a rich source of auxin and each seed is responsible for growth of its own receptacular region to produce normal accessory fruit.
5. Prevention of pre-harvest fruit drop until it can be harvested or prevent premature fall of fruits in citrus and apples.
6. Auxins prevent the sprouting of buds on potato tuber so that they can be stored for longer periods.
7. To increase branching or bushiness in ornamental plants by removing apical buds to eliminate the effect of apical dominance. Gardeners produce thick and impenetrable hedges by periodic pruning (removal of apical shoot tip).
8. Control of leaf and fruit abscission.
9. Herbicides have come into widespread use in agriculture and horticulture as weedicides to kill 'broad-leaved weeds' in cereal grain crops and in lawns, thus virtually eliminating the need for weeding. They have only limited toxicity to plants of the grass family but are extremely effective against most broad-leaved plants.

> Synthetic auxins are used as powerful selective herbicides. Agent orange (code name), a mixture of 2,4-D and the closely related 2,4,5-trichlorophenoxyacetic acid (2, 4, 5-T), was used to defoliate the jungle vegetation during the Vietnam war.

RQ 16.8. Why are potato tubers cut into sections before using as 'seeding material'?

Ans. Potato tubers, if used whole, will produce 'only one' plant from the apical bud. As long as the apical bud is intact, it will prevent the sprouting of other lateral buds through apical dominance. However, when the potato tuber is cut into sections (all the sections still attached), it will produce as many plants as are the number of sections because the effect of apical dominance is eliminated.

> Gardeners produce thick and impenetrable hedges by periodic pruning. This practice enables the lateral buds to produce new branches.

RQ 16.9. In which ways has the phenomenon of apical dominance been exploited in our agricultural practices?

Ans. Apical dominance has many or varied uses in agriculture:

1. Harvesting young tender shoots in tea (*Camellia sinensis*) encourages the development of lateral buds into new tender shoots. These can be again harvested. By using this practice, the growers can have 3 to 4 'flushes' of young shoots in a year.
2. By periodic pruning, growers can have a balanced framework of stem branches, as in coffee. This encourages more production of flowers and fruits.
3. Hedges are pruned, and lawns are mowed periodically to have thick impenetrable fences and a dense carpet of grass, respectively.

In such practices such as 'plucking' in tea or 'pruning' in coffee bushes or hedges in and around the lawn, the effect of apical dominance is eliminated. The apical buds, when removed, would stimulate the growth of lateral buds.

RQ 16.10. List the immediate precursor of IAA, gibberellins, ethylene, cytokinin and brassinosteroids.

Ans. The immediate precursor of the following is:

IAA	Tryptophan
Gibberellins	Mevalonate
Ethylene	Methionine
Cytokinin	Adenine derivatives or mevalonates
Brassinosteroids	Sterols (campesterol)

Jasmonic and salicylic acids are two other putative hormones important in plant defence mechanism.

RQ 16.11. Why synthetic cytokinins are applied to some of the harvested leafy vegetables?

Ans. Synthetic cytokinins are used as spray on leafy vegetables like celery, broccoli and other leafy vegetables to prolong further their shelf-life as they prevent (i) breakdown of chlorophyll, responsible for 'greenness' and (ii) degradation of proteins, nucleic acids, and other macromolecules. Hence, these vegetables retain their freshness and quality over a long time, thus fetching premium prices in the market.

The physiological reasoning is that cytokinins protect membranes against degradation, thus maintaining their integrity, otherwise the hydrolytic enzymes such as proteases, from the vacuole (surrounded by tonoplast) would leak into the cytoplasm and hydrolyse soluble proteins, and proteins of chloroplast and mitochondrial membranes.

Since synthetic cytokinins slow down ageing processes or senescence, they are also employed to enhance the vase-life of cut flowers, e.g., carnations, roses, and such. Cytokinin treatment extends the lifespan of detached leaves by delaying chlorophyll and protein degradation (Richmond and Lang, 1957).

RQ 16.12. Why is it so, that one rotten apple in a barrel can spoil the entire lot?

Ans. The one rotten apple produces a 'volatile gaseous' hormone, ethylene (C_2H_4) which causes degradative changes in the healthy fruits nearby. This leads to their over-ripening or spoilage. They in turn begin to produce ethylene, which affects still other fruits close-by. Thus, a chain reaction continues and more or more ethylene gas is evolved or gets accumulated, leading to spoilage of the entire lot.

This can be overcome by enriching the air with CO_2 during storage.

For this reason, many of the fruits such as tomatoes, bananas, oranges, mangoes, and such like, are shipped in the green stage, in well-aerated crates to stop building up of ethylene but are then exposed to ethylene before reaching the market.

RQ 16.13. Expand the following abbreviations.

Ans.

CCC* (cycocel)	2-Chloroethyl-trimethyl ammonium chloride (cycocel)
DCMU	Dichlorophenyl dimethyl urea
TIBA	2,3,5-Triiodobenzoic acid
2,4,-D	2,4-Dichlorophenoxyacetic acid
2,4,5-T	2,4,5-Trichlorophenoxyacetic acid
IBA	Indole-3-butyric acid
IAA	Indole-3-acetic acid
NAA	Naphthalene acetic acid
2,4-DNP	2,4-Dinitrophenol
IPA	Indole-3-propionic acid
	Indole-3-pyruvic acid
ABA	Abscisic acid
Phosfon-D	2,4-Dichlorobenzyl-tributyl phosphonium chloride

*2-Chloroethyl-trimethyl ammonium chloride (CCC) stops the synthesis of gibberellins (anti-gibberellin).

RQ 16.14. List out examples of: (a) naturally occurring and (b) synthetic inhibitors.

Ans. A large group of chemicals have been discovered that possess inhibitory action. They are categorized as follows:

Naturally occurring	Synthetic
Ethylene	Morphactins
Abscisic acid	Chlorocholine chloride (CCC)
Phenolic acid	AMO 1618*
Phenolic lactones	Maleic hydrazide (MH)

*Ammonium (5-hydroxycarvacryl) trimethyl chloride, piperidine carboxylate.

RQ 16.15. Some fruits like bananas and mangoes are harvested in the green stage and then transported. How is the ripening done by small fruit merchants?

Ans. Fruits like mangoes, bananas, papayas, apples and plums, and so on, are picked up green or in unripened state and are then transported to other areas. There they are subjected to acetylene or ethylene in totally sealed warehouses by placing calcium carbide (CaC_2) or 'masala' in small plastic containers that absorb the moisture in the air to release acetylene (or ethylene) – the ripening hormone that is normally produced by fruits to accelerate the ripening process.

However, the industrial-grade CaC_2 contains arsenic and phosphorus, and its use for ripening fruits is prohibited in many countries under the 'Prevention of Food Adulteration Rules' as it can cause cancer. It also leads to mouth ulceration, gastric irritation, food poisoning and other problems. Some fruit-sellers also immerse fruits in ethephon solution or expose the fruits to gaseous ethylene to accelerate the process of ripening.

The government should formulate stringent measures to punish offenders who are using chemicals for artificially ripening fruits.

RQ 16.16. Name five groups of plant hormones along with their precursors and sites of production.

Ans.

Auxin	Precursor	Site of production
IAA*	Tryptophan	Growing tips of stem.
Cytokinins	Adenine derivatives, and from mevalonate	Primarily in the roots, transported up to the shoots *via* the xylem; abundant in embryos and young fruits.
Gibberellins	Mevalonate	Young leaves near growing tips.
Ethylene	Methionine	Ripening fruits.
ABA	C-Mevalonate	Leaves and roots-transported up through xylem.

*A chlorinated analogue of IAA is known as 4-chloro IAA or 4-chloroindoleacetic acid.

RQ 16.17. List examples of: growth hormones that have antagonistic action.

Ans.
1. Cytokinins are known to promote stomatal opening whereas ABA causes the stomata to close.
2. Cytokinins delay senescence while ethylene triggers senescence.
3. Inhibition of bud growth by ABA can be prevented by simultaneous treatment with gibberellic acid.
4. GA application promotes the induction of α-amylase synthesis while ABA has the opposite effect, i.e., suppresses the expression or induction of α-amylase.
5. Auxins prevent abscission while ethylene triggers abscission i.e., shedding of leaves, flowers and fruits.
6. IAA produced in the apical meristem inhibits the development of axillary buds (a process called apical dominance) while cytokinins counteract the suppression of growth of lateral buds, i.e., they promote bud growth.
7. Exogenous application of cytokinins promotes the development of lateral buds even if the terminal bud is still intact.

RQ 16.18. Millions of plants in a wheat field die at approximately the same time, following the formation of seeds. Explain why?

Ans. Reproduction seems to trigger a series of events that lead to senescence and finally to the death of the plants. During senescence, metabolic activities fall rapidly, reserves are translocated out, and colour changes frequently. It appears that the carbohydrates and proteins of leaves are broken down into sugars and amino acids for onward mobilization as reserve of the seeds (or tubers and rhizomes). Their transport is supposed to be mediated by auxins. Senescence of plants, once initiated as a result of imbalance of growth hormones and nutrients, results in lack of gene activity and enzyme synthesis, and finally autophagy and eventual autolysis.

The final stage is death of the leaf or in the case of annuals or biennials, death of the whole plant. Dead tissue falls to the ground where they undergo decay through microbial action, producing inorganic substances that are cycled back.

RQ 16.19. What is bioassay? Name the plant materials that are principally used for bioassay studies for auxins, gibberellins and cytokinins.

Ans. A Bioassay is the quantitative measurement of biologically active substances, reflected in the amount of their activity under standardized conditions on living organisms or live 'tissue for the known response'. The following materials are chiefly used for bioassay studies:

1. Auxins	(a) Avena coleoptile curvature test (seeds of oat, *Avena sativa*)
2. Gibberellins	(b) (i) Lettuce hypocotyl test (seeds of lettuce, *Lactuca sativa*) (ii) Induction of α-amylase enzyme in barley (*Hordeum vulgare*) aleurone cells by gibberellins
3. Cytokinins	(c) Cotyledon expansion test (seeds of cucumber)

RQ 16.20. What do you mean by the term 'de-novo synthesis'? Give a suitable example to explain the phenomenon.

Ans. 'De-Novo synthesis' of enzyme α-amylase during seed germination of barley is the classical example. Gibberellic acid induces 'de-novo' (a new) synthesis of α-amylase. This has been shown by 'isolating' the 'aleurone layer' of barley seeds and treating it with a dilute solution of gibberellins (Varner and his associates, 1967). Normally gibberellins, produced by the embryo, stimulate the cells of the aleurone layer in the grains of cereals to synthesize α-amylase, proteases and other hydrolases.

RQ 16.21. Discuss briefly the contribution of scientists who played a pioneering role in the discovery and characterization of gibberellins.

Ans. Gibberellins were discovered by a Japanese plant pathologist Kurosawa in 1926 while working on the rice fields where some seedlings grew much taller (foolishly tall) than others – a disease known as 'bakanae disease'. The causal fungus infecting the rice seedlings was named *Gibberella fujikuroi.*

Yabuta (1935) and Yabuta and Sumiki (1938) demonstrated that some substances secreted by the fungus were responsible for this 'spindly' growth. In 1938, they isolated two crystalline substances named gibberellin A and B from the fungus culture filtrate. Today more than 125 different types of gibberellins are known (designated as GA_1 to GA_{125}). The purified chemical compound was named gibberellic acid (Curtis and Cross, 1954).

RQ 16.22. Match the following in column I and II.

Plant hormone	Response
(i) ABA	A. Gaseous or volatile hormone
(ii) Ethylene	B. Stress hormone
(iii) Cytokinin	C. Anti-senescence hormone (delay senescence)
(iv) Ethylene	D. Bolting in rosette plant
(v) Gibberellins	E. The senescence hormone

Ans. (i) B (ii) A (iii) C (iv) E (v) D

RQ 16.23. Why is ABA regarded as a stress hormone?

Ans. Under water stress or water-deficient conditions (such as drought, saline or freezing), the roots respond by enhancing the synthesis of ABA and releasing it into tracheary elements from where it is transported quickly to the leaves. ABA induces stomatal closure in most plant species, hence preventing the loss of water by transpiration.

> As a term, the name abscisic acid is a 'misnomer', as this chemical has nothing much to do with abscission. It induces bud and seed dormancy. However, ABA stimulates the production of seed storage proteins.

RQ 16.24. How is the auxin level in plants prevented from accumulating to excessive levels?

Ans. As the auxin is being continuously produced in plant system, its concentration may become very high or supra-optimal, and that, can have adverse effects. Hence, the plants have developed mechanisms to detoxify or inactivate auxin by either of the three methods.

1. IAA is degraded by light (photo-oxidation). Blue light is most active in photo-oxidation of IAA and is absorbed by two pigments riboflavin and violaxanthin. UV light, X- and gamma-radiations also inactivate IAA.

2. IAA may be oxidized by the enzyme IAA oxidase.

3. IAA is converted to an inactive form by conjugating (or attaching) with various compounds such as aspartic or glutamic acid (forming the peptide derivatives), sugars (glucose and arabinose), polysaccharides and proteins. Auxin glycosyl esters, abundant in seeds, are prime examples of conjugated auxins that are inactive until IAA is released by enzymolysis.

In the conjugated form, IAA is safe from destruction, it can be stored indefinitely in seeds, and it can be transported from cotyledons to the epicotyl during germination. Most of the IAA (up to 80 per cent) in oat seeds is conjugated but this can be disconjugated, releasing free IAA during germination more quickly than its synthesis could.

> Two major ways for IAA destruction are removal of the side group (acetic acid side chain) and oxidation of the five-member indole ring.

RQ 16.25. In which ways are growth retardants commercially useful? Give a few examples.

Ans. Growth retardants or anti-gibberellins are given to potted plants either as a foliar spray or 'soil drench' with the aim of reducing stem elongation that results in shortened and compact plants with dark green leaves. Commercial nurseries can thus produce shorter and compact 'poinsettias', lilies, chrysanthemums and other ornamental plants.

In certain parts of the world, wheat crop undergoes lodging at or near the harvesting time as the plants top become heavy with grains, and fall flat. This can be corrected by spraying plants with anti-gibberellins to produce shorter and stiffer stems, thus preventing lodging. Anti-gibberellins or gibberellin antagonists also have been employed to minimize the need for pruning practices of vegetational cover under power transmission lines.

Examples of synthetic chemicals are AMO-1618, Cycocel (or CCC), B-Nine[9], Phosphon-D and Ancymidol (commercially sold as A-REST). Growth retardants thus imitate the properties of the dwarfing genes, by arresting reactions involved in gibberellin synthesis. Treatment of wheat

[9] B-Nine = N-(Dimethylamino) succinamic acid.

seedlings with Chlormequat or CCC causes short and stiff straw formation, similar to those of Mexican wheat.

RQ 16.26. List four synthetic auxins.

Ans. A number of synthetic compounds having 'auxin-like' activities are known and these do not occur in plants. For example, indole butyric acid (IBA), naphthalene acetic acid (NAA), 2,4-dichlorophenoxyacetic acid (2,4,-D), 2,4,5-trichlorophenoxyacetic acid (2,4,5,-T), and parachlorophenoxyacetic acid (PCPA).

The synthetic auxins are more powerful as compared to the naturally occurring IAA. They are more stable, unlike IAA which is inactivated in the plants once it has accomplished its purpose. IAA can be stored as chemical conjugates such as in the form of glycosyl esters which would release active IAA by the activity of the enzymes.

RQ 16.27. Outline briefly the method to detect and measure the quantity of hormone in the oat coleoptile.

Ans. A 3-day-old seedling of oat is decapitated at the tip and the leaf inside is pulled up. The tip is placed on agar block for few hours. An agar block containing auxin is kept eccentrically (on one side) of the cut stump. Auxin from the agar block is transported lower down the coleoptile and stimulates that side to grow more quickly as compared to the opposite side, thus resulting in curvature. The curvature is measured after 90 min.

Hormone concentration is measured by the degree of bending, by comparing with the degree of angle produced by known IAA concentration (refer to Figure 16.2). The greater the angle, the more is the hormone concentration.

Chapter 17

Ripening, Senescence and Cell Death

Keywords

- Climacteric
- Fibrillins
- Gerontoplasts
- Lipooxygenase
- Lycopene
- Metallothioneins
- Non-climacteric
- Phytoene
- Plastoglobules
- Stay-green mutants

17.1 Fruit Ripening

Fruit ripening is considered to be a terminal phase of development rather than a degenerative process. Fruits are classified into **climacteric** (exhibiting a ripening-associated respiratory burst, often accompanying a spike in ethylene production) or **non-climacteric**. Mature climacteric fruits can be ripened after detaching from the parent plant, a process that can be hastened by exposure to ethylene.

17.1.1 Morphological changes during ripening

(i) Change in fruit color

Fruits such as banana, pepper, tomato and citrus change color during ripening and become brightly colored. When immature, these fruits are green and catabolize chlorophyll on ripening by the identical pathway that operates in senescing leaves.

- The loss of chlorophyll uncovers underlying brightly colored, hydrophobic carotenoids located in chromoplasts. Carotenoids form fibrils, crystals or globules and are associated with specific proteins, **fibrillins.** Fibrillin genes are highly expressed at the stage of fruit ripening, leaf senescence and during the development of floral organs, as well as in response to several types of abiotic stresses. New carotenoids are synthesized *via* the **isoprenoid pathway**. The red carotenoid of tomato and bell pepper is **lycopene** The enzymes involved in lycopene synthesis, become significantly more active during *Capsicum* ripening (Figure 17.1) .

Climacteric fruits: Examples:- Apple, Apricot, Banana, Guava, Kiwifruit, Mango, Papaya, Passion fruit, Peach, Pear, Persimmon, Plum, Sapodilla, Tomato
Non-climacteric fruits: Examples:- Cherry, Cucumber, Grape, Grapefruit, Lemon, Lime, Litchi, Melon, Orange, Pineapple, Pomegranate, Raspberry, Strawberry

Geranylgeranyl diphosphate (2 molecules) $\xrightarrow{\text{Phytoene synthase}}$ Phytoene

Phytoene $\xrightarrow[\text{Zeta-carotene desaturase}]{\text{Phytoene desaturase}}$ Lycopene

Fig. 17.1 Ripening of *Capsicum* fruits. Immature fruits are green and the red carotenoid pigment that appears during the ripening process is lycopene which is synthesized from geranylgeranyl diphosphate, with phytoene as the intermediate.

- Another source of fruit color is the pigmented water-soluble products of **phenylpropanoid metabolism** which accumulate in the central vacuole. Phenylpropanoid pathways originate with the amino acid phenylalanine and are complexed, branched, metabolic sequences with several control points that lead to biosynthesis of a diverse group of phytochemicals, including phenolics, tannins and flavonoids. Fruits like red grapes, cherries, red apples, black currants and strawberries are rich in anthocyanins. The development of anthocyanin coloration in fruits is the by-product of coordinated genes expression for flavonoid biosynthesis under the control of specific **MYB transcription factors**. Red and purple anthocyanins and yellow flavonoids are also responsible for the striking pigments of autumnal foliage in maple trees.

(ii) Change in fruit texture

Fleshy fruits become softer during ripening because of the chemical changes occurring in their cell walls which result in the loss of firmness. Hydrolytic enzymes become active and act on cell wall carbohydrates, de-esterify and depolymerise complex polysaccharides, thereby altering the physical qualities of the wall matrix as well as loosening cell-cell adhesion in soft fruits such as tomato and peach. **Pectins,** that make up more than 50 per cent of the fruit wall, consist of **homogalacturonan** (HG) and **rhamnogalacturonan (RG)** and are esterified with higher molecular weight and natural sugar content in unripe fruits. Although the activity of enzymes **pectin methylesterase (PME)** and **polygalacturonase (PG)** increases during ripening, still, their induction during ripening is not the main determinant of tissue softening. These enzymes work cooperatively with a number of cell wall-modifying enzymes (**xyloglucan endotransglycosylase**, **endo-1,4-β-glucanases** and **expansins**) to bring about fruit softening. The wall-modifying enzymes are believed to restructure the wall by altering the cross-linking network.

(iii) Change in fruit flavours and fragrances

Fruits become sweet during ripening through the import of sugars (sucrose) from the photosynthetic tissues and hydrolysis of starch. **Apoplastic invertase** has a critical role in regulating sugar composition and fruit size in transgenic tomatoes. Variation in sweetness indicates the relative proportions of sucrose and hexoses, which in turn is a consequence of differences in the relative activities of invertase and the sugar-hydrolyzing enzyme, sucrose synthase. The activities of **sucrose synthase, fructokinase** and **ADP-glucose pyrophosphorylase** control the rate of starch accumulation. Enzymes of starch degradation are likely to participate in starch hydrolysis, and starch phosphorylase and amylase have been observed to increase in banana pulp during ripening.

Sourness of the fruits is determined predominantly by organic acids (citric, malic, tartaric) and amino acids (aspartate and glutamate). Astringency is largely determined by **phenolic compounds,** especially condensed tannins or **proanthocyanidins**. Phenolics and pigments (anthocyanins) share a common metabolic origin in the **phenylpropanoid pathway,** and this is often reflected in coordinated regulation of taste and color during fruit ripening.

Increased fragrance of the fruits during ripening process is because of increased production of low molecular volatile organic compounds (such as esters, alcohols, aldehydes and ketones). The biochemical origins of volatiles are diverse and trace back to the pathways of amino acid, fatty acid and carotenoid metabolism. A number of volatiles are oxidation products of lipids' precursors of some aroma compounds, originating in reactions catalyzed by **lipoxygenase.** Genetic or chemical treatments aimed at suppressing ethylene production in climacteric fruits often result in decreased production of aroma compounds.

Ripe fruits are important dietary sources of antioxidants and vitamins, such as carotenoids, vitamin A and folic acid. The biosynthesis and regulation of these nutritionally important compounds in ripening fruits are areas of active biotechnological research.

Fruit Ripening and Climacteric Fruits

Once the fruit (a ripened ovary) has attained its 'maximum size', subtle chemical and physiological changes begin which ultimately cause it to ripen and make it edible. Ripening in certain fruits is closely accompanied by marked increase in the rate of respiration for a short period, followed again by a sharp decline. This period of greatly increased CO_2 output is called '**climacteric**' which signals the onset of irreversible process of degeneration or senescence.

A little ethylene is actually present at all times but this amount increases dramatically, nearly a hundred-fold at the time of climacteric. Following this change, many mature fruits, like a fully grown apple, undergo ripening during which (1) there is disappearance of much of the malic acid (that imparts acidic taste to the 'hard' unripe fruit); (2) complex polysaccharides are degraded into simple sugars; (3) cell walls turn soft because of hydrolytic action of some enzyme such as cellulase and pectinase; (4) chlorophylls are broken down; (5) development of pigmentation; and finally (6) the volatile compounds responsible for flavour and scent are formed. Apple, bananas, avocados, mango and tomatoes are a few examples of climacteric fruits. However, it is possible to keep unripe fruits for extended period during storage by removing the ethylene from the fruits as fast as it is formed by proper ventilation. Ripening can also be prevented by the use of inhibitors of respiration; by high CO_2 or nitrogen concentration or by low temperatures.

Application of ethylene, by contrast, promotes both the climacteric riseand ripening in many fruits like banana, lemons and oranges as well (Figure 17.2) . Tomatoes are generally harvested green and then artificially subjected to ripening by the ethylene treatment. It cannot be applied to the fruits without enclosing them in a chamber. However, a chemical called **Ethrel** (2-chloroethanephosphonic acid) is available in a liquid form and it can be directly sprayed on the plant and after reaching the plant cells, it breaks down releasing ethylene.

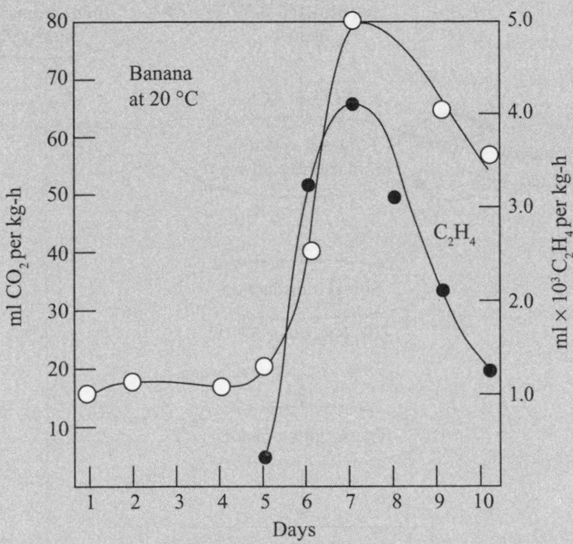

Fig. 17.2 The production of ethylene gas by ripening banana fruit – a climacteric fruit that also produces a sudden spike in the level of CO_2.

17.1.2 Fruit ripening through genetic and hormonal regulation

Based on the genetic and physiological characteristics of mutations affecting tomato fruits, it is possible to arrange genes and processes into a model of ripening that both accounts for experimental observations and defines targets for practical intervention to improve crop yield and post-harvest quality. Mutations in a large number of different genes lead to altered ripening in tomatoes. It is believed that **NOR** (non-ripening), **RIN** (ripening inhibitor) and **CNR** (colorless non-ripening) genes are global regulators that function in the mechanism that directly triggers the ripening syndrome independently of ethylene (Figure 17.3).

Transcription factors encoded by RIN, CNR and NOR genes regulate ripening by activating climacteric ethylene synthesis and are also likely to participate in an ethylene-independent (non-climacteric) pathway. RIN encodes a **MADS-box factor** that influences differential genes expression coding for ethylene biosynthetic enzymes ACC synthase (ACS) and ACC oxidase (ACO) before and after the climacteric. Ethylene biosynthesis is sustained by autocatalysis. **NR** (never-ripe), **GR** (green-ripe) and other ethylene receptor proteins (**ETRs**) promote a cascade of gene activation that initiates and sustains ripening. Light regulates ripening through the activities of **HP** (high-pigment) proteins. Carotenoid accumulation is controlled by UV light, mediated by the gene **HP**. Cryptochrome also functions in light regulation of ripening.

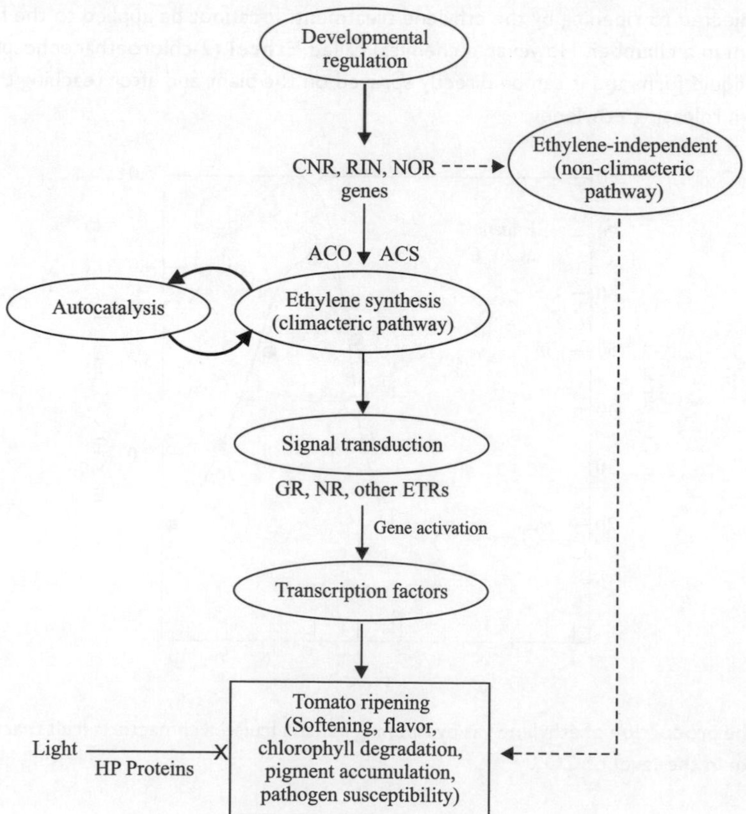

Fig. 17.3 Model explaining the genetic regulation of ripening in tomatoes. ACC-(1-aminocyclopropane-1-carboxylic acid).

17.2 Plant Senescence

Senescence is an energy-dependent, autolytic process that is impacted by the interaction of environmental factors along with genetically regulated developmental programs.

Abscission refers to the detachment of cellular layers that are present at the bases of leaves, floral organs as well as fruits, which permits them to fall easily without inflicting any damage to the plant. In the case of abscission, the region along which cell-cell adhesion will be dissolved; is usually in a predetermined position called the **abscission zone**. Abscission is stimulated by ethylene and inhibited by auxin. Formation of the abscission zone is responsive to stress, abscisic acid and the degree of senescence of the organ to be shed. Under the control of abscission-related transcription factors and protein kinases, genes encoding cell wall-modifying enzymes, particularly cellulase, polygalacturonase, peroxidase and expansin, are turned on in the abscission zone. Stress response factors are also up-regulated, including metallothioneins and pathogenesis-related proteins, which function in defending the exposed surface from attack by pathogens and other environmental variables.

Necrosis represents the death of tissues because of physical damage, poisons (such as herbicides) or other such external agents.

There are three types of plant senescence, based on the level of structural organization of the senescing unit:

(i) Programmed cell death
(ii) Whole plant senescence
(iii) Organ senescence

17.2.1 Programmed cell death

Programmed cell death (PCD) is essentially a feature of normal plant development. PCD is a general term which refers to an active, energy-requiring, genetically determined phase in which metabolism proceeds in an orderly fashion. Most forms of PCD are autolytic i.e. the cell contents are eliminated by autodigestion, which is catalyzed by proteases, nucleases and other lytic enzymes. PCD plays important role in **xylogenesis** (formation of tracheary elements), formation of secretory glands and canals, differentiation of protective surface structures (spines, thorns), generation of perforations and indentations, and formation of aerenchyma in the stems and roots of aquatic plants and development of unisexual flowers.

17.2.1.1 Development and survival of plants

Normal PCD is the general term applied to any process by which plant protoplasm (or sometimes the associated cell wall) is selectively eliminated as part of a genetically determined action. Almost every phase in the life cycle of a plant, from germination through to the development of the next generation of seeds, is influenced in some way by PCD. Responses to various abiotic (such as water, temperature or nutrient deficiency) and biotic stresses (interaction with pathogens, pests, herbivores, pollinators and such like) also often invoke the process of programmed cell death in certain cases. The most extensively studied form of PCD in animals is called **apoptosis**, which is a highly regulated process under the control of specific well-characterized genes and executed *via* defined biochemical pathways. In many cases, plant PCD appears to involve different genes and pathways from apoptosis in animals.

17.2.1.2 Cell viability is maintained during the developmental program leading to death

Two phases of PCD

(i) ***Phase1:*** This phase is reversible. Leaves remain turgid, cells are viable, their organelles and membranes stay intact. Leaf mesophyll cells remain viable until all of the leaf's macromolecules have been recycled and exported to the rest of the plant.

(ii) ***Phase 2:*** This phase is quite rapid, terminal one and is always accompanied by irreversible loss of cell viability.

17.2.1.3 Autolysis: a common form of cell death

An important general characteristic of many kinds of PCD in plants is that they are autolytic i.e. the cell digests itself by lytic enzymes, disposing of cell contents by processes that act on and through the cytoplasm that is being eliminated. Hydrolytic enzymes, including peptidases and nucleases, confined to central vacuole assist in autolysis through an autophagy-type engulfing mechanism or by rupture of vacuolar membrane, thereby releasing their contents into the cytosol. In PCD, endopeptidases (proteases that cut internal peptide bonds in their substrates) play a crucial role. **The ubiquitin-proteasome system** is also important in regulating cell functions through selective protein turnover and the removal of unwanted polypeptides. Chemical inhibitors of proteases block the operation of autophagic pathways.

Autophagy is a particular mechanism of autolysis shared by plants, yeast and other eukaryotic cells. A network of regulatory kinases, responsive to triggers such as nutrient starvation or developmental signals regulate the expression of **autophagy genes (ATGs)**, the activities of ATG proteins, assembly of autophagic vesicles that engulf parts of cytoplasm and the fusion of vesicles with the central vacuole where hydrolysis takes place.

17.2.1.4 Cell death during growth and morphogenesis

PCD contributes to several morphogenetic processes. Some of them are discussed below:

(i) ***Formation of vascular and mechanical tissues:*** PCD is an integral part of **tracheary element (TE)** development (Figure 17.4). **Xylogenesis** (formation of tracheary elements) begins during embryo development and continues throughout the life cycle of the plant. During the final phases of differentiation, tracheary tissues undergo thickening of the secondary cell wall, followed by rupturing of the tonoplast, and disintegration of remaining organelles by autolysis. Finally, all that remains is the cell wall with its characteristic secondary thickenings creating the continuous pipework of water-conducting vascular tissue. An initial dedifferentiation stage is followed by a phase of commitment to redifferentiation, during which there is upregulation of **TE differentiation genes (TEDs)**, including those coding for enymes catalyzing macromolecule degradation and cytoplasmic lysis.The *Zinnia* system has proven to be a good model for xylogenesis in *Arabidopsis* and wood formation in *Populus* trees.

(ii) ***Formation of ducts, glands and channels in plant tissues***: Plant shapes and adaptations for various purposes can be attained by programmed cell death and shedding of parts. For example, *Citrus* peel bears oil glands that arise by a combination of **lysigeny** (loss of cell contents) and **schizogeny** (cell separation). Mucilaginous canals found in the bud scales of *Tilia cordata* arise exclusively by lysigenous cell death. The resinous secretory ducts or channels in the phloem tissues in members of the family Anacardiaceae develop schizogenously. Abscission of plant organs such as, leaves and flowers, and dehiscence of anthers as well as dry fruits are all cell separation events, which result as a consequence of degradation of the middle lamellae of cell walls under the positive and negative regulation of ethylene and auxin, respectively.

(iii) Selective death of cells and tissues: Selective cell death also controls other aspects of vegetative development in plants. Localized PCD occurs during the differentiation of prickles and thorns, and also leads to the development of holes in leaves of swiss cheese plant, *Monstera and Aponogeton*, the aquatic lace plant.

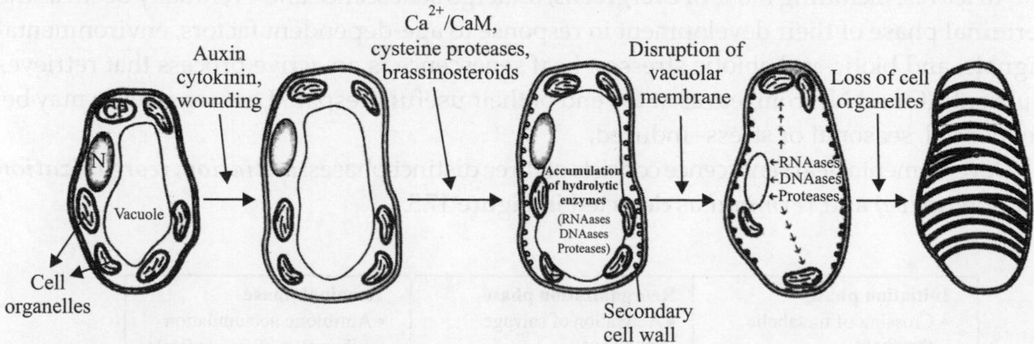

Fig. 17.4 Programmed cell death depicting various stages during differentiation of tracheary elements in mesophyll cells. Major events, such as thickening of secondary cell wall, storage of hydrolytic enzymes within vacuoles accompanied by rupturing of its membrane and loss of cellular contents, are regulated by hormones, calcium ions and wounding.

PCD processes contribute to all stages of reproductive development. Selective death of reproductive structures occurs during the development of unisexual flowers. Petals and sepals undergo senescence. Specific cells undergo senescence and death during gamete and embryo formation. Programmed senescence and death occur during seed development and germination also.

17.2.2 Whole plant senescence

Whole plant senescence covers the death of the entire plant. In general, annuals and biennials reproduce only once before senescing, while perennials can reproduce multiple times before senescing. The whole plant senescence represents an accelerated form of aging where tissues are programmed to fail quickly once certain thresholds are reached. Nutrient or hormonal redistribution from vegetative structures to reproductive sinks may trigger whole plant senescence in monocarpic plants.

17.2.3 Organ senescence

Organ senescence (the senescence of whole leaves, branches, flowers or fruits) occurs at various stages of vegetative and reproductive development including abscission of the organ undergoing senescence.

17.2.3.1 Leaf senescence

Leaf senescence is a vulnerable phase in plant life cycle, when food resources that are being relocated for use in grain filling or tuber development could provide a feast to potential pathogenic species. This may be the reason why plants upregulate defence mechanisms as they trigger off normal senescence, and why some of the senescence hormones, especially jasmonates and salicylates, act as mediators of resistance towards pathogens.

All leaves, including those of evergreens, undergo senescence and eventually death as the terminal phase of their development in response to age-dependent factors, environmental signals, and biotic and abiotic stresses. Leaf senescence is an active process that retrieves nutrients (C and N) from leaves at the end of their useful lifespan. Leaf senescence may be: sequential, seasonal or stress- induced.

Developmental leaf senescence consists of three distinct phases: ***initiation, reorganization (degenerative)*** and ***terminal*** as depicted in Figure 17.5.

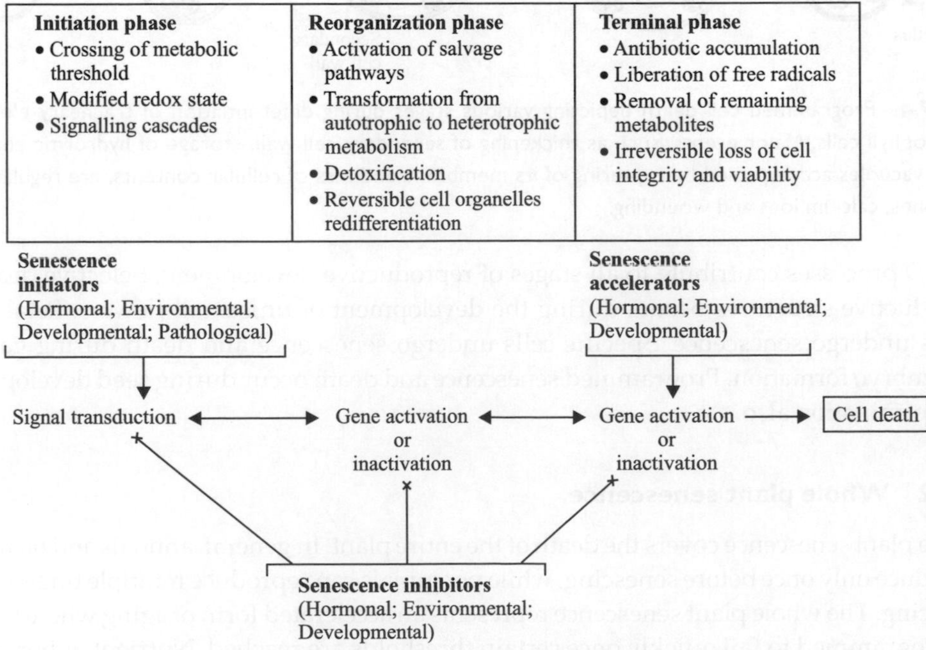

Fig. 17.5 Steps in leaf senescence network from the initiation phase to the terminal phase i.e. cell death.

During the *intiation phase,* the leaf receives developmental and environmental signals that indicate a decline in photosynthesis, and a transition from being a nitrogen sink to a nitrogen source. Transcription of many genes are activated or suppressed in this phase.

During the reorganization phase (*degenerative phase*), autolysis of cellular organelles and macromolecules occurs. The soluble organic nutrients thus produced are remobilized to growing sinks (young leaves, underground storage organs or reproductive structures) *via* phloem. An abscission layer forms during this phase.

During the *terminal phase,* subcellular membranes lose their integrity and separation of cells occurs at the abscission zone (Figure 17.6). The compartmentalisation of biochemical pathways also breaks down during this phase.

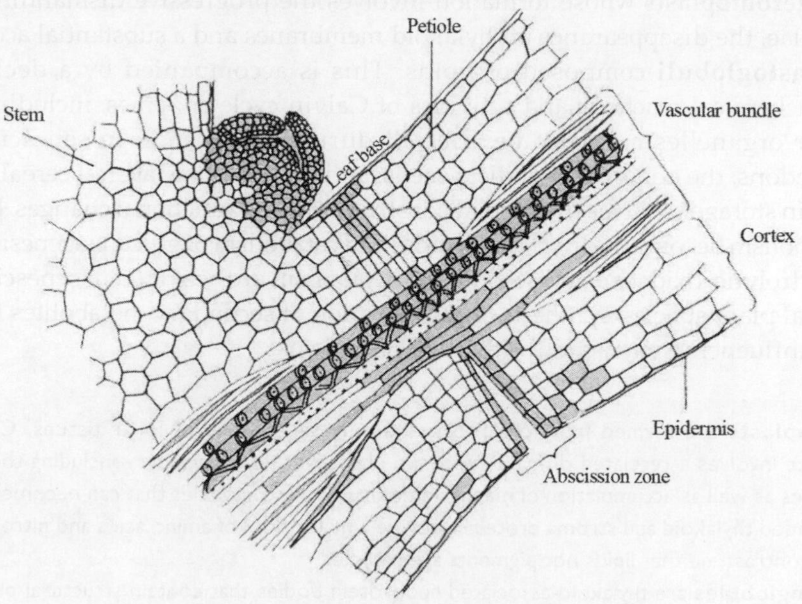

Fig. 17.6 Abscission zone at the base of the petiole of a mature leaf.

Four Phases of the sequence of changes in leaf senescence

Phase 1: Mature leaves, pre-senescent phase

Phase 2: Degradation of chlorophyll and carotenoids
- Anthocyanin accumulation
- Mobilization of N and P
- Metabolism of sugars
- Conversion of chloroplasts into gerontoplasts
- Switching of primary energy sources from chloroplasts towards mitochondria

Phase 3: Depletion of cell contents but metabolism continues
- Less than 5 per cent of original chlorophyll remains

Phase 4: Complete cell death
- Only a few structures like photosystem II are identifiable within residual cell walls

Changes associated with leaf senescence

(i) *Change in leaf colour:* During leaf senescence, the first visible change is in colour. Loss of green color is followed by development of red pigmentation which is directly linked to the regulation of nutrient mobilization from leaf cells. So chlorophyll breakdown and anthocyanin synthesis is a defining feature of senescence. The first step is the release of chlorophyll from the chlorophyll-protein complexes within the thylakoid membrane, mediated by SGR, a factor encoded by the senescence-upregulated gene **STAY-GREEN**. Dissociation of the complex makes the protein nitrogen available for recycling.

(ii) *Change in cell structure and metabolism*: During catabolism, chloroplasts are transformed into **gerontoplasts** whose formation involves the progressive dismantling of grana lamellae, the disappearance of thylakoid membranes and a substantial accumulation of **plastoglobuli** composed of lipids. This is accompanied by a decline in the photochemical reactions and activities of Calvin cycle enzymes, including Rubisco. Other organelles may also be affected during senescence. In senescing storage cotyledons, the endosperm of dicot seeds, and the aleurone layer of cereal grains, the protein storage vacuoles undergo major structural and functional changes. Peroxisome metabolism becomes redirected towards lipid catabolism or gluconeogenesis. Pathways of hydrolytic, oxidative and secondary metabolism are activated in senescing cells. In several plant species, synthesis or modification of secondary metabolites takes place, thus influencing interaction with other organisms.

Gerontoplasts are formed from chloroplasts during senescence of foliar tissues. Chloroplast senescence involves a regulated dismantling of the photosynthetic machinery including the thylakoid membranes as well as accumulation of massive amounts of plastoglobules that can become very large. The degraded thylakoid and stroma proteins provide a major pool of amino acids and nitrogen for the plant. In contrast, neither lipids nor pigments are recycled.

Plastoglobules are thylakoid-associated lipoprotein bodies that contain structural proteins and enzymes employed in the synthesis and accumulation of lipid molecules

(iii) *Degradation of macromolecules*: One of the most important features of senescence is the recycling of macromolecules, including proteins, lipids and nucleic acids to be used in other non-senescent parts of the plant. A major function of senescence and cell death is to regulate the efficient salvage of the N and P locked up in the proteins and nucleic acids of mature tissues and to ensure that these elements are transported to growing plant parts or to storage tissues. Macromolecules must be degraded by hydrolytic enzymes whose activity increases during senescence. Proteases release amino acids that are further metabolized to form amides (glutamine and asparagine), the principal form in which recycled nitrogen is translocated from senescing tissues (Figure 17.7). Nucleases hydrolyze nucleic acids to nucleotides. Phosphatases, which are strongly active during senescence and programmed cell death, liberate phosphate from the nucleotide products, obtained by nuclease action on RNA and DNA. The resulting nucleosides are cleaved into sugars, purines and pyrimidines, which are further catabolized into ammonia and carbon dioxide.

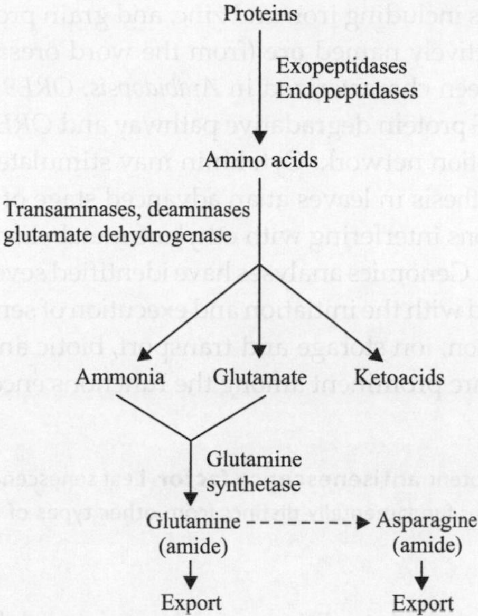

Fig. 17.7 Proteins breakdown to amino acids during leaf senescence. Amino acids are further metabolized to the amides, glutamine and asparagine. Amide nitrogen is the major form in which recycled protein nitrogen is exported from senescing tissues to developing sinks.

(iv) *Modification of energy and oxidative metabolism*: The state of energy metabolism in plastids, peroxisomes and mitochondria changes markedly during senescence along with cellular redox reactions. **Reactive oxygen species (ROS)** play an important role in senescence. The amount of ROS produced during plant metabolic processes often increases with the age of the tissues. ROS in general, and H_2O_2 in particular, have signalling functions during senescence and their accumulation is under the control of antioxidant metabolites and enzymes. Catalase and ascorbate peroxidase are enzymes that scavenge H_2O_2 and regulate its influence of gene expression and participation in oxidative metabolism.

More than 50 **senescence associated genes (SAGs)** have been discovered till date. One group of SAGs involves genes that encode components of the system e.g., enzymes of the catabolic pathways that are freshly activated during senescence (ubiquitin, proteolytic enzymes, **metallothioneins** - proteins that chelate metal ions). The second group possesses the so-called **stay–green** mutants, characterized by delayed leaf senescence. Some types of stay–green traits are beneficial in agriculture, because they are associated with extended photosynthetic productivity, enhanced crop quality and shelf–life or better stress resistance. But stay-green traits can result in undesirable inefficiencies in nutrient remobilization. SGR, a stay-green gene that functions in the breakdown of chlorophyll-protein complexes, has been isolated by a combination of comparative genetic mapping and functional testing in grasses, pea and *Arabidopsis*.

A similar strategy has identified **a transcription factor, *NAM* (No Apical Meristem),** in wild and domesticated wheats that regulates the onset and progress of leaf senescence, the remobilization of nutrients including iron and zinc, and grain protein content. A number of stay-green mutants, collectively named **ore** (from the word oresara, meaning long-lived in Korean language), have been characterized in *Arabidopsis*. *ORE9* encodes an F-box protein that functions in the UbPS protein degradative pathway and *ORE12-1* is a component of the cytokinin signal transduction network. Cytokinin may stimulate the reversal of yellowing and recovery of photosynthesis in leaves at an advanced stage of senescence. In *Arabidopsis* and other species, mutations interfering with ethylene metabolism and response also result in delayed leaf senescence. Genomics analyses have identified several hundred differentially expressed genes associated with the initiation and execution of senescence. Protein, lipid and carbohydrate remobilization, ion storage and transport, biotic and abiotic stress regulators and transcription factors are prominent among the functions encoded by these SAGs.

Cytokinin is considered a potent **antisenescence factor**. Leaf senescence is potentially reversible at an advanced stage and thus fundamentally distinct from other types of PCD.

17.2.3.2 Effect of environmental conditions on senescence and death

Just like other physiological processes, senescence and PCD in particular also responds to the extreme environmental conditions. Seasonal or other predictive environmental variables can initiate senescence as part of an adaptive strategy.

(i) *Effect of seasons on leaf senescence*: Senescence in many ecosystems of latitudes away from the equator is regulated by photoperiods. Change in photoperiod at different seasons is an important and reliable environmental indicator that prepares a plant to face abiotic stress. In temperate regions, leaf senescence is a principal developmental process. Appropriate timing of leaf senescence is essential if a plant is to maintain a balance of its carbon and nitrogen requirements. Autumnal senescence of temperate deciduous forests is generally initiated in response to decreasing day length. For example, *Aspen* trees always start fall senescence in a coordinated manner approximately at the same date each year, independent of temperature and other environmental factors. Subsequent events are influenced by temperature; chloroplasts are converted to gerontoplasts; photo-oxidative bleaching of chlorophyll followed by anthocyanin stimulation occurs at low temperature which makes the striking display of fall colors in *Aspen*.

MADS-box homeotic transcription factors and phytochrome appear to play regulatory roles in integrating autumnal senescence with growth, flowering and dormancy. Apple, pear and some tree species of the family Rosaceae are exceptions where chilling temperature rather than photoperiod is the major factor to induce foliar senescence.

Besides photoperiod and temperature, high concentrations of sugars may also serve to signal leaf senescence, especially under conditions of low nitrogen availability. Leaf senescence can be accelerated by the sugar depletion and darkness (both related to reduced photosynthetic productivity). Increased levels of the growth regulator cytokinin help in delaying senescence. Besides cytokinins, auxins and gibberellins also suppress senescence, whereas senescence- promoting regulators include ethylene, ABA, jasmonic acid, salicylic acid and brassinosteroids .

(ii) *Complex relationships between programmed senescence, death and aging*: The term 'aging' refers to changes over time. Plants encompass the widest diversity of longevities of all multicellular organisms, with life spans ranging from weeks in ephemeral weeds like *Arabidopsis*, to millennia as exemplified by trees and clonal species. **Monocarpic plants** (which die after flowering once) suffer from a kind of reproductive exhaustion similar to that experienced by semelparous animals. Monocarpic senescence is one kind of survival strategy in which the whole of the vegetative body of the plant is sacrificed in the interests of seed production and propagation of the next generation. In long-lived plants that reproduce serially, the emphasis is on maintenance and survival of the individual. Such a strategy is accompanied by potential susceptibility to accumulated genetic errors and declining physiological integrity

(iii) *Programmed senescence and death are common responses to abiotic stresses* : For example, when plant roots are subjected to flooding, low oxygen (hypoxia or anoxia) sensed by flooding induces the development of **aerenchyma.** Ethylene-signalling pathways regulate PCD in a specific group of root cortical cells, resulting in the development of **schizogenous** or **lysigenous** air spaces, thus improving air flow through the roots. During aerenchyma differentiation, DNA is fragmented, while genes for respiratory, carbohydrate, nitrogen and secondary metabolism are differentially activated. Regulatory systems are also activated, including those for calcium signalling, protein phosphorylation and dephosphorylation, and ethylene biosynthesis resulting in cell wall breakdown associated with increased activity of cellulases.

Research is being carried out in *Arabidopsis* to understand the genetic basis of aerenchyma development. Crop plants, particularly rice and maize in which flooding and poor aeration can reduce yield have been studied in details for aerenchyma development.

(iv) *Senescence and cell death: adaptive and pathological responses to biotic interactions* : PCD is significant in biotic interactions, both those of the beneficial kind represented by attracting pollinators and dispersers, and those of a harmful nature such as attack by pests and pathogens. Pathogenic organisms are classified as **necrotrophs** (which kill the host and feed off its dead tissues) and **biotrophs (**which require host tissues to be alive). The **hypersensitive response** (HR) is a strategy whereby host neutralizes infection by inducing PCD. The mechanism of HR-associated PCD closely resembles autophagy in the role played by vacuolar lytic activities and its sensitivity to chemical inhibitors. ROS and reactive nitrogen species are inducers of HR. Gene-gene interactions in the resistance-avirulance system are essential for triggering PCD. During PCD, the potential vulnerability of senescing tissues to pathogen attack accounts for the prominence of defence genes among SAGs.

The processes of PCD, organ senescence and whole plant senescence differ with respect to the size, cell number and complexity of their senescing units. They also differ with respect to the developmental and environmental signals that trigger them. However, at the cellular level, all of them use the similar genetic pathways for cell autolysis.

Review Questions

RQ 17.1. What are the associated changes observed in leaves during senescence?

Ans. The following changes are observed in leaves during senescence:

(i) Reduction in the level of chlorophyll
(ii) Consequently, a decline in CO_2 fixation
(iii) Disruption in the overall organization of thylakoids of chloroplasts

RQ 17.2. List the various factors that initiate senescence in plants.

Ans. The various factors involved in initiation of senescence in plants, are:

Internal:

(i) Age of the plant
(ii) Degree of flowering
(iii) Time of ripeness of fruits

External:

(i) Length of the photoperiod
(ii) Temperature
(iii) Nutrients

Biotic:

(i) Attack of mites, insects and parasites
(ii) Plucking of leaves

RQ 17.3. Name the enzymes involved in leaf abscission along with their functions.

Ans. The following enzymes are involved in leaf abscission:

(i) Cellulase: hydrolyzes cell wall of the cells in abscission zone
(ii) Polygalacturonase: hydrolyzes pectin which is a major component of middle lamella region of the cell wall shared by adjacent cells

RQ 17.4. Why is ethylene called the senescence hormone? How is ethylene produced?

Ans. Ethylene seems to be a prime controlling agent in many aspects of plant senescence such as:

1. The fading of flowers
2. The ripening of fruits
3. The abscission of leaves.

During senescence of flowers, fruits, and even leaves, the tonoplast or vacuolar membrane becomes slightly leaky so that methionine, the precursor of ethylene, leaks from the vacuole into the cytoplasm where it is converted to ethylene which causes degradative changes.

RQ 17.5. Some fruits like bananas and mangoes are harvested in the green stage and then transported. How is the ripening done by small fruit merchants?

Ans. Fruits like mangoes, bananas, papayas, apples and plums, and such-like, are gathered for the market in green or in unripened states for transporting to other areas. There, they are treated with acetylene or ethylene in totally sealed warehouses by placing calcium carbide (CaC_2) or 'masala' in small plastic containers that absorb the moisture in the air to release acetylene (or ethylene) – the ripening hormone that is normally produced by fruits to accelerate the ripening process.

However, industrial-grade CaC_2 contains arsenic and phosphorus, and so its use for ripening fruits is prohibited in many countries under the 'Prevention of Food Adulteration Rules' as it can cause cancer. It also causes mouth ulcers, gastric irritation, even food poisoning. Some vendors also dip fruits in a solution of ethephon or expose the fruits to ethylene gas to speed up ripening.

The government should formulate stringent measures to punish vendors using such dangerous chemicals to artificially ripen fruits.

RQ 17.6. After seed formation, millions of plants in a wheat field die at approximately the same time. Explain why?

Ans. Reproduction seems to trigger a series of events that lead to senescence and finally to the death of the plants. During senescence, metabolic activities fall rapidly, reserves are translocated out, and colour changes frequently. It appears that the carbohydrates and proteins of leaves are broken down into sugars and amino acids for onward mobilization as reserve food for seeds (or tubers, and rhizomes). Their transport is supposed to be mediated by auxins. Senescence of plants, once initiated as a result of imbalance of growth hormones and nutrients, results in lack of gene activity and enzyme synthesis, and finally autophagy and eventual autolysis.

The final stage is death of the leaf or in the case of annuals or biennials, death of the whole plant. Dead tissue falls to the ground where they undergo decay through microbial action, producing inorganic substances that are cycled back.

Unit IV Physiological Stress and Secondary Metabolites – Their Role in Metabolism

Chapter 18

Abiotic and Biotic Stress

Under both natural and agricultural situations, plants are often subjected to environmental stresses. Stress plays an important role in determining how soil and climate restrict the distribution of plant species. **Stress** is usually defined as a disadvantageous impact on the physiology of a plant, induced upon a sudden shift from optimal environmental condition where homeostasis[1] is maintained to suboptimal level which disturbs the initial homeostatic state. In most cases, stress is measured in terms of plant survival, crop productivity or the primary assimilation processes, which are all co-related to overall growth.

Plant stress can be divided into two categories: **Abiotic** and **Biotic.**

Keywords

- Abiotic stress
- Acclimation
- Adaptation
- Anoxia
- Biotic stress
- Cryoprotectants
- Cytorrhysis
- Exotherm
- Halophytes
- Hardening
- Homeostasis
- LEA proteins

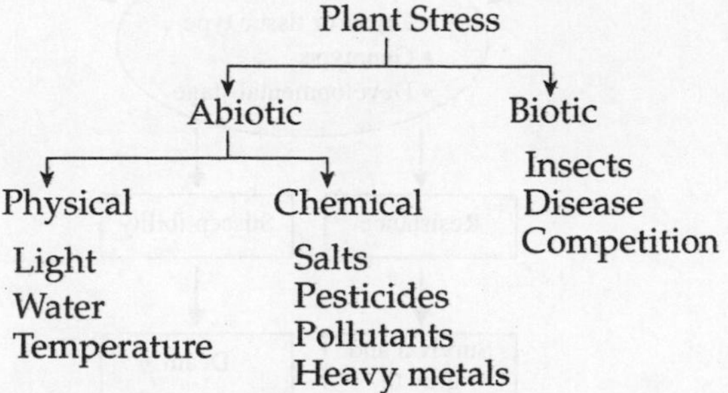

[1] The maintenance of a cellular steady-state far from equilibrium results in an apparently stable condition called **homeostasis**.

18.1 Abiotic Stress

Plants grow and reproduce in hostile environments containing large numbers of abiotic chemical and physical variables, which differ both with time and geographical location. The primary abiotic environmental parameters that affect plant growth are light, water, carbon dioxide, oxygen, soil nutrient content and availability, temperature, salts, and heavy metals. Fluctuations of these abiotic factors generally have negative biochemical and physiological impact on plants. Being fixed, plants are unable to avoid abiotic stress by simply moving to a more suitable environment. Instead, plants have evolved the ability to compensate for stressful conditions by switching over physiological and developmental processes to maintain growth and reproduction.

Responses to abiotic stress depend on the extremity and time duration of the stress, developmental stage, tissue type and interactions between multiple stresses (Figure 18.1). Experiencing stress typically promotes alterations in gene expression and metabolism, and reactions are frequently centred on altered patterns of secondary metabolites. Plants are complex biological systems comprising of thousands of different genes, proteins, regulatory molecules, signalling agents, and chemical compounds that form hundreds of interlinked pathways and networks. Under normal growing conditions, the different biochemical pathways and signalling networks must act in a coordinated manner to balance environmental inputs with the plant's genetic imperative to grow and reproduce. When exposed to unfavourable environmental conditions, this complex interactive system adjusts homeostatically to minimize the negative impacts of stress and maintain metabolic equilibrium.

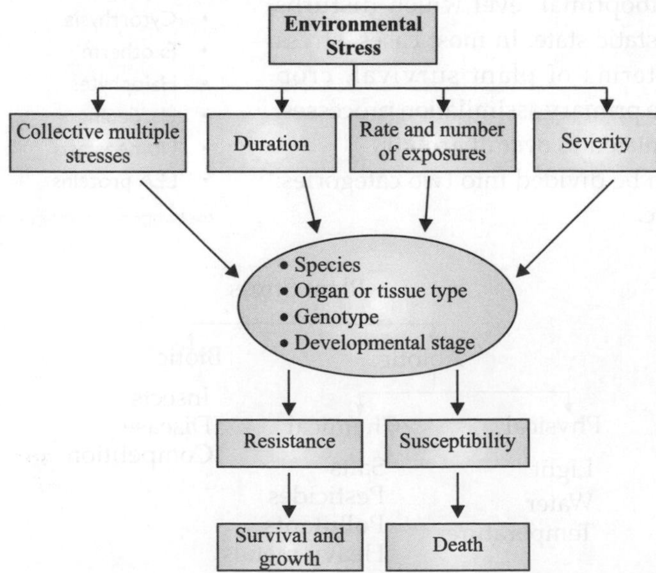

Fig. 18.1 Plants response to environmental stress occurs by several mechanisms. Failure to adjust to acute stress can lead to the plant's death.

The molecular mechanisms controlling plant responses to abiotic stress, including changes in gene expression and cellular signal transduction, have been analyzed extensively, and various **transcription factors** (proteins involved in the process of transcribing DNA into RNA) and **signalling molecules** are believed to play important roles in cellular homeostasis during stressful situations.

Stress-induced alterations in metabolism and development occur because of modified patterns of gene expression. In response to stress conditions, some of the genes show activation or deactivation of expression. Protein products of stress-induced genes normally accumulate in response to unfavourable situations. A stress response is triggered when a plant is able to sense a stress at the cellular level. Stress recognition induces signal transduction pathways that convey information in the individual cells and throughout the plant .Ultimately gene expression changes are incorporated into a response by the entire plant that modifies growth as well as development and that may even affect reproductive potentials. The duration and extremity of the stress determine the extent and timing of the response.

Stress resistance mechanisms have been grouped into two general categories: avoidance **and tolerance mechanisms**. Avoidance mechanisms prevent exposure to stress by ending plants' life cycle during periods of comparatively less stress, whereas tolerance mechanisms allow the plants to tolerate stress by altering their metabolism, which enables them to come to thermodynamic steady state with the stress but not suffer injury (Figure 18.2). Some resistance mechanisms are **constitutive** and active before exposure to stress. In other cases, plants exposed to stress alter their physiology in response, thereby acclimating themselves to an unfavourable environment.

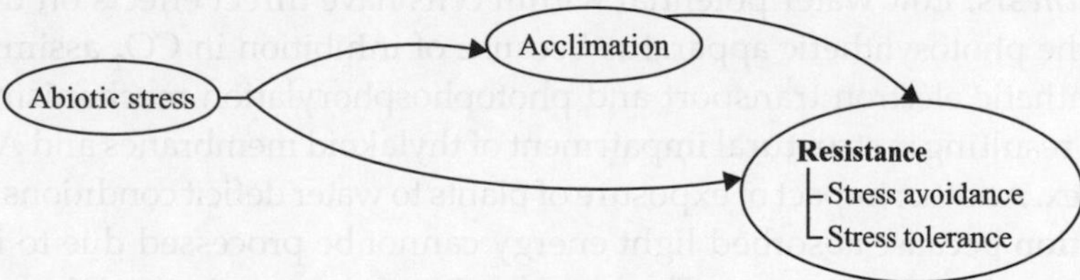

Fig. 18.2 Stress resistance can involve tolerance or avoidance of the stressful environment. Some resistance mechanisms are constitutive and active before exposure to stress, and the plants exposed to stress alter their physiology in response, thereby acclimating themselves to an unfavourable conditions.

Two types of modifications - **adaptation** and **acclimation** are means of achieving tolerance to a particular stress. Adaptation to stress consists of genetic alteration over many generations e.g., morphological and physiological modifications associated with CAM plants. The evolution of adaptive mechanisms in plants to a particular set of environmental conditions generally involves processes that allow avoidance of the potentially damaging effects of these conditions. Acclimation, on the other hand, allows plants to respond to environmental fluctuations. The physiological modifications involved in acclimation are generally reversible and need no genetic alterations. Both genetic adaptation and acclimation can impart complete toleration of extremes to the plants in their abiotic environment. The process of acclimation to a stress is known as **'hardening'** and plants that have potential to acclimate are designated as '**hardy**' species. On the contrary, those plants that have the least capability to acclimate to a particular stress are designated as '**non-hardy**' species.

Plant responses to abiotic stress involve trade-offs between vegetative and reproductive growth, which may differ depending on whether the plant is an annual or perennial.

The most common environmental stresses for plants are those that restrict water availability such as, **drought, salinity** and **low temperature**. The responses of plants to such stress conditions show many similarities in signalling mechanism and in the biochemical and metabolic responses involved.

18.1.1 Water stress

Water stress may develop through either because of excess water (flooding, anoxia) or a water deficit (drought, low water potential)_

18.1.1.1 Water deficit

Stress caused by water deficit is more frequent (often referred to as **drought stress**). The following are the consequences of water deficit:

Membrane damage: Desiccation leads to an increase in solute leakage as a result of shrinkage in the protoplast volume. Water stress also alters the lipid bilayer, thus causing the displacement of membrane proteins. This may result in severe structural and metabolic disorders because of effects on membranes integrity, protein structure, and a general disruption of cellular compartments. In addition to membrane damage, cytoplasmic and organelle activity may undergo significant loss or even complete denaturation.

Photosynthesis: Low water potential within cells have direct effects on the structural integrity of the photosynthetic apparatus because of inhibition in CO_2 assimilation. Both the photosynthetic electron transport and photophosphorylation mechanisms decline in chloroplasts, resulting in structural impairment of thylakoid membranes and ATP synthase (F_0-F_1) complex. Another impact of exposure of plants to water deficit conditions is continued **photoinhibition** because absorbed light energy cannot be processed due to inhibition of photosynthetic electron transport. The specific site of damage is the D1 reaction centre polypeptide of PS II system.

Cellular growth: Cellular activity seems to be the most sensitive response to water stress. The inhibition of cell enlargement is generally accompanied by reduced cell wall synthesis. Cells become smaller in size and lesser leaf development takes place during water stress, resulting in reduced surface area for photosynthetic process. Even moderate drought can decrease the crop yields noticeably. Cell growth and photosynthesis decrease as a consequence of decreasing tissue water potentials.

ABA accumulation: ABA levels begin to increase markedly in leaf tissues and roots followed by closure of stomates and decrease in transpiration rate. Additionally, ABA inhibits shoot growth thus further conserving water and simultaneously root growth appears to be promoted, which could increase the water supply. ABA is released from the cytoplasm of mesophyll cells into the apoplast, from where it can bc transported along with the transpiration stream to the guard cells. ABA disrupts the proton pump operating in the plasma membrane of guard cells and as a consequence, stimulates K^+ efflux from them, leading to loss of turgor and stomatal closure. ABA is associated with a type of early warning system that transmits information regarding soil water potential to the leaves. ABA also appears to be involved as part of a root-to-leaf signal chain that initiates stomatal closure when soil begins to dry out. Drought sends an immediate hydraulic signal, probably a rapid change in xylem tension, that is able to stimulate ABA biosynthesis over long distances, and is rapidly distributed to neighbouring tissues.

ABA synthesized in response to water deficiency is swiftly broken down on rehydration. ABA 8′-hydroxylase, a major enzyme in ABA degradation, gets activated in vascular tissues and stomatal guard cells immediately, after shoot exposure to high humidity. This leads to the synthesis of inactive phaesic acid.

Stomatal response to water deficit: Plants usually respond to water stress by closing their stomata to compensate for the water loss through transpiration. Stomatal closure caused by the direct evaporation of water from the guard cells is known as **'Hydropassive Closure'** and it does not require any metabolic involvement on part of the guard cells. This happens when the rate of water loss from the guard cells exceeds the rate of water movement into guard cells from the surrounding epidermal cells. On the other hand, **'Hydroactive closure'** is dependent on metabolic energy and involves a reversal of the ion fluxes responsible for stomatal opening. It is triggered by reducing water potential in the leaf mesophyll cells and seems to involve ABA. ABA is produced in the cytosol of leaf mesophyll cells and accumulates in the chloroplast.

***Miscellaneous effects*:** Activity of hydrolytic enzymes increases significantly under water stress. Nitrate reductase and phenylammonia lyase enzymes show sharp decline in activity, thus affecting the process of nitrogen fixation. Cytokinin levels decrease in the leaves of some plants even during mildly stressful conditions. Besides this, other physiological processes such as respiration, CO_2 assimilation, and translocation of mineral ions as well as photoassimilates drop to their lowest during extreme stress conditions.

Physiological and cellular responses to water stress

Several environmental factors can cause water deficit. Many drought-resistant plant species adjust their osmotic potential to overcome short-term durations of water stress i.e., **osmotic adjustment,** which results from a net increase in the solute molecules in the plant cell. By lowering plant solute potential, osmotic adjustment can drive root water potential to lesser values compared to soil water potential, thereby permitting water to move from soil to plant down the concentration gradient. Osmotic adjustment by **compatible solutes/osmolytes** and **aquaporins** is known to play an important role in enabling plants to acclimatise to water-deficit or saline habitats.

Osmoprotection

Upon exposure to abiotic stress, many plants accumulate compatible solutes in the form of carbohydrates such as mannitol, trehalose[2], myo-inositol, fructan, galactinol, raffinose

[2] Trehalose, non-reducing disaccharide, is composed of two D-glucose molecules joined by an α-α (1, 1) glycosidic bond. It is also known as tremalose and myose. It is widespread in bacteria, fungi, yeast, plants and invertebrates.

and amino acids (such as proline and glycine betaine). These metabolites not only serve as energy storage but are believed to play other roles also such as, balancing osmotic strength, stabilizing macromolecules, and preserving the membrane. Genetic regulation of the major enzymes associated with the biosynthetic pathways of these carbohydrates can improve abiotic stress tolerance in transgenic plants.

(i) ***Compatible solutes:*** Synthesis and accumulation of compatible solutes are most prevalent in plants, but their distribution varies among species. These solutes are localized in the cytosol as well as the cellular organelles and move across the membrane by specific transporters. Stress can trigger the irreversible synthesis of these compounds (e.g. glycine betaine), alter the balance between their synthesis and catabolism (e.g. proline) or trigger their release from polymeric forms (e.g. release of glucose and fructose from starch and fructans). On removal of stress, these monomers can be repolymerized to facilitate rapid and reversible osmotic adjustment, or they can be metabolized to produce primary metabolites or energy. Compatible solutes are water-soluble and being organic in nature, are non-toxic even at higher concentrations. These possess a neutral charge at physiological pH, either non-ionic or zwitterionic, and are kept out from hydration shells of macromolecules. These allow the cytoplasm to achieve osmotic balance with the vacuole and are believed to play a protective function to minimize the impact of abiotic stress in plants by acting as antioxidants.

- ***Glycine betaine (GB)*** functions as a major osmoprotectant. It is synthesized in chloroplasts under various types of environmental stress, protecting the integrity of thylakoid membranes and thus helps in maintaining photosynthetic efficiency under osmotic stress. It stabilizes the quaternary structures of the PSII protein-pigment under high salinity and maintains the ordered state of membranes at extreme temperatures. Genetic engineering is employed to introduce GB-biosynthetic pathways into non-accumulator species (*Arabidopsis*, rice, tobacco), and is a promising approach to increase abiotic stress tolerance.

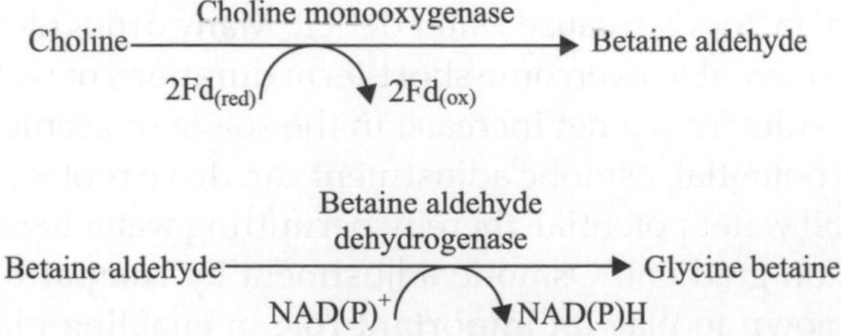

- ***Proline*** accumulates in cytosol, where it supports the stabilization of protein structure, membranes and other subcellular architecture. It also buffers cellular metabolism against fluctuations in redox and pH conditions. On relieving stress, proline rapidly catabolizes to the products which contribute to energy requirements for recovery and repair. Proline is a transcriptional regulator of stress-responsive genes. Proline has received major attention in plant biology because of its accumulation in responses to salt, metal and drought stress. It allows plants to lower their water potential and maintain turgor during dry or saline conditions. Proline metabolism is a target for metabolic engineering of stress tolerance.

Proline: $H_2C—CH_2$, H_2C CH—COOH, N, H (ring)

Sorbitol: H_2COH — HC—OH — HO—CH — HC—OH — HC—OH — H_2COH

Glycine betaine: $(H_3C)_3—N^+—CH_2—COOH$

Proline Sorbitol Glycine betaine

Structure of three compatible solutes involved in osmotic adjustment.

- ***Sorbitol/pinitol /mannitol*** (polyhydric alcohols) behave as scavengers of hydroxyl radicals and also play an important role in moderating the effects of **reactive oxygen species (ROS).** Sucrose accumulation within water-stressed plant species serves an osmoprotective role by protecting cellular components, from damage caused due to dehydration. Yet other plants respond to prolonged water stress by adjusting leaf area, senescence, and abscission of older leaves to reduce leaf area and transpiration.
- ***Oligosaccharides of the raffinose family,*** for example, raffinose and stachyose and their precursor galactinol, play major roles in drought stress tolerance in plants. Raffinose family oligosaccharides (RFOs) are important for reducing the reactive oxygen species that accumulate and trigger oxidative stress under conditions of abiotic stress. Overexpression of the stress-inducible galactinol synthase gene improves drought tolerance by activating the accumulation of galactinol and raffinose in genetically modified *Arabidopsis* plants.

(ii) ***Aquaporins:*** Another possible mechanism for osmotic adjustment is increased flow of water into the cell. The hydrophobic organization of the lipid bilayer offers a great barrier to the free flow of water into the cell and intracellular organelles. However, plasma membranes and tonoplasts can be made more permeable to water by transmembrane water channels, **aquaporins**. Water movement through aquaporins can be regulated quickly, and there is evidence suggesting that these channels may facilitate movement of water in the water-stressed plant tissues and promote fast recovery of turgidity upon watering. The abundance of aquaporin mRNA is correlated with turgor changes in leaves subjected to osmotic stress. High transcript levels[3], enhanced translation, and activation of existing proteins may constitute multiple mechanisms for regulating aquaporin abundance and activity, which may be advantageous for plants coping with water deficit.

LEA proteins

Late embryogenesis-abundant (LEA) proteins are of special interest in the context of stress responses during the plant life cycle. LEA proteins (approx 400 in number) are members of a widespread group of proteins known as **'hydrophilins'** characterized by high hydrophilicity

[3] Transcript level refers to how abundant a transcript is i.e., if a certain mRNA (a transcript) exist in many or few copies.

and a glycine content of more than 6 per cent. LEA proteins generally increase in vegetative tissues experiencing water deficit and these have been shown to protect them against dehydration. It has been observed that overexpression of LEA genes can enhance drought and salinity resistance in plants, bacteria and yeast. Most LEA proteins are 'intrinsically unstructured proteins' (proteins that lack a fixed 3-D structure), existing principally as random coils in solution, which may allow them to prevent enzyme inactivation. LEA proteins may also form a tight network of hydrogen-bonding in desiccating cytoplasm that retains residual water and stabilizes cell structures against osmotic damage. Another hypothesis is that LEA proteins may decrease the collisions between partially denatured proteins, thereby reducing aggregation of exposed hydrophobic domains.

ABA and water deficit

- ABA production is essential for the cellular accumulation of various metabolites and proteins that have protective roles in water deficit resistance. Accumulated ABA induces a number of stress genes, the products of which are important for plant responses and tolerance to dehydration. Endogenous ABA levels, governed by the balance between ABA synthesis and degradation, increase substantially under water deficit or high salinity conditions.
- ABA is synthesized from xanthoxin in the plastids, catalyzed by the enzyme 9-cis-epoxycarotenoid dioxygenase (NCED) (refer to Figure 16.15). NCED is regulated by a multigene family, and the stress-inducible NCED3 gene plays a major role during ABA production under stress situations. In *Arabidopsis*, overexpression of NCED3 increases endogenous ABA levels and improves drought stress tolerance.
- ABA is catabolized either by the oxidative pathway or the sugar conjugation pathway. The oxidative mechanism is induced *via* hydroxylation catalyzed by ABA 8′-hydroxylase (CYP707A family). In the sugar conjugation pathway, ABA is made inactive in sugar-conjugated forms (ABA-glucosyl ester) and is stored in vacuolar sap. Under conditions of dehydration, ABA is set free from the glucosyl ester form by the enzyme β-glucosidase. Regulation of the genes involved in ABA synthesis and catabolism by transgenic technology can improve drought tolerance. Regulation of ABA transport is important for both inter- and intracellular signalling in plants.

Drought stress-inducible genes play a key role during acclimation process (Figure 18.3). The protein products of these genes can be categorized into two groupings: **functional proteins** and **regulatory proteins** (Table 18.1).

Various drought-sensitive genes are induced by exogenous ABA application, including those that encode enzymes of compatible solute metabolism and protective protein synthesis. It also regulates ion fluxes during opening and closing of stomatal guard cells. For example, P5CS1, the gene encoding the proline-synthesizing enzyme Δ^1-pyrroline-5-carboxylate synthetase, is regulated by an ABA-responsive subfamily of bZIP (basic leucine Zipper domain) transcription factors. Role of transcription factors belonging to MYC, MYB, and NAC families has been suggested in ABA regulation of responses to, and tolerance of, drought and salinity stress. Accumulation of many carbohydrate-related compatible solutes, notably oligosaccharides of the raffinose family, occurs *via* an ABA-independent pathway. Genes in this pathway have a drought-responsive element in their promoter regions and are probably regulated by distinct or interacting signal transduction networks.

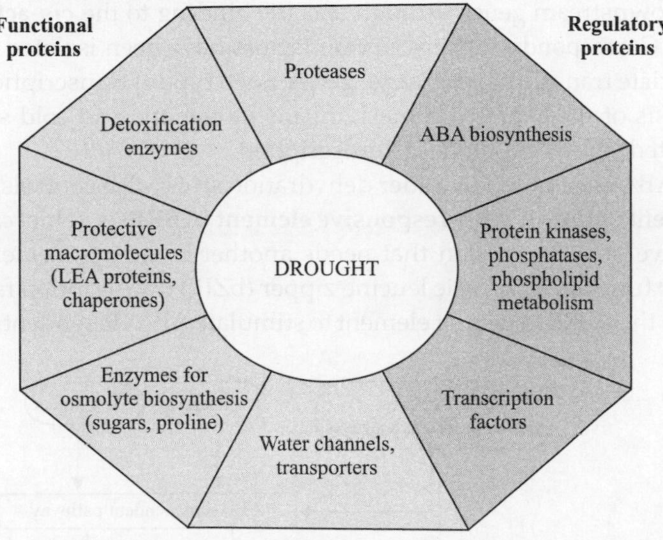

Fig. 18.3 Drought stress-inducible genes perform important functions during acclimation.

Table 18.1 The protein products of drought stress-inducible genes can be classified into two groups.

Proteins	Examples
Functional proteins: (Responsible for executing stress responses)	• Detoxification enzymes and various proteases
	• RNA-binding proteins
	• Protective molecules e.g. LEA proteins, osmotin, antifreeze proteins and chaperones
	• Water channels (aquaporins), and sugar as well as proline transporters
	• Key enzymes for osmolyte (proline, sugars) biosynthesis
Regulatory proteins: (Responsible for signal transduction and control of gene expression)	• Transcription factors
	• Factors involved in post-transcriptional regulation, such as RNA-processing enzymes
	• Protein kinases and protein phosphatases
	• Enzymes of phospholipid metabolism
	• ABA biosynthesis
	• Components of the calmodulin system

Various **transcription factors (TFs)** are important regulatory factors in stress-responsive gene expression (Figure 18.4). Many plant transcription factors, such as AP2/ERF, bZIP, MYB, MYC, Cys2His2 zinc-finger, and NAC, constitute multi-gene families involved in stress-responsive gene expression. Several TFs function as master switches for the expression of various sets of downstream genes through specific binding to the ***cis*-acting elements** in their promoters. Corresponding transcription factors have been isolated that bind to cis-elements and initiate transcription of target genes. Such type of transcription unit is termed '**regulon**'. Analysis of the expression mechanisms of osmotic and cold stress-responsive genes has indicated several regulons in transcription.

Endogenous ABA accumulation under dehydration stress also controls gene expression *via* **ABA-dependent pathway**. **ABA-responsive element (ABRE)** is a chief *cis*-acting element in ABA-responsive gene expression that needs another *cis*-acting element, the **coupling element (CE),** for function. The basic leucine zipper (bZIP) transcription factors AREB and ABF can attach to the ABRE *cis*-acting element to stimulate ABA-dependent gene expression.

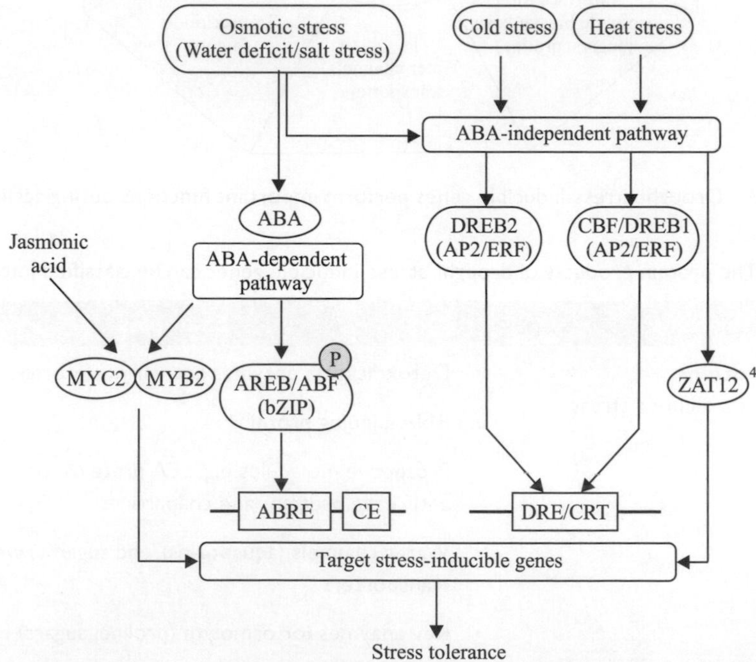

Fig. 18.4 Role of transcription factors and *cis*-acting elements in plant responses to osmotic, heat, and cold stress. ABA-dependent pathway: ABRE is an important *cis*-acting element in ABA-responsive gene expression that requires coupling element for function. Transcription factors (AREB/ABF, MYC2 and MYB2) play important part in this pathway. ABA-independent pathway: DRE/CRT is a major *cis*-acting element and CBF/DREB1 and DREB2 transcription factors participate in this pathway.

CBF; C-Repeat binding factor

DREB; Dehydration response element binding factor.

[4] ZAT12, the zinc-finger protein, responds to a large number of abiotic and biotic stresses. ZAT12 is thought to be involved in cold and oxidative stress signalling in *Arabidopsis*.

Water stresses cause considerable loss of agricultural productivity, making it desirable to develop drought-tolerant plants by conventional or transgenic strategies.

ABA-independent pathways can also influence the abiotic stress response and DREB2 transcription factors. The promoters of genes such as RD29A and COR15, which are activated by dehydration, high salinity, and low temperature, contain two chief *cis*-acting elements, **ABRE** and **DRE (dehydration responsive element)/CRT (C-Repeat)**, both of which are associated with stress-inducible gene expression. Two transcription factors of the AP2/ERF family, CBF/DREB1 and DREB2, bind to the DRE/CRT element. Overexpression of CBF/DREB1 in transgenic plants increases resistance to drought, freezing, and salinity stress, suggesting that these proteins contribute to the development of cold stress tolerance without post-translational modification.

Constitutive expression of transcription factors has great potential for the development of drought-resistant crops, for example, involvement of SNAC1 and ZmNF-YB2 transcription factors in improvement of drought resistance in transgenic rice and maize.

18.1.1.2 Flooding

Flooding stress, most commonly known as **oxygen stress**, deprives plants of oxygen (**hypoxia/anoxia**) and affects respiratory processes (fermentation, ethanol formation, Pasteur effect), thus depressing crop yields. Oxygen deficit causes a transition from aerobic to anaerobic metabolism in terms of metabolic response followed by changes in gene expression. Anoxia (or hypoxia)-induced injury is a result of acidification of cytosol which, in turn, leads to reduced protein synthesis, mitochondrial destruction, retardation of cell division and elongation, inhibition of ion transport and root meristem cell death.

Flooding-tolerant plants can overcome cytosolic acidosis and continuously synthesize ATP by stimulating ethanolic fermentation. Wetland plants, which are adapted to long-term flooding, possess anatomical, morphological and physiological features (such as differentiation of aerenchyma, adventitious roots, a thickened root hypodermis that reduces O_2 loss to the anaerobic soil and formation of lenticels and pneumatophores) that permit survival in water-logged soils.

Role of aerenchyma in the growth of rice and other plants undergoing waterlogging

Flooding is the chief abiotic stress usually, and is harmful to almost all the higher plants, resulting in their decreased productivity. The major damaging effect to plants in flooded soil is an insufficient availability of O_2 to the submerged tissues. Additionally, the deprivation of oxygen in the soil leads to transition from aerobic to anaerobic microbial activity, which results in reduction of oxidized compounds and the generation of phytotoxic ionic forms of Mn^{2+}, Fe^{2+}, and S^{2-}.

Flood-tolerant crops develop morphological, anatomical, and metabolic adaptations in response to waterlogging. Rice and many wetland plants show elongation of the shoot during submergence – an **escape strategy** (or **avoidance strategy**), enabling plants to grow taller, in a growth response driven by the plant hormones like gibberellins and ethylene. This keeps the leaves above water to resume photosynthetic activity by raising their shoots above the surface of water.

An important metabolic response to flooding is the induction of ethylene synthesis.

Ethylene is synthesized in two steps:

(i) The first step is the production of 1-aminocyclopropane-1-carboxylate (ACC) from methionine, catalyzed by ACC synthase (ACS) enzyme.
(ii) The second step involves ethylene production from ACC, catalyzed by ACC oxidase (ACO) enzyme.

Methionine
(1) ↓ ACC-synthase
AOC 1-aminocyclopropane-1-carboxylic
(2) ↓ ACC-oxidase
Ethylene

Steps in the biosynthesis of ethylene

Ethylene is not only evolved in such stress conditions but has also been reported to be an important signalling molecule during flooding, regulating stress-related morphological and anatomical responses such as **aerenchyma** development which allows exchange of gases between the shoot and the roots (Figure 18.5). In response to waterlogging, rice and other aquatic plants develop aerenchyma, a tissue composed of a network of interconnected gas conducting channels which supplies plant roots with oxygen under hypoxic conditions. Aerenchyma tissue is formed firstly by the collapse and programmed death of some cells in the cortex of the root to produce air-filled cavities or spaces (**lysigenous type**), and secondly through destruction of pectic compounds in the middle lamellar region, leading to separation of the cells (**schizogenous type**).

In view of oxygen deficiency, root cells generate highly reactive molecules known as **reactive oxygen species (ROS)**. The calcium ions concentration also increases before aerenchyma formation. ROS and calcium ions are involved in reactions catalyzed by enzymes termed as **respiratory burst oxidase homologs (RBOHs)** and **calcium-dependent protein kinases (CDPKs)**. Ca^{2+} ions stimulate the activity of CDPK enzymes which, in turn, activate RBOH enzymes, and the activated RBOH generates a kind of ROS known as **superoxide**. The genes coding for these enzymes are highly expressed in rice roots under waterlogging.

ROS are by-products of different metabolic pathways and are generated both enzymatically, or, non-enzymatically. **Non-enzymatic ROS** generation can take place in the chloroplasts and mitochondria through electron transport chain. In both cell organelles, the occasional spilling of electrons to O_2 during electron transport system can lead to the partial reduction of O_2, generating $\mathbf{O_2^-}$, which produces more reactive ROS. **Enzymatic ROS** generation can happen at several regions within the cell, including peroxisomes, cell walls, plasma membrane, and apoplast. Another source of enzymatic ROS is plasma membrane-bound NADPH oxidase, or respiratory burst oxidase homologues (RBOHs). RBOHs convert extracellular oxygen to O_2^-, using cytoplasmic NADPH.

However, plants overcome excess of ROS generation with an equally efficient scavenging system that comprises of ROS-scavenging enzymes like superoxide dismutase, ascorbate peroxidase (APX), catalase (CAT), and compounds such as glutathione, ascorbic acid as well as α- tocopherol.

Rice crops can grow very well in flooded soils, where many other plants will die off. So flooding of rice paddies is an important measure to eradicate weeds in rice fields. However, even rice plants can incur yield loss or die if the water level is too deep for too long a period of time.

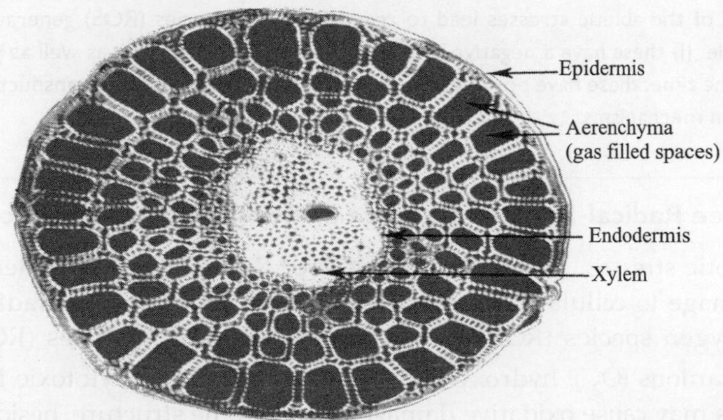

Fig. 18.5 Transverse section of an aquatic plant showing aerenchyma (gas-filled spaces) in the cortex (*Courtesy*: Mr Andrei Savitsky, Cherkassy, Ukraine).

18.1.1.3 Reactive oxygen species (ROS)

A number of abiotic stresses lead to the accumulation of ROS in cells. ROS play a dual role during abiotic stress (Figure 18.6). On one hand, ROS have a negative effect on plant growth, development, and yield. On the other hand, ROS accumulation has a positive effect on cells by stimulating signal transduction pathways that induce acclimation mechanisms. These, in turn, counteract the negative effects of stress, including ROS accumulation.

Reactive oxygen is a substrate in a number of important biochemical reactions, including the formation of lignin. ROS function as enzyme substrates and products, signalling components, redox regulators and harmful oxidizers. These also participate in responses to biotic stress.

Receptor proteins and transcription factors sense ROS status and trigger the signal transduction networks, which in turn result in acclimatory changes in gene expression. More than 150 genes have been identified in the ROS regulatory network, coding for a range of ROS-scavenging and ROS-producing proteins with a high degree of functional redundancy.

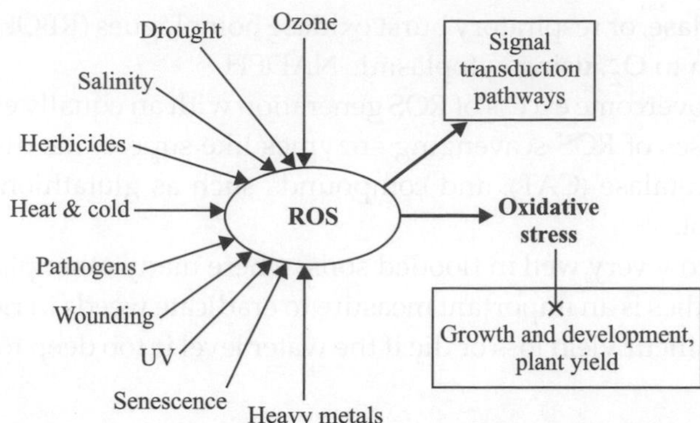

Fig. 18.6 Most of the abiotic stresses lead to reactive oxygen species (ROS) generation in plant cells. ROS play dual role: (i) these have a negative impact on growth, development, as well as yield of the plants (ii) but, at the same time, these have positive effect on cells by stimulating signal transduction pathways that induce acclimation mechanisms.

Free Radical-Induced Damage or Reactive Oxygen Species

During abiotic stresses, a large number of 'Free Radicals' are generated that cause intense damage to cellular components. The most important free radicals are the reactive oxygen species (ROS) or reactive oxygen intermediates (ROI), such as superoxide anions (O_2^{*-}), hydroxyl free radicals OH^{*-} and phytotoxic H_2O_2. These free radicals may cause oxidative damage to membrane structure, besides affecting vital molecules such as enzymes, proteins, nucleic acids, etc. However, the plants have evolved protective or defensive mechanisms to scavenge these free radicals, which consists of antioxidants, e.g., **ascorbate** and **glutathione** and many enzymes, for example, **superoxide dismutase** (SOD), **ascorbate peroxidase, ascorbate reductase** and **glutathione reductase**. These metabolites and enzymes are also induced during stressful conditions. Superoxide radical (O_2^{*-}), with an unpaired electron, is produced when electrons ejected from PSI are acquired by the molecular oxygen instead of its usual acceptor ferredoxin. This happens when CO_2 assimilation is reduced and the turnover of ferredoxin and NADPH is restricted. The superoxide anion is highly reactive towards cellular components and if not metabolized, may cause toxicity. To counteract the accumulation of this radical, the green plants have evolved a mechanism for removing the superoxide through the ubiquitous enzyme 'superoxide dismutase', which converts it into less toxic metabolite hydrogen peroxide (H_2O_2).

$$O_2^- + O_2^- \rightarrow \underset{\text{(Superoxide radical)}}{(2O_2^{*-})} + 2H^+ \rightarrow H_2O_2 + \underset{\text{(Molecular oxygen)}}{O_2}$$

Hydrogen peroxide does not exhibit any free radical properties but being strongly nucleophilic, it can lead to oxidation of –SH groups in proteins. Most plants have in their peroxisomes highly active enzyme catalase which can degrade hydrogen peroxide to water and oxygen. However, the enzyme catalase is either absent or hardly active in cytosol, mitochondria, and chloroplasts. Plants have evolved a more efficient mechanism for scavenging hydrogen peroxide through **'Ascorbate–glutathione cycle'** (also known as the **Asada–Halliwell cycle** after its discoverer). Reduction of hydrogen peroxide is necessary to prevent its reaction with superoxide anions to produce another highly toxic ROS – the hydroxyl radical (OH^*) that can rapidly damage protein. In this cycle, the toxic hydrogen peroxide on combining with ascorbate is reduced by the chloroplastic enzyme, ascorbate peroxidase, to water and monodehydroascorbate (MDHA). The MDHA is reduced back to ascorbate by the enzyme monodehydroascorbate reductase, using NADH or NADPH as electron donors. Under limited supply of NAD(P)H, monodehydroascorbate radicals are converted to ascorbate and dehydroascorbate.

Dehydroascorbate is also converted to ascorbate by reduced glutathione in the presence of enzyme dehydroascorbate reductase (GSH). The oxidised glutathione (GSSH) is, in turn, transformed to reduced glutathione by the enzyme glutathione reductase, using NADPH as reductant (Figure 18.7). Here, the metabolites ascorbate and glutathione and the different enzymes participating in the ascorbate–glutathione mechanism are important in scavenging free radicals.

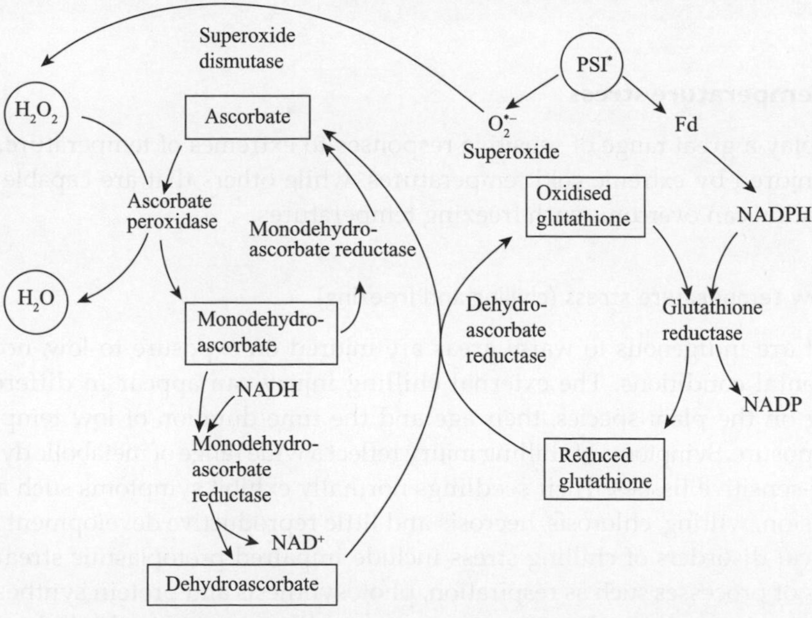

Fig. 18.7 Asada–Halliwell pathway or ascorbate–glutathione cycle.

*Oxygen released during PSII system can also function as alternative acceptor for photosynthetic electron transport system. Even under normal situation, up to 5–10 per cent of the electrons generated by PSI may react with molecular oxygen rather than with ferredoxin/$NADP^+$, producing superoxide radicals (O_2^{*-}). Oxygen can be directly photoreduced by PSI to produce the superoxide free radical (O_2^{*-}) (Mehler reaction as shown on page 558).

Aerobic Rice

Rice is the staple food throughout much of the world, being consumed by nearly 3 billion people. Globally, rice is grown over approximately 149 million hectares with an annual production of 600 million tonnes and the demand is expected to rise with an increase in the population. This will impose a heavy burden on our dwindling fresh water resources. As a solution, the Philippines based IRRI has devised the '**Aerobic Rice Technology**', for water crisis in tropical agriculture. Rice is grown in well-drained, non-puddled and non-irrigated, or non-saturated soils, like wheat, maize and sorghum and such, that grow on dry soils. Aerobic rice is directly seeded into the fields, eliminating the cost of raising nurseries, transplantation and its related impact on labour. Direct seeding also reduces 'seed rates' dramatically. Water requirements can be minimised by lowering water losses caused by seepage, percolation, and evaporation. In a pioneering effort in 2007, India officially released for cultivation, its first drought- tolerant Aerobic Rice Variety **MAS 946-1,** followed by **MAS 26**, developed at the University of Agricultural Sciences, Bangalore . Yields were nearly equal to the traditional irrigated-puddled rice varieties with an average of 5.5–6 tonnes per hectare. Aerobic rice agriculture systems can lower water consumption by as much as 50 per cent. Additionally, aerobic rice emits 80–85 per cent less methane gas (another greenhouse gas) into the atmosphere, thus benefitting the environment. **So, aerobic rice is a better alternative for future climate deterioration under water-deficit conditions,** with surface irrigation provided when necessary.

18.1.2 Temperature stress

Plants display a great range of sensitive responses to extremes of temperature. Some are killed or injured by extreme cold temperatures, while others that are capable of proper acclimatisation can overcome such freezing temperatures.

18.1.2.1 Low temperature stress (chilling and freezing)

Plants that are indigenous to warm areas are injured on exposure to low, non-freezing environmental conditions. The external chilling injury can appear in different forms, depending on the plant species, their age and the time duration of low temperature or chilling exposure. Symptoms of chilling injury reflect a wide range of metabolic dysfunctions in chilling-sensitive tissues. Their seedlings normally exhibit symptoms such as reduced leaf expansion, wilting, chlorosis, necrosis and little reproductive development. The other physiological disorders of chilling stress include impaired protoplasmic streaming, and lower rates of processes such as respiration, photosynthesis and protein synthesis.

One of the immediate plant responses to a chilling stress is the light-dependent retardation of photosynthesis. The chilling-sensitive plants show reversible changes in the physical state of cellular membranes at low temperatures. Their membranes tend to have a higher proportion of unsaturated fatty acids with subsequent, conversion from a fluid to semi-crystalline gel state at a higher temperature as compared to the chilling-resistant plants. The net effect of shift from a fluid membrane to a semi-crystalline state is the disruption of integrity of membrane channels resulting in solute leakage and the impairment of operation of integral proteins making up photosystems and respiratory assemblies. A distinctive reaction to chilling in many plants is the build-up of red anthocyanins, which probably play a defensive role against light/oxidative damage. Chilling may be perceived and transduced into changes in cold responsive (COR) gene expression through temperature-sensitive factors. Among the CORs are genes encoding transcription factors of the ERF family, genes of the ABA-responsive and ABA-independent drought response pathways and genes for antioxidant metabolism.

Freezing stress is frequently exhibited by trees and shrubs growing in north temperate, subarctic and alpine regions. Water freezes first in the apoplastic spaces where there is relatively little dissolved solute. Protoplasmic water will then migrate out of the cells and add to the extracellular crystals. Both of these freezing events are accompanied by a measurable increase in tissue temperature or **exotherm**, due to the heat of fusion released when water crystallizes. A third type of exotherm occurs when ice crystals form intracellularly which kills the tissue. So the freezing injury is largely a result of formation of ice crystals, which in turn lead to dehydration, ionic imbalance and damage to membrane structure and function (Figure 18.8).

Acclimation to freezing is commonly linked to accumulation of sugars (both simple and complex), amino acids and amines. Compatible solutes are effective **cryoprotectants** (i.e. chemicals that prevent low temperature damage).The threshold for chilling injury is associated with the temperature at which lipids of the membrane undergo phase transition, from liquid crystalline to gel. An increase in the amounts of unsaturated fatty acids or sterols causes the membranes to remain functional at low temperature. Another kind of protection against freezing injury involves the synthesis of **'antifreeze proteins'**, which have the unique property of remaining highly hydrated. Freezing sensitivity and susceptibility to pathogens is another example of cross-acclimation between stresses. A number of pathogenesis-related (PR) proteins are known to have antifreeze properties. A distinctive reaction to chilling in most of the plants is the build-up of red anthocyanins (indicative of activation of phenylpropanoid and other secondary metabolic pathways) that play a role in defence against light and oxidative damage.

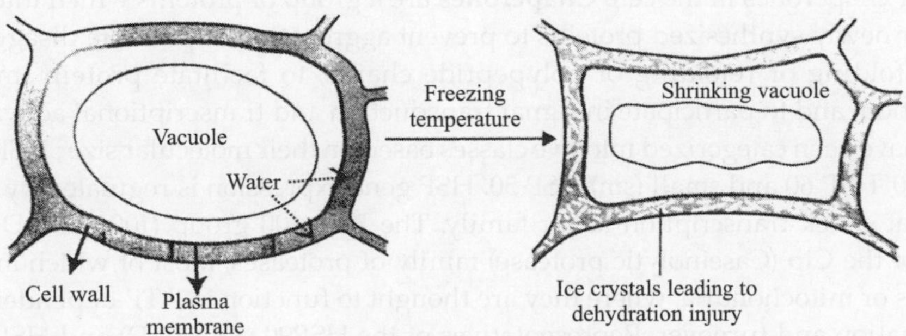

Fig. 18.8 Freezing injury to plant cells. At chilling temperature, ice crystals formed in the apoplastic region or cell wall withdraw water from the cell, which in turn lead to dehydration and damage to membrane structure.

A period of cold acclimation is required to attain maximum freezing tolerance in both woody and herbaceous plant species. Most woody species, including the evergreen conifers attain maximum freezing tolerance through a photoperiod-dependent cessation of growth combined with a low temperature dependent cold acclimation period, thereby inhibiting photosynthesis. In contrast, cold-tolerant herbaceous plants such as wheat, spinach and *Arabidopsis* require active growth and photosynthesis during the cold-acclimation period in order to attain maximum freezing tolerance. These represent two different cold-acclimation mechanisms.

18.1.2.2 High Temperature Stress

Stressful high temperature is a major factor in lowering plant productivity. High temperature stress disrupts plant metabolism through its effects on protein stability and enzymatic reactions, leading to uncoupling of different reactions and the accumulation of toxic intermediates. Heat stress increases membrane fluidity which alters the permeability and catalytic functions of the membrane proteins. Heat stress can destabilize RNA and DNA secondary structures, causing disruption of transcription, translation or RNA processing and turnover. Temperature stress can block protein degradation causing the build-up of protein clumps which interfere with the normal cellular functions. In higher plants, the thylakoid membranes of chloroplasts are most sensitive to high temperature damage, thereby affecting the efficiency of photosynthetic process. Rubisco, Rubisco activase and other photosynthetic enzymes may undergo denaturation at high temperatures. Moreover, exposure to high temperature inhibits the formation of many proteins, including D1 polypeptides of PS II reaction center, thus leading to disruption of D1 repair cycle which is essential to counteract the impacts of chronic photoinhibition.

The membrane of plants adapted to high temperature such as cacti, have a higher amounts of saturated fatty acids, which help to stabilize membranes at high temperatures. Most of the plants subjected briefly to high temperature respond by changes in gene expression. The formation of most of the proteins is suppressed, while the formation of a 'new family' of low molecular weight **heat shock proteins (HSPs)** is induced. HSPs appear to operate as molecular chaperones in the cell. **Chaperones** are a group of proteins which interact with mature or newly synthesized proteins to prevent aggregation or promote disaggregation, to assist folding or refolding of polypeptide chains, to facilitate protein import and translocation, and to participate in signal transduction and transcriptional activation.

HSPs have been categorized into five classes based on their molecular size : HSP 100, HSP 90, HSP 70, HSP 60 and small (sm) HSP 50. HSP gene expression is regulated by members of the heat shock transcription factor family. The HSP 100 group (100–140 kD proteins) consists of the Clp (Caseinolytic protease) family of proteases, most of which are located in plastids or mitochondria, where they are thought to function in ATP-dependent protein disaggregation and turnover. Representatives of the HSP90 (80–94 kD) and HSP70 (69–71 kD), present in cytosol, ER, mitochondria and plastids are ATP-dependent and are essential for normal cell function. Members of the HSP60 group (57–60 kD), generally found in the mitochondrial matrix and chloroplast stroma, are abundant even at normal temperatures. This group includes chaperonin 60, a nuclear-encoded chloroplast protein required for assembly of Rubisco subunits into the functional holoenzyme. The smHSPs (15–30 kD) are not required for normal cell function but participate in the development of thermo-tolerance by promoting ATP-independent protein stabilization and preventing aggregation.

The promoter regions of genes encoding HSPs have the **heat shock element (HSE)**, a conserved DNA sequence that is recognized by members of the **heat shock factor (HSF)** family. Heat stress is required to convert monomeric HSF proteins to the active HSE-binding oligomeric form. Three HSF classes (A, B, C) through their interactions, affect the developmental and stress responses in plant species especially in *Arabidopsis*, tomato and rice. The heat shock response can be detected in tissues experiencing temperatures as little as 50°C above the growth optimum. It defends against damage to organelles, cytoskeleton and membrane function. Manipulation of HSP gene expression not only enhances thermo-tolerance of transgenic plants, but also increases water stress tolerance.

High Night Temperatures and Grain Yield

Recent increases in night temperature has been nearly three times more than the corresponding day temperatures on the surface of the earth. Grain yield has reduced by 10 per cent, with every increase of 1°C minimum temperature. There are evidences of decreased rice yields with high night temperature (HNT) caused by global warming. It is known that plant maintenance respiration shows an increasing trend with increasing temperature and thus, decreases the quantities of photoassimilates available for growth and yield. Rice biomass output was controlled by the balance between net photosynthetic rate and night respiration. HNT has been associated with an increase in respiration rate but reduction in membrane stability and it could potentially decrease biomass output but improve the overall amounts of antioxidants and physiological performance of rice. In addition to this, HNT is also linked to increase in carbon assimilation due to higher nitrogen concentration in the leaves and increased leaf area per unit weight. HNT might impact rice yields due to improved pollen germination and spikelet fertility. Adopting high temperature-resilient varieties is one of the most effective ways to sustain higher yields and stability of rice crop under anticipated climate change. There is need to explore the efficiency of source-sink activity at various levels of heat stress exposure.

18.1.3 Light stress

Although light is essential for photosynthesis, exposure to high light intensities or irradiance may cause damage to photosystems, especially the DI polypeptide of PSII reaction[5] centre which is more sensitive than the rest of the photosynthetic machinery, leading to **photoinhibition** i.e., loss of PSII electron transfer activity which causes significant losses in plant growth and productivity. PSI is rarely damaged, but when it occurs, the damage is irreversible. The extent of photoinhibition depends on the balance between photodamage to the photosystem II (PSII) complex and its repair. In higher plants, the water-splitting photosystem (PSII) complex is located almost exclusively in the appressed regions in grana stacks whereas photosystem (PS I) complex and the enzyme ATP synthase are mostly located

---

[5] Unlike PS II, Pigment system I (PSI) is very rarely damaged, but when it occurs, the damage is practically irreversible.

in the non-appressed areas of grana thylakoids both at the margins as well as on the top and bottom of grana stacks and on the interconnecting single intergranum lamella (frets) whose outer membrane surface face the stroma (refer to Figure 7.3B).

Plants and algae have developed a photoprotective system to continuously repair the photodamaged or photoinhibited PSII through a multistep **DI repair cycle** (Figure 18.9). Functional PSII reaction centres are transformed to damaged PSII ones. When this occurs, PSII is partially disaggregated and DI polypeptide is destroyed by the proteolytic enzymes of the thylakoid. Following this, new DI polypeptide is synthesized *de novo* by the chloroplastic transcriptional and translational machinery. The new DI polypeptide is interpolated into the non-appressed region of stromal thylakoid and a new functional PSII reaction center is recreated. How this damaged PSII complex migrates laterally (lateral heterogeneity) from the granal lamellae stacks to the non-appressed stromal thylakoids for the process of disassembly and reassembly still needs to be researched.

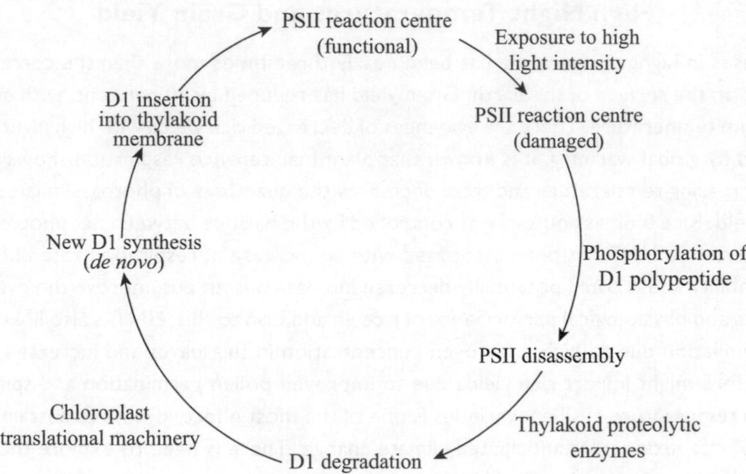

Fig. 18.9 The DI Repair Cycle. The repair cycle of PSII begins with by monomerization of phosphorylated dimeric PSII complex in the granal stacks, followed by dephosphorylation of the core proteins D1, D2, CP43 and CP47.

Plants are able to control the distribution of light energy between PSI and PSII in response to short-term shading and sun-flecking by a redox-regulated phosphorylation of light-harvesting complexes. Over a longer period, shade-adapted plants are able to develop chloroplasts with altered relative numbers of PSII and PSI centers.

Thioredoxin and ferredoxin mediate redox regulation of Calvin-Benson cycle enzymes in response to the excitation state of the thylakoid reactions. Overexcitation of the photosynthetic apparatus can result from production of NADPH and ATP in excess of that required by CO_2 fixation. This can be balanced by a system that links the redox state of ferredoxin, which is reduced by the light energy capture reactions, to that of thioredoxin which in turn regulates the activity of the Calvin cycle enzymes. Ferredoxin-thioredoxin reductase catalyzes the reaction between reduced ferredoxin and thioredoxin. Reduced thioredoxin activates target enzymes by reducing intramolecular disulphide bonds. The Calvin-Benson cycle is also directly regulated by light *via* changes in pH and Mg^{2+} concentration. Reduced thioredoxin formed in the light stimulates the ATP synthase associated with the light capture reactions and the C4 enzyme $NADP^+$-malate dehydrogenase, and inhibits the oxidative pentose phosphate pathway.

18.1.4 Salt stress

There are many regions of the earth where high salinity is a natural part of the environment, such as coastal salt marshes, sea water, inland deserts, shores of inland lakes, and heavily irrigated agricultural land. Salt stress can damage plants at three different levels:

- High salt concentration particularly sodium, will alter the structure of soils (porosity, as well as hydraulic conductance).
- High concentrations of salt are inherently associated with water stress because these causes low soil water potentials, a type of physiological drought that makes it increasingly tough for the plant to absorb water as well as nutrients.
- A third kind of injury includes the toxic effects of specific ions, particularly Na^+ and Cl^-. Excessive Na^+ might result in protein degradation and destabilization of membrane by decreasing the hydration of these macromolecules, leading to enzyme inhibition, or, general metabolic disorders.

The effects of high salinization on plants occur through a two-phase process: a fast response to the high osmotic pressure at the root-soil interface and a slower impact caused by build-up of Na^+ and Cl^- in the foliage. In the osmotic phase, there is reduction in shoot growth coupled with reduced leaf expansion and inhibition of lateral bud formation. The second phase starts with the accumulation of toxic amounts of Na^+ in the leaves, leading to inhibition of photosynthesis and other biosynthetic processes.

The accumulation of compatible solutes and redistribution of ions in response to salt stress enables a plant to acclimate to low water potentials. High salinity triggers greater increase in the endogenous ABA content, followed by principal changes in gene expression and acclimatory and adaptive physiological responses. As with other stresses, salt stress includes changes in the protein synthesis direction, envisaging that new genes may be transcribed. Prior treatment of salt-tolerant cells with ABA also induces new proteins and improves the ability to acclimate to NaCl. An increase in the level of **Osmotin**, a 26 kD polypeptide, is one of the more striking physiological responses to salt and osmotic stress.

Plants inhabiting saline habitats (**halophytes**)[6] have developed many survival mechanisms to withstand strong salt concentrations:

- Most of them are succulents with small, leathery leaves with water storage tissues having low internal water potential (ψ), accomplished by storing unusually high quantities of solutes in the vacuoles of their root cells.
- Most of them produce special types of branched, negatively geotropic roots that come out of the mud surface. These roots are called **pneumatophores** (Figure 18.10) with large lenticels which allow atmospheric oxygen to diffuse into the submerged roots.

[6] Glycophytes are less salt-tolerant plants that are not adapted to salinity.

Fig. 18.10 Pneumatophores of *Avicennia officinalis* (*Courtesy*: Dr Preeti V. Phate).

- Many such plants have specialized 'salt glands' in the leaves through which large quantities of salts are actively secreted[7] (Figure 18.11). Salt excretion not only permits growth in saline habitats but also provides protection against herbivores.

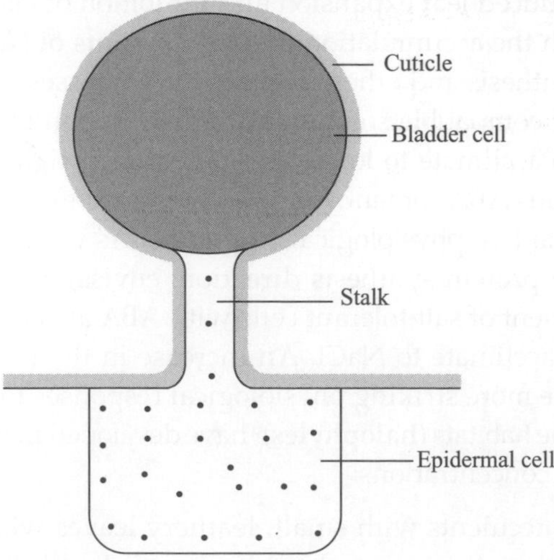

Fig. 18.11 Diagrammatic structure of a salt hair on the leaf surface of *Atriplex spongiosa*.

---

[7] Another saline habitat plant such as the tamarisk tree (*Tamarix*), widely planted in desert areas, expels sodium chloride (NaCl) from salt excreting glands on the leaves. It accumulates in the tissues, at an extraordinary high level of NaCl (2M or 12 per cent).

- Some of the halophytes show vivipary i.e., germination of seeds within the fruit before they are shed from the mother plant (Figure 18.12). Thus the seedlings have a higher vacuole osmotic potential which enables them to establish in soil so saline that it is otherwise physiologically toxic to normal plants.

(a) (b)

Fig. 18.12 A *Rhizophora mucronata* tree (on the left) supported by the arching tangles of aerial roots that are anchored in the mudflat below. The same tree showing viviparous germination (on the right) (*Courtesy*: M S Swaminathan Research Foundation, Chennai, TN, India).

- Halophytes have been found to synthesize the amino acid 'proline', several other amino acids, and galactosyl glycerol and such for ensuring endosmosis.

  Soil salinity is becoming a serious problem worldwide as more and more land is irrigated and fertilized very heavily. Much of the nitrogen that is supplied cannot be absorbed or used by the crop and becomes a serious cause of groundwater pollution. The accumulated nitrate is actually passed on in the 'produce' (seeds) to consumers at toxically high levels.

> The seawater farming project of the M. S. Swaminathan Research Foundation promotes integrated **agri-aqua farming**. Halophytes like *Salicornia, Atriplex,* brown algae, etc., are important sources of nutrition and constitute quality human food as well as animal feed.

Water-deficit and Salinity affect transport across membranes

Plants or cultured cells exposed to high NaCl concentrations tend to accumulate Na^+ as a result of Na^+ influx through **channels and transporters** (Figure 18.13). Ion channels and transporters help maintain ionic homeostasis under conditions of salt stress. Salt treatment increases the activity of Na^+/H^+ antiporters, which then move Na^+ either out of the cellular compartments or into the vacuole. Salt stress triggers changes in Ca^{2+}, which regulate SOS3, which in turn regulates SOS2 protein kinase to activate the plasma membrane SOS1 Na^+/H^+ antiporter. Changes in ABA levels also regulate the NHX1 Na^+/H^+ antiporter and other ion transporters. Other transporters that mediate Na^+ transport into the cell include non-selective cation channels (NSCC) and HKT1/HKT2, the low-affinity cation transporter (LCT1), and K^+ transporters of the KUP/HAK/KT and AKT families.

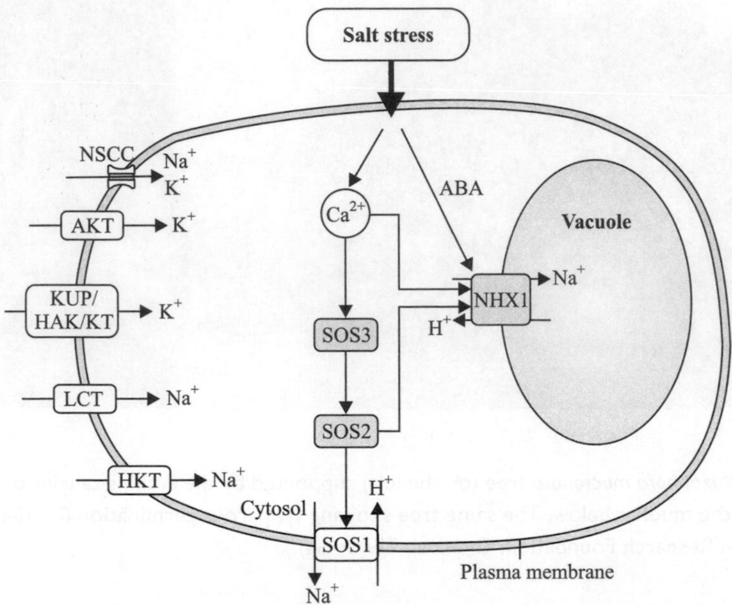

Fig. 18.13 Ion channels and transporters located in the plasma membrane and vacuolar membranes, which help in coping up with salt stress by maintaining ion homeostasis.

SOS (SALT OVERLY SENSITIVE), NHX1 (Na^+/H^+ exchanger1), HKT (High-affinity K^+ transporter), KUP (K^+ uptake permease), HAK (High-affinity K^+ uptake), KT (K^+ transport), AK (*Arabidopsis* K^+ transporter), NSCC (Non-selective cation channels), LCT (Low-affinity cation transporter).

A **salt tolerant gene** known as Asparagine synthetase has been identified using integrative transcriptomic and metabolomics analysis from *Pandanus odorifer.* Now this gene is going to help make rice plants that will grow in seawater of 3.5 molar NaCl. Renowned Chinese agricultural scientist Yuan Longping and his research team have successfully bred more than 200 salt-resistant rice strains in the country's history, yielding 4.5 tonnes/ per hectare of land as compared to the average of 3 metric tonnes of regular rice. If implemented widely, salt-tolerant crops could feed a climate-change ravaged world.

18.1.5 Heavy metal stress

Many heavy metals originating from industrial minefields and sewage treatment plants pollute the environment. Cadmium is a common pollutant of this category. It usually retards plant metabolic activity and also affects human health adversely. Yet, plant selections can be made depending upon their ability to grow on such highly toxic metallic ion concentrations. In many plant species, synthesis of low molecular weight proteins, which are capable of binding with metal ions, is induced in response to heavy metal ions present in the environment. These proteins probably aid the plants in metal accumulation and keeping them far away from the sensitive cellular regions. The uptake of heavy metals such as cadmium (Cd), arsenic (As), and aluminium (Al) by the plant cell can lead to the accumulation of ROS, inhibition of photosynthesis, disruption of membrane structure and ion homeostasis, inhibition of enzymatic reactions, and activation of programmed cell death (PCD). One of the reasons that heavy metals are so toxic is that they can replace essential mineral nutrients such as Ca^{2+} and Mg^{2+}, and disrupt the essential reactions.

18.1.6 Mineral nutrient deficiencies

Mineral nutrient deficiencies can occur even in the presence of an adequate nutrient supply. Most mineral nutrients are available between pH 4.5–6.5 and become unavailable for uptake below or above this range because of insoluble forms. Nutrient stress results in plant metabolic disorders and thus suppression of plant growth and reproduction. Mineral nutrients are components of essential enzymes and building blocks of the cell. For example, an insufficient supply of iron and magnesium results in reduction in heme content, which is essential for chlorophyll and cytochrome biosynthesis and without both of them, energy production in the cell ceases.

18.1.7 Ozone and ultraviolet (UV) stress

Both ozone and UV stress cause suppression of plant growth and reduction of agronomic yields. Ozone enters inside plants through open stomata and is transformed to different types of ROS. These ROS can lead to lipid peroxidation as well as the oxidation of cellular proteins, RNA, and DNA. The toxic effects induce the formation of lesions in leaves that are characteristic of the activation of programmed cell death (PCD). The thinning down of the ozone layer in earth's upper strata reduces the filtering of ultraviolet radiation, thus resulting in increased UV radiation reaching earth's surface. The UV-induced accumulation of ROS causes mutations during DNA replication and activation of PCD, leading to formation of lesions. The type of lesions (leaf chlorosis and tissue necrosis) and the severity of the injuries are dependent on the extent of exposure to ozone and UV light.

Carotenoid biosynthesis also provides a route for dissipating excess excitation energy as heat by pH-dependent quenching of chlorophyll fluorescence *via* the **Xanthophyll Cycle** (refer to Figure 7.7). Phytoene is converted to lycopene in four steps, catalyzed by two desaturases. Lycopene is cyclised to form α-and β-carotenes, precursors of the major plastid xanthophylls lutein, zeaxanthin, violaxanthin and neoxanthin (refer to Figure 7.6).

18.1.8 Developmental and physiological mechanisms that protect plants against abiotic stress

- Plants tend to lower root water potential to continue to absorb water in drying soil.
- Aerenchyma tissue allows O_2 to diffuse down to submerged organs.
- ROS can be detoxified through scavenging pathways that reduce oxidative stress.
- Molecular chaperone proteins protect sensitive proteins and membranes during abiotic stress.
- Prolonged exposure to extreme temperature conditions can alter the membrane lipids composition, thereby allowing plants to maintain membrane fluidity.
- Plants cope with toxic ions through exclusion and internal tolerance mechanisms.
- Plants generate antifreeze proteins to prevent the formation of ice crystals.
- Stomatal closure is prompted by ABA-induced efflux of K^+ and other anions from guard cells into subsidiary cells.
- Plants can modify their leaf architecture and their root-to-shoot biomass ratio to avoid abiotic stress.
- Metabolic shifts enable plants to outlast ephemeral stresses such as flooding or environmental changes from day to night.
- Reversal of stress-response mechanisms must occur in a synchronized manner to avoid production of ROS.
- Agricultural researchers are investigating how plants can sense and acclimatize to stressful situations, and then try to develop crops with increased tolerance.

18.1.9 Stress-sensing mechanisms and activation of signalling pathways in response to abiotic stress in plants

- The initial stress-sensing mechanisms trigger a downstream response that comprises multiple signal transduction pathways (Figure 18.14). These pathways involve many factors such as calcium, protein kinases, protein phosphatases, ROS signalling, stimulation of transcriptional regulators, accumulation of plant hormones, etc. The stress-specific signals that emanate from these mechanisms, in turn, stimulate or repress several networks that may either permit growth and reproduction to continue under stress conditions, or enable the plant to survive the stress until more favourable situations return. Many stress-response pathways share signalling intermediates, allowing these pathways to be interconnected. When plants are exposed to multiple stresses, cross talk can occur between the hormones, secondary messengers, and protein kinases/phosphatases participating in each of the stress pathways (Figure 18.15).

Plants use a number of mechanisms to sense abiotic stress, and each one of these can act individually or collectively to activate downstream signal transduction mechanisms.

- Biochemical sensing involves the presence of specialized proteins that have evolved to sense a particular stress. For example, presence of calcium channels that can sense changes in temperature and alter calcium homeostasis.

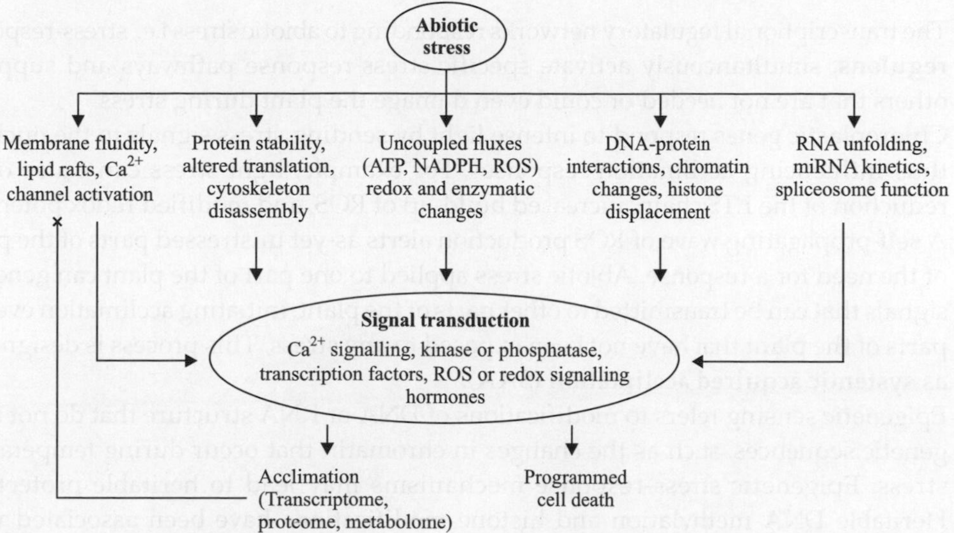

Fig. 18.14 Plants use physical, biophysical, biochemical, metabolic, and epigenetic mechanisms to detect abiotic stresses. A few early events of being subjected to abiotic stress, activates the signal transduction as well as response mechanisms to help the plant cope with such events.

> Possible temperature-sensing mechanisms in acclimation include: membrane fluidity, amounts of critical metabolites, protein configuration, redox status and the state of chromatin.

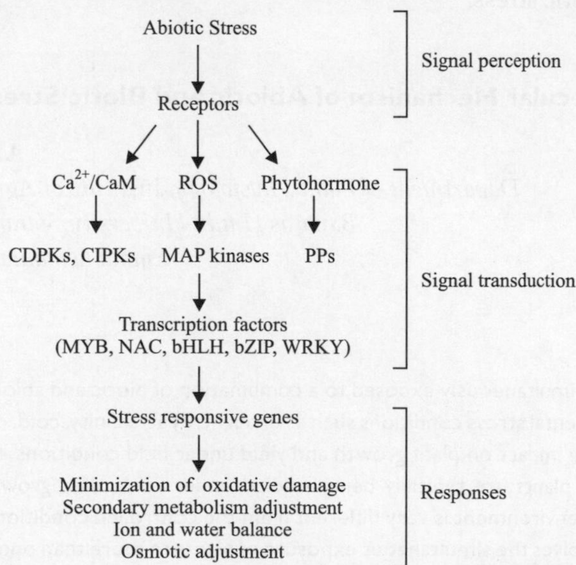

Fig. 18.15 Various signalling mechanisms are initiated in response to abiotic stress. ABA is an important mediator of stress signals. Ca^{2+} and phospholipids (second messengers), and protein phosphorylation contribute significantly in stress signalling pathways. This involves various protein kinases, including MAP kinases, calcium-dependent protein kinases (CDPKs) and CBL interacting protein kinases (CIPKs). Transcriptional regulation is quite essential for the regulation of stress genes.

- The transcriptional regulatory networks responding to abiotic stress i.e., stress-response **regulons**, simultaneously activate specific stress-response pathways and suppress others that are not needed or could even damage the plant during stress.
- Chloroplastic genes respond to intense light by sending stress signals to the nucleus, thus influencing acclimation responses. For example, light stress can cause over-reduction of the ETS chain, increased build-up of ROS, and modified redox potential.
- A self-propagating wave of ROS production alerts as-yet unstressed parts of the plant of the need for a response. Abiotic stress applied to one part of the plant can generate signals that can be transmitted to other parts of the plant, initiating acclimation even in parts of the plant that have not been exposed to any stress. This process is designated as **systemic acquired acclimation (SAA).**
- Epigenetic sensing refers to modifications of DNA or RNA structure that do not alter genetic sequences, such as the changes in chromatin that occur during temperature stress. Epigenetic stress-response mechanisms may lead to heritable protection. Heritable DNA methylation and histone modifications have been associated with specific abiotic stresses. The role of small RNAs in inhibiting protein synthesis and controlling gene expression during different abiotic stresses has also been emphasized.
- Plant hormones act individually or together to regulate abiotic stress response. ABA biosynthesis is among the most quick responses of plants to abiotic stress and is very effective in stomatal closure, resulting in decreased water loss through transpiration under water-stressed situations. The action of auxins, gibberellins, ethylene, jasmonates, salicylic acid, and brassinosteroids, have also contributed significantly towards plants' response to abiotic stress.

Molecular Mechanism of Abiotic and Biotic Stresses

A. Hemantaranjan

Department of Plant Physiology, Institute of Agricultural Sciences,

Banaras Hindu University, Varanasi 221 005, India

Email: hemantaranjan@gmail.com

Introduction

In nature, plants are simultaneously exposed to a combination of biotic and abiotic stresses that limit crop yields. Environmental stress conditions such as drought, heat, salinity, cold, or pathogen infection can have a devastating impact on plant growth and yield under field conditions. However, the effects of these stresses on plants are typically being studied under controlled growth conditions in the laboratory. The field environment is very different from the controlled conditions used in laboratory studies, and often involves the simultaneous exposure of plants to more than one abiotic and/or biotic stress condition, such as a combination of drought and heat, drought and cold, salinity and heat, or any of the major abiotic stresses combined with pathogen infection. Membranes are major targets of environmental stresses where lipids are important membrane components, and changes in their composition may help to maintain membrane integrity and so to preserve cell compartmentation under stress conditions. Though the final cellular response of plants is expected to be specific to stress combinations, the perception and signal transduction events could be operated through some known signalling components. These signalling components include hormone signals, receptors and transcription factors.

It is largely evidenced that the response of plants to a combination of biotic and abiotic stresses is complex involving interaction of various signalling pathways. The two or more stress conditions are exclusive and cannot be directly extrapolated from the response of plants to each of the different stresses applied individually. In addition, the simultaneous occurrence of different stresses results in a high degree of complexity in plant responses, as the responses to the combined stresses are largely controlled by different, and sometimes opposing, signalling pathways that may interact and inhibit each other.

Beyond that, plants exhibit tailored physiological and molecular responses and these tailored responses are suggested to appear only in plants exposed to concurrent stresses. Consequently, this information cannot be inferred from individual stress studies. The majority of laboratory studies on plant stress factors have analyzed each stress in isolation. There is increasing evidence that the response to multiple environmental stresses is distinct from that for individual stresses, and not merely additive.

Categorically, there is necessity for focusing on the response of plants to a combination of different stresses to perceive as to how divergent stress responses are integrated to have their impact on plant growth and physiological traits. Plant responses to different stresses are highly complex and engross changes at the transcriptome, cellular, and physiological levels. Transcriptome studies in recent years have been made in model plants such as tobacco (*Nicotiana tabacum*) and Arabidopsis (*Arabidopsis thaliana*) that have been subjected to multiple abiotic stresses. In the present scenario, evidence supports the view that plants respond to multiple stresses by activating a specific programme of gene expression relating to the exact environmental conditions encountered. To a certain extent, the presence of an abiotic stress can have the effect of reducing or enhancing susceptibility to a biotic pest or pathogen, and vice versa. This interaction between biotic and abiotic stresses is orchestrated by hormone signalling pathways that may induce or antagonize one another, especially that of abscisic acid.

In addition, precision in multiple stress responses is more controlled by a range of molecular mechanisms that act together in a complex regulatory network. The molecular signalling pathways controlling biotic and abiotic stress responses may interact and antagonize one another. Transcription factors, kinase cascades, and reactive oxygen species are key components of this cross-talk, as are heat shock factors and small RNAs. The endeavour is also to characterize the interaction between biotic and abiotic stress responses at a molecular level, focusing on regulatory mechanisms imperative to both pathways. Identifying master regulators that connect both biotic and abiotic stress response pathways is fundamental in providing opportunities for developing broad-spectrum stress-tolerant crop plants.

With the view to explain molecular mechanisms of biotic and abiotic stresses, it is obligatory to be aware with the concise working examples on (i) cross tolerance between biotic and abiotic stress followed by (ii) signalling pathways induced by multiple stress response; (iii) genes identified in the response to concurrent biotic and abiotic stresses; (iv) potential phenotypic response, which have been conversed comprehensively to have a preliminary idea regarding the molecular mechanisms of abiotic and biotic stresses based on recent researches scanned and simplified through unequivocal articles of reputed journals.

Cross-tolerance

With the aim of mitigating varied agricultural limitations, extensive effort has been used up examining the signalling and response pathways of plants to biotic and abiotic stresses. Each stress influences the morphological, metabolic, transcript and protein landscapes of the leaves and other organs in ways that illustrate a high degree of overlap with the responses to extra stresses permitting for cross-tolerance phenomena. There is significant overlap in signalling and response pathways to different biotic and abiotic stresses that include cellular redox status, reactive oxygen species, hormones, protein kinase cascades and calcium gradients as common elements. This overlap in signalling pathways is associated with cross-tolerance phenomena in which exposure to one type of stress enhances plant resistance to other biotic or abiotic stresses (Figure 18.16). Plants co-evolved with an enormous variety of microbial pathogens and insect herbivores under daily and seasonal variations in abiotic environmental conditions. As a result, plant cells display a high capacity to respond to diverse stresses through a flexible and finely balanced response network that involves components such as reduction-oxidation (redox) signalling pathways, stress hormones and growth regulators, as well as calcium and protein kinase cascades. Biotic and abiotic stress responses use common signals, pathways and triggers leading to cross-tolerance phenomena, whereby exposure to one type of stress can activate plant responses that facilitate tolerance to several different types of stress. Often, environmental pressure by abiotic and biotic stress can induce plant resistance. Though, some plants tackled with each stress individually have also been accounted to be more susceptible in contrast to an instantaneous exposure to two different stresses.

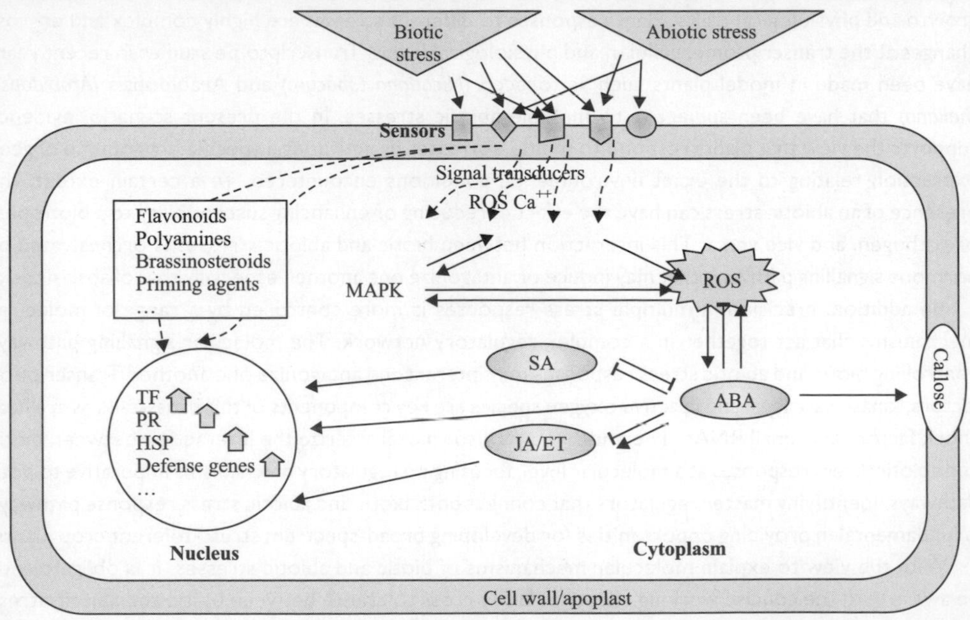

Fig. 18.16 The combination of both stress types leads to an increased accumulation of a large number of signalling compounds expressed as cross-tolerance.

Elements active in cross-tolerance between biotic and abiotic stress are first sensed by the plant cell and the information is transduced to exact downstream-located pathway(s).
Sensors and signal transducers are shared by two types of stressors.
Reactive oxygen species (ROS) and Ca^{2+} play a prominent role as transducers (messengers) and mitogen-activated protein kinases (MAPK) cascades are used by both types of stresses.
MAPKs are centrally positioned in Ca^{2+}-ROS crosstalk as well as in the signal output after exposure to a specific stress.

ROS represent crucial elements in the integration of both stresses during cross-tolerance.
Plant hormone signalling is of utter importance for stress adaptation, while abscisic acid (ABA) is predominantly involved in abiotic stress adaptation.
Salicylic acid (SA) and Jasmonate/ethylene (JA/ET) are more responsible for the plant's reaction to biotic stress.
Huge crosstalk takes place between the various hormonal pathways.
ABA signalling contributes positively to pre-invasion defence, responsible for enhancing callose deposition.
ABA presents a positive interaction with JA/ET signalling.
The activation of SA signalling by pathogen challenge can attenuate ABA responses.
ABA signalling negatively affects signals that trigger systemic acquired resistance, enhancing pathogen spread from the initial site of infection.
The interaction of SA, JA, and ET signalling results in increased resistance to pathogens.
Hormones, secondary metabolites, priming agents, and further chemicals located in the cytoplasm finally up-regulate transcription factors (TF), pathogenesis related (PR) and defence genes, heat shock protein (HSP) genes in addition to genes involved in protection against stress and thus lead to the phenotypic expression known as cross-tolerance.
Arrows: induction; flat-ended lines: repression.

Additionally, certain environmental stresses have the option to incline the plant with the purpose to allow it to act in response quicker and in a resistant way to more challenges. Consequently, cross-tolerance between environmental and biotic stress may influence a positive effect and enhanced resistance in plants and have significant agricultural implications. Intriguingly, abiotic stress regulates the defence mechanisms at the location of pathogen infection as well as in systemic parts, thus ensuring a development of the plant's innate immunity scheme. Certainly, to prepare the plant for the battle, the activation of various detoxifying enzymes, control hormones, signalling pathways, and gene expression are indispensable. Therefore, the defensive response of plants exposed to different stressors is expected to be complex including the interconnection of various signalling pathways regulating numerous metabolic networks.

Signalling pathways induced by multiple stress response

The interaction between abiotic and biotic stress induces complex responses to the different stressors. Under stress, the accumulation of certain metabolites positively affects a plant's response to both stresses and therefore protects it from multiple aggressors. Callose accumulation, changes in ions fluxes, ROS, and phytohormones are the first responses induced to combat the stress and the resulting signal transduction triggers metabolic reprogramming for defence. Understanding complexity of plant defence mechanisms and stress signalling is indispensable in order to minimize the impact of stresses on plant growth, development, and productivity, essentially to maximize their lifespan. Of course, transcription factors and mitogen-activated protein kinases may be particularly important in defining signal specificity as discussed in short in the following captions:

Reactive oxygen species (ROS)

A rapid generation of ROS is observed after stress sensing. One of the major roles of ROS is to serve as signalling molecules in the cells. The production of ROS is fine-modulated by the plant to avoid tissue damage. High levels of ROS lead to cell death, lower levels are mostly responsible to regulate the plant's stress responses. In biotic stress, ROS are mainly involved in signalling. This again might attenuate the oxidative stress caused by abiotic stress. Additionally, the production of ROS can help in cell-to-cell communication by amplifying the signal through the *Respiratory Burst Oxidase Homologue D (RBOHD)*; and can act as a secondary messenger by modifying protein structures and activating defence genes. In *Arabidopsis*, ROS production can be sensed by ROS-sensitive transcription factors leading to the induction of genes participating in the stress responses. ROS are inducers of tolerance by activating stress response-related factors like mitogen-activated protein kinases (MAPKs), transcription factors, antioxidant enzymes, dehydrins, and low-temperature-induced-, heat shock-, and pathogenesis-related proteins.

Mitogen-activated protein kinase (MAPK) cascades

Mitogen-Activated Protein Kinase (MAPK) cascades are activated in response to stress. They control the stress response pathways. MAPKs are highly conserved in all eukaryotes and are responsible for the signal transduction of diverse cellular processes under various abiotic and biotic stress responses, and certain kinases are involved in both kind of stress. In cotton, the kinase *GhMPK6a* negatively regulates both biotic and abiotic stress. MAPK pathways activated by pathogen attack are mediated by salicylic acid (SA), and the resulting expression of *PR* genes induces defence reactions. The *Arabidopsis* protein VIP1 is translocated into the nucleus after phosphorylation by MPK3 and acts as an indirect inducer of *PR1*. MAPK cascades are important in controlling cross-tolerance between stress responses. MPK3 and MPK6 are essential to show full primed defence responses, therefore, these two kinases could be important for mediating tolerance to further stresses. Over-expression of the *OsMPK5* gene and also kinase activity of OsMPK5 induced by ABA contributes to increased abiotic and biotic stress tolerance. MAPK signalling also interacts with ROS and ABA signalling pathways leading to enhanced plant defence and induction of cross-acclimation to both abiotic and biotic stress.

Hormone signalling under stress interaction

Increases in ABA concentration modulate the abiotic stress-regulation network while biotic stress responses are preferentially mediated by antagonism between other stress hormones such as SA and JA/ET. In certain cases, ABA has been shown to accumulate after infection. Recent findings show a positive effect of ABA on biotic stress resistance. This dual effect makes ABA a controversial molecule that can switch from 'good to bad' depending on the environmental conditions (type and timing of the stress). Moreover, under combination of abiotic and biotic stress, ABA mostly acts antagonistically with SA/JA/ethylene inducing a susceptibility of the plant against disease and herbivore attack. ABA and SA have an antagonistic role in plant defence against stressors.

Transcription factors and molecular responses

Transformations in gene expression occur after detection of a given stress, and the reprogramming of the molecular machinery is regulated by the action of transcription factors. The altered expression of certain genes is a key event in helping plants to set up an effective defensive state, and there is convincing evidence that many genes are multifunctional and capable of inducing tolerance in plants for more than one stress. The over-expression of certain transcription factors in plants confronted with cold stress and infection activates cold-responsive *PR* genes, thereby conferring protection against both stressors. The up-regulation of some transcription factors after exposure to abiotic stress leads to an accumulation of PR proteins. The transcription factors C-repeat Binding Factors (*CBF*), Dehydration-Responsive Element-Binding proteins (*DREB*), No Apical meristem, ATAF and Cup-Shaped Cotyledon (*NAC*) have been extensively studied as players of the primary abiotic stress signalling pathways ensuring tolerance under stress. In recent times, it has been proposed that the WHIRLY 1 protein and *REDOX-RESPONSIVE TRANSCRIPTION FACTOR1* (*RRTF1*) could participate in the traffic of communication between plastids and the nucleus. WHIRLY 1 perceives the redox changes in the plastid and carries the information to the nucleus in an NPR1-independent manner.

Many stress combinations lead to phenotypic damage and the defensive response is in accordance with the type of abiotic stress and the pathogens involved. Overall, the complex response of the plant stems from the interplay of specific signalling pathways involved in abiotic and biotic stress. The combination of both stress types leads to an increased accumulation of a large number of signalling compounds that demonstrate cross-tolerance. In this connection, the 'Omics' technology is one of these approaches. Transcriptomics, proteomics, and metabolomics have revealed plant responses under stress along with their underlying mechanisms and point to potential target genes, proteins or metabolites for inducing tolerance and improve plant responses. Much has yet to be known about the 'Omics' characterization of abiotic and biotic stress combinations.

Heat shock proteins and transcription factors

A family of highly conserved genes encodes heat shock proteins (HSPs) across all cells of both, prokaryotes and eukaryotes. These stress response proteins are constitutively expressed in cells under normal conditions, in the absence of environmental stresses, and function as molecular chaperones regulating protein synthesis, folding, assembly, translocation, and degradation. Their expression is increased upon exposure to a wide variety of environmental stimuli to protect cells against stress by preventing protein aggregation and stabilizing misfolded proteins and cellular homeostasis. Current findings reveal that over-expression of OsHSP18.6 augmented tolerance to diverse types of abiotic stresses including heat, drought, salt and cold in rice. In addition, genes encoding HSPs were induced into *Arabidopsis* by necrotrophic fungus *Botrytis cinerea* infection, cold, drought and oxidative stress. Therefore, HSPs may regulate pathogen defence as well as abiotic stress tolerance. Recently, melatonin (N-acetyl-5-methoxytryptamine) has been identified as an important secondary messenger that bestows resistance to heat, salt, drought, cold and pathogens such as *Pseudomonas syringe* pv. tomato in *Arabidopsis*. For this reason, melatonin may remarkably impact crosstalk among stress responses, and control biotic and abiotic stress response as a regulator of HSFs.

MicroRNAs

Only a few studies have been conducted on the role of miRNA in plant responses to both, biotic and abiotic stresses by using Solexa high-throughput sequencing. MicroRNAs (miRNAs) are a class of endogenous tiny non-coding, single-stranded RNAs of approximately 20–24 nucleotides in length. These happen to control the expression of their target genes in plants as key post-transcriptional gene regulators through translational repression or transcript cleavage. miRNAs have been implicated in plant growth and development, cell proliferation and cell death, organogenesis, auxin signalling and also known to perform key roles in response to a wide array of biotic and abiotic stresses. miRNAs were differentially expressed in response to abiotic stresses. miRNA159 and miRNA166 were inversely expressed in response to cold, while they showed the same pattern in response to ABA, mannitol, and NaCl.

Epigenetics and stress response

The term 'epigenetics' describes heritable changes in gene expression patterns that transpire without altering the underlying DNA sequence. Abundant progress has been made in elaborating epigenetic regulation of plant responses to stressors, which established that plants have evolved intricate epigenetic mechanisms in regulating plant responses to environmental factors comprising DNA methylation, histone post-translational modification, chromatin remodeling complexes, small non-coding RNAs (miRNAs and small interfering RNAs, siRNAs), and long non-coding RNAs.

Significant evidence has been collected, indicating that production of ROS, MAP kinase cascades, and hormone signalling are triggered by both, biotic and abiotic environmental stimuli in a non-specific manner and play crucial roles in the response to biotic and abiotic stresses as common components of the immune system. Receptor proteins as a frontline of the immune system are key regulators of crosstalk between biotic and abiotic signalling. Transcription factors as positive and negative regulators of plant defence also play a significant complex role in crosstalk and fine-tuning responses to combined biotic and abiotic stresses.

Genes identified that respond to concurrent biotic and abiotic stresses

In both tobacco (*Nicotiana tabacum*) and *Arabidopsis thaliana*, a combination of drought and heat stress induces a novel programme of gene expression, activating transcripts that are not induced by either stress individually. Correspondingly, microarray analysis has revealed that exposure to multiple biotic stresses (two species of herbivorous insect) elicits a transcriptional response that is distinct from each individual response. Phytohormone abscisic acid (ABA) is produced primarily in response to abiotic stresses and induces a range of downstream processes resulting in stress tolerance. By contrast, the response to biotic stresses is defined by antagonism between the hormones jasmonic acid, salicylic acid, and ethylene. Nevertheless, ABA can also act as a global regulator of stress responses by dominantly suppressing biotic stress signalling pathways. This may facilitate fine-tuning of plant stress responses with a view to focus on the most significant threat and can have both positive and negative effects on defence against pathogens.

Drought stress or ABA treatment can actually increase the susceptibility of *Arabidopsis* to an a virulent strain of *Pseudomonas syringae*, while ABA treatment in tomato (*Solanum lycopersicum*) increases susceptibility to the pathogens *Botrytis cinerea* and *Erwinia chrysanthemi*. ABA-deficient mutants of both *Arabidopsis* and tomato show increased resistance to pathogens. Further, transcription factors and mitogen-activated protein kinases may be particularly important in defining signal specificity, as they frequently control a wide range of downstream events and many are induced by more than one stress.

Genes that differentially regulated between individual and dual stress treatments were identified and selected for further analysis from overrepresented families of transcription factors that were strongly regulated by hormones. Overall, candidate genes with potential roles in controlling the response to multiple stresses were selected and functionally characterized as follows:

(i) *RAPID ALKALINIZATION FACTOR-LIKE8* (*AtRALFL8*) was induced in roots by joint stresses but provided susceptibility to drought stress and nematode infection when overexpressed. Constitutively expressing plants had stunted root systems and extended root hairs.

(ii) Plants may produce signal peptides such as AtRALFL8 to induce cell wall remodeling in response to multiple stresses.

(iii) The methionine homeostasis gene *METHIONINE GAMMA LYASE* (*AtMGL*) was up-regulated by dual stress in leaves, offering resistance to nematodes when overexpressed. It may regulate methionine metabolism under conditions of multiple stresses.

(iv) *AZELAIC ACID INDUCED 1* (AZI 1), involved in defence priming in systemic plant immunity, was noted to be down-regulated in leaves by combined stress and conferred drought susceptibility when overexpressed, potentially as part of abscisic acid-induced repression of pathogen response genes.

Potential phenotypic response

Potential phenotypic response was recorded when plants were exposed to combination of drought stress and pathogen infection. ABA plays positive role in pre-invasive defence. Pathogen infection on already drought stressed plants can either result in plant resistance to the pathogen through drought-induced activation of basal defence or can result in susceptibility due to weakened basal defence. Drought-induced pathogen resistance is presumably due to enhanced induction of anti-microbial PR-proteins activated by drought stress. PR-proteins can protect plants during early stages of pathogen infection. The susceptibility could be attributed to high levels of ABA in drought stressed plants, which can interfere with pathogen-induced plant defence signalling and thereby reduce the expression of defence-related genes.

Exposure of plants to simultaneous drought stress and pathogen infection can result in tolerance to biotic and abiotic stresses due to the inherent ability of plants to induce one and only tailored strategies. Contact of pathogen infected plants to drought stress can result in their tolerance to drought stress through pathogen-induced salicylic acid (SA)-dependent ROS signalling. On the other hand, it can also result in susceptibility of the plants to drought stress due to SA or jasmonic acid (JA)-mediated reduction in responsiveness of plants to ABA. Moreover, increase in ABA concentration before pathogen treatment can either prevent pathogen invasion through stomatal closure or ABA can make plants more susceptible to pathogen by suppressing JA or SA or ethylene-mediated signalling.

Conclusions and outlook

Biotic and abiotic stresses, as a part of the natural ecosystem, seriously impact crop productivity and threaten global food security. Plants in their natural habitats are persistently and simultaneously confronted with a range of biotic (biotrophic and necrotrophic fungi, bacteria, phytoplasmas, oomycetes and nematodes), and non-cellular pathogens. Relatively little information is available on the signalling crosstalk in response to combined biotic/abiotic stresses. Recent evidence highlights the complex nature of interactions between biotic and abiotic stress responses, significant aberrant signalling crosstalk in response to combined stresses and a degree of overlap, but also a unique response to each environmental stimulus. Therefore, it is vital to conduct plant transcriptome analysis in response to combined stresses to identify general or multiple stress- and pathogen-specific genes that confer multiple stress tolerance in plants under climate change. Recent advances in our understanding of the molecular mechanisms of crosstalk in response to biotic and abiotic stresses have been briefly discussed, and should be connected with substantial information already available on the physiological and metabolic changes in plants exposed to individual stresses. Therefore, stress combinations need to be handled as a different state of stress and studied at the laboratory level to adequately explore the interactions. In addition, it also requires a repository of combined stress interactions using past literature to allow users to know what pathogen causes positive or negative interaction on drought on what plant species at what stress level. Last but not the least, key factors identified from these approaches can be used to develop crop plants tolerant to simultaneous stresses.

18.2 Biotic Stress

Biotic stress occurs due to biological factors such as insects and disease, to which a plant may be subjected to during its lifespan. Plants respond to such conditions with alterations in the chemical and physical characteristics of cell walls and synthesis of secondary metabolites. These responses which are collectively known as **hypersensitive reactions;** function to isolate the potentially pathogenic organisms and stop their development and spread across the plant.

Hypersensitive reactions include the following events:

(i) Stimulation of defence-related genes and formation of their products, **pathogenesis-related (PR) proteins**
(ii) Induction of genes that code for enzymes responsible for the synthesis of antimicrobial compounds such as **'phytoalexins and isoflavonoids'** that limit the growth of pathogens
(iii) Alterations in the cell wall composition accompanied by its enhanced strength due to accumulation of **lignin, callose and suberin** in cell walls alongside **hydroxyproline-rich glycoproteins**
(iv) Invaded cells trigger **programmed cell death** which leads to the formation of **necrotic spots** at the site of infection

18.2.1 Beneficial interactions between host plants and microorganisms

In their natural habitats, land plants are colonized by several beneficial microorganisms such as nitrogen-fixing bacteria, mycorrhizal fungi, and rhizobacteria.

1. Symbiotic **nitrogen-fixing rhizobia release nodulation (Nod) factors**, which function as signalling agents when they are in vicinity of legume root. The interaction between specific nod factors and their corresponding receptors is the basis of host-symbiont specificity. Nod factors are lipochitin oligosaccharides that bind to a specific class of **receptor-like kinases (RLKs)** containing N-acetyl glucosamine-binding lysine motifs (LysM) in the extracellular domain. Upon binding to Nod factors, Nod factor receptor (NFR) heterodimer initiates two separate processes:
   - The first involves a signalling pathway that facilitates the infection process itself, and
   - The second involves the activation of a set of genes that regulate the formation of root nodules.

   A second type of receptor, called the **symbiosis-receptor like kinase (SYMRK)** is thought to participate in both the processes. Upon binding to Nod factors, SYMRK is thought to activate subsequent steps shared by these associations, including calcium spiking inside and around the nucleus of the infected epidermal cell, which causes the activation of the core symbiotic genes. The final steps leading to nodulation involve cytokinin signalling mechanism.
2. Arbuscular mycorrhizal associations and nitrogen-fixing symbionts include closely related signalling mechanisms. In leguminous plants, SYMRK and many other genes are required both for legume nodulation and for arbuscular mycorrhiza formation.

Just like the key symbiotic signals (Nod factors) produced by rhizobia, the arbuscular mycorrhizal fungus releases lipochitin oligosaccharides called MYC factors that stimulate the formation of mycorrhiza in several plants. Nitrogen-fixing symbionts and arbuscular mycorrhizas may also involve related receptors.

3. Rhizobacteria can enhance availability of nutrients, initiate branching of roots, and provide protection against pathogens.`A group of **plant growth promoting rhizobacteria (PGPR)** provide several beneficial services to growing plants . For example, volatiles released by bacteria facilitate iron uptake, higher chlorophyll content and photosynthetic efficiency, increased size, alteration of root architecture. PGPR microbes can provide cross-protection against pathogenic organisms by activating the induced systemic resistance pathway and can also control the build-up of harmful soil organisms. In addition, *Pseudomonas aeruginosa* can alleviate the symptoms of biotic and abiotic stress by the release of antibiotics or iron-scavenging siderophores. The plant exerts control over the bacterial population by regulating bacterial quorum sensing signalling pathways through the release of exudates by the roots.

18.2.2 Harmful interactions between plants, pathogens and herbivores

Plants have developed various modes to protect themselves against insect pests and pathogens, mechanical barriers, constitutive chemical defences, and direct as well as indirect inducible defences.

1. **Mechanical barriers** offer the first line of defence against pests and pathogens for several species of plants that include surface structures (thorns, spines, prickles, and trichomes), mineral crystals (silica crystals, phytoliths[8], and calcium oxalate crystals raphides) and thigmonastic (touch-induced) leaf movements in *Mimosa pudica*.
2. **Chemical defence mechanisms** comprise a second line of defence against plant pests and pathogens. Plants can synthesize a wide array of constitutive secondary metabolites and tend to accumulate toxic ones in storage organelles such as vacuoles, or in specialized anatomical structures such as resin ducts, laticifers, or glandular trichomes. Following an attack by herbivores or pathogens, toxins are released and become active at the site of damage, without adversely affecting vital growing areas. Plants tend to store defensive compounds in the vacuoles as non-toxic water soluble conjugates of sugars that are spatially separated from their activating hydrolases.
3. **Inducible defence responses to insect herbivores**
   (i) Rather than producing defensive secondary metabolites continuously, plants can save energy by producing defensive compounds only when triggered by mechanical damage (such as wind or hail) or specific components of insect saliva (**elicitors**). Elicitors derived from insects can activate signalling pathways systemically i.e. throughout the plant, thereby initiating defence responses that can minimize further harmful effects in distal regions of the plant.

[8] Phytoliths are rigid microscopic structures comprised of silica, found in many plant tissues, which persist even after their decay.

(ii) Jasmonic acid (JA) and salicylic acid are rapidly produced in response to insect attack and induces transcription of genes controlling the plant defences. JA functions through a conserved ubiquitin-ligase signalling pathway. JA-induced growth suppression helps in the relocation of nutrients to metabolic pathways regulating defence mechanisms.

(iii) Herbivore damage can induce systemic defences by causing the synthesis of polypeptide signals. For example, **systemin** is released to the apoplast and binds to receptors in undamaged tissues, activating JA synthesis there.

(iv) Other early events associated with insect herbivory are calcium signalling and stimulation of the mitogen-acivated protein kinase (MAPK) pathway. Following activation by an elicitor, increased cytosolic calcium level activates target proteins e.g., calmodulin, and protein kinases, which then activate downstream targets of the signalling pathway i.e. protein phosphorylation and transcriptional activation of stimulus-specific responses.

(v) In addition to polypeptide signals, plants can also project long-distance electrical signals to initiate defensive responses in as-yet undamaged tissues. A group of **glutamate receptor-like (GLR) genes** has been discovered in some plants during herbivory. The electrical signals travel through the vascular system and initiate defence responses *via* JA production. (Figure 18.17).

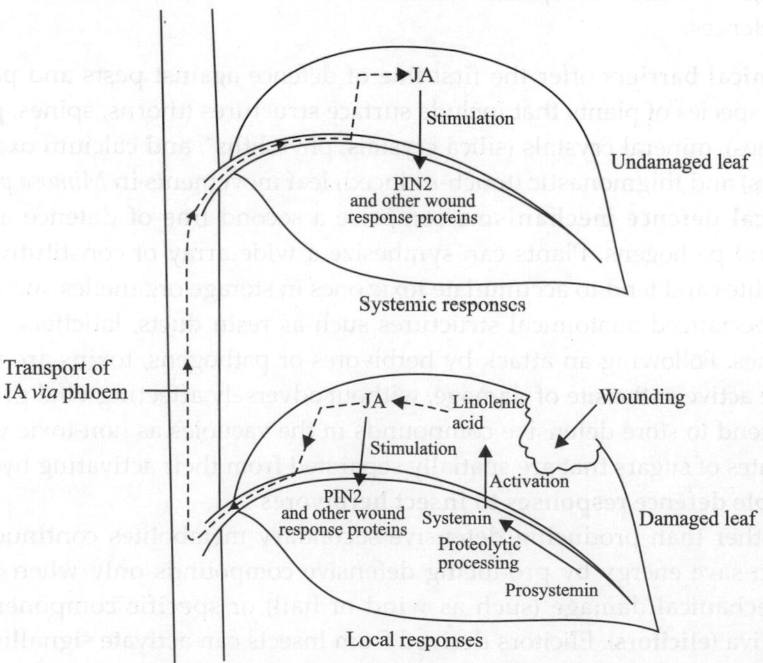

Fig. 18.17 Local and systemic responses to wounding. Wounding promotes proteolytic processing of prosystemin to systemin (a peptide hormone), which induces the synthesis of jasmonic acid (JA) from linolenic acid. JA moves through the phloem and stimulates the transcription of genes encoding proteinase inhibitor (PIN2) and other wound response proteins locally as well as in nearby tissues.

(vi) Plants may release volatile compounds to attract natural enemies of herbivores (predators or parasites) or signal neighbouring plants to initiate defencce mechanisms. In addition, plants also release lipid-derived products, such as green-leaf volatiles (a mixture of 6-C aldehydes, alcohols, and esters) in response to mechanical damage. Moreover, they activate the production of phytoalexins and other anti-microbial compounds and appear to play a major role in the overall defensive strategies of plants. Herbivore-induced volatiles can also function as systemic signals in a plant. Just like several other aspects of plant metabolic and developmental activities, defensive responses to herbivores and pathogens are controlled by circadian rhythms.

18.2.3 Responses to pathogen invasion or plant defences against pathogens

Plant pathogens have evolved various mechanisms to invade their plant hosts and cause disease. These can penetrate through cell walls by secreting lytic enzymes, as well as through stomata, hydathodes, lenticels, and wound sites (caused by insect herbivores). In addition to this, several kinds of viruses are also transmitted by insect herbivores, which function as vectors. Pathogens normally follow any of three attack strategies: **Necrotrophic pathogens** invade their host by secreting **cell wall degrading enzymes (CWDEs)** ; **biotrophic pathogens** survive on substrates supplied by their hosts; **hemibiotrophic pathogens** keep host cells alive but cause extensive tissue damage.

- Pathogens can release a vast array of effector compounds that aid in initial infection. Effectors (i.e., enzymes, toxins, and growth regulators) support their ability to successfully colonize their host and gain nutritional benefits. The effectors of some pathogenic bacteria, such as *Xanthomonas*, are proteins that target the plant cell nucleus and cause marked alterations in gene expression. The **transcription activator-like (TAL)** effectors bind to the host plant DNA and trigger the expression of genes beneficial to the pathogen's growth and subsequent dissemination.
- All plants have **pattern recognition receptors (PRRs)** that set off defensive responses when activated by evolutionarily conserved **microbe-associated molecular patterns (MAMPS)** e.g., flagella, and chitin. Pathogen infection can produce molecular 'warning signals' that are recognized by cell surface pattern recognition receptors.
- Plant resistance genes encode cytosolic receptors that perceive pathogen-derived effector gene products in the cytosol. Binding of an effector gene product to its receptor initiates anti-pathogen signalling pathways.
- In response to infection, many plants produce phytoalexins, secondary metabolites with strong anti-microbial activity.
- Another anti-pathogen defence is the hypersensitive response, in which cells engulfing the infected region die quickly, thereby limiting the spread of infection. The hypersensitive response is normally preceded by quick production of **ROS** and **NO**, which may kill the pathogen directly or aid in cell death.
- A plant that survives local pathogen infection often develops increased resistance to subsequent attack, a phenomenon called **systemic acquired resistance (SAR)**.

- Interactions with non-pathogenic bacteria can stimulate **induced systemic resistance (ISR)**.

18.2.4 Plant defences against other organisms

Nematodes are parasites that move between hosts and induce the formation of feeding structures and galls from vascular plant tissues. In response, plants use defensive signalling pathways similar to those used for pathogen infection. Some plants produce allelopathic secondary metabolites that enable them to outcompete nearby plant species. A few plants are parasitic on other plants. Parasitic plants can be classified into two major groups, **hemiparasites** and **holoparasites**, based on their potential to perform photosynthesis. Parasitic plants use a specialized structure, the **haustorium**, to penetrate their host, grow into the vasculature, and absorb nutrients. Some parasitic plants detect their host by the specific volatile profile that is constitutively released.

Immediate responses of invaded cells

- Generation of ROS
- Synthesis of nitric oxide
- Opening of gated ion channels
- Phosphorylation/dephosphorylation of proteins
- Rearrangement of cytoskeleton
- Hypersensitive cell death
- Gene induction

Local responses and gene activation

- Modifications in the secondary metabolic pathways
- Ceasing of cell cycle
- Production of PR proteins
- Build-up of benzoic as well as salicylic acid
- Synthesis of ethylene and jasmonic acid
- Fortification of cell walls

Systemic responses and gene activation

- β(1,3)-glucanases
- Chitinases
- Peroxidases
- Synthesis of other PR proteins

Plants combat biotic stress by conducting chemical warfare with allelopathic secondary compounds. Secondary metabolites secreted from roots are also used by parasitic plants to locate suitable hosts. Parasitism by nematodes and herbivory by grazing animals can inflict considerable damage on plants. Some species defend themselves by producing spines or hairs. Most plants have a range of local and systemic responses that are invoked by herbivory, predation and wounding. Many of these reactions have features in common with abiotic stress responses, such as ROS production and accumulation of phenolics, tannins and phytoalexins. The synthesis of protease inhibitors, which interfere with invertebrate digestive systems, is a frequent response to herbivore attack.

Wounding can also induce the release of volatile compounds such as, methanol, acetone, formaldehyde and other short-chain carbonyl compounds, terpenoids, phenylpropanoids and derivatives of fatty acids, that act as a signal to different parts of the plant or to nearby plant species in the community. Wounding often stimulates systemic responses in the whole plant as a consequence of the export of signal molecules from a wound site *via* the vascular system. Wounding promotes proteolytic processing of precursor prosystemin (200-amino acids) to form the peptide hormone systemin (18-amino acids). Systemin engages with a receptor present in the plasma membrane and activates the oxylipin pathway of jasmonic acid synthesis from linolenic acid (C 18:3)as well as the transcription of genes encoding a proteinase inhibitor (PIN2). Jasmonic acid moves through the phloem to remote undamaged tissues where it induces the synthesis of PIN2 and other systemic wound response proteins.

Two common features of all stresses are that they: (i) activate or deactivate particular genes or gene families (ii) have a negative impact on photosynthetic productivity. As a result, abiotic and biotic stresses collectively decrease crop productivity.

Review Questions

RQ 18.1. How is it possible for plants growing in temperate and desert habitats to 'sense' and 'react' to an approaching freezing winter or a long dry spell conditions, respectively?

Ans. To survive periods of unfavourable environmental conditions such as frost or drought, plants have devised ways and means by which they are able to 'stop' as well as 'start' physiological processes and have developed a mechanism of sensing (perceiving environmental signals) and reacting to cope with adverse conditions.

Much before the onset of winter-freezing conditions, these temperate plants 'start preparing' themselves by developing features like:

- The scale-covered winter buds. These small thickened, closely appressed scale-like leaves form a protective cap around the growing point. Then scales are also covered by a varnish-like, sticky insulating material or are covered by a tuft of cottony insulating material.
- It is usually accompanied by the senescence and shedding of leaves.
- Below the terminal bud, all leaves have nearly abscised and the leaf scars marking point is well protected by a corky layer.
- The cambial cells become dormant while the parenchyma cells of xylem and phloem have developed very high osmotic concentration that enable them to withstand freezing injury.
- Prior to winter freezing, the inhibitor of IAA appears to be made in the leaves and there is a significant rise in the ABA level which leads to dormancy of buds, and thus reduces the active vegetative growth of a bud.

Winter dormancy persists until it is terminated in spring when there is resumption of growth and ABA is degraded but there is an increase in amounts of various other hormones e.g., gibberellins and cytokinins.

Similarly, in xerophytes, ABA is generated in high concentration during water stress that results in efflux of potassium (K^+) ions from the guard cells to subsidiary cells and simultaneous influx of H^+ ions from subsidiary cells to guard cells. In the absence of potassium malate, the stomata close and this is how the plants are able to conserve water.

Large amounts of ABA are exported from wilted leaves to the rest of the plant, moving through phloem.

RQ 18.2. Explain how deep water or floating rice is able to keep pace with rising flood waters?

Ans. Deep water or floating rice can grow up to a height of about 4.6–6 m. The upper part of the culm keeps floating on the water surface. Submergence of young rice plants during monsoon period triggers an increase in ethylene biosynthesis which elicits internodal elongation that provides a mechanism for the rice plants to keep pace with the rising water level. The extent of internodal elongation can go up to about 'a foot' in a single day.

Ethylene also promotes the formation of aerenchyma in submerged roots and stems.

RQ 18.3. Give at least three examples of growth hormones that are complimentary to each other.

Ans. Quite often, hormones interact together in controlling a developmental response such as:

1. Ethylene and gibberellins interact in the regulation of internode elongation in deep water rice.
2. Ethylene and ABA interact in regulating the process of seed germination.
3. Ethylene as well as auxin interact in the regulation of a characteristic phenotypic growth response known as the 'triple response'.

The three elements of 'triple response' are: hypocotyls become shorter and thicker; roots also become shorter and thicker (stunted); and the apical hook becomes exaggerated. The triple response probably protects delicate structures such as shoot apical meristem from damage as the seedlings force their way out through the soil.

RQ 18.4. Some plants are able to survive if flooded for a long time. Comment.

Ans. During and after rain, soil air is displaced by water, and the roots either have lesser amounts of oxygen (hypoxia) or none at all (anoxia). If flooded for a prolonged period, the roots of 'almost' all species suffer damage. Aquatics, however, survive because of oxygen diffusion from the stems through aerenchyma channels in the cortex or pith-in the petioles of water lilies, cattail (*Typha latifolia*) and rice plants. During flooding, ethylene builds-up in submerged stems and it induces the formation of aerenchyma in certain species.

Hormone levels undergo a change in flooded plants – ethylene level, for example, often increases, while the level of 'gibberellins' and cytokinins usually declines.

RQ 18.5. Explain how have sun-loving plants been able to overcome shade cast by other plants?

Ans. Plants grown in full sunlight have short internodes and petioles but while growing under a foliar canopy they exhibit the classic shade avoidance response, i.e., tending to expend more resources into elongation of stem and petioles and less into expansion of leaves. This 'shade' avoidance response enables the plants to grow out quickly out of the shade and into direct sunlight, thus maximizing the capture of sun's energy for photosynthesis. The sunlight filtering through the vegetation has relatively high far-red content which will be detected by phytochrome B, together with its close relatives phytochromes D and E (not by chlorophyll) and the plants respond back by developing extended internodes and petioles and thus emerge out of the shade cast by the neighbouring plants.

RQ 18.6. In which ways does ABA participate in different kinds of stress resistance?

Ans. ABA appears to be specially participating in many kinds of stress resistance such as:-

- Heat-stressed leaves
- Waterlogged roots
- Chilling
- High salinity

All of these are associated with sudden increases in ABA. When plants begin to wilt, the concentration of ABA in leaf cells increases dramatically from about 20 μg/kg fresh weight to 500 μg/kg, and guard cells close stomatal pores. Large amounts of ABA are exported from wilted leaves to the rest of the plant, moving through phloem.

RQ 18.7. Enumerate the survival strategies developed by the plants of arid regions.

Ans. Plants growing in arid regions have developed extraordinary survival mechanisms by developing many structural and physiological changes.

- Seeds of many desert annuals remain dormant. Once the seeds germinate, they grow rapidly, bloom and complete their life cycle within just a few weeks.
- Desert vegetation is characteristically dwarf, i.e., low growth habits with deep-feeder root systems and restricted leaf areas that help in minimizing the rate of transpiration.
- Transpiration rate is also reduced by structural adaptations such as small, thick, heavily cutinized leaves; relatively small number of sunken stomata; multiple epidermis.[9]
- Mechanisms for folding or rolling of leaves when under water stress.
- Succulent plants store up large amounts of water in special water storage tissues in their leaves or stems.
- The leaves of many xerophytes are covered with copious hairs (hairiness) that reflect light and prevent overheating of leaves (i.e., reducing the 'heat load') – an adaptive mechanism that helps in thermoregulation. The felt of hairs on the leaf surface increases 'boundary layer resistance' which minimizes transpiration.
- Most remarkably, the stomata of some xerophytes (CAM plants) show reverse rhythmic cycles, i.e., open at night time when transpiration rate is low and are closed during day time when conditions are conducive to transpiration.
- Some plants avoid water stress problems by simply shedding their leaves during drought.
- Under water deficit conditions, there is dramatic rise in the level of ABA which leads to stomatal closure.
- Even midday closure of stomata is seen under high temperatures in the leaves because of build-up of CO_2 due to increased respiratory rate, thus helping in conserving as much water as possible. The transpiration in such desert succulent plants is minimal at midday.

[9] Small leaves provide thin boundary layers and result in efficient convective heat transfer. Their temperatures are thus closely coupled to air temperature. Sunken stomata result in high leaf resistance that tends to retard transpiration.

- Many plants cope up with water stress by osmotic adjustment, i.e., accumulation of osmotically active solutes.
- In response to high temperatures, plants synthesize a new group of proteins with relatively high molecular weight (up to 70 kD), the **heat-shock proteins** (HSPs), while the formation of normal proteins declines. The HSPs appear rapidly, constituting at one time a high percentage of the total protein content. The protein synthesis appears to be self-regulatory as it is switched off after 6–8 hours at high temperatures while the synthesis of normal proteins is resumed. The HSPs have an important role in heat tolerance, probably by affording protection to essential enzymes and nucleic acids from getting denatured.

RQ 18.8. Plants survive and even grow in freezing winter. How do they do it?

Ans. Trees and shrubs in temperate, subarctic and alpine regions overcome the effects of winter by undergoing structural and physiological changes. Cold temperate vegetation can, in fact, anticipate and prepare the plants before the onset of winter temperature conditions. Thus, it is an advance warning of impending dangers of the coming winters with near-freezing temperatures. Winter dormancy persists until it is terminated by an appropriate environmental signal that indicates that winter has passed and the plants then begin to grow again when favourable temperatures have returned.

It is believed that ABA accumulates during induction of dormancy in various temperate plants, and its level declines during favourable weather conditions. The temperate plants drop their foliage during chilling winters, leaving behind only more resistant branches with their buds encased in hard scale-like leaves covered with varnish-like material.

- Early in acclimation (cold hardiness), various solutes accumulate in the cells, lowering the osmotic potential and decreasing the likelihood of freezing by lowering the freezing point. The major damage to the cells resulting from freezing comes from the development of ice crystals inside the cells; as these crystals grow, they rupture the cell membranes and kill the cells.
- Cold temperate plants become less brittle because of the greater degree of unsaturation of the membrane lipids, thus lowering their melting points, and permitting them to remain semi-liquid at lower temperature. An increase in the amounts of unsaturated fatty acids or sterols causes the membranes to remain functional at low temperature.

Another kind of protection against freezing injury involves the synthesis of greater quantities and new types of proteins (rich in -SH groups, characteristic of the amino acid cystine), which have the unique property of remaining highly hydrated as much we see in the hydrated flour of baker's dough. This water is virtually unfreezable as it is kept in the vicinity of the protein molecule by forces that prevent ice crystal formation. The more the concentration of such proteins in a cell, the more resistant it will be to freezing. In many of the plants, there is a regular cycle of synthesis of such proteins during winter, and their use as a metabolite in the spring.

> Extracellular (or apoplastic) ice formation does not kill plant cells while intracellular ice disrupts the fine structure of the cell and invariably results in death. Freezing tolerance develops because of 'antifreeze proteins' that inhibit the formation of ice crystals. These proteins as well as sugars (particularly sucrose) protect the cells from freezing by preventing the ice-crystal formation.

RQ 18.9. Enumerate the measures the gardeners or farmers employ to protect their cold-sensitive shrubs.

Ans. The different measures adopted by farmers to protect their orchards from extreme of cold are:

- Use a covering of burlap to (i) protect the plants against excessively low temperatures and (ii) to also shelter the orchards by protecting it against the chilling effects of winds.
- In cold-sensitive orange or peach orchards, frost damage may be averted by employing wind machines which look-like giant aircraft propellers on towers. The machine forces the warmer air of upper levels down to the ground, thus preventing the leaf surfaces from cooling down below freezing by way of thermal emission (radiative heat loss).
- Growers may also light small oil or smudge fires in orchards to warm the air by several degrees, thus minimizing frost damage.

Chapter 19

Secondary Plant Metabolites

Keywords

- Allelopathy
- Nectar guides
- Phytoalexins
- Phytoecdysones
- Primary metabolites
- Secondary metabolites

Plants synthesize a vast array of organic compounds that seemingly have no known role in either assimilation or during growth and development of the organism. These compounds, which are natural products, are called **secondary metabolites** (Figure 19.1). Secondary metabolites are different from primary metabolites (e.g., proteins, lipids, carbohydrates and nucleic acids) because of their restricted distribution in various groups of plants (Figure 19.2). Scientists have long thought that these compounds provide protection to plants from predators and pathogens owing to their toxic effect and repellent nature to herbivores and microbes when studied *in vitro*. Recent studies have shown that the expression of secondary metabolites can be modified by advanced molecular techniques.

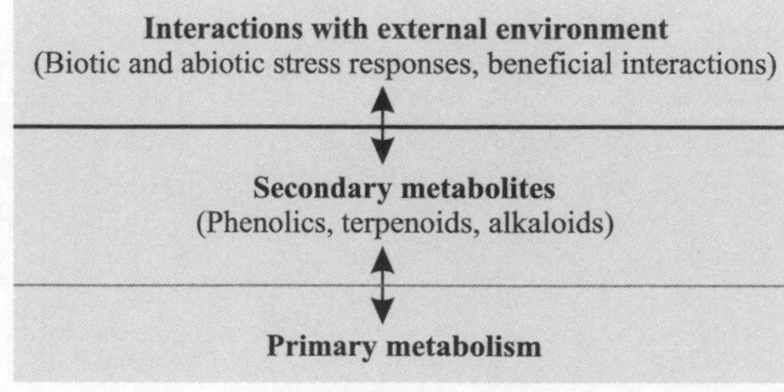

Fig. 19.1 Secondary metabolites (such as terpenoids, phenolics, alkaloids) play defensive role in plants and also act as modulators in response to abiotic stress. Secondary metabolites lie at the interface between primary metabolism and the interactions of organisms with their external environment.

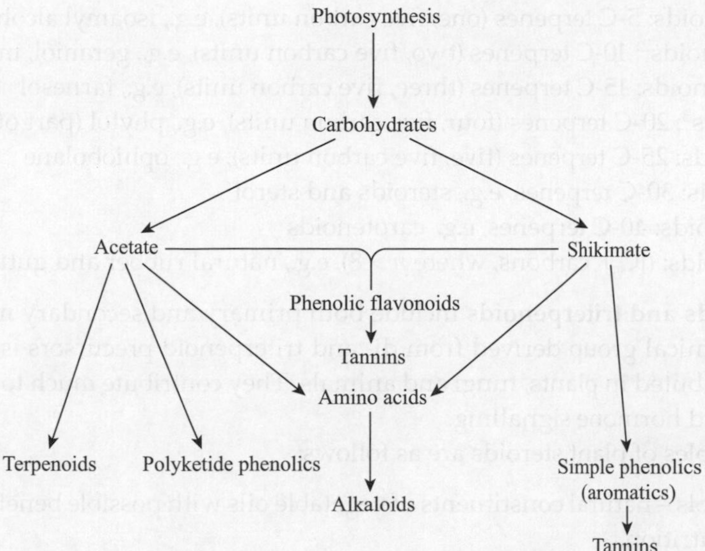

Fig. 19.2 A simplified scheme for the biosynthesis of the several classes of natural products.

Plant secondary metabolites can be classified into four distinct groups, such as **terpenes, phenolics, alkaloids and glycosides.**

19.1 Terpenes (Terpenoids)

Terpenoids take their name from terpenes, the volatile constituents of turpentine (a solvent produced from the distillation of pine tree resin). These include both primary and secondary metabolites. The terpenes represent the largest group of secondary metabolites. The diverse substances belonging to this group are normally insoluble in water and biosynthesized either from acetyl CoA or intermediates of glycolysis. Terpenoids and their derivatives may be considered polymers of isoprene[1] units which consist of the branched five-carbon isoprene skeleton (Figure 19.3). Consequently, terpenoids are often referred to as isoprenoid compounds. The repetitive 5-C structural motif from which terpenoids are built is called the **prenyl group**. Terpenes are grouped by the number of 5-C units they have in their skeleton:

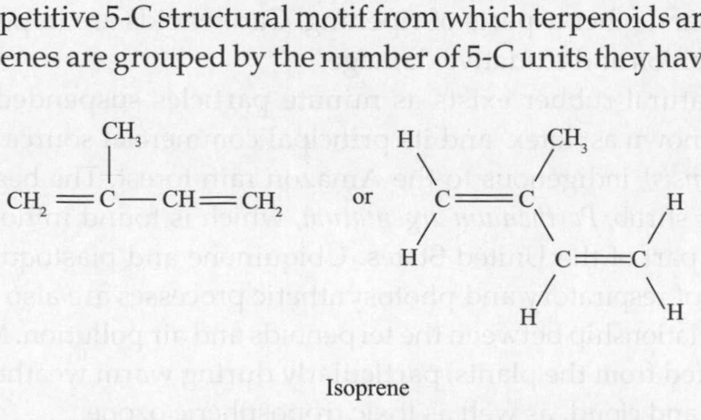

$$CH_2=C(CH_3)-CH=CH_2$$

or

$$(H)(H)C=C(CH_3)-C(H)=CH_2$$

Isoprene

Fig. 19.3 Isoprene, the building blocks of the terpenes.

[1] Isoprene is also a principal reactant in the formation of ozone in the troposphere.

Hemiterpenoids: 5-C terpenes (one, five carbon units), e.g., isoamyl alcohol, tiglic acid
Monoterpenoids[2]: 10-C terpenes (two, five carbon units), e.g., geraniol, menthol
Sesquiterpenoids: 15-C terpenes (three, five carbon units), e.g., farnesol
Diterpenoids[3]: 20-C terpenes (four, five carbon units), e.g., phytol (part of chlorophyll)
Sesterpenoids: 25-C terpenes (five, five carbon units), e.g., ophiobolane
Triterpenoids: 30-C terpenes, e.g., steroids and sterol
Tetraterpenoids: 40-C terpenes, e.g., carotenoids
Polyterpenoids: ($[C_5]_n$ carbons, where $n > 8$), e.g., natural rubber and gutta

Diterpenoids and triterpenoids include both primary and secondary metabolites. An important chemical group derived from di- and triterpenoid precursors is **steroid** which is widely distributed in plants, fungi and animals. They contribute much to the stability of membranes and hormone signalling.

Some examples of plant steroids are as follows:

- Phytosterols – natural constituents of vegetable oils with possible beneficial effects for human nutrition.
- Brassinosteroids – a class of plant hormones.
- Sterols of plant membrane-cholesterol (C_{27}), campesterol (C_{28}), sitosterol and stigmasterol (C_{29}) – all derived from C_{30} triterpenoid precursor.

The **phytoecdysones**, first discovered from the common fern, *Polypodium vulgare*, are a class of plant steroids and have a structural organization similar to the insect moulting hormones. Their ingestion by insects completely impairs moulting and other developmental processes, often with deadly lethal consequences.

The majority of **tetraterpenoids** are **carotenoids,** many of which, especially violaxanthin and lutein, have essential roles in photobiology and oxygen metabolism across the taxonomic range of autotrophs and are therefore **primary metabolites**. **Capsorubin**, a carotenoid secondary metabolite, is responsible for the red colour of the spice paprika. **Apocarotenoids** (e.g., ABA), the products of oxidative cleavage of carotenoids, are widely distributed in living organisms and frequently perform important regulatory, receptor and signalling functions. Carotenoid synthesis is a central feature of fruit development in many species contributing to colour changes and other aspects of ripening. Carotenoids are important in protecting against photoinhibition and oxidative damage.

In the plant, natural rubber exists as minute particles suspended in a milky-white emulsified form known as 'latex' and its principal commercial source is the Para Rubber tree (*Hevea brasiliensis),* indigenous to the Amazon rain forest. The best known source of gutta is the desert shrub, *Parthenium argentatum,* which is found in northern Mexico and the Southwestern part of the United States. Ubiquinone and plastoquinone, the electron transport carriers of respiratory and photosynthetic processes are also **polyterpenoids**.

There is close relationship between the terpenoids and air pollution. Many of the volatile essential oils emitted from the plants, particularly during warm weather, contribute to the formation of haze and cloud, as well as toxic tropospheric ozone.

[2] Menthol is cyclic monoterpenoid.

[3] Gibberellins are cyclic diterpenoids.

Norterpenoids – Terpenoid derivatives that have lost one or more carbons and are no longer C5 multiples. For example: some gibberellins are C19 structures.
Meroterpenoids – These natural products of mixed synthetic origins are partially derived from terpenoids. For example: cytokinins.

Physiological Roles of Terpenes

1. Terpenes are toxins and anti-feedants to many of the plant-feeding insects and mammals; thus, they seem to have defensive roles in the different groups of the plant kingdom, such as diterpenoids of resin from conifer species, thus, helping plants to repel attack by insects.
2. Pyrethrins (monoterpenoids) – originally identified in members of family Asteraceae – are popular components in commercial insecticides due to their negligible toxicity to mammals.
3. Terpenoids often accumulate in specialized structures such as glandular trichomes (on the leaf surface), secretory cavities, resin ducts and blisters. Many plants, such as citrus, mint, basil, peppermint, sage, eucalyptus, produce a mixture of monoterpenes and sesquiterpenes (essential oils), which impart a characteristic aroma to their leaves. Essential oils are well-known for their insect repellent properties.
4. The terpenoid essential oils emitted by the glandular epidermis of petals encourage insect pollination. Specialization of the epidermis is also associated with the formation and excretion of triterpenoid surface waxes, while laticiferous ducts produce certain triterpenoids and rubber.

***Biosynthesis*:** All terpenoids are biosynthesized from simple primary metabolites by a four-step process:

Step 1: Formation of isopentenyl pyrophosphate (IPP) and its isomer form i.e. dimethyl allyl pyrophosphate (DMAP).
Step 2: Synthesis of a series of prenyl pyrophosphate homologs by repetitive additions of IPP.
Step 3: Synthesis of terpenoid skeletons from prenyl pyrophosphate catalyzed by specific terpenoid synthases.
Step 4: Secondary enzymatic modifications of terpenoid skeletons.

Taxol, saponins and cardenolides are examples of bioactive products of step four activities.

The anticancerous drug taxol, obtained from yew (*Taxus* spp.) is a diterpene nucleus. Saponins and cardenolides are powerful bioactive compounds that are toxic to herbivores. The widely used cardioactive drug digitoxigenin is the aglycone moiety derived from foxglove (*Digitalis*) cardenolides.

The Anti-feedant Property of Neem

The anti-feedant property of the Neem tree was put on record for the first time in November 1902 when the newspaper, *The Times of India* reported from Trombay, near Bombay (now Mumbai), 'this morning the trees are almost denuded and the locust, hanging on the twigs and branches, appear to have changed the colour of foliage, for the tops of the tree seem covered with a yellow blight, which on close inspection turns out to be "locusts". They ate almost all the vegetative parts, leaving a bare landscape when they flew away, except for "Neem Trees" with their bitter leaves'.

The Neem or margosa or Indian Lilac Tree (*Azadirachta indica* A.Juss) of the family Meliaceae is a native of India and Myanmar. It is perhaps the most valuable traditional plant in India and has been used here for several thousands of years for multiple purposes. It is now commonly grown in the arid, subtropical and tropical regions of the Southeast Asian Countries, America, Australia and the adjoining South Pacific islands. The tree bears a large crown with scythe-like leaves with a serrated margin. The flowers and fruits develop in axillary clusters. When fully mature, the greenish yellow ellipsoidal drupes consist of a sweet-tasting pulp, enclosing within a seed 1–3 greenish yellow kernels, containing **Azadirachtin** and its homologues. However, the 'anti-feedant property' of Neem was reported to the Western World in November 1959 when a German Entomologist, Professor Dr Heinrich Schmutterer of the Institute of Phytopathology and Applied Zoology, Justus-Liebig University, Giessen (Germany) noticed that it was the 'only one green plant' standing after a swarm of locust swept through the Sudan. Dr Schmutterer has the honour of being 'the Father of the Neem Tree' and he authored a book called *The Neem Tree* with exhaustive details about the tree.

Most parts of the neem tree such as bark, roots, leaves, flowers, seeds, fruits, gum, resin, seed cake, seed oil, fruit pulp as well as its twigs and timber offer solutions to global agricultural and public health problems, whether it be integrated pest management or human health—skin care, cosmetics, personal care, and animal and environmental health, such as afforestation programs, or even as soil enhancer, or in preventing soil erosion, controlling wind and water erosion. Besides this, neem also acts as good manure, fertilizer, soil conditioner, urea coating agent and serves as a shade tree due to its large spreading crown. In agriculture, it has lent itself to many eco-friendly and sustainable-agricultural initiatives. Neem preparations containing Azadirachtin are used in organic farming. Perhaps no other botanical can be better defined as a panacea than the Neem tree. The Neem tree has been researched more intensively than any other tree on this planet and the United Nations has declared the Neem tree 'as the tree of the 21st century'.

Fruiting branches of Neem tree (*Courtesy*: Mr B. K. Basavaraj)

The neem tree provides many useful compounds that are used as pesticides (Figure 19.4). A complex called **tetranortriterpenoid limonoid**, derived from neem seeds, is the major active ingredient in a number of formulations of pesticides because of its known toxic effects on a large variety of insects. Its other homologues include salanin, meliantriol, nimbin, nimbidin, quercetin and gedunin. Azadirachtin behaves as an insect growth regulator, and has several effects on phytophagous insects, and is thought to disrupt insect moulting cycles.

Azadirachtin

Fig. 19.4 Molecular configuration of Azadirachtin

19.2 Phenolic Compounds

Plants produce a vast array of secondary products that possess a phenol grouping – a hydroxyl functional group on an aromatic ring:

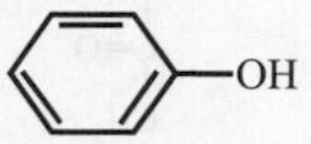

These compounds are categorized as phenolic compounds. Chemically speaking, plant phenolics are a heterogenous group of approximately 10,000 individual compounds: some of which are soluble only in organic solvents, while others are water-soluble carboxylic acids and glycosides and still others are large-sized complex, insoluble polymers (Figure 19.5). Many act as deterrent compounds against herbivores and pathogens. Others provide mechanical support, help in attracting pollinators and fruit dispersers, in absorbing harmful UV radiation or in reducing the growth of nearby competing vegetation (**allelopathy**).

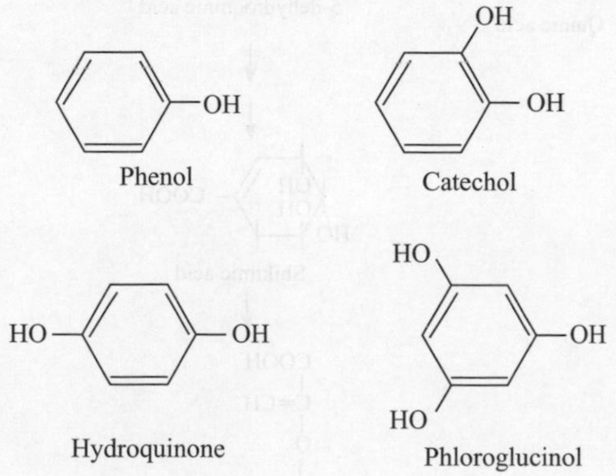

Phenol

Catechol

Hydroquinone

Phloroglucinol

Fig. 19.5 The most common phenolics present in plants are phenol, catechol, hydroquinone, and phloroglucinol.

Plant phenolics are synthesized by two biosynthetic pathways: the **Shikimic acid pathway** (Figure 19.6) and the **Malonic acid pathway** – the former being more prevalent in higher plants. Biosynthesis of most of the secondary phenolic products starts with the deamination of phenylalanine to cinnamic acid, catalyzed by the enzyme, phenylalanine ammonia lyase (PAL). Cinnamic acid appears to be precursor for synthesis of more complex derivatives like **lignin, flavonoids, coumarin and tannins**.

19.2.1 Lignin

After cellulose, the most predominant organic compound in plants is **lignin**- a highly branched polymer of phenylpropanoid groups. Lignin is found in the cell walls of tracheary elements such as tracheids and vessel elements. It generally gets impregnated in the thick secondary wall but can also be found in the primary wall as well as the middle lamella. Besides providing mechanical support, lignin also has a significant defensive role in plants.

It is relatively indigestible to herbivores because of its chemical composition. Being bound to cellulose and protein, lignin further reduces its digestibility. Lignification restricts the growth of pathogens and is the most common response to infection or wounding.

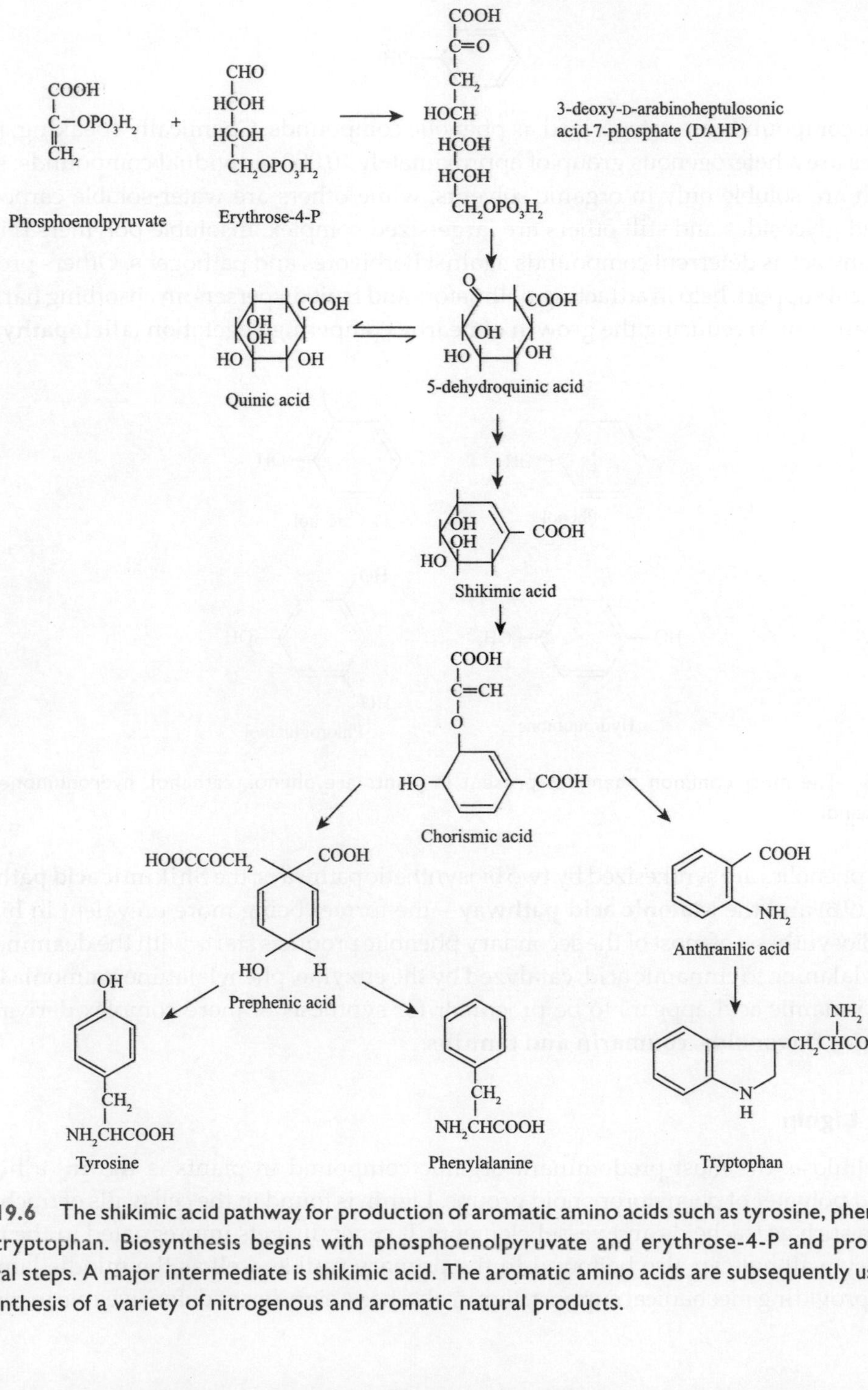

Fig. 19.6 The shikimic acid pathway for production of aromatic amino acids such as tyrosine, phenylalanine and tryptophan. Biosynthesis begins with phosphoenolpyruvate and erythrose-4-P and proceeds *via* several steps. A major intermediate is shikimic acid. The aromatic amino acids are subsequently used in the biosynthesis of a variety of nitrogenous and aromatic natural products.

19.2.2 Flavonoids

Flavonoids are the most widely distributed secondary metabolites with altogether different metabolic functions. Flavonoids are biologically active compounds, and over 6000 different flavonoids have been isolated from plant sources, for example, fruits, vegetables, herbs and non-alcoholic beverages such as tea and red wine. They are the most abundant constituents of human diet, representing two-third of all those consumed. The flavonoids constitute one of the largest groups of plant phenolics with a 15-C carbon skeleton, comprising of two benzene rings linked by a 3-C bridge or connected through a heterocyclic pyrone ring structure.

Therefore, their carbon structural organization can be abbreviated as $C_6 - C_3 - C_6$. The flavonoids are further categorized into nine chief subclasses such as **flavones, flavonols, flavanones, flavanonols, flavanols** or **catechins, isoflavones, anthocyanidins** and **proanthocyanidins,** and **aurones**. Finally, flavonoids with open C ring are known as **chalcones**. The varied groups of flavonoids differ from one another in their oxidation states and the substitution pattern of the carbon ring. They are soluble in water and accumulate in cell vacuolar compartments (Figures 19.7 and 19.8).

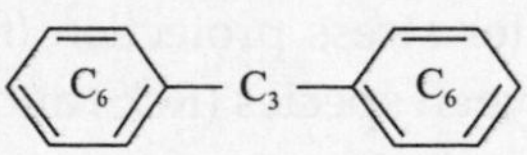

C_6 — C_3 — C_6

Fig. 19.7 Basic structure of flavonoids.

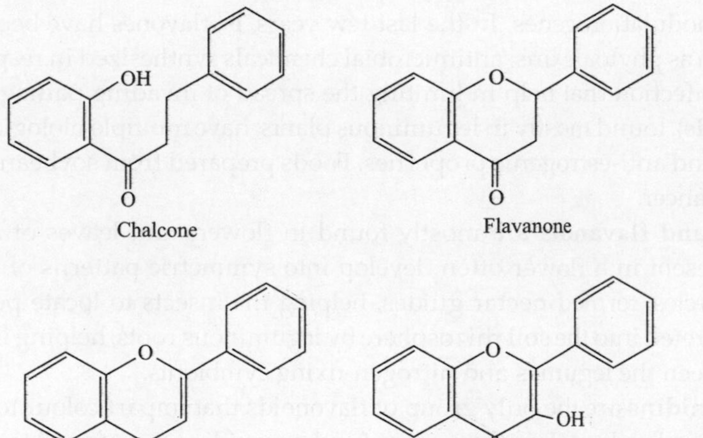

OH O Chalcone

O O Flavanone

O O Flavone

O OH O Flavonol

+ O OH Anthocyanidin

Fig. 19.8 The flavonoids constitute one of the largest groups of plant phenolics – with 15 carbons organized in two aromatic rings linked with each other *via* a 3-C bridge. The varied groups of flavonoids are different from one another in their degree of oxidation and the substitution pattern of the carbon ring.

Flavonoids are widely used in a variety of nutraceuticals, cosmetics, and pharmaceutical applications by virtue of their antioxidant, anti-inflammatory, anti-microbial, anti-carcinogenic, anti-mutagenic properties coupled with their capacity to modulate key cellular functions.

Flavonoids have varied biological roles, such as:

- As **phytoalexins** (provide protection against herbivores and pathogens)
- Signalling during nodulation (in symbiont recognition between rhizobia and host roots and regulation of nodulation)
- Regulation of male fertility
- Regulation of plant growth and development as modulators of polar auxin transport
- Impart colour variation to flowers, thus serving as visual signals for attracting insects, birds as well as animals for pollination and dispersal of seeds
- Offer stress protection (functioning as scavengers of free radicals such as reactive oxygen species (ROS) and chelating metallic compounds that produce ROS
- Flavonoids are now much appreciated as sunscreen pigments that protect the photosynthetic apparatus of leaves against UV-B radiations (280–320 nm) damage

More specific functions have been assigned to certain flavones and isoflavones as inducers of rhizobial nodulation genes. In the last few years, isoflavones have become popular for their function as phytoalexins, antimicrobial chemicals synthesized in response to bacterial and fungal infection that help in limiting the spread of invading pathogens. **Isoflavones (isoflavanoids)**, found mostly in leguminous plants, have multiple biological functions like insecticidal and anti-estrogenic properties. Foods prepared from soybeans are responsible for curbing cancer.

Flavones and flavanols are mostly found in flowers and leaves of all green plants. Flavanols present in a flower often develop into symmetric patterns of stripes, spots, or concentric circles, termed nectar guides, helping the insects to locate pollen and nectar. These are secreted into the soil rhizosphere by leguminous roots, helping in the recognition process between the legumes and nitrogen-fixing symbionts.

Anthocyanidins are the only group of flavonoids that impart colour to plants (all other flavonoids are colourless). The glycosides of anthocyanidins are referred to as **anthocyanins**, which are a prevalent class of pigmented flavonoids, imparting brilliant red, pink, purple, and blue colours to fruits, vegetables, flowers and grains; especially black soybean which is an excellent source of anthocyanins. Anthocyanins are the most important water-soluble pigments in plants, after chlorophyll, amongst the plant pigments that are visible to the human eye. Its pH has a marked impact on the colour of an anthocyanin solution. Anthocyanins undergo reversible structural transformations with change in pH, which are followed by dramatic changes in colour. As a result of their amphoteric nature, anthocyanins behave somewhat like pH indicators. Below pH 3, anthocyanin solutions exhibit their most intense red coloration. On increasing pH of such solutions, their red colour normally fades to the point where they appear colourless in the pH range of 4–5. Further increase in pH gives rise to purple and blue anthocyanin solutions.

In plants, anthocyanins perform various roles:

- Serve as attractants for pollination and seed dispersal,
- Provide protection against the harmful effects of UV irradiation
- Provide plants with antiviral and antimicrobial activities.

Polyphenol oxidase (PPO) enzyme is responsible for anthocyanin degradation. On mechanical damage, decompartmentalization of phenolic compounds and PPO occurs and the resulting contact between these two leads to the formation of brown pigments, affecting marketability, sensory quality and nutritional value. Oxygen can be detrimental on anthocyanins through direct/indirect oxidation mechanisms, wherein oxidized constituents of the medium are capable of reacting with anthocyanins to yield colourless or brown pigmented products.

***Biosynthesis*:** The flavonoids are synthesized from phenylalanine and three molecules of malonic acid to form chalconaringenin from which virtually all other flavonoids skeletons are derived.

Anthocyanins – pemium functional superfood supplements

The demand for anthocyanins in the healthcare sector and in personal care products has witnessed a great increase in the last few years. With better understanding about their chemistry and biochemical significance, the market for anthocyanins has mushroomed globally with applications that range from **food and beverages to pharmaceuticals, and personal care.**

Anthocyanins as antioxidants: The antioxidant potential is mainly analyzed based on the ability to readily absorb and quench free radicals and other reactive oxygen species. Anthocyanins have been found to be great lipophilic radical scavengers which help them to retard chain oxidation reactions of lipids and thereby enabling them to prevent cell damage caused by oxidative stress. Cyanidin has been found to have the highest antioxidant potential, measured in terms of **Oxygen radical absorbing capacity (ORAC);** followed by malvidin, peonidin, and delphinidin. Anthocyanins as antioxidants with possible roles as **neuroprotectant** and **anti-cancer drug**, are gaining prominence for applications in pharmaceutical products. Their anti-oxidant properties help in improving skin appearance, which makes them preferred ingredients in personal care products.

Anthocyanins as bio-colourants: Anthocyanins are making fresh inroads in the food colourant industry as several synthetic food colourants, including orange II, fast red, amaranth, auramine and rhodamine face a ban. Due to their bright and vivid colors, these are the preferred natural food colours for the international food and beverage industry. The pH has a significant effect on the colour of anthocyanin solutions. Because of the color changes with pH, use of anthocyanins has typically been restricted to acidic foods. **Co-pigmentation** i.e. interaction of anthocyanins with compounds such as phenolic compounds, flavonoids, alkaloids, and amino acids is of immense value as a natural tool to improve the colour of anthocyanin-rich food products.

Stilbenes are a class of C_6–C_2–C_6 compounds, that exhibit a chemical defensive role, primarily preventing the spread of fungal invasion of heartwood. These are one of the many phenylpropanoid derived products that are continuously deposited into the heartwood (or duramen), central wood core of trees after the lignification of cells. These are synthesized by the sequential condensation of three molecules of malonyl-Co A with either Cinnamoyl-Co A or *p*-Coumaroyl-Co A.

19.2.3 Coumarins

Over 300 **coumarins** have been discovered and isolated from more than 800 species of plants, belonging to more than 70 angiospermic families, for example, in tonka bean, lavender, liquorice, strawberries, cherries, cassia, cinnamon, sweet clover (a forage legume) and grasses (like sweet grass and vanilla grass). They include the widespread class of lactones called **benzopyranones**. The two simplest fragrant chemical compounds are – coumarin and 7-hydroxycoumarin (umbelliferone), resulting from the reaction of cinnamic acid and *p*-coumaric acid respectively, followed by a ring closure. Some other naturally occurring coumarin derivatives are aesculetin (6,7-dihydroxycoumarin), herniarin (7-methoxycoumarin) and psoralen.

Physiological roles: As a group, coumarins are known for their role as antimicrobial agents, feeding deterrents, and germination inhibitors. Scopoletin is perhaps the most commonly occurring coumarin in flowering plant species, and is often present in the seed coats where it is suspected to inhibit germination, and therefore it must be leached out before sowing. While coumarin itself is only mildly toxic, its derivative 'dicoumarol' is a powerful anticoagulant present in mouldy hay or sweet clover silage where it can lead to fatal haemorrhage in cattle by retarding vitamin K synthesis (i.e. antagonist), an essential cofactor required in blood clotting (Figure 19.9).

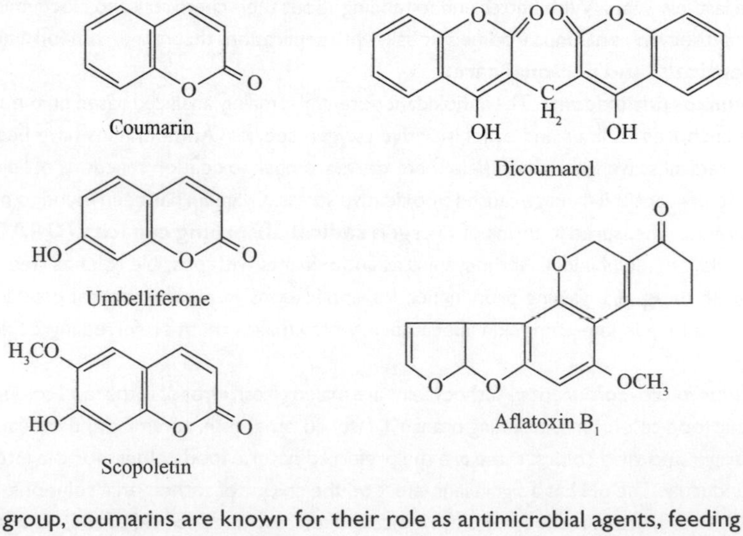

Coumarin

Dicoumarol

Umbelliferone

Aflatoxin B_1

Scopoletin

Fig. 19.9 As a group, coumarins are known for their role as antimicrobial agents, feeding deterrents and germination inhibitors.

Coumarin is a component of Bergamot oil, a volatile oil used for flavouring pipe tabacco, tea, perfumes, fabric conditioner and other products. Coumarin gives the freshly mown hay its characteristic sweet odour, but its bitter taste discourages animals from eating such plants. Besides this, coumarin has blood thinning, antifungal and anti- tumour activities but can be toxic when used in higher dosage for a prolonged period. Warfarin is a synthetic coumarin that is used as a very potent rodenticide. Aflatoxin B_1 (a mycotoxin synthesized by the fungus, *Aspergillus flavus*), commonly found in livestock feed, moist peanut, and maize (the latter when damaged by European corn borer), is among the most carcinogenic of naturally occurring compounds, and can impair immune systems and damage livers and kidneys.

19.2.4 Tannins

Tannins are polymerization products of phenolic compounds. These are responsible for the brown coloration of oak leaves in the fall as well as discoloration of injured plant tissues. The defensive roles of most of the tannins are because of their toxic nature, which is normally due to their potential to bind proteins non-specifically. Plant tannins also function as defences against microorganisms. Tannins are general toxins that significantly reduce growth and affect survival of many herbivores when ingested. Additionally, tannins function as feeding repellents to a wide variety of animals. Frequently, immature fruits have very high tannin contents that are mainly concentrated in their outer cell layers.

> Interestingly, tannins are one of the components in a variety of fruits and drinks such as coffee, tea as well as red wine. Recently, tannins in red wine[4] have been reported to block the formation of endothelin-1, a signalling molecule that makes blood vessels constrict.

The two types of tannins are: condensed and hydrolyzable.[4]

Condensed tannins are the products synthesized through polymerization of flavonoid units and are frequently present in woody plants (Figure 19.10a).

Hydrolysable tannins, on the other hand, are heterogenous polymers composed of gallic acid and simple sugars (Figure 19.10b).

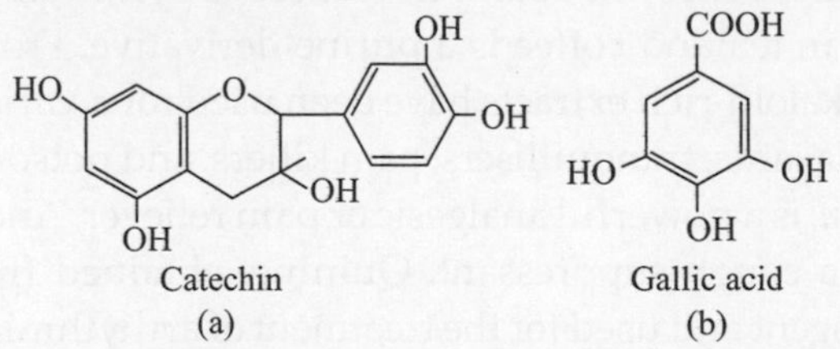

Fig. 19.10 (a) Structure of condensed tannins and (b) hydrolyzable tannins – the former are largely polymeric condensation products of catechin while in hydrolyzable tannins, gallic acid is the primary component.

These are typical of the defensive compounds found in leaves, fruits, pods and galls of a range of dicot species.

Many healthy tissues have tannins, which are frequently synthesized when injury causes cellular disruption. So tannins may function as a protective device to prevent infection of injured tissues.

---

[4] Tannins protect wine from bacterial attack but begin to precipitate with time. Thus, the tannin content of wine decreases with age, thereby improving the taste of the wine.

19.3 Alkaloids

Alkaloids are a diverse class of naturally occurring heterocyclic nitrogen-containing organic compounds with significant pharmacological properties and medicinal applications. Members of the alkaloid family share some principal characters: they are soluble in water; they possess at least one nitrogen atom; most of them are colourless, non-volatile crystalline solids; they tend to impart a bitter taste; they are generally alkaline in nature, combining with acid to form water-soluble salts. Alkaloids name generally end in the suffix-ine. Well-known alkaloids include **morphine, quinine, ephedrine, strychnine, reserpine, caffeine, atropine, cocaine, colchicine** and **nicotine.**

Over 12,000 different kinds of alkaloids have been discovered from more than 4000 plants, chiefly flowering plants. They are synthesized by plants and are found in the leaf, bark, seed, or other parts. A few alkaloids have been reported to occur in the animal kingdom such as the New World beaver (*Castor canadensis*). A few fungi also produce them, for example, ergot alkaloids (from *Claviceps purpurea*).

Most of the alkaloids now play a role as defences against predators due to their toxicity and deterrent properties. The use of alkaloids can be traced back to the ancient times but the scientific study of these chemicals had to await the growth of organic chemistry. These alkaloids have much varied and powerful physiological effects on humans as well as animals, most often by interfering with neurotransmitters. The first alkaloid to be isolated and crystallised in 1804 was morphine, a potent alkaloid obtained from poppy (*Papaver somniferum*).

***Physiological roles*:** Alkaloids are derived from some common amino acids e.g., ornithine, arginine, tyrosine, tryptophan, lysine, and phenylalanine. Alkaloids may be present constitutively or the plant may be induced to synthesize them in response to attack by pathogens. The alkaloid nicotine found in tobacco is synthesized from nicotinic acid, while caffeine present in tea and coffee is a purine derivative. From ancient times to the present, alkaloids or alkaloid-rich extracts have been used for a variety of pharmacological functions, as muscle relaxants, tranquillisers, pain killers, and poisons. Morphine, obtained from *Papaver somniferum*, is a powerful analgesic or pain reliever. Another alkaloid, codeine, is commonly used as a cough-suppressant. Quinine, obtained from *Cinchona* spp., is a powerful antimalarial agent and used for the treatment of arrhythmia, (irregular rhythms of the heart). Rauwolfia alkaloids (obtained from *Rauvolfia serpentina*) are used as hypotensive agent and used as tranquiliser for treating insanity (Figure 19.11). Atropine, derived from *Atropa belladonna*, is used for ophthalmic studies. Two alkaloids, vincristine and vinblastine (from *Catharanthus roseus*) are commonly used in the treatment of many forms of cancers. Ergonovine, derived from the fungus *Claviceps purpurea*, is used for reducing uterine haemorrhage after childbirth. Ephedrine, from *Ephedra* spp., relieves discomforts during common cold, sinusitis, hay fever and asthamatic attacks. Nicotine, derived from tobacco (*Nicotiana tabacum* and *N. rustica*) has a calming or sedative action on users. Many alkaloids such as cocaine possess anaesthetic properties. Cocaine (from *Erythroxylum coca*) is very potent anaesthetic. In the Western Hemisphere, the Azetecs and other native peoples have been using the peyote cactus or mescal buttons (*Lophophora williamsii*) containing mescaline for its hallucinogenic properties.

Vinblastine (R=Me)
Vincristine (R=CHO)

(a)

Ajmalicine

Reserpine $R = OC$

Rescinnamine $R = OC$

Ajmaline

(b)

Codeine

Morphine

(c)

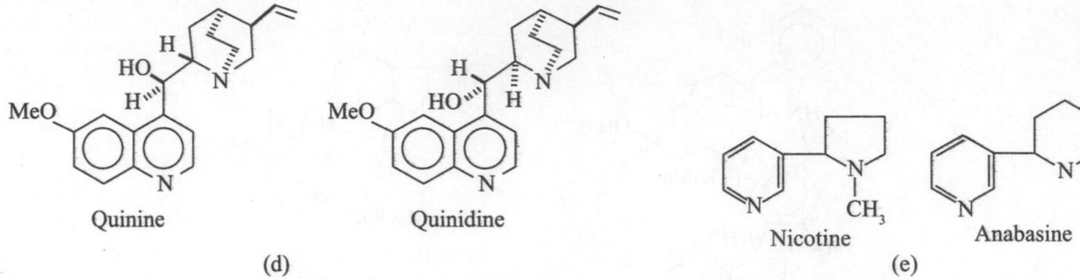

Quinine Quinidine

(d)

Nicotine Anabasine

(e)

Fig. 19.11 Structure of major alkaloids having pharmacological properties, derived from *(a) Catharanthus roseus*, (b) *Rauvolfia serpentina*, (c) *Papaver somniferum*, (d) *Cinchona calisaya*, and (e) *Nicotiana tabacum*.

The alkaloid tubocurarine is the active component in arrow poison, curare (obtained from *Chondrodendron tomentosum*) has been utilized by the aborigines or tribals in many countries to kill animals. Strychnine (from *Strychnos toxifera* and others) is another powerful arrow poison of South American natives. In ancient Greece, the most famous hemlock poisoning (*Conium maculatum*) occurred 399 BC when the great Athenian philosopher Socrates was convicted of impiety or lack of reverence towards gods and for corrupting the youth. He was condemned to death and was made to drink poisonous hemlock.

19.4 Glycosides

A glycoside compound has a carbohydrate moiety (sugar) bound to another functional group by a glycosidic bond. The sugar group represents the '**glycone**' while the non-sugar group as the '**aglycone**' or the 'genin' component of the glycoside. The linkage between glycone and aglycone is called glycosidic linkage. The glycone and aglycone portion can be separated from each other chemically *via* hydrolysis in the presence of acid. Several enzymes are known that can form and break glycosidic bonds.

Glycosides are categorized based on the following:

1. Chemical characteristic of the aglycone part
2. By the type of glycosidic bond, whether it is positioned at 'upper or lower' level of the cyclic sugar molecule, that is, α-glycoside or β-glycoside
3. The therapeutic activity of the aglycone part.

Depending upon the chemical nature of the aglycone moiety or derivative present in the glycosides they are called: alcoholic glycosides; anthraquinone glycosides; phenol glycosides; steroid glycosides; flavonoid glycosides; coumarin glycosides; furano-coumarin glycosides; cyanogenic glycosides; thioglycosides; saponin glycosides (closely related to steroids) and steviol glycosides, etc.

Broadly speaking, on the basis of pharmacological attributes, glycosides are classified into three groups, namely: i) the **cardiac glycosides** or **cardenolides**, ii) the **cyanogenic glycosides**, and iii) the **saponins**. A fourth family, the **glucosinolates** (technically not a glycoside) can be added to the list as they have a similar structure.

19.4.1 Cardiac glycosides or cardenolides

The cardenolides or cardiac glycosides are an important class of naturally occurring drugs where the range between lethal and therapeutic dose is quite narrow. Therefore, they must be used with utmost caution. They are widely distributed in more than 200 species of flowering plants, belonging to 55 genera and 12 families. Perhaps the best known is the common foxglove plant which was first used to treat dropsy (unwanted accumulation of water in the body cavities), and now much valued as a medicine in congested heart problems and arrhythmia (irregular rhythm of heart beat), to strengthen the heart muscles by increasing the contraction rate, improve circulation and lower blood pressure, and connected issues. The foxglove plant refers to *Digitalis purpurea* and *D. lanata*. The active glycosides of the leaves of *D. purpurea* are digitoxin, gitoxin and gitalin while in *D. lanata* gitalin is replaced by digoxin, the other two remain the same.

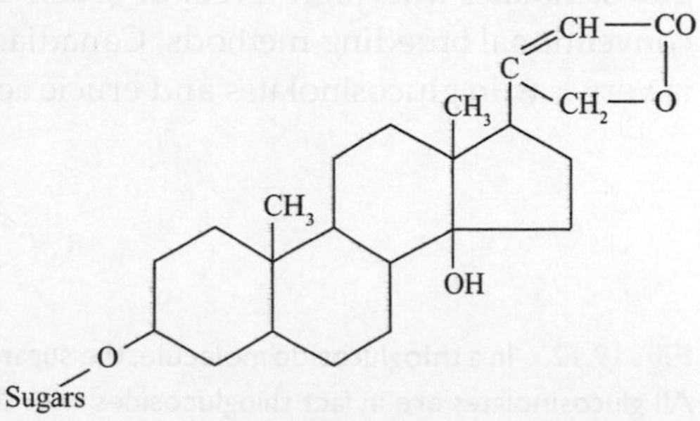

CH—CO
C
CH_3 CH_2—O
CH_3
OH
O
Sugars

Digitoxin

Some other examples of cardiac glycosides are species of *Strophanthus* and *Scilla*. Another very interesting source of cardenolides are the milkweeds, particularly those belonging to the genus *Calotropis* and *Asclepias* that produce milky-white, cardenolides-rich latex. They are also the chief host for ovipositing monarch butterflies. The larvae live on the leaves of milkweed and suck the toxic steroids (cardenolides). These cardenolides act as a defence against the predators. When a bird (such as blue jays) or other predator attempts to feed both on the larvae and adult monarchs, the bad taste of accumulated toxic compounds force them to vomit out. The predator very wisely learns to avoid preying on monarchs, at least for some time.

The cardenolides-rich seed extracts of the woody *Strophanthus* vines have been used by African hunters to poison the tip of arrows and spears and blow darts.

19.4.2 Thioglycosides

As the name suggests, these compounds contain sulphur in a S-glycosidic bond (Figure 19.12). Examples include sinigrin (or potassium myronate), present in black mustard (*Brassica nigra*), and sinalbin, contained in white mustard (*B. alba*). The seeds of both species are virtually without any physiological activity, but on hydrolysis with the enzyme myrosin (a thioglucosidase) they yield dextrose and the essential oil of mustard which is responsible for its pungency. Like the cyanogenic glycosides, thioglycosides or glucosinolates are separated spatially from the hydrolytic enzymes so that the oil of mustard is produced only when the seeds are crushed or disintegrated, allowing the enzyme and substrate to come in direct contact. The distinctive flavour deter or repel some of the herbivores, thus acting as a defence protection while in some others these glucosinolates or mustard oils act as attractant for insects and further stimulate feeding and ovipositing. Rape seed oil, obtained principally from *Brassica napus*, is important vegetable oil, but the presence of glucosinolates and high levels of erucic acid impart to it an undesirable taste. By using conventional breeding methods, Canadian scientists have produced a Canola variety that is very low in glucosinolates and erucic acid.

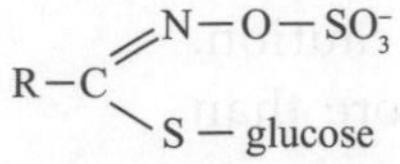

$$R-C\begin{matrix}\nearrow N-O-SO_3^- \\ \searrow S-\text{glucose}\end{matrix}$$

Fig. 19.12 In a thioglucoside molecule, the sugar is attached to the remaining molecule *via* a sulphur atom. All glucosinolates are in fact thioglucosides with the same basic structure.

19.4.3 Cyanogenic glycosides

Cyanogenic glycosides have been found in over 200 different species of plants, including dicots, monocots, gymnosperms, and ferns. They are essentially, an aglycone with a cyanide group attached. They are made through the conversion of amino acids to oximes, which are then glycosylated. In the cell, the cyanogenic glycosides are contained in separate compartments from the enzymes which break them down. Upon rupture of the cells by herbivore damage, the contents of the compartments mix and the glycosides are degraded to liberate hydrogen cyanide (HCN) which acts as a potent protection for plants against both herbivores and pathogens.

Cyanogenic glycosides are found in the seeds (and wilted leaves) of a number of plant species of the family Rosaceae such as peaches, apples, apricots, cherries, plums, and crab apple. Amygdalin derived from bitter almond is a typical example of cyanogenic glycosides (Figure 19.13). Flax seed (*Linum usitatissimum*), a popular health food, is another multipurpose crop that contains cyanogenic compounds but these are lost during roasting. Cassava or manioc (*Manihot esculenta* of the family Euphorbiaceae), another food plant of South Africa and South America, also contains cyanogenic glycosides in their tubers but these are destroyed by boiling, roasting, expression or fermentation.

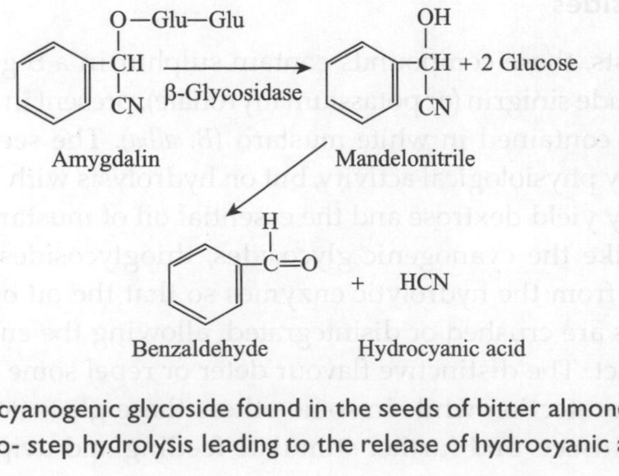

O—Glu—Glu | CH | CN (Amygdalin) —β-Glycosidase→ OH | CH | CN (Mandelonitrile) + 2 Glucose → H | C=O (Benzaldehyde) + HCN (Hydrocyanic acid)

Fig. 19.13 Amygdalin, a cyanogenic glycoside found in the seeds of bitter almond and other members of Rosaceae, undergoes a two- step hydrolysis leading to the release of hydrocyanic acid (HCN).

19.4.4 Saponin glycosides

Saponins are a special group of complex chemical molecules with foaming characteristics. Their name is derived from soapwort plant (*Saponaria* spp.), the roots of which have been used since historic times as a soap. Saponins consist of polycyclic aglycones linked to one or more carbohydrate side chains. The aglycone portion of the saponins (or sapogenin) is either a steroid (C_{27}) or a triterpene (C_{30}). Saponins have been extracted from nearly 100 different angiospermic families, mainly Sapindaceae, Aceraceae, Hippocastanaceae, Cucurbitaceae, and Araliaceae. In the animal kingdom, saponins are found in most sea cucumbers and starfish. However, the commercial supplies are extracted, mainly from the soapbark tree, *Quillaja saponaria* and *Yucca schidigera*.

Steroidal saponins, for example, in wild yam (*Dioscorea* spp.) and ginseng (*Panax quinquefolia* and *P. ginseng*) are widely used in medicines. The saponin glycyrrhizin, found in liquorice (*Glycyrrhiza glabra*), has expectorant, corticosteroid and anti-inflammatory properties. They are sometimes used as a sweetener and flavour-additives in foods and cigarettes.

Saponins have a bitter taste, may function as anti-feedants, thus protecting the plant against many pathogenic fungi. The root epidermal cells of oats (*Avena sativa*) produce a triterpenoid saponin, avenacin A-1 that protects against fungal invasion. Alfalfa saponins (from *Medicago sativa*) can cause digestive problems and bloating. Saponins are phytonutrients, found in many food sources such as vegetables, beans, peas, soybean. Some saponins (e.g., obtained from oat and spinach) may increase nutrient absorption. Because of their detergent properties, saponins are popular in cosmetic formulations such as shampoos, facial cleansers, and cosmetic creams. Saponins from *Yucca* and *Quillaja* are used in some beverages, such as beers, to produce stable foam.

A special mention may be made of Steviol glycoside, i.e., stevia plant (*Stevia rebaudiana*) containing sweet glycoside that possesses as much as 40–300 times sweeter than even sweet-tasting sucrose. Two primary glycosides, stevioside and rebaudioside A are utilized as natural sweetener in place of normal sugar in many countries. Steviol is the aglycone part of such glycosides.

19.5 Role of Secondary Metabolites in Plant Defences

19.5.1 Plant defences against pathogens

- A common defence against pathogens (bacterial/fungal) is the **hypersensitive response**, where cells surrounding the infection site die quickly, thereby limiting the spread of the pathogen. Nitric oxide (NO) and reactive oxygen species (ROS) contribute to host cell death or directly kill the pathogen.
- In response to pathogen attack, many plant species react by synthesizing **lignin** or **callose** polymers that block its spread by acting as barriers.
- Certain cell wall proteins get modified following the pathogen attack. This strengthens the cell walls near the site of infection, thereby increasing their resistance to microbial digestion.
- Infected host plants may synthesize hydrolysing enzymes that damage the cell walls of fungal pathogens. These hydrolytic enzymes are comprised of a group of proteins

known as **pathogenesis-related (PR) proteins** such as chitinases, nucleases, and proteases.

- Many plants produce **phytoalexins,** a vast group of secondary metabolites with strong antimicrobial action, in response to infection.
- Plants that survive a pathogenic infection at one place, generally develop greater resistance to subsequent attacks in different parts of the plant – a phenomenon known as **systemic acquired resistance (SAR)**. SAR is transmitted from the infection site to other regions of the plant. The amount of salicylic acid and its methyl ester increase significantly during this process and lead to the production of PR proteins.
- Exposure of plants to non-pathogenic microbes can initiate **induced systemic resistance (ISR)** through a process, mediated by ethylene and jasmonic acid.

19.5.2 Induced plant defences against insect herbivores

Plant defensive strategies against insect herbivores can be classified into two groups:

(i) **Constitutive defences:** These always exist within the plant and are generally species-specific. They may occur as stored compounds, conjugated compounds, or precursors of active compounds.

(ii) **Induced defences:** These are initiated in response to mechanical damage induced by herbivores or by specific components of insect saliva, called elicitors. Jasmonic acid (JA) level rises significantly in response to herbivore attack. Jasmonic acid is activated by conjugation with amino acid, isoleucine and triggers the transcription of key genes responsible for the synthesis of defensive proteins such as lectins, protease inhibitors as well as toxic secondary metabolites.

Herbivore damage can also induce systemic defences by causing the synthesis of polypeptide signals such as **systemin**. Systemin binds to cell surface receptors in the phloem and activates the synthesis of jasmonic acid as well as the transcription of genes encoding proteinase inhibitor (PIN2) and other systemic wound response proteins (refer to Figure 18.17).

Review Questions

RQ 19.1. When plant tissues are injured, it is often that a brown colouration rapidly develops. How would you prevent such tanning from occurring?

Ans. Upon injury, plant tissues rapidly develop a brown colouration because of the polymerization of quinones present in them. In the laboratory it is most common to manipulate the tissue under anaerobic conditions or add an agent that will chelate the copper in the polyphenol oxidases, preventing the reaction from occurring. However in the kitchen, vegetables prone to tanning can be placed under water after cutting; or vinegar might be added.

RQ 19.2. Why is it so that secondary metabolites, originally designed by plants to deter pests or attract pollinators, have commercial value to humans?

Ans. Humans have used many of the secondary metabolites of plant origin for medical, culinary, or other purposes. It has been found that 50,000 to 100,000 secondary compounds are found within the plant kingdom, and only a few thousand of these have been identified so far.

Secondary metabolites are often obtained by modification of amino acids and related compounds to synthesize alkaloids, or through specialized transformations i.e., the shikimic acid pathway (phenolics) and mevalonic acid pathway (terpenoids). For example, lignin (a component of secondary cell walls) is synthesized *via* the shikimic acid pathway. Lignin being hard to digest and also toxic to many predators, it provides protection to plants from herbivores.

RQ 19.3. What additional metabolic pathways are demonstrated by plants, and why are they necessary?

Ans. There are many additional processes in plants, leading toward growth, development, reproduction and survival. Many of these employ intermediate steps, but they can't proceed without the main processes such as photosynthesis and respiration. The essential compounds produced from such pathways include sugar phosphates, amino acids, nucleotides, nucleic acids, proteins, fatty acids, oils, waxes, chlorophylls, carotenoids and cytochromes.

RQ 19.4. 'Secondary metabolism is not so essential for normal plant growth and development, yet plants can't do without them'. Comment.

Ans. Metabolic processes that are not needed for normal plant growth and development are called **secondary metabolism.** Although not essential, several products from this pathway help plants to survive and exist under special situations. These products impart to the plant unique colours, aromas, poisons and provide other chemicals that may attract or deter other organisms, allowing them to draw upon environmental synergies as well as to survive threats.

RQ 19.5. Distinguish between primary metabolites and secondary metabolites.

Ans. Primary metabolites and secondary metabolites differ in several ways:

Primary metabolites	Secondary metabolites
1. These are found throughout the plant kingdom.	These have restricted distribution within the plant kingdom i.e highly species–specific.
2. These are essential chemicals, e.g., proteins, carbohydrates, lipids, nucleic acid, etc.	These are non-essential chemicals, often toxic, anti-nutritional, allergenic, bad-tasting and therefore harmful for human consumption e.g., terpenoids, phenolics, alkaloids, and so on.
3. These metabolites directly participate in core processes of growth and development.	These serve various ecological roles in plants: i. provide protection to plants against ingestion by herbivores or infection by pathogens of microbial origin. ii. serve as attractants for pollinating agents and seed-dispersal animals. iii. act as agencies of plant–plant competition and plant–microbe interactions. iv. play adaptive and acclimation roles to environmental stresses.
4. These are generally a component of the essential molecular organization or function of the cell.	These are natural products, which generally occur in relatively low quantities and are important as drugs, poisons, flavours and industrial products.

Unit V Crop Physiology – An Innovative Approach

Chapter 20

Agricultural Implications of Plant Physiology

The science of Plant Physiology deals with various life processes occurring both within cells, and between cells and the environment. The former includes metabolic pathways like photosynthesis, respiration, nitrogen and fat metabolism, translocation of food materials and growth regulators and similar processes; while the latter encompasses diffusion of gases, absorption of water and minerals, ascent of sap and transpiration and such. Understanding the physiological requirements of different crops is the key to successful agriculture. A deep understanding of water, nutritional and other edaphic and climatic parameters controlling the crop physiological functions is essential for optimizing productivity for the sustainable development of our exploding population. To emphasise this, we will examine the pre-green revolution era when we had tall varieties of wheat and rice with a tendency towards lodging when heavily fertilised. Using the **Norin 10 dwarfing genes** in wheat, and the **Dee-geo-woo-gen genes** in rice, agricultural scientists were able to change the morphological architecture (the Ideotype concept) by developing high-yielding dwarf and semi-dwarf varieties. But their success depended upon the crop's water, nitrogenous fertilizer and other edapho-climatic requirements which were optimised through the concerted efforts of plant physiologists, agronomists and others. This is the success story of the **Green Revolution** in the early 1960s.

Plant Physiology is a rapidly advancing field of botany and serves as the foundation for the numerous advances in **agriculture** (including horticulture), **environmental sciences, floriculture, agronomy, plant pathology, forestry**, and **pharmacology**. Adequate knowledge and understanding of crop physiology is the key to success in optimizing farm productivity in terms of yield and quality. In the case of crops, their genetic make-up and environmental factors play a crucial role in manipulating their growth. Once the breeders understand the physiology of crops they are trying to handle, it becomes easy for them to deal with the problems for selecting varieties for higher yield and to cope up with environmental stresses as well as insect invasions, etc.

The scope of plant physiology covers the issues per the headings given:

20.1 Role of Plant Hormones

***Auxins*:** Compounds exhibiting activity similar to IAA such as Indole-3-butyric acid (IBA) and α-Naphthaleneacetic acid (NAA) are generally marketed as **rooting hormones** and are widely employed in vegetative propagation i.e., multiplication of plants from stem and leaf cuttings by stimulation of adventitious root formation e.g. in economically useful plants, such as sugarcane and forest trees. Stem cuttings are put in the preparations of synthetic auxins mixed with talcum powder/lanolin paste before allowing them to grow in sandy soil to promote root initiation (rate of formation as well as number of roots). The synthetic auxins are favoured over the naturally occurring ones in commercial applications because these are cost-effective and have greater chemical stability.

In 1946, the introduction of two synthetic auxins- 2,4-dichlorophenoxyacetic acid (2,4-D) and 4-chlorophenoxyacetic acid (4-CPA) changed our approach to agriculture. Both of them can be used as herbicides to eliminate unwanted plants in the fields. 2,4-D is being used as the principal component of **weed-and-feed** mixtures in the lawns and also for controlling broadleaved weeds in cereal crops. 4-CPA has been used to increase flowering and fruit set in tomatoes, while NAA is used to promote flowering in pineapples as well as to prevent pre-harvest fruit drop in apples and pears. Pruning of the apical bud is now widely used in ornamental plants, fruit-yielding crops, tea cultivation and garden hedges to allow growth of lateral buds. Synthetic auxins are widely used for prolonging dormancy in potato tubers and for producing parthenocarpic fruits.

***Gibberellins*:** The enzyme α-amylase activity induced by gibberellins (in barley aleurone layer of its endosperm) has resulted in their extensive use in the **malting industry** to accelerate production of malt. One principal commercial use of GAs is in the production of the 'Thompson Seedless' variety of table grapes. Treatment with gibberellins at the onset of flowering stimulates elongation of the floral axis, thus thinning the cluster and allowing for the development of fungus-free, large grapes with better market value. GAs have also been helpful in breaking dormancy, increasing germination, stimulating early seedling emergence and uniform crop emergence in seed potatoes, and growth in fruit crops such as grapes, apples, peach, and cherry. Besides, gibberellins increase yield in cotton and delay ripening of lemons.

Several **anti-gibberellins** such as Alar (B-nine), AMO-1618, Ancymidol (A-REST), Cycocel (CCC), and Phosphon-D function as growth retardants and have found to be useful for the horticultural industry. These are commercially employed for the production of ornamental plants like lilies, chrysanthemums, poinsettias and other ornamentals where they reduce stem elongation, resulting in short and compact plants with dark green foliage. CCC is used for flowering in cotton, and thus increases its yields. The plants treated with anti-gibberellins help in produing the shorter and stiffer stem that prevents lodging at harvesting time, thus increasing yield especially in wheat, oat and several strains of rye. Growth retardants such as maleic hydrazide prevent sprouting in potato, onion and certain root crops.

***Cytokinins*:** The cytokinin to auxin ratio determines shoot and root initiation in callus tissues cultures as well as the growth of axillary buds. This relationship is the basis for regenerating large numbers of plants by **micropropagation**. Micropropagation can also be an effective means to produce virus- and pathogen-free propagules commercially, particularly in orchids of the genus *Cymbidium*, potato, lilies, tulips and other species. Micropropagation has also been used extensively in the production of forest tree species and in many of the temperate fruits like apple, peach, and pear. **Mericloning** reduces the time required to create disease-free, desirable strains of crop plants and especially forest trees.

The capability of cytokinins in breaking dormancy, increased branching, delaying of senescence, extending shelf life of lettuce, tomatoes, apples and others, and causing cell division have proven to be of great value to plant physiologists. Moreover, cytokinins also regulate source-sink relations leading to improvement in yield parameters.

Ethylene: Understanding fruit development and ripening is important for agriculturists to increase yield, nutritional value, and quality; while maintaining post-harvest life. Scientists are trying various methods for controlling **fruit ripening** either by uniform ripening (such as in bananas) or by delaying the ripening process in apples and other climacteric fruits thus extending their marketable window (period). For example, apples are now available in the stores throughout the whole year because of cold-storage facilities with a combination of low temperature, low ambient oxygen (1–3%), and high ambient CO_2 (upto 8 per cent). These factors reduce ethylene production as well as the respiratory activities of the stored products. Unripe bananas are harvested, treated with fungicide, packed in cartons and shipped worldwide. At their destinations, bananas are kept in temperature-controlled rooms and exposed to small amounts of ethylene gas to initiate ripening. This ensures that fruits at different stages of maturity will ripen at the same time, making them easier to market.

In elite varieties of tomato, the ***rin* mutation** (refer to section 17.1.2) is widely used in the heterozygous form to slow the rate of ripening and extend shelf life (*Flavr Savr* tomato) but these fruits are often deficient in optimal levels of flavour, aroma, and other qualitative characteristics. A more effective approach would be to target individual ripening processes, for instance by extending the shelf life of fruits by slowing softening in the absence of detrimental effects on colour and flavour. Now with the help of tomato genome sequencing, scientists can identify the genes underlying complex factors controlling individual aspects of fruit quality.

More recent advances in our understanding of signal transduction in plants have come from *Arabidopsis thaliana*. The application of this knowledge may have practical benefits, for example in areas such as drought tolerance.

20.2 Photoperiodism, Vernalization and Dormancy

Understanding of the principles of growth are now being applied in the fields of agriculture and horticulture. Discoveries of photoperiodism and vernalization have enabled us to grow and make certain wanted plants flower in their off-seasons by suitably reducing or increasing the photoperiod or by giving low temperature treatment to winter varieties and biennial plants. The evolution of both the phenomena precisely regulates flowering so that it occurs at the optimal time for reproductive success. Synchronized flowering favours crossbreeding and helps ensure seed production under favourable conditions.

Dr Lee Hickey, a plant geneticist at the University of Queensland in Australia along with his team has been working on **speed breeding** crops under controlled light and temperature conditions to send plant growth into overdrive. This has enabled them to harvest seeds and to start growing the next generation of crops sooner. Crops, such as wheat, barley, chickpeas and canola, are made to flower early by exposure to blue and red LED lights for 22 hours a day at temperatures between 62 – 72° F resulting in as many as six generations in a year as compared to one or two by traditional methods.

Knowledge of the causes and mechanisms governing the dormancy of seeds enables farmers to manage such dormancy in order to obtain another quick crop.

20.3 Phloem Transport

Girdling is practised in certain fruit bearing plants to divert the translocated food material to the growing fruits for getting large-sized fruits. The knowledge of **photosynthate allocation and partitioning** to achieve higher yield is the prime area of research now-a-days in the agricultural crops. These two processes must be coordinated in the whole plant so that increased transport of photosynthates to sink (edible) tissues does not occur at the expense of other ongoing important physiological processes. Another way to improve crop yield is to retain photosynthates, which are normally lost by the plants.

The control of carbon exchange from source to sink is a big challenge for plant breeders for enhanced yield of crop plants. The increase in carbon exchange to the sink organs (i.e. such as tubers and seeds) can be achieved by regulating metabolism either at the source end or at the sink end. Several developmental and environmental factors affect the functioning of source and sink in determining crop yield.

20.4 Mineral Nutrition

Research on mineral nutrition of plants is central to improving modern agricultural practices and environmental protection, as well as for understanding plant ecological interactions in natural ecosystems. High agricultural productivity depends on fertilizers application with essential mineral nutrients. In fact, yields of most crops are directly proportionate to the amount of fertilizers absorbed by them from the soil. World food demand increased significantly between 1960 and 1990; necessitating the use of higher NPK fertilizers, whose consumption rose from 30 million metric tons in 1960 to 143 million metric tons in 1990. Over the next decade, consumption of these NPK fertilizers stabilised as they were used more judiciously in an attempt to offset rising costs. During the past few years, however, annual consumption has climbed to 180 million metric tons. An elemental analysis of soil and plant makes it possible to apply the correct type and dose of fertilizer and thus judicious application. Adding suitable micronutrients along with NPK fertilizers to the soil may also correct a pre-existing deficiency and improve the quality of field crops. For example, many acidic, sandy soils in humid regions are micronutrient-deficient typically as in boron, copper, zinc, manganese, molybdenum, or iron, and can thus gain from micronutrient supplements.

Fertilizers, when used along with **enhancers** give better results in terms of yield. The enhancers/additives involving micronutrients (Ni, Co, Mo) are being used to boost soil microbial activity, enhanced water and nutrient flow and improved nitrogen use efficiency as these are essential cofactors for key enzymes, including nitrogenase, thus leading to more amino acid production, eventually paving way fro more efficient and larger plant growth.

Enhancement products improve plant start-up, as well as plant and root health, water and nutrient uptake, and stress and drought mitigation. Regardless of the fertilizer applied, the **4R principle- right source, right time, right place, right rate** - should always be considered to improve agricultural productivity and minimize environmental impact.

Problems of soil acidity can be solved either by adding organic matter or treating the soil with lime to improve fertility or oxygen levels in plants. The application of green manure to the soil is a sustainable and affordable alternative to chemical fertilizers and enhances nutrient cycling while building-up organic matter. So organic farming is the way to go and a way to offset climate change.

Concerted efforts of soil scientists, hydrologists, microbiologists, ecologists as well as plant physiologists would be more meaningful in studies of mineral nutrition and complex soil-plant-atmosphere relationships. Mineral nutrition studies play crucial role in maintaining soil health, especially the micro-flora i.e. bacteria, fungi and symbiotic associations with blue green algae and *Azolla* in rice fields (***Azolla* and Duck association during rice cultivation** refer to chapter 8). Plant breeders may use methods of selection for crop improvement such as its protein, oil, carbohydrate contents and various other substances of nutritive importance and recommend specific application of organic or inorganic salts using knowledge of plant physiology.

Hydroponics[1], the practice of growing plants in soil-less medium and with their roots submerged in mineral nutrient solution. Hydroponic systems are engineered as a highly space and resource efficient forms of farming as compared to traditional agricultural practices, providing higher yield and economic returns. These systems are not affected by weather, wild animals and any other external biotic and abiotic factors. Such systems also make less and efficient utilization of water. These are now being used commercially for the production of many greenhouse and indoor crops, such as tomato (*Solanum lycopersicum*) and cucumber (*Cucumis sativus*) and other leafy greens like lettuce, basil, spinach, coriander, amaranth, fenugreek, etc.

Mycorrhizae modify the plant root system and influence plant mineral nutrient acquisition. Arbuscular mycorrhizal fungi are known to be important in the uptake of nutrients such as phosphate, zinc and nitrogen to some extent. As fertilizers are becoming increasingly expensive, harnessing arbuscular mycorrhizal symbiosis to optimize crop nutrition will depend on understanding how the symbiotic partners interact to influence nutrient acquisition.

Phytoremediation techniques can be used to detoxify metal-contaminated soil by means of biological activity of **hyperaccumulator plants.** Over 450 plant species have been identified as hyper-accumulators. Hyper-accumulators are of great interest because of their potential application in bioremediation i.e. the reclamation of contaminated soil by removal of toxic materials, including trace elements (such as Zn, Co, Cd, Ni, Cu, Mn), metalloids (As) and non-metals (Se).

---

[1] **To grow 1 kilogram of tomatoes through traditional methods of farming requires 400 litres of water. With hydroponics, this comes down to 70 litres of water, and only 20 litres for aeroponics (the process of growing plants in the air or misty environment. Aeroponics, however, requires high initial investment for temperature controlled spaces, which may not be feasible for the common man, leave alone urban farmers.**

20.5 Reproductive Development

Plant reproduction is a central process in the plant's life cycle. Flowering, pollination, and fertilization culminate in the formation of the seeds, which ensure the propagation of the species but are also the main output of agricultural production. A strict control of flowering time ensures the formation of as many seeds as possible, which is critical for the competitive advantage of a plant in its natural habitat, as well as in an agricultural setting. The importance of plant reproduction is demonstrated by the agricultural importance of fruit and seed crops, with cereals contributing as much as 80–90 per cent of the world's calorie supply.

20.6 Fruit Preservation

Plant physiology has led to the development of many useful processes for food technology; food preservation and food processing such as pickling, reducing the water content and increasing the viscosity of tomato pulp for the manufacture of tomato ketchup, and increasing the starch content of potatoes.

20.7 Nitrate Assimilation

As expected, atmospheric levels of CO_2 will double during this century, CO_2 inhibition of shoot nitrate assimilation will increasingly affect plant-nutrient relations. Food quality of C3 crops such as wheat has already suffered deterioration and will decline even more during the coming decades. Breeding crops for enhanced root nitrate and ammonium assimilation has the potential to mitigate such drops in food quality.

20.8 Plant Secondary Metabolites (Natural Products)

Metabolic engineering of natural product pathways is the future of **pharmaceutical biotechnology**. Chemists have long been interested in secondary metabolites for their potential uses as dyes, polymers, fibres, glues, oils, waxes, flavours, fragrances, drugs, insecticides, and herbicides. Today it is recognized that natural products have important ecological roles in plants. Photosyhthetic pigments, carotenoids, have become very popular as **neutraceuticals** (foods with health benefits) and are believed to help prevent cancer, heart disease, and macular degeneration.

Anthocyanin levels in fruits can be manipulated by transgenic approaches (by over-expressing transcription factors that control the biosynthesis of these compounds in *Antirrhinum*), even to the extent of introducing high levels of these compounds in tomato flesh, where they do not normally occur. A greater understanding of the molecular determinants of other aspects of fruit development, such as the production of volatiles, will presumably offer further possibilities to improve fruit quality.

Terpenoids play important roles in human society and are of nutritional and pharmaceutical significance. Many of the terpenoids that are in high demand as flavours, fragrances, pigments, or drugs are produced in low amounts by plants. Therefore, enthusiasm has grown for creating transgenic plant lines capable of producing increased levels of valuable terpenoids. Recent successes in cloning genes that encode enzymes of terpenoid biosynthesis have provided the necessary tools for this effort. The best known example of terpenoid engineering in plants is the creation of '**golden rice**', a variety of rice with increased β-carotene in its kernel. β-carotene (a precursor of vitamin A) is scarce in the staple crops of many developing countries, causing a number of ailments, especially in children. Although rice does not synthesize β-carotene, it does produce a precursor, geranylgeranyl pyrophosphate (GGPP), to β-carotene. However, rice lacks the enzymes necessary to convert GGPP to β-carotene (provitamin A) . A team of Swiss and German scientists have succeeded in genetically modifying rice by adding the four alien genes that code the missing enzymes that are needed to convert the precursor GGPP found in rice to β-carotene. The result is a variety of rice known as 'golden rice', so named because of the colour of β-carotene that accumulates in the rice grains. Splitting of β-carotene yields 2 molecules of Vitamin A. Genes for four enzymes have been identified through recombinant gene technology. They are: *Phytoene synthase, Phytoene desaturase, G-carotene desaturase, and Lycopene β-cyclase* and can now be synthesized.

Golden Rice

Golden rice was the result of collaboration between Professor Ingo Potrykus of the Swiss Federal Institute of Technology and Professor Peter Beyer of the University of Freiburg, Germany. Their work included two genes from daffodil plants and a third from a bacterium. A plant microbe was used to ferry its genes into plant cell. In 2018, Australia, Canada, New Zealand, and the US approved of 'golden rice' for cultivation. The Philippines approved the cultivation of Golden Rice in 2019 whereas Bangladesh is still considering application for approval.

A genetically modified (GM) '**Super-banana**' backed by the Bill and Melinda Gate Foundation is enriched with α- and β-carotene (making it orange rather than cream-coloured) and could save the lives of hundreds of thousands children in underdeveloped countries to combat vitamin-A deficiency. High in potassium and low in sodium, these bananas lower blood pressure and protect against heart attacks and strokes.

In another example, plant terpenoid genes were used to engineer microbes to produce artemisinic acid, a precursor to the antimalarial sesquiterpene artimisinin.[2] Reduction of cyanogenic glycoside levels in cassava (*Manihot esculenta*), barley (*Hordeum vulgare*), and sorghum (*Sorghum bicolor*) may be accomplished through genetic engineering to make them suitable for human consumption, especially in rural populations.

20.9 Strategies Employed by Plant Breeders to Eradicate Weeds in Crop Fields

(i) Introduction of **glyphosate-resistant crops** have helped farmers to destroy weeds selectively without affecting the main crop. A glyphosate spray blocks the Shikimic acid pathway followed by disruption of synthesis of aromatic amino acids and their derivatives, thus eliminating weeds from the fields.

[2] For the discovery of artemisinin, a new class of antimalarial drug, China's Tu Youyou was awarded the Noble Prize in Medicine (2015).

(ii) **Photosynthetic inhibitors** such as DCMU, atrazine, paraquet, DBMIB, behave as commercial herbicides and generally control annual and perennial broad-leaved weeds as well as grasses. They bind to specific sites in the thylakoid membrane of the chloroplast, block electron transport chain resulting in interruption of the photosynthesis in the susceptible plants .

(iii) Removal of parasitic weeds in crop fields by initiation of germination of their seeds in the absence of any host crop (**suicide germination**), results in loss of seed bank of parasitic plants. For example, *Striga hermonthica* (a genus from Orobanchaceae) parasitic on cereal crops, causes huge crop losses, particularly in Africa. Planting non-host crops in the fields is an effective strategy to induce suicide germination.

20.10 Genetic Engineering of Crops

- **Amino acid biosynthesis pathways,** unique to plants, are goals for fundamental and applied research. Our major aim is to understand the genes that control the assimilation of inorganic nitrogen into amino acids and the factors that control the synthesis of secondary compounds from amino acids in plants. Research on amino acid biosynthetic pathways and their manipulation in transgenic plants can enhance crop resistance to osmotic stress, and improve composition of food protein as well as production of pharmaceutically useful compounds.
- **Proline metabolism** is a way forward for metabolic engineering of stress tolerance. Water stress is a primary reason for considerable loss of agricultural productivity, making it desirable to develop drought-tolerant plants by conventional or transgenic strategies. Studies of genetically engineered plants that accumulate high levels of proline have provided clear evidence of the ability of proline to confer increased osmotic tolerance. Proline also scavenges reactive oxygen species (ROS), which are generated under stressful conditions, and thus an additional benefit of proline in stress situations.
- Production of biomass may be enhanced by **engineering photorespiration**. A solution to the current food and energy shortages depends on the degree to which land plants can be developed for greater CO_2 assimilation. When O_2 outcompetes CO_2, the Rubisco behaves as oxygenase and reduces the amount of carbon entering the Calvin cycle. Therefore to understand how to engineer leaf cells for improving the photosynthetic efficiency, plant physiologists are tackling various features of the photorespiratory pathway, from modification of the active site of Rubisco to the introduction of parallel photorespiratory pathways by genetic engineering.
- As we know, the photorespiratory cycle is essential for land plants, an attractive possibility is to incorporate different mechanisms for retrieving the carbon atoms of 2-phosphoglycolate. One approach introduces a **bacterial glycolate catabolic pathway** into chloroplasts of land plants (*Arabidopsis*). This reduces the continuous exchange of photorespiratory metabolites through peroxisomes and mitochondria, releasing photorespired CO_2 in the chloroplasts, where it can be directly fixed. Chloroplasts of the transgenic plants have an entirely functional photorespiratory cycle while additionally accommodating the bacterial enzymes: glycolate dehydrogenase, tartronic semialdehyde synthase, and tartronic semialdehyde reductase. The engineered plants grow faster, have increased biomass, and contain higher levels of soluble sugars.

- Use of **antitranspirants** offers great potential in commercial floriculture to maximize the shelf-life of horticultural crops/herbaceous plants/ornamental plants and enhance their aesthetic characteristics. A biologically active form of ABA (S-ABA) as spray has been used successfully to enhance longevity of *Chrysanthemum, Petunia, Tagetes, Aster* and *Impatiens* species during shipping and retailing.

20.11 Developing Crops with Increased Tolerance to Stress (Biotic and Abiotic)

This strategy can prevent annual losses of billions of dollars to agricultural production. Traditional breeding in combination with genetic engineering, has resulted in the identification and successful use of selected genes responsible for tolerance of salt stress in wheat and rice, boron; aluminium stress in wheat, sorghum, and barley; and anaerobic stress in rice.

20.12 Biofortification for Selecting and Developing Crop Cultivars Denser in Micronutrients

Biofortification is a new type of nutrition intervention to address the global problem of micronutrient deficiencies of iron, zinc, iodine, selenium and vitamin A (also known as **hidden hunger**) because people suffering from this type of under-nutrition appear healthy, but are in fact more vulnerable to illness and infection. Both plant breeders and biotechnologists are adopting three major strategies of biofortifications to eradicate hidden hunger by using options like: 1. **conventional plant breeding** or Mendelian methods, 2. **agronomic biofortification** (e.g., increasing micronutrient levels in crops *via* foliar application or supplying these deficient nutrients along with irrigation water), and 3. use of **transgenic crops** or **GM crops** - the latter method of genetic engineering is more precise and it takes less time but there are great public concerns about their bio-safety, environmental safety, unfavourable impact on biodiversity, and long-term impact on human and animal health. Until such concerns are adequately addressed and approved by all nations, work will have to be limited to the first two approaches.

We have had many successes in this regard. To begin with, a research team from the Indira Gandhi Agricultural University (IGAU), Raipur, had developed a new rice variety rich in zinc named **Chattisgarh Zinc Rice-1** after a region called Chattisgarh in India. Next, with the support of IRRI, Manila, Philippines, the Bangladesh Rice Research Institute was able to develop a high-zinc rice variety **BRRI - dhan 62** which has 20–22 parts per million (ppm) of zinc as compared to earlier varieties of rice having just 14–16 ppm. This variety was developed using Mendelian breeding methods. Finally, cultivation of an **iron-rich pearl millet** was commercialised in India and a few hundred acres have been brought under this crop.

The Consultative Group of the International Agricultural Research (CGIAR) has undertaken the initiative of bio-fortifying seven food crops, including wheat, rice, pearl millet, and maize through its initiative - *HarvestPlus*. Dr Howarth Bouis (US) founded

HarvestPlus at the International Food Policy Research Institute, a major NGO in the development of biofortified crops, primarily using conventional breeding techniques. Three scientists from the International Potato Centre (known by its Spanish acronym CIP) - Dr Maria Andrade (Cape Verdes), Dr Robert Mwanga (Uganda) and Dr Jan Low of the United States and Howarth Bouis also from the US were awarded the 2016 World Food Prize for their achievement in developing the single most successful example of micronutrient and vitamin A enriched, **orange-fleshed sweet potato** (OFSP) which has been released in more than 40 countries. This variety can prevent Vitamin A deficiency, and was also found to be disease resistant, drought-tolerant and high-yielding, making it easier to grow in the variable soils and climatic conditions found in Sub-Saharan Africa. This has resulted in great health benefits for millions of people suffering from micronutrient deficiency.

Efforts are on to develop high-micronutrient varieties of Cassava (*Manihot esculenta)* and leguminous crops too. The Bill and Melinda Gate Foundation is also investing lot of funds for encouraging work on other crops. Professor M. S. Swaminathan, Founder Chairman of the M S Swaminathan Research Foundation, in Chennai (India) has advocated that **Drumstick** (*Moringa oleifera*), sweet potatoes, nutritious millets, besides fruits and vegetables should be cultivated and consumed in India to overcome malnutrition and undernutrition; and, we need to introduce agricultural remedies for nutritional maladies. *Moringa* is a superfood because its leaves provide four times the vitamin A in carrots; four times the calcium in milk ; three times the potassium in bananas; seven times the vitamin C in oranges; and twice the protein in milk . The fact that *Moringa* is easily cultivable makes it a sustainable remedy for malnutrition.

Unsaturated and Monosaturated Fatty Acids are Considered Healthier than Saturated Fatty Acids

In 1996, the first genetically modified soybean was introduced to the US market by Monsanto, and since then it has been grown extensively in USA, Canada, Brazil, Argentina, Paraguay, South Africa, Uruguay, Bolivia, Mexico, Chile and Costa Rica. Scientists at the major chemical giant DuPont Pioneer have created a transgenic soybean (*Glycine max*) line whose oil contains up to 85 per cent oleic acid (a monosaturated fat) as compared to the 25 per cent that is characteristic of the conventional variety.

Native refined canola or mustard oil (*Brassica napus*) contains high levels of both glucosinolates and erucic acid that tend to impart a characteristically undesirable taste to the oil, and has been linked to cardiovascular diseases. By using the traditional plant breeding technologies, Keith Downey and Baldur R. Stefansson were the the first two scientists to create a new variety 'Canola' with less than 2 per cent erucic acid and low glucosinolates content. However, herbicide- tolerant GM canola, a cousin of canola, produced through genetic engineering, has a very low level of erucic acid and glucosinolates, and it has been under cultivation in the US, Canada, and Australia for over a decade, and as such does not present any health issues to the consumers and without any harmful effect on the environment. GM mustard has undergone rigorous testing for all bio-safety characters and for its environmental impact. GM mustard will revolutionise the oil crop productivity and will greatly contribute to alleviating cooking oil shortages in the country.

20.13 Genetic Engineering of Lipids

Improvement of edible oil quality by metabolic engineering is a major objective of plant breeders. Extensive research was done in the 1950s to produce cultivars of *Brassica napus*, now known as **canola**, with low erucic acid content. Oil from genetically engineered soybean variety is similar in composition to olive oil and has improved health benefits due to reduction in omega-6 fatty acids and improved oxidative stability. Molecular genetical approaches have been used to increase seed oil yields. Oil palm (*Elaeis guineensis*), which produces 4000 litres of oil per hectare per year with low inputs of agrochemicals, can be genetically engineered, and it may be viable to produce biodiesel from a renewable source at a cost that is competitive with petroleum. Similarly high lauric-content rapeseed can be successfully used in oilseed engineering.

20.14 Engineering Cell Wall-degrading Enzymes into Growing Plants to Deliver Cost-effective Bioethanol Technology

Two recent biological techniques can be used to convert biomass to ligno-cellulosic ethanol :

(i) engineering cell wall-degrading enzyme into growing plants
(ii) use of the gene-timed expression technique.

The plant cell wall represents the most abundant renewable biomass resource for lignocellulosic ethanol production. Engineering cell wall-degrading enzymes (cellulases - endoglucanases, exoglucanases and β-glucosidase) into growing plants provides a promising solution for converting tough plant biomass to biofuels.

To conclude, agriculture advances are vital to meeting the food demands of billions of peoples. Not only as population grows but also in the face of increasing individual consumption and as production comes under pressure from climate change.

Plant Physiology offers one of the most striking examples of the symbiotic relationship that exists between fundamental principles and applied research. We owe much of the green revolution in agriculture to the knowledge of plant physiology.

Unit VI Breakthroughs in Plant Physiology

Chapter 21

Seminal Contributions of Plant Physiologists

Thomas Graham

It was Thomas Graham who coined the term 'colloid' in 1861 for substances that cannot diffuse through parchment membranes, to distinguish such matter from crystalloid (true solution). He also coined the terms 'sol' and 'gel'.

He discovered the principle of 'Graham's Law of Diffusion': that the relative rate of diffusion of different gases is inversely proportional to the square roots of their densities. The rate of diffusion of hydrogen is four times more than that of oxygen.

$$\frac{r_1}{r_2} = \sqrt{\frac{d_2}{d_1}}$$

$$\frac{rh}{ro} = \sqrt{\frac{do}{dh}} = \sqrt{\frac{16}{1}} = 4$$

He also discovered 'The law of independent diffusion of gases', i.e., the diffusion of different gases is controlled by their 'own diffusion pressure gradient' irrespective of diffusion pressure gradient of other gases.

He devised **Graham's dialyser** by which a colloid can be separated from crystalloid by using a parchment membrane or collodion, e.g., starch (colloid), sodium chloride or glucose (crystalloid). The process is known as **'dialysis'**.

Ernst Münch

In 1930, Ernst Münch, the German plant physiologist, introduced the concept of **apoplastic** (cell wall plus intercellular spaces, representing non-living continuum) and **symplastic** (through the cytoplasm and connecting plasmodesmata) pathways of water and ion (both anion and cation) transport.

Before that in 1926, he had developed the model of phloem transport, currently favoured by most plant physiologists, Münch's 'pressure-flow model', is also called 'mass flow hypothesis'. The pressure flow hypothesis is the simplest and most widely accepted mechanism for long-distance assimilate transport. The solutes are translocated 'en masse' through the sieve tubes from the 'source' (the supply end) to the 'sink' (the consumption end) along a turgor pressure gradient.

Again in 1930, by using the girdling technique, he showed that phloem transport can take place in both upward and downward directions (i.e., bidirectional transport).

Paul J. Kramer

An American plant physiologist well-known for his researches on **water-plant relations**, he authored the famous book '*Water Relations of Plants* (1983) - Academic Press.

Paul J. Kramer, one of the deans of the water relations of plants, carried out researches in diverse fields, including physiology of woody plants.

He was a solid supporter of the neglected theory of Renner that two mechanisms are involved in the absorption of water. This propounds that roots function as osmometers under mesophytic and lowly transpiring conditions as prevailing in the early morning before sunrise, in the evening and at night time, thus leading to the development of positive root pressure that pushes the water up from below, and is accompanied by guttation. Under xeric and highly transpiring conditions, though, water is pulled up through the roots and stem passively, by forces developing in the transpiring cells (passive absorption) and here osmotic movement of water is negligible. Passive absorption accounts for about 98 per cent of the total water absorbed.

Paul J. Kramer, one of the deans of the water relations of plants, carried out researches in diverse fields, including physiology of woody plants.

He proposed, in 1937, that the rate of absorption of water is approximately equal to the rate of transpiration.

Kramer along with Kozlowski gave great support to the cohesion-adhesion hypothesis, originally proposed by Dixon and Joly in 1894. They studied dendrographic measurement of the diameter variations in tree trunks and found that the diameter of the stem decreased during day time when transpiration is high but increased during periods of low transpiration, indicating that water is under tension in transpiring plants.

H. F. Thut

By using leafy twigs of different kinds of woody plants, both angiosperms and gymnosperms, Dr H. Thut at Cornell University, Ithaca (USA), demonstrated in 1932, a rise of mercury above atmospheric pressure, ranging up to 101.8 cm when the atmospheric pressure can support a column of mercury 76 cm high in a vacuum. Perhaps this was the strongest evidence in support of the cohesion hypothesis.

Later in 1941, in a classroom demonstration using 'hemlock', he reported a rise of mercury (Hg) of the order of 118 cm.

One atmosphere supports a column of mercury (Hg) 76 cm high in a vacuum (or a barometric height of 76 cm Hg), so that the corresponding column of water would be about 34 ft (76 × 13.6 = 1033.6 cm or about 34 ft). It will not be possible for water to rise beyond that, and if water is to be transported to a height of 400 ft, this would require a pressure of about 12 atm (400/34 = 12.2 atm).

Dennis R. Hoagland

A pioneer in the study of plant mineral nutrition, and famed for Hoagland's solution (which he developed in 1920 before modifying it to what would be known as Arnon and Hoagland's solution) he established the famous 'Division of Plant Nutrition' at Berkeley, California, USA. Among his well-known students were Daniel I. Arnon, Perry R. Stout, T. C. Broyer, Emanuel Epstein and C. E. Hagen.

By using radioactive potassium (^{42}K), he provided conclusive evidence that the upward movement of salts takes place in the xylem tissue and from where the 'lateral translocation' of salts into phloem occurs (Hoagland and Stout, 1939). He demonstrated unequivocally that both anions and cations accumulate within plants against the concentration gradient (i.e., through active transport). Analysis of ion accumulation in the cell sap of *Nitella clavata* and *Valonia macrophysa* has shown that the accumulation of salts or salt absorption mechanism is both selective and also against the concentration gradient. Furthermore, he concluded that salt absorption and accumulation is inhibited at low temperatures, low oxygen tension, metabolic inhibitors and so on, indicating thereby that the ion accumulation that occurs in plants requires metabolic energy.

This has been further corroborated by the works of Hober (1945) and Raven (1967) on *Nitella translucens, Chara australis* and *Hydrodictyon africanum*. The active absorption concept of salt uptake is supported by scientists like Street, Lundegardh, and Epstein.

A. Hemantaranjan

Hemantaranjan was born in Chapra, India, on 30 November 1954 and it was while studying for M.Sc.in Botany, at the University of Bihar at Muzaffarpur, that he specialized in Plant Physiology under the aegis of Professor K. K. Jha, a doctorate from Cambridge University. Hemantaranjan then went on to a PhD himself, in Botany-Plant Physiology, at the Banaras Hindu University in 1982.

Specialising as one of the pioneers in physiology and biochemistry of micronutrients zinc and boron in crop plants; he was granted prestigious UGC and CSIR Fellowships and was awarded the 'Research Scientist' Award of the Government of

India for his work on the role of zinc in delaying leaf senescence by improved auxin and polyamine production; boron in pollen germination and reproduction.

Hemantaranjan has served many roles as teacher as he has risen through the ranks of faculty - culminating in a stint as Head of the Department (2007–2010); at the BHU. In the course of his teaching career, he has been the guide for at least 13 PhD and 65 M.Sc students. He has also been a prolific writer, with over 127 research articles (highly cited & H-indexed); 30 review articles and book chapters, and over 30 books to his credit.

Besides his expertise on iron, zinc and boron nutrition, he established the significance of salicylic acid, brassinosteroids and paclobutrazole's significance in mitigating abiotic stresses, such as drought, salinity, terminal heat and flooding stresses in chickpea, pea, wheat and mungbean. Importantly, he has been able to furnish explanations for factors underlying altered physiological and phytochemical mechanisms.

His work has included discoveries of the legume-rhizobium relationship and complexities for efficient biological nitrogen fixation; enhanced nodular carbon fixing enzyme RuBP carboxylase in French bean and soybean; mitigation techniques at molecular levels on wheat terminal heat stress, chickpea drought stress and Field pea salinity stress.

Hemantaranjan was first accorded global recognition in 1987 when he delivered three lectures on iron nutrition research in the USA, on invitation. This was followed by the publication of five international research articles serially in US journals. Then, he published two international books: 'Advancements in Iron Nutrition Research' and 'Advancements in Micronutrient Research' in 1995 and 1996. Thereafter he was appointed the *Editor-in-Chief*, International Treatise Series on ***Advances in Plant Physiology*** (1998–2019: editing over 18 *Volumes*.

His awards have been many and in addition to the Research Scientist Award of the UGC; he has been Fellow, Indian Society for Plant Physiology; and, Japanese Society for Plant Physiologists. He has won the 'Shiksha Rattan Award' in 2010; the Agricultural Excellence Award in 2013 and a Life Time Achievement Award (2014); besides holding prominent posts on a number of important Plant Physiology forums across the globe.

Receiving invites from the world over, Hemantaranjan is prominent on the lecture circuit, addressing and chairing gatherings of scientists and organising conferences.

Emanuel Epstein

German born but educated in later years at Davis and Berkeley Campuses of the University of California, USA, Emanuel Epstein joined Dennis R. Hoagland's famous 'Division of Plant Nutrition' group at Berkeley where he pioneered the technique of 'excised barley-roots' in 1950s and 1960s about how roots absorb mineral salts. 'Low-salt roots' were used in ion-uptake studies. Radioactive isotopes of nutrient elements were employed.

He demonstrated that absorption of solutes is specific and selective. He also contributed to the carrier concept of ion transport across the membranes.

Along with Laggett in 1956, he reported that back diffusion (**desorption**) of absorbed ions from outer space in plants is possible but the ions that have been actively transported into inner space cannot diffuse back during desorption.

Together with Hagen (1952) reported that during movement – potassium (K^+), caesium (Cs), rubidium (Rb) that are in the same column of the periodic table compete with each other for the same carrier or binding site, whereas lithium (Li) and sodium (Na) do not. This was supported by Epstein (1978) during his interaction of ion studies.

He is an authority on 'salt-tolerance' in crops. In 1972, he wrote the famous book on *'Mineral Nutrition of Plants: Principles and Perspectives'*, New York: Wiley.

J. D. Sayre

The so-called 'classical theory' of stomatal movement was formulated in 1926 by J. D. Sayre. According to him, removal of CO_2 during photosynthesis leads to increased pH (alkalinity or basicity) that favours conversion of starch into 'soluble sugars', thereby increasing the osmotic concentration of guard cells and withdrawal of water from subsidiary cells. Increased turgidity in turn leads to opening of stomata.

On the other hand, accumulation of CO_2 in the guard cells at night time leads to lowering of pH (i.e., acidity) which promotes conversion of sugars into 'insoluble' starch, followed by loss of turgor pressure of guard cells and hence closing of stomata.

$$\text{Starch} + \text{P}_i \underset{\text{pH 5}}{\overset{\text{pH 7}}{\underset{\text{Phosphorylase}}{\rightleftharpoons}}} \text{Glucose-1-phosphate}$$

G. W. Scarth

According to him (1930–31), it is carbon dioxide, through the formation of carbonic acid (H_2CO_3), that is the main controlling force determining the pH changes in guard cells. During day time, CO_2 after entering the leaves is instantaneously fixed during photosynthesis. This lowers CO_2 levels and so lowers the production of H_2CO_3. The consequent lower availability of H^+ ions leads to alkalinity (pH 8–9) which favours the conversion of starch into 'soluble sugars'.

$$\underset{\text{(insoluble)}}{\text{Starch} + \text{P}_i} \underset{\text{pH 5}}{\overset{\text{pH 7}}{\underset{\text{Starch phosphorylase}}{\rightleftharpoons}}} \underset{\text{(soluble)}}{\text{Glucose-1-phosphate}}$$

At night time, there is 'build-up' of CO_2 due to non-photosynthetic activity and respiration that goes on at night, thus resulting in greater production of H_2CO_3; leading to more production of H^+ ions (i.e., acidity) that promotes conversion of glucose-1-phosphate to 'insoluble starch' that ultimately leads to stomatal closure.

Yin and Tung (1948) reported the presence of enzyme phosphorylase in guard cells where it carries out the interconversion of starch to sugars.

F. C. Steward

Steward was a Cornell University (USA) plant physiologist well-known for the following works.

He was the author or editor of:

(i) *Plants at Work.*
(ii) *About Plants -Topics in Plant Biology* (with A. D. Krikorian).
(iii) *Plant Physiology - A Treatise* (1960–72), Vols. 1–6, edited by F. C. Steward (Academic Press).

In 1958, he and his co-workers were able to produce a whole plant from a single phloem cell of a carrot (*Daucus carota*) root, complete with flowers and roots, on a basal medium under aseptic conditions, thus confirming German botanist Gottlieb Haberlandt's hypothesis of **'totipotency'**, i.e., each cell has all the genetic information for the whole plant.

He extended the 'starch-sugar' hypothesis in 1964 that glucose-1-P is converted to glucose and inorganic phosphate - both of which are osmotically active and play an important role in the opening of stomata (Figure 21.1).

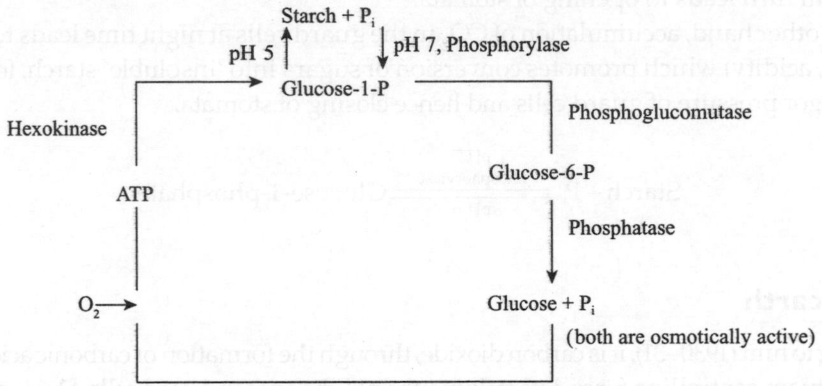

Fig. 21.1 Steward's starch-sugar model

Steward and his co-workers reported greatly increased growth of carrot root tissue by adding coconut milk to the culture medium.

Jacob Levitt

In 1967, J. Levitt modified the 'classical theory' by proposing that carbonic acid is 'too weak an acid' to bring about pH changes in the guard cells where conversion of 'starch-sugar' (or *vice versa*) occurs during day or night.

Removal of CO_2 during photosynthesis would lead to 'low CO_2 concentration' in the guard cells that favours 'decarboxylation' (breakdown of organic acids), that would result in fewer H^+ ions, i.e., alkalinity or high pH. Under alkaline pH, starch is converted into soluble sugar that would lead to low ψ, thus withdrawing water from adjacent subsidiary cells and leave the stomata open.

On the other hand, during night time, there is a build-up of CO_2 in the guard cells. High CO_2 concentration in the guard cells promotes 'carboxylation' (formation of organic acids) which leads to low pH (acidic reaction) that would favour conversion of sugar to 'insoluble starch', followed by outward movement of water (exosmosis) into subsidiary cells. Thus, the stomata close down.

Proposed in 1974, 'the Proton Transport Theory' of photoactive stomatal opening, combining the good points of Scarth's classical pH theory and active K^+ ion transport theory.

Levitt authored a book titled *'Introduction to Plant Physiology'* (1969). C. V. Mosby Company, St. Louis, Tokyo; and another titled *'Responses of Plants to Environmental Stresses'* (Vols. I and II, 1972) Academic Press .

Franz Knoop

In 1904, Franz Knoop first studied the process of β-oxidation during fatty acid degradation in plants concluding that during this process, 2-carbon fragments are successively released in the form of acetyl CoA.

In 1911, he demonstrated that ammonia was incorporated into glutamic acid by the activity of enzyme glutamic/glutamate dehydrogenase.

His work should not be confused with that of W. Knop (a German physiologist) who in 1868 employed a water culture method for studying the essentiality of minerals for plant nutrition. He devised a nutrient solution which still bears his name.

Hugo De Vries

Dutch botanist, Hugo De Vries devised the incipient plasmolytic method for determining the osmotic pressure of plant cells.

In 1885, he proposed the 'Protoplasm Streaming Hypothesis' for translocation of food (or solutes) materials. The rapid streaming movement (cyclosis) of protoplasm is presumed to carry along with the enclosed solute particles from one end of the sieve tube to the other.

Well-known for the famous *'The Mutation Theory of Evolution*. It is he who coined the term 'tonoplast' for the vacuolar membrane.

J. C. Bose

Although basically a physicist, Jagdish Chandra Bose is regarded as the father of plant physiology in India.

He devised several instruments, including sensitive mechanical recorders for measuring growth (Bose's crescograph which magnify growth 1000–10,000 times), the movement of leaves', stem contraction and responses to mechanical and electrical stimuli.

He also devised a 'bubbler' with automatic recorder for measuring photosynthesis in *Hydrilla* - an aquatic plant.

Based upon his extensive investigations (between 1907 and 1914), he proposed that the transfer of stimulus in a sensitive plant or touch-me-not (*Mimosa pudica*) was in the form of some electrical impulse (action potential) just like the nervous impulse in animals. This action potential travels very fast at velocities of 2 cm/s.

An advocate of the 'vital theory' of ascent of sap, and in 1923, Sir J. C. Bose proposed that the ascent of sap is due to the pulsatory activity of the cortical cells of the stem just outside the endodermis (for details see chapters 2 and 22).

Stephen Hales

Stephen Hales is regarded as the father of plant physiology (1677–1761) in the English speaking world.

In his book *'Vegetable Staticks'* (1727), he first described a method of demonstrating transpiration in plants which when enclosed under a glass vessel produced liquid water that condensed on the vessel walls (it was long before photosynthesis was known). He also devised methods for measuring the rate of transpiration such as 'weighing method' and Lysimeter and the Gravimetric method (i.e., weighing a potted plant in the beginning and at the end of a prescribed time period).

He coined the term 'root pressure' in 1727, and described the method to record root pressure by means of a mercury manometer attached to the stump of a cut stem, and in 1727, he suggested that plants derive their nutrition partly from air, and that light participated somehow in the process of photosynthesis.

Otto Renner

It was Otto Renner in Germany who coined the terms 'active absorption' and 'passive absorption' for water uptake in 1912 1nd 1915 respectively. During active absorption, the forces responsible for the intake or uptake of water develop in the living cells of roots while in passive absorption, the forces are originating in the transpiring cells of the leaf, and the root cells play only a secondary role, acting as a 'passive' absorbing organs.

His work provided ample support to Dixon and Joly's 'cohesion-adhesion' hypothesis for the ascent of sap.

He also coined the term 'Suction force' (Saugkraft) or 'Suction pressure' (SP) - earlier known as diffusion pressure deficit. Nowadays the term 'water potential' is used.

Frederick Sanger

In 1953, the same year in which Watson and Crick cracked or unravelled the structure of DNA, Frederick Sanger, a British biochemist, succeeded in determining the amino acid sequence of insulin - a small protein hormone with a molecular weight of 5733 D and is made up of 51 amino acids (the first protein to be sequenced). He was awarded the Nobel Prize for this discovery in 1958.

The 'Chain termination' technique, developed initially by Sanger, is currently used for DNA sequencing, and earned him his 'second' Nobel Prize (1980).

Emil Fischer

Emil Fischer, a great German organic chemist, proposed in 1894 'the Lock and Key hypothesis' to explain the mode of enzyme action (i.e., 'enzyme-substrate complex').

$$E + S \underset{K_2}{\overset{K_1}{\rightleftharpoons}} ES \xrightarrow{K_3} E + \text{Products}$$

For a 'key' to work it must be provided with the right 'lock', and so is the case with enzymes and substrates. In this analogous situation, some transient reactions between an enzyme and a substrate must occur, however briefly.

Also Fischer proposed for the first time that proteins are produced by the condensation of amino acids forming dipeptides, tripeptides, and such through peptide linkages or bonds (-CONH). He succeeded in linking together 18 amino acid molecules (15 of glycine and 3 of leucine) and thus producing a 'synthetic' polypeptide.

$$\underset{\underset{NH_2}{|}}{CH_2}-CO\boxed{OH + H}-\overset{\overset{CH_2-COOH}{|}}{\underset{\underset{H}{|}}{N}} \longrightarrow \underset{\underset{NH_2}{|}}{CH_2}-\boxed{CO-NH}\overset{CH_2-COOH}{} + H_2O$$

peptide bond

D. Koshland

In 1958, D. Koshland propounded the 'Induced-fit model' to explain the enzyme action. It envisages a flexible active site structure. The active site is not very rigid and its configuration changes according to the substrate configuration. When the substrate fits into the active site of the enzyme, the geometry of enzyme is altered. For enzyme action to take place, the proper orientation is induced by the substrate. It causes a change in the geometry so that a perfect fitting with the substrate is achieved.

In 1963, Koshland further proposed that an enzyme basically consists of 'four' categories of amino acids (Figure 6.2):

(i) Catalytic residues (catalytic site).
(ii) Binding residues (binding site)
(Both (i) and (ii) residues together form the active site).
(iii) Structural residues.
(iv) Non-essential residues.

C. B. van Niel

The Dutch-born microbiologist, Cornelis B. van Niel, then a graduate student at Stanford University, USA, in 1931 first proposed that oxygen comes from water and not from carbon dioxide in green plants as was previously thought.

He developed the following generalized equation for green plants and photosynthetic bacteria.

Elemental sulphur in photosynthetic bacteria

$$CO_2 + 2H_2A \xrightarrow{\text{Light}} CH_2O + 2A + H_2O$$

O_2 in green plants

For photosynthetic bacteria:

$$CO_2 + 2H_2S \xrightarrow{\text{Light}} [CH_2O] + H_2O + 2S$$

The globules of sulphur accumulate inside the bacterial cell wall. The purple sulphur bacteria reduce CO_2 to carbohydrates during photosynthesis but do not evolve oxygen.

In short, van Niel proposed that water, not CO_2, was the source of O_2 in green plants. This 'revolutionary' proposal was confirmed by the works of Robert (or Robin) Hill (1937) and Samuel Ruben and Martin D. Kamen (1941). This was a great 'breakthrough' in the history of photosynthesis.

Otto Warburg

The German biologist Otto Warburg, 'a Nobel Laureate', is credited with many important discoveries in the field of plant physiology that include:

His experiments with intermittent light established that when the light is supplied in short intense flashes separated by dark periods, the photosynthetic yield per second was higher than in the case of application of continuous light of the same intensity. Photosynthesis involves two steps: one is a faster photochemical (or light) reaction and the second is a slower chemical (or temperature-sensitive dark) reaction.

The inhibitory action of oxygen on photosynthesis was first observed in 1920 and is called the 'Warburg effect'. Warburg used *Chlorella* as basic system for studying photosynthesis.

In 1934, Warburg proposed the nomenclature for two coenzymes: Coenzyme I or diphosphopyridine nucleotide (DPN - now referred to as NAD) and Coenzyme II or triphosphopyridine nucleotide (TPN - now referred to as NADP).

Warburg and his co-workers in 1935 conceptualised an alternative pathway of respiration, known as pentose phosphate pathway or HMP pathway or direct oxidation pathway.

He designed the 'Warburg Constant Volume Respirometer' for measuring the rate of O_2 uptake and CO_2 release by plant tissues.

In the early 1940s, he found that chloride ions had a stimulating effect on the Hill's reaction, probably by facilitating the release of O_2 from OH^- ions.

And finally, he conceptualised the quantum requirement or efficiency (the light energy required to reduce one molecule of CO_2 to carbohydrates). He calculated that 4 quanta are required for the fixation of one molecule of carbon dioxide.[1]

F. F. Blackman

F. F. Blackman, an English physiologist, is well-known for the following discoveries:

Based upon Q_{10} studies, F. F. Blackman postulated the existence of at least two reactions - one is an exceedingly rapid photochemical or Light reaction (light dependent and temperature independent), and the second is a relatively slower thermochemical or Dark reaction (temperature dependent and light independent):

$$A \xrightarrow[\text{reaction}]{\text{Light}} B \xrightarrow[\text{reaction}]{\text{Dark}} C$$

[1] It is now generally believed that at least 8–12 quanta are required to reduce one molecule of CO_2 (or produce one molecule of O_2). The requirement of at least 8 quanta for each molecule of O_2 produced (or CO_2 reduced) is justified because it takes 4 (e^- + H^+) to reduce one CO_2 to carbohydrate level, and each electron transfered requires 2 quanta.

In 1905, he enunciated the 'Law of Limiting Factors' in photosynthesis which states - 'when a process is conditioned as to its rapidity by a number of separate factors, the rate of the process is limited by the pace of the slowest factor'. The 'slowest factor' means a factor which is present in quantity or intensity that is lesser than what is required for the process.

The Dark reaction (fixation of CO_2 into carbohydrates) is also known as Blackman's reaction.

Daniel I. Arnon

Daniel Arnon's many contributions to plant physiology have included:

Coining the terms 'cyclic and non-cyclic photophosphorylation'[2].

Accomplishing the complete light reaction with isolated and intact chloroplasts from the leaves, i.e., the formation of ATP (in cyclic photophosphorylation) and NADPH and the evolution of oxygen during non-cyclic photophosphorylation without any corresponding reduction of CO_2.

Along with James Bassham, Arnon devised methods for extracting chloroplasts from leaves without breaking the chloroplast envelope and was able to carry out the light as well as dark reactions for fixation of CO_2 in photosynthesis.

Along with Perry Stout in 1939, the two collaborators laid down three criteria for the essentiality of minerals. By then, Arnon and Stout had already assisted Dennis Hoagland at the University of California, Berkeley, USA, to develop a nutrient solution in 1938 – the widely used Hoagland's solution. Later in 1940, they developed an improved modification which came to be known as the 'Arnon and Hoagland's solution'.

Was co-editor together with Leonard Machlis *'The First Annual Review of Plant Physiology'* in 1950.

Bessel Kok

In 1970, Bessel Kok of the Martin Marietta Laboratories in Baltimore (USA) proposed a model called 'Oxygen-Evolving Complex or Clock' for the photosynthetic splitting of H_2O (photolysis). As the name implies, the OEC is responsible for splitting (oxidation) of water and consequent evolution of molecular oxygen. The OEC is located on the lumenal side of thylakoid membrane and is bound to the D1 and D2 proteins of PS II reaction centre and functions to stabilize the Mn^{++} cluster. It is also bound to chloride ions.

The oxidation of two molecules of water generates one molecule of oxygen, four protons and four electrons:

$$2H_2O \xrightarrow{\text{4 photons}} O_2 + 4H^+ + 4e^-$$

Not all the electrons released from water molecules reach PS II (P 680^+) at the same time.

According to Kok and his co-workers, the OEC system can exist in five transient states, S_0

[2] A third type of photophosphorylation called 'pseudocyclic' was first demonstrated by A. H. Mehler in 1951, where ATP is produced in the presence of FMN, vitamin K and oxygen (in the absence of $NADP^+$ and CO_2).

to S_4 which represent different oxidation states of Mn atoms which are known to catalyze electron transfer reaction. Successive photon or quantum (of extremely short flashes, i.e., short enough to excite essentially one electron at a time) advance the system from S_0 to S_1 to S_2 and then to the next S_3, etc., until state S_4 is reached. This S_4 state is unstable and reacts with a pair of water molecules (**two**), giving rise to electrons, protons and molecular oxygen (Figure 21.2).

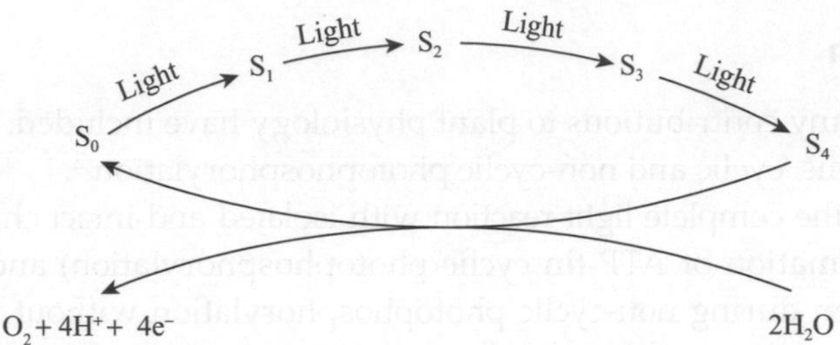

Fig. 21.2 The oxygen-evolving complex or clock.

A series of short flashes is given to a sample of dark-adapted chloroplasts. Little or no oxygen is produced on the 'first **two** flashes' and the maximum production of O_2 occurs on the **third** flash with secondary maxima occurring on every **fourth** flash.

From S_4 it goes back to S_0, making it possible for the cycle to begin again. The OEC, however, releases four electrons simultaneously but PS II can accept only one electron at a time. After one flash of light (equivalent to one photon) P 680 (Chl a 680) becomes P 680^+ (or Chl a 680^+), each excitation of P 680 is followed by withdrawal of one electron from the Mn^{++} cluster.

Together with A. T. Jagendorf, in 1963, he edited a book on *'Photosynthetic Mechanism of Green Plants'*.

N. E. Tolbert

N. E. Tolbert and his colleagues at Michigan State University, USA, formulated the steps involved in the glycolate pathway (i.e., photorespiration) that involves three cell organelles: chloroplasts, peroxisomes and mitochondria.

André Jagendorf

A. T. Jagendorf and his co-workers (1967) carried out an elegant experiment that provided the most convincing evidence for the Mitchell's chemiosmotic theory (Figure 21.3).

They equilibrated the chloroplast lamellar membranes (grana) in a pH 4 buffer. The grana were then transferred to a medium of pH 8, and ADP and P_i were added shortly afterward. This manipulation generates a pH gradient of 4 units across the thylakoid membranes (pH 4 inside the grana and pH 8 outside). Significant amounts of ATP were produced from ADP and P_i in the '**dark**'.

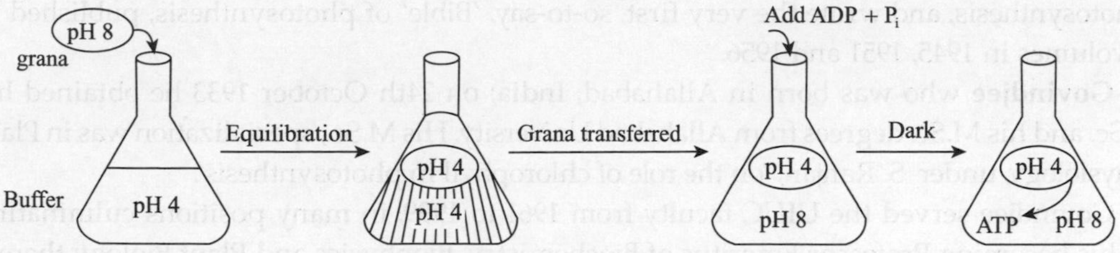

Fig. 21.3 Jagendorf's acid bath experiment.

This experiment provided strong evidence in favour of Mitchell's chemiosmotic theory – that pH gradients and electrical potentials across the membrane are a source of energy for ATP synthesis.

Jagendorf was co-editor with B. Kok in 1963, on the book on *Photosynthetic Mechanisms of Green Plants*.

Robert Emerson

Emerson is famous for the following discoveries in the field of photosynthesis:

(a) Red drop or long red wave-length 'drop off'.

(b) The Emerson enhancement effect.

(c) His findings about the 'quantum requirement or efficiency' for photosynthesis. In 1939, he calculated that 8 quanta are required for the fixation of one molecule of CO_2 (or to release one O_2 molecule).

(d) Emerson along with Arnold concluded that light is absorbed **not** by independent pigment molecules, but rather by clusters of chlorophyll and accessory pigment molecules - known as photosystems (PSII and PSI). A photosystem thus consists of two closely linked components (1) an 'antenna complex' of hundreds of pigment molecules that gather photons, and feed the captured light energy to the reaction centre and (2) a 'reaction centre' consisting of one or more Chl *a* molecules in a matrix of proteins.

R. G. Govindjee[3]

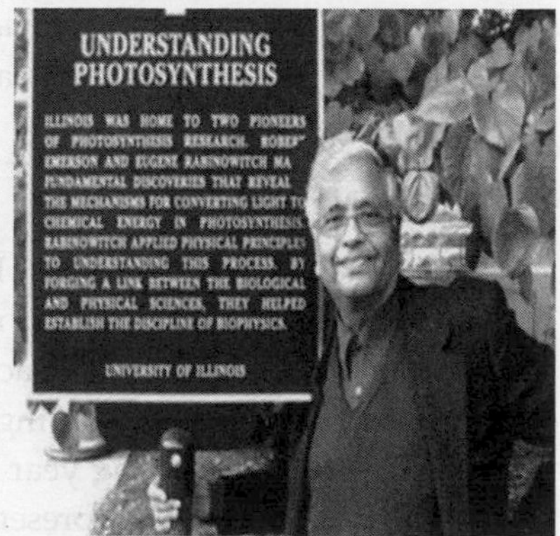

Hailing from India originally, Govindjee has been on the faculty of the University of Illinois at Urbana-Champaign (UIUC), where he earned his PhD in Biophysics, in 1960, specialising in *'Photosynthesis'*. He was mentored by two of the greatest giants in that field: Robert Emerson the discoverer of the 'photosynthetic unit', of two photosystems and two light reactions in photosynthesis; and, Eugene Rabinowitch, the genius, who integrated physics, chemistry and biology of

[3] His last name, Asthana, was dropped by his father as a protest against the caste system in India.

photosynthesis, and wrote the very first, so-to-say, 'Bible' of photosynthesis, published in 3 volumes in 1945, 1951 and 1956.

Govindjee who was born in Allahabad, India; on 24th October 1933 he obtained his B.Sc. and his M.Sc. degrees from Allahabad University. His M.Sc., specialization was in Plant Physiology, under S. Ranjan, 'on the role of chlorophyll in photosynthesis'.

Govindjee served the UIUC faculty from 1961 to 1999, in many positions culminating in his becoming Professor Emeritus of Biochemistry, Biophysics and Plant Biology there.

He is a leading authority on the *Basics of Photosynthesis*, the engine for life that provides food, biofuel and the very oxygen that we breathe. The focus of his research has been on how plants, algae and cyanobacteria convert light energy into chemical energy, using the two systems (Photosystem I and Photosystem II), oxidizing water to oxygen and producing the reducing power, needed to convert carbon dioxide to sugars.

Govindjee's discoveries include:

(a) that Photosystems I and II are run by different forms of chlorophyll a;
(b) that light energy is converted into chemical energy in both the photosystems within picoseconds (10^{-12} s); and,
(c) that CO_2 (in the form of bicarbonate) is required not only for making sugars, but for electron and proton transfer between Photosystem I and II.

His work has been recognised and he has been honoured world-wide. He is Fellow of the American Association of Advancement of Science; Fellow and Lifetime member of the National Academy of Sciences (India); the first recipient of the Lifetime Achievement Award of the Rebeiz Foundation for Basic Biology, 2006; recipient of the Communication Award of the International Society of Photosynthesis Research, 2007; and Lifetime Achievement Award of the College of Liberal Arts & Sciences, University of Illinois at Urbana, 2008.

He has been President of the American Society for Photobiology (1980–1981) and Honorary President of the 2004 International Photosynthesis Congress (Montréal, Canada).

Govindjee is known to the world through his wonderful and popular 1969 book 'Photosynthesis' (available for free on his website: http://www.life.illinois.edu/govindjee); for his Scientific American articles on 'Photosynthesis' (published in 1965, 1974, and 1990,; and for the Series *Advances in Photosynthesis and Respiration* (Springer); the latest volume (40) deals with how plants, algae and cyanobacteria protect themselves against excess light.

Fay Bendall

Fay Bendall together with Robert Hill (original discoverer of the Hill reaction) proposed in 1960 the 'Z-Scheme' for the light reaction in photosynthesis. The Hill and Bendall model involved two photochemical reactions operating in a series - one serving to oxidise the cytochromes and the other serving to reduce them.

Duysens in the following year confirmed the model by Hill and Bendall. that has been the catalyst that has led to our present fairly comprehensive understanding of electron transport and oxygen evolution in the light reaction of photosynthesis.

Louis Duysens

Louis Duysens of the Netherlands (Duysens et al. 1961) found that when a sample of a red alga was exposed to long-wavelength light, the cytochrome became largely oxidised. If shorter wavelength light was also present, the effect was partially 'reversed' (antagonistic effects). Chloroplasts contain cytochromes (iron-containing proteins) that function as intermediate electron carriers in photosynthesis.

Melvin Calvin

Melvin Calvin was awarded the Nobel Prize[4] in Chemistry in 1961, after he and Andrew Benson of the University of California at Berkeley (USA) had come up with an elucidation of the C3 pathway using radioactive (carbon 14 based) CO_2 as a tracer.

Melvin E. Calvin, an American chemist, famous for discovering the Calvin cycle, along with Andrew Benson and James Bassham for which he was awarded the 1961 Nobel Prize. His researches on alternative sources of energy have opened-up new vistas for future investigations (*Courtesy*: Professor Melvin Calvin).

The Calvin-Benson cycle, for CO_2 fixation or the dark reaction, as it came to be known, involved three steps:

(i) carboxylation,
(ii) reduction of phosphoglyceric acid (PGA), and
(iii) regeneration of ribulose bisphosphate (RuBP)-CO_2 acceptor.

Calvin along with Wolken, also proposed the molecular model of the chloroplast lamellae, where, the porphyrin 'head' of chlorophyll extends into the aqueous protein layer while the 'phytol' tail extends into the lipid layer.

He was the founder Director of the Laboratory of Chemical Biodynamics, while simultaneously holding the position of Associate Director of the Berkeley Radiation Laboratory. Later he worked on 'Alternative Sources of Energy' (biofuels), with special focus on *Euphorbia lathyris* although he also studied other species. He passed away on 8th January 1997.

Albert Szent-Györgyi

In 1935, the Hungarian biologist, Albert Szent-Györgyi studied respiration in isolated pigeon-breast muscles which have a very fast respiration. Its by-products are carbon dioxide and water. Furthermore, he found that if succinic, fumaric, or oxaloacetic acids were added to muscle

---

[4] The second Nobel Prize for photosynthetic research was awarded to Johann Deisenhofer, Hartmut Michel, and Robert Huber of Germany for their work on crystallization of light-harvesting complex from a purple photosynthetic bacterium and the determination of its (bacterial reaction center) structure by X-ray crystallographic analysis (1988).

preparations, a large amount of citric acid would be produced. This idea helped Hans Krebs, an English biochemist, to work out the exact sequence of reactions occurring during aerobic respiration for which he was awarded the Nobel Prize in 1954.

Hans A. Krebs

In 1937, H. A. Krebs proposed the Krebs cycle (or the citric acid cycle - as citric acid is an important intermediate) during aerobic respiration. It is also called the tricarboxylic acid (TCA) cycle because citric and isocitric acids have three carboxyl groups. For this outstanding discovery, the British biochemist was awarded the Nobel Prize in medicine in 1954.

Together with H. L. Kornberg in 1957, he worked out the mechanism of conversion of fats to carbohydrates (glyoxylate cycle) in the bacterium *Pseudomonas*. Later, the different reactions (and enzymes) of the β-oxidation were discovered by Harry Beevers and his associates. The cycle operates in special cell organelles called glyoxysomes, abundant in germinating fatty seeds. The two unique enzymes of glyoxysomes are 'isocitrate lyase' and 'malate synthase'.

And, together with K. Henseleit, the duo discovered the 'urea cycle' which is the conversion of ammonia (which is toxic beyond a certain level), to the less toxic urea.This is a cyclical process, with ornithine being regenerated. The chief site of the urea cycle is the liver.

Harry Beevers

Harry Beevers was associated with Purdue University and the University of California, Santa Cruz, in the USA. In the early 1950s, he had worked out the details of the 'pentose phosphate pathway' or 'Warburg-Dickens pathway'

The glyoxylate cycle was first studied in plants by Beevers in 1967 while he was investigating how castor bean (*Ricinus communis*) could convert fats to sugars during germination.

The discovery of glyoxysomes in castor bean and much of information about the reactions and enzymes of the β-oxidation is due to the work of Beevers and his colleagues. Acetyl CoA produced during oxidative degradation of the fatty acids is metabolized in the glyoxysome *via* the glyoxylate cycle that operates through **two** unique enzymes: isocitrate lyase and malate synthase. The succinate so produced is transported to the mitochondria where it gets converted to oxaloacetate. Later in the cytosol, oxaloacetate is finally converted to glucose *via* 'gluconeogenesis'. However, this glyoxylate cycle in the germinating fatty seeds ceases as soon as the fat resources have been exhausted. This cycle does not operate in seeds that store starch.

He authored the book: *'Respiratory Metabolism in Plants'* in 1961.

Frits W. Went

While working as a graduate student in his father's laboratory at the University of Utrecht in the Netherlands, Frits Went conducted experiments that could be considered those of the epoch. These were on coleoptiles of oats (*Avena sativa*). He removed coleoptile tips and placed them on 'agar blocks' to allow the growth substance to diffuse into the agar. When agar block, containing the diffusible substance, was placed directly on to the decapitated coleoptile stump, the growth (elongation) was normal, i.e., vertical. However, when these agar blocks were placed asymmetrically or unequally, curvature would result – the basis for the famous '*Avena* Curvature Bioassay Test'. The angle of curvature can be used as a bioassay for auxin.

Frits W. Went, then a young Dutch graduate student, was able to confirm that grass coleoptile tips produce significant quantities of diffusible substance (auxin) that controls the growth and development. Avena curvature bioassay test, developed by him, was the first use of a bioassay to quantify a plant growth hormone.

In 1926, according to Went, unidirectional light causes lateral redistribution of the endogenous auxin which due to polar transport diffuse more on the shaded side than the illuminated half (Cholodny-Went Model).

Having established the first phytotron at California Institute of Technology, Pasadena, California, USA, he spent the later part of his career working on 'desert ecology' at the Desert Biology Laboratory at the University of Nevada (USA). He died on 1st May 1990.

F. Kögl

In 1934, the Dutch chemist F. Kögl and A. J. Haagen-Smit isolated auxin from human urine and identified it as heteroauxin or indoleacetic acid (IAA) - now known to be naturally occurring, universally present hormone in higher plants.

Initially, he and his associates isolated two more compounds - auxin A (or auxentriolic acid) and auxin B or auxenolinic acid (non-indole compounds).

R. Gane

In 1934, R. Gane showed that ripening fruits produced ethylene gas and it was then proposed that ethylene was a hormone-natural product of plant metabolism.

K. V. Thimann

It is the studies of K. V. Thimann and his associates that have established the fact that apical dominance is the result of the interactions of hormones, largely auxins and cytokinins (as cytokinins can overcome the auxin inhibition of lateral buds.).

In 1934, Thimann and Skoog announced their discovery that auxin is synthesized in the apical meristem and is then translocated downwards (basipetal transport) where it inhibits the growth of lateral buds (i.e., apical dominance). In 1935, he isolated IAA from *Rhizopus* culture filtrate and in 1951 he ideated the 'non-osmotic mechanism' of water uptake in plants.

Thimann's works include: the book '*Hormone Action in the Whole Life of Plants*' published by the University of Massachusetts Press, USA, in 1977; the book '*Plant Biology and Its Relation to Human Affairs'* (John Wiley, New York) which he co-authored with J. H. Langenheim; and in 1980, he edited a book, *'Senescence in Plants'* published by CRC Press, Boca Raton, Florida.

Gottlieb Haberlandt

It was German botanist, Gottlieb Haberlandt who proposed in 1902 that all living plant cells are totipotent, i.e., each cell possesses the full genetic potential of the organism. The importance of plant tissue and cell culture in studying morphogenesis was first anticipated by him. This 'totipotent hypothesis' was confirmed by F. C. Steward at Cornell University (USA) who was able to regenerate plants from 'isolated' secondary phloem tissue of carrot roots when grown in a flask of liquid growth medium.

Also in 1902, Haberlandt along with Němec of Czechoslovakia proposed that movable or mobile solid bodies statoliths (or amyloplasts) present within cells called 'statocytes' of root caps perceive geotropic stimulus.

In 1913, he demonstrated that an unknown compound (now called cytokinins) present in the extract from phloem cells of different plants could induce cell division (cytokinesis) in the non-dividing parenchymatous tissues of the potato tuber where it causes cork-cambium formation and wound healing.

That very year, he further postulated that the injured cells release a 'wound hormone' which causes the adjacent uninjured cells to become meristematic and continue dividing till the wound is healed up. Later this active substance was named 'traumatic acid'.

He proposed that the rapid transfer of a stimulus over long distances takes place through the 'conducting elements' of the sensitive plant (*Mimosa pudica*). The response occurs when turgid cells lose water.

Johannes van Overbeek

In the early 1940s, American physiologist Johannes. van Overbeek and his associates discovered that the growth in culture of embryos of Jimson weed (*Datura stramonium*) was greatly enhanced by the addition of milky endosperm from immature coconut or coconut milk (the liquid endosperm of the coconut seed). This potent growth factor(s) promoted the growth of isolated plant tissues and organs in a test tube. Later these growth factors would be called 'cytokinins'.

He and his co-workers then reported that the growth of small, free-floating aquatic plant duckweed (*Lemna*) is reduced by the addition of one part per billion of ABA to the culture medium.

And in 1966, he outlined the sequence of events in barely seed germination.

Folke Skoog

During the early 1950s, F. Skoog and his colleagues at the University of Wisconsin reported that 'coconut milk' and yeast stimulated the growth or cell division of tobacco tissue cultures in the presence of auxin.

He and his co-worker Carlos O. Miller (then his postdoctoral student) discovered that 'herring sperm DNA' (the sperm of a fish) stimulated tobacco cells to divide and grow. Surprisingly, they found that 'freshly prepared DNA samples' of herring sperm DNA did not promote cell division. However, the autoclaved herring sperm DNA (heated at 15 pound pressure for 30 min) had a powerful stimulating effect on cell division. It was then logically surmised or concluded that the cell division stimulant must be a breakdown product of DNA (a derivative of purine-adenine). They named the new growth substance kinetin (6-furfurylaminopurine). Collectively, these are called 'cytokinins.'

The word cytokinins are derived from cytokinesis – a term for cell division.

D. S. Letham

In 1964, D. S. Letham of Australia and his co-workers isolated a cytokinin from corn or maize (*Zea mays*) seeds, naming it 'Zeatin'. It was the first naturally occurring cytokinins isolated from plant tissues.

F. T. Addicott

In 1964, F. T. Addicott and his associates at the University of California at Davis (USA) reported the discovery in leaves and fruits (cotton) of a substance which was responsible for the premature fall (or abscission) of fruits from the plant. They named the substance **abscission** II (identical chemically to **dormin**) which was finally named abscisic acid in 1967.

P. F. Wareing

In 1965, the English physiologist P. F. Wareing and his co-workers (University College of Wales at Aberystwyth, UK) isolated a substance from the leaves of birch, *Betula*, English sycamore, and other trees that caused winter 'bud dormancy'. They named it '**dormin**'. It had the identical structure as found in abscission II, discovered by Addicott. Later the name abscisic acid was given in 1967 by both groups of researchers led by Addicott and Wareing.

The level of ABA in the dormant buds remains very high throughout the winter. However, on the onset of spring, level of ABA decreases while the levels of auxin, gibberellins and cytokinins begin to rise. The dormant buds are activated and stem begins to grow, thereby unfolding leaves.

Together with I. D. J. Phillips, Wareing authored a book titled *'The Control of Growth and Differentiation in Plants'* (1978) - Pergamon Press, Oxford, England.

J. E. Varner

It was in 1955 that Varner and Webster discovered the process of formation of amides – glutamine and asparagine (amidation) from glutamic acid and aspartic acid, respectively, by glutamine synthetase and asparagine synthetase (see Chapter 10). Glutamine and asparagine are the two most abundant amides found in plants, and are transported from the regions of their formation to areas where amino acids and proteins are being synthesized.

In 1965, Varner and his group proposed that gibberellins activated (or caused de-repression) genes producing mRNA which in turn produce new enzymes that lead to morphogenetic changes.

And, in 1967, Varner and his colleague P. Tuan-Hua Ho reported that exogenous application of gibberellins stimulated the synthesis of 'hydrolytic enzymes' in the aleurone layer of germinating cereal grains such as barley. These enzymes are secreted into the endosperm tissue where they bring about hydrolysis of starch into sugars, and proteins into amino acids. The sugars, amino acids, and nucleic acids are absorbed by the scutellum (cotyledon) and are transported to the shoot and roots for further growth.

They further showed that actinomycin D (RNA synthesis inhibitor) will prevent α-amylase production in the aleurone tissue. Whereby, they suggested that gibberellins seem to increase DNA transcription and translation - both responsible for protein (or enzyme) synthesis.

In 1976 he would edit a book along with J. Bonner; titled: '*Plant Biochemistry*', published by the Academic Press.

L. G. Paleg

L. G. Paleg in the late 1950s first suggested the role of gibberellins in the mobilization of the reserve food material in the endosperm of cereal grains such as barley. The application of gibberellins to 'embryonectomized' (embryo-less) barley seeds with aleurone intact results in the appearance and secretion of amylases as well as protease enzymes that will hydrolyse the reserve food materials. α- and β-amylases together digest starch into glucose while protease brings about breakdown of proteins. The sugars, amino acids and other products are absorbed from the endosperm by the scutellum, and then transported to the growing regions of the embryo where they are utilized for further growth.

It is believed that gibberellins activate certain genes, resulting in the synthesis of messenger RNA molecules specific for hydrolytic enzymes. Such studies on the mechanism of gibberellins activity were carried out simultaneously by researchers in Japan, Australia, and the United States.

L. H. Flint

In 1935, L. H. Flint (of the USDA, Beltsville, Maryland) and E. D. McAlister (of the Smithsonian Institute Washington, D.C.), discovered that the germination of seeds of 'Grand Rapids' variety of lettuce (*Lactuca sativa*) is not only promoted by red light (660–690 nm) but is also inhibited by far-red light (at about 730 nm) of the spectrum. Sequential light exposure (e.g., 100 or more alternate red and far-red light) proved that the last exposure to light determined the germination response (Table 21.1). If the last exposure given is red, then seed germination occurs but if it is far-red, the seeds fail to germinate or there was reduced germination (Figure 21.4).

Table 21.1 Reversal of seed germination by far-red radiation. (If the last irradiation is red, the seeds germinate)

Irradiation	Germination
R	Yes
R-FR	No
R-FR-R	Yes
R-FR-R-FR	No
R-FR-R-FR-R	Yes
R-FR-R-FR-R-FR	No
R-FR-R-FR-R-FR-R	Yes

This discovery led to the hypothesis that the 'light-absorbing pigment' could be 'pushed' (or 'tailored') in one direction by red light, and in other direction by far-red light.

This reversal of seed germination with red (R) and far-red (FR) light was later shown in 1950s by Borthwick and Hendricks - two USDA researchers working at Beltsville, Maryland where Garner and Allard had discovered the phenomenon of photoperiodism. In 1959, they discovered a new pigment that existed in 2 states, a red-light absorbing form (660 nm) and a far-red light absorbing form (730 nm). They named this pigment '**phytochrome**'.

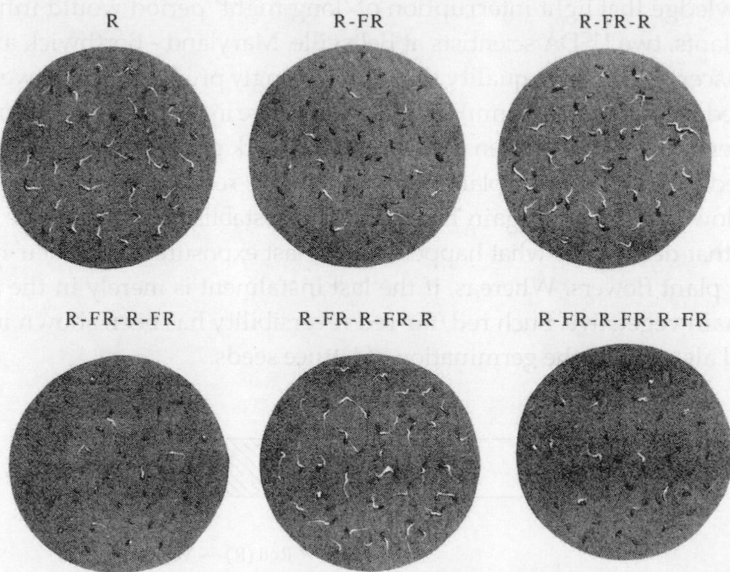

Fig. 21.4 The germination of lettuce seeds is promoted by red light (R) but is reversed by far-red light (FR). This typical photoreversible response is controlled by phytochrome.

Michael H. Chailakhyan

M. H. Chailakhyan, a well-known Russian plant physiologist, proposed the concept of a flowering hormone in 1936. He named this hypothetical hormone 'Florigen' - it is hypothetical because it has not been isolated and chemically identified.

This concept of a flowering hormone was based primarily on the fact that the hormone was produced and released by the leaves (sites of perception) and is then transported through the phloem to the apical and lateral meristems (sites of reaction or expression). In some cases, only a single leaf perceives the stimulus from where it is passed onto the meristems (the stimulus can be transmitted even across a graft-union).

Florigen was supposed to be a complex of two types of chemical substances, 'gibberellins' and 'anthesins' (that actually exist) - the former is essential for the formation of growth of stems while 'anthesins' are known to stimulate or initiate flowering.

The florigen model has been in use since 1940s, but it was first replaced by the 'Nutrient Diversion hypothesis' and now even this has given way to the 'Multifactorial hypothesis', according to which a number of factors, such as promoters, inhibitors, phytohormones, and nutrients that are present at the shoot apex at appropriate times and in appropriate concentrations, are responsible for flowering. Further he found that cultured shoots of *Perilla* could be induced to form flower primordia by the application of kinetin.

H. A. Borthwick

With the knowledge that light interruption of 'long-night' period would inhibit flowering in short-day plants, two USDA scientists at Beltsville, Maryland - Borthwick and Hendricks attempted to ascertain the light quality that most strongly promoted the flowering response. They discovered that red light (660 nm) is the most effective in preventing flowering in short-day plants. However, if the red light given in the middle of dark period is immediately followed by a flash of far-red (730 nm), then the plant flowers. If the far-red is again followed by exposure to red light, flowering is once again inhibited. They established that it is the last treatment of light given that determines what happens. If the last exposure is at the far-red part of the spectrum, the plant flowers. Whereas, if the last instalment is merely in the red part, then the plants remain vegetative. Such red/far-red reversibility has been shown in a number of SD plants, and also during the germination of lettuce seeds.

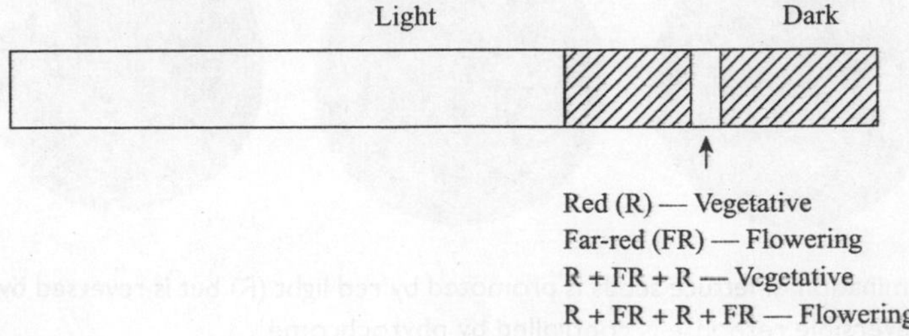

The dark period should be continuous. Red light given in the middle of dark period inhibits flowering while the far-red reverses the effect of red light and flowering occurs.

They concluded that the light-absorbing pigment **phytochrome** existed in two forms: a red-light absorbing form, Pr (660 nm) and another far-red light-absorbing form, Pfr (730 nm). Phytochrome is synthesized in the dark and on receiving red light (660 nm), it is converted from Pr (red light-absorbing form) to Pfr (far-red absorbing form) - the two forms being interconvertible. The Pr form is biologically active and induces a response.

W. W. Garner and H. A. Allard

In 1920, while working in the U.S. Department of Agriculture in Maryland, Garner and Allard discovered that flowering of plants can be controlled by regulating the daily daylight length (photoperiod), and the flowering response of plants to such photoperiods. 'Photoperiodism' is a term coined by them.

They noticed that the Maryland Mammoth variety of tobacco (*Nicotiana tabacum*) failed to flower during summer but showed profuse vegetative growth. The same variety when grown in a greenhouse during the winter produced flowers and fruits. Further they concluded that it was the short day length which controlled flowering in these plants. They called this plant a 'short-day plant'. Other examples of short day plants include soybean and cosmos.

However, there are other plants such as spinach, radish, lettuce that remained vegetative under short days but flowered when exposed to long days (long-day plants). Recognising this, Garner and Allard classified plants into three classes according to their photoperiodic response: short-day plants (SDP), long-say plants (LDP) and day-neutral plants.

The photoperiodic signal is perceived by the 'leaves', for which, as we now know, the pigment 'phytochrome' is the principal photoreceptor.

Hans Möhr

A German biologist, plant physiologist and an authority in the field of photomorphogenesis, Hans Mohr and his associates propounded that the process involves two important stages, namely: *pattern specification* (during which cells and tissues develop the ability to react to light) and *pattern realization*, during which period the light-dependent activities occur.

Based on his researches with *Sinapis alba* seedlings, he concluded that phytochrome control activation and deactivation of specific genes that code for specific mRNA molecules which in turn code for specific enzymes (new proteins) that bring about the biochemcal changes. This was in contrast to the earlier theory of the fast membrane-permeability effect.

In 1966 he investigated the phytochrome-related biosynthesis of anthocyanins that is typical of HIR (high irradiation response). All the while, he researched extensively on the physiological, biochemical and molecular basis of photomorphogenesis, biological signal reaction and nitrate assimilation, and then went on to build 'a centre of research for plant developmental biology.'

Working with the famed Bernhard Hassenstein, the duo introduced structural reforms and built the famous Botanical Institute at Freiburg where he was Professor of Botany.

He did his post-doctoral work at the Beltsville Center of the US department of Agriculture with the working group of Sterling Hendricks. He authored the books: '*Lectures on Photomorphogenesis* (1972), Berlin, Springer-Verlag; and *A Comprehensive Textbook of Plant Physiology* (1969); translations of which are available world-wide.

Chapter 22

The Status and Development of Plant Physiology in India

Plant physiology as we know, deals with different life processes operating within the cell, and interactions between the cell and the environment based upon physical, chemical and biological concepts. All these make it highly dynamic and exacting in nature. The science of plant physiology is never static but always changing as new facts are discovered and fresh concepts are developed. With new instrumentation and advances in the knowledge of working of the cell and the discovery of structure of DNA by Watson and Crick (1953), plant physiology has become first increasingly biochemical and then molecular.

A knowledge of plant physiology is essential for different fields of applied botany, whether agronomy, floriculture, forestry, horticulture, landscape gardening, plant breeding, plant pathology, or pharmacognosy. All these applied courses depend upon plant physiology for information regarding how plants grow and develop.

Until the forties, laboratories in England and Germany dominated the scientific scene and many Indian scientists began their career in these countries. However, after the Second World War, during which much of Europe was completely devastated, the focus of major research activity shifted to USA and Canada where new schools were established and many of our scientists went over there to enrich their expertise.

Compared to other parts of the world in India, the discipline of plant physiology has not received the attention it ought to have deserved but still some of our scientists earned distinction at the international level. The undisputed Indian pioneer in experimental research on plants was J.C. Bose (1858–1937), who was knighted by the British in recognition of his contributions to scientific endeavour; entitling him to the use of 'Sir' before his name. Basically a physicist, he was widely acclaimed for his discoveries on radio waves and wireless. But in his later years, and particularly after retirement (which to most people spells the end of one's career), Bose was also greatly interested in plant life and became hugely involved in research on plants. In fact, in

Sir J. C. Bose – the father of plant physiology in India

1917, he founded an autonomous institute, the Bose Institute (now among the pioneering research centres in India).

His work is summarized in several volumes, published by the Longman Green and Company during 1902 to 1929 and are listed as under:

(*i*) *Response in the Living and Non-Living* (1906), (*ii*) *Comparative Electrophysiology* (1907), (*iii*) *Researches on Irritability of Plants* (1913), (*iv*) *Physiology of Photosynthesis* (1923), (*v*) *The Nervous Mechanism of Plants* (1928), and (*vi*) *Growth and Tropic Movement of Plants.* Additionally, he published articles regularly in the proceedings of the Royal Society and the Philosophical Transactions - both published from London. And so, very aptly, Professor (Sir) J. C. Bose is regarded as the 'father of plant physiology' in our country.

Another Indian pioneer, also to be knighted by the British was Sir J. C. Ghosh, at the Chemistry Department at Dhaka (also spelled a Dacca) University (now in Bangladesh). Ghose pioneered research on aspects of plant chemistry that included one of the first protocols for the estimation of chlorophyll pigments, employing spectrophotometers of the earliest design (together with S. B. Sen Gupta).

Around this time, P. K. De of the same department made pioneering investigations, reporting fixation of nitrogen in rice fields and ascribed it to the presence of blue-green algae. Around 1915, R. S. Inamdar (1890–1968) was the first Indian to have undertaken advanced research in plant physiology under F.F. Blackman of Cambridge (renowned for enunciating **'the Law of Limiting Factors'**). Among the other well known Indians associated with him at Cambridge were Birbal Sahni and Sir C. D. Deshmukh, who became the Finance Minister in Nehru's Cabinet and also the Vice-Chancellor of Delhi University in early sixties. In 1919, Inamdar was appointed as the first Professor and Head of the new Department of Botany at Banaras University.

In the twenties, R. H. Dastur (1896–1961) undertook advanced research in plant physiology with W. Stiles in Reading (U.K.). After returning to India, he worked at the Royal Institute of Science in Bombay (now Mumbai) on a variety of problems in cotton physiology and photosynthesis. One of his remarkable discoveries was the effect of blue and red lights that led him to propose that in photosynthesis there might be two photochemical 'stages' instead of one (this is much before the discovery of the red drop and the Emerson enhancement effect [*the original paper is reproduced in Sinha and Bose's - Classical Research Papers in Plant Physiology from India*]). However, in his later life he devoted his time totally to the 'physiology of cotton'.

In the early thirties, P. Parija (1896–1978) worked in Cambridge with F. F. Blackman. There, his work was on respiration of fruits like apple, leading to the discovery of the **'climacteric'**. On his return, Parija founded a school of plant physiology at the Ravenshaw College at Cuttack in Odisha (which later became a part of Utkal University). Later, in an ICAR project, Parija worked on the eradication of water hyacinth or the terror of Bengal (*Eichhornia crassipes*) from rivers and water pools.

R. Ranjan (1899–1969), who also worked with Blackman in Cambridge initially but later went on to his doctorate in Tolouse in France, founded a school of plant physiology at Allahabad where he researched on possible light effects on respiration and on nitrogen metabolism. Ranjan trained, in the early fifties, outstanding students of the likes of Ravinder Kaur (now Kaur-Sawhney, who initially worked with J. Beale in Los Angeles and later collaborated with A Galston at Yale University, contributing much to the physiology

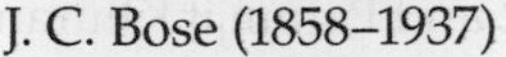

J. C. Bose (1858–1937) R. S. Inamdar (1890–1968) R. H. Dastur (1896–1961)

P. Parija (1896–1978) S. Ranjan (1899–1969) P. Maheshwari (1904–1966)

of polyamines), Govindjee (who joined R. Emerson but completed his PhD work with E. Rabinowitch), and M. M. Lalorya (who did post doctoral work with K. V. Thimann at Harvard). In later years, Ranjan's interests were on the influence of light on respiration, relations of catalase to respiratory activity, glycine metabolism, and role of hormones in dormancy.

In the seventies, there was a major expansion of research programme in plant physiology at the IARI with the strong support of Drs B. P. Pal and M. S. Swaminathan (both FRS) who were then heading ICAR.

Over the decades, several contemporary schools of plant physiology have been established in different parts of India and are listed below:

Kalyani (Professor S. P. Sen); ***Varanasi*** (Professors H. D. Kumar, E. R. S. Talapasayi and Drs (Mrs) Bandana Bose, J. P. Srivastava and A. Hemantaranjan of Department of Plant Physiology, Institute of Agricultural Sciences, BHU). ***Ahmedabad*** (Late Professor J. J. Chinoy); ***Lucknow*** (Professor S. C. Agarwala - who previously worked with E. J. Hewitt in UK); ***Aligarh*** (Professor M. M. R. Afridi); ***Indore*** (Professor M. M. Laloraya, a pupil of late Professor S. Ranjan); ***Delhi, IARI*** (Drs G. S. Sirohi, R. D. Asana - who worked with Gregory in UK, S. K. Sinha, Y. P. Abrol, S. C. Bhargava, G. C. Srivastava, U. K. Sen-Gupta, M. C. Ghildiyal, S. N. Bhardwaj, P. S. Deshmukh; ***Delhi University*** (Late Professor P. Maheshwari, Professors S. C. Maheshwari, H. Y. Mohan Ram, R. C. Sachar, S. K. Sawhney – pupil of Professor K. K. Nanda, K. R. Shivanna, S. S. Bhojwani, N. S. Rangaswamy, P. S. Ganapathy, S. C. Bhatla); ***Tirupati*** (Professor V. S. Rama Das); ***Chandigarh*** (Professor K. K.

Nanda); ***PAU, Ludhiana*** (Professors C. P. Malik, A. K. Srivastava, O. S. Singh, Gurbux Singh, N. C. Bhattacharaya, Rangeel Singh, J. S. Bhatia); ***JNU, Delhi*** (Professors G. S. Singhal, P. K. Mohanty, Sipra Guha-Mukherjee, S. K. Sopory, A. Dutta, K. Upadhyay); ***Madurai*** (A. Gnanam, a student of A. Jagendorf); ***Bombay, TIFR*** (Professors O. Siddiqi, M. M. Johri who worked with J. Varner); ***HAU, Hisar*** (Professors O. P. Garg, H. N. Krishnamoorthy, Randhir Singh); ***Garhwal*** (Professor A. N. Purohit); Vadodra (earlier ***Baroda***) (Professor A. R. Mehta); ***Sri Nagar*** (M. Farooq); ***National Biotechnology Center at Delhi*** (Professors V. L. Chopra, R. P. Sharma, K. R. Koundal, K. C. Bansal who did post-doctorate work with Bogorad at Harvard); ***Delhi University - South Campus*** (Professors D. Pental, A. Pradhan, M. Rajam in the ***Department of Genetics***, and Professors A. K. Tyagi, J. P. and Paramjit Khurana and A. Grover in the ***Department of Plant Moleculer Biology***); ***ICGEB, Delhi*** (Dr K. K. Tewari); ***Rajasthan Universities (Jodhpur, Udaipur and Jaipur)*** (Professors H. C. Arya, Pushpa Khanna, N. S. Shekhawat, K. Ramawat and S. L. Kothari).

However, we have tried to have highlighted here the seminal contributions of some of our pioneers in plant physiology that had an impact on science:—

Brief biographies of Indian pioneers who have made seminal contributions to the field of plant physiology now follow:

S. M. SIRCAR (1908–1978)

Having worked with Professor F. C. Gregory at the Imperial College of Science and Technology (London), on the relationship between respiration and nitrogen metabolism in potato tuber; Sourindra Mohan Sircar, took over as Acting Head of the Department of Biology at Dacca (now Dhaka) University (1945–47) before a brief stay at the Central College of Agriculture at IARI, 1947–48, (then affiliated to Delhi University).

Sircar returned to Calcutta (now Kolkata) to become the Head of the Botany Department. Toward the later part of his career, he became the Director of the Bose Institute (among the pioneering institutions in India that was founded by Sir J. C. Bose).

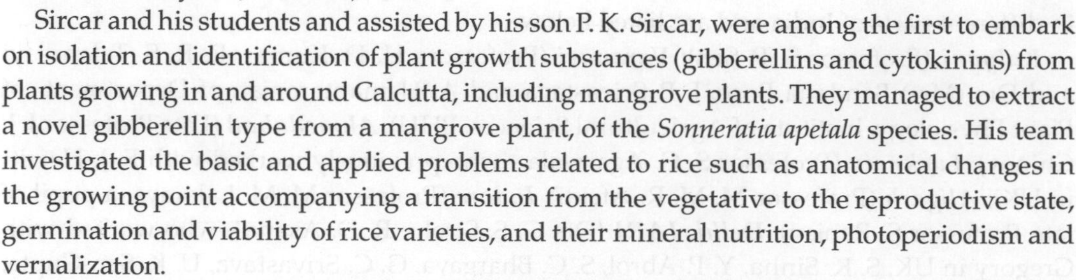

Sircar and his students and assisted by his son P. K. Sircar, were among the first to embark on isolation and identification of plant growth substances (gibberellins and cytokinins) from plants growing in and around Calcutta, including mangrove plants. They managed to extract a novel gibberellin type from a mangrove plant, of the *Sonneratia apetala* species. His team investigated the basic and applied problems related to rice such as anatomical changes in the growing point accompanying a transition from the vegetative to the reproductive state, germination and viability of rice varieties, and their mineral nutrition, photoperiodism and vernalization.

His other interests included biochemical changes in rice and mungbean seed germination and their control mechanism. In 1971, he published his book *'Plant Hormone Research in India'*, ICAR, New Delhi.

R. D. ASANA (1908–1999)

A former student of R. H. Dastur, R. D. Asana also worked with Gregory in London. While returning from the UK, he visited the famous Botanical Laboratory at Utrecht (in the Netherlands) where the father and son team of F. A. F. C. and F. W. Went worked. There his work was devoted to 'Avena test' for estimation of auxins. Upon return to India, he joined IARI in 1946 and teamed up with J. J. Chinoy to study the growth and flowering of various varieties of wheat to identify the best one suited to local conditions.

After Chinoy moved to Delhi University in 1947 as Reader in the newly established Botany Department, Asana became the first Head of the Division of Plant Physiology, a position carved out from the former Division of Botany.

Asana work elucidated the physiological basis of yield, and outlined 'the ideotype concept of wheat' for improving crop productivity under rainfed conditions. Its main features were: that ears should have a large number of spikelets; that ears should be branched (allowing large number of grains to develop); the length of the ear bearing internode (forming a substantial proportion of the length of the stem); about seven leaves on the main axis, preferably horizontally disposed; substantial root growth at a depth of third and fourth of soil horizon (probably timing with the ear development); appearance of ear at such a time that would allow grains to develop for at least 5 weeks at a mean maximum temperature of 25°C.

Thus Dr Asana laid the foundation of crop physiology research at IARI and after his retirement in 1970, Dr G. S. Sirohi took over as Head of the Division of Plant Physiology (and built upon Asana's contributions through his focus on sophisticated instruments). Along with M. N. Sarin, Asana wrote a book *Crop Physiology in India*, published by ICAR, New Delhi.

G. S. SIROHI (1928–)

Awarded double PhDs in 'Crop Science' (Delhi University, 1951– 55) and 'Biological Clock' (University of California, Los Angeles, USA, 1955–58); he was Professor and Head at the Division of Plant Physiology at Indian Agricultural Research Institute, New Delhi (1971–88) and established a centre of excellence for research and teaching plant physiology.

He was also the first Indian to have conducted experiments on 'biological clock' at the South Pole (1960–61) in the snow tunnels, at an altitude of 10,000 ft., using soybean and hamsters as plant and animal material respectively. In recognition of his services at Antarctica, a place was named 'Sirohi Point' by the US Board for Geographic names. Later, he would assist Dr Z. A. Quasim to finalise the first Indian Expedition to Antarctica (1981).

He prepared an invaluable report on 'Photosynthesis and Crop Productivity under tropical environments' (1980).

Sirohi has contributed immensely to the promotion and popularisation of plant physiology and was chairman of the 'Apex Body' of the Journal of Indian Society for Plant Physiology. He was honoured with a 'Life-time Achievement Award' by the President of India. His other awards include the J. J. Chinoy gold medal for excellent research work on the subject of the 'Physiology of Flowering'; and the R. D. Asana gold medal for distinguished work in the field of 'Stress Physiology'.

He published a book called *Glossary of Plant Physiology* (G. S. R. Murti, G. S. Sirohi and K. K. Upreti, 2004).

J. J. CHINOY (1909–1978)

After training under Professors R. H. Dastur and F. G. Gregory at the Imperial College of Science and Technology in London, he became Professor and then Head, Department of Life Sciences, Ahmedabad.

He worked for the most on the synthesis, utilisation and effect of ascorbic acid in different plant processes. He made the discovery of considerable increase in the ascorbic acid content of growing points during a transition from the vegetative to reproductive phase. Investigating the role of ascorbic acid at molecular and sub-molecular levels of differentiation, he found that:

- Exogenous application of ascorbic acid increased α-amylase activity in starchy, oily and proteinaceous seeds.
- Pre-soaking of the seeds in ascorbic acid prior to sowing increased growth and yield.

He studied the effect of drought on the development of the plant and found that drought, at any early stage of the life cycle of the plant, produces beneficial effects on growth and yield characters while drought at later stages proves harmful or deleterious.

K. K. NANDA (1918–1983)

His career started with Delhi University, but after a short stint at the Forest Research Institute in Dehradun, he moved to Punjab University, Chandigarh where as Professor and Head, Department of Botany, he built and led a strong school of Plant Physiology.

His core work was in investigating the physiology of adventitious root formation in stem cuttings. His findings include: that some inhibitors of nucleic acid synthesis, protein synthesis and oxidative phosphorylation enhance rooting i.e., the rooting pressure is normally repressed by some endogenous inhibitors. He reported the existence of circadium rhythms in the flowering response to Biloxi soybean and the concepts of a biological clock in *Impatiens balsamina* which enabled them to measure the passage of time quite accurately. His other interests were forest tree physiology and effects of temperature and light on growth, development and yield on diverse crops.

He was awarded the Rafi Ahmed Kidwai Memorial Prize (1972–73), J. J. Chinoy gold medals (1969, 1974), and the Birbal Sahni Medal (1977) for his outstanding contributions to the study of plant physiology.

S. K. SINHA (1934–2002)

Double-doctorate Sinha was awarded his first PhD from Agra University under I. M. Rao for mineral nutrition studies in barley and linseed with reference to P and N, and pH interactions. His second was awarded when, under E. Cossins at Edmonton University, Alberta, Canada, he worked on amino acid biosynthesis in plants. Later, he was appointed Professor of Plant Physiology at the Water Technology Centre, IARI (Delhi). Other fields covered in the course of his research include dryland agriculture, drought tolerance, senescence, irrigation and energy studies in agriculture.

He was the Secretary General of first International Congress of Plant Physiology in New Delhi (1988).

He collaborated with the Professor M. S. Swaminathan in producing the book *Global Aspects of Food Production*, published by IRRI and Tycooly International. Collaborating with Salil Bose (JNU), the two of them compiled *Classical Research Papers in Plant Physiology*, published by the Society for Physiology and Biochemistry on the occasion of the International Congress of Plant Physiology, New Delhi (1988).

S. C. AGARWALA (1909–1983)

Was Head, Department of Botany, Lucknow University where he built up a strong school of plant physiology. His group carried out work on the roles of trace elements. He worked with Professor E. J. Hewitt in the UK on the effect of molybdenum deficiency on cauliflower plant growth, and nitrogen supply in nitrate and ammonia forms. Molybdenum stress resulted in decreased activity of nitrate reductase.

His group surveyed the soils on the Gangetic plains for any mineral deficiency. His group reported positive correlation of chlorophyll and catalase content with iron, and suggested a common porphyrin-like precursor for both of these compounds. They deduced evidence that excessive supply of cobalt, nickel and chromium stimulates the synthesis of catalase in barley seedlings.

S. C. MAHESHWARI (1933–2019)

Like his father the late Panchanan Maheshwari, a specialist and world leader in plant embryology, S. C. Maheshwari also carved a niche for himself in the field and is internationally renowned. He was Professor and Head, Department of Botany, University of Delhi having worked with such authorities as A. Galston, J. Bonner and interacted with H. Borthwick, S. Hendricks and W. S. Hilman.

Together with Professor H. Y. Mohan Ram, he built-up tissue culture facilities at Delhi University, and jointly shared teaching plant physiology. Together with Sipra-Guha, the duo discovered the anther culture technique for production of haploids[1] which holds great promise in producing homozygous lines in plant breeding in such short spans, unlike the conventional methods. For this outstanding contribution to anther culture pioneering studies, he was awarded the Shanti Swarup Bhatnagar prize. Along with his students, his group has done pioneering studies in India on the isolation and characterization of plant growth hormones, especially cytokinins.

Among his other research interests are acetylcholine metabolism in plants, circadium rhythms, phytochrome, growth regulators (novel cytokinins, cAMP), physiology of flowering (especially duckweed, *Lemna minor*), nitrate reductase activity and RNA polymerase activity in nuclei and chloroplasts (together with R. S. Bandurski).

Liberal grants from ICAR, DST and DBT, have enabled him and his co-workers to establish the unit for 'Plant Cell and Molecular Biology' taken over by Delhi University and upgraded in the new South Campus of the University of Delhi as a 'Department of Plant Molecular Biology'.

H. Y. MOHAN RAM (1930–2018)

Worked with F. C. Steward on morphogenesis at the Cornell University (USA) and interacted with the late Professor J. P. Nitsch and Dr H. Harda in France. He was Professor and Head, Department of Botany, University of Delhi.

His wide research interests include the effects of growth regulators on crop plants, physiology of sex expression and flowering, tissue culture and morphogenesis. He reported the induction of male flower formation in genetically female plants of *Cannabis sativa* by exogenous application of gibberellins while ethephon (ethylene) spray enhanced the number of pistillate flowers in male plants. In general, female flowers and female plants have higher endogenous levels of auxin and ethylene while male flowers and male plants have a higher level of gibberellins. In fact it is the balance between growth

[1] Haploid plants are valuable in agriculture as the time-frame for obtaining homozygous pure lines required for breeding purposes is reduced from about 20 years to a few weeks!

hormones that controls the phenomenon of the sex expression. He also researched the effect of cytokinins on flowering in aquatic plant (*Ceratophyllum demursum*).

R. C. SACHAR (1929–)

Professor R. C. Sachar moved from IARI, Delhi back to Delhi University where he initiated promising work in plant biochemistry. His significant contributions include work in the area of molecular mechanism of gene regulation by the phytohormone GA_3. He has enriched our knowledge concerning the phytohormonal regulation of RNA processing enzymes namely poly (A) polymerase and poly $(A)^+$ RNA in wheat aleurones and excised wheat embryos. He was the first to report the presence of poly $(A)^+$ RNA in tobacco leaf homogenate and was able to isolate the enzyme system responsible for its synthesis. He and his associates demonstrated the pivotal role of phosphorylation in the regulation of RuBPcase in moss and spinach.

The research group under his leadership provided evidence that the polyphenol oxidase system in wheat is composed of two separate enzymes, monophenolase and oligophenolase. This team also provided evidence that messengers for O-diphenolase are long lived and pre-exist in the dry seeds. They have also shown that cAMP does stimulate RNase activity in germinating cowpea seeds.

Unit VII Some Experimental Exercises

Chapter 23

Experimental Exercises

I. Plant–Water Relations

Experiment I.I

To determine the osmotic potential of any given plant material by using the incipient plasmolytic method. Also, to determine the osmotic potential of an unknown solution.

Requirements

Plant material : The leaves of *Tradescantia spathacea*[1]/*Commelina*/*Zebrina*
Family : Commelinaceae
Chemicals : Sucrose solution (1 molal), distilled water, unknown sucrose solution
Glass apparatus : Petri dishes, glass slides, cover slips, beakers, pipettes (1 ml, 10 ml)
Miscellaneous : Microscope, forceps, needle, stop watch, graph paper

Principle

An actively metabolizing cell comprises of a permeable cell wall and semipermeable plasma membrane and tonoplast. The bulk of the space in a cell is occupied by the vacuole, which is a storehouse of minerals, osmotic substances like sugars, amino acids, organic acids and secondary metabolites like anthocyanin, and so on. Under normal, fully distended conditions, cells are said to be 'turgid'. When such cells are shifted to a hypertonic solution, the movement of water occurs from a region of high water potential to a region of low water potential resulting in the loss of turgor. This is followed by shrinking of the cytoplasm away from the cell wall, i.e., **Plasmolysis**. The stage when there is initial pulling away of the cell

---

[1] The leaves of *Tradescantia spathacea* (earlier known as *Rhoeo discolor*) are selected for this experiment because these contain anthocyanin pigments in their vacuolar sap. Anthocyanin pigments are water-soluble and respond to changes in pH values of the external medium.

membrane from the cell wall and the turgor pressure becomes zero, is known as 'incipient plasmolysis'. Plasmolysed cells can be deplasmolysed by placing them in hypotonic solutions. In hypotonic solutions, turgor pressure develops, which presses the membrane against the cell wall. The rigid cell wall exerts an equal and opposite pressure termed as wall pressure and the cell is said to be turgid.

Water potential is the algebraic sum of three components:

$$\underset{\text{Water potential}}{\Psi_w} = \underset{(-)}{\overset{\text{Osmotic potential}}{\Psi_\pi}} + \underset{(+)}{\overset{\text{Pressure potential}}{\Psi_p}} + \underset{(-)}{\overset{\text{Matric potential}}{\Psi_m}}$$

Procedure

Prepare a series of sucrose solutions ranging from 0.15 to 0.35 m, using 1 m sucrose as stock solution (Dilution Table). Take out epidermal peels from the zone near central midrib portion of lower surface of the leaf by applying differential or unequal pressure. Place 3 peels in each petri dish after a time gap of 5 minutes. Wait for 10 minutes[2] or so, mount the peels on the slides in their respective solutions and observe under microscope (6×, 40×). Count the total number of pigmented cells visible in the microscopic field and also the number of cells that had undergone partial or complete plasmolysis. Plot a graph between the percentage cells plasmolysed against the sucrose concentrations and from the graph find out the concentration at which 50 per cent of the cells are plasmolysed - the stage of incipient plasmolysis. Calculate the osmotic potential of the cells by the following formula:

$$\text{Osmotic potential} = -\text{CRT}$$

whereas C = molal concentration at incipient plasmolysis
R = gas constant
T = absolute temperature in °K

A negative sign is given to the calculated value since solutions have a lower water potential than pure water. This method assumes that at incipient plasmolysis, the turgor pressure is zero.

Dilution table

Stock solution = 1 molal (m) sucrose [342 g in 1000 ml of distilled water]

Concentration of sucrose solution (m)	Amount of 1 m sucrose (ml)	Amount of solvent distilled water (ml)	Total volume (ml)
Control	–	10.0	10
0.15	1.5	8.5	10
0.20	2.0	8.0	10
0.22	2.2	7.8	10

[2] Adjust the time limit by keeping the peels in 0.3 m solution for 10 min (trial run) and if there is 100 per cent plasmolysis, then reduce the time limit to 5 min or so.

Concentration of sucrose solution (m)	Amount of 1 m sucrose (ml)	Amount of solvent distilled water (ml)	Total volume (ml)
0.24	2.4	7.6	10
0.26	2.6	7.4	10
0.30	3.0	7.0	10
0.35	3.5	6.5	10

Observation table

Sucrose concentration (m)	No. of pigmented cells examined (mean)	No. of cells plasmolysed (mean)	Degree of plasmolysis (%)
Control	34 32 } 32 30	0 0 } 0 0	0
0.15	35 33 } 34 34	3 5 } 4.3 5	12.74
0.20	40 39 } 39.6 40	10 9 } 9 8	22.69
0.22	41 40 } 39 36	14 10 } 10.6 8	27.35
0.24	35 35 } 33.6 31	12 11 } 11 10	32.67
0.26	32 29 } 30.3 30	14 11 } 12 11	39.56
0.30	33 32 } 32.6 33	18 20 } 18.6 18	57.14
0.35	33 33 } 32.3 31	24 27 } 25.3 25	78.35
Unknown solution X	29 27 } 28 28	14 13 } 13.3 13	47.62

Initial room temperature = 33 °C

Final room temperature = 34 °C

$$\text{Mean temperature} = \frac{33\,°C + 34\,°C}{2} = \frac{67\,°C}{2} = 33.5\,°C$$

$\therefore$ Absolute temperature = 273 + 33.5 °C = 306.5 °K

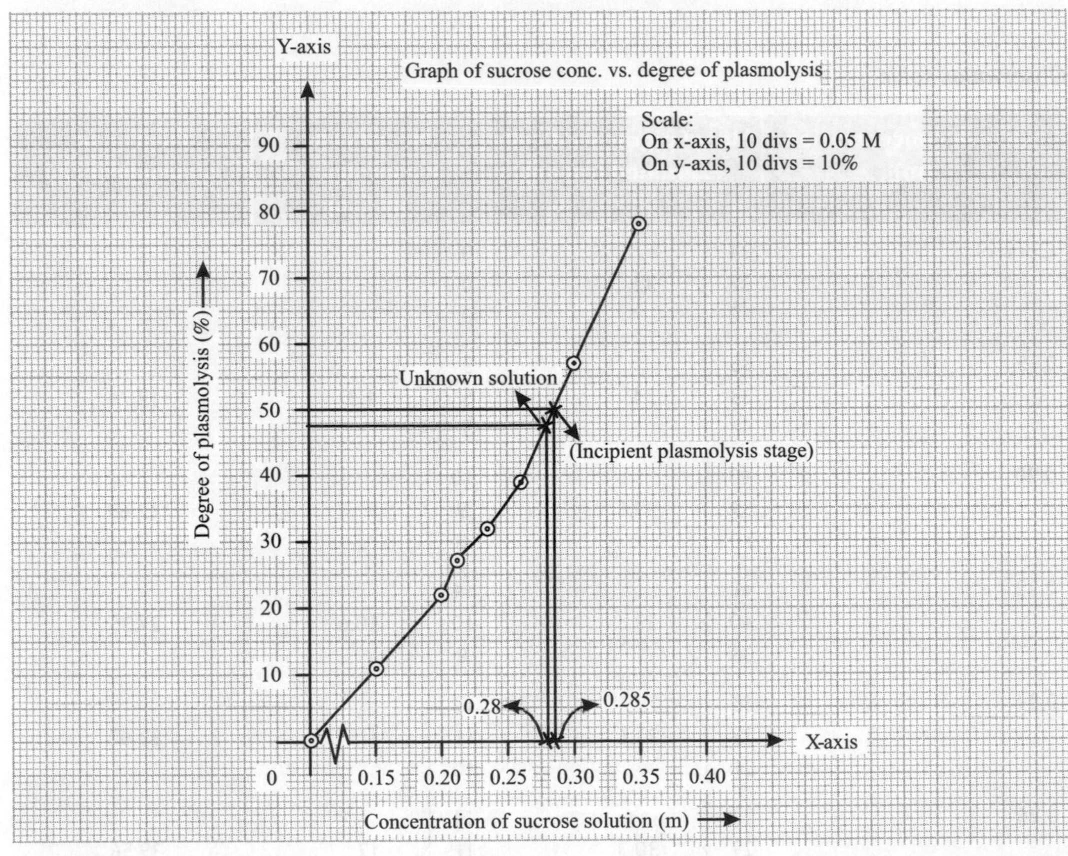

Calculations

I. For given plant material

C = isotonic concentration = 0.285 m (from the graph)

R = gas constant = 0.082 L atm $K^{-1}mol^{-1}$

T = absolute temperature = 306.5 °K

$\psi_\pi = -CRT$

$= -0.285 \times 0.082 \times 306.5$

$\psi_\pi = -7.163$ atm

II. For the given unknown solution X,

$$C_x = 0.28 \text{ m (from the graph)}$$
$$R = 0.082 \text{ L atm K}^{-1}\text{mol}^{-1}$$
$$T = 306.5\,^{\circ}\text{K}$$
$$\psi_\pi = -C_x RT$$
$$= -0.28 \times 0.082 \times 306.5$$
$$\psi_\pi = -7.04 \text{ atm}$$

Result and Discussion

Plasmolysis cannot be observed in the control (distilled water). With an increase in sucrose concentration, the number of plasmolysed cells increases because the external medium becomes hypertonic with respect to cell sap. As plasmolysis proceeds, the cytoplasm along with cell membrane starts shrinking. A space is created between the cell wall and the cell membrane, which gets filled up with the plasmolysing solution (external sucrose solution and water from the cell sap). The sucrose concentration where 50 per cent of the cells are plasmolysed, i.e., isotonic concentration, is determined from the graph. The concentration of an unknown solution can also be determined from the graph and used for calculating osmotic potential of that solution.

Precautions

1. The peels should be one-celled thick and free from green tissue, and air bubbles.
2. All cells of the peel should invariably contain anthocyanin pigment.
3. Before use, keep the peels fully immersed in water to avoid the entry of air bubbles.
4. The peels should be examined after a fixed duration of time, the time to be adjusted before starting the experiment (trial run for fixing time period may be done).
5. Mounting of the peels should be done in their respective concentrations, and not in water to prevent deplasmolysis.
6. The solutions should be carefully prepared.
7. The petri dishes containing different sucrose solutions should be covered to prevent evaporation.
8. Even if the cells are slightly plasmolysed, they should be categorised as having undergone plasmolysis.
9. Observations should be made under high power (6×, 40×) of the microscope.

Peels should not be stained for observations because stains are usually acidic, and may change the state of the cells or may kill them.

Experiment 1.2

To determine the water potential of any given plant material by using the gravimetric (weight) method.

Requirements

Plant material : Potato (*Solanum tuberosum*)
Family : Solanaceae
Chemicals : Sucrose solution (1 m), distilled water
Glass apparatus : Test tubes, glass rod, pipettes (1 ml, 10 ml), a large petri dish
Miscellaneous : Cork borer, electric balance, blotting sheet, aluminium foil, stop-watch, blunt forceps, graph paper.

Principle

In plant-water relations, water potential (ψ_w) is generally referred to as chemical potential of water. The chemical potential of pure water is arbitrarily set at zero. Any addition of water with a solute establishes a potential that is less than that of pure water and is expressed as a negative number. Water potential of a cell is a measure of the tendency for water to enter the cell when placed in it. The determination rests upon the fact if the cell is placed in a solution having lower water potential (hypertonic solution), it will lose water and hence decrease in weight and volume. On the other hand, if the cell is kept in a solution having higher water potential (hypotonic solution) or pure water, it will absorb water and hence increase in weight and volume. Water moves from high water potential to low water potential (energetically downhill) or from less negative water potential to more negative water potential.

$$\psi_w = \psi_\pi + \psi_p + \psi_m$$

At isotonic concentration,

$$\psi_m = \text{negligible}, \psi_p = 0$$

$$\therefore \quad \psi_w = \psi_\pi = -\text{CRT}$$

This method involves the placement of pre-weighed plant tissue into a graded series of solutions of sucrose of known concentration or at known osmotic potential. A representative sampling of tissue is incubated for a predetermined time in the solutions, removed and reweighed. The weight gain or loss is plotted against the water potential of each solution. When the points are connected, the intercept at the abscissa (through zero) represents the water potential of the tissue, with zero weight gain or loss. The water potential of the solution corresponding to the intercept point is equal to that of the tissue.

Procedure

Prepare sucrose solutions of different concentrations ranging from 0.15 to 0.4 m (Dilution Table). Using the cork borer, take out cylindrical pieces from a large potato tuber and keep them in the folds of moist blotting sheet. Cut these pieces into 1 cm size each. Note down the initial weight of 3 potato cylinders after blotting them dry and immerse in 0.15 m sucrose solution. Repeat the same procedure for other concentrations. After 1 hour, remove the potato cylinders from the sucrose solution using blunt forceps, blot them dry and note down the final weight.

Plot the values of sucrose concentrations against change in weight (gain/loss) on the graph paper and determine the isotonic concentration. Record the initial and final temperature and calculate the mean to find out absolute temperature.

Dilution table

Stock solution = 1 m sucrose (342 g sucrose in 1000 ml of distilled water)

Concentration of sucrose solution (m)	Amount of 1 m sucrose solution (ml)	Amount of distilled water (ml)	Total volume (ml)
Control	–	10.0	10
0.15	1.5	8.5	10
0.20	2.0	8.0	10
0.25	2.5	7.5	10
0.30	3.0	7.0	10
0.35	3.5	6.5	10
0.40	4.0	6.0	10

Observation table

Conc. of sucrose solution (m)	Initial wt. of cylinders (w_1) (g)	Final wt. of cylinders (w_2) (g)	Change in weight ($w_2 - w_1$) (g)	Inference
Control	3.345	4.110	+0.765	Endosmosis
0.15	3.360	3.612	+0.252	Endosmosis
0.20	3.425	3.640	+0.215	Endosmosis
0.25	3.430	3.515	+0.085	Endosmosis
0.30	3.497	3.430	−0.067	Exosmosis
0.35	3.440	3.270	−0.17	Exosmosis
0.40	3.504	3.160	−0.344	Exosmosis

Initial room temperature = 37 °C
Final room temperature = 37 °C
Mean temperature = 37 °C
Absolute temperature = (37 °C + 273) K = 310 °K

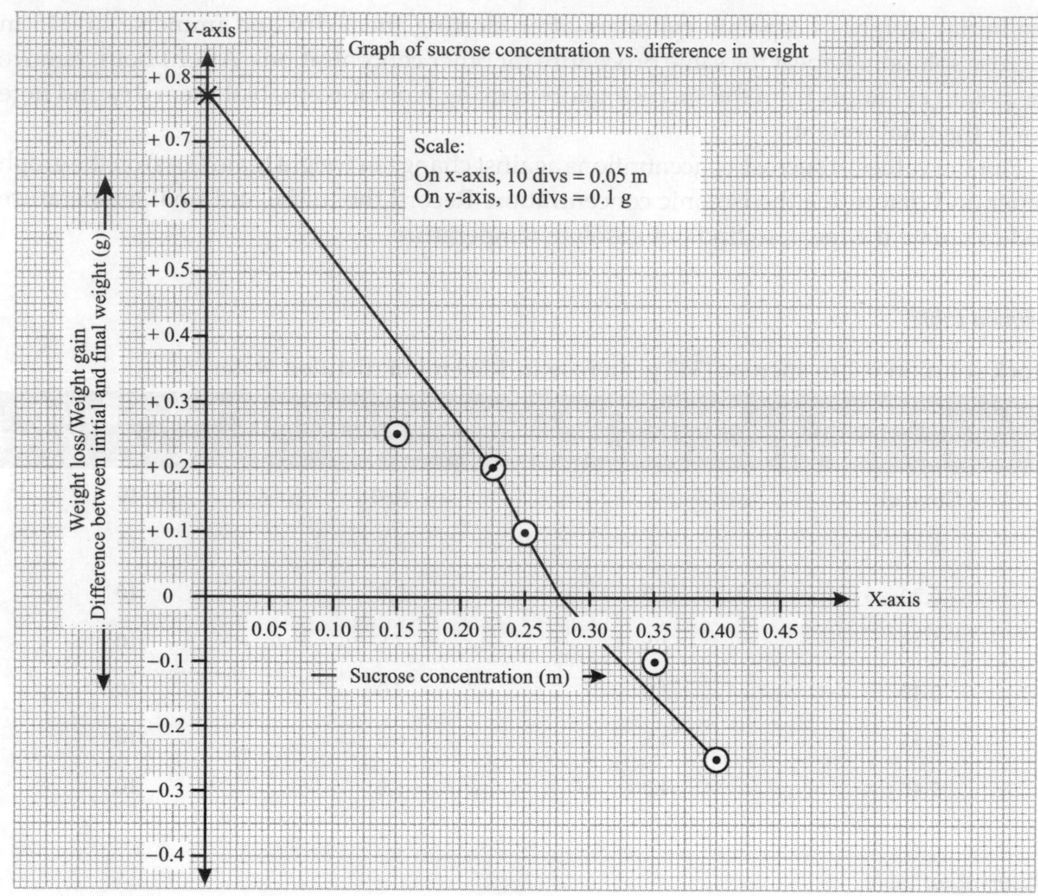

Calculations

Isotonic concentration $C = 0.28$ m (from the graph)

$R = 0.082$ L atm $K^{-1}mol^{-1}$

$T =$ absolute temperature $= 310\,°K$

$$\psi_w = -CRT$$

$$= -0.28 \times 0.082 \times 310$$

$$\psi_w = -7.118 \text{ atm}$$

Result and Discussion

From the graph based on observations, it is observed that the intercept at the abscissa is 0.28 m. Hence 0.28 m is the isotonic concentration where there is no gain or loss in weight of the potato cylinders. Below 0.28 m, gain in weight of potato cylinders is observed because of endosmosis, i.e., water potential of the surrounding sucrose solution is more than that of cell sap (hypotonic solution). Beyond 0.28 m, loss in weight of potato cylinders occurs because of exosmosis, i.e., water potential of the surrounding sucrose solution is less than that of cell sap (hypertonic solution).

Precautions

1. A large sized potato tuber should be used for experimentation.
2. All the cylindrical pieces of uniform size should be taken out from the same potato tuber (as these are at the same osmotic level).
3. Potato peel (periderm) should be removed as it is impervious to water.
4. Test tubes should be covered with aluminum foil to prevent evaporation of sucrose solution.
5. The cylinders should be blotted dry before weighing them.
6. Three potato cylinders for each concentration should be weighed together and completely immersed in their respective sucrose solution.
7. Blunt forceps should be used to transfer the cylinders into and out of the solution.
8. The treatment should be given for a fixed duration of time; say for 60 minutes.
9. The solutions should be prepared very carefully.

Potato cylinders should not be kept in water but in the folds of a moist blotting sheet until used and are therefore not exposed to air.

Experiment 1.3

To determine the water potential of any given plant material using Chardakov's falling drop method or density gradient method.

Requirements

Plant material	: A large sized potato tuber (*Solanum tuberosum*)
Family	: Solanaceae
Chemicals	: Sucrose solution (1 m), methylene blue (0.2 per cent), distilled water
Glass apparatus	: Test tubes (12), pipettes (1 ml, 10 ml), a large petri dish
Miscellaneous	: Test tube stand, cork borer, blotting sheet, blunt forceps, dropper, aluminum foil, stop watch

Principle

The credit for developing the 'falling drop method' for the determination of water potential goes to the Russian scientist, V. S. Chardakov (1948). The change in density of the external solution (hypotonic, hypertonic, isotonic) forms the basis of the falling drop method.

In this method, two graded series of sucrose solutions are placed in test tubes, set up in duplicate. Homogenous plant tissue is placed into each test tube of one of the series – 'Test Series'. Only a drop of methylene blue is mixed into each solution of the second series – 'Control Series'. Plant tissue is not added to the control series and the methylene blue dye does not appreciably change the osmotic potential.

The concentration of isotonic sucrose solution is determined from the experiment and is used to calculate water potential of the given plant material in accordance with the equation

$$\psi = -CRT$$

Procedure

Prepare two graded series of sucrose solutions of concentrations (0.15 to 0.4 m) as control series and test series (Dilution Table). Take out cylindrical pieces from a large-sized potato tuber with the help of the cork borer, trim them to 1 cm size and preserve in moist blotting sheet.

Test series: It contains potato cylinders (3 in each tube) in various concentrations of sucrose solution.

Control series: It contains a drop of methylene blue in various concentrations of sucrose solution.

After 60 minutes of incubation, remove the potato cylinders from the test series using blunt forceps. A drop of the respective control series solution is then introduced carefully below the surface (more or less in the centre) of its corresponding test solution. Observe the movement of the drop. Repeat the procedure using the test series and control series solutions of all other concentrations (Figure E1.1).

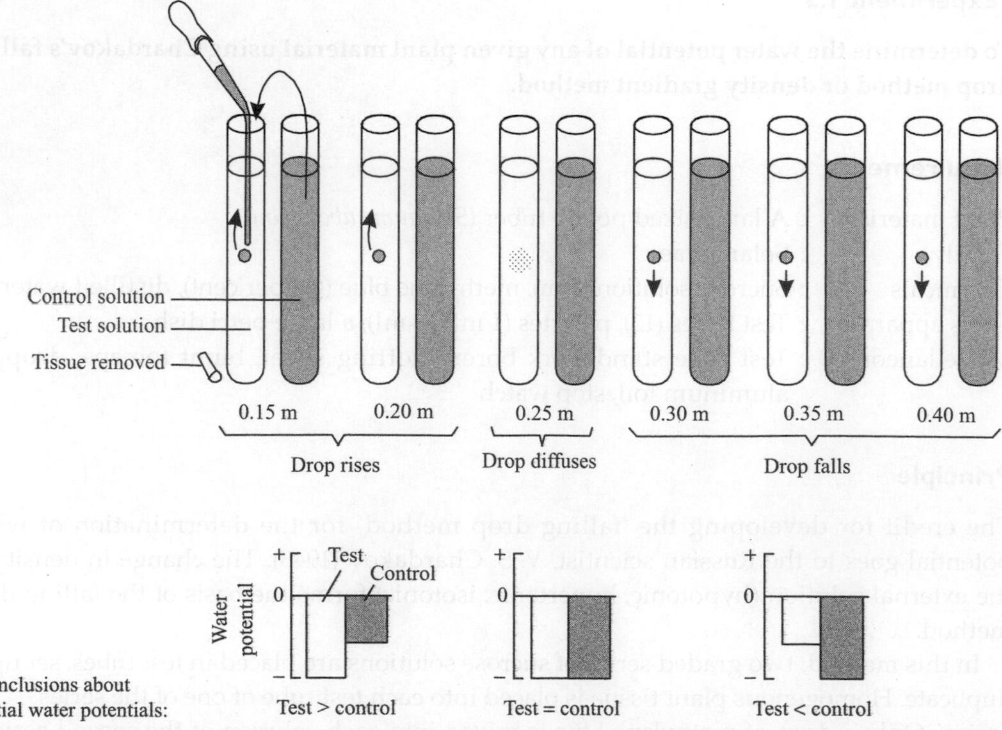

Fig. E1.1 Chardakov's falling drop method. Drop rising in the test solution indicates test solution became denser than its counterpart control solution. Water diffused out of test solution into tissue which initially had a more negative ψ_w. The reverse is true when drop falls in the test solution. Diffusion of drop in the test solution indicates no change, hence water potentials were equal initially.

Dilution table

Stock solution = 1 m sucrose (342 g sucrose in 1000 ml of distilled water)

Concentration of sucrose solution (m)	Amount of 1 m sucrose solution (ml)	Amount of distilled water (ml)	Total volume (ml)
0.15	1.5	8.5	10
0.20	2.0	8.0	10
0.25	2.5	7.5	10
0.30	3.0	7.0	10
0.35	3.5	6.5	10
0.40	4.0	6.0	10

Observations and calculations

Initial room temperature = 34 °C
Final room temperature = 35 °C
Mean temperature = 34.5 °C
Absolute temperature = (34.5° + 273) K = 307.5 °K

Observation table

Concentration of sucrose solution (m)	Movement of drop	Change in density of external solution	Inference	
			Nature of sucrose solution	ψ
0.15	Upward	More dense	Hypotonic	$\psi_E > \psi_T$
0.20	Upward	More dense	Hypotonic	$\psi_E > \psi_T$
0.25	Stationary	No change	Isotonic	$\psi_E = \psi_T$
0.30	Downward	Less dense	Hypertonic	$\psi_E < \psi_T$
0.35	Downward	Less dense	Hypertonic	$\psi_E < \psi_T$
0.40	Downward	Less dense	Hypertonic	$\psi_E < \psi_T$

$$\left[\begin{array}{l}\psi_E = \text{Water potential of external solution} \\ \psi_T = \text{Water potential of plant tissue}\end{array}\right]$$

Isotonic concentration, C = 0.25 m (from the observation table)
Gas constant, R = 0.082 L atm $K^{-1}mol^{-1}$
Absolute temperature, T = 307.5 K

$$\begin{aligned}\text{Water potential of given plant material} &= -CRT \\ &= -0.25 \times 0.082 \times 307.5 \\ \psi_T &= -6.304 \text{ atm}\end{aligned}$$

Result and Discussion

In 0.15 and 0.2 m sucrose solutions, a drop of methylene blue of control solution 'rises upwards' in the corresponding test solution. This means that the external solution is more 'denser' or more hypotonic than the potato tissue. In other words, the water potential of external solution (ψ_E) is more than that of the tissue (ψ_T).

$$\therefore \psi_E > \psi_T$$

In the 0.25 m sucrose solution, a drop of methylene blue from the control solution remains 'stationary' or diffuses evenly into the test solution. This means that density of external solution is equal to that of potato tissue. In other words, the two solutions are isotonic.

$$\therefore \psi_E = \psi_T$$

In 0.3 m, 0.35 m and 0.4 m sucrose solutions, a drop of methylene blue from the control solution 'sinks' to the bottom of the test solution. This means that the external solution is 'less dense' or hypertonic compared to that of the potato tissue. In other words, the water potential of the external solution is less than that of the tissue.

$$\therefore \psi_E < \psi_T$$

Using isotonic concentration, C = 0.25 m, water potential of the potato tuber is found to be –6.304 atm.

Precautions

1. A large-sized potato tuber should be selected for the experiment and all the cylinders should be taken from the same tuber and should be of uniform size.
2. Cylinders must not be exposed to air but kept within the folds of the moist blotting sheet.
3. Cylinders should be blotted dry before use.
4. Test tubes should be kept covered to prevent evaporation.
5. Blunt forceps should be used to handle the potato cylinders to avoid any injury to them.
6. Only a drop of methylene blue should be added to the solutions of the control series.
7. A dropper of with a very fine nozzle should be selected to transfer a drop of methylene blue from the control series to the test series.
8. The test solution should not be disturbed in any way and the movement of the drop must be carefully observed.

The actual time of incubation should not exceed 1 hour; it should be just long enough for osmosis to occur and change the concentration of each solution in the test series. The attainment of equilibrium is not necessary.

Questions for Viva Voce

1. What is the difference between a molar and a molal solution?
2. What occupies the space between the cell wall and the cytoplasm in a plasmolysed cell?
3. Where is anthocyanin pigment located in the cell? Does it change colour in response to different pH solutions?
4. What is the chemical nature of plasma membrane? Which is the most accepted model to explain the structure of plasma membrane?
5. Why is sucrose preferred over other solutes?
6. What modifications should be brought in the experiment if onion peels are used instead of *Rhoeo* peel?
7. What will happen if the peels are kept in alcohol before putting in sucrose solution? Will the cells show plasmolysis or not?
8. Why is mounting of peels not done in water?
9. Why is 50 per cent plasmolysis taken as the stage of incipient plasmolysis?
10. Which instrument is used for measuring water potential in (i) cut stems (ii) soil.
11. What happens when animal cells are placed in a hypotonic or a hypertonic solution?
12. Suggest alternative materials for conducting the experiment.

Experiment 1.4

To study the effects of temperature on the permeability of cell membranes.

Requirements

Plant material	: Beet root (*Beta vulgaris*)
Family	: Chenopodiaceae
Chemicals	: Distilled water
Glass apparatus	: Boiling or test tubes, petri dish, glass rod, pipette (10 ml), a thermometer
Miscellaneous	: Cork borer, test tube holders, water baths (5), blunt forceps, blade, graph paper, scale, photoelectric colorimeter or spectrophotometer

Principle

Beet root is a source of beet sugar as well as betacyanin pigments. Betacyanin pigments are water-soluble and confined to vacuolar sap.

Membrane permeability is affected by several factors like temperature, organic solvents, pH, radiations and electric current. Temperature affects membrane permeability by interacting with the protein part of cell membrane. At higher temperatures, proteins get denatured and membrane permeability is altered resulting in release of betacyanin pigments from within the cell into its immediate environment/ medium. Moreover, high temperature induces phase transition of the phospholipid bilayer from a highly ordered gel-like substance to a more mobile fluid state over a short temperature range.

Procedure

Remove cylinders of beetroot (each of 1 cm length) using the cork borer. Thoroughly wash the beet cylinders for 10 minutes in several lots of distilled water and then transfer two of them into 10 ml distilled water in boiling tube maintained at 35 °C for exactly 5 min. Repeat the same procedure for 50 °C, 65 °C, 80 °C and 100 °C. Keep one boiling tube at room temperature, as a control.

After 5 minutes, gently mix the contents of the tubes by inverting, decant, and either arrange according to intensity of colour obtained, or measure spectrophotometrically at 540 nm.

Results of absorbance vs. temperature are presented as a graph.

Observation table

Reaction time = 5 min

S. No.	Temperature (°C)	Absorbance (540 nm)	Transmission (%)
1.	17 (Room temp.)	0	100
2.	35	0.01	94
3.	50	0.13	74
4.	65	0.65	23
5.	80	0.50	32
6.	100	0.40	40

Result and Discussion

With an increase in temperature, the absorbance of the solution increases. Denaturation of proteins at higher temperatures results in distortion of the membrane and its permeability properties, leading to release of betacyanin pigments in the external medium. Absorbance at 80 °C or 100 °C may be even less because of breakdown of betacyanin pigment. As long as the tissue remains healthy and viable, its pigments, sugars and salts may be retained within the tissue cylinder. After treatment with heat, the solutes from within the cell diffuse into the external solution at a much faster rate because of an increase in kinetic energy of the pigment molecules.

Precautions

1. All beetroot cylinders must be removed from the same tuber as these are at the same osmotic level and tuber should be healthy.
2. Cylinders should be of uniform size and colour intensity, devoid of any white pithy specks.
3. Amount of distilled water and reaction time should be constant for all treatments. However, reaction time can vary depending upon the plant material.
4. Blunt forceps should be used for transferring the beetroot cylinders into and out of the boiling tubes otherwise these would get damaged.
5. Cylinders should be cut under water and thoroughly washed to remove the pigments sticking to the damaged surface.

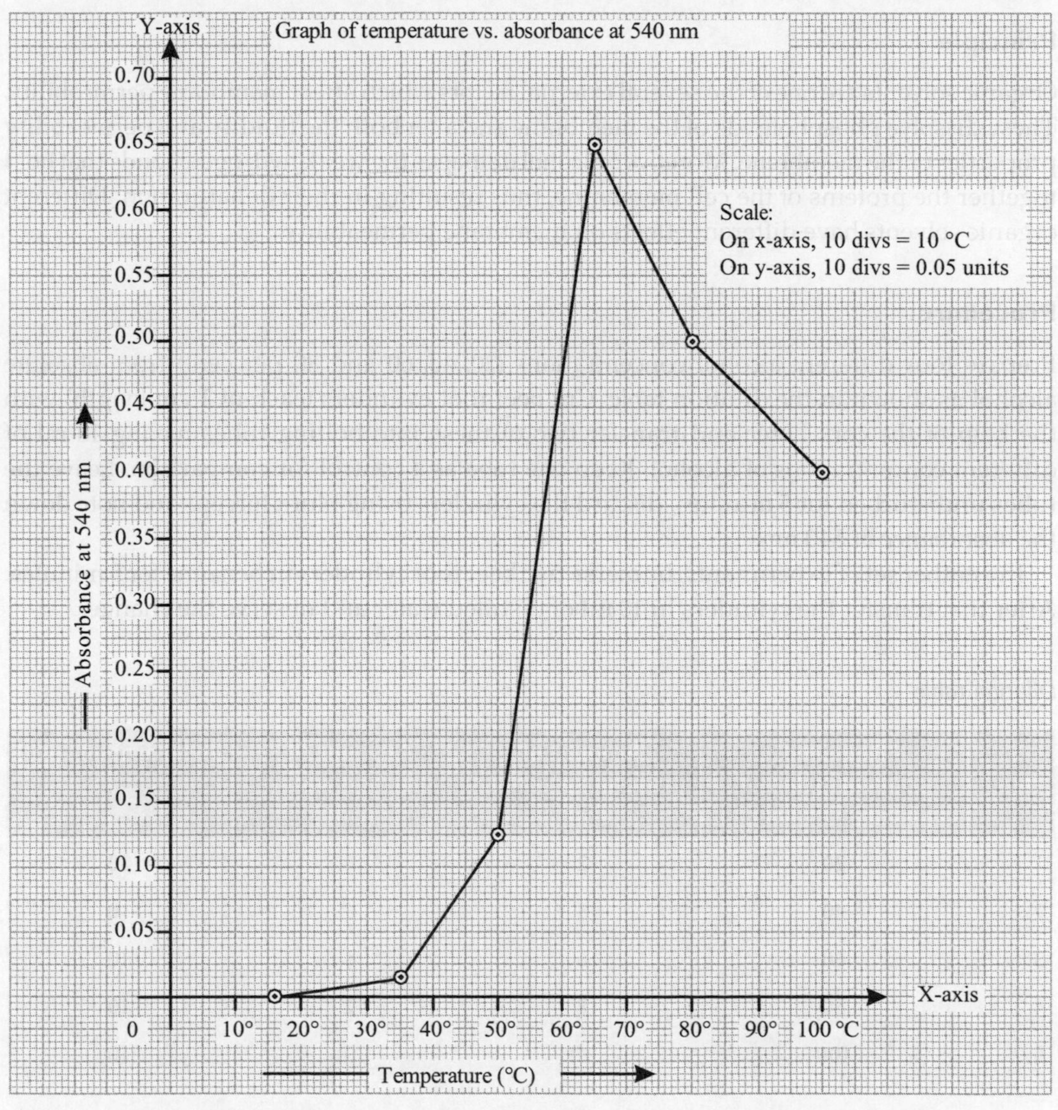

Experiment 1.5

To study the effect of organic solvents on the permeability of cell membranes.

Requirements

Plant material : Beet root (*Beta vulgaris*)
Family : Chenopodiaceae
Chemicals : Organic solvents (absolute alcohol or acetone, benzene), distilled water
Glass apparatus : Boiling tubes, petri dishes, glass rod
Miscellaneous : Cork borer, blunt forceps, blade, scale, graph paper, photoelectric colorimeter or spectrophotometer

Principle

Organic solvents are one of the several external factors which affect membrane permeability. These dissolve phospholipids of the membrane and alter both the fluidity and permeability properties of the membrane. Moreover, alcohol damages the weak hydrogen bonds that hold together the proteins of the cell membrane, thus resulting in their denaturation. Different organic solvents have different effects on membrane permeability.

Procedure

Prepare four different alcohol concentrations (20, 40, 60 and 80 per cent) from absolute alcohol as shown in the dilution table. Cut beet root cylinders (each of 1 cm length) with the help of the cork borer and transfer two of these into boiling tubes having 10 ml of different concentrations of alcohol. Keep one tube as a control for comparison. Leave the tubes undisturbed for 5 minutes, discard the cylinders using blunt forceps and note down the absorbance at 540 nm.

In another set of experiments, immerse two beet root cylinders in benzene for 5 minutes. After that transfer the cylinders in 10 ml of distilled water and observe the change.

Dilution table

S. No.	Required concentration of absolute alcohol (%)	Amount of absolute alcohol (ml)	Amount of distilled water (ml)	Total amount (ml)
1.	20	2	8	10
2.	40	4	6	10
3.	60	6	4	10
4.	80	8	2	10
5.	100	10	0	10

Observation table

Reaction time = 5 min

S. No.	Concentration of absolute alcohol (%)	Absorbance (540 nm)	Transmission (%)
1.	Control (distilled H_2O)	0	100
2.	20	0.005	99
3.	40	0.02	95
4.	60	0.13	73
5.	80	0.75	18
6.	100	0.23	63

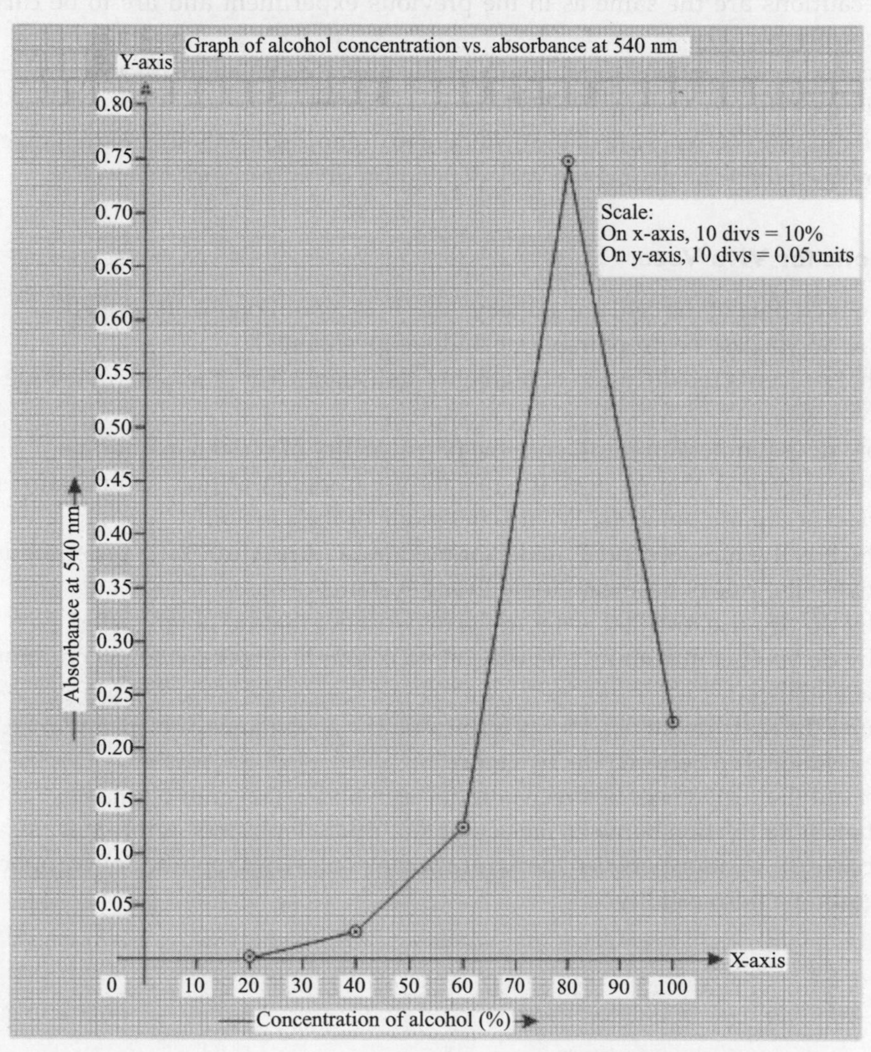

Result and Discussion

With an increase in concentration of organic solvent (alcohol), absorbance of the solution increases because of more and more release of betacyanin pigments. Alcohol at higher concentrations damages the membranes and alters their permeability by dissolving their phospholipid components. At 100 per cent alcohol, the solution shows less absorbance as compared to 80 per cent alcohol as less water is available in the external medium and betacyanin pigments are water-soluble.

In another set, benzene solution remains colourless. But when beet root cylinders are transferred from benzene solution to distilled water, because of loss in permeability of the membrane, water immediately turns dark pink.

Precautions

1–4. Precautions are the same as in the previous experiment and are to be carried out accordingly.

5. Alcohol concentrations should be carefully prepared using distilled water.
6. All tubes should contain an equal amount of solution.
7. The duration of exposure of the cylinders to organic solvents should remain constant. However, the time period can vary depending upon the plant material.

Questions for Viva-Voce

1. In which part of the cell, are anthocyanin/betacyanin pigments located?
2. Why do you use beetroot in permeability experiments?
3. What prevents the diffusion of pigments and sugars from the beet root while it is still growing?
4. How does temperature influence the permeability of plasma membrane?
5. What chemical change does the membrane undergo at high temperatures and in different organic solvents? What is the mode of their action?
6. Which is the most accepted model for membrane structure? Who proposed it?
7. What is the role of permeability in living systems?
8. Why do you use distilled water in the experiment instead of tap water?
9. Why do you get maximum intensity of red colour at 60 per cent or 80 per cent alcohol concentration?
10. Name the instrument used for determining the absorbance or transmittance. What is the relationship between the two terms?
11. List out the differences between anthocyanin and betacyanin pigments.
12. What is the relation between molecular size and membrane permeability?
13. What is the relation between the nature of molecules (hydrophilic or hydrophobic) and membrane permeability?

Physiological Set-Up

1. To Demonstrate Dialysis

It was first described by Thomas Graham in 1861.[3] Dialysis is a technique of separation of solutes of a mixture on the basis of molecular weight differences using a semi-permeable membrane. The solute mixture, usually glucose and starch, is placed into dialysis bag (prepared from cellophane or collodion, i.e., cellulose nitrate) and the bag is sealed at both the ends. The bag is then immersed in an aqueous medium having iodine (Figure E 1.2). Molecules small enough to pass through the bag (salts, water, iodine, monosaccharides, etc.) tend to move into or out of the dialysis bag in the direction of decreasing concentration, therefore displaying diffusion. Larger molecules (proteins, polysaccharides, and such) that have dimensions significantly greater than the pore diameter are retained inside the dialysis bag. However, high molecular weight solutes do differentially penetrate membranes having large pore sizes.

Graham's discovery and what he developed has become the basis for modern-day dialysis which traces back to his study of colloids. This work led to his ability to separate colloids from crystalloids using a so-called 'dialyzer', the precursor of today's dialysis machine.

In medicine, haemodialysis is a method for removing waste products such as creatinine and urea, as well as free water from the blood when the kidneys are in renal failure. The haemodialysis machine pumps the patient's blood and the dialysate through the dialyzer with the two flows running on different sides of semi-permeable membranes. Osmotic pressure causes the waste matter in blood to permeate into the dialysate which carries them away.

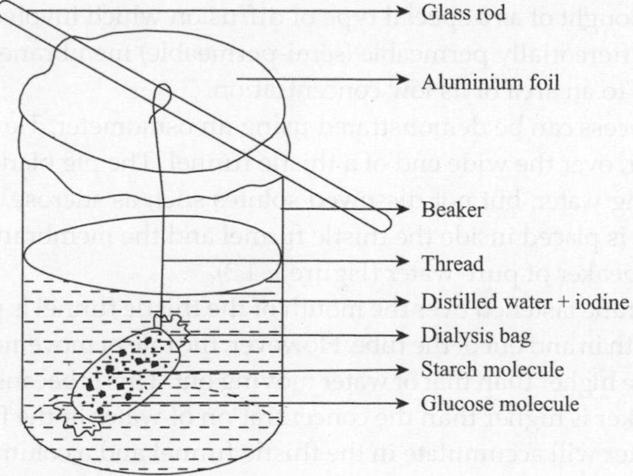

Fig. E1.2 An experimental setup for dialysis

---

3 Thomas Graham (1805–1869), a Scottish chemist, is best remembered today for 'Graham's law' for diffusion of gas, besides pioneering work in dialysis. Graham is also known as founder of colloidal chemistry.

Dialysis thus provides an 'artificial replacement' for lost kidney function in people with renal failure.

Precautions

1. The ends of the dialysis bag should be tightly sealed to prevent any leakage of solution.
2. The dialysis bag should be fully immersed in a beaker containing water.
3. The dialysis bag should be washed thoroughly with sufficient water before suspending it in the beaker.

Questions

1. Who introduced the technique of dialysis?
2. What is the chemical composition of the dialysis bag?
3. Name a part of the cell which simulates the dialysis bag.
4. How would you test the presence of glucose in the outer aqueous solution?
5. If a mixture of starch and NaCl is used inside the dialysis bag, how would you confirm the presence of NaCl in the aqueous solution?
6. What results do you expect, if crude amylase extract is added to the dialysis bag?
7. How can the process of separation be speeded up?
8. What is meant by ultrafiltration?
9. Write about the application of dialysis in medicine

2. To Demonstrate Osmosis

Osmosis may be thought of as a special type of diffusion which involves the movement of water through a differentially permeable (semi-permeable) membrane from an area of its high concentration to an area of its low concentration.

The osmotic process can be demonstrated using an osmometer. Tie a piece of material, such as pig bladder, over the wide end of a thistle funnel. The pig bladder is differentially permeable, allowing water, but not dissolved solutes such as sucrose, to pass through. A solution of sucrose is placed inside the thistle funnel and the membrane end of the funnel is submerged in a beaker of pure water (Figure E 1.3).

Since the membrane fastened over the mouth of the thistle funnel is permeable to water, water will move both in and out of the tube. However, the rate of movement of water moving into the tube will be higher than that of water moving out. This is because the concentration of water in the beaker is higher than the concentration of water in the funnel. Under those circumstances, water will accumulate in the thistle funnel and a column of water will rise in the tube. As a result, the sucrose solution becomes more and more diluted and the rate of water moving into the funnel slows down accordingly, i.e., there is less and less difference between the concentration of water in the beaker and in the funnel.

However, as the column of water builds up in the funnel, it exerts a hydrostatic pressure on the membrane in the downward direction. When this pressure becomes equal to osmotic entry of water from the beaker, osmosis ceases.

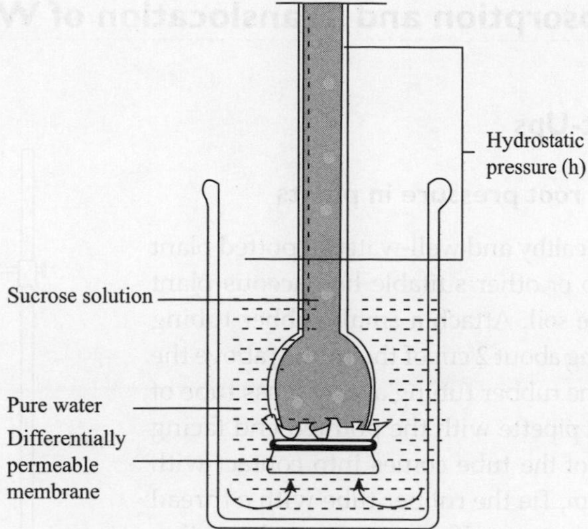

Fig. E1.3 Demonstration of osmosis.

Questions

1. In what way is osmosis similar to diffusion? Mention one distinguishing feature between the two.
2. Give other examples of artificial semi-permeable membranes.
3. What will happen if pig's bladder (sheep's bladder) membrane is replaced with a (i) rubber sheet; (ii) thick sheet of writing paper?
4. In the living cell, which membranes behave as semi-permeable membranes?
5. Vacuolated plant cells behave as osmometers'. Comment.
6. Explain the significance of osmosis in plants

2. Absorption and Translocation of Water

Physiological Set-Ups

1. To demonstrate root pressure in plants

Cut off the stem of a healthy and well-watered potted plant of geranium, tobacco or other suitable herbaceous plant about 5 cm above the soil. Attach a small rubber tubing into the stump, leaving about 2 cm of the tubing above the cut end. Insert into the rubber tubing a long glass tube of proper diameter or a pipette with the pointed end facing upward till the end of the tube comes into contact with the cut end of the stem. Tie the rubber tube with a thread without damaging the stump. If necessary, apply sealing wax around the rubber tube to make the connection air-tight. Support the tubes and the pipette with a stand (Figure E2.1).

Within a short time, there will be a rise of xylem sap in the glass tube. Measure at regular intervals with the help of manometer the amount of liquid exuded from the stump.

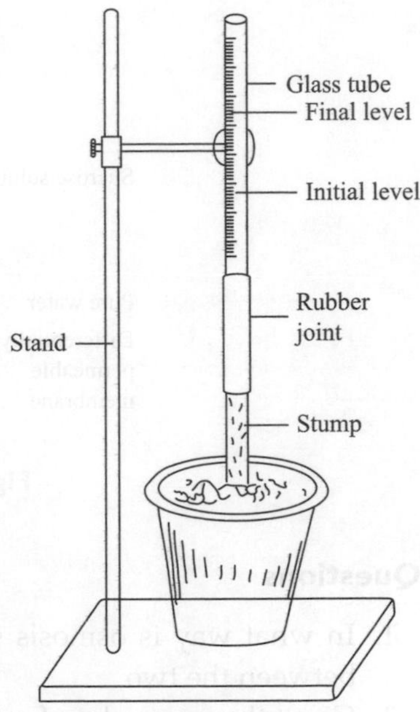

Fig. E2.1 Demonstration of root pressure.

Precaution

Keep the plant well-watered so that the rise of the xylem sap is well maintained. (Alternative plant materials - Tomato, Balsam, *Bryophyllum*.)

Question

1. Who coined the term 'root pressure'?
2. Comment upon root pressure and its significance.
3. 'Root pressure is seldom exhibited by an actively transpiring plant' - Explain.
4. 'Root pressure may account for the rise of water in certain plants under certain conditions' – Justify.

2. To demonstrate that the amount of water absorbed by the roots is equal to the amount of water transpired

Take a large glass bottle, provided with a graduated side tube. Fill the apparatus with water and fix a well developed twig of a mesophyte in an air-tight way into the bottle so that its roots are in water. Cover the surface of water in the side tube with an oil drop to prevent direct evaporation (Figure E2.2). Wrap black cloth around the bottle to exclude light from the roots. Mark the water level in the side tube. Weigh the whole apparatus and place it in diffuse light in the laboratory. Reweigh at hourly intervals. The loss in weight will give you the rate of water transpired. Also measure the fall in the level of water in the side tube (1 ml = 1 gram). Hourly readings give you the rate of water absorption by the plant.

The two rates will be approximately equal, though not identical.

Carry out similar determinations by placing the set-up (i) in direct sunlight in the open (ii) in front of an electric fan so that transpiration is increased (iii) in humid air after covering the apparatus with a bell jar lined with wet blotting paper. In each case, compare the rates of water absorption and water loss.

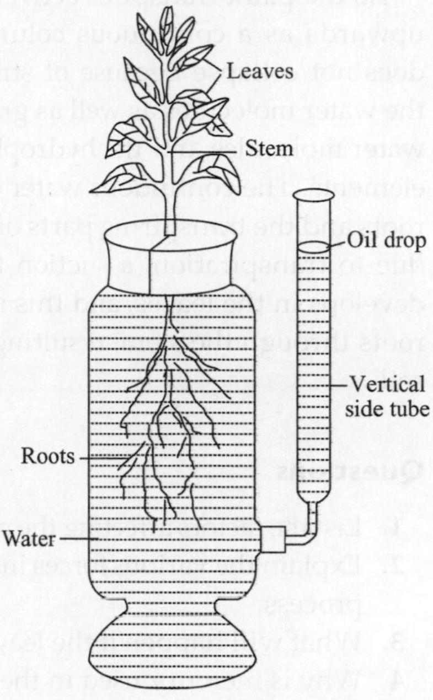

Fig. E2.2 Demonstration of comparison of the amount of water transpired with the amount of water absorbed by the roots.

Precautions

1. Connections should be made airtight - the water level will not fall even if the apparatus is inverted.
2. There should be no air bubbles inside the apparatus.
3. The roots should be well washed to remove any adhering soil, otherwise the water will become turbid.
4. A drop of oil should be added in the graduated side tube to prevent evaporation of water.

Questions

1. What result do you expect if an experimental twig is replaced by a xerophytic twig?
2. Why is it necessary to add a drop of oil in the vertical side tube?
3. What change do you observe after an hour or so in the (i) level of water in the side tube (ii) weight of the entire apparatus? What does it indicate?

3. To demonstrate suction due to transpiration

The set-up (simple potometer) includes a hollow glass tube with one end submerged in a trough containing mercury and the other end fitted with the shoot of an actively transpiring plant (such as guava, sunflower or geranium), under airtight conditions (Figure E2.3). Rise in the level of mercury column after some time period (30 minutes or so) indicates suction due to transpiration.

$$\text{Water lifting power of transpiration} = \pi r^2 h \times d$$

where r= radius of the capillary tube
h= rise of mercury level (height)
d= relative density of mercury (13.6 g)

As the plant transpires actively in nature, water is lifted upwards as a continuous column. Such a water column does not collapse because of strong cohesive force among the water molecules as well as great adhesive force between water molecules and the hydrophilic walls of the tracheary elements. The continuous water column exists between the roots and the transpiring parts of the plant (i.e. leaves). Thus due to transpiration, a suction force or transpiration pull develops in the leaves, and this is transmitted below to the roots through the stem, resulting in water uptake from the soil.

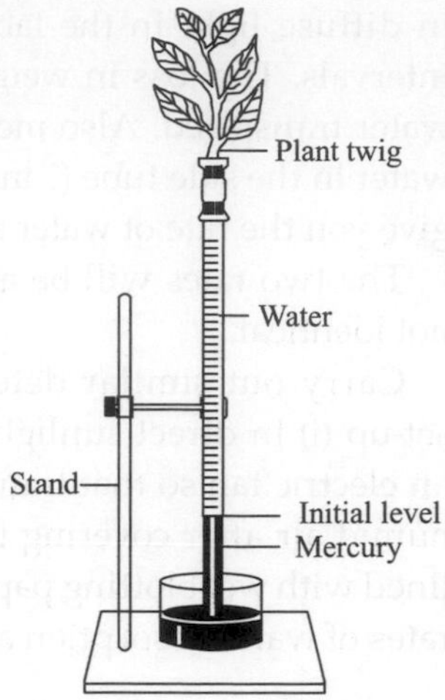

Fig. E2.3 Demonstration of suction due to transpiration.

Questions

1. List the factors affecting the process being demonstrated.
2. Explain the various forces involved in this physiological process.
3. What will happen if the leaves are smeared with grease or wax?
4. Why is mercury used in the set-up?
5. Why does mercury-water move as an unbroken column?
6. What will happen if the set-up is shifted in the humid environment?
7. Which plant hormone regulates stomatal closure during water deficit conditions?
8. Can 'guttation' be observed in the twig used in the set-up?
9. How would you differentiate between samples of guttation water and transpired water?
10. What results do you expect, if the plant material in the set-up is replaced by (i) wet sponge (ii) dry sponge (iii) unglazed porous pot.
11. Explain the significance of the process.

4. To demonstrate that ascent of sap takes place through the xylem by (i) Eosin method, (ii) Ringing method, (iii) Blocking of xylem vessels with grease

(i) Eosin method. Wash the roots of a small potted plant, e.g., Balsam or cabbage leaf carefully with tap water and insert them into a test tube filled with eosin solution. Plug the mouth of the tube with cotton and keep it fixed to a stand (Figure E2.4).

After a few hours, you will observe that xylem strands of the roots, stem and leaves, stain red. This indicates that ascent of sap occurs through xylem strands.

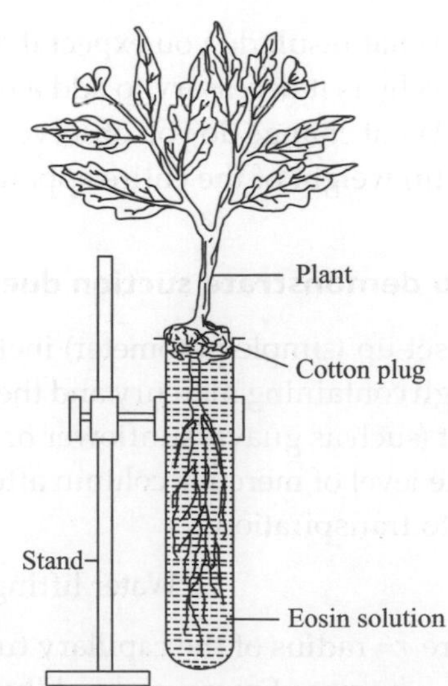

Fig. E2.4 Demonstration of movement of water through xylem by eosin method.

(ii) Ringing method. Take two woody twigs of the plant. Remove a ring of bark (all the tissues from epidermis up to phloem) about an inch in length without causing injury to the cambium below from the first twig (Figure E2.5). Then remove xylem elements from another twig without injuring the cortex and the phloem (transport in the xylem or phloem and cortex can be blocked by applying wax or grease to these elements).

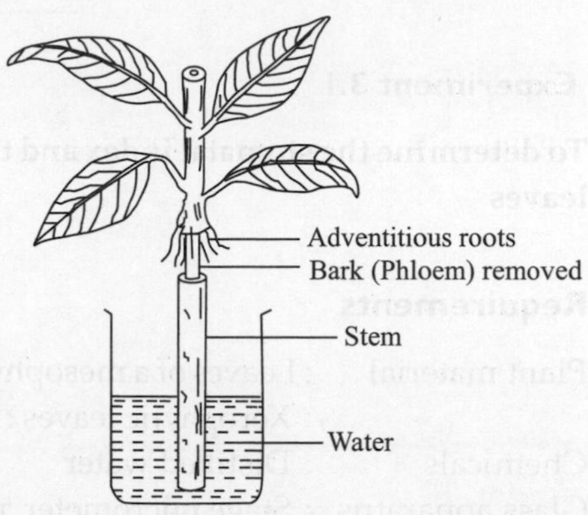

Fig. E2.5 The ringing or girdling experiment.

The ringing (removal of tissues) should always be done while the twigs are under water. Immerse the twigs in separate beakers containing water. Allow them to stand for a few hours.

In the first twig, the leaves are turgid even after ringing because of continued transport of water whereas in the second twig, leaves show wilting because water does not reach to them to keep them turgid.

This demonstrates that ascent of sap occurs through xylem.

(iii) Blocking of xylem vessels with grease. Cut two woody stems with about equal number of leaves on them. Place the cut end of one stem in melted paraffin for about an hour.

Take out the stem and allow the paraffin to solidify. Make a fresh cut near the end of the shoot so that the cell walls are exposed. Place both the stems in beaker containing water. The shoot which has blocked xylem vessels will wilt, whereas the other shoot will not wilt.

Questions

1. What do you understand by 'ascent of sap'? How does the sap rise in the plants and what is its path?
2. Discuss the various theories that have been put forward to explain the mechanism of ascent of sap in plant.
3. Which is the most accepted theory for ascent of sap in plants?
4. Comment on the resistances faced by water column during its rise in tall plants.
5. Trace the path of molecules of water from the time it enters the root hair until it escapes as water vapour through the stomata.
6. Suggest a substitute for the eosin solution used in the set-up.
7. Who introduced the 'girdling technique' and when?
8. Name the tissues responsible for 'long-distance' and 'short-distance' transport of water and solutes in plants.

3. Transpiration

Experiment 3.1

To determine the stomatal index and the stomatal frequency in mesophyte and xerophyte leaves

Requirements

Plant material : Leaves of a mesophytic plant : *Withania somnifera or Crinum or Pancratium*
: Xerophytic leaves : *Bryophyllum*
Chemicals : Distilled water
Glass apparatus : Stage micrometer, petri dishes, slides, cover slips
Miscellaneous : Compound microscope, blade, blotting paper, brush, needle

Principle

The epidermal surface of a leaf bears a large number of pores called stomata. The stomata are microscopic and are bordered by two specialised epidermal cells called guard cells, which control the opening and closing of the stomata. The surface of a leaf, depending upon the species, may contain anywhere from 1000 to 60,000 stomata per square centimeter. Stomata are more frequently found on the undersurface of leaves (hypostomatous), but in many species they are found on both the surfaces (amphistomatous). In certain floating aquatic leaves, the stomata occur only on the upper surface (epistomatous). Submerged leaves normally do not have stomata.

Stomata play a major role in the physiology of plants by regulating gaseous exchange. In general, they are open in the light and closed in the dark.

Stomatal index (%) $$I = \frac{S}{E + S} \times 100$$

S = Number of stomata per unit area.

E = Number of epidermal cells in the same area.

Under a given set of conditions, a species tends to form a definite proportion of stomatal indices. It is believed that high humidity tends to reduce the proportion of stomata formed and aquatic plants have low stomatal indices. Nutritional conditions also affect the stomatal index.

$$\text{Stomatal frequency (per cm}^2\text{)} = \frac{\text{Number of stomata in a given microscopic field}}{\text{Area of microscopic field}}$$

Area of microscopic field = πr^2 [r= radius of the microscopic field]

Salisbury concluded that stomatal frequency differs with environmental conditions, and is related to the availability of water.

Micrometry is a technique that is used to determine dimensions of any microscopic material. Micrometers are of two types : stage and ocular micrometer.

(i) Stage micrometer[4]: It is in the form of a glass slide with markings etched which can measure up to an accuracy of 100th part of a mm, i.e., 100 divisions = 1 mm. The stage micrometer is placed on the stage of the microscope and is used to calibrate the ocular micrometer and also to calculate area of microscopic field.

(ii) Ocular micrometer: It is in the form of a circular glass disc with 100 equidistant divisions but of unknown dimensions. It is placed between the ocular and the objective lenses. Since the ocular micrometer is not calibrated, it is done using the stage micrometer.

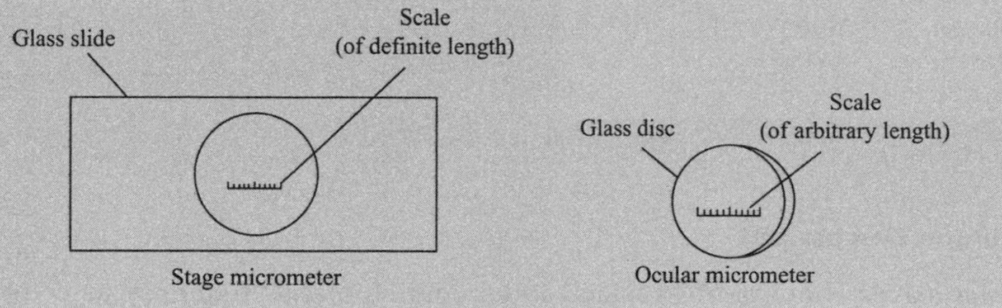

Procedure

Remove several epidermal peels from the lower surface of the leaves of *Withania* and *Bryophyllum.* Stain them with dilute safranin, mount in glycerine and observe under high power (6×, 40×) of the compound microscope.

Count the number of stomata and number of epidermal cells in the microscopic field. Record three concordant readings. Calculate the 'stomatal index' and 'stomatal frequency' using the appropriate formulae.

Calculate the area of the microscopic field using the stage micrometer.

Observations and calculations

Number of divisions of the stage micrometer in the microscopic field = 51

∴ Diameter of the microscopic field = 51 × .01 mm = 0.51 mm [1 stage division = .01 mm]

$$\therefore \text{Radius}(r)\text{ of microscopic field} = \frac{0.51}{2}\text{mm}$$

$$\therefore \text{Area of microscopic field} = \pi r^2$$

$$= \frac{22}{7} \times \frac{0.51}{2} \times \frac{0.51}{2} = 0.204\,\text{mm}^2 \text{ or } 0.00204\,\text{cm}^2$$

[4] Stage micrometers are also known as objective micrometers.

Observation table

Plant material	Number of stomata		Number of epidermal cells		Stomatal index (%)	Stomatal frequency (/cm^2)
	Number	Mean	Number	Mean		
Withania somnifera	(i) 58		(i) 173			
	(ii) 51	55	(ii) 163	164	25.114	26960.78
	(iii) 56		(iii) 156			
Bryophyllum	(i) 23		(i) 123			
	(ii) 18	20	(ii) 94	113	15.037	9803.92
	(iii) 20		(iii) 123			

Result and Discussion

Withania leaves show more stomatal index and stomatal frequency as compared to *Bryophyllum* leaves. The latter being succulent have fewer number of stomata to reduce the rate of transpiration.

Precautions

1. Peels should be taken from the same leaf by applying unequal pressure.
2. Avoid counting of stomata near edges of leaf strips (border effect).
3. Peels should be one-celled thick and free from green mesophyll tissue.
4. Magnification should remain the same throughout the experiment.
5. Peels should be immediately kept in water to prevent the entrance of air bubbles.

Note:

(i) 'Stomatal index' and 'Stomatal frequency' can be calculated for upper and lower surfaces of the given leaf separately for comparison.

(ii) A mesophyte and a xerophyte leaf can be replaced by a dicot (e.g. bean) and a monocot (e.g. barley) leaf also.

Experiment 3.2

To determine the percentage of leaf area open through stomata

Requirements

Plant material : As in previous experiment
Chemicals : 1 per cent KOH, distilled water
Glass apparatus : Stage micrometer, ocular micrometer, petri dishes, slides, cover slips
Miscellaneous : Compound microscope, blade, blotting paper, dropper, brush, needle

Principle

The stomatal pores can be thought of as ports of exchange between the external environment and the interior of the leaf. Therefore, the physical factors influencing the diffusion of water vapour through these pores are important in the study of transpiration. When fully opened, the stomatal pore may measure from 3 to 12 μm across and from 10 to 40 mm in length. The surface of a leaf may contain approximately 1000 - 60,000 stomata per square cm. As large as these figures are, the stomatal pores are so small that they occupy, when fully opened, only 1 or 2 per cent of the total leaf surface. The diffusion of water vapours through these pores often exceeds 50 per cent of that evaporating from a free water surface.

Percentage leaf area open through stomata (%) = Stomatal frequency × Area of open stomata × 100

where Area of open stomata, $A = \frac{\pi}{4} \times \text{length} \times \text{breadth}\,(\text{of stomatal aperture})$

Procedure

Calibrate ocular micrometer using stage micrometer.

Remove several pieces of epidermis from the lower surface of the leaves and plunge instantly into 1 per cent KOH solution for 15 minutes. Measure the length and breadth of stomatal aperture using ocular micrometer. Calculate the area of open stomatal aperture and the percentage of leaf area open through stomata using the formulae mentioned above.

Compute the pore area as a percentage of total area. Calculate total leaf area using graph paper.

Observations and calculations

(a) Calibration of ocular micrometer:

5th division of ocular micrometer coincides with 2nd division of stage micrometer.

$$\therefore \text{1 division of ocular micrometer} = \frac{2}{5} \times 0.1\,\text{mm} = 0.004\,\text{mm or } 0.0004\,\text{cm}$$

(b) $$\text{Area of microscopic field} = \pi r^2$$
$$= 0.00204\,\text{cm}^2\,(\text{from the previous experiment})^5$$

(c) $$\text{Stomatal frequency} = \left[\begin{array}{l} \textit{Withania}\text{: } 26960.78/\text{cm}^2 \\ \\ \textit{Bryophyllum}\text{: } 9803.92/\text{cm}^2 \end{array}\right](\text{from the previous experiment})^5$$

[5] Assuming that both the experiments (1, 2) are done on a particular day, using the same microscopic magnification, otherwise calibration should be done again.

Plant material	Length (L) of stomatal aperture (cm)	Breadth (B) of stomatal aperture (cm)	Area $\left(\frac{\pi}{4}\times L\times B\right)$ of stomatal aperture (cm^2)	% leaf area open through stomata = area of stomatal aperture × stomatal frequency × 100
Withania somnifera	5 div ×	1 div ×	0.00000063	
	0.0004 cm	0.0004 cm	6.3×10^{-7} cm^2	1.7%
Bryophyllum	5 div ×	1 div ×	0.00000063	0.6%
	0.0004cm	0.0004 cm	6.3×10^{-7} cm^2	

Result and Discussion

Withania has a greater percentage of leaf area open through stomata (1.7 per cent) as compared to *Bryophyllum* (0.6 per cent).

Variation in the size of the stomatal aperture occurs not only in different species but also in different leaves of the same plant and in the same leaf at different points.

Precautions

1–5. Same as in previous experiment.

6. The breadth of the stomatal pore should be taken at the point of maximum width, which can be obtained by placing the peels in KOH solution for 15 minutes.

While measuring length and breadth of stomatal pore, the peels should not be stained as stain would kill the cells which would lead to closure of stomata.

Experiment 3.3

To study the effect of different light intensities on the rate of transpiration using Ganong's Potometer

Requirements

Plant material : Guava (*Psidium guajava*)[6]
Family : Myrtaceae
Glass apparatus : Ganong's potometer, beaker, large sized petri dish
Miscellaneous : Luxmeter, stop watch, table lamp, clamp stand, graph paper, blade, rubber cork, grease

[6] Alternative plant materials: *Tecoma*, *Polyalthia*.

Principle

The process of water loss from the plants in the form of water vapour is called 'Transpiration'. The loss of water may occur through stomata (stomatal) or lenticels (lenticular) or through cuticle on stem (cuticular). The amount of water loss through lenticular and cuticular transpiration is insignificant as compared to the amount of water lost through stomatal transpiration.

Factors affecting rate of transpiration: In fact, any factor which affects stomatal movement is bound to affect the rate of transpiration. Environmental factors like light, wind, humidity of the air, temperature and available soil water will affect the steepness of the water potential gradient between the internal and external atmosphere of the leaf and therefore, affect the rate of transpiration.

Light: Generally, the stomata of a leaf are open when exposed to light and remain still open under continuous light unless some other factor becomes limiting. When darkness returns, the stomata close. CAM plants are an exception as they exhibit inverted stomatal behaviour.

Light occupies a prominent position among the factors influencing transpiration and an increase in light intensity increases the transpiration rate. The rate of transpiration can be interpreted in relation to 'transpiration flux.'

$$\text{Transpiration flux} = \frac{\text{Magnitude of the driving force}}{\text{Resistance in the pathway}}$$

$$= \frac{C_l - C_a}{R_s + R_a}$$

C_l = vapour concentration of substomatal cavity of the leaf
C_a = vapour concentration of air
R_s = resistance offered by stomata
R_a = resistance offered by boundary layer, i.e., air

Rate of transpiration is inversely related to resistance, i.e., the greater the resistance in the pathway, the lower would be the rate of transpiration.

Resistance depends upon different variables, such as stomatal aperture and stomatal frequency.

The rate of transpiration can be conveniently measured with help of **Ganong's potometer** (Figure E3.1).

A twig of guava, tecoma, neem or some other suitable plant is sealed in Ganong's potometer. Before the rate of transpiration can be measured, the entire apparatus is filled with water so that no air spaces are present. This may be accomplished by manipulating the stopcock, which controls the flow of water into the vessel from the reservoir. An air bubble is then introduced into the capillary tube. As transpiration proceeds, the air bubble will move along the capillary tube, giving a measure of the rate of transpiration.

The potometer method is ideal for observing the effects of different environmental factors (temperature, light, air movement) on transpiration rates. However, its reliability is limited because it actually measures water absorption rather than transpiration; under certain circumstances the two can vary considerably.

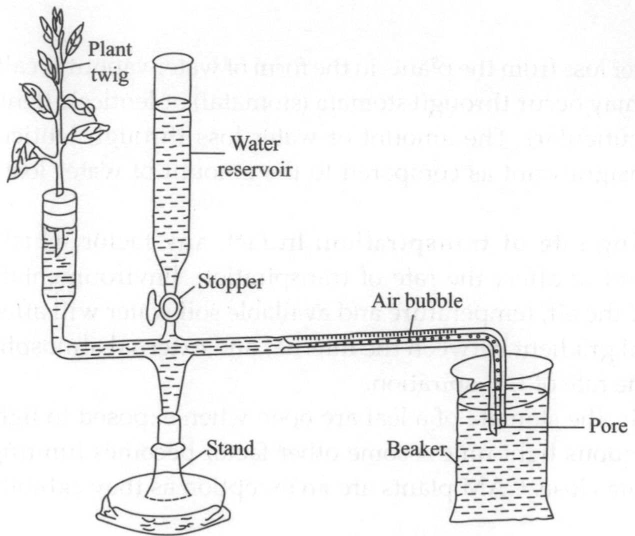

Fig. E3.1 A Ganong's potometer.

Procedure

Set up the Ganong's potometer as described earlier. For determining the effect of different light intensities, the rate of transpiration in terms of time taken by an introduced air bubble to move over a fixed distance of 2 cm is measured in (a) diffuse sunlight (b) diffuse sunlight + 1 lamp (c) diffuse sunlight + 2 lamps (one on either side)

Note down the light intensity using a luxmeter.

Observation table

S. No.	Environmental factor	Light intensity (lux)	Distance travelled by air bubble (D) (cm)	Time taken (T) (s)	Mean value	Rate of transpiration $\frac{D}{T}$ × 60 (cm/min)
1.	Diffuse sunlight		(a) 2	53		$\frac{2}{53} \times 60 = 2.26$
		496	(b) 2	50	53	
			(c) 2	56		
2.	Diffuse sunlight + 1 lamp		(a) 2	45		$\frac{2}{43} \times 60 = 2.79$
		550	(b) 2	43	43	
			(c) 2	41		
3.	Diffuse sunlight + 2 lamps		(a) 2	23		$\frac{2}{25} \times 60 = 4.8$
		634	(b) 2	27	25	
			(c) 2	25		

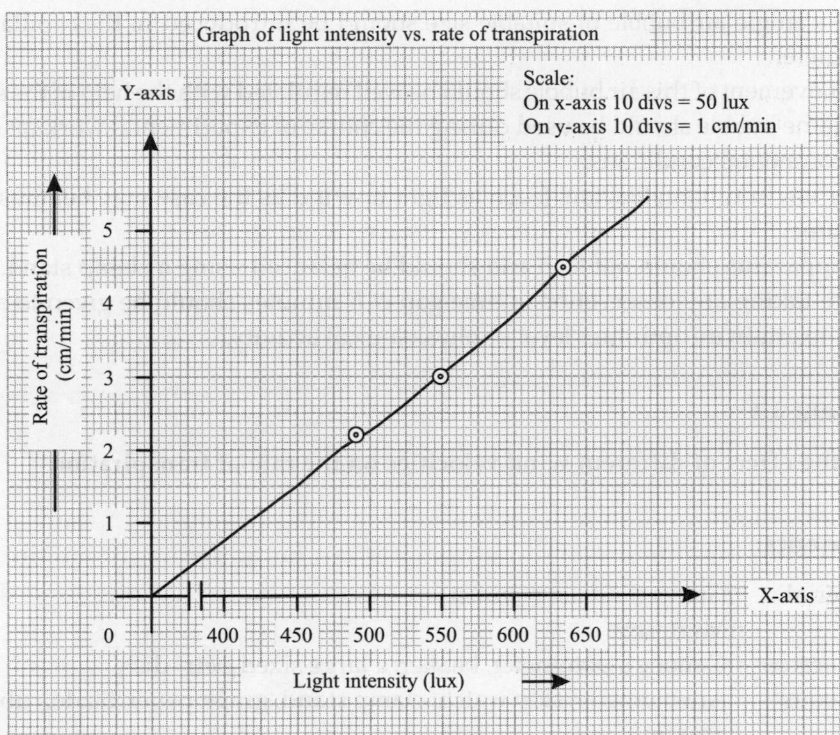

Result and Discussion

With an increase in light intensity, the time taken by the bubble to move a distance of 2 cm decreases; indicating increase in rate of transpiration. Increase in light intensity leads to increase in pH. A high pH is accompanied by a decrease in starch (osmotically inactive) and an increase in reducing sugars (osmotically active), resulting in an increase in turgor of guard cells, causing the stomata to open and hence an increase in the transpiration rate.

In fact, in this experiment, one simply measures the rate of absorption of water or indirectly the rate of transpiration, as the absorption of water is approximately equal to the rate of transpiration.

Precautions

1. The Ganong's potometer should be thoroughly washed, using chromic acid.
2. The diameter of the leafy twig should be such that it fits tightly in the cork - *neither* too tight (otherwise the bark will peel off) *nor* too loose (otherwise the apparatus will not be airtight).
3. The twig should be given an oblique cut while being kept in water.
4. The apparatus should be made air tight, using grease. The reservoir should be completely filled with water.
5. Before applying grease, the water sticking to the surface of the cork must be blotted or removed otherwise adhesion between the cork and the grease will not be effective.

6. Only a single air bubble of suitable size should be introduced in the capillary of the potometer.
7. The movement of this air bubble should be well regulated with the help of the stopcock.
8. The same bubble should be used during the course of experimental work for different environmental conditions.
9. Other environmental factors (such as light or wind as the case may be) must remain the same.
10. The apparatus should not lean and should be balanced using a clamp stand.
11. Before taking any observation, a time gap of 5 minutes should be given to allow the plant to attain its optimum level of physiological activity.

Experiment 3.4

To study the effect of different wind velocities on the rate of transpiration

Requirements

Plant material	: Guava (*Psidium guajava*)
Family	: Myrtaceae
Glass apparatus	: Ganong's potometer, beaker, a large sized petri dish
Miscellaneous	: Anemometer, stop watch, clamp stand, graph paper, blade, rubber cork, grease

Principle

Wind is an important factor regulating the rate of transpiration.

Air in the immediate vicinity of a transpiring leaf becomes more and more concentrated with water vapour. Under these circumstances, the vapour pressure gradient is lowered and the rate of transpiration decreases. However, if the water vapour concentrating in the area of the leaf is dispersed by wind, transpiration will again increase.

Increase in transpiration as a result of wind is not proportional to the wind velocity. It has been shown by several investigators that when plants are suddenly exposed to wind, there is a sharp increase in the rate of transpiration, followed by a gradual falling off of this increase, indicating that the effect of wind on transpiration may be rather complex.

It is realized that wind blowing over an evaporating surface would have a significant cooling effect, a circumstance which would lower the vapour pressure gradient and thus the rate of transpiration. In addition, winds of high velocity may possibly cause stomatal closure.

> It appears, therefore, that the increase in transpiration caused by the wind is actually the sum total of positive and negative influences.

Procedure

Set up the Ganong's potometer as described in the previous experiment.

For measuring the effect of different wind velocities on the rate of transpiration, experiment is carried out under diffuse sunlight conditions and the time taken by the introduced air bubble to move a distance of 2 cm is recorded for various wind conditions: (a) fan off (b) fan on low speed (c) fan on high speed.

Wind velocity is measured using anemometer.

Observation table

Distance travelled by the bubble = 2 cm

S. No.	Environmental factor	Wind velocity (m/s)	Time taken for movement of bubble	Mean time (s)	Rate of transpiration (cm/min)
1.	Diffuse sunlight	0.0	(a) 30 (b) 32 (c) 32	31.3	$\frac{2}{31.3} \times 60 = 3.78$
2.	Diffuse sunlight + fan (low speed)	0.60	(a) 22 (b) 21 (c) 20	21	$\frac{2}{21} \times 60 = 5.7$
3.	Diffuse sunlight + fan (high speed)	0.80	(a) 20 (b) 21 (c) 18	19.6	$\frac{2}{19.6} \times 60 = 6.12$

*Rate of transpiration can be better expressed in cm/h.

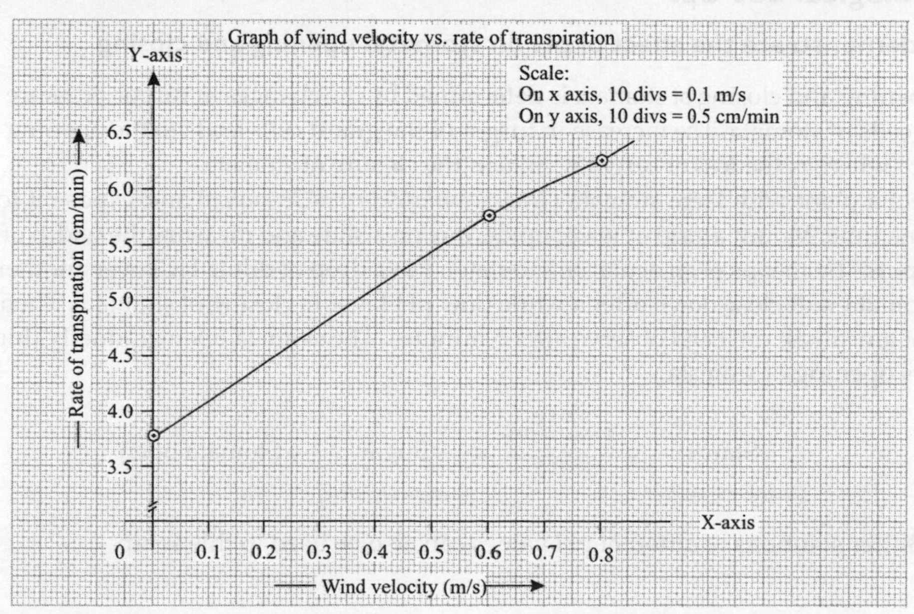

Result and Discussion

With an increase in wind velocity, there is increase in rate of transpiration. However, it may be followed by a decline on further increase in wind velocity. This is due to two reasons:

(a) Strong winds create a water imbalance in the guard cells because of increased transpiration, thus leading to stomatal closure.

(b) Strong winds exert a cooling effect and thus lower the vapour pressure gradient and in turn the rate of transpiration.

Precautions

The precautions here, are the same as in previous experiment (Experiment 3.3).

Note: Alternate procedure: The distance moved by the bubble can be also recorded at fixed intervals of 2 minutes.

Questions for Viva Voce

1. What are the limitations of using Ganong's potometer to study rate of transpiration under certain conditions?
2. Why is an oblique cut given to the twig? Why is it given under the water?
3. Why is it important to wait for a few minutes before repeating the experiment under different environmental conditions?
4. What will happen if the apparatus is kept (i) in the dark (ii) under a high-speed fan.
5. Describe the significance of transpiration.
6. Explain the difference between guttation and transpiration.

Physiological Set-Ups

1. To demonstrate the effect of pH on stomatal opening and closing

The opening and closing of stomata is determined by the changes in turgor pressure of the guard cells of stomata. The increase in turgor of the guard cells leads to the opening of the stomata whereas the loss of turgor results in the closure of stomata.

Select a plant with large stomata (*Tradescantia/Zebrina/Geranium*) and expose it to light for some time. Mount a strip of lower epidermis on a slide and measure the width of some of the pores by means of an ocular micrometer. Put a drop of dilute HCl over another peel. The stomata will close down after a while (low pH favours closing of stomata). Then put a drop of any alkali on it and observe after 5 minutes. The stomata open gradually (high pH favours opening of stomata.)

$$\underset{\text{(osmotically inactive)}}{\text{Starch}} \xrightleftharpoons[\text{pH 5}]{\text{pH 7} \atop \text{Phosphorylase}} \underset{\text{(osmotically active)}}{\text{Sugar}} \rightarrow \text{Stomata open}$$

Note: Buffer solutions of different pH values can also be used in this setup.

Questions

1. How does increase and decrease in pH affect this phenomenon?
2. Explain the effect of K^+ ions on this process.
3. What will happen if the cells are placed in pure water?
4. Name the phytohormone involved in this process. What is its mode of action?
5. What change would occur in the water potential of guard cells on addition of K^+ and malate ions?
6. Can NaCl be replaced by KCl?
7. What factors other than pH and hormones can affect this process?
8. Name one growth hormone that inhibits this process.
9. Explain briefly the role of (i) K^+ ions (ii) Light in the mechanism being demonstrated.
10. List two peculiar structural features of guard cells that govern the physiological process.

2. To compare the relative rates of transpiration from both the surfaces of a plant (mesophyte and xerophyte) using $CoCl_2$ paper method

Select well-watered potted plants of Guava (mesophyte) and Oleander[7] (xerophyte). Place dried $CoCl_2$ paper strips (filter paper strips dipped in slightly acidic 3 per cent solution of cobalt chloride), one each on the upper and lower leaf surfaces. Press those closer to leaf surfaces by glass slides and clip the slides together to make the apparatus air tight. (Figure E3.2). When exposed to a transpiring leaf surface, the cobalt chloride treated paper will gradually change from blue to pink. Note down the time taken by the $CoCl_2$ paper to change its colour.

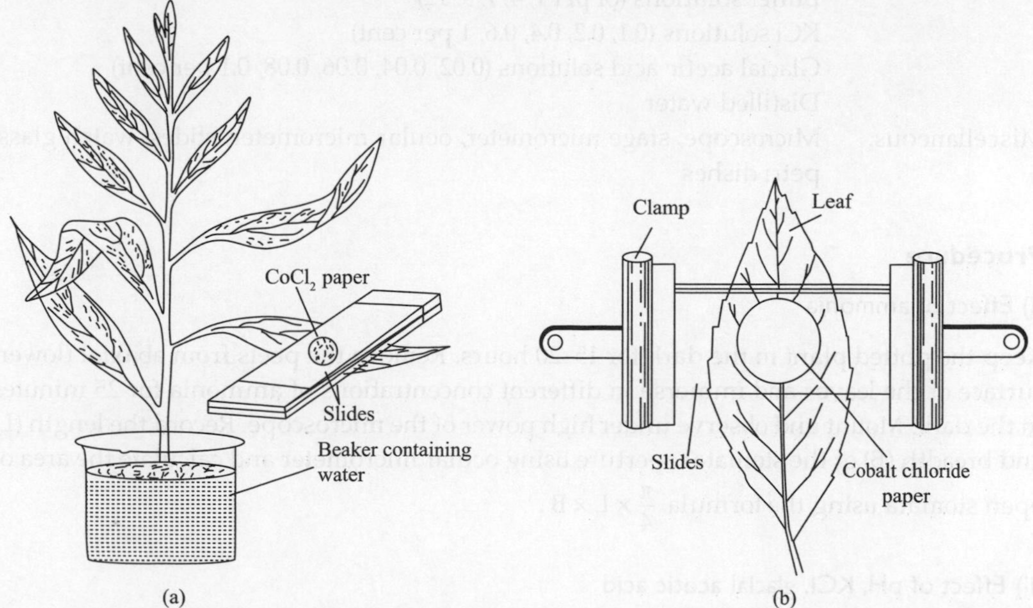

Fig. E3.2 Comparison of transpiration from the upper and the lower surface of a leaf by (a) cobalt chloride paper method (b) close-up view.

[7] Oleander (*Nerium indicum*) has a heavily cuticularized multiple epidermis and stomata present in crypts, closely guarded by trichomes.

The rate of colour change of $CoCl_2$ paper is indicative of the rate of transpiration.

Note:

1. $CoCl_2$ paper strips should be stored in a desiccator to avoid exposure to humidity.
2. Paper strips should be of uniform blue intensity.

Questions

1. List two environmental factors affecting this process.
2. Why do you use $CoCl_2$ paper?
3. What is so special about the paper due to which it changes its colour?
4. List two important precautions.
5. Explain the underlying principle of the experiment.

Project

To study the effect of NH_3, different pH values, KCl, glacial acetic acid and light on stomatal movements.

Requirements

Plant Material : *Crinum* (potted plant)

Reagents : Ammonia solutions (5, 10, 15, 20, 25 per cent)
Buffer solutions (of pH 1, 4, 7, 9, 12)
KCl solutions (0.1, 0.2, 0.4, 0.6, 1 per cent)
Glacial acetic acid solutions (0.02, 0.04, 0.06, 0.08, 0.1 per cent)
Distilled water

Miscellaneous: Microscope, stage micrometer, ocular micrometer, slides, watch glass, petri dishes

Procedure

(i) Effect of ammonia

Keep the potted plant in the dark for 15–20 hours. Remove the peels from abaxial (lower) surface of the leaves and immerse in different concentrations of ammonia for 25 minutes in the dark. Mount and observe under high power of the microscope. Record the length (L) and breadth (B) of the stomatal aperture using ocular micrometer and calculate the area of open stomata using the formula $\frac{\pi}{4} \times L \times B$.

(ii) Effect of pH, KCl, glacial acetic acid

Keep the potted plant in the daylight for 5 to 6 hours. Remove the peels from the abaxial surface of the leaves and immerse in (a) buffer solutions of different pH values (b) KCl solutions and (c) glacial acetic acid solutions of different concentrations. Keep the peels for 25 minutes, mount and observe under high power of the microscope. Record the length and breadth of the stomatal aperture and calculate the area of open stomata.

Stomata can be made to open or close 'at will' by the application of various chemicals. High pH (alkaline medium) favours opening of stomata whereas low pH (acidic medium) favours closing of stomata.

Note: The peel should be mounted in the same very solution in which it is immersed.

4. Mineral Nutrition

Experiment 4.1

Estimation of water, dry matter and ash contents in a variety of plant material of differing water content.

Requirements

Plant material	: Seedlings, fresh leaves, woody stem, fruits and aquatic plants
Chemicals	: Concentrated NH_4NO_2, alcohol
Glass apparatus	: Watch glasses, petri dishes, desiccator with anhydrous $CaCl_2$, crucible
Miscellaneous	: Bunsen burner, muffle furnace, tripod stand, crucible tong, electric balance, hot air oven

Theory

Water content in plants is generally very high. Since all cellular metabolic processes take place in an aqueous medium, cell walls and protoplast must maintain a certain water level to avoid adverse effects. Water content varies from plant to plant and organ to organ of the same plant. It may range from 10 per cent in dry seeds to more than 80 per cent in aquatic plants.

The dry matter contains both organic as well as inorganic compounds found in plants. Like water content, dry matter also varies from plants to plants, organs to organs and tissues to tissues. When strongly heated, organic compounds get oxidised by atmospheric oxygen and are lost in the form of gases whereas the inorganic compounds are changed into different oxides - collectively known as ash. Ash contains various macronutrients and micronutrients.

Procedure

(a) Estimation of water content and dry matter: Weigh 10 g of plant material in a clean petri dish whose empty weight has been noted. Dry the material at 105 °C for 3 to 4 hours. Cool the material in the dessicator and note down the weight. Repeat the drying, cooling and weighing process till a constant weight is achieved. Calculate water and dry matter content as given below:

$$\text{Water content} = \text{Fresh weight} - \text{Dry weight}$$

$$\text{Dry matter content} = \text{Fresh weight} - \text{Water}$$

Results: Calculate the water and dry matter contents in terms of percentages of fresh matter. Compare the water and dry matter contents of different plants and plant organs.

(b) Estimation of mineral elements (ash): Weigh 1 - 2 grams of dried plant material in a pre-weighed ashing crucible. Heat the material in a muffle furnace at 550 ° C for 4 hours, till it turns to ash. When the material is converted to white ash powder, cool the crucible in a dessicator and weigh the crucible again. The difference in weight gives the amount of ash.

Alternatively, ignite the plant material in a pre-weighed crucible over a bunsen burner by first heating gently and then strongly. Add occasionally, a few drops of concentrated NH_4NO_2 solution (oxidising agent) into the crucible. If clumps are formed, add a few drops of alcohol and pulverise the clumps with the help of a glass rod. Continue heating till a white powdered ash is formed. Cool the crucible in the dessicator and note down its weight. Calculate the amount of ash formed.

Results: Calculate the ash content in terms of the percentage of dry matter.

Experiment 4.2

Making qualitative assessments of some of the macroelements in plant ash.

Requirements

Plant material	: Seeds, leaves, stem, fruits and tubers – which should be fresh
Chemicals	: 5 per cent solutions (w/v) of KH_2PO_4, K_2HPO_4, $MgSO_4$, K_2SO_4, $CaCl_2$ and KSCN separately, 15 per cent (v/v) $HClO_4$, 10 per cent (w/v) $BaCl_2$, 50 per cent (v/v) H_2SO_4, ammonium molybdate reagent, magnesium reagent, Diphenylamine, concentrated sulphuric acid, and distilled water.
Glass apparatus	: Containers, drop bottles, crucible, slides, pipettes and burettes
Miscellaneous	: Spoon/spatulas, balances/weighing equipment, crucible tongs, muffle furnace

Preparation of Reagents

(i) **Ammonium molybdate reagent.** Dissolve 7.0 grams ammonium molybdate in 50 ml of warm distilled water. Filter the solution and pour 50 ml concentrated HNO_3 (Sp. gr. 1.42) into the filtrate. Then slowly pour the resulting solution into 100 ml distilled water. Store it in a drop bottle. **Do not touch** with hand, it being corrosive.

(ii) **Magnesium reagent.** Dissolve 0.1 gram of Na_3PO_4 and 30 grams of NH_4Cl in 50 ml distilled water. Add 2.5 ml concentrated NH_4OH and dilute to 100 ml with distilled water. Store it in a drop bottle.

(iii) **Diphenylamine-concentrated H_2SO_4 reagent.** Dissolve 1 gram diphenylamine in 100 ml concentrated H_2SO_4; store in a drop bottle.

Dry Ashing. Take 50 gram of fresh plant material such as seeds, tubers, stems, or leaves in a pre-weighed container. Allow the material to dry out in an oven at 100–105 °C for 24 hours. Then weigh the container and the sample again, and calculate the dry weight of sample.

Grind the material and transfer it to a crucible. After noting the weight, put the crucible into a muffle furnace at 550 °C for 2 hours or until a constant weight is obtained. The constant weight indicates that there are no further reactions taking place and that all this material has turned to ash. After ashing, cool and reweigh the crucible and contents of the crucible.

Detection of macroelements in plant ash: Dissolve the ash in 4 ml of distilled water and use this solution in the following.

Phosphorus:

(a) Put a drop of 2 per cent KH_2PO_4 and a drop of ammonium molybdate reagent at some distance from the former on a clean slide. Draw one of the drops into the other with the edge of a coverslip and wait for several minutes. Observe the yellow colour and form of the crystals formed. Repeat the same procedure for one drop of ash solution. Compare the two and derive the conclusion.

(b) Put sections on a slide in a solution of ammonium molybdate in HNO_3 (1.0 gram ammonium molybdate in 12 ml HNO_3). Small yellow drops with a black border changing into crystals, show the presence of phosphorus. Wash in dilute HCl and add a drop of 1 per cent phenylhydrazine hydrochloride. A deep blue colour appears.

Potassium: Place a drop of K_2HPO_4 solution (5 per cent) on a slide and next to it a drop of 15 per cent $HClO_4$. Following the same method as for phosphorus, mix the drops and look for the formation of $KClO_4$ crystals. After noting and recording their characteristic shape and colour, perform the same test on a drop of the ash solution. Record the result.

Magnesium: Test a drop of 5 per cent $MgSO_4$ with a drop of magnesium reagent in the same manner as above and observe ammonium-magnesium-phosphate crystals. Test the ash solution and note the result.

Sulphur: Put a drop of 5 per cent K_2SO_4 and a drop of 10 per cent $BaCl_2$ on a slide. After mixing the drops, note the crystals of $BaSO_4$ which indicate the presence of sulphur. Repeat the test for the ash solution.

Calcium:

(a) Test a drop of 5 per cent $CaCl_2$ with a drop of 50 per cent H_2SO_4. Observe the formation of $CaSO_4$ crystals. Test the ash solution in the same way.

(b) Put sections into hot 3 per cent ammonium oxalate on a slide. Heat gently for a minute and wash with water. Calcium oxalate crystals appear if calcium is present.

Nitrate: To a clean glass spot slide, add 5 drops of the ash solution. Add a drop of 1 per cent diphenylamine-H_2SO_4 reagent and note the development of a blue colour which indicates the presence of nitrates.

Iron: Add 5 drops of the ash solution and 3 drops of 5 per cent KSCN to a clean spot slide. The formation of a red colour indicates the presence of iron.

Add a drop of 2 per cent $K_4Fe_6(CN)_6$ and a drop of 5 per cent HCl to a drop of liquid from the water-ash mixture on a slide. A blue colour indicates ferric iron.

Iron: Remove chlorophyll from a leaf with the aid of alcohol. Cut sections and place them on a slide in a 2 per cent solution of potassium ferrocyanide. After 15 minutes add a drop of HCl. Wash with distilled water after sometime. A dark blue colour indicates the presence of iron. Use a clean razor for sectioning and handle sections with glass.

Manganese: Place a drop of $MnCl_2$ solution in the centre of a small filter paper. Add a drop of 1 per cent KOH to the spot and a drop of benzidine reagent (Dissolve 0.05 g of benzidine, free base or hydrochloride in 10 ml acetic acid and add distilled water to make 100 ml of solution). A blue colour develops. Repeat the test with a drop of the water-ash mixture.

Project

To demonstrate hydroponics

Grow some seedlings of maize or oat for 10 or 12 days. Take seven large glass jars, each fitted with a split cork. Clean them thoroughly with hot water and finally rinse with distilled water. Prepare Sachs' or Knop's normal nutrient solution (Table 1) or Hoagland's solution (Table 2).

Table 1 Sach's and Knop's nutrient solutions.

Sach's solution (1860)		Knop's solution (1865)	
salts	Grams per litre	Salts	Grams per litre
KNO_3	1.00	$Ca(NO_3)_2$	0.8
$Ca_3(PO_4)_2$	0.50	KNO_3	0.2
$MgSO_4 \cdot 7H_2O$	0.50	KH_2PO_4	0.2
$CaSO_4$	0.50	$MgSO_4 \cdot 7H_2O$	0.2
NaC1	0.25	$FePO_4$	0.1
$FeSO_4$	Trace		

Table 2 Hoagland's Solution modified by Arnon, 1940.

Salts	Grams per litre	Salts	Grams per litre
KNO_3	1.02	$CuSO_4 \cdot 5H_2O$	0.08
$Ca(NO_3)_2$	0.49	$ZnSO_4 \cdot 7H_2O$	0.22
$NH_4H_2PO_4$	0.23	$H_2MoO_4 \cdot H_2O$	0.09
$MgSO_4 \cdot 7H_2O$	0.49	$FeSO_4 \cdot 7H_2O$ 0.5%	0.6 ml
H_3BO_3	2.86	Tartaric acid 0.4%	0.6 ml
$MnCl_2 \cdot 4H_2O$	1.81		

Fill one jar with this solution, the second with the nutrient solution minus nitrate, the third containing no phosphate, the fourth containing no potassium salt, the fifth containing no calcium salt, the sixth containing no iron and the seventh containing no magnesium salt.

Select seedlings which are nearly of the same size. Fix a few seedlings between each of the split corks and fit the corks into the jar, so that roots are in liquid (Figure E4.1).

The seedlings should be supported with the non-absorbent cotton wool packed into the hole. Place the plant in warm bright conditions. Wrap each jar with black paper to prevent the growth of algae. Renew the liquid every other day.

Compare the relative growth and general health of the plant, over a period of one month or more, e.g., leaf development, colour of the leaves, growth of the stem and flower and fruit formation.

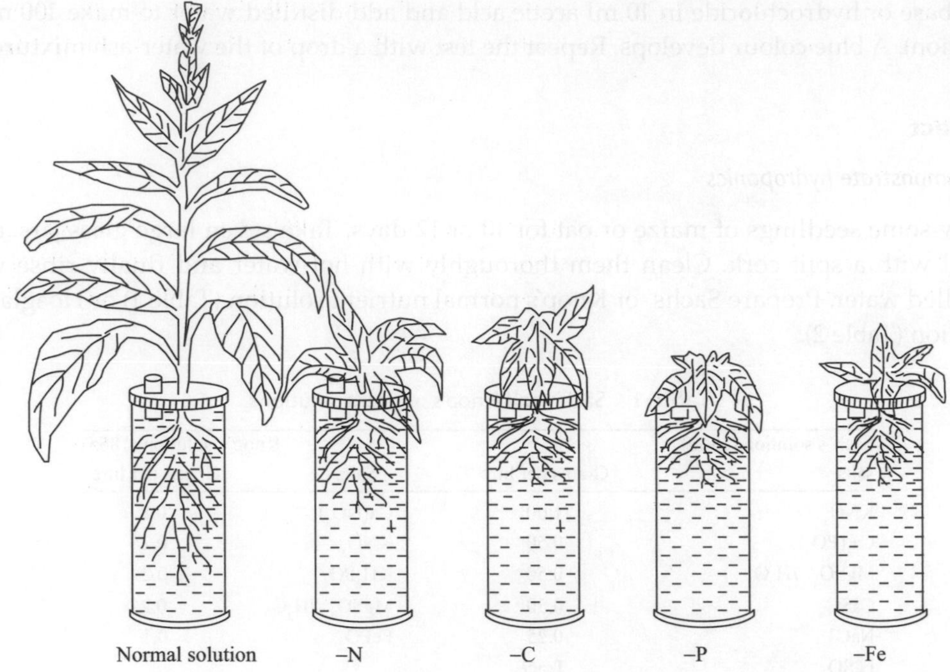

Fig. E4.1 The connection of nutrients with growth and health

The solution for the different jars may be prepared from Sachs' solution as under:

1. For calcium deficiency, use potassium chloride and sodium phosphate for sulphate and calcium phosphate, respectively.
2. For nitrogen deficiency, use potassium chloride for potassium nitrate.
3. For phosphorus deficiency, use calcium nitrate for calcium phosphate.
4. For potassium deficiency, use sodium nitrate for potassium nitrate.
5. For iron deficiency, omit ferrous sulphate.
6. For magnesium deficiency, use potassium sulphate for magnesium sulphate.

5. Enzymes

Experiment 5.1

To demonstrate the activity of the following enzymes—Urease, Invertase, Catalase, Peroxidase, Tyrosinase (Polyphenol oxidase), Succinate dehydrogenase, and Amylase.
Study the following properties of enzymes, using any one of them:

(i) Proteinaceous nature
(ii) Specificity
(iii) Thermolability
(iv) Effects of non-competitive inhibitors

Requirements

Urease
Plant material : Watermelon seeds
Chemicals : 2 per cent urea, distilled water, phenolphthalein

Invertase
Plant material : Yeast pellets
Chemicals : 1 per cent sucrose, Fehling's solution (A and B) or Benedict's reagent, distilled water

Catalase
Plant material : Potato tuber
Chemicals : 3 per cent H_2O_2, distilled water

Peroxidase
Plant material : Potato tuber
Chemicals : 3 per cent H_2O_2, distilled water, 1 per cent phenol, oil

Tyrosinase/Polyphenol oxidase
Plant material : Potato tuber
Chemicals : distilled water, 1 per cent phenol, oil

Succinate dehydrogenase
Plant material : Potato tuber
Chemicals : 0.2 M sodium succinate, 0.2 M sodium azide 0.02 per cent DCPIP (2, 6-dichlorophenolindophenol)

Amylase
Plant material : Germinating wheat grains
Chemicals : 1 per cent starch, dilute iodine

Glass apparatus : Test tubes, pipettes (1 ml, 10 ml), measuring cylinder, beaker, pestle and mortar

Miscellaneous : Test tubes stand, test tube holder, water bath at 37 °C, spirit lamp, blotting paper, tripod stand, funnel, muslin cloth, marker pen

Details of location and nature of various enzymes have been listed in the following table.

Enzyme	Source		Location	Nature of enzyme
	Common name	Scientific name		
Urease	Watermelon (seeds)	*Citrullus lanatus*	Cytoplasm	Simple
Invertase*	Yeast (pellets)	*Saccharomyces cerevisiae*	Cytoplasm	Simple
Catalase	Potato (tuber)	*Solanum tuberosum*	Peroxisome	Conjugated (Fe-cofactor)
Peroxidase	"	"	"	"
Tyrosinase (Polyphenol oxidase)	"	"	"	Conjugated (Cu-cofactor)
Succinate dehydrogenase	"	"	Mitochondrial inner membrane	Conjugated (FAD as prosthetic group)
Amylase	Wheat (germinating grains)	*Triticum aestivum*	Aleurone layer of endosperm	Simple

*Alternative names for invertase - sucrase, β-fructosidase, saccharase.

Principle

Enzymes are proteins synthesized by living cells and are essential to all living organisms as they catalyze the biochemical reactions required to maintain the cell. An enzyme-catalyzed reaction is thought to proceed through the formation of any enzyme-substrate (ES) complex, the simplified equation being,

$$E + S \rightleftharpoons ES \longrightarrow E + P$$

where, E is the enzyme, S is the substrate and P represents the product of the reaction. Enzymes are active in extremely small amounts. Enzyme-catalyzed reactions are reversible and enzymes do not alter the equilibrium constant. A remarkable property of enzymes is their specificity, i.e., enzymes will usually catalyze a narrow range of reactions or in many cases just one reaction. The lock and key model proposed by Emil Fischer (1894) explains the specific nature of the enzymes. Because of their proteinaceous nature, enzymes may undergo denaturation by extreme temperature, pH, inhibitors, poisons, and a number of such factors.

Enzymes have catalytic properties because they lower the activation energy of a reaction. It is believed that the formation of an ES complex somehow strains the bonds of the substrate so that it becomes more reactive and gets converted to product.

Many enzymes are simple proteins and require no additional factors to show activity. Some enzymes, however, require the presence of additional factors for their activity (holoenzymes). The protein part is called apoenzyme and the non-protein part is called the cofactor. Cofactors may be loosely bound to the enzyme (coenzyme) or firmly bound (prosthetic group). In many cases, metallic cations are needed to activate the enzymes.

The activity of an enzyme may be assayed by measuring either (a) the increase of the concentration of the reaction products or (b) the decrease of the concentration of substrate. It is preferable, wherever possible, to measure the increase in the concentration of reaction products since the substrate may be broken down non-enzymatically to yield other products.

Procedure

Prepare the crude enzyme extracts (except amylase) from their respective sources.

Crush the plant materials using pestle and mortar in distilled water/buffer solution (pH 7.0) (1:5). Allow the contents to stand for 15 minutes and filter through muslin cloth. The filtrate is a crude extract of enzymes.

Amylase extract: Crush germinated wheat seeds along with a pinch of moist sand using distilled water (1:5). Then add 1 per cent NaCl to activate the enzyme. Allow the contents to stand for 15 minutes, followed by centrifugation at 5,000 rpm for 10 minutes.

5 g plant material + 25 ml of distilled water/buffer solution
↓
Crush using pestle and mortar
↓
Leave undisturbed for 15–20 min
↓
Filter through muslin cloth
↓
Crude enzyme extract

Observations

1. Enzyme : Urease (5 ml)[8]
   Substrate : 2 per cent urea (2 ml)
   Indicator : Phenolphthalein (4 drops)
   Reaction time : 10 min

$$\underset{\text{Urea}}{(H_2N)_2C{=}O} + H_2O \xrightarrow{\text{Urease}} CO_2 + 2NH_3 \xrightarrow{H_2O} \underset{\text{Ammonium hydroxide (alkaline)}}{NH_4OH}$$

[8] Amounts of enzyme/substrate/indicator as well as reaction time for all the experiments related to enzymes can be subjected to change based on the environmental/physiological factors.

S. No.	Nature of enzyme	Observations		Inference
		Initial	Final	
1.	Active	White	Pink	Phenolphthalein gives pink colour in the alkaline medium.
2.	Denatured (boiled)	White	White	No alkaline medium as NH_3 is not formed.
3.	Control* (distilled water)	Colourless	Colourless	Water does not react with substrate urea, so no reaction.

* Control = without enzyme (distilled water is used in place of enzyme).

2. Enzyme : Invertase (5 ml)
   Substrate : 1 per cent sucrose (2 ml)
   Indicator : Fehling's solution A and B (1 : 1) or Benedict's reagent (2 ml)
   Reaction time : 5 min

$$\underset{\text{Sucrose}}{C_{12}H_{22}O_{11}} + H_2O \xrightarrow{\text{Invertase}} \underbrace{\underset{\text{Glucose}}{C_6H_{12}O_6} + \underset{\text{Fructose}}{C_6H_{12}O_6}}_{\text{(Reducing sugars)}}$$

$$\underset{\text{Glucose}}{\begin{array}{c} CHO \\ | \\ H - C - OH \\ | \\ HO - C - H \\ | \\ H - C - OH \\ | \\ H - C - OH \\ | \\ CH_2OH \end{array}} + 2Cu^{2+} + 4OH^- \xrightarrow{\text{Boil}} \underset{\text{Gluconic acid}}{\begin{array}{c} COOH \\ | \\ H - C - OH \\ | \\ HO - C - H \\ | \\ H - C - OH \\ | \\ H - C - OH \\ | \\ CH_2OH \end{array}} + \underset{\substack{\text{Cuprous oxide} \\ \text{(brick red ppt.)}}}{Cu_2O} + 2H_2O$$

S. No.	Nature of enzyme	Observations		Inference
		Initial	Final	
1.	Active	Dark blue	Brick red precipitates on boiling for 5 min	Cuprous oxide gives brick red precipitates.
2.	Denatured	Dark blue	Blue	Cu_2O not formed, so no brick red precipitates.
3.	Control	Dark blue	Blue	No reaction due to the absence of enzyme.

3. Enzyme : Catalase (5 ml)
   Substrate : 3 per cent H_2O_2 (2 ml)
   Indicator : None
   Reaction time : 10 min

$$2H_2O_2 \xrightarrow{\text{Catalase}} 2H_2O + O_2\uparrow$$

Hydrogen peroxide — Oxygen

S. No.	Nature of enzyme	Observations		Inference
		Initial	Final	
1.	Active	No froth	Froth	Froth formation because of evolution of O_2 gas.
2.	Denatured	No froth	No froth	No froth because O_2 gas is not evolved.
3.	Control	No froth	No froth	No froth because reaction mixture lacks the enzyme.

4. Enzyme : Peroxidase (5 ml)
   Substrate : 3 per cent H_2O_2 (2 ml)
   Indicator : 1 per cent phenol (1 ml)
   Reaction time : 10 min

$$\text{Phenol} + H_2O_2 \xrightarrow{\text{Peroxidase}} \text{Quinone} + H_2O$$

S. No.	Nature of enzyme	Observations		Inference
		Initial	Final	
1.	Active (+ oil drop)	Colourless	Brown colour	Phenol is reduced to quinone, so brown colour of quinone formed.
2.	Denatured (+ oil drop)	Colourless	Colourless	No quinones formed, so no brown colour.
3.	Control (+ oil drop)	Colourless	Colourless	No enzyme, so no reaction.

[Oil drop is added to block the passage of atmospheric oxygen]

5. Enzyme : Tyrosinase/Polyphenol oxidase (5 ml)
   Substrate : Atmospheric O_2
   Indicator : 1 per cent phenol (1 ml)
   Reaction time : 10 min

$$\text{Phenol} + O_2 \xrightarrow{\text{Polyphenol oxidase}} \text{Quinone} + H_2O_2$$

S. No.	Nature of enzyme	Observations		Inference
		Initial	Final	
1.	Active	Colourless	Brown colour	Phenol is reduced to quinone, so brown colour of quinone obtained.
2.	Denatured	Colourless	Colourless	Quinones are not formed, so no change in colour.
3.	Control	Colourless	Colourless	Enzyme absent, so no reaction.
4.	Active (+ oil drop)	Colourless	Colourless	Oil blocks atmospheric O_2, so no reaction.

6. Enzyme : Succinate dehydrogenase (5 ml)
   Substrate : 0.2 M sodium succinate + 0.2 M sodium azide (1:1)
   Indicator : 2, 6-DCPIP (4 drops)
   Reaction time : 10 min

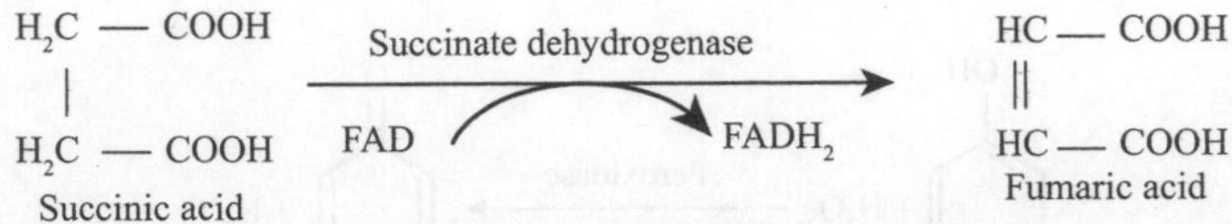

S. No.	Nature of enzyme	Observations		Inference
		Initial	Final	
1.	Active	Blue colour	Colourless	DCPIP is blue in the oxidized state but colourless in reduced form.
2.	Denatured	Blue colour	Blue colour	Denatured enzyme does not act on the substrate.
3.	Control	Blue colour	Blue colour	No enzyme, so no reaction.

7. Enzyme : Amylase (5 ml)
   Substrate : 1 per cent Starch (2 ml)
   Indicator : Iodine (4 drops)
   Reaction : 10 min

$$\text{Starch (glycogen)} \xrightarrow{\alpha\text{-amylase}} \text{Dextrins} \xrightarrow{\beta\text{-amylase}} \text{Maltose} \xrightarrow{\text{Maltase}} \text{Glucose}$$

S. No.	Nature of enzyme	Observations		Inference
		Initial	Final	
1.	Active	Blue colour	Blue colour disappears	Amylase converts starch to reducing sugars which do not give blue colour with iodine.
2.	Boiled (denatured)	Blue colour	Blue colour	Denatured enzyme does not act on starch.
3.	Control	Blue colour	Blue colour	No enzyme, so no reaction.

Result and Discussion

1. **Urease:** Urease converts urea to ammonia which combines with water to form alkaline ammonium hydroxide which thus gives a pink colour with the indicator phenolphthalein.
2. **Invertase:** Invertase, also known as saccharase, sucrase or β-fructosidase, catalyzes the hydrolysis of sucrose (a non-reducing sugar) to yield glucose and fructose (both reducing sugars).

   $$\text{Sucrose} + H_2O \rightarrow \text{Glucose} + \text{Fructose}$$

   The enzyme is present in the root and shoot systems of higher plants, as well as in microorganisms. The reaction catalyzed by invertase can be assayed by measuring the amounts of reducing sugars formed.
3. **Catalase:** The enzyme catalase will catalyze the decomposition of hydrogen peroxide:

   $$2H_2O_2 \rightarrow 2H_2O + O_2$$

   This is an oxidation–reduction reaction in which one molecule of hydrogen peroxide is oxidized to yield oxygen and the other molecule is reduced to form water. Catalase occurs widely in higher plant tissues. It detoxifies any hydrogen peroxide that may arise in plant cells.
4. **Peroxidase:** Peroxidase is an oxido-reductase and iron-containing porphyrin. It oxidises phenol to quinone, which is a brown coloured compound, in the presence of H_2O_2.
5. **Tyrosinase (Polyphenol oxidase):** Tyrosinase is a metallo-protein that contains copper. It oxidises colourless phenolic compounds to quinones, which polymerize further, and form tannins. Source of oxygen for this oxidation reaction is the atmosphere. If supply of oxygen from the atmosphere is cut off, the phenols would not be oxidised.
6. **Succinate dehydrogenase:** SDH is an oxido-reductase enzyme capable of oxidizing succinic acid to fumaric acid in the inner mitochondrial membrane. Flavin adenine dinucleotide (FAD) acts as a prosthetic group of the enzyme. Sodium succinate, substrate for SDH, gets oxidized and meanwhile DCPIP gets reduced (DCPIP is blue in the oxidised form and colourless in reduced form). Sodium azide is used to stop the transfer of electrons from succinic acid to oxygen *via* cytochromes.
7. **Amylase:** Amylase enzyme is connected with polysaccharide metabolism. α-amylase hydrolyses α-1,4 glycosidic bonds in polyglycans (amylose and amylopectin) and results in the appearance of dextrins, maltose and small amounts of glucose. β-amylase will also hydrolyse α-1,4 bonds. In contrast to the α-amylase, this enzyme cleaves the penultimate bond from the non-reducing end group of the substrate, thus releasing one maltose molecule after another. This continues until the enzyme comes to a branch point where its action ceases. A good source of the amylases is barley seeds. In ungerminated seeds, β-amylase is the major enzyme whilst in germinating seeds, α-amylase is predominant.

Properties of the Enzymes

1. **Proteinaceous nature:** To confirm the protein nature of any enzyme, perform the following tests as given in table below.

(a) Xanthoproteic test
(b) Biuret test
(c) Millon's test

Proteinaceous nature of enzyme (e.g. Urease)

S. No.	Procedure	Observations	Inference
1.	*Xanthoproteic test:* 2 ml of enzyme + 1 ml of conc. HNO_3. Boil, cool and add NH_4OH in excess	Colourless to yellow and then orange colour	Protein present
2.	*Biuret test:* 2 ml of enzyme + 2 ml of biuret reagent (1 ml of 1% $CuSO_4$ + 1 ml of 40% NaOH)	Bluish violet coloration	Protein confirmed
3.	*Millon's test:* 2 ml of enzyme + 2 ml of Millon's reagent. Boil, cool and add $NaNO_2$	Pink flocculation	Protein confirmed

2. **Specificity:** Enzymes are highly specific in their action. Specificity is of two types - *substrate specificity* and *group specificity*. In substrate specificity, enzymes are so specific for their substrates that they will not even recognize their geometric stereoisomer. In group specificity, enzymes accept any of whole group of substrates as long as they possess some common structural feature.

Specificity of enzyme (Urease)
(Urease enzyme is specific for urea substrate only)

S. No.	Enzyme (5 ml)	Substrate (2 ml)	Indicator	Observations		Inference
				Initial	Final	
1.	Urease	2% Urea	Phenolphthalein (4 drops)	White colour	Pink colour	Phenolphthalein gives pink colour in the alkaline medium.
2.	Urease	3% H_2O_2	–	White colour	No froth	Urease does not react with H_2O_2, so no O_2 evolution and no froth.
3.	Urease	1% Sucrose	Fehling's solution (A + B) (2 ml)	Blue colour	On boiling, no brick red ppt	Urease does not react with sucrose, so no reducing sugars formed and that is why no brick red ppt.

3. **Thermolability:** Enzymes are highly sensitive to change in temperature. Gradual increases in temperature increase the kinetic energy of both enzyme and substrate molecules resulting in more frequent collisions, thereby increasing the likelihood of correct substrate binding. However, further increases in temperature are counterproductive because the enzyme molecules begin to denature. Hydrogen bonds

break, hydrophobic interactions change and the structural integrity of the active site is disrupted causing a loss of activity.

Thermolability of enzyme (Urease)
(Urease enzyme shows maximum activity at 35 °C)

S. No.	Enzyme (5 ml)	Substrate (2 ml)	Indicator (4 drops)	Temp. (°C)	Observation		Inference
					Initial	Final	
1.	Urease	2% urea	Phenolphthalein	0	White colour	White colour	Enzyme gets inactivated at low temperature.
2.	Urease	2% urea	Phenolphthalein	35	White colour	Pink colour	Optimum temperature for enzyme.
3.	Urease	2% urea	Phenolphthalein	80	White colour	White colour	Enzyme gets denatured at high temperature.

4. **Effect of inhibitors on enzyme activity:** Inhibitors affect the enzyme activity, thereby reducing (or sometimes even inhibiting) the reaction rate with the desired substrate.

    Inhibitors may be reversible or irreversible. The two most common forms of reversible inhibitors are called competitive inhibitors and noncompetitive inhibitors. The former bind to the active site of the enzyme whereas the latter bind to the enzyme surface at a location other than the active site. An irreversible inhibitor binds to the enzyme covalently, causing an irrevocable loss of catalytic activity.

Effect of non-competitive inhibitors/heavy metals on enzyme (Urease)

S. No.	Enzyme (5 ml)	Inhibitor (1 ml)	Substrate (2 ml)	Indicator (4 drops)	Observation		Inference
					Initial	Final	
1.	Urease	1% $HgCl_2$	2% urea	Phenolphthalein	White colour	White colour	Inhibitors bind to enzyme at site other than the active site and alter its structure, so no reaction.
2.	Urease	1% $AgNO_3$	2% urea	Phenolphthalein	White colour	White colour	
3.	Urease (control)	Distilled water (to equalize the volume)	2% urea	Phenolphthalein	White colour	Pink colour	In the absence of inhibitor, urease acts on urea.

Precautions

1. Glass apparatus must be scrupulously clean.
2. Distilled water should be used for rinsing glass apparatus and for making up all reaction solutions and reagents.
3. Enzyme and substrate should be kept in water bath at 35 °C prior to and after mixing, if the room temperature is low.
4. The amount of substrate, enzyme and the indicator should be kept constant.
5. Enzyme extract should be freshly prepared.
6. Before use, the enzyme extract should be boiled for 5–7 minutes to denature it completely, followed by cooling at room temperature before being added to substrate.
7. Reaction time for enzyme action should be equal.

Care must be taken to keep enzymes ice-cold before use, as in the isolated form they are unstable.

Experiment 5.2

To study the effect of temperature on the enzyme activity of Urease.

Requirements

Plant material : Watermelon seeds (*Citrullus lanatus*)
Family Cucurbitaceae
Chemicals : 2 per cent urea (substrate), phenolphthalein (indicator), 0.02N HCl, distilled water.
Glassware : Test tubes, beaker, pipettes, burette, conical flask, boiling tube, white tile, funnel, thermometers
Miscellaneous : Test tube stand, burette stand, pestle and mortar, water baths (35, 45, 65, 70, 80 °C), muslin cloth, graph paper, tripod stand.

Principle

All enzyme catalyzed reactions are influenced by temperature due to their proteinaceous nature. Increase in temperature results in an increase of reaction rate because of (i) increase in kinetic energy of the interacting molecules; (ii) increase in frequency of collision among the enzyme and substrate molecules. At elevated temperature, these collisions become strong enough to overcome the attraction of hydrogen bonds which are relatively weak. As the hydrogen bonds are broken, enzymes gradually unfold and lose their 3D structure as well as their secondary and tertiary structures. This collapse of protein structure is known as **denaturation**. Till about 40 °C, the temperature coefficient of enzyme-catalyzed reactions is usually higher (Q_{10}= 2), i.e., for every 10 °C rise in temperature; the rate of reaction becomes double.

However, the accelerating effect of temperature increase is countered above 50 °C by denaturing effects on the enzyme protein. However, some of the bacteria living in hot springs are active at even 85 °C.

Procedure

Prepare the crude extract of urease enzyme. To 5 ml of enzyme in a test tube, add 2 drops of phenolphthalein and keep in water bath at 35 °C. Keep another test tube containg 3 ml of 2 per cent urea in the same water bath. When the contents of the two tubes attain the temperature of the water bath (to be taken as zero time), mix them up and keep for 10 minutes at the same temperature for the reaction to proceed. Transfer the contents to a conical flask and titrate against 0.02N HCl till the pink colour disappears. Record the initial and final volume of 0.02N HCl in the burette.

Repeat the same experiment at 45, 65, 70 and 80 °C along with control (at room temperature). Plot a graph of the rate of reaction (volume of HCl used) against temperature.

Observation table

Amount of enzyme : 5 ml   Amount of substrate : 3 ml
Amount of indicator : 2 drops   Reaction time : 10 min

S. No.	Temperature (°C)	0.02N HCl in burette		Vol. of HCl used (ml)
		Initial reading (ml)	Final reading (ml)	
1.	25 (Room temp.)	0	12.5	12.5
2.	35	0	17.6	17.6
3.	45	0	15.0	15.0
4.	65	0	11.5	11.5
5.	70	0	8.0	8.0
6.	80	0	0.0	0.0

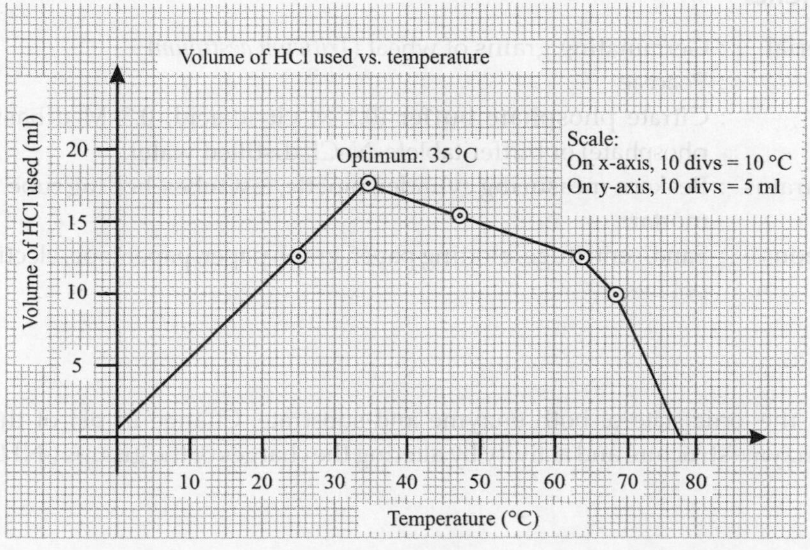

Result and Discussion

From the bell-shaped graph, it is observed that the activity of enzyme is optimum at 35 °C, i.e., when the enzyme is more active, a greater amount of urea is converted to NH_3 which in turn generates more NH_4OH. Thus the pink colour is more intense using phenolphthalein, indicating greater alkalinity. And so, more HCl is required to neutralize this alkalinity. Gradual increase in the temperature increases the kinetic energy of both enzyme and the substrate molecules. Beyond 35 °C, the reaction rate declines and ranging up to 80 °C, it counteracts the positive effect of increased temperature.

Precautions

1. All the glass apparatus should be thoroughly clean.
2. Equal amounts of substrate, enzyme and indicator should be used in all the temperature treatments.
3. The measurements should be accurately taken.
4. The reaction time should be constant.
5. Lower meniscus should be noted down in the burette.
6. Mixing the enzyme and the substrate should be done when both of them are at same temperature.
7. The burette containing HCl should be vertical.
8. The colour in the conical flask should be examined/measured against a white background.
9. A pipette used for any particular solution should never be used for any other.

Experiment 5.3

To study the effect of pH on enzyme activity (Amylase) by cavity tile method.

Requirements

Plant material : Germinating grains of wheat (*Triticum aestivum*)
Family : Poaceae
Chemicals : Citrate phosphate buffer (0.1 M citric acid, 0.3 M dibasic sodium phosphate) or buffer tablets, NaCl, distilled water.
Glass apparatus : Beakers, measuring cylinder, pipette, test tubes, boiling tube, pestle and mortar.
Miscellaneous : Sand, centrifuge, blotting paper, graph paper, paper tape, electric balance, pH meter.

Principle

pH affects the ionic state of both enzyme and substrate. Change in pH of the medium modifies enzyme structure and activity by altering the charges of the groups ($-NH_2$, -COOH) carried on amino acid side chains. These groups are capable of releasing and accepting protons and getting converted to ionized or charged forms.

$$NH_2 \xrightarrow{+H^+} NH_3^+$$

$$COOH \xrightarrow{-H^+} COO^-$$

Each of these groups, depending upon their location in a protein molecule, undergoes conversion from an uncharged to a charged form at a particular pH. Changes of this type affect the charge of the groups holding an enzyme in their final 3D shape and also the activity of the charged groups that may have been introduced into the active site to fit the reactants, products or the transitional state of reaction catalyzed by the enzyme. Each enzyme has got a fixed optimum pH at which charge taken by its amino acid side group is exactly 'correct' and the enzyme is most efficient in speeding the rate of its specific biochemical reaction. On either side of this optimum pH, charges of these groups change; the resulting alterations in the folding pattern of enzyme and the charge of active site causes the rate of reaction to drop off.

$$Enz^- + SH^+ \xrightarrow{\text{optimum pH}} Enz-SH \text{ complex}$$

At acidic pH, enzyme gets protonated and therefore, substrate cannot fit properly

$$Enz^- + H^+ \xrightarrow{\text{low pH}} Enz-H$$

At alkaline pH, substrate ionizes and loses its positive charge

$$SH^+ \xrightarrow{\text{high pH}} S + H^+$$

Since the only forms which interact are Enz^- and SH^+, extreme pH on either side will lower the effective concentration of Enz^- and SH^+, thus lowering the reaction velocity.

Procedure

Prepare citrate-phosphate buffers of different pH values as shown in buffer preparation table.

Using germinating wheat grains, prepare a crude extract of amylase enzyme. Add 2 ml of enzyme in one test tube and in another, 1 ml of starch + 3 ml buffer (pH 3.8). Maintain at 35 °C in a water bath and mix up the contents. Add 8 drops of dilute iodine in each cavity of the cavity tile. Then, in each cavity, add 2–3 drops of reaction mixture at regular intervals of time (30 seconds) and observe the change in colour till there is no further change in the colour of iodine. This is called the **achromic point**.

Repeat the process for all other buffer solutions.

Buffer preparation table

A = 0.1 M stock solution of citric acid

B = 0.3 M stock solution of dibasic sodium phosphate

Citrate phosphate buffer = X ml A + Y ml B. Add equal amount of distilled water.

S. No.	pH	X ml of A solution	Y ml of B solution	Distilled water (ml)
1.	3.8	3.23	1.77	5
2.	4.2	2.94	2.06	5
3.	4.6	2.67	2.33	5
4.	5.0	2.43	2.57	5
5.	5.4	2.22	2.78	5
6.	5.8	1.97	3.03	5

Prepare at least 10 ml of a few buffer solutions of different pH values.

Observation table

Amount of enzyme = 2 ml
Amount of starch = 1 ml
Amount of buffer = 3 ml
Amount of iodine = 8 drops

S. No.	pH	Time for decolourisation, t (min)	$x = 1/t$ (min^{-1})
1.	3.8	5	0.20
2.	4.2	4	0.25
3.	4.6	6	0.17
4.	5.0	8	0.12
5.	5.4	12	0.08
6.	5.8	18	0.06

Result and Discussion

From the graph, it is observed that the enzyme shows maximum activity at a pH of 4.2, i.e., the achromic point is 4.2. At this achromic point, when iodine is added to the mixture of enzyme and substrate, it does not turn blue in colour as the substrate is completely digested. Every enzyme has a particular pH at which the activity is maximum, above and below which its activity decreases, resulting in a bell-shaped curve when activity is plotted. The lesser the time required to reach its achromic point, the more active is the enzyme. Hence, the effect of pH on enzyme activity is inversely proportional to the time taken to reach the achromic point.

Precautions

1. Enzyme and buffer solution should be freshly prepared.

2. Iodine should be diluted 50 times.
3. Quantity of enzyme, substrate and iodine should be same in all the cases.
4. Temperature of enzyme and substrate should be same before mixing.

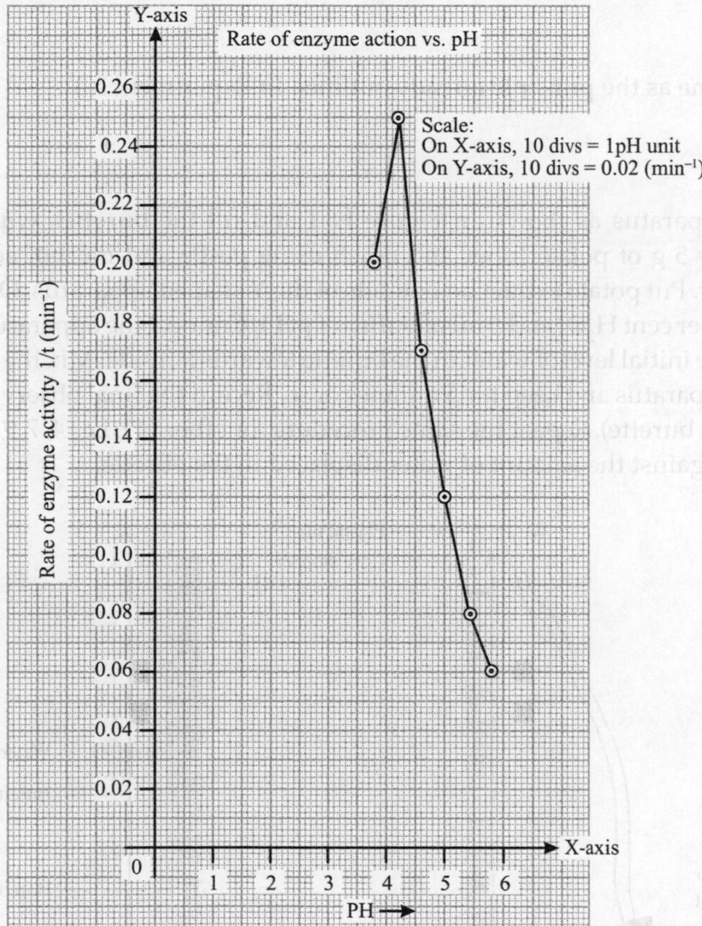

To obtain the enzyme extract, the mixture should be centrifuged at 5000 rpm for 10 minutes to remove the starch (as wheat grains contain both enzyme and starch)

Experiment 5.4

To study the effect of pH on catalase enzyme, using Y-shaped apparatus.

Requirements

Plant material : Potato tuber (*Solanum tuberosum*)
Family : Solanaceae

Chemicals : 3 per cent H_2O_2, buffer solutions (of pH 1, 4, 7, 9, 12), distilled water.
Glass apparatus : Y-shaped apparatus, burettes (2), beaker, measuring cylinder.
Miscellaneous : Clamp stands (2)

Principle

This is the same as the principle already outlined in Experiment 5.3.

Procedure

Set up the apparatus as shown in Figure E5.1 and fill the burettes with water. Weigh approximately 5 g of potato tuber and crush using pestle and mortar, adding 1–2 ml of distilled water. Put potato extract on one side of the Y-shaped apparatus. On the other side, add 5 ml of 3 per cent H_2O_2 and 5 ml of buffer of pH 1. Connect the apparatus to the burette. Note down the initial level of water in the burette towards left side. Mix the contents of both the sides of apparatus and keep for 5 minutes or so. Record the final observation (final level of water in the burette). Repeat the same procedure for other pH, i.e., 4, 7, 9, 12. Plot a graph between pH against the amount of water displaced in the burette.

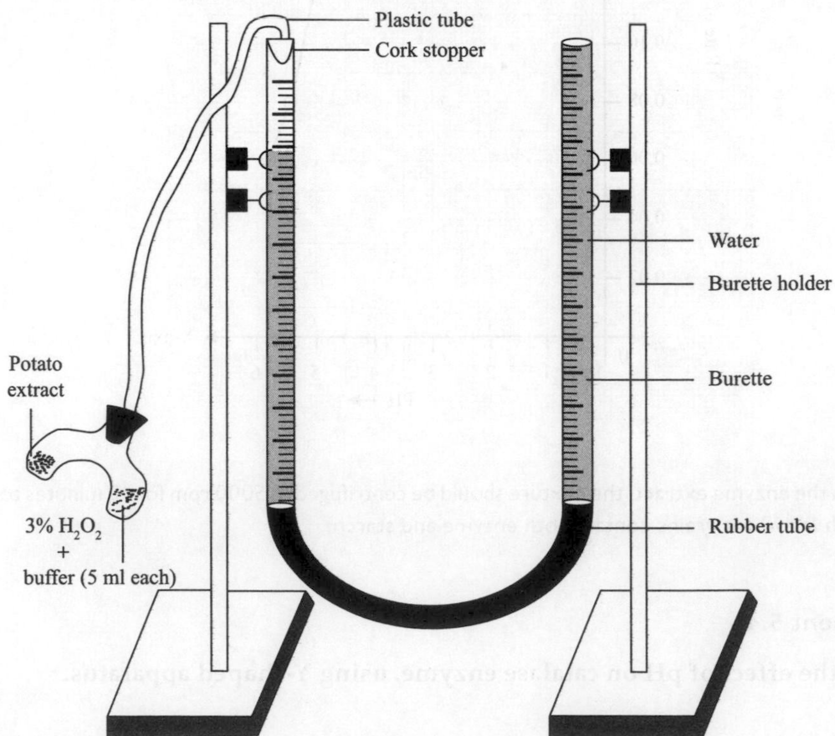

Fig. E5.1 Estimation of catalase activity by Y-shaped apparatus.

Observation table

S. No	pH of buffer solution (5 ml)	Substrate (5 ml)	Enzyme extract (5 g)	Burette readings		Volume of water displaced (ml)
				Initial (ml)	Final (ml)	
1.	1	3% H_2O_2	Potato tuber	0	2.0	2
2.	4	3% H_2O_2	"	5	8.0	3.0
3.	7	3% H_2O_2	"	10	20.0	10.0
4.	9	3% H_2O_2	"	2	10.5	8.5
5.	12	3% H_2O_2	"	12	16.5	4.5

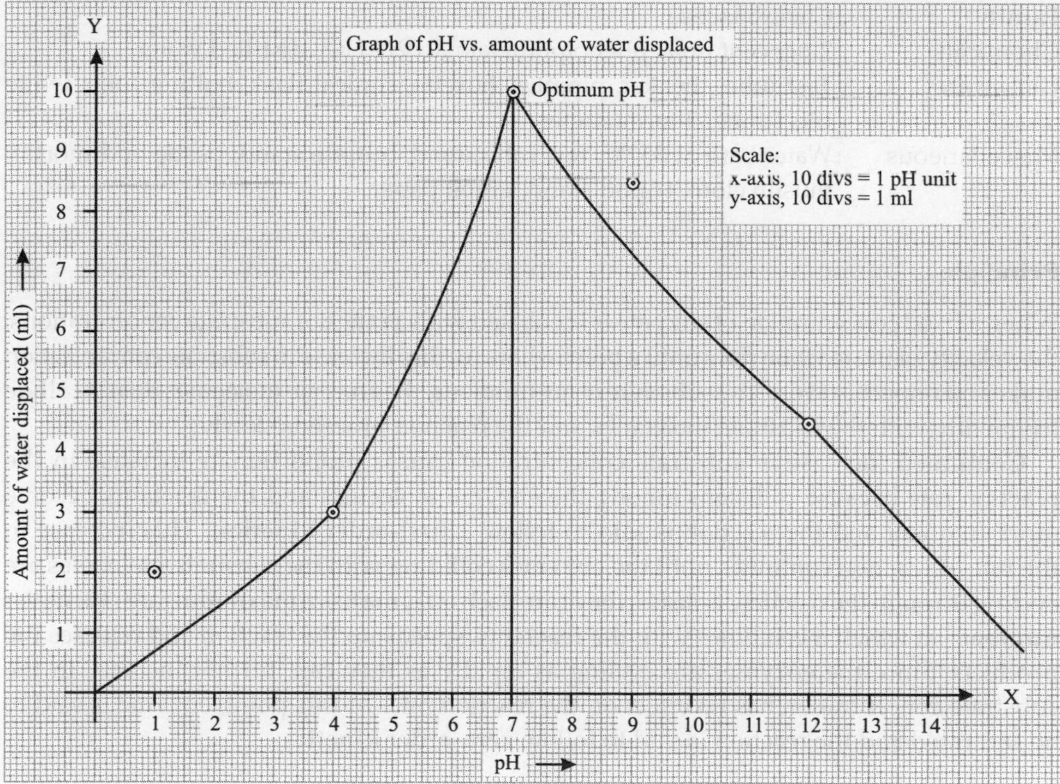

Result and Discussion

The amount of water displaced in the burette indicates the amount of O_2 evolved which in turn depends upon the catalase activity. In this experiment, the maximum activity of enzyme was recorded at pH 7.0, where water displaced was 10.0 ml. Enzyme activity showed a decline above and below pH 7.0. So the graph obtained was bell-shaped.

Precautions

1. Enzyme extract and buffer solution should be freshly prepared.
2. The initial reading should be noted carefully.
3. Reaction time should be constant in all the buffer solutions.
4. Note the initial level of water in burette after closing the mouth of the Y-shaped apparatus with stopper but before mixing of contents.

Experiment 5.5

To study the effect of substrate concentration on enzyme activity

Requirements

Plant material : Watermelon seeds (*Citrullus lanatus*)
Family : Cucurbitaceae
Chemicals : 2 per cent urea, phenolphthalein, 0.02N HCl, distilled water
Glassware : Test tubes, beaker, pipettes, burette, conical flask, boiling tube, white tile, funnel, thermometer
Miscellaneous : Water bath at 35 °C, test tube stand, burette stand, pestle and mortar, muslin cloth, tripod stand, graph paper.

Principle

When a fixed concentration of enzyme is incubated with a range of substrate concentrations, the velocity of the reaction rises until a point where further increase in substrate concentration makes no impact. At very low substrate concentrations, the enzyme molecules do not collide frequently with the substrate molecules and the reaction proceeds slowly.

Continued increase in substrate concentration causes proportionate increase in the reaction rate because collisions among the enzyme molecules and the reactants become more frequent and the reaction is said to follow a **first-order kinetics**.[9] This increase in the rate of reaction continues until the enzymes colliding with the substrate molecules are cycling through the reaction as fast as they can. Each enzyme has a maximum rate - **turnover number** - at which it combines with the reactants, speeds up the reaction and releases the products. Since the enzymes are cycling as rapidly as they can, further increase in the substrate concentration has no effect on the reaction rate. At this point, enzyme is said to be saturated and the reaction is said to follow **zero-order kinetics**.

Procedure

Prepare the crude urease enzyme extract from watermelon seeds. Using a 2 per cent urea (stock solution) and distilled water, prepare a range of substrate concentrations (0.1, 0.2, 0.4, 0.8 and 1 per cent). Keep six sets of two test tubes each in water bath at 35 °C.

[9] The study of the rates of conversion of substrates to products is called 'Kinetics'.

Test tube A: 4 ml substrate + 1 drop of phenolphthalein
Test tube B: 1 ml enzyme

After the temperature in both the tubes have stabilised, mix the contents and record the time as zero time. Keep the tubes back in the water bath for another 10 minutes and titrate the pink coloured reaction mixture against 0.02N HCl in burette. Disappearance of the pink colour indicates the end point. Note down the volume of HCl used in all the sets.

Observation table

Amount of enzyme = 1 ml
Amount of substrate = 4 ml
Amount of indicator = 1 drop
Reaction time = 10 min

S. No.	Conc. of substrate, i.e., urea, %	Burette readings		Amount of 0.02N HCl used (ml)
		Initial reading (ml)	Final reading (ml)	
1.	0.1	0	3.4	3.4
2.	0.2	0	6.4	6.4
3.	0.4	0	14.7	14.7
4.	0.8	0	27.9	27.9
5.	1	0	34.7	34.7
6.	2	0	36.4	36.4

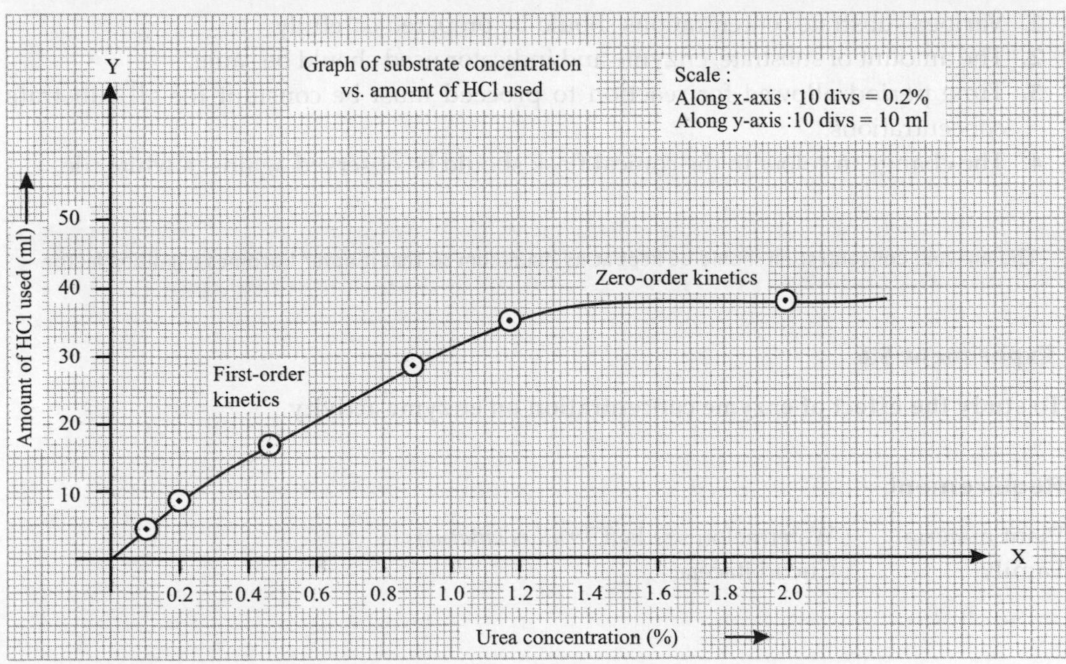

Result and Discussion

The reaction of urea with urease, using phenolphthalein as an indicator will produce a pink colour, which is neutralised by titrating against 0.02N HCl. The amount of HCl used in neutralising indicates the activity of an enzyme.

As the concentration of substrate increases, the volume of HCl used for reaching the end point also increases, i.e., more of pink colour or more of NH_3 is released. So the enzyme activity increases proportionately with increasing substrate concentrations (first-order kinetics). This happens till 1 per cent urea concentration is reached. A further increase in urea concentration to 2 per cent brings about nearly nil increase in enzyme activity which, thus, becomes constant and independent of substrate concentration (zero-order kinetics). Because all the active sites of enzyme molecules have been saturated and so the addition of substrate does not affect enzyme activity. Rather than an increase in enzyme activity, a higher concentration of substrate may retard activity due to accumulation of end products.

The hyperbolic graph thus obtained, follows the Michaelis–Menten equation.

$$V_0 = \frac{V_{max}[S]}{K_m + [S]}$$

The limiting velocity is called the maximum velocity (V_{max}). The substrate concentration required to give half the maximum velocity ($V_{max}/2$) is known as **Michaelis constant** (K_m). This is a measure of the affinity of the enzyme for the substrate. The smaller the value of K_m, the greater is this affinity.

Precautions

1. Various substrate concentrations should be prepared carefully.
2. The amount of substrate, enzyme and indicator used should be same.
3. Time period allowed for reaction to proceed must be constant for all substrate concentrations.
4. The change in colour in the conical flask should be observed against a white tile.

The enzyme concentration should be kept at its maximum so that it never becomes a limiting factor.

Experiment 5.6

To study the effect of enzyme concentration on enzyme activity

Requirement

Plant material : Watermelon seeds (*Citrullus lanatus*)
Family : Cucurbitaceae

Chemicals : 2 per cent urea, phenolphthalein, 0.02N HCl, distilled water.

Glass apparatus : Test tubes, beaker, pipettes, burette, conical flask, boiling tube, white tile, funnel

Miscellaneous : Test tube stand, burette stand, pestle and mortar, muslin cloth, graph paper, tripod stand, water bath (35 °C)

Principle

There are two general mechanisms involved in enzyme catalysis. First, the presence of enzyme increases the likelihood that the potentially reacting molecular species will encounter each other with the required orientations in space. This occurs because the enzyme has high affinity for the substrates and temporarily binds with them. Second, the formation of temporary bonds (non-covalent) between the enzyme and substrate forces a redistribution of electrons within the substrate molecules, which imposes a strain upon specific covalent bonds within the substrate molecules, culminating in bond breakage (**substrate activation**). The net effect of this is to greatly increase the percentage of molecules in the population which are sufficiently reactive towards each other.

> The enzyme greatly increases the rate at which equilibrium is achieved but does not alter the respective equilibrium concentrations of the various molecular species.

An increase in enzyme concentration is expected to increase the reaction rate. However, this increase is only till a certain level after which the reaction rate becomes constant. Because when all the active sites of the enzyme are occupied by the substrate molecules, the substrate concentration becomes limiting. There is no further increase in reaction rate with increase in enzyme concentration because of unoccupied active sites of the enzyme molecules.

The effect of enzyme on reaction rate can be estimated by considering the turnover number of an enzyme.

Procedure

Prepare a crude extract of urease enzyme (taken as 1 per cent). Dilute it to 0.1, 0.2, 0.4, 0.6 and 0.8 per cent by using pipettes for measuring out accurate amounts of enzyme and distilled water.

Keep the test tubes containing 5 ml of substrate + 2 drops of phenolphthalein and 5 ml of enzyme separately in water bath at 35 °C. Repeat the same for other concentrations of enzymes as well. After attainment of water bath temperature, mix the contents of the two tubes (zero time) and again keep at 35 °C for 10 minutes. Immediately transfer the contents into a conical flask, titrate against 0.02N HCl in burette and mark the end point that is signaled by disappearance of pink colour. Record the initial and final readings in the burette. Plot the graph between enzyme concentration vs. rate of reaction (volume of HCl used in the burette).

Observation table

Amount of enzyme = 5 ml
Amount of substrate = 5 ml
Amount of indicator = 2 drops
Stock solution of enzyme = 1 per cent (hypothetical)
Reaction time = 10 min

S. No.	Concentration of enzyme (%)	0.02N HCl in burette		Vol. of HCl used (ml)
		Initial reading (ml)	Final reading (ml)	
1.	0.1	1.0	2.1	1.1
2.	0.2	2.1	4.3	2.2
3.	0.4	3.0	7.3	4.3
4.	0.6	7.3	12.5	5.2
5.	0.8	12.1	18.1	6.0
6.	1.0	0	6.1	6.1

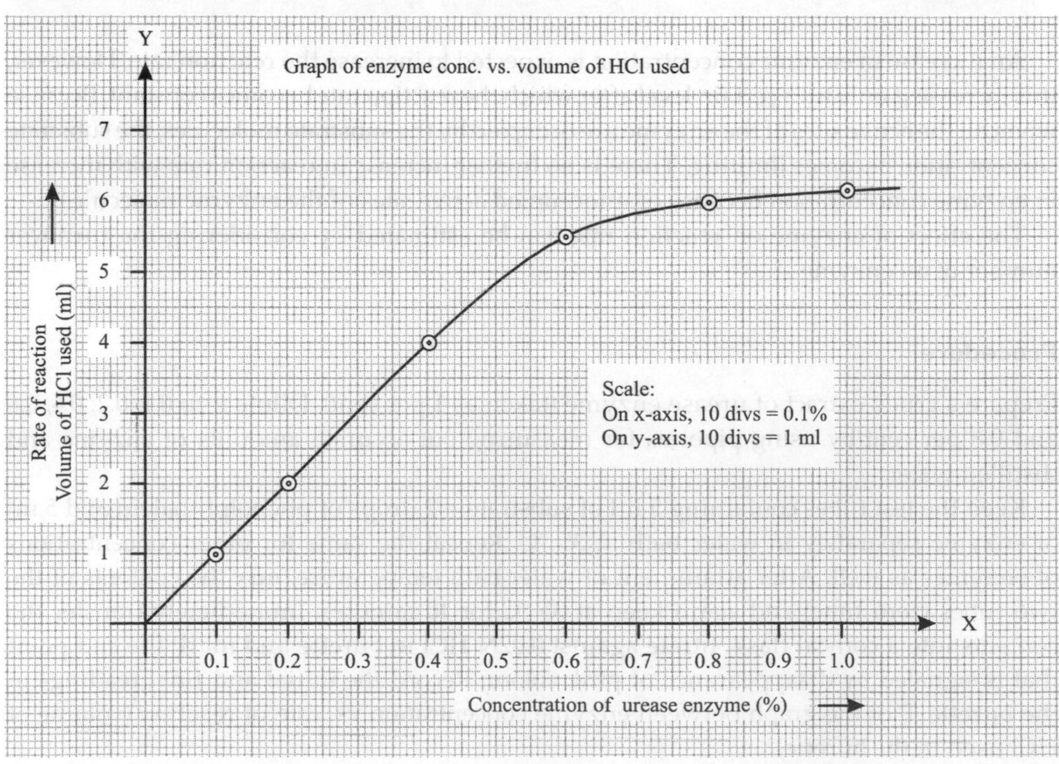

Result and Discussion

With an increase in enzyme concentration from 0.1 to 0.8 per cent the volume of HCl used for titration increases correspondingly (volume of HCl used is an indicator of enzyme activity). The more the enzyme concentration, the greater is the amount of NH_3 produced and more of HCl is required to neutralize it. However, beyond 0.8 per cent the enzyme activity becomes stabilised.

Thus, the graph obtained is regular hyperbola. Initially, reaction rate is directly proportional to the concentration of enzyme (**first-order kinetics**) followed by stabilization of reaction rate (**zero-order kinetics**), i.e., reaction rate is independent of enzyme concentration, and indicative of the fact that the enzyme is working at maximum efficiency.

Precautions

1. The glassware should be cleaned thoroughly.
2. The amount of substrate, enzyme and indicator used should be same.
3. Before mixing, the enzyme and substrate should be maintained at the same temperature.
4. Reaction time should be constant.
5. There should be no mixing of pipettes.
6. The change in colour in the conical flask (end point) should be carefully observed against a white background.
7. The burette should be in a vertical position.
8. The substrate concentration should always be kept in excess.

Questions for Viva Voce

1. Why are enzymes specific in nature?
2. Why do enzymes show very little activity at very low or very high temperatures or pH values?
3. How do enzymes bring about catalytic functions?
4. Cite examples of competitive and non-competitive inhibition.
5. How does catalase differ from peroxidase in its function? Where are they located in the cell?
6. What does effervescence produced during catalase activity indicate?
7. Where do you find succinate dehydrogenase in the cell? Name the coenzyme associated with it.
8. Why do we use phenolphthalein or lime water to find out urease activity?
9. Why can't we use phenolphthalein for finding out the optimal pH for urease enzyme?
10. Where and why do we use 2, 6-dichlorophenolindophenol (DCPIP)?
11. Define (i) turnover number of enzyme (ii) K_m.
12. What is the role of catalase in plant tissues?
13. Does H_2O_2 occur naturally in plant tissues? How is it formed?
14. Is there any similarity between the response of catalase activity to pH and the effect of pH on proteins? Explain.

15. **What are the products formed by the enzymatic hydrolysis of sucrose by invertase? Write the reaction.**

16. **Explain the chemical reaction involved during the treatment of reducing sugars with Benedict's/Fehling's reagent.**

17. **How is sucrose synthesized in plants?**

18. **Write the chemical equation showing complete hydrolysis of starch to glucose by amylases.**

19. **Why should the potato extract be used immediately after extraction?**

6. Photosynthesis

Experiment 6.1

To separate and identify chloroplast pigments by paper chromatography.

Requirements

Plant material : Fresh leaves of spinach (*Spinacea oleracea*) *or Amaranthus or Chenopodium*
Chemicals : 80 per cent acetone, petroleum ether, acetone/benzene
Glass apparatus : Capillary tube, funnel, chromatography jar or 100 ml measuring cylinder
Miscellaneous : Pestle and mortar, Whatman paper, hair dryer, ruler, muslin cloth, sand

Principle

The separation of solutes is based on **liquid-liquid partitioning** in the paper chromatography technique. It is based on the differences in solute partition between two immiscible phases - a stationary aqueous phase tightly bound to the cellulose fibres of the paper and a mobile organic phase passing through the paper by capillary action. The more soluble a solute is in the mobile phase, the farther the molecules will travel along the paper. The migration rate of a substance may be expressed according to its R_f value.

$$R_f = \frac{\text{Distance travelled by solute}}{\text{Distance travelled by solvent front}}$$

Procedure

Grind 5 grams freshly collected leaves[10] of spinach with a little sand and 20 ml of 80 per cent acetone. Filter through 3 or 4 folds of muslin cloth to get a clear extract. Pour 10 ml of organic solvent (Petroleum ether: acetone/benzene, 9:1) in a clean, dry chromatography jar and allow the internal atmosphere to equilibrate for 10 minutes. Cut the chromatographic paper to a desired size and draw a line with pencil about 2 cm away from the bottom. Load the pigment extract with the help of a capillary jet and let it dry completely. Repeat the application of extract until the spot is dark-green in colour (15–20 times). The spot should be as small as possible for better resolution. Suspend the chromatographic paper in organic solvent in the jar and cover with lid (Figure E6.1A). Remove the paper after the solvent moves almost to the other end.

Mark the solvent front and dry the chromatogram.

Result and Discussion

Observe the separation of pigments on the paper. The pigments are arranged in the following sequence from top (solvent front) to bottom: Orange-yellow carotenes, one or more yellow bands of xanthophylls, blue-green chlorophyll *a* and yellow-green chlorophyll *b*.

[10] For better resolution of the pigments, an extract of dried leaves (in powder form) can be used for the experiment.

Mark the spots with a pencil since the colours will fade away quickly. Calculate the R_f value of each pigment and identify by comparing with those of original values.

The mobility of pigments on the paper depends upon their degree of solubility in the solvent system (the mobile phase) and the affinity to the chromatographic paper (stationary phase) which is made of pure cellulose fibre.

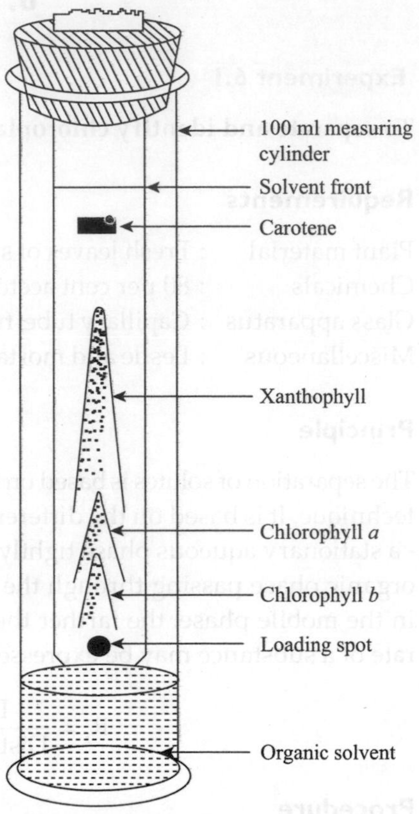

Fig. E6.1A Separation of leaf pigments by paper chromatography.

Precautions

1. The size of the chromatography paper should be adjusted according to the size of the jar.
2. Fine capillary should be used for loading the leaf extract.
3. Cold air should be used for drying the spot at the time of loading.
4. Jar should be completely saturated with the vapours of the organic solvent before starting the experiment for efficient capillary action of the solvent.
5. The chromatography paper should be suspended vertically. It should neither touch the side walls nor the bottom of the jar.
6. Mark the solvent front when the paper is already wet.
7. Keep the chromatogram in the folds of black paper to prevent photooxidation of pigments.

Experiment 6.2

To separate and identify chloroplast pigments by thin layer chromatography (TLC).

Requirements

Plant material	: Fresh spinach leaves (*Spinacea oleracea*)
Family	: Chenopodiaceae
Chemicals	: Silica gel-G, 80 per cent acetone, chloroform, petroleum ether, methanol
Glass apparatus	: Stoppered flask, glass plate (5 cm × 20 cm)
Miscellaneous	: Gel applicator, mounting board, desiccator, fan, chromatography jar, spectrophotometer, homogeniser, rotary evaporator.

Principle

TLC is a relatively new technique for the separation of plant pigments. In this method, a thin layer of an adsorbent is coated on a glass plate and the sample is loaded on this coating and developed in a chromatography jar containing a small amount of solvent, by the ascending method.

The significant advantages of this method over paper chromatography lie in the fact that the TLC method is much faster (20–40 minutes for most applications) than other chromatography and requires smaller quantities of sample. It is far more sensitive than other methods and the separation is noticeably much sharper. As long as there is a difference in the adsorption properties of the fractions involved, this method can be successfully used to separate any compound by choosing the right combination of adsorbent and solvent.

The compounds on the chromatogram can be identified on the basis of their R_f values.

Procedure

Extract chloroplast pigments from spinach leaves by grinding approximately 25 grams of fresh leaves in 40 ml of acetone in a homogenizer. When the acetone has turned deep green in colour, filter the preparation through muslin cloth. Evaporate the acetone extract in a rotary evaporator and dissolve the dried pigments in 10 ml of chloroform.

TLC plate: Prepare a slurry of silica gel-G, by vigorously shaking 30 grams of silica with 60 ml of distilled water in a stoppered 250 ml flask. The slurry must be prepared quickly and used immediately. Using the special mounting board and the applicator provided, coat a glass plate with the slurry of silica gel. The applicator should be set to apply a layer of adsorbent 250 μm thick. Place the coated plate in a drying rack and store in a desiccator until it is time for the sample to be chromatographed.

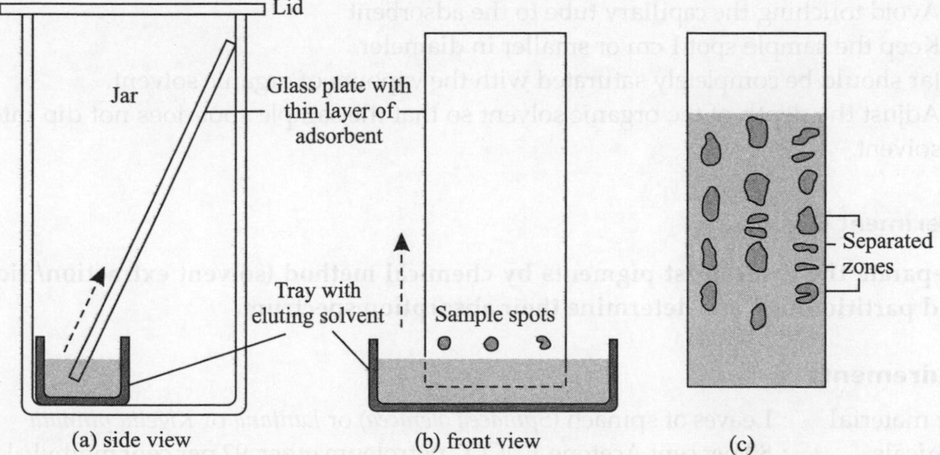

Fig. E6.1B Thin-layer chromatography (a) and (b) Glass plate containing thin layer of adsorbent and applied samples is immersed in tray containing shallow layer of solvent. Capillary action causes solvent to ascend though adsorbent (arrows), which separates samples spots into series of zones (c).

Apply a small aliquot of the sample using a very fine capillary tube in the centre of a line drawn leaving 1.5 cm space from the end. Keep the sample spot 1 cm or smaller in diameter. A cool air fan may be used to hasten the evaporation of the solvent if repeated application of dilute extract is necessary. When the sample spot is completely dry, place the plate in a chromatography jar containing organic solvent (petroleum ether: methanol, 98:2, v/v). Observe the differential migration of various pigments up the chromatogram. When the solvent has advanced 15 cm above the origin, remove the chromatogram from the jar and allow it to dry (Figure E6.1B).

Plant pigment	R_f value	Colour
Chlorophyll *b*	0.17	Yellow-green
Chlorophyll *a*	0.23	Blue-green
Xanthophyll	0.35	Yellow
Carotene	0.96	Yellow-orange

Results

Make a scaled diagram of the chromatogram, showing all the pigments which are visible. Identify as many of the pigments as possible. Calculate the R_f value of each pigment. Then remove bands of silica gel containing the individual pigments, using a clean razor blade. Place each sample of silica gel containing a single pigment in a centrifuge tube and add 5 ml of acetone. Agitate the tube to facilitate the elution of pigment off the gel. Centrifuge the tubes to sediment the gel. Then pipette off the supernatant from each tube, and transfer the pigment solution from each tube. Adjust the concentration of each pigment solution and determine the absorption spectra of the pigments by using a spectrophotometer (400–720 nm) and plot the absorption spectra.

Precautions

1. Slurry of silica gel should be prepared quickly and used immediately.
2. Store the TLC plates in a dessiccator to avoid moisture.
3. Avoid touching the capillary tube to the adsorbent.
4. Keep the sample spot 1 cm or smaller in diameter.
5. Jar should be completely saturated with the vapours of organic solvent.
6. Adjust the depth of the organic solvent so that the sample spot does not dip into the solvent.

Experiment 6.3

To separate the chloroplast pigments by chemical method (solvent extraction/ liquid-liquid partitioning), and determine their absorption spectrum.

Requirements

Plant material : Leaves of spinach (*Spinacea oleracea*) or *Lantana* or *Kigelia pinnata*
Chemicals : 80 per cent Acetone, $CaCO_3$, petroleum ether, 92 per cent methyl alcohol (Leishman solution), 30 per cent methanolic KOH, diethyl ether.
Glass apparatus : Separating funnel, test tubes, beaker, funnel, measuring cylinder.
Miscellaneous : Electric balance, tripod stand, test tube stand, muslin cloth, sand, spectrophotometer, pestle and mortar.

Principle

The most important molecules in the conversion of light energy to chemical energy during the process of photosynthesis are the pigments that exist within chloroplasts of plants.

Pigments involved in trapping light energy can be divided into two categories: vital pigments (Chl *a*) and accessory pigments (Chl *b*, carotene, xanthophyll). The accessory pigments capture solar energy and transfer it to Chl *a* by an inductive resonance process.

Besides paper chromatography and thin layer chromatography, the solvent extraction (liquid-liquid partitioning) method is also used to separate chloroplast pigments. This works on the principle that the 'pigments exhibit differential solubility in different solvents'.

During the initial stage of extraction, however, acetone is used since it has greater penetrating power into the leaf tissues for removal of pigments. Essentially solvent extraction is a method of separating a compound from the solution by shaking it with other solvent in which it is more soluble. The solvent and the solution should be immiscible with each other and form two distinct layers (Figure E6.2).

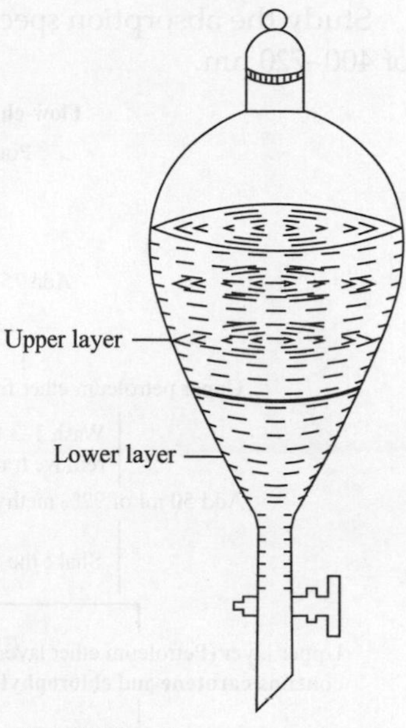

Fig. E6.2 A separating funnel.

One important step of the solvent extraction method is that the solvent (acetone[11] or petroleum ether[12]) is always added 'in lots' and not in one go. This act can be justified by the 'Partition Law' according to which it is better to repeat the process of extraction with smaller amounts of the solvent.

Procedure

Crush 20 grams of spinach leaves (about 5 grams of dry leaf powder) in 40 ml of 80 per cent acetone (in 4 lots of 10 ml each) using pestle and mortar along with a pinch of sand which acts as an abrasive material. Add also a little bit of $CaCO_3$ while crushing as it helps in neutralization of acids present in cell vacuoles and also prevents chlorophyll destabilisation. Filter the extract through muslin cloth and transfer in a separating funnel. Proceed for separation of chloroplast pigments as shown in the flow-chart 1.

Result and Discussion

The liquid-liquid partitioning method is used for separation of pigments based on their differential affinities for different solvents. These pigments can be separated by dissolving them out in different solvents like petroleum ether, acetone, methanol, diethyl ether, and many more.

Polar pigments (Chl *b* and xanthophyll) are soluble in polar solvents such as methanol. Non-polar pigments (Chl *a* and carotene) are soluble in non-polar solvents such as petroleum ether. Methanolic KOH solution saponifies the phytol chain of chlorophyll molecules making them water-soluble.

---

[11] Acetone - Less non-polar and water-soluble.

[12] Petroleum ether - Highly non-polar and water-insoluble.

Study the absorption spectrum of the pigments using spectrophotometer over a range of 400–720 nm.

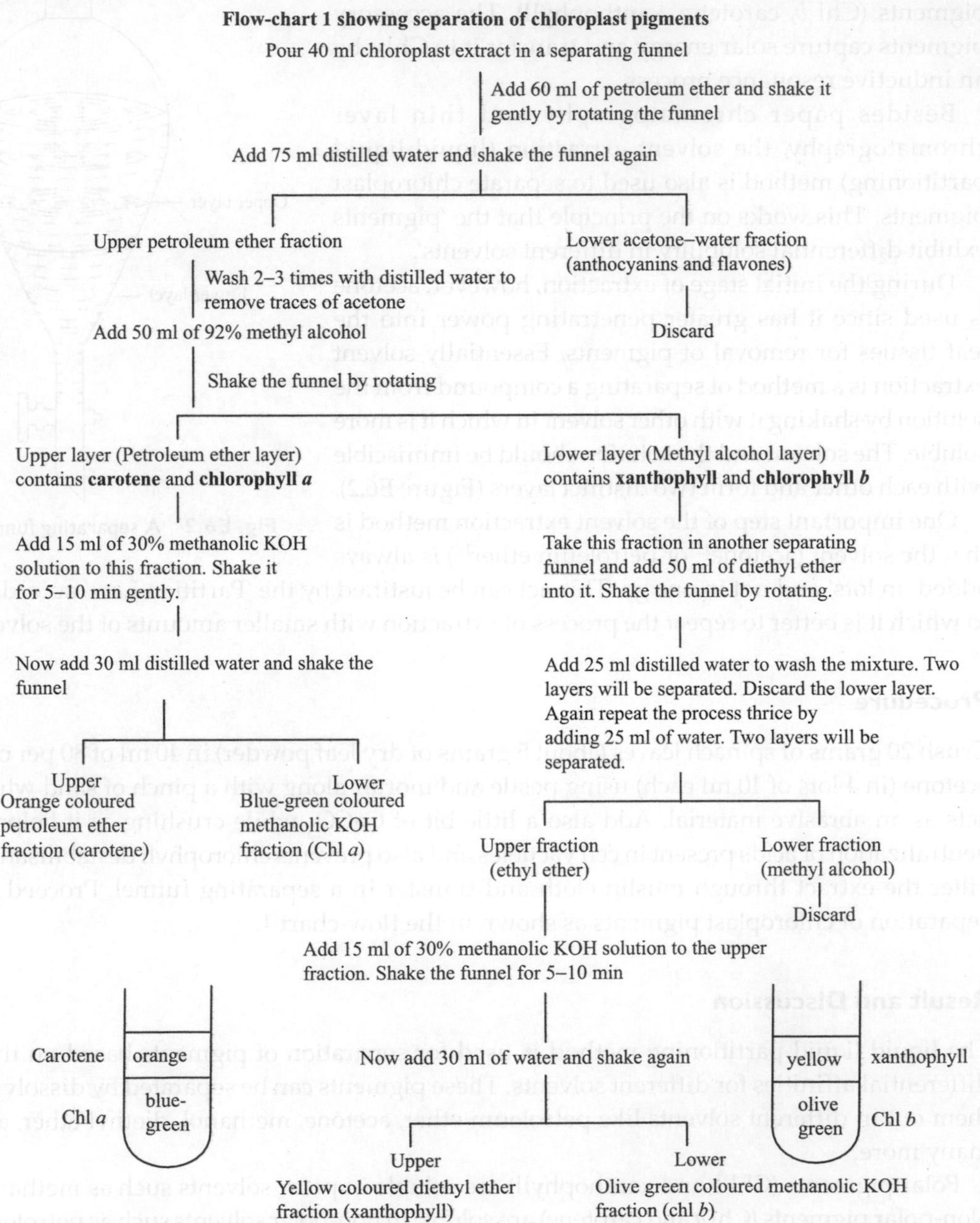

Flow-chart 1 showing separation of chloroplast pigments
Pour 40 ml chloroplast extract in a separating funnel
Add 60 ml of petroleum ether and shake it gently by rotating the funnel
Add 75 ml distilled water and shake the funnel again
Upper petroleum ether fraction
Lower acetone–water fraction (anthocyanins and flavones)
Wash 2–3 times with distilled water to remove traces of acetone
Discard
Add 50 ml of 92% methyl alcohol
Shake the funnel by rotating
Upper layer (Petroleum ether layer) contains carotene and chlorophyll a
Lower layer (Methyl alcohol layer) contains xanthophyll and chlorophyll b
Add 15 ml of 30% methanolic KOH solution to this fraction. Shake it for 5–10 min gently.
Take this fraction in another separating funnel and add 50 ml of diethyl ether into it. Shake the funnel by rotating.
Now add 30 ml distilled water and shake the funnel
Add 25 ml distilled water to wash the mixture. Two layers will be separated. Discard the lower layer. Again repeat the process thrice by adding 25 ml of water. Two layers will be separated.
Upper
Orange coloured petroleum ether fraction (carotene)
Lower
Blue-green coloured methanolic KOH fraction (Chl a)
Upper fraction (ethyl ether)
Lower fraction (methyl alcohol)
Discard
Add 15 ml of 30% methanolic KOH solution to the upper fraction. Shake the funnel for 5–10 min
Carotene
orange
Chl a
blue-green
Now add 30 ml of water and shake again
yellow
xanthophyll
olive green
Chl b
Upper
Yellow coloured diethyl ether fraction (xanthophyll)
Lower
Olive green coloured methanolic KOH fraction (chl b)

Note: Carotenes and xanthophylls are more stable in alkaline medium.

Precautions

1. The central mid rib part of the spinach leaves must be removed as it contains a lot of water.

2. Leaf lamina should be cut apart into small pieces before crushing.
3. Leaves should be crushed gently to avoid damage to chloroplasts.
4. Organic solvents should be added in lots.
5. Only muslin cloth, not the filter paper, should be used for filtration, otherwise pigments get adsorbed on the surface of filter paper.
6. Separating funnel should be gently rotated while adding the solvent to avoid precipitation and formation of emulsion.
7. The stopper should be periodically removed to allow the vapours of organic solvent to escape, otherwise stopper be blown-off due to pressure created inside the funnel.
8. Repeated rinsing of the upper petroleum ether layer (with dissolved pigments) with distilled water is necessary to remove even the last traces of acetone.
9. Dry tubes should be used for pigment separation.
10. Addition of methanol should be followed by vigorous shaking.
11. Methanol being poisonous, its vapours should not be inhaled.

$CaCO_3$[13] should be added to neutralize the acids present in cell vacuoles. Otherwise the acidic medium will lead to replacement of Mg^{2+} ion present in with chlorophyll with H^+, thus converting it into pheophytin.

Experiment 6.4

To demonstrate the Hill reaction using isolated chloroplasts and study the effect of different light intensities.

Requirements

Plant material	: Fresh green leaves of spinach (*Spinacea oleracea*)
Family	: Chenopodiaceae
Chemicals	: Phosphate buffer (0.2 M Na_2HPO_4, 0.2 M KH_2PO_4) 1 M sucrose, 0.2 M KCl 0.1 per cent DCPIP (2, 6-Dichlorophenolindophenol)
Glass apparatus	: Beaker, pipettes (1 ml, 10 ml), test tubes, funnel, glass rod, measuring cylinder
Miscellaneous	: Pestle and mortar, refrigerator, ice-bucket, muslin cloth, centrifuge, centrifuge tubes, test tube stand, lux meter, table lamps, black paper, graph paper, gas burner, aluminium foil.

Principle

Photosynthesis is an anabolic process during which simpler compounds like CO_2 and water are converted into complex organic compounds. It involves light reaction (Hill reaction) as well as dark reaction (Blackman's reaction).

[13] However, excess of $CaCO_3$ should be avoided to prevent the formation of emulsion.

During light reaction of photosynthesis, hydrogen atoms of water molecules are transferred to the natural electron acceptor $NADP^+$, thus resulting in the formation of NADPH + H^+ which is used to reduce carbon dioxide to form carbohydrates. In 1937, Robert Hill demonstrated that the isolated chloroplasts generated oxygen when illuminated in the presence of suitable electron acceptors (such as ferricyanide) without a simultaneous reduction of CO_2. Such a reaction is known as **Hill reaction.**

In the given experiment, Hill reaction can be demonstrated using 2,6-dichlorophenolindophenol (DCPIP), the most frequently used artificial electron acceptor in the laboratory. This is blue coloured in the oxidized form and turns colourless in the reduced form ($DCPIPH_2$). Photolysis of water by isolated chloroplasts is accompanied by reduction of DCPIP to $DCPIPH_2$. This test indicates that photosynthesis is an oxidation-reduction process (oxidation of water and reduction of DCPIP/NADP).

1. Overall equation of photosynthesis

$$6CO_2 + 12H_2O \rightarrow C_6H_{12}O_6 + 6H_2O + 6O_2\uparrow$$

2. van Niel proposed following equations

$$CO_2 + 2H_2S \rightarrow (CH_2O) + H_2O + 2S \text{ (for photosynthetic bacteria)}$$

$$CO_2 + 2H_2O \rightarrow (CH_2O) + H_2O + O_2 \text{ (for higher plants)}$$

$$CO_2 + 2H_2A \rightarrow (CH_2O) + H_2O + 2A \text{ (generalized equation)}$$

3. Robert Hill

$$2H_2O + 2A \rightarrow 2AH_2 + O_2 \text{ (A = Hill's oxidant or hydrogen acceptor)}$$

4. Biological system

$$2H_2O + 2NADP \rightarrow 2NADPH_2 + O_2\uparrow$$
$$(-0.32\ E°')$$

5. *In vitro* conditions (experimental conditions)

$$2H_2O + 2DCPIP \longrightarrow 2DCPIPH_2 + O_2\uparrow$$
$$(-0.32\ E°')$$

Oxidized (blue) → Reduced (colourless)

$$\text{2,6-DCPIP (oxidized-blue)} + 2H^+ \xrightarrow[\text{hv}]{\text{Chloroplast} + H_2O} \text{2,6-DCPIPH}_2\text{ (reduced-colourless)} + 1/2\ O_2$$

(Structures: HO–C₆H₂Cl₂–N=C₆H₄=O + 2H⁺ → HO–C₆H₂Cl₂–NH–C₆H₄–OH + 1/2 O₂)

2, 6-DCPIP (oxidized-blue)

2, 6-$DCPIPH_2$ (reduced-colourless)

Procedure

Homogenize 20 grams of spinach leaves (without midrib) in 50 ml of extraction mixture using pestle and mortar and filter through muslin cloth. Centrifuge the filtrate at 1000 rpm for 3 minutes. Discard the pellet but centrifuge the supernatant once again at 4000 rpm for 12 minutes. Discard the supernatant but keep the second pellet that now results. Suspend this pellet in 10 ml of extraction mixture (chloroplast suspension) and keep at freezing temperature. Perform the experiment as shown in Observation Table 1 (S. No. 1 → 4). Study the effect of two light intensities on Hill reaction as mentioned in Observation Table 2 (S. No. 5 → 7).

Preparation of Extraction Mixture

S. No.	Molarity of stock solution	Required molarity of stock solution	Volume of stock solution (ml)
1.	1 M Sucrose	0.33 M	60
2.	0.2 M Na_2HPO_4	0.05 M	50
3.	0.2 M KH_2PO_4	0.05 M	50
4.	0.2 M KCl	0.04 M	40
		Total	200
			(pH 6.5–6.8)

Role of Different Components of Extraction Mixture

- Sucrose – an osmoticum, isotonic, non-ionizable compound
- Na_2HPO_4 } Buffering agent
- KH_2PO_4 } Buffering agent
- KCl – source of Cl^- ions for photolysis of H_2O

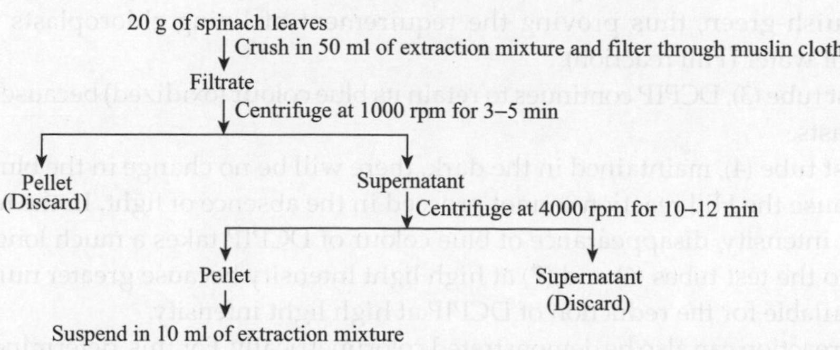
Flow-sheet for
preparation of chloroplast suspension
20 g of spinach leaves
Crush in 50 ml of extraction mixture and filter through muslin cloth
Filtrate
Centrifuge at 1000 rpm for 3–5 min
Pellet
(Discard)
Supernatant
Centrifuge at 4000 rpm for 10–12 min
Pellet
Supernatant
(Discard)
Suspend in 10 ml of extraction mixture

Observation table 1

S. No.	Light condition	Volume of reaction* mixture (ml)	Volume of chloroplast		Volume of dye (DCPIP) (ml)	Total volume (ml)	Inference	
			Living	Boiled			Initial	Final
1.	Light	8	1 ml	-	1	10	Bluish-green	Greenish
2.	Light	8	-	1 ml	1	10	Bluish-green	Bluish-green
3.	Light	8	-	-	1	10	Blue	Blue
4.	Dark	8	1 ml	-	1	10	Bluish-green	Bluish-green

(In tube number 3, 1 ml of distilled water is added in place of chloroplast suspension).
* Extraction mixture is now designated as reaction mixture.

Observation table 2: Effect of different light intensities on Hill's reaction

S. No.	Light intensity	Time taken for decolourisation of DCPIP (min)
5.	Control (450 lux)	28
6.	1 lamp (912 lux)	20
7.	2 lamps (1760 lux)	10

Result and Discussion

In the test tube (1) completely exposed to the light, the blue colour of DCPIP will gradually disappear due to the reduction of DCPIP to $DCPIPH_2$ as a result of photolysis of water and electron transfer by chloroplasts.

In the test tube (2) containing denatured/boiled chloroplast suspension, the solution remains bluish-green, thus proving the requirement of living chloroplasts as sites of photolysis of water (Hill reaction).

In the test tube (3), DCPIP continues to retain its blue colour (oxidized) because of absence of chloroplasts.

In the test tube (4), maintained in the dark, there will be no change in the blue colour of DCPIP because the Hill reaction cannot proceed in the absence of light. In the test tube (5) at low light intensity, disappearance of blue colour of DCPIP takes a much longer time as compared to the test tubes (6) and (7) at high light intensity because greater number of H^+ ions are available for the reduction of DCPIP at high light intensity.

The Hill reaction can also be demonstrated colorimetrically. For this, determine the initial absorbance of the mixture at 620 nm. Record the decrease in absorbance at an interval of every 1 or 2 minutes after exposing the mixture to the light. Draw a graph showing the decrease in absorbance with time, which is a function of rate of the Hill reaction.

Precautions

1. Spinach leaves should be fresh and without any midrib portion as they contain excessive water.
2. Plant material, chemicals, glass apparatus should be pre-chilled before the experiment.
3. Extraction of chloroplastic pigments must be conducted at subfreezing temperature to prevent denaturation of chloroplastic enzymes.
4. The amount of DCPIP and reaction mixture should remain constant in all the test tubes.
5. Filtration of chloroplast solution should be done using muslin cloth and not filter paper.
6. DCPIP amount should be sufficient enough to give a distinct bluish-green colour to the chloroplast suspension.
7. Centrifugation and decantation should be done carefully to remove starch grains or heavier fractions before removing the chloroplast.

The pH of the extraction mixture should be maintained between 6.5 and 6.8; otherwise in an acidic medium, the Mg of chlorophyll is replaced by H, forming pheophytin.

Experiment 6.5

To study the effect of different light intensities on the rate of photosynthesis.

Requirements

Plant material	: *Hydrilla verticillata*
Family	: Hydrocharitaceae
Chemicals	: Filtered pond water, sodium bicarbonate
Glass apparatus	: Measuring cylinder (100 ml), glass rod, beaker, pipette, a large petri dish
Miscellaneous	: Meter rod, blade, thread, stop watch, tally counter, lux meter, table lamps, electric balance, graph paper

Principle

Photosynthesis is an anabolic process and can be defined as the formation of carbohydrates from CO_2 and H_2O by illuminated green cells with O_2 and H_2O being the by-products.

$$6CO_2 + 12H_2O \xrightarrow[\text{Chlorophyll}]{\text{Light}} \underset{\text{Carbohydrate}}{C_6H_{12}O_6} + 6O_2 + 6H_2O$$

Various factors that affect the process of photosynthesis are:

External factors: CO_2, O_2, light, temperature, amount of available water.

Internal factors: Amount of chlorophyll, protoplasmic factors, accumulation of end products, leaf area, genetic control.

According to the concept of three cardinal points introduced by Sachs (1880), there is a minimum, optimum and maximum for each factor in relation to photosynthesis.

For example, any species has a minimum temperature below which no photosynthesis takes place, an optimum temperature at which the highest rate takes place, and a maximum temperature above which photosynthesis ceases altogether.

Effect of light on photosynthesis: Light affects photosynthesis through its intensity, quality and duration. A direct relationship between the rate of photosynthesis and light intensity is shown at the lower light intensities. As the intensity of light is increased, however, there is a decline of the photosynthetic rate because of some other limiting factor or the destructive effects of high light intensity (photo-oxidation). Also, the point of saturation may have been achieved, at which point the rate of photosynthesis becomes stationary.

Bubble-counting is one method used for measurement of photosynthesis in which the rate of evolution of bubbles by a submerged aquatic plant is considered to be proportional to the rate of apparent photosynthesis.

Procedure

Cut a healthy or actively growing twig of *Hydrilla,* about 15 cm length (with apex intact), under water. Place it in a measuring cylinder containing 100 ml of filtered pond water. With the help of a glass rod, tie the twig loosely so that when placed in water it stays submerged. The cut end of the twig should face upwards whereas the intact shoot apex faces downwards.

When the twig starts producing bubbles uniformly, count the number of bubbles given out from the cut end at an interval of one minute, using a tally counter. Repeat such counting at least thrice. Repeat the bubble counting under different intensities of light achieved by varying the distance between the light source and the plant material.

Allow the plant to be exposed to each intensity for 5 minutes before starting the count. Record the light intensity using a lux meter.

$$NaHCO_3 \rightleftharpoons Na^+ + HCO_3^-$$

$$H_2O \longrightarrow H^+ + OH^-$$

$$Na^+ + OH^- \longrightarrow NaOH$$

$$H^+ + HCO_3^- \longrightarrow \underbrace{H_2CO_3}_{\text{Carbonic acid}} \longrightarrow \underbrace{H_2O + CO_2}_{\text{Both act as reactants in photosynthesis}}$$

Observation table 2

Measuring cylinder = 100 ml pond water + 0.1 g $NaHCO_3$

Environmental factor	Light intensity (lux)	Oxygen bubbles per min	Mean
Diffuse sunlight	325	i) 2 ii) 1 iii) 1	1.33 (= 1)
Diffuse sunlight + 1 lamp	500	i) 2 ii) 2 iii) 2	2
Diffuse sunlight + 2 lamps	710	i) 5 ii) 4 iii) 5	4.66 (= 5)

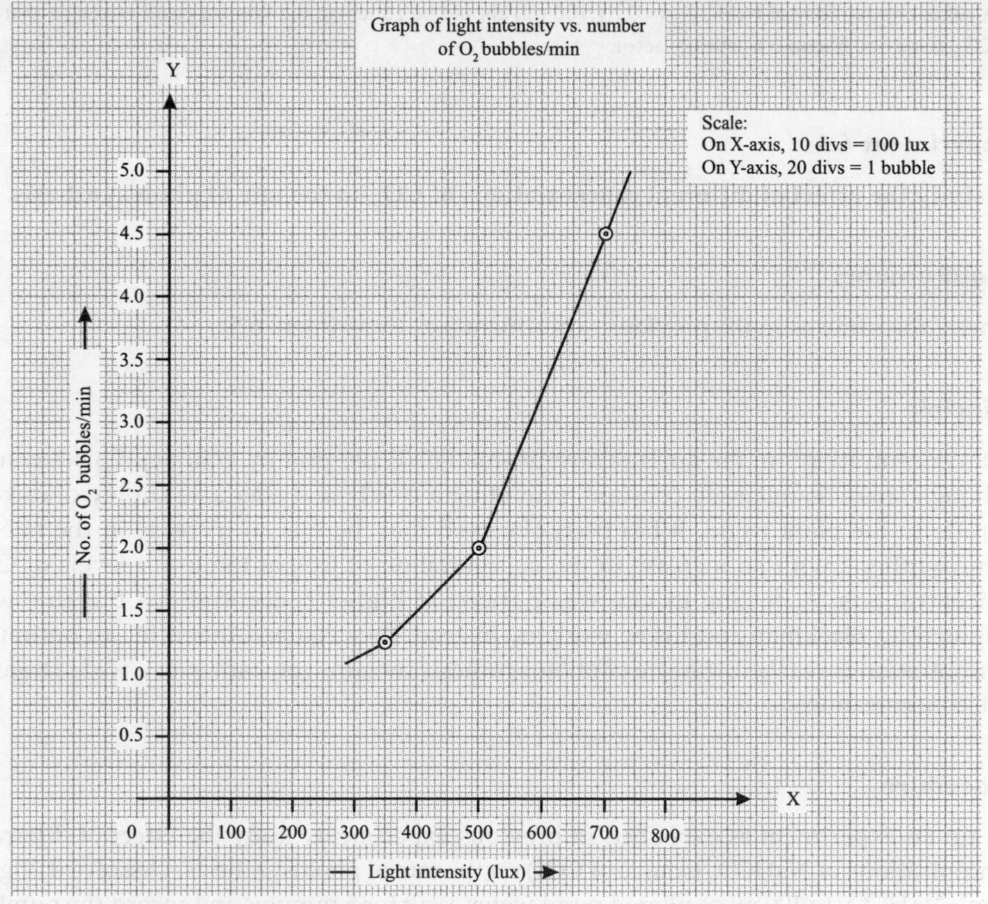

Result and Discussion

With an increase in light intensity, the rate of photosynthesis goes on increasing till some other factor (CO_2 or temperature) becomes limiting.

Precautions

1. *Hydrilla* twig should be healthy, dark green, injury-free and with an intact shoot apex.
2. Oblique cut should be given to the twig, under water.
3. Twig should not touch the side walls of the cylinder and may be supported by using a glass rod.
4. The same twig should be used throughout the experiment so that size of the oxygen bubbles remains constant.
5. There should be free movement of oxygen bubbles, i.e., they should not be allowed to accumulate at the cut end of the twig.
6. The twig should be allowed to be exposed to each light intensity for 3–5 minutes before starting the bubble count.

A sufficient quantity of $NaHCO_3$ (≈ 0.1 g/100 ml) should be added to the filtered pond water so that CO_2 does not become a limiting factor.

Experiment 6.6

To study the effect of different concentrations of CO_2 on the rate of photosynthesis.

Requirements

Exactly the same as in the previous experiment (i.e., Experiment 6.5).

Principle

CO_2 concentration[14] in the air is relatively small (0.03 per cent by volume). This amount, although small, is relatively constant, providing a steady and adequate supply of carbon dioxide to the plant world. It has been observed that there is an increase in the rate of photosynthesis with an increase in the concentration of CO_2 at the lower levels of CO_2 concentrations. A decline in rate was noticed at higher concentrations but this was disputed by Blackman and his co-workers. Instead, they claimed that after an optimum concentration is reached, the rate of photosynthesis remains constant over a wide range of CO_2 concentrations.

Procedure

Set up an experiment as explained in previous experiment. Study the effect of different concentrations of $NaHCO_3$ (source of CO_2) on the rate of photosynthesis. Dissolve 0.01 g or 10 mg $NaHCO_3$ to 100 ml of filtered pond water. Wait for 5 minutes and record the number of oxygen bubbles evolved per minute (3 concordant readings).

[14] The level of CO_2 – a greenhouse gas, has gone up to 385 ppm (by volume 0.0385 per cent) in recent times.

Sol. A : .01 per cent - 0.01 g dissolved per 100 ml of pond water
Sol. B : 0.1 per cent - 0.09 g + Sol. A (0.1 g/100 ml)
Sol. C : 1 per cent - 0.9 g + Sol. B (1 g/100 ml)

To prepare solution B, take out 5 ml of solution A, add 0.09 g of $NaHCO_3$, dissolve and then put the solution back into A. Wait for 5 minutes and note down the observations. Similarly, to prepare solution C, take out 5 ml of solution B, add 0.9 g of $NaHCO_3$, dissolve and then put the solution back into B. Record the observations per minute.

Observation table 3

Fixed environmental factor : diffuse sunlight + 1 lamp (at a distance of 30 cm)

S.No.	Concentration of $NaHCO_3$ (%)	O_2 bubbles evolved per min	Mean
1	0.01	i) 2 ii) 2 iii) 2	2
2	0.1	i) 2 ii) 3 iii) 3	3
3	1	i) 4 ii) 5 iii) 5	5

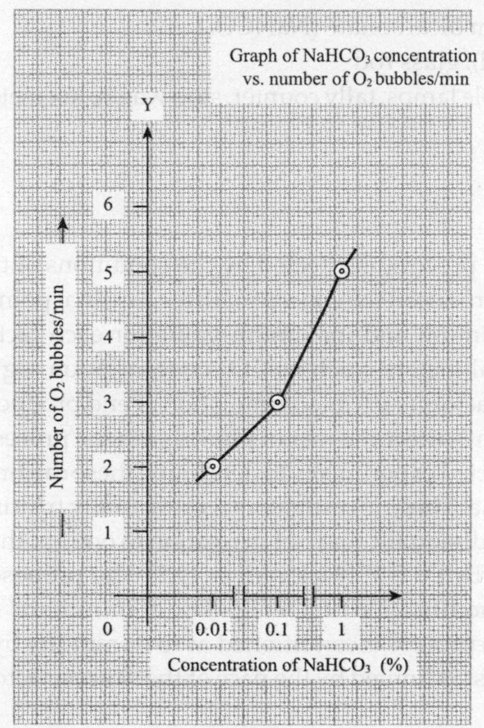

Graph of $NaHCO_3$ concentration vs. number of O_2 bubbles/min

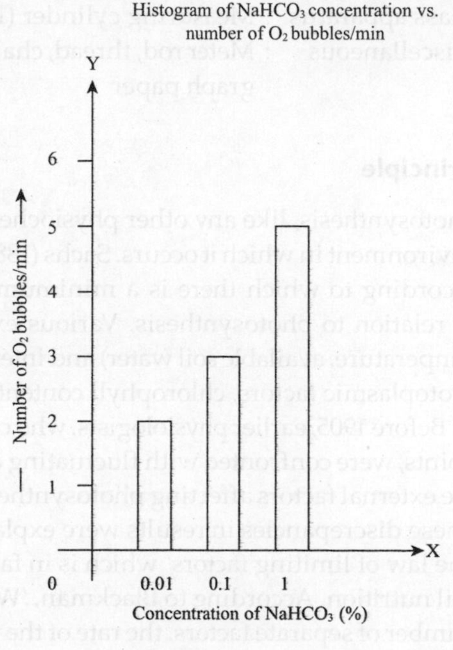

Histogram of $NaHCO_3$ concentration vs. number of O_2 bubbles/min

Result and Discussion

With an increase in CO_2 concentration, the rate of photosynthesis increases till some other factor (light/temperature) becomes limiting.

Precautions

1. Sufficient light intensity should be provided to the experimental set-up so that it does not become a limiting factor.
2. The twig should be allowed to remain in each $NaHCO_3$ concentration for 3 to 5 minutes before starting the bubble count.

> It is desirable not to add different concentrations of $NaHCO_3$ directly into the solution in the cylinder but should be first dissolved in 5 ml of solution withdrawn from the cylinder, using a pipette and then added back.

Experiment 6.7

To demonstrate Blackman's law of limiting factors.

Requirements

Plant material	: *Hydrilla verticillata*
Family	: Hydrocharitaceae
Chemicals	: Filtered pond water, sodium bicarbonate ($NaHCO_3$)
Glass apparatus	: Measuring cylinder (100 ml), glass rod
Miscellaneous	: Meter rod, thread, chalk, table lamps, tally counter, stop watch, lux meter, graph paper

Principle

Photosynthesis, like any other physiochemical process, is affected by the conditions of the environment in which it occurs. Sachs (1880) introduced the concept of three cardinal points, according to which there is a minimum, optimum and maximum value for each factor in relation to photosynthesis. Various external factors (such as CO_2 concentration, light, temperature, available soil water) and internal factors (such as accumulation of end products, protoplasmic factors, chlorophyll content of leaf) affect the rate of photosynthetic process.

Before 1905, earlier physiologists, who conducted experiments to determine these cardinal points, were confronted with fluctuating optima. These investigators were actually treating the external factors affecting photosynthesis individually and not in relation to one another. These discrepancies in results were explained by Dr F. F. Blackman in 1905 who proposed 'the law of limiting factors' which is in fact a modification of 'Liebig's law of minimum' on soil nutrition. According to Blackman, 'When a process is conditioned as to its rapidity by a number of separate factors, the rate of the process is limited by the pace of the slowest factor.'

Hydrilla twig is used as an experimental material because it is a submerged aquatic plant and has lot of aerenchyma tissue. It is capable of accumulating CO_2 for photosynthesis during the day and O_2 for respiration during the night. Visible O_2 bubbles are formed as the plant utilizes CO_2 from the surrounding solution (water + $NaHCO_3$) and carries out photosynthesis. The number of bubbles formed per minute is indicative of the rate of photosynthesis.

$NaHCO_3$ is used as CO_2 source for the plant (and not Na_2CO_3) because it is an unstable compound which decomposes in water to form CO_2 whereas Na_2CO_3 is a stable compound, not dissociating easily and cannot be used as an appropriate CO_2 source.

Procedure

Select a healthy, uninjured *Hydrilla* twig and give an oblique cut under water at its lower end. With the help of a meter rod, draw a line on the table. Mark a point in the centre as zero and on either side of it, mark points at an interval of 10 cm, upto 100 cm. Tie the twig loosely with a thread on the glass rod with apex facing downwards and cut end pointing upwards. Put the glass rod with the attached twig in a measuring cylinder containing 100 ml of filtered pond water and 0.02 per cent $NaHCO_3$ (20 mg $NaHCO_3$/100 ml) (Figure E6.3).

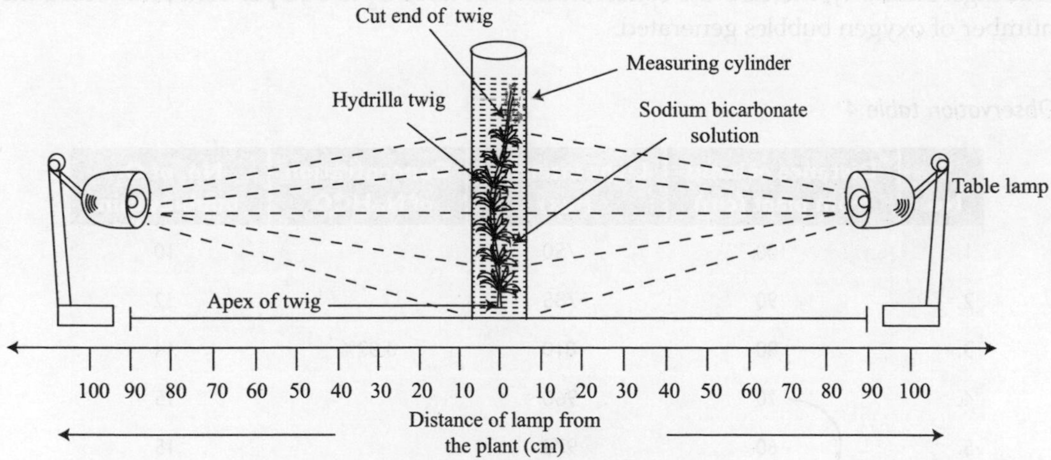

Fig. E6.3 Design of experimental set-up

$$NaHCO_3 \rightleftharpoons Na^+ + HCO_3^-$$

$$H_2O \longrightarrow H^+ + OH^-$$

$$Na^+ + OH^- \longrightarrow NaOH$$

$$H^+ + HCO_3^- \longrightarrow H_2CO_3 \text{ (Carbonic acid)}$$

$$H_2CO_3 \longrightarrow \underbrace{H_2O + CO_2\uparrow}_{\text{Both act as reactants in photosynthesis}}$$

Preparation of $NaHCO_3$ Solution

A. 0.02 per cent $NaHCO_3$: 0.02 g (20 mg) dissolved per 100 ml of pond water
B. 0.04 per cent $NaHCO_3$: 0.04 g (40 mg)/100 ml of pond water [i.e. Sol. A + 0.02 g $NaHCO_3$]
C. 0.06 per cent $NaHCO_3$: 0.06 g (60 mg)/100 ml of pond water [i.e. Sol. B + 0.02 g $NaHCO_3$]

Place the measuring cylinder at 'zero' mark of the line and position two table lamps at 100 cm distance on either side of the cylinder with their light directed towards plant material. Measure the light intensity at point zero with the help of the lux meter. Wait for 5 minutes to allow the plant to attain its optimum physiological level of activity. Count the number of oxygen bubbles coming out per minute, using tally counter.

On either side of the plant, reduce the distance of the two lamps from the cylinder containing plant by 10 cm, wait for 5 minutes and record the number of bubbles produced per minute. Keep on reducing the distance between lamps and the plant while recording the number of bubbles produced per minute till two nearly concordant readings of the number of bubbles produced per minute are obtained. At this point of time, pipette out a small amount of solution from the cylinder, dissolve 20 mg of $NaHCO_3$ more in it to increase $NaHCO_3$ concentration in the solution to 0.04 per cent and record the number of O_2 bubbles generated by placing the lamps at a distance farther of the two which gave concordant readings. Similarly, increase the concentration of $NaHCO_3$ to 0.06 per cent and record the number of oxygen bubbles generated.

Observation table 4

S. No.	Distance of lamp from plant (cm)	Light intensity (lux)	Concentration of $NaHCO_3$	No. of O_2 bubbles/min
1.	100	750		10
2.	90	785		12
3.	80	810	0.02%	14
4.	70	960		15
5.	60	978		15
6.	70	960		18
7.	60	978		20
8.	50	1008	0.04%	22
9.	40	1025		23
10.	30	1048		23
11.	40	1025		26
12.	30	1048		28
13.	20	1081	0.06%	28

Result and Discussion

As the distance of the two lamps from the plant is reduced, the available light intensity to the plant increases, thus increasing number of O_2 bubbles produced per minute. At 0.02 per cent $NaHCO_3$, when light intensity is increased from 750 to 978 lux, the saturation point of O_2 evolution is achieved, i.e., the rate of photosynthesis has been stabilised now, as CO_2 concentration acts as a limiting factor. When CO_2 concentration is increased from 0.02 per cent to 0.04 per cent to 0.06 per cent, the rate of photosynthesis is again found to increase with increasing light intensity till CO_2 concentration becomes 'limiting' again.

The graph between number of O_2 bubbles released per minute and light intensity shows that the rate of photosynthesis levels off as CO_2 concentration becomes limiting factor. This rate of stabilisation is found to be gradual and not abrupt as proposed by Blackman.

Precautions

1. The plant selected should be healthy, actively photosynthesising and uninjured.
2. The plant apex should be intact.
3. Give an oblique cut to the plant under water, using a sharp blade. The cut should be smooth, otherwise resistance is provided to free bubble movement by uneven or rough ends.
4. The plant should be tied to the glass rod loosely.
5. Additional amounts of $NaHCO_3$ should be added only when photosynthetic yield gets stabilised.
6. $NaHCO_3$ should not be added directly to the cylinder otherwise the set-up is disturbed.
7. A gap of 5 minutes should be given for attaining physiological stability whenever the set-up is shifted to new environmental conditions.

It is preferable to move the light source rather than the 'plant'. Because the bubbling rate is at times very sensitive to jarring or any slight disturbance to the plant.

Drawbacks of the Experiment

The principle of limiting factors could not be accepted in the simple form by most workers because when rate of photosynthesis is plotted against quantitative increase in some factors, a smooth curve is obtained, whereas according to the law, an abrupt transition should take place. It may be because two or more factors may be limiting simultaneously or all the chloroplast in the photosynthetic tissue may not be uniformly exposed to same conditions simultaneously. It is virtually impossible to vary only one factor at a time to affect only one physiological process or to determine the optimum quantity of a given factor.

However, though its strict application is avoided, Blackman's law of limiting factors provides a good approximation for interpreting the effect of various factors on the rate of photosynthesis.

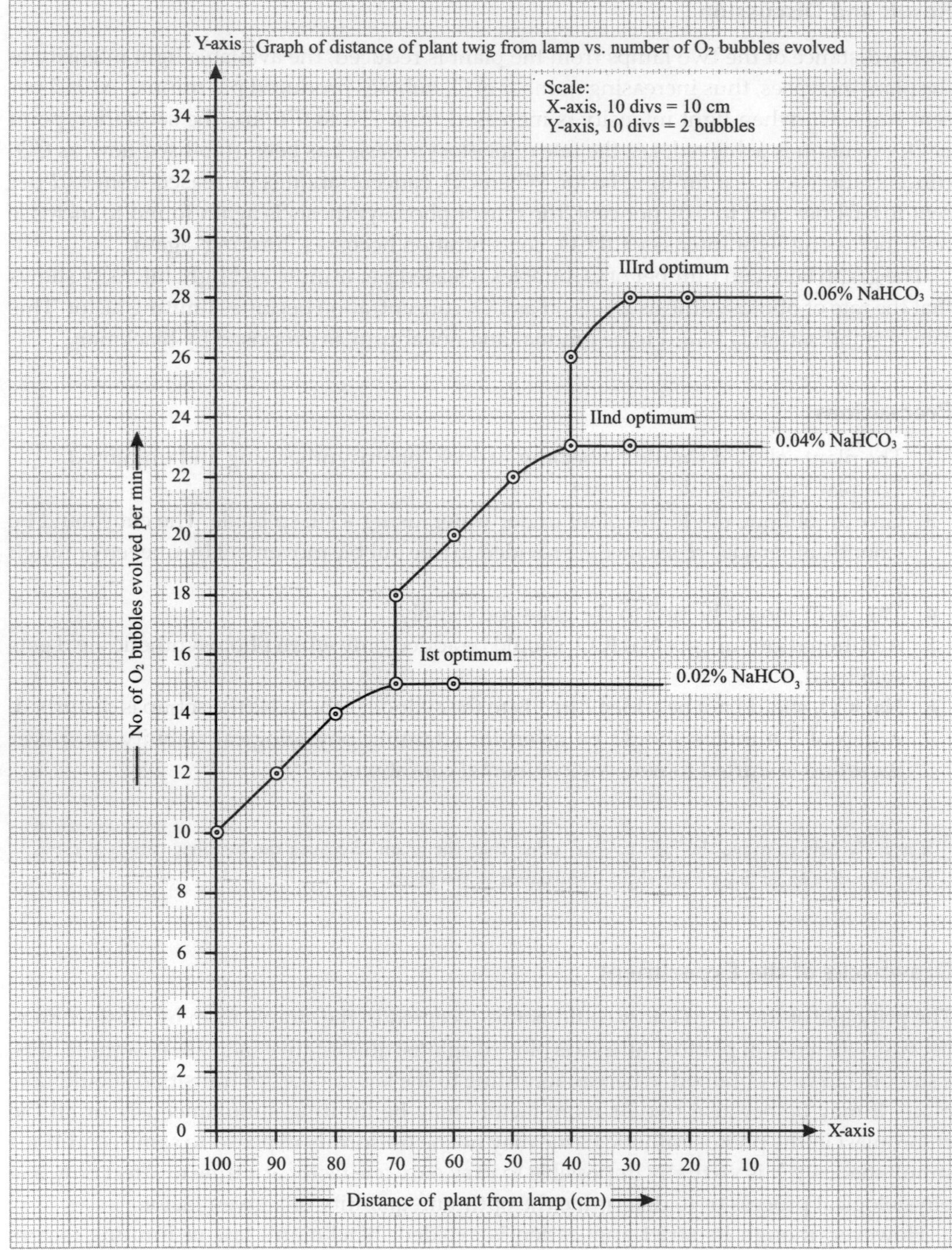

Y-axis
Graph of distance of plant twig from lamp vs. number of O_2 bubbles evolved
Scale:
X-axis, 10 divs = 10 cm
Y-axis, 10 divs = 2 bubbles
34
32
30
28
26
24
22
20
18
16
14
12
10
8
6
4
2
0
IIIrd optimum
0.06% $NaHCO_3$
IInd optimum
0.04% $NaHCO_3$
Ist optimum
0.02% $NaHCO_3$
No. of O_2 bubbles evolved per min
X-axis
100
90
80
70
60
50
40
30
20
10
Distance of plant from lamp (cm)

Questions for Viva Voce

Chromatography

1. **Who discovered the technique of paper chromatography?**
2. **What will happen if the organic solvent is replaced by water or some non-polar solvent?**
3. **How does the nature of interphase affect the R_f value?**
4. **Can the solutes be separated on ordinary writing paper or filter paper?**
5. **Why does the alcoholic solution of chloroplastic pigments appear red in transmitted light?**
6. **Which instrument is used to study the absorption spectrum of pigments?**
7. **How can you obtain greater resolution of solutes using the technique of paper chromatography?**
8. **What is the basic principle involved in the method of paper chromatography?**
9. **How could you make this method quantitative?**
10. **Explain the principle of TLC. What is the merit of this method of analysis?**
11. **What other substances can be detected by paper chromatography?**

Hill reaction

1. **What is the Hill reaction? Explain by giving an equation.**
2. **Explain the role of different components of the reaction mixture.**
3. **Why is chloroplastic extraction done in an ice-bucket?**
4. **What is the role of Mn^{2+} and Cl in photosynthesis?**
5. **Why do you use DCPIP? Can it be substituted by blue ink? If not so; why?**
6. **Name a natural hydrogen accepter.**
7. **Why do you keep the pH of the reaction mixture between 6.5 and 6.8?**
8. **Why does Chl *a* exist in more than one form?**
9. **What is the source of the hydrogen that reduces the dye in this experiment?**
10. **What is the relation between the Hill reaction and the overall reaction of photosynthesis in the intact green plant?**

Blackman's laws of limiting factors

1. **Why do you select *Hydrilla* as an experimental material?**
2. **Explain the role of $NaHCO_3$ is this experiment. Can it be substituted by Na_2CO_3?**
3. **Why is it necessary to take pond water for this experiment?**

Physiological Set-Ups

I. To demonstrate fluorescence in alcoholic solution of chlorophyll

When a pure alcoholic solution of chlorophyll (spinach leaf extract) is subjected to a beam of light, the pigments when viewed in reflected light appear red.

This phenomenon is called **fluorescence** (Figure E6.4).

A pigment molecule (chlorophyll) is able to absorb light of a specific wavelength in the visible range. Electrons present in the chlorophyll molecule can remain in an excited state only for a short period, 10^{-9} seconds or so. The excitation energy can be completely lost in the form of heat as the electrons move back to ground state or by a combination of heat loss and fluorescence.

Fluorescence can thus be defined as light production accompanying rapid decay of excited electrons. Greater the fluorescence, lesser will be the rate of photosynthesis. Chlorophyll molecules fluoresce red since the 'emitted photon' contains less energy than the 'absorbed photon' and that falls in the longer wavelength region (red region) of the electromagnetic spectrum.

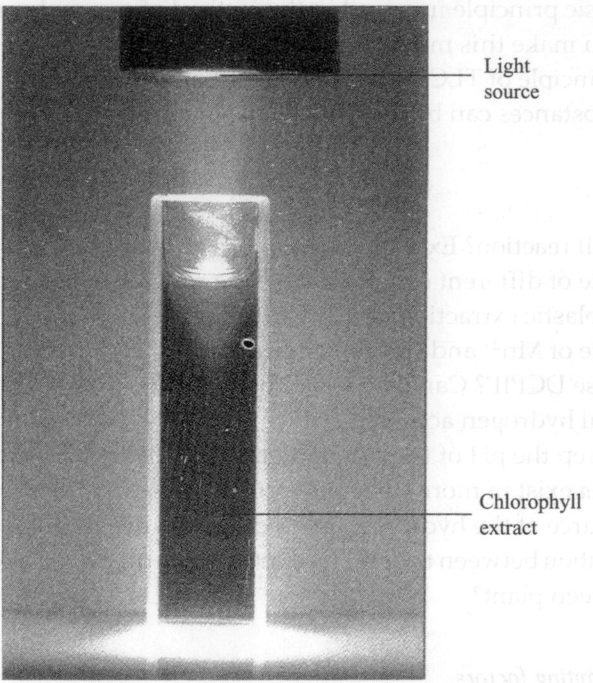

Fig. E6.4 Demonstration of fluorescence.

Fluorescence occurs in conditions such as:

(i) Requirement for photosynthesis, i.e., reactants are not available.
(ii) Electron transport carriers are not available.
(iii) Pigment molecules are in a dilute solution and thus, not in close proximity.

Questions

1. Why does fluorescence not occur in green leaves?
2. What is the biological significance of this process?

3. Mention the conditions leading to fluorescence.
4. Will you still be able to make the similar observations if an intact leaf is kept in place of the extract?
5. Would it make any difference if the pigment solution is replaced by fast green solution?
6. Distinguish between fluorescence and phosphorescence.

2. To demonstrate that CO_2, water, light and chlorophyll are essential for photosynthesis (Moll's half-leaf experiment)

Destarch a potted plant by keeping it in the dark for 3 days. Insert one of the leaves of this plant in a bottle through a split cork. The bottle is partly filled with a strong solution of caustic potash (KOH) (Figure E6.5). Keep the apparatus in a warm well-lit room near the window so that photosynthesis can take place freely. Leave for 2–3 days and then perform the iodine test for starch.

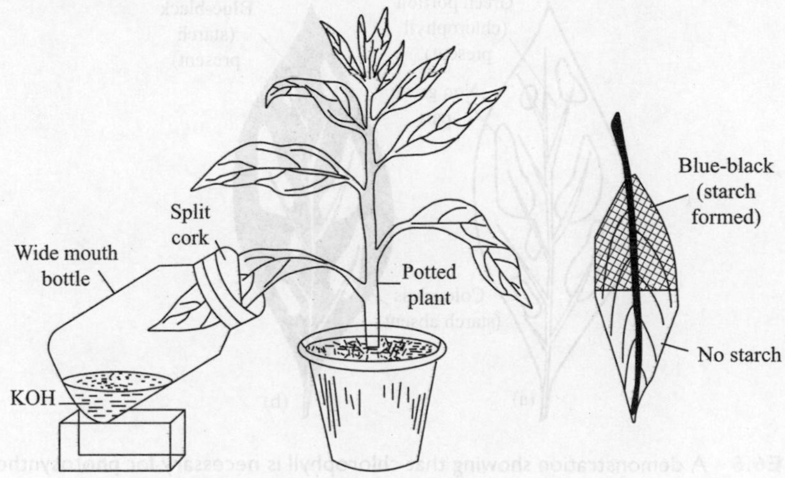

Fig. E6.5. Moll's half leaf experiment.

You will notice that the portion of the leaf inside the bottle shows a negative test result. The portion of the leaf between the two halves of the split cork also shows a negative result. But the portion of the leaf outside the cork and bottle tests positively.

This proves that photosynthesis requires CO_2, light, water and chlorophyll, without any of which the process cannot continue.

Questions

1. Why do you use KOH inside the bottle?
2. What result do you expect if another experiment is conducted without keeping KOH in the bottle?
3. How can you test the presence of starch in the part of the leaf outside the bottle?

3. To demonstrate that chlorophyll is necessary for photosynthesis

Pluck a variegated leaf of a *Coleus* or *Croton* plant kept in sunlight. Test this leaf for starch by iodine solution. You will notice that only those parts of leaf turn blue which were originally green. The yellow parts or white parts of leaf show negative test. It indicates that chlorophyll is necessary for photosynthesis (Figure E6.6).

Questions

1. Suggest some alternative plant materials for this experiment.
2. Classify the chloroplastic pigments.
3. Differentiate between Chl *a* and Chl *b*.
4. Besides chlorophyll, what are the other pigments that participate in photosynthesis?
5. Name the instrument used to study the absorption spectrum of the plant pigments.

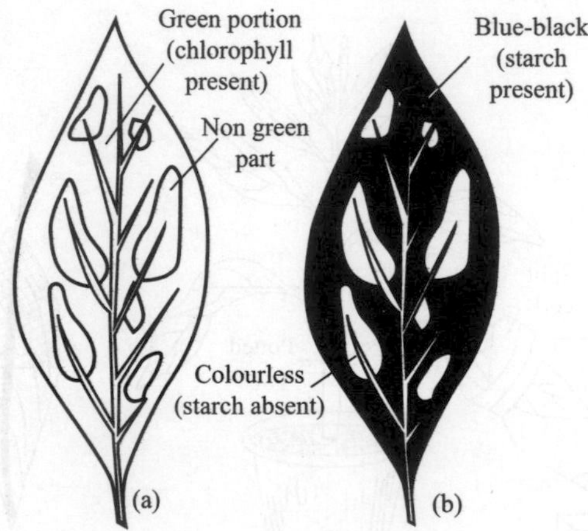

Fig. E6.6 A demonstration showing that chlorophyll is necessary for photosynthesis.

4. To demonstrate that O_2 is evolved during photosynthesis

Take a beaker filled with filtered pond water. Insert the twigs of *Hydrilla* in the funnel in such a way that the cut ends are firmly introduced in the tube of funnel but spreading in its wide mouth. Place the funnel fully immersed in water and then invert a test tube filled with water over the stem of the funnel. Keep this apparatus in sunlight for a few hours and observe (Figure E6.7).

Bubbles will come out through the stem of the funnel indicating the evolution of a gas. The oxygen gas will be collected in the tube after replacing water. When the test tube is completely filled with the gas, remove it from the funnel and test the gas with the help of either a burning match stick or by pyrogallol solution. The burning match stick will burn brighter and more quickly and the gas will be dissolved in the pyrogallol solution after which it will not affect the burning match stick further.

Both the tests indicate that oxygen is evolved during photosynthesis due to photolysis of water.

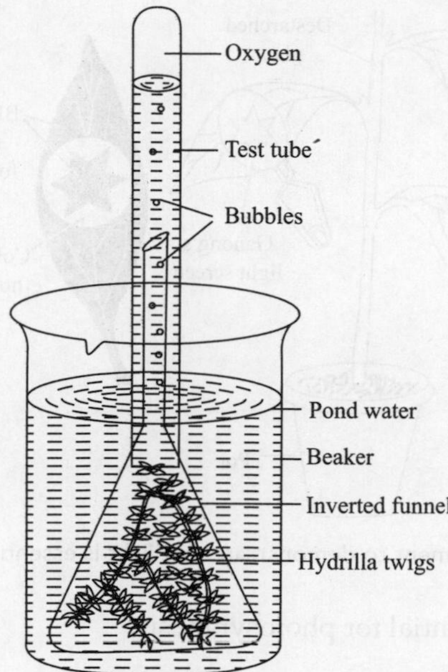

Fig. E6.7 Demonstration of evolution of O_2 during photosynthesis.

Questions

1. Why do you use *Hydrilla* in the experiment? Can it be replaced by a mesophytic or a xerophytic plant?
2. If (i) boiled water (ii) soda water is used instead of pond water, what effect it will have on the experimental set-up?
3. How will the rate of photosynthesis be affected if some toxic or anaesthetic substance is added to H_2O?
4. Why do you add $NaHCO_3$ in the pond water? Can this be substituted by Na_2CO_3?
5. Which test confirms the evolution of oxygen in photosynthesis?
6. Write a balanced chemical equation for photosynthesis in plants.
7. What is the source of oxygen released during this process?
8. List some of the external and internal factors affecting this physiological process.
9. Distinguish between photosynthesis in plants and bacteria.
10. Define the greenhouse effect.

5. To demonstrate that light is essential for photosynthesis

Place a potted plant in dark for 3 days so that leaves become free from starch. Fix a Ganong's light screen to one of the leaves. Keep the plant in bright sunlight. After a few hours, detach the leaf and boil it in 70 per cent alcohol till it becomes decolorized. Stain the leaf with iodine solution. The leaf turns blue only in those parts which were exposed to light. The portions which were covered with the light screen will not show blue stain (Figure E6.8).

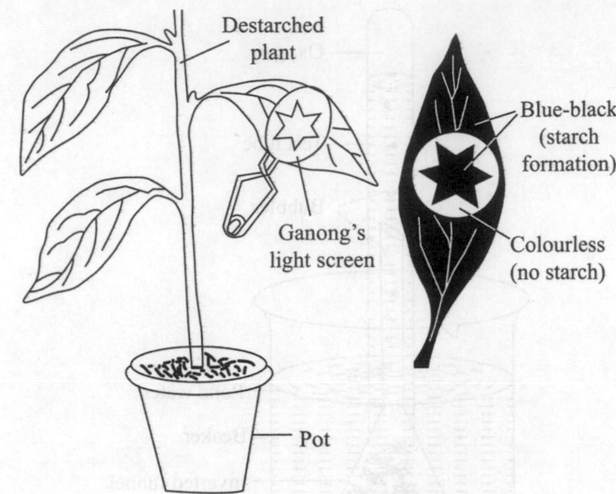

Fig. E6.8 An experiment to demonstrate that light is essential for photosynthesis.

It proves that light is essential for photosynthesis.

Questions

1. Describe the role of light in photosynthesis.
2. Mention the wavelengths of visible light which are most effective in photosynthesis.
3. Can you suggest any other method to demonstrate this experiment?

7. Respiration

Experiment 7.1

To compare the rates of respiration in different plant materials/different organs of a plant

Requirements

Plant material : *Ageratum conyzoides*
Family : Asteraceae
Chemicals : Brine solution (saturated solution of NaCl), KOH pellets
Glass apparatus : Buchner's flasks (3 sets), measuring cylinder (1000 ml), beakers (250 ml), ignition tubes (3)
Miscellaneous : Electric balance, pestle and mortar, clamp stands, black paper

Principle

Cellular respiration consists of a series of interdependent pathways by which carbohydrate and other molecules are oxidised for the purpose of retrieving the energy stored in photosynthesis and to obtain the carbon skeletons that serve as precursors for other molecules used in the growth and maintenance of the cell. As a general rule, rate of respiration is a reflection of 'metabolic demands'. Younger plants, organs or tissues respire more rapidly than older plants, organs or tissues. The rapid rate of respiration during early stages of growth is presumably related to synthetic requirements of rapidly dividing and enlarging cells. As the plant or organ ages and approaches maturity, growth and its associated metabolic demands decline. However, many organs, especially leaves and some fruits, experience a transient rise in respiration, called a **'climacteric'**, that marks the onset of senescence.

Respiratory rate also depends upon the protoplasmic factors, i.e., amount of protoplasm and state of its activity. Amount of active protoplasm is highest in the younger cells, especially in actively growing regions. Moreover, the growing regions require a much larger amount of energy than the older ones and this is supplied by the higher rate of respiration. Dry seeds and spores have the lowest respiration rates. This is because certain changes in the protoplasm especially desiccation shut off metabolism.

Other factors being constant, the rate of respiration of a given tissue is governed by the available amount of respiratory substrate, sucrose.

Green leaves starved in the dark and etiolated leaves show a fall in the rate of respiration. The rate of respiration increases rapidly, if the leaves are placed in sucrose solution. Similarly, green leaves show an increase in the rate of respiration after a period of illumination on account of the increase in the amount of respiratory material as a result of photosynthesis. However, high concentrations of sugars have an adverse effect on respiration, probably due to osmotic effects.

Besides age and metabolic state, the respiratory rate of whole plants and organs varies widely with environmental conditions; light, temperature, gaseous composition of the atmosphere, salts and mechanical stimuli.

Procedure

Weigh equal amounts (7 -10 grams) of three organs of the same plant, i.e., leaves, roots and flower buds. Set up 3 apparatus consisting of Buchner's flask with a rubber cork, an ignition tube containing a thick paste of 10 - 12 pellets and a beaker containing brine solution (Figure E7.1). Place the three plant materials/organs in separate Buchner's flasks and make the apparatus airtight, while the graduated side tube is immersed in the measuring cylinder containing brine solution. Cover the apparatus having green leaves with black paper to avoid photosynthesis. Remove the graduated side tube from the cylinder and shift it to a beaker containing brine solution and note down the initial level. Later on, note down the rise in level in all the three set-ups at 5, 10 and 15 minute intervals.

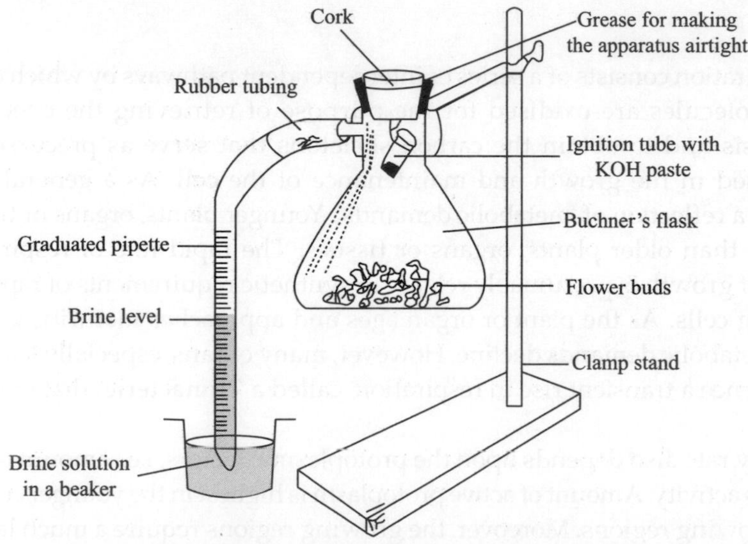

Fig. E7.1 An experiment to determine the rate of respiration.

Observation table

Plant material	Rise in level in ml (*d*)				Rate of respiration*
	$t \rightarrow$ Initial	5 min	10 min	15 min	d/t (ml/h)
Roots	4.5	4.3	4.1	3.9	0.6
Leaves	4.5	4.2	4.0	3.8	0.7
Flower buds	4.3	3.8	3.2	2.8	1.5

* Rate of respiration can also be expressed as ml/h/g of plant tissue.

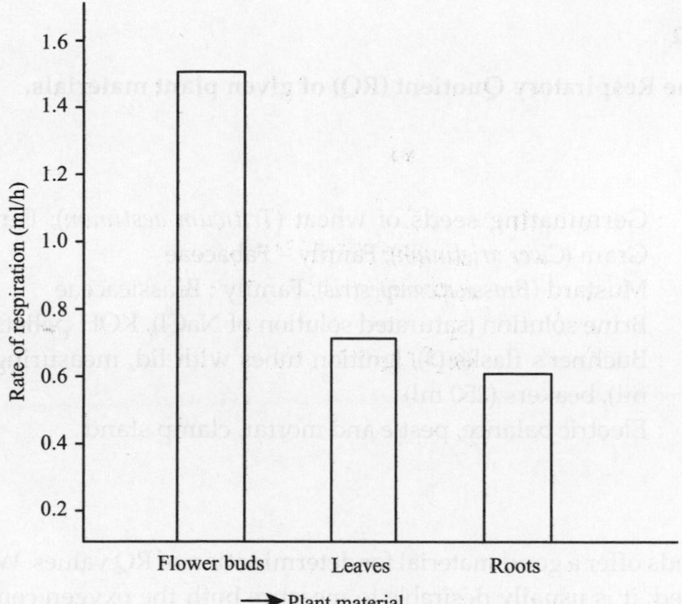

Histogram representation showing the rate of respiration

Result and Discussion

KOH in the ignition tube absorbs CO_2 released during respiration of plant parts. Hence, a vacuum is created in the flask causing brine level to rise. The rise in level of brine in the side tube after 15 minutes is taken as the measure of the rate of respiration per hour. Compare the values obtained for the three plant organs.

Respiration rate is found to be at a maximum in the flower buds followed by leaves and then the roots.

Flower buds > leaves > roots

Flower buds comprising of actively dividing cells show a very high respiratory activity due to increased energy demand. Leaves are also rich in photosynthetic tissue with a high level of respiratory substrate available. Mature roots, because of low metabolic rates, probably show less respiration as compared to flower buds and leaves.

Precautions

1. The amount of the plant materials should remain same in all the three sets.
2. The ignition tube containing KOH should remain open throughout the experiment.
3. The apparatus should be vertical and airtight.
4. The graduated side tube should not touch the walls or bottom of the beaker.
5. Plant material should not be washed before the start of an experiment.

The apparatus containing green leaves should be covered with black paper to cut off photosynthetic activity; hence not affecting the amount of CO_2 given out.

Experiment 7.2

To determine the Respiratory Quotient (RQ) of given plant materials.

Requirements

Plant material : Germinating seeds of wheat (*Triticum aestivum*); Family – Poaceae, Gram (*Cicer arietinum*); Family – Fabaceae
Mustard (*Brassica campestris*); Family : Brassicaceae
Chemicals : Brine solution (saturated solution of NaCl), KOH pellets
Glass apparatus : Buchner's flasks (3), ignition tubes with lid, measuring cylinder (1000 ml), beakers (250 ml)
Miscellaneous : Electric balance, pestle and mortar, clamp stand

Principle

Germinating seeds offer a good material for determination of RQ values. When respiration is being measured, it is usually desirable to measure both the oxygen consumed and the CO_2 evolved. The ratio of moles of CO_2 produced to moles of O_2 consumed is called the **Respiratory Quotient** (RQ).

$$RQ = \frac{\text{moles of } CO_2 \text{ evolved}}{\text{moles of } O_2 \text{ consumed}}$$

RQ value varies with the nature of the respiratory substrate: carbohydrates = 1; Proteins/ fats = less than 1 (as these are highly reduced and require more oxygen for oxidation); organic acids = more than 1 (as these are highly oxidized and require less oxygen for oxidation).

(For chemical equations and importance of RQ, refer to the Chapter 9).

Procedure

Take 15 to 20 grams of well germinated seeds in the Buchner's flask. Suspend inside the flask an ignition tube with lid having thick KOH paste. Dip the graduated side tube in a measuring cylinder filled with brine solution and allow this to rise considerably (I). Make the apparatus airtight. Gently remove the graduated tube from the cylinder and transfer it into a beaker filled with brine solution. Note down the level of brine solution in the graduated tube (F_1) after 15 minutes.

Shake the apparatus to remove the lid of ignition tube and after 5 minutes, note down the final level (F_2). Repeat the experiment using other plant materials and compare the RQ values obtained.

Calculations

Initial reading in pipette = I
First final reading before dislodging the cover-glass = F_1
Amount of oxygen consumed in excess = $I - F_1$

Second final reading after dismantling the cover-glass = F_2
Amount of CO_2 released during the course of experimentation = $F_1 - F_2$

$$RQ = \frac{\text{Amount of } CO_2 \text{ evolved}}{\text{Amount of } O_2 \text{ consumed}} = \frac{F_1 - F_2}{(F_1 - F_2) + (I - F_1)}$$

Brassica campestris
I = 3.1 ml
F_1 = 1.0 ml (Level moves up)
F_2 = 0.4 ml (Level moves up)

$$RQ = \frac{1.0 - 0.4}{(1.0 - 0.4) + (3.1 - 1.0)} = \frac{0.6}{0.6 + 2.1} = \frac{0.6}{2.7} = 0.2$$

Triticum aestivum
I = 2.0 ml
F_1 = 2.0 ml (Level remains the same)
F_2 = 1.5 ml (Level moves up)

$$RQ = \frac{2.0 - 1.5}{(2.0 - 1.5) + (2.0 - 2.0)} = \frac{0.5}{0.5} = 1.0$$

Cicer arietinum
I = 4.3 ml
F_1 = 2.8 ml (Level moves up)
F_2 = 1.6 ml (Level moves up)

$$RQ = \frac{2.8 - 1.6}{(2.8 - 1.6) + (4.3 - 2.8)} = \frac{1.2}{1.2 + 1.5} = \frac{1.2}{2.7} = 0.4$$

Plant material	RQ value whether <, =, > 1	Nature of respiratory material
Brassica	$0.2 \therefore RQ < 1$	Fats
Triticum	$1.0 \therefore RQ = 1$	Carbohydrates
Cicer	$0.4 \therefore RQ < 1$	Proteins

Theoretically, the value of the respiratory quotient in fatty and proteinaceous material should be about 0.7 and 0.5, respectively. But, actually it is not so because the seeds generally contain both proteins and fats together.

Result and Discussion

There are three possibilities regarding brine level in the graduated tube.

(a) ***The level moves up***: This is when more of O_2 is being consumed during respiration by the germinated seeds as compared to amount of CO_2 released. A vacuum is created in the flask which results in upward movement of brine level in the graduated tube. This is observed in the case of *Brassica* and *Cicer*. RQ values of these materials as expected, are found to be <1, since amount of CO_2 released is less than amount of oxygen consumed.

(b) ***The level remains the same***: This is when amount of CO_2 released during respiration is balanced by the amount of O_2 consumed. Thus, no vacuum develops in the flask and the brine level in the graduated tube remains stable.

This is observed in the case of germinating seeds of wheat. RQ value comes out to be 1, since amount of CO_2 released is equal to amount of O_2 absorbed.

(c) ***The level moves down***: Though no plant material was found to exhibit this, the situation could have arisen in case amount of CO_2 released is more than amount of O_2 absorbed during respiration. Increased CO_2 level in the flask exerts a pressure on the brine column causing the level to fall. RQ values for such plant materials (*Sedum, Crassula, Bryophyllum*) is greater than 1, since amount of O_2 absorbed is less than amount of CO_2 released.

After the cover-glass has been dislodged, KOH absorbs CO_2 released during respiration by germinating seeds. Hence a vacuum is created in the flask causing brine level in graduated tube to rise.

Precautions

1. Same amount of germinating seeds should be used in all the set-ups.
2. The seeds must be well germinated so that respiration rate is high (radicles should have emerged out).
3. Seeds must be blotted dry in the folds of a blotting paper as the moisture may absorb CO_2.
4. Number of KOH pellets should be same for all plant materials and a concentrated paste should be made.
5. The lid of the ignition tube should be placed lightly to facilitate its easy dislodging on slight jerking of apparatus when required.
6. The set-up should be vertical and airtight.
7. Before making the apparatus airtight, the graduated tube should be completely immersed in brine solution[15] in a measuring cylinder so that the level can be brought somewhere in the middle.
8. No trickling of brine solution should occur once the apparatus is airtight.
9. The graduated tube should not touch the bottom of the beaker.

Mercury can also be used to indicate the level as CO_2 is completely insoluble in mercury. But it is rarely used because it is expensive.

[15] Brine solutions are generally used in the experiment since CO_2 is insoluble in it. However, CO_2 is soluble in water.

Questions for Viva Voce

1. Why are germinated seeds used instead of dry ones in the experiment?
2. Why do you use the brine solution?
3. Why do you use a strong solution of KOH inside the ignition tube?
4. Can boiled seeds/seeds pre-treated with chemical be used for the experiment?
5. Define RQ. In which way it is useful in knowing the chemical composition of respiratory substrate?

Physiological Set-Ups

I. To demonstrate that CO_2 is evolved during aerobic respiration by conical flask method

Germinating seeds (of gram, wheat, or mustard) are placed in a conical flask fitted with a cork. One end of a narrow glass tube which is bent at two places is inserted into the conical flask through a hole in the cork, while the other end is dipped in saline water contained in a beaker. A small glass tube containing KOH solution is suspended into the conical flask by means of a thread (Figure E7.2).

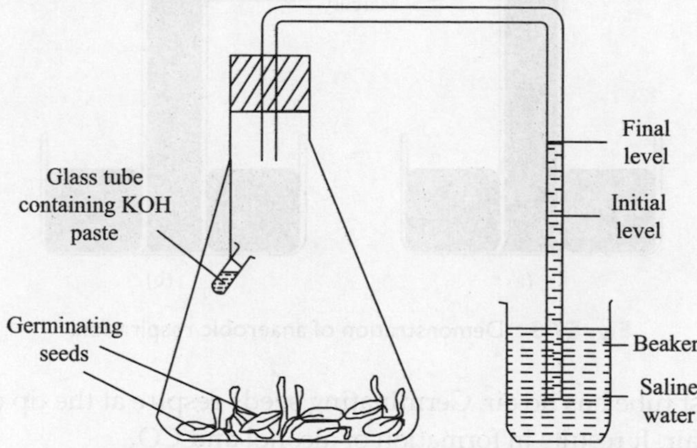

Fig. E7.2 An experiment to demonstrate the evolution of CO_2 during aerobic respiration.

After some time, the CO_2 produced during aerobic respiration of germinating seeds will be absorbed by the KOH solution and the level of water will rise in the glass tube.

Questions

1. Describe briefly the major steps in the mechanism of aerobic respiration.
2. List the factors which increase the rate of respiration of a given plant material.
3. Name four distinct processes in plants which require the energy of respiration.
4. Land animals need lungs for inhaling and exhaling air. How do the plants manage without them? Explain.
5. Write the balanced chemical equation for the complete oxidation of a glucose molecule.

6. What precaution would you take if green plant is used instead as the respiratory material?
7. Mention the role of KOH kept in the glass tube.
8. Suggest some other alternative of saline solution in the beaker.

2. To demonstrate anaerobic respiration

A test tube is completely filled with mercury and inverted over a petri dish filled half with mercury. Germinating seeds (soaked gram seeds) are introduced into the tube by forceps through its open end. The seeds rise to the top of the tube (Figure E7.3). The apparatus is allowed to remain as such for a few hours. The level of mercury falls down. Then KOH crystals are inserted into the tube. The mercury begins to rise again.

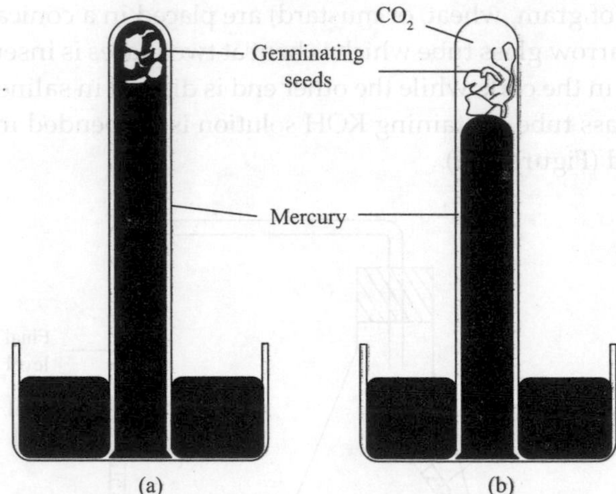

Fig. E7.3 Demonstration of anaerobic respiration.

Initially, the test tube has no air. Germinating seeds respire at the tip of the tube in the absence of air. It results in formation of alcohol and CO_2.

$$\underset{\text{Glucose}}{C_6H_{12}O_6} \rightarrow \underset{\text{Carbon dioxide}}{2CO_2} + \underset{\text{Ethyl alcohol}}{2C_2H_5OH}$$

CO_2 produced pushes down the mercury column. On the introduction of KOH crystals, this CO_2 is absorbed. Thus, mercury rises once again.

Questions

1. In which way is anaerobic respiration useful to plants?
2. What is the difference between aerobic and anaerobic respiration?
3. In what way does the anaerobic respiration resemble alcoholic fermentation?
4. Explain the mechanism of glucose breakdown in anaerobic respiration.
5. Give an account of the energy-releasing process during anaerobic respiration.
6. How will you test the presence of carbon dioxide gas in the inverted test tube?

3. To demonstrate the process of fermentation

Prepare a 10 per cent solution of glucose and add a small quantity of baker's yeast to it. Pour the mixture into Kuhne's fermentation vessel. The complete upright tube and half of the bent bulb is filled. The open end of the bent bulb is plugged with a stopper. The apparatus is allowed to stand for a few hours (Figure E7.4).

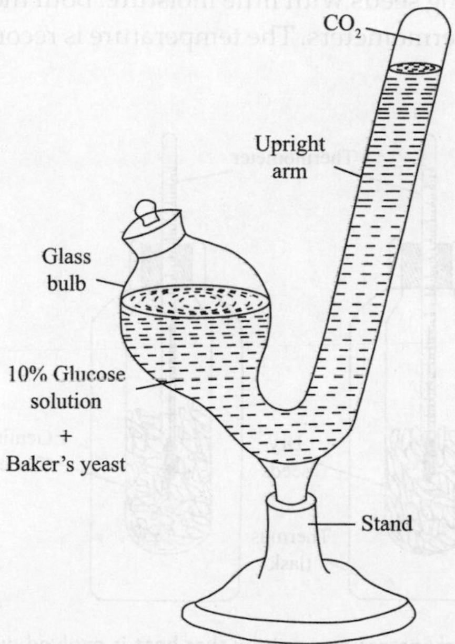

Fig. E7.4. Kuhne's fermentation vessel.

Carbon dioxide gas begins to collect in the upright arm of the vessel with a consequent fall in the level of solution. CO_2 gas can be tested by introducing caustic potash in the vessel. The smell of alcohol can be detected from the remaining solution. (For chemical equations, refer to Chapter 9).

Questions

1. What would happen if the contents of the tube were pre-boiled?
2. What occupies the space at the tip of the tube?
3. What is the scientific name of yeast?
4. List the factors affecting the phenomenon.
5. What will happen if the set-up is kept at (i) 4 °C and (ii) 100 °C?
6. Write about the usefulness of the technique.
7. Which gas is formed during fermentation?
8. Write the chemical equation for the reaction of the alkali with the gas collected in the upright arm.
9. Name other kinds of fermentation carried out by microorganisms.
10. Where does fermentation occur within a cell?

11. Name the two enzymes responsible for fermentation.
12. How does fermentation differ from aerobic respiration?

4. To demonstrate that energy is released in the form of heat during respiration

Introduce some ungerminated dry seeds in one of the thermos flask and in the other, the same amount of germinating seeds with little moisture. Both the flasks are corked airtight with the cork fitted with thermometers. The temperature is recorded before and after some time (Figure E7.5).

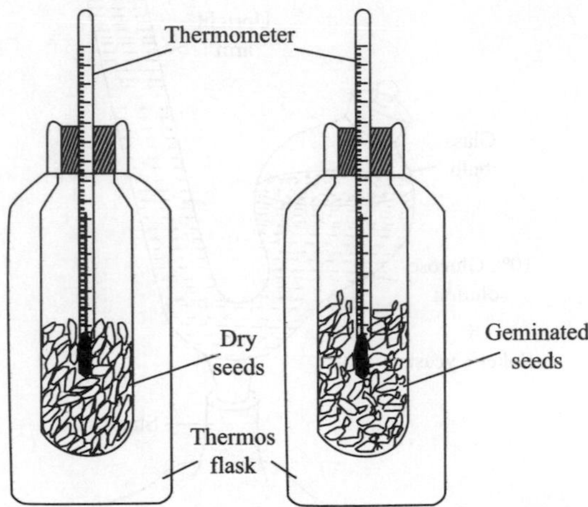

Fig. E7.5 A demonstration to show that heat is evolved during respiration.

The thermometer of the thermos flask containing the germinating seeds records a marked rise in temperature because during respiration some untrapped energy is released in the form of heat.

Questions

1. What difference would it make if boiled seeds are used in another thermos flask?
2. List two important precautions which you will observe while setting up the apparatus.

5. To demonstrate root respiration

Place a small rooted plant (e.g., *Tagetes*) with the intact root in a bottle or flask containing water made slightly alkaline with dilute NaOH solution and coloured red with phenolphthalein. Prepare a second flask to serve as control, stopper tightly and leave without a plant. Allow both flasks to stand in diffuse light and examine the solution after some time.

Observe that the control flask does not show change in the colour while the one with the roots becomes colourless. The respiring roots release CO_2, which reacts with water to form carbonic acid (H_2CO_3).

$$CO_2 + H_2O \rightarrow H_2CO_3$$

Carbonic acid neutralizes the NaOH present in the flask and the alkalinity of the solution starts decreasing, causing the red colour to fade (phenolphthalein is colourless in the neutral medium).

$$H_2CO_3 + 2NaOH \rightarrow Na_2CO_3 + 2H_2O$$

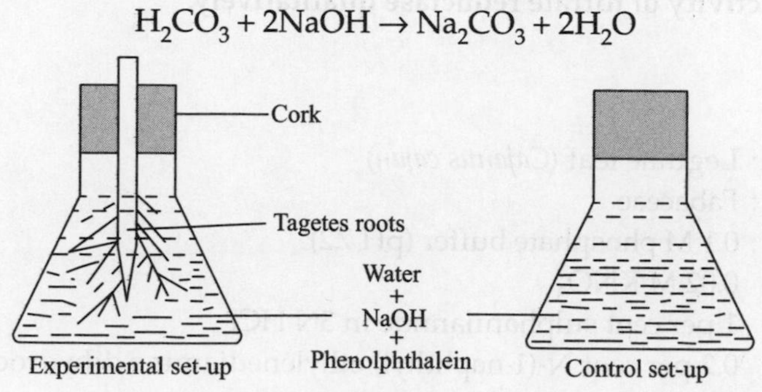

Questions

1. Explain, with the use of equations, the physiological mechanism involved in this set-up.
2. What would happen if the solution in the flask contains a buffer solution of pH 7.8?
3. What will happen if a drop of HCl is added to the experimental flask?
4. What results do you expect, if the roots in the experimental flask are pre-boiled?
5. Will there be any change if the set-up is shifted to darkness?
6. In which way is root respiration helpful in the nutrient uptake from the soil? Explain.

8. Nitrogen Metabolism

Experiment 8.1

To estimate the activity of nitrate reductase qualitatively.

Requirements

Plant material : Legume leaf (*Cajanus cajan*)
Family : Fabaceae
Chemicals : 0.1 M phosphate buffer (pH 7.2)
0.02 M KNO_3
1 per cent sulphanilamide in 3N HCl
0.2 per cent N-(1-naphthyl) ethylenediamine dihydrochloride (NEDD), distilled water, sodium nitrite
Glass apparatus : Test tubes
Miscellaneous : Test tube stand, scissors, vacuum pump, water bath at 33 °C, photoelectric colorimeter or spectrophotometer

Principle

The assimilatory reduction of nitrate by plants is a fundamental biological process in which a highly oxidised form of inorganic nitrogen is reduced to nitrite and then to ammonia.

$$NO_3^- + 2H^+ + 2e^- \rightarrow NO_2^- + H_2O$$

Nitrate reductase is a ubiquitous enzyme found in both prokaryotic and eukaryotic cells.[16] It has been most intensively studied in *Neurospora* by A. Nason and H. I. Evans (1953). Nitrate reductase is cytosolic, inducible and a highly regulated enzyme. It has long been recognized that both NO_3^- and light are required for its maximum activity. Its activity can also be reversibly regulated by red and far-red light, indicating control by the phytochrome system. Nitrate reductase is metalloflavoprotein in nature consisting of NADH, FAD and molybdenum.

Procedure

Wash 0.3 grams fresh legume leaf and cut into small pieces. Add 3 ml of phosphate buffer and 3 ml of KNO_3 to it. Incubate at 33 °C for 1 h in the dark. Take out 0.2 ml aliquot in a test tube and terminate the reaction by adding 1 ml of sulphanilamide and 1 ml of NEDD. Make the volume upto 6 ml using distilled water. Keep in dark for 20 minutes. Read absorbance at 540 nm. Prepare a standard curve with sodium nitrite (5–50 µg).

[16] In plants, nitrate reductase is generally found in the roots and leaves of legumes.

Protocol (*in vivo* method of Jaworski, 1971)

0.3 g fresh plant material (legume leaf, e.g., *Cajanus cajan*)

↓

Wash and cut into small pieces

↓

Add 3 ml phosphate buffer and 3 ml KNO_3 to the plant material in the test tube

↓

Evacuate air by vacuum pump till the plant material settles down

↓

Shake and incubate at 33 °C in water bath for 1 h in the dark

↓

Keep in hot water bath for 5 minutes to stop the reaction

↓

Take out 0.2 ml aliquot; to this add 1 ml sulphanilamide and 1 ml NEDD

↓

Make the volume to 6 ml by adding 3.8 ml distilled water

↓

Keep tubes in dark for 20 minutes

↓

Read absorbance at 540 nm using photoelectric colorimeter or spectrophotometer.

Result and Discussion

Nitrate reductase activity is measured by colorimetric determination of nitrite produced when nitrite diazotizes with sulphanilamide at pH 1.0. The *p*-diazonium sulphanilamide thus formed subsequently reacts with N-(1-naphthyl)ethylenediamine dihydrochloride to form a pink, azo dye with an absorption maximum at 540 nm. The pink colour (diazo complex) thus obtained by nitrate is stable for 2 - 3 hours. Activity is expressed as micromole nitrite produced per minute per mg protein or per gram fresh tissue.

NH_2 — SO_2NH_2 Sulphanilamide (SAN) $\xrightarrow[H^+]{NO_2^-}$ $N \equiv N^+$ — SO_2NH_2 *p*-diazonium sulphanilamide + HN— NH_2 *N*-(1-Naphthyl)ethylene -diamine dihydrochloride (NEDD) → HN— NH_2 — N=N — SO_2NH_2 Azo chromophore (λ_{max} = 543 nm)

Precaution

Fresh leaves used for the experiment should be torn into small pieces.

Questions for Viva Voce

1. Where does nitrate reduction occur in the plant cell?
2. Write the probable sequence in the reduction of nitrate to the amino acids in plant tissues.
3. What is the source of energy for the reduction of nitrate?
4. Would you expect nitrate or ammonium nitrogen to be more rapidly incorporated into plant proteins? Why?

Experiment 8.2

To separate and identify amino acids from a given mixture by paper chromatography and thin layer chromatography.

Requirements

Chemicals : Amino acids (proline, glycine, leucine in N/10 HCl), ninhydrin solution (0.2 per cent in acetone); Organic solvent[17] (butanol: acetic acid:distilled water, 3:1:1); Silica gel-G

Glass apparatus : Chromatography jar with lid, glass plate, capillary tube or microsyringe

Miscellaneous : Whatman paper (No. 1), atomizer, TLC applicator, hair-dryer

Principle

For details, refer to experiment 6.1 and 6.2.

Procedure

Cut a sheet of Whatman No. 1 paper to a convenient size (for TLC, take a glass plate coated with a thin layer of silica gel-G). Draw a line 2.5 cm from the edge of the paper/ glass plate. Load the samples using fine capillary tubes, keeping the diameter of each spot as small as possible (less than 1 cm). Repeat the loading process 5 or 6 times. After each application, dry the first spot with the help of a hair-dryer.

Meanwhile, saturate the chromatography jar with the vapours of the organic solvent. Lower the chromatography paper/spotted plate into the solvent, making sure the solvent is below the origin and replace the lid. When the chromatogram has run for the desired

[17] Role of components of organic solvent
(i) Butanol - Mobile phase (non-polar)
(ii) Acetic acid - Buffering action
(iii) Distilled water - Hydrated interphase

time (allow the solvent to run about 2/3rd of the length of paper/ plate), remove it from the jar and mark the position of the solvent front. Dry the chromatogram with a hair-dryer. The spots are located by spraying the paper/plate with the ninhydrin solution using an atomizer. Dry the paper/plate for 5 minutes at room temperature followed by at 100 °C in an oven for 2–3 minutes.

Observation table

Solvent front = 9.7 cm

Solute or amino acid	Distance travelled by solute (cm)	$R_f = \frac{\text{Distance travelled by solute}}{\text{Distance travelled by solvent}}$
Proline	5.1	$\frac{5.1}{9.7} = 0.52$
Leucine	7.9	$\frac{7.9}{9.7} = 0.81$
Glycine	3.4	$\frac{3.4}{9.7} = 0.35$
Mixture	(i) 8.0	$\frac{8.0}{9.7} = 0.82$
	(ii) 3.4	$\frac{3.4}{9.7} = 0.35$

Result

The given mixture of amino acids contains leucine and glycine.

Glycine: $CH_2(NH_2)COOH$

Leucine: $H_3C—HC(CH_3)—H_2C—CH(NH_3)—COOH$

Proline: $CH_2—CH_2$, CH_2, $CH—COOH$, NH (ring)

Glycine Leucine Proline

Discussion: Ninhydrin (triketohydrindene hydrate) reacts with α-amino acids when the pH is between 4 and 8, to yield a purple coloured complex. Not all amino acids give exactly the same intensity of colour. Proline (or hydroxyproline) gives a yellow colour. The migration of a substance relative to the solvent front is a characteristic of the substance. This is called the R_f value.

$$R_f = \frac{\text{Distance moved by the solute}}{\text{Distance moved by the solvent front}}$$

Identification of the spots may be achieved by comparing their R_f values with those of unknown compounds run on the same chromatogram. Molecular weight of the amino acids and the length of the side chains of the amino acids form a basis for their separation by paper chromatography/thin-layer chromatography.

$$\alpha\text{-Amino acid} + \text{Ninhydrin (oxidized)} \rightarrow \alpha\text{-Imino acid} + \text{Ninhydrin (reduced)}$$

$$\alpha\text{-Imino acid} \xrightarrow[\text{Hydrolysis}]{\text{Decorboxylation}} RCHO + CO_2 + NH_3$$

$$\underset{\text{(oxidized)}}{\text{Ninhydrin}} + NH_3 + \underset{\text{(reduced)}}{\text{Ninhydrin}} \longrightarrow \text{Blue complex}$$

$$RCHNH_2{\cdot}COOH + \text{Ninhydrin (oxidized)} \longrightarrow RC{=}NHCOOH + \text{Ninhydrin (reduced)} + H_2O$$

Ninhydrin (oxidized) Ninhydrin (reduced)

$$RC = NHCOOH + H_2O \longrightarrow RCHO + CO_2 + NH_3$$

$$\text{Ninhydrin (oxidized)} + NH_3 + \text{Ninhydrin (reduced)} \longrightarrow \text{Blue product (Ruheman's complex)} + 3H_2O$$

Ninhydrin (oxidized) Ninhydrin (reduced) Blue product (Ruheman's complex)

Precautions

1. The organic solvent should not touch the spots on the chromatography paper or glass plate, while being transferred into the jar.
2. Capillary tube should be fine and spot should be kept small for maximum resolution.
3. Jar should be completely saturated with the vapours of the organic solvent prior to the experiment.
4. Chromatography paper should not touch the side walls or bottom of the jar.
5. Apparatus once set up, should not be disturbed.
6. Ninhydrin solution should be prepared fresh.
7. Solvent front should be marked when the paper or plate is still wet.

Handle the chromatography sheet very carefully until developed as otherwise amino acids from fingers will contaminate it. It is advisable to hold the chromatographic paper between a fold of filter paper piece or to be kept in a notebook.

Experiment 8.3

To prepare a standard curve for proteins using the Biuret method.

Requirements

Chemicals : Bovine serum albumin also known as BSA (20 mg/ml)
Biuret reagent
[3 g $CuSO_4 . 5H_2O$ + 9 g sodium potassium tartarate in 500 ml of 0.2 mol/L of NaOH. Add 5 g of KI → make upto 1 L with 0.2 mol/L of NaOH]
BSA samples of unknown concentrations

Glass apparatus : Test tubes, pipettes (1 ml, 10 ml), beaker

Miscellaneous : Test tube stand, water bath at 37 °C, graph paper, spectrophotometer or photoelectric colorimeter, vortex mixer

Principle

Proteins[18], one of the important macromolecules found in all living cells, are polymers of different amino acids. There are about 20 different amino acids which make these proteins. In proteins, the amino acids are connected by peptide bonds.

Various methods such as Biuret, UV, Lowry, dye binding and fluorescamine are available to determine the concentration of proteins in biological samples. Each of these methods vary in their sensitivity and hence their applicability.

Biuret method: It is a widely employed colour reaction of peptides and proteins - producing colours not associated with free amino acids. Such colours are of excellent intensity for measurement by photospectrometers. The name of the test comes from the compound known as **biuret** which gives a typical positive reaction for the test.

$CONH_2$
|
NH
|
$CONH_2$

Biuret

Treatment of a peptide or protein with Cu^{2+} and alkali yields a purple Cu^{2+}-peptide complex, which can be measured quantitatively in a spectrophotometer.

Procedure

Prepare BSA solutions of various concentrations using 20 mg/ml BSA as the stock solution. Add 3 ml of biuret reagent to 2 ml of BSA solution. Mix and warm at 37 °C for 10 minutes. Cool and read the absorbance at 540 nm using tube 1 to zero the instrument. Plot your readings against the BSA concentration on a graph paper.

---

[18] The term 'protein' was coined by J. J. Berzelius in 1838 and derived from Greek word 'proteinous' which means of primary importance.

Dilution table

Stock solution of BSA = 20 mg/ml
Total volume of the solution = 1 ml

$$\text{Dilution} = \frac{\text{Required concentration}}{\text{Given concentration}} \times \text{Total volume of the solution}$$

S. No.	Conc. of BSA (mg/ml)	Amount of BSA (ml)	Amount of distilled H_2O (ml)
1	0	0	1
2	2	0.1	0.9
3	4	0.2	0.8
4	6	0.3	0.7
5	8	0.4	0.6
6	10	0.5	0.5
7	12	0.6	0.4
8	14	0.7	0.3
9	16	0.8	0.2
10	18	0.9	0.1
11	20	1.0	0

Observation table

S. No.	Conc. of BSA (mg/ml)	Absorbance at 540 nm
1	0	0.0
2	2	0.07
3	4	0.08
4	6	0.09
5	8	0.12
6	10	0.15
7	12	0.17
8	14	0.22
9	16	0.25
10	18	0.28
11	20	0.30
12(A)	Unknown	0.16
12(B)	Unknown	0.28

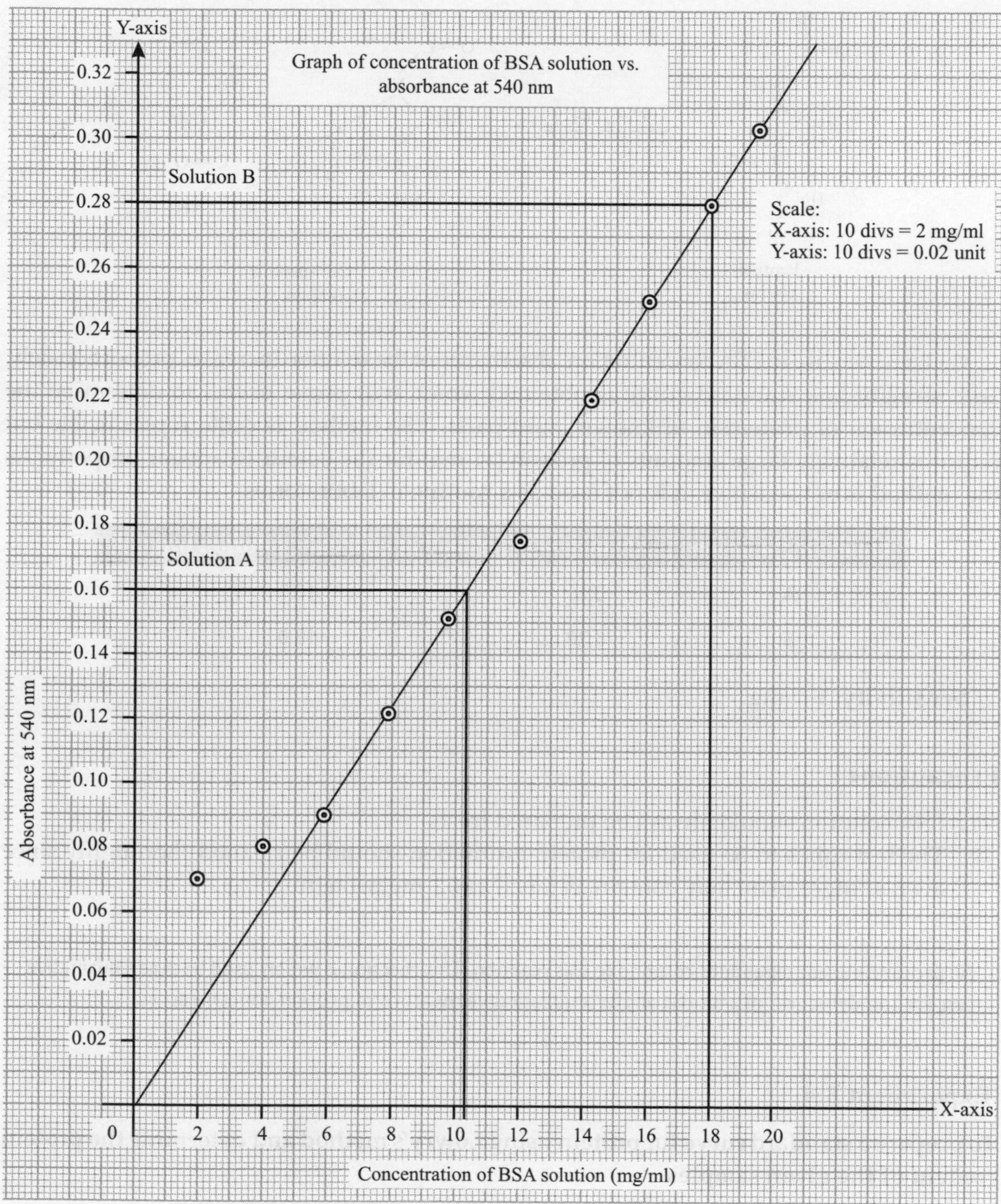

Result and Discussion

Peptides which have two or more peptide bonds will give a violet-purple colour when a small amount of copper sulphate is added to a strongly alkaline solution of protein. The colour produced is probably due to the formation of a coordination complex between a copper atom and four of the nitrogen atoms of the peptide bonds in proteins.

$2\ RHC$ (Peptide chains) $\xrightarrow{Cu^{2+},\ OH^-}$ Cu^{2+}-peptide complex

Peptide chains

Cu^{2+}-peptide complex

With an increase in BSA concentration, absorbance of the solution increases proportionately. So a straight-line standard curve is obtained which will prove useful to determine the concentration of protein in unknown plant parts. The method is quite reproducible but requires relatively large amounts of protein (>1 mg) for the colour to appear properly. As the method is not sensitive, it is not used as a routine in laboratories

Unknown solution	Absorbance	Conc. of unknown solution (mg/ml)
A	0.16	10.2
B	0.28	18.0

Sensitivity: 1–5 mg.

Precautions

1. BSA solution should always be prepared fresh.
2. The test tubes to be used for protein estimation should be cleaned thoroughly since even a finger print may contain as much as 0.1 mg protein.

All proteins except dipeptides respond to this reaction. This reaction is widely used both as a qualitative test for the detection of proteins and as a quantitative measure of protein concentration.

Experiment 8.4

To prepare a standard curve for protein using Lowry's method and determine the protein content of the given samples.

Requirements

Chemicals : Bovine serum albumin[19] (0.1 per cent)
Reagent A: 2 per cent Na_2CO_3 in 0.1N NaOH

[19] The BSA solution can be replaced with egg albumin as well.

Reagent B: 0.5 per cent $CuSO_4$ in 1 per cent sodium potassium tartrate
Reagent C: 50 ml of reagent A + 1 ml of reagent B
Folin-Ciocalteau reagent (AR)
Protein samples of unknown concentration (solutions A and B), distilled water

Glass apparatus : Test tubes, beaker, pipettes (1 ml, 10 ml)
Miscellaneous : Test tube stand, aluminium foil, colorimeter, graph, vortex mixer

Principle

Lowry's method is the most widely used method to determine the protein content in the biological investigations in all laboratories.

In this method, the colour development relies –

(i) On the formation of a copper-protein complex as in biuret reaction.
(ii) On the reduction of phosphomolybdate and phosphotungstate anions present in Folin-Ciocalteau phenol reagent by the tyrosine and tryptophan residues of the proteins, to heteropolymolybdenum blue and tungsten blue, respectively, which give a blue coloured complex with a λ_{max} of 660 nm/750 nm.
(iii) Cu^{2+} also acts as a catalyst in the reduction reaction. The intensity of the colour mainly depends on the aromatic amino acid content of a protein. Approximately, 75 per cent reduction occurs due to the copper-protein complex with tyrosine and 25 per cent with tryptophan residue.

Procedure

Prepare 8 tubes containing increasing amounts of BSA (0, 25, 50, 100, 150, 200, 250, 300 μg/ml) in a total volume of 1 ml. Meanwhile, prepare two tubes with unknown samples. Add 5 ml of alkaline reagent C to each tube, shake and let it stand for 10 minutes at room temperature. Add 0.5 ml of 1N Folin's reagent and mix immediately using a vortex mixer. The blue colour is measured at 660 nm after 30 minutes at room temperature.

Plot the values of A_{660} vs. concentration of BSA. Estimate the protein concentration of the unknown samples using the standard curve.

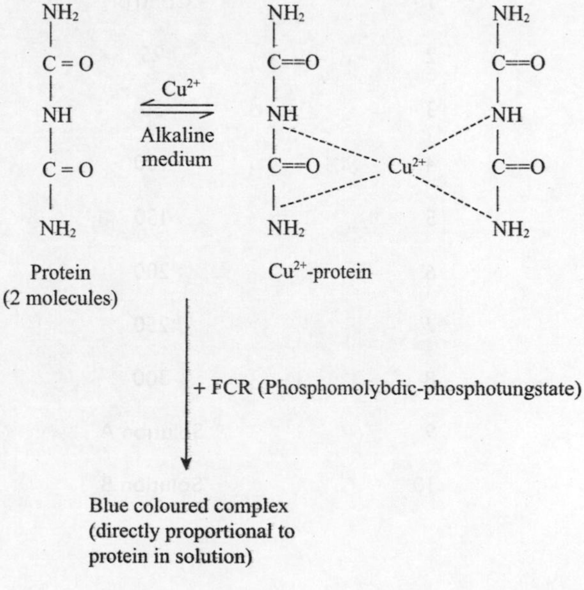

NH₂–C=O–NH–C=O–NH₂
Protein (2 molecules)
Cu^{2+}, Alkaline medium
Cu^{2+}-protein
+ FCR (Phosphomolybdic-phosphotungstate)
Blue coloured complex (directly proportional to protein in solution)

Dilution table

Stock solution: 0.1 per cent BSA (0.1 g BSA/100 ml of distilled water)

Concentration of BSA solution (μg/ml)	Amount of stock solution (ml)	Amount of distilled water (ml)
Control	0	10
25	0.25	9.75
50	0.50	9.50
100	1.00	9.00
150	1.50	8.50
200	2.00	8.00
250	2.50	7.50
300	3.00	7.00

Observation table

S. No.	Concentration of BSA solution (μg/ml)	Absorbance at 660 nm
1	Control	0.0
2	25	0.06
3	50	0.15
4	100	0.30
5	150	0.47
6	200	0.59
7	250	0.75
8	300	0.90
9	Solution A	0.35
10	Solution B	0.47

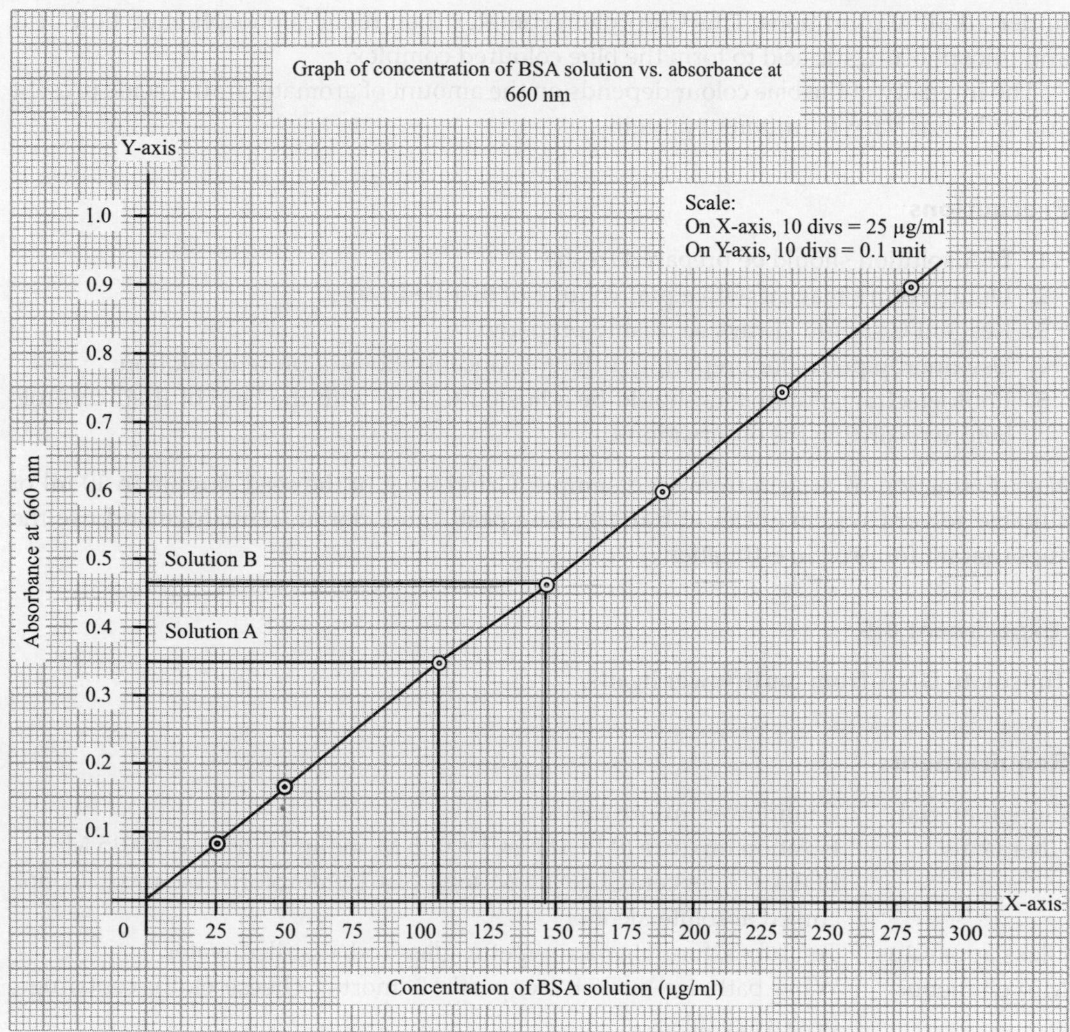

Result and Discussion

With increasing BSA concentrations, the absorbance of the solution increases. A straight-line standard curve is obtained on the graph paper which is further used for estimating protein content in the unknown samples.

Unknown solution	Absorbance	Concentration of BSA (µg/ml)
A	0.35	110
B	0.47	145

All proteins containing aromatic amino acids (tyrosine and tryptophan) react with Folin's reagent at alkaline pH (>7) and reduce it to a blue coloured complex. This is due to the formation of Cu-protein complex when peptide bonds of proteins combine with $CuSO_4$

under alkaline conditions. This Cu-protein complex then reduces phosphomolybdic acid and phosphotungstic acid to form the blue coloured complex.

The intensity of the blue colour depends on the amount of aromatic amino acids present and will thus vary for different proteins.

Precautions

1. BSA solution should be prepared fresh.
2. Dilute Folin's reagent with equal volume of distilled water.
3. Reagent C should always be freshly prepared.
4. Use micropipettes for preparing different concentrations of protein solution.
5. Photoelectric colorimeters should be adjusted to zero with the help of their controls before recording the observations.

Note: Extraction of protein from seed sample: Grind 0.5 g of the seed sample in 10 ml of suitable solvent system (water or buffer) using pestle and mortar, centrifuge and use the supernatant for protein estimation.

Experiment 8.5

To isolate DNA from cauliflower heads.

Requirements

Plant material	: Cauliflower inflorescence
Chemicals	: Isolation mixture (0.15 M NaCl + 0.15 M sodium citrate + 0.0015 M EDTA + 5 per cent SDS[20]), 1 M solid NaCl, chloroform, isoamyl alcohol, chilled absolute alcohol
Glass apparatus	: Test tubes, funnel, micropipette, glass rod, distilled water.
Miscellaneous	: Water bath, ice-bucket, ice, pestle & mortar, cheese cloth, centrifuge, centrifuge tubes

Theory

Deoxyribonucleic acid (DNA) is a genetic material which contains all the information required for metabolism, reproduction and hereditary characters of an organism. DNA contains coded sequences that specify the structure of ribonucleic acid (RNA) and proteins. DNA is organized into units called genes, each of which is a code for a particular RNA or protein sequence. The study of genes is essential to learn about biological structure and function, evolution, diseases and many other aspects of living systems. To study genes in detail, DNA must be isolated and purified from cells of interest.

DNA was first isolated by the Swiss physician Friedrich Miescher who, in 1869, discovered a microscopic substance, 'nuclein', in the pus of discarded surgical bandages. In 1920s, P. A.

[20] SDS – Sodium dodecyl sulphate.

Levene concluded that DNA molecule consists of 'nucleotides' and each nucleotide is made up of three components - nitrogenous base (Purines: adenine and guanine, Pyrimidines: cytosine, thiamine and uracil), phosphate group and five- carbon sugar (deoxyribose in DNA; ribose in RNA) (refer to Chapter 10). In 1953, Rosalind Franklin, after studies based on X-ray crystallography (diffraction), suggested that the DNA molecule is either helical or cork screw in shape. Finally, in 1953, James Watson and Francis Crick prepared an acceptable model for the structure of DNA.

Procedure

DNA isolation procedures involve disruption of cells by manually homogenizing in a homogenization medium. Ice-cold ethanol/isopropanol is added to this homogenate to precipitate the DNA. The DNA can then be isolated by slowly stirring a glass rod in the homogenate literally spooling the DNA on to the glass rod. DNA spooling is an excellent demonstration method since an impressive tangle of pure DNA is clearly visible.

The protocol is as follows:

50 g of finely chopped inflorescence

↓

Add 100 ml of isolation mixture

↓

Incubate at 60 °C for 5 min

↓

Immediately plunge in ice bucket (at 4 °C)

↓

Grind it into slurry at 4 °C

↓

Filter through single-layered cheese cloth

↓

Centrifuge the filtrate at 5000 rpm for 10 min

↓

To the supernatant, add 1 M solid NaCl and equal
volume of chloroform and isoamyl alcohol in the ratio of 20:1

↓

Centrifuge at 10,000 rpm for 30 min

↓

3 layers obtained

↓

Take out upper aqueous layer using micropipette
Discard middle and bottom layer

↓

To this layer, add chilled absolute alcohol drop-by-drop
almost double the amount of aqueous solution

↓

DNA fibres are spooled out on a glass rod

↓

DNA is dissolved in distilled water for further experiment

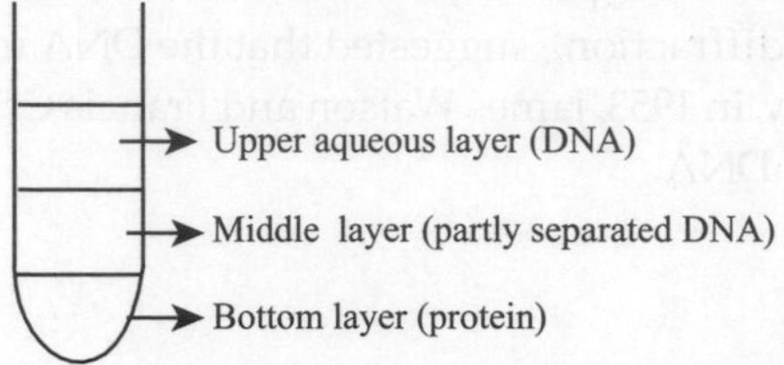

Note:

(i) DNA content in cauliflower is very high, even higher than DNAase.

(ii) Role of various components used in the experiment:
EDTA - Chelating agent: binds to Ca and Mg ions and inactivates enzyme DNAase.
SDS - Dissolves plasma and nuclear membrane.
Chloroform and isoamyl alcohol - Precipitate protein.
Absolute alcohol - Precipitates DNA from the solution.

Applications

The isolation of DNA is one of the most commonly used procedures in many areas of bacterial genetics, molecular biology and biochemistry. Purified DNA is required for many applications such as studying DNA structure and chemistry, examining DNA-protein interactions, carrying out DNA hybridizations, sequencing or PCR, performing various genetic studies or gene cloning.

Precautions

1. The experiment must be conducted in cold conditions to prevent nucleic acid hydrolysis.
2. After centrifugation, three layers should not be mixed up and the topmost layer which contains the DNA should be separated very carefully using a micropipette.
3. Absolute alcohol should be chilled and added drop by drop to the topmost layer.
4. Amount of 1 M solid NaCl should be calculated according to the volume of supernatant.

Isolation of DNA from leaves of higher plants

10 g of fresh leaves (from a single plant)

↓

Grind in a mortar and pestle in liquid nitrogen until a fine powder is formed

↓

Transfer the powder to 50 ml centrifuge tube containing 20 ml of extraction buffer

↓

Incubate the samples at 60 °C water bath for 15 minutes and break up the clumps, if any, with a glass rod

↓

Shake the sample for 5–10 minutes at the speed of 80–100 rpm

↓

Add 5 M cold potassium acetate (5 ml) and shake for 5 min

↓

Add 5 ml of chloroform and isoamyl alcohol mixture (24:1) and shake for 10–15 minutes

↓

Centrifuge the sample for 20–25 minutes at a speed of 3,000 rpm

↓

Sieve the sample through 4–5 layers of cheese cloth

↓

Transfer the solution to new tube and again add 5 ml of chloroform and isoamyl alcohol mixture (24:1) and shake for 10–15 minutes

↓

Collect the aqueous phase (top) into new 50 ml tube and add equal amount of isopropanol and invert until DNA comes together

↓

With the help of glass rod (with hook), take out DNA and place it in a beaker containing 70 per cent ethanol

↓

Dry the DNA with soft tissue paper and dissolve in 5 ml of TE (Tris-EDTA) buffer, pH 8.0

↓

Add RNase and incubate for 5 minutes at room temperature

↓

Add 1/10 volume of 3 M sodium acetate (pH 6.8) and 2 volume of ethanol

↓

Hook out the DNA, wash with 70 per cent ethanol and dissolve in TE buffer

↓

Centrifuge to remove undissolved material

DNA isolation from *E. coli*

1.5 ml of overnight grown culture in a microfuge tube

↓

Centrifuge at 5,000 rpm for 5 minutes to obtain the pellet

↓

Discard the supernatant and again add 1.5 ml of culture to the microfuge tube

↓

Centrifuge at 5,000 rpm for 5 minutes to obtain the pellet

↓

Discard the supernatant and add 200 µl of GTE mix. Break the pellet by knuckles

↓

Incubate at room temperature for 5 minutes

↓

Add 400 µl of 1 per cent SDS and keep on ice for 5 minutes

↓

Add 60 µl of 5 mM NaCl (1/10th volume) and 700 µl of isopropanol

↓

Invert, mix and incubate at room temperature for 5 minutes.
DNA threads are visible

↓

Centrifuge at 10,000 rpm for 20 minutes

↓

Decant and add 300 µl of 70 per cent ethanol

↓

Centrifuge it at 10,000 rpm for 20 minutes

↓

Decant the supernatant and air dry

↓

Dissolve in 100 µl of distilled water

↓

Centrifuge at 10,000 rpm for 5 minutes

↓

Take the supernatant in a fresh microfuge tube and subject it to gel electrophoresis.

Note: GTE mix is a mixture of glucose, tris and EDTA.

Component	Working (mM)	Stock (M)
Glucose	50	1
Tris	50	1
EDTA	10	0.5

Projects

Project 8.1

Track changes in the amino acid content of fruits on ripening using the chromatographic technique

Requirements

Apple juice, tomato juice, ethanol; 0.2 per cent ninhydrin, organic solvent (ethanol: water: 0.88 per cent ammonia - 80:10:10), centrifuge, chromatography paper.

Procedure

Extract juice from the chosen fruits and centrifuge or filter to rid particulate material. Spot the juice on the chromatography paper together with marker amino acids. (The juices may be examined for free amino acids by precipitating the protein with ethanol. Treat 1 ml of juice with 2 ml of ethanol and centrifuge to get clear supernatant.)

Identify the number of different amino acids present and try to estimate the relative amounts by the size of the individual spots.

Note: 'Changes in amino acid content in developing seeds' can be estimated at different stages of development (grains and legumes can be chosen for this project.)

Project 8.2

Determination of nitrite reductase activity (NiR)

Requirements

Leaves of any plant material, 0.2M potassium phosphate buffer (pH 7.0), 5 mM sodium nitrite (in water), 1.5 mM methyl viologen, dithionite reagent (40 mg each of sodium dithionite and sodium hydrogen carbonate in 10 ml of distilled water – freshly prepared)

Procedure

Grind 7 g of leaves in two volumes of 20 mM potassium phosphate buffer, containing 5 mM cysteine HCI and 0.05 mM EDTA for 90 seconds. Strain through two layers of muslin cloth, add charcoal (0.5 g/8 ml), stir, centrifuge at 5000 g and use the supernatant for enzyme activity. To 0.2 ml–1 ml of supernatant, add 0.2 ml phosphate buffer, 0.1 ml sodium nitrite, and 0.1 ml methyl viologen and incubate at 30°C. Start the reaction by adding 0.2 ml of the dithionite reagent and incubate for 10 minutes. Stop the reaction by vigorously shaking the mixture until the diothionite is completely oxidized and the dye colour has disappeared . Determine the amount of nitrite in the reaction mixture by the procedure given for nitrate reductase activity,using the blank (without the enzyme and the boiled enzyme). Express the specific activity of nitrite reductase as μ moles of nitrite removed/min/mg protein.

9. Lipid Metabolism

Experiment 9.1

To test the presence of fats in the given plant material by microchemical tests.

Requirements

Plant material : Corn or linseed or olive or castor or mustard oil (fats)
Chemicals : Butyric, oleic, propionic, stearic acids (fatty acids)
Organic solvents (ethanol, diethyl ether), soaps, dilute NaOH, alcoholic solution of Sudan III (IV)
Glass apparatus : Test tubes
Miscellaneous : Filter paper, gas burner

Tests

1. **Solubility test:** Take small samples of each of the fatty acids and fats and test their solubilities in ethanol, diethyl ether and water.
   All fats and oils are slightly soluble in cold ethanol and very soluble in hot ethanol. Liquid fats are insoluble in water and form emulsions.
2. **Grease spot test:** Place a drop of each of the liquid fatty acids and fats on a filter paper and allow them to dry. A translucent greasy spot will be visible around each.
3. **Emulsification:** Shake a drop of olive oil or any other oil with 5 ml of water in a test tube. A temporary emulsion is formed because the fat breaks into tiny globules which are finely dispersed in the water. The emulsion with water is unstable but can be stabilised by the addition of soaps.
4. **Soap formation:** Heat a little stearic acid or any other fatty acid with dilute alkali (NaOH). Observe the formation of a soapy solution.
5. **Sudan test:** Shake a little groundnut or mustard oil with an alcoholic solution of Sudan III or Sudan IV (prepared by dissolving 0.5 g in 100 ml of 70 per cent alcohol). The oil will be stained red.

   or

   Boil the aqueous suspension containing oil (obtained by crushing oilseeds, such as groundnut or castor) in a test tube. Once a translucent layer appears on the surface, pour a little Sudan III and shake it lightly. A red colour in the upper layer indicates the presence of fat.

Experiment 9.2

To demonstrate the activity of enzyme lipase.

Requirements

Plant material : Germinated seeds of cotton (*Gossypium hirsutum*) or mustard (*Brassica campestris*) or castor (*Ricinus communis*)

Chemicals : Toluene, oil emulsion, phenolphthalein, NaOH (4 per cent)
Glass apparatus : Pestle and mortar, funnel, test tubes
Miscellaneous : Muslin cloth, tripod stand

Principle

Lipases are a group of enzymes that catalyze the hydrolysis of the esters of glycerol:

$$\begin{array}{l} CH_2O.CO.R^1 \\ | \\ CHO.CO.R^2 + H_2O \\ | \\ CH_2O.CO.R^3 \\ \text{Triglyceride} \end{array} \xrightarrow{\text{Lipase}} \begin{array}{l} CH_2OH \\ | \\ CHOH \\ | \\ CH_2OH \\ \text{Glycerol} \end{array} + \underbrace{R^1COOH + R^2COOH + R^3COOH}_{\text{Fatty acids}}$$

There are a variety of lipases and they may show some specificity for one or the other of the three fatty acids linked in the triglycerides. Thus more than one lipase may be required for the complete hydrolysis of any one triglyceride. Lipases are found in dormant and germinating seeds which contain fatty oils as food reserves. Castor seeds are a very rich source of the lipases.

Procedure

Grind 5 grams of germinated seeds of cotton or mustard (for castor oil seeds, remove the shells from six seeds). Add just sufficient water until the paste is of uniform consistency. Adjust the total volume to 25 ml with distilled water. Put a drop of toluene while grinding to prevent fermentation. Allow the contents to stand for 15 minutes and filter through muslin cloth. Centrifuge the filtrate at 3,500 rpm for 15 minutes to obtain the clear supernatant.

Observation table

S. No	State of enzyme (5 ml)	Substrate (2 ml)	Indicator	Observations		Inference
				Initial	Final	
1.	Active	Oil emulsion	Phenolphthalein + 4% NaOH	Pink colour	Colourless	Lipase converts oil emulsion into fatty acids and glycerol. Initial pink colour due to alkaline phenolphthalein disappears due to presence of fatty acids.
2.	Boiled	Oil emulsion	Phenolphthalein + 4% NaOH	Pink colour	Pink colour	Denatured enzyme does not convert oil into fatty acids.
3.	Control (distilled water)	Oil emulsion	Phenolphthalein + 4% NaOH	Pink colour	Pink colour	No enzyme, so no reaction.

Result and Discussion

Lipases act on lipids (oil) by splitting them into glycerol and fatty acids. Initially, the indicator phenolphthalein gives pink colour due to alkaline substrate. As the enzyme acts on oil emulsion and converts it into fatty acids and glycerol, neutral conditions are created and the pink colour disappears.

Precautions

1. Oil emulsion should not be turbid.
2. Equal quantities of enzyme, substrate and indicator should be added in all test tubes.

Questions for Viva Voce

1. Write the generalized chemical equation for the reaction catalyzed by lipase.
2. From what respiratory intermediates are the components of a fat synthesized?
3. What is the physiological function of fat in seeds?

10. Phloem Transport

Experiment 10.1

To separate sugars by paper chromatography.

Requirements

Plant material : Fruit juices

Chemicals : 1 per cent solution of known sugars (glucose, fructose, lactose, maltose, galactose, xylose, mannose, ribose, arabinose, rhamnose)

- Solvents (any one)
  - (a) *t*-butanol:glacial acetic acid:water (3:1:1) or (4:1:5)
  - (b) 96 per cent ethanol:water (7:3)
  - (c) Isopropanol:water (4:1)
  - (d) Isopropanol:pyridine:water:acetic acid (8:8:4:1)
  - (e) Water saturated phenol:1 per cent ammonia
- Spraying reagents (any one)
  - (a) Alkaline permanganate
    1 g $KMnO_4$ + 2 g Na_2CO_3 in 100 ml distilled water
  - (b) Ammoniacal silver nitrate
    Equal volumes of NH_4OH and saturated solution of $AgNO_3$ and dilute to give a final concentration of 0.3 M
  - (c) Aniline-diphenylamine reagent
    - (i) 5 volumes of 1 per cent aniline
    - (ii) 5 volumes of 1 per cent diphenylamine in acetone with 1 volume of 85 per cent phosphoric acid and heating for a short period at 100 °C
  - (d) Resorcinol reagent
    1 per cent ethanolic solution of resorcinol in 0.2N HCl
  - (e) Benzidine reagent
    (0.5 g benzidine, 10 ml acetic acid, 10 ml of 45 per cent trichloroacetic acid and 100 ml of 95 per cent ethanol)

Glass apparatus : Beakers, funnel, chromatography jar, separatory funnel, fine capillary tube

Miscellaneous : Mortar & pestle, centrifuge, filter paper, Whatman No. 1 filter paper, hot air oven, atomiser (spraying bottle)

Principle

Refer to experiment 6.1.

Procedure

Place sufficient solvent (any one) into the chromatography jar and immediately cover and allow the jar to be saturated with the solvent. Select a sheet of Whatman No. 1 chromatography paper and mark several spots on the paper with a pencil, leaving 2 cm space from the lower edge. Load 2–3 drops of fruit juice on one of the spots and 2–3 drops of known sugar solutions on the other spots with the help of fine capillary tube. Make sure that spots are allowed to dry before the application of the next drop. Place the chromatography paper in the jar, close it with the lid and allow the solvent to flow for about 35–45 minutes. Remove the paper and immediately mark the position of the solvent front with a pencil. Dry the paper and spray with the spraying reagents (anyone). Again dry in hot air oven for 5–10 minutes and the reducing sugars will appear as brown spots. The colours are stable for some weeks if kept in the dark and away from acid vapours.

Carefully encircle the position of each spot with a pencil.

Result and Discussion

With the help of the position of spots of known sugars, identify the sugars present in fruit juices and calculate their R_f values. Speed of migration of sugars (in paper chromatography as well as TLC) decreases in the following order:

Pentoses > hexoses > disaccharides > trisaccharides

Pentose or Hexose $\xrightarrow{\text{Acid}}$ Furfural or hydroxymethyl furfural

+ Diphenylamine (C₆H₅–NH–C₆H₅)

↓

Coloured complex

Precautions

Refer to experiment 6.1.

Experiment 10.2

To separate and identify the sugars by thin layer chromatography (TLC).

Requirements

Plant material : Fruit juices

Chemicals : Silica gel G, Boric acid[21] (0.1 M)

[21] Boric acid increases the polarity of silica.

3 per cent standard sugar solution (e.g., lactose, ribose, xylose, fructose, glucose, sucrose, rhamnose)
Solvents (any one)

- *n*-butanol:glacial acetic acid:water (60:30:10)
- Ethyl acetate:isopropanol:water:pyridine (26:14:7:2)
- *n*-butanol:pyridine:water (6:4:3)

Developing reagent: 1.5 ml aniline + 1.6 g phthalic acid in 100 ml of 95 per cent acetone

Glass apparatus : Glass plates, glass with lid, beakers, capillary tubes
Miscellaneous : Hair-dryer, oven, atomizer

Principle

Refer to experiment 6.2.

Procedure

Prepare TLC plates using 1 part of silica gel G and 2 parts of 0.1 M boric acid. Load the spots of fruit juices and standard sugars separately on the TLC plate. Place the plate in a glass jar saturated with solvent mixture (any one) and leave it till the solvent has ascended 3/4th of the distance from the place of origin. Dry the plates in a stream of cold air and locate the sugars by spraying with developing reagent. Heat the plate at 105 °C for 5 minutes.

Results

Note the colour of the standard sugars and calculate their R_f values. Also observe the colour of the spots developed from the fruit juices to determine their identity. The sugar spots appear as yellow brown in the blue background of the plate.

Precautions

Refer to experiment 6.2.

11. Dormancy, Germination and Flowering

Project 11.1

To demonstrate various aspects of seed germination.

(i) Seed viability test

Requirements

Plant material : Dry seeds of *Zea mays, Cucumis, Helianthus, Phaseolus*
Chemicals : 2, 3, 5-Triphenyl tetrazolium chloride solution (1 per cent),
Other : Petri dishes, filter paper, incubator

Procedure

Soak 50 seeds overnight in water. Cut them into half by a sharp scalpel and place in a petri dish with the cut surface in contact with filter paper soaked in sufficient amount of tetrazolium salt. Incubate at 20–25 °C in the dark. Repeat the experiment using seeds which have been previously heated at 100 °C for at least 30 minutes.

Determine after 24 hours, the percentage of seeds that have turned red due to the accumulation of formazan. Note that the seeds which are killed by heating, will fail to turn red due to the lack of dehydrogenase activity. The dead seeds cannot reduce such dyes because they do not respire while the viable seeds respire and therefore can reduce colourless tetrazolium dyes into highly coloured compounds.

$+2H^+ + 2e^- \longrightarrow$ (From aerobic respiration) $+ HCl$

2, 3, 5-Triphenyl tetrazolium chloride
(Soluble and colourless)
(The dotted line represents the position from where the ring breaks)

Triphenyl formazan (insoluble and red)

(ii) Water uptake by germinating seeds

(a) *Demonstration of increase in seed volume due to imbibition*

(Imbibition is a purely physical phenomenon)

Requirements

Plant material : Dry pea seeds
Others : Measuring cylinder (250 ml)

Procedure

Fill a measuring cylinder with dry seeds upto 100 ml mark. Pour enough water into the cylinder. Change water frequently to prevent fermentation. Repeat the Experiment with seeds which have been killed by previously heating at 100 °C for about an hour.

Record the increase in seed volume (of living as well as dead cells) in regular time interval, every 3 h in the first few hours and every 12–24 h later. Represent the result graphically.

(b) *Demonstration of imbibition pressure*

Requirements

Plant material : Dry pea seeds
Others : Test tubes, cork, thread

Procedure

Fill a test tube with dry pea seeds and water. Close the mouth of the tube with a cork. Within few hours enough pressure will be built inside the seeds to break open the test tube.

(c) *Effect of glucose/NaCl on the uptake of water by imbibing seeds*

Requirements

Plant material : Dry pea seeds
Chemical : Glucose/NaCl solution (0.25, 0.5, 1.0 and 2.0 M)
Others : Balance, petri dishes, test tubes, pipettes, filter paper

Procedure

Soak about 10 grams of dry seeds each in 10 ml of glucose/NaCl solutions of different concentrations. Keep a control simultaneously by soaking the same amount of seeds in distilled water. After 24 hours remove the seeds, dry the surface off seeds with filter paper and weigh them again.

Express water uptake by seeds in terms of percentage of original dry weight of seeds and present the results graphically with per cent increase in weight against the concentrations of glucose/NaCl solution.

Water uptake by imbibing seeds is inhibited by osmotic media.

Fresh weight of the seeds = 10 g

Concentration of glucose/NaCl (M)	Dry weight of the seeds (g)	Difference in weight (g) (amount of water uptake)
0.25		
0.5		
1.0		
2.0		

(d) *Role of reserve food material in the uptake of water by imbibing seeds*

Requirements

Plant material : Seeds of maize, sunflower and pea
Others : Balance, petri dishes, filter paper

Procedure

Place 10–15 g of dry seeds in petri dishes containing excess of water. After 48 h remove the seeds and wipe off water from the surface of seeds with a dry filter paper and weigh them.

Calculate the difference between the weight of seeds before and after soaking in water. Express increase in mass in terms of percentage of the original weight of the seeds.

Determine how the nature of reserve food (carbohydrate or protein or fat) influences the water uptake.

(e) *Pre-requisite of oxygen for seed germination*

Requirements

Plant material : Seeds of wheat (*Triticum aestivum*) or oat (*Avena sativa*)
Others : Petri dishes

Procedure

Soak seeds in water for 24 h and transfer them to two petri dishes. In one of the dishes, add just enough water so that seeds are only half submerged in water. In the second dish, put enough water to completely submerge the seeds and add a thin layer of liquid paraffin on the top of water to cut off oxygen supply from the air (Figure E11.1).

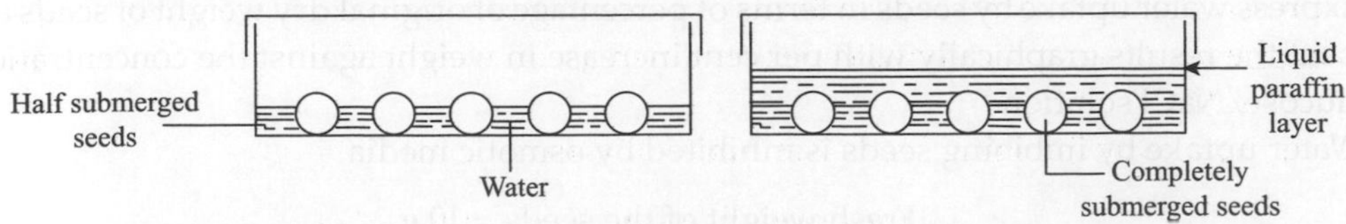

Fig. E11.1 An experiment showing pre-requisite of oxygen for seed germination.

The half submerged seeds will germinate readily, whereas the completely submerged seeds fail to germinate because of lack of oxygen.

Project 11.2

To study the effect of gibberellins on α-amylase in cereal grains.

Requirements

Plant material : Dehusked barley grains
Reagents : 1 per cent sodium hypochlorite (NaOCl)
Stock solution of gibberellic acid (GA_3) - 2 µm
Streptomycin (0.1 per cent in 1 mM citrate-phosphate buffer, pH 4.8)
Starch solution (6.7 g soluble starch + 8.16 g KH_2 PO_4 dissolved in 1000 ml distilled water)
Iodine reagent (0.6 g KI + 60 mg I_2 dissolved in 1000 ml of 50 mM HCl)
Miscellaneous : Sand, incubator, small stoppered glass tubes, mechanical shaker, petri dishes, pipettes (10, 1, 0.1 ml), micropipettes, spectrophotometer, forceps, razor blade

Procedure

Cut 60 dehusked barley grains transversely with a sharp razor blade at the middle. Collect the apical embryo free halves of the grains in a petri dish. Place 10 grain halves with embryo in another petri dish. Incubate the grain halves in 1 per cent NaOCl solution for 15–20 minutes. Discard NaOCl solution and wash the seeds with enough sterile distilled water at least three times. Sterilize seven petri dishes, each containing 100 grams of sand in an oven at 170 °C for 1 hour. After the sand and petri dishes have cooled to room temperature; transfer 10 grain halves with embryos to one of the petri dishes, and 10 embryo-free grain halves each to the remaining petri dishes. Add 20 ml sterile distilled water to each petri dish. After 24 hours, transfer grain halves from petri dishes to six sterilized stoppered glass tubes containing the following solutions.

Tubes	GA_3 solution (ml)	Streptomycin solution (ml)	Distilled water (ml)
1	0.0	1.0	1.0
2	1.0	1.0	0.0
3	0.5	1.0	0.5
4	0.1	1.0	0.9
5	0.05	1.0	0.95
6	0.01	1.0	0.99

Transfer 10 embryo-free grains each into tube numbers 1 to 6. Incubate the tubes at 25–30 °C for 24 hours. If available, shake the tubes using a mechanical shaker. In a cuvette of spectrophotometer, mix 0.1 ml of incubation medium from each tube with 1 ml of starch solution and 0.9 ml distilled water. Exactly after 10 minutes, add 1 ml of iodine reagent. Shake well and measure the absorbance at 620 nm. Decrease in absorbance is a measure of α-amylase activity. Draw a graph showing the relative absorbance against the gibberellin concentration. Lesser the absorbance, higher is the α-amylase activity.

Project 11.3

Mobilization of starch during seed germination by amylases.

Requirements

Plant material : Maize seeds (soaked in water for 2–3 days)
Chemicals : Iodine solution (0.5–1.0 g KI in 100 ml of distilled water + 1 g of iodine crystals, shaken well).
Starch - agar solution (2 g agar + 0.1 g soluble starch in 100 ml water)
Miscellaneous : Beakers, petri dishes, razor blade, filter paper

Procedure

Mix 0.1 gram soluble starch in 100 ml distilled water. Heat gently with constant stirring, till a clear solution is obtained. Add 2 grams agar powder and warm the mixture till agar is dissolved and a clear viscous medium is produced. Pour a few mm thick layer of medium into several petri dishes. After the medium solidifies and is cooled to room temperature, cut a few dry and soaked seeds transversely into two halves with a sharp razor blade (Figure E11.2a). Place them on petri dishes with cut surface in contact with the agar medium (Figure E11.2b). After 1 or 2 days, remove the seeds and pour iodine solution on the agar medium.

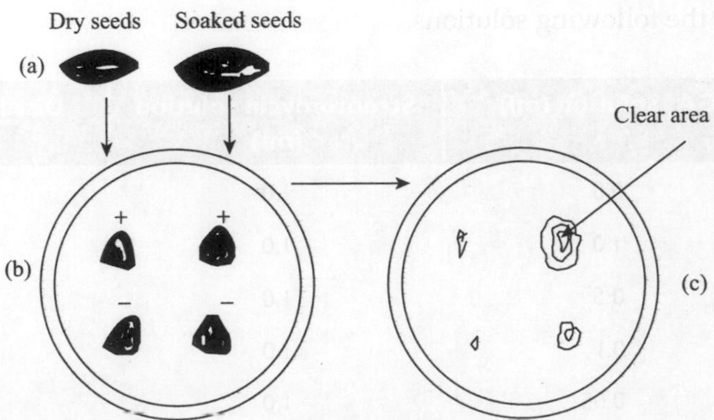

Fig. E11.2 The sequence of extracellular breakdown of starch, by amylases, during the germination of cereal grains. +/- indicate grain-halves with and without embryos.

The agar medium on which dry seeds are placed will stain uniformly blue showing that starch has remained unchanged. Similarly, the spots on agar medium where seed-halves without embryo are placed will also stain blue. But around the spots on agar medium where soaked seed-halves with embryo are placed, a clear area will appear showing that starch is hydrolysed in these regions due to the secretion of amylases by embryo (Figure E11.2c).

Project 11.4

To show the inhibitory effect of various fruit juices on seed germination.

Requirements

Plant material : *Brassica* seeds
Other : Tomato juice, orange juice, distilled water, petri dishes

Procedure

Extract juice of tomatoes and oranges (≈250 grams) and dilute it as shown in the table. Soak the filter paper with 10 ml each of pure juice and diluted juices in different petri dishes. Add 25 seeds of *Brassica* in each petri dish and keep them in dark at 27 °C. If required, add another 10 ml of respective dilution after 2 days. Note down the number of seeds germinated in each petri dish after 24 hours till all the seeds are germinated in control petri dish.

Dilution table

Dilution of juice	Juice (ml)	Distilled water (ml)	Total volume (ml)
Pure	25	0	25
1:2	15	30	45
1:3	10	30	40
1:4	5	20	25
1:5	5	25	30
Control	0	25	25

Number of seeds in each petridish = 25

Dilution of juice	Number of seeds germinated				
	18 h	24 h	48 h	72 h	96 h
Pure					
1:2					
1:3					

Dilution of juice	Number of seeds germinated				
	18 h	24 h	48 h	72 h	96 h
1:4					
1: 5					
Control					

Represent your results graphically - number of seeds germinated vs. time of germination.

Fruit juices contain germination inhibitors (probably a mixture of ABA, caffeic acid, ferulic acid, coumarin, *para* ascorbic acid, phenols, *trans* cinnamic acid, alkaloids, essential oils, etc.) which tend to prolong the germination process. Dilution of fruit juices results in better germination percentage of the seeds.

Project 11.5

Photomorphogenetic effects of light on the development of seedlings (comparison of light-grown and etiolated seedlings).

Requirements

Plant material : Seeds of white mustard (*Sinapis alba*) or radish (*Raphanus sativus*) or pea (*Pisum sativum*)

Chemicals : Acetone, extraction medium: *n*-propanol - 32 per cent HCl - H_2O (18:1:81 V/V)

Miscellaneous : Plastic boxes, filter paper, centrifuge, glass vials, Buchner's funnel, spectrophotometer

Procedure

Soak a lot of 25 seeds and sow them in a plastic box of suitable size lined with 4 layers of filter paper moistened with adequate amount of distilled water. Place one lot of seeds in the dark and expose the other lot to white light. After 4–5 days, determine the differences in the following growth parameters between the light- and dark-grown seedlings:

(i) Length of hypocotyl
(ii) Size and fresh weight of cotyledons
(iii) Differentiation of stomata in the lower epidermis of cotyledons

Estimate the chlorophyll content in dark and light grown seedlings described on page 810 (Project 13.1).

Select 20 seedlings with uniform hypocotyl length, cut off cotyledons at the base of the lamina. Transfer the excised cotyledons to glass vials containing 10 ml of extraction medium. Estimate the anthocyanin content in 20 pairs of cotyledons as shown in experiment 15.1.

Compare the different parameters of growth and development in light and dark-grown seedlings. Draw diagrams of both the types revealing their morphological and anatomical differences.

Parameters	Light-grown seedlings		Etiolated seedlings (dark-grown seedlings)	
	Day 4	Day 8	Day 4	Day 8
Length of hypocotyl				
Fresh weight of cotyledons				
Number of stomata				
Chlorophyll content				
Anthocyanin content				

Project 11.6

Effect of red and far-red light (phytochrome) on seed germination

Requirements

Plant material : Seeds of *Lactuca sativa* var. Grand rapids (incubate dry seeds at 35–37 °C for 3 days before experiment)

Miscellaneous : Petri dishes, filter paper, light sources

Procedure

Place at equidistances 100 seeds each in 6 petri dishes containing three layers of filter paper moistened with 5 ml distilled water. Since the seeds become very sensitive to light after imbibition, carry out this experiment in a dark room under green safe light. Cover petri dishes with dark paper or cloth and transfer quickly to the dark. After 3 hours incubation in the dark at room temperature, expose the seeds to red- and far-red light as shown below and return them to the dark. The far-red and red light treatments should follow immediately after one another.

S. No.	Treatment	Number of seeds germinated
1.	2 minutes red → dark	–
2.	2 minutes far-red → dark	–
3.	2 minutes red + 2 min far-red → dark	–
4.	2 minutes red + 2 minutes far-red + 2 minutes red → dark	–
5.	2 minutes far-red + 2 minutes red → dark	–
6.	Continuous dark (control)	–

After 48 hours, count the number of seeds germinated. The light effect depends upon the light quality of last exposure. The red light promotes seed germination and the effect of red light is reversed by far-red and *vice versa* (Figure E11.3).

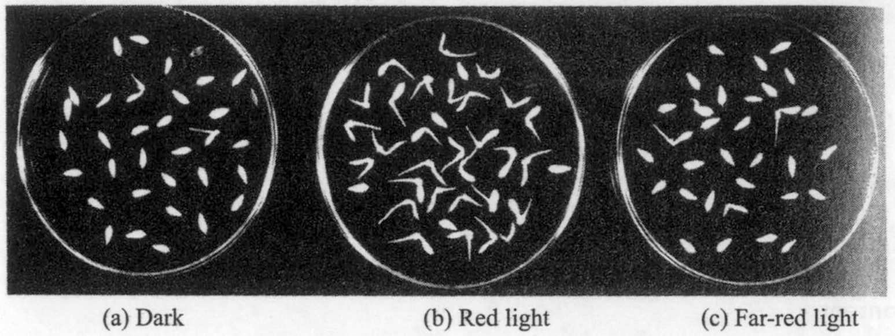

(a) Dark (b) Red light (c) Far-red light

Fig. E11.3 Seed germination in (a) dark (b) red light (c) far-red light.

Light sources

(i) *Green safe light*: Filter the light from a table lamp equipped with a 15 W fluorescent tube through a layer of dark green cellophane.
(ii) *Red light*: Use a table lamp equipped with a red or white incandescent bulb (40 W) or a cool fluorescent tube or a red fluorescent tube. Filter the light through 2 layers of red cellophane.
(iii) *Far-red light*: Use a table lamp equipped with a 40 or 60 W incandescent bulb. Filter the light with a combination of 4 layers of blue, 1 layer of green and 2 layers of red cellophane.

Project 11.7

Effect of light (phytochrome) and kinetin on seed germination.

Requirements

Plant material : Seeds of *Lactuca sativa* var. Grand rapids (incubate dry seeds at 35–37 °C for 3 days before experiment), summer squash (*Cucurbita pepo*)
Chemicals : Kinetin solutions (1, 2, 5, 10 ppm)
Miscellaneous : Petri dishes, filter paper

Procedure

Sow two batches of 50 seeds each of *Lactuca* and *Cucurbita* in petri dishes lined with filter paper moistened with 10 ml of test solutions. Place one set of seeds under diffuse light in the laboratory and place another set in the dark. Similarly, place 50 *Lactuca* seeds each in petri dishes containing different concentrations of kinetin solution (0, 1, 2, 5 and 10 ppm) and place them in the dark. Count the number of seeds germinated after 4–5 days.

Plant material	Number of seeds germinated						
	Light	Dark	Kinetin (ppm)				
			0	1	2	5	10
Lactuca sativa							
Cucurbita pepo							

Calculate percentage of seeds germinated in each case.

Percentage of seed germination will be higher in light in case of *Lactuca,* whereas in *Cucurbita* the seed germination will be higher in the dark. Observe if kinetin can replace light in promoting seed germination in *Lactuca.*

Note: Seeds of *Lactuca sativa* can be replaced with *Nicotiana tabacum* also.

Project 11.8

Demonstrating breaking seed dormancy by scarification (chemical as well as mechanical).

Requirements

Plant material : Seeds of *Luffa cylindrica* or *Cassia floribunda* or *Leucaena leucocephala*
Chemicals : Concentrated H_2SO_4, absolute alcohol, 1 per cent NaOCl, distilled water
Miscellaneous : Petri dishes, test tubes, blotting paper, scissors, scalpel, needle, sand paper.

Procedure

Take 8 sets of 10 seeds each of *Luffa cylindrica* and sterilize them in NaOCl.

- Keep first set of seeds in petri dish lined with moist blotting paper - *Control.*
- Treat two sets of seeds with conc. H_2SO_4 for 15 and 30 seconds and another two sets with alcohol for 15 and 30 seconds - *Chemical scarification.*

  Wash them 3–4 times with distilled water and transfer to petri dishes lined with layers of blotting paper soaked in distilled water. Keep in the incubator at 25 °C for 4 days.
- Scratch another set of seeds either with a needle or scalpel or sand paper at the hilar end - *Mechanical scarification.* Keep them for germination for 4 days.

Also record the initial and final weights of all sets of seeds on day 0 and 4. Find out percentage increase in dry weight and tabulate your results for discussion.

Total number of seeds in each set = 10

S. No.	Treatment	Duration (s)	Number of seeds germinated			
			Day 1	Day 2	Day 3	Day 4
1.	Control	–				
2.	Conc. H_2SO_4	15				
3.	Conc. H_2SO_4	30				
4.	Alcohol	15				
5.	Alcohol	30				
6.	Needle (mechanical)	–				
7.	Scalpel (mechanical)	–				
8.	Sand paper (mechanical)	–				

Dormancy of the seeds caused due to hard seed coat can be removed by mechanical or chemical scarification. However, the duration of the treatment varies in different species. Even the concentration of chemicals can be changed depending on the species.

Note:

(i) Sulphuric acid should be used with extreme caution. Vinegar is safer but less effective and can be used for species that do not have an extremely hard seed coat.

(ii) Another method is hot water scarification. Bring water to a boil, remove the beaker from the gas and place the seeds into the water. Allow the seeds to soak until the water cools to room temperature. Remove the seeds from the water and keep for germination.

(iii) Following scarification, seeds should be dull in appearance, but not cracked as to damage the embryo. Scarified seeds should be planted as soon as possible after treatment.

Project 11.9

Demonstrating the breakdown of seed dormancy by stratification (chilling treatment).

Requirements

Freshly harvested seeds of wild plants such as species of *Chenopodium, Atriplex, Oenothera* Petri dishes, refrigerator

Procedure

Place the dormant seeds in petri dishes lined with moist filter paper and then store them in a refrigerator (3–5 °C) for 0, 12, 24, 48 and 72 hours. Add more distilled water to compensate for the water loss by evaporation during storage in the refrigerator. After such chilling treatment for different durations, transfer the seeds to a location at room temperature or an incubator (20–25 °C).

Find out the rate of seed germination under different treatments after a week of transferring the seeds from the chilling temperature to the normal temperature. Determine the optimal duration of chilling treatment for proper seed germination.

Total number of seeds = 20

Period of treatment at low temperature (h)	Number of seeds germinated						
	Day 1	Day 2	Day 3	Day 4	Day 5	Day 6	Day 7
0							
12							
24							
48							
72							

12. Plant Hormones and Signal Transduction

Project 12.1

1. Effect of IAA on *Avena* coleoptile curvature (Bioassay for auxin).

The *Avena curvature test* developed by F. W. Went (1926) was the first bioassay for auxin and probably the best one.

The measurement of auxin activity by the *Avena* curvature test depends upon the strict, rapid polar transport of auxin in the *Avena* coleoptile. Because of this property, auxin applied to one side of the coleoptile will diffuse down that side rapidly. The differential growth that is caused by the transport of auxin down only one side of the coleoptile produces a curvature that is proportional, within limits, to the amount of auxin applied.

Requirements

Plant material : Oat (*Avena sativa*) seedlings

Chemicals : IAA solutions (prepare a stock solution of 2 mM by dissolving IAA in a few drops of absolute alcohol and subsequently adding required amount of distilled water. Prepare solutions of different concentrations - 200, 20, 2, 0.2, and 0.02 μM - by further dilution of stock solution), agar powder

Others : Light source, petri dishes, glass vials, sand, sharp razor, forceps, slide projector.

Procedure

Pre-soak seeds in water for 24 hours and incubate them in petri dishes containing 5 ml of distilled water at 25–30 °C for another 24 hours. Transfer the seeds with 2–3 mm long coleoptiles to glass vials containing sand and just enough water to wet the sand. Plant the seeds at a 45° angle in the sand, the embryo facing downward, so that the coleoptiles grow vertically. Place the glass vials in the dark in an incubator maintained at 25–30 °C for 24 hours. Inconvenient elongation, of the first internode may be reduced by exposing the seedlings 2 days after germination to 2–4 hours of red light. After a height of 15–30 mm has been reached, remove 1 mm of the apical tip of the coleoptile, thus removing the natural source of auxin.

Concentration of auxin in the agar block (μM)	Angle of coleoptile curvature
200	
20	
2	
0.2	
0.02	

A second decapitation of 2–4 mm is necessary after a 3-hour period to remove tissue that has regenerated and produces auxin. Gently pull out the primary leaf with a forceps so that it is broken at the base and comes up a few millimeters out of the coleoptile. This serves as the vertical support (tip of the primary leaf) for the agar block that will be placed on the coleoptile.

Place agar blocks containing different concentrations of auxin on one side of the severed end of the coleoptile. After 90 minutes, the shadows of the seedlings are projected onto a strip of bromide paper and photographed (or illuminate the coleoptiles from the side with a slide projector in such a way that the shadow of the coleoptiles falls on a drawing paper held closely to the coleoptile. Draw the outline of coleoptile on the paper with a pencil.)

Use the shadow graph (photograph) to measure the angle of curvature.

Draw a graph showing the relationship between the angle of coleoptile curvature and the concentration of auxin used in the agar blocks. Within a certain range of concentrations of IAA, there is a linear relationship between the concentration and the amount of curvature obtained (Refer to Chapter 16).

Preparation of Agar Blocks

Dissolve 1 gram agar powder in 100 ml distilled water in boiling water bath. Cool till it forms a viscous fluid (do not let it solidify). Pour the melted agar on a clean microscopic slide to a thickness of about 2 mm and allow it to solidify. Place the slide over a graph paper and prepare agar blocks of 2 mm × 2 mm by cutting with a razor blade. Place 10 such agar blocks in a petri dish containing 1–2 ml of IAA solution of different concentrations. Allow the agar to imbibe the hormone for at least 2 hours before use.

Note: Select *Avena* seedlings with straight coleoptiles for the experiment.

Project 12.2

To demonstrate the effect of IAA on apical dominance.

Requirements

Plant material : Kidney bean plant (*Phaseolus vulgaris*)
Other : 1 per cent IAA-lanolin paste, lanolin paste, razor blade

Procedure

Soak the dwarf or runner bean seeds in water for 24 hours and plant in soil for 2–3 weeks. Select three bean plants of identical growth and age. Leave one plant undisturbed (control), decapitate the other two plants using razor blade. To the stump (cut apex) of one of these, apply sufficient amount of 1 per cent IAA-lanolin paste and plain lanolin paste to the other. Observe the results with respect to the forcing out of lateral bud every two weeks and record.

Measure the growth of the axillary buds of each series (of decapitated and intact plants) daily and compare the results.

IAA-Lanolin Paste

Dissolve the required amount of crystalline IAA in a little absolute ethanol. Add this to the appropriate quantity of warm hydrous lanolin. Stir vigorously to ensure a uniform mixture.

Note: Perform the experiment in triplicate to get uniform results. Within a few days, in the intact plants, the growth of axillary buds remains inhibited (effect of apical dominance) whereas the axillary buds in decapitated plants begin to grow and develop.

Project 12.3

To study the effects of auxin on rooting.

Requirements

Plant material : (a) Uniform sized stem cuttings of *Salix* sp. or Rose or *Ficus* (15 cm × 1 cm)
(b) 2–3 weeks old bean plants (*Phaseolus vulgaris*) with well developed primary leaves.

Chemicals : IAA solution (1 mg/ml) - dissolve 100 mg IAA in a few drops of ethanol and make a final volume of 100 ml by adding distilled water. Store in a refrigerator till use.

Procedure

(a) ***Rooting in stem cuttings***: Prepare uniform sized stem cuttings of *Salix* sp. Dip them in auxin, i.e., IAA solutions of different concentrations (0, 0.1, 1, 10, 100 µg/ml). After 2 hours of hormone treatment, plant the stem cuttings in a bed of moist sand. Place under diffuse light in the laboratory at room temperature. Add water to the sand from time to time to keep it moist throughout. For each hormone concentration, use five replicates.

Parameters	Concentration of auxin (µg/ml)				
	0	0.1	1	10	100
(i) Average number of elongated roots per plant					
(ii) Average number of primordia per plant					
(iii) Average total number of roots per plant					
(iv) Average root length (mm)					

(b) ***Rooting in isolated leaves***: Soak bean seeds overnight in water. Sow the seeds on a flat bed of moist sand or vermiculite. Detach 5 well-developed primary leaves from 2 to 3 weeks old plants. Dip about 10 mm of the leaf petiole into IAA solutions of different concentrations. After 2 hours, transfer the detached leaves to petri dishes lined with moist filter paper. Cover the petri dishes and place them under diffuse light at room temperature. Add water regularly to keep the filter paper moist throughout. Use five replicates for each IAA concentration.

Observe the root formation at the base of stem cuttings and leaf petioles treated with different concentrations of auxin after about 3 weeks. Count the number of roots formed in each stem cutting and excised leaf. Find the average number of roots formed as a result of treatment. Draw a graph showing the relation between hormone concentration and the number of roots. Also determine the time of initiation of rooting under different IAA concentration.

Note: Auxins like indole-3-butyric acid (IBA) and naphthalene acetic acid (NAA) can also be used in place of IAA. Rootex powder is used commercially to increase rooting in stem cuttings.

Project 12.4

To demonstrate the herbicidal effects of auxins.

Requirements

Plant mterial : Dicot seeds - cucumber (*Cucumis sativus*)
Monocot seeds - wheat (*Triticum aestivum*)
Chemicals : Auxin (2,4-D) - 100 ppm (100 mg/L). Dilute to 10, 1, 0.7, 0.5, 0.1 mg/L,
Miscellaneous : Petri dishes, pipettes.

Procedure

Take 14 sterilized petri dishes (two for each treatment including control, i.e., distilled water). Place filter paper at the bottom of each petri dish and add 25 seeds of cucumber and wheat in them. Then add 10 ml of auxin (2,4-D) solution in the respective petri dishes. Keep the petri dishes in the laboratory at 25–27 °C for about a week. Note down the germination percentage and the root length daily for a week and plot the values against the number of days.

Total number of seeds = 25

Conc. of 2,4-D (mg/L)	Number of seeds germinated						
	Day 1	Day 2	Day 3	Day 4	Day 5	Day 6	Day 7
Control							
100							
10							
1							
0.7							
0.5							
0.1							

Similarly, draw an observation table for (a) the root length of cucumber seedlings (b) the number of seeds germinated in wheat; and (c) the root length of wheat seedlings.

High concentrations of 2,4-D inhibits germination of seeds in both dicots and monocots. Hence high concentrations of auxin (2,4-D) show herbicidal effects. 2,4-D is the most widely used herbicide in the world for the control of broad-leaf weeds.

Besides 2,4-D, the most widely used synthetic auxins in weed control at the present time are 2,4,5-trichlorophenoxyacetic acid (2,4,5-T), 2-methyl-4-chlorophenoxyacetic acid (MCPA) or mixtures of these. IAA is relatively ineffective as a weed-killer.

Project 12.5

To demonstrate the process of bolting by exogenous application of gibberellic acid (GA_3)

or

Induction of internode elongation in rosette plants by gibberellins.

Requirements

4-week-old potted plants of lettuce (*Lactuca sativa*), GA_3 solution (0, 0.1, 1, 5 and 10 mg/L), distilled water

Procedure

Select 15 suitable plants of equal size. Divide them into 5 groups, each consisting of three plants. Expose the shoot apex of the plants by carefully shifting the leaves. Apply a drop of GA_3 solution of different concentrations on the shoot apex with the help of cotton swab. For each hormone concentration, use three test plants. Repeat the application of hormone every third day.

Follow the plant growth for 2–3 weeks and determine the length of the elongated stem of each plant using a meter rod.

Draw a graph/histogram showing the relation between the length of the stem and the concentration of GA_3 applied.

Note: Support the growing stem of the plants with a wooden stick, if necessary.

Stock solution of GA_3= 100 mg/L

Dilution table

S. No.	Concentration of GA_3 (mg/L)	Amount of stock solution (ml)	Amount of distilled water (ml)	Total volume (ml)
1	0 (Control)			
2	0.1			
3	1			
4	5			
5	10			

S. No.	Time (days)	Height of plant (cm)				
		Control	0.1 mg/L	1 mg/L	5 mg/L	10 mg/L
1	0					
2	2					
3	4					
4	6					
5	8					
6	10					
7	12					
8	14					
9	16					

Project 12.6

To study the effect of different concentrations of nitric oxide (NO) on seed germination.

Requirements

Plant material : Seeds of *Brassica campestris, Triticum* sp., *Lactuca sativa* cv. Grand Rapids, *Vigna radiata*

Chemicals : Sodium nitroprusside (SNP), 70 per cent ethanol

Miscellaneous : Petri plates, incubator, white fluorescent lamps

Procedure

Wash the seeds of different plants in distilled water and immerse in 70 per cent ethanol for one minute. After five washes in sterile distilled water, germinate in different petri plates on filter paper soaked in distilled water (control) and different concentrations of SNP (0–10 mM) and maintained at 25 ± 2 °C for 2 days in dark, followed by a 14 hour photoperiod provided by white fluorescent lamps.

Count the number of germinated seeds at 24, 48, 72 and 120 hours after keeping in light. Find out the concentration of SNP showing the highest percentage of seed germination. The higher levels (>1 mM) completely inhibit the germination process.

Note: Petri plates with the seeds treated with SNP should be completely sealed with paraffin wax.

13. Ripening, Senescence and Cell Death

Project 13.1

To study the effect of cytokinin on delaying senescence and chlorophyll retention.

Requirements

Plant material : 10–15 days old wheat seedlings (*Triticum aestivum*), leaves of tobacco, sunflower, cocklebur, garden nasturtium

Chemicals : Kinetin solution (100, 10, 1, 0.1 0.01 μM), 80 per cent ethanol or acetone

Miscellaneous : Petri dishes or coupling jars, pipettes, Buchner funnel, photoelectric colorimeter

Procedure

Detach 60–80 leaves from healthy 15 day old wheat plants or any other plant material and place them in a beaker with cut ends immersed in distilled water. Allow the leaves to remain in continuous fluorescent light for about 48 hours to initiate the process of aging. Take 15 clean, sterile petri dishes to make up three sets. The average of their readings will constitute the results for both treatment as well as control. Transfer 5 of the aging leaves from the beaker to each petri dish, cover and place them under fluorescent light. Determine the total chlorophyll content (mg/g of tissue) of the leaves remaining in the beaker. Observe the colour change of the leaves on days 4, 8 and 12. Also determine the total chlorophyll content of leaves on days 4, 8 and 12 for each treatment of kinetin.

Determination of total chlorophyll content

Determine the chlorophyll content of 1 gram of leaf material by cutting it into small pieces and placing them in 15 ml of 80 per cent ethanol in a test tube. Heat the tube in a boiling water bath for 2–5 minutes to extract chlorophyll. Cool the tube in the dark and adjust the volume of the solution to 15 ml by adding more of 80 per cent ethanol. Read absorbance of the solution at 665 nm against 80 per cent ethanol solvent.

Note:

(i) Petri dishes should be kept covered to prevent evaporation.
(ii) Kinetin solutions must be replaced regularly during the entire course of experiment to avoid drying of the leaves.
(iii) Petioles should be removed at the time of chlorophyll estimation as they contain a lot of water.
(iv) Equal amounts of leaves should be taken each time for the estimation of chlorophyll.

Concentration of Kinetin solution (μM)	Total chlorophyll content (A_{665})			
	Day 0	Day 4	Day 8	Day 12
100				
10				
1				
0.1				
0.01				
0 (control)				

The excised leaves treated with kinetin remain green for a longer period and the amount of chlorophyll retained is proportional to the hormone concentration. Plot relative absorbance as a function of chlorophyll content of leaves against the kinetin concentration.

Project 13.2

Acceleration of chlorophyll degradation by abscisic acid (ABA).

Requirements

Plant material : 7–10 days old green seedlings of oat (*Avena sativa*) or wheat (*Triticum aestivum*) or mature and aging leaves of tobacco, sunflower or other thin-leaved species
Chemicals : ABA solution (0, 0.005, 0.05, 0.5 and 5 mg/L), 80 per cent ethanol
Miscellaneous : Petri dishes, Buchner funnel with suction, spectrophotometer

Procedure

Raise oat or wheat seedlings in the light for 7–10 days at room temperature. Prepare 3 cm long sections of the primary leaves by cutting about 15 mm below the tip of the leaves. Place them in a petri dish containing a layer of moist filter paper. Transfer randomly, 10 leaf sections each to a petri dish containing 25 ml of ABA solution of different concentration. Maintain the petri dishes at room temperature under diffuse light.

Extract and measure total chlorophyll content in 10 leaf sections. The total chlorophyll content obtained initially will provide a reference for comparison with chlorophyll content of leaf sections treated with different concentration of ABA. Determine the chlorophyll content in leaf sections after 4 days of ABA treatment. Draw a graph showing the relation between chlorophyll degradation and the concentration of ABA. Express chlorophyll content considering the initial chlorophyll content before treating with ABA as 100 per cent (relative absorbance is taken as a measure of chlorophyll content).

Concentration of ABA (mg/l)	Total chlorophyll content (A_{665})
0	
0.005	
0.05	
0.5	
5.0	

Project 13.3

Induction of leaf abscission by ethylene.

Requirements

Plant material : 3-week-old plants of *Phaseolus vulgaris*
Chemicals : Ethrel (50 ppm), benzyladenine (50 μM)
Miscellaneous : Conical flasks

Procedure

Select out 20 uniform-sized plants. Cut-off the roots from the plants and place 5 plants each in a flask with the cut ends dipping in the following solutions: (i) Distilled water; (ii) ethrel solution; (iii) benzyladenine solution; (iv) benzyladenine + ethrel solution.

Over the next several days, follow the leaf fall by counting the number of leaves shed by the plants under different treatments. Also make longitudinal sections of the petioles and study the anatomical structure of the abscission layer.

Ethylene accelerates leaf fall and benzyladenine delays it.

Project 13.4

Effect of ethylene on leaf abscission and fading of flowers.

Requirements

Plant material: Holly or *Bougainvillea* twigs, ripe banana and apple
Miscellaneous: Conical flasks (3), glass jars (3)

Procedure

Place three similar twigs in conical flasks having water under glass jars for a week. At the same time, place a ripe apple and a banana under the two jars separately. Ethylene produced by the apple and banana will cause abscission of the English Holly leaves and fading of flowers. Count the total number of leaves and abscised leaves on each twig for about a week.

Ethylene apparently can trigger its own production.

14. Abiotic and Biotic Stress

Project 14.1

Biochemical analysis to study the effect of abiotic stresses on seed germination.

Requirements

Plant material : Seeds of wheat (*Triticum aestivum*), rice (*Oryza sativa*), mustard (*Brassica juncea*),

Chemicals : Sodium hypochlorite (1 per cent), Polyethylene glycol 400, sodium chloride,

Others : Petri plates, B.O.D. incubator, refrigerator.

Procedure

Wash and sterilize the seeds using sodium hypochlorite. Soak them in distilled water at room temperature overnight. Place 20–30 seeds in perti plates lined with filter paper and treat them as follows:

- For **drought (water stress)**, use various concentrations (0–10 per cent) of polyethylene glycol 400 (PEG 400).
- For **salt stress**, use different concentrations (0–500 mM) of sodium chloride.
- For **low temperature stress**, place petri plates containing seeds at (i) room temperature, (ii) 4 °C (in refrigerator), and, (iii) −20 °C (in freezer).
- For **high temperature stress**, place petri plates containing seeds in incubator at 40 and 50 °C respectively.

Allow the seeds to germinate for 6–7 days and record the observations.

(For biochemical analysis, i.e., enzyme assays, HPLC and proteomic analysis, i.e., protein gels, prepare the extracts from stress-treated seeds and use for analysis).

Seed germination being the primary step in plant growth is quite vulnerable to abiotic stresses resulting in major loss of crop productivity worldwide.

Project 14.2

Estimation of proline.

Requirements

Plant tissue, acid ninhydrin reagent (1.25 g ninhydrin in a mixture of warm 30 ml glacial acetic acid and 20 ml of 6M phosphoric acid, pH 1.0. Mix vigorously to dissolve. Store at 4°C and use within 24 hours), 3 per cent aqueous sulphosalicyclic acid, glacial acetic acid, toluene, standard proline solution (range 0.1–36 μmole).

Procedure

Homogenize 0.5 g plant tissue with 10 ml of sulphosalicylic acid in a pestle and mortar and filter through Whatman No. 2 filter paper. Repeat the extraction and pool the filtrates. Add 2 ml each of glacial acetic acid and ninhydrin reagent to 2 ml of filtrate. Keep in boiling water bath for 1 hour and terminate reaction by placing on ice bath. Add 4 ml of toluene and shake vigorously for 20–30 seconds. Aspirate the toluene layer and warm to room temperature. Measure the absorbance of red colour obtained at 520 nm against a reagent blank. The colour is stable for at least one hour.

Calculate the amount of proline in the sample using a standard curve prepared from pure proline and express on fresh weight basis of sample.

$$\mu\text{moles of proline/g tissue} = \frac{\mu\text{g proline/ml} \times \text{ml toluene}}{115.5} \times \frac{5}{\text{g sample}}$$

(Molecular weight of proline is 115.5)

Project 14.3

Estimation of superoxide dismutase (SOD) activity in salt-stressed seedlings of *vigna radiata*.

The enzyme superoxide dismutase (SOD) is present in all subcellular compartments of aerobic organisms susceptible to oxidative stress. SOD out competes damaging reactions of superoxide protecting the cell from superoxide toxicity. It catalyzes the dismutation of superoxide radical ($O_2^{\bullet -}$) into hydrogen peroxide (H_2O_2) and oxygen as per the following reaction:

$$O_2^{\bullet -} + 2H^+ \xrightarrow{\text{SOD}} H_2O_2 + O_2$$

The assay is based on formation of blue coloured formazone by nitro-blue tetrazolium and $O_2^{\bullet -}$ radical, which absorbs at 560 nm and the SOD enzyme decreases this absorbance due to the reduction in the formation of $O_2^{\bullet -}$ radical by the enzyme.

Requirements

Seeds of *Vigna radiata*, Nacl, Fluorescent white light lamps, shaker equipped with transparent tubes holder, spectrophotometer, NaCl.

Reaction mixture

- 0.1 M phosphate buffer (pH 7.5) : Mix 16 ml of 0.2 M KH_2PO_4 with 84 ml of 0.2 M K_2HPO_4 and add 100 ml of distilled water.
- Riboflavin 24 M : Dissolve 9 mg of riboflavin in 100 ml of distilled water.
- Methionine 150 mM : Dissolve 0.56 g of methionine in 25 ml of distilled water.
- Na_2EDTA 1.2 mM : Dissolve 17.2 mg of Na_2EDTA in 25 ml of distilled water.
- p-Nitro blue tetrazolium (NBT) 840 M : Dissolve 17.2 mg of NBT in 25 ml of distilled water.

Procedure

Prepare enzyme extract of 7-day old germinated seedlings in 10 ml phosphate buffer and centrifuge in refrigerated centrifuge for 15 min at 15,000xg at 4°C. Germination was done in distilled water, 50, 75, and 100 mM Nacl.

Prepare 3 ml of reaction mixture for each treatment and one control (without enzyme) as per the following table.

Reaction mixture for *Superoxide dismutase* (SOD) activity assay

Initial concentration	Quantity
0.1 M phosphate buffer pH 7.5	1.95 ml
Riboflavin 24 µM	250 µl
Methionine 150 mM	250 µl
Na_2EDTA 1.2 mM	250 µl
NBT 840 µM	250 µl
Extract	50 µl
Total volume	3 ml

Place the four tubes on shaker and incubate at 25°C in the fluorescent light for 15 minutes. Keep another similar set of four tubes in the dark at 25°C.

After incubation period, measure absorbance at 560 nm using respective dark-incubated samples as reference for test samples.

Results

The colour in the tubes incubated under light turns yellow-purplish from yellow. Maximum absorbance is seen in control and minimum in 100mM NaCl.

If superoxides are present, NBT will be reduced to formazan and form blue colour. In salt-stressed seedlings, more SOD is produced (maximum in 100 mM) to overcome the high salt stress. So less superoxides are freely available to reduce NBT (SOD has scavenged the superoxide radicals) and SOD inhibits the formation of the blue formazan. The formation of blue formazan is inversely proportional to the activity of the enzyme SOD.

Project 14.4

Estimation of ascorbate peroxidase (APX) activity.

Ascorbate peroxidases belong to scavenging enzymes that detoxify peroxides (i.e., H_2O_2) using ascorbate as substrate. These catalyze the transfer of electrons from ascorbate to peroxide, producing dehydroascorbate and water as products.

$$\text{Ascorbate } C_6H_8O_6 + H_2O_2 \xrightarrow{\text{APX}} \text{Dehydroascorbate } C_6H_6O_6 + 2\,H_2O$$

Method for its activity assay base on spectrophotometrical monitoring of of the rate of decomposition of H_2O_2 by ascorbate peroxidase in the presence of ascorbate as hydrogen donor, results in dehydroascorbate production.

Requirements

Freshly and partially purified seedlings extract, UV spectrophotometer, 0.1M phosphate buffer (pH 7.0), 1.2 mM Na_2EDTA (11.2 mg in 25 ml distilled water), 35 mM H_2O_2 (prepare fresh 80 µl of 30 per cent H_2O_2 in 19.92 ml distilled water), 15 mM ascorbic acid (66 mg in 25 ml distilled water).

Procedure

Prepare 3 ml reaction mixture in spectrophotometric cuvette as per the following table and monitor ascorbate oxidation i.e. change in absorbance at 290 nm for 500 seconds at 25°C. Make at least three independent repetitions.

Reaction mixture for APX activity assay

Initial concentration	Quantity
0.1 M phosphate buffer pH 7	2.4 ml
EDTA 1.2 mM	250 µl
H_2O_2 35 mM	50 µl
Ascorbic acid 15 mM	100 µl
Extract	200 µl
Total volume	3 ml

Project 14.5

Estimation of catalase (CAT) activity.

Catalase enzyme catalyzes the dismutation of hydrogen peroxide into water and oxygen. This enzyme is important for the removal of H_2O_2 generated in peroxisomes by oxidases involved in β-oxidation of fatty acids, glyoxysomes (the glyoxylate cycle) and purine catabolism.

$$2\,H_2O_2 \xrightarrow{CAT} 2H_2O + O_2$$

Requirements

Freshly and partially purified seedlings extract, 0.1 M phosphate buffer, 264 mM H_2O_2 (add 300 µl of 30 per cent H_2O_2 solution to 9.7 ml of distilled water), UV spectrophotometer.

Procedure

Prepare 3 ml reaction mixture in spectrophotometric cuvette as per the following table. Measure change in absorbance at 240 nm for 100 seconds at 25°C. Make at least three independent repetitions per extract.

Reaction mixture for CAT activity assay

Initial concentration	Quantity
0.1 M phosphate buffer pH7	1.95 ml
H_2O_2 264 mM	100 μl
Extract	50 μl
Total Volume	3 ml

15. Secondary Plant Metabolites

Experiment 15.1

To extract anthocyanin pigments from two plant sources and study the effect of pH on them.

Requirements

Plant material	: Flower petals of *Hibiscus rosa-sinensis* Leaves of *Tradescantia*
Chemicals	: Extraction medium – *n*-propanol – 32 per cent $HCl:H_2O$ (18:1:81 v/v) NaOH, HCl
Glass apparatus	: Glass vials with stopper, beaker, pipette (10 ml), test tubes, measuring cylinder (50 ml)
Miscellaneous	: Water bath, centrifuge, photoelectric colorimeter, carbon paper, test tube holder

Principle

Anthocyanins are the major determinant pigments of flower colour. Most red, purple and blue plant pigments found in angiosperm flowers are due to anthocyanins. Anthocyanins are water-soluble and localized in vacuoles. The colour of anthocyanins depends upon the acidity of the cell sap. Several different anthocyanins exist in angiosperm flowers. All anthocyanins have a basic structure of anthocyanidine but they differ from one another in the attachment of sugars to the anthocyanidine. Anthocyanins belong to a major group of the most important pigments involved in floral coloration, the flavonoids. In addition to flowers, anthocyanins are also found in leaves, fruits and young seedlings. In young seedlings, anthocyanins protect the plant against the harmful UV radiation. Anthocyanins may aid as attractants of pollinators. Sometimes they may also serve as inhibitors of bacterial growth or a means to classification by taxonomists.

Synthesis of anthocyanins takes place at low temperature, nutrient stress and UV light sensitive conditions.

Procedure

Transfer 500 mg of plant material to glass vials containing 20 ml of extraction medium. Stopper the glass vials and heat for 5 minutes in a boiling water bath. Allow them to stand for 3 hours in the dark for complete extraction of anthocyanins. Centrifuge the extract to obtain a clear supernatant. Measure absorbance at 535 and 650 nm respectively. Calculate the absorbance due to anthocyanins using the following formula.

$$A_{anth} = A_{535} - 2.2\ A_{650}$$

Observation table

Plant source	A_{535}	A_{650}	A_{anth}
Hibiscus rosa-sinensis	0.61	0.02	0.57
Tradescantia	0.27	0.33	0.45

Compare anthocyanin contents in different plant materials using the relative absorbance as a measure of anthocyanin content. Divide the supernatant into two parts. Add a few drops of NaOH and HCl to each of this supernatant and observe the change in colour.

Result and Discussion

Anthocyanins are sensitive to pH. At neutral pH, these are violet coloured but change to red and blue colour at acidic and basic pH, respectively.

Precautions

1. The extraction mixture should be freshly prepared.
2. The supernatant obtained after centrifugation should be absolutely clear.
3. The glass vials should be quickly covered with carbon paper after taking out from water bath.

Experiment 15.2

Estimation of caffeine content in tea leaves.

Requirements

Plant material : Black tea leaves (*Camellia sinensis*)
Chemicals : Chloroform, 10 per cent ammonium hydroxide, absolute alcohol
Glass apparatus : Conical flask, mortar and pestle, beaker
Miscellaneous : Water bath, filter paper

Theory

Caffeine is a group of alkaloids commonly found in tea and coffee. Depending upon the quality, tea contains 3–5 per cent caffeine. Caffeine in tea leaves can be extracted in chloroform.

Procedure

Take 5 grams of tea leaves in an Erlenmeyer flask containing 100 ml chloroform. Add 5 ml of 10 per cent ammonium hydroxide and shake the flask intermittently for 2 hours.

Now filter the mixture through a filter paper. Keep the extract on boiling water bath till it comes to dryness. Add 3 ml of absolute alcohol and again boil to dryness. With the crude extract still in the water bath, add 10 ml of 30 per cent ethyl alcohol and stir vigorously. This will precipitate the chlorophyll, fat and waxes. Filter the solution and rinse the beaker with 10 ml of 30 per cent ethyl alcohol. Filter again and add the filtrate to the first filtrate.

Now evaporate the combined filtrate to dryness on a water bath. Weigh the nearly pure caffeine residues.

Flow-Sheet

5 g of tea leaves in 100 ml chloroform

↓

Add 5 ml of 10 per cent ammonium hydroxide

↓

Shake the flask for 2 h

Filter and dry the extract on boiling water bath

Add 3 ml of absolute alcohol and boil to dryness

Add 10 ml of 30 per cent ethyl alcohol and stir vigorously

Filter and evaporate the filtrate to dryness

Pure caffeine residues obtained

Result: Calculate the percentage of caffeine in the tea, using the following formula:

$$\% \text{ Caffeine} = \frac{W_1}{W_2} \times 100$$

[W_1 is the weight of the extracted caffeine and W_2 is the weight of the tea leaves.]

Experiment 15.3

Estimation of total alkaloids in tobacco leaves.

Requirements

Plant material : Dry tobacco leaves (*Nicotiana tabacum*)

Chemicals : Starch, petroleum ether (60–80 °C), 0.1N HCl, 0.1N NaOH, methyl red indicator, barium hydroxide

Glass apparatus : Mortar and pestle, Soxhlet apparatus, thimble, separating funnel, volumetric flask (50 ml), pipette (10 ml), burette (50 ml)
Miscellaneous : Water bath

Principle

Alkaloids are a heterogeneous group of secondary plant products. Alkaloids possess a nitrogen-containing heterocyclic ring system and are mostly alkaline in nature. These are generally amino acid derivatives and some of them are derivatives of purine nucleotides (e.g., caffeine). Approximately 3000 alkaloids are known so far. Common examples are nicotine, caffeine, colchicine, atropine, hyoscyamine, codeine, cocaine, quinine, ephedrine, etc.

Procedure

Mix 5 grams of finely powdered dry tobacco leaves with 1 gram of barium hydroxide powder, in a mortar. Add a few ml of water and continue mixing till a thick paste is formed. Dry the paste to a fine powder by mixing it with sufficient amount of starch powder. Transfer the powder quantitatively into a thimble of a Soxhlet apparatus (Figure E15.1). Extract alkaloid for about 3 hours. Separate the alkaloid extracted from petroleum ether solvent by evaporation in a rotatory evaporator. To the resulting alkaloid extract, add 50 ml of 0.1N HCl and shake well to dissolve the alkaloid. Titrate 20 ml of the solution against 0.1N NaOH using a drop of methyl red as indicator to determine the amount of HCl left after being neutralized by the alkaloids. Assume that 1 ml of 0.1N HCl is equivalent to 0.01629 g nicotine.

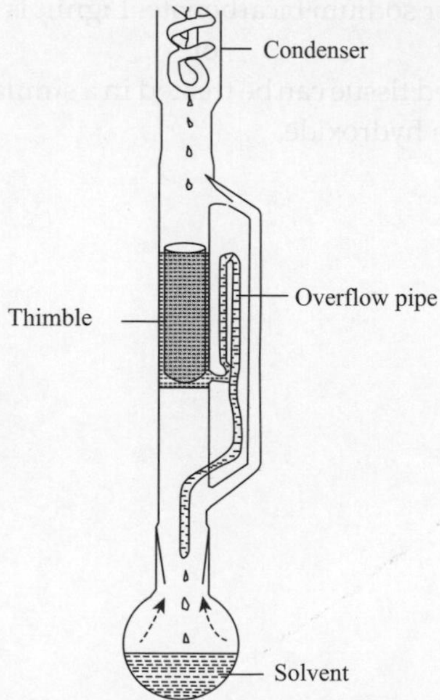

Fig. E15.1 Soxhlet extractor.

Results

Express alkaloid content (equivalent to nicotine) in terms of percentage in dry matter of tobacco leaf.

Physiological Set-Up

To demonstrate the presence of tannins and lignin in the given plant materials.

Tannins

Mount thin sections of tea leaves or rose stem with 10 per cent aqueous solution of ferric chloride. The cells containing tannins turn black.

Lignin

(i) Add a few drops of acidified phloroglucinol (a saturated solution of phloroglucinol in 18 per cent HCl) to jute fibres or coir fibres or a newspaper strip or a match-stick handle. Lignin in the cell walls stains cherry red.

(ii) Treat jute fibres with a solution of aniline hydrochloride or aniline sulphate. The fibres turn bright yellow in colour.

(iii) Add 1 per cent $KMnO_4$ solution to the lignin containing plant material kept in a watch-glass. Keep it for 5 minutes (Maule test). After thorough washing in water, place it in dilute HCl (2 per cent) for a minute. Wash it again and keep in a solution of 2 per cent ammonium hydroxide or sodium bicarbonate. Lignin is stained deep red to brown.

or

Sections with lignified tissue can be treated in a similar manner and then mounted in 2 per cent ammonium hydroxide.

Glossary

Abiotic factors: Physical, chemical and other non-living components of the environment such as light, temperature, moisture, nutrients and other edaphic factors.

Abscission: The shedding of leaves, flowers, and fruits from the living plant is called as abscission. It occurs in a region called the abscission zone.

Abscisic acid (ABA): ABA, often termed the 'stress hormone', is induced by many different environmental stresses, including drought, salinity, freezing, chilling, wounding, and hypoxia (low oxygen). Both bud dormancy and seed or fruit dormancy are caused by ABA (but reversed by the application of gibberellic acid). 'Abscisic acid' is a misnomer for this compound, for it has little to do with abscission. ABA stimulates the induction of storage protein synthesis and the closing of stomata. It is widely regarded as a growth inhibitor, found in dormant buds and fruits.

Acclimation: The capacity of a plant to adjust to and survive a stress.

AcetylCoA: A compound consisting of coenzymeA and acetyl group derived from pyruvate. It is the substrate for fatty acid synthesis and the Krebs cycle, and the product of fatty acid oxidation and other catabolic processes.

Activation energy: The extra energy required to destabilize existing chemical bonds and initiate a chemical reaction is called 'activation energy'. Catalysts reduce the activation energy and so increase the rate of reactions. Catalysts work by reducing the amount of energy which reactant molecules must possess before they can undergo a chemical change.

Active site: The portion of the enzyme where the substrate fits very precisely into a groove or pocket on the protein surface. The active site also has exactly the correct array of charged or uncharged (or hydrophilic and hydrophobic areas) on its binding surface.

Active transport: The transport of ions or molecules across a cellular membrane against a concentration gradient (from a region of low concentration to a region of high concentration) with the help of membrane protein, and energy is supplied by ATP.

Acyl carrier protein (ACP): The protein to which acyl chains are covalently linked during chain elongation by the fatty acid synthase complex.

Acyltransferases: Enzymes that catalyze the transfer of an acyl group from one molecule (usually an acyl CoA) to another.

Adenosine diphosphate (ADP): A compound made up of adenine, five carbon sugar ribose and two phosphate groups. It is involved in the mobilization of energy for cellular functions.

Adenosine triphosphate (ATP): ATP is the energy currency of all cells. Cells use their supply of ATP to power all energy-requiring processes they carry out. The ATP molecule is often referred to as a 'coiled spring', the phosphates straining away from one another. ATP has three components: the first is a five-carbon sugar, ribose; the second component is adenine-composed of two carbon-nitrogen rings; and the third component is a triphosphate group (three phosphates linked in a chain by high energy bonds-represented by a squiggle ~). Its decomposition into ADP and phosphate results in the release of energy needed for cellular functions.

A–P ~ P ~ P (terminal phosphate)

The high-energy phosphate bond is symbolically represented by the expression ~P while the low-energy P group is represented by –P.

Adhesion: The force of attraction between water and the hydrophilic cell walls of treachery elements. It prevents the collapse of water column when it is under tension (or negative pressure).

ADP-glucose: A sugar nucleotide formed by reaction of glucose-1-phosphate and ATP, catalyzed by ADP-glucose pyrophosphorylase; it is the substrate for starch synthesis.

Aerial spraying: Application of pesticide or fertilizer in the form of spray by using aeroplane or helicopter, with the purpose of covering a vast area in a short time.

Aerobic rice technology: Rice is grown in well-drained, non-puddled, and non-irrigated or non-saturated soils, just like wheat, maize, and sorghum and such. It is rice growing in aerobic soil, with the use of external inputs such as supplementary irrigation, fertilizers and aiming at high yields. Aerobic rice is the solution to future climate change under drought stress conditions with lesser methane gas (greenhouse gas) emission.

Aeroponics: It is a method for growing plants with their roots bathed in the nutrient mist (a cloud of moisture in the air).

Agent orange: A mixture of 2, 4-D and 2, 4, 5-T was used by the U.S. military to defoliate the forests in parts of South Vietnam. It is a herbicide.

Allelopathic: Chemicals leached out of one plant may inhibit the growth and seed germination of other species, thus disrupting the ecosystem underneath, e.g., *Eucalyptus* leaves (Blue gum).

Amphibolic pathway: The major function of Krebs cycle (TCA cycle) is clearly catabolism but at the same time it is involved in a variety of anabolic processes such as transamination reactions that reversibly convert α-keto intermediates into amino acids alanine, aspartic acid, glutamic acid from pyruvate, oxaloacetate and α-ketoglutrate respectively. As the TCA cycle can function both in a catabolic mode and as a source of precursors for anabolic pathways, it is called an amphibolic pathway (from the Greek amphi, meaning 'both').

Amphipathic: Molecules with distinct hydrophobic and hydrophilic regions, e.g., phospholipids.

Amphiprotic or Amphoteric compounds: Compounds that can act both as an acid and as a base. Amino acids are compounds with the properties of both acids and bases. Every amino acid contains at least one carboxyl (–COOH) group and one or more amino ($–NH_2$) groups. That is why amino acids (or proteins) behave as acids towards alkalies and as bases towards acids or they can react with acids and bases to form salts.

Amylopectin: It is a fraction of starch molecules; multibranched polymer of α (1→4) linked glucose containing occasional α (1→6) linkages at branch points.

Amyloplast: A colourless plastid in which starch is synthesized. Starch is a polysaccharide built up from glucose units joined together by a different form of linkage, catalyzed by 'starch synthase'.

Amylose: It is a fraction of starch molecules, consisting of a long chain of α (1→4) linked glucose residues, i.e., in between the first and fourth carbons.

Anammox: (An abbreviation for anaerobic ammonium oxidation) It is an important microbial process of the nitrogen cycle which is cost-effective and sustainable way of removing ammonium from effluents and ammonia from waste gas. This reaction oxidizes ammonium to nitrogen gas using nitrite as the electron acceptor under anoxic conditions and is carried out by anaerobic bacteria discovered in waste-water sludge .

Anaplerotic reaction: The Citric Acid or Krebs cycle will cease to operate unless new oxaloacetate is formed because acetyl CoA cannot enter the cycle unless it condenses with oxaloacetate, which is formed by the carboxylation of pyruvate, in a reaction catalyzed by pyruvate carboxylase.

$$\text{Pyruvate} + CO_2 + \text{ATP} + H_2O \rightarrow \text{Oxaloacetate} + \text{ADP} + \text{Pi} + 2\text{H}$$

Oxaloacetate is both intermediate in gluconeogenesis and Krebs cycle. The carboxylation of pyruvate is an example of an anaplerotic reaction (from the Greek, to 'fill up'). Intermediates drawn off from TCA cycle for biosynthesis are replenished by the formation of oxaloacetate from pyruvate.

Anion or salt respiration: The increase in respiration when a tissue or plant is transferred from water to salt solution.

Anoxia (or anaerobiosis): When the soil air is completely dispelled or displaced by water as it happens during prolonged flooding. The soil becomes depleted or deprived of oxygen. Anoxia inhibits root growth, respiration of root cells and finally reduces nutrient uptake from the soil as it needs metabolic energy.

Anoxygenic photosynthesis: Bacterial photosynthesis that does not use water as an electron donor and does not evolve oxygen as one of the end product.

Antagonism: It refers to two ions or chemical compounds that produce responses which are opposite to one another.

Antennae (or subsidiary) pigments: These accessory or antennae pigments consist of chl b, carotene and xanthophyll. After absorbing or gathering light at appropriate wavelengths, they transfer energy to the reaction centres of PS II and PS I through inductive resonance.

Anthocyanins: These are water soluble pigments, occurring in the vacuole. Many flower and fruit colour, including pinks, reds, blues, and purples are due to anthocyanins. The floral colour of many plants are pH dependent, being blue when the cell sap is alkaline and red when the cell sap is acidic. Examples are beet root, red cabbage, floral bracts of *Euphorbia pulcherrima*.

Antifreeze proteins: Proteins that inhibit the nucleation of ice crystals.

Antioxidants: Compounds that can scavenge reactive oxygen species by becoming oxidized themselves.

Antiport: Two ion species or solutes (molecules) move in opposite direction through the membrane. The ions during cotransport are moving against the electrochemical gradient (i.e., active transport), requiring energy input.

Antitranspirant: Refers to a group of chemicals or materials that retard or inhibit transpiration, e.g., PMA, silicone oil. These help to conserve water.

Apical dominance: The inhibitory effect of a terminal bud upon lateral or axillary bud development.

Apoplast: The intercellular spaces and cell walls of a plant; all the volume of a plant that is not occupied by protoplasm.

Aquaporins: The plasma membrane and tonoplast contain water channel proteins which facilitate the movement of water across cell membranes. Water passes relatively freely across the lipid bilayer of cell membrane, but the aquaporins permit water to diffuse much more rapidly through the plasma membrane into the cell, and then through the tonoplast into the vacuole.

Ascorbate-glutathione cycle: The metabolic reaction cycle that converts excessive hydrogen peroxide to water in chloroplasts.

Autoradiography (radioautography): A technique whereby the location of radioactive material is determined by the use of photographic film. This process is especially useful in tracing different steps involved in a pathway.

Auxins: Hormones involved in cell elongation, apical dominance, and rooting, among other processes.

Assimilatory power: The two compounds ATP (energy currency of the cell) and NADPH (a biological reductant) together constitute the assimilatory power of the cell.

Bacteriochlorophyll: Light-harvesting pigment involved in bacterial photosynthesis.

Bacteroids: Differentiated forms of rhizobial bacteria in the nodule cells of a host plant.

Ballast elements: Minerals like aluminium and silicon are called ballast elements because they are generally present in large amounts in the soil, although the plants can grow normally without them.

Biliproteins: The biliproteins (or bile pigments) are linear tetrapyrroles found in the photosynthetic tissues of many algae. They include the well-known pigments phycocyanin (blue) and phycoerythrin (red) of cyanophyceae (cyanobacteria) and rhodophyceae respectively.

Bioassay: As plant hormones occur in minute quantities, it is not always possible to detect and estimate them by chemical methods. Therefore they are estimated by using some live plants or plant parts. Such a technique of measuring the activity or response of hormones by living organism is known as bioassay.

Bioenergetics: The study of energy flow through living organism. There are two forms of energy, one that is available to perform work (free energy or Gibbs free energy in honour of the English physicist J.W. Gibbs) and the other that is not available (entropy).

Biofortification: A new type of nutrition intervention to address the global problem of micronutrient deficiencies (also known as hidden hunger) of iron, zinc, iodine, selenium and vitamin A.

Biogeochemical cycle: The flow of chemical elements and compounds between living organisms and the physical environment. For example, nitrogen cycle, phosphorus cycle, sulphur cycle, iron cycle, carbon cycle, oxygen cycle, and water cycle.

Biological clock: An internal biological timing system within the organism.

Bleeding: It is the exudation of liquid water or sap from injured parts of the plant. The exudation of sap from sugar maple and toddy palm are two examples.

Bolting: It refers to elongation of the shoot apex, usually just prior to flower formation. Treatment with gibberellin will cause the rosette to elongate and bolt prematurely, e.g., *Agave.*

Blue-light responses: Plant developmental responses triggered by blue light; the photoreceptor is a small protein called phototropin.

Bone meal: A fertilizer made of finely ground dried animal bones. It contains 20–24 per cent P_2O_5. It has at least 3 per cent nitrogen too.

Bud scale: A small, specialized leaf, usually waxy or corky, that protects an unopened bud.

Buffer: Any substance that absorbs or releases protons (H^+) so that the pH of the solution remains stable even when acid or base is added.

Brassinosteroid: A family of polyhydroxy steroid. Compounds showing hormone-like activity, involved in leaf morphogenesis, root and stem growth, and vascular differentiation. Discovered in a mutant form of *Arabidopsis thaliana* where the plants develop abnormally. Normally it has large leaves in rosette, followed by bolting, i.e., producing the flowering stalks with long internodes. They regulate many plant processes, including cell expansion, cell division, and vascular development.

Calcareous soil: Soil containing sufficient calcium carbonate (often with magnesium carbonate); effervescence visibly when treated with cold 0.1N HCl.

Calcium-pumping ATPase: This operates in the cell wall where it regulates the intracellular calcium concentration. The excess calcium is transported out from the cytosol to the walls where it helps in 'stiffening.'

Callose: Carbohydrate deposited on the sieve-plates of sieve tubes.

Calmodulin: The principal calcium receptor in plant and animal cells. It is a highly conserved, ubiquitous protein that is known to stimulate several classes of enzymes, including NAD^+ kinases, protein kinases and Ca^{2+}-ATPases. NAD kinase catalyzes the phosphorylation of NAD to NADP in the presence of ATP.

Carbon partitioning: The distribution of photosynthetic carbon and associated energy throughout the plant.

Carotenoids: A class of lipid soluble accessory pigments in chloroplasts and chromoplast. These are orange yellow pigments, consisting of carotene (C_{40} H_{56} – a hydrocarbon) and xanthophyll ($C_{40}H_{56}O_2$). These accessory pigments gather light and pass it on to chlorophyll a molecules. They protect chlorophyll molecule from being destroyed by strong light (photooxidation). Flower colours in the yellow, orange, and red range are often due to carotenoids, present in plastids of petal cell.

Carrier proteins: They actually bind the transported molecule on one side of the membrane and release it on the other side.

Catabolism: Reactions in a living organism which involve breaking down complex organic molecules into simple ones. These reactions are typically 'exergonic'. The aerobic respiration of glucose is a classic example.

Cavitation: A process of rapid formation of bubbles in the xylem.

Cellulase: An enzyme that breaks down cellulose into its monosaccharide units.

CF_0–CF_1 Complex: Part of ATP synthase. CF_0 is a proton channel and CF_1 is a set of enzymes that phosphorylates ADP to ATP.

Channel proteins: These are transmembrane proteins that facilitate ion movement across cellular membranes. Channel proteins may exist in two different configurations – open and closed. In the open confirmation, the ions are permitted to pass through the channel, i.e., the gate is 'open' and the ions are free to diffuse *or* 'closed' to ion flow. Opening is stimulated by a number of stimuli such as voltage, light, hormones or ion themselves. The channel protein is believed to contain a sensor that is able to perceive the appropriate stimulus. Through these channels (or pores) the solutes diffuse down their electrochemical or chemical gradient.

Chaperones: A group of proteins which interact with mature or newly synthesized proteins to assist folding or refolding of polypeptide chains, to facilitate protein import and translocation, and to participate in signal transduction and transcriptional activation.

Chlorosis: A physiological disorder in green plants characterized by yellowing of plant tissues due to loss of chlorophyll or lack of chlorophyll synthesis. The yellowing may be generalized all over the entire plant or restricted to the leaves or isolated between some leaf veins, etc. (interveinal chlorosis).

Circadian: Having an approximately 24 h rhythm.

Climacteric fruits: They ripen slowly as they mature, but in the final stages, several developmental changes occur rapidly because of sudden burst (or increase) of ethylene which leads to rapid completion of maturation of the fruits. During the final stages, starches are converted to sugars; cell walls break down and soften; flavours and aromas develop; and colour changes do take place. A little ethylene is present in the initial stages.

Climacteric rise: In many fruits the ethylene gas release during the ripening process is coincident with a sharp rise in the respiratory rate, i.e., CO_2 evolution.

Coenzymes: These are larger 'organic molecules' which bind reversibly to an enzyme. They are regenerated and can therefore be reused. Examples are: NAD, NADP, FAD and CoA.

Cofactors: These are mostly ions (especially divalent cations) or small molecules which normally bind reversibly to an enzyme and can thus easily be removed by dialysis. Some promote enzyme activity as they are components of the active site, but many are not. Examples are Ca^{2+}, Zn^{2+}, Mg^{2+}, Cl^{-}, AMP.

Cohesion: The mutual attractive forces between the water molecules.

Coleoptile (pronounced coal–ee–op–tile): The outermost sheath-like structure which covers the young shoot or plumule of grass seedlings. Oat (*Avena sativa*) coleoptiles are the organ studied most often and show a strong positive phototropic response.

Competitive inhibitor: A small molecule that inhibits an enzyme by binding to the active site.

Compost: Partially decayed organic matter used in gardening and farming to enrich the soil and to increase the water-holding capacity.

Conjugated proteins: Comprised mostly of globular proteins that have a non-protein component. If the compound is a tightly bound organic compound, it is called a 'prosthetic group'. If it is a small detachable molecule or ion it is usually called a 'cofactor'.

Constitutive enzymes: These enzymes always tend to be present such as respiratory enzymes. The glycolytic enzymes are constitutive.

Cotransport: Carrier proteins which transfer two different ions or solutes simultaneously. Here both the ions are transported in the same direction (symport) or in opposite direction (antiport).

Coupled reaction: The endergonic reaction in a living system may take place at the expense of free energy released by the exergonic reactions. Since the two kinds are energetically linked, they are called coupled reaction.

Cryptochrome: A putative blue or UV-A light receptor. These photoreceptors detect and mediate responses to light in the blue and ultraviolet regions of the spectrum. They are made up of apoproteins and chromophores that absorb light. Unlike phytochrome, the chromophore of cryptochromes is attached noncovalently to the apoprotein. The chromophore of cryptochrome is possibly flavin adenine dinucleotide (FAD). None has yet been identified which makes the 'crypto' part of the word especially appropriate.

Cyanide-resistant respiration: Synonym for thermogenic respiration.

Cyclic photophosphorylation: In cyclic photophosphorylation, the electron which was ejected from the chlorophyll molecule travels 'in a cycle' through electron transport carriers and ultimately returns to the 'same' chlorophyll molecule from which it was expelled, i.e., the electron donor and electron acceptor is the same chlorophyll molecule. In *Chromatium* (photosynthetic bacteria) it is the only photosynthetic reaction, producing ATP.

Cytochrome: A protein molecule with the characteristic molecular structure having a porphyrin ring. In contrast to haemoglobin, the central iron atom of the porphyrin ring is easily oxidized and reduced. It is the primary electron transport carrier during oxidative phosphorylation occurring in mitochondria. It exists in several forms, cyt b, c, a and a_3.

Cytokinins: Compounds that stimulate cell division in plants. For the most part they are derivatives of purine (adenine) bases. Kinetin is one example. They have an anti-senescence effect. They are mainly synthesized in the roots from where they migrate upward and interact in various ways. All the cytokinins, both natural and synthetic, are derivatives of adenine.

De-etiolation: When the seedling shoot emerges out from the soil surface, and is exposed to light, its development 'switches' from skotomorphogenesis (dark grown) to photomorphogenesis (light grown). This process is known as *de-etiolation.*

DELLA proteins: A class of nuclear proteins, or transcription factors, that function as repressors in gibberellin signalling.

Dendrograph: A devise to measure diameter variations in tree trunk. There is a decrease in diameter of the stem during the day when there is high rate of transpiration, and increase in diameter during periods of low transpiration. This was designed by D.T. MacDougal of the Carnegie Institute of Washington.

Depolarized: It is referred to a membrane where the distribution of charged ions is approximately the same on both sides, so that the potential across the membrane is 'zero'.

Derepression: Removal of a repressor molecule from the gene, facilitating transcription.

Development: It is the sum of growth and differentiation, i.e., the sum of 'all' the changes that occur in an organism during its life cycle – right from germination of the seeds onward to vegetative growth, maturation, flowering, seed formation and finally senescence. It also applies equally well to cells, tissues and organs. Development is thus an orderly sequence of events controlled by a precise set of genes turning on and off at every stage.

Devernalization: The chilling treatment given during vernalization, if immediately subjected to high temperature, will not result in flowering. This reversal is referred to as devernalization.

Die-Back: Killing of shoot apex or tip-dying, caused by copper deficiency.

Dialysis: It is a method by which pure colloidal solution can be separated from a true solution. Examples are collodion or parchment membrane. They are used to separate starch (colloidal solution) and salt (NaCl).

Differentially or selectively permeable membrane: A membrane which allows some substances to pass through it more readily than others or restricting the passage of others. Examples are all living biological membranes.

Differentiation: It refers to qualitative changes in cells and cell constituents, i.e., the formation of subcellular structures such as organelles; formation of specialized cells such as conductive, protective, storage tissues; specialized cell constituents (such as terpenes, amino acids, latex, cuticle or pigments, etc.); and the development at the organ level (such as leaves, flowers and fruits).

Diffusion Pressure deficit (or Suction pressure): It is the net force with which water is drawn into the cell, and is equal to the difference between the osmotic pressure of the cell sap and the wall pressure (the latter being equal to the turgor pressure).

$$DPD = OP - TP$$

or

the difference between the diffusion pressure of the solution and its solvent at a particular temperature and atmospheric conditions. If the solution is more concentrated its DPD goes up but it decreases with the dilution of the solution (*or* DPD α concentration of the solution).

Disulphide bond: A linkage between the sulphur atoms of two different amino acids in a protein.

Dormancy: It is temporary cessation of growth and suppression of metabolism as it happens during winter and drought periods.

Drought: A period during which the soil contains little or no water (physical drought) or it is rendered unavailable (physiological drought or dryness).

Ectomycorrhiza: Symbiotic associations between fungi and plant roots in which the fungal hyphae surround the roots as a sheath.

Electrochemical potential: It is a measure of the energy status of an ion (analogous to ψ) which includes both concentration (chemical) and electrical potentials.

Electrogenic: Whenever an ion moves into or out of a cell membranes (plasma membrane or tonoplast) unbalanced by a counter ion of opposite charge, a voltage difference is created across the membrane. Such type of transport is called **'electrogenic'.** It is quite common in living cells, being mediated by a carrier called an **electrogenic pump,** requiring energy (i.e., active transport). Electrogenic transport occurs against electrochemical gradient. This process requires energy which is often supplied by hydrolysis of ATP. In plants, H^+–pumping ATPase is the primary electrogenic pump or transporter in both the plasma membrane and the vacuolar membrane (tonoplast).

Electromagnetic spectrum of light: Visible light which is composed of a wavelength spectrum from 390 to 760 nm with violet, indigo, blue, green, yellow, orange and red monochromatic light.

Embden-Meyerhof-Parnas (EMP) pathway: The conversion of glucose to pyruvic acid is known as EMP pathway, named after three German physiologists. This is also known as glycolysis or glycolytic pathway. The reactions of glycolysis occur in the cytoplasm and do not need the presence of oxygen.

Embolism: An obstruction caused by the formation of large gas bubbles in the xylem.

Endergonic: It refers to energy-consuming reaction.

Endergonic reaction: In such reaction, the products of the reaction contain more free energy than the reactants and the extra energy must be supplied for the reaction to proceed. The ΔG is positive. Such reactions do not proceed spontaneously.

Endomycorrhiza: Symbiotic associations between fungi and plant roots in which the fungal mycelium penetrate the roots, forming vesicles and arbuscules in the root cells.

Energy: It is defined as the capacity to do work (really the ability to perform work), which may be mechanical, electrical, osmotic, or chemical. As the energy is converted and work proceeds, the matter becomes less structured and less orderly, and the randomness (entropy) becomes greater.

Entropy: It represents a measure of the disorder of a system. The second law of thermodynamics states that disorder (entropy) in the universe is increasing. It had held all the potential energy ever since it was formed 10–20 billion years ago. It has become progressively more disorganized ever since. And with every energy exchange the amount of entropy is increasing. However, at absolute zero, all molecular motion ceases, entropy is zero. Above absolute zero (–273.18 °C, or 0 °K), molecules are free to move about, that is, the more random, less ordered or chaotic the system is. Hence greater will be their entropy.

Epinasty: The increased growth on the upper surface of a plant organ or part (especially leaves) causing it to bend downward.

Essential amino acids: Amino acids that cannot be synthesized by the human body, e.g., isoleucine, leucine, lysine, methionine, phenylalanine, threonine, tryptophan, and valine.

Etiolation: A phenomenon developing in green plants when grown in dark. The seedlings show abnormally elongated internodes (spindly); leaves fail to develop normally and they remain pale yellow or white in colour because they fail to develop chlorophyll.

Eutrophication: Nutrient enrichment leading to a rapid increase in algae or plant growth in an aquatic system.

Exergonic: It pertains to a reaction that gives off energy (energy yielding) and occurs spontaneously. The potential energy at the end of the reaction is less than the potential energy at the start of the reaction. In many cases, an exergonic chemical reaction is also an exothermic reaction – that is, it gives off heat and thus has a negative ΔH. The complete oxidation of a molecule of glucose to CO_2 plus water yields 673 Kcal:

$$C_6H_{12}O_6 + 6O_2 \rightarrow 6CO_2 + 6H_2O + 673/686 \text{ Kcal (Heat released)}$$

$$\Delta H = -673 \text{ Kcal/mole}$$

Exergonic reaction: In such reactions the products contain less energy than the reactants and the excess energy is released. In these reactions, the ΔG is negative. Such reactions tend to proceed spontaneously.

Facilitated diffusion: It is the passive transport of ions or solutes (down the substance's electrochemical gradient) with the assistance of a carrier protein.

Field capacity: It is a state when the soil is fully wetted or saturated, after draining under gravity. Its water potential is essentially 'zero.'

First Law of Thermodynamics: No matter is either lost or gained during ordinary chemical reactions. The sum total of matter and energy in the universe remains constant.

Floral meristem identity genes: These genes code for transcription factors essential for the early induction of floral organ identity genes. The major floral meristem identity genes in *Arabidopsis* are LFY and APETELA1 (AP1).

Floral organ identity genes: These genes are associated with the direct regulation of floral organ identity.

Florigen: A hypothetical plant hormone produced and released by leaves and is then moved to the apex, where it stimulates flowering.

Fluence: The amount of red light required to induce phytochrome effects. Phytochrome responses are classified into very low fluence (VLF), low fluence (LF), and high fluence (HF).

Fluid mosaic: Cell membranes are approximately half phospholipid and half proteins. The large protein molecules float in a sea of lipids, forming a mosaic structure.

Free radicals: These are atoms or molecules with one or two unpaired electrons, e.g., chlorine atoms, Cl, or superoxide (negatively charged oxygen molecules), O_2^-. They are highly reactive.

Freezing (frost) injury: Plant tissues are killed or severely injured when they are exposed to temperature low enough to cause ice formation within them.

Geotropism: The orientation of stem and roots in response to the force of gravity is known as geotropism.

Gibberellins: A phytohormone first extracted from the fungus *Gibberella* that caused hyperelongation (or foolishly tall growth) in shoots of rice – stimulated both by cell division and cell elongation. Several structural variations of gibberellins are known.

Gibbs free energy (G): It represents the amount of energy that is available to accomplish work.

Gluconeogenesis: The conversion of PEP to glucose by a reversal of glycolysis. It takes place in the cytosol where oxaloacetate (derived from the glyoxylate cycle) is decarboxylated by the enzyme phosphoenolpyruvate carboxykinase (PEPCK) to form PEP (phosphoenolpyruvate) or the production of sugars from PEP is frequently called gluconeogenesis.

Glycolipid: Lipid molecules with sugar attached.

Glycolysis: The metabolic pathway by which glucose is broken down to pyruvic acid. Synonym: Embden–Meyerhof–Parnas pathway.

Glycoprotein: A protein with sugar attached; often the sugar occurs in short chains (less than ten sugar molecules long).

Glyoxylate cycle: The cycle was worked out largely by H. L. Kornberg and Hans Krebs. However, H. Beevers (USA) demonstrated that Glyoxylate cycle occurs in germinating fatty seeds (castor bean). It is now known to occur in many other bacteria, yeasts, molds and higher plants. It is completed in glyoxysomes, mitochondria and cytosol.

Glyoxysome: A single membrane-bound cell organelle, similar in structure to the peroxisome, where the fatty acid undergoes β-oxidation, the fatty acid chain is cleaved at every second carbon, resulting in the formation of acetylCoA. Glyoxysomes are present only in those tissues which convert fats into carbohydrates (sucrose). The glyoxysome are now known to contain all the necessary enzymes for β-oxidation.

Guttation: Oozing out of water from vein endings. It occurs at night time or in the early hours of morning when positive root pressure forces the fluid out of vein endings, through a special structure called 'water stomata' or 'hydathodes'.

Gravitational potential: Gravity causes water to move downward. The effect of gravity on water potential depends upon the height (h) of the water above the reference state water, the density of water (ρw), and the acceleration due to gravity (g). In symbols,

$$f\,(\text{gravity}) = \rho_w gh$$

For transport over a short vertical height (say, less than 5 or 10 m) or between adjacent cells, the gravitational potential is small and is commonly omitted, giving

$$\psi = \psi_\pi + \psi_p + \psi_m$$

Growth: It is irreversible increase in size or volume. The permanent increase in dry matter is interpreted as growth.

Growth retardants (or anti-gibberellins): These are synthetic chemicals that block gibberellin biosynthesis. Thus the growth of stems can be reduced or inhibited. Examples are AMO-1618, cycocel (CCC), phosphon-D.

Half-life: The amount of time required for half the atoms of a given radioactive sample to decay to a more stable form.

Halophytes: Plants growing in highly salty soils.

H^+-ATPase: A membrane protein (enzyme) that hydrolyzes ATP and uses the released energy to drive proton movement across the membrane.

Hardening: Increase in the tolerance of a plant tissue to stress or adverse conditions.

Heat Shock Proteins (HSPs): Proteins that accumulate in response to heat stress.

Herbicides: Chemical substances used in eradication of weeds.

Hexokinases: The enzymes that catalyze the phosphorylation of a hexose (six-carbon sugar), using ATP, to form the hexose phosphate.

Heterocyst: A specialized cell of certain cyanobacteria, usually larger and thick-walled; this cell excludes oxygen and allows the anaerobic nitrogenase activity.

Hydrolases (or hydrolytic enzymes): A group of enzymes that cause the split of complex organic compounds, e.g., amylases that convert starch into sugars; proteases that breakdown proteins to amino acids; nucleases that hydrolyse nucleic acids to nucleotide and finally amino acids, and lipases that split fats into glycerol and fatty acids.

Hydroponics: The growing of plants in watery solutions of essential nutrients without soil (i.e., soil-less growth).

Hyperaccumulators: These are the plants capable of growing in water with very high concentrations of metals by absorbing those through their roots, and concentrating extremely high levels of metals in their tissues.

Hypoxia: When in the rhizosphere, there is lesser amounts of oxygen as it occurs during and after rain or flooding.

In-*Vitro* : In glass or in test tubes, outside the organism.

In-*Vivo*: In a living organism.

Incipient plasmolysis: It is defined as the point at which 50 per cent of the cells are plasmolysed. At this point the turgor pressure is zero.

Indole-3-acetic acid (IAA): It is synthesized in stem tips and migrate basipetally to the zone of elongation.

Inducible enzymes: Enzymes are produced in response to particular conditions, such as the presence of an appropriate substrate, e.g., nitrate reductase, nitrite reductase.

Intermediary metabolism: Connected reaction sequences in cells that represent the major pathways, common to cells in many biological organisms.

Inulin: It is a polyfructose, occurring as a storage product in tubers of *Dahlia* and roots of Jerusalem artichoke (*Helianthus tuberosus*).

Invertase: This enzyme brings about irreversible hydrolysis of sucrose to free glucose and fructose.

$$\text{Sucrose} + H_2O \rightarrow \text{Glucose} + \text{fructose.}$$

This enzyme is more important for slow-growing and mature cells, providing glucose and fructose for respiration.

Invertase (acid): This enzyme is found associated with cell walls and the vacuole, and has a pH optimum near 5. Acid invertase has an important function during mobilization of sucrose in sugarcane (*Saccharum officinarum*) or sugar beets (*Beta vulgaris*) where it is stored in the vacuoles of specialized storage cells of the stem or root.

Invertase (alkaline): This enzyme has a pH optimum near 7.5. Like sucrose synthase, it appears to be localized in the cytosol.

Ion antagonism: The antagonistic effect of one ion on the function of other is called ion antagonism.

Isocitrate lyase: In the glyoxylate cycle this enzyme catalyzes the conversion of isocitrate to glyoxylate and succinate.

Isoelectric point: The pH value at which the net charge on a molecule in a solution is zero. At this pH, amino acids exist almost entirely in the zwitterion state with positive and negative groups equally ionized.

Isomers: Compounds with the same molecular formula but different spatial arrangements of atoms.

Isozymes (or Isoenzymes): The identical reaction occurring in different pathways are catalyzed by different enzymes, differing slightly in molecular configuration. Each enzyme is coded for by a different set of genes. Examples are malic dehydrogenases, and enzymes controlling glyoxylate and TCA cycles where some of the reaction are common. These are enzymes with similar catalytic function but have different structures and catalytic parameters, and are encoded by different genes.

Jasmonic acid: The compound shows hormone-like activity, being involved in defence against animals and fungi. In response to fungal attack (or chewing of plants by animals), somehow jasmonic acid is produced from linolenic acid – a membrane lipid. It is a volatile phytohormone (known as oxylipins), mainly involved in plant growth regulation and defence (or in signalling a response to biotic challenges).

Karrikins: These are a chemically defined family of plant growth regulators discovered in smoke from burning plant material. Karrikins are potent in breaking seed dormancy in a large number of species adapted to environments that regularly experience fire and smoke.

Kinetic energy: It is the energy of motion (or is actively engaged in doing work).

Kinetin: A compound that simulates cell division in plants. It is a derivative of purine (adenine) bases. It was first prepared by Carlos Miller by heat treatment (autoclaving) of DNA and was named 6-furfurylaminopurine. Kinetin is a classical example of a cytokinin.

Kranz anatomy: A specialized leaf anatomy found in C_4 plants in which the vascular bundle is surrounded by bundle sheath cells.

Late embryogenesis abundant (LEA) proteins: These proteins are members of a widespread group of proteins known as 'hydrophilins' and were initially discovered accumulating late in embryogenesis of cotton seeds. These generally increase in vegetative tissues experiencing osmotic stresses or desiccation and protect them against dehydration.

Leghaemoglobin: This protein, which imparts a pink colour to central region of the nodule, is produced partly by the bacteroid (producing the haeme portion) and partly by the host plant (the globin portion).

Light compensation point: It is the light intensity where the rate of carbon dioxide assimilation in photosynthesis equals the rate of loss of CO_2 during respiration, i.e., the volume of CO_2 being released in respiration is equal to the volume of CO_2 being consumed in photosynthesis.

Lipase: The enzyme that catalyzes the hydrolysis of triglyceride (lipids) to glycerol and fatty acids.

Lipid bilayer: The basic structure of biological membranes. Two lipid layers are organized in such a way that their hydrocarbon tails face each other and the polar (charged) head groups of both lipid layers are on either side of the membrane.

Lipids: A class of compounds that are hydrophobic and water insoluble, e.g., fats, oils, waxes.

Little Leaf Disease: Leaves remain quite small in size. It is caused by Zn deficiency.

Long day plant: A plant that flowers when the length of day exceeds some critical value.

Lysigeny: The localized disruption of plant cells to form a cavity, surrounded by remnants of the broken cells, in which secretions accumulate. For example, oil cavities in the leaves of citrus trees.

Lysosomes: These are membrane bound organelles, containing enzymes that have lytic (breakdown) activity. Examples are proteases, nucleases, lipase. As long as the lysosome is intact, these enzymes cannot reach the cell's contents. But following injury, they may rupture and release their enzyme. These are found in animal cells.

Macronutrients: A mineral element that is required in relatively large amounts (usually 500 ppm or above in the plant) for the growth of plants, e.g., C, H, O, Ca, Mg, K, P, S and N.

Malate synthase: In the glyoxylate cycle, the enzyme that catalyzes the condensation of acetyl CoA and glyoxylate to produce malate.

Maltases: enzymes that catalyze the hydrolysis of maltose to glucose.

Matric potential: It is a component of water potential, developing as a result of hydrophilic matrix such as cell walls and protoplasm. It has a negative value. It refers to decrease in the water potential (the free energy) because of affinity of water molecules to colloidal substance (such as protoplasm and cell walls) to which water molecules are adsorbed.

Mericloning: Regeneration of a large numbers of plants by micropropagation. It is an effective means to reduce the time required to create disease-free, desirable strains of crop plants.

Metabolic engineering: It is the practice of optimizing genetic and regulatory processes within cells to increase the cells' production of a certain substance.

Metabolomes: The complete set of small-molecule chemicals found within a biological sample e.g. a cell, cellular organelle, organ, tissue extract, biofluid or an entire organism.

Metabolons: These are multienzyme protein complexes, composed of enzymes catalyzing sequential reactions in a metabolic pathway.

Michaelis constant (Km): It is defined as the concentration of substrate necessary to produce half the maximum velocity. A small Km means that only a small substrate concentration is needed to attain maximum velocity. The smaller the Km, the greater the affinity between the enzyme and substrate.

Micronutrient (minor element or trace element): A mineral element that is needed in small amounts (usually less than 50 ppm) for the growth of plants, e.g., B, Cl, Cu, Fe, Mn, Mo, Zn.

MicroRNAs (miRNAs): A class of endogenous tiny non-coding, single-stranded RNAs of approximately 20–24 nucleotides in length, which play major role in response to biotic and abiotic stresses. These control the expression of their target genes in plants as key post-transcriptional gene regulators through translational repression or transcript cleavage.

Middle lamella: The cementing layer of pectic material between primary cell walls.

Minimal medium: A medium containing only those elements absolutely essential for the growth of a particular organism.

Molal solution: One gram molecular weight of a substance is dissolved in exactly 1000 ml (1 L) of water. Thus the final volume of the solution becomes more than 1 L.

Molar solution: A gram molecular weight of a substance is dissolved in water to make the total volume up to 1 L.

Motor cells: Cells that swell and shrink in plant organs capable of repeated reversible movement, such as insect traps and petioles of leaves that undergo sleep movement.

Mucigel: The mucilaginous, slimy material secreted by root caps and root hairs. Mucigel is rich in carbohydrates and amino acids, which promote rapid growth of soil bacteria around the root tip. These microbes are known to help release nutrients from the soil matrix. Mucigel lubricates passage of the roots through the soils.

Muriate of potash: Commercial potassium chloride salt containing 60 per cent K_2O and is readily available to plants.

Mycorrhizae (singular mycorrhiza): Fungi that form a symbiotic relationship with roots, usually beneficial to each, providing phosphorus to the plant; ecotomycorrhizae – A type in which the fungi invade only the outermost cells of the root; and endomycorrhizae – A type where fungi invade all cells of the root cortex.

NAD (nicotinamide adenine dinucleotide): A coenzyme that acts as an electron acceptor, particularly important in respiration (An alternate name for DPN).

NADP (nicotinamide adenine dinucleotide phosphate): A coenzyme that acts as an electron acceptor, particularly important in photosynthesis. An alternate name for TPN.

Necrosis: It refers to dead or dying tissue along the leaf margins and tip of the leaves, sometimes occurring in spots.

Nitrate reductase: The enzyme that catalyzes reduction of nitrate (NO_3^-) to nitrite (NO_2^-).

Nitrate transporters: Membrane proteins that allow the movement of nitrate across the membrane.

Nitrite reductase: The enzyme that catalyzes reduction of nitrite (NO_2^-) to ammonia (NH_3).

Nitrogen-fixation (N_2-fixation): The process by which molecular nitrogen ($N \equiv N$) is reduced to ammonia.

Nitrogenase complex: The enzyme system that catalyzes conversion of molecular nitrogen (N_2) to ammonia (NH_3).

Nod (nodulation) factors: Signal molecules produced by rhizobia during initiation of nodules or tubercles on the roots of leguminous plants.

Nodules (or tubercles): Swellings on the roots of certain leguminous plants where nitrogen-fixing bacteria reside in a symbiotic relationship.

Noncyclic photophosphorylation: In this scheme of light reaction, the electrons expelled from PS II pass through a series of electron transport carriers, before entering into PS I where they are used up in the reduction of NADP to NADPH. The electron do not return to the chlorophyll molecule from which they were ejected, i.e., the electron donor and electron acceptors are different chlorophyll molecules.

Nonclimacteric fruits: Fruits that ripen slowly and steadily, without a sudden burst of CO_2 or ethylene.

Nonpolar molecule: A molecule that does not carry even a partial charge anywhere.

Nutrient depletion zone: It is the zone in the rhizosphere from which nutrients are readily absorbed by the root system. Mycorrhizal fungi assist in the uptake of minerals by extending their mycelia beyond the depletion zone.

Oleosomes (or Spherosomes): Storage lipids are deposited as oil droplets. These are normally found in storage cells of cotyledons or endosperm. These are also called oil bodies.

Osmoprotectants (Compatible solutes): These are small, organic molecules with neutral charge and low toxicity at high concentrations that act as osmolytes and help plants survive extreme osmotic (water deficit) stress.

Osmosis: The movement of water through a selectively permeable membrane from a region of high water potential (low solute concentration) to a region of lower water potential (higher solute concentration).

Osmotic pressure: The pressure of a solution that can develop when the solution is separated from pure water by a selectively permeable membrane.

Osmoregulator: Any substance, either organic or inorganic, that controls the solute potential of a solution.

Osmotic Potential: The change in free energy brought about by the dissolved solutes in the vacuole. The osmotic potential of a cell (or solution) is directly related to the concentration of dissolved solutes. As the concentration increases, the osmotic potential becomes 'more negative'. Osmotic potential (ψ_π) and osmotic pressure values are numerically equal but the former carries a negative sign, expressed in bars, while the latter has a positive sign and expressed in atmosphere units. The osmotic potential of pure water is 0 (zero).

Oxidation: The loss of electrons is called oxidation, regardless of whether hydrogen, oxygen, or other atoms are involved. Oxygen is the usual final acceptor, or oxidizing agent, but oxidation can occur that does not involve oxygen *per se*.

Oxygen debt: The term 'oxygen debt' is often used to describe the amount of oxygen needed to convert lactic acid back into glucose. More precisely it is the amount of O_2 needed during the oxidation of fatty acids to provide ATP for the conversion of lactate to glucose

$$2C_3H_6O_3(\text{lactate}) \xrightarrow[\text{6 ATP} \;\to\; \text{6 ADP + 6Pi}]{} C_6H_{12}O_6\,(\text{glucose})$$

Lactic acid is toxic and must ultimately be removed before the level in the body or human system may become dangerously high.

α-oxidation (process): It is an important process of fatty acid breakdown in plants. During this process 2-carbons are removed at a time at each step, forming acetyl CoA. This was first studied by F. Knoop (1904). The fatty acid chain is cleaved at every second carbon, releasing acetyl CoA. The cycle was worked out largely by H. L. Kornberg at Oxford University (together with Hans Krebs) in *Pseudomonas.*

β-oxidation (process): This breakdown of fatty acids into chunks of acetyl CoA occurring in mitochondria of plant and animals is called β-oxidation and the chain becomes progressively shorter. This pathways is sometimes called the fatty acid oxidation spiral. The process involves three steps:

(i) Transporting the fatty acid from the cytosol through the relatively impermeable inner mitochondrial membrane, using a carrier molecule.
(ii) Activation of the fatty acid by ATP and CoA.
(iii) Intermediate reactions involving cracking off a two-carbon fragment, (i.e., between the α and β carbons).

The step (iii) is repeated until the entire fatty acid has been reduced into chunks of acetyl CoA which enters the Krebs cycle.

Oxidation–reduction (or redox) reactions: Reactions involving the transfer of electrons from one molecule to another. An atom or molecule that loses electrons is oxidized, and the one that gains electrons is reduced, these two kinds of reactions always occur simultaneously.

Oxidative phosphorylation: The synthesis of ATP in mitochondria linked to electron transfer through the ET chain from NADH to oxygen. The ET chain consists mostly of haeme-containing proteins called cytochromes. It is so called as oxygen is consumed in the terminal step, forming water.

Pacemaker enzymes: The pacemaker enzyme, also known as the rate-limiting step in the pathway, is the enzyme that has the slowest reaction rate.

Passive transport: The transport of ions or solutes down a concentration gradient or an electrochemical gradient. Examples of this type are simple diffusion and facilitated diffusion. Most passive transport requires carrier protein to facilitate the passage of ions.

Pasteur effect: A 19th century observation by Louis Pasteur that glucose utilization is decreased in the presence of oxygen in yeast metabolism.

Patatin: A storage protein of potato tubers.

PEP carboxykinase: The enzyme that catalyzes ATP-dependent conversion of oxaloacetate to phosphoenol pyruvate (PEP) and CO_2 during gluconeogenesis and in the bundle sheath cells of some C4 plants.

Peribacteroid membrane: The membrane surrounding the groups of bacteroid, and is derived from the host cell plasma membrane.

Perlite: It is non-nutritive, coarsely ground glassy volcanic rock.

Phenylpropanoids: A class of secondary metabolite derived from phenylalanine.

Pheophytin: Loss of the magnesium ion from chlorophyll results in the formation of a non-green product, 'pheophytin'. It is readily produced during chlorophyll extraction under acidic condition.

Phosphoenolpyruvate (PEP) carboxylase: The primary cytosolic carboxylating enzyme in C4 plants that catalyzes addition of CO_2 to phosphoenolpyruvate to form oxaloacetate. This enzyme uses the hydrated form of CO_2 which is HCO_3^- (bicarbonate ions). Its activity is not affected by the concentration of O_2.

$$\text{PEP} \xrightarrow[\text{PEP carboxylase}]{HCO_3^- \searrow \nearrow Pi} \text{Oxaloacetate}$$

Phospholipids: A phosphorylated lipid with two fatty acids attached to the glycerol backbone, with the third carbon of glycerol linked to phosphorus containing molecule. They are important components of cellular membrane, since their phosphate ends are water soluble, and their fatty acid ends fat soluble.

Photochemical reaction: A photochemical reaction is light dependent and temperature independent. For a purely photochemical reaction the value of Q_{10} is approximately 1.

Photomorphogenesis: It is defined as growth and development, directly dependent on light but not related to photosynthesis.

Photoperiodism: It refers to the response of plants to the length of the day light period (photoperiod) with respect to flowering. The phenomenon of photoperiodism was discovered by two American scientists Garner and Allard (1920) while working at the USDA at Beltsville, Maryland.

Photophosphorylation: The synthesis of ATP by the phosphorylation of ADP in the presence of light energy is called photophosphorylation.

$$ADP + Pi \rightarrow ATP$$

Photorespiration: The 'light stimulated' respiration in green plants is called photorespiration. It is a wasteful process, yielding *neither* ATP *nor* NADH and in some plants as much as 50 per cent of the carbon fixed in photosynthesis may be reoxidized to CO_2. This occurs in C3 plants when $O_2 > CO_2$, then Rubisco has oxygenase activity.

Photosystem I: The second part of the Z-scheme in which a chlorophyll a absorbs most effectively at 700 nm.

Photosystem II: The first part of the Z-scheme of light reaction in which a chlorophyll a molecule absorbs most effectively at 680 nm.

Phototropins: A family of photoreceptors responsible for blue light activated phototropism, and also mediate the photoreception of blue light in guard cells, thus controlling the opening of stomata. It is a small protein with two flavin mononucleotide attached as chromophore. It is a component of a cell's plasma membrane and becomes phosphorylated when exposed to blue light.

Phototaxis: The movement of an entire organism or organelle in response to light, e.g., plastid movement in *Mougeotia*.

Phototropism: The orientation of stems and roots in response to unilateral illumination is called phototropism.

Phycobilins: The accessory pigments of cyanobacteria and red algae. Phycocyanin absorbs blue light while phycoerythrin absorbs red light.

Phycobilisomes: Groups of pigmented cells containing phycocyanin and phycoerythrin.

Phycocyanin: A blue photosynthetic pigment found in cyanobacteria and red algae.

Phycoerythrin: A red photosynthetic pigment found in cyanobacteria and rhodophyceae (red algae).

Phytoalexins: These are antifungal compounds formed in plants in response to infection by a fungus, i.e., these are formed *only* when the host cells come into contact with the parasite.

Phytochrome: These are homodimeric proteins, consisting of two identical polypeptide chains (*apoproteins*), each is covalently linked to 'chromophore' which is a phytochromobilin – a tetrapyrrole. The chromophore captures a light signal and undergoes isomerization, resulting in a conformational change in the protein. It is unique as it exists in two forms that are photoreversible.

The form P_{660} (or Pr) absorbs maximally at 660 nm; converting the pigment to far-red absorbing form that absorbs strongly at 730 nm – converting it back to the red absorbing form. Pfr is believed to be an active form of the pigment that is capable of initiating a wide range of morphogenetic responses.

Phytoferritin: A protein molecule that binds, and stores iron, mostly found in plastids.

Phytoremediation: It refers to the technologies that use living plants to clean up soil, air, and water contaminated with hazardous contaminants.

Phytosiderophore: Plant compounds that chelate iron in the soil and facilitate its uptake.

Phytotron: It is generally a building complex where plants can be grown in rooms under controlled conditions of artificial light, temperature and humidity to investigate the influence of a particular environmental factor or their interaction. The first phytotron was established by Dr Fritz Went in California Institute of Technology, Pasadena, California, USA.

P_{680}: The reaction centre of photosystem II.

P_{700}: The reaction centre of photosystem I, believed to be a chlorophyll with absorption peak at 700 nm.

Plant hormone: It is an organic compound produced in one part of a plant and then transported to another part where in very low concentration it exerts a physiological response.

Plasmodesmata (singular, plasmodesma): These are cytoplasmic bridges that penetrate the plasma membrane and cell wall, and connect the cytoplasm of one cell to the next. They allow the free exchange of small molecules dissolved in the cytosol but prevent organellar movement between cells.

Plasmolysis: The shrinkage of cytoplasm from the cell wall under the influence of hypertonic solution is called plasmolysis.

Plastoquinone: It is one of the carriers that transfers protons (H^+ ions) as well as electrons, moving the protons vectorially from the exterior of thylakoid to the interior, resulting in build-up of proton concentration inside the thylakoid.

Polar molecule: A molecule that has a partial positive charge at one site and a partial negative charge at another site, like the two poles of a magnet. Water is one of the most polar molecules known.

Polarized: A term applied to a cell membrane across which an electric potential exists, due to an unequal distribution of charged ions on the two sides of the membrane.

Polyamines: These are strongly basic molecules of low molecular weight and are found to be essential for growth and development, affecting the processes of mitosis and meiosis.

Polyprotic acid: A polyprotic acid contains more than one proton, each with a different dissociation constant. Phosphorus occurs in the soil solution as forms of the polyprotic H_3PO_4 (phosphoric acid).

Potential energy: It represents the stored energy. The energy is stored as potential energy in the covalent bonds between atoms in the sugar molecules. Much of the work that living organisms carry out involves transforming potential energy to kinetic energy.

Pressure bomb method: A technique employed for determining the ψ of leaf cells. In this method a freshly cut leafy shoot is placed within an enclosed chamber, leaving its cut portion outside, and pressure will be applied inside the chamber by using nitrogen to force xylem sap to appear at the exposed cut end of the shoot. This pressure so applied represents the water potential of leaf cells. It was discovered by P.F. Scholander and his coworkers in 1966.

Pressure potential: It refers to the hydrostatic pressure developing as a result of osmotic entry of water into the vacuole that pushes the cytoplasm against the cell wall. Pressure potential is usually 'positive in sign', i.e., it increases the water potential of a cell.

Principle of limiting factor (Liebig's Law of minimum): It states that when a process is conditioned as to its rapidity by a number of separate factors, the rate of the process is limited by the factor which is at the minimum (applicable to soil nutrition and photosynthesis).

Programmed cell death (PCD): It is the death of a cell in any form, mediated by an intracellular program. PCD in plants is a crucial component of development and defence mechanisms.

Proteasomes: These are protein complexes which degrade unwanted or damaged proteins by proteolysis. The proteosome is one of the major degradation machineries in eukaryotic cells, designed to carry out selective, efficient and processive hydrolysis.

Proteome: The entire set of proteins that is, expressed by a genome, cell, tissue or an organism at a certain time.

Proton Pump: An integral membrane protein that moves protons across a plasma or organellar membrane, thereby creating a difference in (i.e., a gradient of) both pH and electrical charge across the membrane and tending to establish an electrochemical potential.

Prosthetic groups: These are non-protein groups tightly bound to proteins. It is not normally possible to remove them without disrupting the structure of the enzyme. Examples include the 'porphyrins' with a metal ion (Fe^{2+}) at the centre, 'cytochromes' and haemoglobin.

Proteasome: A cytosolic protein complex in which proteins enter and are degraded to small peptides.

Proton motive force (or Δp): The electro-chemical H^+ gradient is often termed as proton motive force, representing the difference in electrochemical potential between protons inside and outside the cell. This term was first introduced by Peter Mitchell, the British Nobel Laureate.

Pumps: Carrier mediated active transport systems that move ions or molecules against a chemical or electrochemical gradient are often known as 'Pumps'. H^+-pumping ATPase is the primary electrogenic transport in both the plasma membrane and the tonoplast. Light-stimulated H^+-pumping ATPase occurs in the stomatal guard cells where it regulates stomatal movement.

Quiescent centre: At the centre of the root meristem, there is a region called the 'quiescent' centre where there is little or no mitotic activity. This seems to function as a 'reserve bank' of cells, which can undergo division if the normal dividing cells are somehow damaged. The cells on either side of the quiescent centre contribute new cells to the root cap, which is continuously worn away as roots push through the soil.

Radioautography: A technique used for identification of radioactive compounds on a chromatogram. This is done by exposing the chromatogram to sensitized photographic film which will then develop spots where radioactive substances are present.

Reactive oxygen species (ROS): During abiotic stresses, a large number of free radicals are generated which cause oxidative damage to cellular membrane. For example, superoxide anions, hydroxyl free radicals and phytotoxic hydrogen peroxide.

Redox potential: It refers to electron-donating ability. The redox potential ($E°'$) is measured in volts, varies with pH. Hence, during oxidative phosphorylation or photophosphorylation, different biologically important chemicals are arranged in a sequence according to their ability to donate electrons.

Reduction: It consists of the addition of electrons from a compound. Once the electron is transferred, a proton may also follow, the net result being the addition of hydrogen during reduction and its removal during oxidation. The gain of electrons by a molecule constitutes reduction, regardless of whether hydrogen, oxygen, or other atoms are involved.

Redox reactions: Many biologically important reactions involve the gain or loss of electrons. These are called redox reactions (reduction-oxidation reactions). The substance which gains an electron is said to be reduced while the other compound that loses one is said to be oxidized. In cells, the movement of electron is usually accompanied by the movement of proton (hydrogen ion), thus in most cases reduction means the gain of hydrogen, and oxidation the loss of hydrogen. Therefore redox reactions in cells normally involve hydrogen (electron) carriers.

Thus NADH can pass hydrogen to oxygen, forming water, but water cannot spontaneously pass hydrogen back to NAD to re-form $NADH_2$

$$\underset{\text{reduced}}{NADH} + \underset{\text{oxidized}}{\tfrac{1}{2} O_2} \longrightarrow \underset{\text{oxidized}}{NAD^+} + \underset{\text{reduced}}{H_2O}$$

Resonance transfer: The passage of energy from antennae pigments (carotenoids) to chlorophyll, and from one chlorophyll molecule to another and finally to the Reaction Center of PS I or PS II.

Respiratory crisis: Just before pollination, the tissues of the spadix of skunk cabbage (*Symplocarpus foetidus*) undergo a surge in oxygen consumption, called 'respiratory crisis'. This process is due almost entirely to an increase in alternative pathway respiration that elevates the temperature of the spadix by as much as 10°C above ambient.

Reversed Stomatal Rhythm: In CAM plants, the stomata open at night and close during daytime. By opening stomata when it is cool and when the atmosphere is generally more humid, water loss is considerably reduced.

Rhizodermis: The outermost region consisting of the root epidermis is often referred to as the rhizodermis.

Rhizosphere: A region of the soil surrounding the immediate root surface. The root-microbe interactions may be beneficial in the inorganic nutrition of the plants or may become pathogenic and cause injury (soil borne diseases).

Rhizotrons: A laboratory with glass-walled subterranean chambers to study the root growth while the aerial parts are exposed to natural field conditions. Rhizotron is derived from two Greek words *rhizos*, meaning root, and *tron*, meaning a device for studying construction of subterranean walkways, called rhizotrons, equipped with windows through which the growth and development of roots could be observed.

Ribonuclease: An enzyme catalyzing the breakdown of RNA.

Ribozymes: Certain reactions involving RNA molecules appear to be catalyzed in cells by 'RNA itself'. This was discovered by Tom Cech and his colleagues at the University of Colorado in 1981. It now seems at least possible that RNA may have evolved first and catalyzed the formation of the first proteins. It was found that the ribosomal RNA has a segment which catalyzes its own removal.

Ribulose-1, 5-bisphosphate carboxylase/oxygenase (Rubisco): The primary carboxylation enzyme that catalyzes the reaction between CO_2 and RuBP to form two molecules of PGA. Rubisco uses CO_2 as its substrate, unlike HCO_3^- (the bicarbonate ions) as in C4 plants. Its activity is affected by the presence of concentration of oxygen, in contrast to PEP carboxylase.

Root Pressure: The hydrostatic pressure developed due to the accumulation of water absorbed by the roots and is controlled by the living activity of the root cells. It ceases when the root cells are killed by poisons or high temperature.

Salicylic acid: A compound similar in structure to aspirin (acetyl salicylic acid). It plays a central role both in defence responses around the site of infection and in triggering systemic acquired resistance (SAR). A phytohormone that affects plant growth and development; it also serves as an endogenous signal in a plant's defence against pathogen, especially viruses.

Salt glands: In some plants, the excess salts are secreted onto the leaf surface through special salt glands, e.g., the tamarisk bush, and sea-lavender (*Limonium latifolium*) taste salty when licked. The leaves are encrusted with salts.

Schizogeny: The localized separation of plant cells to form a cavity, surrounded by the intact cells, in which secretions accumulate. For example, resin canals in some coniferous plant species and oil ducts in aniseed fruits.

Secondary messengers: A variety of small, mobile molecules that relay information from the receptor to the biochemical machinery inside the cells. These messengers tend to amplify the original signal by initiating a cascade of biochemical events, e.g., G-proteins; protein kinases; c-AMP (cyclic adenosine monophosphate), calcium (Ca^{2+}) ions, and inositol triphosphate.

Secondary metabolites: Organic compounds synthesized by the plant that are not immediately essential for growth, development or reproduction.

Seismonastic: The shock movements in response to touch, blow, etc., and are caused by the differential loss of turgor on the two sides of the pulvinus, e.g., sensitive plant or touch-me-not (*Mimosa pudica*).

Seleniferous soils: Soils rich in selenium. *Astragalus* and *Stanleya* are two major indicator plants of such soils.

Semipermeable membrane: A membrane which permits the passage of pure solvent molecules (water) through it but does not allow the passage of solutes.

Senescence: Mature leaves and fruits rapidly senesce and die. Senescence is characterized by the breakdown of protein, nucleic acids, and other macromolecules. The ageing process is usually characterized by the loss of some functional activity or capacity, including reproduction.

Short day plant: A plant that flowers when the length of the day (or photoperiod) is shorter (or does not exceed) than some critical value.

Sigmoid Growth Curve (or Grand Period Curve): The S-shaped, or sigmoid curve, is typical of growth of all organs, plants and population of plants. It consists of at least four distinct components: initially growth is slow (*Lag Phase*), then gains speed (*Log Phase*), eventually slows down (*Decreasing Growth Rate*) and finally comes to a halt or ceases growth (*Stationary Phase*), and size is stabilized.

Signal perception: All developmental stimuli have three basic sequences in common: signal perception, signal transduction and response. Signal perception requires a receptor molecule, which is normally a protein. Receptors are known for red and blue light, the hormone auxin and ethylene.

Signal transduction: Following signal perception there occurs a diverse array of biochemical or metabolic events that ultimately determines the cell response. This often complex web of interacting pathway is referred to as 'signal transduction'.

Sink organs: Organs or tissues to which photoassimilate is transported.

Skotomorphogenesis (*Skotos in Greek for darkness*): Dark grown seedlings exhibit elongated hypocotyl, apical hook, closed or folded cotyledons with apical meristem inside.

Spectrophotometer: It is an instrument which measures the wavelength of light absorbed by a substance. The wavelengths vary for different chemicals, e.g., NAD and NADP have an absorption peak at 340 nm.

Speed breeding: The breeding of crops under controlled light and temperature conditions to send plant growth into overdrive, thus enabling quick harvesting of seeds and growing the next generation of crops sooner.

Spherosomes: These are thought to be equivalent to the lysosomes of animal cells. They contain hydrolytic enzymes, such as proteases (for breakdown of proteins), nucleases (breakdown of DNA and RNA), lipases (that digest fats), and phosphatases. These are single membrane cellular organelles where triglycerides in most seeds are stored in the cytoplasm of either cotyledon or endosperm.

Starch-branching enzyme: An enzyme that catalyzes the introduction of α-1, 6 linkages into α-1, 4-linked glucose chain during the synthesis of amylopectin.

Statocyte: A cell which is gravity sensitive. It contains statoliths.

Substrate-level phosphorylation: The formation of ATP by the direct transfer of a phosphate group to ADP from another phosphorylated molecule.

Sucrose phosphate phosphatase: The enzyme that catalyzes conversion of sucrose phosphate to sucrose.

Sucrose phosphate synthase: The enzyme that catalyzes synthesis of sucrose phosphate from fructose 6-phosphate and UDP-glucose.

Sucrose synthase: It is the primary enzyme that degrades sucrose in starch-storage organs (for example, developing seeds and potato tubers) or in rapidly growing tissue, that are converting translocated sucrose to cell-wall polysaccharides. It catalyzes cleavage of sucrose to form UDP-glucose and fructose.

Suicide germination: Initiation of germination of seeds of parasitic weeds in the absence of any host crop in crop fields.

Surface tension: The water molecules on the surface are attracted downward and sideways by other water molecules (cohesion), never upwards by molecule of gases in the air. Hence they are subjected to unequal forces of attraction and are held to form an elastic membrane. Therefore the liquid–air interface behaves as an elastic skin that contracts to a minimum area. This property of liquid is called surface tension.

Superoxide dismutase (SOD): It converts superoxide radicals to hydrogen peroxide and molecular oxygen.

$$O_2^{*-} + O_2^{*-} \xrightarrow{+2H^+} H_2O_2 + O_2$$

Symport: The two ions or molecules (solute) are transported in the same direction through the membrane. This is an example of 'cotransport' and the molecules move against the electrochemical gradient (active transport). Two ions may be transported in the same direction (symport) or in the opposite direction to one another (antiport).

Temperature coefficient (Q_{10}): The effect of temperature on a thermochemical reaction is normally expressed mathematically by the Q_{10} which tells us how a 10°C rise in temperature affects the rate of reaction.

$$Q_{10} = \frac{\text{rate of reaction}(X+10)°\text{C}}{\text{rate of reaction at } X°\text{C}}$$

The rate of reaction in a purely biochemical reaction is 2 or 3 but for a photochemical reaction $Q_{10} = 1$

Temporary wilting: Wilting condition of the plant due to loss of water which would recover from wilted condition during the night without addition of any water.

The first law of thermodynamics: According to this law 'energy can be changed from one form to another but cannot be created or destroyed. To be more precise, in all energy exchanges and conversions, the total energy of the system and its surroundings after the conversion is equal to the total energy before the conversion. The potential energy of reactants (the initial state) is equal to the potential energy of products (the final state) plus the energy released in the process or reaction.

The log phase: The phase of most rapid growth in a sigmoid curve.

The second law of thermodynamics: In all energy exchanges and conversions, if no energy leaves or enters the system under study, the potential energy of the final state will always be less than the potential energy of the initial state. A process in which the potential energy of the final state is less than that of the initial state is one that releases energy, i.e., exergonic ('energy output').

The sodium-potassium pump: The protein channel called as the 'sodium-potassium pump' transports sodium (Na^+) and potassium (K^+) ions across the cell membrane. For every three Na^+ that are transported out of the cell, two K^+ ions are transported into the cell. This pump is fuelled by ATP. Most animal cells have a low internal concentration of Na^+ (but high concentration of K^+) relative to the surroundings.

Thermal emission: It is the physical loss of heat energy from the leaves into the air. It accounts for the dissipation of all absorbed radiant energy.

Thermochemical reaction: A chemical reaction which is temperature dependent and unaffected by light. For ordinary chemical reactions the temperature coefficient (Q_{10}) is at least 2.

Thermodynamics: The science that deals with energy changes in chemical and physical processes.

Thermogenic respiration: Respiration in which electron transport is uncoupled from ATP synthesis, so heat is liberated. Synonym: cyanide resistant respiration.

Thigmotropism: The growth movements of plant organs in response to the stimulus of contact or friction. It is commonly seen in plants which climb by means of tendrils, e.g., cucurbitaceous plants, leaf tip of *Gloriosa*, leaf stalks of *Clematis*, leaves (*Lathyrus aphaca*), leaflets (*Pisum sativum*).

Thin-layer chromatography (TLC): A technique used for separating molecules based on size and polarity. A thin layer of adsorbent material, such as silica gel or cellulose, is coated on a plate. A liquid phase, consisting of a solution of the molecules to be separated, is drawn up the plate by capillary action.

Turgor pressure: It is a hydrostatic pressure developing in the cell vacuole that presses the cytoplasm against the cell wall.

Transcriptome: The full range of mRNA molecules expressed by an organism.

Transfer cells: There are specialized transporting cells involved in rapid short distance transfer of material. They have highly convoluted inner walls (labyrinthine wall) that take up and carry a variety of ions, molecules, and metabolite.

Translocation (of food materials): The transport of photoassimilates over long distance from the leaves (*source*) to the roots (*sinks*) in known as translocation of food materials.

Transpiration ratio: The ratio of the mass of water transpired to the mass of dry matter produced.

Tricarboxylic acid cycle (TCA cycle): Synonym for Citric Acid cycle and Krebs cycle.

Triple response: A seedling's response to ethylene, i.e., inhibition of root and hypocotyl elongation; exaggerated tightening of the apical hook; swelling of the hypocotyl. This was discovered by the Russian physiologist Dimitry N. Neljubow. Furthermore, in response to ethylene, there is inhibited stem elongation, increased stem thickening and a horizontal growth habit.

Turnover number: The number of molecules of substrate converted per minute by one molecule of an enzyme to the end products is called the turnover number. The turnover numbers of enzymes vary greatly, ranging from one hundred to over three million.

Ubiquitin: A small protein that is linked to other proteins targeted for degradation by proteasomes.

Ubiquitination: The process of attachment of ubiquitin to a protein.

Uniport: The transport of *only* one ion (or solute) from one side of the membrane to the other *via* transport proteins.

UV RESISTANCE LOCUS 8 (UVR8): A UV-B-sensing protein (280–315 nm) found in plants, and is associated with initiation of the plant stress response.

Vermiculite: It is a silicate mineral of the mica family. It expands on heating to produce a lightweight product that is highly water retentive and is commonly used as a mulch. It is an inert substance, containing no plant nutrient.

Vernalin: The chilling stimulus for floral induction which is transmitted from shoot apex to other tissues. This hypothetical vernalization stimulus was named 'vernalin' in 1937 by George Melchers of Germany. It is thought to be a hormone but it has not been extracted and purified. Gibberellins are believed to be involved in vernalization.

Vernalization: The cold or chilling treatment (0–2 °C for about a week) given for early flower induction. This was named by the famous (or infamous) Russian botainst T.D. Lysenko.

Vicillins: A group of storage proteins of legume seeds.

Vivipary: The germination of a seed while still attached to the mother plants. It is a characteristic feature of halophytes. A precocious germination 'before' the embryo reaches maturity or the seed is released from the fruit.

Voltage-gated channels: Ion channels in a membrane that open and close in response to the changes in membrane potential.

Wall Pressure: The cell wall, being rigid, exerts an equal and opposite counter pressure on the protoplasm against the cell vacuole. Wall pressure is equal in magnitude but opposite in direction to turgor pressure at a given time.

White Bud: Chlorosis affecting young leaves as well as buds so that the latter are whitish instead of greenish colour.

Water potential (pronounced 'sigh' and has the symbol ψ): It is an expression of the energy status of water within cells or other parts of a transport system. It is expressed by the Greek letter psi (ψ) and is measured in bars. Water will flow in the energetically downhill direction, that is, from high ψ to low ψ. It is an algebraic sum of osmotic potential, pressure potential and matric potential.

$$\psi = \psi_\pi + \psi_p + \psi_m$$

Wilting: Under water stress or deficit condition the leaves and young stem tips droop down due to the loss of turgor of their cells, becoming flaccid. There is a dramatic increase in the level of ABA and large amounts of it are exported from the wilted leaves to the rest of the plants moving through phloem where it leads to stomatal closure - thus helping to conserve as much water as possible – an adaptation to xeric mode of life. Unlike plasmolysis, 'the cytoplasm never ever shrinks from the cell wall.'

Xanthophyll cycle: It is a mechanism which provides protection to plants against oxidative stress. The xanthophyll cycle involves the enzymatic removal of epoxy groups from xanthophyll (violaxanthin, antheraxanthin) to produce de-epoxidised xanthophyll (zeaxanthin).

X-ray crystallography: A technique which enables the three dimensional structure of crystalline molecules to be deduced from the pattern produced when X-rays are passed through them such as enzymes, DNA, etc.

Xenobiotics: These are compounds not produced naturally in organisms and include drugs, food additives, and environmental pollutants

Xerophyte: A plant adapted to xeric (or desert) condition.

Xylogenesis: It is a complex developmental process culminating in programmed cell death as a truly terminal differentiation event.

Z-scheme (for photosynthesis): During the light reaction of photosynthesis, the electrons from PS II move *'uphill'* and are captured by the electron acceptor (Q) from where they move *downhill* through electron carriers and enter into PS I. From PS I, the electrons are boosted *'uphill'* and are captured by a strong reductant X from where they are transferred *downhill* to ferredoxin and NADP, forming an English letter Z. That is why it is known as Z-scheme.

Zeatin: It was the first naturally occurring cytokinin isolated from the immature endosperm of corn (*Zea mays*). It was discovered independently by two groups of workers in two different countries – Carlos Miller in the United States and D.S. Letham in Australia.

Zymase: It is an iron-containing multienzyme complex, originally discovered in yeast (*Saccharomyces*). It decomposes glucose into ethyl alcohol and CO_2. Glycolase and carboxylases are well-known components of this enzyme aggregate.

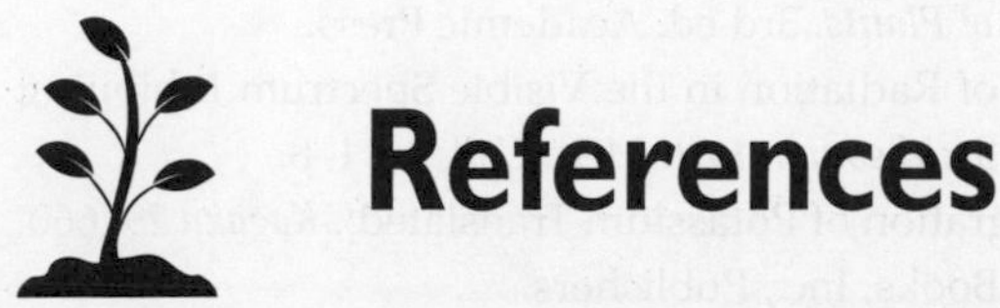

References

Adams, P., J. J. W. Baker, and G. E. Allen. 1970. *The Study of Botany*. California. London: Addison Wesley Publishing Company.

Albert, B. et al. 2019. *Essential Cell Biology*. 5th ed. W. W. Norton.

Arnon, Daniel. I. 1966. 'The Role of Light in Photosynthesis'. *Sci. Am.* 203 (5): 104–18.

Bajracharya, D. 1999. *Experiments in Plant Physiology – A Laboratory Manual*. New Delhi: Narosa Publishing House.

Bandurski, R. S. 1965. 'Biological Reduction of Sulfate and Nitrate'. In *Plant Biochemistry* J. Bonner and J. E. Varner, (edited) New York: Academic Press, Inc.

Bassham, J. A. 1962. 'The Path of Carbon in Photosynthesis'. *Sci. Am.* 206 (6): 88–100.

Berg, J. M. et al. 2019. *Biochemistry*. 9th ed. Macmillan.

Bhaskar, A., and V. G. Vidhya. 2009. *Enzyme Technology*. Chennai, India: MJP Publishers.

Bhatla, S. C. and M. A. Lal. 2018. *Plant Physiology,* Development and Metabolism. Singapore: Springer.

Bidlack, J. E., and S. H. Jansky. 2011. *Stern's Introductory Plant Biology*. 12th ed. New York: McGraw-Hill Companies, Inc.

Bidwell, R. G. S. 1979. *Plant Physiology*. 2nd ed. New York: Macmillan Publishing Co., Inc.

Blackman, F. F. 1905. 'Optima and Limiting Factors'. *Ann. Bot*. 19: 281–95.

Buchanan, B., W. Gruissem, and R. Jones. 2015. *Biochemistry and Molecular Biology of Plants*. 12th ed. John Wiley & Sons. Ltd.

Calvin, M. 1956. 'The Photosynthetic Carbon Cycle'. *J. Am. Chem. Soc.* 78: 1895.

Calvin, M., and A. A. Benson. 1948. 'The Path of Carbon in Photosynthesis'. *Science* 107: 476–80.

Campbell, N. A. et al. 2018. *Biology*. 11th ed. Francisco: Pearson Benjamin Cummings.

Cech, T. R. 1986. 'RNA as an Enzyme'. *Sci. Am.* 255 (5): 76–84.

Cooper, G. M. 2019. *The Cell: A Moleculer Approach*. 8th ed. Sinauer/OUP.

Curtis, O. F., and D. G. Clark. 1950. *An Introduction to Plant Physiology*. New York: McGraw-Hill Book Company, Inc.

Dennis, D. T. et al. 1997. *Plant Metabolism*. New York: Addison Wesley/Longman.

Devlin, R. M. 2017. *Outline of Plant Physiology*. India: MedTech.

Devlin, R. M., F. H. Witham, and D. F. Blaydes. 2017. *Exercises in Plant Physiology*. 2nd ed. India: MedTech.

Duysens, L. N. M., J. Amesz, and B. M. Kemp. 1961. 'Two Photochemical Systems in Photosynthesis'. *Nature* 190: 510–11.

Elliot, W. Y. 2009. *Biochemistry and Molecular Biology*. 4th ed. London: Oxford Publishers.

Fischer, R. A. 1968. 'Stomatal Opening-Role of Potassium Uptake by Guard Cells'. *Science* 160: 784–85.

Fischer, R. A., and T. C. Hsiao. 1968. 'Stomatal Opening in Isolated Epidermal Strips of *Vicia faba* II. Response to KCl Concentration and Role of Potassium Absorption'. *Plant Physiol.* 43: 1953–58.

Fitter, A., and R. Hay. 2012. *Environmental Physiology of Plants*. 3rd ed. Academic Press.

Flint, L. H., and E. D. McAlister. 1935. 'Wavelength of Radiation in the Visible Spectrum Inhibiting the Germination of Light Sensitive Lettuce Seed'. *Smithsonian Inst. Misc. Coll.* 96: 1–8.

Fujino, M. 1959. 'Stomatal Movement and Active Migration of Potassium Translated'. *Kagaku* 29: 660.

Galston, A. W. 1981. *Green Wisdom*. New York: Basic Books, Inc., Publishers.

Galston, A. W., P. I. Davies, and R. L. Satter. 1980. *The Life of the Green Plant*. 3rd ed. Englewood Cliffs, New Jersey: Prentice-Hall, Inc.

Gane, R. 1934. 'Production of Ethylene by Some Ripening Fruits'. *Nature* 134: 1008.

Garner, W. W., and H. A. Allard. 1920. 'Effect of the Relative Length of the Day and Night and Other Factors of the Environment on Growth and Reproduction in Plants.' *J. Agric. Res.* 18: 553–603.

Govindjee, R. G., and E. I. Rabinowitch. 1960. 'Two Forms of Chlorophyll a *In Vivo* with Two Distinct Photochemical Functions'. *Science* 132: 355.

Govindjee, R. G., and R. Govindjee. 1974. 'The Absorption of Light in Photosynthesis'. *Scientific American* 231: 68–82.

Greulach, V. A. 1973. *Plant Function and Structure*. New York: Macmillan Publishing Co., Inc.

Greulach, V. A., and J. E. Adams. 1967. *Plants – An Introduction to Modern Botany*. New York. London: John Wiley and Sons, Inc.

Guha, S., and S. C. Maheshwari. 1966. 'Cell Division and Differentiation of Embryos in the Pollen Grains of *Datura In Vitro*. *Nature* (London), 212: 97–8.

———. 1967. 'Development of Embryoids from Pollen Grains of *Datura In Vitro*'. *Phytomorphology* 17: 454–61.

Hardin, J. and G. Beotoni. 2018. *Becker's World of the Cell*. Pearson.

Hartmann, H. T., W. J. Flocker, and A. M., Kofranek. 1981. *Plant Science-Growth, Development and Utilization of Cultivated Plants*. New Jersey: Prentice-Hall, Inc.

Hatch, M. D., and C. R. Slack. 1966. 'Photosynthesis by Sugarcane Leaves. A New Carboxylation Reaction and Pathway of Sugar Formation'. *Biochem. J.* 101: 103–11.

Haupt, W. 1982 'Light-Mediated Movement of Chloroplasts'. *Annu. Rev. Plant Physiol.* 33: 205–33.

Heath, O. V. S., and B. Orchard. 1957. 'Midday Closure of Stomata'. *Nature* (London), 180: 181.

Hess, D. 1975. *Plant Physiology*. New York: Springer–Verlag.

Hill, R. 1937. 'Oxygen Evolved by Isolated Chloroplasts'. *Nature* 139: 881–82.

Hill, R., and F. Bendall. 1960. 'Function of the Two Cytochrome Components in Chloroplasts – A Working Hypothesis'. *Nature* 186: 136–37.

Hopkins, W. G., and N. P. A. Hüner. 2009. *Introduction to Plant Physiology*. 4th ed. London. Ontario: John Wiley and Sons, Inc.

Humble, G. D., and K. Raschke. 1971. 'Stomatal Opening Quantitatively Related to Potassium Transport'. Evidence from Electron Probe Analysis. *Plant Physiol.* 48: 447–53.

Jagendorf, A. T. 1967. 'Acid–Base Transitions and Phosphorylation by Chloroplasts.' *Fed. Proc. Am. Soc. Exp. Biol.* 26: 1361–69.

Kochhar, P. L., and H. N. Krishnamoorthy. 1989. *A Text Book of Plant Physiology*. Delhi, Lucknow: Atma Ram and Sons.

Kochhar, S. L., and Sukhbir Kaur Gujral, 2012. *Comprehensive Practical Plant Physiology*. Delhi: Macmillian Publishers India Ltd.

Kornberg, H. L., and H. A. Krebs. 1957. 'Synthesis of Cell Constituents from C2-Units by a Modified Tricarboxylic Acid Cycle'. *Nature* 179: 988.

Kortschak, H. P., C. E. Hartt, and G. O. Burr. 1965. 'Carbon dioxide Fixation in Sugarcane Leaves.' *Plant. Physiol.* 40: 209–13.

Kramer, P. J. 1937. 'The Relation Between Rate of Transpiration and Rate of Absorption of Water in Plants.' *Am. J. Bot.* 24: 10.

Krebs, H. A. 1970. 'The History of the Tricarboxylic Acid Cycle'. *Perspect. Biol. Med.* 14: 154–70.

Levitt. J. 1956. *The Hardiness of Plants*. New York: Academic Press, Inc.

———. 1967. 'The Mechanism of Stomatal Action'. *Planta* 74: 101–8. (It gives an account of Levitt's modified classical theory).

———. 1969. *Introduction of Plant Physiology*. Saint Louis: The C.V. Mosby Company.

———. 1974. 'Mechanism of Stomatal Movements – Once More'. *Protoplasma* 82 (1–2): 1–17.

Machlis, L., and J. G. Torrey. 1956. *Plants in Action*. San Francisco: W.H. Freeman and Company.

Maheshwari, S. C. 2003. 'A Rise of Experimental Plant Physiology in India–A Personal View'. Souvenir: *2nd International Congress of Plant Physiology*, New Delhi, India. 1–13.

Mauseth, J. D. 2019. *Botany–An Introduction to plant Biology*. 6th ed. Boston: Jones and Bartlett Publishers.

Meidner, H. 1984. *Class Experiments in Plant Physiology*. London: George Allen and Unwin.

Meyer, B. S., D. B. Anderson, and R. H. Böhning. 1960. *Introduction to Plant Physiology*. Princeton, N.J.: D. Van Nostrand Company, Inc.

Mitchell, P. 1961. 'Coupling of Phosphorylation to Electron and Hydrogen Transfer by a Chemiosmotic Type of Mechanism. *Nature* 191: 144–48.

Mothes, K., and L. Engelbrecht. 1961. 'Kinetin and Its Role in Nitrogen Metabolism'. In *Proc. Int. Bot. Cong.*, 9th Cong., Montreal, Canada 2: 996. Toronto: University of Toronto Press.

Narwal, S. S. et al. 2009. *Plant Biochemistry*. Studium Press, LLC.

Nelson, D. L., and M. M. Cox. 2017. *Lehninger Principles of Biochemistry*. 7th ed. Machmillan Higher Education.

Nickelsen, Kärin, and Govindjee. 2011. *The Maximum Quantum Yield Controversy: Otto Warburg and the 'Midwest-Gang'*. Bern Studies in the History and Philosophy of Science, University of Bern. Switzerland: Institut für Philoshie.

Noggle, G. R., and G. J. Fritz. 1986. *Introductory Plant Physiology*. 2nd ed. New Delhi: Prentice-Hall of India.

Northington, D. K., and J. R. Goodin. 1984. *The Botanical World*. St. Louis. Toronto: Times Mirror/ Mosby College Publishing.

Ochoa, S., and W. Vishniac. 1952. 'Carboxylation Reactions and Photosynthesis'. *Science* 115: 297.

Ochs, R. S. 2014. *Biochemistry*. Jones and Bartlett Learning.

Pandey, S. N., and B. K. Sinha. 2006. *Plant Physiology*. 4th ed. New Delhi: Vikas Publishing House Pvt. Ltd.

Park, R. B., and Biggins. 1964. 'Quantasome: Size and Composition'. *Science* 144: 1009–11.

Plummer, D. T. 1990. *An Introduction to Practical Biochemistry*. 3rd ed. New Delhi. New York: Tata McGraw-Hill Publishing Company Limited.

Rabinowitch, E. I. 1948. 'Photosynthesis'. *Sci. Am.* 179 (2): 24–35.

Rabinowitch, E. I., and R. G. Govindjee. 1965. 'The Role of Chlorophyll in Photosynthesis'. *Sci. Am.* 213 (1): 74–83.

Raven, P. H., R. F. Evert, and S. E. Eichhorn. 2019. *Biology of Plants*. 12th ed. McGraw-Hill.

Richmond, A., and A. Lang. 1957. 'Effect of Kinetin on Protein Content and Survival of Detached *Xanthium* Leaves'. *Science* 125: 650–51.

Ridge, I. 1991. *Plant Physiology*. Milton Keynes: Hodder and Stoughton.

Ruben, S., and M.D. Kamen. 1940a. 'Photosynthesis with Radioactive Carbon. IV. Molecular Weight of the Intermediate Products and a Tentative Theory of Photosynthesis'. *J. Am. Chem. Soc.* 62: 3451.

———. 1940b. 'Radioactive Carbon in the Study of Respiration in Heterotrophic Systems'. *Proc. Natl. Acad. Sci. USA*. 26: 418.

Russell, J., H. Oughan, H. Thomas, and S. Waaland. 2013. *The Molecular Life of Plants*: John Wiley & Sons Ltd.

Sadasivam, S. and A. Manikam. 1992. *Biochemical Methods for Agricultural Sciences*. India: Wiley Eastern Limited.

Salisbury, F. B., and C. W. Ross. 1992. *Plant Physiology*. 4th ed. California: Wadsworth Publishing Co. Ltd.

Sawada, K. 1912. 'Disease of Agricultural Products in Japan'. *Formosan. Agr. Rev.* 36: 10.

Sawada, K., and Kurosawa. 1924. 'On the Prevention of the Bakanae Disease of Rice'. *Exp. Sta. Bull. Formosa.* 21: 1.

Sawhney, S. K., and R. Singh. 2009. *Introductory Practical Biochemistry*. India: Narosa Publishing House.

Sayre, J. D. 1926. 'Physiology of the Stomata of *Rumex patientia*'. *Ohio. J. Sci.* 26: 233.

Scarth, G. W. 1932. 'Mechanism of Action of Light and Other Factors on Stomatal Movement'. *Plant Physiol*. 7: 481–504.

Scholander, P. F., H. T. Hammel, and E. D. Bradstreet. 1965. 'Sap Pressure in Vascular Plants'. *Science* 148: 339–46.

Scientific American-Resource Library Readings in the *Life Sciences*. Vols. 1–7. San Francisco, California: W. H. Freeman and Company.

Scott, P. 2008. *Physiology and Behaviour of Plants*. London: John Wiley and Sons, Ltd.

Sengar, R. S. S. Gupta, and A. K. Sharma. 2011. *Laboratory Manual on Biotechnology*. India. N. R. Book Distributor.

Sheeler, P., and D. E. Bianchi. 1987. *Cell and Molecular Biology*. 3rd ed. New York: John Wiley and Sons, Inc.

Siegelman, H. W., and E. M. Firir. 1964. 'Purification of Phytochrome from Oat Seedlings'. *Biochemistry* 3: 418.

Siegelman, H. W., and W. L. Butler. 1965. 'Properties of Phytochrome'. *Ann. Rev. Plant Physiol.* 16: 383.

Singer, S. J., and G. L. Nicolson. 1972. 'The Fluid Mosaic Model of the Structure of Cell Membranes'. *Science* 175: 720–31.

Sinha, R. K. 2007. *Modern Plant Physiology*. New Delhi: Narosa Publishing House.

Skoog, F., and C. O. Miller. 1957. 'Chemical Regulation of Growth and Organ Formation in Plant Tissues Cultured *In vivo*'. *Symp. Soc. Exp. Biol.* 11: 118.

Slatyer, R. O., and S. A. Taylor. 1960. 'Terminology in Plant and Soil Water Relations'. *Nature*. 187: 922–24.

Smith, A. M. et al. 2010. *Plant Biology*. New York: Garland Science.

Srivastava, H. N. 2000. *Plant Physiology*. New Delhi: Pradeep Publications.

Steward, F. C. 1967. *Plants At Work*. New York: Addison–Wesley Publishing Company.

Stryer, L. 1995. *Biochemistry*, 3rd ed. San Francisco: W. H. Freeman and Company.

Sumner, J. B. 1926. 'The Isolation and Crystallization of the Enzyme Urease'. *J. Biol. Chem.* 69: 435.

Taiz, L., E. Zeiger, I. M. Moller, and A. Murphy. 2015. *Plant Physiology and Development*. 6th ed. USA: Sinauer Associates Inc. Publishers.

Tanada, T. 1968. 'A Rapid Photoreversible Response of Barley Root Tips in the Presence of 3-Indole Acetic Acid'. *Proc. Natl. Acad. Sci. USA*. 59: 376–79.

Thaine, R. 1961. 'Transcellular Strands and Particle Movement in Mature Sieve Tubes'. *Nature* 192: 72.

Thimmaiah, S. K. 2009. *Standard Methods of Biochemical Analysis*. New Delhi: Kalyani Publishers.

Thut, H. F. 1932. 'Demonstrating the Lifting Power of Transpiration'. *Am. J. Bot.* 19: 358.

Ting, I. P. 1982. *Plant Physiology*. California, U.S.A.: Addison-Wesley Publishing Company.

Ting, I., and W. Loomis. 1963. 'Diffusion Through Stomates'. *Am. J. Bot.* 50: 866.

Tolbert, N. E. 1981. 'Metabolic Pathways in Peroxisomes and Glyoxysomes'. *Ann. Rev. Biochem.* 50: 133–57.

Tyree, M. T. 1997. 'The Cohesion-Tension Theory of Sap Ascent: Current Controversies'. *J. Exp. Bot.* 1753–56.

Van Niel, C. B. 1931. 'On the Morphology and Physiology of the Purple and Green Sulphur Bacteria'. *Arch. Mikrobiol*. 3: 1.

Verma, V. 2007. *Textbook of Plant Physiology*. New Delhi: Ane Books India.

Voet, D., and J. G. Voet. 2019. *Fundamentals of Biochemistry*. 5th ed. New York: John Wiley Sons Inc.

Wareing, P. F., and I. D. J. Philips. 1978. *The Control of Growth and Differentiation in Plants*. 2nd ed. New York: Pergamon Press.

Watson, J. D., and F. H. C. Crick. 1953. 'Molecular Structure of Nucleic Acids'. *Nature* 171: 737–38.

Weier, T. E., C. R. Stocking, M. G. Barbour, and T. L. Rost. 1970. *Botany*. 6th ed. New York: John Wiley and Sons.

Wilkins, M. 1963. 'Molecular Configuration of Nucleic Acids'. *Science* 140: 1941–63.

———. 1988. *Plant Watching*. London: Macmillan.

Wilkins, Malcolm B. ed. 1984. *Advanced Plant Physiology.* London: ELBS, Longman.

Yin, H. C., and Y. T. Tung. 1948. 'Phosphorylase in Guard Cells'. *Science* 108: 87.

Zelitch, I. 1963. 'The Control and Mechanism of Stomata and Water Relations in Plants'. *Conn. Agric. Expt. Sta. Bull.* 664: 18–42.

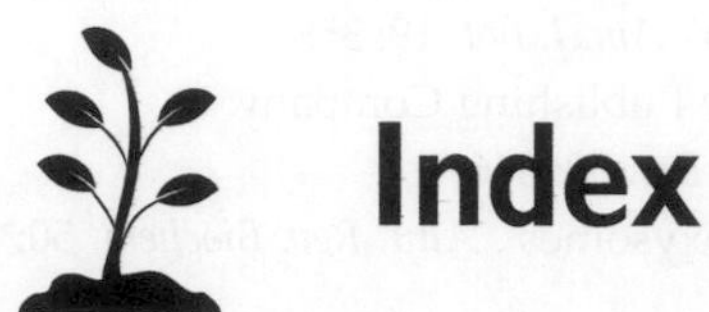

Index

Colour Plates

Picture on the Jacket Cover

Young maturing capsule of opium poppy (*Papaver somniferum*) – the latex oozing out from incisions is rich in alkaloids such as morphine, codeine, etc. (*Courtesy*: Mr Sandesh Parashar)

Unit I

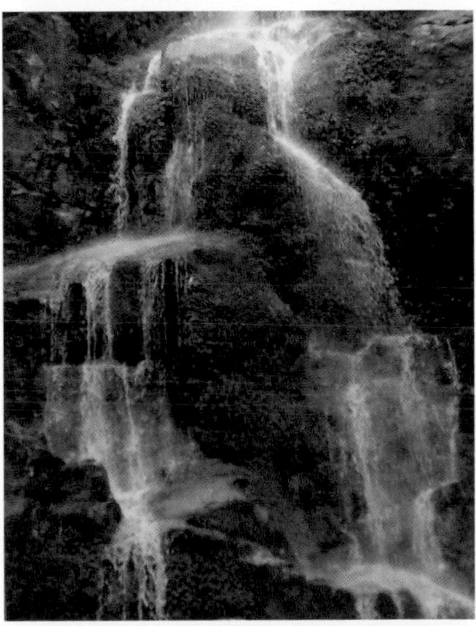

A waterfall in pachmarhi, MP, India. The water potential (ψ) of water upstream is greater than the ψ of water downstream by virtue of its position in the gravitational field, thus the water will move downhill (*Courtesy*: Mr Satya Prakash Mahapatra).

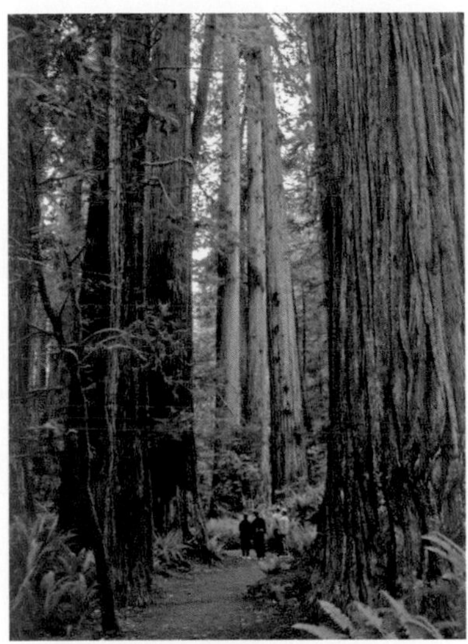

A forest of Redwood tree, *Sequoia sempervirens* in California, USA (*Courtesy*: Redwood National and State Park (RNSP) California.

A bed of *Rhoeo spathacea* showing purple undersurface of the leaves (*Courtesy*: Mr R. Thilakan Makkiseril).

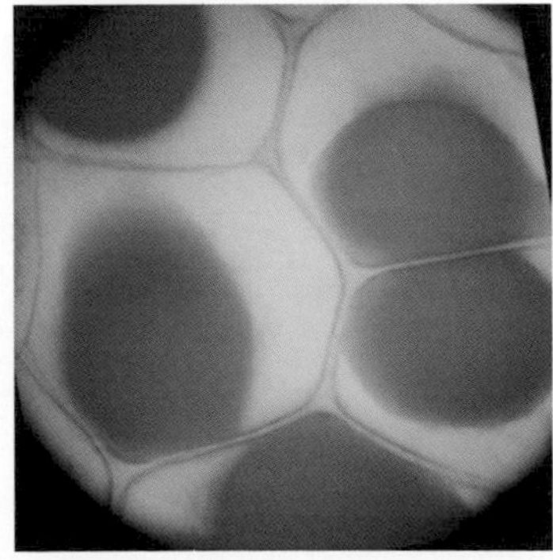

Plasmolysed cells of *Rhoeo* peel (*Courtesy*: Miss Madhumati Yashwant Shinde, Ichalkaranji, Maharashtra).

Unit II

A fruiting pineapple (*Ananas comosus*) – a source of protein-digesting enzyme bromelian. An important agricultural crop showing CAM pathway (*Courtesy*: Mr. R. Thilakan Makkiseril).

A closeup of Pitcher plant, *Nepenthes khasiana* (*Courtesy*: Dr C. T. Chandra Lekha).

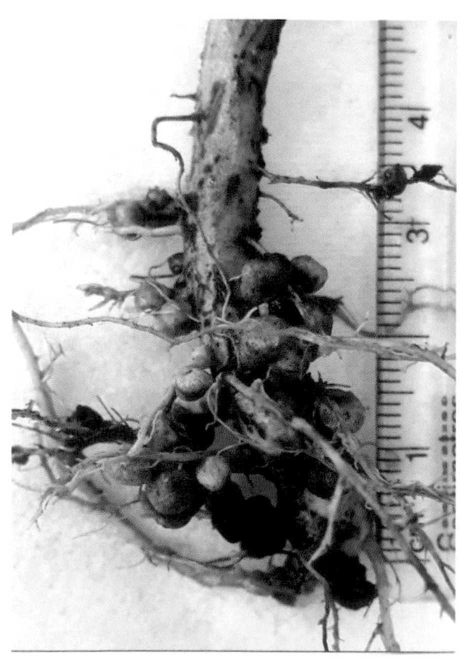

Nodulated roots with tubercles or nodules – a site of nitrogen fixation (*Courtesy*: Dr Surendera Singh, Bhartpur, Rajasthan).

A flowering and fruiting branch of castor bean (*Ricinus communis*) – a plant material where the Glyoxylate pathway was studied (*Courtesy*: Dr B. K. Mukherjee).

Unit III

A closeup of the flower of tobacco (*Nicotiana tabacum*) where photoperiodic phenomenon was discovered (*Courtesy*: Dr C. T. Chandra Lekha).

A closeup of flowering shoot of Touch-me-not (*Mimosa pudica*). The leaves exemplify seismonastic movements in response to mechanical touch or shock (*Courtesy*: Mr Sandesh Parashar).

A closeup of flowering and fruiting shoot of cocklebur (*Xanthium strumarium*) – a short day plant (SDP) where photoperiodic studies were conducted (*Courtesy*: Dr Sushain Babu).

A closeup of a flowering head (capitulum) of sunflower (*Helianthus annuus*) showing solar-tracking movements (Heliotropic movements) (*Courtesy*: Mr B. K. Basavaraj).

Unit IV

A closeup of a flowering shoot of tea (*Camellia sinensis*) – the leaves are a source of polyphenols with antioxidant properties (*Courtesy*: Dr Rabiul Hasan Dollar, Chapainawabganj, Bangladesh).

A closeup of flower of Madagascar periwinkle (*Catharanthus roseus*) – the leaves are a source of cancer-protective compounds such as vincristine and vinblastine (*Courtesy*: Mrs Jyotirmayi Parija).

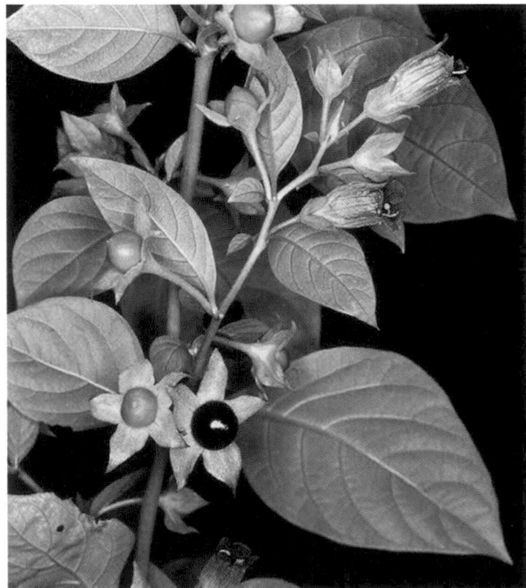

Flowering and fruiting branches of Belladonna or Deadly Nightshade plant (*Atropa belladonna*) – the leaves are a source of alkaloids like atropine, hyoscyamine and scopolamine (*Courtesy*: Dr Prabir Ranjan Sur, BSI Kolkata).

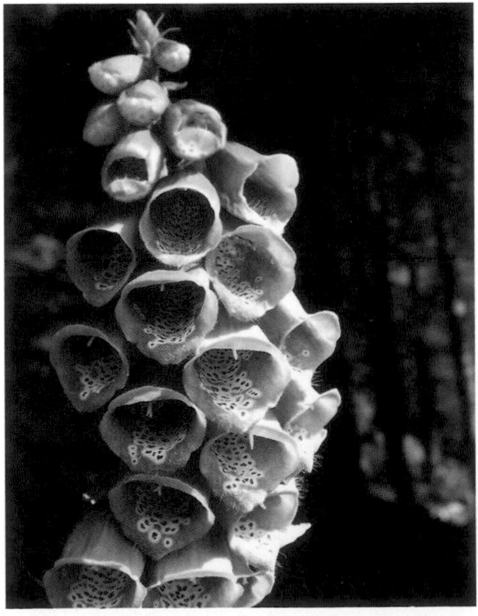

A closeup of the inflorescence of common foxglove (*Digitalis purpurea*) – the leaves are a source of cardiotonic glycosides containing digitoxin, gitoxin and digoxin (*Courtesy*: Dr Manju Chaudhary).